USED
W9-AYG-386

THERMODYNAMICS

THERMODYNAMICS

Fourth Edition

$h = u + Pv$

Kenneth Wark

Associate Professor of Mechanical Engineering
Purdue University

McGraw-Hill Book Company

New York St. Louis San Francisco Auckland Bogotá Hamburg
Johannesburg London Madrid Mexico Montreal New Delhi
Panama Paris São Paulo Singapore Sydney Tokyo Toronto

This book was set in Times Roman by Santype-Byrd.
The editors were Rodger H. Klas and J. W. Maisel;
the production supervisor was Leroy A. Young.
New drawings were done by ECL Art Associates, Inc.
The cover was designed by Robin Hessel.
R. R. Donnelley & Sons Company was printer and binder.

THERMODYNAMICS

4567890 DOCDOC 8987654

ISBN 0-07-068284-4

Library of Congress Cataloging in Publication Data

Wark, Kenneth, date
Thermodynamics.

Includes bibliographies and index.
1. Thermodynamics. I. Title.
QC311.W3 1983 536′.7 82-10031
ISBN 0-07-068284-4 AACR2

CONTENTS

PREFACE

This edition retains the format of a dual system of units found in the third edition. Example problems, assignment problems at the end of each chapter, and data in the appendices are presented in both metric (SI) and English engineering (USCS) units. In general, problems are stated in one set or the other, but units are not mixed. Consequently, the text may be used exclusively in SI or USCS units. In addition, both sets of units may be emphasized by the proper choice of problem assignments. The problems for each chapter are grouped into specific topics and are identified by an appropriate title. This should make it easier for an instructor or student to choose problems for study. As in the preceding edition, any example problem, assignment problem, or table in the appendix which employs metric data is designated by the symbol M after the number. Metric pressure units in this edition include both the bar and the megapascal, while metric specific volume data appear primarily in either cubic centimeters per gram or cubic meters per kilogram.

As in preceding editions, the introductory concepts of the second law may be approached from either a macroscopic or microscopic viewpoint in this edition. If a statistical introduction is desired, in place of the macroscopic approach, the first five chapters should be followed by Chap. 9. An alternative is to use Chap. 9 as a supplement to Chap. 6 if desired.

On the basis of classroom experience a number of basic changes have been made in the general layout of the text. Subject matter has been carefully examined for changes which might aid in the clarification of material. In some instances this has led to the elimination or addition of a substantial amount of material within a section. Chapters 9, 12, and 13 of the third edition have remained essentially intact, while Chap. 10 on the applications of quantum-statistical mechanics has been entirely removed. This latter material is probably better covered by a separate course in statistical thermodynamics for those with a specific need for information in this area. Chapter 10 in this edition now involves transient flow analysis formerly presented at the end of Chap. 5. Chapter 8, on the concept of availability, has been completely rewritten and tested in the classroom on the senior level. Problems involving the application of availability concepts to engineering systems appear in later chapters.

All other chapters have been modestly or extensively revised. The ideal-gas material of Chap. 1 in the third edition now has been incorporated into the closed-system energy analysis of ideal gases in Chap. 3. The state postulate material formerly in Chap. 3 is now combined with the introduction to the first law for closed systems presented in Chap. 2. Introductory second law concepts involving Carnot heat engines and refrigerators which appeared early in Chap. 7 have been moved back to the end of Chap. 6. Chapter 17 has been split into two chapters, with the material on vapor power cycles remaining in this chapter, and that on refrigeration devices forming a new Chap. 18. Finally, based on items removed from other chapters and some new discussions, a new Chap. 19 covers some alternative and innovative energy conversion systems of current interest. This chapter is not all-inclusive, but simply contains discussions of some approaches to energy conservation and usage which are heavily based on thermodynamic analysis.

The problems carried over from the third edition have been carefully examined for their appropriateness. Data in problems have been altered, and problems have been dropped and others added. The large number of problems available should allow an instructor to select appropriate problems without repetition over several years.

The author is indebted to the students and faculty of the Mechanical Engineering School at Purdue University for their helpful input over the years. A special debt of gratitude goes to Dr. Peter E. Liley of this school, who has provided resource information for new and revised thermodynamic data which appear in the text.

Kenneth Wark

CHAPTER

ONE

BASIC CONCEPTS AND DEFINITIONS

1-1 THE NATURE OF THERMODYNAMICS

Thermodynamics is a science which comprises the study of energy transformations and of the relationships among the various physical qualities, or properties, of substances which are affected by these transformations. Predictions of the physical properties of substances can be carried out either by analyzing the large-scale (gross) behavior of a substance or by statistically averaging the behavior of the individual particles which make up the substance. A thermodynamic analysis which is undertaken without recourse to the nature of the individual particles and their interactions falls into the domain of *classical thermodynamics.* This is the macroscopic viewpoint toward matter and its interactions, where the overall, large-scale effect is the focal point of interest. It requires no hypothesis on the detailed structure of matter on the atomic scale; consequently, the general laws of classical thermodynamics are not subject to change as new knowledge on the nature of matter is uncovered.

Another approach to the study of thermodynamic properties and energy relationships is based on the statistical behavior of large groups of individual particles. This method, founded on a microscopic viewpoint, is called *statistical thermodynamics.* It combines the computational techniques of statistical mechanics with the findings of quantum theory. The dual purpose of statistical thermodynamics is to predict and interpret the macroscopic characteristics of matter in equilibrium situations from their microscopic origins. It postulates that the values of macroscopic properties (such as pressure, temperature, and density, among others), which we measure directly or calculate from other measurements, merely reflect some sort of statistical average of the behavior of a tremendous number of particles. In present technology, where substances are employed under extreme ranges of temperature and pressure, the predictive methods of statistical thermodynamics are extremely important. In addition, the interpretive role of the microscopic approach is of value. For example, the interpretation in a microscopic context of a thermodynamic property called entropy frequently is helpful in increasing a student's understanding of this concept. This viewpoint of entropy is considered in Chap. 9. This theory also was helpful in the modern development of new, direct energy-conversion methods, such as thermionics and thermoelectrics.

Another independent approach to the analysis of particle behavior is the field of *kinetic theory*. This subject is the study of particle behavior based normally on newtonian mechanics, and requires a detailed knowledge of the interactions between particles. It is extremely useful in developing relations for the transport properties of rate processes, such as viscosity, thermal conductivity, and diffusion coefficients. Since it does not take into account the quantization of energy, it is not successful in predicting thermodynamic properties except in limiting cases. A third microscopic approach is *information theory*. Its applications to thermodynamics has been developed since the 1950s. It has made some important contributions to the interpretation of the macroscopic state of matter in terms of microscopic behavior.

Emphasis will be placed upon the macroscopic viewpoint in this text, with the statistical viewpoint playing a supporting role. There are several reasons for this. First of all, the solution of a great majority of thermodynamic problems requires an analysis only in terms of macroscopic variables. An understanding of the underlying mechanisms for the process in terms of particle behavior would not be necessary in these cases, nor especially beneficial. Second, classical thermodynamics is an easier, more direct approach to the solution of engineering problems, and often has fewer mathematical complications. It is somewhat more free of abstractions. Finally, it may be pointed out that macroscopic analyses in a number of instances have demonstrated what major area could be studied profitably by the methods of statistical mechanics.

1-2 ENERGY SOURCES AND CONVERSION EFFICIENCIES

One measure of a country's standard of living is gross national product (GNP). Worldwide data indicate that GNP is closely related to the energy consumption per capita in a particular country. Hence, the use of energy by industrialized countries is an important factor in their continued growth. In addition, the desire of underdeveloped nations to drastically improve their standard of living will lead to an increase in energy usage throughout the world, even if the growth rate of energy usage in developed countries were to approach zero. As a result, a great deal of activity must occur in the energy field to maintain the average standard of living in industrialized countries and to increase the standard of living in underdeveloped nations.

In terms of fossil fuels, we need to discover and develop additional sources of oil and gas. Offshore drilling and the further development of underground reserves in Alaska are two options open to the United States. The vast coal reserves in the United States must also be utilized, consistent with environmental concerns. Another major possibility is to continue developing alternative energy concerns. Nuclear reactors continue to provide an increasing share of the electric power production in many countries. One concern in this area, however, is environmental safety. Unfortunately, a second concern exists in the lack of a plentiful supply of ^{235}U which is used in conventional reactors. If nuclear fuel is to remain an important source of energy, it may be necessary to develop breeder reactors which will produce additional fuel as well as power during their operation. Several countries have breeder reactors under active investigation; commercial operation of such reactors may be close at hand. The use of passive and active solar energy units is undergoing tremendous growth, while commercial power production from steam produced in geothermal sources beneath the ground is available on a relatively small scale. In addition, wind and tidal power is under active investigation, as well as the use of the temperature difference between the surface and deeper layers of the oceans as a potential source of power production.

A third possibility for maintaining the standard of living in an industrialized country is to cut its growth rate of energy usage. Such action does not imply a decline in living standards, but it does require at least two other things. One of these is the reallocation of energy usage, which involves establishing the products and services that are really necessary for maintaining the health and welfare of a nation. This action is primarily a political one, and leads to allocation of energy on a priority basis. A second requirement is the serious move that must be made to cut the wasteful use of energy in industry, in transportation, and in residential and commercial applications. In 1980 the allocation of energy to these three basic users in the United States was estimated to be 37, 34, and 29 percent, respectively. The amount of the total allocated energy which went to a useful purpose was around 46 percent. Hence, over one-half of the energy was wasted. Moves to reduce energy losses are already in evidence in a number of areas. The use of greater insulation in walls and attics of homes and businesses, and the improvement in average miles per gallon for gasoline-powered automobiles are but two examples of energy conservation.

The availability of cheap energy from fossil fuels before the early 1970s encouraged the development of devices which were inefficient in the conversion of energy to more useful forms. With increasing cost and decreasing supply of conventional fossil fuels, it is imperative that engineers look seriously at increasing the efficiency of energy-conversion devices. By efficiency or effectiveness we generally mean the ratio of the desired output to the required input. That is, it is a measure of what is accomplished compared to what it costs.

One important efficiency term for cyclic energy conversion devices which transform thermal energy (heat) into work output is the *thermal* efficiency. It is defined as the ratio of the net work output (which is desired) to the heat input (which is costly). Hence,

$$\eta_{\text{thermal}} \equiv \frac{\text{net work output}}{\text{heat input}} \tag{1-1}$$

This definition of thermal efficiency leads to a value which normally lies between 0 and 1. In most instances the ratio in Eq. (1-1) is multiplied by 100, and the thermal efficiency is expressed as a percent. The thermal efficiencies of conventional modern devices are startlingly low. Table 1-1 lists some values for mobile and stationary equipment. One of the larger consumers of fuel in the United States, the spark ignition automotive engine, has essentially the lowest efficiency. In city driving the conversion efficiency probably lies between 10 and 15 percent. The most efficient cyclic device for the production of power is the large steam plant operating on fossil fuels. Even here the efficiency barely exceeds 40 percent. The thermal efficiency of a nuclear-fuel plant is well below this value. There are means of increasing the overall efficiency of a steam power plant, but such changes increase the cost of production of electricity. Eventually the consumer must decide whether the increase in goods and services brought about by expenditure of energy per capita is worth the cost.

The data of Table 1-1 reflect the efficiencies of cyclic devices which convert chemical energy (essentially in the form of fossil fuels) into mechanical energy.

Table 1-1 Typical thermal efficiencies, in percent, of some power-producing devices

Type	Conditions	Efficiency
Automobile		
Spark	Optimum	25
Ignition, gasoline	Steady 60 mi/h	18
	Steady 30 m/h	12
Truck		
Compression	Full load	35
Ignition, diesel	Half load	31
Locomotive		
Diesel		30
Gas turbine (100 hp)		
Without regeneration	Optimum	12
With regeneration	Optimum	16
Gas turbine (>7500 kW)		
Without regeneration	Optimum	25
With regeneration	Optimum	34
Steam power plant (>350,000 kW)	Optimum	41

Table 1-2 Approximate converter efficiencies, in percent, of some well-known devices

Type	Efficiency
Chemical to thermal	
Home furnace	70
Chemical to electrical	
Storage battery	70
Dry cell battery	90
Fuel cells	60
Electrical to radiant	
Incandescent lamp	7
Fluorescent lamp	21
Sodium vapor lamp	40
Electrical to mechanical	
Electric motor	90
Chemical to kinetic	
Rockets	45
Jet engines	40
Potential to mechanical	
Hydraulic turbine	95
Radiant to electrical	
Solar cell	12

Table 1-2 lists the approximate efficiencies of some other types of well-known energy converters. These converters do not operate in a cyclic manner, and have appreciably higher efficiencies. The home furnace is an excellent example of a device where considerable savings in energy consumption might be made. Although only 30 percent of the chemical energy input is lost (as hot gas up the chimney) if the furnace combustion process is optimized, a large number of home furnaces in this country are not checked frequently enough to maintain optimum firing conditions. With dirty fuel nozzles and incorrect air supply, for example, the conversion efficiency can be relatively poor. It has been estimated that the average oil- and gas-heated home may be operating only at the 50 percent level. The significant difference in converter efficiency between a fluorescent and an incandescent lamp should also be noted. The fuel cell listed operates somewhat like a battery, with one major difference. The reactants responsible for the production of electricity from chemical energy are fed continuously to the fuel cell, and then the products of the reaction are continuously withdrawn. Unlike a battery, then, the fuel cell theoretically has an unlimited life of operation. The development of the fuel cell for commercial production of electricity is a spin-off of the space age. Fuel cells provide the main source of internal power for space capsules. One modern converter—not listed in Table 1-2—which may ultimately play a role in large-scale power production is the MHD (magnetohydrodynamic) cycle. An MHD unit generates electricity by passing a very high-temperature gas (called a plasma) through a magnetic field. The high-temperature, ionized gas could be produced through the use of chemical, nuclear, or solar sources. The gas emitted from an MHD unit is still quite hot. Hence the exit gas can be used further as a thermal source for a conventional steam power plant. A similar situation exists for a combined cycle involving a gas turbine–steam power plant combination. In this case the hot gas emitted from the gas turbine is used as the thermal source for the steam power plant. These latter two methods are just two of the possible methods under investigation for increasing the overall conversion efficiency for large electrical power generation units. Figure 1-1 shows schematic diagrams of four conventional or potential energy-conversion systems.

1-3 INTRODUCTION TO THE LAWS OF THERMODYNAMICS

This text presents five laws or postulates which govern the study of energy transformations and the relationships among thermodynamic properties. Two of these—the first and second laws—deal with energy, directly or indirectly. Consequently they are of fundamental importance in engineering studies of energy transformations and usage. The remaining three statements—the zeroth law, the third law, and the state postulate—relate to thermodynamic properties. They are important, since the first and second laws lead to equations which involve properties of matter. These latter three statements aid in the evaluation of such properties. In an engineering context, then, the zeroth and third laws and the state postulate are means to an end and not an end in themselves.

The first law of thermodynamics, in its ultimate useful form, is a conservation of energy statement. When energy is transferred from one region to another or changes form within a system, the total quantity of energy is constant. (In this text we shall not consider nuclear transformations of mass to energy.) Most persons have some degree of familiarity with this principle on the basis of their living experience. In addition, restrictive forms of the law appear in many introductory chemistry and physics texts. One of the purposes of this text is to organize these previous experiences into a more general and logical pattern.

The second law of thermodynamics has many ramifications with respect to engineering processes. Among others, it determines the direction of change toward equilibrium for a particular system under a given set of constraints. The major importance of the second law to society today is this: The first law deals with the *quantity* of energy in terms of a conservation rule. The second law deals with the *quality* of energy. It is essentially a nonconservation rule. To speak of the quality of energy may be surprising, since quality implies that some forms of energy are more useful to society than others. However, such an idea is directly connected with the discussion of energy and societal needs presented in Sec. 1-2,

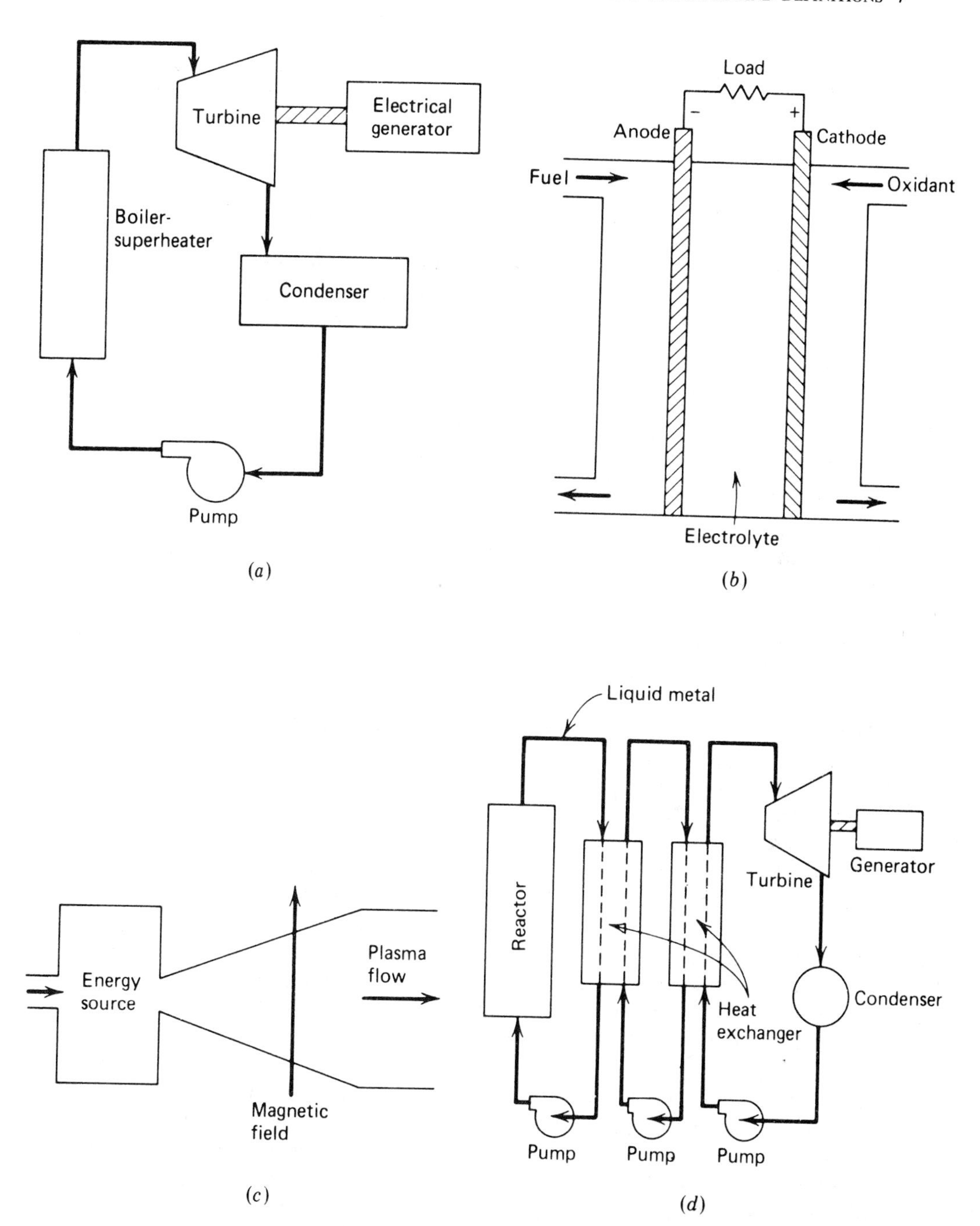

Figure 1-1 Schematic diagram of some energy-conversion devices. (*a*) Conventional steam power plant; (*b*) fuel cell; (*c*) magnetohydrodynamic generator (MHD); (*d*) breeder reactor cycle.

which stressed the need to optimize the conversion, transmission, and consumption of energy. The concept of optimization of energy usage implies that there are better and worse ways of using energy. It is the potential of various forms of energy to do *useful work* for humanity that determines the quality of those forms. The second law places some restrictions on the transformation of some forms of energy to a more useful type. In addition, we are all familiar with the notion that certain "losses" are associated with energy usage. The presence of friction in a process denotes to most people a loss in performance. Thus certain features of an energy process leads to a "degradation" of energy or, again, a loss in quality. The second law enables the engineer to measure this degradation or change in quality in quantitative terms. That is, the second law provides standards of performance against which actual (real) processes may be compared, in terms of energy usage. If society is to practice conservation in the conversion and consumption of energy, such standards are vitally important. In Sec. 1-2 the values of the thermal efficiency of a number of conventional power-producing devices were tabulated. The values range from roughly 10 to 40 percent. Apparently some energy-conversion devices are much better for converting chemical, solar, or nuclear energy to mechanical work than others. However, an equally important point is whether the value of 40 percent is good, in absolute terms. If a perfect device would yield 45 or 50 percent conversion, then the actual device looks pretty good. On the other hand, if a perfect device is capable of 90 to 95 percent conversion, the 40 percent converter doesn't look too attractive and the 10 percent converter is an extremely poor device. This is one example where the second law provides that standard of performance. Thus, in terms of energy usage, the second law of thermodynamics plays an increasingly important role in the analysis of new processes and in the improvement of old ones.

1-4 THERMODYNAMIC SYSTEM, PROPERTY, AND STATE

There are a number of terms used in the study of thermodynamics that are usually introduced to the student in branches of chemistry, physics, and other engineering sciences. Many of these words also are in common usage by the general populace. At this point, and throughout the text, we shall take the opportunity to review concepts and redefine words that should already be a part of the vocabulary of the reader. These concepts and terms are extremely important in thermodynamic studies, hence their meaning must be clear. In some cases it may be necessary to clarify a concept solely because the meaning has been corrupted by sloppy common usage. For example, the word "heat" typically falls into this classification.

A macroscopic thermodynamic *system* is any three-dimensional region of space which is bounded by one or more arbitrary geometric surfaces. The bounding surfaces may be real or imaginary and may be at rest or in motion. The boundary may change its size or shape. The region of physical space which lies outside the arbitrarily selected boundaries of the system is called the *surroundings* or the *environment*. In its usual context the term "surroundings" is restricted to that specific localized region which interacts in some fashion with the system and hence has a detectable influence on the system.

Two examples of macroscopic systems and the representative boundaries are shown in Fig. 1-2. Part (*a*) of the figure illustrates the flow of a fluid through a pipe or duct. The dashed line is a two-dimensional representation of one possible choice of a boundary, fixed in space, which outlines the region of space for which a thermodynamic study is to be made. The inside surface of the pipe may be selected as part of the boundary, and it represents a real obstruction to the flow of matter. However, it is noted that a portion of the boundary is imaginary; i.e., there is no real surface which marks the position of the boundary at the open ends. These latter boundaries are selected arbitrarily for accounting purposes, and they have no real effect or significance in the actual physical process. Thus we see that it is not necessary for any or all of the boundary to be physically distinguishable when making thermodynamic analyses. Nevertheless, it is extremely important to establish the boundaries of a system clearly before beginning any form of analysis.

In Fig. 1-2*b* a piston-cylinder assembly is shown. Again the selected boundary is given by the dashed line. In this particular case the boundary is physically well established, for the envelope lies just inside the walls of the cylinder and the piston. For this example it should be noted that the shape of the boundary will change when the position of the piston is altered, if the boundary is selected to surround the fluid within the assembly. The change in shape or position of the boundary is always permissible as long as such changes are recognized in subsequent calculations.

The analysis of thermodynamic processes includes the study of the transfer of mass and energy across the boundaries of a system and the effect of these interactions on the state of the system. When the transfer of mass across the bounding surface of a system is prohibited, the system is called a *closed* system. Although

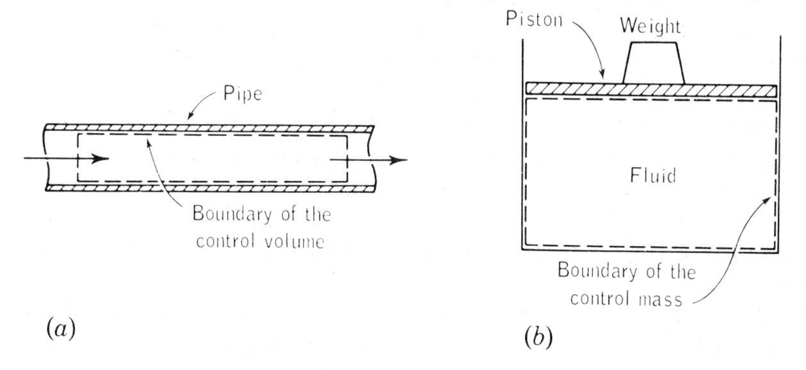

Figure 1-2 The boundaries of two typical thermodynamic systems. (*a*) Pipe flow; (*b*) piston-cylinder device.

the quantity of matter is fixed in a closed system, energy may be allowed to cross its boundaries in the form of work and heat. The matter may also change in chemical composition within the boundaries. The quantity of the matter under observation frequently is called the *control mass*. The control mass within the piston-cylinder assembly in Fig. 1-2*b* is in a closed system. It is appropriate to define an *open* system as a system for which mass is permitted to cross the selected boundaries, as well as energy in the form of heat and work. The region of space described in Fig. 1-2*a* would be analyzed as an open system.

A *property* is any characteristic of a system which can in principle be specified by describing an operation or test to which the system is to be subjected. As a result of the particular operation or test, a value can be assigned to the property under investigation. Properties are defined macroscopically without recourse either to a specific molecular model of matter or to a statistical evaluation of the microscopic behavior of the particles within the system. Examples include pressure, temperature, mass, volume, density, electrical conductivity, acoustic velocity, coefficient of expansion, and many others.

The *state* of a macroscopic system is the condition of the system characterized by the values of its properties. Attention in this text will be directed toward what are known as equilibrium states. The word "equilibrium" is being used in its generally accepted context—the equality of forces, or the state of balance. In future discussion the term "state" will refer to an equilibrium state unless otherwise noted. The concept of equilibrium is an important one, since it is only in an equilibrium state that the thermodynamic properties have any real meaning. By definition:

> A system is in thermodynamic equilibrium if it is not capable of a finite, spontaneous change to another state without a finite change in the state of the surroundings.

There are many types of equilibrium, all of which must be met to fulfill the condition of thermodynamic equilibrium. Included among these are thermal, mechanical, phase, and chemical equilibriums. Thermal and mechanical equilibriums are related to the concepts of temperature and pressure, respectively. Phase equilibrium deals with the lack of a tendency for the net transfer of one or more chemical species from one phase to another, when the phases are in contact. A mixture of substances is in chemical equilibrium if there is no tendency for a net chemical reaction to occur.

A *process* is any transformation of a system from one equilibrium state to another. A complete description of a process typically involves specification of the initial and final equilibrium states, the path (if identifiable), and the interactions which take place across the boundaries of the system during the process. *Path* in thermodynamics refers to the specification of a series of states through which the system passes. There are several processes in thermodynamics of special interest. These are processes during which one property remains constant, and they are designated by the prefix iso- before the property. Table 1-3 is a list of the more important ones found in this text. The last two properties in this list

Table 1-3 A list of special constant-property processes

Property held constant	Name of process
Temperature	Isothermal
Pressure	Isobaric (isopiestic)
Volume	Isometric (isovolumic)
Entropy	Isentropic
Enthalpy	Isenthalpic

may be unfamiliar to the student. They will be described in detail later in the text. A *cyclic* process is defined as a process with identical end states. The change in the value of any property for a cyclic process is zero.

Any property of a thermodynamic system has a fixed value in a given equilibrium state, regardless of how the system arrives at that state. Therefore the change that occurs in the value of a property when a system is altered from one equilibrium state to another is always the same. This is true regardless of the method used to bring about a change between the two end states. The converse of this statement is equally true. If a measured quantity always has the same value between two given states, that quantity is a measure in the change in a property.

The uniqueness of a property value for a given state can be described mathematically in the following manner. The integral of an *exact differential* dY is given by

$$\int_1^2 dY = Y_2 - Y_1 = \Delta Y \tag{1-2}$$

Thus the value of the integral depends solely on the initial and final states. But the change in the value of a property likewise depends only on the end states. Hence the differential change dY in a property Y is an exact differential. Throughout this text the infinitesimal variation of a property will be signified by the differential symbol d preceding the property symbol. For example, the infinitesimal change in the pressure P of a system is given by dP. The finite change in a property is denoted by the symbol Δ (capital delta), for example, ΔP. The change in a property value ΔY always represents the final value minus the initial value. This convention must be kept in mind.

Properties are classified as either extensive or intensive. Consider a system arbitrarily divided into a group of subsystems. A property is *extensive* if its value for the whole system is the sum of its values for the various subsystems or parts. If a system is divided into n subsystems, then the extensive property Y for the whole system is given by

$$Y_{\text{system}} = \sum_{i=1}^{n} Y_i$$

where Y_i is the property value for the ith subsystem. Examples of extensive properties include volume V, energy E, and quantity of electric charge Q_e. Unlike extensive properties, *intensive* properties have values which are independent of the extent or mass of the system. If a single-phase system in equilibrium is divided arbitrarily into a set of subsystems, the value of a given intensive property will be the same for each macroscopic subsystem. Thus intensive properties have the same value at every small region or point within a system at equilibrium. Examples of intensive properties include temperature, pressure, density, velocity, and concentration. When an extensive property is divided by the total mass of a system, the resulting property is called a *specific* property. For systems in equilibrium a specific property is an intensive property of the system. Examples include specific volume ($v = V/m$) and specific energy ($e = E/m$). Capital letters will be used, generally, to denote extensive properties (with mass m a major exception). Lowercase letters will be used for intensive (specific) properties, the most notable exceptions being pressure P and temperature T. These latter two properties are always intensive.

1-5 DIMENSIONS AND UNITS

Dimensions are those names which are used to characterize physical quantities. Common examples of dimensions include length L, time t, force F, mass m, electric charge Q_e, and temperature T. In engineering analysis any equation which relates physical quantities must be dimensionally homogeneous. Dimensional homogeneity requires that the dimensions of terms on one side of an equation equal those on the other side. Such homogeneity also is retained by an equation during any subsequent mathematical operation, and thus is a powerful tool for checking the internal consistency of an equation.

In order to make numerical computations involving physical quantities, there is the additional requirement that units, as well as the dimensions, be homogeneous. *Units* are those arbitrary magnitudes and names assigned to dimensions which are adopted as standard for measurements. For example, the primary dimension of length may be measured in units of feet, miles, centimeters, angstroms, etc. These are all arbitrary lengths which may be related to each other in terms of unit conversion factors, or unitary constants. Unit conversion factors include 12 in = 1 ft and 60 s = 1 min. A comparable way of writing unit conversion factors is

$$\frac{12 \text{ in}}{1 \text{ ft}} = 1 \qquad \frac{60 \text{ s}}{1 \text{ min}} = 1$$

Terms in equations can always be multiplied by unit conversion factors, since it is always permissible to multiply by unity. A given physical quality may often be measured in several sets of units; it is therefore important to have available tables of common unit conversion factors. These are presented in most engineering and scientific handbooks, such as Marks's handbook (mechanical engineering),

Perry's handbook (chemical engineering), the Handbook of Chemistry & Physics, and others. A brief tabulation of unit conversion factors useful for thermodynamic problems met in this text is presented in Tables A-1 and A-1M, in the Appendix.

A number of systems of units have been developed over the years. However, in this text we shall consider only two: the SI and the United States Customary system (USCS).

a The Système Internationale (International System)

The fundamental system of units chosen for scientific work all over the world is the Système Internationale, which is usually abbreviated as SI. The SI employs seven primary dimensions: mass, length, time, temperature, electric current, luminous intensity, and amount of substance. Table 1-4 lists the SI units which are standards for these dimensions. All these units are defined operationally. The SI unit of length is the meter (m), and is defined as 1,650,763.73 wavelengths of the orange-red line of emission from krypton-86 atoms in vacuum. (Previously the accepted standard was a distance marked on a platinum-iridium bar maintained under prescribed conditions in Sèvres, France.) The unit of time is the second (s), and it is defined as the duration of 9,192,631,770 $\pm$ 20 cycles of a specified transition within the cesium atom. The previous legal standard for time was the mean solar second, which corresponds to 1/86,400 of the mean solar day. The SI unit of mass is the kilogram (kg), and it is represented by the mass of a platinum-iridium cylinder also preserved in Sèvres, France, at the International Bureau of Weights and Measures. The unit of temperature is the kelvin (K), but in this text it will be known as the °K. The origin of this unit is discussed more fully in Sec. 1-7. There are also precise operational definitions of the ampere (A), the candela (cd), and the mole (mol).

A mole is the amount of substance containing the same number of particles as there are atoms in 0.012 kg of the pure carbon nuclide ^{12}C. It has heretofore been frequently used as a unit for mass. For example, the gram-mole used by chemists is that amount of mass which contains as many atoms or molecules of a substance as the number of atoms in 0.012 kg (12 g) of carbon-12. A kilogram-

Table 1-4 SI and USCS base units

Physical quantity	SI unit and symbol	USCS unit and symbol
Mass	kilogram (kg)	pound-mass (lb_m)
Length	meter (m)	foot (ft)
Time	second (s)	second (s)
Temperature	kelvin (K)	Rankine (°R)
Electric current	ampere (A)	ampere (A)
Luminous intensity	candela (cd)	candle
Amount of substance	mole (mol)	
Force		pound-force (lb_f)

mole of a substance is 1000 times as large as a gram-mole. For example, a kilogram-mole of diatomic oxygen O_2 contains 32 kg of oxygen.

All other units in SI are secondary ones, and are derivable in terms of these seven primary or base units. The SI unit of force is the newton (N), and it is derived from Newton's second law, $F = ma$. On the basis of this equation, a force of one newton accelerates one kilogram of mass at one meter per second per second. Since $1\text{ N} = 1\text{ kg} \times 1\text{ m/s}^2$, then

$$1\text{ N} \cdot \text{s}^2/\text{kg} \cdot \text{m} = 1 \tag{1-3}$$

Equation (1-3) is a unit conversion factor which relates the derived force unit to the primary mass, length, and time units in the SI. It is useful in converting other secondary units, which are partially defined in terms of force, into a set of units containing only the kilogram, the meter, and the second. For example, the SI unit for pressure is the pascal (Pa), which is defined in force and length units as 1 N/m^2. Employing Eq. (1-3) we find that

$$1\text{ Pa} = 1\text{ N/m}^2 \times \frac{\text{kg} \cdot \text{m}}{1\text{ N} \cdot \text{s}^2} = 1\text{ kg/m} \cdot \text{s}^2$$

As a result, the secondary unit of pressure, the pascal, can be expressed in terms of its primary equivalent, $1\text{ kg/m} \cdot \text{s}^2$. Table 1-5 lists some SI secondary or derived units in terms of the primary or base units found in Table 1-4.

Special decimal multiples are designated by certain prefixes in SI units. Table 1-6 lists the values of the multiplier, the name of the prefix, and the symbol for the prefix. Although multiples of 10^3 are used in the SI system, other multiples are sometimes employed in specific fields.

A practical example of the use of the unit conversion factor relating force to mass is the conversion of a set of primary units into a secondary unit. Consider the concept of linear kinetic energy, which is derived from Newton's second law. The familiar result is the term $mV^2/2$, or simply $V^2/2$ for the kinetic energy per unit mass. In primary units this latter quantity is expressed as m^2/s^2. However, such usage is unconventional, since the secondary unit for energy in the SI is the joule (J). Conversion to the joule is accomplished through the use of Eq. (1-3). For linear kinetic energy per unit mass

$$\frac{V^2}{2} = \text{m}^2/\text{s}^2 \times 1\text{ N} \cdot \text{s}^2/\text{kg} \cdot \text{m} = \text{N} \cdot \text{m/kg} = \text{J/kg}$$

Since it is more customary to evaluate energy in terms of kJ/kg or J/g, the following conversion is appropriate:

$$\text{KE} = \frac{V^2}{2}\text{ m}^2/\text{s}^2 \times 1\text{ N} \cdot \text{s}^2/\text{kg} \cdot \text{m} \times \frac{\text{kJ}}{10^3\text{ N} \cdot \text{m}} = \frac{V^2}{2000}\text{ kJ/kg} \quad (\text{or J/g})$$

The gravitational potential energy relative to the earth's surface is given by gz. If g is in m/s^2 and the elevation relative to some datum is measured in meters, then the potential energy will be expressed in $\text{N} \cdot \text{m/kg}$ or J/kg. Consequently, the value of gz must be divided by 1000, much as in the case of kinetic energy, if the energy is to be in units of kJ/kg or J/g. That is,

$$PE = gz\ \text{m}^2/\text{s}^2 \times 1\ \text{N} \cdot \text{s}^2/\text{kg} \cdot \text{m} \times \frac{\text{kJ}}{10^3\ \text{N} \cdot \text{m}} = \frac{gz}{1000}\ \text{kJ/kg} \quad (\text{or J/g})$$

Recall that the value of g on earth's surface is roughly 9.8 m/s^2.

Table 1-5 Some SI and USCS derived units

SI (metric)			
Physical quantity	Unit	Symbol	Definition
Force	newton	N	$1\ \text{kg} \cdot \text{m/s}^2$
Pressure	pascal	Pa	$1\ \text{kg/m} \cdot \text{s}^2 (= 1\ \text{N/m}^2)$
	bar	bar	$10^5\ \text{kg/m} \cdot \text{s}^2 (= 10^5\ \text{N/m}^2)$
Energy	joule	J	$1\ \text{kg} \cdot \text{m}^2/\text{s}^2\ (= 1\ \text{N} \cdot \text{m})$
Power	watt	W	$1\ \text{kg} \cdot \text{m}^2/\text{s}^3\ (= 1\ \text{J/s})$
Electric quantity	coulomb	C	$1\ \text{A} \cdot \text{s}$
Electric potential difference	volt	V	$1\ \text{kg} \cdot \text{m}^2/\text{A} \cdot \text{s}^3\ (= 1\ \text{A} \cdot \Omega)$
Electric resistance	ohm	Ω	$1\ \text{kg} \cdot \text{m}^2/\text{A}^2 \cdot \text{s}^3\ (= 1\ \text{V/A})$
Electric capacitance	farad	F	$1\ \text{A}^2 \cdot \text{s}^4/\text{kg} \cdot \text{m}^2\ (= 1\ \text{C/V})$

USCS (engineering)			
Physical quantity	Unit	Symbol	Definition
Force	pound-force	lb_f	$32.174\ \text{lb}_m \cdot \text{ft/s}^2$
Pressure	atmosphere	atm	$68{,}087\ \text{lb}_m/\text{ft} \cdot \text{s}^2$ $(= 14.696\ \text{lb}_f/\text{in}^2)$
Energy	foot-pound-force	$\text{ft} \cdot \text{lb}_f$	$32.174\ \text{lb}_m \cdot \text{ft}^2/\text{s}^2$
Power	foot-pound-force/second	$\text{ft} \cdot \text{lb}_f/\text{s}$	$32.174\ \text{lb}_m \cdot \text{ft}^2/\text{s}^3$ $(= 1.82 \times 10^{-3}\ \text{hp})$

Table 1-6 Names and symbols for common multipliers of SI units

Multiplier	Prefix	Symbol
10^9	giga	G
10^6	mega	M
10^3	kilo	k
10^{-1}	deci	d
10^{-2}	centi	c
10^{-3}	milli	m
10^{-6}	micro	μ
10^{-9}	nano	n

In conjunction with Newton's second law it is important to note that *weight* always refers to a force. When it is stated that a body weighs a given amount, what we really mean is that this is the force with which the body is attracted toward another body, such as the earth or moon. The acceleration of gravity varies with the distance between two bodies (such as the height of a body above the earth's surface). Thus the weight of a body varies with elevation, while the mass of a body is constant with elevation.

Example 1-1M The weight of a piece of metal is 100.0 N at a location where the local acceleration of gravity g is 10.60 m/s^2. What is the mass of the metal, in kilograms, and what is the weight of the metal on the surface of the moon where $g = 1.67$ m/s^2?

SOLUTION In this case we write Newton's second law as $F = mg$, since $a = g$. Hence,

$$m = \frac{F}{g} = \frac{100.0\ \text{N}}{10.60\ \text{m/s}^2} = 9.434\ \frac{\text{N}}{\text{m/s}^2} = 9.434\ \text{kg}$$

The mass of the piece of metal will remain the same regardless of its location. The weight will change, however, with a change in gravitational acceleration. Equating weight to force on the surface of the moon,

$$\text{Weight} = F_{\text{moon}} = mg = 9.434\ \text{kg} \times 1.67\ \text{m/s}^2 = 15.8\ \text{N}$$

Although the mass is the same at the two locations, the weight is significantly different.

b The United States Customary System of Units

An important system of units employed in the United States is the USCS. Unfortunately, it introduces two points of confusion. First of all, pound is used to denote both a unit of force (lb_f) and a unit of mass (lb_m). If the symbol "lb" is found in this text without a subscript, it should always be taken as a mass unit. Second, both force and mass are chosen as primary dimensions. Consequently there are four primary dimensions to be used in conjunction with Newton's second law, rather than the three (mass, length, and time) used in the SI. Independent of Newton's second law, the standard gravitational force is defined such that a force of one pound (lb_f) will accelerate a mass of one pound (lb_m) at a rate of 32.1740 ft/s^2. Consequently we must write Newton's second law in the form

$$F = \frac{ma}{g_c}$$

when we use the USCS units, where g_c is a proportionality constant. Substituting the definition of a pound force into this modified form of Newton's law, we find that

$$1\ \text{lb}_f = \frac{1\ \text{lb}_m (32.174\ \text{ft/s}^2)}{g_c}$$

or that

$$g_c = 32.174\ \text{lb}_m \cdot \text{ft/lb}_f \cdot \text{s}^2$$

Hence, in order to use the USCS units, Newton's second law must be written as

$$F = \frac{ma}{g_c} = \frac{ma}{32.174} \tag{1-4}$$

If, in this equation, the mass is expressed in lb_m and the acceleration in ft/s^2, then the value of the force will be in lb_f. Also keep in mind that weight is a force.

Example 1-1 The weight of a piece of metal is 220.5 lb_f at a location where the local acceleration of gravity g is 30.50 ft/s^2. What is the mass of the metal, in lb_m, and what is the weight of the metal on the surface of the moon, where $g = 5.48$ ft/s^2?

SOLUTION In this case we write Newton's second law as $F = mg/g_c$, since $a = g$. Hence,

$$m = \frac{g_c F}{g} = \frac{32.174 \text{ lb}_m \text{ft/lb}_f \cdot \text{s}^2 \times 220.5 \text{ lb}_f}{30.50 \text{ ft/s}^2} = 232.6 \text{ lb}_m$$

The mass of the piece of metal will remain the same regardless of its location. Its weight will change, however, with a change in gravitational acceleration. Equating weight to force on the surface of the moon,

$$\text{Weight} = F_{\text{moon}} = \frac{mg}{g_c} = \frac{232.6 \text{ lb}_m \times 5.48 \text{ ft/s}^2}{32.174 \text{ lb}_m \cdot \text{ft/lb}_f \cdot \text{s}^2} = 39.6 \text{ lb}_f$$

Although the mass is the same at the two locations, the weight is significantly different.

In the USCS we can also use the mole as a unit of mass. In this case it is called the pound-mole (lb-mol). Since the definitions of a mole is the same as noted earlier for the SI units, a pound-mole of diatomic oxygen will contain 32.0 lb_m. There are approximately 454 g in a pound-mass. Hence a pound-mole is equivalent to 454 g · mol or 0.454 kg · mol.

The definition of a pound force presented earlier allows us to develop a unit conversion factor relating the pound mass to the pound force. Since

$$1 \text{ lb}_f \equiv 32.174 \text{ lb}_m \cdot \text{ft/s}^2$$

it follows that

$$32.174 \text{ lb}_m \cdot \text{ft/lb}_f \cdot \text{s}^2 = 1 \tag{1-5}$$

Because the term on the left is equal to unity, it can be inserted into any equation without altering the validity of the equation. This unit conversion factor is useful when it is necessary to convert a particular set of units into a more convenient or conventional set. For instance, consider the linear kinetic energy term, $mV^2/2$. If the mass is given in lb_m and the velocity in ft/s (and neglecting the factor 2 which is unitless), then in the USCS, or English engineering system,

$$mV^2 = \text{lb}_m \times \text{ft}^2/\text{s}^2 = \text{lb}_m \cdot \text{ft}^2/\text{s}^2$$

Although the above units for mV^2 are correct, the more customary units for kinetic energy (KE) are ft · lb_f. We achieve the more desirable units by using Eq. (1-5) in the following manner.

$$\text{KE} = \frac{mV^2}{2} \times \text{lb}_m \cdot \text{ft}^2/\text{s}^2 \times \frac{\text{lb}_f \cdot \text{s}^2}{32.174 \text{ lb}_m \cdot \text{ft}} = \frac{mV^2}{2(32.174)} \text{ ft} \cdot \text{lb}_f$$

Thus, when the primary USCS units given in Table 1-5 are used to evaluate the linear kinetic energy, the resulting value must be divided by the unit conversion factor given by Eq. (1-5) if units of $\text{ft} \cdot \text{lb}_f$ are desired.

The kinetic energy quantity is not the only term which frequently requires the use of the unit conversion factor in Eq. (1-5), or g_c, in its evaluation. The gravitational potential energy mgz likewise requires a unit conversion when USCS units are used and $\text{ft} \cdot \text{lb}_f$ are the desired units. There are other relationships in the engineering literature which typically require the unit conversion factor given by Eq. (1-5). In this textbook the quantity g_c is omitted from all equations. One must check the units of all terms in an equation, and determine which ones might require some type of unit conversion in order to get all terms in the equation into a consistent set of units.

1-6 DENSITY, SPECIFIC WEIGHT, SPECIFIC VOLUME, AND PRESSURE

Macroscopic thermodynamics deals with systems containing a very large number of particles. It is convenient to consider matter within such systems as a continuum. That is, matter is regarded as distributed continuously throughout space. From a continuum viewpoint it is acceptable to speak of properties at a point. The *density* ρ in the vicinity of a point within a system is defined as the mass per unit volume, m/V. The *specific weight* w of a substance is defined as the weight per unit volume. The relation between the specific weight of a body and its density may be obtained from Newton's second law, $F = ma$. If both sides of this equation are divided by the volume of a substance, then $F/V = (m/V)a$, or

$$w = \rho g \tag{1-6a}$$

where g is the local acceleration of gravity. When the USCS of units is used, this equation must be modified to

$$w = \frac{\rho g}{g_c} = \frac{\rho g}{32.17} \tag{1-6b}$$

The specific weight of a substance is not truly a property of the substance, since the value of w depends upon the local acceleration of gravity. The *specific gravity* of a substance is defined as the ratio of its density to that of water at a specified temperature. The density of water at room temperature is close to 1.00 g/cm^3 (kg/L) or 62.3 lb_m/ft^3.

The *specific volume* v is defined as the reciprocal of the density. Thus $v = 1/\rho = V/m =$ volume/mass. The mass units are normally pounds or kilograms. In certain circumstances, however, it is useful to employ the mole as the unit of mass. This may be either a pound-mole or a gram-mole. The number of moles N of a substance is defined as $N = m/M$, where M is the molar mass. The *moler mass* is the mass of a substance which is numerically equal to its molecular weight in any mass unit expressed as mass per mole. For example, the molar mass of helium is 4.003 g/g · mol or 4.003 lb/lb · mol. Thus any property which is partially defined in terms of the mass of the system may be expressed in two ways. As an example, the density might be a mass density or a molar density. (They differ from each other only in that their ratio is a constant.) No special symbol will be used to differentiate between these two ways of expressing properties in terms of mass units. They will be distinguished either by context and usage or by a definite comment as to which is being employed.

The *pressure* P is defined as the normal force per unit area acting on some real or imaginary boundary. Normal forces in static equilibrium will always be considered as compressive; hence the pressure is a positive quantity. In static systems the pressure is uniform in all directions around the vicinity of an elemental volume of fluid. It is well known, however, that the pressure may vary throughout the system for a fluid in the presence of a gravitational field. A familiar example is the variation of the pressure with depth in the common swimming pool.

Pressure differences are frequently measured in terms of the height of a liquid column. Consider the situation shown in Fig. 1-3*a*. A tube containing a liquid is connected to a tank which is filled with a gas at a uniform pressure P_1. A pressure P_2 exists outside the tank, and this pressure is also applied to the top of the liquid column. The pressure differential $P_1 - P_2$ can be determined from a knowledge of the height Δz of the liquid column. The differential height dz shown in Fig. 1-3*a* is shown as a fluid element in Fig. 1-3*b*. Three forces act on the fluid element in the z direction. Two of these are normal compressive forces, and the third is the weight of the element in the gravitational field g. On the basis of a force balance for the static situation,

$$\rho(A\,dz)g + (P + dP)A = PA$$

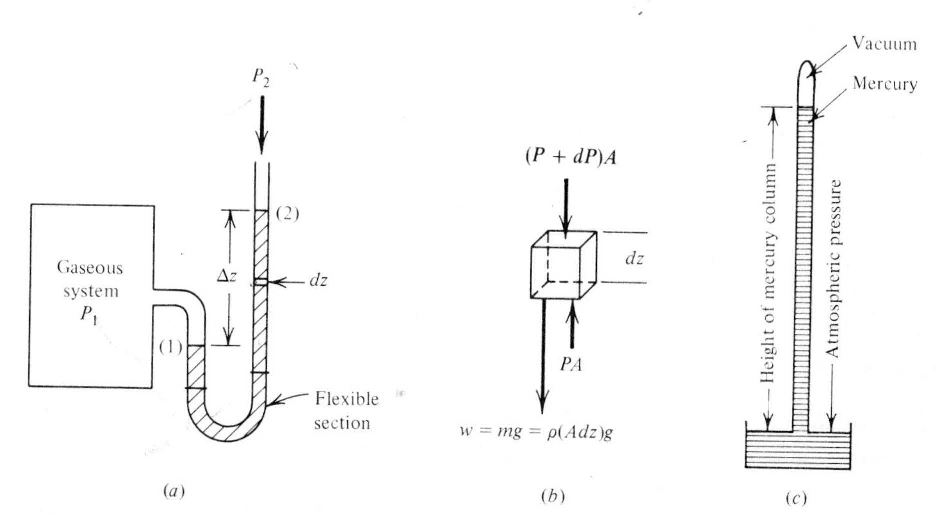

Figure 1-3 (*a*) Measurement of a pressure differential in terms of a height of a liquid column; (*b*) force balance on a fluid element within the liquid column; (*c*) device for measuring the barometric pressure.

where A is the cross-sectional area normal to the z direction. This equation reduces to $dP = -\rho g\, dz$. If the gravitational acceleration g and the density of the fluid ρ are assumed to be constant, then integration of this expression shows that

$$\Delta P = P_2 - P_1 = -\rho g\, \Delta z = -w\, \Delta z \tag{1-7}$$

The negative sign results from the convention that the height z is measured as positive upward, whereas P decreases in this direction. When English engineering units are employed, the term w must be evaluated by means of Eq. (1-6b) if ΔP is to be expressed as force per unit area.

The actual pressure at a given position in a system is called its *absolute* pressure. The modifying adjective is necessary since, experimentally, most pressure-measuring devices indicate what is known as gage or vacuum pressure. Absolute pressures must be used in thermodynamic relationships. A positive *gage* pressure is the difference between the absolute pressure and the pressure exerted by the atmosphere at that location. That is,

$$P_{\text{gage}} = P_{\text{abs}} - P_{\text{atm}} \tag{1-8}$$

A negative gage pressure (which occurs when the atmospheric pressure is greater than the absolute pressure) is called a *vacuum.* In order to make a vacuum measurement a positive value it is defined as

$$P_{\text{vacuum}} = P_{\text{atm}} - P_{\text{abs}} \tag{1-9}$$

Note that positive and negative gage pressures are pressure differentials. For pressures below and slightly above atmospheric this pressure differential is frequently measured by the method shown in Fig. 1-3a. The device is known as a manometer, and the liquid it contains may be mercury, alcohol, water, oil, or some other fluid. The pressure differences expressed by Eqs. (1-8) and (1-9) can be evaluated by using Eq. (1-7). The absolute pressure within a system is then determined by combining the evaluation of the pressure difference with an independent measurement of the atmospheric pressure. Figure 1-4 illustrates the relations among the various types of pressures.

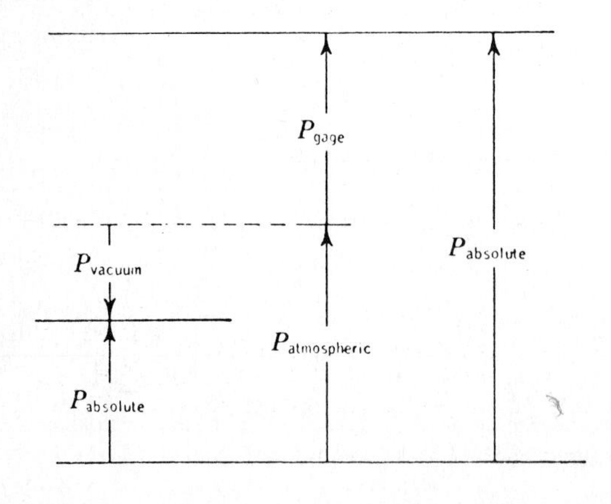

Figure 1-4 The relationships among the absolute, atmospheric, gage, and vacuum pressures.

Atmospheric pressure is commonly spoken of as the *barometric* pressure. Its value is not fixed, but varies with time and location on the earth. A sketch of a barometer for measuring the barometric or atmospheric pressure is shown in Fig. 1-3*c*. One reference value for pressure frequently used is the *standard atmosphere*. It is defined as the pressure produced by a column of mercury exactly 760 mm in height at 273.15°K or 0°C(32°F) and under standard gravitational acceleration. Some equivalent values in other units are as follows:

$$1 \text{ standard atmosphere (atm)} = \begin{cases} 14.696 \text{ lb}_f/\text{in}^2 \text{ (psi)} \\ 29.92 \text{ inHg at } 0^\circ \text{ C} \\ 1.013 \times 10^5 \text{ N/m}^2 = 1.013 \text{ bar} \end{cases}$$

In order to distinguish between absolute and gage pressures in this text, it should be assumed that data are absolute values unless denoted explicitly or implicitly as gage values. In the technical literature the letters *a* and *g* are frequently added to abbreviations in order to signify the difference. For example, the absolute and gage pressures in pounds per square inch are denoted by the terms "psia" and "psig," respectively. In some circumstances the symbols "lb/in^2 gage" or "mbar gage" may be used.

As previously mentioned, the standard pressure unit in SI is the pascal (Pa), which is defined as 1 N/m^2. For tabulating or reporting data for substances, the kilopascal (kPa) or megapascal (MPa) is commonly used, since the pascal is a relatively small pressure unit. In this text the *bar* is used most frequently as the SI pressure unit and is equal to 10^5 N/m^2. As noted above, 1 bar is slightly smaller than 1 atm.

Example 1-2M A barometer reads 735 mmHg. Determine the atmospheric pressure in millibars by employing Eq. (1-7).

SOLUTION The use of Eq. (1-7) requires data on ρ and g. The density of mercury at 0°C is 13.595 g/cm^3, and we shall assume this value is relatively independent of temperature. Since the location is not specified, we shall also assume that g is the standard value of 9.807 m/s^2. With reference to Fig. 1-3*c*, for a barometer P_1 is atmospheric pressure and P_2 is a complete vacuum, that is, $P_2 = 0$. On the basis of Eq. (1-7),

$$\begin{aligned} P_{\text{atm}} = P_1 &= 13.595 \text{ g/cm}^3 \times 9.807 \text{ m/s}^2 \times 735 \text{ mm} \times 10^{-3} \text{ m/mm} \\ &\quad \times 10^6 \text{ cm}^3/\text{m}^3 \times 1 \text{ N} \cdot \text{s}^2/\text{kg} \cdot \text{m} \times 10^{-3} \text{ kg/g} \\ &= 0.980 \times 10^5 \text{ N/m}^2 = 0.980 \text{ bar} = 980 \text{ mbar} \end{aligned}$$

This value is slightly less than a standard atmosphere, which is 1013 mbar.

Example 1-2 A barometer reads 28.90 inHg. Determine the atmospheric pressure in lb$_f$/in^2 (psi) by employing Eq. (1-7).

SOLUTION The use of Eq. (1-7) requires data on ρ and g. The density of mercury at 32°F (which is assumed to be constant) is 13.595 g/cm^3. Since the location is not known, we shall assume that g is the standard value of 32.174 ft/s^2. With reference to Fig. 1-3*c*, for a barometer P_1 is atmospheric peressure and P_2 is zero, that is, a vacuum. First, on the basis of Table A-1,

$$\rho = 13.595 \text{ g/cm}^3 \times (1 \text{ lb/ft}^3/0.01602 \text{ g/cm}^3) = 849 \text{ lb/ft}^3$$

Application of Eq. (1-7) then yields

$$P_{\text{atm}} = 849\ \text{lb/ft}^3 \times 32.174\ \text{ft/s}^2 \times 28.90\ \text{in}$$
$$\times (\text{ft}^3/1728\ \text{in}^3) \times (\text{lb}_f \cdot \text{s}^2/32.174\ \text{lb}_m \cdot \text{ft})$$
$$= 14.20\ \text{lb}_f/\text{in}^2$$

Example 1-3M If the barometer reads 735 mmHg, determine what absolute pressure, in bars, is equivalent to a vacuum of 280 mmHg within a system. Neglect effect of temperature on the density of mercury.

SOLUTION A vacuum reading is the difference between the barometric or atmospheric pressure and the absolute pressure, on the basis of Eq. (1-9). In this case the absolute pressure within the system is equal to 735 − 280, or 455 mmHg. From Table A-1M it is noted that 760 mmHg = 1 atm = 1.013 bar at 0°C. Hence

$$P_{\text{abs}} = 455\ \text{mmHg} \times 1.013\ \text{bar}/760\ \text{mmHg} = 0.606\ \text{bar}$$

Example 1-3 If the barometer reads 29.1 inHg, determine what absolute pressure in psia is equivalent to a vacuum of 11 inHg within a system.

SOLUTION Since a vacuum reading is the difference between the barometric or atmospheric pressure and the absolute pressure, the absolute pressure within the system is equal to 29.1 − 11.0, or 18.1 inHg abs. From Table A-1 it is noted that 1 inHg is equal to 0.491 psi at 32°F. Assuming that the specific gravity of mercury is constant, we find that

$$P_{\text{abs}} = 18.1\ \text{inHg} \times 0.491\ \text{psia/inHg} = 8.9\ \text{psia}$$

Example 1-4M A pressure gage reads 1.60 bars when the barometer has a reading of 755 mmHg. Determine the absolute pressure of the system, in bars.

SOLUTION The absolute pressure is the sum of the gage pressure and the barometric pressure. Using the conversion factors from Table A-1M,

$$P_{\text{bar}} = 755\ \text{mmHg} \times 1.013\ \text{bars}/760\ \text{mmHg} = 1.01\ \text{bars}$$

$$P_{\text{abs}} = P_{\text{gage}} + P_{\text{bar}} = 1.60 + 1.01 = 2.61\ \text{bars}$$

This is equivalent to 2.58 atm.

Example 1-4 A pressure gage reads 23.0 psi when the barometer has a reading of 28.9 inHg. Determine the absolute pressure of the system.

SOLUTION The absolute pressure is the sum of the gage pressure and the barometric pressure. Therefore

$$P_{\text{bar}} = 28.9\ \text{inHg} \times 0.491\ \text{psi/inHg} = 14.1\ \text{psi}$$

$$P_{\text{abs}} = P_{\text{gage}} + P_{\text{bar}} = 23.0 + 14.1 = 37.1\ \text{psia}$$

1-7 TEMPERATURE AND THE ZEROTH LAW

The concept of an absolute temperature is tied in intimately with the second law of thermodynamics. Hence a discussion along these lines must be delayed until later. In this section we shall discuss some important characteristics of temperature. Consider two systems X and Y which initially are individually in equilibrium. If they are brought into contact through a common rigid boundary, there are two possible outcomes with respect to the final states of the systems. One possibility is for the states of X and Y to remain unchanged macroscopically. As a second possibility, both systems will be observed to undergo a change of state until each reaches a new equilibrium state. These changes of state are due to an interaction between X and Y. When two systems which are isolated from the local surroundings undergo no further change of state even when in contact through a common rigid boundary, they are said to be in *thermal equilibrium*. It is important to note that both systems change their state as they proceed toward thermal equilibrium.

It is a matter of experience that the value of a single property is sufficient to determine whether systems will be in thermal equilibrium when placed in contact through a common rigid boundary. This property is the *temperature* T. If an interaction occurs, the two systems involved are said to be of unequal temperatures. Such an interaction will continue until the temperatures of the two systems become equal and thermal equilibrium prevails.

Temperature is a property of great importance in thermodynamics, and its value can be obtained easily by indirect measurement with calibrated instruments. The temperature of a system is determined by bringing a second body, a thermometer, into contact with the system and allowing thermal equilibrium to be reached. The value of the temperature is found by measuring some temperature-dependent property of the thermometer. Any such property is called a *thermometric* property. Commonly used properties of materials employed in temperature sensing devices include:

1. Volume of gases, liquids, and solids
2. Pressure of gases at constant volume
3. Electric resistance of solids
4. Electromotive force of two dissimilar solids
5. Intensity of radiation (at high temperatures)
6. Magnetic effects (at extremely low temperatures)

On the basis of experimental observation, it is found that when each of two systems is in thermal equilibrium with a third system, they also will be in thermal equilibrium with each other. This statement is a thermodynamic postulate known as the *zeroth* law. Although it is a statement of common experience, it is not derivable from other laws or definitions. Historically the first and second laws were already numbered before the zeroth law was recognized as an independent postulate. Therefore it is called the zeroth law since in any logical presentation of thermodynamics it precedes the development of the first and second laws.

This law is of importance in the field of thermometry and in the establishment of empirical temperature scales. In practice, the third system in the zeroth law is a thermometer. It is brought into thermal equilibrium with a set of temperature standards and is calibrated. At some later time the thermometer is equilibrated with a system at an unknown temperature, and a value is determined. If thermal equilibrium exists during the calibration process and during the test of the system, the temperature of the system must be the same as that set by the calibration standards, on the basis of the zeroth law.

The absolute temperature scale used by scientists and engineers in the SI is the Kelvin scale. Since 1954 it has been recommended by an international conference that a reference state value of 273.16 be assigned on the Kelvin temperature scale to that state where solid, liquid, and gaseous water all coexist in equilibrium. A state in which three phases of a pure substance coexist in equilibrium is called a triple state of the substance. The triple state of water is 0.01°K higher than the freezing or ice point of water. Thus water freezes at 273.15°K. The Celsius temperature scale, in °C (formerly called the centrigrade scale), is related to the Kelvin scale by the relation

$$T_C = T_K - 273.15 \tag{1-10}$$

Two additional temperature scales are in common usage in this country—the Rankine and the Fahrenheit scale. The temperature in degrees Rankine (°R) is defined arbitrarily as 1.8 times the temperature in kelvins. Consequently,

$$T_R = 1.8T_K \tag{1-11}$$

In terms of temperature intervals, one sees that

$$1°\text{K} = 1.8°\text{R}$$

The triple-state temperature of water is, therefore, 491.69°R. The Fahrenheit scale T_F (°F) is defined as

$$T_F = T_R - 459.67 \tag{1-12}$$

It is fairly common to round off the last number of Eqs. (1-10) and (1-12) to 273 and 460, respectively, when making engineering calculations. A comparison of these four temperature scales is shown in Fig. 1-5. Familiarity with the interrelationships of these different scales is essential in carrying out thermodynamic calculations. In thermodynamic relationships a symbol T for temperature always implies °K or °R unless specifically stated in terms of the other two scales.

1-8 CONSTANT-VOLUME GAS THERMOMETER

Among the thermometric properties employed in the measurement of temperature is the pressure of a gas maintained at constant volume. On the basis of the experimental data shown in Fig. 1-6 we see that all gases exhibit the same value of Pv at a given temperature if the pressure is extremely low. Therefore the quantity Pv can be used as a thermometric property to measure the temperature, regardless of the nature of the gas. At extremely low pressures, Pv will vary

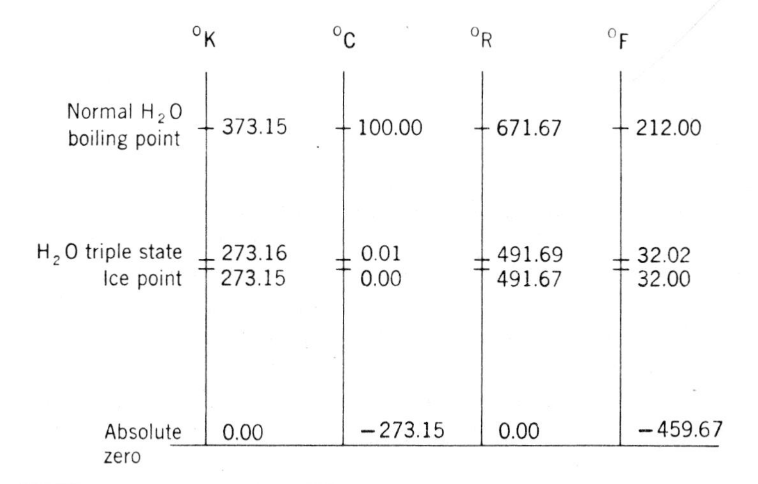

Figure 1-5 Comparison of temperature scales.

linearly with T to a high degree of accuracy. If a thermometer contains a low-pressure gas, then when the thermometer is placed in contact with two systems at different temperatures, we should find that in the limit as $P \to 0$,

$$\frac{T}{T^*} = \frac{Pv}{(Pv)^*}$$

where the asterisk value will be used to represent a reference state. If the gas in

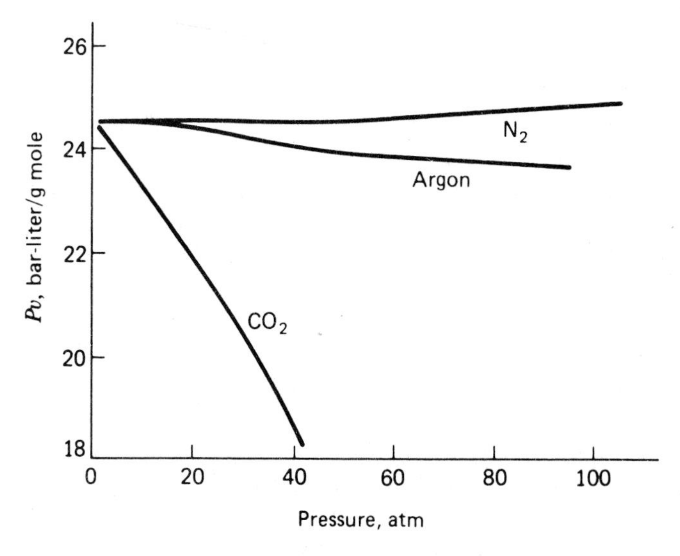

Figure 1-6 Experimental data showing the variation of Pv with pressure at a given temperature for several gases.

the thermometer is maintained at constant volume, the above equation reduces to

$$\frac{T}{T^*} = \frac{P}{P^*} \tag{1-13}$$

That is, when one brings a constant-volume-gas thermometer into contact with a system of interest, and then in contact with a system at a reference state, the ratio of the measured gas presssures enable one to evaluate the temperature of the actual system relative to an assigned value T^* at some reference state. If T^* is chosen to be 273.16, then the Kelvin temperature T_K at any other state as measured by a constant-volume-gas thermometer is given by

$$T_K = 273.16 \frac{P}{P^*} \tag{1-14}$$

Hence the temperature of a given state of a substance may be determined by measuring the value of P when the gas thermometer is in equilibrium with the substance and also measuring P^* when the thermometer is in equilibrium with the triple state of water.

The operation of a constant-volume-gas thermometer is illustrated in Fig. 1-7. The tube on the left is raised or lowered until the mercury level on the right side of the U-tube is at the indicated mark. Then the height z of the mercury column is a measure of the gage pressure of the gas within the thermometer bulb. The height z will vary as the temperature of the gas within the bulb changes.

Example 1-5M A constant-volume-gas thermometer containing nitrogen is brought into contact with a system of unknown temperature and then into contact with a system maintained at the triple state of water. The mercury column attached to the device has readings of 59.2 and

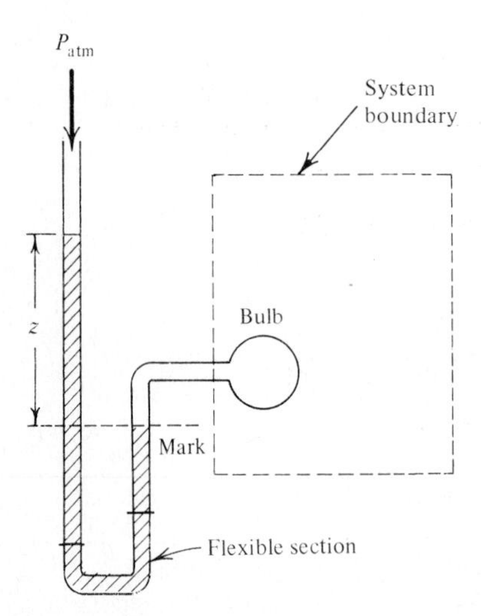

Figure 1-7 Constant-volume gas thermometer.

2.28 cm, respectively, for the two systems. If the barometric pressure is 960 mbar (96.0 kPa), what is the unknown temperature, in Kelvins, if $g = 9.806$ m/s^2?

SOLUTION The unknown temperature is found by applying Eq. (1–14) to the situation. The values of P and P^* must be found in absolute units. The values of the absolute pressures are found, in turn, from Eq. (1-8),

$$P_{\text{abs}} = P_{\text{gage}} + P_{\text{atm}}$$

The gage pressure due to Δz may be determined from Eq. (1-7). Thus a number of basic principles are involved here. Applying Eq. (1-7) with suitable conversion factors, we find that the gage pressures for the two cases are

$$P_{\text{gage}} \text{ (unknown temperature)} = 790 \text{ mbar } (79.0 \text{ kPa})$$

$$P_{\text{gage}} \text{ (triple-state temperature)} = 30.4 \text{ mbar } (3.04 \text{ kPa})$$

Therefore, the absolute pressures in the two cases are

$$P_{\text{abs}} \text{ (unknown temperature)} = 790 + 960 = 1750 \text{ mbar } (175 \text{ kPa})$$

$$P_{\text{abs}} \text{ (triple-state temperature)} = 30.4 + 960 = 990.4 \text{ mbar } (99.04 \text{ kPa})$$

Equation (1-14) then yields for the unknown absolute temperature,

$$T = 273.16 \frac{P}{P^*} = 273.16\left(\frac{1750}{990}\right) = 483°\text{K } (210°\text{C})$$

In the above calculation the specific gravity of mercury is taken to be 13.6.

Example 1-5 A constant-volume-gas thermometer containing nitrogen gas is brought into contact with a system of unknown temperature and then into contact with a system maintained at the triple state of water. The mercury column attached to the device has readings of 20.1 and 1.6 in, respectively, for the two systems. If the barometric pressure is 29.75 inHg, what is the unknown temperature, in degrees Fahrenheit, if $g = 32.0$ ft/s^2 ?

SOLUTION The unknown temperature is found by applying Eq. (1-14) to the situation. The values of P and P^* must be found in absolute units. The values of the absolute pressures are found, in turn, from Eq. (1-8)

$$P_{\text{abs}} = P_{\text{gage}} + P_{\text{atm}}$$

The gage pressure due to Δz may be determined from Eq. (1-7). Thus a number of basic principles are involved here. Applying Eq. (1-7) with suitable conversion factors, we find that

$$P_{\text{gage}} \text{ (unknown temperature)} = 9.82 \text{ psia} = 20.0 \text{ inHg}$$

$$P_{\text{gage}} \text{ (triple-state temperature)} = 0.78 \text{ psia} = 1.59 \text{ inHg}$$

Therefore, the absolute pressures in the two cases are

$$P_{\text{abs}} \text{ (unknown temperature)} = 20.0 + 29.75 = 49.75 \text{ inHg}$$

$$P_{\text{abs}} \text{ (triple-state temperature)} = 1.59 + 29.75 = 31.34 \text{ inHg}$$

Equation (1-14) then yields for the unknown absolute temperature,

$$T = 491.69 \frac{P}{P^*} = 491.69\left(\frac{49.75}{31.34}\right) = 780°\text{R} = 320°\text{F}$$

The value of 491.69 is the Rankine temperature at the triple state of water. In the above calculation the specific gravity of mercury is taken to be 13.6.

REFERENCES

Burghardt, M. D.; "Engineering Thermodynamics with Applications," Harper & Row, New York, 1982.
Kestin, J.: "A Course in Thermodynamics," Blaisdell, Waltham, Mass., 1966.
Obert, E. F., and R. A. Gaggioli: "Thermodynamics," McGraw-Hill, New York, 1963.
Zemansky, M. W.: "Heat and Thermodynamics," 5th ed., McGraw-Hill, New York, 1968.

PROBLEMS (METRIC)

Converter efficiencies

1-1M An automobile running steadily at 90 km/h for a period of 1.1 h consumes 11.5 L of gasoline. The energy content of the gasoline is 44,000 kJ/kg and its density is 0.75 g/cm^3. The power delivered to the wheels of the car is 28 kW.

(*a*) Determine the thermal efficiency of the engine, in percent, assuming the entire energy content of the fuel is released by combustion.

(*b*) Determine the fuel economy, in km/L.

1-2M Consider Prob. 1-1M, except that the power output is 30 kW and the fuel consumption is 17 L over a period of 1.53 h.

(*a*) Determine the thermal efficiency.

(*b*) Determine the fuel economy, in km/L.

1-3M A small industrial power plant generates 11,600 kg/h of steam, and has an output of 3800 kW. The plant consumes 1450 kg/h of coal, which has an energy content of 29,000 kJ/kg.

(*a*) Determine the overall plant thermal efficiency, in percent.

(*b*) If the energy added to the steam in the steam-generating unit is 2950 kJ/kg, what is the efficiency of the generating unit, in percent?

1-4M A steam power plant generates 180,000 kg/h of steam, and has an output of 55,000 kW of power. The plant consumes 19,500 kg/h of coal, which has an energy content of 30,000 kJ/kg.

(*a*) Determine the overall plant thermal efficiency, in percent.

(*b*) If the energy added to the steam in the steam-generating unit of the cycle is 2680 kJ/kg, what is the efficiency of the generating unit, in percent?

1-5M A home furnace uses 3.4 L/h of fuel oil, which has a density of 0.88 g/cm^3 and an energy content of 46,400 kJ/kg. The furnace is rated at 95,000 kJ/h of heat input to the room air. Determine the chemical to thermal converter efficiency, in percent.

1-6M Consider Prob. 1-5M, except that the fuel oil usage is 4.3 L/h and the furnace is rated at 124,000 kJ/h of heat input. Determine the chemical-to-thermal converter efficiency, in percent.

1-7M A jet engine consumes 9.00 kg/min of fuel which has a heating value of 41,800 kJ/kg. The velocity of the gas leaving the exit nozzle of the engine is 490 m/s. The ratio of air supplied to fuel consumed on a mass basis is 125 : 1.

(*a*) Determine the converter efficiency of chemical energy to kinetic energy.

(*b*) Determine the equivalent power output of the exit gas stream, in kilowatts.

1-8M Consider Prob. 1-7M, except that the jet engine consumes 0.30 kg/s of fuel and the exit gas velocity is 340 m/s.

(*a*) Determine the converter efficiency of chemical energy to kinetic energy.

(*b*) Determine the equivalent power output of the exit gas stream, in kilowatts.

1-9M Consider an air compressor which is rated as a 4475-kW unit and is driven by a natural-gas engine. Fuel efficient engine-compressor units have a fuel rating as low as 9050 kJ/h per kilowatt required. Assume a unit operates 8600 h/year, and that the heating or energy content of the fuel is 37,000 kJ/SCM (standard cubic meter).

(*a*) Determine the energy efficiency of the engine, that is, the percent of the energy input which is used to drive the compressor.

(*b*) If the cost of fuel is $150 per thousand standard cubic meters, determine the cost of operation over a one-year period.

(*c*) If another company offers a unit with a fuel rating of 9500 kJ/kW · h, determine the fuel savings, in dollars, if one uses the 9050 kJ/kW · h unit for a one-year period.

1-10M The chemical energy released in a hydrogen-oxygen fuel cell is 15,850 kJ/kg of fuel, and the electrical power produced is 1.0 kW. If the flow rate of fuel to the converter is 0.37 kg/h, determine the converter efficiency, in percent.

Intensive and extensive properties

1-11M Specify whether the following properties are intensive or extensive.

(*a*) Mass, (*b*), weight, (*c*) volume, (*d*) velocity, (*e*) density, (*f*) energy, (*g*) specific weight, (*h*) strain, (*i*) molar density, (*j*) mass concentration, (*k*) mole fraction, (*l*) pressure, (*m*) temperature, (*n*) surface area, (*o*) elevation, (*p*) potential energy, (*q*) stress.

1-12M Three cubic meters of air at 25°C and 1 bar have a mass of 3.51 kg.

(*a*) List the values of three intensive and two extensive properties for this system.

(*b*) If local gravity g is 9.65 m/s^2, evaluate another intensive property, the specific weight.

1-13M Four cubic meters of water at 25°C and 1 bar have a mass of 3990 kg.

(*a*) List the values of two extensive and three intensive properties of the system.

(*b*) If the local gravity g for the system is 9.7 m/s^2, evaluate the specific weight.

Force, mass, density, and specific weight

1-14M A body which has a mass of 7.0 kg is accelerated at a rate of 3.0 m/s^2. What total force is required if (*a*) the body is moving along a horizontal frictionless plane, and (*b*) the body is moving vertically upward at a location where local gravity is 9.45 m/s^2?

1-15M The acceleration of gravity as a function of elevation above sea level at 45° latitude is given by $g = 9.807 - 3.32 \times 10^{-6}\, z$, where g is in m/s^2 and z is in meters. Find the height, in kilometers, above sea level where the weight of a man will have decreased by (*a*) 1 percent, (*b*) 2 percent, and (*c*) 4 percent.

1-16M A mass of 2 kg is subjected to a vertical force of 25 N. The local gravity g is 9.60 m/s^2 and frictional effects are neglected. Determine the acceleration of the mass if the external vertical force is (*a*) downward, and (*b*) upward.

1-17M The density of a certain liquid is 0.80 g/cm^3. Determine the specific weight, in N/m^3, where local gravity g is (*a*) 2.50 m/s^2 and (*b*) 9.50 m/s^2.

1-18M On the surface of the moon where local gravity g is 1.67 m/s^2, 4.4 kg of a gas occupy a volume of 1.2 m^3. Determine (*a*) the specific volume of the gas, in m^3/kg, (*b*) the density, in g/cm^3, and (*c*) the specific weight, in N/m^3.

Pressure

1-19M Determine the pressure equivalent to 1 bar in terms of meters of a column of liquid at room temperature, where the liquid is (*a*) water, and (*b*) ethyl alcohol. The specific gravity of ethyl alcohol is 0.789.

1-20M If the density of mercury is 13.59 g/cm^3, determine the height, in centimeters, of a liquid column which would be equivalent to 1 bar.

1-21M The gage pressure within a system is equivalent to a height of 75 cm of a fluid with a specific gravity of 0.75. If the barometric pressure is 0.980 bar, compute the absolute pressure within the chamber, in millibars.

1-22M If the barometric pressure is 930 mbar, convert (*a*) an absolute pressure of 2.30 bars to a gage reading in bars, (*b*) a vacuum reading of 500 mbar to an absolute value in bars, (*c*) 0.70 bar absolute to millibars vacuum, and (*d*) a gage reading of 1.30 bars to kilopascals.

1-23M If the barometric pressure is 1020 mbar (102.0 kPa), convert (*a*) an absolute pressure of 1.70 bars to a gage reading in bars, (*b*) a vacuum reading of 600 mbar to an absolute value, in bars and kilopascals, (*c*) an absolute pressure of 0.60 bar (60 kPa) to millibars vacuum, and (*d*) a gage reading of 2.20 bars to an absolute pressure, in kilopascals.

1-24M A tank is divided by a rigid wall into two sections as shown in Fig. P1-24M. Pressure gage B reads 1.75 bars, and gage C reads 1.10 bars. If the barometric pressure is 0.97 bar, determine the reading on gage A, in bars.

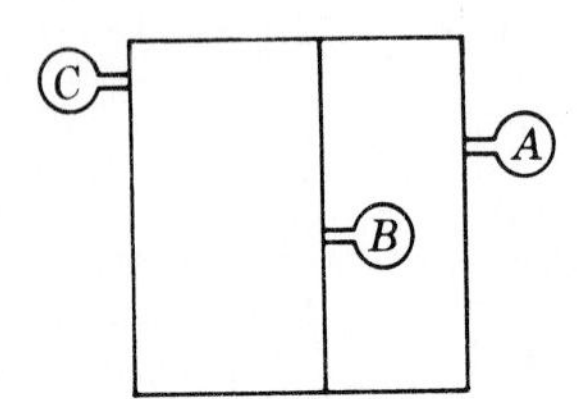

Figure P1-24M

1-25 Consider Prob. 1-24M, except that gage C reads 4.50 bars and gage B reads 2.20 bars. Determine the reading of gage A and convert this reading to an absolute value.

1-26M Consider Prob. 1-24M, except that gage C is a vacuum reading of 240 mbar and gage B reads 0.440 bar. Determine the reading on gage A, in bars.

1-27M A mountain climber carries a barometer which reads 950 mbar at the base point of his ascent. During his climb he takes three additional readings which are (*a*) 894 mbar, (*b*) 846 mbar, and (*c*) 765 mbar. Estimate the vertical distance, in meters, he has climbed from the base point if the average air density is assumed to be 1.20 kg/m^3. Neglect the effect of altitude on local gravity.

1-28M Determine the pressure exerted on a skin diver who has descended to (*a*) 10 m, and (*b*) 20 m below the surface of the sea if the barometric pressure is 0.96 bar at sea level and the specific gravity of seawater is 1.03 in this region of the ocean.

1-29M If the atmosphere is assumed to be isothermal at 20°C and follows the relationship $Pv = RT$ (an ideal gas), compute the pressure, in bars, and the density, in kg/m^3, at (*a*) 2500 m, and (*b*) 1000 m above sea level. The pressure and density at sea level are taken to be 1 bar and 1.19 kg/m^3, respectively.

Constant-volume gas thermometer

1-30M A constant-volume-gas thermometer is brought into contact with a system of unknown temperature T and then into contact with the triple state of water. The mercury column attached to the thermometer has readings of -10.7 and -15.5 cm, respectively. Determine the unknown temperature, in kelvins. The barometric pressure is 980 mbar (98.0 kPa), and the specific gravity of mercury is 13.6.

1-31M Consider Prob. 1-30M, except that the reading for the unknown temperature is $+10.7$ cm instead of -10.7 cm. Find the unknown temperature.

1-32M A constant-volume-gas thermometer is brought into contact with a system of unknown temperature T and then into contact with the triple state of water. The mercury column attached to the thermometer has readings of 22.2 and 2.2 cm, respectively. The barometric pressure is 975 mbar (97.5 kPa), and the specific gravity of mercury is 13.6. Find the value of the unknown temperature in kelvins.

1-33M Consider Prob. 1-32M, except that the reading for the triple state is -12.6 cm, and the reading for the unknown system is 39.6 cm. Determine the temperature, in kelvins.

1-34M A constant-volume-gas thermometer is used to measure the temperature of a system. The mass of the gas within the thermometer is varied, and the values of P and P^* are measured. On the basis of the following two sets of data for (*a*) system A and (*b*) system B, determine the value of the unknown temperature, in kelvins, for each case.

(*a*) System A	P, mbar	1200	1000	800	600	400	200
	P/P^*	1.89	1.81	1.74	1.68	1.63	1.59
(*b*) System B	P, kPa	60.0	50.0	40.0	30.0	20.0	10.0
	P/P^*	1.52	1.46	1.41	1.37	1.34	1.32

PROBLEMS (USCS)

Converter efficiencies

1-1 An automobile running steadily at 55 mi/h for a period of 1.3 h consumes 4.0 gal of gasoline. The energy content of the gasoline is 19,000 Btu/lb, and its density is 47 lb/ft^3. The power delivered to the wheels of the car is 40 hp.

(*a*) Determine the thermal efficiency of the engine, in percent, assuming the entire energy content of the fuel is released by combustion.

(*b*) Determine the fuel economy, in mi/gal.

1-2 Consider Prob. 1-1, except that the power output is 35 hp and the fuel consumed is 2.2 gal over a period of 0.73 h.

(*a*) Determine the thermal efficiency.

(*b*) Determine the fuel economy, in mi/gal.

1-3 A small industrial power plant generates 25,000 lb/h of steam, and has an output of 3800 kW of power. The plant consumes 3200 lb/h of coal, which has a heating value of 12,500 Btu/lb.

(*a*) Determine the overall plant thermal efficiency, in percent.

(*b*) If the energy added to the steam in the generating unit is 1150 Btu/lb, what is the efficiency of the steam-generating unit, in percent?

1-4 A steam power plant generates 400,000 lb/h of steam, and has an output of 55,000 kW of power. The plant consumes 42,800 lb/h of coal, which has a heating value of 13,000 Btu/lb.

(*a*) Determine the overall thermal efficiency of the plant, in percent.

(*b*) If the energy added to the steam in the steam-generating unit is 1150 Btu/lb, what is the efficiency of the generating unit, in percent?

1-5 A home furnace uses 0.90 gal/h of fuel oil, which has a density of 55 lb/ft^3 and a heating value of 20,000 Btu/lb. The furnace is rated at 90,000 Btu/h of heat input to the room air, which is heated by the furnace. Determine the chemical-to-thermal conversion efficiency, in percent.

1-6 Consider Prob. 1-5, except that the fuel oil usage is 1.2 gal/h and the furnace is rated at 110,000 Btu/h of heat input. Determine the chemical-to-thermal converter efficiency, in percent.

1-7 A jet engine consumes 2.0 lb/min of fuel which has a heating value of 18,000 Btu/lb. The velocity of the gas leaving the exit nozzle of the engine is 1200 ft/s. The ratio of air supplied to fuel consumed on a mass basis is 125 : 1.

(*a*) Determine the converter efficiency of chemical energy to kinetic energy.

(*b*) Determine the equivalent power output of the exit gas stream, in horsepower.

1-8 Consider Prob. 1-7, except that the jet engine consumes 0.50 lb/s of fuel and the exit gas velocity is 950 ft/s.

(*a*) Determine the converter efficiency of chemical energy to kinetic energy.

(*b*) Determine the equivalent power output of the exit gas stream, in horsepower.

1-9 Consider an air compressor which is rated as a 6000-hp unit, and is driven by a natural-gas engine. Fuel efficient engine-compressor units have a fuel rating as low as 6400 Btu/h per horsepower required. Assume a unit operates 8600 h/yr, and that the heating or energy content of the fuel is 1000 Btu/SCF (standard cubic foot).

(*a*) Determine the energy efficiency of the device, that is, the percent of the energy input which is used to drive the compressor.

(*b*) If the cost of fuel is $4.00 per thousand cubic feet, determine the cost of operation over a one-year period.

(*c*) If another company offers a unit with a fuel rating of 6700 Btu/hp · h, determine the fuel savings, in dollars, if one uses the 6400 Btu/hp · h unit for a one-year period.

1-10 The chemical energy released in a hydrogen-oxygen fuel cell is 6830 Btu/lb of fuel and the electrical power produced is 1.2 kW. If the flow rate of fuel to the converter is 0.98 lb/h, determine the converter efficiency, in percent.

Intensive and extensive properties

1-11 Specify whether the following properties are intensive or extensive.

(*a*) Mass, (*b*) weight, (*c*) volume, (*d*) velocity, (*e*) density, (*f*) energy, (*g*) specific weight, (*h*) strain, (*i*) molar density, (*j*) mass concentration, (*k*) mole fraction, (*l*) pressure, (*m*) temperature, (*n*) surface area, (*o*) elevation, (*p*) potential energy, and (*q*) stress.

1-12 Two cubic feet of air at 70°F and 14.6 psia have a mass of 0.149 lb.

(*a*) List the values of three intensive and two extensive properties.

(*b*) If the local gravity g is 31.2 ft/s^2, evaluate the specific weight.

1-13 Three cubic feet of water at 60°F and 14.7 psia have a mass of 187 lb.

(*a*) List the values of two extensive and three intensive properties of the system.

(*b*) If the local gravity g is 30.8 ft/s^2, evaluate the specific weight.

Force, mass, density, and specific weight

1-14 A body which has a mass of 20 lb$_m$ is acceleterated at a rate of 10 ft/s^2. What total force is necessary if (*a*) the body is moving along a horizontal frictionless plane, and (*b*) the body is moving vertically upward at a location where the local effect of gravity is 31.0 ft/s^2?

1-15 The acceleration of gravity as a function of elevation above sea level at 45° latitude is given by $g = 32.17 - 3.32 \times 10^{-6}z$, where g is in ft/s^2 and z is in ft. Find the height, in miles, above sea level where the weight of a man will have decreased by (*a*) 1 percent, and (*b*) 2 percent.

1-16 A mass of 5 lb$_m$ is subjected to an external vertical force of 7 lb$_f$. The local gravity g is 31.1 ft/s^2, and frictional effects are to be neglected. Determine the acceleration of the mass if the external vertical force is (*a*) downward, and (*b*) upward.

1-17 The density of a certain liquid is 50 lb$_m$/ft^3. Determine the specific weight, in lb$_f$/ft^3, where (*a*) g is 8.05 ft/s^2 and (*b*) g is 30.0 ft/s^2.

1-18 On the surface of the moon, where the local acceleration of gravity g is 5.47 ft/s^2, 8 lb$_m$ of a gas occupy a volume of 40 ft^3. Determine (*a*) the specific volume of the gas, in ft^3/lb$_m$, (*b*) the density, and (*c*) the specific weight, in lb$_f$/ft^3.

Pressure

1-19 Determine the pressure equivalent to 1 atm in terms of feet of a column of liquid at room temperature, where the liquid is (*a*) water, and (*b*) ethyl alcohol. The specific gravity of ethyl alcohol is 0.789.

1-20 If the specific gravity of mercury is 13.59, determine the height, in inches, of a liquid column which would be equivalent to 1 atm. Assume $g = 32.2$ ft/s^2.

1-21 The gage pressure within a system is equivalent to a height of 24 in of a fluid with a specific gravity of 0.75. If the barometric pressure is 29.5 inHg, compute the absolute pressure within the chamber, in psia.

1-22 If the barometric pressure is 30.15 inHg, convert (*a*) 35 psia to psig, (*b*) 20 inHg vacuum to inHg absolute and to psia, (*c*) 10 psia to inHg vacuum, and (*d*) 20 inHg gage to psia.

1-23 If the barometric pessure is 29.90 inHg, convert (*a*) an absolute pressure of 27.0 psia to psig, (*b*) a vacuum reading of 24.0 inHg to an absolute value, in inHg and psia, (*c*) an absolute pressure of 12.0 psia to inHg vacuum, and (*d*) a gage reading of 14.0 inHg to an absolute pressure, in psia.

1-24 A tank is divided by a rigid wall into two sections as shown in Fig. P1.24. Pressure gage *B* reads 24.8 psig and gage *C* reads 15 psig. If the barometric pressure is 29.5 inHg, determine the reading on gage *A*, in psig.

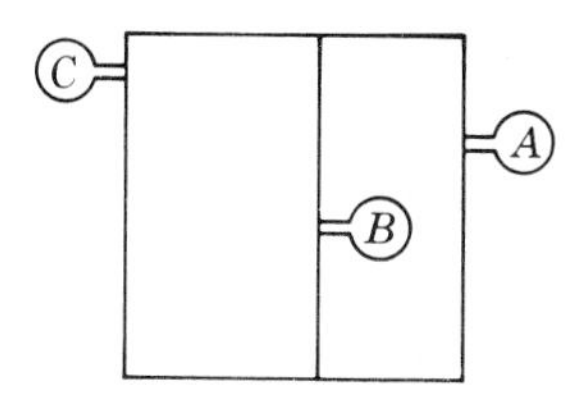

Figure P1-24

1-25 Consider Prob. 1-24, except that gage *C* reads 60 psig and gage *B* reads 35 psig. Determine the reading of gage *A* and convert this reading to an absolute value.

1-26 Consider Prob. 1-24, except that gage *C* has a vacuum reading of 10 psig, and gage *B* reads 18 psig. Determine the reading on gage *A*, in psig.

1-27 A mountain climber carries a barometer which reads 30.10 inHg at the base point of his ascent. During his climb he takes three additional readings which are (*a*) 29.05 inHg, (*b*) 27.74 inHg, and (*c*) 26.28 inHg. Estimate the vertical distance he has climbed from the base point, in feet, if the average air density is taken to be 0.074 lb/ft^3. Neglect the effect of altitude on local gravity.

1-28 Determine the pressue exerted on a skin diver who has descended to (*a*) 35 ft, and (*b*) 65 ft below the surface of the sea if the barometric pressure is 14.5 psia at sea level and the specific gravity of seawater is 1.03 in this region of the ocean.

1-29 If the atmospheric air is assumed to be isothermal at 60°F and follows the relationship $Pv = RT$ (an ideal gas), compute the pressure, in psia, and the density, in lb_m/ft^3, at (*a*) 7000 ft, and (*b*) 3000 ft above sea level. The pressure and density at sea level ate taken to be 14.7 psia and 0.077 lb_m/ft^3, respectively.

Constant-volume gas thermometer

1-30 A constant-volume-gas thermometer is brought into contact with a system of unknown temperature T and then into contact with the triple state of water. The mercury column attached to the thermometer has readings of -4.20 and -6.10 in, respectively. Determine the unknown temperature, in degrees Rankine. The barometric pressure is 29.20 inHg, and the specific gravity of mercury is 13.6.

1-31 Consider Prob. 1-30, except that the reading for the unknown temperature is $+4.20$ in, instead of -4.20 in. Find the unknown temperature.

1-32 A constant-volume-gas thermometer is brought into contact with a system of unknown temperature T and then into contact with the triple state of water. The mercury column attached to the thermometer has readings of 22.2 and 2.2 in, respectively. The barometric pressure is 29.80 inHg, and the specific gravity of mercury is 13.6. Find the value of the unknown temperature, in degrees Rankine.

1-33 Consider Prob. 1-32, except that the reading for the triple state is -2.6 in, and the reading for the unknown system is 19.6 in. Determine the temperature, in degrees Rankine.

1-34 A constant-volume-gas thermometer is used to measure the temperature of a system. The mass of the gas within the thermometer is varied, and the values of P and P^* are measured. On the basis of the following two sets of data for (*a*) system A, and (*b*) system B, determine the value of the unknown temperature, in degrees Rankine, for each case.

(*a*) System A

P, atm	1.20	1.00	0.80	0.60	0.40	0.20
P/P^*	1.89	1.81	1.74	1.68	1.63	1.59

(*b*) System B

P, inHg	60.0	50.0	40.0	30.0	20.0	10.0
P/P^*	1.42	1.36	1.31	1.27	1.24	1.22

CHAPTER

TWO

THE FIRST LAW OF THERMODYNAMICS AND THE STATE POSTULATE

2-1 INTRODUCTION

As stated earlier, the first law of thermodynamics is concerned with the study of energy transformations. In basic mechanics a few forms of energy, such as gravitational potential and linear kinetic energy, are examined. In electromagnetics other forms of energy, associated with electric and magnetic fields, are introduced. To the chemist the study of the energy associated with atomic and nuclear binding forces is extremely important. Thermodynamics relates these and other forms of energy and describes the change in energy of various types of systems in terms of interactions at the boundaries of the system. Thus thermodynamics is an all-encompassing field which ties together these many diverse areas, which are usually studied independently of one another. The basic laws of thermodynamics are, then, extremely broad in their scope. One of the most important laws of thermodynamics leads to a general conservation of energy principle. The law on which this conservation principle is based is called the first law of thermodynamics.

2-2 THE CONCEPT OF WORK AND THE ADIABATIC PROCESS

The concept of work is usually introduced in the study of mechanics. Mechanical work is defined as the product of a force F and the displacement s of the force when both are measured in the same direction. The expression for a differential quantity of work δW which results from a differential displacement ds is given by

$$\delta W = F\,ds = \mathbf{F} \cdot d\mathbf{s}$$

The last term in this equation is the vector notation for work, which is a scalar quantity. The total work for a finite displacement is obtained from the integration of $F\,ds$. This will require a functional relationship between F and s, since F usually will not be a constant.

In thermodynamics it is convenient to define the concept of work as follows: *Work* is an interaction between a system and its surroundings, and is done by a system if the sole effect external to the boundaries of the system could have been the raising of a weight. This definition leads to a broader interpretation of work than the purely mechanical one. In actual circumstances, effects may appear in the surroundings which do not have associated with them a recognizable force moving through a distance. Nevertheless, the real test is whether or not the interaction could have been carried out so that the sole external effect would have been the raising of a weight.

Consider the following example, illustrated in Fig. 2-1*a*. A closed system contains a pulley-weight system which is attached to a shaft which extends through the boundary of the system. Externally the shaft is attached to a paddle-wheel which rotates in a fluid as the weight is lowered within the system. Is the interaction at the boundary to be called work, in the light of our general definition? Note that the end result of the interaction is a "heating" of the fluid; that is, its temperature increases. One should not infer, however, that the interaction is a heat interaction until the operational test for work has been made. Figure 2-1*b* shows that the paddle-wheel apparatus could be replaced by another pulley-weight system, and the sole effect external to the system boundary would have been the raising of a weight. Hence the interaction at the boundary in this case must be called work. One should not use the result of an interaction as an

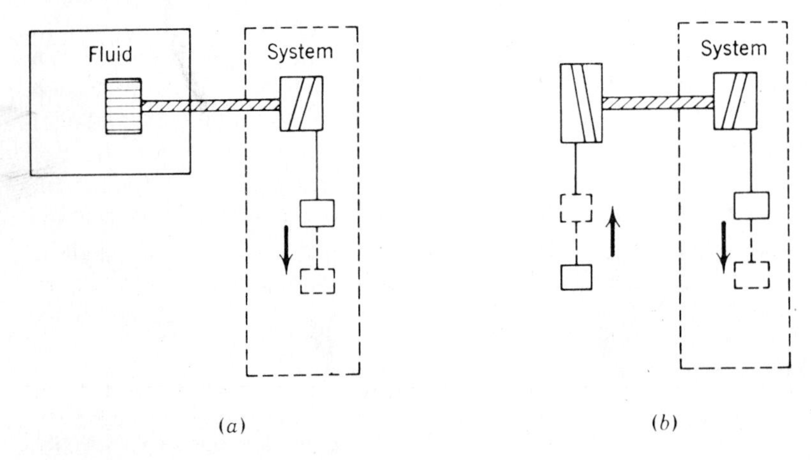

Figure 2-1 Illustration of paddle-wheel work.

intuitive measure of the nature of the interaction itself. Carefully note that a temperature rise of a system does not always imply that a heat interaction has occurred.

The value of a work interaction must necessarily be positive with respect to one system and negative with respect to the other. The sign convention adopted in this edition of the text is that work done *on* a system is positive. Likewise, work done *by* a system is given a negative number. This is a natural convention, in that quantities added are positive and quantities removed are negative.† Occasionally the sign on a work quantity will be omitted, and its direction indicated by assigning a subscript "in" or "out" to the symbol for work. The time rate at which work is done on or by a system is defined as *power*. Power is a scalar (nondirectional) quantity.

Any process which involves only work interactions is defined as an *adiabatic* process. In conjunction with this definition, an adiabatic boundary or surface is one which prevents or excludes all interactions except those which can be classified as work effects. Hence, an adiabatic boundary is one which thermally insulates a system, that is, prevents heat interactions.

2-3 WORK AS RELATED TO MECHANICAL AND ELECTROSTATIC FORMS OF ENERGY

A combination of the concept of work with Newton's second law of motion leads to the derivation of mechanical forms of energy. Consider the adiabatic change in position of a body of mass m in a gravitational field. The work required to bring about this change of state is obtained by integrating the expression $\delta W = F\,ds$. The force acting on the body is given by $F = mg$, where g is the local acceleration of gravity. Hence,

$$\delta W_{\text{ad}} = mg\,dz$$

where the distance s has been replaced by the coordinate z in the direction of the gravitational field. The subscript "ad" stands for adiabatic. If one assumes that the value of g is constant over the given change in elevation, then integration of the above equation leads to

$$W_{\text{ad}} = mg(z_2 - z_1)$$

Thus the work associated with an adiabatic change in position of a system is solely a function of the initial and final positions. In this restricted case, the work required is independent of the path. The quantity mzg is defined as the gravi-

† The sign convention that work done on a system is positive is not a universal convention. There are a number of books in various fields where the author employs the opposite sign convention. The reader must remain aware of the fact that a number of practicing engineers and scientists use the more traditional sign convention, namely, that work done by a system is positive.

tational potential-energy function, and we note in this special case that

$$\Delta(\text{gravitational potential energy}) = \Delta(\text{PE}) = W_{\text{ad}}$$

The above equation relates the adiabatic work done on a system to the change in the gravitational potential-energy function. The actual value of the function is arbitrary, a zero value of the potential-energy function being dependent upon the selection of some elevation where z is taken to be zero.

In a similar manner a system may undergo a change in velocity while other properties of the system, including its elevation, remain constant. The acceleration may be carried out adiabatically, and the work required is again measured by $\delta W = F\,ds$. According to Newton's second law the external force which is used to accelerate the system is given by $ma = m(dV/dt)$. For a differential time interval dt the distance ds moved by the system is simply $V\,dt$. Therefore

$$\delta W_{\text{ad}} = m\frac{dV}{dt}(V\,dt) = mV\,dV$$

Integration of the above expression yields

$$W_{\text{ad}} = \frac{m}{2}(V_2^2 - V_1^2)$$

The quantity $mV^2/2$ is called the linear, or translational, kinetic-energy function KE. Any change in its value is independent of the path of the process and can be measured by the adiabatic work required to bring about the change in velocity. It is convenient to define the kinetic energy as zero when the velocity is zero relative to some frame of reference.

Recall also from elementary physics that the adiabatic work required to move an electric charge through an electrostatic field from one position to another is independent of the path. Consequently, one can define a new function, the electrostatic potential-energy function, such that the adiabatic work done on moving a charge through an electrostatic field is measured by the change in the electrostatic potential-energy function. Symbolically, we may write that $\mathbf{F} = Q_e\mathbf{E}$ and $\delta W = \mathbf{F} \cdot d\mathbf{s}$. Therefore

$$W_{\text{ad}} = Q_e \int_i^f \mathbf{E} \cdot d\mathbf{s} = \Delta(\text{electrostatic potential energy})$$

where Q_e = charge
$\mathbf{E}$ = electrostatic field
$\mathbf{s}$ = distance

The three cases cited above have a number of common features, which are summarized below.

1. The work done on a system in each case has been equated to the change in a specific potential function, and each of the potential functions has been given the name of a specific form of energy.

2. Each form of energy has a value at a given state which is arbitrary, relative to some reference state.
3. Each relationship between work and a change in energy is derived for adiabatic conditions. That is, work is the only type of interaction allowed during the change of state. If these processes were nonadiabatic, properties of the system other than the elevation, velocity, or position would change during the process. In such cases the simple relationships derived above would not generally be valid.

This review of basic concepts from physics reveals that the concept of work frequently is related to specific forms of a quantity called energy. However, the relationships are valid only under adiabatic conditions. We shall next examine the nature of adiabatic changes of state for other types of processes than those discussed above.

2-4 THE FIRST LAW OF THERMODYNAMICS

In the preceding section we considered three special cases of adiabatic processes involving work interactions. In each separate case the value of the work interaction was independent of the path between given end states. These cases represented, however, processes with quite different overall changes of state. As a next step, we shall investigate the effect of different types of work interactions on a given system which all lead to the same change of state. To illustrate the point, consider a constant-volume closed system which undergoes a change of state between two given equilibrium states. Two different methods could be employed experimentally to carry out the specified change of state. Process A (see Fig. 2-2a) is carried out by allowing a paddle-wheel to rotate in the fluid within the system.

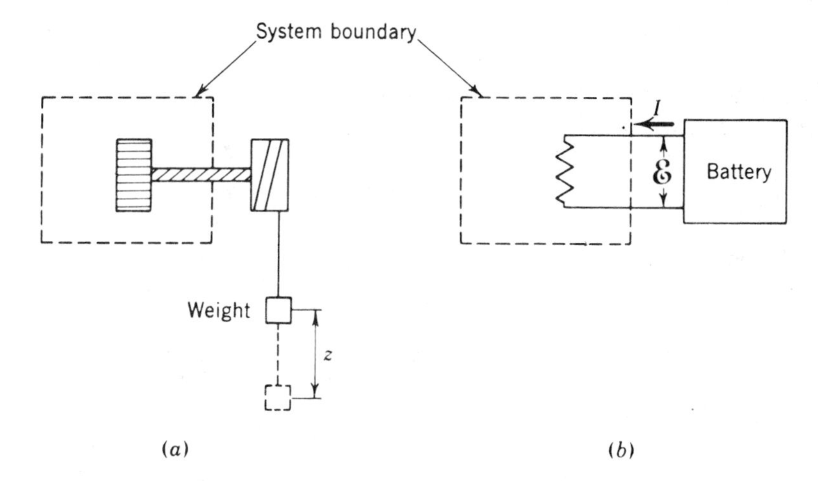

Figure 2-2 Two different work interactions on the same adiabatic, closed system.

If the rotation occurs by dropping a weight, as shown in the figure, then the work done on the system can be measured in terms of the quantity $mg\ \Delta z$, where m is the mass of the weight used and Δz is the distance through which the mass moves in the direction of the gravitational field. In process B (see Fig. 2-2*b*) an electrical resistor has been placed within the fluid, and it is connected through the boundaries of the system to an external battery. The amount of electrical work required is the product of the voltage times the charge delivered, or, in an equivalent manner, it can be represented by ξIt, where ξ is the cell voltage, I is the current, and t is the time.

Both processes, A and B, start at the same initial state and end at the same final state. In addition, one might also consider a process C where both paddle-wheel and electrical work are used to bring about the required change of state. It would be found that the work required is always the same for the described adiabatic processes between the same equilibrium states of the closed system. Thus one would find that the value of $mg\ \Delta z$ for process A would be identical to the value of ξIt for process B, within experimental accuracy, if each process were adiabatic. For process C the sum of the combination of paddle-wheel and electrical work would be the same as for the other two processes if we again require an adiabatic process.

On the basis of experimental evidence of this type, which began with the work of Joule in the middle of the nineteenth century, it is possible to make the following broad assertion. This postulate, called the *first law of thermodynamics*, states the following:

> When a closed system is altered adiabatically, the total work associated with the change of state is the same for all possible processes between the two given equilibrium states.

Stated another way, the value of the work carried out on or by a closed adiabatic system is dependent solely upon the end states of the process. This postulate is made regardless of the type of work interaction involved in the process, the type of process, and the nature of the closed system.

The first law of thermodynamics, together with those relationships derivable from it, are so well established that one does not question its validity when properly applied to an extensive range of scientific and engineering problems. Not only is it applicable to simple adiabatic processes such as those which involve paddle-wheel or electrical work in a closed system, but it may also be applied to systems which experience electromagnetic, surface tension, shear, and gravitational field effects. In an expanded formulation it can be applied to closed systems undergoing heat interactions as well as work interactions. Finally, the above generalization can be extended to open systems as well, where mass transfer is permitted across the boundaries of the system.

2-5 THE NATURE OF E, THE TOTAL ENERGY

It has been postulated that the work done is the same for all adiabatic processes which connect two equilibrium states of a closed system. This statement, the first law of thermodynamics, leads to a general definition of the change in the energy of a closed system between equilibrium states. Recall that the change in the value of a property is fixed by the end states and is independent of the process. If this is true, then specifying the end states will fix the change in values of all properties, regardless of the process between the two specified states. Consequently, it follows logically that any quantity which is fixed by the end states for all processes between those end states must be a measure of the change in the value of a property. Since the adiabatic work is solely a function of the end states for a closed-system process, the quantity of adiabatic work defines or measures the change in a new property function. We shall call this new function the energy E (or total energy) of a closed system.

As a consequence of the first law of thermodynamics we may write

$$W_{\text{ad}} \equiv E_2 - E_1 = \Delta E \tag{2-1}$$

where 1 and 2 indicate the initial and final states, and W is the work carried out for any adiabatic process between those two states. This equation is an operational definition of the change in the energy of a closed system between two equilibrium states. Equation (2-1) is a fundamental relation which enables one to evaluate the energy of a set of states in terms of adiabatic work interactions. The actual value of E can be established only by arbitrarily assigning a value for a particular substance at a reference state. However, this same procedure was necessary for the mechanical and electrostatic forms of energy discussed earlier.

There are numerous forms of energy which constitute the total energy of a macroscopic system. In addition to the gravitational and electrostatic potential energies discussed already with respect to gravitational and electric force fields, a potential energy related to magnetic fields must also be considered. Beside linear kinetic energy related to translational motion of the system as a whole with respect to a reference frame, there is angular or rotational energy of a rigid body. Solids under anisotropic stress have a so-called elastic potential energy, or strain energy, which arises from the work done on a solid material to overcome intermolecular forces while changing its shape. In certain other cases one may wish to acknowledge the energy stored in the surface layer of a liquid, as in the study of interfacial effects between liquid layers.

Even in the absence of the effects of external force fields, motion (linear or rotational), anisotropic shear, and surface tension, a change in the energy of another important type may occur as the result of a work interaction. In the preceding section adiabatic work was added to a closed system in the forms of paddle-wheel work and electrical work dissipated in a resistor (recall Fig. 2-2). In both cases the system underwent the same change of state, but none of the forms of energy outlined above were affected. What was affected in these latter cases was a form of energy called *internal energy*, and it is designated by the symbol U (or u on a unit-mass basis). We have seen that kinetic and potential energies may be associated with macroscopic systems. These same forms of energy may be ascribed to the individual particles of a substance as well. In the gas phase, for instance, particles move at random around the containing vessel as a result of interactions with the walls or with other particles. Based on their velocities relative to the boundaries of the system, there is a translational (linear) kinetic energy associated with each particle. The particles may also have rotational or vibratory motions. Moreover, the particles have a potential energy which arises from the concept of intermolecular forces between the particles of the substance. These forces include the gravitational forces due to mass attraction, electrostatic forces between charges on the particles, nuclear binding forces, etc. The actual nature of these forces on the microscopic level is unimportant to us, since we are interested solely in the gross behavior of a large group of particles. Although internal energy has been described in terms of microscopic characteristics, it is not necessary to acknowledge the particle concept of matter in order to define or evaluate the internal energy. The change in its value can always be established by the quantity of adiabatic work required to change the state of the system. That is, in accordance with Eq. (2-1), when other forms of energy are not affected, $W_{ad} = \Delta U$.

In summary, the total energy E of a system may be broken down into contributions from a number of sources. In general we may write

$$\text{Total energy} = \text{internal energy} + \text{kinetic energy} + \text{gravitational potential energy} + \text{electrostatic energy} + \text{magnetic energy} + \text{strain energy} + \text{surface energy} + \cdots \tag{2-2}$$

The proper inclusion of one or more of the terms on the right side of this expression in a thermodynamic analysis depends primarily upon the type of system under analysis.

2-6 A CONSERVATION OF ENERGY PRINCIPLE FOR CLOSED SYSTEMS

The first law of thermodynamics has led to an operational definition of energy. The change in the energy of control mass is equal to the work done on or by the system during an adiabatic process. There are many forms of energy and work which must be considered in the analysis and synthesis of engineering systems. In addition, there is another type of interaction which cannot be classifed as a work effect. Consider the stirring of a fluid within an adiabatic, constant-volume system. It is experimentally known that the same change of state could also be accomplished by bringing the walls in contact with a high-temperature source. The effect which has occurred in the latter case falls into the category of a heat interaction. An adiabatic surface prevents heat transfer. In the case above, the quantity of energy Q transferred in the heat interaction is equal to the change in the energy of the control mass. That is, $Q = \Delta E$. Heat and work are the sole mechanisms by which energy is exchanged with a closed system.

Consider a closed system undergoing a process during which both heat and work effects are present. In this case, the work done will not equal the change in the energy of the system, since the process is nonadiabatic. The difference between the energy change of the system and the work done on the system is defined as a measure of the heat interaction that took place during the process. Mathematically this relationship is expressed by

$$Q \equiv (E_2 - E_1) - W \tag{2-3}$$

or

$$Q + W = \Delta E \tag{2-4}$$

Equation (2-4) is a conservation of energy statement for a closed system or control mass.† Note that heat transfer *into* a system is taken as *positive*.

From Eq. (2-4) it is seen that the sum of Q and W is unique between given end states, because ΔE is fixed in value by the end states; however, the actual values of Q and W depend upon the nature of the process between the given end states. The differentials of quantities which depend upon the path of a process (such as Q and W) are known as *inexact* differentials. To emphasize the need for a specification of the path, these quantities will be preceded by the symbol δ (lowercase delta) for infinitesimal variations in the state of the system (rather than the usual differential symbol, d). As a result, Eq. (2-4) for a differential change of state is written as

$$\delta Q + \delta W = dE \tag{2-5}$$

From the comparison between Eqs. (2-4) and (2-5), note that the integration of an inexact differential does not lead to the use of the symbol Δ. For inexact quantities such as work and heat one does not speak of the change in the quantity (such as one does for energy ΔE), but rather its overall value for the process in question. Note also that since the value of a work interaction (and a heat interaction) is a function of the path, the evaluation of δW (or δQ) around a cycle should not necessarily lead to zero as it does for the cyclic evaluation of dE around any path. Finally, it cannot be overemphasized that both work and heat are transient phenomena which occur across the boundary of the system. Thermodynamically, these phenomena are interactions which cease to exist once the process has ended.

Frequently it is convenient to analyze closed systems on a unit-mass basis. If the heat transfer per unit mass q is defined by $q = Q/m$, and the work per unit mass w by $w = W/m$, then the conservation of energy principle for a differential change of state is written as

$$\delta q + \delta w = de \tag{2-6}$$

For a finite change of state this becomes

$$q + w = \Delta e \tag{2-7}$$

Equations such as these are important in the thermodynamic analysis of closed systems.

† As just pointed out in the previous footnote, some books use the more traditional sign convention that work done by a system is positive. In such cases the positive sign between Q and W in Eq. (2-4) would be replaced by a negative sign.

In order to use Eqs. (2-4) and (2-7) in engineering analysis, it is necessary to independently evaluate heat-transfer and work quantities and changes in specific forms of energy of the system. A brief description of heat-transfer calculations is presented in the next paragraph. Sections 2-7 through 2-10 are devoted to the evaluation of specific types of work interactions, while the evaluation of the internal energy change for various classes of substances, along with other thermodynamic properties, is covered in Chaps. 3 and 4.

The concept of heat is defined operationally in the preceding section as a measure of the difference between the change in the energy of a closed system and the work done on the system. Nevertheless, it is a common experience to think of heat transfer as an energy-transfer process which occurs by virtue of a temperature difference between two systems. This is a calorimetric definition of heat transfer. In actual practice the engineer evaluates Q by adopting mechanisms for various types of heat-transfer phenomena. These mechanisms are known as conduction and radiation, in conjunction with convective transfer by fluid motion. They involve temperature as the driving force, and are described mathematically in textbooks and courses in heat transfer. Unfortunately, formal course work in the independent field of heat transfer usually follows an introductory course in thermodynamics. This lack of knowledge of independent heat-transfer calculations places a restriction on thermodynamic problems developed in this text. With regard to Q in such problems, there are three possiblilites:

1. It is the unknown in the problem.
2. It is essentially zero and is neglected.
3. It is a known quantity stated in the problem.

With respect to possibility 2, this occurs when the system is deliberately insulated or the temperature difference between the system and its environment is reasonably small, thus making the process adiabatic. When Q is given, it is implied that independent heat-transfer calculations have been carried out with the resulting reported data.

Example 2-1M The total energy of a closed system increases by 55.0 kJ during a process as work is done on the system in the amount of 100.0 kJ. How much heat is transferred during the process, and is it added or removed from the system?

SOLUTION The basic starting place in the analysis is the conservation of energy principle represented by Eq. (2-4), $Q + W = \Delta E$. Since work done on a system is positive by convention, the substitution of values for W and ΔE leads to

$$Q + W = \Delta E$$

$$Q + (+100.0) = +55.0$$

or

$$Q = +55.0 - 100.0 = -45.0 \text{ kJ}$$

Note that the answer is -45.0 kJ, and not just -45.0. Answers to engineering problems have units, and they *must* be stated specifically. The negative sign on the answer indicates that 45.0 kJ of energy in the form of heat must be removed during the process. Another way of expressing the answer is

$$Q_{out} = 45.0 \text{ kJ}$$

When the subscript "out" or "in" is placed on Q (or W), the sign no longer is necessary.

Example 2-1 The stored energy of a closed system increases by 55 Btu during a process as work is done on the system in the amount of 77,800 ft · lb_f. How much heat is transferred during the process, and is it added or removed from the system?

SOLUTION The basic starting place in the analysis is the conservation of energy principle represented by Eq. (2-4), $Q + W = \Delta E$. Application of this equation to the system of interest requires that the units of energy for Q, W, and ΔE must be the same. If we choose to work in Btu, the amount of work done on the system becomes

$$W = 77{,}800 \text{ ft} \cdot \text{lb}_f \frac{1 \text{ Btu}}{778 \text{ ft} \cdot \text{lb}_f} = 100 \text{ Btu}$$

where the conversion factors for the units found in Table A-1 have been rounded off to three significant figures. We are now in a position to substitute values into Eq. (2-4). Since work done on a system by convention is taken as positive, the substitution of values for W and ΔE leads to

$$Q + W = \Delta E$$

$$Q + (+100) = +55$$

or

$$Q = +55 \text{ Btu} - 100 \text{ Btu} = -45 \text{ Btu}$$

The negative sign on the answer indicates that 45 Btu of energy in the form of heat must be removed during the process.

2-7 QUASISTATIC PROCESSES

Consider a system in thermodynamic equilibrium. An external constraint or force on the system which has kept the system in equilibrium now undergoes a finite change in value, so that a finite unbalanced force acts on the system. This places the system in a state of nonequilibrium. As a result, a number of nonuniformities might show up within the system. For example, gradients of temperature and pressure may occur within the system and the fluid itself may be in some type of turbulent motion. When a finite unbalanced force is present, a system changes through a series of nonequilibrium states toward a final equilibrium state. Since the properties of such a system cannot be described during a nonequilibrium process, certain intermediate information on the system during the process is missing. Nevertheless, various overall effects can still be predicted, even though a detailed description is not possible. However, there are other thermodynamic situations where a detailed accounting of the state of the system during a process is important.

In order for the thermodynamic state of a system to be known at every instant during a process, finite unbalanced forces cannot exist. However, some type of imbalance must exist if a change of state is to occur. We can conceive of an *idealized* process during which an external force is only slightly different from the opposing force within the system at a given time. Mathematically speaking, the unbalanced force is differential in size. As a result, during such a process the system internally will be infinitesimally close to a state of equilibrium at all times. Any process carried out in this idealized manner is called *quasistatic*. As a result, the thermodynamic properties of the system are defined and measurable at all times. Although a quasistatic process is an idealization, many actual processes approximate quasistatic conditions closely. This is true since the time required for many substances to reach internal equilibrium is short compared to the time required for the overall change of state.

It is frequently necessary in engineering analysis for an actual process to be modeled as a quasistatic one. Only under this constraint can a mathematical model be developed for a particular phenomenon or process. In addition, it becomes possible to plot the path of a quasistatic process on diagrams where the coordinates represent thermodynamic properties. Such diagrams are extremely helpful in analyzing design problems. When plotting such diagrams, it is convenient to distinguish between equilibrium (quasistatic) and nonequilibrium processes. On property diagrams it is common practice to denote quasistatic processes by a solid or continuous line, while a nonequilibrium process is represented by a dashed line between the given end states. The discontinuous nature of the dashed line serves to emphasize the fact that the values of the thermodynamic coordinates are not known between the initial and final states. Hence the position of the dashed line is arbitrary, and is not a true indication of property values during the nonequilibrium change of state. The first use of process lines on property diagrams begins in Sec. 2-8.

2-8 QUASISTATIC EXPANSION AND COMPRESSION WORK

As an example of the use of the quasistatic process to calculate the value of work effects, it is interesting to examine the mechanical work done when the volume of a closed system, or control mass, is changed. This common type of mechanical work is usually called expansion or compression work, or simply boundary work. In some instances it is termed the $P\,dV$ work, for reasons which will become apparent in a moment. An example of this type of work is illustrated by a substance enclosed in a piston-cylinder assembly as shown in Fig. 2-3*a*. The system boundaries are indicated by the dashed line in the diagram. The cross-sectional area of the piston is A, and the pressure at the initial equilibrium state is P, throughout. From basic mechanics the expression for the work done on a system is again given by

$$\delta W = \mathbf{F} \cdot d\mathbf{s} = F\,ds$$

where F is the force exerted by the surroundings on the system and ds is the infinitesimal displacement of the system in the direction of F. The force of the

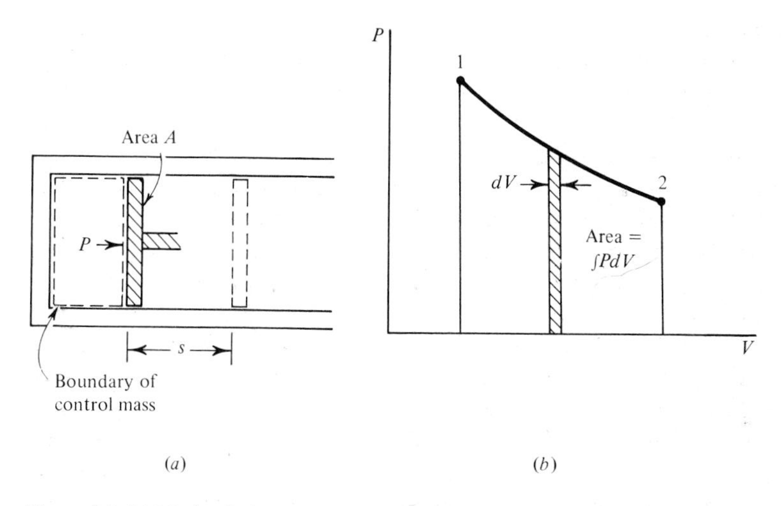

Figure 2-3 (*a*) Mechanical work associated with moving the boundary of a piston-cylinder device; (*b*) area representation of boundary work for a quasistatic process.

piston on the system is simply PA, and the process is one of compression because we are considering work done on the system. For the process,

$$\delta W = PA\, ds$$

Since ds is positive by convention, but dV is negative during the compression process, then $dV = -A\, ds$. Therefore the differential quantity of work done when a substance undergoes a compression or expansion process is

$$\delta W = -P\, dV \tag{2-8}$$

The total boundary work done on or by the substance during a finite change in volume is the sum of the $P\, dV$ terms for each differential change in volume. Mathematically, this is expressed by the relation

$$W_{\text{boundary}} = -\int_{V_i}^{V_f} P\, dV \tag{2-9}$$

where V_i and V_f are the initial and final volumes of the substance. The pressure is expressed in absolute units. It should be noted that the integral of δW is simply W, and not ΔW. A work interaction is associated with a process and is a function of the path selected.

A plot of a quasistatic process on pressure-volume coordinates is quite useful for describing graphically the boundary work of a process involving a closed system. From integral calculus it is known that the area beneath the curve which represents the quasistatic path of the process is equal to the integral of $P\, dV$ on a pressure-volume diagram. A typical diagram for the evaluation of boundary work is shown in Fig. 2-3*b*. The cross-hatched area on the diagram represents the work done by the gas within the cylinder when the volume is increased by the amount

dV. The entire area beneath the curve from point 1 to point 2 represents the total work done when the gas expands from state 1 to state 2. An indefinitely large number of paths could be drawn between these same two end states, any number of which could be quasistatic. In general, the area beneath each of these paths will be different. This merely emphasizes the fact that work is a path function, and, unlike the change in the value of a property, is not solely dependent upon the end states of the process. (Only in the special case of adiabatic processes does the value of work interactions become independent of the path.)

The integration of the equation for the boundary work requires a knowledge of the functional relationship between P and V. This can be found from experimental data or from knowledge of the specific type of process involved and a representative equation of state. An equation of state is an equation which relates P functionally to V and other independent variables. This latter point is best illustrated by examples.

$P\,dV$ Work for an Isothermal Gas System

The thermodynamic relationship between P and V for gases typically is presented in terms of an equation of state relating P, V, and T. This P-V-T relationship, based on experimental measurements, can be used to determine the boundary work done on or by the gas as it undergoes a quasistatic change of state. This requires expressing P explicitly as a function of V and T, and then substituting this functional relationship into Eq. (2-9). The general method is illustrated below.

Example 2-2M Two kilograms of a gas in a piston-cylinder device at 27°C and 0.040 m^3 are compressed isothermally to 0.020 m^3. The P-V-T equation of state for the gas is given in this case by $PV = mRT[1 + (a/V)]$, where R has a value of 0.140 kJ/(kg)(°K) and a is 0.010 m^3. Determine the minimum work of compression, in kilojoules, for the gas.

Solution The minimum work of compression requires that the process be quasistatic and is found from the integral of $-P\,dV$. When the expression for P from the equation of state is substituted into this integral, then

$$W_T = -\int_1^2 P\,dV = -\int_1^2 \frac{mRT}{V}\left(1 + \frac{a}{V}\right) dV$$

$$= -mRT \ln \frac{V_2}{V_1} - mRTa\left(\frac{1}{V_1} - \frac{1}{V_2}\right)$$

where the subscript T of W indicates an isothermal process. Substitution of the known data leads to

$$W_T = -2 \text{ kg} \times 0.140 \text{ kJ/(kg)(°K)} \times 300\text{°K} \times \ln \tfrac{0.020}{0.040}$$

$$-2 \text{ kg} \times 0.140 \text{ kJ/(kg)(°K)} \times 300\text{°K} \times 0.010 \text{ m}^3\left(\tfrac{1}{0.040} - \tfrac{1}{0.020}\right) \text{m}^{-3}$$

$$= 58.2 + 21.0 = 79.2 \text{ kJ}$$

The answer is positive, because the process is one of compression. When a work interaction is to be expressed per unit mass, the lowercase letter w is used. In this example, on a unit-mass basis, $w = 39.6$ kJ/kg.

Example 2-2 Two pounds of a gas in a piston-cylinder device at 80°F and 0.40 ft^3 are compressed isothermally to 0.20 ft^3. The P-V-T equation of state for the gas is given in this case by $PV = mRT[1 + (a/V)]$, where R has a value of 25.0 ft · lb$_f$/(lb$_m$)(°R) and a is 0.10 ft^3. Determine the minimum work of compression, in ft · lb$_f$, for the process.

SOLUTION The minimum work of compression requires that the process be quasistatic and is found from the integral of $-P\,dV$. When the expression for P from the equation of state is substituted into this integral, then

$$W_T = -\int_1^2 P\,dV = -\int_1^2 \frac{mRT}{V}\left(1 + \frac{a}{V}\right) dV$$

$$= -mRT \ln \frac{V_2}{V_1} - mRTa\left(\frac{1}{V_1} - \frac{1}{V_2}\right)$$

where the subscript T on W indicates an isothermal process. Substitution of the known data leads to

$$W_T = -2\ \text{lb}_m \times 25.0\ \text{ft} \cdot \text{lb}_f/(\text{lb}_m)(°\text{R}) \times 540\ °\text{R} \times \ln \tfrac{0.20}{0.40}$$

$$-2\ \text{lb}_m \times 25.0\ \text{ft} \cdot \text{lb}_f/(\text{lb}_m)(°\text{R}) \times 540\ °\text{R} \times 0.10\ \text{ft}^3 \times \left(\tfrac{1}{0.40} - \tfrac{1}{0.20}\right) \text{ft}^{-3}$$

$$= 18{,}720 + 6750 = 25{,}470\ \text{ft} \cdot \text{lb}_f$$

The answer is positive, because the process is one of compression. When a work interaction is to be expressed per unit mass, the lowercase letter w is used. For this example, on a unit-mass basis, $w = 12{,}735$ ft · lb$_f$/lb$_m$.

Boundary Work on Solid and Liquid Phases

The integral of $P\,dV$ may be used to evaluate the quasistatic expansion or compression work for any phase of interest. It is particularly appropriate in this form for the boundary work associated with gaseous systems, since the functional relation between P and v may be established by direct experimental measurements. For solids and liquids, the variation of pressure with volume is more frequently expressed indirectly in terms of a property called the *isothermal coefficient of compressibility* K_T. This property is defined as

$$K_T = -\frac{1}{V}\left(\frac{\partial V}{\partial P}\right)_T = -\frac{1}{v}\left(\frac{\partial v}{\partial P}\right)_T \tag{2-10}$$

It is an experimental observation that $(\partial v/\partial P)_T$ is always negative for all phases of matter. The negative sign in the definition of K_T is introduced so that tabulated values of the isothermal compressibility are always positive. It is apparent from the defining equation that K_T will always be given in reciprocal pressure units (for example, atm^{-1}).

To accommodate the use of K_T in the evaluation of boundary work, it is necessary to replace dv in the integral of $P\,dv$ by an equivalent quantity. On the basis that v is solely a function of T and P, that is, $v = v(T, P)$, then the total differential is written as

$$dv = \left(\frac{\partial v}{\partial T}\right)_P dT + \left(\frac{\partial v}{\partial P}\right)_T dP \tag{2-11}$$

For isothermal processes ($dT = 0$) Eq. (2-11) reduces to $dv = (\partial v/\partial P)_T \, dP = -vK_T \, dP$. Therefore, the isothermal work of compression per unit mass is given by

$$w_T = -\int_1^2 P \, dv = \int_1^2 vK_T P \, dP$$

The coefficient K_T and the specific volume v frequently are nearly constant for liquids and solids over considerable ranges of temperature and pressure. Under this condition integration of the above expression yields

$$w_T = \frac{vK_T}{2}(P_2^2 - P_1^2) \tag{2-12}$$

This equation provides a means for estimating the work required to compress a solid or liquid under isothermal conditions. Table 2-1 lists K_T values for solid copper and liquid water as a function of temperature. Note that K_T is extremely small with the value in both cases being on the order of 10^{-6} bar^{-1}.

Example 2-3M Water is compressed from 1 to 100 bar while the temperature is maintained at 20°C. Estimate the work required on the closed system, in joules.

SOLUTION Equation (2-12) is applicable if we assume that both K_T and v are reasonably independent of pressure. The value of K_T is found in Table 2-1 to be 45.90×10^{-6} bar^{-1}. The density of water at 20°C is close to 1.0 g/cm^3. Hence

$$w_T = \tfrac{1}{2} \times 1 \text{ cm}^3/\text{g} \times 45.90 \times 10^{-6} \text{ bar}^{-1} \times [100^2 - 1^2] \text{ bar}^2$$

$$= 0.230 \text{ cm}^3 \cdot \text{bar/g} \times 10^5 \text{ N/bar} \cdot \text{m}^2 \times (\text{m}^3/10^6 \text{ cm}^3) = 0.0230 \text{ J/g}$$

This is a small quantity of energy.

Example 2-3 Water is compressed from 15 to 1500 psia while the temperature is maintained at 100°F. Estimate the work required on the closed system in ft · lb_f/lb_m.

SOLUTION Equation (2-12) is applicable if we assume that both K_T and v are reasonably independent of pressure. On the basis of Table 2-1, the value of K_T is roughly 3.06×10^{-6} psi^{-1}.

Table 2-1 The isothermal compressibility K_T for solid copper and liquid water as a function of temperature, in units of bar^{-1}

Copper							
Temperature, °K	100	150	200	250	300	500	800
$K_T \times 10^6$	0.721	0.734	0.749	0.763	0.778	0.839	0.922
Water							
Temperature, °C	0	10	20	30	40	60	80
$K_T \times 10^6$	50.89	47.81	45.90	44.77	44.24	44.50	46.15

The specific volume of water is essentially 0.01613 ft^3/lb. Hence

$$w_T = \frac{1}{2} \times 0.01613 \text{ ft}^3/\text{lb}_m \times 3.06 \times 10^{-6} \text{ in}^2/\text{lb}_f \times [1500^2 - 15^2] \frac{\text{lb}_f^2}{\text{in}^4}$$

$$= 0.0555 \text{ lb}_f \cdot \text{ft}^3/\text{lb}_m \cdot \text{in}^2 = 8.00 \text{ ft} \cdot \text{lb}_f/\text{lb}_m$$

Further Considerations of $P\,dV$ Work

Consider now the evaluation of the boundary work for a closed system which undergoes a cyclic, quasistatic process. One possible Pv diagram for the control mass is shown in Fig. 2-4. The path of the cyclic process is from state 1 to states 2, 3, and 4, and finally back to the initial state. For those processes in which the final volume is greater than the initial volume, it is seen that work is done by the system on the surroundings. Hence, for paths 1–2 and 2–3, the work interaction is done by the system. In a similar fashion it will be found that the surroundings do work on the system for path 4–1, since ΔV is negative. It is apparent that no boundary work is present during path 3–4, because the volume is constant. As a result, the net work done by the system is represented by the area enclosed by the complete cyclic path, and in this case it is a negative value. If the cycle were carried out in the opposite direction, the net work would still be shown by the cross-hatched area in the figure, only in this case it would actually be done by the surroundings on the system. Finally, it should be noted that, in general, the area enclosed by any cycle will not be zero. The change in all properties will be zero for the cycle, but the net work will be finite.

It has been noted that, in order to evaluate the integral of $P\,dV$, the variation of P in terms of V must be known. If the pressure varied erratically over the volume of the substance, it would be highly unlikely that any simple functional relationship between P and V existed. Consider a piston-cylinder assembly filled with a gas; the surfaces of the piston and cylinder are frictionless. If the piston is suddenly accelerated inwardly, the effective pressure at the piston face becomes greater than it would have been if the piston had been moved gradually. From a molecular viewpoint, the gas particles are not able to move away from the approaching piston face fast enough. Consequently, they pile up near the piston and offer a greater resistance to the movement of the piston. As a result, more

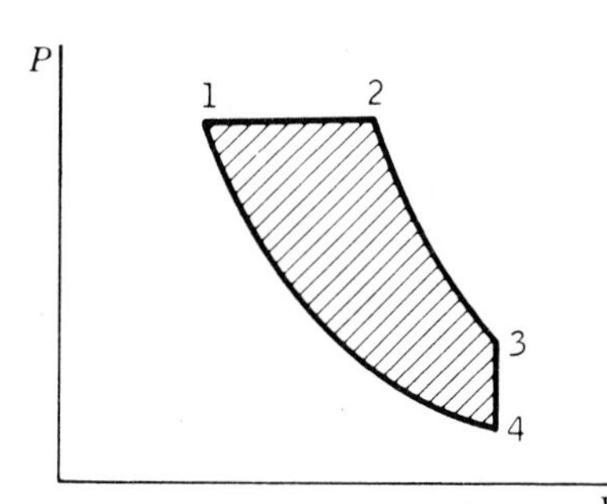

Figure 2-4 Net boundary work during a cyclic process for a closed system.

work is required to compress a gas rapidly than for the corresponding quasistatic process, all other factors being equal. By similar reasoning it would be found that less expansion work is done by the system when the substance is in nonequilibrium, compared with that for a quasistatic expansion. In the case of either nonequilibrium expansion or compression, the value of the work interaction must be measured directly. It could not be calculated from Eq. (2-9) because there is no way of relating P and V thermodynamically for nonequilibrium changes of state. Although the equation for boundary work was derived for volume changes produced by a piston, it should be apparent that this equation applies to any volume change which is carried out quasistatically.

2-9 OTHER QUASISTATIC FORMS OF WORK

There are a number of other quasistatic forms of work developed in basic physics and chemistry, in addition to boundary work. A brief summary of several of these is presented in this section.

1 Work of a Reversible Chemical Cell

A galvanic cell is a device for converting chemical energy into electrical energy by means of a controlled chemical reaction. A reversible cell is one for which the reaction is capable of proceeding in either direction. In order to operate the cell so that it remains close to equilibrium, a slightly lower electric potential must oppose the potential developed by the cell. This could be accomplished by placing a potentiometer in the external circuit which counterbalances the developed potential of the cell. Under these circumstances the ideal potential difference developed is called the electromotive force (emf) and is represented by $\mathcal{E}$.

The electrical work delivered by the cell for the passage of a differential quantity of charge dQ_e is found in the following way. For a potential difference $\mathcal{E}$, a current I performs work at the rate $\dot{W} = \delta W/dt = \mathcal{E}I$. Also, $I = dQ_e/dt$. Combination of these two expressions yields, for the electrical work,

$$\delta W_{\text{rev cell}} = \mathcal{E}\, dQ_e \tag{2-13}$$

The value of dQ_e is taken to be negative when the cell is discharged. Consequently δW will be negative during discharge of the cell, in accordance with our sign convention on work.

If the charge is removed at a finite rate, the potential developed at the terminals of the cell is less than the emf of the cell, which is the maximum potential of the cell. This loss in potential is due to the presence of a finite electric resistance internal to the cell, as shown in Fig. 2-5. The current I causes a voltage drop $\Delta\mathcal{E} = RI$. Therefore, on discharging, $\mathcal{E}_{\text{ext}} = \mathcal{E}_{\text{emf}} - RI$. Upon charging, the applied voltage $\mathcal{E}_{\text{ext}}$ must be greater than the emf by the amount RI. Therefore, on charging, $\mathcal{E}_{\text{ext}} = \mathcal{E}_{\text{emf}} + RI$. Hence the maximum work output and the minimum

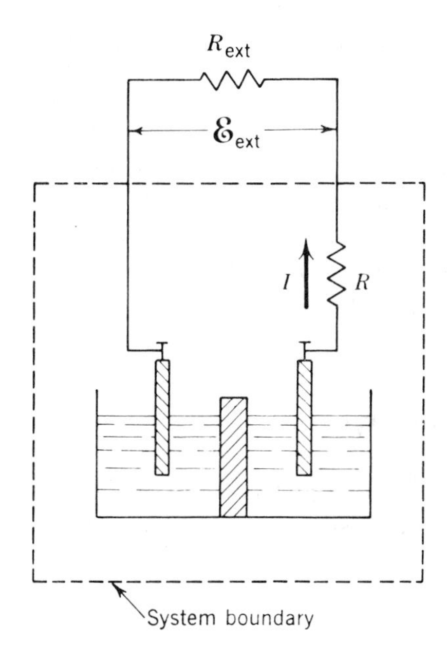

Figure 2-5 Effect of internal resistance on chemical-cell performance.

work input for a reversible cell require quasistatic processes in both directions. For finite rates of charging or discharging, the work associated with a chemical cell would be given by $\delta W_{\text{chem cell}} = \mathcal{E}_{\text{ext}}\, dQ_e$. However, in this case the work must be measured rather than calculated, since $\mathcal{E}_{\text{ext}}$ is not a thermodynamic variable and no unique relation between $\mathcal{E}_{\text{ext}}$ and Q_e exists.

2 Work in Stretching a Liquid Surface

In heterogeneous systems where gas and liquid (or two liquid phases) may be in contact, the phenomenon of surface tension occurs. For a molecule well inside the liquid phase no resultant force is experienced as it moves around in the fluid because of intermolecular attractions, there being an equal number of molecules immediately surrounding it. However, a molecule near the interface experiences a resultant force directed back into the liquid since the forces of attraction due to gas molecules above it are quite weak compared to those of the liquid phase. Hence additional work must be done on any molecule which is brought from the interior of the liquid to the surface as the surface area is increased. Molecules at the surface may be considered to acquire a surface potential energy over that of the molecules in the interior. This additional energy for the liquid as a whole obviously would be proportional to the surface area between the phases. To increase the surface area of a liquid, then, requires work equal to the increase in the surface potential energy. The quantity of work required is defined in terms of a thermostatic property of the system called the surface tension. The surface tension γ of a liquid phase (with respect to the other phase) is defined as the work per unit change in area required to increase the surface area. Consequently, the change in surface energy of a liquid for a differential area change is given by

$$\delta W_{\text{surf tension}} = \gamma\, dA \tag{2-14}$$

The surface tension is given in dimensions of energy per length squared, or force per unit length. Typical values are about 50×10^{-5} N/cm, or 3×10^{-3} lb_f/ft.

3 Work Done on Elastic Solids

In order to change the length of a spring or wire in tension or compression, it is necessary to exert a force F which alters the length L. The equation for the differential work required to alter the length from L to $L + dL$ follows directly from the mechanical definition of work, namely,

$$\delta W_{\text{elastic}} = F\, dL \tag{2-15}$$

This expression requires a functional relation between F and L in order to carry out the integration. The force F on a linear elastic spring is commonly related to the displacement x of the spring by the relation $F = k_s x$, where k_s is the spring constant and x is the change in the length of the spring from its unstressed state or condition. That is,

$$F_s = k_s(L_s - L_{s,0}) \tag{2-16}$$

where $L_{s,0}$ is the unstressed length of the spring and F_s is the force on the spring when its length is L_s. In this format Eq. (2-15) becomes

$$\delta W_{\text{spring}} = F_s\, dL_s \tag{2-17}$$

Substitution of Eq. (2-16) into Eq. (2-17) and subsequent integration will yield the work required to compress or stretch an elastic spring. The result is

$$W_{\text{spring}} = k_s(x_2^2 - x_1^2)/2 \tag{2-18}$$

where $x_2 = L_2 - L_0$ and $x_1 = L_1 - L_0$.

It is sometimes more convenient to express the work done on an elastic solid bar or wire in terms of the stress σ and the strain ϵ. The stress in the axial direction is defined by $\sigma = F/A$, where A is the cross-sectional area of the solid. The strain for a differential change in length is given by $d\epsilon = dL/L$, as shown in Fig. 2-6. Substitution of these expressions into Eq. (2-15) yields

$$\delta W_{\text{elastic}} = (\sigma A)(L\, d\epsilon) = V_0\, \sigma\, d\epsilon \tag{2-19}$$

where V_0 is the initial volume of the solid. If the work on the elastic solid is done isothermally, then the relation between stress and strain is given by Young's isothermal modulus, $E_T = \sigma/\epsilon$. When an elastic deformation occurs (the Hooke's law region of a solid), then E_T is a constant.

4 Work of Polarization and Magnetization

It may be shown from the theory of electromagnetic fields that work is done on a substance contained within an electric or magnetic field when the field is altered.

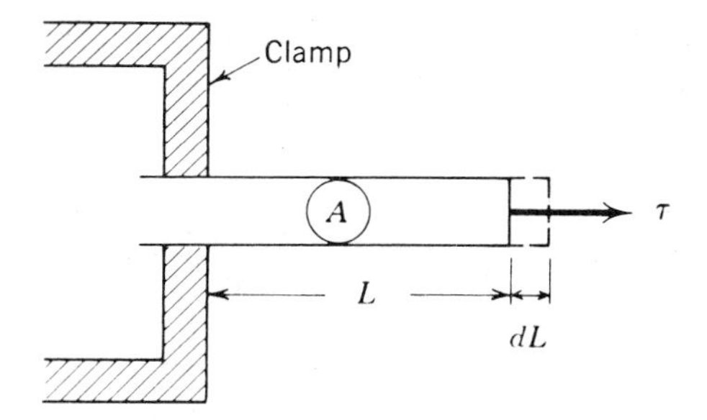

Figure 2-6 Stretching a wire.

For a dielctric material which lies within an electric field, the work supplied externally to increase the polarization of the dielectric is given by

$$\delta W_{\text{polarization}} = V\mathbf{E} \cdot d\mathbf{P} \tag{2-20}$$

where V is the volume, **E** is the electric field strength, and **P** is the polarization, or electric dipole moment, of the dielectric. In this equation the boldface type emphasizes that **E** and **P** are both vector quantities, and as such should be multiplied together in accordance with the rules of vector analysis. A similar equation for the work done in increasing the magnetization of a substance due to a change in the magnetic field is expressed by

$$\delta W_{\text{magnetization}} = V\mu_0 \mathbf{H} \cdot d\mathbf{M} \tag{2-21}$$

where **H** = magnetic field strength
M = magnetization per unit volume
μ_0 = permeability of free space
V = volume

Additional terms may be added to the above equations for the work required to change either the electric field or magnetic field of free space.

2-10 GENERALIZED QUASISTATIC WORK INTERACTIONS

In the preceding sections a number of work interactions were introduced. It is convenient to refer to all types of quasistatic work interactions as a product of a force and a displacement, even though the factors within a given work expression may not bring to mind physical forces and displacements. The intensive property that appears in the equations representing work will be called a *generalized force*, and will be symbolized by F_k. In an analogous fashion the extensive property found in these equations will be called a *generalized displacement*, represented by X_k. The subscript k simply refers to the kth type of a work interaction. The sum of the quasistatic work effects of these types is given by

$$\begin{aligned}\delta W_{\text{tot}} &= \sum_k F_k \, dX_k \\ &= -P\,dV + F\,dL + \gamma\,dA + V\mu_0 \mathbf{H} \cdot d\mathbf{M} + \cdots \end{aligned} \tag{2-22}$$

Table 2-2 Generalized quasistatic work interactions

System	Generalized force F_k (intensive)	Generalized displacement X_k (extensive)	Equations for work
Linear mechanical	F	s	$F\ ds$
Elastic	F (or σ)	L (or ϵ)	$F\ dL$ (or $V\sigma\ d\epsilon$)
Electrostatic and reversible cell	$\mathcal{E}$	Q_e	$\mathcal{E}\ dQ_e$
Surface	γ	A	$\gamma\ dA$
Capacitor	E	P	$V\mathbf{E}\cdot d\mathbf{P}$
Magnetic	H	M	$V\mu_0\,\mathbf{H}\cdot d\mathbf{M}$
Boundary	P	V	$-P\ dV$

In accordance with the general definition of work given earlier, the work $F_k\ dX_k$ (done by the system), when integrated over the path of the quasistatic process, is equivalent solely to the raising of a weight external to the system boundaries.

It is a matter of experience when various work effects should be taken into account. In some instances their contribution may be so small as to be neglected. Table 2-2 shows various work interactions in terms of the intensive and extensive factors, as well as the overall differential expressions. This table emphasizes the fact that most generalized forces do not have the dimensions of force, nor do the generalized displacements have the dimensions of length. Nevertheless, the product of the dimensions for any corresponding pair of a generalized force and displacement has the dimensions of work, i.e., energy.

2-11 NONQUASISTATIC FORMS OF WORK

Some important forms of work interactions were summarized in Table 2-2. All these interactions can be expressed by an equation of the form $\delta W = F\ dX$. If a process is quasistatic, then F and X are properties of the system and some functional relationship exists between F and X which permits integration of the expression for work. Quasistatic work interactions have the following characteristics:

1. The value of F_k depends only upon the state of the system and is independent of the direction of change of X. The change in X can be either positive or negative.
2. Theoretically a system can be returned to its initial state after a quasistatic work interaction merely by reversing the direction of the original work effect.
3. The value of F_k remains finite as dX approaches zero. In each equilibrium state F_k has a fixed and finite value.

Consider, for example, the boundary work done by a piston-cylinder assembly.

For a given initial equilibrium state with a pressure P, the magnitude of P in the expression $P\,dV$ is the same whether V increases or decreases. After a differential increase in V, the system can be returned to its initial state by reversing the direction of the piston movement. Whether for a finite or differential change of state, the value of P remains finite and well defined as long as the process is quasistatic.

There are other important work interactions which do not fall into the quasistatic class. These nonquasistatic forms of work have a number of characteristics which differentiate them from quasistatic interactions. Among these are:

1. The force F depends upon the rate of change of state.
2. The work interaction is unidirectional, and the effect cannot in practice be undone by reversing the original process.
3. The force F approaches zero as the rate of change of state approaches zero. Consequently the work approaches zero for a differential change of state.

A typical example of a nonquasistatic-type work interaction is the case of a fluid undergoing shear deformation due to the rotation of a paddle-wheel or its equivalent. The shear is proportional to the rate of fluid deformation, and as the rate goes to zero, the shear force does likewise. It should be noted that paddle-wheel work may be done on a closed system, but the reverse process is not possible. Thus the interaction is unidirectional. Another example of a unidirectional, nonquasistatic work interaction is the passage of an electric current through a resistor which lies within the system boundaries. The electrical work done on the system is due to some external voltage source, which causes current to pass along conductors connected with the resistor. The usual result of such an interaction is a rise in the temperature of the substance within the system. The reverse process, the spontaneous removal of energy from the system which causes a current to flow through the circuit, never occurs. In cases of both paddle-wheel work and electrical-resistor work, the fluid within the system usually is not in equilibrium during the process. Since the process is nonquasistatic, the work must be measured external to the boundaries of the system. In comparison, a quasistatic form of work is evaluated in terms of changes within the system.

In terms of paddle-wheel work, for example, the rotational mechanical work is evaluated in terms of the external torque τ transmitted by a rotating shaft. To review the development of the expression relating torque to work, consider the apparatus shown in Fig. 2-7. An external force F acts at a distance r from the center of a shaft. This external force could be represented by a weight attached to the pulley system, for example. The work required to move the force through a differential distance ds is given by $F\,ds$. From the definition of torque we may write that

$$\tau = Fr$$

Also, by definition the angle $d\theta$ in radians is

$$ds = r\,d\theta$$

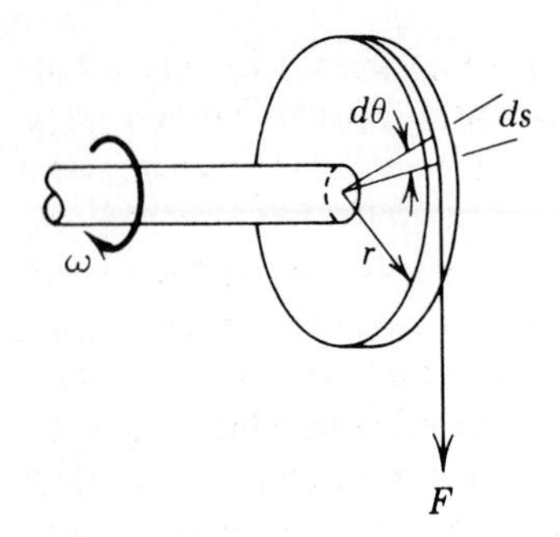

Figure 2-7 Schematic showing relationship of torque and angular deflection to rotational mechanical work.

Hence the relation between rotational mechanical work and torque is simply

$$\delta W = F\,ds = \frac{\tau}{r}(r\,d\theta) = \tau\,d\theta \tag{2-23}$$

If a constant torque is applied, then

$$W = \tau\theta \tag{2-24}$$

where θ must be expressed in radians, and not degrees. The power associated with rotational work is

$$\dot{P} = \frac{\delta W}{dt} = \tau\frac{d\theta}{dt} = \tau\omega \tag{2-25}$$

where ω is the angular velocity in radians per unit time.

The evaluation of the electrical work dissipated within a resistor likewise depends upon external measurements. As noted earlier for the reversible chemical cell, the rate of doing electrical work (or the electrical power) is

$$\text{Power} = \frac{\delta W}{dt} = VI \tag{2-26}$$

When a current of one ampere (A) passes through an electrostatic potential of one volt (V), the power required is defined as one watt (W), which is also 1 J/s. Upon rearrangement, Eq. (2-26) yields the differential work required to pass a current I through a voltage potential V over a differential time period. That is,

$$\delta W_{\text{electrostatic}} = VI\,dt \tag{2-27}$$

Knowledge of current, voltage, and time relations permits the integration of this expression.

2-12 THE STATE POSTULATE

An important objective of thermodynamics is to ascertain the functional relationships among certain macroscopic properties of a system at equilibrium. In order to develop these relations, the number of independent variables necessary to fix or establish the state of the system under given conditions must be known. This information is provided by a basic law of thermodynamics known as the state postulate.

Thermodynamics directs its attention to a major class of properties called *intrinsic* or *thermostatic* properties. This class of properties arises solely from particular characteristics of the mass within the boundaries of the system. That is, intrinsic properties characterize an average of some type of molecular or microscopic behavior of the particles of the substance within the system. Common examples of intrinsic or thermostatic properties include pressure, temperature, specific volume, and internal energy. Because intrinsic properties are all characteristics of molecular behavior, it is reasonable to expect them to be functionally related. By functionally related, we mean the following. Consider the general situation where some dependent property y_0 is a function of n other intensive properties. Mathematically, this means that $y_0 = f(y_1, y_2, \ldots, y_n)$. Once we have selected values for all n independent properties, the value of the dependent property y_0 is fixed. As a simple example, it is known that pure water boils at 100°C or 212°F if the pressure is adjusted to 1 atm. If the pressure is lowered, such as on a mountain top, the water boils at a lower temperature. Symbolically for this two-phase system, we write that $T = f(P)$.

We seek a general rule that allows us to determine the number of independently variable properties n for any prescribed physical situation. Experience has shown that the number of independent variables depends upon the nature of the physical situation of interest. On the basis of experience, the following statement, or *state* postulate, may be presented.

> The number of independent intrinsic properties required to fix the state of a substance is equal to one plus the number of possibly relevant quasistatic work modes.

By "relevant work modes" we mean that they have an appreciable effect on the state of the substance if they are altered during a process. Objects on earth, for example, are usually under the influence of the natural gravitational, electric, and magnetic force fields of earth. However, their effects on the outcome of most processes are often negligible. Consequently we ignore their effect when determining the number of independent variables. However, if a strong electric field is applied to a substance containing electric dipoles, we certainly expect its state to be a strong function of the electric field strength. In this case the applied electric field would need to be considered an independent variable. Nonquasistatic work modes are never relevant, in terms of the state postulate.

The state postulate applies to any system, regardless of the amount of mass within the system. However, as a matter of convenience we usually relate by some means only intensive values of intrinsic properties. That is, the functional relationship among intrinsic properties is usually expressed in terms of their values on a unit-mass basis. The general rule does not change for this situation. The number of intensive intrinsic properties is still equal to the number of relevant quasistatic work modes plus one.

2-13 SIMPLE SYSTEMS

The state postulate asserts that the number of independent thermostatic properties of a substance is related to the number of relevant quasistatic work modes that possibly could alter the state of the substance. When one examines past and present-day engineering and scientific studies, one thing becomes apparent. Seldom does one encounter a system where more than one of the quasistatic work modes listed in Table 2-2 is used to alter the state of the substance. That is, most problem areas involve compressible systems, elastic systems, electrostatic systems, etc., but seldom does one find a combination of these effects influencing the state of the system.

On the basis of this observation, it is convenient to classify various systems as simple systems. By definition, a *simple* system is one for which only one quasistatic work mode might possibly affect the state of the system. On the basis of the state postulate asserted above, the following statement may be made as a state postulate for simple substances:

> The equilibrium state of a simple, homogeneous substance is fixed by specifying the values of any two independent, intrinsic properties.

Once the two independent properties are fixed in value, the values of all remaining intrinsic or thermostatic properties are also fixed. Thus only two intrinsic properties may be independently varied for any simple substance. The state postulate for simple systems is extremely important, since a wide range of scientific and engineering problems involves systems which meet the conditions of a simple system.

The various types of simple systems may be differentiated form each other by speaking of simple compressible systems, simple elastic systems, simple magnetic systems, etc. A *simple compressible* system is defined as one for which the only quasistatic work interaction is boundary ($P\,dV$) work. For such a substance the effects of capillarity, anisotropic stress, and external force fields are negligible. From a practical viewpoint this means that the system is not influenced by these effects, even though they may be present to some degree. The absence of capillarity, for example, implies that volume effects are much more significant than interfacial surface effects. Due to their importance in engineering studies, property relations for simple compressible substances will be stressed in this and the following chapter.

Any two independent, intrinsic properties are sufficient to fix the intensive state of a simple substance. Thus the functional relationship between a set of intensive properties is given by

$$y_0 = y(y_1, y_2) \tag{2-28}$$

where the quantity (y_1, y_2) represents a set of two independent variables. This equation expresses some, as yet, unknown relation between two independent, intrinsic, intensive properties and a third dependent property. One of the major tasks of thermodynamics is to develop from theoretical or experimental considerations some explicit expressions for these relationships. Expressions such as Eq. (2-28) are known as *equations of state* because they relate properties at a given state of a substance.

The state principle, based on experimental evidence, enables one to predict the number of independent, intrinsic properties required to fix the equilibrium state of a system. On the basis of this principle we may begin to develop various techniques—analytical, graphical, and tabular—for relating and evaluating intrinsic properties. The most important properties to begin examining in terms of their functional relationships are pressure P, temperature T, specific volume v, and internal energy u. In addition, we shall find it necessary to define and relate three other properties of interest—the enthalpy h, the specific heat at constant pressure c_p, and the specific heat at constant volume c_v.

2-14 THE CONSERVATION OF ENERGY PRINCIPLE FOR SIMPLE, COMPRESSIBLE CLOSED SYSTEMS

The conservation of energy equation for closed systems (control masses) can be expanded into the following form on the basis of Eqs. (2-2) and (2-4).

$$Q + W = \Delta E$$
$$= \Delta U + \Delta KE + \Delta PE + \Delta(\text{electrostatic energy}) + \cdots$$

where the terms on the right include many possible forms of energy associated with the system. It must also be remembered that the work term W may include several types of work interactions that occur during the specified process. If, in addition to the effects of magnetic and electric fields and surface tension, we elect also to neglect the effects of motion on a substance, the above equation reduces to

$$Q + W = \Delta U \tag{2-29}$$

Other useful forms of Eq. (2-29) are

$$\delta Q + \delta W = dU \tag{2-30}$$

$$q + w = \Delta u \tag{2-31}$$

and

$$\delta q + \delta w = du \tag{2-32}$$

Equations (2-29) to (2-32) apply to simple compressible closed systems which are stationary. The last two of these equations are applied to a substance of unit mass.

In addition to the internal energy U of a substance, we shall find it convenient to employ in energy balances another intrinsic property called the enthalpy. It is defined by the relation

$$H = U + PV \tag{2-33}$$

or, per unit mass, by

$$h \equiv u + Pv \tag{2-34}$$

Since u, P, and v are properties, h also must be a property. In many cases the enthalpy function has no specific physical interpretation, although it does have the dimension of energy. In a number of thermodynamic equations the quantities U and Pv often appear together. Therefore it is advantageous to write the symbol h in these equations, rather than $u + Pv$. As a result it would also be helpful to tabulate numerical values of the enthalpy along with other property correlations, just as a labor-saving device. In reality, we shall find that the enthalpy function appears in tables and charts more frequently than the internal-energy function. However, the enthalpy h should not be thought of as a specific form of energy, but simply as a defined quantity which is useful in the solution of engineering and scientific problems.

The enthalpy function is a convenient property when analyzing volume changes under the condition of constant pressure. Since the volume of the control mass is altered, a work interaction is associated with the process. In order to evaluate the work done in moving the boundary, it is necessary to assume that the process is carried out quasistatically. On this basis the work is given by the integral of $-P\,dV$. Equation (2-30) then becomes

$$\delta Q - P\,dV = dU$$

or

$$\delta Q = dU + P\,dV \tag{a}$$

Now, since $H = U + PV$, for a process at constant pressure it is seen that

$$dH = d(U + PV) = dU + d(PV) = dU + P\,dV \tag{b}$$

Substitution of Eq. (b) into Eq. (a) leads to an interesting result under the restrictions imposed: that is,

$$\delta Q_P = dH \tag{c}$$

This relation indicates that the quantity of heat transferred to or from a simple compressible system during a quasistatic, constant-pressure process is equal to the change in the enthalpy of the control mass within the boundaries of the system. A knowledge of the enthalpies at the initial and final states would lead to a rapid evaluation of the heat interaction required to bring about the required change of state.

In general, other work interactions may occur during the constant-pressure change of state, in addition to boundary work. For example, if the system were a fluid it might be stirred with a paddle-wheel. As another possibility, electrical work might be performed on an electrical resistor within the system. In order to make a broader statement of the special energy Eq. (c), we may write that

$$\delta Q_P + \delta W' - P\,dV = dU \tag{2-35}$$

or

$$\delta Q_P + \delta W' = dH \tag{2-36}$$

where $\delta W'$ in these two equations includes all forms of work other than boundary work. These equations are still restricted to quasistatic, constant-pressure processes.

From the equations developed in this section for simple compressible stationary systems it is apparent that information on the internal energy and the enthalpy over a broad range of conditions will be required in order to analyze and synthesize engineering systems of interest. It will be shown in subsequent chapters that the methods of presenting or evaluating u and h depend upon the class or phase of the substance under study.

REFERENCES

Reif, F.: "Fundamentals of Statistical and Thermal Physics," McGraw-Hill, New York, 1968.

Reynolds, W. C., and H. C. Perkins: "Engineering Thermodynamics," 2d ed., McGraw-Hill, New York, 1977.

Zemansky, M. W., and G. C. Van Ness: "Basic Engineering Thermodynamics," 2d ed., McGraw-Hill, New York, 1974.

PROBLEMS (METRIC)

Kinetic and potential energy

2-1M Determine the work required, in kilojoules, to accelerate a 0.98-kg body from a velocity of (*a*) 100 to 200 m/s and (*b*) 80 to 180 m/s.

2-2M It requires 60 kJ of work to raise a body 90 m in the earth's gravitational field where the local acceleration of gravity is 9.55 m/s^2.

(*a*) Find the mass of the body, in kilograms.

(*b*) If the initial gravitational potential energy of the body was 20 kJ with respect to the earth's surface, determine the final elevation of the body above this surface, in meters.

2-3M The acceleration of gravity as a function of elevation above sea level is given by $g = 9.807 - 3.32 \times 10^{-6}z$, where g is in m/s^2 and z is in meters. A satellite with a mass of 240 kg is boosted 400 km above the earth's surface. What work is required, in kilojoules?

2-4M Two hundred kilojoules of work are used to change the kinetic energy of a body by 100 N · m/g.

(*a*) Determine the mass of the body, in kilograms.

(*b*) If the kinetic energy relative to some reference is 40 kJ, determine the final velocity relative to the same reference, in m/s.

General first law analysis

2-5M For each of the following cases of a process involving a closed system, fill in the missing data.

	Q	W	E_1	E_2	ΔE		Q	W	E_1	E_2	ΔE
(*a*)	24	−13		−8		(*d*)	18		26		10
(*b*)	−8			62	−18	(*e*)	−8	13	28		
(*c*)		17	−12		20	(*f*)		−11		7	−12

2-6M A closed system is changed from state 1 to state 2 along path x. It is then returned to the initial state along either of two different paths, y and z. Values of Q and W along the three different paths are listed below for two different systems, A and B. Determine the missing data for Q and W for (*a*) system A, and (*b*) system B.

System A

Path	Process	Q	W
x	1 to 2	10	
y	2 to 1	-7	4
z	2 to 1		8

System B

Path	Process	Q	W
x	1 to 2		-7
y	2 to 1	-4	9
z	2 to 1	10	

Boundary work

2-7M During an expansion process for a closed system, the volume of a gas changes from 0.01 to 0.03 m^3. The pressure changes during the quasistatic process according to the relation $P = -4000V + 240$, where the units of P and V are kilopascals and cubic meters, respectively.

(*a*) Find by integration the work done by the gas, in joules.

(*b*) Check the answer by geometric considerations of the process on a PV diagram.

(*c*) Find the useful work output if the constant frictional force between piston and cylinder is 600 N and the piston area is 0.04 m^2.

2-8M A gas is compressed from a state of 1 bar, 0.30 m^3, to a final state of 4 bars. The process equation relating P and V is $P = aV + b$, where $a = -15$ bars/m^3. Compute by integration the necessary work, in kilojoules, and sketch the process path on PV coordinates.

2-9M A gas is confined in a frictionless piston-cylinder device surrounded by the atmosphere. Initially, the pressure of the gas is 1400 kPa and the volume is 0.030 m^3. If the gas expands to a final volume of 0.060 m^3 and the piston area is 0.1 m^2, calculate the work done, in newton-meters, against the force F along the shaft connected to the piston. The atmospheric pressure is 100 kPa. Assume the processes connecting the end states are of the following types: (*a*) the pressure is constant, (*b*) the product PV is a constant, and (*c*) the product PV^2 is a constant. Compare the different quasistatic processes by use of a PV diagram. $W = W_g - W_a$

2-10M One-fifth kilogram of a gas is contained in a piston-cylinder assembly at initial conditions of 0.02 m^3 and 7 bars. The gas is allowed to expand to a final volume of 0.050 m^3. Determine the amount of work done, in joules, for the following processes:

(*a*) The pressure is constant.

(*b*) The product PV is a constant.

(*c*) The product PV^2 is a constant.

Compare the different quasistatic processes by use of a PV diagram.

2-11M In many real processes it is found that gases at low pressures fulfill the relationship $PV^n = c$, where the exponent n and c are constants. If n is taken as 1.3, determine the work done, in newton-meters, by air in a piston-cylinder arrangement when it expands quasistatically from 0.01 m^3 and 3 bars to a final pressure of 1 bar.

2-12M A quantity of sulfur dioxide (SO_2) gas is expanded in a piston-cylinder device. The following experimental data were taken:

P, bars	3.45	2.75	2.07	1.38	0.69
v, m^3/kg	0.125	0.150	0.187	0.268	0.474

(*a*) Estimate the work of expansion done in kilojoules, per kilogram of SO_2.

(*b*) If the friction between the cylinder and piston was equivalent to $0.15P$, what was the work of expansion appearing in the surroundings, in kJ/kg?

2-13M A piston-cylinder device contains a gas which undergoes a quasistatic process for which the pressure-volume relationship is given by $P = (10.8/V^2) + 3.5$. In this equation the pressure is in bars and the volume is in cubic meters. The initial volume is 1 m^3 and the final volume is 2 m^3.

(*a*) Sketch the process to approximate scale on a PV diagram.

(*b*) Determine the units of the quantity 10.8 in the equation.

(*c*) Calculate the work done, in kilojoules, and state explicitly whether the work is *into* or *out* of the system.

2-14M One kilogram of a gas with a molar mass of 35 is compressed isothermally (constant temperature) at 77°C from a volume of 0.05 m^3 to a volume of 0.025 m^3. The PvT relationship for the gas is given by $Pv = RT[1 + (c/v^2)]$, where $c = 2.0 \text{ m}^6/(\text{kg} \cdot \text{mol})^2$.

(*a*) Compute the work done on the gas, in newton-meters, if quasistatic.

(*b*) If $c = 0$, would the work required be greater than, equal to, or less than that calculated in part (*a*)? Defend your answer in terms of two process lines on a Pv diagram.

2-15M One kilogram of a gas with a molar mass of 60 is compressed at a constant temperature of 27°C from a volume of 0.12 to 0.04 m^3. The PvT relationship for the gas is given by $Pv = RT[1 + (b/v)]$, where b is 0.012 m^3/kg.

(*a*) Determine the quasistatic work required, in newton-meters.

(*b*) Find the useful work output if the friction force between piston and cylinder is 1000 N and the piston moves 0.5 m.

2-16M A gas with a molar mass of 46.0 is compressed from 0.08 to 0.04 m^3. The process equation relating P and V is given by $P = 0.1V^{-2} + 80$, where P is in kilopascals and V is in cubic meters.

(*a*) Determine the required work of compression, in kilojoules.

(*b*) Plot the process roughly to scale on a PV diagram, and indicate the area representation of the work input.

2-17M A gas is compressed quasistatically in a piston-cylinder device from an initial state of 2.5 bars, 0.4 m^3, to a final state of 0.2 m^3 in accordance with the following relations: (*a*) $P(V)^{1/2}$ = constant, (*b*) PV = constant, and (*c*) $Pv(\ln v)$ = constant. Determine the work required for the various processes, in kilojoules.

2-18M A gas expands from 1 to 3 m^3 at a constant temperature of 17°C. The PvT relationship for the gas is given by $[P + (a/v^2)]v = R_u T$, where $a = 15 \text{ bar m}^6/(\text{kg} \cdot \text{mol})^2$. Compute the work done quasistatically by the gas, in newton-meters, if the system contains 0.3 kg · mol.

2-19M Compute the work required to compress copper isothermally from 1 to 500 bars at (*a*) 300°K, and (*b*) 500°K. Assume the density is 8.90 g/cm^3 at both temperatures. Answer in kJ/kg.

2-20M The pressure on 1 kg of water is increased isothermally and quasistatically from 1 bar to 1000 bars. The density of water is 1.0 g/cm^3. Estimate the work required, in kilojoules, if the temperature is (*a*) 20°C, and (*b*) 50°C.

2-21M The isothermal compressibility for liquid water at 60°C is given by $K_T = 0.125/v(P + 2740)$, where K_T is in bar^{-1}, P is in bars, and v is in cm^3/g. Determine the work required, in joules, to compress 2.5 kg of water isothermally from 1 bar to 600 bars.

2-22M Determine the work required, in joules, to compress 20 cm^3 of liquid mercury at a constant temperature of 0°C from a pressure of 1 bar to (*a*) 500 bars, (*b*) 1000 bars, and (*c*) 1500 bars. The isothermal compressibility of mercury at the given temperature is given by $K_T = 3.9 \times 10^{-6} - 1.0 \times 10^{-10}P$, where K_T is in bar^{-1}, and P is in bars. The density of mercury may be taken as 13.6 g/cm^3.

Elastic work

2-23M The relation between the tension τ, and length L, and the temperature T for an elastic substance is given by $\tau = KT(x - x^{-2})$. In this expression K is a constant, $x = L/L_0$, and L_0 is the value of L at zero tension and is solely a function of temperature. Determine the work required to stretch such a substance isothermally and quasistatically from $L = L_0$ to (*a*) $L = 1.2L_0$, (*b*) $L = 1.4L_0$, and (*c*) $L = 1.5L_0$.

2-24M (*a*) Show that the work required to stretch an initially unstressed wire within the elastic region is given by $W = 0.5ALE\epsilon^2$, where L is the initial length, A is the cross-sectional area, and Young's modulus $E = \sigma/\epsilon$.

(*b*) What is the work required, in newton-meters, to increase the length of an unstressed steel bar from 10.00 to 10.01 m, if $E = 2.07 \times 10^7$ N/cm^2 and $A = 0.3$ cm^2?

2-25M In Prob. 2-24M use the same values of E and A, but stretch the 10-m wire until the force on the bar is (*a*) 8000 N, and (*b*) 50,000 N.

2-26M An elastic spring with a spring constant k_s of 100 N/cm has a length of 20 cm when not under compression. Determine the work required to compress the spring from (*a*) 20 to 19 cm, and (*b*) 19 to 18.5 cm in length.

2-27M An elastic linear spring with a spring constant of 144 N/cm is compressed from an initial unconstrained length to a final length of 6 cm. If the work required on the spring is (*a*) 6.48 J, and (*b*) 2.88 J, determine the initial length of the spring.

2-28M An elastic linear spring requires 12.6 J of work to compress it from 11 to (*a*) 9 cm, and (*b*) 10 cm. Find the value of the spring constant k_s, in N/cm if the spring initially is unstressed.

2-29M An elastic linear spring with a free length of 15 cm is compressed by a work input of (*a*) 20 J, and (*b*) 4.0 J. If the spring constant k_s is 80 N/cm, determine the final length of the spring.

Electric polarization and magnetization

2-30M A parallel-plate condenser in a dc circuit is charged by slowly increasing the voltage across the condenser from 0 to 110 V. Under this condition the voltage V and the charge Q_e are related by the equation $Q_e = kV$, where k is the capacitance of the condenser. Calculate the work, in joules, necessary to charge the condenser if $k = 2.0 \times 10^{-5}$ C/V.

2-31M With respect to the general discussion in Prob. 2-30M, a parallel-plate condenser with a capacitance of 1×10^{-5} C/V has a 10-V dc potential slowly impressed across it and maintained until the condenser is charged.

(*a*) What is the final charge on the condenser?

(*b*) How much work was done on the condenser to charge it?

2-32M A parallel-plate capacitor is quasistatically charged at room temperature to a potential of 100 V. The capacitor plates are 7 cm square and their separation is 1 mm. The dielectric equation of state for the air between the plates is given by $P = 4.75 \times 10^{-15}E$, where P is in C/m^2, and E is in V/m. Determine the work, in joules, required to polarize the air.

2-33M Curie's law for paramagnetic substances is given by the relation $M = CH/T$, where C is a constant. For a quasistatic, isothermal change of state, show that the work done per unit volume to change the magnetization is given by

$$w = \mu_0 \frac{T}{2C}(M_f^2 - M_i^2) = \mu_0 \frac{C}{2T}(H_f^2 - H_i^2)$$

Other work interactions

2-34M The drive shaft of an automobile rotates at 3000 revolutions per minute (rpm) and transmits (*a*) 75 kW, and (*b*) 90 kW of power from the engine to the rear wheels. Compute the torque developed by the engine, in newton-meters.

2-35M A torque of 150 N · m is associated with a shaft rotating at (*a*) 1500 rpm, and (*b*) 2500 rpm. Determine the power transmitted in each case, in kilowatts.

2-36M A drive shaft delivers (*a*) 80 kW when the torque is 500 N · m, and (*b*) 60 kW when the torque is 120 N · m. Determine the shaft rotational speed, in rpm.

2-37M A dc motor draws a current of 50 A at 24 V. The torque applied to the shaft is (*a*) 10.5 N · m at 1000 rpm, and (*b*) 6.8 N · m at 1500 rpm. What is the rate of heat transfer to or from the motor, in kJ/h? Assume the process is steady state.

2-38M An electric potential of 115 V is impressed on a resistor such that a current of 9 A passes through the resistor over a period of (*a*) 2 min, and (*b*) 5 min. Find the amount of electrical work, in kilojoules, done in each case.

2-39M A 12-V battery is used to pass a current of (*a*) 1.5 A, and (*b*) 4 A through an external resistance for a period of 15 s. Find the amount of electrical work, in kilojoules, done by the battery in each case.

2-40M A 12-V storage battery delivers a current of 10 A for 0.20 h. What is the heat transfer, in kilojoules, if the energy of the battery decreases by 94 kJ?

2-41M A 12-V battery is charged by supplying a current of 5 A for 40 min. During the charging period a heat loss of 27 kJ occurs from the battery. Find the change in the energy stored in the battery during the specified time period.

General energy considerations

2-42M A piston-cylinder assembly contains a gas which undergoes a series of quasistatic processes which make up a cycle. The processes are as follows: 1–2, adiabatic compression; 2–3, constant pressure; 3–4, adiabatic expansion; 4–1, constant volume. The table below gives data at the beginning and end of each process.

System A					System B				
State	*P*, bars	*V*, cm³	*T*, °C	*U*, kJ	State	*P*, kPa	*V*, cm³	*T*, °K	*U*, kJ
1	0.95	5700	20	1.47	1	110	500	300	0.137
2	23.9	570	465	3.67	2	950	125	650	0.305
3	23.9	1710	1940	11.02	3	950	250	1300	0.659
4	4.45	5700	1095	6.79	4	390	500	1060	0.522

For (*a*) system A, and (*b*) system B, roughly sketch the cycle on *PV* coordinates and determine the work and heat interactions, in kilojoules, for each of the four processes.

2-43M An insulated piston-cylinder assembly containing a fluid has a stirring device operated externally. The piston is frictionless, and the force holding it against the fluid is due to standard atmospheric pressure and a coil spring. The spring constant is 7200 N/m. The stirring device is turned 1000 revolutions with an average torque of 0.68 N · m. As a result, the 0.3-m-diameter piston moves outward 0.10 m. Find the change in the internal energy of the fluid, in kilojoules, if $F_{s,i} = 0$.

2-44M A piston-cylinder assembly is equipped with a paddle-wheel driven by an external motor and is filled with a gas. The walls of the cylinder, the piston, and the paddle-wheel and its shaft are all made of a material which is a very good insulator. The system contains 50 g of gas. Initially the gas is in state 1 (see table). The paddle-wheel is then operated, but the piston is allowed to move to keep the pressure constant. Then the paddle-wheel is stopped, and the system is found to be in state 2. There is very little friction between the piston and cylinder walls. Determine the energy transfer in joules along the paddle-wheel shaft.

State	*P*, bars	*v*, cm³/g	*u*, J/g	*h*, J/g
1	35	7.11	22.75	47.64
2	35	19.16	97.63	164.69

2-45M A heavily insulated piston-cylinder device contains a gas which is initially at 6 bars and 177°C and occupies 0.05 m^3. The gas undergoes a quasistatic process according to the equation $PV^2 =$ constant. The final pressure is 1.5 bars. Determine (*a*) the work done, in newton-meters, and (*b*) the change in internal energy, in kilojoules.

2-46M A closed cylinder with its axis vertical is fitted with a piston in its upper end. The piston is loaded by a weight so that a constant pressure of 3 bars is exerted on the 0.8 kg of gas within the cylinder. The gas decreases in volume from 0.1 to 0.03 m^3, and the internal energy decreases by 60 kJ/kg. If the process is quasistatic, determine (*a*) the work done on or by the gas, in kilojoules, (*b*) the amount of heat added or removed, in kilojoules, and (*c*) the enthalpy change, in kJ/kg.

2-47M A piston-cylinder device is maintained at a constant pressure of 5 bars and contains 1.4 kg of a gas. During a process the heat transfer out is 50 kJ, while the volume changes from 0.15 to 0.09 m^3. Find the change in the internal energy in kJ/kg.

2-48M A vertical piston-cylinder assembly contains a gas which is compressed by a frictionless piston weighing 3000 N. During an interval of time, a paddle-wheel within the cylinder does 6800 N · m of work on the gas. If the heat transfer out of the gas is 10 kJ and the change in the internal energy is −1 kJ, determine the distance the piston moves, in meters. The area of the piston is 52 cm^2 and the atmospheric pressure acting on the outside of the piston is 1.0 bar.

PROBLEMS (USCS)

Kinetic and potential energy

2-1 Determine the work required, in ft · lb_f, to accelerate a 6.44 lb_m-body from (*a*) a velocity of 100 to 300 ft/s, and (*b*) a velocity of 50 to 150 ft/s.

2-2 It requires 50,000 ft · lb_f of work to raise a body 280 ft in the earth's gravitational field where the local acceleration of gravity is 31.9 ft/s^2.

(*a*) Find the mass of the body in lb_m.

(*b*) If the initial gravitational potential energy of the body was 15,000 ft · lb_f with respect to the earth's surface, determine the final elevation of the body above this surface, in feet.

2-3 The acceleration of gravity as a function of elevation above sea level is given by $g = 32.17 - 3.32 \times 10^{-6}z$, where g is in ft/s^2 and z is in feet. A satellite with a mass of 520 lb is boosted 240 mi above the earth's surface. What work is required, in ft · lb_f?

2-4 Work in the amount of 160,000 ft · lb_f is used to change the kinetic energy of a body by 40,000 ft · lb_f/lb_m.

(*a*) Determine the mass of the body, in lb_m.

(*b*) If the initial kinetic energy of the body relative to some reference is 30,000 ft · lb_f, find the final velocity relative to the same reference, in ft/s.

General first law analysis

2-5 For each of the following cases of a process involving a closed system, fill in the blank spaces.

	Q	W	E_1	E_2	ΔE		Q	W	E_1	E_2	ΔE
(*a*)	25	−10		−10		(*d*)	20		27		10
(*b*)	−10			65	−20	(*e*)	−9	12	29		
(*c*)		15	−10		20	(*f*)		−10		6	−10

2-6 A closed system undergoes a cycle made up of three processes. Fill in the missing data in the tables below for (*a*) system A, and (*b*) system B.

System A

Process	Q	W	ΔE
1–2		0	50
2–3	0	−40	
3–1	−30		

System B

Process	Q	W	ΔE
1–2	50	0	
2–3	0	40	
3–1		−20	

Boundary work

2-7 During an expansion process for a closed system, the volume changes from 0.10 to 0.30 ft^3. The pressure of the gas changes during the quasistatic process according to the relation $P = -4.0V + 2.4$, where the units of P and V are atmospheres and cubic feet, respectively.

(*a*) Find by integration the work done by the gas, in ft · lb$_f$.

(*b*) Check your answer by geometric considerations of the process on a PV diagram.

(*c*) Find the useful work output if the constant frictional force between piston and cylinder is 140 lb$_f$ and the piston area is 0.40 ft^2.

2-8 A gas is compressed from a state of 15 psia, 0.5 ft^3, to a final state of 60 psia. The process equation relating P and V is $P = aV + b$, where $a = -150$ psia/ft^3. Compute by integration the necessary work input in ft · lb$_f$ for the quasistatic process, and sketch the process path on PV coordinates.

2-9 A gas is confined in a frictionless piston-cylinder device surrounded by the atmosphere. Initially, the pressure of the gas is 200 psia and the volume is 1.0 ft^3. If the gas expands quasistatically to a final volume of 2.0 ft^3 and the piston area is 1.50 ft^2, calculate the work done, in ft · lb$_f$, against the force F along the shaft connected to the piston. The atmospheric pressure is 1 atm. Assume the processes connecting the end states are of the following types: (*a*) the pressure is constant, (*b*) the product PV is a constant, and (*c*) the product PV^2 is a constant. Compare the different quasistatic processes by use of a PV diagram.

2-10 Two-tenths pound of a gas is contained in a piston-cylinder assembly at initial conditions of 0.2 ft^3 and 100 psia. The gas is allowed to expand to a final volume of 0.5 ft^3. Determine the amount of work done, in ft · lb$_f$, for the following processes:

(*a*) The pressure remains constant.

(*b*) The product PV is a constant.

(*c*) The product PV^2 is a constant.

Compare the different quasistatic processes by the use of a PV diagram.

2-11 In many real processes it is found that gases at low pressures fulfill the relationship $PV^n = c$, where n and c are constants. If n is taken as 1.3, determine the work done by air, in ft · lb$_f$, in a piston-cylinder arrangement when it expands quasistatically from 0.2 ft^3 and 60 psia to a final pressure of 20 psia.

2-12 A quantity of sulfur dioxide (SO_2) gas is expanded in a cylinder equipped with a piston. The following data are taken:

P, psia	50	40	30	20	10
v, ft^3/lb	2.0	2.4	3.0	4.3	7.6

(*a*) Calculate the work of expansion done per pound of SO_2.

(*b*) If the friction between the cylinder and piston was equivalent to $0.2P$, what was the work of expansion appearing in the surroundings per pound of sulfur dioxide?

2-13 A piston-cylinder device is used to compress a gas quasistatically from 1.0 to 0.50 ft^3. The process equation relating P and V is given by $P = 720V^{-2} + 1440$, where P is in lb_f/ft^2 and V is in cubic feet. (*a*) Sketch the process to approximate scale on a PV diagram. (*b*) Determine the units on the quantity 720 in the equation. (*c*) Calculate the work required, in $ft \cdot lb_f$.

2-14 One pound of a gas with a molar mass of 30 is compressed isothermally (constant temperature) at 140°F from a volume of 2 ft^3 to a volume of 1 ft^3. The PvT relationship for the gas is given by $Pv = RT[1 + (c/v^2)]$, where $c = 450\ ft^6/(lb \cdot mol)^2$.

(*a*) Compute the work done on the gas in $ft \cdot lb_f/lb_m$ if the process is quasistatic.

(*b*) If $c = 0$, would the work required be greater than, equal to, or less than that calculated in part (*a*)? Defend your answer in terms of two process lines on a Pv diagram.

2-15 One-tenth pound of a gas with a molar mass of 60 is compressed at a constant temperature of 140°F from a volume of 0.2 to 0.1 ft^3. The PvT relationship for the gas is given by $Pv = RT[1 + (b/v)]$, where b is 0.2 ft^3/lb_m.

(*a*) Determine the quasistatic work required in $ft \cdot lb_f/lb_m$.

(*b*) If $b = 0$, would the work input be greater than, equal to, or less than that of part (*a*)? Defend your answer in terms of a Pv diagram.

(*c*) What would be the answer to part (*b*) if the constant b were originally negative, rather than positive?

2-16 A gas with a molar mass of 46 is compressed from 1.0 to 0.5 ft^3. The process equation relating P and V is given by $P = 0.2V^{-2} + 1.2$, where P is expressed in atmospheres and V is in cubic feet.

(*a*) Determine the required work of compression, in $ft \cdot lb_f$.

(*b*) Plot the process roughly to scale on a PV diagram, and indicate the area representation of the work input.

2-17 A gas is compressed in a piston-cylinder device quasistatically from an initial state of 40 psia, 0.6 ft^3, to a final state of 0.3 ft^3, in accordance with the following process relations: (*a*) $P(V)^{1/2}$ = constant, (*b*) PV = constant, and (*c*) $PV(\ln V)$ = constant. Determine the work required for the various processes, in $ft \cdot lb_f$.

2-18 A gas expands from 2 to 4 ft^3 at a constant temperature of 140°F. The PvT relationship for the gas is given by $[P + (a/v^2)]v = R_u T$, where $a = 350\ atm \cdot ft^6/(lb \cdot mol)^2$. Compute the work done quasistatically by the gas, in $ft \cdot lb_f$, if the system contains 0.2 lb · mol.

2-19 Compute the work required to compress copper isothermally from 1 to 500 atm at (*a*) 540°R, and (*b*) 900°R. Assume the density of copper is 555 lb/ft^3 at both temperatures. Answer in ft lb_f/lb_m.

2-20 The pressure on 1 lb of water is increased quasistatically and isothermally from 1 to 1000 atm. The density of water is 62.3 lb/ft^3. Estimate the work required, in $ft \cdot lb_f$, if the temperature is (*a*) 68°F, and (*b*) 122°F.

2-21 The isothermal compressibility for liquid water at 140°F is given by $K_T = (2.01 \times 10^{-3})/v(P + 2700)$, where K_T is in atm^{-1}, P is in atm, and v is in ft^3/lb. Determine the work, in $ft \cdot lb_f$, required to compress 4.5 lb of water isothermally from 1 to 600 atm.

2-22 Determine the work, in $ft \cdot lb_f$, required to compress 1.0 in^3 of liquid mercury at a constant temperature of 32°F from a pressure of 1 atm to (*a*) 500 atm, (*b*) 1000 atm, and (*c*) 1500 atm. The isothermal compressibility of mercury at the given temperature is given by $K_T = 3.9 \times 10^{-6} - 1.0 \times 10^{-10}P$, where K_T is in atm^{-1} and P is in atm. The density of mercury may be taken as 850 lb/ft^3.

Elastic work

2-23 The relation between the tension τ, the length L, and the temperature T for an elastic substance is given by $\tau = KT(x - x^{-2})$. In this expression K is a constant, $x = L/L_0$, and L_0 is the value of length L at zero tension and is solely a function of the temperature. Determine the work required to stretch such a substance isothermally from $L = L_0$ to (*a*) $L = 1.1L_0$, (*b*) $L = 1.2L_0$, and (*c*) $L = 1.3L_0$.

2-24 (*a*) Show that the work required to stretch an initially unstressed wire within the elastic region is given by $W = 0.5\ ALE\epsilon^2$, where L is the initial length, A is the cross-sectional area, and Young's modulus $E = \sigma/\epsilon$.

(*b*) What is the work required, in $\text{ft} \cdot \text{lb}_f$, to increase the length of an unstressed steel wire from 20.00 to 20.01 ft, if $E = 3 \times 10^7\ \text{lb}_f/\text{in}^2$ and $A = 0.10\ \text{in}^2$?

2-25 In Prob. 2-24 use the same values of E and A, but stretch the 20-ft wire until the force on the wire is (*a*) 2000 lb_f, and (*b*) 10,000 lb_f.

2-26 An elastic spring with a spring constant k_s of 100 lb_f/in has a length of 10 in when not under compression. Determine the work required to compress the length of the spring from (*a*) 10 to 9 in, and (*b*) 9 to 8 in.

2-27 An elastic spring with a spring constant k_s of 72 lb_f/in is compressed from an initial unconstrained length to a final length of 3 in. If the work required on the linear spring is (*a*) 54 $\text{ft} \cdot \text{lb}_f$, and (*b*) 81 $\text{ft} \cdot \text{lb}_f$, determine the initial length of the spring.

2-28 An elastic linear spring requires 252 $\text{ft} \cdot \text{lb}_f$ of work on it to compress it from 12 to (*a*) 8 in, and (*b*) 9 in. Find the value of the spring constant k_s, in lb_f/in, if the spring initially is unstressed.

2-29 An elastic linear spring with a free length of 8 in is compressed by a work input of (*a*) 28 $\text{ft} \cdot \text{lb}_f$, and (*b*) 14 $\text{ft} \cdot \text{lb}_f$. If the spring constant k_s is 48 lb_f/in, determine the final length of the spring.

Electric polarization and magnetization

2-30 A parallel-plate condenser in a dc circuit is charged by slowly increasing the voltage across the condenser from 0 to 110 V. Under this condition the voltage V and the charge Q_e are related by the equation $Q_e = kV$, where k is the capacitance of the condenser. Calculate the work, in joules, necessary to charge the condenser if $k = 2.0 \times 10^{-5}\text{F}$ (1 F = 1 C/V).

2-31 A parallel-plate condenser with a capacitance of 1×10^{-5} F has a 10-V dc potential slowly impressed across it and maintained until the condenser is charged.

(*a*) What is the final charge on the condenser?

(*b*) How much work was done on the condenser to charge it?

2-32 A parallel-plate capacitor is quasistatically charged at room temperature to a potential of 100 V. The capacitor plates are 4 in square and their separation is 0.040 in. The dielectric equation of state for the air between the plates is given by $P = 4.75 \times 10^{15}E$, where P is in C/m^2 and E is in V/m. Determine the work, in joules, required to polarize the air.

2-33 Curie's law for paramagnetic substances is given by the relation $M = CH/T$, where C is a constant. If a paramagnetic substance undergoes a quasistatic, isothermal change of state, show that the work done by unit volume to change the magnetization is given by

$$w = \mu_0 \frac{T}{2C}(M_f^2 - M_i^2) = \mu_0 \frac{C}{2T}(H_f^2 - H_i^2)$$

Other work interactions

2-34 The drive shaft of an automobile rotates at 3000 revolutions per minute (rpm) and transmits (*a*) 90 hp, and (*b*) 120 hp from the engine to the rear wheels. Compute the torque developed by the engine, in $\text{lb}_f \cdot \text{ft}$.

2-35 A torque of 150 $\text{lb}_f \cdot \text{ft}$ is associated with a shaft rotating at (*a*) 1500 rpm, and (*b*) 2500 rpm. Determine the power transmitted in each case, in horsepower.

2-36 A drive shaft delivers (*a*) 50 hp when the torque is 150 $\text{lb}_f \cdot \text{ft}$, and (*b*) 40 hp when the torque is 120 $\text{lb}_f \cdot \text{ft}$. Determine the shaft rotational speed, in rpm.

2-37 A dc motor draws a current of 60 A at 24 V. The torque applied to the shaft is (*a*) 110 $\text{lb}_f \cdot \text{in}$ at 1000 rpm, and (*b*) 90 $\text{lb}_f \cdot \text{in}$ at 1200 rpm. What is the rate of heat transfer to or from the motor, in Btu/h?

2-38 An electric potential of 110 V is impressed on a resistor such that a current of 8 A passes through the resistor over a period of (*a*) 3 min, and (*b*) 4 min. Find the amount of electrical work, in Btu, done in each case.

2-39 A 12-V battery is used to pass a current of (*a*) 0.75 A, and (*b*) 3.5 A through an external resistance for a period of 24 s. Find the amount of electrical work, in Btu, done in each case.

2-40 A 12-V storage battery delivers a current of 10 A for 0.20 h. What is the heat transfer if the energy of the battery decreases by 98 Btu?

2-41 A 12-V battery is charged by supplying a current of 5 A for 40 min. During the charging period a heat loss of 26 Btu occurs from the battery. Find the change in the energy stored in the battery during the specified time period.

2-42 An experimental energy converter has a heat input of 80,000 Btu/h and a work input of 1.2 hp. The converter produces an electric power output of 18 kW. Calculate the change in the energy of the converter, in Btu, over a time period of 4 min.

General energy considerations

2-43 A piston-cylinder assembly contains a gas which undergoes a series of quasistatic processes which make up a cycle. The processes are as follows: 1–2, adiabatic compressions; 2–3, constant pressure; 3–4, adiabatic expansion; 4–1, constant volume. The table below gives data at the beginning and end of the various processes.

System A					System B				
State	P, psia	v, ft^3	T, °F	U, Btu	State	P, psia	V, ft^3	T, °R	U, Btu
1	14	0.20	70	1.39	1	16	0.100	540	0.736
2	352	0.02	870	3.48	2	140	0.025	1180	1.635
3	352	0.06	3530	10.45	3	140	0.050	2360	3.540
4	65	0.20	2000	6.44	4	58	0.100	1950	2.860

For (*a*) system A, and (*b*) system B, roughly sketch the cycle on PV coordinates and determine the work and heat interactions, in Btu, for each of the four processes.

2-44 An insulated piston-cylinder assembly containing a fluid has a stirring device operated externally. The piston is frictionless, and the force holding it against the fluid is due to standard atmospheric pressure and a coil spring. The spring constant is 500 lb_f/ft. The stirring device is turned 10,000 revolutions with an average torque of 0.50 $\text{lb}_f \cdot$ ft. As a result, the 2.0-ft-diameter piston moves outward 2 ft. Find the change in the internal energy of the fluid, in Btu, if $F_{s,i} = 0$.

2-45 A piston-cylinder assembly is equipped with a paddle-wheel driven by an external motor and is filled with a gas. The assembly is heavily insulated, and the paddle-wheel is made of a material which is a poor conductor of heat. The system contains 0.1 lb of gas. Initially the gas is in state 1 (see table). The paddle-wheel is then operated, but the piston is allowed to move to keep the pressure constant. When the paddle-wheel is stopped, the system is found to be in state 2. There is very little friction between the piston and cylinder walls.

State	P, atm	v, in^3/lb	u, Btu/lb	h, Btu/lb
1	35	197	9.80	20.65
2	35	531	42.06	71.34

Determine the energy transfer in the form of paddle-wheel work, in ft $\cdot$ lb_f.

2-46 A heavily insulated piston-cylinder device contains a gas which is initially at 100 psia and

350°F, and occupies 1 ft^3. The gas undergoes a quasistatic process according to the equation $PV^2 =$ constant. The final pressure is 25 psia. Determine (*a*) the work done, in ft · lb_f, and (*b*) the change in internal energy, in Btu.

2-47 A closed cylinder with its axis vertical is fitted with a piston in its upper end. The piston is loaded by a weight so that a constant pressure of 50 psia is exerted on the 2 lb of gas within the cylinder. The gas decreases in volume from 3 to 1 ft^3, and the internal energy decreases by 30 Btu/lb. Assuming the process to be quasistatic, determine (*a*) the work done on or by the gas, in ft · lb, (*b*) the amount of heat added or removed, in Btu, and (*c*) the enthalpy change, in Btu/lb.

2-48 A piston-cylinder device maintained at a constant pressure of 80 psia contains 3 lb of a gas. During a cooling process the heat transfer out is 50 Btu, while the volume changes from 5 to 3 ft^3. Find the change in the internal energy, in Btu/lb.

2-49 A vertical piston-cylinder assembly contains a gas which is compressed by a frictionless piston weighing 684 lb_f. During an interval of time, a paddle-wheel within the cylinder does 5000 ft · lb_f of work on the gas. If the heat transfer out of the gas is 10 Btu and the change in the internal energy is −1.0 Btu, determine the distance the piston moves, in feet. The area of the piston is 8.0 in^2 and the atmospheric pressure acting on the outside of the piston is 14.5 psia.

CHAPTER

THREE

IDEAL GAS AND SPECIFIC HEAT

3-1 INTRODUCTION

In the preceding chapter a conservation of energy equation was developed for simple compressible substances contained within a closed, stationary system. The only quasistatic work interaction permitted under these conditions is boundary, or $P\,dV$, work. However, the state of the substance may also be altered by nonquasistatic work interactions as well as heat effects. As noted in Sec. 2-14, any heat and work interactions are related to changes in the internal energy or enthalpy of the substance. In general, the substance could be in any one of three phases—solid, liquid, or gas—or in several phases simultaneously. Thus, methods of evaluating U and H for various classes of substances is needed in order to complete the energy analysis of a closed-system process. In this chapter we shall restrict ourselves to gaseous substances which can be modeled as an ideal gas system. This model is quite accurate for a large number of engineering systems.

3-2 IDEAL-GAS EQUATION OF STATE

Of particular interest in thermodynamics are equations which relate the variables P, v, and T. On the basis of experimental work carried out originally by Boyle, Charles, and Gay-Lussac, the PvT behavior of many gases at low pressures and moderate temperatures can be approximated quite well by the *ideal (perfect)-gas equation* of state, namely,

$$PV = NR_u T \tag{3-1}$$

or

$$P\bar{v} = R_u T \tag{3-2}$$

where N = number of moles of a gas
$\bar{v}$ = specific volume on a molar basis
R_u = a universal gas constant
(The bar over the v is used only in this section to denote the specific volume on a mole basis. In future sections the units on v should be clear from the context of the equation or discussion.) The values of R_u in several sets of units are

$$R_u = \begin{cases} 0.08314 \text{ bar} \cdot \text{m}^3/(\text{kg} \cdot \text{mol})(^\circ\text{K}) \\ 8.314 \text{ kJ}/(\text{kg} \cdot \text{mol})(^\circ\text{K}) \\ 1545 \text{ ft} \cdot \text{lb}_f/(\text{lb} \cdot \text{mol})(^\circ\text{R}) \\ 0.730 \text{ atm} \cdot \text{ft}^3/(\text{lb} \cdot \text{mol})(^\circ\text{R}) \\ 1.986 \text{ Btu}/(\text{lb} \cdot \text{mol})(^\circ\text{R}) \end{cases}$$

These values are also tabulated in Table A-1 in the appendix.

The ideal-gas equation is frequently used with mass units such as pounds and grams instead of pound-moles and gram-moles. In such cases one uses a specific gas constant R in the equation instead of the universal value R_u. It is to be noted that

$$R \equiv \frac{R_u}{M}$$

where M is the molar mass. Values of M are given for some elements and common substances in Table A-2. The equivalent forms of the ideal-gas equation become

$$Pv = \frac{R_u T}{M} = RT \qquad PV = mRT \qquad P = \rho RT \qquad PV = \frac{mR_u T}{M} \tag{3-3}$$

where v = specific volume on a mass basis
ρ = density
m = mass of system
Since R depends upon the molar mass of a substance, its value is different for each substance, even when expressed in the same set of units. For example, if the average molar mass of air is taken as 28.97 (or roughly 29), then R for air in the SI is 0.287 kJ/(kg)(°K). The R value for hydrogen in the same set of units is 4.12 kJ/(kg)(°K). It should be carefully noted that the temperature in the ideal-gas equation of state is always expressed in either kelvins or degrees Rankine.

Example 3-1M Nitrogen gas N_2 at a pressure of 1.40 bars is maintained at a temperature of 27°C. Determine the specific volume, in m³/kg, if it is reasonable to assume that the gas behaves ideally.

SOLUTION From Eq. 3-3 it is seen that $v = R_u T/PM$. The absolute temperature is $273 + 27 = 300$°K, the molar mass of nitrogen from Table A-2 is 28.01 kg/kg · mol, and a convenient value of R_u is 0.08314 bar · m³/(kg · mol)(°K). Substitution of these values leads to

$$v = \frac{0.08314 \text{ bar} \cdot \text{m}^3}{(\text{kg} \cdot \text{mol})(°\text{K})} \frac{300°\text{K}}{1.40 \text{ bars}} \frac{\text{kg} \cdot \text{mol}}{28.01 \text{ kg}} = 0.636 \text{ m}^3/\text{kg}$$

Example 3-1 Nitrogen gas at a pressure of 20 psia is maintained at a temperature of 80°F. Determine the specific volume, in ft³/lb$_m$, if it is reasonable to assume that the gas behaves ideally.

SOLUTION From Eq. (3-3) it is seen that $v = R_u T/PM$. Substitution of the above value for P and T, and use of 10.73 psia · ft³/(lb · mol)(°R) as a proper value for R_u and 28 lb$_m$/lb · mol for M, leads to

$$v = \frac{10.73 \text{ psia} \cdot \text{ft}^3}{(\text{lb} \cdot \text{mol})(°\text{R})} \frac{80 + 460 \text{ °R}}{20 \text{ psia}} \frac{\text{lb} \cdot \text{mol}}{28 \text{ lb}_m}$$

$$= 10.35 \text{ ft}^3/\text{lb}_m$$

In some calculations involving ideal gases it is not necessary to know the value of the gas constant. We shall find that there are a number of thermodynamic relations which require a knowledge of ratios of a given property rather than a knowledge of the actual values of the property. For example, information on the value of v_2/v_1 might be needed. It is easily seen that for a given ideal gas

$$\frac{v_2}{v_1} = \frac{RT_2/P_2}{RT_1/P_1} = \frac{T_2 P_1}{T_1 P_2}$$

To calculate the individual values of v_2 and v_1 directly from the ideal-gas equation and then take the ratio of the two values involves considerably more work than using the above relationship directly. In addition, there is the possibility of mathematical error or errors in the units required when separate calculations are made. Therefore one must keep in mind such shortcut methods for calculations involving ideal gases as that illustrated by the above example.

If a gas were ideal at all pressures, at constant temperature the quantity Pv would be a constant, independent of pressure. A plot of Pv versus P at a given temperature would be a horizontal line. Typical experimental data on such a plot are shown in Fig. 1-6. Note that nitrogen gas approximates ideal behavior over a wide range of pressures, since the line drawn through experimental data is fairly horizontal out to at least 30 atm. Argon gas begins to deviate after about 10 atm. Carbon dioxide, on the other hand, apparently is nonideal at all pressures. Therefore, the quantity Pv for gases generally is a function of both P and T. However, as the pressure is lowered, the Pv product approaches the same value, regardless of the nature of the gas. That is, the limiting value of Pv at zero pressure is the same for all gases at the same temperature. Similar plots would result for data at other temperatures, except that the value of the zero pressure intercept differs for every new temperature. Consequently, at zero pressure all gases behave according to the equation $P\bar{v} = R_u T$. As the pressure is raised, some gases deviate from ideal-gas behavior faster than others. Apparently the ideal-gas equation would be considerably in error for CO_2 at the temperature used for Fig. 1-6 if the pressure exceeded more than several atmospheres.

As a generalization, the ideal-gas equation of state is only an approximation at best, strictly valid only at zero pressure. Nevertheless, for monatomic and diatomic gases the ideal-gas equation is usually a good approximation up to pressures of 10 to 20 atm at room temperature and above, for errors in accuracy not exceeding several percent. The range of validity of the ideal-gas equation of state for any particular gas depends upon the degree of accuracy desired. In Chap. 4 we shall reexamine the usefulness of this basic equation in the light of further experimental data. A simple method for correcting the equation to give a more accurate PvT relationship for nonideal gases will be shown then.

1 PvT Surface of an Ideal Gas

One of the interesting consequences of an ideal-gas model is that the equilibrium states of such a gas can be represented by a fairly simple surface in a rectangular coordinate system. If we plot the equation $Pv = RT$ on a PvT coordinate system, then a surface is generated which has the general shape shown in the center of Fig. 3-1. Constant-temperature lines (isotherms) along the surface will appear as hyperbolas because Pv is a constant in this case. Conditions of constant pressure or constant volume are represented by straight lines on the surface. Figure 3-1 also shows the PT and Pv projections of the PvT surface of an ideal gas. The hyperbolic nature of constant-temperature lines on the Pv plane and the straight-line nature of constant-volume lines on the PT plane are clearly shown. Although not shown, the surface can also be projected onto the Tv plane. In Chap. 2 we used the Pv plane to exhibit and compare the work associated with various quasistatic processes involving boundary work. Boundary work in that case had an area representation on the Pv plane. The next subsection illustrates this same point for a special ideal-gas process.

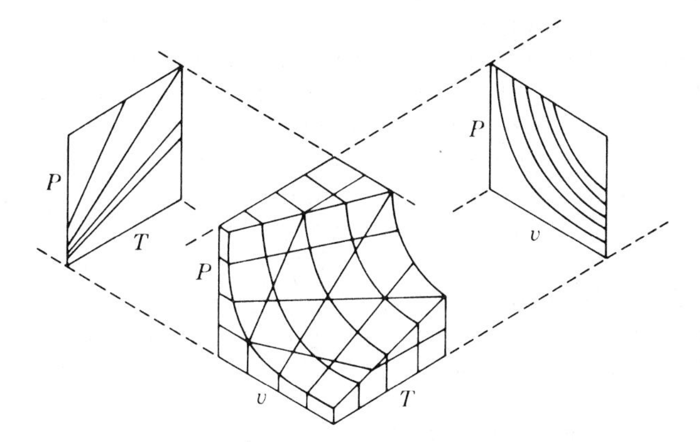

Figure 3-1 PvT surface and the PT and Pv projections for ideal-gas behavior.

2 $P\,dV$ Work for an Isothermal Ideal-Gas System

Consider a piston-cylinder device that contains a gas which fulfills the condition of an ideal gas, that is, $PV = NR_u T$. The gas is allowed to expand quasistatically and isothermally from state 1 to state 2. The work done by the gas on the piston face is given by the integral of $-P\,dV$. Noting that T is a constant by definition of an isothermal process, we find that

$$W = -\int_{V_1}^{V_2} P\,dV = -\int_{V_1}^{V_2} \frac{NR_u T}{V}\,dV = -NR_u T \ln \frac{V_2}{V_1}$$

By measurement of the temperature of the process and the initial and final volumes, the maximum work done by the system can easily be calculated.

Example 3-2M Two kilograms of nitrogen gas at 27°C and 1.5 bars are compressed isothermally to 3.0 bars. Determine the minimum work of compression, in kilojoules, for the gas.

SOLUTION The minimum work of compression is found when the process is quasistatic. In this case the boundary work is given by the integral of $-P\,dV$. Since the maximum pressure on the diatomic gas is only 3.0 bars, the assumption that nitrogen is an ideal gas under these conditions is a reasonably good one. It is shown above for an isothermal process that the boundary work for an ideal gas is given by $-NR_u T \ln (V_2/V_1)$. For an ideal gas, also, $P_1V_1/T_1 = P_2V_2/T_2$. For an isothermal process of an ideal gas, then, $V_2/V_1 = P_1/P_2$. Consequently, using R_u from Table A-1M and the molar mass from Table A-2M,

$$W = -NR_u T \ln (V_2/V_1) = NR_u T \ln (P_2/P_1)$$

$$= 2/28 \text{ kg} \cdot \text{mol} \times 8.314 \text{ kJ/(kg} \cdot \text{mol)(°K)} \times 300\text{°K} \times \ln (3.0/1.5)$$

$$= 0.0714(8.314)(300)(0.693) = 123 \text{ kJ}$$

The answer is expected to be positive, since the process is one of compression and work in is positive. The answer could also be expressed as $W_{in} = 123$ kJ.

Example 3-2 Two pounds of nitrogen gas at 80°F and 20 psia are compressed isothermally to 40 psia. Determine the minimum work of compression, in ft · lb_f, for the gas.

SOLUTION The minimum work of compression is found when the process is quasistatic, in which case the boundary work is given by the integral of $-P\,dV$. Since the pressure never exceeds 40 psia, the assumption can be made that diatomic nitrogen is an ideal gas during the process. It is shown above for an isothermal process of an ideal gas that the boundary work is given by $-NR_u T \ln V_2/V_1$. Also, for an ideal gas, $P_1V_1/T_1 = P_2V_2/T_2$. Under isothermal conditions, then, V_2/V_1 may be replaced by P_1/P_2. Consequently, on the basis of data from Tables A-1 and A-2,

$$W = -NR_u T \ln V_2/V_1 = NR_u T \ln (P_2/P_1)$$

$$= 2/28 \text{ lb} \cdot \text{mol} \times 1545 \text{ ft} \cdot \text{lb}_f/(\text{lb} \cdot \text{mol})(\text{°R}) \times 540\text{°R} \times \ln (40/20)$$

$$= 0.0714(1545)(540)(0.693) = 41{,}300 \text{ ft} \cdot \text{lb}_f$$

This answer is positive, since the process is one of compression. The answer could also be expressed as $W_{in} = 41{,}300$ ft · lb_f.

The value of the work required in the above examples could be represented by an area under a hyperbolic line which is an isotherm on the Pv plane for an ideal gas.

3-3 SPECIFIC HEATS c_v AND c_p

The conservation of energy principle developed in Sec. 2-14 requires the knowledge of the internal energy of simple substances. In some special cases we shall find that the enthalpy function h, defined as $u + Pv$, is a convenient function to employ in conjunction with this conservation principle. Since the internal energy and the enthalpy of substances are not directly measurable, it is necessary to develop equations for these properties in terms of other measurable properties, such as P, v, and T. For simple compressible systems the internal energy u is a function of two other intensive, intrinsic properties. It is advantageous to select the temperatures and the specific volume as the independent variables. If $u = u(T, v)$, we can write for the total differential of u that

$$du = \left(\frac{\partial u}{\partial T}\right)_v dT + \left(\frac{\partial u}{\partial v}\right)_T dv \tag{3-4}$$

The first partial derivative on the right is defined as c_v, the specific heat at constant volume. That is,

$$c_v \equiv \left(\frac{\partial u}{\partial T}\right)_v \tag{3-5}$$

Since the specific internal energy u may be expressed either on a unit-mass basis or a mole basis, values of c_v found in the literature may be expressed also on a unit-mass or mole basis. That is, c_v is commonly tabulated in units of both kilograms (or pounds) and kilogram-moles (or pound-moles).

Evaluation of the enthalpy function h is usually made in terms of temperature and pressure as the independent variables. By letting $h = h(T, P)$ we may then write that

$$dh = \left(\frac{\partial h}{\partial T}\right)_P dT + \left(\frac{\partial h}{\partial P}\right)_T dP \tag{3-6}$$

The first partial derivative on the right is defined as the specific heat at constant pressure c_p. Then we have, in a fashion analogous to Eq. (3-5),

$$c_p \equiv \left(\frac{\partial h}{\partial T}\right)_P \tag{3-7}$$

Similar to the comments made with respect to c_v, the values of c_p commonly are tabulated on either a unit-mass or mole basis. Thus we find that derivatives of properties with respect to other properties are state functions of the system. The specific heats c_v and c_p are two of the most important derivative-type properties in thermodynamics.

The dimensions on the values of the specific heats defined above are in terms of energy/(mass)(temperature difference). It is important to note that the temperature quantity in the denominator involves a change in temperature, and not the value of the temperature itself. Consequently, the specific heat may be expressed in terms of either kelvins or degrees Celsius. The symbols °K and °C for these two temperature scales have the same significance in this special case, where only differences in temperature are of interest. The same reasoning obviously applies to degrees Rankine versus degrees Fahrenheit. Common units for specific heats in the SI are kJ/(kg)(°C)[or kJ/(kg)(°K)] and kJ/(kg · mol)(°C). In the USCS the units employed most frequently are Btu/(lb_m)(°F)[or Btu/(lb_m)(°R)] or Btu/(lb · mol)(°F).

3-4 INTERNAL ENERGY, ENTHALPY, AND SPECIFIC-HEAT RELATIONS FOR IDEAL GASES

When gases are at sufficiently low pressures, their PvT behavior is approximated very closely by the relationship $Pv = RT$. In order to make suitable energy balances for processes involving ideal gases, it is necessary to evaluate internal-energy and enthalpy changes for these gases. The internal-energy change for any simple compressible substance is given by Eq. (3-4). When this equation is combined with Eq. (3-5), we find that in general

$$du = c_v \, dT + \left(\frac{\partial u}{\partial v}\right)_T dv \tag{3-8}$$

The second coefficient, $(\partial u/\partial v)_T$, is a measure of the change in the internal energy of a substance as the volume is altered at constant temperature. From microscopic considerations it can be argued that the internal energy of an ideal gas should not be a function of the volume of the system. This result is confirmed by macroscopic measurements. As early as the middle of the nineteenth century, Joule carried out a series of experiments which indirectly indicated that the internal energy of gases at low pressures was essentially a function of the temperature only. The fact that $(\partial u/\partial v)_T$ is approximately zero at low pressures for gases is sometimes called Joule's law. Consequently, for substances which approximate ideal-gas behavior, the coefficient of the last term in Eq. (3-8) may be taken as zero. Thus, for ideal gases, we may write

$$du = c_v \, dT \qquad \text{for ideal gases} \tag{3-9}$$

for *all* processes, whether constant-volume or not. Hence the internal energy of an ideal gas, unlike that of real gases, is a function of only one independent variable, the temperature. Equation (3-9) is an excellent approximation for many gases up to pressures of several hundred pounds per square inch. This equation is misleading, however, in one respect. The presence of the term c_v in the equation often leads one to misconstrue the proper use of the equation. If a gas behaves essentially as an ideal gas, this expression for du is valid for all processes, regardless of its path. The use of the equation is not restricted to constant-volume processes. The simplicity of the relationship is a consequence of the special nature of an ideal gas. Since the specific heat at constant volume is defined for simple compressible substances as $(\partial u/\partial T)_v$, the value of c_v must also be solely a function of temperature for ideal gases.

Integration of Eq. (3-9) for any finite process involving an ideal gas leads to

$$\Delta u = \int c_v \, dT \qquad \text{for ideal gases} \tag{3-10}$$

The right-hand side can be evaluated once the empirical data for c_v as a function of temperature are measured.

The extension of these results to the property enthalpy is straightforward. By definition, $h = u + Pv$, and for an ideal gas, $Pv = RT$. Thus we may write

$$dh = du + d(Pv) \qquad \text{and} \qquad d(Pv) = d(RT) = R\,dT$$

The change in the enthalpy for an ideal gas then becomes

$$dh = du + R\,dT \tag{3-11}$$

The terms on the right-hand side of Eq. (3-11) are solely a function of temperature for an ideal gas. Consequently, the enthalpy of a hypothetical ideal gas is also only temperature-dependent. Likewise, the c_p values for ideal gases are a function of temperature only.

To evaluate enthalpy changes of ideal gases, one uses the basic equation represented by Eq. (3-6). When this latter equation is combined with Eq. (3-7) for c_p, we find that a general expression for dh for any simple compressible substance is

$$dh = c_p\,dT + \left(\frac{\partial h}{\partial P}\right)_T dP \tag{3-12}$$

Since the enthalpy of an ideal gas is solely a function of temperature, the above equation reduces to

$$dh = c_p\,dT \qquad \text{for ideal gases} \tag{3-13}$$

or

$$\Delta h = \int c_p\,dT \qquad \text{for ideal gases} \tag{3-14}$$

This set of equations is valid for *all* processes of an ideal gas, and is not restricted to constant-pressure processes.

A special relationship between c_p and c_v for ideal gases is obtained by substituting Eqs. (3-9) and (3-13) into Eq. (3-11). This yields

$$c_p\,dT = c_v\,dT + R\,dT$$

or

$$c_p - c_v = R \qquad \text{for ideal gases} \tag{3-15}$$

This simple relationship between c_p and c_v for an ideal gas is an important one, since a knowledge of either c_p or c_v allows the other one to be calculated by the above equation. When the specific heats are given as molar values, the value of R in this equation is R_u, the universal gas constant.

The integration of Eqs. (3-10) and (3-14) for Δu and Δh requires a knowledge of the specific-heat variation with temperature at low pressures. The following section discusses the general behavior of the specific heats of gases in terms of their order of magnitude, as well as their functional dependence on both temperature and pressure.

3-5 SPECIFIC HEAT VARIATION WITH TEMPERATURE

The molar specific heats at constant pressure of some common gases are illustrated in Fig. 3-2 as a function of temperature. The data have been extrapolated to zero pressure and are known as zero-pressure or ideal-gas specific heats. The symbols $c_{p,0}$ and $c_{v,0}$ are used frequently to signify values at this state of very low pressure. In Fig. 3-2 it is seen that a monatomic gas such as argon has a value of $c_{p,0}$ which is very close to 5.0 Btu/(lb · mol)(°F) over the entire range of temperature. This value, which is 20.8 kJ/(kg · mol)(°C), is characteristic of all monatomic gases. On the basis of Eq. (3-15) it is found that $c_{v,0}$ for monatomic gases must be approximately 12.5 kJ/(kg · mol)(°C) or 3.0 Btu/(lb · mol)(°F) over a wide range of temperature. As shown in the figure, molecules with two or more atoms do not have a constant value of c_p. These more complex molecules exhibit an increase in c_p with increasing temperature at low pressures. The c_p values of the diatomic gases shown in Fig. 3-2 increase as much as 25 percent over a range of 0 to 2000°F (0 to 1100°C). Values of the constant-pressure and constant-volume specific heats are tabulated, in Tables A-4M and A-4, as a function of temperature for a few common gases at essentially zero pressure. The values of c_v and c_p for monatomic gases quoted above are also repeated in these tables.

In order to integrate Eqs. (3-10) and (3-14) for Δu and Δh, we generally need equations which relate c_v and c_p to temperature. Specific-heat data as a function of temperature are measured directly or evaluated from theory based on a molecular model of matter. Reasonably accurate algebraic equations may then be

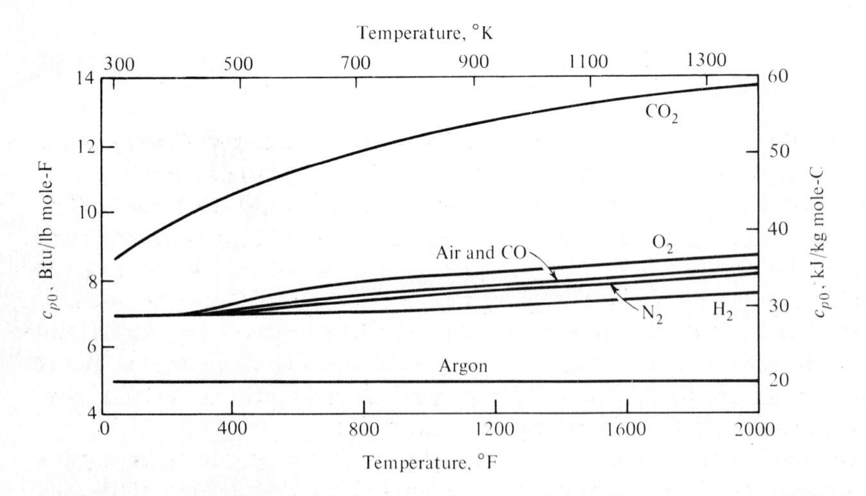

Figure 3-2 Values of $c_{p,0}$ for seven common gases. (*Based on data from NBS Circular 564, 1955.*)

fitted to the experimental or calculated data. Equations for zero-pressure specific-heat data based on spectroscopic measurements are given in Tables A-4M and A-4 for a number of common gases. The units to be employed in the equations must be specified by the author of the equations. In many tables of specific-heat equations the range of temperatures over which an equation is valid will be listed. Also, the maximum percent error of an equation, when used within the prescribed range of temperatures, is frequently noted. The algebraic format of such equations is purely arbitrary.

On the basis of the information on specific-heat data presented above, we are now in a position to evaluate internal-energy and enthalpy changes for ideal gases. In recapitulation, for ideal gases,

$$\Delta u = \int c_v \, dT \tag{3-10}$$

and

$$\Delta h = \int c_p \, dT \tag{3-14}$$

With the use of the types of equations presented in Tables A-4M and A-4, direct integration of these two equations is possible. Such integrations normally would be carried out for each set of temperature limits of interest. This method of evaluation is acceptable when only a few integrations are anticipated. However, this type of computation is highly undesirable for design purposes if many repetitive calculations are required for specific gases. Consequently the internal-energy and enthalpy data for many gases have been extensively tabulated over small temperature increments. These tables are the result of the integration of accurate specific-heat data and they greatly simplify the evaluation of Δu and Δh. In order to obtain the values of u and h shown, the integration process is based on the following modification of Eqs. (3-10) and (3-14), namely,

$$u = u_{\text{ref}} + \int_{T_{\text{ref}}}^{T} c_v \, dT \qquad \text{and} \qquad h = h_{\text{ref}} + \int_{T_{\text{ref}}}^{T} c_p \, dT \tag{3-16}$$

That is, to obtain any u or h value at a given T, an arbitrary reference value u_{ref} or h_{ref} is set at a reference temperature T_{ref}. In the ideal-gas tables in the appendix the artibrary reference value of zero enthalpy is chosen to be at 0°K. The h and u data for air in Table A-5M are in kJ/kg (or J/g), while the same properties in Table A-5 are in Btu/lb. Tables A-6M to A-11M list the ideal-gas enthalpies and internal energies of the gases N_2, O_2, CO, CO_2, H_2O, and H_2, and these h and u data are *molar* values, that is, kJ/kg · mol (or J/g · mol). Data for these six gases in USCS units appear in Tables A-6 to A-11, and the molar units in this case are Btu/lb · mol. Other properties that appear in these tables, in addition to h and u, will be introduced in a later chapter.

When the temperature interval is relatively small, the specific heats of gases are nearly constant. This can be confirmed by noting data over a several hundred degree interval in either Table A-4M or A-4. Frequently it is convenient in such cases to assume a constant c_P or c_v value, so that Eqs. (3-10) and (3-14) reduce to

$$\Delta u = c_{v,\,\text{av}}\,\Delta T \tag{3-17}$$

and

$$\Delta h = c_{p,\,\text{av}}\,\Delta T \tag{3-18}$$

where $c_{v,\,\text{av}}$ and $c_{p,\,\text{av}}$ in these equations are arithmetic average values for the given temperature interval. This method is especially useful when only tabular specific-heat data are available. The example below illustrates and compares various methods of evaluating the enthalpy change of an ideal gas.

Example 3-3M Find the change in enthalpy of 1 kg of air which is heated at low pressure from 300 to 500°K by use of (*a*) empirical specific-heat data, (*b*) average specific-heat data, and (*c*) Table A-5M, the table for air.

SOLUTION (*a*) The constant-pressure specific heat of air given in Table A-4M is

$$c_p = 30.26 - 9.154 \times 10^{-3}T + 20.50 \times 10^{-6}T^2 - 7.832 \times 10^{-9}T^3$$

where c_p is measured in kJ/(kg · mol)(°K) and T is in kelvins. If the air is assumed to behave as an ideal gas, then $\Delta h = \int c_p\,dT$. Substitution of the above equation for c_p into the expression for Δh, and subsequent integration, leads to

$$\Delta h = 30.26(T_2 - T_1) - 4.577 \times 10^{-3}(T_2^2 - T_1^2) + 6.834 \times 10^{-6}(T_2^3 - T_1^3)$$

$$- 1.958 \times 10^{-9}(T_2^4 - T_1^4)$$

$$= 6052 - 732 + 670 - 107 = 5883 \text{ kJ/kg} \cdot \text{mol}$$

Since 1 kg · mol of air contains approximately 29 kg,

$$\Delta h = 203.0 \text{ kJ/kg}$$

Hence the enthalpy change for 1 kg is 203.0 kJ.

(*b*) On the basis of an average specific heat, we note from Eq. (3-18) that $\Delta h = c_{P,\,\text{av}}\,\Delta T$. At 300 and 500°K, from Table A-4M, we note c_p values of 1.005 and 1.029 kJ/(kg)(°K), respectively. Using the arithmetic average, we find that

$$\Delta h = 1.017(200) = 203.4 \text{ kJ/kg}$$

This answer, based on a linear variation of c_p with temperature over the given temperature range, differs by roughly 0.2 percent from the integrated value. This close approximation is not unexpected, since the temperature interval is reasonably small.

(*c*) From Table A-5M, the listing of properties for dry air, the h values are read directly. Thus,

$$\Delta h = h_2 - h_1 = 503.02 - 300.19 = 202.8 \text{ kJ/kg}$$

All three answers are substantially the same. This problem illustrates the point that the use of tabular data is much preferred over direct integrations, which are time-consuming. For desk calculations tabular data are extremely useful. In the absence of tabular data and specific-heat equations, use of constant or average specific-heat data lead to reasonably accurate answers if the temperature range is small. In fact, in this particular problem, the use of c_P at 300°K (the initial temperature) of 1.005 kJ/(kg)(°C) leads to a Δh equal to 201 kJ/kg, which is very close to the other values.

Example 3-3 Find the change in enthalpy of 1 lb of air which is heated at low pressure from 100 to 500°F by use of (*a*) empirical specific-heat data, (*b*) average specific-heat data, and (*c*) Table A-5, the table for air.

SOLUTION (*a*) The constant-pressure specific heat of air given in Table A-4 is

$$c_p = 7.229 - 1.215 \times 10^{-3}T + 1.511 \times 10^{-6}T^2 - 0.321 \times 10^{-9}T^3$$

where c_p is measured in Btu/(lb · mol)(°R) and T is in degrees Rankine. If the air is assumed to behave as an ideal gas, then $\Delta h = \int c_p\, dT$. Substitution of the above equation for c_p into the expression for Δh, and subsequent integration, yields

$$\Delta h = 7.229(T_2 - T_1) - 0.608 \times 10^{-3}(T_2^2 - T_1^2) + 5.037 \times 10^{-6}(T_2^3 - T_1^3)$$
$$- 0.803 \times 10^{-9}(T_2^4 - T_1^4)$$

where $T_2 = 500°\text{F} = 960°\text{R}$ and $T_1 = 100°\text{F} = 560°\text{F}$. Upon evaluation,

$$\Delta h = 2892 - 369 + 357 - 60 = 2820 \text{ Btu/lb} \cdot \text{mol}$$

Since 1 lb · mol of air contains approximately 29.0 lb,

$$\Delta h = 97.3 \text{ Btu/lb}$$

(*b*) On the basis of an average specific heat we shall employ Eq. (3-18). At 100 and 500°F, from Table A-4, we note c_p values of 0.240 and 0.248 Btu/(lb)(°F), respectively. Using the arithmetic average of c_p leads to

$$\Delta h = c_{p,\text{ av}}\,\Delta T = 0.244(400) = 97.6 \text{ Btu/lb}$$

The use of an arithmetic value of c_p for the temperature range leads to roughly a 0.3 percent difference, when compared to the integration method. This small difference is not unexpected, since the temperature range is relatively small.

(*c*) The values of h at the two temperatures may be read directly from Table A-5, the listing of properties for dry air. Thus,

$$\Delta h = 231.06 - 133.86 = 97.2 \text{ Btu/lb}$$

Note that all three methods give answers which are in substantial agreement. However, the problem illustrates the point that the use of tabular data is much preferred over repeated direct integrations, which are time-consuming. For desk calculations, tabular data are extremely useful. In the absence of tabular data and specific-heat equations, use of constant or average specific-heat data lead to reasonably accurate results if the temperature range is small. In fact, in this particular problem, the use of the initial c_p value at 100°F of 0.240 Btu/(lb)(°F) leads to a Δh of 96.0 Btu/lb, which is only in error by 2 percent.

3-6 PROBLEM-SOLVING TECHNIQUES

Thermodynamics is a science which aids the engineer in the design of processes and equipment which are useful to mankind. Engineering design involves a great deal of problem solving. Hence, it is important for a student early in his or her academic career to begin to acquire good habits with respect to problem-solving procedures. There are certain steps and procedures which are fairly common to most engineering analysis. A general methodology for problem solving involves the following major points.

1. An engineering problem usually begins as a written or verbal statement. The information contained in this statement must be translated into sketches, diagrams, and symbols. Students frequently attempt to put numbers into equations without ever determining the nature of the problem.

2. Sketches of the system, with the appropriate system boundaries indicated, are of great value in approaching a problem in a consistent manner. Input data on the state of the system and on heat and work interactions should be indicated in appropriate places.
3. Process diagrams, such as a Pv diagram, are extremely helpful as an aid in picturing the initial and final states and the path of the process. Such diagrams are useful sometimes in determining what type of tabular data may be needed eventually in the numerical solution.
4. Idealizations or assumptions should be listed that might be necessary in order to solve the problem. For example, is the process necessarily quasistatic, or is the fluid an ideal gas?
5. When one makes use of a relation such as the ideal-gas relation, one has selected an *equation of state*. Equations of state relate properties at a given equilibrium state, but give no information on how these properties vary during a process.
6. The work statement of the problem may indicate a special nature of the path of the process, such as isothermal, constant pressure, adiabatic, etc. These word statements of how things happen are extremely important and are known as process relations or *process equations*. Make use of this information.
7. Determine what energy interactions are important, and recognize the sign conventions on these terms.
8. Fundamentals should be applied at this point. For example, write a suitable *basic* energy balance. Indicate which terms are zero or negligible.
9. Complete the solution. Watch the units used in the various equations, so that they are consistent. Is the numerical answer reasonable in the light of your common experience? All problems in this text of an engineering nature have been designed to give answers which are consistent with common practice.

Note that you may wish to check a number of things before ever attempting the numerical solution suggested in item 9. Not all of these items may be necessary in a given problem, and no specific order of the items is implied. However, a student's difficulty with a given problem usually arises because one or more of these items have been neglected. This checklist will be employed in the problem examples which follow. The student should look upon the problem examples as a major source for acquiring an understanding of the fundamentals of thermodynamics. In numerical problems it should be assumed that all input data are accurate to three significant figures, even though they frequently will be reported to less than this.

3-7 ENERGY ANALYSIS OF CLOSED IDEAL-GAS SYSTEMS

The preceding sections have developed some basic relationships among the properties of an ideal gas. The most useful of these include

$$Pv = RT \tag{3-3}$$

$$du = c_v \, dT \tag{3-9}$$

$$dh = c_p \, dT \tag{3-13}$$

and

$$c_p - c_v = R \tag{3-15}$$

In addition, c_p and c_v are solely functions of temperature. For monatomic and diatomic gases these equations can be used up to pressures of several hundred psia and above with reasonable accuracy if the temperature is room temperature and above. If the temperature interval is small (several hundred degrees or less), the internal-energy and enthalpy changes can be estimated accurately by assuming constant specific heats. In this case $\Delta u = c_v \Delta T$ and $\Delta h = c_p \Delta T$. For some gases the equations for du and dh have been integrated, using accurate specific-heat data, and the results have been presented in tabular form.

With the availability of data for u, h, c_p, and c_v, we are in a position to employ the conservation of energy principle to closed systems containing gases at relatively low pressures. For simple, compressible substances in a stationary, closed system, recall that in general

$$Q + W = \Delta U \tag{2-29}$$

or, on a unit-mass basis,

$$q + w = \Delta u \tag{2-31}$$

Before proceeding, however, it is important to recall the characteristics of a Pv diagram for an ideal gas. This diagram will be a useful problem-solving device. Figure 3-3 shows a general Pv diagram for an ideal gas. Two isotherms (constant-temperature lines) are shown as hyperbolas, where $T_2 > T_1$. The change in temperature along the constant-pressure path ab is also the same along the constant-volume path cb. Recall that u and h are solely functions of temperature. Hence the curve along which a and c lie is also a constant-internal-energy and a constant-enthalpy line, u_1 and h_1. The curve which contains b must also represent

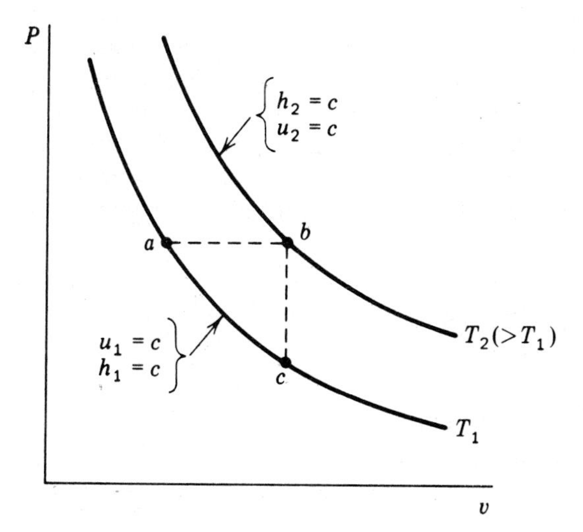

Figure 3-3 General Pv plot for an ideal gas, showing position of isotherms.

u_2 and h_2, as well as T_2. Consequently, the five important properties—P, v, T, u, and h—are all easily represented on the Pv diagram of an ideal gas.

The following examples illustrate the application of the problem-solving procedure discussed in Sec. 3-6 to ideal-gas systems. The use of the Pv diagram is also demonstrated.

Example 3-4M One kilogram of air is compressed slowly in a piston-cylinder assembly from 1 bar, 290°K, to a final pressure of 6 bars. During the process, heat is exchanged with the surroundings at a rate sufficient to make the process isothermal. Determine (*a*) the change in internal energy of the air, (*b*) the work interaction, in kJ/kg, and (*c*) the quantity of heat transferred, in kJ/kg.

SOLUTION (*a*) The air is considered a simple compressible system which undergoes a constant temperature (isothermal) process. The path of the process is shown on the Pv diagram of Fig. 3-4, and the system boundary is shown in the sketch of the system. Since the system contains 1 kg of gas, the basic energy equation can be written as

$$q + w = \Delta u$$

Since air is composed primarily of diatomic gases, it behaves as an ideal gas at 290°K (17°C) even at a pressure of 6 bars. The internal energy of an ideal gas is only a function of temperature. Since the temperature is constant (process equation), the change in internal energy is zero.

(*b*) On the basis of part (*a*), the energy equation reduces to $q = -w$. If either of the terms can be evaluated, then the other automatically is found. We have not had any specific equations for computing heat-transfer quantities independently, and therefore we must first evaluate w. From the description of the process the only work interaction is boundary work, and it is given by the integral of $-P\,dv$. P and v are functionally related by the ideal-gas equation of state $Pv = RT$ if we assume that the process is quasistatic. Hence,

$$w = -\int P\,dv = -\int RT\,\frac{dv}{v} = -RT\ln\frac{v_2}{v_1} = RT\ln\frac{P_2}{P_1}$$

The gas constant R is found in Table A-1M. Substitution of the proper data yields

$$w = \frac{8.314}{29}\ \text{kJ/(kg)(°K)} \times 290\text{°K} \times \ln\frac{6}{1} = 149\ \text{kJ/kg}$$

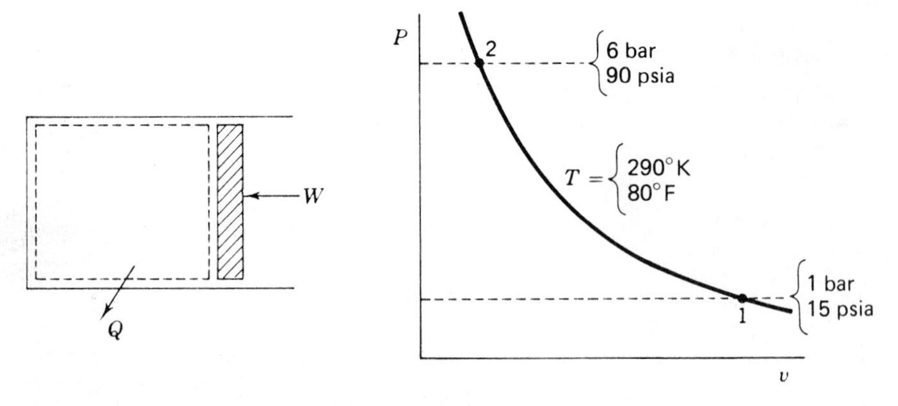

Figure 3-4 Schematic of system and a Pv diagram of the process for Examples 3-4M and 3-4.

The positive sign on w indicates that the work is done on the system.

(c) By equating the heat transfer to the negative of the work interaction we note that q equals -149 kJ/kg, which is out of the system.

Example 3-4 One pound of air is compressed slowly in a piston-cylinder assembly from 15 psia, 80°F, to a final pressure of 90 psia. During the process heat is exchanged with the atmosphere at a rate sufficient to make the process isothermal. Determine (a) the change in the internal energy of the gas, (b) the work interaction in ft · lb_f/lb_m, and (c) the quantity of heat transferred in Btu/lb.

SOLUTION (a) The air is a simple system which undergoes a constant-temperature (isothermal) process. The path of the process is shown on the Pv diagram of Fig. 3-4, and the system boundary is shown in the sketch of the system. Since the system contains 1 lb of gas, the basic energy equation can be written as

$$q + w = \Delta u$$

Even at a pressure of 90 psia air behaves like an ideal gas at this temperature. The internal energy of an ideal gas is only a function of the temperature. Since the temperature is constant, the change in the internal energy is zero.

(b) On the basis of part (a), the energy equation reduces to $q = -w$. If either of the terms can be calculated, then the other automatically is obtained. We have not had any specific equations or methods introduced for computing heat-transfer quantities independently, and therefore we must tackle the evaluation of w. The only work interaction is boundary work, as given by the integral of $-P\,dv$. P and v are functionally related by the ideal-gas equation $Pv = RT$ if we assume that the process is quasistatic. Hence,

$$w = -\int P\,dv = -\int RT\frac{dv}{v} = -RT\ln\frac{v_2}{v_1} = RT\ln\frac{P_2}{P_1}$$

Substitution of proper data yields

$$w = \frac{1545}{29}\ \text{ft}\cdot\text{lb}_f/(\text{lb}_m)(°\text{R}) \times 540°\text{R} \times \ln\frac{90}{15} = 51{,}500\ \text{ft}\cdot\text{lb}_f/\text{lb}_m$$

(c) By equating the heat transfer to the work interaction and converting to the desired units, we find that

$$q = -w = -\frac{51{,}500}{778} = -66.2\ \text{Btu/lb}$$

The positive value of w indicates that work is done on the system, and the negative value of q indicates that heat is removed from the gas.

Example 3-5M An ideal gas has a constant-pressure specific heat of 2.20 kJ/(kg)(°C) and a molar mass of 16.04. Eight kilograms of the gas are heated from 17 to 187°C at constant volume. Determine (a) the work done by the gas, (b) the change in enthalpy of the gas, in kilojoules, and (c) the heat transferred, in kilojoules.

SOLUTION (a) The system is constant-volume; therefore no boundary work is done on the simple compressible substance. There is no other work interaction described in the problem; hence $W = 0$. The fact that the boundary work is zero is confirmed by noting that the area under the process curve on the Pv diagram in Fig. 3-5 is zero.

(b) The change in enthalpy of an ideal gas is given by $dH = mc_p\,dT$. There is only a single value of c_p given, so we will assume it is constant. The enthalpy change becomes

$$\Delta H = mc_p\,\Delta T = 8\ \text{kg} \times 2.20\ \text{kJ/(kg)(°C)} \times (187 - 17)°\text{C}$$

$$= 2990\ \text{kJ}$$

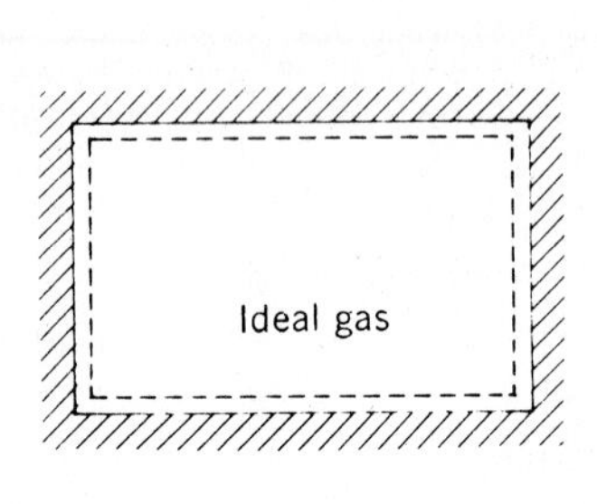

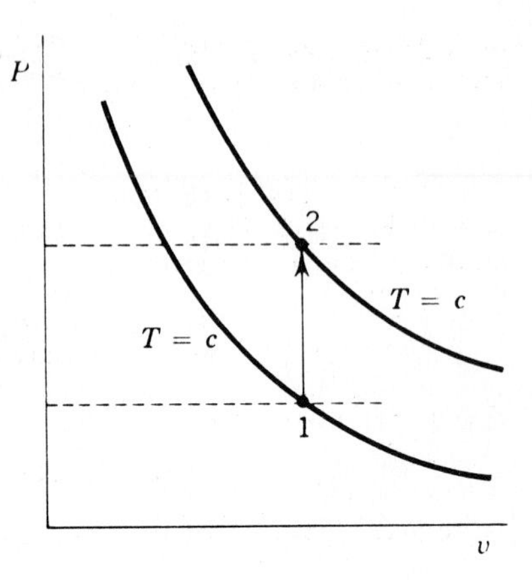

Figure 3-5 Schematic of system and a Pv diagram of the process for Examples 3-5M and 3-5.

It is important to note that the enthalpy change in this problem is just a number, and has no physical significance.

(c) The quantity of heat transfer is found from the conservation of energy principle. Basically, $Q + W = \Delta U$. Because $W = 0$, $Q = \Delta U$. The internal energy of an ideal gas with a constant specific heat is given by $\Delta U = mc_v \, \Delta T$. Therefore,

$$Q = mc_v \, \Delta T$$

The value of c_v is not given in the statement of the problem. However, it is related to c_p for an ideal gas by the relation $c_p - c_v = R$. Use of the universal gas constant from Table A-1M and the molar mass of the gas enables one to compute c_v.

$$c_v = c_p - \frac{R_u}{M} = 2.20 - \frac{8.314}{16.04} = 1.68 \text{ kJ/(kg)(°C)}$$

The quantity of heat transferred becomes

$$Q = 8 \text{ kg} \times 1.68 \text{ kJ/(kg)(°C)} \times (187 - 17)\text{°C} = 2285 \text{ kJ}$$

The positive value shows that heat is added to the system.

Example 3-5 An ideal gas has a constant-pressure specific heat of 0.526 Btu/(lb)(°F) and a molar mass of 16.04. Fifteen pounds of the gas are heated from 45 to 345°F at constant volume. Determine (*a*) the work done by the gas, (*b*) the change in enthalpy of the gas, and (*c*) the heat transferred.

SOLUTION (*a*) The system boundary is rigid; therefore, no boundary work is done. There is no other work interaction described in the problem; hence $W = 0$. Note that the area under the process curve on a Pv diagram in Fig. 3-5 is zero.

(*b*) The change in enthalpy of an ideal gas is given by $dH = mc_p \, dT$. There is only a single value of c_p given, and so we will assume it is constant. The enthalpy change becomes

$$\Delta H = mc_p \, \Delta T = 15 \text{ lb} \times 0.526 \text{ Btu/(lb)(°F)} \times (345 - 45)\text{°F}$$

$$= 2370 \text{ Btu}$$

The enthalpy change in this problem is just a number, and has no physical significance.

(*c*) The quantity of heat transfer is found from the conservation of energy principle. Basically, $Q + W = \Delta U$. Because $W = 0$, $Q = \Delta U$. The internal energy of an ideal gas with constant specific heats is given by $\Delta U = mc_v \, \Delta T$. Therefore,

$$Q = mc_v \, \Delta T$$

The value of c_v is not given in the statement of the problem. However, it is related to c_p for an ideal gas by the relation $c_p - c_v = R$. Use of the universal gas constant and the molar mass of the gas enables one to compute c_v.

$$c_v = c_p - \frac{R_u}{M} = 0.526 - \frac{1.986}{16.04} = 0.526 - 0.124 = 0.402 \text{ Btu/(lb)(°F)}$$

The quantity of heat transferred becomes

$$Q = 15(0.402)(345 - 45) = 1810 \text{ Btu}$$

The positive value shows that heat was added to the system.

Example 3-6M A vertical piston-cylinder assembly which initially has a volume of 0.1 m^3 is filled with 0.1 kg of nitrogen gas. The piston is weighted so that the pressure on the diatomic nitrogen is always maintained at 1.15 bars. Heat transfer is allowed to take place until the volume is 75 percent of its initial value. Determine the quantity and direction of the heat transfer, in kilojoules, and the final temperature of the nitrogen at equilibrium, in degrees Kelvin.

SOLUTION The system, or control mass, is taken as the mass of nitrogen within the piston-cylinder. A schematic diagram of the system and a process diagram on a Pv plot are shown in Fig. 3-6. Since field effects, etc., are absent, we assume the system to be a simple compressible one. Under this assumption the conservation of energy principle reduces to Eq. (2-30); namely,

$$\delta Q + \delta W = dU$$

Since the volume of the control mass is altered, boundary work is associated with the process. In order to evaluate this form of work, it is necessary to assume that the process is carried out

cons T no ΔU

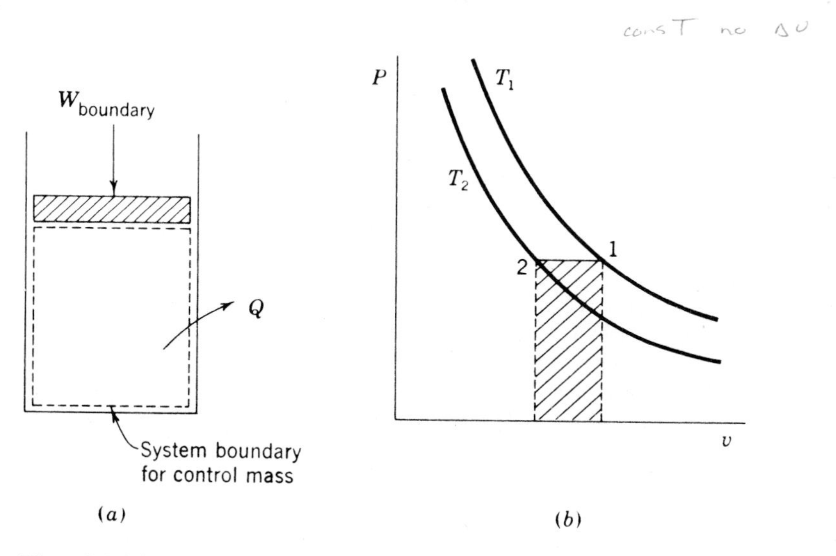

Figure 3-6 Schematic of system and a Pv diagram of the process for Examples 3-6M and 3-6.

quasistatically. On this basis the work is given by the integral of $-P\,dV$. The energy equation becomes

$$\delta Q - P\,dV = dU$$

or

$$\delta Q = dU + P\,dV = dH$$

The replacement of $dU + P\,dV$ by dH is possible since the process is one of constant pressure.

The enthalpy change can be determined by noting that at 1.15 bars, which is just slightly above atmospheric pressure, nitrogen gas behaves essentially as an ideal gas. For an ideal gas, $dH = mc_p\,dT$. Hence the heat transfer for the proposed process is evaluated by

$$\delta Q = dH = mc_p\,dT$$

or

$$Q = \int mc_p\,dT$$

Integration of the right-hand side of this result depends upon the functional relation between c_p and T. If the temperature change is not too great, an average value for c_p can be chosen and the integration carried out quickly. If the temperature change is large, then tabular data would be appropriate.

In order to obtain the initial temperature T_i of the gas, the ideal-gas equation of state is used.

$$T_i = \frac{P_i V_i}{NR_u} = 1.15 \text{ bars} \times 0.1 \text{ m}^3 \times \frac{28}{0.1} \frac{1}{\text{kg} \cdot \text{mol}} \times \frac{1}{0.08314} (\text{kg} \cdot \text{mol})(^\circ\text{K})/\text{bar} \cdot \text{m}^3$$

$$= 387^\circ\text{K}$$

Application of the ideal-gas relation for a constant-pressure process also shows that $T_f = T_i(V_f/V_i)$. Thus the final temperature T_f is

$$T_f = 387 \frac{0.75}{1.0} = 290^\circ\text{K}$$

In the small range of temperatures from T_i to T_f, Table A-4M indicates that c_p is nearly constant at 1.041 kJ/(kg)(°C). Consequently, the quantity of heat transferred is

$$Q = mc_{p,\text{av}}(T_f - T_i) = 0.1 \text{ kg} \times 1.041 \text{ kJ/(kg)(}^\circ\text{C)} \times (290 - 387)^\circ\text{C}$$

$$= -10.1 \text{ kJ}$$

The negative sign reveals that heat must be removed from the system in order to reduce the volume to 75 percent of its initial value at constant pressure.

Alternative solutions and discussion Since the conservation of energy principle under the stated restrictions reduces to $Q = \Delta H = m(h_f - h_i)$, the numerical answer can also be obtained by using the gas table for nitrogen, Table A-6M, at the calculated temperatures. This table indicates values of roughly 11,260 and 8432 kJ/kg · mol for the initial and final enthalpies, respectively, when based on linear interpolation. Hence the value of the heat transfer is

$$Q = 0.1 \text{ kg} \times \tfrac{1}{28} \text{ kg} \cdot \text{mol/kg} \times (8{,}432 - 11{,}260) \text{ kJ/kg} \cdot \text{mol}$$

$$= -10.1 \text{ kJ}$$

Good agreement with the result based on an average c_p value is obtained because the variation of c_p with temperature is negligible.

Another alternative solution is to base the calculation on

$$Q = \Delta U + P\,\Delta V$$

without a further combination of the last two terms into ΔH. The change in ΔU can be found either by $c_{v,\text{av}}\,\Delta T$ or from u data in the nitrogen table. Since ΔV is known explicitly, $P\,\Delta V$ can be computed directly. The reader may wish to confirm that ΔU equals -7.2 kJ by either method, and that $P\,\Delta V$ equals -2.9 kJ. The work required is shown by the cross-hatched area on the Pv plot, and is into the system. Again the overall answer is -10.1 kJ. This latter method is somewhat more descriptive, since it shows explicitly the sources of the energy which contributed to Q. The ΔH computation is faster, but shows less detailed information on the overall process.

Example 3-6 A vertical piston-cylinder assembly which initially has a volume of 1.0 ft^3 is filled with 0.1 lb$_m$ of nitrogen. The piston is weighted so that the pressure on the nitrogen is always maintained at 20 psia. Heat transfer is allowed to take place until the volume is 90 percent of its initial value. Determine the quantity and direction of the heat transfer and the final temperature of the nitrogen at equilibrium.

SOLUTION The system, or control mass, is taken as the mass of nitrogen within the confines of the piston cylinder. A schematic diagram of the system and a process diagram on a Pv plot are shown in Fig. 3-6. The system can be treated as a simple compressible system, since the effects of magnetic and electric fields, etc., are absent. Under this assumption the conservation of energy principle reduces to Eq. (2-30); namely,

$$\delta Q + \delta W = dU$$

Since the volume of the control mass is altered, a work interaction is associated with the process. In order to evaluate the work done in moving the boundary, it is necessary to assume that the process is carried out quasistatically. On this basis the work is given by the integral of $-P\,dV$. The energy equation then becomes

$$\delta Q - P\,dV = dU$$

or

$$\delta Q = dU + P\,dV = dH$$

The enthalpy change can be determined by noting that at 20 psia, that is, just slightly above atmospheric pressure, nitrogen gas behaves essentially as an ideal gas. For an ideal gas, $dH = mc_p\,dT$. Hence the heat transfer for the proposed process can be evaluated by

$$\delta Q = dH = mc_p\,dT$$

or

$$Q = \int mc_p\,dT$$

Integration of the right-hand side of this result depends upon the functional relation between c_p and T. If the temperature change is not too great, an average value for c_p can be chosen and the integration carried out quickly. We have noted already that the specific heats of diatomic gases do not change rapidly with temperature at about room temperature and above. In order to obtain the initial temperature of the gas, the ideal-gas equation of state is used.

$$T_i = \frac{P_i V_i}{NR_u} = 20 \times 144\ \text{lb}_f/\text{ft}^2 \times 1.0\ \text{ft}^3 \times \frac{28}{0.1}\,\frac{1}{\text{lb}\cdot\text{mol}} \times \frac{1}{1545}\,(\text{lb}\cdot\text{mol})(^\circ\text{R})/\text{ft}\cdot\text{lb}_f$$

$$= 522^\circ\text{R} = 62^\circ\text{F}$$

Application of the ideal-gas relation for a constant-pressure process also shows that, since $P_i V_i/T_i = P_f V_f/T_f$, then $T_f = T_i(V_f/V_i)$. Thus the final temperature is

$$T_f = 522\,\frac{0.9}{1.0} = 470^\circ\text{R} = 10^\circ\text{F}$$

In the range of temperatures from T_i to T_f the specific heat at constant pressure is nearly constant at 0.248 Btu/(lb$_m$)(°F). Consequently, the quantity of heat transfer is

$$Q = \int mc_p \, dT = mc_p(T_f - T_i) = 0.1(0.248)(470 - 522) = -1.29 \text{ Btu}$$

The negative sign reveals that heat must be removed from the control mass in order to reduce the volume to 90 percent of its initial value at constant pressure.

ALTERNATIVE SOLUTIONS AND DISCUSSION Since the conservation of energy principle for the control mass under the stated restrictions reduces to $Q = \Delta H = m(h_f - h_i)$, the numerical answer can also be obtained by using the gas table for nitrogen, Table A-6, at the calculated temperatures. This table indicates values of 3625 and 3264 Btu/lb · mol for the initial and final enthalpies, respectively, when based on linear interpolation. Hence the value of the heat transfer is

$$Q = 0.1(3264 - 3625)(\tfrac{1}{28}) = -1.29 \text{ Btu}$$

Good agreement with the result based on an average c_p value is obtained because the variation of c_p with temperature is negligible.

Another alternative solution is to base the calculation on the relation

$$Q = \Delta U + P\,\Delta V$$

without a further combination of the last two terms into ΔH. The change in ΔU can be found either by $c_{v,\text{av}}\, T$ or from u data in the nitrogen table. Since ΔV is known from the input data, $P\,\Delta V$ can be computed directly. The reader may wish to confirm that ΔU equals -0.92 Btu by either method, and that $P\,\Delta V$ equals -0.37 Btu. The work required is shown as the cross-hatched area on the Pv diagram, and is into the system. Again, the overall answer is -1.29 Btu. This latter method is somewhat slower than using the ΔH method, but it does show the individual sources of energy which contribute to Q.

Example 3-7M A rigid insulated tank is divided into two equal volumes by a partition. Initially, 1 kg of a gas is introduced into one side of the partitioned tank, and the other side remains evacuated. In this equilibrium state the pressure and temperature are 2 bars and 100°C, respectively. The partition is then pulled out, and the gas is allowed to expand into the entire tank. Determine the final pressure and temperature at equilibrium if the gas is ideal.

SOLUTION The boundaries of the closed system will be selected to lie just inside the walls of the entire rigid, insulated tank, including the section which is initially evacuated (see Fig. 3-7). The system may be idealized as a simple compressible one. For such a system the energy equation is

$$Q + W = \Delta U$$

The boundary work associated with a rigid tank is zero, and all other work effects are absent. In addition, Q is zero for the insulated boundary. Because Q and W are zero for this process, we may write for the finite change of state that

$$\Delta U = 0$$

Consequently, the expansion of the gas in this case is one of constant internal energy; that is, U_2 equals U_1.

For an ideal gas the internal energy is solely a function of the temperature for a given equilibrium state. Since the internal energy is constant, the initial and final temperatures of the ideal gas undergoing this "free expansion" must be the same. Hence T_2 is 100°C. For an ideal gas at constant temperature the ideal-gas equation of state leads to the relation $P_2 = P_1(V_1/V_2)$. Since the final volume is twice the initial volume, the final pressure is one-half the original value, or 1 bar.

The process is shown on the PV diagram. The path between states 1 and 2 is drawn as a

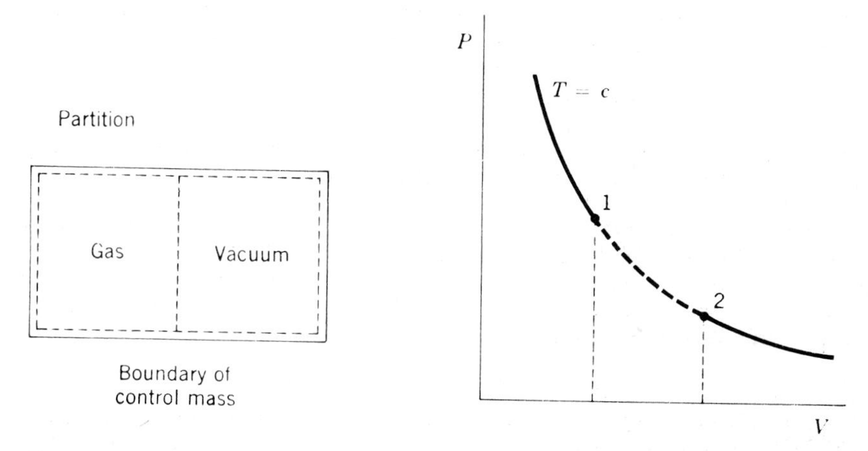

Figure 3-7 Schematic of system and a PV diagram of the process for Examples 3-7M and 3-7.

dashed line, since the process is nonquasistatic and the temperature is not defined except at the equilibrium end states.

Example 3-7 A rigid insulated tank is divided into two equal volumes by a partition. Initially, 1 lb of a gas is introduced into one side of the partitioned tank, and the other side remains evacuated. In this equilibrium state the properties of the gas are measured and recorded. The partition is then pulled out, and the gas is allowed to expand into the entire tank. Determine the final pressure and temperature at equilibrium if the gas is ideal and initially at 15 psia and 600°F.

SOLUTION The boundaries of the closed system will be selected to lie just inside the walls of the entire rigid, insulated tank, including the section which is originally evacuated (see Fig. 3-7). The system may be idealized as a simple one. In the absence of all other work effects, the rigid insulated tank assures that the system is an isolated one. For a simple system of fixed composition the energy equation is

$$Q + W = \Delta U$$

However, because Q and W are zero for this process, we may write for the finite change of state that

$$\Delta U = 0$$

Consequently, the expansion of the gas in this case is one of constant internal energy; that is, U_2 equals U_1.

For an ideal gas it has been shown that the internal energy is solely a function of the temperature for a given equilibrium state. Since the internal energy is constant, the initial and final temperatures of the ideal gas undergoing this "free expansion" must be the same. Hence T_2 is 600°F. For an ideal gas at constant temperature the ideal-gas equation leads to the relation $P_2 = P_1(V_1/V_2)$. Since the final volume is twice the initial volume, the final pressure is one-half the original value, or 7.5 psia.

The process is shown on the PV diagram. The path is drawn as a dashed line, since the process is nonquasistatic and the temperature is not defined except at the equilibrium end states.

REFERENCES

Burghardt, M. D.: "Engineering Thermodynamics with Applications," Harper & Row, New York, 1982.

Joint Army, Navy, and Air Force (JANAF) Thermochemical Tables, NSRDS-NBS-37, June 1971.

Keenan, J. H., and J. Kaye: "Gas Tables," Wiley, New York, 1945.

Reynolds, W. C., and H. C. Perkins: "Engineering Thermodynamics," 2d ed., McGraw-Hill, New York, 1977.

Zemansky, M. W., and H. C. Van Ness: "Basic Engineering Thermodynamics," 2d ed., McGraw-Hill, New York, 1974.

PROBLEMS (METRIC)

Ideal gas

3-1M A balloon is filled with methane gas (CH_4) at 20°C and 1 bar until the volume is (*a*) 42.5, and (*b*) 26.4 m^3. Calculate the mass, in kilograms, of methane under ideal-gas conditions.

3-2M Fifty liters of carbon monoxide are maintained at (*a*) 1.3 bars and 27°C, and (*b*) 2.1 bars and 127°C. Determine the mass of the ideal gas, in kilograms.

3-3M Propane (C_3H_8), in the amount of 1500 kg, is to be stored in a gas reservoir at (*a*) 12°C and 3 bars, and (*b*) 42°C and 450 kPa. The molar mass of propane is 44.0. What is the necessary volume of the reservoir, in cubic meters?

3-4M One-half kilogram of helium is placed in a rigid tank 0.5 m^3 in volume. If the temperature is (*a*) 27°C, and (*b*) 112°C, and the barometric pressure is 1.0 bar, what would be the reading, in bars, of a pressure gage which is attached to the tank?

3-5M Determine the reading, in kilopascals, on a pressure gage attached to a tank of hydrogen at (*a*) 22°C, and (*b*) 122°C when the density of the gas is 0.0798 kg/m^3. The barometric pressure is 98 kPa.

3-6M One kilogram of argon gas ($M = 40$) is placed in a rigid tank with a volume and pressure of (*a*) 0.60 m^3 and 1.40 bars, and (*b*) 0.45 m^3 and 0.170 MPa. If argon behaves as an ideal gas, what is the temperature in degrees Celsius?

3-7M One kilogram of air is maintained in (*a*) a 0.6-m^3 tank at a gage reading of 500 mbar, and (*b*) a 0.35-m^3 tank at a gage reading of 258 kPa. If the atmospheric (barometric) pressure is 970 millibars (97 kPa), what is the temperature of the ideal gas, in degrees Celsius?

3-8M A rigid tank has a volume of 3.0 m^3. It contains a gas having a molar mass of 30 at 8 bars and 47°C. Gas leaks out of the vessel until the pressure is 3 bars at 27°C. What volume is occupied by the gas which escaped if it is at 1.0 bar and 22°C?

3-9M A tank contains carbon dioxide at 5 bars and 30°C. A leak occurs in the tank which is not detected until the pressure has dropped to (*a*) 4.2 bars, and (*b*) 3.4 bars. If the temperature of the gas at the time of detection of the leak is 20°C, determine the mass of carbon dioxide that has leaked out if the original mass was 25 kg.

3-10M A tank contains helium at 6 bars and 40°C. A kilogram of the gas is removed, which causes the pressure and temperature to change to (*a*) 4 bars and 30°C, and (*b*) 3.4 bars and 10°C. Determine the volume of the tank, in liters.

3-11M Two tanks A and B, are connected by suitable pipes through a valve, which initially is closed. Tank A initially contains 0.3 m^3 of nitrogen at 5 bars and 60°C, and tank B is evacuated. The valve is then opened, and nitrogen flows into tank B until the pressure in this tank reaches 1.5 bars and the temperature is 27°C. As a result, the pressure in tank A drops to 4 bars and the temperature changes to 50°C. Determine the volume of tank B, in cubic meters.

Isothermal boundary work

3-12M Nitrogen gas initially at 1.2 bars and 30°C is compressed isothermally in a piston-cylinder device to 3.0 bars.

(*a*) Compute the work required by some outside source and show the area representation on a Pv diagram. Express w in J/g.

(*b*) Now consider that atmospheric air at 0.97 bar acts on the back of the piston, in addition to the connecting rod. Compute the net work required by the outside source in this case, for the same change in state of the nitrogen, and the Pv area representation.

(*c*) Finally, consider in addition to part (*b*) that friction acts between the piston and cylinder and that the frictional resistance is equivalent to an effective resistive pressure of 0.50 bar which is constant throughout the piston travel. Determine in this case the net work that must be delivered along the connecting rod in N · m/g. Show the area representation of this final net work on a Pv diagram.

3-13M A certain quantity of work is needed to reduce quasistatically the volume of an ideal gas to one-half of its initial value at a constant temperature T_1. At what temperature T_2 will the same amount of work reduce the volume isothermally to a quarter of the same initial volume if (*a*) $T_1 = 250°C$, (*b*) $T_1 = 350°C$, and (*c*) $T_1 = 400°C$? Report the answer in degrees kelvin.

3-14M Consider Prob. 3-12M, except that the gas is carbon dioxide, the atmospheric pressure is 96.0 kPa, and the frictional resistance is equivalent to a resistive pressure of 40 kPa.

3-15M An ideal gas initially in a piston-cylinder device at 1.5 bars and 0.03 m^3 is first heated at constant pressure until the volume is doubled. It is then allowed to expand isothermally until the volume is again doubled. Determine the total work done by the gas, in kJ/kg · mol, and plot the quasistatic processes on a PV diagram. The initial gas temperature is (*a*) 500°K, and (*b*) 300°K.

3-16M A piston-cylinder device contains 0.09 m^3 of an ideal gas at 1.2 bars and 20°C. The gas is compressed to a final pressure of (*a*) 3.6 bars, and (*b*) 2.8 bars, and a temperature of 20°C.

(1) Determine the minimum work of compression, in kilojoules, if the quasistatic process is isothermal.

(2) Consider now that a process occurs between the same two end states as in part (1) such that it follows a straight line on PV coordinates. Determine the work of compression for this process, and compare the two processes on a PV diagram.

Enthalpy and internal energy changes from specific heat data

3-17M The temperature of oxygen is increased from (*a*) 300 to 500°K, (*b*) 300 to 700°K, and (*c*) 400 to 800°K.

(1) On the basis of the equation for c_p as a function of temperature at zero pressure for oxygen (Table A-4M), calculate the change in enthalpy, in kJ/kg · mol.

(2) Compare this answer to that obtained by use of h data from Table A-7M.

(3) What percent error will be introduced by using the arithmetic-mean specific heat c_p from Table A-4M to evaluate Δh between the given temperature limits?

3-18M Consider Prob. 3-17M, except that the gas is carbon dioxide.

3-19M The temperature of oxygen is increased from (*a*) 300 to 600°K, and (*b*) 300 to 800°K.

(1) On the basis of the empirical equation for c_p of oxygen (Table A-4M), calculate the internal energy change, in kJ/kg · mol, by integration techniques.

(2) Compare this answer with that obtained by use of u data from Table A-7M.

(3) What percent error will be introduced by using the arithmetic-mean specific heat c_v from Table A-4M to evaluate Δu, compared to that in part 1?

3-20M Using the specific-heat data cited for monatomic gases in Table A-4M, calculate the change in internal energy and enthalpy of (*a*) helium for the temperature change from 100 to 600°C, and (*b*) argon for the temperature change from 300 to 700°K, in kJ/kg.

Energy analysis at constant temperature

3-21M One-half kilogram of air is compressed isothermally and quasistatically in a closed system from 1 bar, 20°C to a final pressure of (*a*) 6 bars, and (*b*) 2.4 bars. Determine (1) the internal energy change and (2) the heat transfer, in kilojoules.

3-22M Carbon dioxide is expanded isothermally and quasistatically in a closed system from initial conditions of 1.3 bars, 150°C, and 0.1 m^3 to a final volume of (*a*) 0.2 m^3, and (*b*) 0.3 m^3. Calculate the magnitude and direction of any necessary heat transfer, in kilojoules.

3-23M A piston-cylinder device contains 0.12 kg of air initially at 200 kPa and 123°C. During a quasistatic, isothermal process, heat is removed in the amount of (*a*) 20 kJ, and (*b*) 15 kJ, and electrical work is carried out on the system in the volume of 1.75 W · h. Determine the ratio of the final volume to the initial volume.

3-24M In a closed system 0.1 kg of argon ($M = 40$) expands quasistatically and isothermally from 2 bars, 325 K until (*a*) the volume is doubled and 42.0 kJ/kg of heat is added, and (*b*) the volume is tripled and 68.0 kJ/kg of heat is added. If during the process a 12-V battery is operated for 20 s, determine the constant current supplied, in amperes, to a resistor within the system. Neglect energy storage in the resistor.

3-25M Air is contained within a piston-cylinder device at initial conditions of 4 bars and 0.01 m^3. A paddle-wheel within the gas is turned under the conditions of (*a*) an applied torque of 500 N · cm for 4000 revolutions, and (*b*) an applied torque of 750 N · cm for 2400 revolutions. During the process the temperature remains constant while the volume is doubled. Determine the magnitude, in kilojoules, and the direction of any heat transfer.

Energy analysis at constant pressure

3-26M A system consists of 0.88 kg of carbon dioxide within a constant-pressure piston-cylinder device which is frictionless. The initial conditions are 7 bars and 17°C. Heat is added until the volume is doubled. On the basis of data in Table A-9M determine (*a*) the change in internal energy, in kilojoules, (*b*) the change in enthalpy, in kilojoules, (*c*) the work performed by or on the gas, in newton-meters, and (*d*) the quantity of heat supplied, in kilojoules.

3-27M A piston-cylinder device which initially has a volume of 0.1 m^3 contains 0.014 kg of hydrogen at 210 kPa. Heat is transferred until the final volume is (*a*) 0.082, and (*b*) 0.075 m^3. If the quasistatic process occurs at constant pressure, determine (1) the final temperature, in degrees Celsius, (2) the heat transfer, in kilojoules, based on data from Table A-11M, and (3) the heat transfer based on specific-heat data from Table A-4M.

3-28M Air is quasistatically heated at constant pressure in a piston-cylinder device from 2 bars, 0.06 m^3, and 47°C to a final volume of 0.09 m^3. Determine the heat transfer, in kilojoules, by using (*a*) average c_v data, and (*b*) average c_p data from Table A-4M.

3-29M One-hundredth of a kilogram-mole of helium is contained within an adiabatic piston-cylinder device and is maintained at a constant pressure of 2 bars. A paddle-wheel is operated within the gas until the volume of the gas has increased by 25 percent. Compute (*a*) the quantity of paddle-wheel work required, in newton-meters, and (*b*) the net work done on the system, if the piston is frictionless and the initial temperature of the gas is 20°C.

3-30M One-tenth cubic meter of nitrogen at 1 bar and 25°C is contained in a piston-cylinder device. A paddle-wheel within the cylinder is turned until (*a*) 23,700 N · m, and (*b*) 17,850 N · m of energy have been added. If the process is quasistatic, adiabatic, and at constant pressure, determine the final temperature of the gas in degrees Celsius. Neglect energy storage in the paddle-wheel, and use data from Table A-6M.

3-31M A piston-cylinder device maintains nitrogen at a constant pressure of 200 kPa. The initial volume and temperature are 0.10 m^3 and 20°C, respectively. Electrical work is performed on the system by allowing 2 A at 12 V to pass through a resistor for 20 min. As a result, the volume increases to (*a*) 0.14 m^3, and (*b*) 0.16 m^3. Determine (1) the net work for the process, (2) the change in internal energy of the gas, using specific heat data, and (3) the magnitude and direction of any heat transfer, all in kilojoules. Neglect energy storage in the resistor.

3-32M An adiabatic piston-cylinder device contains 0.3 kg of air initially at 300°K and 3 bars. Energy is added to the gas by passing a current of 5 A through a resistor within the system for a time period of 30 s. The gas expands quasistatically at constant pressure until the volume is doubled. Determine the size of the resistor, in ohms. Recall that the power dissipated in a resistor is I^2R.

Energy analysis at constant volume

3-33M Air is contained in a rigid vessel with insulated walls. A paddle-wheel of negligible mass in the vessel is driven by an external motor. During a process, the temperature rises from 27°C to (*a*) 127°C, and (*b*) 227°C. Determine the work done, in kJ/kg, if (1) average specific-heat data are used from Table A-4M, and (2) data from Table A-5M are used.

3-34M Nitrogen contained in a rigid tank at 27°C is heated until the pressure increases by a factor of (*a*) 1.5, and (*b*) 2. Determine the heat transfer required, in kJ/kg, by using (1) average specific-heat data from Table A-4M, and (2) data from Table A-6M.

3-35M A rigid tank with a volume of 1.0 m^3 contains oxygen at 0.20 MPa and 20°C. A paddle-wheel within the gas undergoes 160 revolutions with an applied torque of 43.2 N · m. The final temperature is 40°C. Determine (*a*) the change in the internal energy of the gas, in kilojoules, using specific-heat data, and (*b*) the magnitude, in kilojoules, and the direction of any heat transfer. Neglect energy storage in the paddle-wheel.

3-36M A rigid tank contains nitrogen gas initially at 100 kPa and 17°C. A paddle-wheel within the tank is rotated by an external source which provides a torque of 11.0 N · m for 100 revolutions of the shaft until the final pressure is (*a*) 120 kPa, and (*b*) 130 kPa. During the process a heat loss of 1.0 kJ occurs. Determine (1) the mass within the tank, in kilograms, and (2) the volume of the tank, in cubic meters. Neglect energy storage in the paddle-wheel.

3-37M A rigid tank initially contains 0.80 g of air at 295°K and 1.5 bar. An electric resistor within the tank is energized by passing a current of 0.6 A for a period of 30 s from a 12.0-V source. At the same time a heat loss of 126 J occurs.

(*a*) Determine the final temperature of the gas, in degrees kelvin.
(*b*) Find the final pressure in bars.

Specific heat evaluation

3-38M One-tenth of a kilogram of an ideal gas is enclosed in a rigid tank at a pressure of 1.2 bars and a temperature of 30°C. A paddle-wheel within the tank does 520 N · m of work on the substance, and 810 J of heat are added at the same time. The temperature of the gas, which has a molecular weight of 48, rises 25°C during the process. Compute the average specific heat c_v for the gas, in kJ/(kg)(°C).

3-39M Ten grams of an ideal gas having a molar mass of 32 undergoes a quasistatic expansion at constant pressure from 1.3 bars, 20°C, to 80°C. During the process 550 J of heat are added to the closed system. Compute the average value of c_v for the gas, in kJ/(kg)(°C).

3-40M One-tenth of a kilogram of an ideal gas (mol wt = 40) expands quasistatically at constant pressure in a closed system from 1 bar, 40°C to 150°C, and 5200 J of heat are added. Determine the average value of c_v, in kJ/(kg)(°C).

3-41M One-half kilogram of a gas is contained within a vessel made of rigid walls which do not absorb or transmit heat. Work of 10,000 N · m is done on the gas by rotating a paddle-wheel of negligible mass within the system. In addition, 15,000 J are supplied as electric energy to a resistance heater of negligible mass located in the system. During the process, the temperature of the gas increases 50°C. Determine an average c_v value for the gas, in kJ/(kg)(°C).

Additional problems

3-42M Electric energy is supplied to a mass of 224 g of cadmium at a constant rate of 17 W and at a constant pressure. The cadmium is thermally insulated, and temperature readings are taken at certain time intervals as follows:

Time, s	Temperature, °C	Time, s	Temperature, °C
0	39	285	137
15	45	345	155
45	57	405	172.3
105	80	465	191
165	100	525	208
225	119.2		

Draw a graph with time as abscissa and temperature as ordinate. The atomic mass of cadmium is 112. Estimate c_p values at (*a*) 39°C, (*b*) 100°C, and (*c*) 200°C.

3-43M A rigid, insulated tank with a total volume of 3.0 m^3 is divided in half by an insulated, rigid partition. Both sides of the tank contain an ideal monatomic gas. On one side the initial temperature and pressure are 200°C and 0.40 bar, and on the other side the values are 40°C and 1.0 bar. The internal partition is broken and complete mixing occurs. Determine (*a*) the final equilibrium temperature, degrees Celsius, and (*b*) the final pressure, in bar.

3-44M In a tank 0.60 kg of nitrogen are stored at 2 bars and 50°C. Attached to this tank through a suitable valve is a second tank which is 0.50 m^3 in volume and completely evacuated. Both tanks are thoroughly insulated. If the valve is opened and equilibrium is allowed to be reached, determine the final pressure, in bars.

3-45M In a tank, 1 kg of oxygen is stored at 1 bar and 60°C. Attached to this tank through a suitable valve is a second tank which contains 1 kg of oxygen at 2 bars and 7°C. Both tanks are insulated. If the valve is opened and equilibrium is allowed to be reached, determine the final pressure, in bars.

3-46M One-tenth cubic meter of air is contained in a piston-cylinder assembly at 1.5 bar and 27°C. It is first heated at constant volume (process 1–2) until the pressure has doubled. Then it is expanded at constant pressure (process 2–3) until the volume has doubled. Find the total heat added, in kilojoules, using the air table.

3-47M Carbon monoxide gas is contained within a piston-cylinder device at 1 bar and 27°C.

(*a*) In process *A* the gas is heated at constant volume until the pressure has doubled. It is then expanded at constant pressure until the volume is three times its initial value.

(*b*) In process *B* the same gas in the same initial state is first expanded at constant pressure until the volume has tripled, then the gas is heated at constant volume until it reaches the same final pressure as in process *A*.

For these two overall processes determine the net heat effect, the net work effect, and the changes in internal energy, in kJ/kg. Compare the results for the two processes.

3-48M A piston-cylinder arrangement contains 1 kg of air initially at 2 bars and 77°C. Two processes are involved: a constant-volume process followed by a constant-pressure process. During the first process, heat is added in the amount of 50,670 J. During the second process, which follows the first, heat is added at constant pressure until the volume is 2.0 m^3. If the processes are quasistatic, evaluate the total work of expansion, in kilojoules, for the overall system change.

3-49M An ideal gas with a molar mass of 50 is contained within a piston-cylinder device initially at 20°C and 0.24 m^3/kg. It undergoes an isothermal process to state 2, where the specific volume is 0.12 m^3/kg. It is then expanded at constant pressure to state 3, where the specific volume is 0.36 m^3/kg. Finally, it is returned to its initial state along a path which is a straight line on Pv coordinates.

(*a*) Sketch the cyclic process on a Pv diagram.

(*b*) Determine the values of P_2 and P_3, in bars.

(*c*) Calculate the net work of the cycle, in newton-meters per gram.

PROBLEMS (USCS)

Ideal gas

3-1 A balloon is filled with methane gas (CH_4) at 70°F and 30.0 inHg until the volume is (*a*) 1500, and (*b*) 900 ft^3. Calculate the mass, in pounds, of methane under ideal-gas conditions.

3-2 Two cubic feet of carbon monoxide are maintained at (*a*) 20 psia and 85°F, and (*b*) 30 psia and 260°F. Determine the mass of the ideal gas, in pounds.

3-3 One-hundred pounds of methane (CH_4) are to be stored in a gas reservoir at (*a*) 55°F and 200 psia, and (*b*) 160°F and 250 psia. The molar mass of methane is 16.0. What is the necessary volume of the reservoir, in cubic feet?

3-4 One pound of helium is placed in a rigid tank 15 ft^3 in volume. If the temperature is (*a*) 170°F, and (*b*) 70°F, and the barometric pressure is 30.5 inHg, what would be the reading, in psi, of a pressure gage which is attached to the tank?

3-5 Determine the reading, in psig, on a pressure gage attached to a tank of hydrogen at (*a*) 73°F, and (*b*) 140°F when the density of the gas is 0.00498 lb/ft^3. The barometric pressure is 29 inHg.

3-6 One pound of argon gas ($M = 40$) is placed in a rigid tank with a volume and pressure of (*a*) 10.0 ft^3 and 20 psia, and (*b*) 7.5 ft^3 and 26 psia. If argon behaves as an ideal gas, what is the temperature in degrees Fahrenheit?

3-7 One pound of air is maintained in (*a*) a 10.0-ft^3 tank at a gage reading of 7.0 psig, and (*b*) a 9.0-ft^3 tank at a gage reading of 4 psig. If the atmospheric (barometric) pressure is 14.5 psia, what is the temperature of the ideal gas, in degrees Fahrenheit?

3-8 A rigid tank has a volume of 100 ft^3. It contains a gas having a molar mass of 30 at 100 psia and 100°F. Gas leaks out of the vessel until the pressure is 50 psia at 80°F. What volume, in cubic feet, is occupied by the gas which escaped if it is at 14.6 psia and 70°F?

3-9 A tank contains carbon dioxide at 80 psia and 100°F. A leak occurs in the tank which is not detected until the pressure has dropped to (*a*) 60 psia, and (*b*) 45 psia. If the temperature of the gas at the time of detection of the leak is 70°F, determine the mass of carbon dioxide that has leaked out if the original mass was 60 lb.

3-10 A tank contains helium at 6 atm and 100°F. Two pounds of the gas are removed, which causes the pressure and temperature to change to (*a*) 4 atm and 80°F, and (*b*) 3.4 atm and 60°F. Determine the volume of the tank, in cubic feet.

3-11 Two tanks A and B are connected by suitable pipes through a valve, which initially is closed. Tank A initially contains 10 ft^3 of nitrogen at 100 psia and 140°F, and tank B is evacuated. The valve is then opened, and nitrogen flows into tank B until the pressure in this tank reaches 20 psia and the temperature is 60°F. As a result, the pressure in tank A drops to 60 psia and the temperature changes to 110°F. Determine the volume of tank B, in cubic feet.

Isothermal boundary work

3-12 Nitrogen gas initially at 18 psia and 80°F is compressed isothermally in a piston-cylinder device to 45 psia.

(*a*) Compute the work required in ft · lb_f/lb_m by some outside source and show the area representation on a Pv diagram.

(*b*) Now consider that atmospheric air at 14.5 psia acts on the back of the piston, in addition to the connecting rod. Compute the net work required by the outside source in this case, for the same change in state of the nitrogen and the Pv area representation.

(*c*) Finally, consider in addition to part (*b*) that friction acts between the piston and cylinder and that the frictional resistance is equivalent to an effective resistive pressure of 7.0 psia which is constant throughout the piston travel. Determine in this case the net work that must be delivered along the connecting rod. Show the area representation of this final net work on a Pv diagram.

3-13 Consider Prob. 3-12, except that the gas is carbon dioxide, the atmospheric pressure is 14.4 psia, and the frictional resistance is equivalent to a resistive pressure of 6 psia.

3-14 A certain quantity of work is needed to reduce quasistatically the volume of an ideal gas to one-half its initial value at a constant temperature T_1. At what temperature T_2 will the same amount of work reduce the volume isothermally to a quarter of the same initial volume if (*a*) $T_1 = 500°F$, (*b*) $T_1 = 600°F$, and (*c*) $T_1 = 700°F$? Report the answer in °F.

3-15 An ideal gas initially in a piston-cylinder device at 20 psia and 1.0 ft^3 is first heated at constant pressure until the volume is doubled. It is then allowed to expand isothermally until the volume is again doubled. Determine the total work done by the gas, in ft · lb$_f$/lb · mol, and plot the quasistatic processes on a *PV* diagram. The initial gas temperature is (*a*) 40°F, and (*b*) 140°F.

3-16 A piston-cylinder device contains 3 ft^3 of an ideal gas at 20 psia and 60°F. The gas is compressed to a final pressure of (*a*) 60 psia, and (*b*) 50 psia, and a temperature of 60°F. (1) Determine the minimum work of compression, in ft · lb$_f$, if the quasistatic process is isothermal. (2) Consider now that a process occurs between the same two end states, as in part (1), such that it follows a straight line on *PV* coordinates. Determine the work of compression for this process, and compare the two processes on a *PV* diagram.

Enthalpy and internal energy changes from specific heat data

3-17 The temperature of oxygen is increased from (*a*) 100 to 800°F, (*b*) 200 to 1000°F, and (*c*) 300 to 1200°F. (1) On the basis of the equation in Table A-4 for c_p as a function of temperature at zero pressure for oxygen, calculate the change in enthalpy, in Btu/lb · mol. (2) Compare this answer to that obtained by use of *h* data from Table A-7. (3) What percent error will be introduced by using the arithmetic-mean specific heat from Table A-4 to evaluate Δh between the given temperature limits, compared to that in part 2?

3-18 Consider Prob. 3-17, except that the gas is carbon dioxide.

3-19 The temperature of oxygen is increased from (*a*) 100 to 800°F, and (*b*) 200 to 1000°F. (1) On the basis of the empirical equation for c_p of oxygen (Table A-4), calculate the internal energy change in Btu/lb · mol by integration techniques. (2) Compare this answer with that obtained by use of *u* data from Table A-7. (3) What percent error will be introduced by using the arithmetic mean specific heat c_v from Table A-4 to evaluate Δu, compared to that in part 1?

3-20 Using the specific-heat data cited for monatomic gases in Table A-4, calculate the change in internal energy and enthalpy of (*a*) helium for the temperature change from 100 to 1000°F, and (*b*) argon for the temperature change from 80 to 800°F, in Btu/lb.

Energy analysis at constant temperature

3-21 One-half pound of air is compressed isothermally and quasistatically in a closed system from 15 psia, 60°F to a final pressure of (*a*) 90 psia, and (*b*) 36 psia. Determine (1) the internal energy change and (2) the heat transfer, in Btu.

3-22 Carbon dioxide is expanded isothermally and quasistatically in a closed system from initial conditions of 20 psia, 340°F, and 2.0 ft^3 to a final volume of (*a*) 4.0 ft^3, and (*b*) 3.0 ft^3. Calculate the magnitude and direction of any necessary heat transfer, in Btu.

3-23 A piston-cylinder device contains 0.27 lb of air initially at 35 psia and 240°F. During a quasistatic, isothermal process, heat is removed in the amount of (*a*) 20 Btu, and (*b*) 15 Btu, and electrical work is carried out on the system in the amount of 1.76 W · h. Determine the ratio of the final to the initial volume.

3-24 In a closed system 0.1 lb of argon ($M = 40$) expands quasistatically and isothermally from 30 psia, 140°F until (*a*) the volume is doubled and 18.5 Btu/lb of heat is added, and (*b*) the volume is tripled and 28.5 Btu/lb of heat is added. If during the process a 12-V battery is operated for 20 s, determine the constant current supplied, in amperes, to a resistor within the system. Neglect energy storage in the resistor.

3-25 Air is contained within a piston-cylinder device at initial conditions of 4 atm and 0.35 ft^3. A paddle-wheel within the gas is turned under the conditions of (*a*) an applied torque of 3.5 $lb_f \cdot ft$ for 4000 revolutions, and (*b*) an applied torque of 5.3 $lb_f \cdot ft$ for 2400 revolutions. During the process the temperature remains constant while the volume is doubled. Determine the magnitude, in Btu, and the direction of any heat transfer.

Energy analysis at constant pressure

3-26 A system consists of 0.88 lb of carbon dioxide within a constant-pressure piston-cylinder device which is frictionless. The initial conditions are 100 psia and 40°F. Heat is added until the volume is doubled. On the basis of data in Table A-9 determine (*a*) the change in internal energy, in Btu, (*b*) the change in enthalpy, in Btu, (*c*) the work performed by or on the system, in $ft \cdot lb_f$, and (*d*) the quantity of heat supplied, in Btu.

3-27 A piston-cylinder device which initially has a volume of 1 ft^3 contains 0.010 lb of hydrogen at 30 psia. Heat is transferred until the final volume is (*a*) 0.82, and (*b*) 0.88 m^3. If the quasistatic process occurs at constant pressure, determine (1) the final temperature, in degrees Fahrenheit, (2) the heat transfer, in Btu, based on data from Table A-11, and (3) the heat transfer, in Btu, based on specific-heat data from Table A-4.

3-28 Air is quasistatically heated at constant pressure in a piston-cylinder device from 30 psia, 2.0 ft^3, and 120°F to a final volume of 3.0 ft^3. Determine the heat transfer, in Btu, by using (*a*) average c_v data, and (*b*) average c_p data from Table A-4.

3-29 One-hundredth of a pound-mole of helium is contained within an adiabatic piston-cylinder assembly and is maintained at a constant pressure of 30 psia. A paddle-wheel is operated within the cylinder until the volume of the gas has increased by 25 percent. Compute (*a*) the quantity of paddle-wheel work, in $ft \cdot lb_f$, and (*b*) the net work done on the system, if the piston is frictionless and the initial temperature of the gas is 60°F.

3-30 A cubic foot of nitrogen at 15 psia and 80°F is contained in a piston-cylinder device. A paddle-wheel within the cylinder is turned until (*a*) 7070 $ft \cdot lb_f$, and (*b*) 5650 $ft \cdot lb_f$ of energy have been added. If the process is quasistatic, adiabatic, and at constant pressure, determine the final temperature of the gas, in degrees Fahrenheit. Neglect energy storage in the paddle-wheel, and use data from Table A-6.

3-31 A piston-cylinder device maintains nitrogen at a constant pressure of 30 psia. The initial volume and temperature are 0.35 ft^3 and 60°F, respectively. Electrical work is performed on the system by allowing 2 A at 12 V to pass through a resistor for 2.0 min. As a result, the volume increases to (*a*) 0.50 ft^3, and (*b*) 0.60 ft^3. Determine (1) the net work for the process, (2) the change in internal energy of the gas, using specific heat data, and (3) the magnitude and direction of any heat transfer, all in Btu. Neglect energy storage in the resistor.

3-32 An adiabatic piston-cylinder device contains 0.66 lb of air initially at 80°F and 3 atm. Energy is added to the gas by passing a current of 5 A through a resistor within the system for a time period of 30 s. The gas expands quasistatically at a constant pressure until the volume is doubled. Neglecting energy storage in the resistor, determine the size of the resistor, in ohms. Recall that power dissipated in a resistor is I^2R.

Energy analysis at constant volume

3-33 Air is contained in a rigid vessel with insulated walls. A paddle-wheel of negligible mass in the vessel is driven by an external motor. During a process, the temperature rises from 60°F to (*a*) 160°F, and (*b*) 200°F. Determine the work done, in Btu/lb, if (1) average specific-heat data are used from Table A-4, and (2) data from Table A-5 are used.

3-34 Nitrogen contained in a rigid tank at 100°F is heated until the pressure increases by a factor of (*a*) 1.5, and (*b*) 2. Determine the heat transfer required, in Btu/lb, by using (1) average specific-heat data from Table A-4, and (2) data from Table A-6.

3-35 A rigid tank with a volume of 4.0 ft^3 contains oxygen at 30 psia and 70°F. A paddle-wheel within the gas undergoes 160 revolutions with an applied torque of 4.70 $lb_f \cdot ft$. The final temperature is 120°F. Determine (*a*) the change in the internal energy of the gas, in Btu, using specific-heat data, and (*b*) the magnitude, in Btu, and the direction of any heat transfer. Neglect energy storage in the paddle-wheel.

3-36 A rigid tank contains nitrogen gas initially at 15 psia and 40°F. A paddle-wheel within the tank is rotated by an external source which provides a torque of 8.0 $lb_f \cdot ft$ for 100 revolutions of the shaft until the final pressure is (*a*) 18 psia, and (*b*) 17.4 psia. During the process a heat loss of 1.0 Btu occurs. Determine (1) the mass within the tank, in pounds, and (2) the volume of the tank, in cubic feet. Neglect energy storage in the paddle-wheel.

3-37 A rigid tank contains 0.23 lb · mol of nitrogen at 3 atm and 100°F. An electric resistor within the tank is energized by passing a current of 12 A for a period of 50 s from a 120-V source. If, during this process, 125 Btu of heat are removed, determine (*a*) the final temperature, in degrees Fahrenheit, and (*b*) the final pressure, in atm.

Specific-heat evaluation

3-38 One-tenth pound of an ideal gas is enclosed in a rigid tank at a pressure of 18 psia and a temperature of 80°F. A paddle-wheel within the tank does 390 ft · lb_f of work on the substance, and 0.77 Btu of heat is added at the same time. If the temperature of the gas, which has a molecular mass of 48, rises 80°F during the process, compute the average specific heat c_v for the gas, in Btu/(lb)(°F).

3-39 Ten grams of an ideal gas having a molar mass of 32 undergoes a quasistatic expansion at constant pressure from 20 psia, 65°F, to 170°F. During the process 0.51 Btu of heat are added to the closed system. Compute the average value of c_v for the gas, in Btu/(lb_m)(°F).

3-40 Two-tenths pound of an ideal gas (mol wt = 40) expands quasistatically at constant pressure in a closed system from 15 psia, 140°F, to 340°F, and 4.9 Btu of heat is added. Determine the average value of c_v, in Btu/(lb)(°F).

3-41 Ten pounds of a gas are contained within a vessel made of rigid walls which do not absorb or transmit heat. Work of 77,800 ft · lb_f is done on the gas by rotating a paddle-wheel of negligible mass within the system. In addition, 150 Btu are supplied as electric energy to a resistance heater of negligible mass located in the system. During the process, the temperature of the gas increases 100°F. Determine an average c_v value for the gas, in Btu/(lb)(°F).

Additional problems

3-42 A rigid, insulated cylinder is divided into two parts by an uninsulated, frictionless piston which initially is held in a fixed position. One part of the cylinder contains 1 lb of a gas at 60 psia and 80°F, while the other part contains 1 lb of the same gas at 20 psia and 80°F. The piston is then released, and equilibrium is established between the two parts.

(*a*) What is the final equilibrium temperature, in degrees Fahrenheit?

(*b*) What is the final equilibrium pressure, in psia? Assume that the specific heats, c_v and c_p, are constant.

3-43 A rigid, insulated tank with a total volume of 3.0 ft^3 is divided in half by an insulated, rigid partition. Both sides of the tank contain an ideal monatomic gas. On one side the initial temperature and pressure are 400°F and 1.2 atm, respectively, and on the other side the values are 120°F and 2.2 atm. The internal partition is broken and complete mixing occurs. Determine (*a*) the final equilibrium temperature, degrees Fahrenheit, and (*b*) the final pressure, in atm.

3-44 In a tank 0.130 lb of nitrogen is stored at 30 psia and 140°F. Attached to this tank through a suitable valve is a second tank which is 2.0 ft^3 in volume and completely evacuated. Both tanks are thoroughly insulated. If the valve is opened and equilibrium is allowed to be reached, determine the final pressure, in psia.

3-45 Two insulated tanks are connected by a valve. One tank contains 10 lb_m of an ideal gas at 100°F, and the other tank contains 20 lb_m of the same gas at 200°F. The valve is opened, and the gases allowed to mix until equilibrium is reached. All specific heats may be regarded as constant.

(*a*) Calculate the mixture temperature, in degrees Fahrenheit.

(*b*) Calculate the total enthalpy change for the two-tank system. Show that this answer is the same regardless of the mass and the temperature of the gas in each tank initially.

3-46 A piston-cylinder machine contains air initially at 20 psia, 240°F, and 0.5 ft^3. The piston moves slowly and with negligible friction until the pressure rises to 100 psia. The process is described by the equation $V = 0.6 - 0.005P$, where V is in cubic feet and P is in psia.

(*a*) What is the final temperature?

(*b*) What mass of air is present?

(*c*) Determine the work done, in $ft \cdot lb_f$.

(*d*) Determine the heat transfer, in Btu.

3-47 Carbon monoxide gas is contained within a piston-cylinder device at 15 psia and 60°F.

(*a*) In process A the gas is heated at constant volume until the pressure has doubled. It is then expanded at constant pressure until the volume is three times its initial value.

(*b*) In process B the same gas in the same initial state is first expanded at constant pressure until the volume has tripled; then the gas is heated at constant volume until it reaches the same final pressure as in process A.

For these two processes determine the net heat effect, the net work effect, and the change in internal energy, all in Btu/lb. Compare the results for the two processes.

3-48 Air initially at 30 psia and 1 ft^3 and with a mass of 0.1 lb expands at constant pressure to a volume at 3 ft^3. It then changes state at constant volume to a pressure of 15 psia. If the processes are quasistatic, find the total work done, in Btu, the total heat transferred, in Btu, and the overall change in internal energy, in Btu.

3-49 One cubic foot of air is contained in a piston-cylinder assembly at 20 psia and 40°F. It is first heated at constant volume (process 1–2) until the pressure has doubled. Then it is expanded at constant pressure (process 2–3) until the volume has doubled. Find the total heat added, in Btu, using the air table (Table A-5).

CHAPTER

FOUR

PROPERTIES OF A PURE, SIMPLE COMPRESSIBLE SUBSTANCE

4-1 THE PvT SURFACE

In Chap. 2 the state postulate indicated that any intensive, intrinsic property of a simple, compressible substance is solely a function of two other intrinsic properties. That is, $y_1 = y(y_2, y_3)$, where y, in general, is any intrinsic property. A specific example of this situation was examined in Chap. 3, namely, the ideal gas. The relationships among the properties P, v, T, u, h, c_v, and c_p were developed from theoretical and experimental considerations. We now wish to examine the relationships among these properties for nonideal gases, liquids, solids, and two-phase systems of a pure substance. Experimental data reveal a consistent pattern in the behavior of simple compressible substances in the solid, liquid, and gas phases. It is on this consistency that we wish to focus our attention.

From a mathematical viewpoint, any equation involving two independent variables (such as y_2 and y_3) can be represented in a rectangular, three-dimensional space as a surface. Consequently, the equilibrium states of any simple, compressible substance can be represented as a surface in space, where the geometric coordinates are intrinsic properties of interest. The PvT surface is extremely important because it clearly exhibits the basic structure of matter in a general fashion. The PvT surface of an ideal gas is presented in Fig. 3-1. More generally, the PvT diagram for a substance which contracts on freezing is shown in Fig. 4-1. (Note that water is an exception to this rule.) It should be emphasized that this figure is not to scale. The solid, liquid, and gas phases appear as surfaces. In addition to these single-phase conditions, the process of a phase change is well known. Two phases are coexistent during any phase change, e.g.,

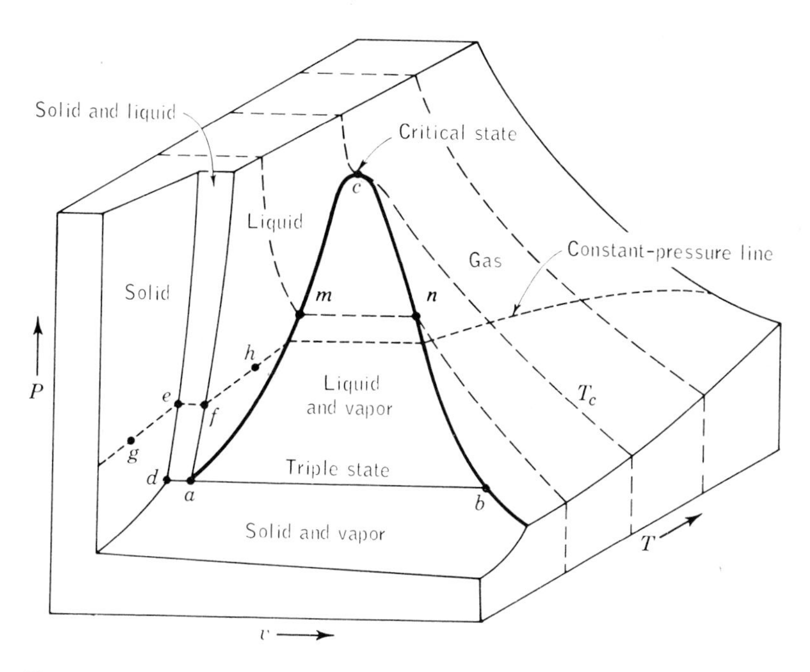

Figure 4-1 *PvT* surface for a substance which contracts on freezing.

melting, vaporization, or sublimation. (Sublimation is the transformation of a solid directly into a gas.) The single-phase regions on the surface are necessarily separated by two-phase regions. For example, the solid- and liquid-phase regions are seen to be separated by another surface which represents a two-phase mixture of solid and liquid. The surface of liquid-vapor and solid-vapor mixtures are also shown. (The words vapor and gas are used interchangeably at this point.)

Any state represented by a point lying on a line separating a single-phase region from a two-phase region in Fig. 4-1 is known as a *saturation* state. These are states where a discontinuity occurs in the isotherms (constant-temperature lines), such as states *m* and *n* in the figure. It is significant to note that a change of phase occurs without a change of pressure or temperature. The curved line which separates the liquid region from the liquid-vapor region, line *a-m-c*, is referred to as the liquid-saturation line. Any state represented by a point on this line between *a* and *c* is known as a saturated-liquid state. Similarly, the states represented on the curve *c-n-b* are saturated-vapor states.

Since the figure is not to scale and no values are listed on the coordinates, we must rely on actual experimental data to describe more quantitatively the general trends for the majority of substances. The spacing between particles (and hence the volume) undergoes only a very small change during a solid-liquid phase transformation. This volume change is given, for example, by the horizontal distance along the constant-pressure line connecting points *e* and *f* in Fig. 4-1. Since the volume change is small compared with the specific volume of either single phase at that pressure, the solid-liquid region would be quite thin if the diagram were to scale.

In the transition from the liquid to the gas phase, the change in volume generally is very much greater. This would be more readily apparent if the volume axis were more nearly to scale. It has been greatly foreshortened. For most substances the change in specific volume between states a and b in Fig. 4-1 would be thousands of times greater than the change between states a and d. It is important to notice, however, that as one moves up the liquid-vapor surface to higher pressures (and temperatures), the volume change for the process of boiling becomes smaller, and eventually becomes nonexistent. Beyond certain pressure-temperature conditions the process of vaporization (or the inverse process of condensation) cannot occur. The state which is the limit beyond which a liquid-vapor transformation is not possible is called the *critical state.* On a PvT diagram it appears as a point on the general surface. Associated with it are certain property values which are commonly signified by the subscript c. The three properties of present interest will be denoted by P_c, v_c, and T_c at the critical state.

A substance which is at a temperature higher than its critical temperature will not be capable of undergoing condensation to the liquid phase, no matter how high a pressure is exerted. Note further in Fig. 4-1 that the liquid and gas (vapor) phases merge into each other in the region above the critical state. All known substances exhibit this behavior. The existence of the critical state demonstrates that the distinction between liquid and gas phases is not clear-cut, if not impossible, in certain situations. When the pressure is greater than the critical pressure, the state is frequently referred to as a supercritical state. Many familiar substances have fairly high critical pressures, but critical temperatures which are below normal atmospheric conditions. For example, the critical pressures of hydrogen and oxygen are approximately 13.1 and 50.3 bars (190 and 730 psia), respectively, and the corresponding critical temperatures are -240 and -118°C (-400 and -181°F). The critical temperature of carbon dioxide is 31°C (88°F), whereas that of water is quite high, 374°C (705°F). Hence no generalization can be made for the range of critical temperatures commonly found. The critical pressures of most common substances, however, are above 1 atm. A limited list of critical data for some common substances is reported in Tables A-3M and A-3. It might be noted that a few modern steam-power plants operate so that the water during part of its cycle is in states beyond the critical state. In this particular case this means that water is not only at temperatures above 650 K or 705°F, but also at pressures above 220 atm.

Lines ge and fh in Fig. 4-1 indicate that, at a given pressure, the volume of the solid or liquid phase changes very little as the temperature is increased. This is true when compared with even the small volume change for the solid-liquid phase transition. This is in general agreement with the physical notion that solids and liquids are observed not to undergo any significant volume change during heating or cooling.

An additional unique state of matter may be noted from Fig. 4-1. This is represented by the line parallel to the Pv plane marked as the *triple* state. As the term implies, this is a state in which three phases coexist in equilibrium. The triple state in this figure is for equilibrium between a solid, a liquid, and a gas phase. For water, this state appears at 0.0061 atm and 0.01°C (32.02°F). Recall that the triple state of water is used as the reference point for establishing the Kelvin temperature scale. The triple state of water is assigned a temperature value of 273.16°K. As another example, carbon dioxide exists in these three phases simultaneously at roughly 5 atm and −57°C (−70°F). If a piece of solid carbon dioxide (dry ice) is placed in the open atmosphere, it is noticed that the solid goes directly into the gas or vapor phase. The reason is that at 1 atm, carbon dioxide is considerably below the minimum pressure at which its liquid phase can exist; consequently, it sublimes at this pressure. Paradichlorobenzene (used in moth crystals) and iodine crystals are two other common examples of substances undergoing a direct solid-vapor phase change under atmospheric pressure conditions. Some triple-state data appear in Table 4-1.

Water is an anomalous substance in that it expands upon freezing. Thus the specific volume of the solid phase is greater than that of the liquid phase. The PvT surface modified to take this feature into account is shown in Fig. 4-2.

4-2 THE PRESSURE-TEMPERATURE DIAGRAM

Three-dimensional diagrams for the equilibrium states of simple systems are extremely useful in introducing the general relationships between the three phases of matter normally under consideration. The relation of two-phase regions to the single-phase regions is clearly shown, as well as the significance of the critical and triple states of matter. Nevertheless, it will be found to be more convenient in the thermodynamic analysis of simple, compressible systems to work with two-dimensional diagrams. All two-dimensional diagrams may be thought of simply as projections of a three-dimensional surface. For example, the surface presented in Fig. 4-1 may be projected on a PT, Pv, or Tv plane (as was done for an ideal gas in Fig. 3-1). Only the PT and Pv projections will be explored in any depth at the present time.

A projection of a PvT surface upon the PT plane is commonly termed a phase diagram. It is an empirical fact that both the temperature and the pressure

Table 4-1 Triple-state data

Substance	T, °K	P, atm	T, °F
Helium 4 (λ point)	2.17	0.050	−456
Hydrogen, H_2	13.84	0.070	−435
Oxygen, O_2	54.36	0.0015	−362
Nitrogen, N_2	63.18	0.124	−346
Ammonia, NH_3	195.40	0.061	−108
Carbon dioxide, CO_2	216.55	5.10	−70
Water, H_2O	273.16	0.006	32

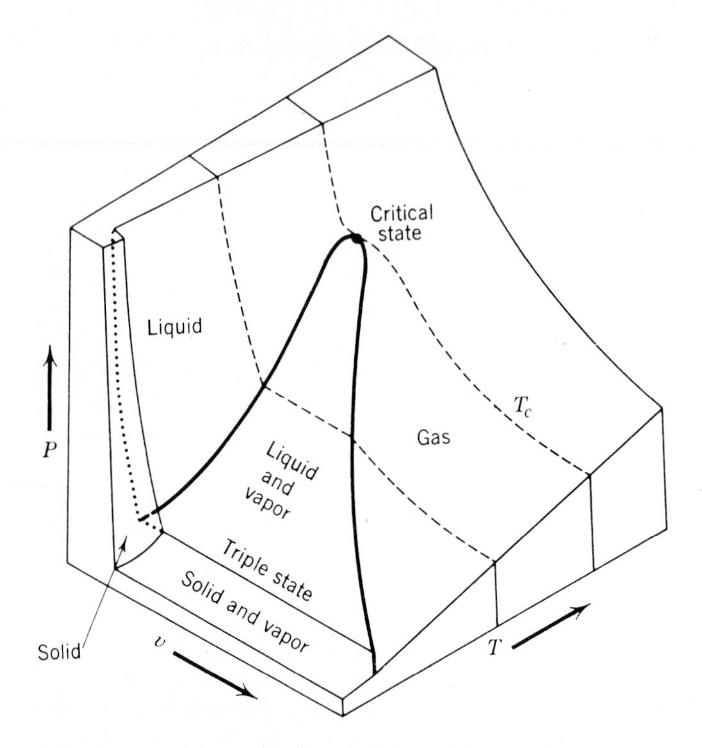

Figure 4-2 *PvT* surface for a substance which expands on freezing.

remain constant during a phase change. This is well known for the boiling of water or melting of ice. Consequently, the surfaces in Fig. 4-1 which represent two phases in equilibrium are parallel to the v axis. Hence these surfaces appear as lines when projected onto the PT plane. A typical pressure-temperature plot (based on the general characteristics of Fig. 4-1) is presented in Fig. 4-3. This figure carries the same restriction as before, i.e., it is for a substance which contracts on freezing.

The liquid-vapor surface which appears as a line on the PT diagram is called the liquid-vapor saturation line, since in this case the saturated-liquid and saturated-vapor lines project on top of each other. It is also known as the vaporization curve. Similarly, the solid-liquid and solid-vapor surfaces of a PvT diagram are shown as the freezing (or melting) curve and the sublimation curve, respectively, in Fig. 4-3. These are also known as the solid-liquid and the solid-vapor saturation lines. The critical and triple states are designated by the points marked c and t. Whereas two-phase systems appear as lines, single-phase systems are represented by areas. The dashed line in the figure represents the fusion (freezing) curve for a substance such as water which expands on freezing. All other specific properties, except the volume, decrease during the freezing of water. Note that for ice an increase in pressure decreases the freezing-point temperature.

The single-phase liquid and gas regions, represented by labeled areas in Fig.

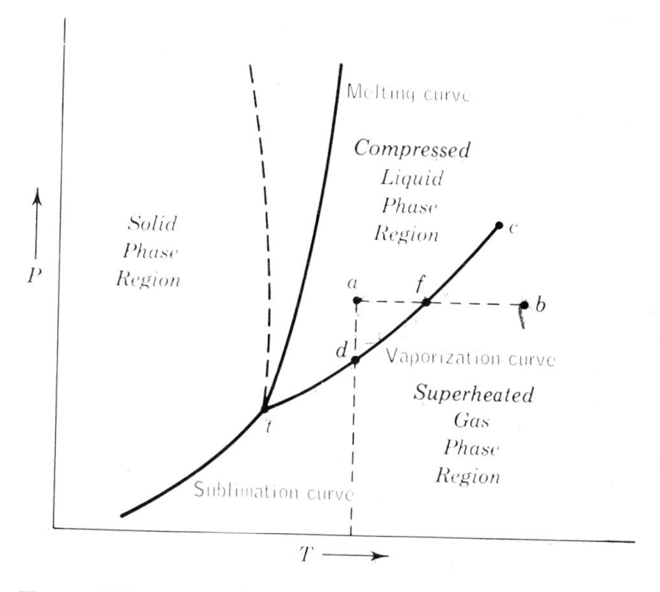

Figure 4-3 Phase (PT) diagram for a substance that contracts on freezing.

4-3, are sometimes given special names. It is seen from the figure that, for any state in the area marked liquid (such as point a), the temperature T_a is below the saturation temperature T_f for the same value of pressure. Such a state is said to be that of a *subcooled* liquid, since it may be achieved by cooling the liquid below its saturation temperature at a given pressure. On the other hand, the pressure P_a of state a is above the saturation pressure P_d for the same temperature. Hence state a is also called the state of a *compressed* liquid, since it may be achieved by compressing the liquid above its saturation pressure at a given temperature. The terms "subcooled" and "compressed" are thus synonymous. In a similar fashion, if the temperature T_b of a substance at state b is above the vapor-liquid saturation temperature T_f for a given pressure, the substance is in a *superheated-vapor* state. The process of superheating generally is defined as one for which the temperature of a vapor is increased at constant pressure.

It is apparent from Fig. 4-3 that for each saturation pressure there is only one saturation temperature. For a liquid-vapor system the saturation pressure is known as the *vapor* pressure. Although there are two independent intensive properties for a single phase of a simple, compressible substance, this number is reduced to one for a single-component, two-phase system. If the pressure is fixed for this latter system, the temperature is fixed and cannot be varied independently. All other intensive properties of each of the two phases in equilibrium are also fixed when one intensive property such as the pressure is fixed. This change in the number of independent variables from two to one has a major influence on the method of presenting tabular property data for single-phase versus two-phase systems. In the case of a three-phase system of a single component there are no independent variables. Thus the intensive properties at the triple state cannot be varied arbitrarily, but are limited to a single set of values for a given compound. There is only one pressure and one temperature for a substance at its triple state. This is true also for all other intensive properties of the three phases. Hence the triple state of single-component systems is especially well defined in a thermodynamic sense, and is easily reproduced. It is this latter fact that makes the triple state of water an excellent reference state for the establishment of a temperature scale.

4-3 THE PRESSURE-VOLUME DIAGRAM

Some of the fundamental characteristics of the two-phase surfaces that appear in Fig. 4-1 are masked by the projection of these surfaces onto the PT plane. Hence the projection of the PvT surface onto the Pv plane is of interest. Again, use is made of Fig. 4-1, which is valid for substances which contract on freezing. The resultant projection onto a plane parallel to the Pv axes is shown in Fig. 4-4. Both single- and two-phase regions appear as areas on this new diagram. The saturated-liquid line represents the states of the substance such that any further infinitesimal addition of energy to the substance at constant pressure will change

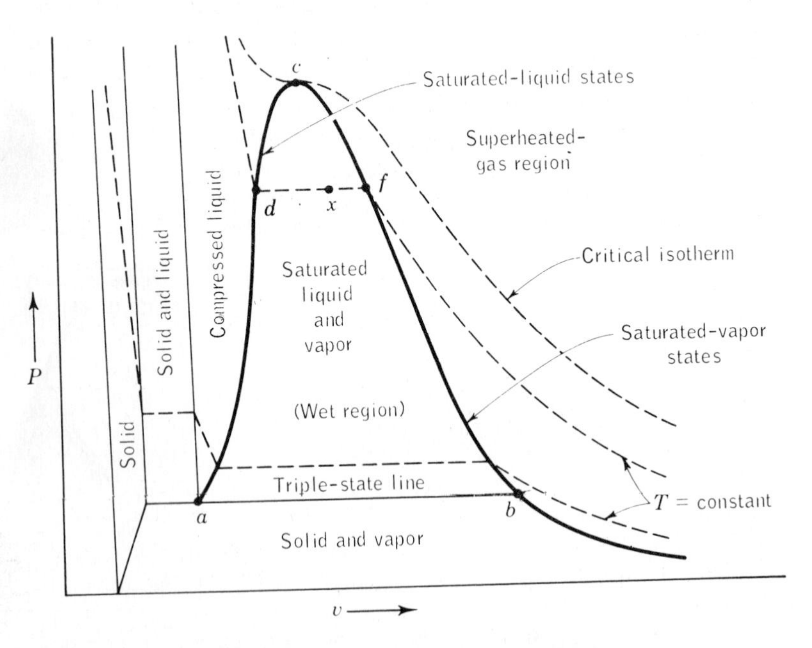

Figure 4-4 Pv diagram for a substance which contracts on freezing.

a small fraction of the liquid into vapor. Similarly, removal of energy from the substance at any state which lies on the saturated-vapor line results in partial condensation of the vapor, whereas addition of energy makes the vapor superheated. The two-phase region labeled saturated liquid and vapor, which lies between the saturated-liquid and saturated-vapor lines, is commonly called the *wet region*, or the wet dome. The state at the top of the wet region indicated by point c is again the critical state.

A state represented by a point in the liquid-vapor region (wet region), such as x in Fig. 4-5, is a mixture of saturated liquid and saturated vapor. The specific volumes of these two phases must be points on the saturation line, such as points d and f in the figure. Hence the specific volume of state x simply represents the *average* property value for the two phases in equilibrium. In order to find this average value of v (or any other specific property of a liquid-vapor mixture, such as u and h), we need to know the proportions of vapor and liquid in the saturated vapor-liquid mixture. To achieve this we define the *quality*, usually represented by the symbol x, as the fraction of the total mixture which is vapor, based on mass (or weight). That is,

$$\text{Quality} = x = \frac{m_{\text{vapor}}}{m_{\text{total}}} = \frac{m_g}{m_g + m_f} \tag{4-1}$$

In this equation the subscript g applies to the saturated vapor state, while subscript f denotes the saturated liquid state. It is common to speak of quality also as a percentage, in which case the value of x defined above is multiplied by 100. For example, a system composed of a saturated liquid alone may be referred to as having a quality of 0 percent ($x = 0.00$), and a saturated vapor alone has a quality of 100 percent ($x = 1.00$). A saturated vapor alone is frequently called a dry saturated vapor. The word *dry* in this case indicates there is no liquid in the saturated state. The quality is limited to values between zero and unity (or 0 to 100 percent). The use of the term is restricted solely to saturated liquid-vapor mixtures. Later in this chapter we shall develop equations for computing intensive properties of two-phase mixtures. The quality is an important parameter in such computations.

Extension of the saturated-vapor line below the triple-state line is valid. If energy were now removed, the saturated vapor represented by points along this lower portion of the saturated-vapor curve would condense out into a solid phase rather than a liquid one. The solid, solid-liquid, and compressed-liquid regions may be considered similarly. One final comment is appropriate to the Pv diagram shown in Fig. 4-4. It has been noted earlier that other families of curves representing various properties could be shown on any of the two-dimensional or three-dimensional diagrams. For this reason a family of constant-temperature lines is also shown on the diagram. The general trends shown by these lines can be verified by referring back to the PvT diagram of Fig. 4-1. It is important for the student to have some general concept of the position of constant-temperature lines on a Pv diagram since this diagram will be used in future problem analysis.

4-4 TABLES OF PROPERTIES OF PURE SUBSTANCES

The preceding sections were presented as a brief introduction to the *PvT* behavior of simple compressible systems containing a pure substance. In addition to the acquisition of the necessary nomenclature required for future discussions, a general qualitative feeling for the *PvT* characteristics of substances has been gained. Some numerical values were also cited to establish the range of interest for some of the various thermodynamic variables. However, in order to carry out quantitative calculations, it is necessary to investigate the more established methods of assembling or storing physical data so that they are readily accessible and in a convenient format for calculation purposes. Such data are primarily either obtained directly from experimental measurements or calculated from equations which employ the preceding experimental measurements.

The relationships among thermodynamic data are frequently presented in the form of tables. Tabular data are listed at convenient increments of the independent variables. For our initial study the following intensive, intrinsic properties will be of primary interest: pressure P, specific volume v, temperature T, specific internal energy u, specific enthalpy h, and specific entropy s. (The property entropy has not yet been defined. For the present discussion one is asked to accept its existence without proof. It will not be used to any extent until Chap. 7.)

a Superheat Tables

In a single-phase region, such as the superheat region, two intensive properties are required to fix or identify the equilibrium state. Variables such as v, u, h, and s are usually tabulated in superheat tables as a function of P and T, since the latter are conveniently measurable properties. The format of a superheat table is readily apparent if we refer again to a *PT* diagram. Figure 4-5 shows the superheated vapor region of a *PT* diagram divided into a grid. The lines of the grid represent

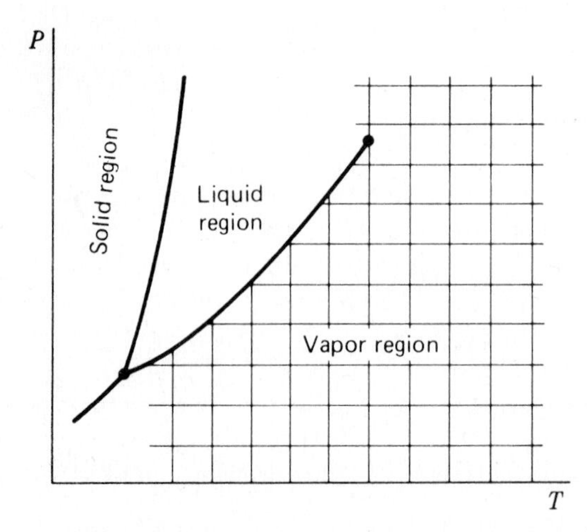

Figure 4-5 Sketch of *PT* diagram indicating method of tabulating superheat data.

integral values of the pressure and the temperature. A superheat table then reports values of v, h, and s (and frequently u) at the grid points of the figure. The values of u, h, and s are completely arbitrary, each being based on an assigned value at some reference state.

Table 4-2 illustrates one possible format for tabulation of data in the superheat region. A more complete compilation of superheated steam data in metric units is found in Table A-14M, while Table A-14 presents data in USCS units. The format of both tables is identical. The units for specific properties are usually listed at the top of a table, under the heading. With respect to Table 4-2 and Fig. 4-5, note that the data begin with saturated *vapor* data and then proceed at a

Table 4-2 Properties of superheated steam (H_2O)

Metric: v, cm^3/g; u, kJ/kg; h, kJ/kg: s, kJ/(kg)(°K)

Temperature, °C	v	u	h	s
	1.0 bar (99.63°C)			
Sat.	1694.0	2506.1	2675.5	7.3594
100	1696.0	2506.7	2676.2	7.3614
120	1793.0	2537.3	2716.6	7.4668
160	1984.0	2597.8	2796.2	7.6597
200	2172.0	2658.1	2875.3	7.8343
	10.0 bars (179.91°C)			
Sat.	194.4	2583.6	2778.1	6.5865
200	206.0	2621.9	2827.9	6.6940
240	227.5	2692.9	2920.4	6.8817
280	248.0	2760.2	3008.2	7.0465
320	267.8	2826.1	3093.9	7.1962

Source: Abstracted from Table A-14M.

USCS: v, ft^3/lb; u, Btu/lb; h, Btu/lb; s, Btu/(lb)(°R)

Temperature, °F	v	u	h	s
	60 psia (292.7°F)			
Sat.	7.17	1098.3	1178.0	1.6444
300	7.26	1101.3	1181.9	1.6496
400	8.35	1140.8	1233.5	1.7134
500	9.40	1178.6	1283.0	1.7678
600	10.43	1216.3	1332.1	1.8165

Source: Abstracted from Table A-14.

given pressure to higher integral temperatures. Some superheat tables in the literature will not, however, carry the data down to the saturated vapor state. The temperature of the saturated vapor data is indicated by the value in parentheses after the pressure value. For example, from Table 4-2, at 1.0-bar pressure the saturation temperature is 99.63°C and the specific volume is 1694 cm^3/g.

Many scientific and engineering problems involve states of matter which do not fall on the grid of data available for that substance. Interpolation of data then becomes necessary. The intervals for the matrix of data found in unabridged superheat tables are usually chosen so that *linear* interpolation leads to reasonable accuracy. Although the superheat tables in the appendix are condensed versions of the original works, we shall still assume linear interpolation is valid. For professional use one should always employ the complete tables as they appear in the literature, unless high accuracy is not desired. Superheat tables for refrigerant 12 (dichlorodifluoromethane) in metric and engineering units appear as Tables A-18M and A-18. Superheat tables for diatomic nitrogen (N_2) appear as Tables A-31M and A-27. The following examples illustrate the use of superheat tables.

Example 4-1M Determine the internal energy of superheated water vapor at (*a*) 1.0 bar and 110°C, and (*b*) 6 bars and 220°C.

SOLUTION (*a*) The value of u at 1 bar and 110°C can be obtained from Table 4-2 or the more complete data of Table A-14M. With reference to either of these tables, we find that at 1 bar

$$u = 2537.3 \text{ kJ/kg at } 120°\text{C}$$

$$u = 2506.7 \text{ kJ/kg at } 100°\text{C}$$

Linear interpolation between these two values leads to an internal energy value of 2522.0 kJ/kg at 110°C.

(*b*) In order to find the value of u at 6 bars and 220°C we must turn to Table A-14M in the appendix. In Table A-14M we find data only at 5 and 7 bars and only at 200 and 240°C. To find the desired value of u we must rely on double interpolation. By linearly interpolating, first, with respect to temperature,

At 5 bars and 220°C: $u = 2675.2$ kJ/kg

At 7 bars and 220°C: $u = 2668.3$ kJ/kg

Consequently, at 6 bars and 220°C the internal energy is the average of these two latter values, or approximately 2671.8 kJ/kg.

Example 4-1 Determine the enthalpy of superheated water vapor at (*a*) 60 psia and 450°F, and (*b*) 110 psia and 550°F.

SOLUTION (*a*) The value of h at 60 psia and 450°F can be obtained from Table 4-2 or Table A-14. With reference to these tables, we find that at 60 psia

$$h = 1233.5 \text{ Btu/lb} \quad \text{at } 400°\text{F}$$

$$h = 1283.0 \text{ Btu/lb} \quad \text{at } 500°\text{F}$$

Linear interpolation between these two values leads to an enthalpy value of 1258.2 Btu/lb at 450°F and 60 psi.

(*b*) In Table A-14 data are reported only at 100 and 120 psia, and only at 500 and 600°F. To find the desired value at 110 psia and 550°F we must rely on double interpolation. By linearly interpolating first with respect to temperature,

At 100 psia and 550°F: $h = 1304.2$ Btu/lb

At 120 psia and 550°F: $h = 1302.4$ Btu/lb

Consequently, at 110 psia and 550°F the enthalpy is the average of these latter two values, or approximately 1303.3 Btu/lb.

Example 4-2M Determine the pressure of superheated refrigerant 12 at a state of 40°C and an enthalpy of (*a*) 214.44 kJ/kg, and (*b*) 214.70 kJ/kg.

SOLUTION This problem illustrates the point that *any* two independent properties fix the state of a simple, compressible substance. In this case we seek P as a function of T and h.

(*a*) The superheat data for refrigerant 12 are tabulated in Table A-18M, with the temperature running from top to bottom at any pressure. Starting at 40°C and the lowest pressure (0.6 bar), we note that the enthalpy starts at 216.77 kJ/kg and decreases as the pressure increases for the same temperature. If we continue to read to higher pressures at 40°C, we finally find that $h = 214.44$ kJ/kg at 2.4 bars.

(*b*) The value of 214.70 kJ/kg for the enthalpy at 40°C lies between 2.0 and 2.4 bars. By linear interpolation we estimate that the pressure corresponding to an enthalpy of 214.70 kJ/kg at 40°C is 2.2 bars.

Example 4-2 Determine (*a*) the temperature of superheated refrigerant 12 at a state of 50 psia and an enthalpy of 90.953 Btu/lb, and (*b*) the pressure at a state of 180°F and an internal energy of 93.800 Btu/lb.

SOLUTION This problem illustrates the point that *any* two independent properties fix the state of a simple, compressible substance.

(*a*) In this case, we seek T as a function of P and h. The superheat data for refrigerant 12 are found in Table A-18, with the temperature running from the top to bottom for any pressure. At 50 psia the saturated-vapor value is 81.249 Btu/lb, and the enthalpy increases with increasing temperature. The desired value of 90.953 Btu/lb for the enthalpy is found at 100°F.

(*b*) In this case, we seek P as a function of T and u. Starting at 180°F and the lowest pressure (10 psia), we note that the internal energy starts at 94.367 Btu/lb and decreases as the pressure increases for the same temperature. If we continue to read u at higher pressures and 180°F, we finally find that $u = 93.800$ Btu/lb at 40 psia.

The preceding examples have illustrated the use of the superheat tables for determining property values of several substances. In general, superheat tables in the literature are arranged like those found in the appendix, although you will find that the position of the temperature and pressure coordinates are frequently reversed. Although interpolation is frequently necessary in order to determine an accurate value of a property, the technique is straightforward. Whether u data are reported in a given table is a matter of choice.

b Saturation Tables

A saturation table lists specific property values (such as v, u, and h) for the saturated liquid and saturated vapor states. Recall from Sec. 4-3 that properties for these two phases are denoted by the subscripts f and g, respectively. Thus, for any extensive property Y (such as V, U, and H), the specific property y in the saturated liquid or saturated vapor state is denoted by y_f or y_g, accordingly. For a unit mass of a mixture of the two phases, a specific property value y_x is determined by adding the contributions of the two phases. Recall [see Eq. (4-1)] that the quality x is defined as the ratio of the mass of vapor to the total mass of liquid plus vapor. Hence for each unit mass of mixture the contribution by the vapor (gas) phase is xy_g, while that by the liquid phase is $(1 - x)y_f$. Consequently, the expression for y_x (per unit mass of mixture) can be written as

$$y_x = (1 - x)y_f + xy_g \tag{4-2}$$

Alternatively, if we designate the difference between the saturated-vapor and saturated-liquid intensive properties by the symbol y_{fg}, that is

$$y_{fg} \equiv y_g - y_f$$

then on the basis of Eq. (4-2) we may also write

$$y_x = y_f + x(y_g - y_f) = y_f + xy_{fg} \tag{4-3}$$

Equations (4-2) and (4-3) are equivalent. These two equations emphasize the fact that a state in the wet region represents only a hypothetical state which has properties that are average characteristics of both phases. The actual state is the state of the two separate phases which are in contact.

Rearrangement of Eq. (4-3) gives

$$x = \frac{y_x - y_f}{y_{fg}}$$

If this particular equation is applied to the specific volume, we may write $x = (v_x - v_f)/v_{fg}$. This result is useful with respect to the Pv diagram in Fig. 4-5. It is seen that the quality of any state of a liquid-vapor mixture is given by the ratio of horizontal distances within the two-phase region on the diagram. A state of 50 percent quality lies midway between the saturated-liquid and saturated-vapor lines along any constant-pressure line. The above equations are valid for any specific property of a two-phase mixture, such as volume, internal energy, enthalpy, etc.

Specific properties of a two-phase mixture are readily calculated from equations such as (4-2) and (4-3). These equations require a knowledge of the property data for the saturation states of the two phases involved and the quality of the mixtures. Only one intensive property is required to identify the intensive state of the two phases in equilibrium when it is known that a saturated mixture exists. Again it is preferable to tabulate the various specific properties of each phase of the mixture against either the pressure or the temperature. We have already noted, especially by the use of Fig. 4-3, that fixing the saturation temperature automatically fixes the saturation pressure, and vice versa, for two-phase systems. In many instances two saturation tables may be given, one with temperature as the independent variable and the other with pressure as the independent variable. In these tables the independent variable usually appears only in integral values.

Table 4-3 Properties of saturated liquid and vapor for water

Metric: v, cm^3/g; u, kJ/kg; h, kJ/kg; s, kJ/(kg)(°K)

		Specific volume		Enthalpy		Entropy	
Temperature, °C T	Pressure, bars P	Sat. liquid v_f	Sat. vapor v_g	Sat. liquid h_f	Sat. vapor h_g	Sat. liquid s_f	Sat. vapor s_g
20	0.02339	1.0018	57791	83.96	2538.1	0.2966	8.6672
40	0.07384	1.0078	19523	167.57	2574.3	0.5725	8.2570
60	0.1994	1.0172	7671	251.13	2609.6	0.8312	7.9096
80	0.4739	1.0291	3407	334.91	2643.7	1.0753	7.6122
100	1.014	1.0435	1673	419.04	2676.1	1.3069	7.3549

Source: Abstracted from Table A-12M.

USCS: v, ft^3/lb; u, Btu/lb; h, Btu/lb; s, Btu/(lb)(°R); T, °F; P, psia

T	P	v_f	v_g	h_f	h_g	s_f	s_g
60	0.2563	0.01604	1207	28.08	1087.7	0.0555	2.0943
70	0.3632	0.01605	867.7	38.09	1092.0	0.0746	2.0642
80	0.5073	0.01607	632.8	48.09	1096.4	0.0933	2.0356
90	0.6988	0.01610	467.7	58.07	1100.7	0.1117	2.0083
100	0.9503	0.01613	350.0	68.05	1105.0	0.1296	1.9822

Source: Abstracted from Table A-12.

Although this may appear as a duplication of data, it is quite convenient when interpolation of either T or P is required.

Table 4-3 is an abridgment of the properties of saturated water as a function of integral values of temperature, in both metric and USCS engineering units. More complete data for saturated water are given in Tables A-12M and A-12 as a function of temperature, and in Tables A-13M and A-13 as a function of pressure. Saturation tables for refrigerant 12 appear as Tables A-16M, A-16, A-17M, and A-17. In some of these tables there are columns of data listed as h_{fg} and s_{fg}. As noted earlier, the difference between saturated-vapor and saturated-liquid properties is symbolized by the subscript fg. The quantity h_{fg} is called the *enthalpy of vaporization*, or the latent heat of vaporization. It represents the quantity of energy required to vaporize a unit mass of a saturated liquid at a given temperature or pressure. Hence, it is an important thermodynamic property. In these tables s_{fg} is the entropy of vaporization. Both h_{fg} and s_{fg} become zero at the critical state.

The following examples illustrate data retrieval from the various saturation tables for water and refrigerant 12. Saturated data for diatomic nitrogen (N_2) appear in Tables A-30M and A-26.

Example 4-3M Determine the volume change when 1 g of saturated liquid water is completely vaporized at (*a*) 100°C, and (*b*) 300°C.

SOLUTION The volume change during vaporization is given by $v_{fg} = v_g - v_f$.

(*a*) These values can be found for 100°C from either Table 4-3 or Table A-12M. The result is

$$v_{fg} = 1673 - 1.04 = 1672 \text{ cm}^3/\text{g}$$

(*b*) At 300°C the data are found only in Table A-12M in the appendix. At this state,

$$v_{fg} = 21.67 - 1.40 = 20.27 \text{ cm}^3/\text{g}$$

This problem demonstrates the vast change in v_{fg} as the temperature approaches the critical state, which in this case is 374°C. The volume change differs by over a factor of 80. These data indicate that the saturated-vapor line on Fig. 4-4 is considerably out of scale. To agree with the above data, this line should fall much more slowly with decreasing pressure. This trend is typical of most substances.

Example 4-3 Determine the volume change when 1 lb of saturated liquid water is completely vaporized at (*a*) 100°F, and (*b*) 600°F.

SOLUTION The volume change during vaporization is given by $v_{fg} = v_g - v_f$.

(*a*) These values for 100°F can be found from either Table 4-3 or Table A-12. The result is

$$v_{fg} = 350.0 - 0.016 = 350 \text{ ft}^3/\text{lb}$$

(*b*) At 600°F the data are found only in Table A-12 in the appendix. From this table,

$$v_{fg} = 0.2677 - 0.0236 = 0.2441 \text{ ft}^3/\text{lb}$$

These data demonstrate the significant change in v_{fg} as the temperature approaches the critical state, which in this case is 705°F. The volume change differs by over a factor of 1000. The data indicate that the saturated-vapor line on Fig. 4-4 is considerably out of position. To account for the wide variation in v_{fg} with temperature (or pressure), the saturated-vapor line should fall much more slowly with decreasing pressure. This trend is typical of most substances.

Example 4-4M Two kilograms of water substance at 200°C are contained in a 0.2-m^3 vessel. Determine (*a*) the pressure, (*b*) the enthalpy in kJ/kg, and (*c*) the mass and volume of the vapor within the vessel.

SOLUTION (*a*) On the basis of the mass and volume data, the overall specific volume is found to be 0.1 m^3/kg or 100 cm^3/g. From Table A-12M, at 200°C, one notes that the water must be a liquid-vapor mixture since v is greater than v_f of 1.1565 cm^3/g, but less than v_g of 127.4 cm^3/g at this temperature. Thus the pressure must be the corresponding saturation vapor pressure at 200°C. This is 15.54 bars, found in column two of Table A-12M.

(*b*) The specific enthalpy of the mixture may be found from a form of Eq. (4-3), namely,

$$h_x = h_f + xh_{fg} = h_f + x(h_g - h_f)$$

However, this equation requires a knowledge of the quality. This may be determined from information on v_x, which is 100 cm^3/g. Note that another form of Eq. (4-3) is

$$v_x = v_f + x(v_g - v_f)$$

Substitution of the values for v_f and v_g at 200°C from Table A-12M leads to

$$100 = 1.1565 + x(127.4 - 1.1565)$$

$$x = \frac{98.84}{126.2} = 0.783$$

The specific enthalpy, then, is found to be

$$h_x = 852.45 + 0.783(1940.7) = 2372 \text{ kJ/kg}$$

(c) The mass of the vapor is 0.783(2) = 1.57 kg. Hence, the volume occupied by the vapor is

$$V_g = m_g v_g = 1570 \text{ g} \times 127.4 \text{ cm}^3/\text{g} = 199{,}470 \text{ cm}^3 = 0.1995 \text{ m}^3$$

Thus the vapor fills nearly all the volume of the vessel.

Example 4-4 Five pounds of water substance at 400°F are contained in a 7.0-ft³ vessel. Determine (a) the pressure, (b) the enthalpy, in Btu/lb, and (c) the mass and volume of the vapor within the vessel.

SOLUTION (a) From the given data on the total volume and mass we find the specific volume to be $\frac{7}{5}$ or 1.4 ft³/lb. From Table A-12, at 400°F, it is found that v is greater than v_f of 0.01864 ft³/lb but less than v_g of 1.866 ft³/lb. Hence, the water must be a liquid-vapor mixture. Thus, the pressure must be the corresponding saturation vapor pressure at 400°F. This is 247.1 psia, which is found in column two of Table A-12.

(b) The specific enthalpy of the mixture may be found from a form of Eq. (4-3), namely,

$$h_x = h_f + xh_{fg} = h_f + x(h_g - h_f)$$

The quantities h_f and h_{fg} may be found from the saturation tables, but the quality x is unknown. This latter value is determined from the value of v_x, which we found to be 1.4 ft³/lb. We may write Eq. (4-3) as

$$v_x = v_f + x(v_g - v_f)$$

Substitution of the values of v_f and v_g at 400°F from Table A-12 leads to

$$1.4 = 0.01864 + x(1.866 - 0.01864)$$

$$x = \frac{1.38}{1.847} = 0.747$$

With a knowledge of the quality, the specific enthalpy is now found to be

$$h_x = 375.1 + 0.747(826.8) = 992.7 \text{ Btu/lb}$$

(c) The mass of the vapor, based on the quality, is simply 0.747(5) = 3.735 lb. Hence the volume occupied by the vapor is

$$V_g = m_g v_g = 3.735 \text{ lb} \times 1.866 \text{ ft}^3/\text{lb} = 6.97 \text{ ft}^3$$

Thus the vapor fills nearly all the volume of the vessel.

Example 4-5M Three kilograms of saturated liquid water is contained in a constant-pressure system at 5 bars. Energy is added to the fluid until it has a quality of 60 percent. Determine (a) the initial temperature, (b) the final pressure and temperature, and (c) the volume and enthalpy changes of the water substance.

SOLUTION (a) In a saturation state there is a unique temperature for a given pressure. From the saturation pressure table, Table A-13M, for water we find the saturation temperature corresponding to 5 bars to be 151.9°C.

(b) Since the fluid is not completely vaporized, the pressure and temperature remain equal to their initial values of 5 bars and 151.9°C.

(c) The volume and enthalpy changes are computed from the relations $\Delta V = m(v_2 - v_1)$ and $\Delta H = m(h_2 - h_1)$, where 1 and 2 represent the initial and final states. The initial specific volume and specific enthalpy are read directly from Table A-13M, in terms of v_f and h_f. These values are

$$v_1 = v_f = 1.0926\ \text{cm}^3/\text{g} \qquad \text{and} \qquad h_1 = h_f = 640.23\ \text{kJ/kg}$$

The values of v_2 and h_2 must be calculated on a basis of a liquid-vapor mixture of 60 percent quality. Hence

$$v_2 = v_f + xv_{fg} = 1.09 + 0.60(374.9 - 1.1) = 225.4\ \text{cm}^3/\text{g}$$

$$h_2 = h_f + xh_{fg} = 640.2 + 0.60(2108.5) = 1905\ \text{kJ/kg}$$

Consequently,

$$\Delta V = 3000\ \text{g} \times (225.4 - 1.1)\ \text{cm}^3/\text{g} = 672{,}900\ \text{cm}^3 = .0673\ \text{m}^3$$

$$\Delta H = 3.0\ \text{kg} \times (1905 - 640)\ \text{kJ/kg} = 3800\ \text{kJ}$$

The enthalpy change in this case is equal to the energy added to the system.

Example 4-5 Three pounds of saturated liquid water are contained in a constant-pressure system at 30 psia. Energy is added to the fluid until it has a quality of 70 percent. Determine (*a*) the initial temperature, (*b*) the final pressure and temperature, and (*c*) the volume and enthalpy changes of the water substance.

SOLUTION (*a*) In a saturation state there is only one saturation temperature for a given pressure. From the saturation pressure table, Table A-13, for water the saturation temperature corresponding to 30 psia is 250.34°F.

(*b*) Since the fluid is not completely vaporized, the pressure and temperature remain equal to the initial values of 30 psia and 250.34°F.

(*c*) The volume and enthalpy changes are computed from the relations, $\Delta V = m(v_2 - v_1)$ and $\Delta H = m(h_2 - h_1)$, where 1 and 2 represent the initial and final states. The initial specific volume and specific enthalpy are read directly from Table A-13, in terms of v_f and h_f. These values are

$$v_1 = v_f = 0.017\ \text{ft}^3/\text{lb} \qquad \text{and} \qquad h_1 = h_f = 218.93\ \text{Btu/lb}$$

The values of v_2 and h_2 must be calculated on a basis of a liquid-vapor mixture of 70 percent quality. Hence

$$v_2 = v_f + xv_{fg} = 0.017 + 0.70(13.75 - 0.017) = 9.63\ \text{ft}^3/\text{lb}$$

$$h_2 = h_f + xh_{fg} = 218.9 + 0.70(945.4) = 880.7\ \text{Btu/lb}$$

Consequently,

$$\Delta V = 3\ \text{lb} \times (9.63 - 0.02)\ \text{ft}^3/\text{lb} = 28.8\ \text{ft}^3$$

$$\Delta H = 3\ \text{lb} \times (880.7 - 218.9)\ \text{Btu/lb} = 1985\ \text{Btu}$$

The enthalpy change in this case is equal to the energy added to the system during the process.

When searching the literature for property values of a substance, one usually finds both saturation and superheat tables. On the basis of some given property data, it frequently is difficult for those unfamiliar with the tables to know which of these two tables contains the desired information. For example, if the enthalpy and temperature of refrigerant 12 are 212.25 kJ/kg and 40°C, what is the corresponding pressure? When temperature (or pressure) and another property value (such as v, u, h, or s) are given, the best method usually is to check in the saturation tables first. By using Table A-16M (saturation-temperature table for refrigerant 12) it is found that $h_f = 74.59$ kJ/kg and $h_g = 203.20$ kJ/kg at 40°C. If the given value had been between 74.59 and 203.2, the substance would be a two-phase mixture, and its quality could be calculated for the saturation pressure of 9.6066 bars. Since the actual value of h is 212.25 kJ/kg, which is greater than h_g, the state we seek must be in the superheat region. Now, using Table A-18M, we find that the pressure corresponding to the given h and T values is 4.0 bars. (The reader should confirm this value of the pressure.) As a general rule, examination of data in the saturation tables first will shorten the time required to find other property values if the condition of the substance is not known.

c Compressed- or Subcooled-liquid Table

There is not a great deal of tabular data for compressed or subcooled liquids in the literature. However, since water is used as the fluid in fossil-fuel power plants, considerable data are available for this substance in the liquid region. The data shown in Table 4-4 are taken from more extensive compilations found in Tables A-15M and A-15. The first row in each set of data is the saturated-liquid data at that temperature. The variation of properties of a compressed liquid with pressure is seen to be slight. By using data for water in the saturated-liquid state of 80°C and 0.4739 bar, for example, as an approximation for the compressed-liquid state of 80°C and 100 bars, errors of 0.45, 0.68, 2.3, and 0.61 percent are made in the values of v, u, h, and s, respectively.

When compressed-liquid data based on experimental values are available, they should be used in engineering calculations. However, in the frequent absence of such data the above comparison indicates a general approximation rule. Compressed-liquid data in most cases can be approximated closely by using the property values of the saturated liquid state at the given *temperature*. This simply

Table 4-4 Properties of compressed liquid water

Metric: data at 80°C				
P, bars	v, cm^3/g	u, kJ/kg	h, kJ/kg	s, kJ/(kg)(°K)
0.474 (sat.)	1.0291	334.86	334.91	1.0753
50	1.0268	333.72	338.85	1.0720
100	1.0245	332.59	342.83	1.0688
USCS: data at 150°F				
P, psia	v, ft^3/lb	u, Btu/lb	h, Btu/lb	s, Bru/(lb)(°R)
3.722 (sat.)	0.01634	117.95	117.96	0.2150
500	0.01632	117.66	119.17	0.2146
1000	0.01629	117.38	120.40	0.2141

Source: Abstracted from Tables A-15M and A-15.

implies that compressed-liquid data are more temperature-dependent than pressure-dependent. If we designate a compressed-liquid state by the subscript c, then, in general, $y_c \approx y_f$ at the given temperature, where y is any specific property.

Example 4-6M Find the change in the specific internal energy of water for a change of state from 40°C, 25 bars, to 80°C, 75 bars by (*a*) use of the compressed-liquid table, and (*b*) use of the approximation rule using saturated data.

SOLUTION (*a*) The data for compressed liquid water appear in Table A-15M. From this table we determine that

$$u_2 - u_1 = 333.15 - 167.25 = 165.90 \text{ kJ/kg}$$

(*b*) As an approximation method, the values of u_f are used at the specified temperatures, and the pressure data are ignored. Use of Table A-12M leads to

$$u_2 - u_1 = 334.86 - 167.56 = 167.30 \text{ kJ/kg}$$

In the absence of a compressed-liquid table the use of a saturation-temperature table produces an answer which is 0.84 percent in error.

Example 4-6 Find the change in the specific enthalpy of water for a change of state from 50°F, 500 psia, to 100°F, 1500 psia, by (*a*) use of the compressed-liquid table, and (*b*) use of the approximation rule using saturated data.

SOLUTION (*a*) The data for compressed liquid water appear in Table A-15. From this table we determine that

$$h_2 - h_1 = 71.99 - 19.50 = 52.49 \text{ Btu/lb}$$

(*b*) As an approximation method, the values of h_f are used at the specified temperatures, and the pressure data are ignored. Use of Table A-12 leads to

$$h_2 - h_1 = 68.05 - 18.06 = 49.99 \text{ Btu/lb}$$

In the absence of a compressed-liquid table the use of a saturation-temperature table produces an answer which is 4.76 percent in error. A much smaller percent error would occur for changes in v, u, and s for the same initial and final states.

d Representations of Pv and PT

Process diagrams are an important aid in problem solving. As we use tabular data in the solution of problems involving the conservation of energy principle, it will be helpful to employ Pv and PT diagrams in the overall analysis. The following example lists some processes which are then plotted on these two diagrams. In some cases it is necessary to examine tabular data in order to find the position of the end states.

Example 4-7M Plot the following processes on Pv and PT diagrams.

(*a*) Superheated vapor is cooled at constant pressure until liquid just begins to form.

(*b*) A liquid-vapor mixture with a quality of 60 percent is heated at constant volume until its quality is 100 percent.

(*c*) A liquid-vapor mixture of water with a quality of 50 percent is heated at a constant temperature of 200°C until its volume is 4.67 times the initial volume.

(*d*) Refrigerant 12 at 8 bars is a saturated liquid. It is heated at constant pressure until its enthalpy has increased by a factor of 2.

SOLUTION The Pv and PT diagrams for processes (*a*) and (*b*) are fairly self-evident, as shown in Fig. 4-6. For process (*c*) the final state is not known. It is fixed, however, by the temperature and the specific volume. For the initial state,

$$v_1 = v_f + xv_{fg} = 1.16 + 0.5(127.4 - 1.16) = 64.3 \text{ cm}^3/\text{g}$$

Since the final volume is 4.67 times the initial value,

$$v_2 = 4.67(64.3) = 300.3 \text{ cm}^3/\text{g}$$

AT 200°C the value of v_g is 127.4 cm³/g. Thus the final state is located far into the superheat region at 200°C. From Table A-14M it is found that $v = 299.9$ cm³/g at 200°C and 7.0 bars. Hence the final pressure is close to 7.0 bars, while the initial saturation pressure is 15.54 bars. The final state of process (*d*) is determined by the enthalpy at 8 bars. The initial enthalpy, from Table A-17M, is found to be 67.30 kJ/kg. The final enthalpy, then, is twice this, or $h_2 =$ 134.6 kJ/kg. This value is less than h_g at 8 bars; therefore, the final state is a wet mixture. To find the final quality,

$$h_2 = 134.6 = h_f + xh_{fg} = 67.3 + x(133.33)$$

$$x = \frac{67.3}{133.33} = 0.505$$

Hence the final state of process (*d*) lies roughly halfway across the wet region at 8 bars.

Example 4-7 Plot the following processes on Pv and PT diagrams.

(*a*) Superheated vapor is cooled at constant pressure until liquid just begins to form.

(*b*) A liquid-vapor mixture with a quality of 60 percent is heated at constant volume until its quality is 100 percent.

(*c*) A liquid-vapor mixture of water with a quality of 50 percent is heated at a constant temperature of 400°F until its volume is 3.16 times the initial value.

(*d*) Refrigerant 12 at 100 psia is a saturated liquid. It is heated at constant pressure until its enthalpy has increased by a factor of 2.

SOLUTION The Pv and PT diagrams for processes (*a*) and (*b*) are fairly self-evident, as shown in Fig. 4-6. For process (*c*) the final state is not known. It is fixed, however, by the temperature and the specific volume. For the initial state,

$$v_1 = v_f + xv_{fg} = 0.019 + 0.5(1.866) = 0.952 \text{ ft}^3/\text{lb}$$

Since the final volume is 3.16 times the initial value,

$$v_2 = 3.16(0.952) = 3.008 \text{ ft}^3/\text{lb}$$

At 400°F the value of v_g is 1.866 ft³/lb. Thus the final state is located in the superheat region. From Table A-14 it is found that $v = 3.007$ ft³/lb at 400°F and 160 psia, which is the final state. The final state of process (*d*) is determined by the final enthalpy at 100 psia. The initial enthalpy, from Table A-17, is 26.542 Btu/lb. Since this value is less than h_g at 100 psia, the final state is a wet mixture. To find the final quality,

$$h_2 = 2(26.542) = h_f + xh_{fg} = 26.542 + x(58.809)$$

$$x_2 = \frac{26.542}{58.809} = 0.451$$

Hence the final state of process (*d*) lies roughly halfway across the wet region at 100 psia.

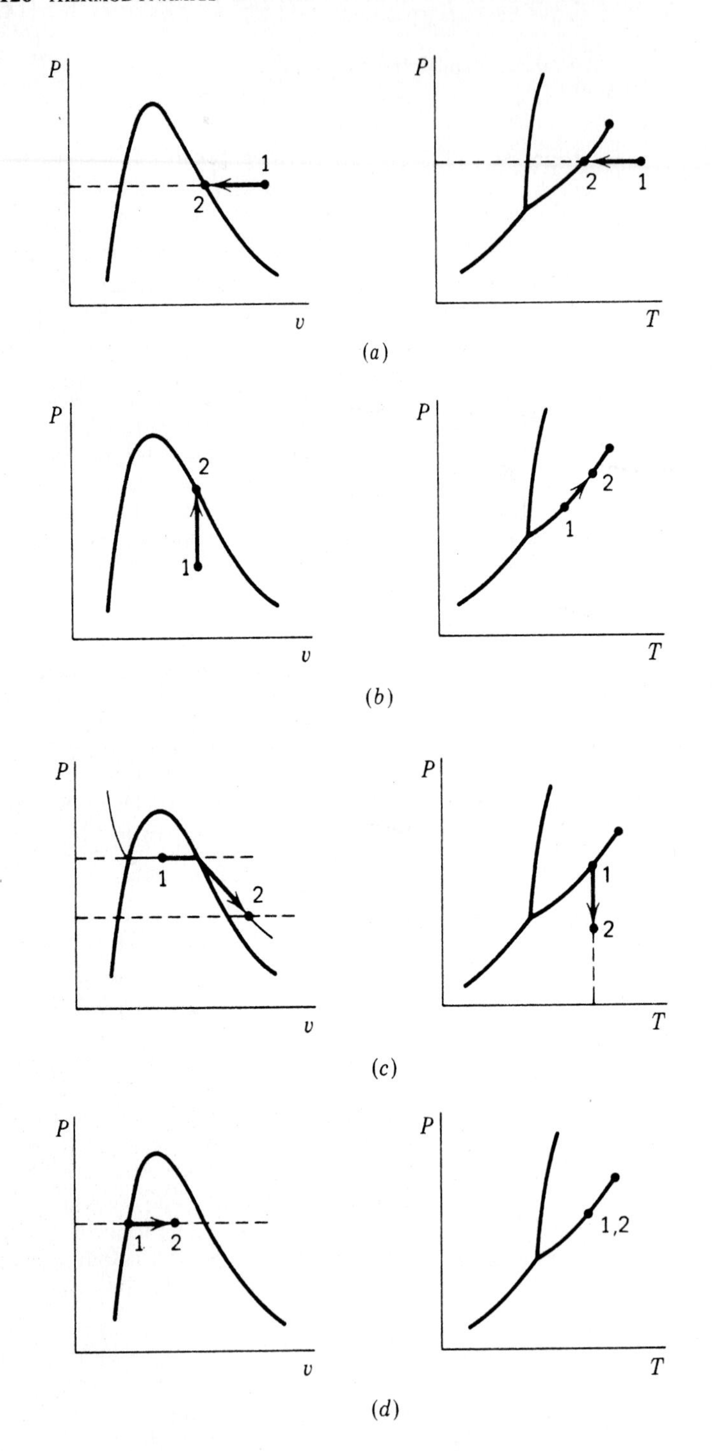

Figure 4-6 Pv and PT diagrams for processes in Examples 4-7M and 4-7.

4-5 TABULAR DATA AND CLOSED-SYSTEM ENERGY ANALYSIS

The preceding section described the format used in the general literature for presenting basic property data in tabular form for simple compressible substances. This discussion covered the superheat, compressed-liquid, and saturation states. We are now in a position to employ these tables in the solution of problems involving the conservation of energy principle for closed systems. Before proceeding, the reader may wish to review the major points introduced in Sec. 3-6 with respect to problem-solving techniques. Among these techniques is the use of property diagrams, which were also discussed in the preceding section. The following examples illustrate the general approach.

a Use of Superheat Data

Example 4-8M One kilogram of water substance is maintained in a weighted piston-cylinder assembly at 30 bars and 240°C. The substance is slowly heated at constant pressure until the temperature reaches 320°C. Determine (*a*) the work required to raise the weighted piston, and (*b*) the required heat input, in kJ/kg.

SOLUTION The system is the mass of water within the assembly. The initial and final states are determined by first checking the saturation table, A-13M. At 30-bar pressure the saturation temperature is 233.9°C. Since the system temperature is always above this value, the fluid is superheated vapor throughout. However, the initial state lies close to the saturation line. A Pv diagram is shown in Fig. 4-7*a*, which illustrates the path of the quasistatic heat-addition process.

(*a*) For a slow heat addition, the work during the quasistatic process is given by the integral of $-P\,dv$. Because the pressure is constant, the work effect is given simply by $-P(v_2 - v_1)$. Taking the specific volume values from the superheat table, Table A-14M, we find that

$$w = -30 \text{ bar} \times (85.0 - 68.2) \text{ cm}^3/\text{g} \times 10^{-1} \text{ N} \cdot \text{m/cm}^3 \cdot \text{bar}$$
$$= -50.4 \text{ N} \cdot \text{m/g} = -50.4 \text{ kJ/kg}$$

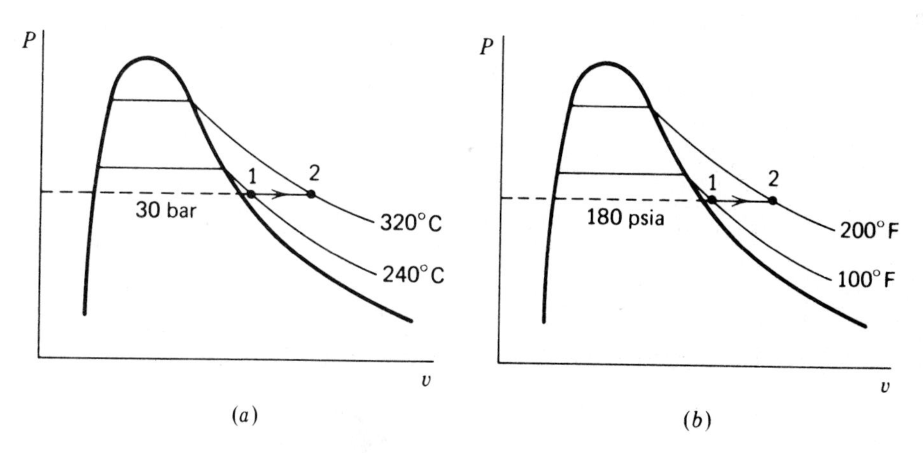

Figure 4-7 Pv diagrams for processes discussed in (*a*) Example 4-8M, and (*b*) Example 4-8.

The negative sign indicates that work is done by the system on the surroundings.

(*b*) The heat transfer is given by the relation, $q = \Delta u - w$. Since $w = -P\,\Delta v$, the heat transfer for this quasistatic, constant-pressure process is also given by $q = \Delta h$. Again making use of Table A-14M, the heat transfer is

$$q = h_2 - h_1 = 3043.4 - 2824.3 = 219.1 \text{ kJ/kg}$$

The positive value indicates that heat is transferred into the system. It would be equally correct in this case to evaluate Δu from the tables and subtract from it the value of w found in part *a*. If this is done, $q = u_2 - u_1 - w = 2788.4 - 2619.7 - (-50.4) = 219.1$ kJ/kg, as before.

If we assume that water vapor is an ideal gas under these conditions, then we can use $c_{p,0}$ data for the given temperature range to calculate Δh. An average $c_{p,0}$ value for water between 240 and 320°C is approximately 1.984 kJ/(kg)(°C). Hence, for an ideal gas,

$$q = \Delta h = c_{p,0}(T_2 - T_1) = 1.984(80) = 158.7 \text{ kJ/kg}$$

This value constitutes an error of over 27 percent when compared to the result based on tabular (experimental) data. Thus, great care must be taken when analyzing gaseous systems because in many cases the assumption of an ideal gas leads to considerable error.

Example 4-8 One pound of water substance is maintained in a weighted piston-cylinder assembly at 450 psia and 500°F. The substance is slowly heated at constant pressure until the temperature reaches 600°F. Determine (*a*) the work required to raise the weighted piston, and (*b*) the required heat input, in Btu/lb.

SOLUTION The system is the mass of water within the assembly. The initial and final states are determined by first checking the saturation table (Table A-13). At 450 psia the saturation temperature is 456.4°F. Since the system temperature is always above this value, the fluid is a superheated vapor throughout. However, the initial state lies close to the saturation line. A Pv diagram is shown in Fig. 4-7*b*, which illustrates the path of the quasistatic heat-addition process.

(*a*) For a slow heat addition, the work during the quasistatic process is given by the integral of $-P\,dv$. Because the pressure is constant, the work interaction is given simply by $-P(v_2 - v_1)$. Taking the specific volume values from the superheat table, A-14, we find that

$$w = -450(144) \text{ lb}_f/\text{ft}^2 \times (1.300 - 1.123) \text{ ft}^3/\text{lb}_m$$

$$= -11{,}470 \text{ ft} \cdot \text{lb}_f/\text{lb}_m$$

The negative sign indicates that work is done by the system on the surroundings.

(*b*) The heat transfer is given by the relation $q = \Delta u - w$. Since $w = -P\,\Delta v$, the heat transfer for this quasistatic, constant-pressure process is also given by $q = \Delta h$. Again making use of Table A-14, the heat transfer is

$$q = h_2 - h_1 = 1302.5 - 1238.5 = 64.0 \text{ Btu/lb}$$

The positive value indicates that heat is transferred into the system. It would be equally correct in this case to evaluate Δu from the tables, and subtract from it the value of w found in part *a*. If this is done, $q = u_2 - u_1 - w = 1194.3 - 1145.1 - (-11{,}470/778) = 64.0$ Btu/lb, as before.

If we assume that water vapor is an ideal gas under these conditions, then we can use $c_{p,0}$ data for the given temperature range to calculate Δh. An average $c_{p,0}$ value for water between 500 and 600°F is approximately 0.476 Btu/(lb)(°F). Hence, for an ideal gas,

$$q = \Delta h = c_{p,0}\,\Delta T = 0.476(100) = 47.6 \text{ Btu/lb}$$

This value constitutes an error of over 25 percent when compared to the result based on tabular (experimental) data. Thus, great care must be taken when analyzing gaseous systems because in many cases the assumption of an ideal gas leads to considerable error.

Example 4-9M Refrigerant 12 is contained within a rigid tank at an initial state of 2.8 bars and 100°C. Heat is removed until the pressure drops to 2.4 bars. During the process a paddle-wheel within the tank is turned with a constant torque of 6 N · m for 30 revolutions. If the system contains 0.1 kg, compute the required heat transfer, in kilojoules.

SOLUTION Considering the refrigerant within the tank as the system, we write the energy balance in the form

$$q + w = \Delta u = u_f - u_i$$

where f and i stand for the final and initial states. In the absence of boundary work (because of the rigid tank),

$$q = u_f - u_i - w_{\text{paddle}}$$

The saturation temperature at 2.8 bars, from Table A-17M, is −2.93°C. Hence the initial state of 100°C is well into the superheat region. From the superheat table (Table A-18M), we find that

$$u_i = 228.89 \text{ kJ/kg} \qquad \text{and} \qquad v_i = 89.24 \text{ cm}^3/\text{g}$$

The value of v_i is extremely important since, for a rigid system, $v_f = v_i$. Consequently the final state has a pressure of 2.4 bars and a specific volume of 89.24 cm³/g. At 2.4 bars, from the saturated data of Table A-17M, we find that v_g is only 60.76 cm³/g. Thus the final state is also in the superheat region. Looking up the 2.4-bar data in Table A-18M, one sees that a v value of 89.24 cm³/g corresponds very closely to a state of 50°C. Therefore, $u_f = 199.51$ kJ/kg. The Pv diagram shows the vertical path above the saturated vapor line (see Fig. 4-8a).

To complete the analysis we need a value for the paddle-wheel work. Based on the discussion in Chap. 2, rotational mechanical work is given by $\tau\theta$ when τ is a constant; θ must be expressed in radians. Hence,

$$W_{\text{paddle}} = 6 \text{ N} \cdot \text{m} \times 30(2\pi) = 1131 \text{ N} \cdot \text{m}$$

Substitution of the proper values into the energy equation yields

$$q = (199.51 - 228.89) \text{ kJ/kg} + \frac{-1131 \text{ N} \cdot \text{m}}{100 \text{ g}}$$

$$= -29.38 - 11.31 = -40.69 \text{ kJ/kg}$$

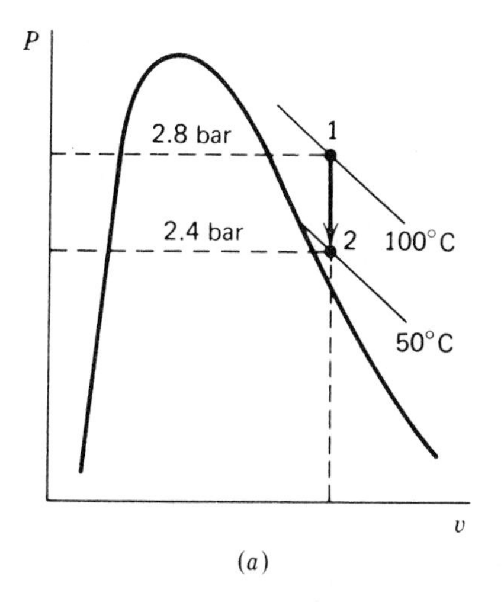

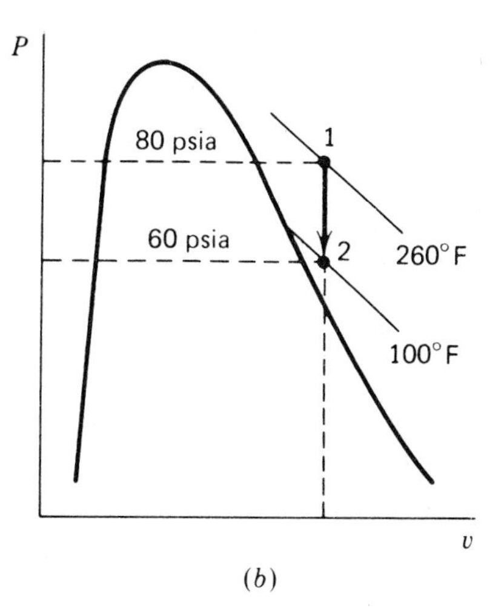

Figure 4-8 Pv diagrams for processes discussed in (a) Example 4-9M, and (b) Example 4-9.

Finally,

$$Q = mq = 0.1 \text{ kg} \times -40.69 \text{ kJ/kg} = -4.07 \text{ kJ}$$

The total heat interaction must account not only for a decrease in the internal energy of the system, but also for the energy added in the form of paddle-wheel work.

Example 4-9 Refrigerant 12 is contained within a rigid tank at an initial state of 80 psia and 260°F. Heat is removed until the pressure drops to 60 psia. During the process a paddle-wheel within the tank is turned with a constant torque of 3.0 $\text{lb}_f \cdot \text{ft}$ for 600 revolutions. If the system contains 1.2 lb, compute the required heat transfer, in Btu.

SOLUTION Considering the refrigerant within the tank as the system, we write the energy balance in the form

$$q + w = \Delta u = u_f - u_i$$

where f and i stand for the final and initial states. In the absence of boundary work (because of the rigid tank),

$$q = u_f - u_i - w_{\text{paddle}}$$

The saturation temperature at 80 psia, from Table A-17, is 66.21°F. Hence the initial state of 260°F is well into the superheat region. From the superheat table (Table A-18), we find that

$$u_i = 104.919 \qquad \text{and} \qquad v_i = 0.7654 \text{ ft}^3\text{/lb}$$

The value of v_i is extremely important since, for a rigid system, $v_f = v_i$. Consequently the final state has a pressure of 60 psia and a specific volume of 0.7654 ft^3/lb. At 60 psia, from the saturated data of Table A-17, we find that v_g is only 0.6701 ft^3/lb. Thus the final state also is in the superheat region. Looking up the 60-psia data in Table A-18, one sees that a v value of 0.7654 ft^3/lb corresponds very closely to a state of 100°F. Therefore, $u_f = 82.024$ Btu/lb. The Pv diagram shows the vertical path above the saturated-vapor line (see Fig. 4-8*b*).

To complete the analysis we need a value for the paddle-wheel work. Based on the discussion in Chap. 2, rotational mechanical work is given by $\tau\theta$ when τ is a constant; θ must be expressed in radians. Hence,

$$W_{\text{paddle}} = 3 \text{ lb}_f \cdot \text{ft} \times 600(2\pi) = 11{,}310 \text{ ft} \cdot \text{lb}_f$$

Substitution of the proper values into the energy equation yields

$$q = (82.024 - 104.919) \text{ Btu/lb} - \frac{(11{,}310/778) \text{ Btu}}{1.2 \text{ lb}}$$

$$= -22.895 - 12.1 = -35.0 \text{ Btu/lb}$$

Finally,

$$Q = mq = 1.2 \text{ lb} \times -35.0 \text{ Btu/lb} = -42.0 \text{ Btu}$$

b Use of Saturation data

Example 4-10M Refrigerant 12 is condensed in a closed system from an initial state of 6 bars and 60°C to a final state of saturated liquid at the same pressure.

(*a*) Determine the value of the work interaction, in $\text{N} \cdot \text{m/g}$.

(*b*) Compute the value of the heat interaction, in kJ/kg.

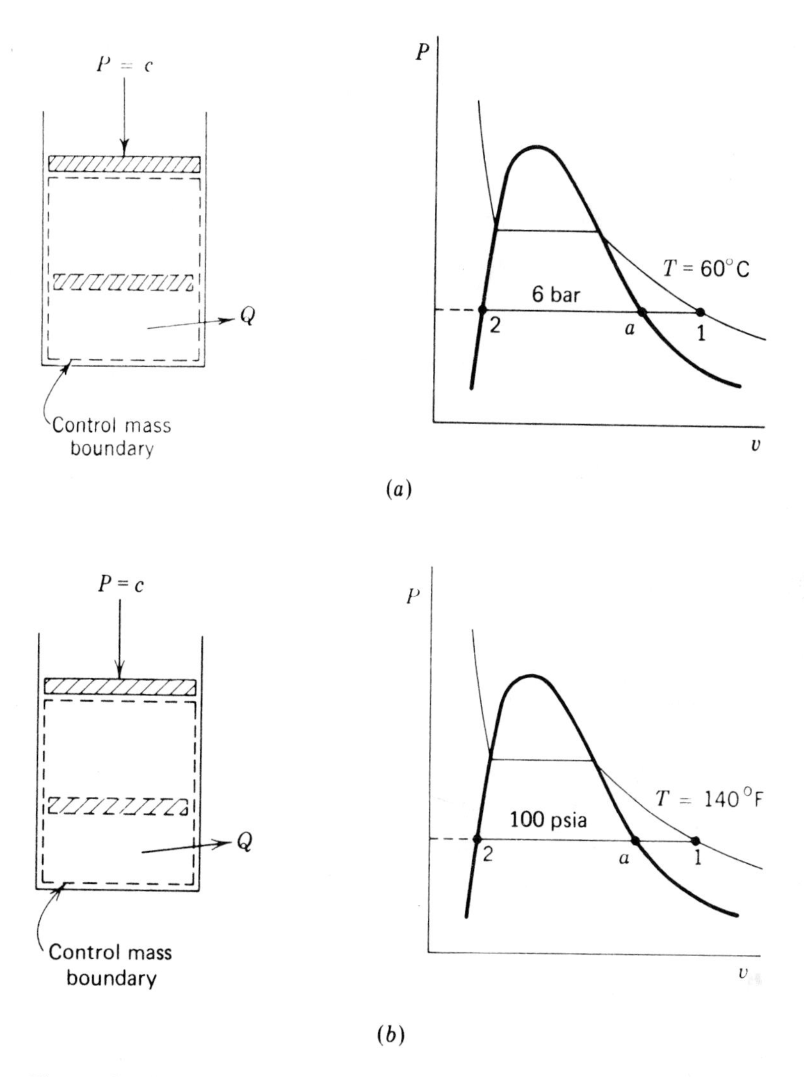

Figure 4-9 Schematic sketches and *Pv* diagrams for processes represented in (*a*) Example 4-10M, and (*b*) Example 4-10.

SOLUTION (*a*) A sketch of the equipment and a *Pv* diagram are shown in Fig. 4-9*a*. A piston-cylinder assembly is used to keep the fluid at a constant pressure at 6 bars. At 6 bars the saturation temperature is 22.0°C (Table A-17M). Therefore the initial state of 60°C is a superheated one, as shown on the *Pv* diagram. The only work interaction is boundary work, and it is given by the integral of $-P\,dv$ if we assume that the process is quasistatic. In addition, the pressure is held constant. Consequently a knowledge of the initial and final specific volumes is sufficient to evaluate w. From the superheat table (Table A-18M), $v_1 = 34.89\ \text{cm}^3/\text{g}$, and from the pressure-saturation table (Table A-17M), $v_2 = v_f = 0.7566\ \text{cm}^3/\text{g}$ at 6 bars. Hence,

$$w = -P(v_2 - v_1) = -6\ \text{bars} \times (0.7566 - 34.89)\ \text{cm}^3/\text{g} \times 10^{-1}\ \text{N}\cdot\text{m/cm}^3\cdot\text{bar} = 20.5\ \text{N}\cdot\text{m/g}$$

(*b*) For a simple, compressible substance the conservation of energy principle reduces to

$q + w = \Delta u$. The value of q can be found directly from this equation because the internal-energy change is fixed by the end states and these are known. From Table A-18M, $u_1 = 202.34$ kJ/kg and from Table A-17M, $u_2 = u_f = 56.35$ kJ/kg. Therefore,

$$q = u_2 - u_1 - w = 56.35 - 202.34 - 20.5 = -166.5 \text{ kJ/kg}$$

The answer is negative, as expected, since heat must be removed in order to condense the vapor.

Since the process is at constant pressure with boundary work and assumed quasistatic, the energy equation can be written as $q = \Delta h$. By recognizing this special form of the energy balance one does not need to calculate w and Δu separately. For this particular problem, $q = h_2 - h_1 = 56.80 - 223.27 = -166.5$ kJ/kg, as before. It might be pointed out that the heat removed between states a and 2 shown on the Pv diagram is equal to $-h_{fg}$, the negative of the enthalpy of vaporization. Since our energy equation shows that $\Delta h = q$ for a liquid-vapor phase change, it is easy to see why h_{fg} is also called the latent "heat" of vaporization.

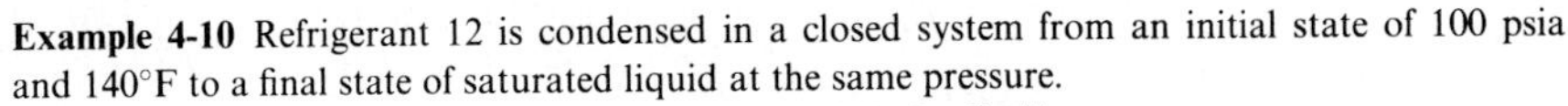

Example 4-10 Refrigerant 12 is condensed in a closed system from an initial state of 100 psia and 140°F to a final state of saturated liquid at the same pressure.

(*a*) Determine the value of the work interaction, in ft · lb_f/lb_m.

(*b*) Compute the value of the heat interaction, in Btu/lb_m.

SOLUTION (*a*) A sketch of the equipment and a process diagram are shown in Fig. 4-9*b*. A piston-cylinder assembly is used to keep the system at a constant pressure of 100 psia. At 100 psia the saturation temperature of refrigerant 12 is 80.76°F (Table A-17). Therefore the initial state of 100°F is a superheated one, as shown on the Pv diagram. The only work interaction is boundary work, and it is given by the intergral of $-P\,dV$ if we assume that the process is quasistatic. In addition, the pressure is held constant. Consequently a knowledge of the initial and final specific volumes is sufficient to evaluate w. From the superheat table (Table A-18), $v_1 = 0.4788$ ft³/lb, and from the pressure saturation table (Table A-17), $v_2 = v_f = 0.0123$ ft³/lb at 100 psia. Hence,

$$w = -P(v_2 - v_1) = -100(144)\ \text{lb}_f/\text{ft}^2 \times (0.0123 - 0.4788)\ \text{ft}^3/\text{lb}_m$$

$$= 6{,}710\ \text{ft} \cdot \text{lb}_f/\text{lb}_m$$

(*b*) For a simple, compressible substance the conservation of energy principle reduces to $q + w = \Delta u$. The value of q can be found directly from this equation because the internal-energy change is fixed by the end states and these are known. From Table A-18, $u_1 = 86.647$ Btu/lb and from Table A-17, $u_2 = u_f = 26.31$ Btu/lb. Therefore,

$$q = u_2 - u_1 - w = 26.31 - 86.65 - \frac{6{,}710}{778} = -69.0 \text{ Btu/lb}$$

The answer is negative, as expected, since heat must be removed in order to condense the vapor.

Since the process is constant pressure with boundary work and assumed quasistatic, the energy equation can be written as $q = \Delta h$. (See Sec. 2-14.) By recognizing this special form of the energy balance one does not need to calculate w and Δu separately. For this particular problem, $q = h_2 - h_1 = 26.54 - 95.507 = -69.0$ Btu/lb, as before. It might be pointed out that the heat removed between states a and 2 shown on the Pv diagram is equal to $-h_{fg}$, the negative of the enthalpy of vaporizations. Since our energy balance shows that $\Delta h = q$ for a liquid-vapor phase change, it is easy to see why h_{fg} is also called the latent "heat" of vaporization.

Example 4-11M One-tenth of a kilogram of water at 3 bars and 76.3 percent quality is contained in a rigid tank which is thermally insulated. A paddle-wheel inside the tank is turned by an external motor until the substance is a saturated vapor. Determine the work necessary to complete the process, and the final pressure and temperature of the water.

SOLUTION The boundaries of the control mass are chosen to lie just inside the tank. The fluid is assumed to be a simple, compressible substance, and during the process the water is in fluid shear and is not in equilibrium. However, for this process a knowledge of the end states only is sufficient. For this system the energy equation again is

$$Q + W = \Delta U$$

No heat transfer or boundary work occurs because of the restrictions on the system (insulated and rigid). Work is done on the system, though, by the action of the paddle-wheel, so that the above equation reduces to

$$W_{\text{paddle}} = \Delta U = U_2 - U_1 = m(u_2 - u_1)$$

The work requirements are determined, then, from an evaluation of the initial and final specific internal energies. The initial value of u is found from data in the saturation-pressure table (Table A-13M), in conjunction with the equations for intensive properties expressed in terms of the quality. That is, for 3 bars and $x = 0.763$, we find that

$$u_1 = u_f(1 - x) + u_g x = 561.2(0.237) + 2543.6(0.763)$$
$$= 2074 \text{ kJ/kg}$$

To find the final state one recognizes that the final specific volume is the same as the initial value, since the tank is rigid and closed. The initial specific volume is

$$v_1 = v_f(1 - x) + v_g x = 1.073(0.237) + 605.8(0.763)$$
$$= 462.5 \text{ cm}^3/\text{g}$$

Thus the final state is a saturated vapor with a specific volume of 462.5 cm^3/g. From the accompanying Pv diagram (see Fig. 4-10) it is noted that in order to proceed at constant v from the wet region at 3 bars and 76.3 percent quality to the state of a saturated vapor, the final pressure must be greater than 3 bars. From the saturation-pressure table (Table A-13M), for steam it is found that a value of $v_g = 462.5$ cm^3/g corresponds to a pressure of 4 bars and a temperature of 143.6°C.

At this final state the specific internal energy is 2553.6 kJ/kg. The paddle-wheel work that must be supplied, then, is

$$W_{\text{paddle}} = m(u_2 - u_1) = 0.1 \text{ kg} \times (2553.6 - 2074) \text{ kJ/kg}$$
$$= 48.0 \text{ kJ}$$

The value of the paddle-wheel work is positive, since the work is done on the system by the rotating shaft.

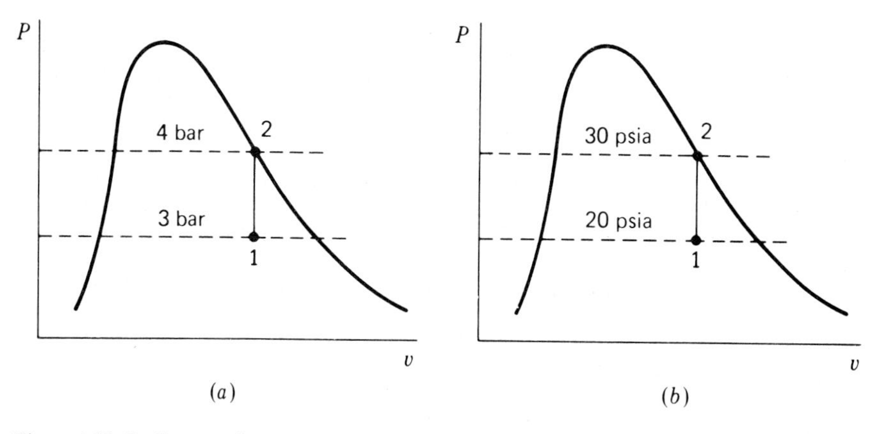

Figure 4-10 Pv diagrams for processes discussed in (*a*) Example 4-11M, and (*b*) Example 4-11.

Example 4-11 One-tenth of a pound of steam at 20 psia and 68 percent quality is contained in a rigid tank which is thermally insulated. A paddle-wheel inside the tank is turned by an external motor until the substance is a saturated vapor. Determine the work necessary to complete the process and the final pressure and temperature of the steam.

SOLUTION The boundaries of the control mass are chosen to lie just inside the tank. In the initial and final equilibrium states the system can be assumed to be a simple system. During the process the water is in fluid shear and is not in equilibrium. However, we shall find that, for this process, a knowledge of the end states only is sufficient. For a simple, compressible system the energy equation is again

$$Q + W = \Delta U$$

No heat transfer or boundary work occurs, because of the restrictions on the system (insulated and rigid). Work is done on the system, though, by the action of the paddle-wheel, so that the above equation reduces to

$$W_{\text{paddle}} = \Delta U = U_2 - U_1 = m(u_2 - u_1)$$

The work requirements are determined, then, from an evaluation of the initial and final sensible internal energies. The initial value of u is found by employing data from the steam table (Table A-13), in conjunction with the equations developed earlier for intensive properties expressed in terms of the quality. That is, for 20 psia and $x = 0.68$, we find that

$$u_x = u_f(1 - x) + u_g x = 196.2(0.32) + (1082.0)(0.68) = 798.2 \text{ Btu/lb}$$

Since the tank is rigid and closed, the final specific volume is the same as the initial value. The initial specific volume is

$$v_1 = v_x = v_f(1 - x) + v_g x = 0.017(0.32) + 20.09(0.68) = 13.67 \text{ ft}^3\text{/lb}$$

Thus the final state is a state of saturated vapor where the specific volume is 13.67 ft^3/lb. From the accompanying Pv diagram (Fig. 4-10) it is noted that, in order to proceed from the wet region at 20 psia and 68 percent quality to the state of a saturated vapor, the final pressure must be above 20 psia. From the saturation table for steam (Table A-13), it is found that the value of v_g corresponds closely to this at a pressure of 30 psia and a temperature of 250°F. At this state the sensible internal energy is 1088.0 Btu/lb. The paddle-wheel work that must be supplied, then, is

$$\begin{aligned} W_{\text{paddle}} &= (1088.0 - 798.2)(0.1) \\ &= 29.0 \text{ Btu} \\ &= 22{,}600 \text{ ft} \cdot \text{lb}_f \end{aligned}$$

The value of the paddle-wheel work must be positive because of our choice of a sign convention for work interactions.

Example 4-12M A piston-cylinder assembly with an initial volume of 0.01 m^3 is filled with saturated refrigerant 12 vapor at 16°C. The substance is compressed until a state of 9 bars and 60°C is reached. During the compression process the heat loss amounts to 0.4 kJ. Compute the boundary work required in kilojoules.

SOLUTION The conservation of energy principle for the simple, compressible substance can be written in the form

$$W = \Delta U - Q = m(u_2 - u_1) - Q$$

The mass of the system is found from the basic relation, $m = V/v$, when evaluated at the initial state. From Table A-16M we find that $v_1 = v_g = 34.42\ \text{cm}^3/\text{g}$ at 16°C. Therefore,

$$m = \frac{V}{v} = \frac{0.01\ \text{m}^3}{34.42\ \text{cm}^3/\text{g}} \times 10^6\ \text{cm}^3/\text{m}^3 = 291\ \text{g} = 0.291\ \text{kg}$$

The value of u_1 is u_g at 16°C, or 176.78 kJ/kg. The saturation temperature at 9 bars, from Table A-17M, is only 37.37°C. Thus at 9 bars and 60°C the substance is superheated. From Table A-18M we find that $u_2 = 199.56$ kJ/kg.

Substitution of the given and acquired data yields

$$W = m(u_2 - u_1) - Q = 0.291\ \text{kg} \times (199.56 - 176.78)\ \text{kJ/kg} + 0.4\ \text{kJ}$$
$$= 6.63 + 0.4 = 7.03\ \text{kJ}$$

The heat loss in this case is roughly only 6 percent of the work required. The accompanying Pv sketch of the process (Fig. 4-11) shows the decrease in volume as the vapor is compressed and becomes superheated. Since P as a function of v is not known, the work cannot be evaluated by the integral of $-P\,dV$.

Example 4-12 A piston-cylinder assembly with an initial volume of 600 in^3 is filled with saturated refrigerant 12 vapor at 60°F. The substance is compressed until a state of 120 psia and 140°F is reached. During the compression process the heat loss amounts to 0.5 Btu. Compute the work required, in ft · lb$_f$.

SOLUTION The conservation of energy principle for the simple, compressible substance can be written as

$$W = \Delta U - Q = m(u_2 - u_1) - Q$$

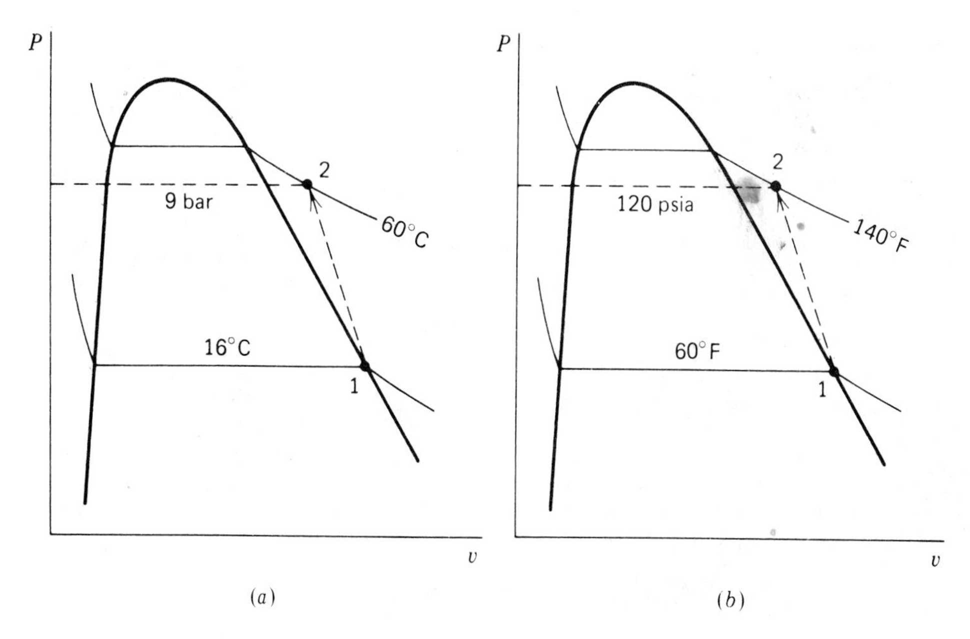

Figure 4-11 Pv diagrams for processes discussed in (a) Example 4-12M, and (b) Example 4-12.

The mass of the system is found from the basic relation, $m = V/v$, when evaluated at the initial state. From Table A-16 we find that $v_1 = v_g = 0.5584\ \text{ft}^3/\text{lb}$ at 60°F. Therefore,

$$m = \frac{V}{v} = \frac{600\ \text{in}^3}{0.5584\ \text{ft}^3/\text{lb}} \times \frac{1\ \text{ft}^3}{1278\ \text{in}^3} = 0.622\ \text{lb}_m$$

The value of u_1 is u_g at 60°F, or 75.92 Btu/lb. The saturation temperature at 120 psia, from Table A-17, is only 93.29°F. Thus at 120 psia and 140°F the substance is superheated vapor. From Table A-18, then, we find that $u_2 = 86.098$ Btu/lb.

Substitution of the given and acquired data yields

$$\begin{aligned} W &= m(u_2 - u_1) - Q = 0.622\ \text{lb} \times (86.098 - 75.92)\ \text{Btu/lb} + 0.5\ \text{Btu} \\ &= 6.33 + 0.5 = 6.83\ \text{Btu} = 5310\ \text{ft} \cdot \text{lb}_f \end{aligned}$$

The heat loss in this case is roughly only 9 percent of the work required. The accompanying Pv sketch of the process (Fig. 4-11) shows the decrease in volume as the vapor is compressed and becomes superheated. Since P as a function of v is not known, the work cannot be evaluated by the integral of $-P\,dV$.

c Use of Compressed-liquid Data

Example 4-13M Liquid water in a compressed-liquid state of 75 bars and 40°C is heated quasistatically at constant pressure until it becomes a saturated liquid. Compute the heat input required, in kJ/kg.

Solution The path of the process is shown on the accompanying Pv diagram (Fig. 4-12), where the region close to the saturated-liquid line has been greatly enlarged. At constant pressure the quasistatic work is given by $-P\,\Delta v$. Hence the heat input becomes

$$q = \Delta u - w = \Delta u + P\,\Delta v = \Delta h$$

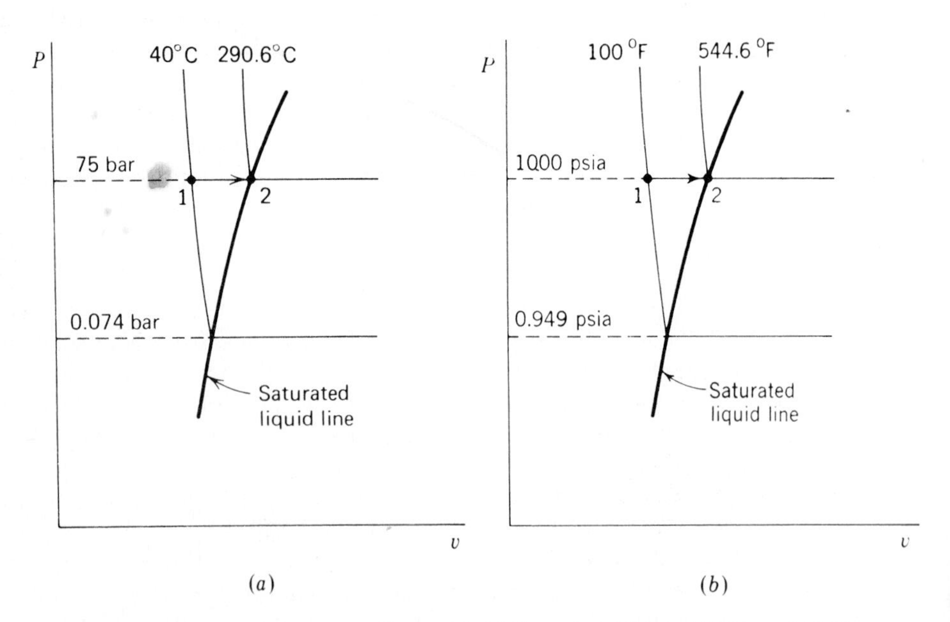

Figure 4-12 Pv diagrams for processes discussed in (*a*) Example 4-13M, and (*b*) Example 4-13.

The initial enthalpy at 75 bars and 40°C is found from Table A-15M to be 174.18 kJ/kg. In the final saturated-liquid state at 75 bars the temperature is 290.59°C and the enthalpy is 1292.2 kJ/kg (see Table A-15M). Consequently,

$$q = h_2 - h_1 = 1292.2 - 174.2 = 1118.0 \text{ kJ/kg}$$

This large value for q is due to the fact that the temperature changes by over 250°C during the process.

Example 4-13 Liquid water in a compressed-liquid state of 1000 psia and 100°F is heated quasistatically at constant pressure until it becomes a saturated liquid. Compute the heat input required, in Btu/lb.

SOLUTION The path of the process is shown on the accompanying Pv diagram (Fig. 4-12), where the region close to the saturated-liquid line has been greatly enlarged. At constant pressure, the quasistatic work is given by $P\,\Delta v$. Hence the heat input becomes

$$q = \Delta u + w = \Delta u + P\,\Delta v = \Delta h$$

The initial enthalpy at 1000 psia and 100°F is found from Table A-15 to be 70.68 Btu/lb. In the final saturated-liquid state at 1000 psia the temperature is 544.61°F and the enthalpy is 542.4 Btu/lb. Consequently,

$$q = 542.4 - 70.7 = 471.7 \text{ Btu/lb}$$

4-6 THE COMPRESSIBILITY FACTOR AND CORRESPONDING STATES

Unless the pressure is reasonably low and the temperature relatively high, gases do not exhibit a PvT behavior which can be represented accurately by the ideal-gas equation, $Pv = RT$. One method of retaining a similar format for the PvT relationship of nonideal gases and still retain reasonable accuracy is through the use of a compressibility-factor correction. The departure from ideal-gas behavior may be characterized by a compressibility factor Z, which is defined as

$$Z \equiv \frac{Pv}{RT} \tag{4-4}$$

Since RT/P is the ideal-gas specific volume (v_{ideal}) of a gas, the compressibility factor may be considered a measure of the ratio of the actual specific volume to the ideal-gas specific volume. That is, $Z = v_{\text{actual}}/v_{\text{ideal}}$. For the hypothetical ideal gas the compressibility factor is unity, but for actual gases it can be either less than or greater than unity. Hence, the compressibility factor measures the deviation of an actual gas from ideal-gas behavior.

The specific volume of any gas could be estimated at any desired pressure and temperature if the compressibility factor were known for the gas as a function of P and T, since $v = ZRT/P$. This information is found through application of what is known as the *principle of corresponding states*. The principle predicts that the Z factor for all gases is approximately the same when the gases have the same reduced pressure and temperature. The reduced pressure P_R and reduced temperature T_R are defined by

$$P_R \equiv \frac{P}{P_c} \quad \text{and} \quad T_R \equiv \frac{T}{T_c} \tag{4-5}$$

Absolute pressures, and temperatures in kelvins or degrees Rankine, must be used in these equations. Thus the critical pressure and temperature of a substance are used to define a reduced state. The validity of such a principle must be based on experimental evidence. When reduced isotherms T_R are plotted on a ZP_R chart, the average deviation of experimental data for a number of gases is somewhat less than 5 percent. Figure 4-13 shows a correlation of actual data for 10 gases for a limited number of reduced isotherms. When the best curves are fitted to all the data, a more complete plot, such as Fig. A-24M in the appendix, results. This generalized compressibility chart is for P_R values from 0 to 1.0 and T_R values from 0.60 to 5.0. Once the chart has been drawn from data for a limited number of substances, it is assumed to be generally applicable to all gases. As we mentioned above, the generalized chart is only an approximation. However, it is remarkably good for many engineering design problems. The main virtue of the generalized compressibility chart is that it requires only a knowledge of critical pressures and temperatures to predict approximately the specific volume of a real gas. It must be emphasized that the generalized compressibility chart should not

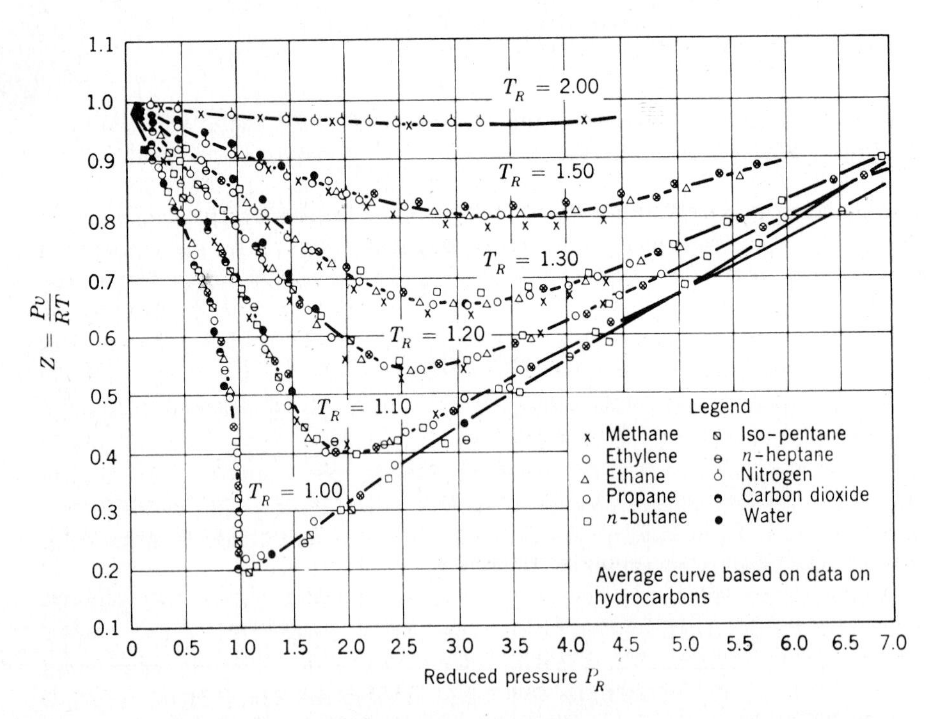

Figure 4-13 Experimental data correlation for a generalized Z chart. [Gour-Jen Su: Modified Law of Corresponding States, *Ind. Eng. Chem.* (Intern. Edition), **38**:803 (1946).]

be used as a substitute for accurate experimental PvT data. The major role of a generalized compressibility chart is to provide reasonable estimates of PvT behavior in the absence of accurate measurements.

A few general characteristics of the compressibility charts should be noted:

1. For reduced temperatures greater than 2.5, the value of Z is greater than unity for all pressures. Under these circumstances the actual volume will always be greater than the ideal-gas volume for the same pressure and temperature.
2. For reduced temperatures below 2.5, the reduced isotherms go through a minimum at fairly low reduced pressures. In this region the volume is less than the ideal-gas volume, and the deviation from ideal-gas behavior is substantial.
3. When P_R is greater than 10, the deviation from ideal-gas behavior can approach several hundred percent.
4. In the limit, of course, as P_R approaches zero, the value of Z approaches unity for all values of the reduced temperature.

The compressibility factor can also be found when vT data or vP data are given. For correlation purposes it has been found best to use a pseudocritical volume in the definition of the reduced volume, rather than the actual critical volume. If we define a pseudocritical volume v'_c by the quantity RT_c/P_c, then the pseudoreduced volume v'_R is

$$v'_R = \frac{vP_c}{RT_c} \tag{4-6}$$

Note that again only a knowledge of T_c and P_c is required. Lines of constant v'_R are also shown in Fig. A-24M, with values running from 0.70 to 8.00.

From the critical data shown in Tables A-3M and A-3, one finds that P_c generally is larger than 30 atm. Standard atmospheric pressure then corresponds to a reduced pressure of about 0.03 or less for most substances. From Fig. A-24M, one sees that the maximum deviation from the ideal-gas law for any gas at 1 atm, regardless of temperature, will be 5 percent or less. When T_R is 1.0 or greater at 1 atm, the deviation will be 1 percent or less. On the other hand, the effect of the critical temperature can be quite pronounced. Consider water, for example, with a critical temperature of 647°K (1165°R). At temperatures from 150 to 550°C (300 to 1000°F) the reduced temperatures range roughly from 0.70 to 1.25. Figure A-24M indicates that water vapor would deviate greatly from ideal-gas behavior in this temperature range, especially as P_R approaches unity. A P_R value of 1.0 for water corresponds to a pressure of 218 atm.

If one considers nitrogen or argon instead of water, a different situation exists. The critical temperatures of these two gases are 126 and 151°K, respectively. Consequently, for room temperature or above, the reduced temperature would always be greater than 2. For such values the compressibility factor is close to unity and is relatively independent of pressure. The assumption of ideal-gas behavior is quite good in these cases (± 2 percent). Therefore, critical data are extremely useful in estimating whether or not a given gaseous substance will approach ideal-gas behavior at a specified temperature and pressure. In addition, a generalized compressibility chart allows one to predict fairly accurately the deviation from ideality.

Example 4-14M Determine the specific volume of water vapor, in cm^3/g, at 200 bars and 520°C by (*a*) the ideal-gas equation of state, (*b*) the principle of corresponding states, and (*c*) the experimental value in the superheat table.

SOLUTION (*a*) On the basis of the ideal-gas equation, $v = RT/P$. The gas constant R, based on Table A-1, is

$$R = \frac{R_u}{M} = \frac{0.08314 \text{ bar} \cdot \text{m}^3/(\text{kg} \cdot \text{mol})(°\text{K})}{18 \text{ kg/kg} \cdot \text{mol}} \times \frac{10^6 \text{ cm}^3}{\text{m}^3} \times \frac{\text{kg}}{10^3 \text{ g}}$$

$$= 4.62 \text{ bar} \cdot \text{cm}^3/(\text{g})(°\text{K})$$

Hence,

$$v = \frac{RT}{P} = 4.62 \text{ bar} \cdot \text{cm}^3/(\text{g})(°\text{K}) \times \frac{793°\text{K}}{200 \text{ bars}} = 18.3 \text{ cm}^3/\text{g}$$

(*b*) The specific volume based on the principle of corresponding states is given by $v_{\text{actual}} = Zv_{\text{ideal}}$. The Z factor is found by computing T_R and P_R and then using Fig. A-24M to find Z. The critical data are found in Table A-3M. Based on these data

$$T_R = \frac{793°\text{K}}{647°\text{K}} = 1.23 \qquad P_R = \frac{200 \text{ bars}}{218.3 \text{ atm}} \times \frac{0.987 \text{ atm}}{1 \text{ bar}} = 0.904$$

From Fig. A-24M the Z value is approximately 0.83. Therefore

$$v = 0.83(18.3) = 15.2 \text{ cm}^3/\text{g}$$

(*c*) The tabulated value, based on experimental data, is found in Table A-14M to be 15.51 cm^3/g. In comparison to the tabulated value, the ideal-gas equation is in error by nearly 20 percent. However, the corresponding states principle leads to an error of only about 2 percent. This chosen state is typical of superheater conditions in large, modern, steam power plants. Based on its Z factor, water vapor in this state is far from behaving as an ideal gas.

Example 4-14 Determine the specific volume of water vapor, in ft^3/lb, at 1000°F and 3000 psia by (*a*) the ideal-gas equation of state, (*b*) the principle of corresponding states, and (*c*) the experimental value in the superheat table.

SOLUTION On the basis of the ideal-gas equation, $v = RT/P$. The gas constant R, based on Table A-1, is

$$R = \frac{R_u}{M} = \frac{10.73 \text{ psi} \cdot \text{ft}^3/(\text{lb} \cdot \text{mol})(°\text{R})}{18 \text{ lb/lb} \cdot \text{mol}} = 0.596 \text{ psi} \cdot \text{ft}^3/(\text{lb})(°\text{R})$$

Hence,

$$v = \frac{RT}{P} = 0.596 \text{ psi} \cdot \text{ft}^3/(\text{lb})(°\text{R}) \times \frac{1460°\text{R}}{3000 \text{ psi}} = 0.290 \text{ ft}^3/\text{lb}$$

(*b*) The specific volume based on the principle of corresponding states is given by $v_{\text{actual}} = Zv_{\text{ideal}}$. The Z factor is found by computing T_R and P_R and then using Fig. A-24M to find Z.

The critical data are found in Table A-3. Based on these data,

$$T_R = \frac{1460°\text{R}}{1165°\text{R}} = 1.25 \qquad P_R = \frac{3000 \text{ psi}}{3204 \text{ psi}} = 0.936$$

From Fig. A-24M, the Z value is approximately 0.83. Therefore, $v = 0.83(0.290) = 0.241$ ft³/lb.

(*c*) The tabulated value, based on experimental data, is found in Table A-14 to be 0.2485 ft³/lb. In comparison to the tabulated value, the ideal-gas equation is in error by nearly 17 percent. However, the corresponding states principle leads to an error of only about 3 percent. This chosen state is typical of superheater conditions in large modern steam power plants. Based on its Z factor, water vapor in this state is far from behaving as an ideal gas.

The preceding examples illustrate that PvT data can be estimated to a fair accuracy by employing the principle of corresponding states. It should be emphasized, however, that the Z chart should never be used if experimental data are available.

Example 4-15M Ethane gas (C_2H_6) is placed in a rigid tank at a pressure of 34.2 bars and a specific volume of 0.0208 m³/kg. It is heated until the pressure reaches 46.4 bars. Estimate the temperature change for the process in kelvins, based on the generalized Z chart.

SOLUTION The temperature change can be estimated by finding the values of T_R for the initial and final states. These values are found from a knowledge of P_R and v'_R for each state. The critical pressure and temperature of ethane are given in Table A-3M as 48.8 bars and 305 K, respectively. The molar mass of ethane is 30, hence the reduced properties are:

$$v'_R = \frac{vP_c}{RT_c} = \frac{0.0204(30)(48.8)}{0.08314(305)} = 1.20$$

$$P_{R,1} = \frac{P_1}{P_c} = \frac{34.2}{48.8} = 0.70 \qquad \text{and} \qquad P_{R,2} = \frac{P_2}{P_c} = \frac{46.4}{48.8} = 0.95$$

Because the volume is fixed, the process follows a constant v'_R line of 1.20 from a P_R value of 0.70 to one of 0.95. From Fig. A-24M, we find that $T_{R,1} = 1.07$ and $T_{R,2} = 1.33$, approximately. Hence,

$$T_2 = T_c T_{R,2} = 305(1.33) = 406°\text{K} \qquad \text{and} \qquad T_1 = T_c T_{R,1} = 305(1.07) = 326°\text{K}$$

Therefore, the temperature change equals (406 − 326) or 80°K.

Example 4-15 Ethane gas (C_2H_6) is placed in a rigid tank at a pressure of 31.3 atm and a specific volume of 0.333 ft³/lb. It is heated until the pressure reaches 45.8 atm. Estimate the temperature change for the process, in degrees Fahrenheit, based on the generalized Z chart.

SOLUTION The temperature change can be estimated by finding the values of T_R for the initial and final states. These values are found from a knowledge of P_R and v'_R for each state. The critical pressure and temperature of ethane are given in Table A-3 as 48.2 atm and 550°R, respectively. The molar mass of ethane is 30, hence the reduced properties are:

$$v'_R = \frac{vP_c}{RT_c} = \frac{0.333(30)(48.2)}{0.730\,(550)} = 1.20$$

$$P_{R,1} = \frac{P_1}{P_c} = \frac{31.3}{48.2} = 0.649 \qquad \text{and} \qquad P_{R,2} = \frac{P_2}{P_c} = \frac{45.8}{48.2} = 0.95$$

Because the volume is fixed, the process follows a constant v'_R line of 1.20 from a P_R value of 0.649 to one of 0.95. From Fig. A-24M, we find that $T_{R,1} = 1.02$ and $T_{R,2} = 1.33$, approximately. Hence,

$$T_2 = T_c T_{R,2} = 550(1.33) = 732°\text{R} \qquad \text{and} \qquad T_1 = T_c T_{R,1} = 550(1.02) = 561°\text{R}$$

Therefore, the temperature change equals (732 − 561) or 171°F.

4-7 PROPERTY RELATIONS FOR INCOMPRESSIBLE SUBSTANCES

On the basis of the PvT surface for simple compressible substances (see Fig. 4-1), it is noted that for many solids and liquids there are wide regions on the PvT surface of equilibrium states where the variation in the specific volume is negligible. Therefore, the assumption that the specific volume (and the density) is constant in the region of interest is often a good approximation to reality, and it leads to no serious error in computations. The equation of state for these two phases often can be represented then by

$$v = \text{constant} \qquad \text{or} \qquad \rho = \text{constant} \tag{4-7}$$

By definition, a substance of constant density is said to be *incompressible*. Geometrically, on a PvT surface, an incompressible phase would be represented by a plane which is perpendicular to the v axis. The boundary or $P\,dV$ work associated with the change of state of an incompressible substance must always be zero. However, for a simple, compressible substance the only quasistatic work interaction permitted is boundary work. Consequently the only ways to change the internal energy of a simple, incompressible material are by a heat interaction or by nonquasistatic work interactions. For example, stirring of an incompressible liquid by means of a paddle-wheel would be permissible. We noted in Sec. 3-3 that the internal energy of a simple substance can be expressed as $u = u(T, v)$. The total differential becomes

$$du = \left(\frac{\partial u}{\partial T}\right)_v dT + \left(\frac{\partial u}{\partial v}\right)_T dv \tag{3-4}$$

The first partial derivative is c_v. In addition, dv is zero for an incompressible substance. Therefore, we may write for a simple, incompressible material that

$$du = c_v\,dT \qquad \text{incompressible} \tag{4-8}$$

That is, the internal energy of an incompressible substance is solely a function of temperature. Based on the definition of c_v, it is also clear that c_v for an incompressible material is solely a function of the temperature. Hence,

$$u_2 - u_1 = \int_1^2 c_v\,dT \qquad \text{incompressible} \tag{4-9}$$

While the internal energy of an incompressible material depends only on the temperature, this is not true of the enthalpy of an incompressible substance.

From the definition of the enthalpy function $h = u + Pv$, we see that

$$h_2 - h_1 = u_2 - u_1 + v(P_2 - P_1) \qquad \text{incompressible} \tag{4-10}$$

Hence the enthalpy of an incompressible substance is a function of both the temperature and the pressure.

If Eq. (4-10) is written in differential form, then $dh_{\text{inc}} = du + v\,dP = c_v\,dT + v\,dP$. On the basis of this result the variation of h with respect to T at constant pressure for an incompressible substance is given by $(\partial h/\partial T)_{P,\,\text{inc}} = c_v$. But, by definition for any substance, $(\partial h/\partial T)_P = c_p$. Thus, for incompressible substances we realize that

$$c_p = c_v = c \tag{4-11}$$

For an incompressible substance we need not distinguish between c_p and c_v, and therefore both can be represented by the symbol c. Consequently Eqs. (4-9) and (4-10) can be written as

$$u_2 - u_1 = \int_1^2 c\,dT \tag{4-12}$$

and

$$h_2 - h_1 = \int_1^2 c\,dT + v(P_2 - P_1) \tag{4-13}$$

for materials that can be approximated as incompressible.

Equation (4-12) supports an approximation introduced in Sec. 4-5 regarding compressed- or subcooled-liquid data. Because subcooled liquids are essentially incompressible, the internal energy of subcooled liquids according to Eq. (4-12) should be only temperature-dependent. For a given temperature, the integral of $c\,dT$ must be zero. Therefore u_2 of the subcooled liquid would equal u_1 of the saturated liquid at the same temperature. This equality is not true for the enthalpy of a subcooled liquid, since the term $v(P_2 - P_1)$ is present in Eq. (4-13). However, if state 2 represents the subcooled state and state 1 is the saturated liquid state at the same temperature, then Eq. (4-13) becomes

$$h\ (\text{subcooled at } T, P) = h_{f,T} + v_{f,T}(P - P_{\text{sat}}) \tag{4-14}$$

where P_{sat} is the saturation pressure at the given temperature. The value of v is usually quite small, so that the contribution of the last term in Eq. (4-14) is small compared to h_f for the saturated liquid. However, it does increase the accuracy of estimating h in the subcooled or compressed liquid state when experimental data are not available.

The equations developed in this section are valid if a substance is incompressible. It must be remembered, however, that incompressibility is an idealization, much like that of an ideal gas. Hence these equations yield values which are approximations to real behavior. Appreciable errors may result if the equations are used for regions of the PvT surface where the assumption of constant density is not appropriate.

Example 4-16M Determine the values of u and h for subcooled water at 20°C and 1 bar on the basis of an incompressible fluid.

SOLUTION The accompanying PT diagram (Fig. 4-14) shows state 2 lying in the subcooled- (or compressed-) liquid region at 20°C and 1 bar. We shall use Eq. (4-12) to evaluate u at state 2 on the diagram in terms of state 1 on the saturation line at the same temperature. Since path 1–2 is at constant temperature, the integral in Eq. (4-12) is zero, and $u_2 = u_1 = u_f$ at 20°C. From the saturated data in Table A-12M we find

$$u_2 = u_f \text{ at } 20°\text{C} = 83.95 \text{ kJ/kg}$$

The enthalpy at state 2 on the diagram is approximated by employing Eq. (4-13) for the assumed incompressible fluid. Again, the evaluation is along the constant-temperature path. As a result, Eq. (4-13) reduces to Eq. (4-14) and

$$h_2 = h_1 + v(P_2 - P_1)$$

The values of v, P_1, and h_1 to use are the saturated-liquid values at 20°C from Table A-12M. Consequently

$$h_2 = 83.96 \text{ kJ/kg} + 1.0018 \text{ cm}^3\text{/g} \times (1 - 0.0234) \text{ bar} \times \frac{\text{J}}{10 \text{ cm}^3 \cdot \text{bar}} \times 1000 \text{ g/kg}$$

$$= 83.96 + 0.10 = 84.06 \text{ kJ/kg}$$

Note that the enthalpy in the subcooled region in this case differs by only roughly 0.1 percent from the saturated-liquid value at the same temperature.

The enthalpy at state 2 could also be found by selecting path 3–2 at a constant pressure of 1 bar. The enthalpy at state 3 is h_f at 1 bar, which is found in Table A-13M to be 417.46 J/g. Equation (4-13) in this case requires knowledge of c. The specific heat c of water between 20 and 99.6°C is relatively constant (see Table A-19M) at 4.18 J/(g)(°C). Hence

$$h_2 - h_3 = c(T_1 - T_3) = 4.18(20 - 99.63) = -332.7 \text{ kJ/kg}$$

$$h_2 = 417.46 - 333.71 = 84.29 \text{ kJ/kg}$$

This method gives about the same answer. However, its accuracy is somewhat in doubt since the use of four significant figures in the calculation is questionable.

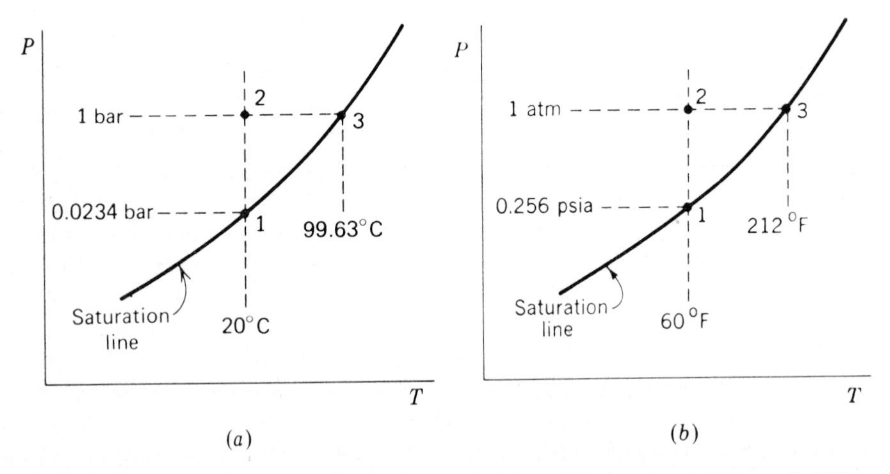

Figure 4-14 PT diagrams for processes discussed in (*a*) Example 4-16M, and (*b*) Example 4-16.

Example 4-16 Determine the values of u and h for subcooled water at 60°F and 1 atm on the basis of an incompressible fluid.

SOLUTION The accompanying PT diagram (Fig. 4-14) shows state 2 lying in the subcooled- (or compressed-) liquid region at 60°F and 1 atm. Equation (4-12) is used to evaluate u at state 2 on the diagram in terms of state 1 on the saturation line at the same temperature. Since path 1–2 is a constant-temperature path, the integral in Eq. (4-12) is zero, and $u_2 = u_1 = u_f$ at 60°F. From the saturated data in Table A-12 we find

$$u_2 = u_f \text{ at } 60°\text{F} = 28.08 \text{ Btu/lb}$$

The enthalpy at state 2 on the diagram is approximated by employing Eq. (4-13) for the assumed incompressible fluid. Again, the evaluation is along the constant temperature path. As a result, Eq. (4-13) reduces to Eq. (4-14) and

$$h_2 = h_1 + v(P_2 - P_1)$$

The values of v_1, P_1, and h_1 to use are the saturated-liquid values at 60°F from Table A-12. Consequently,

$$h_2 = 28.08 \text{ Btu/lb} + 0.01604 \text{ ft}^3/\text{lb}_m \times (14.70 - 0.256)(144) \text{ lb}_f/\text{ft}^2 \times \frac{\text{Btu}}{778 \text{ ft} \cdot \text{lb}_f}$$

$$= 28.08 + 0.04 = 28.12 \text{ Btu/lb}_m$$

Note that the enthalpy in the subcooled region in this case differs only slightly from the saturated-liquid value at the same temperature.

The enthalpy at state 2 could also be found by selecting path 3–2 at a constant pressure of 1 atm. The enthalpy at state 3 is h_f at 1 atm, which is found in Table A-13 to be 180.15 Btu/lb. Equation (4-13) in this case requires knowledge of c. The specific heat c of water between 60 and 212°F is relatively constant (see Table A-19) at 1.0 Btu/(lb)(°F). Hence

$$h_2 - h_3 = c(T_2 - T_3) = 1.0(60 - 212) = -152.0 \text{ Btu/lb}$$

$$h_2 = 180.15 - 152.0 = 28.15 \text{ Btu/lb}$$

This method gives about the same answer. However, its accuracy is somewhat in doubt due to the lack of accuracy for the value of c.

Equations (4-12) and (4-13) for the internal-energy and enthalpy changes of an incompressible substance require the use of specific-heat data for solids and liquids. The specific heat of a simple compressible substance theoretically is a function of two independent intensive variables, such as temperature and pressure. Experimental data indicate that the specific heats of solids are primarily a function of the temperature only. This small dependency on pressure is also generally exhibited by the specific heats of liquids. On the other hand, the variation of the specific heats of solids with temperature usually is quite large, whereas liquids are little affected. Data for various solids are tabulated in Tables A-19M and A-19.

The variation of c_p from the solid to the liquid phase for water is shown in Fig. 4-15. The rapid rise in the specific heat with temperature is noted for the solid phase (ice) up to the melting point. The value of c_p then doubles as the substance becomes a liquid. The c_p of liquid water remains relatively constant up to the boiling point at 1 atm. The relative temperature-independence of the

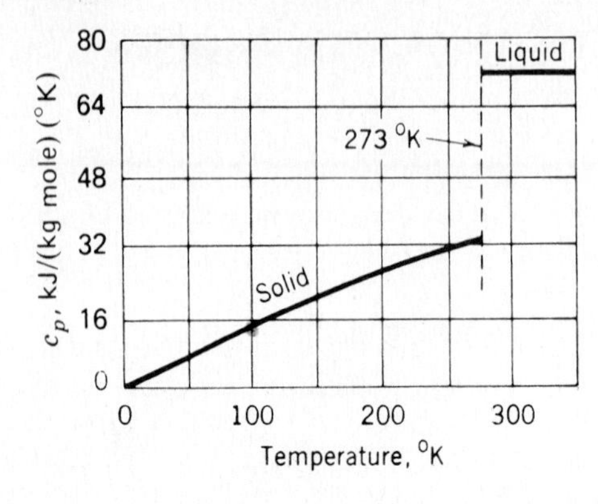

Figure 4-15 Variation of c_p of water with temperature.

specific heats of liquids is quite common. The specific heats of some common liquids are found in Tables A-19M and A-19, where the small variation with temperature again may be noted. Although not indicated, the effect of pressure is slight. The value of c_p for liquid water undergoes a change of less than 3 percent at 35°C (100°F) when the pressure is raised from 1 to 400 atm.

REFERENCES

Baumeister, T.: "Marks' Standard Handbook for Mechanical Engineers," 8th ed., McGraw-Hill, New York, 1978.

Keenan, J. H., et al.: "Thermodynamic Properties of Steam," Wiley, New York, 1969.

Perry, R. H., C. H. Chilton, and S. D. Kirkpatrick (Eds.): "Chemical Engineers' Handbook," 5th ed., McGraw-Hill, New York, 1973.

Zemansky, M. W.: "Heat and Thermodynamics," 5th ed., McGraw-Hill, New York, 1968.

PROBLEMS (METRIC)

PT and *Tv* diagrams

4-1M Sketch a *Tv* diagram which includes the compressed-liquid, wet, and superheated-vapor regions of a substance. Include lines of constant pressure which pass through all three regions.

4-2M On a phase diagram (*PT*) for water, show approximately the location of constant-volume lines in the three single-phase regions.

Superheat and saturation data

4-3M Fill in the data omitted in the table, for water substance.

	Pressure, bars	Temperature, °C	Specific volume, cm^3/g	Enthalpy, kJ/kg	Internal energy, kJ/kg	Quality (if appropriate)
(*a*)		200	127.4			
(*b*)	20	400				
(*c*)		100		2300		
(*d*)	60		20.0			
(*e*)	75	140				
(*f*)	30	400				
(*g*)	10					0.90
(*h*)		190		2786.4		
(*i*)		100			2000	
(*j*)		110		461.3		
(*k*)	2.5				535.1	

4-4M Determine the required data for water for the following specified conditions: (*a*) the pressure and specific volume of saturated liquid at 20°C, (*b*) the temperature and enthalpy of saturated vapor at 9 bars, (*c*) the specific volume and internal energy at 10 bars and 280°C, (*d*) the temperature and specific volume at 8 bars and a quality of 80 percent, (*e*) the specific volume and enthalpy at 100°C and 100 bars, (*f*) the pressure and enthalpy at 150°C and 70 percent quality, (*g*) the temperature and internal energy at 15 bars and an enthalpy of 2899.3 kJ/kg, (*h*) the quality and specific volume at 200°C and an enthalpy of 1822.8 kJ/kg, (*i*) the internal energy and specific volume at 140°C and an enthalpy of 2733.9 kJ/kg, (*j*) the pressure and enthalpy at 280°C and an internal energy of 2760.2 kJ/kg, and (*k*) the temperature and specific volume at 200 bars and an enthalpy of 434.06 kJ/kg.

4-5M Complete the following table of properties of refrigerant 12.

	T, °C	P, bars	x, %	v, cm^3/g	h, kJ/kg	u, kJ/kg
(*a*)		10.0		17.44		
(*b*)	30					187.71
(*c*)		8.0			134.0	
(*d*)	44				204.54	
(*e*)		4.0		0.7299		
(*f*)	60	6.0				
(*g*)	−15		70.0			
(*h*)	16					50.67
(*i*)		5.0			203.96	
(*j*)	36			0.7880		
(*k*)		10.0				164.2

4-6M Determine the required data for refrigerant 12 at the following specified conditions: (*a*) the pressure and specific volume of saturated liquid at 24°C, (*b*) the temperature and enthalpy of saturated vapor at 6 bars, (*c*) the specific volume and internal energy at 7.0 bars and 60°C, (*d*) the temperature and specific volume at 3.2 bars and a quality of 30 percent, (*e*) the approximate specific volume and enthalpy at 4°C and 12 bars, (*f*) the pressure and enthalpy at −15°C and 20 percent quality, (*g*) the temperature and internal energy at 9.0 bars and an enthalpy of 211.92 kJ/kg, (*h*) the quality and specific volume at 44°C and an enthalpy of 166.8 kJ/kg, (*i*) the internal energy and specific volume at 20°C and an enthalpy of 195.78 kJ/kg, (*j*) the pressure and enthalpy at 40°C and an internal energy of 190.13 kJ/kg, and (*k*) the approximate enthalpy and specific volume at 10 bars and 20°C.

4-7M A piston-cylinder device contains 1.0 kg of liquid water and 0.01 kg of water vapor (steam) in equilibrium at 7 bars. The pressure is held constant and heat is added until the temperature reaches 280°C.

(*a*) What is the initial temperature of the mixture?
(*b*) Determine the change in volume, in cubic meters.
(*c*) Plot the process on a Pv diagram.

4-8M Water substance is contained in a piston-cylinder device initially at 10 bars and 267.8 cm^3/g. It is compressed at constant pressure until it becomes a saturated vapor.

(*a*) What is the initial temperature in °C?
(*b*) What is the final temperature?
(*c*) If the process is quasistatic, how much work input is required in kJ/kg?
(*d*) If the original volume is 1000 cm^3, determine the internal energy change of the fluid, in kilojoules.

4-9M Use the ideal-gas equation to calculate the specific volume of steam at the states listed below, and compare your answer with data from the steam tables.

(*a*) 0.06 bar, 200°C
(*b*) 10 bars, 200°C
(*c*) 10 bars, 600°C
(*d*) 40 bars, 280°C
(*e*) 40 bars, 740°C

4-10M Initially, a 0.2-m^3 rigid tank contains saturated water vapor at (*a*) 2.0 bars and (*b*) 5.0 bars. Heat transfer from the substance results in a drop in pressure to 1 bar. From the final equilibrium state determine (1) the temperature, in degrees Celsius, (2) the ratio of the mass of liquid to the mass of vapor, and (3) the quality. Also show the process on a Pv diagram.

4-11M A wet mixture of water substance is maintained in a rigid tank at (*a*) 1 bar, and (*b*) 60°C. The system is heated until the final state is the critical state. Determine (1) the quality of the initial mixture, and (2) the initial ratio of volume of vapor to liquid.

4-12M Consider Prob. 4-11M, except the substance is refrigerant 12.

4-13M Determine the internal energy of 0.1 m^3 of refrigerant 12 at 0°C if it is known that the specific volume is (*a*) 35.0 cm^3/g, and (*b*) 45.0 cm^3/g in that state.

4-14M A closed system contains a mixture of liquid and vapor water at 200°C. The internal energy of the saturated liquid is (*a*) 10 percent, and (*b*) 5 percent of the total internal energy of the system. What is the quality of the mixture?

4-15M Water vapor at 20 bars and 280°C is cooled at constant volume until the pressure is (*a*) 10 bars, (*b*) 5 bars, and (*c*) 2 bars. Determine the quality in the final state and sketch the process on a Pv diagram.

4-16M A wet mixture of water substance undergoes a process such that the enthalpy is maintained constant. Determine the quality of the initial mixture, in percent, if (*a*) the initial pressure is 20 bars and the final state is 1 bar and 120°C, and (*b*) the initial pressure is 7 bars and the final state is 0.7 bar and 100°C.

4-17M Refrigerant 12 at a pressure of 0.50 MPa has a specific volume of 25.0 cm^3/g (state 1). It expands at constant temperature until the pressure falls to 0.28 MPa (state 2). Finally, the fluid is cooled at constant pressure until it becomes a saturated vapor (state 3).

(*a*) Determine the change in specific volume between states 1 and 2, and states 1 and 3.
(*b*) Determine the specific internal energy change, in kJ/kg, between states 1 and 2.
(*c*) Determine the specific enthalpy change, in kJ/kg, between state 2 and 3.
(*d*) Show the path of the process on a Pv diagram.

4-18M Dry saturated steam at 20 bars (state 1) is contained within a piston-cylinder device which initially has a volume of 0.03 m^3. The steam is cooled at constant volume until the temperature reaches 200°C (state 2). The system is then expanded isothermally until the volume in state 3 is twice the initial value.

(*a*) Determine the pressure at state 2.
(*b*) Determine the pressure at state 3.
(*c*) Determine the quality or temperature of superheat at state 3.
(*d*) Determine the change in the internal energy for the two processes $u_2 - u_1$ and $u_3 - u_2$.
(*e*) Plot the two processes on a *Pv* diagram.

4-19M A cylinder having an initial volume of 2 m^3 initially contains steam at 10 bars and 200°C (state 1). The vessel is cooled at constant temperature until the volume is 56.4 percent of the initial volume (state 2). The constant-temperature process is followed by a constant-volume process which ends with a pressure in the cylinder of 30 bars (state 3).

(*a*) Determine the pressure in bars and the enthalpy, in kJ/kg, at state 2.
(*b*) Determine the temperature, in degrees Celsius, and the enthalpy at state 3.
(*c*) Sketch the two processes on a *Pv* diagram, with respect to the vapor dome.

4-20M Steam initially at 30 bars and 400°C (state 1) is cooled at constant volume to a temperature of 200°C (state 2). The fluid is then further cooled at constant temperature to a saturated liquid (state 3). Ascertain (*a*) the final pressure in bars, (*b*) the quality after the constant-volume process, (*c*) the overall specific-volume change, in cm^3/g, (*d*) the change in specific internal energy, in kJ/kg, between states 2 and 3. Finally, sketch the processes on a *Pv* diagram.

Compressed-liquid data

4-21M Saturated liquid water at 80°C is compressed at constant temperature to (*a*) 50 bars, (*b*) 100 bars, and (*c*) 200 bars. Determine the enthalpy change on the basis of experimental compressed-liquid data. In each case, what would be the answer if saturated data were used as an approximation?

4-22M Saturated liquid water at 40°C is compressed to 140°C and (*a*) 50 bars, (*b*) 100 bars, and (*c*) 200 bars. Determine the enthalpy change on the basis of compressed-liquid data. In each case, approximate the answer also on the basis of saturated data. What percent error is involved in the second answer?

4-23M Consider Prob. 4-22M, except find the change in the specific volume instead.

4-24M Consider Prob. 4-22M, except find the change in the internal energy instead.

Process sketches on *Pv* diagrams

4-25M Steam is compressed isothermally from 1.5 bars 200°C, to a final specific volume of (*a*) 250 cm^3/g, and (*b*) 100 cm^3/g. Sketch the processes on a *Pv* diagram, relative to the saturation line.

4-26M Refrigerant 12 undergoes a change of state at constant pressure from 3.2 bars, 20°C, to (*a*) a final temperature of 60°C, and (*b*) a final specific volume of 30 cm^3/g. Sketch the processes on a *Pv* diagram, relative to the saturation line.

4-27M Steam undergoes a change of state at constant pressure from 20 bars and 280°C to (*a*) a final temperature of 400°C, and (*b*) a final enthalpy of 2400 kJ/kg. Sketch the processes on a *Pv* diagram, relative to the saturation line.

4-28M Refrigerant 12 undergoes a change of state at constant volume from 6 bars and 60°C to (*a*) a final pressure of 5 bars, and (*b*) a final temperature of 100°C. Sketch the processes on a *Pv* diagram, relative to the saturation line.

4-29M Steam undergoes a change of state at constant volume from 15 bars and 162.7 cm^3/g to (*a*) a final temperature of 360°C, and (*b*) a final pressure of 8 bars. Sketch the processes on a *Pv* diagram, relative to the saturation line.

4-30M Refrigerant 12 undergoes a change of state at constant temperature from 60°C and 9 cm^3/g to (*a*) a final internal energy of 94.43 kJ/kg, and (*b*) a final pressure of 7 bars. Sketch the processes on a *Pv* diagram, relative to the saturation line.

Energy analysis, saturation and superheat data

4-31M Refrigerant 12 is condensed from an initial state of 6 bars and 100°C to a final state of saturated liquid at the same pressure. Determine (*a*) the work interaction, and (*b*) the heat interaction, both in kJ/kg. Plot the process on a *Pv* diagram.

4-32M One-tenth kilogram of water substance at a pressure of 3 bars occupies a volume of 0.0303 m^3 in a piston-cylinder device weighted to maintain constant pressure. Heat is added in the amount of 122 kJ. Find (*a*) the final temperature, in degrees Celsius, and (*b*) the work output, in kilojoules. Also, sketch the path on a *Pv* diagram, relative to the saturation line.

4-33M A piston-cylinder device initially contains 0.20 kg of refrigerant 12 at 8 bars and 60°C. The fluid is compressed at constant pressure until its volume is 60 percent of its initial value. Determine (*a*) the work done, in kilojoules, and (*b*) the heat transferred, in kilojoules. Also, plot the process on a *Pv* diagram.

4-34M One and one-half kilograms of dry saturated steam at 3 bars are contained in a piston-cylinder device. Heat is added in the amount of 600 kJ, and a 96-volt dc electrical source supplies current to a resistor within the vapor for a period of 8 min. If the final temperature is 400°C and the pressure remains constant, determine the current required, in amperes. Neglect energy storage in the resistor.

4-35M One kilogram of dry, saturated steam at 5 bars is contained in a piston-cylinder assembly. Heat is added in the amount of 225 kJ, and some electrical work is done by passing a current of 1.5 A through a resistor in the fluid for 0.5 h. If the final temperature of the steam is (*a*) 280°C, and (*b*) 400°C and the process is constant pressure, determine the necessary voltage, in volts, of the battery which supplied the potential for the current. Neglect energy storage in the resistor.

4-36M A closed container maintained at 1-bar pressure is subdivided into two sections by an insulated partition. One section contains 10 g of ice at 0°C, while the other contains dry, saturated steam at 100°C. Determine the amount of steam present if, on removing the partition, it is just sufficient to melt all the ice. (The enthalpy of melting of water is 335 kJ/kg.)

4-37M Refrigerant 12 is contained in a rigid tank initially at 2 bars, a quality of 50.5 percent, and a volume of 0.10 m^3. Heat is added until the pressure reaches 5 bars. Determine (*a*) the mass within the system, in kilograms, and (*b*) the quantity of heat added, in kilojoules. Also sketch the process path on a *Pv* diagram.

4-38M A rigid tank contains 12.0 kg of refrigerant 12 at 6 bars and 80°C. A paddle-wheel within the tank adds energy at a constant torque of 150 Nm for 1200 revolutions. At the same time, the system is cooled to a final temperature of 12°C. Determine (*a*) the final internal energy, in kilojoules, and (*b*) the direction and magnitude of the heat transfer, in kilojoules. Also sketch the process on a *Pv* diagram, relative to the saturation line. Neglect energy storage in the paddle-wheel.

4-39M A closed, rigid tank contains 0.5 kg of dry, saturated steam at 4 bars. Heat is added in the amount of 50 kJ, and some work is done by means of a paddle-wheel until the steam is at 7 bars. Calculate the work required, in kilojoules.

4-40M A rigid vessel with a volume of 0.05 m^3 is initially filled with saturated steam at 1 bar. The contents are cooled to 75°C.

(*a*) Sketch the process on *Pv* coordinates with respect to the saturation line.
(*b*) What is the final pressure, in bars?
(*c*) Find the heat transferred from the steam, in kilojoules.

4-41M Water substance is contained in a piston-cylinder device at 10 bars and 240°C. It is compressed isothermally until its quality is 50 percent. The work required to carry this out is 180 kJ/kg.

(*a*) What is the final pressure in bars?
(*b*) Determine the magnitude, in kJ/kg, and the direction of any heat transfer.
(*c*) Sketch the process path relative to the saturation line on a *Pv* diagram.

4-42M One kilogram of water substance initially at 10 bars and 200°C is altered isothermally ($T = c$) until its volume is 40 percent of the initial value. During the process the boundary work is 170 kJ/kg, and paddle-wheel work in the amount of 79 N·m/g also takes place.

(*a*) Determine the magnitude, in kilojoules, and direction of any heat transfer.
(*b*) Sketch the process on a *Pv* diagram, relative to the saturation line.

4-43M One-tenth kilogram of refrigerant 12 initially is a wet mixture with a quality of 50 percent at 40°C. It expands isothermally to a pressure of 6 bars. The measured work output due to the expansion is 17 N · m/g.

(*a*) Determine the magnitude, in kilojoules, and direction of any heat transfer.
(*b*) Sketch the process on a *Pv* diagram, relative to the saturation line.

4-44M Two kilograms of water substance are contained in a piston-cylinder device at 320°C. The substance undergoes a constant temperature process with the volume changing from 0.02 to 0.17 m^3. The measured work output is 889 kJ. Determine (*a*) the final pressure, in bars, and (*b*) the heat transfer, in kilojoules. Also sketch the process on a *Pv* diagram, relative to the saturation line.

4-45M A piston-cylinder device contains refrigerant 12 initially at 2.8 bars and 40°C, and the volume is 0.1 m^3. The piston is kept stationary and there is heat transfer to the gas until its pressure rises to 3.2 bars. Then additional heat transfer occurs from the gas during a process in which the volume varies but the pressure is constant. This latter process ends when the temperature reaches 50°C. Assume the processes are quasistatic and find (*a*) the mass in the system, in kilograms, (*b*) the heat transfer, in kilojoules, during the constant-volume process, and (*c*) the heat transfer for the constant-pressure process, in kilojoules.

4-46M Saturated water vapor at 8 bars (state 1) is heated at constant volume until its pressure is 10 bars (state 2). It is then expanded isothermally to 7 bars (state 3) by adding 80 kJ/kg of heat. Finally, it is cooled to the saturation temperature by a constant-pressure process (state 4). Determine the value of the work interaction for (*a*) process 1–2, (*b*) process 2–3, (*c*) process 3–4, and (*d*) the overall process, all in kJ/kg. Sketch this series of processes on a *Pv* diagram, relative to the saturation line.

4-47M A system having an initial volume of 2.0 m^3 is filled with steam at 30 bars and 400°C (state 1). The system is cooled at constant volume to 200°C (state 2). The first process is followed by a constant-temperature process ending with saturated liquid water (state 3). Find the total heat transfer required, in kilojoules, and its direction. Sketch the two processes on a *Pv* diagram relative to the saturation line.

Energy analysis, compressed-liquid data

4-48M One kilogram of water at 100 bars and 40°C is heated at constant pressure to a temperature of (*a*) 100°C, and (*b*) 180°C. Determine the heat required using experimental compressed liquid data, in kilojoules. Repeat the calculation using saturation data as an approximation. What percent error is involved with the second method?

4-49M A closed container maintained at 25 bars is subdivided into two sections by an insulated partition. One section contains 0.50 kg of water at 20°C, while the other contains saturated steam. Determine the amount of steam present if, on removing the partition, the final state of the system is (*a*) a saturated liquid, and (*b*) a wet mixture with 20 percent quality.

Compressibility factor

4-50M Water vapor exists at 400°C and at (*a*) 140 bars, and (*b*) 180 bars. Determine for these pressures the specific volume, in cm^3/g, based on (1) the ideal-gas equation, (2) the corresponding states principle, and (3) the tables of superheat data.

4-51M Refrigerant 12 is maintained at (*a*) 50°C and 12 bars, and (*b*) 60°C and 14 bars. Determine for these states the specific volume, in cm^3/g, based on (1) the ideal-gas equation, (2) the principle of corresponding states, and (3) experimental data in the tables.

4-52M Determine the pressure, in bars, of water vapor at 360°C and 30.89 cm^3/g on the basis of (*a*) the ideal-gas equation, (*b*) the principle of corresponding states, and (*c*) the experimental value.

4-53M Refrigerant 12 at a temperature of 120°C has a specific volume of 14.61 cm^3/g. Determine the pressure of the gas on the basis of (*a*) the ideal-gas equation, (*b*) the corresponding states principle, and (*c*) the experimental value in the tables.

4-54M Determine the temperature, in degrees Celsius, of refrigerant 12 at 14 bars and 14.25 cm^3/g on the basis of (*a*) the ideal-gas equation, (b) the corresponding states principle, and (*c*) the experimental value.

4-55M Steam, initially at 160 bars and 440°C, expands isothermally until its volume is doubled. Determine the final pressure, in bars, if (*a*) the ideal-gas equation applies, (*b*) the corresponding states principle is used, and (*c*) the steam table is used.

4-56M Propane (C_3H_8), initially at 97°C and 29.9 bars, is heated at constant pressure until the temperature reaches 210°C. Estimate the change in volume, in m^3/kg, if (*a*) the ideal-gas equation is used, and (*b*) the corresponding states principle is used.

4-57M Air at 15 bars and 120°K is heated in a rigid tank until the temperature reaches 213°K. Estimate the final pressure using (*a*) the ideal-gas equation, and (*b*) the corresponding states principle.

Incompressible substances

4-58M A 22-kg mass of copper is dropped into an insulated tank which contains water at 20.0°C and 1 bar. The initial temperature of the copper is 60°C and the final equilibrum temperature of the copper and water at 1 bar is (*a*) 20.80°C, and (*b*) 21.24°C. Estimate the initial volume of water used, in cubic meters.

4-59M An unknown mass of copper at 77°C is dropped into an insulated tank which contains 0.2 m^3 of water at 30°C. If the equilibrium final temperature is (*a*) 31°C, and (*b*) 32°C, find the mass of copper, in kilograms.

4-60M Consider three liquids *A*, *B*, and *C* which have specific heats at constant pressure of 2.0, 4.0, and 7.0 kJ/(kg)(°C), respectively. The three liquids occupy three compartments of an insulated tank, but the walls separating the compartments conduct heat. Initially, one section contains 5 kg of *A* at 100°C, and the other compartments contain 2 kg of *B* and 1 kg of *C*, respectively, each initially at 50°C. Find the temperature of the three fluids at thermal equilibrium, if the overall process is one of constant pressure.

4-61M Consider Prob. 4-60M, except the initial temperatures of *A*, *B*, and *C* are 20°C, 60°C, and 100°C, respectively. Find the final equilibrium temperature, in degrees Celsius.

4-62M A 2 kg mass of lead −100°C is brought into contact with an unknown mass of aluminum initially at 0°C. If the final equilibrium temperature of the two metals is (*a*) −60°C, and (*b*) −40°C, determine the mass of aluminum, in kilograms.

4-63M Consider water at 220°C and (*a*) 75 bars, (*b*) 100 bars, and (c) 150 bars. Use Eq. (4-14) to estimate the enthalpy at these states, in kJ/kg, and compare the answer to the tabulated value in the compressed liquid table.

4-64M Consider Prob. 4-22M. Evaluate the enthalpy change when the final pressure is (*a*) 50 bars, (*b*) 100 bars, and (*c*) 200 bars on the basis that the fluid is an incompressible substance. What percent error is involved in assuming an incompressible substance, in comparison to compressed-liquid data?

4-65M Consider Prob. 4-24M. Evaluate the internal energy change when the final pressure is (*a*) 50 bars, (*b*) 100 bars, and (*c*) 200 bars on the basis that the fluid is an incompressible substance. What percent error is involved in assuming an incompressible fluid, in comparison to compressed-liquid data?

4-66M A 4-kW resistance heater is placed in a 0.10 m^3 container filled with water at 20°C and 1 bar. The heater is allowed to operate for 20 min. The mass of the heater is 2.0 kg and its specific heat is 0.45 kJ(kg)(°C). If the container is insulated, determine the temperature of the water and heater element in the final state, in degrees Celsius.

4-67M A paddle-wheel driven by a 240-W motor is used to agitate liquid water in a closed, insulated tank maintained at 1 bar. The initial temperature is 20°C. Determine the temperature rise of the contents for (*a*) a time period of 0.50 h if the mass of water is 5.0 kg, and the mass and specific heat of the paddle-wheel are 0.3 kg and 0.90 kJ/(kg)(°C), respectively, and (*b*) a time period of 0.80 h if the mass of water is 6 kg, and the mass and specific heat of the paddle-wheel are 0.4 kg and 0.96 kJ/(kg)(°C), respectively.

PROBLEMS (USCS)

PT and Tv diagrams

4-1 Sketch a Tv diagram which includes the compressed-liquid, wet, and superheated-vapor regions of a substance. Include lines of constant pressure which pass through all three regions.

4-2 On a PT diagram for water, show approximately the location of constant-volume lines in the three single-phase regions.

Superheat and saturation data

4-3 Fill in the data omitted in the table, for water substance.

	Pressure, psia	Temperature, °F	Specific volume, ft³/lb	Enthalpy, Btu/lb	Internal energy, Btu/lb	Quality (if appropriate)
(*a*)	300			1203.9		
(*b*)	500	700				
(*c*)		240		952		
(*d*)	400		0.700			
(*e*)	1500	200				
(*f*)	180				1190	
(*g*)	250					0.90
(*h*)		300	6.472			
(*i*)		200			1000	
(*j*)		350		321.8		
(*k*)	80				282	

4-4 Determine the required data for water for the following specified conditions: (*a*) the pressure and specific volume of saturated liquid at 150°F, (*b*) the temperature and enthalpy of saturated vapor at 80 psia, (*c*) the specific volume and internal energy at 140 psia and 500°F, (*d*) the temperature and specific volume at 100 psia and a quality of 80 percent, (*e*) the specific volume and enthalpy at 100°F and 1500 psia, (*f*) the pressure and enthalpy at 300°F and 70 percent quality, (*g*) the temperature and internal energy at 200 psia and an enthalpy of 1268.8 Btu/lb, (*h*) the quality and specific volume at 370°F and an enthalpy of 770 Btu/lb, (*i*) the internal energy and specific volume at 240°F and an enthalpy of 1160.7 Btu/lb, (*j*) the pressure and enthalpy at 500°F and an internal energy of 1172.7 Btu/lb, and (*k*) the temperature and specific volume at 2000 psia and an enthalpy of 73.3 Btu/lb.

4-5 Complete the following table of properties of refrigerant 12.

	T, °F	P, psia	x, %	v, ft³/lb	h, Btu/lb	u, Btu/lb
(*a*)		90		0.4514		
(*b*)	120					85.369
(*c*)		80			60.0	
(*d*)	100				87.03	
(*e*)		120		0.0126		
(*f*)	160	100				
(*g*)	20		70.0			
(*h*)	50					19.51
(*i*)		80			89.64	
(*j*)	110			0.01292		
(*k*)		160				80.2

4-6 Determine the required data for refrigerant 12 at the following specified conditions: (*a*) the pressure and specific volume of saturated liquid at 80°F, (*b*) the temperature and enthalpy of saturated vapor at 160 psia, (*c*) the specific volume and internal energy at 80 psia and 140°F, (*d*) the temperature and specific volume at 40 psia and a quality of 30 percent, (*e*) the approximate specific volume and enthalpy at 20°F and 60 psia, (*f*) the pressure and enthalpy at 10°F and 25 percent quality, (*g*) the temperature and internal energy at 100 psia and an enthalpy of 98.884 Btu/lb, (*h*) the quality and specific volume at 80°F and an enthalpy of 67.6 Btu/lb, (*i*) the internal energy and specific volume at 100°F and an enthalpy of 87.03 Btu/lb, (*j*) the pressure and enthlapy at 140°F and an internal energy of 85.516 Btu/lb, and (*k*) the approximate enthalpy and specific volume at 100 psia and 50°F.

4-7 A piston-cylinder device contains 3.0 lb of liquid water and 0.03 lb of water vapor (steam) in equilibrium at 200 psia. The pressure is held constant and heat is added until the temperature reaches 500°F.

(*a*) What is the initial temperature of the mixture?
(*b*) Determine the change in volume in cubic feet.
(*c*) Plot the process on a *Pv* diagram.

4-8 Water substance is contained in a piston-cylinder device initially at 120 psia and 4.36 ft^3/lb. It is compressed at constant pressure until it becomes a saturated vapor.

(*a*) What is the initial temperature in degrees Fahrenheit?
(*b*) What is the final temperature?
(*c*) If the process is quasistatic, how much work input is required, in Btu/lb?
(*d*) If the original volume is 100 in^3, determine the internal energy change of the fluid, in Btu.

4-9 Use the ideal-gas equation to calculate the specific volume of steam at the states listed below, and compare your answer with data from the steam tables.

(*a*) 1 psia, 300°F
(*b*) 200 psia, 400°F
(*c*) 200 psia, 1000°F
(*d*) 2000 psia, 650°F
(*e*) 2000 psia, 1600°F

4-10 Initially, a 1.0 ft^3 rigid tank contains saturated water vapor at (*a*) 25 psia, and (*b*) 50 psia. Heat transfer from the substance results in a drop in pressure to 15 psia. For the final equilibrium state determine (1) the temperature, in degrees Fahrenheit, (2) the ratio of the mass of liquid to the mass of vapor, and (3) the quality. Also show the process on a *Pv* diagram.

4-11 A wet mixture of water substance is maintained in a rigid tank at (*a*) 15 psia, and (*b*) 200°F. The system is heated until the final state is the critical state. Determine (1) the quality of the intial mixture, and (2) the initial ratio of volume of vapor to liquid.

4-12 Reconsider Prob. 4-11, except the substance is refrigerant 12.

4-13 Determine the internal energy of 1.0 ft^3 of refrigerant 12 at 30°F, in Btu, if it is known that the specific volume is (*a*) 0.70 ft^3/lb, and (*b*) 0.80 ft^3/lb in that state.

4-14 A closed system contains a mixture of liquid and vapor water at 500°F. The internal energy of the saturated liquid is (*a*) 10 percent, and (*b*) 5 percent of the total internal energy of the system. What is the quality of the mixture?

4-15 Water vapor at 300 psia and 450°F is cooled at constant volume until the pressure is (*a*) 100 psia, (*b*) 50 psia, and (*c*) 30 psia. Determine the quality in the final state and sketch the process on a *Pv* diagram.

4-16 A wet mixture of water substance undergoes a process such that the enthalpy is maintained constant. Determine the quality of the initial mixture, in percent, if (*a*) the initial pressure is 450 psia and the final state is 14.7 psia and 250°F, and (*b*) the initial pressure is 500 psia and the final state is 10 psia and 200°F.

4-17 Refrigerant 12 at a pressure of 120 psia has a specific volume of 0.25 ft^3/lb (state 1). It expands at constant temperature until the pressure falls to 50 psia (state 2). Finally, the fluid is cooled at constant pressure until it becomes a saturated vapor (state 3).

(*a*) Determine the change in specific volume between states 1 and 2, and states 1 and 3.
(*b*) Determine the specific internal energy change, in Btu/lb, between states 1 and 2.
(*c*) Determine the specific enthalpy change, in Btu/lb, between states 2 and 3.
(*d*) Show the path of the process on a *Pv* diagram.

4-18 Dry saturated steam at 120 psia (state 1) is contained within a piston-cylinder assembly which initially has a volume of 1 ft^3. The steam is cooled at constant volume until the temperature reaches 300°F (state 2). The system is then expanded isothermally until the volume in state 3 is twice the initial value. Determine (*a*) the pressure at state 2, (*b*) the pressure at state 3, (*c*) the quality or superheat temperature at state 3, and (*d*) the change in the internal energy for the two processes $u_2 - u_1$ and $u_3 - u_2$. Plot the two processes on a *Pv* diagram.

4-19 A cylinder having an initial volume of 2 ft^3 contains steam at 140 psia and 400°F (state 1). The cylinder is cooled at constant temperature until the volume is 37.5 percent of the initial volume (state 2). The constant-temperature process is followed by a constant-volume process which ends with the pressure in the cylinder at 450 psia (state 3).
(*a*) Determine the temperature, in degrees Fahrenheit, and the enthalpy, in Btu/lb, at state 2.
(*b*) Determine the temperature, in degrees Fahrenheit, and the enthalpy at state 3.
(*c*) Sketch the two processes on a *Pv* diagram with respect to the wet region.

4-20 Steam initially at 40 psia and 600°F (state 1) is cooled at constant volume to a pressure of 15 psia (state 2). The fluid is then further cooled at constant temperature to a saturated liquid (state 3). Ascertain (*a*) the final pressure in psia, (*b*) the quality after the constant volume process, (*c*) the overall specific volume change in ft^3/lb, and (*d*) the change in specific internal energy, in Btu/lb, between states 2 and 3. Finally, sketch the processes on a *Pv* diagram.

Compressed liquid data

4-21 Saturated liquid water at 150°F is compressed isothermally to (*a*) 500 psia, (*b*) 1500 psia, and (*c*) 3000 psia. Determine the enthalpy change on the basis of experimental compressed-liquid data. In each case what would the answer be if saturated data were used as an approximation?

4-22 Saturated liquid water at 200°F is compressed to 300°F and (*a*) 500 psia, (*b*) 1500 psia, and (*c*) 3000 psia. Determine the enthalpy change on the basis of compressed-liquid data. In each case also approximate the answer on the basis of saturated data. What percent error is involved in the second answer?

4-23 Consider Prob. 4-22, except find the change in specific volume instead.

4-24 Consider Prob. 4-22, except find the change in internal energy instead.

Process sketches on *Pv* diagrams

4-25 Steam is compressed isothermally from 300°F, 40 psia, to a final specific volume of (*a*) 4.0 ft^3/lb, and (*b*) 7 ft^3/lb. Sketch the processes on a *Pv* diagram, relative to the saturation line.

4-26 Refrigerant 12 undergoes a change of state at constant pressure from 50 psia, 60°F, to (*a*) a final temperature of 100°F, and (*b*) a final specific volume of 0.60 ft^3/lb. Sketch the processes on a *Pv* diagram, relative to the saturation line.

4-27 Steam undergoes a change of state at constant pressure from 400 psia and 500°F to (*a*) a final temperature of 700°F, and (b) a final enthalpy of 1100 Btu/lb. Sketch the processes on a *Pv* diagram, relative to the saturation line.

4-28 Refrigerant 12 undergoes a change of state at constant volume from 100 psia and 140°F to (*a*) a final pressure of 60 psia, and (*b*) a final temperature of 180°F. Sketch the processes on a *Pv* diagram, relative to the saturation line.

4-29 Steam undergoes a change of state at constant volume from 200 psia and 2.724 ft^3/lb to (*a*) a final temperature of 300°F, and (*b*) a final pressure of 170 psia. Sketch the processes on a *Pv* diagram, relative to the saturation line.

4-30 Refrigerant 12 undergoes a change of state at constant temperature from 100°F and 0.15 ft^3/lb to (*a*) a final internal energy of 79.5 Btu/lb, and (*b*) a final pressure of 70 psia. Sketch the processes on a *Pv* diagram, relative to the saturation line.

Energy analysis, saturation, and superheat data

4-31 Refrigerant 12 is condensed from an initial state of 100 psia and 140°F to a final state of saturated liquid at the same pressure. Determine (*a*) the work interaction, in ft · lb_f/lb_m, and (*b*) the heat interaction, in Btu/lb. Plot the process on a *Pv* diagram.

4-32 A piston-cylinder device with an initial volume of 0.10 ft^3 contains water substance initially at 160 psia and 50 percent quality. Heat is added in the amount of 35.6 Btu while the pressure remains constant. Determine (*a*) the mass of the system, in pounds, and (*b*) the final temperature, in degrees Fahrenheit. Also sketch the process on a *Pv* diagram.

4-33 A piston-cylinder device initially contains 0.40 lb of refrigerant 12 at 120 psia and 140°F. The fluid is compressed at constant pressure until its volume is 60 percent of its initial value. Determine (*a*) the work done, and (*b*) the heat transferred, in Btu. Also plot the process on a *Pv* diagram.

4-34 Three pounds of dry, saturated steam at 40 psia are contained in a piston cylinder. Heat is added in the amount of 600 Btu, and some work is done on the steam by means of a paddle-wheel until a temperature of 800°F is reached. If the pressure remains constant at 40 psia, determine the amount of constant torque applied, in lb_f · ft, to the paddle-wheel for 5000 revolutions. Neglect energy storage in the paddle-wheel.

4-35 One pound of dry, saturated steam at 40 psia is contained in a piston-cylinder assembly. Heat is added in the amount of 92 Btu, and some electrical work is done on the fluid by passing a steady current of 1.5 A through a resistor in the fluid for a period of 0.5 h. If the final temperature of the steam is 700°F and the process was constant pressure, determine the necessary voltage of the battery which supplied the potential for the flow of electricity, in volts. Neglect energy storage in the resistor.

4-36 A closed container maintained at 1-atm pressure is subdivided into two sections by an insulated partition. One section contains 0.02 lb of ice at 32°F, while the other section contains saturated water vapor at 212°F. Determine the amount of steam present if, on removing the partition, it is just sufficient to melt all the ice. (The enthalpy of melting of water is 144 Btu/lb.)

4-37 Refrigerant 12 is contained in a rigid tank initially at 30 psia, a quality of 46.8 percent, and a volume of 3.0 ft^3. Heat is added until the pressure reaches 80 psia. Determine (*a*) the mass within the system, in pounds, and (*b*) the quantity of heat added, in Btu. Also sketch the process path on a *Pv* diagram.

4-38 A rigid tank contains 25.0 lb of refrigerant 12 at 80 psia and 180°F. A paddle-wheel within the tank adds energy at a constant torque of 120 ft · lb_f for 1200 revolutions. At the same time, the system is cooled to a final temperature of 40°F. Determine (*a*) the final internal energy, in Btu, and (*b*) the direction and magnitude of the heat transfer, in Btu. Also sketch the process on a *Pv* diagram relative to the saturation line. Neglect energy storage in the paddle-wheel.

4-39 A closed rigid tank contains 2 lb of dry, saturated steam at 60 psia. Heat is added in the amount of 100 Btu, and some work is done on the substance by means of a paddle-wheel until the steam is at 100 psia. Calculate the work done, in Btu.

4-40 A rigid steel vessel having a volume of 2 ft^3 is initially filled with saturated steam at 14.7 psia. The vessel is sealed and chilled to 150°F.

(*a*) Sketch the process on *Pv* coordinates with respect to the saturated-vapor line.
(*b*) What is the final pressure, in psia?
(*c*) Determine the heat transferred from the steam, in Btu.

4-41 Water substance is contained in a piston-cylinder device at 160 psia and 500°F. It is compressed isothermally until its quality is 50 percent. The work required to carry this out is 210,000 ft · lb_f/lb_m.

(*a*) What is the final pressure, in psia?
(*b*) Determine the magnitude, in Btu/lb, and the direction of any heat transfer.
(*c*) Sketch the process path, relative to the saturation line on a *Pv* diagram.

4-42 One pound of water substance initially at 140 psia and 400°F is altered isothermally ($T = c$)

until its volume is 40 percent of the initial value. During the process the boundary work is 65,000 ft · lb_f/lb_m, and paddle-wheel work in the amount of 30,000 ft · lb_f/lb_m also takes place.

(*a*) Determine the magnitude, in Btu, and direction of any heat transfer.

(*b*) Sketch the process on a *Pv* diagram, relative to the saturation line.

4-43 One-tenth pound of refrigerant 12 initially is a wet mixture with a quality of 50 percent at 100°F. It expands isothermally to a pressure of 90 psia. The measured work output due to the expansion is 5400 ft · lb_f/lb_m.

(*a*) Determine the magnitude, in Btu, and direction of any heat transfer.

(*b*) Sketch the process on a *Pv* diagram.

4-44 Four pounds of water substance are contained in a piston-cylinder device at 500°F. The substance undergoes a constant-temperature process with the volume changing from 1.40 to 8.60 ft^3. The measured work output is 675 Btu. Determine (*a*) the final pressure, in psia, and (*b*) the heat transfer, in Btu. Also sketch the process on a *Pv* diagram, relative to the saturation line.

4-45 A piston-cylinder machine contains refrigerant 12 initially at 30 psia and 20°F, and the volume is 1 ft^3. The piston is kept stationary and there is heat transfer to the gas until its pressure rises to 40 psia. Then additional heat transfer occurs from the gas during a process in which the volume varies but the pressure is constant. This latter process terminates when the temperature reaches 100°F. Assume that the processes are quasistatic and determine (*a*) the mass present, (*b*) the heat transfer for the constant-volume process, and (*c*) the heat transfer for the constant-pressure process, in Btu.

4-46 Saturated water vapor at 120 psia (state 1) is heated at constant volume until its pressure is 165 psia (state 2). It is then expanded isothermally to 100 psia (state 3) by adding 61 Btu/lb of heat. Finally, it is cooled to the saturation temperature by a constant-pressure process (state 4). Determine the value of the work interaction for (*a*) process 1-2, (*b*) process 2-3, (*c*) process 3-4, and (*d*) the overall process, all in Btu/lb. Sketch this series of processes on a *Pv* diagram, relative to the saturation line.

4-47 A system having an initial volume of 5.0 ft^3 is filled with steam at 450 psia and 700°F (state 1). The system is cooled at constant volume to 400°F (state 2). The first process is followed by a constant-temperature process ending with saturated liquid water (state 3). Find the total heat transfer required, in Btu, and its direction. Sketch the two processes on a *Pv* diagram, relative to the saturation line.

Energy analysis, compressed-liquid data

4-48 One pound of water at 1500 psia and 50°F is heated at constant pressure to a temperature of (*a*) 150°F, and (*b*) 300°F. Determine the heat required, in Btu, using experimental compressed-liquid data. Repeat the calculation using saturation data as an approximation. What percent error is involved with the second method?

4-49 A closed container maintained at 500 psia is subdivided into two sections by an insulated partition. One section contains 0.8 lb of water at 50°F, while the other contains saturated steam. Determine the amount of steam present if, on removing the partition, the final state of the system is (*a*) a saturated liquid, and (*b*) a wet mixture with 20 percent quality.

Compressibility factor

4-50 Water vapor exists at 700°F and at (*a*) 1800 psia, and (*b*) 2500 psia. Determine for these pressures the specific volume, in ft^3/lb, based on (1) the ideal-gas equation, (2) the corresponding states principle, and (3) the tables of superheat data.

4-51 Refrigerant 12 is maintained at (*a*) 140°F and 180 psia, and (*b*) 160°F and 200 psia. Determine for these states the specific volume, in ft^3/lb, based on (1) the ideal-gas equation, (2) the principle of corresponding states, and (3) experimental data in the tables.

4-52 Determine the pressure, in psia, of water vapor at 700°F and 0.491 ft^3/lb on the basis of (*a*) the ideal-gas equation, (*b*) the principle of corresponding states, and (*c*) the experimental value.

4-53 Refrigerant 12 at a temperature of 240°F has a specific volume of 0.1676 ft^3/lb. Determine the pressure of the gas, in psia, on the basis of (*a*) the ideal-gas equation, (*b*) the corresponding states principle, and (*c*) the experimental value in the tables.

4-54 Determine the temperature, in degrees Fahrenheit, of refrigerant 12 at 200 psia and 0.2486 ft^3/lb on the basis of (*a*) the ideal-gas equation, (*b*) the corresponding states principle, and (*c*) the experimental value.

4-55 Steam, initially at 1600 psia and 900°F, is compressed isothermally to one-half its initial volume. Determine the final pressure, in psia, if (*a*) the ideal-gas equation applies, (*b*) the corresponding states principle is used, and (*c*) the steam table is used.

4-56 Propane (C_3H_8), initially at 205°F and 29.5 atm, is heated at constant pressure until the temperature reaches 405°F. Estimate the change in volume, in ft^3/lb, if (*a*) the ideal-gas equation is used, and (*b*) the corresponding states principle is used.

4-57 Carbon dioxide at 64.6 atm and 640°F is cooled in a rigid tank until the temperature reaches 115°F. Estimate the final pressure, in atm, using (*a*) the ideal-gas equation, and (*b*) the corresponding states principle.

Incompressible substances

4-58 A 40-lb mass of copper is dropped into an insulated tank which contains water at 70.0°F and 1 atm. The initial temperature of the copper is 150°F and the final equilibrium temperature for the copper and water at 1 bar is (*a*) 71.60°F, and (*b*) 72.80°F. Estimate the initial volume of water used, in cubic feet.

4-59 An unknown mass of copper at 170°F is dropped into an insulated tank which contains 7 ft^3 of water at 90°F. If the final equilibrium temperature is (*a*) 91.60°F, and (*b*) 93.5°F, find the mass of copper in pounds.

4-60 Consider three liquids *A*, *B*, and *C* which have specific heats at constant pressure of 0.3, 0.5, and 0.8 Btu/(lb)(°F), respectively. The three liquids occupy three compartments of tank, and the walls separating the compartments conduct heat. Initially, one section contains 10 lb of *A* at 200°F, and the other compartments contain 5 lb of *B* and 2 lb of *C*, respectively, each initially at 100°F. Calculate the temperature of the three fluids at thermal equilibrium, assuming the tank is adiabatic and the process is one of constant pressure.

4-61 Consider Prob. 4-60, except that the initial temperatures of *A*, *B*, and *C* are 80°F, 140°F, and 210°F, respectively. Find the final equilibrium temperature, in degrees Fahrenheit.

4-62 A 4-lb mass of lead at −150°F is brought into thermal contact with an unknown mass of aluminum initially at 30°F. If the final equilibrium temperature of the two metals is (*a*) −70°F, and (*b*) −40°F, determine the mass of aluminum, in pounds.

4-63 Consider water at 300°F and (*a*) 1000 psia, (*b*) 1500 psia, and (*c*) 2000 psia. Use Eq. (4-14) to estimate the enthalpy at these states, in Btu/lb, and compare the answer to the tabulated value in the compressed liquid table.

4-64 Consider Prob. 4-22. Evaluate the enthalpy change when the final pressure is (*a*) 500 psia, (*b*) 1500 psia, and (*c*) 3000 psia, assuming the fluid is incompressible. What percent error is involved in assuming an incompressible fluid, in comparison to compressed-liquid data?

4-65 Consider Prob. 4-24. Evaluate the internal energy change when the final pressure is (*a*) 500 psia, (*b*) 1500 psia, and (*c*) 3000 psia, assuming the fluid is incompressible. What percent error is involved in assuming an incompressible fluid, in comparison to compressed-liquid data?

4-66 A 1-kW resistance heater is placed in a 1.0 ft^3 container filled with water at 70°F and 1 atm. The heater is allowed to operate for 20 min. The mass of the heater is 6.0 lb and its specific heat is 0.12 Btu/(lb)(°F). If the container is insulated, determine the temperature of the water and heater element in the final state, in degrees Fahrenheit.

4-67 A paddle-wheel driven by a 240-W motor is used to agitate liquid water in a closed, insulated tank maintained at 1 atm. The initial temperature is 60°F. Determine the temperature rise of the contents for (*a*) a time period of 0.50 h if the mass of water is 12.0 lb, and the mass and specific heat of the paddle-wheel are 0.6 lb and 0.22 Btu/(lb)(°F), respectively, and (*b*) a time period of 0.80 h if the mass of water is 14 lb, and the mass and specific heat of the paddle-wheel are 0.9 lb and 0.24 Btu/(lb)(°F), respectively.

CHAPTER

FIVE

CONTROL-VOLUME ENERGY ANALYSIS

In the two preceding chapters we have developed the basic conservation of energy equations for closed systems, or control masses, and in addition have become familiar with a number of basic relationships among properties. In this chapter we shall develop the basic energy equations for open-system analyses. In terms of the control volume, we must account for energy transported across various parts of the boundary due to mass transfer, as well as account for heat and work interactions. Again, we shall find that knowledge of intrinsic properties of matter is extremely important. The uses of and restrictions on the various equations are of prime interest, but the general methodology for analyzing engineering systems in terms of the energy equations will also be emphasized.

5-1 IDEALIZATIONS FOR STEADY-STATE CONTROL-VOLUME ANALYSIS

The first law of thermodynamics resulted from experimental observations on closed systems. This principle generally may be applied to any control mass which can be physically identified throughout a given process. However, a large number of engineering problems involve open systems where matter flows continuously in and out of a defined region of space. In most cases it is physically impossible to trace a given control mass, much less measure the heat and work interactions that might occur to the unit of mass. For example, consider an element of fluid passing through a pump. During its passage the velocity, pressure, and other macroscopic characteristics of the element are continually changing. It is impractical to attempt to measure all the interactions that occur between this fluid element and its neighboring elements. Furthermore, in this case we are not interested in the behavior of an individual element, but rather in the average behavior of all the fluid elements. Thus the analysis of flow processes will require an important modification in one's viewpoint of the system. Fortunately, the results developed for a control mass can be extended properly to any general open system. We shall find it appropriate also to include a conservation of mass principle in terms of an open system.

The analysis of flow processes begins by selecting a region of space called a *control volume*. The boundary of the control volume may be in part a well-defined physical barrier, or all of it may be an imaginary envelope. The establishment of the boundary, or *control surface*, is an important first step in the analysis of any open system. In our initial development the control volume will be fixed in size and shape, as well as fixed in position relative to the observer. An energy balance on the control volume necessitates not only measurements of heat and work interactions, but also an accounting for the energy carried into or out of the control volume by mass transferred across the control surface. Such an energy accounting requires a knowledge of the state of matter as it passes across the control surface. This implies that the conditions of static equilibrium are closely met in the region of the control surface. These conditions are fulfilled if the properties of the fluid vary in a continuous manner across the control surface. This requirement does not prevent the properties at a given open boundary from changing with time, as they must in unsteady-state processes. Nevertheless, it means that the flow is sufficiently well behaved so that its properties are known with adequate accuracy at the boundaries. Since a flow system seems quite divorced from the conditions of static equilibrium, several pertinent comments need to be made to clarify the situation.

When we consider open systems, a problem arises with regard to the significance of a property value at the control surface. Since the properties of an element of mass crossing the control surface of the control volume may be changing rapidly as it passes the surface, there is a tendency to question what is meant by a property in this case, and what is its relation to properties defined in thermostatic equilibrium. Are the results of the state postulate still valid? In most cases the answer to this last question is yes, if the physical situation fulfills the conditions for what is frequently called *local equilibrium*. By local equilibrium we mean to imply that the change in any property between two small adjacent fluid elements is very small compared with the average value of the property in this region. That is, $dy/y \ll 1$, where y is some intensive property. If an incremental volume at the boundary is selected, the microscopic characteristics of the particles entering the volume are only slightly different from those leaving it. Hence the unit of volume is so close to static equilibrium that the general results of this latter static condition may be applied to the dynamic situation of open-flow systems. Local equilibrium for open systems is somewhat analogous to quasi-static changes in closed systems. The region of interest is extremely close to equilibrium, so that errors in assuming true equilibrium are negligible. Local equilibrium does not exist, for example, in a differential element within a shock wave. Here the criterion that dy/y is very small is violated. In fact, the element is so far from equilibrium that the normal concept of macroscopic properties is meaningless, and the macroscopic-property variations are essentially discontinuous in this region.

There is a wide variety of engineering problems which may be analyzed from the control-volume viewpoint. It is meaningful at this point if we restrict ourselves to those flow systems which fall under the classification of a steady state. A *steady-state* condition for a control volume is one such that the properties at a given position within or at the boundaries of the control volume are invariant with time. The properties of interest include intrinsic ones, such as temperature, pressure, density, and the specific internal energy, as well as those which refer to the overall flow stream, such as velocity. It is the invariance of these properties with respect to time at the control surface where mass interactions occur that is of special importance. Keep in mind, however, that the state of the fluid does change as it passes through the control volume. The major purpose of some flow devices is simply to alter the state of the fluid.

In addition to steady state, we shall also restrict ourselves at the present time to one-dimensional-flow problems. By *one-dimensional flow* we mean that the properties are constant across any area normal to the flow, and thus vary only in the direction of flow. Such an idealized condition is an approximation to actual behavior. For example, the velocity associated with a flow stream usually varies from zero at the wall to a maximum value at the relative center of the flow channel at any given cross section. Thus two- or three-dimensional configuration should be considered. In most text problems, however, we shall assume the velocity to be constant at any cross section.

5-2 CONSERVATION OF MASS PRINCIPLE FOR A CONTROL VOLUME IN STEADY STATE

An equation which represents a conservation of mass principle for an open system in terms of a control volume may be developed by referral to Fig. 5-1. A control mass (CM) will be followed through the control volume (CV) over a time

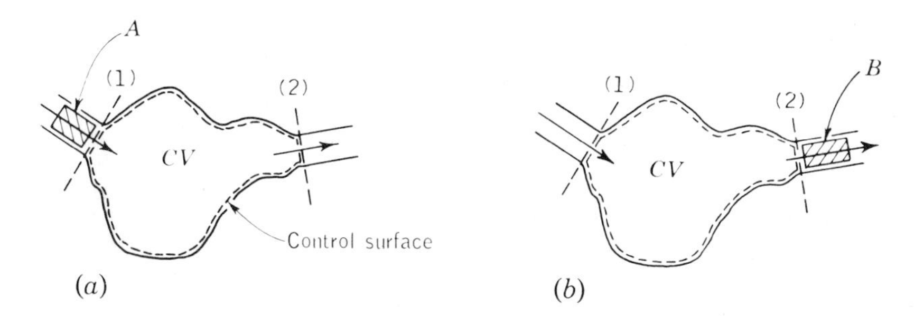

Figure 5-1 The development of conservation principles for a control volume analysis. (*a*) Control mass at time t; (*b*) control mass at time $t + \Delta t$.

interval dt. The control volume is assumed to be of fixed size and shape. At time t we shall consider that mass which occupies the control-volume region marked CV and the small region outside of the inlet 1 indicated by the symbol A. Thus the control mass at time t may be written as

$$m_{CM,t} = m_A + m_{CV,t}$$

The region A is chosen so that after a time period dt all the mass initially in region A has now passed into the control volume. However, in this same time period part of the control mass initially in the control volume has passed into region B which lies just outside the outlet cross section 2. Hence at time $t + dt$, the control mass lies within regions CV and B, that is,

$$m_{CM,t+dt} = m_{CV,t+dt} + m_B$$

By definition of the system under analysis, the control mass at times t and $t + dt$ must be identical. Likewise, since the control volume is a fixed region of space with invariant density at any position within it, $m_{CV,t} = m_{CV,t+dt}$. Consequently, the above two equations show that $m_A = m_B$. That is, under steady-state conditions the masses entering and leaving the control volume in a given time interval must be equal.

The quantities m_A and m_B can be expressed in terms of properties of the fluid and the geometry of the control volume. If regions A and B are made extremely small, the masses within these small regions A and B will lie extremely close to the cross sections 1 and 2. Therefore we may express the quantities m_A and m_B in terms of the properties associated with the control surface at these cross sections. The volume of region A (assuming it to be of constant cross-sectional area) is simply equal to its cross-sectional area A_1 times its length $V_1\,dt$, or $V_1 A_1\,dt$, where V_1 is the average velocity under one-dimensional-flow conditions. Since the density of the fluid in region A is ρ_1 to a close approximation, the mass m_A within region A is given by $\rho_1 V_1 A_1\,dt$. A similar equation is valid for region B at cross section 2. Equating these two expressions, one finds that $\rho_1 V_1 A_1\,dt = \rho_2 V_2 A_2\,dt$, or that

$$\rho_1 V_1 A_1 = \rho_2 V_2 A_2 \tag{5-1}$$

This equation is frequently spoken of as the *continuity-of-flow equation* or simply the *continuity equation* for one-dimensional steady flow. It is appropriate to define the quantity ρAV as the mass rate of flow or the *mass flow* rate, and represent it by the symbol $\dot{m}$. Hence

$$\dot{m} \equiv \frac{\delta m}{dt} = \rho AV \tag{5-2}$$

and

$$\dot{m}_1 = \dot{m}_2 \tag{5-3}$$

where the subscripts 1 and 2 again refer to inlet and outlet. An equivalent form of

Eq. (5-1) which frequently is quite useful is

$$\left(\frac{VA}{v}\right)_{\text{in}} = \left(\frac{VA}{v}\right)_{\text{out}} = \text{constant} \tag{5-4}$$

where v is the specific volume. For the more general case where there is more than one inlet or outlet, the conservation of mass principle for a control volume under steady-state conditions would be given by

$$\sum \dot{m}_{\text{in}} = \sum \dot{m}_{\text{out}} \tag{5-5}$$

or

$$\sum_{\text{in}} (\rho VA) = \sum_{\text{out}} (\rho VA) \tag{5-6}$$

The summation signs indicate that the mass flow rates must be summed over all appropriate inlets and outlets. An equivalent expression similar to Eq. (5-4) would also be valid. These equations imply that under steady-state conditions there can be no net accumulation of mass within a control volume; hence the sum of the constant rates of flow into the control volume must equal the sum of the constant flow rates out. Steady state then implies steady flow. Before turning to the development of the conservation of energy equation for steady-state flow processes, short examples of the use of the continuity equation are presented.

Example 5-1M Refrigerant 12 at 8 bars and 50°C flows through a pipe of 1.50 cm internal diameter at a rate of 2 kg/min. Compute the velocity of the fluid, in m/s.

Solution At 8 bars the saturation temperature from Table A-17M is found to be 32.74°C. Hence the fluid at 50°C is superheated. From Table A-18M the specific volume at 8 bars and 50°C is given as 24.07 cm^3/g. The cross-sectional area of the pipe, normal to the flow velocity, is

$$A = \frac{\pi D^2}{4} = \frac{\pi(1.5)^2}{4} = 1.77\ \text{cm}^2$$

Employing Eq. (5-2), we find that

$$V = \frac{\dot{m}}{\rho A} = \frac{\dot{m}v}{A} = \frac{2\ \text{kg/min} \times 24.07\ \text{cm}^3/\text{g}}{1.77\ \text{cm}^2} \times 10^3\ \text{g/kg}$$

$$= 27{,}200\ \text{cm/min} = 272\ \text{m/min} = 4.53\ \text{m/s}$$

Velocities in the range of 4 to 40 m/s are quite common for the flow of gases through pipes in commercial processes.

Example 5-1 Refrigerant 12 at 120 psia and 140°F flows through a pipe of 0.5-in internal diameter at a rate of 4 lb/min. Compute the velocity of the fluid, in ft/s.

Solution The saturation temperature of 120 psia from Table A-17 is 93.29°F. Thus the fluid at 140°F is superheated. From Table A-18 the specific volume at 120 psia and 140°F is given as 0.3890 ft^3/lb. The cross-sectional area of the pipe, normal to the flow velocity, is

$$A = \frac{\pi D^2}{4} = \frac{\pi(0.5)^2}{4(144)} = 0.001363\ \text{ft}^2$$

Employimg Eq. (5-2), we find that

$$V = \frac{\dot{m}}{\rho A} = \frac{\dot{m}v}{A} = \frac{4 \text{ lb/min} \times 0.3890 \text{ ft}^3/\text{lb}}{0.001363 \text{ ft}^2} = 1140 \text{ ft/min}$$

$$= 19.0 \text{ ft/s}$$

Velocities in the range of 10 to 100 ft/s are quite common for the flow of gases through pipes in commercial processes.

Example 5-2M Water at 20°C flows through a 1.50-cm rubber hose at a rate of 10 L/min. Compute the velocity of the fluid in m/s and the mass flow rate, in kg/min.

SOLUTION The velocity is found by dividing the volume flow rate by the cross-sectional area. The value of A was found in the preceding example to be 1.77 cm^2. Thus,

$$V = \frac{\text{volume rate}}{\text{area}} = \frac{10 \text{ L/min}}{1.77 \text{ cm}^2} \times \frac{10^3 \text{ cm}^3}{1 \text{ L}}$$

$$= 5650 \text{ cm/min} = 56.5 \text{ m/min} = 0.942 \text{ m/s}$$

The saturation pressure of water at 20°C from Table A-12M is 0.023 bar. Although the pressure is not given, it normally would be above atmospheric conditions. Hence the water is subcooled or compressed. We can approximate the specific volume of the fluid by using the saturated-liquid value at the given temperature. From Table A-12M this is given as 1.002 cm3/g. As a result the mass flow rate becomes

$$\dot{m} = \rho A V = \frac{\text{volume rate}}{v} = \frac{10 \text{ L/min}}{1.002 \text{ cm}^3/g} \times \frac{10^3 \text{ cm}^3}{1 \text{ L}}$$

$$= 9980 \text{ g/min} = 9.98 \text{ kg/min}$$

Example 5-2 Water at 60°F flows through a $\frac{1}{2}$-in rubber hose at a rate of 3 gal/min. Compute the velocity of the fluid, in ft/s, and the mass flow rate, in lb/min.

SOLUTION There are 7.48 gal in 1 ft^3. Thus the volume rate is 3/7.48, or 0.401 ft^3/min. Using the saturated-liquid data in Table A-12 as an approximation, we find the specific volume of the fluid at 60°F to be 0.01604 ft^3/lb. The cross-sectional area of the hose is 0.001363 ft^2, as found in the preceding example. Hence the velocity is

$$V = \frac{\text{volume rate}}{\text{area}} = \frac{0.401}{0.001363(60)} = 4.9 \text{ ft/s}$$

The mass flow rate becomes

$$\dot{m} = \rho A V = \frac{\text{volume rate}}{v} = \frac{0.401}{0.01604} = 25.0 \text{ lb/min}$$

Note that although the velocity of the liquid in Example 5-2M (or Example 5-2) is much less than the gas velocity found in Example 5-1M (or Example 5-1) for the same area of flow, the mass flow rate of liquid is much greater due to the larger density. Velocities from 0.3 to 3 m/s (or 1 to 10 ft/s) are quite common for liquids in pipe-flow applications.

5-3 CONSERVATION OF ENERGY PRINCIPLE FOR A CONTROL VOLUME IN STEADY STATE

Because all properties of the fluid within the control volume are invariant with time at a given position during steady-state operation, the energy associated with the control volume must remain constant. This requires that the sum of the rates of energy transfer into and out of the control volume must add up to zero. The modes of energy transfer include that due to the bulk movement of mass across the control surface, as well as heat and work interactions. Because mass flow rates are constant in steady-state processes, the rate of convection of energy associated with mass crossing every open boundary is also constant. In view of this, the rates of heat and work interactions must also remain constant in steady-state processes.

Considering again the control-mass approach employed in Fig. 5-1, we may write for the control mass that

$$\delta Q + \delta W = dE_{\text{CM}} \tag{5-7}$$

The energy associated with the control mass at time t is given by

$$E_{\text{CM},\,t} = E_A + E_{\text{CV},\,t}$$

Similarly, at time $t + dt$ we find that

$$E_{\text{CM},\,t+dt} = E_{\text{CV},\,t+dt} + E_B$$

But E_{CV} is the same at times t and $t + dt$. Consequently, the change in energy of the control mass over the time period dt is simply

$$dE_{\text{CM}} = E_{\text{CM},\,t+dt} - E_{\text{CM},\,t} = E_B - E_A$$

Again choosing regions A and B to be quite small, the energy E_A associated with the mass in region A is equal to the energy per unit mass e_1 times the mass m_A of the region. But the mass of region A has already been seen to be equal to $\rho_1 A_1 V_1\,dt$. Hence $E_A = e_1 \rho_1 A_1 V_1\,dt$ and a similar equation holds for region B in the vicinity of cross section 2. Substitution of these relations for E_A and E_B into the above expression for dE_{CM} yields

$$\begin{aligned} dE_{\text{CM}} &= e_2 \rho_2 A_2 V_2\,dt - e_1 \rho_1 A_1 V_1\,dt \\ &= e_2 \dot{m}_2\,dt - e_1 \dot{m}_1\,dt \end{aligned} \tag{a}$$

Substitution of Eq. (*a*) into Eq. (5-7) leads to

$$\delta Q + \delta W = e_2 \dot{m}_2\,dt - e_1 \dot{m}_1\,dt \tag{b}$$

The energy e per unit mass represents the sum of the internal energy, the kinetic energy, and the potential energy due to conservative force fields.

At this point it is necessary to recognize that the term δW in Eq. (*b*) represents several distinct types of work interactions. One important form of work associated with open systems is called *shaft work.* This arises from the tangential shearing motions of the fluid at the boundary of a blade on a rotating shaft. Shaft work is found in the operation of a turbine, compressor, pump, or fan. This quantity of work can be into or out of the control volume, as the situation dictates. In addition, we must account for the work required to push mass into or out of the control volume. Refer again to Fig. 5-1. In time interval dt, region A will pass through section 1, and region B will pass across section 2. The fluid pressure behind region A pushes the mass within region A into the control volume. In the same time interval the mass in region B must push the fluid ahead of it out of the way. Work δW_1 is done on the mass of region A, while the mass of region B does work δW_2 on the surroundings.

First consider the work δW_1 necessary to push the mass in region A into the control volume. The force exerted by the fluid on the mass in A is simply $P_1 A_1$. This force is exerted through a distance $V_1\, dt$. Hence the required work is $P_1 A_1 V_1\, dt$. Similarly, the work δW_2 done by the fluid within the control volume as it displaces fluid into region B is given by $P_2 A_2 V_2\, dt$. Recalling that $\dot{m} = \rho A V$, we can alter the equations for δW_1 and δW_2 into the following forms.

$$\delta W_1 = P_1 A_1 V_1\, dt = P_1 v_1 \dot{m}_1\, dt \tag{c}$$

$$\delta W_2 = P_2 A_2 V_2\, dt = P_2 v_2 \dot{m}_2\, dt \tag{d}$$

This type of work associated with control-volume analysis is frequently called *flow work* or *displacement work.*

It is now appropriate to substitute the various forms of work interactions into Eq. (*b*). Thus,

$$\delta Q + \delta W_{\text{shaft}} + (P_1 v_1 \dot{m}_1\, dt) + (-P_2 v_2 \dot{m}_2\, dt) = e_2 \dot{m}_2\, dt - e_1 \dot{m}_1\, dt \tag{e}$$

The minus sign on the displacement-work term for region B is necessary due to the sign convention adopted for work interactions. Because regions A and B are assumed to be negligible in size compared to the control volume, the heat and shaft-work interactions represent interactions solely with respect to the control volume. Collecting terms, we find that

$$\delta Q + \delta W_{\text{shaft}} = (e_2 + P_2 v_2)\dot{m}_2\, dt - (e_1 + P_1 v_1)\dot{m}_1\, dt \tag{f}$$

Finally, let us define the rate of doing work (or the power) $\dot{w}$ and the rate of heat transfer $\dot{Q}$ as

$$\dot{W}_{\text{shaft}} \equiv \frac{\delta W_{\text{shaft}}}{dt} \quad \text{and} \quad \dot{Q} \equiv \frac{\delta Q}{dt} \tag{g}$$

Then Eq. (*f*) becomes, after dividing by dt,

$$\dot{Q} + \dot{W}_{\text{shaft}} = (e + Pv)_{\text{out}}\, \dot{m} - (e + Pv)_{\text{in}}\, \dot{m} \tag{5-8}$$

The subscripts have been left off the mass flow rates because under steady-state conditions with one inlet and one outlet, $\dot{m}_1 = \dot{m}_2$. Thus, the subscript on $\dot{m}$ is not needed. Also, note that all the quantities on the right side of the equation are evaluated at the control surface where the mass crosses the boundary.

If more than one inlet or outlet exists, the conservation of energy principle for a control volume on a rate basis must be written as

$$\dot{Q} + \dot{W}_{shaft} = \sum_{out} (e + Pv)\dot{m} - \sum_{in} (e + Pv)\dot{m} \tag{5-9}$$

In Eq. (5-9) it must be remembered that the mass flow rates $\dot{m}$ at each open boundary are not usually equal. However, their values are related by Eq. (5-5); that is,

$$\sum_{in} \dot{m} = \sum_{out} \dot{m} \tag{5-5}$$

The analysis of steady-state, steady-flow systems will frequently require the use of both the conservation of energy and conservation of mass principles outlined above.

It is convenient to have a conservation of energy principle on a unit-mass basis for control volumes which have a single inlet and outlet. Since $\dot{m} = \delta m/dt$, Eq. (f) can be written in the form

$$\delta Q + \delta W_{shaft} = (e + Pv)_2 \, dm - (e + Pv)_1 \, dm$$

Integration of this equation over a time period during which a unit mass enters and another unit mass leaves, the control volume yields

$$q + w_{shaft} = (e + Pv)_2 - (e + Pv)_1 \tag{5-10}$$

where the subscripts 1 and 2 again refer to the inlet and outlet cross sections. Equations (5-8), (5-9), and (5-10) are extremely important statements of the conservation of energy principle for a control volume under steady-state conditions.

5-4 SPECIAL CONSERVATION EQUATIONS

The application of Eqs. (5-8), (5-9), and (5-10) requires the evaluation of the energy e of a unit mass of fluid. If we restrict ourselves to simple, compressible substances the energy e includes the internal energy, the linear kinetic energy, and the gravitational potential energy of the flow stream. In this case:

$$e = u + \frac{V^2}{2} + gz \tag{5-11}$$

Under this restriction Eq. (5-10) for the conservation of energy per unit mass of a steady-flow process becomes

$$q + w_{shaft} = \left(u + Pv + \frac{V^2}{2} + gz\right)_{out} - \left(u + Pv + \frac{V^2}{2} + gz\right)_{in} \tag{5-12}$$

At this point an important reason for introducing the enthalpy function h is seen. At every control surface where mass transfer occurs, the sum of the internal energy and the flow work, $u + Pv$, is associated with each unit of mass crossing the boundary. Therefore, we may write the conservation of energy principle as

$$q + w_{\text{shaft}} = \left(h + \frac{V^2}{2} + gz\right)_{\text{out}} - \left(h + \frac{V^2}{2} + gz\right)_{\text{in}} \tag{5-13}$$

It should be apparent that the datum points for the velocities and the elevations and the reference state for the enthalpy (or internal energy) must be the same for the inlet and outlet. By rearranging and grouping common terms, the steady-state energy balance is often found written as

$$q + w_{\text{shaft}} = (h_2 - h_1) + \frac{V_2^2 - V_1^2}{2} + g(z_2 - z_1) \tag{5-14}$$

Subscript 1 designates the inlet; the subscript 2 designates the outlet.

On a rate basis Eq. (5-9) becomes

$$\dot{Q} + \dot{W}_{\text{shaft}} = \sum_{\text{out}} \left(h + \frac{V^2}{2} + gz\right)\dot{m} - \sum_{\text{in}} \left(h + \frac{V^2}{2} + gz\right)\dot{m} \tag{5-15}$$

It must be remembered that the $\dot{m}$ quantities in the summations are usually different at each open boundary. When a control volume has only one inlet and one outlet, the rate equation is

$$\dot{Q} + \dot{W}_{\text{shaft}} = \left[(h_2 - h_1) + \frac{V_2^2 - V_1^2}{2} + g(z_2 - z_1)\right]\dot{m} \tag{5-16}$$

Although the enthalpy function appears in the steady-state energy balance, it should not be considered a specific form of energy. It is simply a convenient sum of two quantities ($u + Pv$) which appear together in the energy balance.

Since open systems are common in engineering practice, the student must be familiar with the use of the two conservation laws presented in this section—those of mass and energy. In order to familiarize the student with the general method, a number of examples are presented in Sec. 5-5. These problems will attempt to give a feeling of orders of magnitude of various terms, the handling of units, and the importance of various terms in the energy equation when applied to definite types of engineering equipment. Confidence in the use of the control-volume approach to the solution of open-system problems will come only with experience. These solutions require not only the use of one or more of the conservation equations, but also relationships among the properties of matter. Thus we again rely on information from earlier chapters.

5-5 ENGINEERING APPLICATIONS INVOLVING STEADY-STATE SYSTEMS

The following examples illustrate the use of the various forms of the conservation of energy equation for a number of control-volume analyses. At this point it is important for the reader to be reminded of a few of the major points that need to be considered in any thermodynamic analysis. These include:

1. Select a suitable control volume for analysis, and sketch the system, indicating the appropriate boundaries.
2. Determine what energy interactions are important, and recognize certain sign conventions on such terms.
3. Write a suitable energy balance for the chosen system.
4. Obtain physical data for the substance under study. Is an equation of state available (such as $Pv = RT$), or must graphical and/or tabular data be employed? What are other property relations for the substance?
5. Determine the path of the process between the initial and final states. Is it isothermal, constant-pressure, quasistatic, adiabatic, etc.?
6. What other idealizations or assumptions are necessary to complete the solution? Is kinetic energy negligible, etc.?
7. Draw suitable diagram(s) for the process, as an aid in picturing the overall problem.
8. Complete the solution for the required item(s) on the basis of the information supplied. Check units in each equation used.

Two of the items in the above list are extremely important. The first item suggests sketching the system under analysis. On the sketch one should indicate all the relevant energy-transfer terms consistent with the boundaries selected for the system. Sketches help the engineer approach a problem in a straightforward and consistent manner. Equally important is the process diagram. By process diagram we mean a drawing of the process on thermodynamic coordinates, such as a Pv plane. The process diagram is especially meaningful in terms of establishing the initial and final states of the process. The value of the process diagram will become more apparent as more properties are introduced, and as problems become more complex. Although the process diagram is listed as item 7, it is frequently one of the first things that is done in a thermodynamic analysis.

Before proceeding with the analysis of some specific steady-state devices, it is appropriate to examine the relative magnitude of kinetic energy and potential energy terms. This will be done for both SI and USCS units.

a Kinetic energy and potential energy in SI units The linear kinetic energy of a unit mass is $V^2/2$. Since it is customary to evaluate energy terms in J/g and kJ/kg, then,

$$\mathrm{KE} = \frac{V^2}{2}\ \mathrm{m^2/s^2} \times \mathrm{N \cdot s^2/kg \cdot m} \times \frac{\mathrm{kg}}{10^3\mathrm{g}} = \frac{V^2}{2000}\ \mathrm{kJ/kg\ (or\ J/g)}$$

For many engineering applications the values of various energy terms are usually at least 1 kJ/kg in magnitude, and more commonly in the range of 10 to 100 kJ/kg. For a KE of 1 kJ/kg, the above equation requires a velocity of $2000^{1/2}$, or 45 m/s, and a kinetic energy of 10 kJ/kg requires a velocity of 140 m/s. The significant point is that a relatively high velocity is required if the kinetic energy is to be important relative to other terms in an energy balance.

The gravitational potential energy relative to earth's surface is given by gz. The value of g on earth's surface is roughly 9.8 m/s^2. For a potential energy of 1 kJ/kg, the elevation required is

$$z = \frac{\text{PE}}{g} = 1 \text{ kJ/kg} \times \frac{\text{s}^2}{9.8 \text{ m}} \times \text{kg} \cdot \text{m/N} \cdot \text{s}^2 \times 1000 \text{ N} \cdot \text{m/kJ} = 102 \text{ m}$$

For 10 kJ/kg, the height requirement obviously is 1020 m. Since the flow in many industrial devices seldom has an elevation change of this magnitude, the gravitation potential energy change is frequently negligible. Some care must be taken, however, about neglecting the potential energy change when pumping liquids. This will be illustrated by a numerical example later.

b Kinetic energy and potential energy in USCS units The linear kinetic energy of a unit mass is $V^2/2$. If the velocity is expressed in ft/s, then the kinetic energy will be expressed in ft $\cdot$ lb$_f$/lb$_m$ if the term is divided by the unit conversion factor, 32.2 lb$_m$ $\cdot$ ft/lb$_f$ $\cdot$ s^2 = 1. In conjunction with tabular data and property equations, we shall find it useful to express energy in Btu/lb. Hence,

$$\text{KE} = \frac{V^2}{2} \text{ ft}^2/\text{s}^2 \times \frac{\text{lb}_f \cdot \text{s}^2}{32.2 \text{ lb}_m \cdot \text{ft}} \times \frac{\text{Btu}}{778 \text{ ft} \cdot \text{lb}_f} \approx \frac{V^2}{50{,}000} \text{ Btu/lb}_m$$

where the constant 50,000 is approximate. For many engineering applications the values of various energy terms are usually at least 1 to 10 Btu/lb, with some values greater than 100 Btu/lb. For a kinetic energy of 1 Btu/lb, the above equation requires a velocity of $50{,}000^{1/2}$, or 225 ft/s. This is a reasonably high velocity for gas flow in commercial applications. Thus relatively high velocities are required if the kinetic energy of a flow stream is to be important relative to other energy quantities.

The gravitational potential energy relative to earth's surface is given by gz. The value of g on the earth's surface is roughly 32.2 ft/s^2. For a potential energy of 1 Btu/lb, the elevation required is

$$z = \frac{\text{PE}}{g} = 1 \text{ Btu/lb}_m \times \frac{\text{s}^2}{32.2 \text{ ft}} \times 32.2 \text{ lb}_m \cdot \text{ft/lb}_f \cdot \text{s}^2 \times 778 \text{ ft} \cdot \text{lb}_f/\text{Btu}$$

$$= 778 \text{ ft}$$

Since the flow in many industrial processes seldom has an elevation change of this magnitude, the gravitational potential energy change frequently is negligible. Some care must be taken, however, about neglecting this term when pumping liquids through modest elevation changes. This will be illustrated later.

1 Nozzles and Diffusers

In many steady-flow processes there is a need to either increase or decrease the velocity of a flow stream. A device which increases the velocity (and hence the kinetic energy) of a fluid at the expense of a pressure drop in the direction of flow is called a *nozzle.* A *diffuser* is a device for increasing the pressure of a flow stream at the expense of a decrease in velocity. These defining conditions apply for both subsonic and supersonic flow. Figure 5-2 is a set of sketches for the general shapes of a nozzle and a diffuser under the conditions of subsonic and supersonic flow. Note that a nozzle is a converging passage for subsonic flow, whereas the passage is diverging for supersonic flow. The opposite conditions hold for a diffuser. We are not in a position to explain the reason for these differences at this point, theoretically; so they should be accepted as a matter of experience. One outgrowth of this which is extremely important in rocket design is that a converging-diverging nozzle must be used if a fluid is to be accelerated from subsonic to supersonic velocities. The actual shapes of these devices in practice are not as shown, but will probably have curved surfaces in the direction of flow. The true shape is unimportant for the overall energy balances that we wish to make. Since both these devices are merely ducts, it is apparent that no shaft work is involved, and the change in potential energy, if any, would be negligible under most conditions. For a simple, compressible substance the steady-flow energy equation per unit mass reduces to

$$q = \frac{V_2^2 - V_1^2}{2} + h_2 - h_1$$

For nozzles, the exit pressure may be imposed or controlled externally. In many cases the heat transfer per unit mass may be quite small compared to the kinetic-energy and enthalpy changes. In some cases the duct may be deliberately insulated. Even without insulation, the fluid velocity may be so large that there is not time enough for the fluid to come to thermal equilibrium with the walls. Also the surface area of the duct is frequently relatively small for effective heat transfer. Thus in many applications the assumptions of an adiabatic process is a good approximation for nozzles and diffusers.

For adiabatic nozzles and diffusers the energy equation reduces to

$$-\Delta h = \Delta \text{KE}$$

or

$$-\Delta u - \Delta(Pv) = \Delta \text{KE}$$

The latter equation emphasizes the fact that the kinetic-energy change is due to two effects, namely, a change in the internal energy of the fluid and a change in

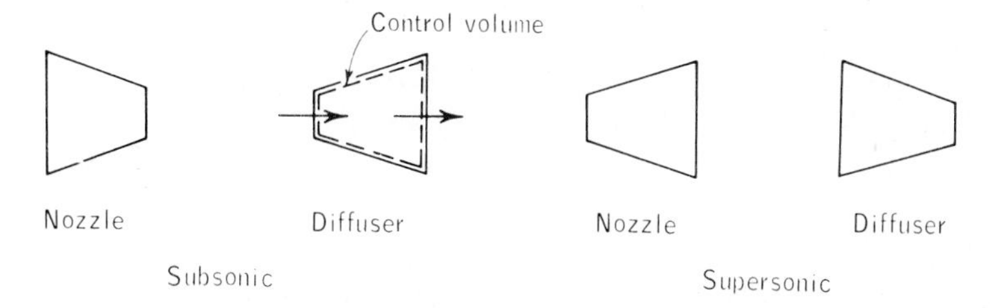

Figure 5-2 General shapes of nozzles and diffusers for subsonic and supersonic flow.

displacement work during the process. Although the use of the enthalpy function is a real convenience, one must not forget the underlying factors which contribute to its change. If we assume, for the purpose of a qualitative illustration, that the fluid is an ideal gas in the nozzle or diffuser, then the last equation can be written as

$$c_v(T_1 - T_2) + R(T_1 - T_2) = \Delta\text{KE}$$

Recall that for gases, $c_v > R$, but they are the same order of magnitude. For example, as an approximation for air, $c_v/R = 5/2$. Hence the internal-energy change for air is about twice that of the change in displacement work as the gas is accelerated or decelerated.

In the application of the conservation of energy principle to a nozzle or diffuser, the specification of the inlet conditions is insufficient to determine the outlet state of a fluid. One outlet property or a specification of the type of process is required to fix the final state. The continuity equation may also be useful because of the variable-area nature of these devices.

Example 5-3M Water vapor enters a subsonic diffuser at a pressure of 0.7 bar, a temperature of 160°C, and a velocity of 180 m/s. The inlet to the diffuser is 100 cm². During passage through the diffuser the fluid velocity is reduced to 60 m/s, the pressure increases to 1.0 bars, and 0.6 J/g of heat are transferred to the surroundings. Determine (*a*) the final temperature, (*b*) the mass flow rate, and (*c*) the outlet area in cm².

SOLUTION (*a*) The final temperature is a function of the final pressure, which is 1.0 bars, and one other property. The continuity-of-flow equation is insufficient to help at this point, since both A_2 and v_2 are unknown. The energy balance for this steady-state device on a unit-mass basis is, since $w = 0$,

$$q = h_2 - h_1 + \frac{V_2^2 - V_1^2}{2}$$

The potential-energy change is assumed to be zero. It is seen that sufficient information is given to evaluate all the quantities except h_2. From the steam table in the superheat region (Table A-14M), it is found that $h_1 = 2798.2$ kJ/kg. Upon substituting the known values, we find that

$$-0.6 = h_2 - 2798.2 + \frac{60^2 - 180^2}{2(1000)}$$

or

$$h_2 = 2798.2 - 0.6 + 14.4 = 2812.0 \text{ kJ/kg}$$

Knowledge of P_2 and h_2 establishes the values of all other properties, such as the temperature. The enthalpy of saturated vapor at 1.0 bar is given in Table A-13M as 2675.5 kJ/kg. Since $h_2 = 2812.0$ kJ/kg, the final state is in the superheat region. From the 1.0-bar set of data in Table A-14M, the calculated value of h_2 corresponds to a temperature of 168°C.

(*b*) The mass flow rate is found from the continuity equation.

$$\dot{m} = \frac{V_1 A_1}{v_1} = \frac{180 \text{ m/s} \times 100 \text{ cm}^2}{2841 \text{ cm}^3/\text{g}} \times 100 \text{ cm/m} = 634 \text{ g/s} = 0.634 \text{ kg/s}$$

(*c*) The outlet area is found also by application of the continuity equation. From the superheated steam table A-14M, the value of v_2 is roughly 2022 cm³/g by interpolation. Hence

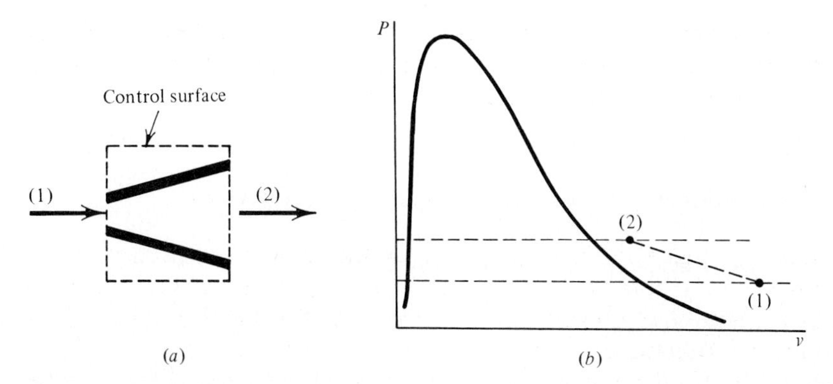

Figure 5-3 (*a*) Control volume schematic, and (*b*) a pressure-volume diagram showing pressure increase in a diffuser.

the final area is

$$A_2 = \frac{A_1 V_1 v_2}{V_2 v_1} = 100 \text{ cm}^2 \times \frac{180}{60} \times \frac{2022}{2841} = 214 \text{ cm}^2$$

Figure 5-3 illustrates the process on a Pv diagram, which is not to scale.

Example 5-3 Water vapor enters a subsonic diffuser at a pressure of 14.7 psia, a temperature of 300°F, and a velocity of 500 ft/s. The inlet to the diffuser is 0.1 ft^2. During passage through the steady-flow device the fluid velocity is reduced to 175 ft/s, the pressure increases to 20 psia, and 0.7 Btu/lb of heat are transferred to the surroundings. Determine (*a*) the final temperature in degrees Fahrenheit, (*b*) the mass flow rate, and (*c*) the outlet area, in square feet.

SOLUTION (*a*) Knowledge of two thermostatic properties is required to find the value of the final temperature. Since P_2 is given, one other property value needs to be found. The continuity equation is insufficient to help at this point since both A_2 and v_2 are unknown. The energy balance for this device on a unit-mass basis is, since $w = 0$,

$$q = h_2 - h_1 + \frac{V_2^2 - V_1^2}{2}$$

The potential energy change is assumed negligible. It is seen that there is sufficient information given to evaluate all the quantities except h_2. From the superheated steam table A-14, it is found that $h_1 = 1192.6$ Btu/lb. Upon substituting the known values, we find that in units of Btu/lb

$$-0.7 = h_2 - 1192.6 + \frac{175^2 - 500^2}{2(32.2)(778)}$$

or

$$h_2 = 1192.6 - 0.7 + 4.4 = 1196.3 \text{ Btu/lb}$$

Knowledge of P_2 and h_2 determines T_2. The enthalpy of saturated vapor at 20 psia given in Table A-13 is 1156.4 Btu/lb. Since h_2 is much larger than this, the final state is a superheated one. From the 20-psia set of data in Table A-14, the calculated value of h_2 corresponds to a temperature T_2 of 310°F

(*b*) The mass flow rate is found from the continuity equation.

$$\dot{m} = \frac{V_1 A_1}{v_1} = \frac{500 \text{ ft/s} \times 0.1 \text{ ft}^2}{30.52 \text{ ft}^3/\text{lb}} = 1.64 \text{ lb/s}$$

(*c*) The outlet area is also found by application of the continuity equation. From the superheated steam table A-14, the value of v_2 is roughly 22.67 ft³/lb by interpolation. Hence the final area is

$$A_2 = \frac{A_1 V_1 v_2}{V_2 v_1} = 0.1 \text{ ft}^2 \times \frac{500}{175} \times \frac{22.67}{30.52} = 0.212 \text{ ft}^2$$

Figure 5-3*b* illustrates the process on a Pv diagram, which is not to scale.

2 Turbines, Pumps, Compressors, and Fans

A turbine, whether the fluid is a gas or a liquid, is a device in which the fluid does work against some type of blade attached to a rotating shaft. As a result, the device produces work which may be used for some purpose in the surroundings. Pumps, compressors, and fans are devices in which work is done on the fluid which results in an increase in pressure of the fluid. Pumps are usually associated with liquids and compressors and fans are employed for gases. The ratio of outlet-to-inlet pressure across a fan will probably be just slightly above 1, whereas for a compressor the ratio will probably be between 3 and 10. For steady flow through any of these devices the energy equation reduces to

$$q + w = h_2 - h_1 + \frac{V_2^2 - V_1^2}{2}$$

The potential-energy change is normally negligible. Two other terms in this expression require comment. As we have noted already, the inclusion of a heat quantity depends upon the mode of operation. If the device is not insulated, the heat gained or lost by the fluid depends upon such factors as whether or not (1) a large temperature difference exists between the fluid and the surroundings, (2) a small flow-velocity exists, and (3) a large surface area is present. In rotating turbomachinery (axial or centrifugal) velocities can be fairly high, and the heat transfer normally is small compared to the shaft work. In reciprocating devices the heat-transfer effects may be reasonably large. Experience and experimental specifications enable the engineer to estimate the relative importance of heat transfer. As a second point, it is found that the change in kinetic energy is usually quite small in these devices, since the velocities at inlet and outlet are frequently less than several hundred feet per second. There are exceptions, of course. In a steam turbine the exhaust velocity can be quite high, due to the large specific volume of the fluid at the low exhaust pressure. On the basis of the continuity equation, velocities may be kept low by selecting large flow areas. However, this may not be a practical choice. Schematics of an axial-flow turbine and a centrifugal-flow compressor are shown in Fig. 5-4.

In many cases, the steady-flow energy balance for these work-producing or work-absorbing devices becomes

$$w = h_2 - h_1$$

In this approximate solution it is seen that the enthalpy decreases for a turbine and increases in the direction of flow for compressors and pumps.

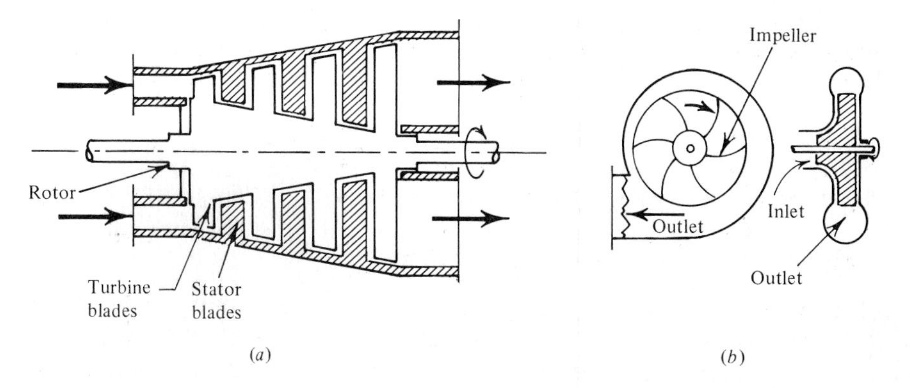

Figure 5-4 Schematics of (*a*) an axial-flow turbine, and (*b*) a centrifugal-flow compressor.

Example 5-4M Air initially at 1 bar and 290°K is compressed to 5 bars and 450°K. The power input to the air under steady-flow conditions is 5 kW, and a heat loss of 5 kJ/kg occurs during the process. If the changes in potential and kinetic energies are neglected, determine the mass flow rate, in kg/min.

SOLUTION The steady-state energy balance given by Eq. (5-14) reduces to

$$q + w_{\text{shaft}} = h_2 - h_1$$

All quantities in this equation are known except the shaft work. Since the inlet and exit temperatures are known, the enthalpies can be read from Table A-5M. Therefore,

$$w_{\text{shaft}} = h_2 - h_1 - q = 451.8 - 290.2 - (-5) = 166.6 \text{ kJ/kg}$$

Power is the product of the mass flow rate and the shaft work. The mass flow rate is determined, then, by

$$\dot{m} = \frac{\text{power}}{w_{\text{shaft}}} = \frac{5 \text{ kW}}{166.6 \text{ kJ/kg}} \times \frac{1 \text{ kJ/s}}{1 \text{ kW}} = 0.030 \text{ kg/s} = 1.80 \text{ kg/min}$$

Example 5-4 Air initially at 15 psia and 60°F is compressed to 75 psia and 400°F. The power input to the air under steady-state conditions is 5 hp, and a heat loss of 4 Btu/lb occurs during the process. If the changes in potential and kinetic energies are neglected, determine the mass flow rate, in lb/min.

SOLUTION The steady-state energy balance given by Eq. (5-14) reduces to

$$q + w_{\text{shaft}} = h_2 - h_1$$

All quantities in this equation are known except the shaft work. The enthalpies are read from Table A-5 in terms of the initial and final temperatures. Therefore,

$$w_{\text{shaft}} = h_2 - h_1 - q = 206.5 - 124.3 - (-4) = 86.2 \text{ Btu/lb}$$

Power is the product of the mass flow rate and the shaft work, that is $\dot{m}w$. The mass flow rate then is

$$\dot{m} = \frac{\text{power}}{w_{\text{shaft}}} = \frac{5 \text{ hp}}{86.2 \text{ Btu/lb}} \times \frac{42.4 \text{ Btu/min}}{1 \text{ hp}} = 2.46 \text{ lb/min}$$

Example 5-5M A steam turbine operates with an inlet condition of 30 bars, 400°C, 160 m/s and an outlet state of a saturated vapor at 0.7 bar with a velocity of 100 m/s. The mass flow rate is 1000 kg/min and the power output is 9300 kW. Determine the magnitude and direction of any heat transfer, in kJ/min.

SOLUTION Since the heat transfer is desired on a rate basis, we shall employ Eq. (5-16) as a suitable energy balance under assumed steady-state conditions. That is,

$$\dot{Q} + \dot{W} = \dot{m}\left(h_2 - h_1 + \frac{V_2^2 - V_1^2}{2}\right)$$

Since the saturation temperature at 30 bars is 233.9°C, the initial state of 400°C is superheated. From the superheat table A-14M, $h_1 = 3230.9$ kJ/kg, and from the saturation pressure table A-13M, $h_2 = h_g = 2660$ kJ/kg. Consequently,

$$\dot{Q} + (-9300)\text{kW} \times \frac{\text{kJ}}{1\ \text{kW} \cdot \text{s}} \times \frac{60\ \text{s}}{\text{min}} = \frac{1000\ \text{kg}}{\text{min}} \times \left[(2660 - 3231) + \frac{100^2 - 160^2}{2(1000)}\right] \text{kJ/kg}$$

$$\dot{Q} = 558{,}000 + 1000(-571 - 8) = -21{,}000\ \text{kJ/min}$$

Note that the heat transfer is from the turbine (since the steam is much hotter than the surroundings) and is only about 4 percent of the work interaction. Also, the kinetic-energy change in only 1.4 percent of the enthalpy change. The process is shown on a Pv diagram in Fig. 5-5.

Example 5-5 A steam turbine operates with an inlet condition of 450 psia, 800°F, 450 ft/s and an outlet state of a saturated vapor at 10 psia with a velocity of 250 ft/s. The mass flow rate is 2000 lb/min and the power output is 12,300 hp. Determine the magnitude and direction of any heat transfer, in Btu/min.

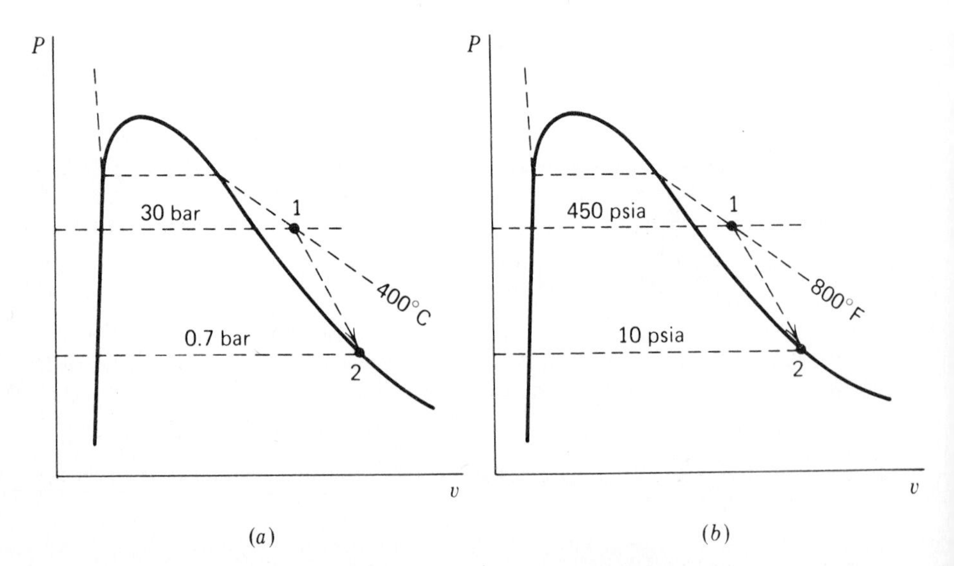

Figure 5-5 The Pv diagrams for processes represented in Examples 5-5M and 5-5.

SOLUTION Since the heat transfer is desired on a rate basis, we shall employ Eq. (5-16) as a suitable energy balance under assumed steady-state conditions. That is,

$$\dot{Q} + \dot{W} = \dot{m}\left(h_2 - h_1 + \frac{V_2^2 - V_1^2}{2}\right)$$

Since the saturation temperature at 450 psia is 456.4°F, the initial state of 800°F is superheated. From Table A-14, $h_1 = 1414.4$ Btu/lb, and from the saturation pressure table A-13, $h_2 = h_g =$ 1143.3 Btu/lb. Consequently,

$$\dot{Q} + (-12{,}300)\text{hp} \times \frac{42.4 \text{ Btu}}{\text{hp} \cdot \text{min}} = \frac{2000 \text{ lb}}{\text{min}} \times \left[(1143.3 - 1414.4) + \frac{250^2 - 450^2}{2(32.2)(778)}\right] \text{Btu/lb}$$

$$\dot{Q} = 521{,}500 + 2000(-271 - 3) = -26{,}500 \text{ Btu/min}$$

Note that the heat transfer is from the turbine (since the steam is much hotter than the surroundings) and is only about 5 percent of the work transfer. Also, the kinetic-energy change is only 1 percent of the enthalpy change. The process is shown on a *Pv* diagram in Fig. 5-5.

3 Throttling Devices

Steady-flow devices such as turbines and nozzles produce a useful effect, such as work output or an increase in kinetic energy of the fluid. These effects are accompanied by a decrease in pressure. There are circumstances in the design of systems where a decrease in pressure is desired, but no other useful effect occurs. This pressure drop is accomplished by inserting in the flow system a component called a throttling device. The main effect of a throttling process is a significant pressure drop without any work interactions or changes in kinetic or potential energy. Flow through a restriction such as a valve or a porous plug or a long capillary tube fulfills the necessary conditions. A throttling valve is shown in Fig. 5-6*a*, and a schematic of the control volume for such devices is shown in Fig. 5-6*b*. By decreasing the cross-sectional area for flow, a greater flow resistance is

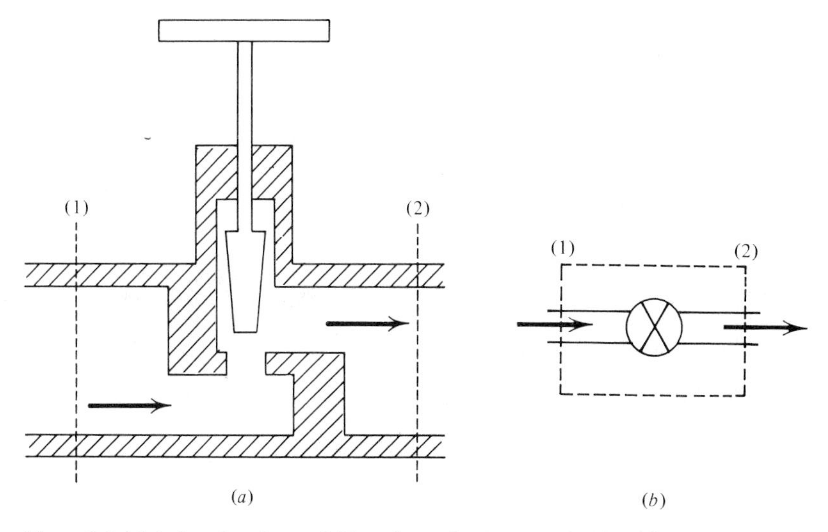

Figure 5-6 (*a*) A throttle valve, and (*b*) a schematic of a control volume for a throttling process.

introduced. For a given mass flow rate, the greater flow resistance leads to a greater pressure drop across the valve.

Although the velocity may be quite high in the region of the restriction, measurements upstream and downstream from the actual valve area will indicate that the change in velocity, and hence the change in kinetic energy, across the restriction is very small. Since the control volume is rigid and no rotating shafts are present, no work interactions are involved, either. The steady-flow energy balance under these restrictions would reduce to $q = h_2 - h_1$. However, in most applications, either the throttling device is insulated or the heat transfer is insignificant. Thus, for a throttling process, the enthalpy change is zero, or

$$h_1 = h_2$$

This statement does not say that the enthalpy is constant for the process, but merely requires that the initial and final enthalpies be the same. The valves in water faucets in the home are examples of throttling devices. These devices are also in common usage in most home refrigeration units. A true throttling process which obeys the above equation is frequently called a Joule-Thomson expansion.

Example 5-6M Refrigerant 12 is throttled from a saturated liquid at 32°C to a final state where the pressure is 2 bars. Determine the final temperature and the physical state of the fluid at the exit.

SOLUTION From the saturation-temperature table A-16M, the enthalpy of saturated liquid at 32°C is 66.57 kJ/kg. This also must be the enthalpy of a unit mass after throttling. From the saturation-pressure table A-17M at 2 bars, the value of h_f is 24.57 kJ/kg and that of h_g is 182.07 kJ/kg. Therefore, the final state must be a mixture of liquid and vapor, and the exit temperature is the corresponding saturation temperature at 2 bars of −12.53°C. The quality in the final state is found by

$$h_2 = h_1 = 66.57 = h_f + xh_{fg} = 24.57 + 157.50x$$

or

$$x = \frac{42.0}{157.5} = 0.267$$

The final mixture is roughly 27 percent vapor and 73 percent liquid. The approximate path of the process in the wet region is sketched in Fig. 5-7.

Example 5-6 Refrigerant 12 is throttled from a saturated liquid at 90°F to a final state where the pressure is 20 psia. Determine the final temperature and the physical state of the fluid at the exit.

SOLUTION From the saturation-temperature table A-16, the enthalpy of a saturated liquid at 90°F is 28.71 Btu/lb. This also must be the enthalpy of a unit mass after throttling. From the saturation-pressure table A-17 at 20 psia, the value of h_f is 6.77 Btu/lb and that of h_g is 76.40 Btu/lb. Consequently, the final state must be a mixture of liquid and vapor, and the exit temperature is the corresponding saturation temperature at 20 psia of −8.13°F. The quality in the final state is found by

$$h_2 = h_1 = 28.71 = h_f + xh_{fg} = 6.77 + 69.63x$$

or

$$x = \frac{21.94}{69.63} = 0.315$$

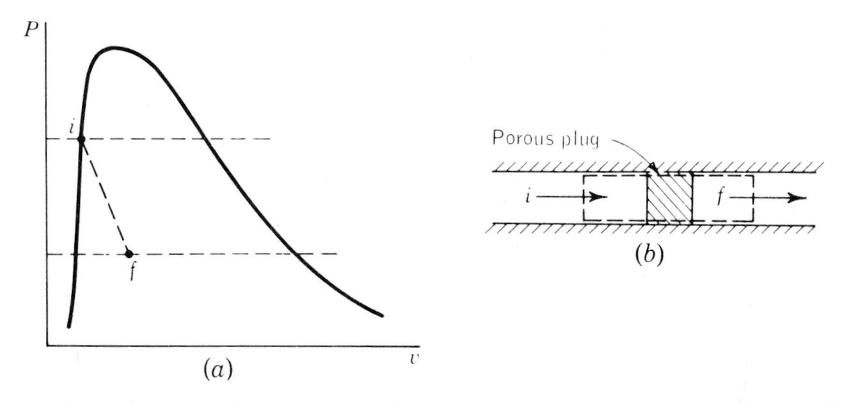

Figure 5-7 Throttling. (*a*) Process representation; (*b*) control volume.

The final mixture is roughly 31 percent vapor and 69 percent liquid. The approximate path of the process in the wet region is sketched in Fig. 5-7.

4 Heat Exchangers

One of the most important steady-flow devices of engineering interest is the heat exchanger. These devices serve two useful purposes. Either they are used to remove (or add) energy from some region of space or they are employed to change deliberately the thermodynamic state of a fluid. The radiator of an automobile is an example of a heat exhanger used for heat removal. Modern gas turbines and electrical generators are frequently cooled internally, and their performance is greatly affected by the heat-transfer process. In steam power plants, heat exchangers are used to remove heat from hot combustion gases and subsequently increase the temperature and enthalpy of the steam in the power cycle. This high-enthalpy fluid is then expanded with a resulting large power output. In the chemical industry, heat exchangers are extremely important in maintaining or attaining certain thermodynamic states as chemical processes are carried out. Modern applications of heat exchangers are numerous.

One of the primary applications of heat exchangers is the exchange of energy between two moving fluids. The changes of kinetic and potential energies are usually negligible and no work interactions are present. The pressure drop through a heat exchanger is usually small, hence an assumption of constant pressure is often quite good as a first approximation. A heat exchanger composed of two concentric pipes is shown in Fig. 5-8. One fluid A flows in the inner pipe, and a second fluid B flows in the annular space between the pipes. Two different equations arise from energy balances on this equipment, depending upon where the system boundaries are selected. In the first case, the control surface will be placed around the entire piece of equipment, as indicated by the dashed line in Fig. 5-8*a*. For this system the steady-flow energy balance on a rate basis [Eq. (5-16)] is applicable. We assume that heat transfer external to the device is zero,

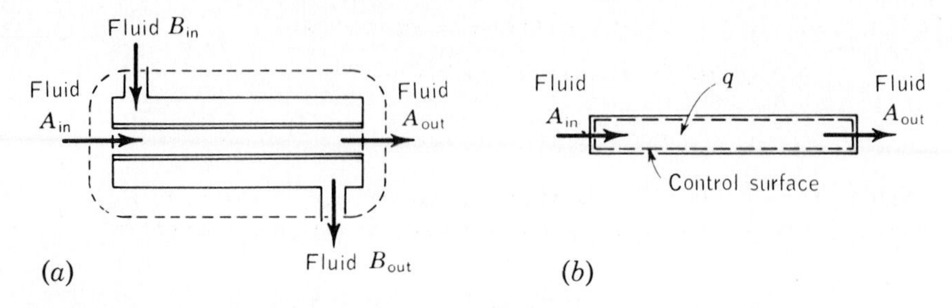

Figure 5-8 Two different control surfaces for a heat exchanger.

and W is zero since no shaft work exists. In addition, the kinetic- and potential-energy changes of the fluid streams usually are negligible. In terms of the notation shown in Fig. 5-8*a*, Eq. (5-16) reduces to

$$\dot{m}_A(h_{A1} - h_{A2}) = \dot{m}_B(h_{B2} - h_{B1})$$

Second, it is sometimes desirable to place a set of boundaries around either of the two fluids. In this case not only is there a change in the enthalpy of the fluid, but a heat-transfer term also appears. For example, if the actual heat transfer in the above case is positive from fluid A to fluid B, then

$$\dot{Q} = \dot{m}_B(h_{B2} - h_{B1})$$

$$-\dot{Q} = \dot{m}_A(h_{A2} - h_{A1})$$

The heat-transfer rates are, of course, identical. Figure 5-8*b* shows the system boundaries in this latter case for fluid A. Properly used, these equations may be applied to specific types of heat-exchanger equipment called boilers, evaporators, and condensers. As the names suggest, in these types of equipment one of the fluids will change phase. Hence the enthalpy of vaporization will be part of the overall enthalpy change for one of the fluids. Several examples of energy balances on heat-exchanger equipment are given below.

Example 5-7M A small nuclear reactor is cooled by passing liquid sodium through it. The liquid sodium leaves the reactor at 2 bars and 400°C. It is cooled to 320°C by passing through a heat exchanger before returning to the reactor. In the heat exchanger, heat is transferred from the liquid sodium to water, which enters the exchanger at 100 bars and 49°C and leaves at the same pressure as a saturated vapor. The mass flow rate of sodium is 10,000 kg/h, and its specific heat is constant at 1.25 J/(g)(°C). Determine the mass flow rate of water evaporated in the heat exchanger, in kg/h, and the heat-transfer rate between the two fluids, in kJ/h.

SOLUTION The mass flow rate can be determined from an energy balance on the entire heat exchanger. It is assumed that the heat loss from the exterior surface of the exchanger is negligible. In terms of the above statement, the energy balance may be written as

$$\dot{m}_{H_2O}(h_{out} - h_{in})_{H_2O} = \dot{m}_{Na}(h_{in} - h_{out})_{Na} = \dot{m}_{Na}\, c_p(T_{in} - T_{out})_{Na}$$

The inlet water stream is a compressed liquid. Its enthalpy is determined from Table A-15M as 176.38kJ/kg. The outlet enthalpy of saturated water vapor at 100 bars is found in Table A-14M to be 2724.7 kJ/kg. Substitution of these values into the above equation yields

$$\dot{m}_{H_2O}(2724.7 - 176.4) = 10{,}000(1.25)(400 - 320)$$

or

$$\dot{m}_{H_2O} = \frac{10{,}000(1.25)(80)}{2548.3} = 392 \text{ kg/h}$$

The heat-transfer rate from the sodium to the water is the value of either the left- or right-hand side of the energy balance. Using the sodium stream to evaluate the heat-transfer rate, one finds that

$$\dot{Q} = \dot{m}_{Na} c_p(T_{in} - T_{out}) = 10{,}000(1.25)(80) = 1.0 \times 10^6 \text{ kJ/h}$$

Example 5-7 A small nuclear reactor is being cooled by passing liquid sodium through it. The liquid sodium leaves the reactor at 25 psia and 725°F. The liquid sodium is cooled to 610°F by passing through a heat exchanger before returning to the reactor. In the heat exchanger, heat is transferred from the liquid sodium to water, which enters the equipment at 1000 psi and 100°F and leaves at the same pressure as a saturated vapor. The mass flow rate of the sodium is 20,000 lb_m/h, and its specific heat is constant at 0.3 Btu/(lb)(°F). Determine the mass flow rate of water evaporated in the heat exchanger, in lb/h, and the heat-transfer rate between the fluids, in Btu/h.

SOLUTION The mass flow rate of water can be determined from an energy balance on the entire heat exchanger. It must be assumed that energy leaves only with the fluid streams, so that heat loss from the exterior surface of the heat exchanger is negligible. In actual practice, the external heat loss could be predicted by heat-transfer calculations, which are independent of thermodynamics. In terms of the above problem, the energy balance may be written as

$$\dot{m}_{H_2O}(h_{out} - h_{in})_{H_2O} = \dot{m}_{Na}(h_{in} - h_{out})_{Na} = \dot{m}_{Na} c_p(T_{in} - T_{out})_{Na}$$

The inlet water stream is a compressed liquid. Its enthalpy is determined from Table A-15 as 70.68 Btu/lb. The outlet enthalpy of saturated water vapor at 1000 psia is found in Table A-13 to be 1192.4 Btu/lb. Substitution of these values into the above equation yields

$$\dot{m}_{H_2O}(1192.4 - 70.7) = 20{,}000(0.3)(725 - 610)$$

or

$$\dot{m}_{H_2O} = \frac{20{,}000(0.3)(115)}{1121.7} = 615 \text{ lb/h}$$

The heat-transfer rate from the sodium to the water is the value of either the left- or right-hand side of the energy balance. Using the sodium stream to evaluate the heat-transfer rate, one finds that

$$\dot{Q} = \dot{m}_{Na} c_p(T_{in} - T_{out}) = 20{,}000(0.3)(115) = 690{,}000 \text{ Btu/h}$$

Example 5-8M In large steam power plants the use of open feedwater heaters is frequently noted. A sketch of an open feedwater heater, which externally appears as a large rigid tank, is shown in Fig. 5-9. This feedwater heater may be classified lossely as a heat exchanger, although

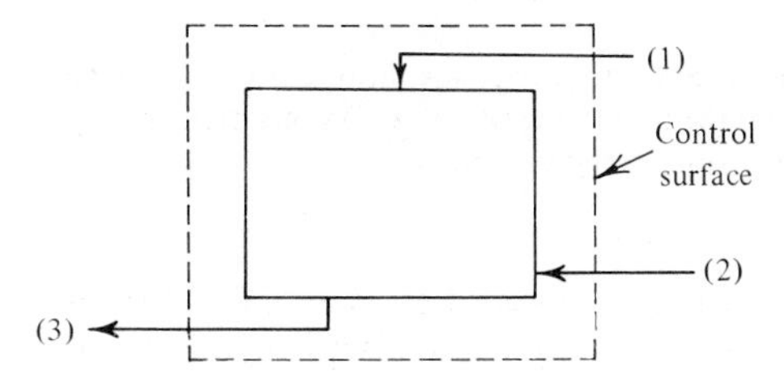

Figure 5-9 Control volume schematic for an open feedwater heater.

in this case the fluid streams come into direct contact inside the heater. Compressed liquid water from a feedwater pump in the power-plant cycle enters at section 1. Superheated steam bled from the turbine in the power cycle is shown entering at section 2. After direct mixing within the heater, the resulting fluid, which is a saturated liquid at the heater pressure, leaves at section 3.

For a control volume with three or more flow streams the steady-flow energy equation is given by

$$\dot{Q} + \dot{W}_{\text{shaft}} = \sum_{\text{out}}\left(h + \frac{V^2}{2} + gz\right)\dot{m} - \sum_{\text{in}}\left(h + \frac{V^2}{2} + gz\right)\dot{m}$$

If the heater is essentially adiabatic and the potential- and kinetic-energy changes of the flow streams are negligible, the steady-flow energy balance on the entire heater becomes

$$\dot{m}_1 h_1 + \dot{m}_2 h_2 = \dot{m}_3 h_3$$

A form of the continuity-of-flow equation also shows that

$$\dot{m}_1 + \dot{m}_1 = \dot{m}_3$$

These two equations are sufficient to determine the state and flow rates at the different sections, in conjunction with other data on the power cycle.

Consider now an open feedwater heater operating at 5 bars. The superheated vapor is bled off the turbine at this pressure and 200°C. The compressed liquid coming from the pump is at 40°C. For every kilogram of steam bled from the turbine per unit time, how many kilograms of compressed liquid enter through section 1?

Solution The first task is to evaluate the enthalpies at the three sections. At 5 bars and 200°C the enthalpy from Table A-14M is 2855.4 kJ/kg. At section 1 the compressed liquid enters at 5 bars and 40°C. Since the pressure is not high, we can estimate the enthalpy by using the saturated liquid value at the given temperature. This is found in Table A-12M to be 167.57 kJ/kg. The fluid leaves at section 3 as a saturated liquid at 5 bars. From Table A-13M, h_3 is 640.23 kJ/kg. If we let $\dot{m}_2$ be unity, the energy and mass balances lead to

$$2855.4 + \dot{m}_1(167.6) = \dot{m}_3(640.2)$$

or

$$2855.4 + \dot{m}_1(167.6) = (1 + \dot{m}_1)(640.2)$$

The solution to this expression is

$$\dot{m}_1 = \frac{2885.4 - 640.2}{640.2 - 167.6} = 4.75 \text{ kg/unit time}$$

Thus over 89 percent of the mass entering the feedwater heater is compressed liquid, and the remainder comes from steam bled from the turbine.

Example 5-8 In large steam power plants the use of open feedwater heaters is frequently noted. A sketch of an open feedwater heater, which externally appears as a large rigid tank, is shown in Fig. 5-9. This feedwater heater may be classified loosely as a heat exchanger, although in this case the fluid streams come into direct contact inside the heater. Compressed liquid water from a feedwater pump in the power-plant cycle enters at section 1. Superheated steam bled from the turbine in the power cycle is shown entering at section 2. After direct mixing within the heater, the resulting fluid, which is a saturated liquid at the heater pressure, leaves at section 3.

For a control volume with three or more flow streams the steady-flow energy equation is given by

$$\dot{Q} + \dot{W}_{\text{shaft}} = \sum_{\text{out}}\left(h + \frac{V^2}{2} + gz\right)\dot{m} - \sum_{\text{in}}\left(h + \frac{V^2}{2} + gz\right)\dot{m}$$

If the heater is essentially adiabatic and the potential- and kinetic-energy changes of the flow streams are negligible, the steady-flow energy balance on the entire heater becomes

$$\dot{m}_1 h_1 + \dot{m}_2 h_2 = \dot{m}_3 h_3$$

A form of the continuity-of-flow equation also shows that

$$\dot{m}_1 + \dot{m}_2 = \dot{m}_3$$

These two equations are sufficient to determine the state and flow rates at the different sections, in conjunction with other data on the power cycle.

Consider now an open feedwater heater operating at 80 psia. The superheated vapor is bled off the turbine at this pressure and 400°F. The compressed liquid coming from the pump is at 100°F. For every pound of steam bled from the turbine per unit time, how many pounds of compressed liquid enter section 1?

SOLUTION The first task is to evaluate the enthalpies at the three sections. At 80 psia and 400°F the enthalpy from Table A-14 is 1230.6 Btu/lb. At section 1 the compressed liquid enters at 80 psia and 100°F. Since the pressure is not high, we can estimate the enthalpy by using the saturated liquid value at the given temperature. This is found in Table A-12 to be 68.1 Btu/lb. The fluid leaves at section 3 as a saturated liquid at 80 psia. From Table A-13, h_3 is 282.2 Btu/lb. If we let $\dot{m}_2$ be unity, the energy and mass balances lead to

$$1230.6 + \dot{m}_1(68.1) = \dot{m}_3(282.2)$$

or

$$1230.6 + \dot{m}_1(68.1) = (1 + \dot{m}_1)(282.2)$$

The solution to this equation is

$$\dot{m}_1 = \frac{1230.6 - 282.2}{282.2 - 68.2} = 4.44 \text{ lb/unit time}$$

Thus a little over 80 percent of the mass entering the feedwater heater per unit time is compressed liquid, and the remainder comes from steam bled from the turbine.

5 Pipe Flow

The transport of fluids, either gaseous or liquid, in pipes or ducts within a system or between systems is of primary importance in numerous engineering analyses. Pipes for a given system may have different diameters at different sections. In addition, the fluid may undergo a considerable elevation change, which is not the case with other pieces of standard equipment. The pipes or ducts may be uninsulated, with heat transfer occurring between the fluid and the environment. If the piping system does not include any work input or output device, then the steady-flow energy equation reduces to

$$q = \Delta h + \Delta \mathrm{KE} + \Delta \mathrm{PE}$$

In the case of liquid flow which is assumed to be incompressible

$$\Delta h = \Delta u + v\,\Delta P \qquad \text{and} \qquad \Delta u = c\,\Delta T$$

In this case the above equation becomes

$$q = c\,\Delta T + v\,\Delta P + \Delta \mathrm{KE} + \Delta \mathrm{PE} \qquad \text{incompressible}$$

Finally, when the effects of heat transfer and fluid temperature change are negligible, we find that

$$v\Delta P + \Delta\text{KE} + \Delta\text{PE} = 0 \qquad \text{incompressible; } q = 0;\ \Delta T = 0$$

or

$$v(P_2 - P_1) + \frac{V_2^2 - V_1^2}{2} + g(z_2 - z_1) = 0 \tag{5-17}$$

This latter equation is known as the Bernoulli equation in elementary fluid mechanics.

Example 5-9M At a certain position in a piping system a solution with a specific gravity of 1.50 passes through an 8-cm pipe with a velocity of 1.2 m/s. At some position downstream the elevation of the pipe has increased 15.0 m and the pipe size has been reduced to 5 cm. The temperature of the fluid is assumed constant at 30°C, and a heat loss of 25 N · m/kg occurs. Determine the change in pressure, in bars and megapascals.

Solution The inlet and exit positions on the control volume are designated as states 1 and 2. The conservation of energy principle for the fluid within the piping system may be written as

$$q = \Delta u + \Delta(Pv) + \Delta\text{KE} + \Delta\text{PE}$$

If we assume the fluid to be incompressible, then $v_1 = v_2$, and Δu is zero because the temperature is constant. As a result the energy equation for the control volume becomes

$$-v\,\Delta P = \Delta\text{KE} + \Delta\text{PE} - q = \frac{V_2^2 - V_1^2}{2} + g(z_2 - z_1) - q$$

The final velocity is found from the continuity equation, since the velocity at state 1 is known. Hence

$$V_2 = V_1\frac{A_1}{A_2} = V_1\left(\frac{D_1}{D_2}\right)^2 = 1.2\left(\frac{8}{5}\right)^2 = 3.1 \text{ m/s}$$

Substitution of the available data into the energy equation, with the assumption that g may be taken to be 9.80 m/s^2, yields

$$-v\,\Delta P = \frac{3.1^2 - 1.2^2}{2} + 15(9.8) - (-25) = 176 \text{ N}\cdot\text{m/kg}$$

The specific volume of water at 30°C is 1.004 cm^3/g if the saturated liquid value is used. Taking into account the specific gravity of the fluid, we find that

$$\Delta P = -176 \text{ N}\cdot\text{m/kg} \times \frac{1.50}{1.004}\text{ g/cm}^3 \times 10^6 \text{ cm}^3/\text{m}^3 \times \frac{\text{bar}\cdot\text{m}^2}{10^5\text{ N}} \times \frac{\text{kg}}{10^3\text{ g}}$$

$$= -2.64 \text{ bars} = -0.264 \text{ MPa}$$

Example 5-9 At a certain position in a piping system a solution with a specific gravity of 1.50 passes through a 3-in pipe with a velocity of 4.0 ft/s. At some position downstream the elevation of the pipe has increased by 50.0 ft and the pipe size has been reduced to 2 in. The temperature of the fluid is assumed constant at 90°F, and a heat loss of 0.013 Btu/lb occurs. Determine the change in pressure, in psi.

Solution The inlet and exit positions on the control volume are designated as states 1 and 2. The conservation of energy principle for the fluid within the piping system may be written as

$$q = \Delta u + \Delta(Pv) + \Delta\text{KE} + \Delta\text{PE}$$

If we assume the fluid to be incompressible, then $v_1 = v_2$, and Δu is zero since the temperature is constant. As a result the energy equation becomes

$$-v\,\Delta P = \Delta\text{KE} + \Delta\text{PE} - q = \frac{V_2^2 - V_1^2}{2} + g(z_2 - z_1) - q$$

The final velocity is found from the continuity equation. Since the velocity at state 1 is known,

$$V_2 = V_1 \frac{A_1}{A_2} = V_1 \left(\frac{D_1}{D_2}\right)^2 = 4.0\left(\frac{3}{2}\right)^2 = 9 \text{ ft/s}$$

With the assumption that g may be taken to be 32.2 ft/s^2, substitution of the available data into the energy equation yields

$$-v\,\Delta P = \frac{9^2 - 4^2}{2(32.2)} + \frac{50(32.2)}{32.2} - (-0.013)(778) = 60.4 \text{ ft} \cdot \text{lb}_f/\text{lb}_m$$

The specific volume of water at 90°F is 0.0161 ft^3/lb if the saturated liquid value is used. Taking into account the specific gravity of the fluid, we find that

$$\Delta P = -60.4 \text{ ft} \cdot \text{lb}_f/\text{lb}_m \times \frac{1.50}{0.0161} \text{lb}_m/\text{ft}^3 \times \frac{\text{ft}^2}{144 \text{ in}^2} = -39.1 \text{ lb}_f/\text{in}^2$$

PROBLEMS (METRIC)

Continuity equations

5-1M Steam enters a steady-flow device at 160 bars and 560°C with a velocity of 80 m/s. At the exit the fluid is a saturated vapor at 0.70 bar and the area is 1000 cm^2. If the mass flow rate is 1000 kg/min, determine (*a*) the inlet area, in square centimeters, and (*b*) the outlet velocity, in m/s.

5-2M Refrigerant 12 enters a steady-state control volume at 5 bars and 100°C where the entrance diameter of the inlet pipe is 0.10 m and the flow velocity is 7.0 m/s. At the exit of the control volume the pressure has reached 0.60 bar and the quality of the fluid is 70 percent. If the exit diameter is 0.20 m, determine (*a*) the mass flow rate, in kg/s, and (*b*) the exit flow velocity, in m/s.

5-3M Steam enters a turbine at 60 bars and 500°C with a velocity of 110 m/s, and leaves as a saturated vapor at 0.30 bar. The turbine inlet pipe has a diameter of 0.60 m and the outlet diameter is 4.5 m. Determine (*a*) the mass flow rate, in kg/h, and (*b*) the exit velocity, in m/s.

5-4M In a steady-flow device 0.50 kg/min of saturated refrigerant 12 vapor at 5 bars enters with a velocity of 4.0 m/s. The exit area is 0.90 cm^2, and the exit temperature and pressure are 60°C and 4.0 bars, respectively. Determine (*a*) the entrance area, in square centimeters, and (*b*) the exit velocity, in m/s.

5-5M Steam enters a turbine at 40 bars and 440°C with a velocity of 100 m/s through an area of 0.50 m^2. It expands to a pressure of 0.30 bars, a quality of 90 percent, and a velocity of 200 m/s. Determine (*a*) the mass flow rate for the steady-state device, in kg/s, and (*b*) the exit area, in square meters.

5-6M A compressor is designed with inlet conditions of 2.4 bars and 0°C for the entering refrigerant 12. A mass flow rate of 2.0 kg/min is to be used. If the inlet velocity of the refrigerant is not to exceed 10 m/s, (*a*) determine the smallest internal diameter of tubing that can be used, in centimeters. (*b*) If the outlet temperature and pressure are 50°C and 8.0 bars, respectively, and the outlet tubing is the same size as the inlet tubing, determine the exit velocity, in m/s.

5-7M Air initially at 0.15 MPa and 80°C flows through an area of 100 cm^2 at a rate of 50 kg/min. Downstream at another position the pressure is 0.25 MPa, the temperature is 100°C, and the velocity is 20 m/s. Determine (*a*) the inlet velocity, in m/s, and (*b*) the outlet area, in square centimeters.

5-8M Air flows through a pipe with a variable cross section. At the pipe inlet the pressure is 6.0 bars, the temperature is 27°C, the area is 35.0 cm^2, and the velocity is 60 m/s. At the pipe exit the conditions are 5.0 bars and 50°C, and the cross-sectional area is 20.0 cm^2. Find (*a*) the mass flow rate, in kg/s, and (*b*) the exit velocity, in m/s.

5-9M Diatomic oxygen enters a control volume in steady state at 18.0 kg/min with a velocity of 20.0 m/s through an area of 0.0080 m^2. The initial temperature is 27°C. The fluid leaves the control volume at 50 m/s and 2.0 bars through an outlet area of 0.0030 m^2. Determine (*a*) the inlet pressure, in bars, and (*b*) the outlet temperature, in kelvins.

5-10M Carbon monoxide passes in steady flow through a control volume with inlet conditions of 3.0 bars, 0.50 m^3/kg, and 80 m/s, and the inlet area is 25 cm^2. At the exit the condtions are 2.4 bars, 157°C, and 50 m/s. Determine (*a*) the mass flow rate, in kg/s, and (*b*) the exit area, in square centimeters.

5-11M Carbon dioxide enters a steady-flow device at 27°C with a velocity of 25 m/s through an area of 4800 cm^2. At the exit of the device, the pressure and temperature are 0.14 MPa and 47°C, respectively, and the gas moves with a velocity of 9 m/s through an area of 7500 cm^2. Determine (*a*) the mass flow rate, in kg/s, and (*b*) the inlet pressure, in MPa. Assume ideal-gas behavior.

5-12M Water at 20°C and 0.20 MPa flows through a garden hose with an inside diameter of 2.50 cm and out of a nozzle with an exit diameter of 0.60 cm. The exit velocity is 6.0 m/s. Determine (*a*) the mass rate of flow, in kg/s, and (*b*) the velocity in the hose, in m/s.

5-13M Water at 15°C flows down a long pipe. At position 1 in the pipe the internal diameter is 20 cm and the velocity is 0.9 m/s. At position 2 downstream from position 1 the velocity is 3.6 m/s. Determine (*a*) the mass flow rate in kg/min, and (*b*) the internal diameter at position 2, in centimeters.

Diffusers

5-14M An adiabatic diffuser is employed to reduce the velocity of an airstream from 250 to 40 m/s. The inlet conditions are 0.1 MPa and 400°C. Determine the required outlet area, in square centimeters, if the mass flow rate is 7 kg/s and the final pressure is 0.12 MPa.

5-15M Air enters a diffuser at 0.7 bar, 57°C, with a velocity of 200 m/s. At the outlet, where the area is 20 percent greater than at the inlet, the pressure is 1.0 bar. Determine the outlet temperature, in degrees Celsius, and the outlet velocity, in m/s, if (*a*) the process is adiabatic, and (*b*) the fluid loses 40 kJ/kg in heat transfer as it passes through.

5-16M The entrance conditions for an adiabatic diffuser operating on air are 0.8 bar, 27°C, and 250 m/s. The exit pressure is 1.0 bar, and A_2/A_1 is 1.4. Compute the exit velocity, in m/s, assuming that the specific heats are constant.

5-17M Steam enters an adiabatic diffuser as a saturated vapor at 110°C with a velocity of 220 m/s. At the exit the pressure and temperature are 1.5 bars and 120°C, respectively. If the exit area is 50 cm^2, determine (*a*) the exit velocity, in m/s, and (*b*) the mass flow rate, in kg/s.

5-18M Refrigerant 12 enters an adiabatic diffuser as a saturated vapor at 32°C with a velocity of 110 m/s. At the exit the pressure and temperature are 8 bars and 40°C, respectively. If the exit area is 50 cm^2, determine (*a*) the exit velocity, in m/s, and (*b*) the mass flow rate, in kg/s.

Nozzles

5-19M Refrigerant 12 enters an adiabatic nozzle at 5 bars and 90 m/s. At the exit the fluid is a saturated vapor at 3.2 bars and has a velocity of 200 m/s. Determine (*a*) the inlet temperature, in degrees Celsius, and (*b*) the mass flow rate, in kg/s, if the exit area is 4.0 cm^2.

5-20M Steam enters a nozzle at 30 bars and 320°C and leaves at 15 bars with a velocity of 535 m/s. The mass flow rate is 8000 kg/h. Neglecting the inlet velocity and considering adiabatic flow, compute (*a*) the exit enthalpy, in kJ/kg, (*b*) the exit temperature, in degrees Celsius, and (*c*) the nozzle exit area, in square centimeters.

5-21M Steam at 10 bars, 200°C, enters an adiabatic nozzle with a velocity of 20 m/s. The exit conditions are 7 bars and 180°C. Determine the ratio of inlet to exit areas, A_1/A_2.

5-22M Air is admitted to an adiabatic nozzle at 3 bars, 200°C, and 50 m/s. The exit conditions are 2 bars, and 150°C. Determine the ratio of the exit area to the entrance area, A_2/A_1.

5-23M Air enters a nozzle at 1.8 bars, 67°C, and 48 m/s. At the outlet the pressure is 1 bar and the velocity is six times its intitial value. If the inlet area is 100 cm^2, determine (*a*) the exit area of the adiabatic nozzle, in square centimeters, and (*b*) the outlet temperature, in degrees Celsius. Use the air table for data.

Turbines

5-24M An adiabatic steam turbine operates with inlet conditions of 120 bars, 480°C, and 100 m/s, and the flow is through an area of 100 cm^2. At the exit the quality is 90 percent at 1 bar, and the velocity is 50 m/s. Determine (*a*) the change in kinetic energy, in kJ/kg, (*b*) the shaft work, in kJ/kg, (*c*) the mass flow rate, in kg/s, and (*d*) the power output, in kilowatts.

5-25M A small gas turbine operating on hydrogen delivers 18 kW. The gas enters the steady-flow device with a velocity of 75 m/s through a cross-sectional area of 0.0020 m^2. The inlet pressure is 2.2 bars and the temperature is 500 K. The velocity change is negligible, the final pressure is 0.8 bar, and the final temperature is 380 K. Compute the heat-transfer rate, in kJ/min.

5-26M Steam flows steadily through a turbine at 20,000 kg/h, entering at 40 bars, 440°C, and leaving at 0.20 bar with 90 percent quality. A heat loss of 20 kJ/kg occurs. The inlet pipe has a 12-cm diameter, and the exhaust section is rectangular with dimensions of 0.6 by 0.7 m. Calculate (*a*) the kinetic-energy change, and (*b*) the power output, in kilowatts.

5-27M A steam turbine develops 10,000 kW of power from a flow rate of 42,000 kg/h. The steam enters at 40 bars and leaves at 0.04 bar with a quality of 92 percent. Assume negligible heat transfer and change in kinetic energy. Calculate (*a*) the inlet temperature in degrees Celsius, and (*b*) the exit area, in square meters, if the exit velocity is 140 m/s.

5-28M Air enters a turbine at 6 bars, 740°K, and 120 m/s. The exit conditions are 1 bar, 450°K, and 220 m/s. A heat loss of 15 kJ/kg occurs, and the inlet area is 4.91 cm^2. Determine (*a*) the kinetic-energy change, in kJ/kg, (*b*) the power output, in kilowatts, and (*c*) the ratio of the inlet- to outlet-pipe diameters.

5-29M Argon gas enters a turbine at 0.35 MPa and 150°C, and leaves the device at 0.15 MPa. A heat loss of 3.7 kJ/kg occurs, and the measured shaft work output is 52.0 kJ/kg. If the kinetic-energy change is negligible, find the outlet temperature, in degrees Celsius.

Compressors

5-30M Carbon dioxide is to be compressed from 0.1 MPa, 310°K to 0.5 MPa, 430°K. The required volume flow rate at inlet conditions is 30 m^3/min. The kinetic-energy change is negligible, but a heat loss of 4.0 kJ/kg occurs. Determine the required power input, in kilowatts, using data from Table A-9M.

5-31M An air compressor handling 300 m^3/min increases the pressure from 1.0 bar to 2.3 bars, and heat is removed at the rate of 1700 kJ/min. The inlet temperature and area are 17°C and 280 cm^2, respectively, and these values for the exit are 137°C and 200 cm^2. Find (*a*) the inlet and exit velocities in m/s, and (*b*) the required power input, in kW.

5-32M A fan receives air at 970 mbar, 20.0°C, and 3 m/s, and discharges it at 1020 mbar, 21.6°C, and 18 m/s. If the flow is adiabatic and 50 m^3/min enters, determine the power input, in kilowatts.

5-33M A steam compressor is supplied with 50 kg/h of saturated vapor at 0.04 bar and discharges at 1.5 bar and 120°C. The power required is measured to be 2.4 kW. What is the rate of heat transfer from the steam, in kJ/min?

5-34M A water-cooled compressor changes the state of refrigerant 12 from a saturated vapor at 1.0 bar to a pressure of 8 bars. The fluid rate is 0.9 kg/min, and the cooling water removes heat at a rate of 140 kJ/min. If the power input is 3.0 kW, determine the exit temperature, in degrees Celsius.

5-35M A compressor steadily inducts 2000 kg/h of refrigerant 12 at 0.6 bar and 0°C through a pipe with an inside diameter of 7 cm. It discharges the gas at 7 bars and 140°C through a 2-cm-diameter pipe. During the process, 40,000 kJ/h are lost as heat to the surroundings. Determine (*a*) the inlet and exit velocities, and (*b*) the power required, in kilowatts.

Throttling devices

5-36M Steam flows through a heavily insulated throttling valve under the following conditions: (*a*) enters at 30 bars, 240°C, and exhausts at 7 bars, (*b*) enters at 280 bars, 480°C, and exits at 200 bars, and (*c*) enters as saturated vapor at 8 bars and exits at 3 bars. Determine the final temperature downstream from the valve at the stated exit pressure, in degrees Celsius.

5-37M Steam is throttled from (*a*) 40 bars to 0.35 bar and 120°C, (*b*) 5 bars to 1 bar and 100°C, and (c) 10 bars to 0.7 bar and 100°C. Determine (1) the quality of the steam entering the throttling process, and (2) the ratio of exit area to inlet area for the device, if the inlet and exit velocities are essentially the same.

5-38M Refrigerant 12 is throttled from (*a*) a saturated liquid at 32°C to a temperature of −20°C, (*b*) a state of 40°C and 10 bars to a temperature of 4°C, and (*c*) a state of 36°C and 9 bars to a temperature of −15°C. Determine (1) the final pressure, in bars, and (2) the final specific volume, in cm^3/g.

5-39M Refrigerant 12 is throttled from (*a*) 48°C and 12 bars to a final pressure of 2.8 bars, and (*b*) a saturated vapor at 16 bars to a final pressure of 1.0 bar. What is the specific volume in the final state, in cm^3/g?

Heat exchangers

5-40M Refrigerant 12 with a mass flow rate of 5 kg/min enters a condenser at 14 bars, 80°C, and leaves at a state of 52°C and 13.8 bars. The coolant in the condenser is (*a*) water which enters at 12°C and leaves at 24°C and 7 bars, and (*b*) air which experiences a 9°C temperature rise at 1.1 bars. Calculate (1) the heat transfer from the refrigerant 12, in kJ/min, and (2) the mass flow rate of coolant required, in kg/min.

5-41M Dry air enters an air conditioning system at 30°C and 0.11 MPa at a volume flow rate of 1.20 m^3/s. The air is cooled by exchanging heat with a stream of refrigerant 12 which enters the heat exchanger at −10°C and a quality of 20 percent. Assume the heat transfer takes place at constant pressure for both flow streams. The refrigerant 12 leaves as a saturated vapor, and 22 kJ/s of heat is removed from the air. Find (*a*) the flow rate of refrigerant 12 required, in kg/s, and (*b*) the temperature of the air leaving the heat exchanger.

5-42M Water at 5 bars and 140°C enters a heat exchanger at a rate of 240 kg/min and leaves at 4.8 bars and 60°C. The water is cooled by passing air through the heat exchanger at an inlet volume flow rate of 1000 m^3/min. The air initially is at 1.10 bars and 25°C, and the exit pressure is 1.05 bars. Determine (*a*) the exit air temperature in degrees Celsius, (*b*) the inlet area for airflow, in m^2, if the inlet air velocity is 25 m/s, and (*c*) the inlet water velocity, in m/s, if the inlet-pipe diameter is 10 cm.

5-43M Steam is condensed on the outside of a heat exchanger tube by passing air through the inside of the tube. The air enters at 1.20 bars, 20°C, and 10 m/s, and exits at 80°C. The steam enters at 3 bars, 200°C, at a mass flow rate of 5 kg/min, and leaves as a saturated liquid. Find (*a*) the mass flow rate of air required, in kg/min, and (*b*) the inlet area of the pipe for airflow, in square meters.

5-44M Steam enters a heat exchanger as a saturated vapor at 5 bars and is condensed on the outside of tubes at constant pressure to a saturated liquid. Compressed liquid water passes through the tubes and experiences a temperature rise change from 15 to 22°C.

(*a*) Determine the ratio of mass flow rate of water to steam required.

(*b*) If the required mass flow rate of steam through the heat exchanger is 2.0 kg/s, the cross-sectional area of a tube is 6.0 cm^2, and the inlet water velocity is 0.8 m/s, determine the total number of tubes required to the nearest integer value.

Mixing processes

5-45M Water substance is fed into a mixing chamber from two different sources. One source delivers steam of 90 percent quality at a rate of 2000 kg/h. The second source delivers steam at a temperature of 280°C and a rate of (*a*) 1750 kg/h, and (*b*) 2790 kg/h. If the mixing process is adiabatic and at a constant pressure of 10 bars, determine the temperature of the mixture at equilibrium downstream from the chamber.

5-46M An open feedwater heater operates at 7 bars. Compressed liquid water at 35°C enters one section. Determine the temperature of steam entering another section if the ratio of the mass flow rate of compressed liquid to superheated vapor is (*a*) 4.52:1, and (*b*) 4.37:1.

5-47M Water is heated in an insulated chamber by mixing it with steam. The water enters at a steady rate of 100 kg/min at 20°C and 3 bars. The steam enters at 320°C and 3 bars, and the mixture leaves the mixing chamber at (*a*) 40°C and 3 bars, and (*b*) 90°C and 3 bars. (1) Determine the amount of steam needed, in kg/min. (2) If the exit area is 25 cm^2, determine the exit velocity, in m/s.

5-48M Refrigerant 12 at 5 bars and 4°C enters a mixing chamber at a steady rate of 2 kg/s. Another source of refrigerant 12 at 6 bars and 50°C is first throttled before it enters the mixing chamber. The mixture leaves the chamber as a saturated liquid at 5 bars. If heat transfer to the chamber occurs at a rate of 4 kJ/s, determine the mass flow rate of refrigerant gas entering the chamber, in kg/s.

Pipe flow

5-49M A steam heating system for a building 150 m high is supplied from boilers 15 m below ground level. Saturated water vapor leaves the boiler at 2.0 bars, and it reaches the 150-m elevation above ground at 1.50 bars. Heat transfer from the supply pipe to the surroundings is 50 kJ/kg. Neglecting kinetic-energy effects, (*a*) find the quality of the steam at the 150-m elevation, and (*b*) find the percent reduction in heat transfer necessary to assure that saturated vapor is present at the 150-m level.

5-50M A long, horizontal pipe with an internal diameter of 5.25 cm carries refrigerant 12. The fluid enters at 3.2 bars with a quality of 40 percent and a velocity of 3 m/s. The fluid leaves the pipe at 2.8 bars and 20°C. Determine (*a*) the mass flow rate, in kg/s, (*b*) the exit velocity, in m/s, and (*c*) the heat-transfer rate, in kJ/s.

5-51M Steam flows through a long, insulated pipe. At one point, where the diameter is 9 cm, the pressure and temperature are 15 bars and 320°C, respectively. Further downstream where the diameter is now 7 cm, the conditions are 10 bars and 280°C. Determine (*a*) the mass flow rate of the steam, in kg/s, and (*b*) the downstream velocity, in m/s.

5-52M In a building, water flows steadily through a series of pipes from 4.0 m below ground level to 120 m above ground level. At the entrance below ground the conditions are 35°C, 0.70 MPa, and 3 m/s, while at the 120-m level the state is 32°C, 0.64 MPa, and 14 m/s. Determine the heat transfer, in kJ/kg, if no shaft work is done. Do not neglect any term which can be calculated. Local g is 9.8 m/s^2.

5-53M Water enters a piping system at 25°C and 7 m/s. Downstream the conditions are 0.20 MPa, 25°C, and 12 m/s, and the elevation is 10.0 m above the inlet. The local gravity is 9.60 m/s^2 and the fluid undergoes a heat loss of 0.010 kJ/kg. If the volume flow rate is 10.0 m^3/min, determine (*a*) the inlet pressure, in MPa, and (*b*) the inlet-pipe diameter, in cm.

5-54M An oil with a specific gravity of 0.90 enters a piping system at 0.240 MPa, 15°C, and 6.0 m/s. Downstream the conditions are 15°C and 4 m/s at an elevation which is 8.0 m below the inlet. The local gravity is 9.65 m/s^2, and the fluid gains 0.0079 kJ/kg as heat as it passes through the system at a mass flow rate of 2000 kg/min. Determine (a) the outlet pressure, in MPa, and (*b*) the outlet-pipe diameter, in cm.

PROBLEMS (USCS)

Continuity of flow

5-1 Steam enters a steady-flow device at 2000 psia and 900°F with a velocity of 250 ft/s. At the exit the fluid is a saturated vapor at 10 psia and the area is 1.1 ft^2. If the mass flow rate is 2000 lb/min, determine (*a*) the inlet area, in square feet, and (*b*) the outlet velocity, in ft/s.

5-2 Refrigerant 12 enters a steady-state control volume at 80 psia and 200°F where the entrance diameter of the inlet pipe is 0.30 ft and the flow velocity is 22 ft/s. At the exit of the control volume the pressure has reached 10 psia and the quality of the fluid is 70 percent. If the exit diameter is 0.60 ft, determine (*a*) the mass flow rate, in lb/s, and (*b*) the exit flow velocity, in ft/s.

5-3 Steam enters a turbine at 1000 psia and 1000°F with a velocity of 220 ft/s, and leaves as a saturated vapor at 1.0 psia. The turbine inlet pipe has a diameter of 1.5 ft, and the outlet diameter is 12 ft. Determine (*a*) the mass flow rate, in lb/h, and (*b*) the exit velocity, in ft/s.

5-4 In a steady-flow device 1.2 lb/min of saturated refrigerant 12 vapor at 70 psia enters with a velocity of 12 ft/s. The exit area is 0.15 in^2, and the exit temperature and pressure are 140°F and 60 psia, respectively. Determine (*a*) the entrance area, in square inches, and (*b*) the exit velocity, in ft/s.

5-5 Steam enters a turbine at 600 psia and 800°F with a velocity of 300 ft/s through an area of 1.5 ft^2. It expands to a pressure of 5 psia, a quality of 90 percent, and a velocity of 560 ft/s. Determine (*a*) the mass flow rate for the steady-state device, in lb/s, and (*b*) the exit area, in square feet.

5-6 A compressor is designed with inlet conditions of 50 psia and 40°F for the entering refrigerant 12. A mass flow rate of 4.0 lb/min is to be used. If the inlet velocity of the refrigerant is not to exceed 30 ft/s, (*a*) what is the smallest internal diameter of tubing that can be used, in inches? (*b*) If the outlet temperature and pressure are 140°F and 160 psia, respectively, and the outlet tubing is the same size as the inlet tubing, determine the exit velocity, in ft/s.

5-7 Air initially at 20 psia and 140°F flows through an area of 0.10 ft^2 at a rate of 100 lb/min. Downstream at another position the pressure is 40 psia, the temperature is 140°F, and the velocity is 50 ft/s. Determine (*a*) the inlet velocity, in ft/s, and (*b*) the outlet area, in square feet.

5-8 Air flows through a pipe with a variable cross section. At the pipe inlet the pressure is 100 psia, the temperature is 80°F, the area is 6.0 in^2, and the velocity is 175 ft/s. At the pipe exit the conditions are 80 psia and 120°F, and the cross-sectional area is 4.0 in^2. Find (*a*) the mass flow rate, in lb/s, and (*b*) the exit velocity, in ft/s.

5-9 Diatomic oxygen enters a control volume in steady state at 28.0 lb/min with a velocity of 60 ft/s through an area of 0.080 ft^2. The initial temperature is 80°F. The fluid leaves the control volume at 150 ft/s and 30 psia through an outlet area of 0.030 ft^2. Determine (*a*) the inlet pressure, in psia, and (*b*) the outlet temperature, in degrees Rankine.

5-10 Carbon monoxide enters a steady-state control volume at 20.0 lb/min through an inlet area of 0.050 ft^2. The inlet pressure is 20 psia and the velocity is 70 ft/s. At the outlet the pressure and temperature are 15.0 psia and 40°F, respectively, and the outlet area is 0.15 ft^2. Determine (*a*) the inlet temperature in degrees Fahrenheit, and (*b*) the outlet velocity, in ft/s.

5-11 Carbon dioxide enters a steady-flow device at 80°F with a velocity of 80 ft/s through an area of 0.050 ft^2. At the exit of the device the pressure and temperature are 20 psia and 120°F, respectively, and the gas moves with a velocity of 27 ft/s through an area of 0.080 ft^2. Determine (*a*) the mass flow rate, in lb/s, and (*b*) the inlet pressure, in psia. Assume ideal-gas behavior.

5-12 Water at 60°F and 20 psia flows through a garden hose with an inside diameter of 1.0 in and out of a nozzle with an exit diameter of 0.25 in. The exit velocity is 20 ft/s. Determine (*a*) the mass rate of flow in lb/s, and (*b*) the velocity in the hose, in ft/s.

5-13 Water at 40°F flows down a long pipe. At position 1 in the pipe the internal diameter is 16 in and the velocity is 2.5 ft/s. At position 2 downstream from position 1 the internal diameter is 8 in. Determine (*a*) the mass flow rate, in lb/h, and (*b*) the fluid velocity at position 2, in ft/s.

Diffusers

5-14 An adiabatic diffuser is employed to reduce the velocity of an airstream from 780 to 120 ft/s. The inlet conditions are 15 psia and 560°F. Determine the required outlet area, in square inches, if the mass flow rate is 15 lb/s and the final pressure is 17.7 psia.

5-15 Air enters a diffuser at 10 psia, 140°F, with a velocity of 800 ft/s. At the outlet, where the area is 28 percent greater than at the inlet, the pressure is 13 psia. Determine the outlet temperature, in degrees Fahrenheit, and the outlet velocity, in ft/s, if (*a*) the process is adiabatic, and (*b*) the fluid loses 2 Btu/lb in heat transfer as it passes through.

5-16 The entrance conditions for an adiabatic diffuser operating on air are 10 psia, 100°F, and 800 ft/s. The exit pressure is 14 psia, and A_2/A_1 is 1.6. Compute the exit velocity, in ft/s, assuming that the specific heats are constant.

5-17 Steam enters an adiabatic diffuser as a saturated vapor at 200°F with a velocity of 1100 ft/s. At the exit the pressure and temperature are 14.7 psia and 250°F, respectively. If the exit area is 8.0 in^2, determine (*a*) the exit velocity, in ft/s, and (*b*) the mass flow rate, in lb/s.

5-18 Refrigerant 12 enters an adiabatic diffuser as a saturated vapor at 80°F with a velocity of 420 ft/s. At the exit the pressure and temperature are 100 psia and 100°F, respectively. If the exit area is 10 in^2, determine (*a*) the exit velocity, in ft/s, and (*b*) the mass flow rate, in lb/s.

Nozzles

5-19 Refrigerant 12 enters an adiabatic nozzle at 80 psia and 100 ft/s. At the exit the fluid is a saturated vapor at 60 psia and has a velocity of 615 ft/s. Determine (*a*) the inlet temperature, in degrees Fahrenheit, and (*b*) the mass flow rate, in lb/s, if the exit area is 0.010 ft^2.

5-20 Steam enters a nozzle at 400 psia and 600°F and leaves at 250 psia with a velocity of 1475 ft/s. The mass flow rate is 18,000 lb/h. Neglecting the inlet velocity and considering adiabatic flow, compute (*a*) the exit enthalpy, in Btu/lb, (*b*) the exit temperature in degrees Fahrenheit, and (*c*) the nozzle exit area, in square feet.

5-21 Steam enters a nozzle at 160 psia and 500°F with a velocity of 210 ft/s. The area of the inlet is 4.53 in^2 and the exit area is 1.20 in^2. At the nozzle exit the conditions are 120 psia and 450°F. Calculate the heat-transfer rate from the steam, in Btu/min.

5-22 Air expands through a nozzle from 25 psia, 200°F, to 15 psia, 80°F. If the inlet velocity is 100 ft/s and the heat loss is 2.0 Btu/lb, find (*a*) the exit velocity, in ft/s, and (*b*) the ratio of the inlet area to the exit area.

5-23 Air enters a nozzle at 25 psia, 140°F, and 100 ft/s. At the outlet the pressure is 15 psia and the velocity is six times its initial value. If the inlet area is 0.10 ft^2, determine (*a*) the exit area of the adiabatic nozzle, in square feet, and (*b*) the outlet temperature, in degrees Fahrenheit. Use the air table for data.

Turbines

5-24 An adiabatic air turbine develops 500 hp. Air enters at 14.7 psia, 70°F, and 400 ft/s through an opening of 1.20 ft^2. The air leaves the power plant at 26 psia and 300°F through a cross-sectional area of 0.80 ft^2. Determine (*a*) the mass flow rate, in lb/s, (*b*) the exit velocity, in ft/s, and (*c*) the rate of any heat transfer, in Btu/h. Use the air table for property data.

5-25 A small gas turbine operating on hydrogen delivers 24 hp. The gas enters the steady-flow device with a velocity of 220 ft/s through a cross-sectional area of 0.020 ft^2. The inlet pressure is 32 psia and the temperature is 440°F. The velocity change is negligible, the final pressure is 12 psia, and the final temperature is 220°F. Compute the heat-transfer rate, in Btu/min.

5-26 Steam flows steadily through a turbine at 300,000 lb/h, entering at 700 psia, 700°F, and leaving at 1 psia as a saturated vapor. A heat loss of 4 Btu/lb occurs. The inlet pipe has a 12-in diameter, and the exhaust section is rectangular with dimensions of 6.0 by 7.0 ft. Calculate (*a*) the kinetic-energy change, in Btu/lb, and (*b*) the power output, in horsepower.

5-27 A steam turbine develops 10,000 hp from a flow rate of 220,000 lb/h. The steam enters at 1000 psia and 800°F and leaves at 200 psia. A heat loss of 17,000 Btu/min occurs and the change in kinetic energy is negligible. Calculate (*a*) the outlet temperature, in degrees Fahrenheit, and (*b*) the exit area, in square feet, if the exit velocity is 440 ft/s.

5-28 Air enters a turbine at 90 psia, 940°F, and 240 ft/s. The exit conditions are 15 psia, 440°F, and 480 ft/s. A heat loss of 6 Btu/lb occurs, and the inlet area is 31.5 in^2. Determine (*a*) the kinetic-energy change, in Btu/lb, (*b*) the power output, in horsepower, and (*c*) the ratio of the inlet- to outlet-pipe diameters.

5-29 Argon gas enters a turbine at 50 psia and 300°F, and leaves the device at 20 psia. A heat loss of 2.0 Btu/lb occurs, and the measured shaft work output is 24.0 Btu/lb. If the kinetic-energy change is negligible, find the outlet temperature in degrees Fahrenheit.

Compressors

5-30 Carbon dioxide is to be compressed from 1 atm, 100°F, to 5 atm, 320°F. The required volume flow rate at inlet conditions is 1000 ft^3/min. The kinetic-energy change is negligible, but a heat loss of 4.0 Btu/lb occurs. Determine the required power input, in kilowatts, using data from Table A-9.

5-31 An air compressor handling 10,000 ft^3/min increases the pressure from 15 psia to 35 psia, and heat is removed at the rate of 750 Btu/min. The inlet temperature and area are 70°F and 0.26 ft^2, respectively, and these values for the exit are 280°F and 0.20 ft^2. Find (*a*) the inlet and exit velocities, in ft/s, and (*b*) the required power input, in horsepower.

5-32 A fan receives air at 14.2 psia, 70.0°F, and 10 ft/s, and discharges it at 14.8 psia, 74.6°F, and 30 ft/s. If the flow is adiabatic and 1800 ft^3/min enters, determine the power input, in horsepower.

5-33 A steam compressor is supplied with 110 lb/h of saturated vapor at 0.60 psia and discharges at 20 psia and 250°F. The power required is measured to be 3.2 hp. Determine the rate of heat transfer from the steam, in Btu/min.

5-34 A water-cooled compressor changes the state of refrigerant 12 from a saturated vapor at 12 psia to a pressure of 120 psia. The fluid rate is 2.0 lb/min, and the cooling water removes heat at a rate of 135 Btu/min. If the power input is 4.0 hp, determine the exit temperature, in degrees Fahrenheit.

5-35 A compressor steadily inducts 5000 lb/h of refrigerant 12 at 5.0 psia and 80°F through a pipe with an inside diameter of 3 in. It discharges the gas at 100 psia and 300°F through a 1-in-diameter pipe. During the process, 75,000 Btu/h are lost as heat to the surroundings. Determine (*a*) the inlet and exit velocities, in ft/s, and (*b*) the power required, in kilowatts.

Throttling devices

5-36 Steam flows through a heavily insulated throttling valve under the following conditions: (*a*) enters at 600 psia, 600°F, and exhausts at 250 psia, (*b*) enters at 4000 psia, 900°F, and exits at 3000 psia, and (*c*) enters as saturated vapor at 100 psia and exits at 40 psia. Determine the final temperature downstream from the valve at the stated exit pressure, in degrees Fahrenheit.

5-37 Steam is throttled from (*a*) 800 psia to 5 psia and 250°F, (*b*) 80 psia to 10 psia and 200°F, and (*c*) 250 psia to 20 psia and 250°F. Determine (1) the quality of the steam entering the throttling process, and (2) the ratio of exit area to inlet area for the device, if the inlet and exit velocities are essentially the same.

5-38 Refrigerant 12 is throttled from (*a*) a saturated liquid at 90°F to a temperature of −10°F, (*b*) a state of 100°F and 135 psia to a temperature of −40°F, and (*c*) a state of 90°F and 120 psia to a temperature of −10°F. Determine (1) the final pressure, in psia, and (2) the final specific volume, in ft^3/lb.

5-39 Refrigerant 12 is throttled from (*a*) 100°F and 135 psia to a final pressure of 30 psia, and (*b*) a saturated vapor at 300 psia to a final pressure of 30 psia. What is the specific volume in the final state, in ft^3/lb?

Heat exchangers

5-40 Refrigerant 12 with a mass flow rate of 10 lb/min enters a condenser at 200 psia, 180°F, and leaves at a state of 120°F and 190 psia. The coolant in the condenser is (*a*) water which enters at 55°F and leaves at 75°F and 90 psia, and (*b*) air which experiences a 15°F temperature rise at 1.1 atm. Calculate (1) the heat transfer from the refrigerant 12, in Btu/min, and (2) the mass flow rate of coolant required, in lb/min.

5-41 Dry air enters an air conditioning system at 100°F and 14.7 psia at a volume flow rate of 30 ft^3/s. The air is cooled by exchanging heat with a stream of refrigerant 12 which enters the heat exchanger at $-30°F$ and a quality of 10 percent. Assume the heat transfer takes place at constant pressure for both flow streams. The refrigerant 12 leaves as a saturated vapor, and 20 Btu/s of heat is removed from the air. Find (*a*) the flow rate of refrigerant 12 required, in lb/s, and (*b*) the temperature of the air leaving the heat exchanger, in degrees Fahrenheit.

5-42 Water at 80 psia and 250°F enters a heat exchanger at a rate of 500 lb/min and leaves at 75 psia and 100°F. The water is cooled by passing air through the heat exchanger at an inlet volume flow rate of 35,400 ft^3/min. The air initially is at 14.8 psia and 80°F, and the exit pressure is 14.6 psia. Determine (*a*) the exit air temperature in degrees Fahrenheit, (*b*) the inlet area for airflow in ft^2, if the inlet air velocity is 100 ft/s, and (*c*) the inlet water velocity, in ft/s, if the inlet-pipe diameter is 4 in.

5-43 Steam is condensed on the outside of a heat exchanger tube by passing air through the inside of the tube. The air enters at 1.20 atm, 80°F, and 30 ft/s, and exits at 180°F. The steam enters at 20 psia, 250°F, at a mass flow rate of 10 lb/min, and leaves as a saturated liquid. Find (*a*) the mass flow rate of air required, in lb/min, and (*b*) the inlet area of the pipe for airflow, in square feet.

5-44 Steam enters a heat exchanger as a saturated vapor at 14.7 psia and is condensed on the outside of tubes at constant pressure to a saturated liquid. Compressed liquid water passes through the tubes and experiences a temperature rise change from 50 to 80°F.

(*a*) Determine the ratio of mass flow rate of water to steam required.

(*b*) If the required mass flow rate of steam through the heat exchanger is 3.0 lb/s, the cross-sectional area of a tube is 1.0 in^2, and the inlet water velocity is 2.4 ft/s, determine the total number of tubes required to the nearest integer value.

Mixing processes

5-45 Water substance is fed into a mixing chamber from two different sources. One source delivers steam of 90 percent quality at a rate of 2100 lb/h. The second source delivers steam at a temperature of 400°F and a rate of (*a*) 4650 lb/h, and (*b*) 7890 lb/h. If the mixing process is adiabatic and at a constant pressure of 100 psia, determine, in degrees Fahrenheit, the temperature of the mixture at equilibrium downstream from the chamber.

5-46 An open feedwater heater operates at 100 psia. Compressed liquid water at 100°F enters one section. Determine the temperature of steam entering another section if the ratio of the mass flow rate of compressed liquid to superheated vapor is (*a*) 4.47:1, and (*b*) 4.69:1.

5-47 Water is heated in an insulated chamber by mixing it with steam. The water enters at a steady rate of 200 lb/min at 60°F and 50 psia. The steam enters at 600°F and 50 psia, and the mixture leaves the mixing chamber at (*a*) 100°F and 48 psia, and (*b*) 200°F and 48 psia. (1) Determine the amount of steam needed, in lb/min. (2) If the exit area is 4.0 in^2, determine the exit velocity, in ft/s.

5-48 Refrigerant 12 at 40 psia and $-40°F$ enters a mixing chamber at a steady rate of 1 lb/s. Another source of refrigerant 12 at 100 psia and 200°F is first throttled before it enters the mixing chamber. The mixture leaves the chamber as a saturated liquid at 40 psia. If heat transfer to the chamber occurs at a rate of 5 Btu/s, determine the mass flow rate of refrigerant gas entering the chamber, in lb/s.

Pipe flow

5-49 A steam heating system for a building 435 ft high is supplied from boilers 30 ft below ground level. Saturated water vapor leaves the boiler at 30 psia, and it reaches the 435-ft elevation above ground at 20 psia. Heat transfer from the supply pipe to the surroundings is 50 Btu/lb. Neglecting kinetic-energy effects, (*a*) find the quality of the steam at the 435-ft elevation, and (*b*) find the percent reduction in heat transfer necessary to assure that saturated vapor is present at the 435-ft level.

5-50 A long, horizontal pipe with an internal diameter of 2.07 in carries refrigerant 12. The fluid enters at 80 psia with a quality of 20 percent and a velocity of 10 ft/s. The fluid leaves the pipe at 50 psia and 80°F. Determine (*a*) the mass flow rate, in lb/s, (*b*) the exit velocity, in ft/s, and (*c*) the heat-transfer rate, in Btu/s.

5-51 Steam flows through a long, insulated pipe. At one point, where the diameter is 4 in, the pressure and temperature are 300 psia and 600°F, respectively. Further downstream where the diameter is now 3 in, the conditions are 250 psia and 550°F. Determine (*a*) the mass flow rate of the steam, in lb/s, and (*b*) the downsteam velocity, in ft/s.

5-52 In a building, water flows steadily through a series of pipes from 10 ft below ground level to 380 ft above ground level. At the entrance below ground the conditions are 100°F, 100 psia, and 10 ft/s while at the 380-ft level the state is 94°F, 80 psia, and 50 ft/s. Determine the heat transfer, in Btu/lb, if no shaft work is done. Do not neglect any term which can be calculated. Local gravity is 32.0 ft/s^2.

5-53 Water enters a piping system at 80°F and 20 ft/s. Downstream the conditions are 20.0 psia, 80°F, and 30 ft/s, and the elevation is 30.0 ft above the inlet. The local gravity is 31.8 ft/s^2 and the fluid undergoes a heat loss of 0.0050 Btu/lb. If the volume flow rate is 250 ft^3/min, determine (*a*) the inlet pressure, in psia, and (*b*) the inlet-pipe diameter, in inches.

5-54 An oil with a specific gravity of 0.90 enters a piping system at 18.0 psia, 60°F, and 15 ft/s. Downstream the conditions are 60°F and 10 ft/s at an elevation which is 20.0 ft below the inlet. The local gravity is 32.0 ft/s^2, and the fluid gains 0.0040 Btu/lb as heat as it passes through the system at a mass flow rate of 2400 lb/min. Determine (*a*) the outlet pressure, in psia, and (*b*) the outlet-pipe diameter in inches.

CHAPTER

SIX

A MACROSCOPIC VIEWPOINT OF THE SECOND LAW

Historically, the study of the second law of thermodynamics was developed by persons such as Carnot (a French engineer), Clausius, and Kelvin in the middle of the nineteenth century. This development was made purely on a macroscopic viewpoint, and it is referred to as the classical approach to the second law. A study based on this viewpoint does not require the existence of an atomic theory of matter.

6-1 INTRODUCTION

The first law of thermodynamics is a conservation law for energy transformations. Regardless of the types of energy involved in processes—thermal, mechanical, electrical, elastic, magnetic, etc.—the change in the energy of a system is equal to the difference between energy input and energy output. All conservation laws are expressed mathematically this way; that is, as equalities. The first law allows free convertibility from one form of energy to another, as long as the overall quantity is conserved. This law, for example, places no restriction on the conversion of work into heat, or on its counterpart—of heat into work. The unrestricted conversion of work into heat is well known to most persons. Frictional effects are frequently associated with mechanical forms of work which result in a temperature rise of the bodies in contact. Subsequent heat transfer in that area tends, then, to distribute the energy away from the region of contact. The work interaction, aided by heat transfer, eventually appears as a change in the internal energy of the materials involved in the process. It was noted in Chap. 2 that electrical work is commonly transformed, through the presence of electric resistance, into a temperature rise of the resistor. If the resistor is uninsulated, heat transfer will then occur to the surroundings. Again, work in some form has ultimately been transformed into internal energy, with heat acting as the intermediate transfer process. Now let us look at the inverse process, the transformation of heat into work.

In nations with a developed or developing technological society, the ability to produce energy in the form of work becomes of prime importance. Work transformations are necessary to transport people and goods, drive machinery, pump liquids, compress gases, and provide energy input to so many other processes that are taken for granted in highly developed societies. Much of the work output in such societies is available in the basic form of electric energy, which is then converted to rotational mechanical work. Although some of this electric energy (work) is produced by hydroelectric power plants, by far the greatest part of it is obtained from fossil or nuclear fuels. These fuels allow the engineer to produce a relatively high-temperature gas or liquid stream which acts as a thermal (heat) source for the production of work. Hence the study of the conversion of heat to work is extremely important, especially in the light of developing shortages of fossil and nuclear fuels.

In Chap. 1 it was pointed out that thermal efficiency is the parameter that is used by engineers to measure the effectiveness of the heat-work conversion process. In general, a reasonable measure of performance of any device is the ratio of the desired output to the required (costly) input. For a heat-to-work converter, the thermal efficiency η_{th} is defined as

$$\eta_{th} \equiv \frac{|W_{net}|}{Q_{in}} \tag{6-1}$$

In the first chapter we noted that the thermal efficiency of common energy-conversion devices typically range from 10 to 40 percent. The important point to note is that the first law places no restriction on this conversion process. A 100 percent conversion is possible in terms of energy conservation alone. One of the contributions of the second law is that it places a restriction on heat-to-work transformations by cyclic devices. The establishment of this restriction is one of the goals of this chapter.

The brief discussion of energy-conversion devices above leads to several other second-law considerations. One of these is the concept that energy has *quality*, as well as quantity. If work is 100 percent convertible to heat, but the reverse situation is not possible (which we must eventually prove), then work is a more valuable form of energy than heat. Although not as obvious, it can also be shown through second-law arguments that heat has quality in terms of the temperature at which it is discharged from a system. The higher the temperature at which heat transfer occurs, the greater the possible energy transformation into work. Thus thermal energy stored at high temperatures generally is more useful to humanity than that available at lower temperatures. (While there is an immense quantity of energy stored in the oceans, for example, its present availability to us for performing useful tasks is quite low.) This implies, in turn, that thermal energy is *degraded* when it is transferred by means of heat transfer from one temperature to a lower one. Other forms of energy degradation include energy transformations due to frictional effects and electric resistance, among others. Such effects are highly undesirable if the use of energy for practical purposes is to be maximized. The second law provides some means of measuring this energy degradation.

There are other well-known phenomena which fall under second-law analysis. Consider the placement of any hot object in a cold environment, or vice versa. It is our experience that objects of different temperatures, when brought into thermal contact, tend to reach thermal equilibrium, or equality of temperature. The first law requires that the energy given up by one object is acquired by the other (assuming no losses elsewhere). However, consider now two objects initially at the same temperature. The first law places no restriction on the possibility that one will become cooler while the other becomes warmer, as long as the energy given up by the cooler object is the same as that gained by the warmer system. As a second example, consider a paddle-wheel that stirs a fluid within an insulated container. The paddle-wheel might be rotated by some pulley-weight mechanism. As a result, the potential energy of the weight decreases and the internal energy of the fluid increases, in light of the conservation of energy principle. However, at some later time we do not expect to see the energy of the fluid decrease and the weight return to its initial position, spontaneously. These and many other possible examples illustrate the fact that processes, of their own accord, have a preferred direction of change, irrespective of the first law. Also, after sufficient time, such processes reach an equilibrium state.

In summary, there are a number of phenomena which cannot be explained by conservation principles of any type. Hence we seek another law which, through its generality, will provide guidelines to the understanding and analysis of diverse effects. Among other considerations, the second law is extremely helpful to the engineer in the following ways:

1. It provides the means of measuring the quality of energy.
2. It establishes the criteria for the "ideal" performance of engineering devices.
3. It determines the direction of change for processes.
4. It establishes the final equilibrium state for spontaneous processes.

Since the second law is used to examine the direction of change of processes, it is expressed mathematically as an inequality. This inequality means that the second law is a nonconservation law. A new intrinsic property, the entropy, will be deduced from the general second-law statement. It is this fundamental property which is not conserved in real processes.

6-2 EQUILIBRIUM AND THE SECOND LAW

The concept of equilibrium has been extremely important in the preceding discussions of properties of matter and the conservation of energy principle. In the analysis of various processes it was either stated or implied that the initial and final states were equilibrium states. This is necessary since the intrinsic properties of matter are commonly defined only for equilibrium states. Recall also that in the evaluation of various work interactions it is necessary to assume a quasistatic process in order to functionally relate the two parameters, such as P and V in boundary work. That is, a series of equilibrium steps must be assumed in order to integrate such expressions as $P\,dV$. The development of the second law of thermodynamics also relies heavily upon the concept of equilibrium states.

By definition, a system is in equilibrium if a further change of state cannot occur unless subject to interactions with the environment. If a system is isolated from the environment, and a sufficient time passes, it is common experience to note no further change in the state of the system. One can then infer that the system is in equilibrium. No further changes of state will occur unless interactions take place across the boundary. If a system is in equilibrium, a finite change in the state of the system requires a permanent and finite change in the state of the environment.

It is our everyday experience that systems of all types tend to reach a state of equilibrium. If a marble is placed halfway down the side of a bowl and released, it is expected that the marble eventually will come to rest at the bottom without any change in the environment. If two systems of different temperatures are placed in contact through a nonadiabatic wall, but they are insulated from the environment, the two systems will reach a final common temperature. They are said to be in thermal equilibrium. When the two plates of a charged capacitor are shorted out, it is soon found that the plates reach a common electric potential. When a gas expands from one tank into another evacuated tank, both of which are isolated from the surroundings, it is soon found that there is a common temperature, pressure, density, etc., throughout the two tanks. If two miscible liquids, such as alcohol and water, are placed in contact by removing a partition which originally separated them, we expect the liquids to mix until a uniform state is reached. Many other examples could be provided. There is no way we can prove that systems left to themselves eventually reach an apparent state of equilibrium. It is simply a matter of common experience. In many cases equilibrium is reached very rapidly, while in other cases the process may be rate-limited, so that equilibrium is attained only after a considerable length of time. Nevertheless, we expect things to reach equilibrium in the absence of interactions between the system of interest and its environment.

In the absence of any proof of this generally found behavior, we must postulate that such behavior is to be expected. The following postulate is one form of the *second law* of thermodynamics.

> Any system having certain specified constraints and having an upper bound in volume can reach from any initial state a stable equilibrium state with no net effect on the environment.

This statement postulates the existence of stable equilibrium states. By constraints we mean those barriers within or outside of the system which restrict its behavior. These include internal partitions, external conservative force fields, rigid impermeable walls, etc.

Before proceeding, it is important to note that the second law is a directional or limiting law. Processes tend to proceed in one direction and may not be reversed, or processes attain a final equilibrium state which is limited in some sense. For example, a marble placed on the side of a bowl rolls to the bottom. One does not observe it spontaneously rolling back up the side once it has come to rest. If a system of high temperature and a system of low temperature are placed in thermal contact, the former does not get hotter and the latter colder. When a chemical reaction occurs, it is quite possible that an equilibrium state is reached before the reaction is complete. That is, a sizable amount of the reactants may still exist, although no further change of state is observed. The system reaches equilibrium, but the extent of the reaction appears to be limited. When we postulate that any system can reach an equilibrium state, in the absence of interactions, we usually imply that one particular equilibrium state will be reached. Thus the approach to equilibrium is quite directional in its nature.

6-3 HEAT ENGINES

We shall now apply the statement of the second law to a special class of devices known as heat engines. A heat engine is a device which operates continuously, or cyclically, and produces work while exchanging heat across its boundaries. The restriction to continuous or cyclic operation implies that the matter within the device is returned to its initial state at regular intervals.

As an example of a cyclic heat engine, consider the piston-cylinder assembly shown in Fig. 6-1*a*. A piston and weight rest on some supporting pegs at position

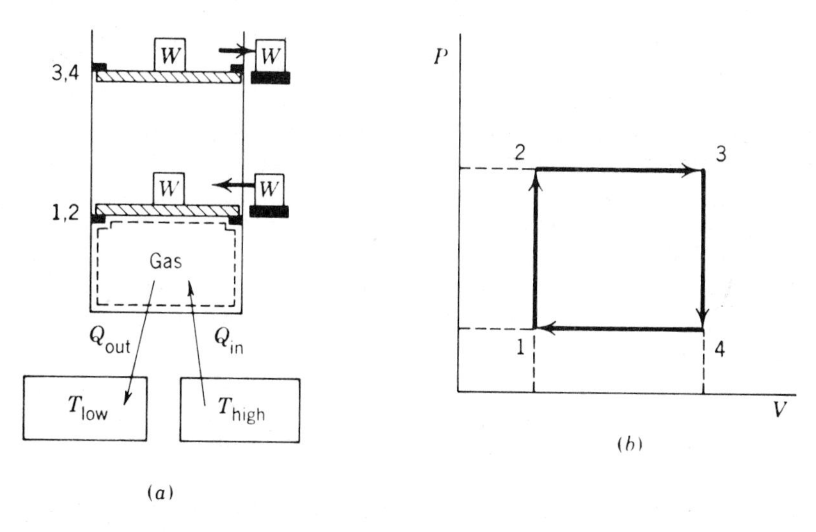

Figure 6-1 A simple, cyclic heat engine.

1, and a gas is contained in the volume below the piston. Initially the pressure of the gas is less than the equivalent pressure exerted by the piston-weight-atmosphere combination above the gas. Heat is then added from a high-temperature source until the gas pressure just balances this opposing pressure. The process to this new state 2 from the original state is shown on a *PV* diagram in Fig. 6-1*b*. If additional heat transfer occurs from the high-temperature source, the gas will expand at constant pressure until the piston hits an upper set of pegs. The heat addition is halted at this point, and the weight is moved horizontally off the piston. The state of the gas is now designated by point 3 on the *PV* diagram. Heat is now rejected from the gas to a low-temperature sink. Until the gas pressure drops so that it just balances the weight of the piston and the atmospheric pressure, the volume does not change. This is state 4. Additional heat removal will lower the piston at constant pressure until the piston reaches its initial position on the lower set of pegs. At this point the cycle has been completed. If another weight were added on at the lower level, the cycle could be repeated. The net work produced by the cycle is measured by the enclosed area on the *PV* diagram. In addition, heat was exchanged between the system and two bodies at different temperatures.

An example of a continuous heat engine is a thermoelectric device such as that shown in Fig. 6-2. Part *a* of the figure illustrates two dissimilar electric conductors with common junctions maintained at two different temperatures. An electric potential is developed as a result of the temperature difference, and work is produced as a result of the current passing through the motor. If the junction temperatures and the load are maintained constant, the device will produce work continuously while heat is exchanged at the two junctions. Figure 6-2*b* illustrates a modern version of a thermoelectric device. The dissimilar electric conductors are semiconductor elements which are placed in series, with alternating junctions

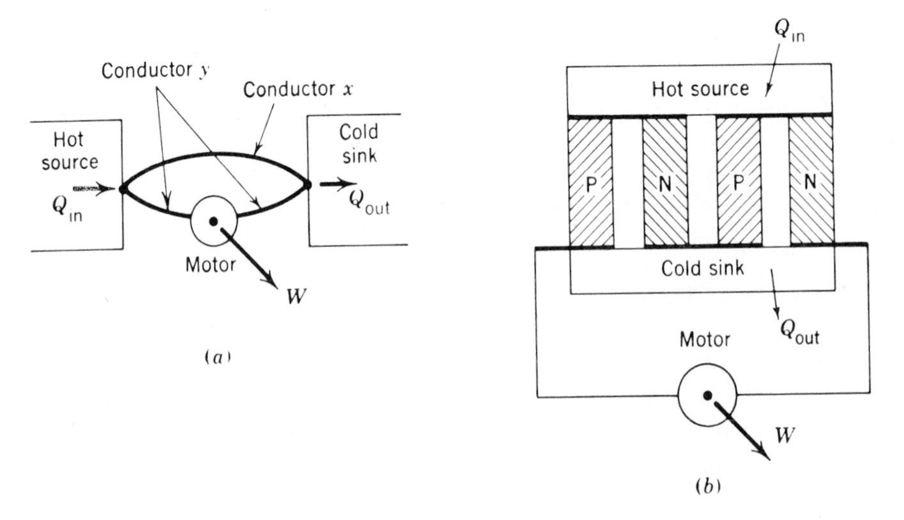

Figure 6-2 A thermoelectric device for power production.

in contact with either a high- or low-temperatue body. Hence the words "cyclic" and "continuous" are descriptive of the nature of processes which are used to convert heat into work.

The two heat-engine devices mentioned above can be represented in a general way by the block diagram of Fig. 6-3. For the heat engine as a closed system, the conservation of energy principle is

$$\sum Q + \sum W = \Delta U$$

When a cyclic heat engine undergoes an integral number of cycles, or for a continuously operating heat engine, the value of ΔU is zero. Thus for either type of heat engine with one heat supply Q_H from a high-temperature source and one heat rejection Q_L to a low-temperature sink we see that

$$Q_H + Q_L + W_{\text{net}} = 0$$

The heat added and removed from the heat engine have different signs by convention. It is convenient to use absolute-value signs in the above energy equation, so that

$$|Q_H| - |Q_L| = -W_{\text{net, eng}}$$

where the net work output by the engine is negative by convention.

The object of a heat engine is to produce work from the energy added as heat to the system. Therefore a reasonable measure of its performance is the thermal efficiency η_{th} of the engine. That is, as previously defined,

$$\eta_{\text{th}} = \frac{|W_{\text{net}}|}{Q_H} \tag{6-1}$$

If we supply 100 units of energy to a heat engine and find that 70 units are rejected, then the net work output is 30 units and the thermal efficiency is 0.30 (or 30 percent). Note that the thermal efficiency will be unity (100 percent) when Q_L is zero. That is, a cyclic heat engine with an efficiency of 100 percent requires no heat rejection to the environment. In the next section we seek to establish

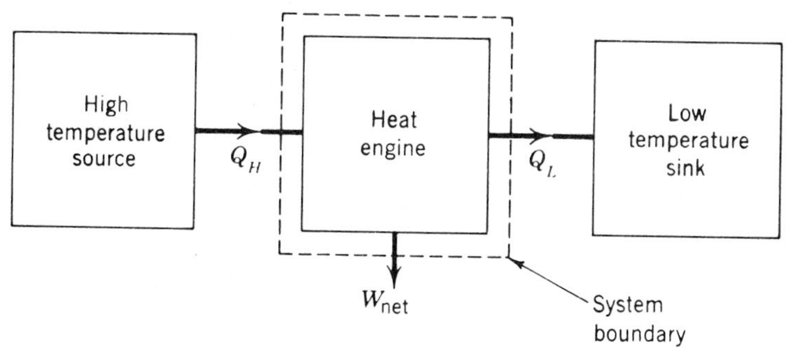

Figure 6-3 Simple schematic of a heat engine.

whether there is theoretically any limit to the value of the thermal efficiency of *any* heat engine.

6-4 PERPETUAL-MOTION MACHINES

It is a matter of experience that the thermal efficiency of all practical heat engines is less than 100 percent. Thus some portion of the heat supplied from a high-temperature source is always rejected to a sink. A basic question with regard to heat engines is whether it is *theoretically* possible to have a device that is 100 percent efficient, even if actual losses due to friction and other factors are neglected. The statement of the second law presented in Sec. 6-2 is sufficient to answer this question. We shall propose the following theorem.

It is not possible to construct a heat engine which produces no other effect than the exchange of heat from a single source initially in an equilibrium state and the production of work.

Hence we propose to prove that no heat engine can be 100 percent efficient. This corollary to the second law is known as the *Kelvin-Planck statement* of the second law.

The proof of the theorem is more easily seen with the aid of the diagram in Fig. 6-4. A hypothetical heat engine of 100 percent thermal efficiency undergoes an integral number of cycles. During the process the engine receives a quantity of heat Q from a source at temperature T, and produces an equivalent amount of work W. The heat engine undergoes a cycle, so that its state is not changed during the process. The source of energy starts at a stable equilibrium state and proceeds to a new state, while the sole effect in the environment is the production of work. The work produced may now be used to further alter the state of the heat source. For example, the heat source might be lifted to a new position or given a finite velocity. As a result of this series of effects, the heat engine is unaltered, the rest of the environment is unchanged, and the heat source has undergone a finite change of state from its original equilibrium state. But, by definition, a system initially in equilibrium cannot undergo a finite change of state without a finite change in its environment. Therefore some part of the proposed process is impossible. The only assumption made in the analysis was

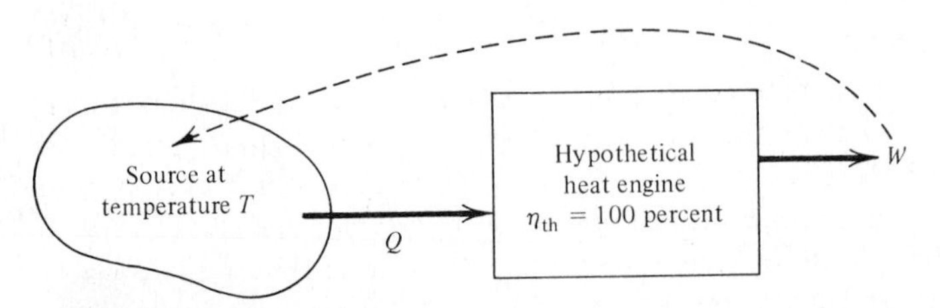

Figure 6-4 Schematic for the proof of the impossibility of a 100 percent efficient heat engine.

that a hypothetical heat engine of 100 percent efficiency could exist. Such an assumption leads to an invalid conclusion. Hence it is not possible to construct a heat engine of 100 percent efficiency if we accept the existence of stable equilibrium states. Therefore all heat engines must reject some portion of the heat supplied to another system (usually called a sink).

Most large fossil- and nuclear-fueled power plants reject heat directly to the environment. Usually the heat is rejected to cooling water which is taken from a river or lake. Since the thermal efficiency of such plants is 40 percent or less, at least 60 percent of the energy released from the fuel must appear in the environment. That portion which is released to the cooling water is what constitutes *thermal pollution* from such plants. For fossil-fueled plants a portion of the 60 percent which is rejected appears as energy in the hot gas stream emitted from the stack of the plant. Nevertheless, approximately one-half of the rejected energy heats the water stream. In a nuclear plant all the rejected energy must appear in the coolant stream, since no combustion gases are associated with such a plant.

By definition, a heat engine which exchanges heat with a single system in a stable state and produces work is called a perpetual-motion machine of the second kind, or a PMM2. It is one of the second kind since it violates the second law of thermodynamics. A device of this type would be extremely useful if it could be constructed, since it could refrigerate a region while simultaneously producing work. The concept of a PMM2 is quite useful in proving subsequent theorems in this chapter. A perpetual-motion machine of the first kind (a PMM1) is a device which creates energy and thus violates the first law of thermodynamics. Any process which creates a PMM1 or a PMM2 is impossible.

6-5 REVERSIBLE AND IRREVERSIBLE PROCESSES

In the preceding section we found that PMM2-type heat engines are impossible. Therefore all heat engines, theoretical and real, must reject a portion of their heat input to a sink, which usually is part of the environment. However, no limiting maximum value has been set on thermal efficiency. Besides showing that a PMM2 is impossible, thermodynamics also provides the engineer with a numerical upper limit on theoretical thermal efficiency. In order to establish the general relationship, however, it is necessary to describe first what is meant by an "ideal" heat engine which would have maximum theoretical efficiency. An ideal heat engine is one which is reversible. Hence the concept of a reversible process must be described in some detail.

In general, a process commencing from an initial equilibrium state is called *reversible* if at any time during the process both the system and environment can be returned to their initial states. The concept of reversibility by its definition requires restorability. But this requirement of restorability is quite strong, applying to *both* the system and its surroundings. Normally, interest is centered during a given process solely on the system. For example, a quasistatic process requires equilibrium conditions within and at the boundaries of a system, but it places no restriction on the effects which occur in the surroundings. However, a reversible process requires something of the environment. It is the nature of a reversible process that all heat and work interactions which occurred across the boundaries during the original (forward) process are equal in magnitude but reversed in direction during the reverse process. Thus no net history is left in the surroundings when the system regains its initial state.

In view of our discussions of work interactions in Chap. 2, any quasistatic form of a work interaction (such as boundary work, electric or magnetic field work, etc.) should be carried out in a series of equilibrium steps. Only in this circumstance will the work output by the system equal the work input for the return path. Consider, as an example, a piston-cylinder assembly which contains a gas. The piston will be taken as *frictionless*, and both the cylinder and the piston are perfect insulators. If the force exerted externally on the gas is decreased by a very small amount, work is performed by the gas on the surroundings. Hypothetically, this amount of work might be stored in the form of the rotational kinetic energy of a flywheel. If the expansion were carried out in a series of small decreases in the external resisting force, the pressure, temperature, and other intrinsic properties of the gas would change uniformly throughout the gas. Hence the work output would be equal to the integral of $P\,dV$, and an equivalent amount of energy would be added to the flywheel. Now the process could be reversed by removing energy from a flywheel through some suitable mechanism. If the gas is now compressed by a series of small increases in the pressure, it can be returned to its initial state. It would be found that the pressure of the gas was exactly the same during the compression as during the expansion for a given position of the piston. Hence the work of compression will equal the work of expansion if the initial and final states of the gas are the same. In addition, all the energy stored in the surroundings (the flywheel) during the expansion will be exactly expended in returning the system to its initial state. Since the system and surroundings are both back in their initial states at the end, the process described above is a reversible one.

The concept of reversibility can also be applied to open systems. As a second example, consider the steady flow of a fluid through an adiabatic frictionless nozzle. The application of the conservation of energy principle shows that the enthalpy of the fluid decreases as the kinetic energy increases. If the nozzle is now followed by an adiabatic, frictionless diffuser, the fluid can be returned to a state identical to that at the entrance of the nozzle. The diffuser increases the enthalpy of the fluid at the expense of a decrease in its kinetic energy. Since the surroundings were not involved, the process described is a reversible one. There are other common examples of reversible processes, and included among these are the ideal pendulum and the pure capacitive-inductive circuit. (See Fig. 6-5.) Both the frictionless pendulum and the ideal capacitive-inductive circuit are forms of an oscillator which undergoes cyclic variations. Another example of a reversible device is an oscillator which interchanges the elastic energy of an ideal spring with the kinetic energy or potential energy of a mass. Other combinations of gravitational potential energy, translational kinetic energy, rotational kinetic

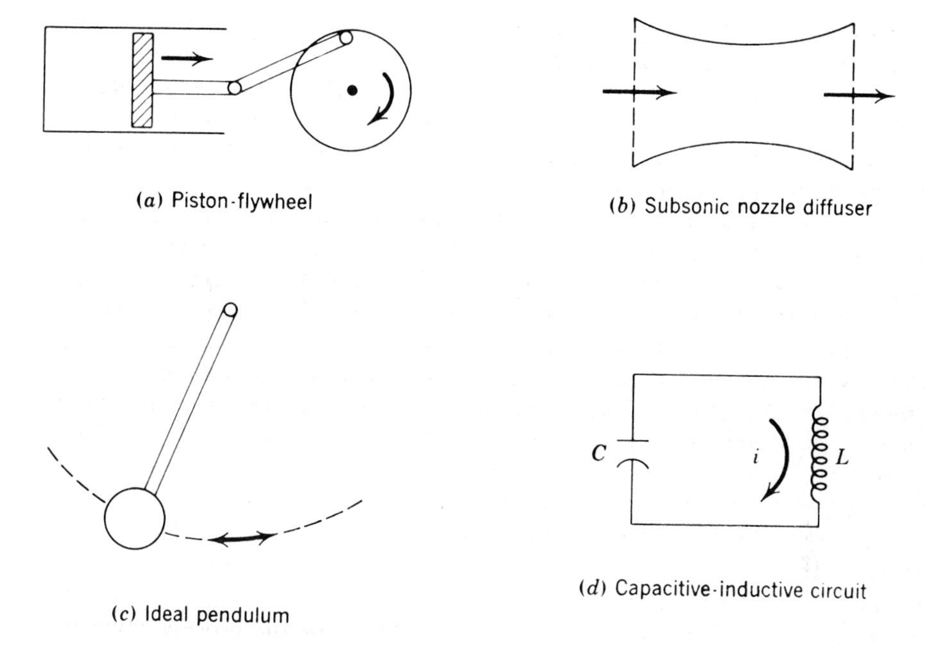

Figure 6-5 Some reversible processes.

energy, elastic energy, magnetic- and electric-field energies, and the frictionless compression and expansion of a gas in a piston cylinder, which would constitute reversible devices, could be considered. Consequently, many of the basic concepts of introductory physics may be employed to devise reversible processes. Among the processes that are frequently idealized as reversible processes are

1. Restrained expansion or compression
2. Frictionless motion
3. Elastic stretching of a solid
4. Electric circuits of zero resistance
5. Polarization and magnetization effects
6. Restrained discharge of a battery

The reversible process is an idealization or fiction. It is a concept which can be approximated very closely at times by actual devices, but never matched. The reason is that the word "ideal" appeared throughout the foregoing discussion. In the analysis of these ideal processes, we required the absence of friction at the bearing surfaces of the piston cylinder, the flywheel, and the pendulum. The pendulum, in addition, could not be acted upon by air resistance. The capacitive-inductive circuit could not contain a resistive element when the current passed through the circuit. Any springs used had to be elastic ones; i.e., they had to follow Hooke's law. In many actual cases the effects of friction, electric resistance, and inelasticity can be substantially reduced, but their complete elimination is usually not found. These effects are frequently called *dissipative* effects, since in all cases a portion of the energy originally in the system is converted or dissipated into a less useful form. Only in the absence of dissipative effects can certain forms of energy be converted into other forms without any apparent loss in the capabilities of the system. In such cases the system is truly reversible.

One other criterion, in addition to the absence of dissipative effects, must be met if an isolated system is to be considered reversible. In the discussion of the piston-cylinder and flywheel combination, it was necessary to carry out the expansion and compression of the gas in a series of equilibrium steps. Otherwise the work output and work input would not be the same, and the process would be irreversible. The requirement that a process be quasistatic, in order to be reversible, is a general one. That is, only infinitesimal unbalanced forces are present during a reversible process.

When a process is such that the system and the surroundings cannot be returned to their initial states, the process is said to be irreversible. We have now seen that irreversibilities arise from two sources:

1. Inherent dissipative effects that stem from the nature of the substance itself. These effects include friction of any type, electrical resistance, magnetic hysteresis, and inelasticity.
2. Absence of mechanical, thermal, or chemical equilibrium during a process, i.e., a nonquasistatic process.

The presence of either class of effects is sufficient to make a process nonideal, or irreversible. Since all actual processes include such effects, the reversible process is a limiting process toward which all actual processes may approach asymptotically in performance, but never succeed in realization. Nevertheless, we shall find reversible processes to be appropriate starting places on which to base engineering design calculations. The usefulness of the concept of the reversible process will become apparent only by repeated application of the concept to different situations in the light of the second law of thermodynamics.

From a practical viewpoint, whether or not a given process is reversible is probably best recognized by ascertaining whether irreversibilities occur during the change of state. As a broad generalization, all irreversibilities have this in common: they leave a history, or an imprint, of their presence. Any system which is returned to its initial state after experiencing an irreversible process will leave a history in the surroundings due to the irreversibilities. A partial list of effects which constitute irreversibilities is presented below. Most of these effects fall into the category of common experience. Irreversibilities include:

1. Electric resistance
2. Inelastic deformation
3. Shock waves
4. Hysteresis effects
5. Viscous flow of a fluid

6. Internal damping of a vibrating system
7. Solid-solid friction
8. Unrestrained expansion of a fluid
9. Fluid flow through valves and porous plugs (throttling)
10. Spontaneous chemical reactions
11. Mixing of dissimilar gases or liquids
12. Osmosis
13. Dissolution of one phase into another phase
14. Mixing of the same two fluids initially at different pressures and temperatures

The foregoing list was made fairly long—not for the purpose of being memorized, although some degree of familiarity is important, but to illustrate two points. First, the common processes that are part of one's experience are all irreversible. Second, these processes cover a diversity of physical and chemical effects.

A gradient within a system, or between two systems, is a source of an irreversibility. Heat transfer is irreversible if its occurrence is due to a finite temperature difference between the system and its surroundings. The transfer of heat is reversible if the temperature difference is made infinitesimally small. In addition, reversibility requires that the quantity of heat transferred on the reversed path be equal in magnitude, but opposite in direction, to that transferred during the original process. This "equal in magnitude, but opposite in sign" concept also applies to any quasistatic work interaction carried out during a reversible process.

We have pointed out that there are either inherent dissipative effects or nonequilibrium effects which make a process irreversible. Absence of these effects in the system and its environment lead to the concept of a *totally reversible* process. In the thermodynamic analysis of engineering systems it frequently is advantageous to consider processes for which irreversibilities are absent within the system of interest, but not necessarily absent from the environment. Such processes are called *internally reversible*. For simple, compressible systems, the only work interaction permitted is quasistatic boundary work if the process is to be internally reversible. Work effects associated with paddle-wheels or electric resistors within the system are not allowed, for obvious reasons. As a more general statement, only quasistatic work interactions, and not nonquasistatic ones, can occur during internally reversible processes. Finally, an *externally reversible* process is one for which irreversibilities may be present within the boundaries of the system of interest, but the surroundings which interact with the system must undergo only reversible changes.

6-6 HEAT AND WORK RESERVOIRS

For the development of further consequences of the second law, it is convenient to define a special heat source or sink known as a heat reservoir. By definition, a heat reservoir is a closed system with the following three characteristics:

1. The only interactions permitted across its boundaries are heat interactions, to the exclusion of work effects.
2. The system equilibrates rapidly, so that it remains essentially in equilibrium during the heat-transfer process.
3. Its temperature remains constant during a finite addition or removal of energy in the form of heat.

Thus energy added or removed from it merely alters its internal energy. The name "heat" reservoir is a misnomer, since it implies that heat is stored in the system. The name really means that this particular system simply acts as a source or sink for the transfer of heat to or from various other systems. Some authors have called this type of reservoir a "thermal energy" reservoir (TER), which might be more appropriate.

A heat reservoir has two important qualities. First, there is no restriction, theoretically, on the phase or chemical composition of the reservoir. Consequently, the only significant property of a heat reservoir is its temperature, which must remain constant. Second, the system changes quasistatically and there are no dissipative effects within the reservoir. Therefore any change in state of a heat reservoir occurs in an *internally reversible* manner.

In practice, a heat reservoir can be achieved in several ways. One method is to have a second system much larger than the system with which it interacts. Any energy in the form of heat added to or removed from the second system will be a small fraction of its total energy. Hence its temperature will tend to remain constant. Large bodies of water such as lakes and oceans, and the atmosphere around the earth, behave essentially as heat reservoirs. Another example of a practical heat reservoir is a two-phase system. Although the ratio of the masses of the two phases will change during heat addition or removal, the temperature will remain fixed as long as both phases coexist.

Another concept of usefulness in thermodynamics is that of a work reservoir, which is called a "mechanical energy" reservoir (MER) by some authors. Similar to a heat reservoir, it is a system which acts as a source or sink for the exchange of work with another system of interest. No heat interactions are associated with it. All motions within a work reservoir are assumed to be frictionless, i.e., nondissipative, and the forces acting across its boundaries are independent of direction or rate of change of the energy-transfer process. Thus a change of state of a work reservoir also occurs in an *internally reversible* manner. Any idealized mechanical, electric, magnetic, or elastic system could operate as a work reservoir. Examples are pulley-weight systems, rotating flywheels, and elastic springs, among others.

6-7 THERMAL EFFICIENCY OF REVERSIBLE AND IRREVERSIBLE ENGINES

Having established the fact that the thermal efficiency of any heat engine is always less than unity, we now turn our attention to a special class of heat engines. With the introduction of the concepts of reversibility and of heat reservoirs, it is possible to classify heat engines either as totally reversible or irreversible. As the name implies, a totally reversible heat engine is one which is free of dissipative or nonequilibrium effects during its operation. These effects must be absent not only within the engine, but also with respect to changes in the environment connected with the operation of the engine. Since a heat engine depends upon heat transfer with at least one source and one sink, it is necessary that all heat interactions be reversible. We consider the transfer of heat to be reversible if the temperature difference between two systems is made infinitesimally small. For a totally reversible heat engine, a temperature difference dT must exist between the working fluid of the engine and the source or sink with which it exchanges energy in the form of heat. If irreversibilities of any kind exist within the engine or result from interactions between the heat engine and its environment, the engine is classified as irreversible.

In this section we shall prove the following theorems regarding the thermal efficiencies of reversible and irreversible heat engines.

1. The efficiency of an irreversible heat engine is always less than the efficiency of a totally reversible heat engine when both are operating between the same heat reservoirs.
2. The efficiencies of two totally reversible heat engines operating between the same heat reservoirs are equal.

These two statements compose what is frequently called Carnot's principle.

The proof of statement 1 is based on the apparatus shown in Fig. 6-6. A totally reversible engine R and an irreversible engine I are operating between the same two heat reservoirs. Both engines receive the same quantity of heat Q_1 during an integral number of cycles for each, and the totally reversible engine

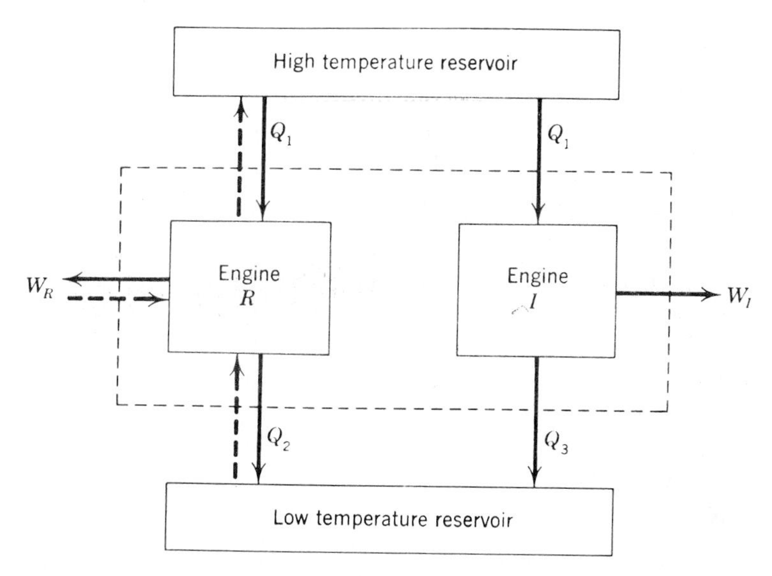

Figure 6-6 Sketch of apparatus for proof that $\eta_I < \eta_R$.

produces work W_R while the irreversible engine produces work W_I. We shall assume that

$$W_I \geq W_R$$

in violation of statement 1. Since engine R is totally reversible, its direction of operation may be reversed. In the reverse direction the magnitudes of Q_1, Q_2, and W_R remain the same, but their signs are changed (as shown by the dashed arrows in Fig. 6-6). The net result of this new operation is that the high-temperature heat reservoir receives no net energy. In addition, the system comprised of engines R and I (shown by the dashed box in the figure) now exchanges a net amount of heat with a single reservoir. If

$$W_I - W_R > 0$$

then a net amount of work will be produced by the composite system. (Recall that W_R is now into the system when it operates in the reverse direction.) This situation of net work production and net heat transfer with a single reservoir is a PMM2, and is not possible. Therefore W_I cannot be greater than W_R.

If, on the other hand,

$$W_I - W_R = 0$$

then from the conservation of energy principle Q_2 equals Q_3 in magnitude, but the quantities are opposite in direction. Hence the low-temperature reservoir in this situation receives no net heat. But this means the process is reversible and engine I is operating just like engine R. Therefore engine I is reversible when $W_I = W_R$. This violates our original assumption, and so W_I cannot equal W_R. Since neither the "greater than" sign nor the "equals" sign is permissible, we have proved statement 1. An irreversible engine is always less efficient than a totally reversible engine when both operate between the same two reservoirs.

The proof of the second statement of Carnot's principle follows immediately from the preceding proof. If the efficiencies of two totally reversible engines were not the same when operating between the same reservoirs, then a PMM2 could be set up by operating one in reverse. This would be true no matter which engine were chosen to be the more efficient. Therefore the only way to avoid setting up a PMM2 is to have the same efficiency for both heat engines.

6-8 THE THERMODYNAMIC TEMPERATURE SCALE

One of the most noteworthy results of Carnot's principle is the statement that all totally reversible heat engines have the same thermal efficiency when operating between the same two heat reservoirs. In the proof of this statement, it is not necessary to specify the nature of the heat engine itself. That is, the construction and manner of operation of the engine does not affect the thermal efficiency if the engine is totally reversible. In addition, the thermal efficiency does not depend upon the nature of the fluid or matter within the engine. The only other possible influence on the thermal efficiency is the nature of the heat reservoirs. But the only significant property of a heat reservoir is its temperature. Therefore the only parameters that fix the thermal efficiency are the temperatures of the two reservoirs.

The very fact that the efficiency is solely dependent on the two reservoir temperatures provides us with a means of establishing an absolute thermodynamic temperature scale. A brief discussion of temperature scales was given in Chap. 1. Various thermometric properties of substances were suggested for use in measuring the temperature. In each case the nature of the substance influences the value of the temperature determined in a measurement. All ideal gases in a constant-volume device give the same reading in the limit as the pressure is made very small within the thermometer. Nevertheless, the reading still depends upon a special class of substances, the ideal gas. The concept of a reversible engine provides us with the chance to establish a thermodynamic temperature scale which is independent of the properties of the substance used for the measurement.

Now consider a totally reversible heat engine operating between two reservoirs at T_H and T_L. According to the Carnot principle we may write that

$$\eta_{th} = f(T_H, T_L)$$

In addition, the definition of the thermal efficiency leads to the relation

$$\eta_{th} \equiv \frac{W}{Q_{in}} = \frac{|Q_H| - |Q_L|}{|Q_H|} = 1 - \frac{|Q_L|}{|Q_H|}$$

Rev

If we now arbitrarily select some $f(T_H, T_L)$, then measurements of Q_H and Q_L for *any* totally reversible engine operating between those reservoirs will provide a relation between T_H and T_L, since

$$f(T_H, T_L) = 1 - \frac{|Q_L|}{|Q_H|}$$

It has been found convenient and useful to select, arbitrarily, the function of T_H and T_L so that

$$\frac{T_H}{T_L} = \frac{|Q_H|}{|Q_L|} \qquad (6\text{-}2)$$

Rev

This relationship helps define an absolute thermodynamic Kelvin temperature scale. Since it only defines a ratio of temperatures, we assign the thermodynamic temperature T^* of a heat reservoir at the triple state of water to be 273.16°K. If we now operate a totally reversibly heat engine between a reservoir at the triple state of water and another reservoir at an unknown temperature T, then this latter temperature is related to T^* by

$$T = 273.16 \frac{Q}{Q^*} \qquad (6\text{-}3)$$

where Q is the heat received by the engine from the reservoir at temperature T, and Q^* is the heat rejected to the sink at 273.16°K.

Although a totally reversible heat engine is an idealization, available extrapolation techniques enable us to achieve fairly good practical results for a temperature scale based upon this concept. In addition, it can be shown that the temperature scale based on the constant-volume, ideal-gas thermometer is compatible with the absolute thermodynamic temperature scale based on a reversible heat engine. That is, they measure equivalent temperatures. In this latter respect, note the similarity between Eqs. (6-3) and (1-14).

6-9 THE CARNOT EFFICIENCY

On the basis of Eq. (6-1) and the first law for a closed system we may write that

$$\eta_{th} = \frac{|W_{net}|}{Q_{in}} = \frac{|Q_{in}| - |Q_{out}|}{|Q_{in}|} = 1 - \frac{|Q_{out}|}{|Q_{in}|}$$

Combination of Eq. (6-2) with this relation yields

$$\eta_{th,\,Carnot} = 1 - \frac{T_L}{T_H} \tag{6-4}$$

The efficiency given by this equation is called the *Carnot* efficiency. According to the Carnot principle, this is the *maximum* efficiency that any heat engine could have when operating between heat reservoirs with temperature T_H and T_L. This is an upper limit, since actual engines are always less efficient than reversible engines. To improve the efficiency of a reversible heat engine it is necessary to raise T_H, lower T_L, or both.

This equation is extremely important with respect to the efficiency data given earlier in Table 1-2. Common power-producing devices have efficiencies ranging from 10 to 40 percent. These values are low relative to 100 percent. However, Eq. (6-4) dictates that efficiencies should not be compared to 100 percent, but to some lower theoretical value. For example, consider a steam power plant with a real efficiency of 40 percent. We shall assume that T_H is 800°K (980°F) and T_L is 300°K (80°F). For a heat engine receiving and rejecting heat at these temperatures, the theoretical thermal efficiency according to Eq. (6-4) would be 62.5 percent. Against this value, an actual value of 40 percent does not appear quite so low.

The Carnot efficiency for heat engines is the first relationship developed from the second law of thermodynamics for measuring the quality of energy. Although the complete conversion of work into heat is possible, the percent conversion of heat into work by a cyclic device will be relatively low. The theoretical highest value is considerably below 100 percent for temperatures commonly employed. Irreversibilities of actual engines lower the thermal efficiency even further. To achieve high thermal efficiencies the second law predicts that T_H should be as high as possible, and irreversibilities should be held to their lowest practical values.

Example 6-1M A totally reversible heat engine operating between fixed temperatures of 900 and 300°K produces 1000 kJ of net work output. Determine the heat input and heat output of the heat engine which operates as a closed system, in kilojoules.

SOLUTION The first and second laws for a heat engine operating as a closed system are

$$Q_{in} - Q_{out} = W_{net,\, out} \quad \text{and} \quad \frac{Q_{in}}{Q_{out}} = \frac{T_{in}}{T_{out}}$$

or

$$Q_H - Q_L = W_{net} \quad \text{and} \quad \frac{Q_H}{Q_L} = \frac{T_H}{T_L}$$

Inserting the known data, we find that

$$Q_H - Q_L = 1000 \text{ kJ} \quad \text{and} \quad \frac{Q_H}{Q_L} = \frac{900}{300}$$

The solution of these two equations yields

$$Q_H = 1500 \text{ kJ} \quad \text{and} \quad Q_L = 500 \text{ kJ}$$

Example 6-1 A totally reversible heat engine operating between fixed temperatures of 1040 and 40°F produces 1000 Btu of net work output. Determine the heat input and heat output of the heat engine which operates as a closed system, in Btu.

SOLUTION The first and second law statements for a heat engine operating as a closed system are

$$Q_{in} - Q_{out} = W_{net,\, out} \quad \text{and} \quad \frac{Q_{in}}{Q_{out}} = \frac{T_{in}}{T_{out}}$$

or

$$Q_H - Q_L = W_{net} \quad \text{and} \quad \frac{Q_H}{Q_L} = \frac{T_H}{T_L}$$

Inserting the known data, we find that

$$Q_H - Q_L = 1000 \quad \text{and} \quad \frac{Q_H}{Q_L} = \frac{1040 + 460}{40 + 460}$$

Note that the temperatures have been converted to the Rankine absolute scale. The solution to these two equations yields

$$Q_H = 1500 \text{ Btu} \quad \text{and} \quad Q_L = 500 \text{ Btu.}$$

6-10 THE CLAUSIUS INEQUALITY

As noted earlier, the second law of thermodynamics is concerned with the directional quality of nature. This was expressed initially by the statement that the stable state of a system can be reached from other states, but a system in a stable state will not spontaneously change to other states. This led to a discussion of heat engines. We have shown that irreversible heat engines are always less efficient than totally reversible heat engines when operating between the same heat reservoirs. Thus we find that the second law leads to expressions involving inequalities. Conservation laws, such as the first law, involve equalities. Directional laws involve inequalities.

Another important inequality in thermodynamics which stems from the second law is known as the Clausius inequality. It states that

The cyclic integral of the quantity $\delta Q/T$ for a closed system is always equal to or less than zero.

That is, when a closed system undergoes a cyclic process, the sum of all the $\delta Q/T$ terms for each increment of the process will add up to zero, or will be less than zero. The proof of this theorem is based on the apparatus shown in Fig. 6-7. A closed system at temperature T has a quantity of heat δQ supplied to it while producing a quantity of work δW. The temperature T is allowed to vary as the state of the system changes. The source of the thermal energy is a heat reservoir at temperature T_R, and $T_R \neq T$. In order to eliminate any source of irreversibility outside of the system, heat is transferred from the reservoir to the system through an intermediary, a totally reversible heat engine R. This engine receives heat δQ_R from the reservoir and supplies heat δQ to the closed system. As a consequence, the engine must deliver a quantity of work δW_E to the environment. During the change of state, the closed system delivers a quantity of work δW. The direction of the energy transfers for the various parts could be that shown or could be in the opposite direction. It is immaterial to the proof. We shall assume that all work effects are delivered to or from work reservoirs, which are reversible in concept. Consequently the only place irreversibilities can occur is within the closed system itself. Thus, changes in the closed system may be internally reversible or internally irreversible.

The proof of the Clausius inequality is based on two first-law and two second-law statements about the process described above and illustrated in Fig. 6-7. The inequality is introduced by first examining the composite of the closed system and the heat engine, shown by the dashed line in the figure. Consider the situation where the closed system undergoes a cycle, while the engine undergoes an integral number of cycles. The composite system then acts like a cyclic device which exchanges heat with a single reservoir. The composite device cannot produce net work output; otherwise, a PMM2 would result. However, there is no restriction against the situation where the work is equal to zero, or net work input is required. Thus, we can restate the second-law requirement on cyclic

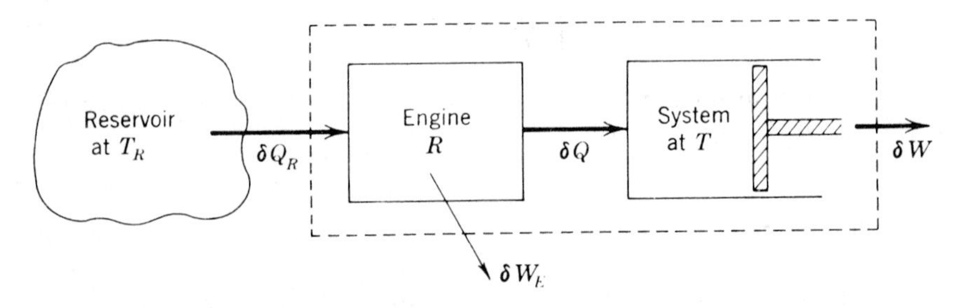

Figure 6-7 Schematic for the proof of the Clausius inequality.

devices that exchange heat with a single reservoir in the following form, using a cyclic integral notation:

$$\oint \delta W_T \geq 0$$

In terms of the composite system described in Fig. 6-7, this notation becomes

$$\oint \delta W_E + \delta W \geq 0 \tag{a}$$

The remaining task is to replace the δW terms by heat and temperature quantities.

First, an energy balance on the heat engine is

$$\delta W_E = -\delta Q_R - \delta Q \tag{b}$$

where δQ is measured relative to the engine. An energy balance on the closed system is

$$\delta W = dU - \delta Q \tag{c}$$

where δQ is measured relative to the closed system. If Eqs. (*b*) and (*c*) are to be compatible, δQ in Eq. (*c*) must be measured, signwise, relative to the heat engine. This is accomplished by writing Eq. (*c*) as

$$\delta W = dU + \delta Q \tag{d}$$

Substitution of Eqs. (*b*) and (*d*) into Eq. (*a*) yields, after canceling common terms,

$$\oint (-\delta Q_R + dU) \geq 0 \tag{e}$$

The cyclic integral of dU is zero, so that term drops out of the above equation. Finally, for a totally reversible heat engine the second law dictates that $\delta Q_R/\delta Q = T_R/T$. Using this expression to replace δQ_R in Eq. (*e*), and noting that T_R is a constant, we find that

$$T_R \oint \frac{\delta Q}{T} \leq 0$$

T_R is always positive. Consequently,

$$\oint \frac{\delta Q}{T} \leq 0 \tag{6-5}$$

This is the Clausius inequality. It applies equally well to a truly cyclic device or a continuous, steady-state device.

6-11 ENTROPY

In the preceding proof of the Clausius inequality, all changes external to the system were made reversible. However, changes within the system could be reversible or irreversible, since the type of process associated with the system was not specified. The possibility of either internally reversible or irreversible processes accounts for the presence of the "equal to" and "less than" signs found in Eq. (6-5). First of all, let us consider that the changes within the system are internally reversible during a specified process. Since the external effects are already specified as reversible, the process is totally reversible. In this case the overall process can be reversed in direction. The magnitudes of heat and work terms will remain the same, with only a change in sign. In particular, for the original direction we may write that

$$\oint \frac{\delta Q}{T} \leq 0$$

If the cyclic path is now reversed, we must write that

$$\oint \frac{\delta Q}{T} \geq 0$$

since the $\delta Q/T$ terms are equal in magnitude but opposite in sign for the reversed direction. However, the greater than sign in this latter inequality violates the Clausius inequality, which must be valid for any cyclic process. Hence the inequality sign cannot apply to a closed system which undergoes an internally reversible, cyclic process. But use of the equality sign is permitted for each direction of the cycle, without any second-law violation. Consequently, for all internally reversible processes involving closed systems

$$\oint \frac{\delta Q}{T} = 0 \qquad \text{internally reversible} \tag{6-6}$$

The presence of internal irreversibilities leads to the inequality

$$\oint \frac{\delta Q}{T} < 0 \qquad \text{internally irreversible} \tag{6-7}$$

In this latter case the presence of dissipative or nonequilibrium effects leads to the inequality. Once these relations are derived they become general in nature. That is, they are valid independent of whether the surroundings are reversible or not. The reason for this is that δQ and T refer solely to the system, and what happens in the surroundings is irrelevant to the evaluation of $\delta Q/T$.

We shall now turn our attention to the internally reversible case, as governed by Eq. (6-6). Consider a closed system undergoing a cyclic process, as shown in Fig. 6-8. The coordinates X and Y represent any two pertinent intrinsic properties, such as P and V. For the internally reversible cycle consisting of paths A and B we may write that

$$\oint \frac{\delta Q}{T} = 0 = \int_1^2 \left(\frac{\delta Q}{T}\right)_A + \int_2^1 \left(\frac{\delta Q}{T}\right)_B$$

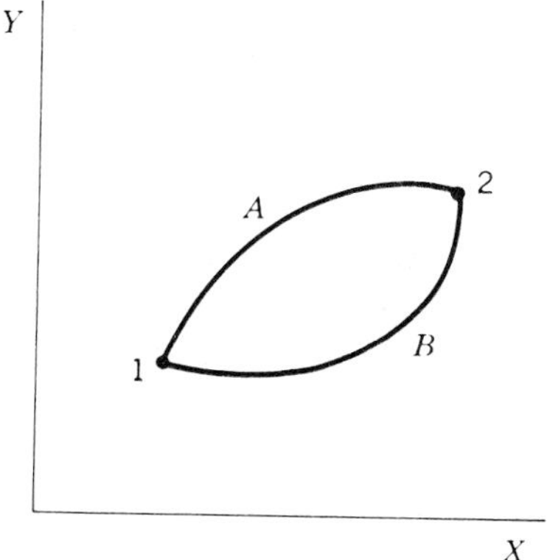

Figure 6-8 Cyclic path for a closed system.

This equation can be rearranged into the form

$$\int_1^2 \left(\frac{\delta Q}{T}\right)_A = \int_1^2 \left(\frac{\delta Q}{T}\right)_B$$

Therefore if the process is internally reversible, the integral of $\delta Q/T$ is the same along paths A and B. Since A and B were chosen arbitrarily, the integral of $\delta Q/T$ should lead to the same value for any internally reversible path between states 1 and 2.

In Chap. 2 we accepted the fact (based on experiment) that the adiabatic work between two given states of a closed system is always the same, regardless of the path chosen. This leads to the definition of the change in energy of a system, that is,

$$W_{\text{adia}} \equiv \Delta E$$

Now we have a similar situation. Between two given end states, the integral of $\delta Q/T$ always has the same value for any internally reversible path between those end states. Therefore this integral defines the change in a new property that is analogous to the change in energy. We shall call this integral a measure of the change in the entropy of the closed system. The function itself will be represented by the symbol S. Hence,

$$\Delta S = S_2 - S_1 \equiv \int \left(\frac{\delta Q}{T}\right)_{\text{int rev}} \tag{6-8}$$

or

$$dS = \left(\frac{\delta Q}{T}\right)_{\text{int rev}} \tag{6-9}$$

If a process is carried out in an internally reversible manner, then only a knowledge of the transient heat effects and of the temperature of the system at each stage of the process is required in order to evaluate ΔS. The restriction of internal reversibility is inherent in the derivation and cannot be removed from this particular equation.

Note that Eq. (6-8) can also be written, on an intensive basis, as

$$s = \int \left(\frac{\delta q}{T}\right)_{\text{int rev}} + s_0$$

This equation shows that the integration of $\delta q/T$ gives the entropy only within an arbitrary constant. As long as we work with pure substances, or nonreacting mixtures, a knowledge of the constant s_0 is unimportant because only a change in the entropy is usually required. Therefore tables and graphs may be constructed for which the base, or reference, value of s is completely arbitrary. For example, in the saturation-temperature tables for water (Tables A-12M and A-12), the specific entropy is chosen to be zero for saturated liquid water at the triple state (0.01°C or 32.02°F). Other reference states are chosen for other substances.

The dimensions of the entropy function are energy/(absolute temperature). Units in SI for the specific entropy are kJ/(kg)(°K) or kJ/(kg · mol)(°K), while USCS units commonly are Btu/(lb)(°R) or Btu/(lb · mol)(°R). Other sets of units may also be found in the literature.

6-12 INCREASE IN ENTROPY PRINCIPLE

It has been demonstrated that the entropy function is related uniquely to the value of $\delta Q/T$ for internally reversible processes. An additional relationship between ΔS and the integral of $\delta Q/T$ can be obtained by considering an internally irreversible process between two given end states of a closed system. Note the cyclic process shown in Fig. 6-8. We shall now retain the statement that path B is internally reversible. However, path A will be defined as an internally irreversible process. For the entire cycle we write that

$$\oint \frac{\delta Q}{T} = \int_1^2 \left(\frac{\delta Q}{T}\right)_A + \int_2^1 \left(\frac{\delta Q}{T}\right)_B = \int_1^2 \left(\frac{\delta Q}{T}\right)_A - \int_1^2 \left(\frac{\delta Q}{T}\right)_B$$

The limits on the integral along path B can be reversed, since in this case the integral represents the change in a property. In addition, for an internally irreversible process, the Clausius inequality requires that

$$\oint \frac{\delta Q}{T} < 0 \tag{6-7}$$

Also, because path B is internally reversible,

$$\int_1^2 \left(\frac{\delta Q}{T}\right)_B = S_2 - S_1$$

Combining these three expressions, we find that

$$\int_1^2 \left(\frac{\delta Q}{T}\right)_A - (S_2 - S_1) < 0$$

Path A is *any* internally irreversible path from 1 to 2. Hence, for internally irreversible processes between equilibrium states

$$S_2 - S_1 > \int \left(\frac{\delta Q}{T}\right)_{\text{irrev}} \tag{6-10}$$

In general, for any type of process involving a closed system,

$$S_2 - S_1 \geq \int \frac{\delta Q}{T} \tag{6-11}$$

or

$$dS \geq \frac{\delta Q}{T} \tag{6-12}$$

For this latter pair of equations, the equality sign applies for internally reversible processes, and the inequality for internally irreversible processes.

From Eq. (6-12) two extremely important thermodynamic relations can be deduced. First, in the absence of heat transfer any closed system must fulfill the condition that

$$dS_{\text{adia}} \geq 0 \tag{6-13}$$

For a finite change of state,

$$\Delta S_{\text{adia}} \geq 0 \tag{6-14}$$

The entropy function always increases in the presence of internal irreversibilities for an adiabatic, closed system. In the limiting case of an internally reversible adiabatic process the entropy will remain constant.

Although Eqs. (6-13) and (6-14) are extremely useful in the analysis of engineering processes, there are a number of processes which are not adiabatic. In such situations these equations are not directly applicable. However, an equivalent inequality in terms of the entropy function can be developed for cases where heat transfer occurs. Rather than studying just the closed system, however, it is necessary to include in the second-law analysis every part of the surroundings which is affected by changes in the system. In effect, we must look at the total changes, and not just at the system change. In practice, these changes outside of the system occur in regions reasonably close to the system. If we enclose in an arbitrary set of boundaries all those parts which are being affected, then no other changes are observed outside of this boundary. Such an enlarged system, which excludes further interactions across its boundaries, is frequently called an *isolated* system. No work or heat interactions or mass transfers occur across the boundaries of an isolated system.

Since no heat or mass transfer takes place across its boundaries, every isolated system fulfills the conditions of being closed and adiabatic. As a result, Eq. (6-13) applies also to an isolated system. Therefore,

$$dS_{\text{isolated system}} \geq 0 \tag{6-15}$$

For a finite change of state,

$$\Delta S_{\text{isolated system}} \geq 0 \tag{6-16}$$

In some textbooks an isolated system is referred to as the "universe."

To emphasize the fact that for nonadiabatic systems we must take into account all parts of physical space that are being affected by the change in the state of a particular system, Eq. (6-16) may be written either as

$$\Delta S_{\text{total}} = \sum \Delta S_{\text{subsystems}} \geq 0 \tag{6-17a}$$

or as

$$\Delta S_{\text{total}} = \Delta S_{\text{system}} + \Delta S_{\text{surroundings}} \geq 0 \tag{6-17b}$$

That is, the algebraic sum of the entropy changes of *all* the subsystems participating in a process is zero or positive. If all the parts or subsystems of an overall system undergo reversible changes, and all interactions between these parts are reversible, then the total entropy of the overall (isolated) system will remain the same. If irreversiblilites occur within any of the parts or are associated with any interaction between two or more of the subsystems, the entropy of the overall system must increase.

Equations (6-13) to (6-17) are the basic governing rules stemming from the second law of thermodynamics. Although other properties and further theorems may be developed which are more useful in certain types of engineering analysis, we begin with the entropy function and the "increase in entropy principles" stated above. Before proceeding, there are several important comments which must be made with respect to these principles.

1. The increase in entropy principles are *directional* statements. They limit the direction in which processes can proceed. A decrease in entropy is not possible for a closed system which is adiabatic, or for a composite of systems which interact among themselves.
2. The entropy function is a *nonconserved* property, and the increase in entropy principles are nonconservation laws. Only in the case of the reversible processes is entropy conserved. Reversible processes do not create entropy.
3. The second law states that all systems, left to themselves, are capable of reaching a state of equilibrium. The law now requires, in addition, that the entropy continually increase as that state of equilibrium is approached. Mathematically, the state of equilibrium must be attained when the entropy function reaches its *maximum* possible value, consistent with the constraints on the system.
4. The increase in entropy principles are intimately connected with the concept of *irreversibility*. It is the presence of irreversibilities that leads to the increase in entropy. We shall find that the greater the magnitude of the irreversibilities, the greater the overall entropy change. Hence the entropy function can be used quantitatively by the engineer to measure dissipative effects or losses associated with any type of process. It may also be used to establish performance criteria.
5. The change in the entropy function, as brought about by irreversibilities, may be used as a measure of the change in the quality of energy during a specified process.

The last two items above are especially important with respect to the analysis of engineering systems. To conserve or make better use of energy supplies, it is necessary to have some measure of the optimum performance of devices. For design purposes, one also would like to be able to measure the change in the quality of energy supplied to competitive processes. The entropy function, or other properties based on this function, is extremely valuable in providing guidelines for these measurements.

6-13 THE ENTROPY CHANGE OF HEAT AND WORK RESERVOIRS

At this point it is appropriate to consider a special application of the integral of $\delta Q/T$. A heat reservoir has been defined as a closed system which undergoes only internally reversible changes at constant temperature, while exchanging heat with another system. As a consequence of being an internally reversible system, the integral of $\delta Q/T$ applies directly to a heat reservoir. The value of the integral is the entropy change of any heat reservoir. Since the temperature is constant, integration leads to

$$\Delta S_{\text{heat reservoir}} = \int \left(\frac{\delta Q}{T}\right)_{\text{int rev}} = \frac{Q}{T} \tag{6-18}$$

The proper sign of Q must be taken into account when numerically evaluating Q/T. When heat is added to a heat reservoir, its entropy must increase, while the removal of heat always decreases the entropy of the reservoir.

The use of property diagrams as an aid in problem solving was emphasized earlier in this text when the conservation of energy principle was applied to closed and open systems. In conjunction with the second law of thermodynamics, it is extremely helpful to be able to plot processes on a diagram for which one of the coordinates is entropy. One of the most fundamental coordinate systems for second-law analysis is the TS diagram. The temperature-entropy diagram is quite useful for processes which occur at constant temperature because the path of the process is a straight line. Heat addition or removal from a heat reservoir is one such example. Rearrangement of Eq. (6-18) shows that

$$Q_{\text{heat reservoir}} = T\,\Delta S \tag{6-19}$$

where T must be on an *absolute* scale. Figure 6-9 illustrates heat addition or removal from a heat reservoir. When the reservoir changes from state 1 to state 2, ΔS is positive and heat is added. For process 2-1, ΔS is negative and heat is removed from the heat reservoir.

A work reservoir is a device which has the capability of delivering or absorbing energy in the form of work in a reversible manner. No heat effect is associated with its operation. The entropy change of a work reservoir is also easily determined. Such devices are described as internally reversible and adiabatic. Under

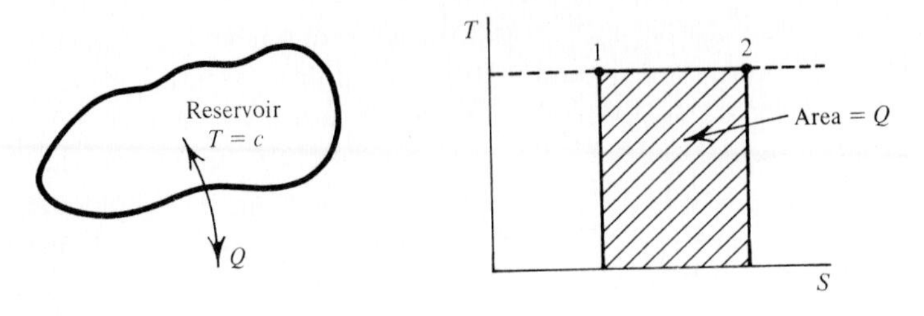

Figure 6-9 Area representation on a TS diagram of heat addition or removal for a heat reservoir.

these conditons the entropy change of reversible work sources or sinks must be zero. That is,

$$\Delta S_{\text{work reservoir}} = \int \left(\frac{\delta Q}{T}\right)_{\text{int rev}} = 0 \tag{6-20}$$

A qualitative explanation for this is as follows. When an idealized pulley-weight or flywheel device is used, for example, only its external state is altered, such as its overall gravitational potential energy or rotational kinetic energy. The intrinsic state of each, represented by variables such as temperature, pressure, or internal energy, is not changed. Since entropy is an intrinsic property, its value for these idealized work storage devices also is not changed.

6-14 THE CARNOT HEAT ENGINE

It has been demonstrated that the maximum thermal efficiency of any heat engine operating between two heat reservoirs is the theoretical Carnot efficiency. It is given by

$$\eta_{\text{Carnot}} = \frac{T_{\text{H}} - T_{\text{L}}}{T_{\text{H}}} = 1 - \frac{T_{\text{L}}}{T_{\text{H}}} \tag{6-4}$$

These are several theoretical cycles, composed of a series of different processes, which have efficiencies equal to the Carnot efficiency. One of the best known of these, which operates in a totally reversible manner between fixed-temperature reservoirs, is called the Carnot cycle. Such an engine can operate as either a closed or open system. Basically, the Carnot cycle requires that the working medium of the engine undergoes a series of four processes. These processes include two isothermal reversible ones and two adiabatic reversible ones.

Consider the following example of a Carnot cycle. Figure 6-10a shows a gas contained within a piston-cylinder device. A heat reservoir at a temperature T_{H} supplies heat to the engine which is at a temperature $T_{\text{H}} - dT$. [See the schematic of the engine in Fig. 6-10b.] Since there is a differential temperature difference between the heat reservoir and the engine, the process is externally reversible. In order to maintain an infinitesimal temperature difference during the

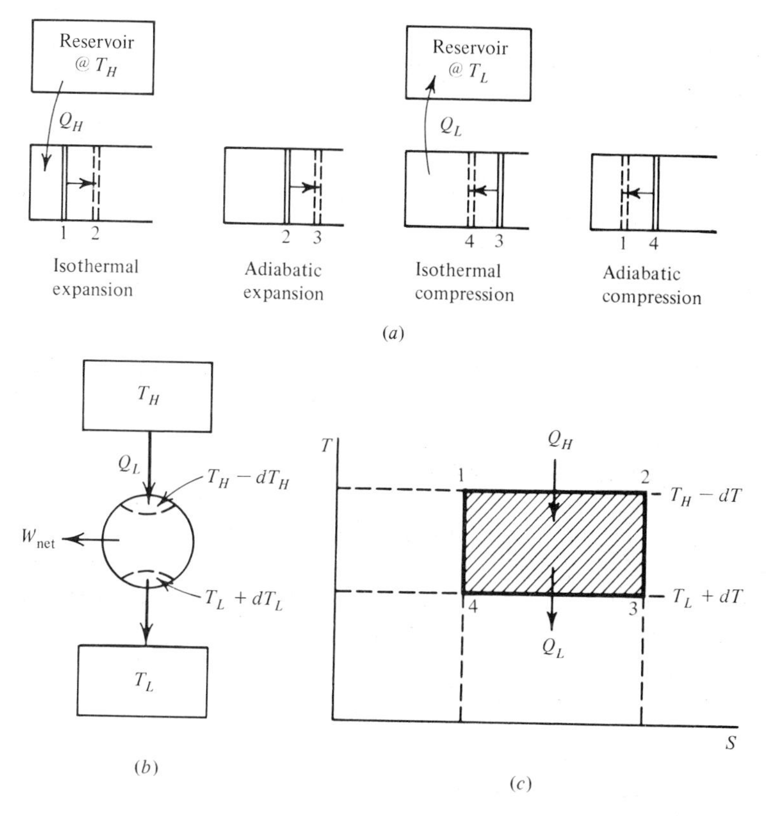

Figure 6-10 Illustrations for a Carnot heat engine. (*a*) Use of a piston-cylinder device; (*b*) schematic of the heat engine; (*c*) a *TS* diagram for a Carnot engine.

heat-addition process, the working medium of the engine is allowed to expand isothermally. This isothermal expansion is shown as process 1-2. The engine produces a net work output during the expansion. From state 2, the working medium is allowed to expand reversibly and adiabatically to a new state 3. During process 2-3 additional work is produced. State 3 corresponds to a temperature of $T_L + dT$. The system is now compressed isothermally to state 4. During the compression process, heat is necessarily removed to maintain the system at $T_L + dT$. The heat is rejected to a heat reservoir at a temperature T_L. Again this process is externally reversible since a temperature difference of dT is maintained across the boundaries of the engine. State 4 is so selected that, by a final reversible, adiabatic compression, the working medium is returned to the initial state. This is process 4-1 in Fig. 6-10*a*. During processes 3-4 and 4-1 work is performed on the system. Overall, heat is exchanged between the working medium of the engine and two fixed-temperature reservoirs, and net work is produced during each complete cycle.

The heat supplied to and rejected from a Carnot engine is easily shown on a

TS diagram. Figure 6-10*c* illustrates this point, where the Carnot cycle appears as a rectangular area. The four states numbered in Fig. 6-10*c* correspond to the same states in Fig. 6-10*a*. For an isothermal reversible process $Q = T\ \Delta S$, as noted in the preceding section. Therefore, the area beneath the horizontal line connecting states 1 and 2 represents the heat supplied Q_H. In a similar fashion the heat rejected Q_L is given by the area beneath the line 3-4. From an energy balance on the heat engine the difference between Q_H and Q_L is the net work produced by the engine during a cycle. Hence the area enclosed by the two isothermal, reversible lines and the two isentropic lines on the *TS* plot in Fig. 6-10*c* also is a measure of the net work output of a Carnot cycle. This figure demonstrates quite well the effect of the values of T_H and T_L on the Carnot efficiency. By either increasing T_H or decreasing T_L the cross-hatched area becomes a larger fraction of the total area beneath line 1-2. Hence the ratio W_{net}/Q_{in} increases under these circumstances.

The term "working medium" has been used in this discussion to emphasize the fact that many different types of systems can undergo a Carnot cycle. Although commonly associated with gases, a Carnot cycle can be executed by liquids, rubber bands (and other elastic media), electric cells, paramagnetic substances, liquid-vapor mixtures, etc. Also note that the use of the Carnot cycle is not restricted to closed systems. If the working medium is a gas or liquid, it may execute a Carnot cycle in a suitable series of steady-flow devices. For example, the isothermal heat-addition and heat-rejection processes are accomplished by heat exchangers. The adiabatic expansion and compression processes require a turbine and a compressor (or pump), respectively.

The Carnot cycle is of particular importance since its thermal efficiency is the maximum value for any heat engine operating between two given heat reservoirs. This efficiency is a theoretical one, because a reversible engine cannot be attained in practice. Nevertheless, the efficiency of a Carnot cycle represents a standard to which all actual heat-engine cycles may be compared. Heat engine operation between two reservoirs can be classified into four groups as follows:

Class I Reversible operation by the second law, $\Delta S = 0$, $\eta_{th} = \eta_{Carnot}$
Class II Irreversible operation by the second law, $\Delta S > 0$, $\eta_{th} < \eta_{Carnot}$
Class III Impossible operation by the second law, $\Delta S < 0$, $\eta_{th} > \eta_{Carnot}$
Class IV Impossible operation by the first law, $Q + W \neq 0$, $W + Q_L > Q_H$

Cited data for a heat engine performance can be used to identify the engine as belonging to one of these four types.

6-15 THE CARNOT REFRIGERATOR AND HEAT PUMP

A Carnot heat-engine cycle has been described as a totally reversible cycle. The engine itself is reversible and heat transfer to or from it occurs through differential temperature differences. Because of its total reversibility, it may truly be operated in the reverse direction. The end result is a device which acts as a refrigerator or heat pump. Heat is absorbed in this case from a low-temperature heat source, and heat is rejected to a high-temperature sink during the cycle. It is essential that work be supplied *to* the device from an external source. This is because of a corollary of the second law known as the *Clausius* statement of the second law, which states:

> It is impossible to operate a cyclic device in such a manner that the sole effect external to the device is the transfer of heat from one heat reservoir to another at a higher temperature.

The Clausius corollary is also an obvious consequence of the increase in entropy principle. Let us assume that a certain cyclic device can bring about the transfer of heat from a cold to a hot body without any further external effects. The entropy change of the device is zero, since it is cyclic. The only other possible entropy change is that of the reservoirs. If H represents the heat reservoir of higher temperature and L the lower one, then for the transfer of a differential quantity of heat δQ,

$$dS_{\text{res}} = \frac{\delta Q}{T_{\text{H}}} - \frac{\delta Q}{T_{\text{L}}} = \delta Q \, \frac{T_{\text{L}} - T_{\text{H}}}{T_{\text{H}} \, T_{\text{L}}}$$

But T_{H} is greater than T_{L}, so that dS of the universe is negative. This is impossible, in view of our second-law statement; consequently, the original premise is incorrect and the Clausius statement is validated.

Figure 6-11 is a schematic of a refrigerator or heat pump operating between fixed-temperature levels. The difference between such a device as a heat pump or a refrigerator is merely one of definition. The purpose of a refrigerator is to

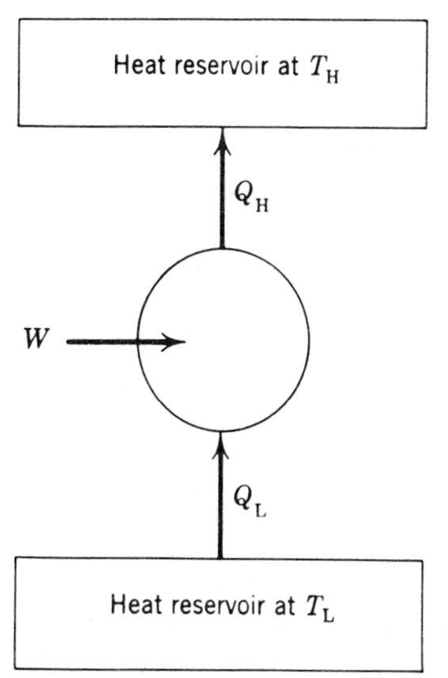

Figure 6-11 Schematic of a Carnot refrigerator or heat pump.

maintain a low-temperature source of finite size at a predetermined temperature by removing heat from it. A heat pump maintains a high-temperature sink at a given level by supplying it with heat taken from a low-temperature source. For example, a home or business office can be supplied with heat taken from the ground or air on a cold winter day. Due to the work input to the reversed engine, the first law dictates that Q_H be larger than Q_L. Even more important is the fact that Q_H can be considerably larger than W_{in}. Hence, in using a heat pump to heat a building, the costly (electrical) work input may be only a fraction of the thermal energy delivered.

The concept of thermal efficiency is not useful when applied to a reversed cycle. When dealing with refrigerators and heat pumps the analogous term usually applied is the coefficient of performance, which is abbreviated COP. In refrigeration processes the object is to remove the maximum quantity of heat from the low-temperature source per unit of net work input. Therefore

$$\text{COP}_{\text{refrig}} \equiv \frac{Q_L}{W_{\text{net}}} = \frac{Q_{\text{in}}}{Q_{\text{out}} - Q_{\text{in}}} \tag{6-21}$$

The purpose of a heat pump is to supply the maximum quantity of heat to a high-temperature sink per unit of required net work input. Hence

$$\text{COP}_{\text{heat pump}} \equiv \frac{-Q_H}{W_{\text{net}}} = \frac{Q_{\text{out}}}{Q_{\text{out}} - Q_{\text{in}}} \tag{6-22}$$

The coefficient of performance may be expressed in terms of temperatures as well as of heat quantities. For heat pumps and refrigerators heat is rejected to the high-temperature sink and heat is removed from a low-temperature source. Hence the entropy change for a differential cycle of the device is

$$dS_{\text{total}} = dS_H + dS_L + dS_{\text{device}} = \frac{\delta Q_H}{T_H} - \frac{\delta Q_L}{T_L} + 0$$

where the signs on the heat quantities have already been assumed. Consequently, only the absolute values of δQ_H and δQ_L should be used in these two equations. The maximum performance of refrigerators and heat pumps, as with heat engines, is attained when dS_{total} is zero. In this circumstance we see that the relationship between temperatures and heat quantities is exactly the same as for heat engines, that is, $\delta Q_H/T_H = \delta Q_L/T_L$, or

$$\frac{|Q_H|}{|Q_L|} = \frac{T_H}{T_L} \tag{6-2}$$

Consequently, for a reversed Carnot cycle, we find that substitution of Eq. (6-2) into Eq. (6-21) yields

$$\text{COP}_{\text{refrig}} = \frac{T_L}{T_H - T_L} \qquad \text{reversed Carnot} \tag{6-23}$$

In a similar manner Eq. (6-22) for a heat pump becomes

$$\text{COP}_{\text{heat pump}} = \frac{T_H}{T_H - T_L} \qquad \text{reversed Carnot} \qquad (6\text{-}24)$$

It should be recognized that, although the limiting value of the thermal efficiency is unity, the coefficient of performance theoretically can be much larger than 1. Figure 6-12 illustrates the variation of $\text{COP}_{\text{refrig}}$ as T_L is changed for a fixed value of the rejection (sink) temperature T_H of 25°C or 77°F. It is noted that the coefficient of performance increases rapidly as the temperature difference between source and sink is decreased. The colder the low-temperature source of a refrigerator must be maintained, the larger the work input must be per unit of heat Q_L removed.

Example 6-2M A Carnot refrigerator is maintaining foodstuffs in a refrigerated area at 2°C by rejecting heat to the atmosphere at 27°C. It is desired to maintain some frozen foods at −17°C with the same sink temperature of 27°C. What percent increase in work input will be required for the frozen-food unit over the refrigerated unit for the same quantity of heat Q_L removed?

Solution For each Carnot refrigerator two equations are valid, namely, $Q_H/Q_L = T_H/T_L$ and $Q_H - Q_L = W$, where the Q and W terms are absolute values. The combination of these equations gives

$$W_{\text{in}} = Q_L\left(\frac{T_H}{T_L} - 1\right)$$

Denoting the work input for the freezer operating between −17 and 27°C by a prime, we find

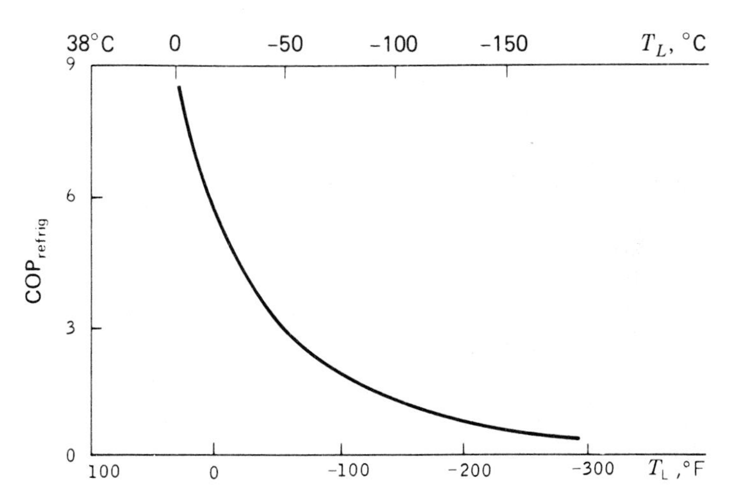

Figure 6-12 Variation of the coefficient of performance with T_L for a fixed value of T_H of 25°C or 77°F.

that

$$W_{\text{in}} = Q_L\left(\frac{300}{275} - 1\right) = 0.0909Q_L$$

and

$$W'_{\text{in}} = Q_L\left(\frac{300}{256} - 1\right) = 0.1719Q_L$$

The percent increase of W' over W is given by

$$\frac{W' - W}{W} = \frac{0.1719Q_L - 0.0909Q_L}{0.0909Q_L} = 0.891 \text{ (or 89.1\%)}$$

Hence, by changing the temperature difference between source and sink from 25 to 44°C, the required theoretical work input is nearly doubled.

Example 6-2 A Carnot refrigerator is maintaining foodstuffs in a refrigerator area at 40°F by rejecting heat to the atmosphere at 80°F. It is desired to maintain some frozen foods at 0°F with the same sink temperature of 80°F. What percent increase in work input will be required for the frozen-food unit over the refrigerated unit for the same quantity of heat removed?

SOLUTION For each Carnot refrigerator two sets of equations are valid, that is, $Q_H/Q_L = T_H/T_L$ and $Q_H - Q_L = W$. The combination of these equations gives

$$W_{\text{in}} = Q_L\left(\frac{T_H}{T_L} - 1\right)$$

Denoting the work input for the freezer operating between 0 and 80°F by a prime, we see that

$$W_{\text{in}} = Q_L(\tfrac{540}{500} - 1) = 0.080Q_L$$

and

$$W'_{\text{in}} = Q_L(\tfrac{540}{460} - 1) = 0.174Q_L$$

The percent increase of W' over W is given by

$$\frac{W' - W}{W} = \frac{0.174Q_L - 0.080Q_L}{0.080Q_L} = 1.175 \text{(or 117.5\%)}$$

Hence, by changing the temperature difference between source and sink from 40 to 80°F, the required work input is more than doubled.

Example 6-3M It is desired to maintain a low-temperature region of $T_{L,R}$ equal to 0°C by removing 1000 kJ/h with a Carnot refrigerator. The energy $Q_{H,R}$ from the refrigerator is transferred to the atmosphere at 22°C (Fig. 6-13*a*). The work to drive the refrigerator is provided by a Carnot heat engine which operates between a supply reservoir $T_{H,E}$ of 282°C and the atmosphere. Determine the heat $Q_{H,E}$ which must be supplied by the high-temperature reservoir to the heat engine, in kJ/h, if all the work output of the heat engine is used to drive the refrigerator.

SOLUTION To solve Carnot heat engine or refrigerator problems, at least three numerical values for the process must be known. In this problem, two values are known for the engine and three values are known for the refrigerator. Hence the refrigerator will be analyzed first. It is immediately seen that

$$Q_{H,R} = Q_{L,R}\left(\frac{T_{H,R}}{T_{L,R}}\right) = 1000\,\frac{22 + 273}{0 + 273} = 1080 \text{ kJ/h}$$

Note that absolute temperatures must be used in the equation. Application of the first law to the refrigerator then yields

$$W_R = W_E = Q_{H,R} - Q_{L,R} = 1080 - 1000 = 80 \text{ kJ/h}$$

Three values are now known for the engine operation. The first law states that

$$W_E = 80 = Q_{\text{H},E} - Q_{\text{L},E}$$

and the second law provides the relation

$$\frac{Q_{\text{H},E}}{Q_{\text{L},E}} = \frac{T_{\text{H},E}}{T_{\text{L},E}} = \frac{282 + 273}{22 + 273} = 1.88$$

Combining the two equations, we find that the required heat supply is

$$Q_{\text{H},E} - 0.532Q_{\text{H},E} = 80$$

$$Q_{\text{H},E} = 171 \text{ kJ/h}$$

Note in this idealized case (no irreversibilities) that a refrigeration effect of 1000 kJ/h is accomplished by a small quantity (171 kJ/h) of heat supplied to the engine.

Example 6-3 It is desired to maintain a low-temperature region of $T_{\text{L},R}$ equal to 0°F by removing 1000 Btu/h with a Carnot refrigerator. The energy $Q_{\text{H},R}$ from the refrigerator is transferred to the atmosphere at 80°F (Fig. 6-13*b*). The work to drive the refrigerator is provided by a Carnot heat engine which operates between a supply reservoir $T_{\text{H},E}$ at 540°F and the atmosphere. Determine the heat $Q_{\text{H},E}$ which must be supplied by the high-temperature reservoir to the heat engine, in Btu/h, if all the work output of the heat engine is used to drive the refrigerator.

SOLUTION To solve Carnot heat engine or refrigerator problems, at least three numerical values for the process must be known. In this problem, two values are known for the engine and three values are known for the refrigerator. Hence the refrigerator will be analyzed first. From the second law it is seen that

$$Q_{\text{H},R} = Q_{\text{L},R}\frac{T_{\text{H},R}}{T_{\text{L},R}} = 1000\,\frac{80 + 460}{0 + 460} = 1174 \text{ Btu/h}$$

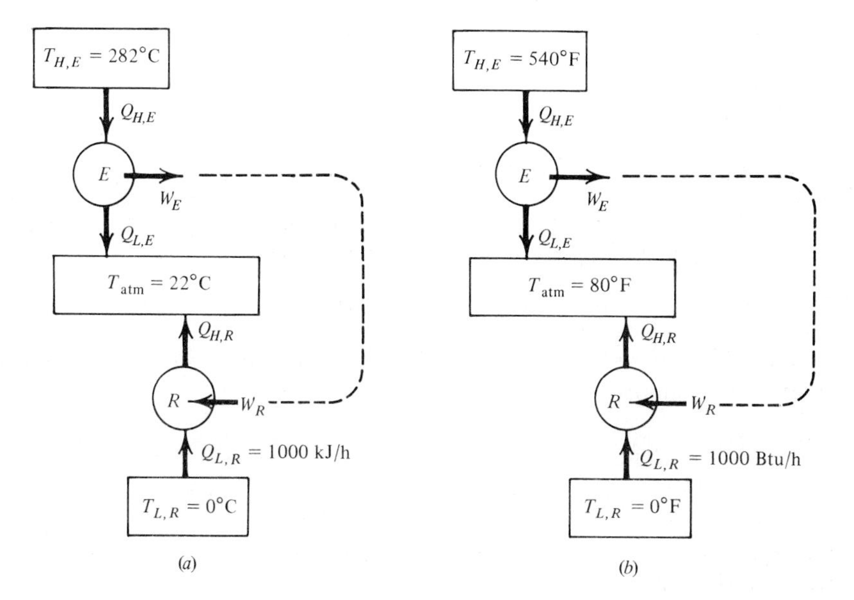

Figure 6-13 Schematics of the processes represented in (*a*) Example 6-3M, and (*b*) Example 6-3.

Note that absolute temperatures must be used in this equation. Application of the first law to the refrigerator then yields

$$W_R = W_E = Q_{H,R} - Q_{L,R} = 1174 - 1000 = 174 \text{ Btu/h}$$

Three values are now known for the engine operation. The first law states that

$$W_E = 174 = Q_{H,E} - Q_{L,E}$$

and the second law provides the relation

$$\frac{Q_{H,E}}{Q_{L,E}} = \frac{T_{H,E}}{T_{L,E}} = \frac{540 + 460}{80 + 460} = 1.852$$

Combining the two equations, we find that the required heat supply is

$$Q_{H,E} - 0.540Q_{H,E} = 174$$

$$Q_{H,E} = 378 \text{ Btu/h}$$

Note in this idealized case (no irreversibilities) that a refrigeration effect of 1000 Btu/h is accomplished by a relatively small quantity (378 Btu/h) of heat supplied to the engine.

One of the important contributions of the second law of thermodynamics to engineering analysis is the establishment of performance standards. In this section the thermal efficiency and the coefficient of performance have been reviewed for cyclic devices operating between heat reservoirs. These theoretical equations give values for the best performance under the stated restrictions. These values for η_{th} and COP are the maximum possible, and may be used as comparative standards for the actual performance of heat engines, heat pumps, and refrigerators.

6-16 EFFECTS OF REVERSIBLE AND IRREVERSIBLE HEAT INTERACTIONS

In an engineering context irreversibilities degrade performance, and the degree or magnitude of the irreversibility is measured by the total entropy change of the interacting systems. An example of this is the effect of irreversible heat transfer on the loss of work potential. Before developing a quantitative analysis of this effect, we shall review the concept of reversible heat transfer in terms of the second law.

Essentially there are two methods of transferring heat reversibly between two systems. One of these is to maintain only an infinitesimal temperature difference between the two systems. To illustrate this, consider the exchange of a quantity of heat Q between two heat reservoirs at temperatures T and $T - \Delta T$. The total entropy change is simply

$$\Delta S_{\text{total}} = \frac{-Q}{T} + \frac{Q}{T - \Delta T} = \frac{Q}{T}\left(\frac{\Delta T}{T - \Delta T}\right)$$

Since ΔT always has to be equal to or less than T, the last term is always equal to or greater than zero, in accordance with the increase in entropy principle. As the two temperatures approach each other, ΔT approaches zero and the total entropy change also approaches zero. In the limit as ΔS_{total} becomes zero, the process by definition becomes reversible. However, we must have a temperature difference in order for heat transfer to occur. We can consider heat transfer to be reversible if the temperature difference between the two systems is made infinitesimally small.

The foregoing discussion indicates that as the temperature difference between two reservoirs (or any two systems in general) increases, the total entropy change increases for the heat exchange process. This point is also illustrated through the use of a TS diagram for such a process. As an example, consider two systems H and L with absolute temperatures T_{H} and T_{L} of 1000 and 500 (in units of either °K or °R). One thousand units of heat are transferred from H to L. The entropy change of H, an internally reversible heat reservoir, is simply Q/T_{H}, which in this case is

$$\Delta S_{\text{H}} = \frac{Q}{T_{\text{H}}} = \frac{-1000}{1000} = -1 \text{ entropy units}$$

Since the reservoir is internally reversible, the values of Q, T_{H}, and S_{H} can be shown on a TS diagram. Figure 6-14 shows the area representation of Q for T_{H} of 1000 and a ΔS of 1 entropy unit.

Now, the heat Q received by L is the same as that given up by H. On the TS diagram of Fig. 6-14 this heat quantity received by L must have the same area as previously shown. But the area must also lie below the temperature line of 500. The cross-hatched area below T_{L} again represents Q. Since in this case $Q = T_{\text{L}}\,\Delta S_{\text{L}}$, and T_{L} is one-half that of T_{H}, ΔS_{L} must be twice as large as ΔS_{H}, or 2 units. It is easily seen that as T_{L} approaches T_{H}, these two cross-hatched areas must approach the same position on the diagram, and the total entropy change must become zero in the limit as T_{L} reaches T_{H}. As T_{L} becomes smaller, ΔS_{L} must become increasingly larger to maintain the same rectangular area on the TS plot. As a consequence ΔS of the composite system of H and L steadily increases. This is a basic example of the increase in entropy principle.

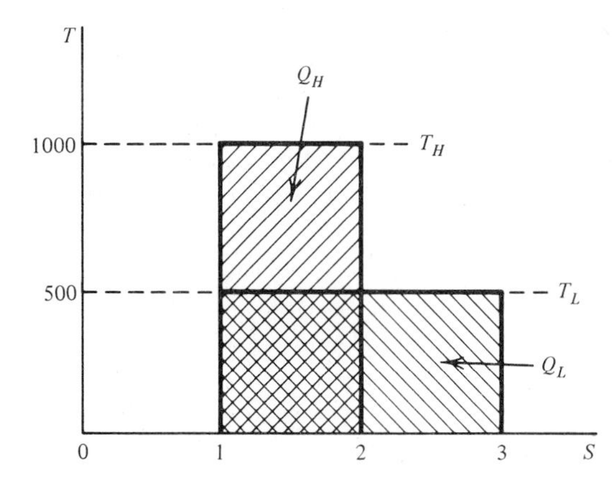

Figure 6-14 Area representation of Q on a TS diagram for an irreversible heat-transfer process.

When the temperature difference between two systems is finite, it is still possible to transfer heat between them in a reversible manner. Based on the discussion in the preceding section, this may be accomplished theoretically by inserting a reversible heat engine between the two systems. If a Carnot engine is used, then

$$\Delta S_H = \frac{Q_H}{T_H} \quad \text{and} \quad \Delta S_L = \frac{Q_L}{T_L}$$

where $Q_H \neq Q_L$. However, for a Carnot engine $Q_H/T_H = Q_L/T_L$. Since the heat-transfer quantities are of opposite sign, the total entropy change is zero, which indicates the reversibility of the process.

If we now compare the two preceding topics—irreversible heat transfer across a finite temperature difference and reversible heat transfer by insertion of a reversible heat engine—then we have one of many possible examples of irreversibilities leading to degradation of energy. When a heat engine is inserted between two systems which have a finite temperature difference between them, work is produced as a result of the heat exchange. For the same ΔT between the systems, direct irreversible heat transfer leads to no work production at all. Consequently, any time direct heat transfer occurs between two systems, the capability of producing some work from the interaction is lost. Since work is a more valuable form of energy in transition than heat, this loss of work capability is equivalent to a degrading or loss of *quality* of the energy involved, where quality is a measure of its potential usefulness.

The preceding discussion indicates that heat transfer through large finite temperature differences should be avoided, if possible, since a large "work potential" is lost. The maximum work output or the work potential obtainable from a quantity of heat Q is given by

$$W_{max} = Q_{in}\,\eta_{Carnot} = Q_{in}\left(1 - \frac{T_L}{T_H}\right) \tag{6-25}$$

where the source and sink for heat transfer are at constant temperature. For most heat engines the environment acts as the sink; therefore, T_L is not a variable under control of the engineer. As T_H increases, there exists the possibility that a larger fraction of Q might be converted into work. That is, a given quantity of heat has a higher potential for conversion into work if it is supplied at a higher temperature to a reversible engine. This implies that heat energy itself has quality as well as quantity. The higher the temperature at which thermal energy is available to the user, the higher its quality at that moment. There is a tremendous amount of energy in the ocean, for example, but its low temperature gives it a low work potential, or quality.

REFERENCES

Burghardt, M. D.: "Engineering Thermodynamics with Applications," Harper & Row, New York, 1982.

Fast, J. D.: "Entropy," McGraw-Hill, New York, 1963.

Hatsopoulos, G. N., and J. H. Keenan: "General Principles of Thermodynamics," Wiley, New York, 1968.
Howerton, M. T.: "Engineering Thermodynamics," Van Nostrand, Princeton, N. J., 1962.
Reif, F.: "Fundamentals of Statistical and Thermal Physics," McGraw-Hill, New York, 1965.

PROBLEMS (METRIC)

Carnot heat engine

6-1M A totally reversible heat engine operates between temperatures of (*a*) 427 and 17°C, and (*b*) 537 and 27°C. Compute the ratio of the heat absorbed from the source to the work output.

6-2M A Carnot engine receives 10 kJ of heat from a reservoir at the normal boiling point of water (100°C) and rejects heat to another reservoir maintained at the triple state of water (0.01°C). Compute the work done, in kilojoules, and the thermal efficiency of the engine.

6-3M At what temperature is heat supplied to a Carnot engine that rejects 1000 kJ/min of heat at 7°C and produces (*a*) 40 kW of power, and (*b*) 50 kW of power?

6-4M A heat engine operates on a Carnot cycle between temperatures of 827 and 17°C. For every kilowatt of net power output, calculate (*a*) the heat supplied and heat rejected, in kJ/h, and (*b*) the thermal efficiency.

6-5M The efficiency of a heat engine operating on a Carnot cycle is 40 percent. A cooling pond at 17°C receives 10^6 J/min of heat from the engine. Determine (*a*) the power output of the engine, in kilowatts, and (*b*) the temperature of the source reservoir, in degrees Celsius.

6-6M A heat engine operating on a Carnot cycle has an efficiency of 60 percent, with 600 kJ/cycle taken from the high-temperature reservoir at 417°C. Calculate (*a*) the sink temperature, in degrees Celsius, and (*b*) the heat rejected to the sink per cycle.

6-7M A heat engine operating on a Carnot cycle has an efficiency of 40 percent and rejects heat to a sink at 25°C. Find (*a*) the net power output, in kilowatts, and (*b*) the temperature of the source in degrees Celsius, if the heat supplied is 4000 kJ/h.

6-8M An ocean thermal-energy-conversion (OTEC) power plant uses the waters' naturally occurring temperature differences.

(*a*) At a given location the surface temperature is 20°C, and at a 700-m depth the value is −5°C. Calculate the maximum net work output from a Carnot cycle per kilojoule of heat addition.

(*b*) Assume the surface water experiences a 5°C decrease during heat addition. Determine the required mass flow rate of seawater required, in kg/s, for a power output from the cycle of 100 MW.

6-9M A geothermal power plant, utilizing an underground source of hot water or steam, receives a supply of water at 160°C.

(*a*) Find the maximum possible thermal efficiency of a cyclic heat engine which uses this source and rejects to the environment at 15°C.

(*b*) Assume the water source experiences a 5°C decrease during heat addition. Determine the required mass flow rate of underground water, in kg/s, for a net power output of 10 MW.

6-10M A reversible heat engine exchanges heat with three reservoirs and produces work in the amount of 400 kJ. Reservoir A has a temperature of 500°K and supplies 1200 kJ to the engine. If reservoirs B and C have temperatures of 400 and 300°K, respectively, how much heat, in kilojoules, does each exchange with the engine and what is the direction of the heat exchanges with respect to the engine?

6-11M Consider Prob. 6-10M, except that the work output is 700 kJ and the temperatures of reservoirs A, B, and C are 600, 400, and 300°K, respectively.

6-12M Two heat engines operating on Carnot cycles are arranged in series. The first engine *A* receives heat at 727°C and rejects heat to a reservoir at temperature *T*. The second engine *B* receives the heat rejected by the first engine, and in turn rejects heat to a reservoir at 7°C. Calculate the temperature T, in degrees Celsius, for the situation where (*a*) the work outputs of the two engines are equal, and (*b*) the efficiencies of the two engines are equal.

6-13M Consider Prob. 6-12M, except that the temperatures of the heat reservoirs are 827 and 127°C.

Clausius inequality

6-14M In a given power cycle the working fluid receives 3150 kJ/kg of heat at an average temperature of 440°C and rejects 1550 kJ/kg of heat to cooling water at 20°C. If there are no other heat interactions to or from the fluid, does the cycle satisfy the Clausius inequality?

6-15M In a given power cycle the working fluid receives 3340 kJ/kg of heat at an average temperature of 480°C and rejects 1220 kJ/kg of heat to the environment at 15°C. In the absence of other heat interactions, does the cycle satisfy the Clausius inequality?

Entropy changes of heat reservoirs

6-16M With reference to Prob. 6-2M, compute the entropy changes of the heat source and heat sink, in kJ/°K.

6-17M With reference to Prob. 6-3M, compute the entropy changes of the heat source and heat sink in kJ/(°K)(min) for part *a* and for part *b*.

6-18M With reference to Prob. 6-6M, compute the entropy changes of the heat source and heat sink in kJ/(°K)(cycle).

6-19M A reversible heat engine receives 1000 kJ of heat from a reservoir at 377°C and rejects heat to another reservoir at 27°C. Compute the entropy changes of the two reservoirs, in kJ/°K.

6-20M A Carnot heat engine operates between 1000 and 300°K. The change in entropy of the supply or source reservoir is 0.50 kJ/°K. Find (*a*) the heat added, and (*b*) the net work output, both in kilojoules.

6-21M A Carnot engine operates between 777 and 7°C. The change in entropy of the sink reservoir is 0.30 kJ/°K. Determine (*a*) the heat added, and (*b*) the net work output, both in kilojoules.

6-22M In a Carnot cycle, heat is added in process 1-2 and rejected in process 3-4 (see Fig. 6-10*a*). For the cycle, $T_1 = 600$°K, $T_4 = 300$°K, and $S_2 - S_1 = 0.10$ kJ/°K.

(*a*) Determine the thermal efficiency of the cycle.

(*b*) Now suppose process 2-3 is not reversible. It is still adiabatic but, as a result of irreversibilities, the entropy increases 0.020 kJ/°K during the process. All other parts of the cycle are assumed to remain reversible. Determine the thermal efficiency in this latter case.

(*c*) Sketch the irreversible cycle on a TS diagram.

6-23M A steady-flow type Carnot engine operates between the temperature limits of 400 and 30°C. During the isothermal-expansion process, 350 kJ of heat are added per kilogram of fluid. However, irreversibilities in the turbine cause an increase in entropy through the adiabatic turbine amounting to 0.04 kJ/(kg)(°K). Calculate the heat rejected and the ratio of the actual work output to the maximum possible work output if the expansion had been reversible. All other parts of the cycle are assumed to be reversible. Sketch the irreversible cycle on a TS diagram.

Classification of heat engine operation

6-24M Four different heat engines are operating between the temperature limits of 1000 and 300°K, and the heat input in each case is 500 kJ/s. Using the heat engine classes I through IV presented in Sec. 6-14, classify the four engines from the data presented below.

(*a*) Engine A, $Q_L = 150$ kJ/s, $W = 350$ kW

(*b*) Engine B, $Q_L = 100$ kJ/s, $W = 400$ kW

(*c*) Engine C, $Q_L = 200$ kJ/s, $W = 350$ kW

(*d*) Engine D, $Q_L = 200$ kJ/s, $W = 300$ kW

6-25M Four different heat engines are operating between the temperature limits of 1200 and 300°K, and the heat rejection in each case is 400 kJ/s. Using the heat engine classes I through IV presented in Sec. 6-14, classify the four engines from the data presented below.

(*a*) Engine *A*, $Q_H = 1500$ kJ/s, $W = 1100$ kW
(*b*) Engine *B*, $Q_H = 1400$ kJ/s, $W = 1050$ kW
(*c*) Engine *C*, $Q_H = 1800$ kJ/s, $W = 1400$ kW
(*d*) Engine *D*, $Q_H = 1600$ kJ/s, $W = 1200$ kW

Carnot refrigerators and heat pumps

6-26M A reversible refrigerating machine absorbs 400 kJ/min from a cold space and requires 3.0 kW to drive it. If the machine is reversed and receives 1600 kJ/min from the hot source, how much power, in kilowatts, does it produce?

6-27M A reversed Carnot engine is to be used to produce ice at 0°C. The heat-rejection temperature is 30°C, and the enthalpy of freezing is 335 kJ/kg. How many kilograms of ice can be formed per hour per kilowatt of power input?

6-28M A ton of refrigeration is defined as a heat-absorption rate of 211 kJ/min from a cold source. It is desired to operate a reversed Carnot cycle between temperature limits of −20 and +35°C so that 8 tons of refrigeration are produced. Calculate (*a*) the number of kilowatts required to operate the cycle, and (*b*) the coefficient of performance.

6-29M A Carnot heat pump is used for heating a building. The outside air at −6°C is the cold reservoir, the building at 26°C is the hot reservoir, and 120,000 kJ/h are required for heating. Find (*a*) the heat taken from the outside, in kJ/h, and (*b*) the power input required, in kilowatts.

6-30M If the thermal efficiency of a Carnot engine is (*a*) 1/6, and (*b*) 1/5, find the coefficient of performance of (1) a Carnot refrigerator, and (2) a Carnot heat pump operating between the same temperature limits.

6-31M A Carnot heat pump operating between temperatures of −7 and 29°C has a power input of 3.5 kW. Determine (*a*) the coefficient of performance, and (*b*) the heat supplied at 29°C, in kJ/s.

6-32M A computer room is to be kept at 20°C. Assume that the computer transfers 1500 kJ/h to the room air. If the environmental temperature is 30°C, what is the minimum power input required for the cooling process?

6-33M A Carnot heat pump is to be used to maintain a home with a heat loss of 80,000 kJ/h at 22°C. The makeup heat is to be supplied from the outside air at −5°C.

(*a*) Find the power input required, in kilowatts.

(*b*) If electricity costs 6.5 cents per kilowatthour, what would be the operating cost for 1 day of continuous operation?

6-34M A heat pump operates on a reversed Carnot cycle, removes heat from a low-temperature source at −15°C, and rejects heat to a sink at 26°C. If electricity costs 5.9 cents per kilowatthour, determine the cost of operation for supplying a home with 50,000 kJ/h.

Combined Carnot heat engines, refrigerators, and heat pumps

6-35M A Carnot heat engine operating between levels of 727 and 27°C is supplied with 500 kJ/cycle. Sixty percent of the work output is used to drive a heat pump which rejects to the surroundings at 27°C. If the heat pump removes 1050 kJ/cycle from a low-temperature reservoir, determine (*a*) the total heat rejected to the surroundings at 27°C, in kJ/cycle, and (*b*) the temperature of the reservoir, in degrees Celsius.

6-36M A Carnot heat engine receives 90 kJ from a reservoir at 627°C. It rejects heat to the environment at 27°C. One-third of its work output is used to drive a Carnot refrigerator. The refrigerator rejects 60 kJ to the environment at 27°C. Find (*a*) the work output of the heat engine, (*b*) the efficiency of the heat engine, (*c*) the temperature of the low-temperature reservoir for the refrigerator in degrees Celsius, and (*d*) the COP of the refrigerator.

6-37M A Carnot heat engine receives heat from a reservoir at T_H in the amount of 800 kJ/min, and rejects heat to the environment at 27°C. The entire net work output of the heat engine is used to drive a Carnot refrigerator which receives heat in the amount of 1000 kJ/min from another heat reservoir at −23°C. The Carnot refrigerator also rejects to the environment at 27°C. Determine (*a*) the work output of the engine, in kJ/min, and (*b*) the temperature T_H of the reservoir supplying heat to the engine.

6-38M A Carnot heat engine operates between temperature levels of 397 and 7°C and rejects 20 kJ/min to the environment. The total net work output of the engine is used to drive a heat pump which is supplied with heat from the environment at 7°C and rejects heat to a home at 40°C. Determine (*a*) the net work delivered by the engine, in kilowatts, (*b*) the heat supplied to the heat pump, in kJ/min, and (*c*) an overall COP for the combined devices, which is defined as the energy rejected to the home divided by the initial energy supplied to the engine.

6-39M A Carnot refrigerator moves heat from a sink at −8°C and rejects heat to the atmosphere at 15°C. The refrigerator is coupled to the output of a Carnot engine which receives heat at 577°C and also rejects to the atmosphere. Determine the ratio of the heat supplied to the engine to the heat removed by the refrigerator.

6-40M A Carnot heat engine is used to drive a Carnot refrigerator. The heat engine receives Q_1 at T_1 and rejects Q_2 at T_2. The refrigerator removes a quantity of heat Q_3 from a source at T_3 and rejects a quantity of heat Q_4 at T_4. Develop an expression for the ratio Q_3/Q_1 in terms of the various temperatures of the heat reservoirs.

Irreversible heat transfer and quality of energy

6-41M One thousand kilojoules of heat are transferred from a heat reservoir at 850°K to a second reservoir at (*a*) 550°K, and (*b*) 330°K. (1) Calculate the entropy change of each reservoir in kJ/°K. (2) Is the sum of the entropy changes of the reservoirs in agreement with the second law?

6-42M One hundred kilojoules of heat are transferred from an isothermal system at (*a*) 550°K, and (*b*) 450°K to the environment at 7°C. (1) Calculate the entropy change of the system and the environment in kJ/°K. (2) Is the sum of the entropy changes of the reservoirs in agreement with the second law?

6-43M Which quantity of heat theoretically has the higher quality, 1000 kJ at 800°K or 3000 kJ at 380°K? The environmental temperature is 7°C.

6-44M Which quantity of heat theoretically has the higher quality, 2000 kJ at 1100°K or 5000 kJ at 530°K? The environmental temperature is 22°C.

6-45M Which quantity of energy theoretically has the higher quality, 1000 kJ of shaft work or 3000 kJ of heat at 500°K? The environmental temperature is 20°C.

6-46M Which quantity of energy theoretically has the higher quality, 530 kJ of shaft work or 1000 kJ of heat at 480°K? The environmental temperature is 27°C.

6-47M Heat is available from a reservoir at (*a*) 1000°K, and (*b*) 800°K. The environmental temperature is 300°K. A quantity of heat is now transferred from the high-temperature reservoir to another reservoir at 600°K. Determine the percent reduction in work potential of the heat quantity due to the irreversible heat transfer to an intermediate temperature of 600°K.

PROBLEMS (USCS)

Carnot heat engines

6-1 A totally reversible heat engine operates between temperatures of 660 and 100°F. Compute the ratio of the heat absorbed from the source to the work output.

6-2 A Carnot engine develops 2 hp and rejects 7500 Btu/h to a sink at 60°F. Determine the source temperature, in degrees Fahrenheit.

6-3 At what temperature is heat supplied to a Carnot engine that rejects 1000 Btu/min of heat at 40°F and produces 50 hp?

6-4 A heat engine operates on a Carnot cycle between temperatures of 1440 and 60°F. For every horsepower of net power output, calculate (*a*) the heat supplied and heat rejected, in Btu/h, and (*b*) the thermal efficiency.

6-5 The efficiency of a heat engine operating on a Carnot cycle is 40 percent. A cooling pond at 80°F receives 700 Btu/min of heat from the engine. Determine (*a*) the power output of the engine, in horsepower, and (*b*) the temperature of the source reservoir, in degrees Fahrenheit.

6-6 A heat engine operating on a Carnot cycle has an efficiency of 48 percent, with 500 Btu/cycle taken from the high-temperature reservoir at 560°F. Calculate (*a*) the sink temperature in degrees Fahrenheit, and (*b*) the heat rejected to the sink, in Btu/cycle.

6-7 A heat engine operating on a Carnot cycle has an efficiency of 40 percent and rejects heat to a sink at 80°F. Find (*a*) the net power output, in horsepower, and (*b*) the temperature of the source, in degrees Fahrenheit, if the heat supplied is 4000 Btu/h.

6-8 An ocean thermal-energy-conversion (OTEC) power plant uses the waters' naturally occurring temperature differences.

(*a*) At a given location the surface temperature is 70°F, and at a 2000-ft depth the value is 30°F. Calculate the maximum net work output from a Carnot cycle per Btu of heat addition.

(*b*) Assume the surface water experiences a 10°F decrease during heat addition. Determine the required mass flow rate of seawater required, in lb/s, for a power output from the cycle of 100 MW.

6-9 A geothermal power plant, utilizing an underground source of hot water or steam, receives a supply of water at 320°F.

(*a*) Find the maximum possible thermal efficiency of a cyclic heat engine which uses this source and rejects to the environment at 60°F.

(*b*) Assume the water source experiences a 10°F decrease during heat addition. Determine the required mass flow rate of underground water, in lb/s, for a net power output of 10 MW.

6-10 A reversible heat engine exchanges heat with three reservoirs and produces work in the amount of 400 Btu. Reservoir A has a temperature of 800°R and supplies 1200 Btu to the engine. If reservoirs B and C have temperatures of 600 and 400°R, respectively, how much heat, in Btu, does each exchange with the engine and what is the direction of the heat exchanges with respect to the engine?

6-11 Same as Prob. 6-10, except that the work output is 700 Btu and the temperatures of reservoirs A, B, and C are 1200, 800, and 600°R, respectively.

6-12 Two heat engines operating on Carnot cycles are arranged in series. The first engine A receives heat at 1200°F and rejects heat to a reservoir at temperature T. The second engine B receives the heat rejected by the first engine, and in turn rejects heat to a reservoir at 60°F. Calculate the temperature T, in degrees Fahrenheit, for the situation where (*a*) the work outputs of the two engines are equal, and (*b*) the efficiencies of the two engines are equal.

6-13 Consider Prob. 6-12, except that the temperatures of the heat reservoirs are 1500 and 100°F.

Clausius inequality

6-14 In a power cycle the working fluid receives 1350 Btu/lb of heat at an average temperature of 820°F and rejects 840 Btu/lb of heat to cooling water at 70°F. If there are no other heat interactions to or from the fluid, does the cycle satisfy the Clausius inequality?

6-15 In a power cycle the working fluid receives 1440 Btu/lb of heat at an average temperature of 900°F and rejects 870 Btu/lb of heat to the environment at 60°F. In the absence of other heat interactions, does the cycle satisfy the Clausius inequality?

Entropy change of heat reservoirs

6-16 With reference to Prob. 6-2, compute the entropy changes of the heat source and heat sink in Btu/(°R)(h).

6-17 With reference to Prob. 6-3, compute the entropy changes of the heat source and sink in Btu/(°R)(min).

6-18 With reference to Prob. 6-6, compute the entropy changes of the heat source and sink in Btu/(°R)(cycle).

6-19 A reversible heat engine receives 1000 Btu of heat from a reservoir at 540°F and rejects heat to another reservoir at 60°F. Compute the entropy change of the two reservoirs, in Btu/°R.

6-20 A Carnot heat engine operates between 1000 and 500°R. The change in entropy of the supply or source reservoir is 0.50 Btu/°R. Find (*a*) the heat added, and (*b*) the net work output, both in Btu.

6-21 A Carnot heat engine operates between 600 and 40°F. The change in entropy of the sink reservoir is 0.50 Btu/°R. Determine (*a*) the heat added, and (*b*) the net work output, both in Btu.

6-22 In a Carnot cycle, heat is added in process 1-2 and rejected in process 3-4 (see Fig. 6-10*a*). For the cycle, $T_1 = 1000$°R, $T_4 = 500$°R, and $S_2 - S_1 = 0.10$ Btu/°R.

(*a*) Determine the thermal efficiency of the cycle.

(*b*) Now suppose process 2-3 is not reversible. It is still adiabatic but, as a result of irreversibilities, the entropy increases 0.020 Btu/°R during the process. All other parts of the cycle are assumed to remain reversible. Determine the thermal efficiency in this latter case.

(*c*) Sketch the irreversible cycle on a *TS* diagram.

6-23 A steady-flow-type Carnot heat engine operates between the temperature limits of 740 and 100°F. During the isothermal expansion process 180 Btu of heat are added per pound of fluid. However, irreversibilities in the turbine cause an increase in entropy through the adiabatic turbine amounting to 0.02 Btu/(lb)(°R). Calculate the heat rejected and the ratio of the actual work output to the maximum possible work output if the expansion had been reversible. All other parts of the cycle are assumed to be reversible. Sketch the irreversible cycle on a *TS* diagram, indicating the areas which represent heat and work.

Classification of heat engine operation

6-24 Four different heat engines are operating between the temperature limits of 1500 and 500°R, and the heat input in each case is 900 Btu/s. Using the heat engine classes I through IV presented in Sec. 6-14, classify the four engines from the data presented below.

(*a*) Engine A, $Q_L = 300$ Btu/s, $W = 600$ Btu/s
(*b*) Engine B, $Q_L = 250$ Btu/s, $W = 650$ Btu/s
(*c*) Engine C, $Q_L = 350$ Btu/s, $W = 600$ Btu/s
(*d*) Engine D, $Q_L = 400$ Btu/s, $W = 500$ Btu/s

6-25 Four different heat engines are operating between the temperature limits of 2000 and 500°R, and the heat rejection in each case is 400 Btu/s. Using the heat engine classes I through IV presented in Sec. 6-14, classify the four engines from the data presented below.

(*a*) Engine A, $Q_H = 1500$ Btu/s, $W = 1100$ Btu/s
(*b*) Engine B, $Q_H = 1400$ Btu/s, $W = 1050$ Btu/s
(*c*) Engine C, $Q_H = 1800$ Btu/s, $W = 1400$ Btu/s
(*d*) Engine D, $Q_H = 1600$ Btu/s, $W = 1200$ Btu/s

Carnot refrigerators and heat pumps

6-26 A reversible refrigerating machine absorbs 200 Btu/min from a cold space and requires 2.0 hp to drive it. If the machine is reversed and receives 800 Btu/min from the hot source, how much power, in horsepower, does it produce?

6-27 A reversed Carnot engine is to be used to produce ice at 32°F. The heat-rejection temperature is 78°F, and the enthalpy of freezing is 144 Btu/lb. How many pounds of ice can be formed per hour per horsepower of power input?

6-28 A ton of refrigeration is defined as a heat-absorption rate of 200 Btu/min from a cold source. It is desired to operate a reversed Carnot cycle between temperature limits of −10 and +85°F so that 8 tons of refrigeration are produced. Calculate (*a*) the horsepower required to operate the cycle, and (*b*) the coefficient of performance.

6-29 A Carnot heat pump is used for heating a building. The outside air at 22°F is the cold reservoir, the building at 76°F is the hot reservoir, and 200,000 Btu/h are required for heating. Find (*a*) the heat taken from the outside, in Btu/h, and (*b*) the power input required, in horsepower.

6-30 If the thermal efficiency of a Carnot engine is (*a*) 1/6, and (*b*) 1/5, find the coefficient of performance of (1) a Carnot refrigerator, and (2) a Carnot heat pump operating between the same temperature limits.

6-31 A Carnot heat pump operating between temperatures of 20 and 80°F has a power input of 4.2 hp. Determine (*a*) the coefficient of performance, and (*b*) the heat supplied at 80°F, in Btu/s.

6-32 A computer room is to be kept at 70°F. Assume that the computer transfers 1500 Btu/h to the room air. If the environmental temperature is 90°F, what is the minimum horsepower input required for the cooling process?

6-33 A Carnot heat pump is to be used to maintain a home with a heat loss of 80,000 Btu/h at 74°F. The makeup heat is to be supplied from the outside air at 20°F.

(*a*) Find the power input required, in horsepower.

(*b*) If electricity costs 6.5 cents per kilowatthour, what would be the operating cost for 1 day of continuous operation?

6-34 A heat pump operates on a reversed Carnot cycle, removes heat from a low-temperature source at −10°F, and rejects heat to a sink at 80°F. If electricity costs 5.9 cents per kilowatthour, determine the cost of operation for supplying a home with 50,000 Btu/h.

Combined Carnot heat engines, refrigerators, and heat pumps

6-35 A Carnot heat engine operating between levels of 1340 and 80°F is supplied with 500 Btu/cycle. Sixty percent of the work output is used to drive a heat pump which rejects to the surroundings at 80°F. If the heat pump removes 1050 Btu/cycle from a low-temperature reservoir, determine (*a*) the total heat rejected to the surroundings at 80°F, in Btu/cycle, and (*b*) the temperature of the reservoir in degrees Fahrenheit.

6-36 A Carnot heat engine receives 90 Btu from a reservoir at 1160°F. It rejects heat to the environment at 60°F. One-third of its work output is used to drive a Carnot refrigerator. The refrigerator rejects 60 Btu to the environment at 60°F. Find (*a*) the work output of the heat engine, in Btu, (*b*) the efficiency of the heat engine, (*c*) the temperature of the low-temperature reservoir for the refrigerator, in degrees Fahrenheit, and (*d*) the COP of the refrigerator.

6-37 A Carnot heat engine receives heat from a reservoir at T_H in the amount of 800 Btu/min, and rejects heat to the environment at 60°F. The entire net work output of the heat engine is used to drive a Carnot refrigerator which receives heat in the amount of 1000 Btu/min from another heat reservoir at −10°F. The Carnot refrigerator also rejects to the environment at 60°F. Determine (*a*) the work output of the engine, in Btu/min, and (*b*) the temperature T_H of the reservoir supplying heat to the engine in degrees Fahrenheit.

6-38 A Carnot heat engine operates between temperature levels of 740 and 20°F and rejects 20 Btu/min to the environment. The total net work output of the engine is used to drive a heat pump which is supplied with heat from the environment at 20°F and rejects heat to a home at 140°F. Determine (*a*) the net work delivered by the engine, in Btu/min, (*b*) the heat supplied to the heat pump, in Btu/min, and (*c*) an overall COP for the combined devices, which is defined as the energy rejected to the home divided by the initial energy supplied to the engine.

6-39 A Carnot refrigerator removes heat from a sink at −20°F and rejects heat to the atmosphere at 50°F. The refrigerator is powered by the output of a Carnot engine which receives heat from a source at 1070°F and also rejects to the atmosphere. Determine the ratio of the heat supplied to the engine to the heat removed by the refrigerator.

6-40 A Carnot heat engine is used to drive a Carnot refrigerator. The heat engine receives Q_1 at T_1 and rejects Q_2 at T_2. The refrigerator removes a quantity of heat Q_3 from a source at T_3 and rejects a large quantity of heat Q_4 at T_4. Develop an expression for the ratio Q_3/Q_1 in terms of the various temperatures of the heat reservoirs.

Irreversible heat transfer and quality of energy

6-41 One thousand Btu of heat are transferred from a heat reservoir at 1040°F to a second reservoir at (*a*) 540°F, and (*b*) 140°F. (1) Calculate the entropy change of each reservoir, in Btu/°R. (2) Is the sum of the entropy changes of the reservoirs compatible with the second law?

6-42 One hundred Btu of heat are transferred from an isothermal system at (*a*) 540°F, and (*b*) 340°F to the environment at 40°F. (1) Calculate the entropy change of the system and the environment, in Btu/°R. (2) Is the sum of the entropy changes in agreement with the second law?

6-43 Which quantity of heat theoretically has the higher quality, 1000 Btu at 1000°F or 3000 Btu at 200°F? The environmental temperature is 40°F.

6-44 Which quantity of heat theoretically has the higher quality, 2000 Btu at 1500°F or 5000 Btu at 500°F? The environmental temperature is 70°F.

6-45 Which quantity of energy theoretically has the higher quality, 1000 Btu of shaft work or 3000 Btu of heat at 500°F? The environmental temperature is 60°F.

6-46 Which quantity of energy theoretically has the higher quality, 40,000 $ft \cdot lb_f$ of shaft work or 100 Btu of heat at 400°F? The environmental temperature is 80°F.

6-47 Heat is available from a reservoir at (*a*) 1800°R, and (*b*) 1500°R. The environmental temperature is 500°R. A quantity of heat is now transferred from the high-temperature reservoir to another reservoir at 1000°R. Determine the percent reduction in work potential of the heat quantity due to the irreversible heat transfer to the intermediate temperature of 1000°R.

CHAPTER

SEVEN

SOME CONSEQUENCES OF THE SECOND LAW

A new intrinsic property—entropy—was introduced in the preceding chapter. It is intimately connected with the second law of thermodynamics, which is concerned with the directional or limiting quality of phenomena and the degradation of performance due to the presence of irreversibilities. These items of directionality and performance will be examined closely in this chapter. In addition, it is necessary to examine in some detail the methods of evaluating entropy changes of common substances.

7-1 THE TEMPERATURE-ENTROPY DIAGRAM

The use of the TS diagram was illustrated in Chap. 6 with regard to the analysis of heat transfer processes from reservoirs, and of Carnot engine cycles. In those cases the source or sink for heat transfer was one of constant temperature. As a result the area representation of Q on a TS plot was a rectangle. In conjunction with second law analyses of various processes and cycles, we now wish to extend the use of the TS diagram to nonisothermal processes.

The basic starting point for nonisothermal processes is the equation for the entropy change of a fixed mass, namely,

$$dS = \left(\frac{\delta Q}{T}\right)_{\text{int rev}} \tag{6-9}$$

This equation can be rearranged to show that

$$\delta Q_{\text{int rev}} = T\,dS \tag{7-1}$$

For a finite change of state this becomes

$$Q_{\text{int rev}} = \int T\,dS \tag{7-2a}$$

and on a unit-mass basis

$$q_{\text{int rev}} = \int T\,dS \tag{7-2b}$$

For a process which is carried out in an internally reversible manner, the heat transferred during the process is represented by an area on a TS diagram. The temperature must be on an absolute scale. Figure 7-1 illustrates this point on a unit-mass basis for some arbitrary internally reversible process. The value of Q may be determined by actual measurement of the area or by integration if the relationship between T and S is known.

Except for special circumstances, the integration of Eq. (7-2) is difficult. However, in many cases a Ts diagram is useful, if only for the qualitative information it provides. In the analysis of simple, compressible systems, sketches of processes on TS coordinate plots are often extremely informative and highly recommended. Hence the general characteristics of a Ts diagram for a specific substance are important. Figure 7-2 illustrates the Ts diagram for CO_2 in a simplified format. Only the characteristic lines in the gas and liquid regions are shown. A more detailed TS diagram for CO_2 which includes all three phase regions is presented in Fig. A-24. To aid in problem solving, the student should become familiar with the placement of constant-volume and constant-pressure lines on a Ts diagram. It should be observed that in the gas-phase region, constant-volume lines have a steeper slope than constant-pressure lines through any given state.

The Pv diagram is extremely useful in the analysis of processes involving boundary ($P\,dv$) work. It is helpful at this point to indicate the position of

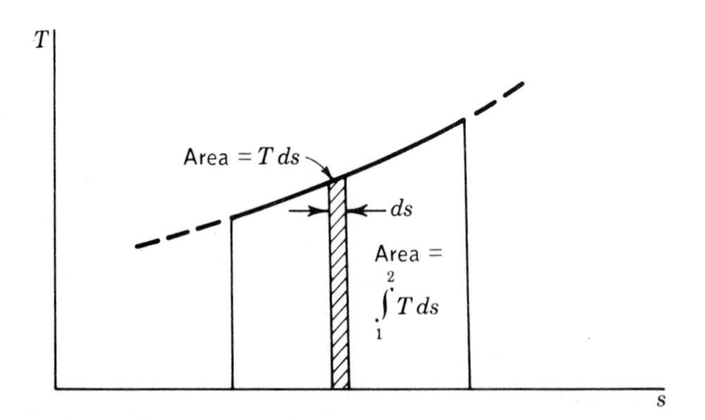

Figure 7-1 Heat transfer representation on a Ts diagram for an internally reversible process with a variation of system temperature.

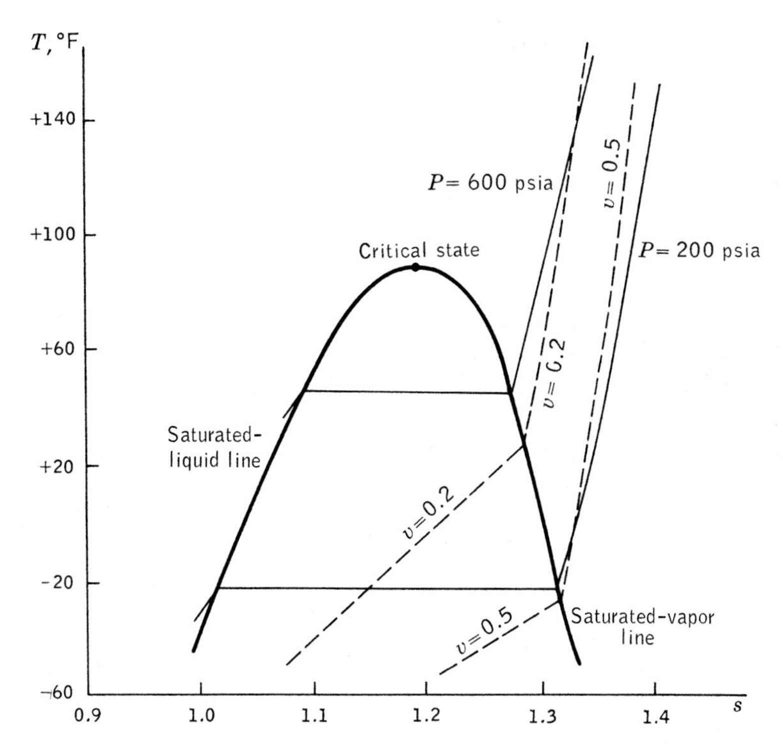

Figure 7-2 A Ts diagram for CO_2.

constant entropy lines on a Pv plot. Figure 7-3 illustrates this situation for the gas phase, and the position of isothermal (constant temperature) lines is also shown for comparison. Note that isentropic (constant entropy) lines have a steeper slope than isothermal lines in the gas region of a Pv diagram.

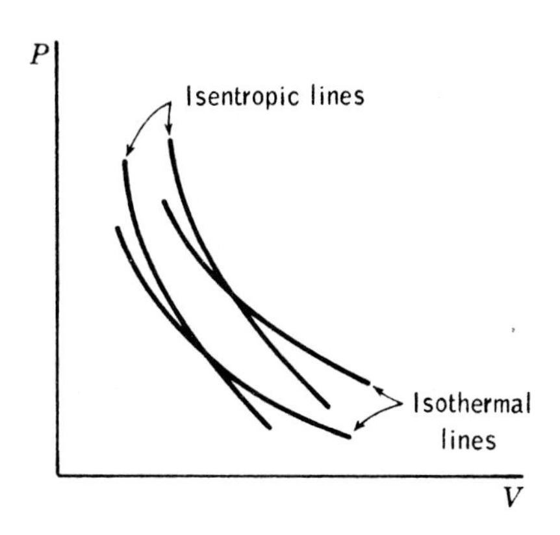

Figure 7-3 Process lines on a pressure-volume diagram.

7-2 THE ENTHALPY-ENTROPY DIAGRAM

In the analysis of steady-flow processes an enthalpy-entropy diagram, as well as a *Ts* plot, is quite useful. The enthalpy is the important property in the steady-flow energy balance, and entropy is the principal property of concern with respect to the second law. Thus the coordinates of an *hs* diagram represent the two major properties of interest in the first- and second-law analysis of open systems. The vertical distance between two states on this diagram is a measure of Δh. The enthalpy change, in turn, is related through the adiabatic steady-flow energy balance to the work and/or kinetic-energy changes for turbines, compressors, nozzles, etc. The horizontal distance Δs between two states is a measure of the degree of irreversibility for an adiabatic process. As a consequence we shall find the *hs* diagram helpful in visualizing process changes for control-volume analyses.

In addition to its use in process visualization, the *hs* diagram is also a means of presenting data. Such data can be read with reasonable accuracy if the diagram is drawn to a suitable size or scale. An *hs* diagram for steam is included in the Appendix as Fig. A-25. A schematic of an *hs* diagram, commonly called a *Mollier* diagram, is shown in Fig. 7-4. On an *hs* plot, constant-pressure lines and constant-temperature lines are straight in the wet (liquid-vapor) region. Lines of constant quality within the wet region lie roughly parallel to the saturated-vapor line. (On some Mollier charts, lines of constant quality are marked as percent-moisture lines. Percent moisture means the percent of liquid in the liquid-vapor mixture. The saturated-vapor line, for example, is a 0 percent–moisture line.)

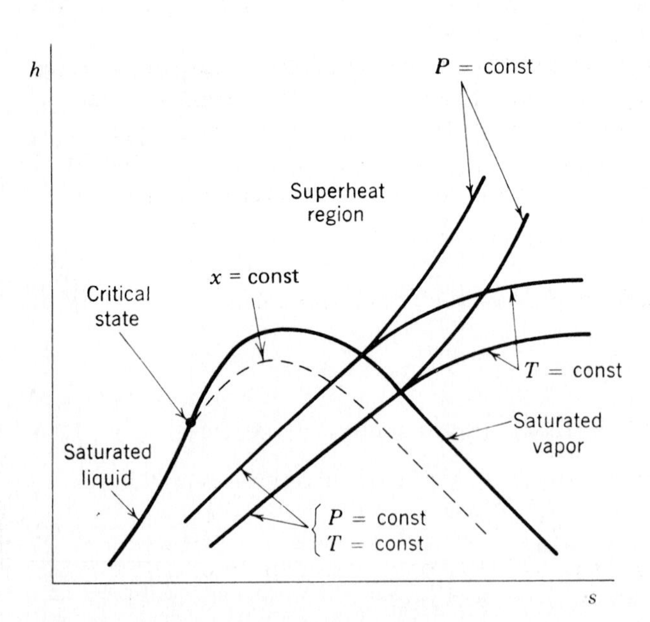

Figure 7-4 Schematic of a Mollier or *hs* diagram.

Another feature of this diagram is that constant-temperature lines become horizontal in the superheat region, which lies on the far right side of the plot. As the pressure is lowered on a gas at constant temperature, the gas behaves more like an ideal gas. For ideal gases the temperature and enthalpy are directly proportional. Hence, at low enough pressures, the temperature lines should lie parallel to the enthalpy lines. As an example, below roughly 40°C or 100°F steam behaves as an ideal gas at all pressures up to the saturation pressure for a given temperature. Finally, note that constant-volume lines are not normally shown on this type of diagram.

It has been pointed out that in certain parts of the superheated-vapor region on an *hs* diagram the temperature and enthalpy lines are parallel. This means that an *hs* and a *Ts* diagram for an ideal gas may be superimposed upon each other. This technique of letting the ordinate of a plot represent both T and h is quite useful for visualization of steady-flow processes. This is especially true when one is analyzing in a qualitative sense the effect of irreversibilities on performance.

7-3 THE $T\ dS$ EQUATIONS

The basic relation for the evaluation of the entropy change of any closed system is the integral of $\delta Q/T$ along an internally reversible path. Under isothermal conditions this integration is straightforward. For heat reservoirs, for example, the entropy change is simply Q/T. When T is a variable, it is necessary to find some functional relationship between δQ and T before integrating. The formal procedure for doing this is outlined in this section.

When a closed system is restricted to internally reversible changes of state, the only work interactions which are permitted are of the quasistatic type discussed in Chap. 2. The sum of all the generalized quasistatic work interactions may be represented by the term $\sum_k F_{k,\text{eq}}\, dX_k$. Therefore the conservation of energy equation for a control mass undergoing internally reversible changes is

$$\delta Q = dU - \sum_k F_{k,\text{eq}}\, dX_k$$

If this relation is substituted into the quantity $\delta Q/T$, we find that

$$dS = \frac{1}{T}\left(dU - \sum_k F_{k,\text{eq}}\, dX_k\right)$$

or
$$T\,dS = dU + P\,dV - \sigma\,d\epsilon - \gamma\,dA - E\,dP - \mu_0 VH\,dM + \cdots \tag{7-3}$$

For a simple, compressible system Eq. (7-3) takes on the abbreviated form

$$T\,dS = dU + P\,dV \tag{7-4}$$

This equation is called the first $T\ dS$ or Gibbsian equation for simple systems. Since $dH = dU + P\,dV + V\,dP$, the preceding equation can be written as

$$T\,dS = dH - V\,dP \tag{7-5}$$

This equation is the second $T\ dS$ or Gibbsian equation for simple, compressible systems.

On a unit-mass basis the $T\ dS$ equations for a simple, compressible substance become

$$T\ ds = du + P\ dv \tag{7-6}$$

and

$$T\ ds = dh - v\ dP \tag{7-7}$$

It is these equations which will be used to evaluate the specific entropy change of several classes of substances.

The set of Eqs. (7-3) to (7-7) was derived on the basis of the process being carried out in an internally reversible manner. However, the integration of these equations will lead to the correct entropy change between two equilibrium states whether the actual process is reversible or irreversible. The change in entropy between two states is independent of the path because entropy is a point function. Integration of the above equations requires a knowledge of the functional relationship among properties. The restrictions, or regions of applicability, for the $T\ dS$ equations are important. They are summarized as follows:

1. They are in general applied to homogeneous systems. They may be applied to heterogeneous systems if the composition of each phase remains fixed and the pressure and temperature are the same for every phase. For example, a two-phase, single-component system would fulfill the requirement over a wide range of temperatures and pressures. However, a two-phase system composed of air could not be used since the composition of each phase differs for each equilibrium temperature.
2. They are valid only for chemically invariant systems. No chemical reactions are permitted within the system boundaries. The equations must be modified to permit their usage for chemically variant systems.
3. They are employed only between equilibrium states.
4. They are equally valid for simple or nonsimple systems as long as the equations are modified to include the proper quasistatic work interactions.

The $T\ dS$ equations are extremely useful, since they allow one to calculate the entropy changes of substances once the functional relationships among the properties involved are known. Shortly we shall examine several special classes of substances for which the integration of the $T\ dS$ equations is quite simple. At the same time we should recognize that the entropy values we find in tables have resulted from the complex integration of equations directly derivable from the $T\ dS$ equations. It is equally important to recognize that these equations are also valid for a mass passing through an open steady-flow system, since we have already seen that the property entropy is solely a function of the intrinsic state of matter. Its value should not be influenced by extrinsic changes of state, such as changes in velocity or position of the mass. The $T\ dS$ equations are extremely valuable in the further development of thermodynamic relationships.

7-4 ENTROPY CHANGES INVOLVING REAL GAS AND SATURATION STATES

In Chap. 4 tables of data representing superheated-vapor, saturation, and compressed-liquid states were introduced. The emphasis then was on the properties P, V, T, u, and h, since we were concerned at that point with first-law analyses. In order to apply the second law to real gases and saturated fluids we must reconsider these tables in the light of the entropy function. There are no simple methods for integrating the $T\,dS$ equations presented in the preceding section for real gases, saturation states, and compressed liquids. The values of s (relative to an arbitrary reference state) in these cases are determined from fairly complex numerical techniques based on experimental P, v, T, and specific-heat data. These data are correct only to within the accuracy permitted by the experimental measurements and numerical computations. Thus reevaluation of tabular data is frequently necessary as improved experimental values become available.

The value of the entropy at a given state is tabulated in exactly the same manner as the properties v, u, and h. In the superheat region, where there are two independent properties, the entropy is tabulated along with the other properties as a function of temperature and pressure. For the saturation states, the values of s_f and s_g are given as a function of either temperature or pressure. For wet mixtures the quality of the mixture must be known in order to ascertain its specific entropy. Finally, if compressed-liquid data are available, the entropy function again is tabulated against the temperature and pressure of the fluid. In the absence of compressed-liquid data, the value of s_c in that region can be estimated by using s_f at the given temperature.

Several examples below illustrate the basic use of tables in terms of the entropy function, in conjunction with the increase in entropy principle. The use of tabulated entropy data will also be shown in numerous later chapters.

Example 7-1M Steam at 40 bars and 280°C is cooled at constant volume to a state of 9 bars. The heat removed from the steam enters the environment at 15°C. Determine (*a*) the entropy change of the steam in kJ/(kg)(°K), and (*b*) the total or overall entropy change for the process in kJ/(kg)(°K). (*c*) Determine if the process is reversible, irreversible, or impossible.

SOLUTION (*a*) The saturation temperature at 40 bars from Table A-13M is 250.4°C. Therefore the initial state is superheated vapor. From Table A-14M the entropy and the specific volume are read directly, and are found to be

$$s_1 = 6.2568 \text{ kJ/(kg)(°K)}$$

$$v_1 = 55.46 \text{ cm}^3\text{/g}$$

The final state is determined or fixed by the value of P_2 and v_2, which is the same as v_1. From Table A-13M, it is found that at 9 bars v_2 lies between v_f and v_g. Consequently, the final state is a wet mixture. The quality of the state is found by

$$v_2 = v_x = v_f + xv_{fg}$$

$$55.46 = 1.12 + x(215.0 - 1.1)$$

$$x = \frac{54.34}{213.9} = 0.254 \text{ (or 25.4\%)}$$

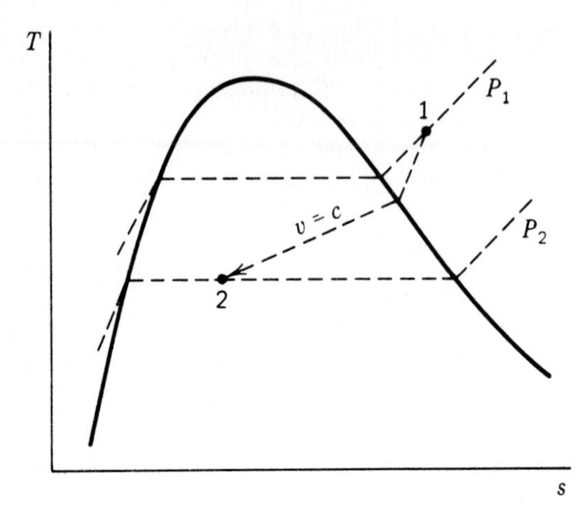

Figure 7-5 Process diagrams for Examples 7-1M and 7-1.

The knowledge of the quality enables us to evaluate the entropy.

$$s_2 = s_f + xs_{fg} = 2.0946 + 0.254(6.6226 - 2.0946) = 3.245 \text{ kJ/(kg)(°K)}$$

Hence the entropy change for the process is

$$\Delta s = s_2 - s_1 = 3.245 - 6.257 = -3.012 \text{ kJ/(kg)(°K)}$$

The process diagram is shown on the Ts plot of Fig. 7-5. The negative sign is not significant, other than to indicate that the gas is cooled during the process.

(*b*) The total entropy change is the sum of that of the system and that of the environment. Because the environment acts like a reservoir of constant temperature, its entropy change is simply Q/T. The heat transfer to the environment is the negative of the heat transfer from the system. This latter value, on a unit-mass basis, is found by $q + w = u_2 - u_1$. The system is one of fixed volume, hence $q = u_2 - u_1$. The internal energy values are found in the tables in the same manner that was used to find s_1 and s_2. From the superheat table for steam $u_1 = 2680.0$ kJ/kg. Knowledge of the quality again enables us to find u_2. That is,

$$u_2 = (1 - x)u_f + xu_g = 0.746(741.83) + 0.254(2580.5) = 1208.9 \text{ kJ/kg}$$

Therefore,

$$q = u_2 - u_1 = 1208.9 - 2680.0 = -1471.1 \text{ kJ/kg}$$

The heat transfer to the environment is the negative of the above value, or 1471.1 kJ/kg. As a result

$$\Delta s_{\text{envir}} = \frac{q}{T} = \frac{1471.1}{273 + 15} = 5.108 \text{ kJ/(kg)(°K)}$$

Finally, the total entropy change for the overall process is

$$\Delta s_{\text{total}} = \Delta s_{\text{system}} + \Delta s_{\text{envir}} = -3.012 + 5.108 = +2.096 \text{ kJ/(kg)(°K)}$$

(*c*) When the sum of the entropy changes for the composite parts is greater than zero, the process is possible, but irreversible. In this particular case the system is internally reversible, but irreversible heat transfer occurs due to the finite temperature difference between the system and the environment.

Example 7-1 Steam at 500 psia and 600°F is cooled at constant volume to a state of 100 psia. The heat removed from the steam enters the environment at 60°F. Determine (*a*) the entropy change of the steam in Btu/(lb)(°R), and (*b*) the total or overall entropy change for the process in Btu/(lb)(°R). (*c*) Determine if the process is reversible, irreversible, or impossible.

SOLUTION (*a*) The saturation temperature at 500 psia is roughly 467°F; therefore, the initial state is a superheated vapor. From Table A-14 the entropy is read directly, and is found to be

$$s_1 = 1.5585 \text{ Btu/(lb)(°R)}$$

The final state is determined by the pressure and the volume. Initially the specific volume is 1.158 ft³/lb. This is also the final specific volume at a pressure of 100 psia. From Table A-13, it is found that at this pressure this value of v lies between v_f and v_g. Consequently the final state is a wet mixture. The quality at this state is found by

$$v_x = v_f + xv_{fg}$$

$$1.158 = 0.01774 + x(4.432 - 0.01774)$$

$$x = \frac{1.14}{4.414} = 0.258$$

The knowledge of the quality enables us to evaluate the entropy.

$$s_2 = s_f + xs_{fg} = 0.4740 + 0.258(1.1286) = 0.7652 \text{ Btu/(lb)(°R)}$$

Here the entropy change is

$$\Delta s = s_2 - s_1 = 0.7652 - 1.5585 = -0.7933 \text{ Btu/(lb)(°R)}$$

The negative sign on the answer is not significant, other than to indicate that the gas is cooled during the process. The process diagram is shown on the accompanying *Ts* plot (Fig. 7-5).

(*b*) The total entropy change is the sum of that of the system and that of the environment. Because the environment acts like a reservoir of constant temperature, its entropy change is simply Q/T. The heat transfer to the environment is the negative of the heat transfer from the system. This latter value, on a unit-mass basis, is found from the basic energy balance for a closed system, namely, $q + w = u_2 - u_1$. The system is one of fixed volume, hence $q = u_2 - u_1$. The internal energy values are found in the tables in the same manner that was used to find s_1 and s_2. From the superheat table for steam, $u_1 = 1191.1$ Btu/lb. Knowledge of the quality in state 2 enables us to calculate u_2.

$$u_2 = (1 - x)(u_f) + xu_g = (0.742)(298.3) + 0.258(1105.8) = 506.6 \text{ Btu/lb}$$

Therefore,

$$q = u_2 - u_1 = 506.6 - 1191.1 = -684.5 \text{ Btu/lb}$$

The heat transfer to the environment is the negative of the above value, or 684.5 Btu/lb. As a result,

$$\Delta s_{\text{envir}} = q/T = \frac{684.5}{460 + 60} = 1.316 \text{ Btu/(lb)(°R)}$$

Finally, the total entropy change for the overall process is

$$\Delta s_{\text{total}} = \Delta s_{\text{system}} + \Delta s_{\text{envir}} = -0.793 + 1.316 = +0.523 \text{ Btu/(lb)(°R)}$$

(*c*) When the sum of the entropy changes for the composite parts is greater than zero, the process is possible, but irreversible. In this particular case the system is internally reversible, but irreversible heat transfer occurs due to the finite temperature difference between the system and the environment.

Example 7-2M One-tenth kilogram of water substance initially at 3 bars and 200°C is contained within a closed system. During a process, a heat removal of 7.70 kJ and a work input of 17,500 N · m occur, resulting in a final pressure of 15 bars. The heat removed appears in the surroundings at 22°C. Determine (*a*) the entropy change of the water substance, in kJ/°K, and (*b*) the total entropy change for the overall process, in kJ/°K.

SOLUTION (*a*) If the steam tables are to be used to evaluate data, a state must be explicitly defined. The initial pressure and temperature are sufficient in this case to fix the initial state. However, only the pressure is known for the final state. Therefore, we need one more property of a simple, compressible substance in order to find the entropy of the final state. Since the values of the energy interactions are known for the process, it is reasonable to suspect that an energy balance on the process will yield more information on the final state. For a closed system,

$$Q + W = \Delta U = m(u_2 - u_1)$$

The only unknown in this expression is the final specific internal energy u_2. The saturation temperature at 3 bars is 133.55°C. Hence the initial temperature of 200°C indicates that the initial state is a superheated vapor. From Table A-14M

$$u_1 = 2650.7 \text{ kJ/kg} \qquad \text{and} \qquad s_1 = 7.3115 \text{ kJ/(kg)(°K)}$$

Use of the energy balance then yields

$$u_2 = u_1 + \frac{Q + W}{m} = 2650.7 + \frac{-7.70 + (17{,}500/1000)}{0.1} = 2748.7 \text{ kJ/kg}$$

The internal energy of a saturated vapor u_g at 15 bars is approximately 2595 kJ/kg. Thus the final state is also a superheated vapor. At 15 bars and a specific internal energy of 2748.7 kJ/kg, we find from Table A-14M that

$$T_2 = 280°\text{C} \qquad \text{and} \qquad s_2 = 6.8381 \text{ kJ/(kg)(°K)}$$

The approximate path of the process on a Ts plot is shown in Fig. 7-6*a*. Consequently the

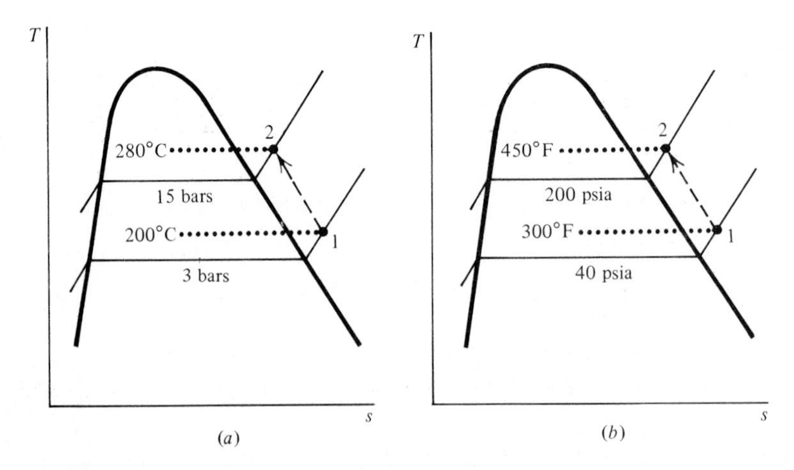

Figure 7-6 Process diagram for (*a*) Example 7-2M, and (*b*) Example 7-2.

entropy change of the steam within the closed system is

$$\Delta S = m\,\Delta s = 0.1 \text{ kg} \times (6.8381 - 7.3115) \text{ kJ/(kg)(°K)}$$
$$= -0.04734 \text{ kJ/°K}$$

The negative sign on the entropy change does not violate the increase in entropy principle, since the closed system is not adiabatic.

(*b*) The total entropy change for the overall process is the sum of the contributions from the system and the surroundings. For the surroundings, which act like a heat reservoir, $\Delta S = Q/T$. Hence

$$\Delta S_{\text{surroundings}} = \frac{Q}{T} = \frac{7.70}{273 + 22} = 0.0261 \text{ kJ/°K}$$

The total entropy change for the overall process becomes

$$\Delta S_{\text{total}} = \Delta S_{\text{system}} + \Delta S_{\text{surroundings}} = -0.04734 + 0.02610 = -0.02124 \text{ kJ/°K}$$

In part *a* the negative sign on ΔS for the system did not constitute a violation of the second law. However, the negative sign on the total entropy change indicates an impossible process. Unlike Example 7-1M, in this example both Q and W were specified. The second law violation is probably due to an incorrect measurement of either Q or W, or both. Whatever the error, this process is not possible as stated.

Example 7-2 One-tenth pound of water substance initially at 40 psia and 300°F is contained within a closed system. During a process a heat removal of 3.40 Btu and a work input of 7.53 Btu occur, resulting in a final pressure of 200 psia. The heat removed appears in the surroundings at 80°F. Determine (*a*) the entropy change of the water substance in Btu/°R, and (*b*) the total entropy change for the overall process in Btu/°R.

SOLUTION (*a*) If the steam tables are to be used to evaluate data, a state must be explicitly defined. The initial pressure and temperature are sufficient to fix the initial state. However, only the pressure is known for the final state. Therefore we need one more property in order to find the entropy of the final state. Since the values of the interactions are known during the process, it is reasonable to suspect that an energy balance on the process will yield more information on the final state.
For a closed system,

$$Q + W = \Delta U = m(u_2 - u_1)$$

The only unknown in this expression is the final specific internal energy u_2. The saturation temperature at 40 psia is roughly 267°F. Hence an initial temperature of 300°F at this pressure indicates that the initial state is a superheated vapor. From the superheat table, Table A-14, we find that

$$u_1 = 1105.1 \text{ Btu/lb} \qquad \text{and} \qquad s_1 = 1.6993 \text{ Btu/(lb)(°R)}$$

Use of the energy balance then yields

$$u_2 = u_1 + \frac{Q + W}{m} = 1105.1 + \frac{-3.40 + 7.53}{0.1} = 1146.4 \text{ Btu/lb}$$

The internal energy of a saturated vapor u_g at 200 psia is approximately 1114 Btu/lb. Thus the final state is also a superheated vapor. At 200 psia and an internal energy of 1146.4 Btu/lb, we find that

$$T_2 = 450\text{°F} \qquad \text{and} \qquad s_2 = 1.5938 \text{ Btu/(lb)(°R)}$$

The approximate path of the process on a *Ts* plot is shown in Fig. 7-6*b*. Consequently the entropy change of the steam within the closed system is

$$\Delta S = m\,\Delta s = 0.1\text{ lb} \times (1.5938 - 1.6993)\text{ Btu/(lb)(°R)}$$

$$= -0.01055\text{ Btu/°R}$$

The negative entropy change found here does not violate the second law, per se, since the system chosen is neither adiabatic or isolated.

(*b*) The total entropy change for the overall process is the sum of the contributions from the system and the surroundings. For the surroundings, which act like a heat reservoir, $\Delta S = Q/T$. Hence

$$\Delta S_{\text{surroundings}} = \frac{Q}{T} = \frac{3.40}{460 + 80} = 0.00630\text{ Btu/°R}$$

The total entropy change for the overall process becomes

$$\Delta S_{\text{total}} = \Delta S_{\text{system}} + \Delta S_{\text{surroundings}} = -0.01055 + 0.00630 = -0.00425\text{ Btu/°R}$$

In part *a* the negative sign on ΔS for the system did not constitute a violation of the second law. However, the negative sign on the total entropy change indicates an impossible process. Unlike Example 7-1, in this example both Q and W were specified. The second law violation is probably due to an incorrect measurement of either Q or W, or both. Whatever the error, this process is not possible as stated.

Example 7-3M Refrigerant 12 initially at 2 bars and 30°C is expanded isothermally and internally reversibly to 1 bar in a steady-flow process. In the absence of work effects, determine the change in kinetic energy, in kJ/kg.

SOLUTION Assuming the potential-energy change to be negligible, the steady-flow energy equation reduces to $q = \Delta h + \Delta\text{KE}$. Since the process is isothermal and reversible, q is given by $T\,\Delta s$. Therefore the change in kinetic energy becomes

$$\Delta\text{KE} = T\,\Delta s - \Delta h = T(s_2 - s_1) - (h_2 - h_1)$$

A sketch of the process is shown in Fig. 7-7 on Pv and Ts diagrams. The saturation temperature is -12.53°C at 2 bars and -30.10°C at 1 bar. Since the process occurs at 30°C, the fluid is in the superheated region throughout the process. From Table A-18M:

$P_1 = 2$ bars	$P_2 = 1$ bar
$T_1 = 30$°C	$T_2 = 30$°C
$s_1 = 0.7978$ kJ/(kg)(°K)	$s_2 = 0.8488$ kJ/(kg)(°K)
$h_1 = 208.60$ kJ/kg	$h_2 = 210.02$ kJ/kg

Substitution of these values into the preceding equation yields

$$\Delta\text{KE} = 303(0.8488 - 0.7978) - (210.02 - 208.60)$$

$$= 15.45 - 1.42$$

$$= 14.03\text{ kJ/kg (or J/g)}$$

Note that in order to maintain the system isothermal, a total of 15.45 kJ/kh of heat must be added to the system. For the heat transfer process

$$q_{\text{system}} = -q_{\text{res}} \qquad \text{or} \qquad T_{\text{system}}\,\Delta S_{\text{system}} = -T_{\text{res}}\,\Delta S_{\text{res}}$$

Because $T_{\text{res}} > T_{\text{system}}$ by necessity, then ΔS of the reservoir is smaller than that of the system, and is also negative in value. Consequently, the sum of ΔS for the system R-12 and for the reservoir is positive. The process, by the second law, is possible but irreversible. The system is internally reversible, but the overall process is externally irreversible.

Example 7-3 Refrigerant 12 initially at 20 psia and 80°F is expanded isothermally and internally reversibly to 15 psia in a steady-flow process. In the absence of work effects, determine the change in kinetic energy, in Btu/lb.

SOULTION Assuming the potential-energy change is negligible, the steady-flow energy balance reduces to $q = \Delta h + \Delta KE$. However, for isothermal, reversible flow, q is given by $T\,\Delta s$. Therefore the change in kinetic energy becomes

$$\Delta KE = T\,\Delta s - \Delta h = T(s_2 - s_1) - (h_2 - h_1)$$

A sketch of the process is shown in Fig. 7-7 on Pv and Ts diagrams. The saturation temperature at 20 psia is -8.13°F and -20.75°F at 15 psia. Since the process occurs at 80°F, the fluid is a superheated vapor throughout the process. From Table A-18:

$P_1 = 20$ psia	$P_2 = 15$ psia
$T_1 = 80$°F	$T_2 = 80$°F
$s_1 = 0.1955$ Btu/(lb)(°R)	$s_2 = 0.2005$ Btu/(lb)(°R)
$h_1 = 89.168$ Btu/lb	$h_2 = 89.383$ Btu/lb

Substitution of these values into the preceding equation yields

$$\begin{aligned}\Delta KE &= 540(0.2005 - 0.1955) - (89.383 - 89.168)\\ &= 2.70 - 0.22\\ &= 2.48 \text{ Btu/lb}\end{aligned}$$

In order to maintain the system isothermal, a total of 2.7 Btu/lb of heat must be added to the

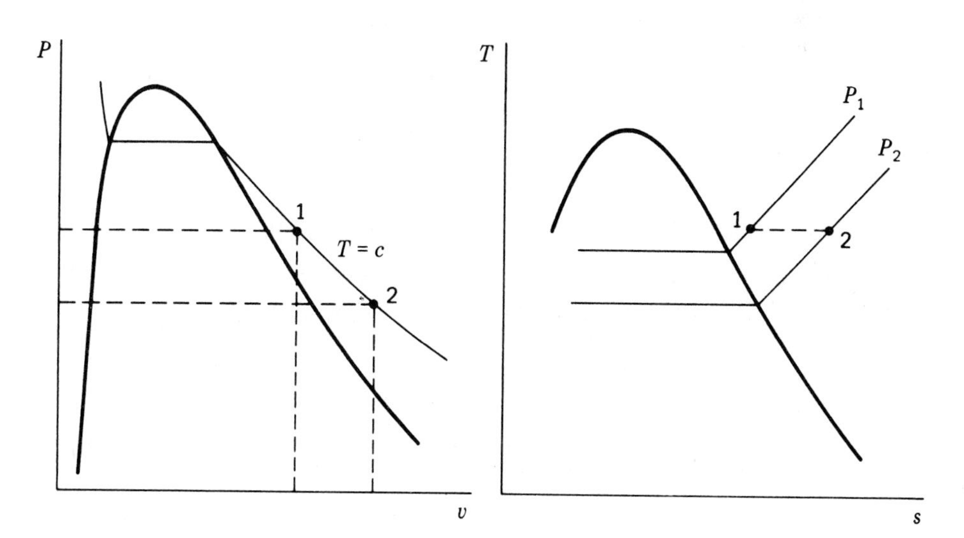

Figure 7-7 The Pv and Ts diagrams for Examples 7-3M and 7-3.

system. For the heat transfer process,

$$q_{\text{system}} = -q_{\text{res}} \qquad \text{or} \qquad T_{\text{system}}\,\Delta S_{\text{system}} = -T_{\text{res}}\,\Delta S_{\text{res}}$$

Because $T_{\text{res}} > T_{\text{system}}$ by necessity, then ΔS of the reservoir is smaller than that of the system, and is also negative in value. Consequently the sum of ΔS for the system R-12 and for the reservoir is positive. The process, by the second law, is possible but irreversible. The system is internally reversible, but the overall process is externally irreversible.

7-5 ENTROPY CHANGES OF IDEAL GASES

The $T\,ds$ equations introduced in Sec. 7-3 allow one to evaluate the entropy changes of substances once the functional relationships among the properties u, h, P, v, T, c_v, and c_p are known. For some substances, such as real gases discussed in the preceding section, these functional relationships are quite involved. Thus complex numerical computations are required in order to set up tables for the entropy function. In some other cases the integration of the $T\,ds$ equations is straightforward. A simple and useful example is the application of the $T\,ds$ equations to systems containing ideal gases. The basic place to start is with the two fundamental relations

$$T\,ds = du + P\,dv \tag{7-6}$$

and

$$T\,ds = dh - v\,dP \tag{7-7}$$

For an ideal gas, $du = c_v\,dT$, $dh = c_p\,dT$, and $Pv = RT$. Therefore

$$ds = \frac{du + P\,dv}{T} = c_v\frac{dT}{T} + R\frac{dv}{v} \tag{7-8}$$

and

$$ds = \frac{dh - v\,dP}{T} = c_p\frac{dT}{T} - R\frac{dP}{P} \tag{7-9}$$

Integration of the last term in these equations requires no further information since R is a constant. We have seen, however, that both c_v and c_p are functions of temperature alone for ideal gases.

a Use of Constant or Arithmetically Averaged Specific-Heat Data

If the temperature range is small, the arithmetically averaged specific heat may be used with negligible error. Table A-4 shows that the variation of c_p and c_v over several hundred degrees is small in many cases. Integration of Eqs. (7-8) and (7-9) for constant specific heats shows that, for ideal gases,

$$\Delta s = c_v \ln\frac{T_2}{T_1} + R\ln\frac{v_2}{v_1} \tag{7-10}$$

and

$$\Delta s = c_p \ln\frac{T_2}{T_1} - R\ln\frac{P_2}{P_1} \tag{7-11}$$

The c_v and c_p value to be used in the above equations may be either the value at the initial temperature or the arithmetically-averaged value over the given temperature range. These equations provide an easy and fairly accurate method for evaluating entropy changes of ideal gases over fairly small temperature ranges.

Example 7-4M Air is compressed from 1 bar, 27°C, to 3.5 bars, 127°C, in a steady-flow system. Determine the entropy change, in kJ/(kg)(°K) or J/(g)(°K).

SOLUTION For the small temperature range under consideration, from 300 to 400°K, the values of c_v and c_p may be considered constant. From Table A-4M it is found that their average values in this range are 0.722 and 1.009 kJ/(kg)(°K), respectively. The value of R needed can be found from the universal values found in Table A-1M, or by recalling that for ideal gases $R = c_p - c_v$. From this latter relation we find that $R = 0.287$ kJ/(kg)(°K). Since both temperature and pressure data are given, Eq. (7-11) is an appropriate equation to use to evaluate Δs. Hence

$$\Delta s = 1.009 \ln \tfrac{400}{300} - 0.287 \ln \tfrac{3.5}{1} = -0.0693 \text{ kJ/(kg)(°K)}$$

The negative answer is not in conflict with the second law, since the process is not stated to be adiabatic. Figure 7-8 shows the process on a Ts diagram. As described in Sec. 7-1, constant-pressure lines run from lower left to upper right on a Ts plot.

Example 7-4 Air is compressed from 15 psia, 80°F, to 50 psia, 240°F, in a steady-flow system. Determine the entropy change per pound.

SOLUTION The accompanying sketch (see Fig. 7-8) shows the process on a Ts diagram. For the small temperature range under consideration, the values of c_v and c_p may be considered constant, and they have average values of 0.173 and 0.241 Btu/(lb)(°F), respectively. Since $R = c_p - c_v$ for ideal gases, R is 0.068 Btu/(lb)(°F). Employing Eq. (7-11), one finds that

$$\Delta s = 0.241 \ln \tfrac{700}{540} - 0.068 \ln \tfrac{50}{15} = -0.0199 \text{ Btu/(lb)(°R)}$$

The negative answer is not in conflict with the increase in entropy principle for a closed system, since the process is not stated to be adiabatic.

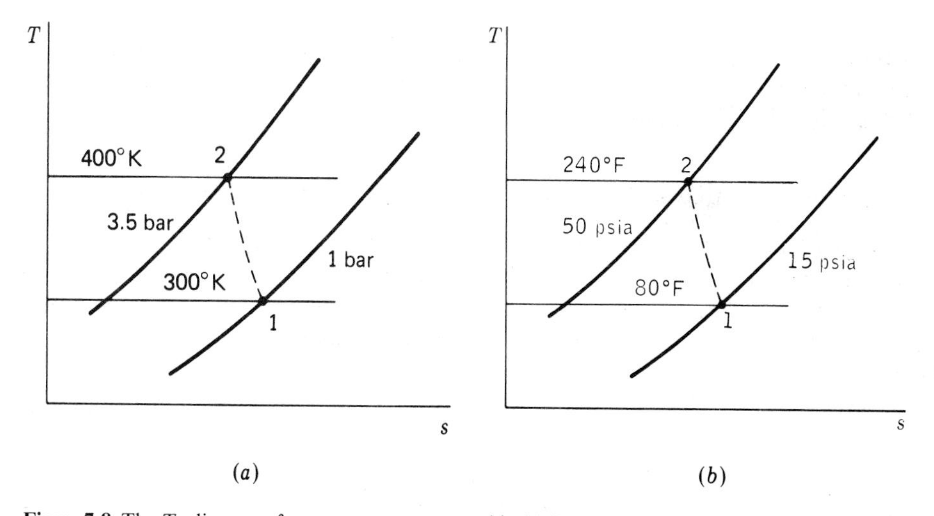

Figure 7-8 The Ts diagrams for processes represented in (a) Example 7-4M, and (b) Example 7-4.

Example 7-5M A rigid, insulated tank contains 1.2 kg of nitrogen gas at 350°K and 1 bar. A paddle-wheel inside the tank is driven by a pulley-weight mechanism. During an experiment 25,000 N · m of work is expended on the gas through the pulley-weight mechanism. Determine the entropy change for the nitrogen, in kJ/°K.

SOLUTION The paddle-wheel work constitutes an internal irreversibility. Nevertheless, the $T\,ds$ equations will still lead to a correct evaluation of ΔS. In oder to use these equations, however, we must determine the final state of the gas. The initial state already is completely specified. Since energy information is given, the application of an energy balance on the system probably will provide additional information on the final state. The tank is insulated and rigid, so boundary work and heat transfer are zero. However, paddle-wheel work is present. Thus,

$$Q + W_{\text{paddle}} = \Delta U = mc_v \,\Delta T$$

$$0 + 25{,}000 \text{ N} \cdot \text{m} = 1200\text{g} \times 0.744 \text{ J/(g)(°C)} \times (T_2 - 350)\text{°K}$$

$$T_2 = 350 + 28.0 = 378\text{°K}$$

A c_v value at the initial temperature is used, since the final temperature is not yet known. Since ΔT is only 28°C, the use of the initial c_v value is quite appropriate in this case.

Since the tank is rigid, the initial and final specific volumes are equal. Therefore Eq. (7-10) is more appropriate to use than Eq. (7-11), because the last term in Eq. (7-10) is zero for the process. As a consequence,

$$\Delta S = m\,\Delta s = mc_v \ln \frac{T_2}{T_1} = 1200(0.744) \ln \frac{378}{350}$$

$$= 68.7 \text{ J/°K} = 0.0687 \text{ kJ/°K}$$

The positive change in entropy is brought about by the irreversible work interaction. Since the process is adiabatic and irreversible, the entropy change must be positive in accordance with the increase in entropy principle for adiabatic, closed systems.

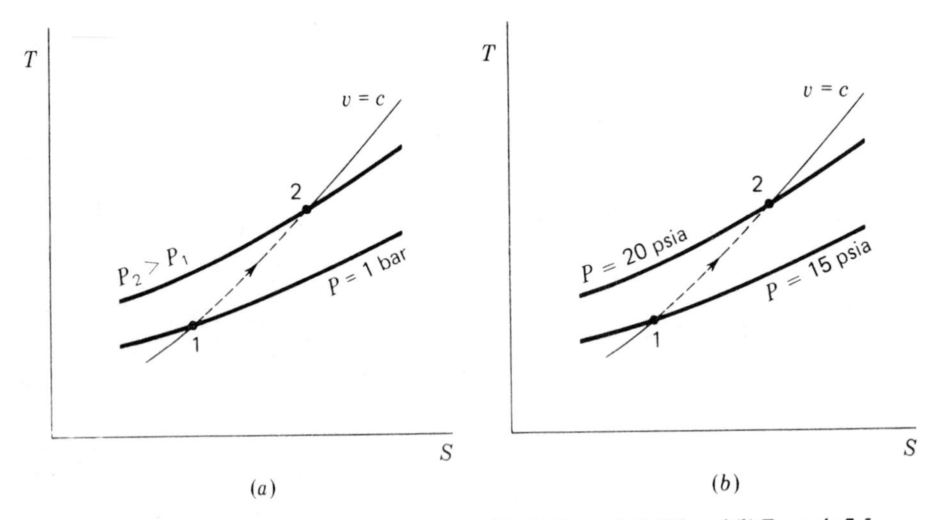

Figure 7-9 The TS diagrams for processes represented in (*a*) Example 7-5M, and (*b*) Example 7-5.

A TS plot of the process is shown in Fig. 7-9*a*. From the ideal-gas equation, $P_2/P_1 = T_2/T_1$. Based on the above data, the pressure increased during the process. We noted in Sec. 7-1 that constant-volume lines have a steeper positive slope than constant-pressure lines on a TS plot. Hence the diagram requires that the entropy change be positive.

Example 7-5 A rigid, insulated tank contains 2 lb of nitrogen gas at 140°F and 15 psia. A paddle-wheel inside the tank is driven by a pulley-weight mechanism. During an experiment 10,000 $\text{ft} \cdot \text{lb}_f$ of work is expended on the gas through the pulley-weight mechanism. Determine the entropy change for the nitrogen, in Btu/°R.

SOLUTION The paddle-wheel work constitutes an internal irreversibility. Nevertheless, the $T\,ds$ equations will still lead to a correct evaluation of ΔS. In order to use these equations, however, the initial and final states must be determined. In this case the initial state already is completely specified, but only the volume of the final state is known. Since energy information is given in the problem statement, the application of an energy balance probably will provide additional information on the final state. The tank is insulated and rigid, so boundary work and heat transfer are zero. However, paddle-wheel work is present. Thus,

$$Q + W_{\text{paddle}} = \Delta U = mc_v\,\Delta T$$

If Btu are the units of each term, then

$$0 + \frac{10{,}000}{778} = 2(0.178)(T_2 - 140)$$

$$T_2 = 140 + 36.1 = 176.1°\text{F} = 636.1°\text{R}$$

A c_v value at the initial temperature, rather than an average value, is used, since the final temperature is not known until the energy balance is made. Since the temperature rise is only 36°F, use of the initial c_v value is appropriate.

Since the tank is rigid, the initial and final specific volumes are equal. Hence the entropy change can be found most easily by using Eq. (7-10) because the last term is zero. Therefore

$$\Delta S = m\,\Delta s = mc_v \ln \frac{T_2}{T_1} = 2(0.178) \ln \frac{636.1}{600}$$

$$= 0.0208 \text{ Btu/°R}$$

The positive change in entropy is brought about by the irreversible work interaction. The entropy change must be positive for this process in accordance with the increase in entropy principle for irreversible, adiabatic closed-system processes.

A TS plot of the process is shown in Fig. 7-9*b*. From the ideal-gas equation, $P_2/P_1 = T_2/T_1$. Therefore the pressure increased during the process. We noted in Sec. 7-1 that constant-volume lines have a steeper slope than constant-pressure lines on a TS plot. Hence the diagram requires that the entropy change be positive.

Examples 7-5M and 7-5 illustrate an interesting point with regard to the increase in entropy principle and nonquasistatic work interactions such as paddle-wheel work. Assume that the reverse process occurs, so that energy is removed from the gas and paddle-wheel work is delivered to the surroundings. For this adiabatic process, the entropy change of the closed system would be equal in magnitude, but opposite in sign, to the original stirring process. But a negative entropy change is not permitted for adiabatic closed systems. Thus processes like paddle-wheel and electrical-resistor work are truly nonreversible.

Example 7-6M A piston-cylinder device maintained at 1.5 bars contains air initially at 500°K. Heat transfer occurs until the air reaches a temperature of 400°K.

(*a*) Determine the specific entropy change of the air within the cylinder in kJ/(kg)(°K).

(*b*) If the heat transfer is between the system and the environment at 27°C, determine the total entropy change for the overall process in kJ/(kg)(°K).

SOLUTION The closed system is chosen to be the air within the assembly. The constant-pressure process is shown as path *A* from state 1 to state 2 on the accompanying *PV* plot (see Fig. 7-10). If we choose Eq. (7-11) to evaluate the entropy change, then the pressure term in the equation drops out, since $P = \text{constant}$. From Table A-4M we find the average c_p value to be 1.021 kJ/(kg)(°K). Therefore

$$\Delta s = c_{p,\,\text{av}} \ln \frac{T_2}{T_1} = 1.021 \ln \frac{400}{500} = -0.228 \text{ kJ/(kg)(°K)}$$

The entropy change is negative since heat is removed during the process in order to lower the temperature.

ALTERNATE SOLUTION Although it appears convenient, it is not necessary to evaluate Δs along path *A* by means of Eq. (7-11). Equation (7-10) involves a temperature term and a specific-volume term. This equation, in reality, represents a path such as *B* shown on the *PV* diagram. The term $c_v \ln (T_2/T_1)$ evaluates the entropy change at constant volume for a change in temperature. In this case this is process 1-*x*. The term $R \ln (v_2/v_1)$ is the change in entropy at constant temperature for a change in volume. This is process *x*-2 of path *B* on the diagram. Physically, starting from state 1, the piston is clamped in position to fix the volume. Heat is then removed until the temperature drops from 500 to 400°K. Then the piston is released and the volume is allowed to decrease isothermally. During this second quasistatic step the pressure increases from P_x to P_2.

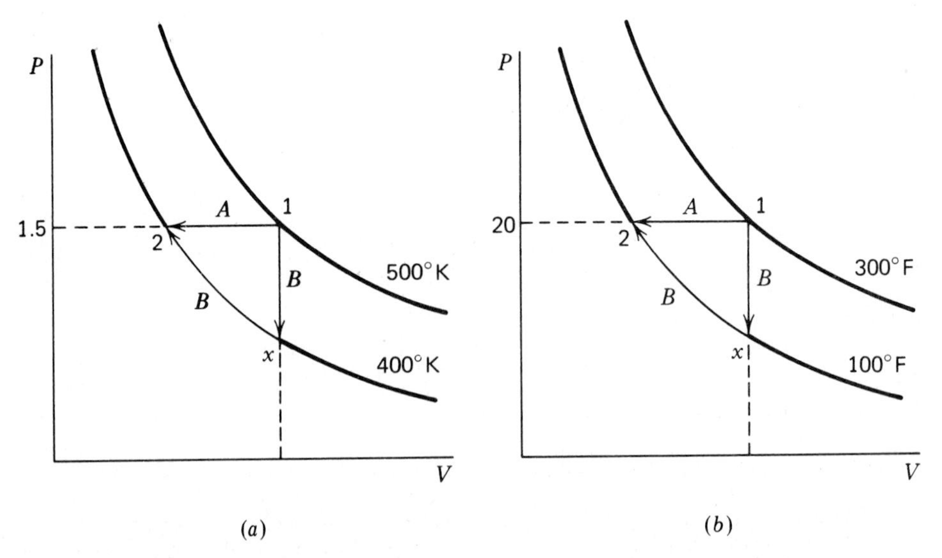

Figure 7-10 The *PV* diagrams for processes discussed in (*a*) Example 7-6M, and (*b*) Example 7-6.

For the process 1-x the volume is constant, and the entropy change for this step is

$$s_x - s_1 = c_{v,\,\mathrm{av}} \ln \frac{T_x}{T_1} = c_{v,\,\mathrm{av}} \ln \frac{T_2}{T_1}$$

$$= 0.734 \ln \tfrac{400}{500} = -0.164 \text{ kJ/(kg)(°K)}$$

For the process x-2 the temperature is constant, and the entropy change for this second step is

$$s_2 - s_x = R \ln \frac{v_2}{v_x} = R \ln \frac{v_2}{v_1}$$

Use of the ideal-gas equation reveals that $v_2/v_1 = P_1 T_2/P_2 T_1 = T_2/T_1$. Consequently,

$$s_2 - s_x = R \ln \frac{T_2}{T_1} = \frac{8.314}{29} \text{ J/(g)(°K)} \times \ln \frac{400}{500}$$

$$= -0.640 \text{ kJ/(kg)(°K)}$$

For the overall process 1-x-2,

$$s_2 - s_1 = -0.164 + (-0.064) = -0.228 \text{ kJ/(kg)(°K)}$$

This answer is the same as that calculated for the direct path 1-2. This is not unexpected. Entropy is a property, and the change in a property depends solely upon the end state. This example does point out, however, that from the evaluation standpoint some paths lead to simpler calculations than others.

(*b*) The total entropy change is the sum of the entropy changes of the air within the piston and of the environment at 27°C. For the environment

$$\Delta s_{\mathrm{envir}} = \left(\frac{Q}{T}\right)_{\mathrm{envir}} = \left(\frac{-Q_{\mathrm{system}}}{T}\right)_{\mathrm{envir}}$$

The heat transfer from the system to the environment is found from an energy balance on the system. For a constant-pressure process

$$q = \Delta u + P\,\Delta v = \Delta h = c_{p,\,\mathrm{av}}\,\Delta T$$

Applying data already used above, we find that for the system

$$q = 1.021 \text{ kJ/(kg)(°K)} \times (400 - 500)\text{°K} = -102.1 \text{ kJ/kg}$$

Therefore,

$$\Delta s_{\mathrm{envir}} = \frac{-(-102.1)}{27 + 273} = +0.340 \text{ kJ/(kg)(°K)}$$

and the total entropy change is

$$\Delta s_{\mathrm{total}} = -0.228 + 0.340 = +0.112 \text{ kJ/(kg)(°K)}$$

Thus the process is possible, but irreversible.

Example 7-6 A piston-cylinder device maintained at 20 psia contains air initially at 300°F. Heat transfer occurs until the air reaches a temperature of 100°F.

(*a*) Determine the specific entropy change of the air within the cylinder in Btu/(lb)(°R).

(*b*) If the heat transfer is between the system and the environment at 40°F, determine the total entropy change for the overall process in Btu/(lb)(°R).

SOLUTION (*a*) The closed system is chosen to be the air within the assembly. The constant-pressure process is shown as path *A* from state 1 to state 2 on the accompanying *PV* plot (see Fig. 7-10). If we choose Eq. (7-11) to evaluate the entropy change, then the pressure term in the equation drops out, since $P =$ constant. From Table A-4 we find the average c_p value to be 0.241 Btu/(lb)(°R). Therefore

$$\Delta s = c_{p,\,\text{av}} \ln \frac{T_2}{T_1} = 0.241 \ln \frac{560}{660} = -0.0737 \text{ Btu/(lb)(°R)}$$

The entropy change is negative since heat is removed during the process in order to lower the temperature.

ALTERNATE SOLUTION Although it appears convenient, it is not necessary to evaluate Δs along path *A* by means of Eq. (7-11). Equation (7-10) involves a temperature term and a specific-volume term. This equation, in reality, represents a path such as *B* shown on the *PV* diagram. The term $c_v \ln (T_2/T_1)$ evaluates the entropy change at constant volume for a change in temperature. In this case this is process 1-*x*. The term $R \ln (v_2/v_1)$ is the change in entropy at constant temperature for a change in volume. This is process *x*-2 of path *B* on the diagram. Physically, starting from state 1, the piston is clamped in position to fix the volume. Heat is then removed until the temperature drops from 760 to 560°R. Then the piston is released and the volume allowed to decrease isothermally. During this second quasistatic step the pressure increases from P_x to P_2.

For the process 1-*x* the volume is constant, and the entropy change for this step is

$$s_x - s_1 = c_{v,\,\text{av}} \ln \frac{T_x}{T_1} = c_{v,\,\text{av}} \ln \frac{T_2}{T_1}$$

$$= 0.173 \ln \tfrac{560}{760} = -0.0528 \text{ Btu/(lb)(°R)}$$

For the process *x*-2 the temperature is constant, and the entropy change for this second step is

$$s_2 - s_x = R \ln \frac{v_2}{v_x} = R \ln \frac{v_2}{v_1}$$

Use of the ideal-gas equation reveals that $v_2/v_1 = P_1 T_2/P_2 T_1 = T_2/T_1$. Consequently,

$$s_2 - s_x = R \ln \frac{T_2}{T_1} = \frac{1.986}{29} \text{ Btu/(lb)(°R)} \times \ln \frac{560}{760}$$

$$= -0.0209 \text{ Btu/(lb)(°R)}$$

For the overall process 1-*x*-2,

$$s_2 - s_1 = -0.0528 + (-0.0209) = -0.0737 \text{ Btu/(lb)(°R)}$$

This answer is the same as that calculated for the direct path 1-2. This is not unexpected. Entropy is a property, and the change in a property depends solely upon the end states. This example does point out, however, that from the evaluation standpoint some paths lead to simpler calculations than others.

(*b*) The total entropy change is the sum of the entropy changes of the air within the piston and of the environment at 40°F. For the environment

$$\Delta s_{\text{envir}} = \left(\frac{Q}{T}\right)_{\text{envir}} = \left(\frac{-Q_{\text{system}}}{T}\right)_{\text{envir}}$$

The heat transfer from the system to the environment is found from an energy balance on the system. For a constant-pressure process

$$q = \Delta u + P\,\Delta v = \Delta h = c_{p,\,\text{av}}\,\Delta T$$

Applying data already used above, we find that for the system

$$q = 0.241 \text{ Btu/(lb)(°F)} \times (100 - 300)\text{°F} = -48.2 \text{ Btu/lb}$$

Therefore,

$$\Delta s_{\text{envir}} = \frac{-(-48.2)}{40 + 460} = 0.964 \text{ Btu/(lb)(°R)}$$

and the total entropy change is

$$\Delta s_{\text{total}} = -0.737 + 0.0964 = +0.0227 \text{ Btu/(lb)(°R)}$$

Thus the process is possible, but irreversible.

Example 7-7M Argon gas is contained in an insulated tank at 25°C and 2 bars. A partition is punctured, and the gas flows into another insulated tank of the same volume. The second portion of the tank is initially evacuated. Determine the entropy change.

SOLUTION Argon may be considered to be an ideal gas. The two portions of the tank, together, are selected as the boundaries of the closed system. Figure 7-11 shows a schematic and a PV diagram. Since the tanks are rigid, the work interactions are zero. The process is adiabatic and highly irreversible. On the basis of the increase in entropy principle we expect a positive change in entropy.

To evaluate Δs we need more information on the final state. Since both Q and W are zero, ΔU is also zero. The internal energy of an ideal gas is solely a function of temperature; consequently, the temperature remains at 25°C during the free expansion. From the ideal-gas equation, $P_1V_1/T_1 = P_2V_2/T_2$. Hence the final pressure is one-half the initial value, or 1 bar. Either Eq. (7-10) or Eq. (7-11) may be used to evaluate Δs since in both cases the temperature term is zero. Choosing the former, we find that

$$\Delta s = R \ln (v_2/v_1) = \frac{8.314}{40} \frac{\text{kJ}}{(\text{kg})(\text{°K})} \times \ln 2 = 0.225 \text{ kJ/(kg)(°K)}$$

This problem illustrates why the restriction of internal reversibility is so important on the integral of $\delta Q/T$. Since Q is zero, the integral of $\delta Q/T$ has no relation to ΔS if the process to which it is applied is internally irreversible.

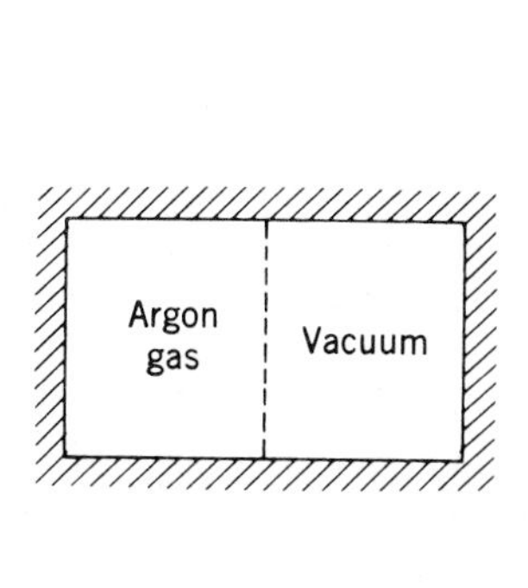

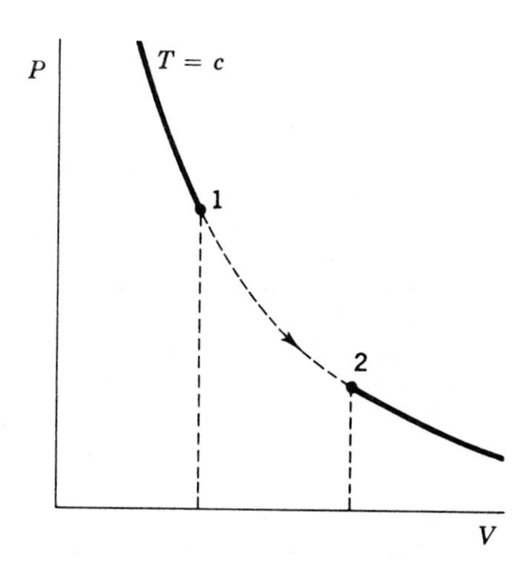

Figure 7-11 Schematic and PV diagram for process discussed in Examples 7-7M and 7-7.

Example 7-7 Argon gas is contained in an insulated tank at 100°F and 30 psia. A partition is punctured, and the gas flows into another insulated tank of the same volume. The second portion of the tank is initially evacuated. Determine the entropy change.

SOLUTION Argon may be considered to be an ideal gas. The two portions of the tank, together, are selected as the boundaries of the closed system. Figure 7-11 shows a schematic and a PV diagram. Since the tanks are rigid, the work interactions are zero. The process is adiabatic and highly irreversible. On the basis of the increase in entropy principle we expect a positive change in entropy.

To evaluate Δs we need more information on the final state. Since both Q and W are zero, ΔU is also zero. The internal energy of an ideal gas is solely a function of temperature; consequently, the temperature remains at 100°F during the free expansion. From the ideal-gas equation, $P_1V_1/T_1 = P_2V_2/T_2$. Hence the final pressure is one-half the initial value, or 15 psia. Either Eq. (7-10) or (7-11) may be used to evaluate Δs since in both cases the temperature term is zero. Choosing the former, we find that

$$\Delta s = R \ln \frac{v_2}{v_1} = \frac{1.986}{40} \text{ Btu/(lb)(°R)} \times \ln 2 = 0.0344 \text{ Btu/(lb)(°R)}$$

This problem illustrates why the restriction of internal reversibility is so important on the integral of $\delta Q/T$. Since Q is zero, the integral of $\delta Q/T$ is zero. However, ΔS is not zero, as shown by our calculation. Thus we see that the integral of $\delta Q/T$ has no relation to ΔS if the process to which it is applied is internally irreversible.

b Use of Integrated Specific-Heat Data

More accurate evaluation of Eqs. (7-8) and (7-9) is attained by the use of analytical expressions for the specific heats as a function of temperature. Table A-4M is an example of such data. By programming equations for specific heat data on a computer, rapid evaluations of Δs may be made. Unless many such calculations are needed, however, it may not be desirable to use a high-speed computer. An alternative is to tabulate the integration of Eqs. (7-8) and (7-9) in some manner similar to the tabulations of h and u in the ideal-gas tables in the Appendix. In devising tables for the accurate evaluation of Δs, Eq. (7-9) is usually used. Integrating the last term of this equation in the usual manner, we note that

$$\Delta s = \int_1^2 c_p \frac{dT}{T} - R \ln \frac{P_2}{P_1} \tag{7-12}$$

Now we shall define a function s^0 such that its value relative to a reference state can be found by

$$s_1^0 - s_{\text{ref}}^0 = \int_{\text{ref}}^1 c_p \frac{dT}{T}$$

The determination of $s_1^0 - s_{\text{ref}}^0$ requires only a knowledge of c_p as a function of T. In a similar manner for another temperature T_2,

$$s_2^0 - s_{\text{ref}}^0 = \int_{\text{ref}}^2 c_p \frac{dT}{T}$$

If we now subtract the first equation for s^0 from the second one,

$$s_2^0 - s_1^0 = \int_1^2 c_p \frac{dT}{T}$$

This is the first term on the right of Eq. (7-12) for the specific entropy. Hence, in general for an ideal gas,

$$s_2 - s_1 = s_2^0 - s_1^0 - R \ln \frac{P_2}{P_1} \tag{7-13}$$

The units of s^0 and R in this equation must be consistent.

Referral to Tables A-5M to A-11M and Tables A-5 to A-11 will show that the function s^0 has been tabulated solely as a function of temperature for these ideal gases. For air the value of s^0 in Table A-5M has units of kJ/(kg)(°K), while in Table A-5 the units are Btu/(lb)(°R). For the remaining gases the metric data have units of kJ/(kg · mol)(°K); the USCS data have units of Btu/(lb · mol)(°R). The superscript on s indicates a standard-state value: the gas is an ideal gas at 1 atm pressure. The following examples illustrate data acquisition from these tables.

Example 7-8M Nitrogen gas maintained at 2 bars is heated from 300°K to (*a*) 500°K, and (*b*) 800°K. Evaluate the entropy change from both sets of temperature increments by means of (1) an average c_p value and (2) the nitrogen table (Table A-6M).

SOLUTION (*a*) Since the pressure in constant, Eq. (7-11) reduces to $\Delta s = c_{p,av} \ln (T_2/T_1)$ and Eq. (7-13) becomes $\Delta s = s_2^0 - s_1^0$. On the basis of data from Tables A-4M and A-6M we find that

$$\Delta s = c_{p,av} \ln \frac{T_2}{T_1} = 1.047(28) \ln \tfrac{500}{300} = 14.98 \text{ kJ/(kg} \cdot \text{mol)(°K)}$$

$$\Delta s = s_2^0 - s_1^0 = 206.630 - 191.682 = 14.95 \text{ kJ/(kg} \cdot \text{mol)(°K)}$$

(*b*) For the second increment of temperature we find that

$$\Delta s = 1.08(28) \ln \tfrac{800}{300} = 29.66 \text{ kJ/(kg} \cdot \text{mol)(°K)}$$

$$\Delta s = 220.907 - 191.682 = 29.23 \text{ kJ/(kg} \cdot \text{mol)(°K)}$$

For the 300 to 500°K interval, use of an average c_p value leads to a 0.2 percent error over the integrated value. For the 300 to 800°K interval, this error increases to 1.5 percent.

Example 7-8 Nitrogen gas maintained at 30 psia is heated from 80°F to (*a*) 400°F, and (*b*) 1000°F. Evaluate the entropy change for both sets of temperature intervals by means of (1) an average c_p value and (2) the nitrogen table (Table A-6).

SOLUTION (*a*) Since the pressure is constant, Eqs. (7-11) and (7-13), respectively, reduce to,

$$\Delta s = c_{p,av} \ln \frac{T_2}{T_1}$$

$$\Delta s = s_2^0 - s_1^0$$

On the basis of data from Tables A-4 and A-6 we find that

$$\Delta s = 0.250(28) \ln \tfrac{860}{540} = 3.258 \text{ Btu/(lb} \cdot \text{mol)(°R)}$$

$$\Delta s = 49.031 - 45.781 = 3.250 \text{ Btu/(lb} \cdot \text{mol)(°R)}$$

(*b*) For the second temperature interval.

$$\Delta s = 0.259(28) \ln \tfrac{1460}{540} = 7.213 \text{ Btu/(lb} \cdot \text{mol)(°R)}$$

$$\Delta s = 52.867 - 45.781 = 7.086 \text{ Btu/(lb} \cdot \text{mol)(°R)}$$

For the 80 to 400°F interval, use of an average c_p value leads to a 0.25 percent error over the integrated value. For the 540 to 1000°F interval, this error increases to 1.8 percent.

When tables of s^0 values are available, they provide a fast and accurate means of evaluating the effect of temperature on the ideal-gas entropy. In addition, the use of average-specific-heat data lead to an increasing error as the temperature interval increases. It must be kept in mind, however, that regardless of the method used to evaluate the $T\,ds$ equations for an ideal gas, the result is equally applicable to a closed system and to a unit mass passing through a steady-state control volume.

7-6 ENTROPY CHANGE OF AN INCOMPRESSIBLE SUBSTANCE

A substance of constant density (or specific volume) was defined in Sec. 4-8 as an incompressible substance. Significantly, the internal energy of such a substance is given by

$$du = c_v\,dT \tag{4-8}$$

and the specific heats at constant volume and constant pressure are equal; that is,

$$c_p = c_v = c \tag{4-11}$$

For any simple, compressible substance we recognize that $T\,ds = du + P\,dv$. However, the last term is zero for an incompressible substance, by definition. Therefore, employing the relations above, we see that the entropy change of an incompressible substance is given by

$$ds = \frac{du}{T} = \frac{c_v\,dT}{T} = \frac{c\,dT}{T} \qquad \text{incompressible} \tag{7-14}$$

In many cases it is reasonable to assume that the specific heat is constant over a small temperature range, in which case we find that

$$s_2 - s_1 = c \ln \frac{T_2}{T_1} \qquad \text{incompressible} \tag{7-15}$$

In those cases where the specific heat c varies significantly with temperature, it would be necessary to integrate Eq. (7-14) by first inserting a functional relationship for c as a function of T.

Example 7-9M Two separated blocks of copper, A and B, have masses of 1 kg and 3 kg and initial temperatures of 100 and 300°C, respectively. They are brought into contact and allowed to come to thermal equilibrium at atmospheric pressure while insulated from the surroundings. Determine the entropy change of each block and the total entropy change.

SOLUTION If the two blocks are taken together as a closed system, then the energy balance on the process is $Q + W = \Delta U$. Because the overall system is insulated from the surroundings, $Q = 0$. In addition, if the blocks are taken to be incompressible, then the volume of the system is constant and $W = P\,\Delta V = 0$. As a result, for the composite system $\Delta U = 0$. There is, however, a change in internal energy for each block, since there is heat transfer internally from block B to block A. Thus $\Delta U_A + \Delta U_B = 0$. For an incompressible substance this may be written as

$$[mc(T_2 - T_1)]_A = -[mc(T_2 - T_1]_B$$

where it has been assumed that the specific heat of each block is constant. If we further assume that the specific heat is the same for each block in the given overall temperature range, then the quantity c may be canceled from the above equation. Substituting in known quantities, we find that the only unknown in the energy balance is T_2. That is,

$$1(T_2 - 100) = -3(T_2 - 300)$$

$$T_2 = 250°\text{C}$$

The entropy change is found by application of Eq. (7-15). Since $\Delta S = m\,\Delta s$,

$$\Delta S = mc \ln \frac{T_2}{T_1}$$

From Table A-19M, the average c_p value in the overall temperature range is 25.9 kJ/(kg·mol)(°K) or 0.407 kJ/(kg)(°K). Hence,

$$\Delta S_A = 1(0.407) \ln \tfrac{523}{373} = 0.138 \text{ kJ/°K}$$

$$\Delta S_B = 3(0.407) \ln \tfrac{523}{573} = -0.111 \text{ kJ/°K}$$

The surroundings are not affected by the process, so the total entropy change is just the sum of A and B.

$$\Delta S_{\text{total}} = 0.138 + (-0.111) = 0.027 \text{ kJ/°K}$$

The increase in entropy is due to the irreversible heat transfer between the two solids. A TS diagram for the process is sketched in Fig. 7-12. The areas under the constant-volume line must be equal, since $Q = \int T\,dS$ for internally reversible processes. In addition, both areas have a common state T_2. Because of the position of the constant-volume line on a TS diagram, the absolute value of the entropy change of B must be less than that of A if the areas are to be equal.

Example 7-9 Two separated blocks of copper, A and B, have masses of 0.1 and 0.3 lb and initial temperatures of 100 and 300°F, respectively. They are brought into contact and allowed to come to thermal equilibrium at atmospheric pressure while insulated from the surroundings. Determine the entropy change of each block and the total entropy change.

SOLUTION If the two blocks are taken together as a closed system, then the energy balance on the process is $Q + W = \Delta U$. Because the overall system is insulated from the surroundings,

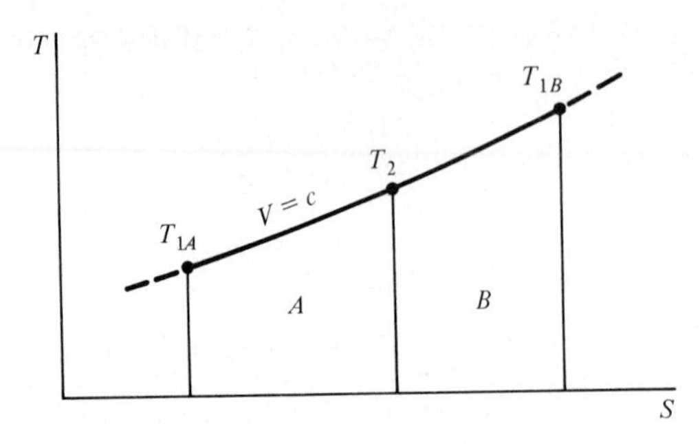

Figure 7-12 Process diagram for Examples 7-9M and 7-9.

$Q = 0$. In addition, if the blocks are taken to be incompressible, then the volume of the system is constant and $W = P\,\Delta V = 0$. As a result, for the composite system $\Delta U = 0$. There is, however, a change in the internal energy for each block, since there is heat transfer internally from block B to block A. Thus $\Delta U_A + \Delta U_B = 0$. For an incompressible substance this may be written as

$$[mc(T_2 - T_1)]_A = -[mc(T_2 - T_1)]_B$$

where it has been assumed that the specific heat of each block is constant. If we further assume that the specific heat is the same for each block in the given overall temperature range, then the quantity c may be canceled from the above equation. Substituting in known quantities, we find that the only unknown in the energy balance is T_2. That is,

$$0.1(T_2 - 100) = -0.3(T_2 - 300)$$

$$T_2 = 250°\text{F}$$

The entropy change is found from Eq. (7-15). Since $\Delta S = m\,\Delta s$,

$$\Delta S = mc \ln \frac{T_2}{T_1}$$

$$\Delta S_A = 0.1(0.0973) \ln \tfrac{710}{560} = 0.00231 \text{ Btu/°R}$$

$$\Delta S_B = 0.3(0.0973) \ln \tfrac{710}{560} = -0.00199 \text{ Btu/°R}$$

The surroundings are not affected by the process, so the total entropy change is just the sum of A and B.

$$\Delta S_{\text{total}} = 0.00231 + (-0.00199) = 0.00032 \text{ Btu/°R}$$

The increase in entropy is due to the irreversible heat transfer between the two solids. A TS diagram for the process is sketched in Fig. 7-12. The areas under the constant-volume line must be equal, since $Q = \int T\,dS$ for internally reversible processes. In addition, both areas have a common state, T_2. Because of the position of the constant-volume line on the TS diagram, the absolute value of the entropy change for B must be less than that of A if the areas are to be equal.

Example 7-10M Saturated liquid water at 40°C is heated and compressed to 80°C and 75 bars. Estimate the entropy change by (*a*) assuming an incompressible fluid, (*b*) using saturation data for the liquid, and (*c*) using the compressed-liquid table.

SOLUTION (*a*) For an incompressible fluid, $\Delta s = c \ln T_2/T_1$. From Table A-19M, the specific heat in this range of temperatures is roughly 4.187 kJ/(kg)(°C). Therefore

$$\Delta s = 4.187 \ln \tfrac{353}{313} = 0.504 \text{ kJ/(kg)(°K)}$$

(*b*) In the absence of compressed-liquid data for state 2, saturated-liquid data may be used at the given temperature. Employing the data in Table A-12M, we find that

$$\Delta s = s_{f,\,80} - s_{f,\,40} = 1.0753 - 0.5725 = 0.5028 \text{ kJ/(kg)(°K)}$$

(*c*) The entropy of state 2 in part *b* is found more accurately by use of the compressed-liquid table A-15M. Hence

$$\Delta s = 1.0704 - 0.5725 = 0.4979 \text{ kJ/(kg)(°K)}$$

This last answer is the most accurate, since it is based directly on experimental data. Use of saturated data leads to a 1 percent error, while the assumption of an incompressible fluid leads to only a 1.2 percent error. Thus, the use of the incompressible-fluid relation for Δs is a reasonable approximation when tabular data are unavailable for the liquid state.

Example 7-10 Saturated liquid water at 32°F is heated and compressed to 200°F and 1000 psia. Determine the entropy change by (*a*) assuming an incompressible fluid, (*b*) using saturated data for the liquid, and (*c*) using the compressed-liquid table.

SOLUTION (*a*) For an incompressible fluid, $\Delta s = c \ln (T_2/T_1)$. From Table A-19, the specific heat of water in this temperature range is roughly 1.0 Btu/(lb)(°F). Therefore

$$\Delta s = 1.0 \ln \tfrac{660}{492} = 0.294 \text{ Btu/(lb)(°R)}$$

(*b*) In the absence of compressed-liquid data for state 2, saturated-liquid data may be used at the given temperature. Employing the data in Table A-12, we find that

$$\Delta s = s_{f,\,200} - s_{f,\,32} = 0.2940 - (-0.0003) = 0.2940 \text{ Btu/(lb)(°R)}$$

(*c*) The entropy of state 2 in part *b* is found more accurately by use of the compressed-liquid table A-15. Hence.

$$\Delta s = 0.29281 - (-0.0003) = 0.2928 \text{ Btu/(lb)(°R)}$$

This last answer is the most accurate since it is based directly on experimental data. Use of saturated data leads to a 0.4 percent error, while the assumption of an incompressible fluid leads to the same percent of error in this case. Thus the use of an incompressible fluid relation for Δs is a reasonable approximation when tabular data are unavailable for the liquid state.

7-7 SOME SECOND-LAW RELATIONSHIPS FOR A CLOSED SYSTEM

In the discussion of quasistatic work interactions for closed systems in Chap. 2, it was pointed out that such interactions lead to the maximim work output and the minimum work input. With the second law as background, we are now in a position to prove the above statement. Some other interesting consequences of the second law with regard to closed systems are discussed in this section.

a Comparison of Reversible and Irreversible Work Interactions

The presence of irreversibilities within a system degrade performance. This is especially important in work-producing and work-absorbing devices. The maximum work output and the minimum work input are obtained by operating the device reversibly. The proof for a closed system is derived by noting we may write two basic equations. In general, for an internally reversible process

$$T\,dS = dU - \delta W_{rev} \tag{7-16}$$

where δW_{rev} would be represented by $-P\,dV$ for a simple, compressible substance. Also, for any process

$$\delta Q_{act} + \delta W_{act} = dU \tag{2-5}$$

Subscripts have been written on the δQ and δW terms to indicate that these are actual values, and apply to either internally reversible or irreversible changes. Consequently, δW_{act} is measured external to the system. That is, the work is the measured input or output. If we use Eq. (2-5) to eliminate dU from Eq. (7-16), then

$$T\,dS = \delta Q_{act} - \delta W_{rev} + \delta W_{act}$$

or

$$dS = \frac{\delta Q_{act}}{T} + \frac{1}{T}(\delta W_{act} - \delta W_{rev}) \tag{7-17}$$

In addition, we have developed from the second law that $dS \geq \delta Q_{act}/T$. Therefore

$$\frac{1}{T}(\delta W_{act} - \delta W_{rev}) = dS - \frac{\delta Q_{act}}{T} \geq 0$$

Since the absolute temperature T is positive by concept, we find that

$$\delta W_{act} \geq \delta W_{rev} \tag{7-18}$$

This inequality is valid regardless of the direction of change of the closed system. That is, it applies whether the work interaction is in or out. Noting that the inequality sign on any of the preceding equations denotes the presence of irreversibilities, we have now shown that irreversibilities within a closed system reduce the work output or increase the work input. To maximize the work output, we must attempt to minimize irreversibilities within a system.

b Entropy Flux and Entropy Production in Closed-System Processes

In Sec. 6-1 it was pointed out that the second law is a nonconservation law, and it is expressed mathematically as an inequality. The increase in entropy principle has been emphasized in the early sections in this chapter. Generally speaking, the presence of irreversibilities anywhere during a process leads to an increase in entropy of the "universe." We are now in a position, however, to examine specifically the sources which contribute to the entropy change of a closed system. It is convenient to rewrite Eq. (7-17) as

$$dS = \frac{\delta Q_{act}}{T} + \frac{1}{T}(\delta W_{act} - \delta W_{rev}) = dS_{flux} + dS_{int\ prod} \tag{7-19}$$

where

$$dS_{\text{flux}} = \frac{\delta Q_{\text{act}}}{T}$$

and
$$dS_{\text{int prod}} \equiv \frac{1}{T}(\delta W_{\text{act}} - \delta W_{\text{rev}})$$

Equation (7-19) states that the change in entropy of a closed system is due to two separate effects. One of these is the transfer of heat to or from the system. The term $\delta Q_{\text{act}}/T$ is called the entropy-flux term. The second effect is the presence of irreversibilities within the system. The second term $dS_{\text{int prod}}$ is usually referred to as the internal production contribution.

A comparison of the possible magnitudes of these two terms is quite interesting. Since the actual heat transfer can be positive, negative, or zero, the entropy-flux term can take on any value. Hence

$$dS_{\text{flux}} = \frac{\delta Q_{\text{act}}}{T} = \text{entropy flux} \lesseqgtr 0 \tag{7-20}$$

On the other hand the second law, in the form of Eq. (7-18), requires that the entropy production due to internal irreversibilities must be equal to or greater than zero. That is,

$$dS_{\text{int prod}} = \frac{1}{T}(\delta W_{\text{act}} - \delta W_{\text{rev}}) = \text{internal entropy production} \geq 0 \tag{7-21}$$

The effect of internal irreversibilities always is to increase the entropy of a given mass. As a consequence of this, the only way to decrease the entropy of a closed system is to transfer heat *from* it.

7-8 SOME SECOND-LAW RELATIONSHIPS FOR A CONTROL VOLUME

In the preceding section the concept of entropy production within a closed system, due to internal irreversibilities, was introduced. This concept will now be applied to a control volume. In addition, a relation for steady-flow mechanical work will be developed.

a Entropy Production and Entropy Flux in Control-Volume Processes

The second law of thermodynamics may be expressed as an increase in entropy principle for adiabatic closed and isolated systems. In each case attention is focused on a particular mass or set of masses within a given boundary. We now wish to extend this principle to a region of space—the control volume. The approach is similar to that used for the derivation of the conservation of energy principle in Chap. 5. Figure 7-13 shows a control mass relative to a control

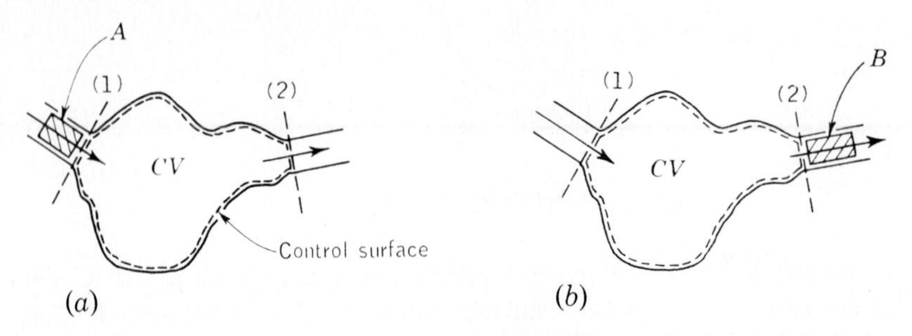

Figure 7-13 Development of an entropy relation for a control volume. (*a*) Control mass at time t; (*b*) control mass at time $t + \Delta t$.

volume at times t and $t + \Delta t$. At time t,

$$S_{\mathrm{CM},t} = S_{\mathrm{CV},t} + S_A \tag{a}$$

and at time $t + \Delta t$,

$$S_{\mathrm{CM},t+\Delta t} = S_{\mathrm{CV},t+\Delta t} + S_B \tag{b}$$

Subtraction of Eq. (*a*) from Eq. (*b*) yields

$$S_{\mathrm{CM},t+\Delta t} - S_{\mathrm{CM},t} = S_{\mathrm{CV},t+\Delta t} - S_{\mathrm{CV},t} + S_B - S_A \tag{c}$$

or

$$\Delta S_{\mathrm{CM}} = \Delta S_{\mathrm{CV}} + s_2 m_2 - s_1 m_1 \tag{d}$$

where $S_B = s_2 m_2$ and $S_A = s_1 m_1$ and the Δ in front of S_{CM} and S_{CV} represents the change over a time period Δt.

The entropy change for the control mass, ΔS_{CM}, in Eq. (*d*) is the sum of the entropy flux contribution, ΔS_{flux}, and the internal entropy production, ΔS_{prod}. Since the temperature varies across the control volume, we are required to sum the quantities Q_i/T_i at each boundary where heat enters or leaves in order to find the entropy flux. The subscript i in this case denotes a location at the control surface of temperature T_i. Hence

$$\Delta S_{\mathrm{flux}} = \sum \frac{Q_i}{T_i}$$

As a result Eq. (*d*) may now be written in the form

$$\sum \frac{Q_i}{T_i} + \Delta S_{\mathrm{prod}} = \Delta S_{\mathrm{CV}} + s_2 m_2 - s_1 m_1$$

or

$$\Delta S_{\mathrm{prod}} = \Delta S_{\mathrm{CV}} + s_2 m_2 - s_1 m_1 - \sum \frac{Qi}{T_i} \geq 0 \tag{7-22}$$

In the above equation ΔS_{prod} refers to the control volume as well as the control mass, since the volumes of A and B can be chosen to be extremely small compared to the control volume. For an infinitesimal time interval dt the above equation for multiple outlets and inlets becomes

$$dS_{\text{prod}} = dS_{\text{CV}} + \sum_{\text{out}} (s\ dm) - \sum_{\text{in}} (s\ dm) - \sum \frac{\delta Q_i}{T_i} \geq 0 \tag{7-23}$$

Finally, on a rate basis for the control volume,

$$\frac{dS_{\text{prod}}}{dt} = \frac{dS_{\text{CV}}}{dt} + \sum_{\text{out}} (s\dot{m}) - \sum_{\text{in}} (s\dot{m}) - \sum \frac{\dot{Q}_i}{T_i} \geq 0 \tag{7-24}$$

In these equations the equality is associated with internally reversible processes and the inequality with internally irreversible processes. For an irreversible process the sum of the terms in the above three equations is positive, and this sum is a measure of the entropy "production" due to the irreversibilities.

Under steady-state conditions, the quantity dS_{CV} (or ΔS_{CV}) must be zero in Eqs. (7-22) to (7-24). In addition, if a process is adiabatic, the sum of the Q_i/T_i terms is zero. When both are valid restrictions, Eq. (7-24) becomes

$$\frac{dS_{\text{prod}}}{dt} = \sum_{\text{out}} (s\dot{m}) - \sum_{\text{in}} (s\dot{m}) \geq 0 \tag{7-25}$$

This expression relates the flow of entropy across the boundaries of an adiabatic, steady-state control volume. When internal irreversibilities are present within the control volume, the sum of the terms on the right side is positive and measures the rate of entropy production, dS_{prod}/dt.

Example 7-11M Consider the open feedwater heater discussed in Example 5-8M. The device is steady-state, with superheated steam (flow stream 1) entering at 1 kg/unit time at 5 bars and 200°C. Subcooled water (flow stream 2) enters at 4.75 kg/unit time at 5 bars and 40°C. Saturated liquid water (flow stream 3) at 5.75 kg/unit time leaves at 5 bars. Determine the rate of entropy production for the irreversible mixing process, in kJ/(°K)(unit time).

SOLUTION Equation (7-25) will be applied to determine the rate of entropy production. The equation may be written as

$$\frac{dS_{\text{prod}}}{dt} = s_3\dot{m}_3 - s_1\dot{m}_1 - s_2\dot{m}_2$$

The specific entropies at states 1, 2, and 3 are found from Tables A-14M, A-12M, and A-13M to be 7.0592, 0.5725, and 1.8607 kJ/(kg)(°K), respectively. Substitution of these values yields

$$\frac{dS_{\text{prod}}}{dt} = 1.8607(5.75) - 7.0592(1) - 0.5725(4.75)$$

$$= 10.70 - 7.06 - 2.72 = 0.92 \text{ kJ/(°K)(unit time)}$$

The entropy production is positive, in accordance with the second law, and is due to the irreversible mixing of flow streams at different temperatures.

Example 7-11 Consider the open feedwater heater discussed in Example 5-8. The device is steady-state, with superheated steam (flow stream 1) entering at 1 lb/unit time at 80 psia and 400°F. Subcooled water (flow stream 2) enters at 4.44 lb/unit time at 80 psia and 100°F.

Saturated liquid water (flow stream 3) at 5.44 lb/unit time leaves at 80 psia. Determine the entropy production of the irreversible mixing process, in Btu/(°R)(unit time).

SOLUTION Equation (7-25) will be applied to determine the rate of entropy production.

$$\frac{dS_{\text{prod}}}{dt} = s_3 m_3 - s_1 m_1 - s_2 m_2$$

The entropies at states 1, 2, and 3 are found from the steam tables to be 1.6790, 0.1295, and 0.4534 Btu/(lb)(°R), respectively. Substitution of these values yields

$$\frac{dS_{\text{prod}}}{dt} = 0.4534(5.44) - 1.6790(1) - 0.1295(4.44)$$

$$= 2.465 - 1.679 - 0.575 = 0.211 \text{ Btu/(°R)(unit time)}$$

The entropy production is positive, and is due to the irreversible mixing of flow streams at different temperatures.

b A Relationship for Reversible, Steady-flow Mechanical Work

The $T\,ds$ equations are also useful in deriving a basic relationship for shaft work associated with a control volume. The conservation of energy principle for a differential mass of a simple, compressible substance passing through a steady-state device is

$$\delta w_{\text{shaft}} = dh + d\text{KE} + d\text{PE} - \delta q$$

If the process is internally reversible, then, for the same unit mass,

$$\delta q = T\,ds = dh - v\,dP \tag{7-7}$$

Substitution of this latter relationship into the energy equation yields

$$\delta w_{\text{shaft}} = dh + d\text{KE} + d\text{PE} - (dh - v\,dP)$$

$$= v\,dP + d\text{KE} + d\text{PE} \tag{7-26}$$

For flow over a finite distance through a steady-state device the frictionless mechanical work associated with a unit mass is

$$w_{\text{sf, rev}} = \int v\,dP + \Delta\text{KE} + \Delta\text{PE} \tag{7-27}$$

If the changes in the kinetic and potential energies are negligible, Eq. (7-27) reduces to

$$w_{\text{sf, rev}} = \int v\,dP \tag{7-28}$$

Although Eqs. (7-27) and (7-28) were developed from an energy balance on a control volume, this approach is not necessary. These expressions can be developed solely from the principles of mechanics. Therefore these equations for reversible, steady-flow mechanical work are independent of an energy analysis on a fluid passing through a control volume. Note that Eqs. (7-27) and (7-28) are useful in a quantitative sense only if the functional relationship between v and P is known. Also, the reader should be careful to distinguish between the nonflow and steady-flow work equations for simple, compressible substances. Confusion often arises from the similarity between $P\,dv$ and $v\,dP$. By sketching a process on a Pv diagram, the student can easily distinguish the difference between these two work expressions in terms of areas on the diagram.

Since Eqs. (7-26) to (7-28) are developed under internally reversible conditions, they should lead to the maximum work output or minimum work input to steady-flow devices such as turbines, compressors, and pumps. By an analysis similar to that given for closed systems in the preceding section, it can be shown that a relation of the type

$$\delta W_{\text{sf, act}} \geq \delta W_{\text{sf, rev}} \tag{7-29}$$

applies to these steady-flow work devices.

7-9 ISENTROPIC PROCESSES

An internally reversible process has a special significance to engineers. As pointed out in the last section, such a process leads to the maximum work output or minimum input for devices such as turbines and compressors. Although it is not as obvious, we shall soon show that internally reversible processes also lead to the best performance of nozzles, diffusers, and other non-work-producing devices. Consequently, the internally reversible process can be used as a standard to which all real processes may be compared, whether applied to closed or open systems. In addition, it was emphasized in Chap. 5 during the discussion of control-volume analysis that many devices are essentially adiabatic, for various reasons. Consequently, the adiabatic, internally reversible process is frequently chosen as a standard to which actual processes may be compared.

The basic relation for the entropy change of any mass of material undergoing an internally reversible process is $dS = \delta Q/T$. It is immediately seen that adiabatic, internally reversible processes must be ones of constant entropy. When the entropy does not change, a process is called *isentropic*. Isentropic processes are used by the engineer in the theoretical analysis of open and closed systems, since such processes are the limit of extrapolation of real adiabatic processes. This situation is illustrated on a Ts diagram in Fig. 7-14 for a process which involves an increase in pressure, for example.

We have already noted that constant-pressure lines have positive slopes on a Ts diagram. If an internally reversible process occurs adiabatically, the final state $2s$ lies directly above the initial state 1. However, if irreversibilities are present, the final state $2a$ must lie to the right of state 1 on the P_2 line; that is, $s_2 > s_1$. This increase in entropy is dictated by the general statement, $dS_{\text{adia}} \geq 0$. (In this discussion and others later in this chapter, the symbol s stands for an isentropic final state, while the symbol a represents an actual final state which occurs due to irreversibilities.) The position of state $2a$ on line P_2 depends upon the extent of

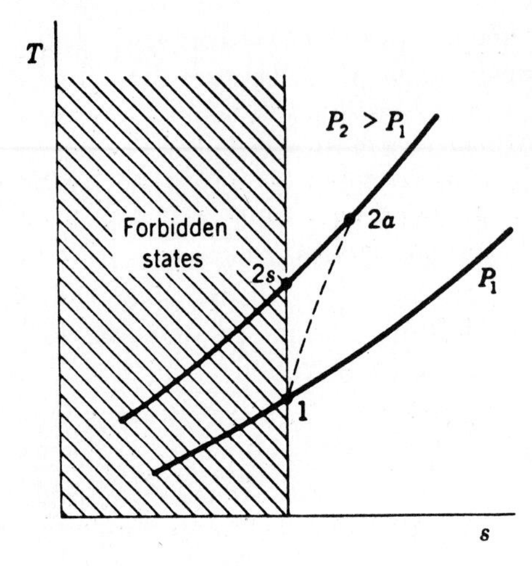

Figure 7-14 Forbidden states for irreversible adiabatic processes.

the irreversibilities. State $2a$ cannot lie to the left of the vertical line through state 1, since all states to the left of this line are forbidden states for an adiabatic process. Consequently, it is seen that the isentropic process is the limiting process as irreversibilities are reduced under adiabatic conditions. As a limiting condition, the isentropic process is a standard of performance against which real processes may be compared.

In previous sections of this chapter, it was shown that special equations can be developed for the entropy change of several specific classes of substances. These special equations may now be used to derive isentropic relations for these same substances.

1 Isentropic Ideal-Gas Relations

In the development of isentropic ideal-gas relations it is important to introduce another intrinsic property, the *specific-heat ratio k.* This ratio is defined as

$$k \equiv \frac{c_p}{c_v} \tag{7-30}$$

Its value for monatomic gases is 1.67 and is essentially constant with temperature. For molecules containing two or more atoms the k value is always less than 1.67 and the values decrease with increasing temperature, as shown by the data in Tables A-4M and A-4. However, from the definition of k and the fact that c_p is always greater than c_v for an ideal gas ($c_p - c_v = R$), k is never less than unity. Many common diatomic gases have a specific-heat ratio of, roughly, 1.4 at room temperature and slightly above. The specific-heat ratio is related to either c_v or c_p by the following relations:

$$c_v = \frac{R}{k-1} \tag{7-31}$$

and
$$c_p = \frac{Rk}{k-1} \tag{7-32}$$

These equations are restricted to ideal gases.

In Sec. 7-5 the $T\,dS$ equations were applied to ideal gases. In the case where the specific heats are assumed constant, or an average value over the given temperature interval is used, the following equations are valid.

$$\Delta s = c_v \ln \frac{T_2}{T_1} + R \ln \frac{v_2}{v_1} \tag{7-10}$$

and
$$\Delta s = c_p \ln \frac{T_2}{T_1} - R \ln \frac{P_2}{P_1} \tag{7-11}$$

For an isentropic process, $\Delta s = 0$. If Eqs. (7-10) and (7-11) are set equal to zero, the following relations result.

$$\frac{T_2}{T_1} = \left(\frac{v_1}{v_2}\right)^{k-1} \qquad \text{isentropic process} \tag{7-33a}$$

and
$$\frac{T_2}{T_1} = \left(\frac{P_2}{P_1}\right)^{(k-1)/k} \qquad \text{isentropic process} \tag{7-34a}$$

The specific-heat ratio k appears in the above equations through the use of Eqs. (7-31) and (7-32). If the ideal-gas equation is substituted into either Eq. (7-33*a*) or Eq. (7-34*a*) so that T is eliminated as a variable, then

$$\frac{P_2}{P_1} = \left(\frac{v_1}{v_2}\right)^{k} \qquad \text{isentropic process} \tag{7-35a}$$

Another way of expressing these relationships among P, v, and T is as follows:

$$T(v)^{k-1} = \text{constant} \tag{7-33b}$$

$$T^k P^{1-k} = \text{constant} \tag{7-34b}$$

$$P(v)^k = \text{constant} \tag{7-35b}$$

Equations (7-33) to (7-35) represent *process* equations for the isentropic change of state for ideal gases. They are fairly accurate when the temperature change during a process does not exceed a few hundred degrees. As a word of warning, note that the three constants in the above set of equations are not equal in value or dimensions.

In Fig. 7-15 isothermal and isentropic lines are plotted on a PV diagram for comparative purposes. Isentropic lines have a larger negative slope than isothermal lines through the same state point on a PV diagram for ideal gases. It is important to note that the above set of relations is valid for any fixed mass of an ideal gas which undergoes a process at constant entropy. The mass may be in a closed system or flowing through a steady-state control volume.

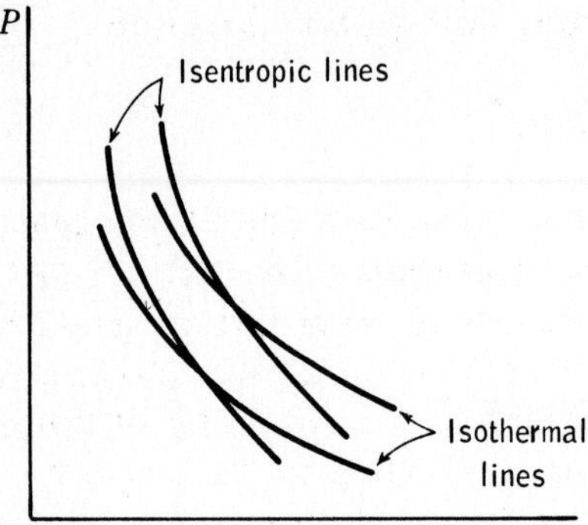

Figure 7-15 Process lines on a pressure-volume line.

Example 7-12M Air enters an adiabatic compressor at 17°C and is compressed through a pressure-ratio of 8.6 : 1. If the compression is assumed to be internally reversible, determine the enthalpy change based on inlet conditions.

SOLUTION The enthalpy change for constant specific heats is given by $\Delta h = c_p(T_2 - T_1)$ for an ideal gas. The final temperature is found from the isentropic relation

$$\frac{T_2}{T_1} = \left(\frac{P_2}{P_1}\right)^{(k-1)/k}$$

At the inlet condition of 17°C, the k value from Table A-4M is found to be 1.40; hence

$$T_2 = 290(8.6)^{0.286} = 290(1.85) = 537°\text{K}$$

Therefore, using the inlet c_p value of 1.005 kJ/(kg)(°C),

$$\Delta h = 1.005(537 - 290) = 248 \text{ kJ/kg}$$

Example 7-2 Air enters an adiabatic compressor at 40°F and is compressed through a pressure ratio of 8.6 : 1. For an internally reversible compression process, determine the enthalpy change based on inlet conditions.

SOLUTION The enthalpy change for a constant c_p value is given by $\Delta h = c_p(T_2 - T_1)$ for an ideal gas. The final temperature is found from the isentropic relation

$$\frac{T_2}{T_1} = \left(\frac{P_2}{P_1}\right)^{(k-1)/k}$$

At the inlet condition of 40°F, the k value from Table A-4 is found to be 1.40; hence

$$T_2 = 500(8.6)^{0.286} = 500(1.85) = 925°\text{R}$$

Therefore, using the inlet c_p value of 0.240 Btu/(lb)(°F),

$$\Delta h = 0.240(925 - 500) = 102 \text{ Btu/lb.}$$

In some isentropic calculations it is necessary to account for the variation of c_p with temperature, if reasonable accuracy is desired. There are two methods by which this may be accomplished. The most fundamental way is based on the direct use of Eq. (7-13). For an isentropic process this equation reduces to

$$0 = s_2^0 - s_1^0 - R \ln \frac{P_2}{P_1}$$

Recall that the quantity $s_2^0 - s_1^0$ is a measure of the integral of $c_p\ dT/T$ between states 1 and 2. Therefore the above equation does account for the variation of specific heat with temperature. Also, we have already seen that s^0 values are tabulated solely as a function of temperature. Such data are available for hundreds of ideal gases, and have been tabulated over a wide range of temperatures in the *JANAF* Tables. (These tables are distributed by the U.S. Government Printing Office.) In this text, the s^0 values for air are given on a unit-mass basis (gram or pound), while for all other gases the values are on a molar basis.

If the initial state is known, then the final state of an isentropic process of an ideal gas is found by using one of the two following equations:

$$s_2^0 = s_1^0 + R \ln \frac{P_2}{P_1} \tag{7-36}$$

and

$$P_2 = P_1 \exp \frac{s_2^0 - s_1^0}{R} \tag{7-37}$$

Equation (7-36) permits the calculation of T_{2s} if P_2 is known, while Eq. (7-37) evaluates P_{2s} if T_2 is known. The subscript s again symbolizes an isentropic state. The use of these expressions is illustrated below.

Example 7-13M Reconsider the data of Example 7-12M. Use variable-specific-heat data to determine the enthalpy change of air when it is compressed through an 8.6 : 1 pressure ratio from an initial temperature of 17°C.

SOLUTION The final temperature for the compression process is found by using Eq. (7-36). In Table A-5M, the value of s_1^0 at 290°K is found to be 1.66802 kJ/(kg)(°K). Substitution of the known data yields

$$s_2^0 = 1.66802 + \tfrac{8.314}{29} \ln 8.6 = 2.285 \text{ kJ/(kg)(°K)}$$

In Table A-5M, this value of s_2^0 is found to lie between 530 and 540°K. By linear interpolation T_{2s} is 533°K. By further interpolation of h data, h_{2s} is 536.8 kJ/kg, and h_1 at 290°K is 290.16 kJ/kg. Hence

$$\Delta h = 536.8 - 290.2 = 246.6 \text{ kJ/kg}$$

This is about 0.6 percent smaller than that obtained from constant-specific-heat data in Example 7-12M. Although the answers in this case differ very little, the above method is quick and always more accurate, since the variation in specific heats is automatically included.

Example 7-13 Reconsider the data of Example 7-12. Use variable-specific-heat data to determine the enthalpy change of air when it is compressed through an 8.6 : 1 pressure ratio from an initial temperature of 40°F.

SOLUTION The final temperature for the compression process is found by using Eq. (7-36). In Table A-5, the value of s_1^0 at 500°R is noted to be 0.58233 Btu/(lb)(°R), while h_1 is 119.48 Btu/lb.

Substitution of the known data into Eq. (7-36) yields

$$s_2^0 = 0.58233 + \tfrac{1.986}{29} \ln 8.6 = 0.7297 \text{ Btu/(lb)(°R)}$$

In Table A-5, this value of s_2^0 falls very close to 920°R. At this temperature, h_2 is 221.18 Btu/lb. Hence

$$\Delta h = 221.1 - 119.5 = 101.7 \text{ Btu/lb}$$

This answer is fairly close to that given by assuming constant specific heats. Nevertheless, the above method is quick and always provides more accuracy, since the variation in specific heats is automatically included.

The use of s^0 data, in conjunction with Eqs. (7-36) and (7-37), provides a general method of evaluating isentropic changes for ideal gases. In some special cases it is useful to extend the method one step further. For example, the preceding method is an iterative (trial and error) process when the volume ratio is known rather than the pressure ratio. This makes it more difficult to evaluate expansion and compression processes in piston-cylinder devices such as automotive engines. Second, for some gases such as air, numerous repetitive calculations may be required for a particular design. In such cases it is helpful, but not essential, to generate a new set of isentropic data.

This set of data is based on Eq. (7-37), namely,

$$P_2 = P_1 \exp \frac{s_2^0 - s_1^0}{R} \tag{7-37}$$

Solving for the pressure ratio under isentropic conditions,

$$\left(\frac{P_2}{P_1}\right)_s = \exp \frac{s_2^0 - s_1^0}{R} = \frac{\exp (s_2^0/R)}{\exp (s_1^0/R)}$$

where the subscript s on (P_2/P_1) emphasizes the isentropic restriction. Note that the quantity $\exp (s^0/R)$ is solely a function of the temperature. It is related mathematically to a function called the *relative pressure* p_r in such a manner that for an isentropic process of an ideal gas,

$$\left(\frac{P_2}{P_1}\right)_s = \frac{p_{r2}}{p_{r1}} \tag{7-38}$$

The values of the relative pressure p_r may be tabulated for a given gas along with u, h, and s^0 as a function of the temperature. Equation (7-38) serves the same purpose that Eq. (7-34*a*) does for ideal gases of constant specific heat.

In some cases it is necessary to work with volume ratios instead of pressure ratios for isentropic processes. A relative volume v_r can be defined in a manner similar to that for the relative pressure. By the use of the ideal-gas equation and Eq. (7-38),

$$\frac{v_2}{v_1} = \frac{P_1 T_2}{P_2 T_1} = \frac{p_{r1} T_2}{p_{r2} T_1} = \frac{T_2}{p_{r2}} \frac{p_{r1}}{T_1} \qquad \text{isentropic process}$$

Note that the quantity T/p_r is solely a function of temperature at a given state.

That is, $(v_2/v_1)_s = f(T_2)/f(T_1)$. This temperature function is related to a property called the *relative volume* v_r in such a manner that for an isentropic process of an ideal gas,

$$\left(\frac{v_2}{v_1}\right)_s = \frac{v_{r2}}{v_{r1}} \tag{7-39}$$

Equation (7-39) for variable specific heats is equivalent to Eq. (7-33*a*) for constant specific heats.

It is convenient, then, to tabulate the functions p_r and v_r versus temperature for ideal gases. This has been done over a wide range of temperatures for a limited number of gases of engineering importance. In this text these two functions appear only in air Tables A-5M and A-5.

Example 7-14M Rework Example 7-13M by using relative-pressure data instead of s^0 data.

SOLUTION In Table A-5M, at 290°K we find p_{r1} to be 1.2311, and h_1 again is 290.16 kJ/kg. The final state is found by

$$P_{r2} = p_{r1}\frac{P_2}{P_1} = 1.2311(8.6) = 10.59$$

In Table A-5M, at 530 and 540°K the values of p_r are 10.37 and 11.10, respectively. By linear interpolation T_{2s} is 533°K, which agrees with the previous calculation, using s^0 data, in Example 7-13M. As a result the enthalpy change again is 246.6 kJ/kg.

Example 7-14 Rework Example 7-13 by using relative-pressure data instead of s^0 data.

SOLUTION In Table A-5, at 500°R we find that p_{r1} is 1.0590. The final state is found by

$$p_{r2} = p_{r1}\frac{P_2}{P_1} = 1.0590(8.6) = 9.106$$

In Table A-5 for air, at 920°R $p_r = 9.102$. Therefore T_{2s} is close to 920°R, in agreement with Example 7-13. As a result the enthalpy change again is 101.7 Btu/lb.

A further example of the use of isentropic relations for ideal gases is in conjunction with equations previously developed for reversible work effects for closed and open systems. For closed-system boundary work

$$w = -\int P\,dv \tag{2-9}$$

and for steady-flow mechanical work

$$w = \int v\,dP + \Delta\text{KE} + \Delta\text{PE} \tag{7-27}$$

In both of these cases a functional relationship between P and v is required. For an isentropic process where the variation of the specific heats with temperature is small,

$$P(v)^k = \text{constant} \tag{7-35b}$$

This latter relation provides a means of evaluating Eqs. (2-9) and (7-27) under adiabatic conditions for ideal gases. By direct substitution of Eq. (7-35*b*) into these expressions followed by integration, it can be shown that

$$w_{\text{nonflow}} = \frac{P_2 v_2 - P_1 v_1}{k - 1} \tag{7-40}$$

$$w_{\text{steady-flow}} = \frac{k(P_2 v_2 - P_1 v_1)}{k - 1} + \Delta\text{KE} + \Delta\text{PE} \tag{7-41}$$

These equations give the maximum work output or minimum work input under adiabatic operation. It can also be shown that these equations are in agreement with the respective energy balances for closed and open systems.

2 Isentropic Relations for Incompressible Substances

For incompressible substances the entropy change is given by Eq. (7-14), namely, $ds = c\ dT/T$. If the specific heat is relatively constant, the integration of this equation leads to Eq. (7-15), that is

$$\Delta s = c \ln \frac{T_2}{T_1} \tag{7-15}$$

For an isentropic process $\Delta s = 0$. Consequently, an isentropic process involving an incompressible substance is also isothermal, so that $T_2 = T_1$. Recall that for an incompressible substance $du = c\ dT$. Therefore $u_2 = u_1$ for the isentropic process also.

The transport of liquids in pipes is of primary importance in numerous engineering designs. A pump may or may not be part of the control volume, and the pipes may have different diameters at different sections of the flow system. In addition, the fluid may undergo a considerable elevation change, which is not the usual case with other pieces of standard equipment. In this section we wish to consider the case where either the pipe is insulated, or the temperature of the fluid is approximately the same as that of the environment. In either situation the heat transfer is essentially zero.

An appropriate conservation of energy principle for pipe flow, based on Eq. (5-12), has the form

$$q + w_{\text{shaft}} = \Delta u + \Delta\ Pv + \Delta\text{KE} + \Delta\text{PE}$$

Liquids flowing through pipes generally can be considered to be incompressible. For incompressible flow $\Delta h = c\ \Delta T + v\ \Delta P$. Therefore, the above equation can be modified to the following form.

$$w_{\text{shaft}} = c\ \Delta T + v\ \Delta P + \Delta\text{KE} + \Delta\text{PE} - q \qquad \text{incompressible flow} \tag{7-42a}$$

For negligible heat transfer, $q = 0$. In addition, for isentropic flow of an incompressible fluid, $\Delta u = 0$. Thus, the energy equation reduces to

$$w_{\text{shaft}} = v\ \Delta P + \Delta\text{KE} + \Delta\text{PE} \qquad \text{incompressible, isentropic} \tag{7-42b}$$

Hence, the work input required to change the state of an incompressible liquid in a piping system under frictionless, adiabatic flow is modeled to be primarily a function of the pressure change, the velocity change, and the elevation change. In most engineering flow devices the energy change due to an elevation change is insignificant. However, for pipe flow in buildings of appreciable height, the change in gravitational potential energy could be a major contributor to the shaft work required for the pumping operation.

If any pump in the piping system is excluded from the control volume, then $w_{\text{shaft}} = 0$. Under this additional restriction the steady-flow energy equation reduces to

$$v\,\Delta P + \Delta\text{KE} + \Delta\text{PE} = 0$$

or

$$v(P_2 - P_1) + \frac{V_2^2 - V_1^2}{2} + g(z_2 - z_1) = 0 \tag{7-42c}$$

This is known as the Bernoulli equation in elementary fluid mechanics. It is applicable to isentropic flow of incompressible fluids through ducts.

Example 7-15M A pump draws a solution with a specific gravity of 1.50 from a storage tank through an 8-cm pipe. The velocity in the suction (inlet) pipe is 1.2 m/s. The open end of the 5-cm discharge pipe is 15 m above the top of the liquid in the storage tank. If the process is assumed to be isentropic and the fluid to be incompressible, determine the power required to the pump in the piping system, in kilowatts.

Solution The top level of the storage tank and the discharge section of the 5-cm pipe are designated as states 1 and 2 (see Fig. 7-16). State a is an intermediate state in the 8-cm pipe, where the velocity is 1.2 m/s. The control volume includes the fluid in the storage tank and within the piping system. Under the stated restrictions, the energy equation is given by Eq. (7-42b), namely,

$$w_{\text{shaft}} = v\,\Delta P + \Delta\text{KE} + \Delta\text{PE}$$

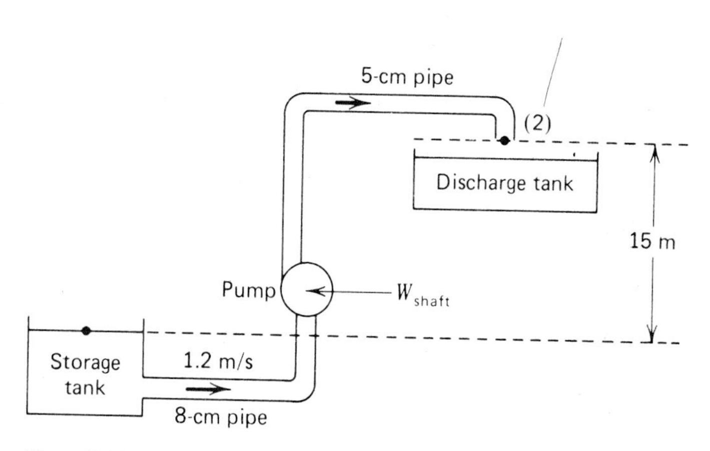

Figure 7-16 Schematic of flow process for Example 7-15M.

For the selected control volume, $P_1 = P_2$, and therefore $v\,\Delta P$ is zero. As a result

$$w_{\text{shaft}} = \frac{V_2^2 - V_1^2}{2} + g(z_2 - z_1)$$

The initial velocity V_1 is zero if we assume no appreciable motion of the fluid at the top of the tank. The final velocity is found from the continuity equation, since the velocity at state a is known. Hence,

$$V_2 = V_a \frac{A_a}{A_2} = V_a \left(\frac{D_a}{D_2}\right)^2 = 1.2\left(\frac{8}{5}\right)^2 = 3.1 \text{ m/s}$$

Substitution of the available data into the energy equation yields

$$w_{\text{shaft}} = \frac{(3.1)^2}{2} + 15(9.8) = 4.8 + 147 = 152 \text{ N}\cdot\text{m/kg}$$

Note that the potential-energy change is responsible for nearly all of the total work required, if g is taken to be 9.8 m/s^2.

The power requirement is obtained from the relation $\dot{W} = \dot{m}w$. The mass flow rate is

$$\dot{m} = \rho A V = (1.50 \times 1.0) \text{ g/cm}^3 \times \frac{\pi(8)^2}{4} \text{ cm}^2 \times 1.2 \text{ m/s} \times 100 \text{ cm/m}$$

$$= 9050 \text{ g/s} = 9.05 \text{ kg/s}$$

In the above calculation the density of water has been taken to be 1.0 g/cm^3. Finally, the power input required is

$$\dot{W} = \dot{m}w = 9.05 \text{ kg/s} \times 152 \text{ N}\cdot\text{m/kg} = 1380 \text{ W} = 1.38 \text{ kW}$$

Example 7-15 A pump draws a solution with a specific gravity of 1.50 from a storage tank through a 3-in pipe. The velocity in the suction (inlet) pipe is 4 ft/s. The open end of the 2-in discharge pipe is 50 ft above the top of the liquid in the storage tank. Assume the process is isentropic and the fluid is incompressible. Determine the power required to the pump in the piping system, in horsepower.

SOLUTION The top level of the storage tank and the discharge section of the 2-in pipe are designated by states 1 and 2 on the accompanying diagram (see Fig. 7-17). State a is an

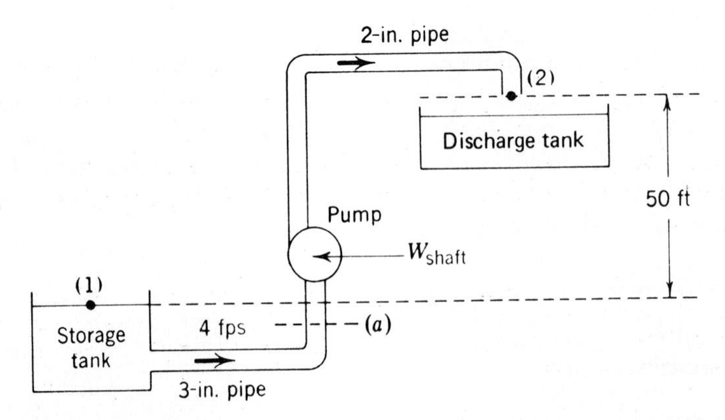

Figure 7-17 Schematic of flow process for Example 7-15.

intermediate state in the 3-in pipe, where the velocity is 4 ft/s. The control volume includes the fluid in the storage tank and within the piping system. Under the stated restrictions the energy equation is given by Eq. (7-42b), namely,

$$w_{\text{shaft}} = v\,\Delta P + \Delta\text{KE} + \Delta\text{PE}$$

For the selected control volume, $P_1 = P_2$, and therefore $v\,\Delta P$ is zero. As a result

$$w_{\text{shaft}} = \frac{V_2^2 - V_1^2}{2} + g(z_2 - z_1)$$

The initial velocity V_1 is zero if we assume no appreciable motion of the fluid at the top of the tank. The final velocity V_2 is found from the continuity equation, since the velocity at state a is known. Hence,

$$V_2 = V_a \frac{A_a}{A_2} = V_a\left(\frac{D_a}{D_2}\right)^2 = 4\left(\frac{3}{2}\right)^2 = 9 \text{ ft/s}$$

Substitution of the available data into the energy equation yields

$$w_{\text{shaft}} = \frac{(9)^2}{2(32.2)} + \frac{50(32.2)}{32.2} = 1.3 + 50 = 51.3 \text{ ft} \cdot \text{lb}_f/\text{lb}_m$$

The local g has been taken to be 32.2 ft/s^2. Note that the potential-energy change is responsible for nearly all of the total work required.

The power requirement is obtained from the relation $\dot{W} = \dot{m}w$. The mass flow rate is

$$\dot{m} = \frac{VA}{v} = 4 \text{ ft/s} \times \frac{\pi(3)^2}{4(144)} \text{ ft}^2 \times \frac{1.50}{0.01605} \text{ lb}_m/\text{ft}^3 = 18.35 \text{ lb}_m/\text{s}$$

The specific volume value of 0.01605 ft^3/lb has been taken from the saturated-liquid-water table at room temperature. Finally, the power input required is

$$\dot{W} = \dot{m}w = 18.35 \text{ lb}_m/\text{s} \times 51.3 \text{ ft} \cdot \text{lb}_f/\text{lb}_m \times \frac{\text{hp} \cdot \text{s}}{550 \text{ ft} \cdot \text{lb}_f} = 1.71 \text{ hp}$$

3 Isentropic Process Evaluation Using Superheat and Saturation Data

There are no special relations for the evaluation of isentropic changes for fluids in the superheat or saturation region, other than $s_1 = s_2$. However, this process information, in conjunction with data on the initial and final states, is usually sufficient. An appropriate use of an energy balance or the continuity-of-flow equation may also be necessary to complete the solution. Several examples of this type of evaluation are provided below.

Example 7-16M Refrigerant 12 passes adiabatically through a nozzle in steady flow until its pressure reaches 2 bars. At the inlet to the nozzle the pressure and temperature are 9 bars and 100°C, respectively. Under reversible-flow conditions in the nozzle, determine the exit velocity, in m/s, if the inlet velocity is small.

SOLUTION If the flow is reversible and adiabatic, it is also isentropic. Thus the inlet and outlet entropy values are equal. The final velocity may be found from an energy balance on the nozzle;

that is, since q and w are zero,

$$h_2 - h_1 = \frac{V_1^2 - V_2^2}{2}$$

V_1 is small and will be neglected. V_2 can be found from the equation if h_2 can be found. From Table A-18M, the value of h_1 is 248.54 kJ/kg and that of s_1 is 0.8190 kJ/(kg)(°K). The final state is determined from the values of P_2 and s_2. At 2 bars the value of s_g is 0.7035 kJ/(kg)(°K), so the final state is superheated vapor also. From Table A-18M, it is found that at 2 bars, $s = 0.8184$ kJ/(kg)(°K) at 40°C. This is close to the value of s_1 which equals s_2. At this state of 40°C and 2 bars, $h_2 = 214.97$ J/g. Substitution of these h values into the energy balance leads to

$$214.97 - 248.54 = -\frac{V_2^2}{2(1000)}$$

where the factor 1000 converts kilograms to grams. Thus

$$V_2 = (67{,}140)^{1/2} = 259 \text{ m/s}$$

Example 7-16 Refrigerant 12 passes adiabatically through a nozzle in steady flow until its pressure reaches 30 psia. At the inlet to the nozzle the pressure is 120 psia and the temperature is 220°F. Under reversible-flow conditions in the nozzle, determine the exit velocity, in ft/s.

SOLUTION If the flow is reversible and adiabatic, it is also isentropic. Thus the inlet and exit entropy values are equal. The final velocity can be found from the energy balance on the nozzle; that is,

$$h_2 - h_1 = \frac{V_1^2 - V_2^2}{2}$$

Since no information is given on the inlet velocity, we shall assume that it is negligible. The final and initial enthalpies can be determined by knowing any two intrinsic properties at those states. From Table A-19, the refrigerant 12 superheat table, it is found that $h_1 = 108.509$ Btu/lb and $s_1 = 0.1993$ Btu/(lb)(°R). The final state is determined from the values of P_2 and s_2. At 30 psia and $s_2 = s_1$, it is found that the final temperature is very close to 120°F. As an approximation, the final enthalpy is then 94.843 Btu/lb. Substitution of these values in the energy balance leads to

$$94.843 - 108.509 = -\frac{V_2^2}{2(32.2)(778)}$$

$$V_2 = 827 \text{ ft/s}$$

Example 7-17M Steam enters a turbine at 30 bars, 500°C, and 70 m/s. It expands through a pressure ratio of 10 : 1 and leaves at a velocity of 140 m/s. If the process is adiabatic and internally reversible, calculate the work output in kJ/kg.

SOLUTION A steady-flow energy balance on the adiabatic turbine shows that

$$w = h_2 - h_1 + \frac{V_2^2 - V_1^2}{2}$$

The potential-energy change, if any, is neglected. Since the velocities are given, the problem is to evaluate $h_2 - h_1$. The value of h_1 is found from knowledge of P_1 and T_1. From steam table A-14M, for the initial state:

$$P_1 = 30 \text{ bars} \qquad h_1 = 3456.5 \text{ kJ/kg}$$

$$T_1 = 500°\text{C} \qquad s_1 = 7.2338 \text{ kJ/(kg)(°K)}$$

The process is adiabatic and reversible and hence isentropic. The specific entropy of a unit mass flowing through the turbine is constant. For the final state, then, the pressure is 3 bars and the entropy is 7.2338 kJ/(kg)(°K). The saturated-vapor entropy at 3 bars is given as 6.9919 kJ/(kg)(°K). Therefore the fluid is slightly superheated on leaving the turbine. Entering Table A-14M again, we find that at 3 bars we must interpolate between 160 and 200°C. Such interpolation shows a turbine outlet temperature of roughly 183°C with an enthalpy of 2830 kJ/kg. Consequently the work done for the isentropic process is

$$w = 2830 - 3456 + \frac{(140)^2 - (70)^2}{2(1000)}$$

$$= -626 + 7.4 = -619 \text{ kJ/kg}$$

The process is illustrated on the Pv diagram in Fig. 7-18. This is the maximum work output for an adiabatic turbine operating under the specified conditions.

Example 7-17 Steam enters a turbine at 600 psia, 900°F, and 250 ft/s. It expands through a pressure ratio of 10 : 1 and leaves at a velocity of 500 ft/s. If the process is adiabatic and internally reversible, calculate the work output per pound of steam.

SOLUTION A steady-flow energy balance on the turbine shows that

$$w = h_2 - h_1 + \frac{V_2^2 - V_1^2}{2}$$

Heat transfer is zero, and the potential-energy change, if any, is negligible. Since the velocities are given, the problem is to evaluate $h_2 - h_1$. Two properties are required to find a third property in the superheat table. From steam table A-4, for the initial state:

$$P_1 = 600 \text{ psia} \qquad h_1 = 1462.9 \text{ Btu/lb}$$

$$T_1 = 900°\text{F} \qquad s_1 = 1.6766 \text{ Btu/(lb)(°R)}$$

The process is adiabatic and reversible and hence isentropic. The specific entropy of each unit mass flowing through the turbine is constant. For the final state, pressure is 60 psia and the entropy is 1.6766 Btu/(lb)(°R). The saturated-vapor entropy at 60 psia (293°F) is given as 1.6443 Btu/(lb)(°R). Therefore the vapor is slightly superheated on leaving the turbine. Entering Table A-14 again, one finds that, at 60 psia and 350°F, s is listed as 1.6830 Btu/(lb)(°R). At 60 psia and 300°F, s is 1.6496 Btu/(lb)(°R). Interpolation shows a turbine outlet temperature of 340°F with an enthalpy of 1203.0 Btu/lb. Consequently the work done per pound for the isentropic process is

$$w = 1203.0 - 1462.9 + \frac{(500)^2 - (250)^2}{2(32.2)(778)}$$

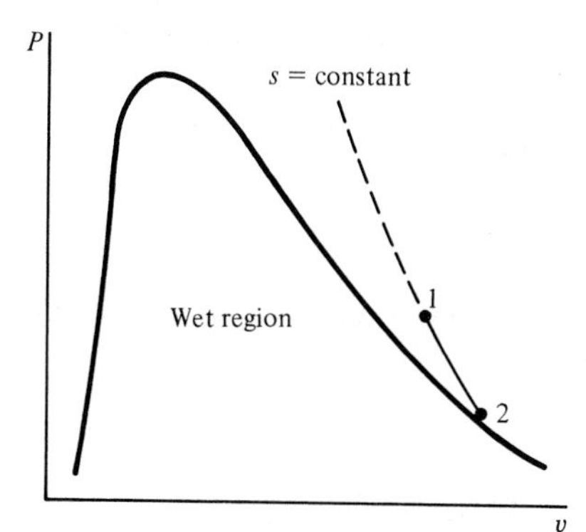

Figure 7-18 A Pv process diagram for Examples 7-17M and 7-17.

or $$w = -256.2 \text{ Btu/lb}$$

The process is illustrated on the accompanying *PV* diagram (Fig. 7-18).

The discussion in this section has centered on isentropic behavior of substances in various closed and open devices. Although no process in practice is isentropic, it is an extremely important concept with respect to devices that are essentially adiabatic. Since internally reversible processes by concept are free of nonequilibrium and dissipative effects, the isentropic performance of a device may be used as a standard to which real behavior is compared. Isentropic conditions lead to a *maximization* or *minimization* of important system variables, such as work effects or final state properties such as velocity, temperature, or pressure.

7-10 EFFICIENCIES OF SOME STEADY-FLOW DEVICES

Dissipative effects inherently accompany the flow of real fluids through steady-flow devices. These irreversibilities downgrade the performance of the device and lead to an increase in the entropy of the fluid. The magnitude of the entropy change is in proportion to the degree of the irreversibility. From an engineering viewpoint it is desirable to have parameters which measure the performance of individual types of equipment in terms of energy degradation.

In order to develop parameters for the comparative performance of specific types of steady-flow devices, we must recognize that the actual flow through many of these devices is approximately adiabatic. Therefore the idealized performance of such equipment occurs when the flow is adiabatic and internally reversible, that is, isentropic. An appropriate measure of their actual performance is a comparison of the actual desired output to the theoretical isentropic output. This ratio is defined as the *adiabatic* or *isentropic* efficiency of the steady-flow device.

The purpose of a turbine is to produce useful work output. Therefore, the isentropic or adiabatic turbine efficiency η_T is defined as

$$\eta_T \equiv \frac{w_a}{w_s} \qquad \text{turbine} \tag{7-43}$$

where the subscripts a and s again represent the actual and the isentropic flows, respectively, for the same initial state and the same final pressure in each case. Under adiabatic conditions and with a negligible potential-energy change, the conservation of energy principle for a unit mass passing through a control volume is

$$w_{\text{shaft}} = \Delta h + \Delta \text{KE} \tag{7-44}$$

When the kinetic-energy change across a turbine is small, the turbine efficiency may be approximated by

$$\eta_T = \frac{(h_1 - h_2)_a}{(h_1 - h_2)_s} \qquad \text{turbine} \tag{7-45}$$

where 1 and 2 represent the inlet and outlet states. The value of η_T for actual machines ranges from 80 to 90 percent.

A nozzle is a flow channel constructed to accelerate the fluid to a higher velocity. Therefore, the adiabatic or isentropic nozzle efficiency η_N is defined as

$$\eta_N \equiv \frac{V_a^2/2}{V_s^2/2} \qquad \text{nozzle} \tag{7-46}$$

where $V_a^2/2$ is the actual kinetic energy at the nozzle exit, and $V_s^2/2$ is the kinetic energy at the nozzle exit for isentropic flow to the same exit pressure. Nozzle efficiencies generally range upward from 90 percent. For converging nozzles used in subsonic flow, efficiencies of 0.95 or higher are quite common. If the inlet nozzle velocity is small compared to V_a or V_s at the exit, then Eq. (7-44) permits the nozzle efficiency to be expressed also by

$$\eta_N = \frac{(h_1 - h_2)_a}{(h_1 - h_2)_s} \qquad \text{nozzle} \tag{7-47}$$

This expression is analogous to Eq. (7-45) for a turbine.

Although turbines typically are insulated to prevent a heat loss from the working fluid, in some cases a deliberate attempt is made to cool the fluid passing through a compressor. For conditions where the heat loss is essentially zero, the adiabatic or isentropic compressor efficiency η_c is defined as the ratio of the isentropic work required to the actual work required for the same initial state and final pressure. That is,

$$\eta_c \equiv \frac{w_s}{w_a} \qquad \text{compressor} \tag{7-48}$$

When the kinetic-energy change across the compressor is negligible, Eq. (7-44) permits us to write Eq. (7-48) as

$$\eta_c = \frac{(h_2 - h_1)_s}{(h_2 - h_1)_a} \qquad \text{compressor} \tag{7-49}$$

where 1 and 2 represent the inlet and outlet states. The value of η_c ranges roughly from 75 to 85 percent for actual gas compressors.

Example 7-18M A small turbine operates on hydrogen gas, initially at 0.523 MPa and 480°K. The expansion pressure ratio is 5.23 : 1. The process is adiabatic, but irreversibilities reduce the work output to 80 percent of the maximum. Neglecting kinetic energy changes, determine (*a*) the maximum work output, in kJ/kg · mol, (*b*) the actual work output, in kJ/kg · mol, and (*c*) the actual final temperature, in kelvins.

SOLUTION (*a*) The first law on the turbine under adiabatic conditions and negligible kinetic- and potential-energy changes is simply $w_{sh} = \Delta h$. The maximum work is obtained from isentropic flow. Hence

$$w_{\text{shaft, max}} = h_{2s} - h_1$$

The initial enthalpy found from Table A-11M is 13,764 kJ/kg mol, and s_1^0 is 144.432 kJ/

(kg · mol)(°K). The final isentropic state is determined by applying the relation

$$s_2^0 = s_1^0 + R \ln \frac{P_2}{P_1}$$

$$= 144.432 + \frac{8.314}{2.018} \ln \frac{1}{5.23} = 137.616 \text{ kJ/(kg} \cdot \text{mol)(°K)}$$

From Table A-11M this value of s_2^0 corresponds very closely to 380°K. At this temperature $h_{2s} = 10{,}843$ kJ/kg · mol. Therefore,

$$w_{\text{shaft, max}} = 10{,}843 - 13{,}764 = -2921 \text{ kJ/kg} \cdot \text{mol}$$

(*b*) The actual work is determined from the adiabatic efficiency of the turbine. Since $\eta_T = w_a/w_s$, then

$$w_{\text{shaft, act}} = \eta_T w_s = 0.80(-2921) = -2337 \text{ kJ/kg} \cdot \text{mol}$$

(*c*) The final actual temperature is found from the final actual enthalpy.

$$h_{2a} = h_1 + w_a = 13{,}764 + (-2337) = 11{,}427 \text{ kJ/kg} \cdot \text{mol}$$

This value of h_{2a} corresponds to a temperature of 400°K in Table A-11M.

Example 7-18 A small turbine operates on hydrogen gas, initially at 63 psia and 400°F. The expansion pressure ratio is 5.30 : 1. The process is adiabatic, but irreversibilities reduce the work output to 80 percent of the maximum. The kinetic-energy change is negligible. Determine (*a*) the maximum work output, in Btu/lb · mol, (*b*) the actual work output, in Btu/lb · mol, and (*c*) the actual final temperature, in degrees Fahrenheit.

SOLUTION (*a*) The first law for the turbine under adiabatic conditions and negligible kinetic- and potential-energy changes is simply $w_{sh} = \Delta h$. The maximum work is obtained from isentropic flow. Hence

$$w_{\text{shaft, max}} = h_{2s} - h_1$$

The initial enthalpy found from Table A-11 is 5886 Btu/lb · mol, and s_1^0 is 34.466 Btu/(lb · mol)(°R). The final isentropic state is determined by applying the relation

$$s_2^0 = s_1^0 + R \ln \frac{P_2}{P_1} = 34.466 + \frac{1.986}{2.018} \ln \frac{1}{5.3} = 32.825 \text{ Btu/(lb} \cdot \text{mol)(°R)}$$

From Table A-11 this value of s_2^0 corresponds very closely to 680°R. At this temperature $h_{2s} = 4631$ Btu/lb · mol. Therefore,

$$w_{\text{shaft, max}} = 4631 - 5886 = -1255 \text{ Btu/lb} \cdot \text{mol}$$

(*b*) The actual work is determined from the adiabatic efficiency of the turbine. Since $\eta_T = w_a/w_s$, then

$$w_{\text{shaft, act}} = \eta_T w_s = 0.80(-1225) = -1004 \text{ Btu/lb} \cdot \text{mol}$$

(*c*) The final actual temperature is found from the final actual enthalpy.

$$h_{2a} = h_1 + w_a = 5886 + (-1004) = 4882 \text{ Btu/lb} \cdot \text{mol}$$

The temperature corresponding to this enthalpy is found by linear interpolation of Table A-11 between temperatures of 700 and 720°R. The result is around 716°R, or 256°F.

7-11 IRREVERSIBLE STEADY-FLOW PROCESSES INVOLVING INCOMPRESSIBLE FLUIDS

One of the major engineering uses of the second law is the prediction of the performance of steady-state flow devices when internal irreversibilities are present. It has been noted in Sec. 7-8 that the performance of adiabatic shaft-work devices is degraded by the presence of irreversibilities. Isentropic processes lead to the maximum work output or minimum work input. It is essential to show that isentropic processes are also standards of performance for adiabatic nonwork devices. In addition, the effects of irreversibilities on changes in various fluid properties need to be established. In this section we shall restrict our analysis to incompressible fluids.

In review, the basic conservation of energy principle for control volumes operating in steady state is

$$q + w_{\text{shaft}} = \Delta h + \Delta \text{KE} + \Delta \text{PE} \tag{5-14}$$

In the following applications the devices are assumed to be adiabatic, and in most cases the effect of gravitational potential-energy changes is assumed negligible. The enthalpy and entropy changes of an incompressible fluid are given by

$$\Delta h = c\,\Delta T + v\,\Delta P \tag{4-13}$$

and

$$\Delta s = c \ln \frac{T_2}{T_1} \tag{7-15}$$

The energy equation for the control volume becomes

$$w_{\text{shaft}} = c\,\Delta T + v\,\Delta P + \Delta \text{KE} \tag{7-50}$$

In addition, the second law for adiabatic processes requires that

$$\Delta S_{\text{adia}} \geq 0 \tag{6-13}$$

Equations (7-15), (7-50), and (6-13) form the basis for the following analysis of various flow devices involving incompressible fluids.

a Turbines

In the analysis of incompressible flow through turbines, it is acceptable as a first approximation to neglect kinetic-energy changes. Hence the energy equation in this case reduces to

$$w_{\text{shaft}} = c\,\Delta T + v\,\Delta P \tag{7-51}$$

In the limiting case of isentropic flow, Eq. (7-15) dictates that $T_2 = T_1$, since $\Delta s = 0$. If the process is isothermal, then the internal-energy term $c\,\Delta T$ is zero. As a result of adiabatic, internally reversible flow, the shaft work is equal to the change in the displacement work. That is,

$$w_{\text{shaft}} = v(P_2 - P_1)$$

Therefore, for isentropic flow all the energy available in the form of displacement work is converted to shaft-work output.

In the case of adiabatic, internally irreversible flow the entropy must increase on the basis of Eq. (6-13). Equation (7-15) now requires that $T_2 > T_1$. The effect of this temperature rise on work output is easily seen if we rearrange Eq. (7-51) to the form

$$v(P_1 - P_2) = -w_{\text{shaft}} + c(T_2 - T_1)$$

The pressure drop $(P_1 - P_2)$ is positive and is fixed for a given problem by the inlet and outlet conditions. The quantity $-w_{\text{shaft}}$ is positive for a turbine, and the internal-energy change is positive due to the temperature rise required by the second law. The two terms on the right of the above equation can take on various positive values, as long as the sum of the two terms equals the fixed value of the displacement work. With irreversibilities present, the change in the displacement work goes to increasing the temperature of the fluid as well as producing shaft-work output. Greater irreversibilities within the control volume (as measured by a larger Δs of the fluid) lead to a larger temperature rise. As a result, on the basis of the above equation, the shaft-work output must decrease. This is an example of mechanical dissipation of energy into a less useful form, an internal-energy increase.

The adiabatic or isentropic efficiency of a turbine was defined in the preceding section by the relation $\eta_T = w_{\text{act}}/w_{\text{isen}}$. For a hydraulic turbine this may be written in the form

$$\eta_T = \frac{c\,\Delta T + v\,\Delta P}{v\,\Delta P} \qquad \text{incompressible flow} \qquad (7\text{-}52)$$

In this equation $c\,\Delta T$ and $v\,\Delta P$ are of opposite sign; hence, the efficiency will be less than unity. The greater the value of ΔT due to irreversibilities, the smaller the turbine efficiency will be.

b Pumps

The basic equations for the analysis of incompressible flow through a pump are exactly the same as for a turbine, namely,

$$\Delta s = c \ln \frac{T_2}{T_1} \qquad (7\text{-}15)$$

and

$$w_{\text{shaft}} = c\,\Delta T + v\,\Delta P \qquad (7\text{-}51)$$

For isentropic flow the temperature remains constant and the shaft work again is related solely to the displacement work. The given shaft-work input goes to increasing the pressure of the fluid, with no other effect.

If the process is adiabatic and internally irreversible, the increase in entropy leads to a rise in temperature of the fluid, as given by Eq. (7-15). The immediate effect on the pressure rise of the fluid is seen from Eq. (7-51). For a given work input, the added energy is split between an increase in internal energy and an increase in displacement work. For the same work input, the pressure rise is greater for the isentropic case than for the irreversible process. On the other hand, for the same pressure rise, the work input will be greater for the irreversible process than for the isentropic case.

The adiabatic or isentropic efficiency η_P of a pump is defined as the ratio of the isentropic work required to the actual work required for the same inlet state and the same final pressure. Thus

$$\eta_P = \frac{w_s}{w_a} = \frac{(h_2 - h_1)_s}{(h_2 - h_1)_a} \qquad \text{pump} \qquad (7\text{-}53)$$

The work terms in Eq. (7-53) may be replaced by the enthalpy terms when the kinetic-energy change is negligible.

Example 7-19M Water enters a pump at 1 bar and 30°C. Shaft work is done on the fluid in the amount of 4500 N · m/kg.

(*a*) Determine the pressure rise if the process is isentropic.

(*b*) Determine the pressure rise if the fluid temperature rises by 0.1°C during the process.

SOLUTION (*a*) For an isentropic process, $w_{\text{shaft}} = v\,\Delta P$ for an incompressible substance. Solving for the pressure rise, we find that

$$\Delta P = \frac{w_{\text{shaft}}}{v} = 4500\text{ N}\cdot\text{m/kg} \times 1\text{ g/cm}^3 \times 10\text{ cm}^3\cdot\text{bar/N}\cdot\text{m} \times \frac{\text{kg}}{1000\text{ g}} = 45\text{ bars}$$

(*b*) For the irreversible process with a temperature rise of 0.1°C, Eq. (7-51) must be employed. Substitution of the proper values yields, when each term is expressed in N · m/g,

$$w_{\text{shaft}} = c\,\Delta T + v\,\Delta P$$

$$\frac{4500}{1000} = 4.18(0.1) + \frac{1.0(P)}{10}$$

$$\Delta P = 10(4.50 - 0.42) = 40.8\text{ bars}$$

When water is compressed from atmospheric conditions to around 40 or 50 bars, every 0.1°C rise in temperature due to irreversibilities results in roughly a 10 percent reduction in the pressure rise under adiabatic flow.

Example 7-19 Water enters a pump at 15 psia and 100°F. Shaft work is done on the fluid in the amount of 1550 ft · lb_f/lb_m.

(*a*) Determine the pressure rise if the process is isentropic.

(*b*) Determine the pressure rise if the fluid temperature rises by only 0.2°F.

SOLUTION (*a*) For an isentropic process, $w_{\text{shaft}} = v\,\Delta P$.
Solving for the pressure rise, we find that

$$\Delta P = \frac{w_{\text{shaft}}}{v} = 1550\text{ ft}\cdot\text{lb}_f/\text{lb}_m \times \frac{\text{lb}_m}{0.01613\text{ ft}^3}\,\frac{\text{ft}^2}{144\text{ in}^2} = 667\text{ psi}$$

(*b*) For the irreversible process with a temperature rise of 0.2°F, Eq. (7-51) must be employed. Substitution of the proper values yields

$$w_{\text{shaft}} = c\,\Delta T + v\,\Delta P$$

$$1550\text{ ft}\cdot\text{lb}_f/\text{lb}_m = 1.0\text{ Btu/(lb}_m\text{)(°F)} \times 0.2\text{°F} \times 778\text{ ft}\cdot\text{lb}_f/\text{Btu} + 0.01613\text{ ft}^3/\text{lb}_m \times \Delta P$$

$$\Delta P = \frac{1550 - 155}{0.01613} = (667 - 66.7)\text{ psi} = 600\text{ psi}$$

The results of this calculation are quite revealing. When water is compressed from atmospheric conditions to around 600 to 700 psia, every 0.2°F rise in temperature due to irreversibilities results in a 10 percent reduction in the pressure rise under adiabatic conditions.

c Nozzles

For flow through adiabatic nozzles, the basic equations for an incompressible fluid are

$$-\Delta h = \Delta\text{KE}$$

or

$$(u_1 - u_2) + v(P_1 - P_2) = \Delta\text{KE} \tag{7-54}$$

and

$$s_2 - s_1 = c \ln \frac{T_2}{T_1} \tag{7-15}$$

Under the conditions of isentropic flow the temperature change is zero. Hence the internal-energy change is also zero, and the energy equation for the control volume reduces to

$$v(P_1 - P_2) = \Delta\text{KE}$$

This result simply states that the desired kinetic-energy change through the nozzle is due solely to the change in displacement work if the flow is incompressible and isentropic.

In the case of internal irreversibilities, which lead to a rise in temperature of the incompressible fluid, the change in kinetic energy is expressed by a rearranged form of Eq. (7-54), i.e.,

$$v(P_1 - P_2) = \Delta\text{KE} + (u_2 - u_1)$$
$$= \Delta\text{KE} + c(T_2 - T_1)$$

We see that the increase in internal energy decreases the kinetic-energy change for a given inlet and outlet pressure. Dissipation of some of the displacement work increases the internal energy of the fluid. As a result, the kinetic-energy change for irreversible flow is less than that for isentropic flow. Note that this will require a different exit area for the irreversible nozzle, as demanded by continuity considerations for this case of fixed-pressure drop. The adiabatic efficiency of a nozzle in this situation was defined in the preceding section by the relation $\eta_N = (V_{2a}^2/2)/(V_{2s}^2/2)$. For the flow of an incompressible fluid through a nozzle this becomes (for $V_1 \approx 0$),

$$\eta_N = \frac{c(T_1 - T_2) + v(P_1 - P_2)}{v(P_1 - P_2)} \qquad \text{incompressible flow} \qquad (7\text{-}55)$$

Note that the quantities $c(T_1 - T_2)$ and $v(P_1 - P_2)$ are of opposite sign.

If the nozzle is of fixed geometry, then the kinetic-energy term is the same for both reversible and irreversible flow. In this case either the inlet pressure must be increased or the outlet pressure decreased for irreversible flow, to make up for the positive internal-energy change.

Example 7-20M Liquid water enters a nozzle at 1.244 bar, 30°C, and with a negligible velocity. The fluid expands to 1.0 bar.

(*a*) Find the maximum exit velocity under adiabatic conditions, in m/s.

(*b*) If the adiabatic efficiency is 90 percent, determine the temperature rise in the fluid, in degrees Celsius, and the final actual velocity, in m/s.

SOLUTION (*a*) The maximum exit velocity is attained under isentropic flow. For isentropic flow of an incompressible fluid, $T_1 = T_2$, and the energy equation for the control volume is

$$\frac{V_2^2}{2} = v(P_1 - P_2)$$

Therefore, when both terms are expressed in N · m/g and the pressure in bars,

$$\frac{V_{2s}^2}{2(1000)} = 1.004(1.244 - 1.0)\frac{10^5}{10^6} = 0.0245$$

$$V_{2s} = 7.0 \text{ m/s}$$

(*b*) Equation (7-54) enables us to determine the temperature rise. Using data from part *a*, we find that

$$0.90 = \frac{-4.184\ \Delta T + 0.0245}{0.0245}$$

$$\Delta T = 5.9 \times 10^{-4}\ ^\circ\text{C}$$

The actual exit velocity is found from Eq. (7-46), namely, $\eta_N = V_{2a}^2/V_{2s}^2$.

$$V_{2a} = [0.90(7)^2]^{1/2} = 6.64 \text{ m/s}$$

Although the temperature rise is immeasurable, there is a significant change in the exit velocity due to irreversibilities. This change in velocity will require a change in the exit area.

Example 7-20 Liquid water enters a nozzle at 17.67 psia, 100°F, and with negligible velocity. The fluid expands to 15.0 psia.

(*a*) Find the maximum exit velocity under adiabatic conditions, in ft/s.

(*b*) If the adiabatic efficiency is 90 percent, determine the temperature rise in the fluid, in degrees Fahrenheit, and the final actual velocity, in ft/s.

SOLUTION (*a*) The maximum exit velocity is attained under isentropic flow. For isentropic flow of an incompressible fluid, $T_1 = T_2$, and the energy equation for the control volume is

$$\frac{V_{2s}^2}{2} = v(P_1 - P_2)$$

Therefore, when both terms are expressed in ft · lb_f/lb_m,

$$\frac{V_{2s}^2}{2(32.2)} = 0.01613(17.67 - 15.0)(144) = 6.202$$

$$V_{2s} = 20.0 \text{ ft/s}$$

(*b*) Equation (7-54) enables us to determine the temperature rise. Using data from part *a* and assuming the specific heat of water is 1.0 Btu/(lb)(°F), we find that

$$0.90 = \frac{-1.0(\Delta T)(778) + 6.202}{6.202}$$

$$\Delta T = 8.0 \times 10^{-4} \text{ °F}$$

The actual exit velocity is found from Eq. (7-46), namely, $\eta_N = V_{2a}^2/V_{2s}^2$.

$$V_{2a} = [0.90(20)^2]^{1/2} = 19.0 \text{ ft/s}$$

Although the temperature rise is immeasurable, there is a 5 percent change in the exit velocity due to irreversibilities. This change in velocity will require a change in the design exit area for the irreversible nozzle, as compared to the area of the reversible nozzle.

d Throttling Devices

A throttling device has been described earlier as a flow restriction which leads to a drop in pressure of the fluid. No interactions occur to the fluid, and kinetic- and potential-energy changes are negligible. Therefore $h_1 = h_2$ if the inlet and outlet are selected to be relatively far upstream and downstream from the flow restriction. No proof of the drop in pressure across these devices has yet been shown. The proof stems from the second law. Splitting the enthalpy function back into its two contributions of internal energy and displacement work, we see that the statement $h_1 = h_2$ can be written as

$$u_2 - u_1 = v(P_1 - P_2)$$

or

$$c(T_2 - T_1) = v(P_1 - P_2)$$

For incompressible fluids,

$$\Delta s = c \ln \frac{T_2}{T_1}$$

If the flow is assumed to be isentropic, then $T_2 = T_1$ on the basis of the Δs expression. The energy relation then requires that $P_1 = P_2$. Hence no properties change in the direction of flow, and flow itself cannot occur. The flow must be irreversible. In this case $T_2 > T_1$ since the entropy increases. The energy relation now requires that $P_2 < P_1$, which is the result we expected but had not yet proved. The flow accelerates and the pressure increases as the fluid reaches the restriction. At the outlet of the restriction the fluid decelerates to the original velocity, so that the original enthalpy is recovered. However, the original pressure is not recovered. This is offset by a rise in temperature of the fluid.

e Pipe Flow

The basic equations for incompressible flow through piping systems are

$$w_{\text{shaft}} = c\ \Delta T + v\ \Delta P + \Delta\text{KE} + \Delta\text{PE} - q \tag{7-42a}$$

and

$$s_2 - s_1 = c \ln \frac{T_2}{T_1} \tag{7-15}$$

If we restrict ourselves to adiabatic flow, then Eq. (7-42*a*) becomes

$$w_{\text{shaft}} = c\ \Delta T + v\ \Delta P + \Delta\text{KE} + \Delta\text{PE} \tag{7-56}$$

and if the piping system does not include a pump within the control volume, then

$$c\ \Delta T + v\ \Delta P + \Delta\text{KE} + \Delta\text{PE} = 0 \tag{7-57}$$

The isentropic case was discussed in Sec. 7-9. As shown by Eq. (7-15) above, the temperature remains constant for isentropic flow of an incompressible fluid. Thus the internal-energy change, given by $c\ \Delta T$, is zero.

If the process is adiabatic and internally irreversible, so that Δs is positive, then Eq. (7-15) dictates that the temperature increases in the direction of flow. The effect of the temperature rise on other properties of the flow system is found by writing Eq. (7-57) in the form

$$v(P_1 - P_2) = c(T_2 - T_1) + \Delta\text{KE} + \Delta\text{PE}$$

The changes in kinetic and potential energy are fixed by the geometry and mass flow rate of the flow system. Therefore any change in temperature of the fluid due to internal irreversibilities requires an increase in the pressure drop from inlet to outlet.

REFERENCES

Fast, J. D.: "Entropy," McGraw-Hill, New York, 1963.

Joint Army, Navy and Air Force (JANAF) Thermochemical Tables, NSRDS-NBS-37, June, 1971.

Keenan, J. H., and J. Kaye: "Gas Tables," New York, 1945.

PROBLEMS (METRIC)

Entropy tabulations for superheat and saturation states

7-1M Find the specific entropy, in kJ/(kg)(°K), for water substance in parts *a* through *j* listed in Prob. 4-3M.

7-2M Find the specific entropy, in kJ/(kg)(°K), for water substance in parts *a* through *k* listed in Prob. 4-4M.

7-3M Find the specific entropy, in kJ/(kg)(°K), for refrigerant 12 in parts *a* through *k* listed in Prob. 4-5M.

7-4M Find the specific entropy, in kJ/(kg)(°K), for refrigerant 12 in parts *a* through *k* listed in Prob. 4-6M.

Second-law analysis for superheat and saturation states

7-5M Refrigerant 12 at 2.8 bars and 60°C is compressed in a closed system to 14 bars. The process is isothermal and internally reversible, and the environmental temperature is 25°C. Determine (*a*) the heat transfer and the work done, in kJ/kg, and (*b*) the total entropy change for the overall process, in kJ/(kg)(°K). Sketch the process on a *Ts* diagram.

7-6M One-tenth kilogram of steam at 3.0 bars and 200°C is compressed isothermally to 15 bars in a piston-cylinder device. During this process 20,000 J of work is done on the gas, and heat is lost to the environment which is at 30°C. Determine (*a*) the heat transfer, in kJ, (*b*) the entropy change of the steam, in kJ/°K, and (*c*) the total entropy change for the overall process, in kJ/°K. Is the process reversible, irreversible, or impossible?

7-7M Steam originally at 60 bars, 500°C, is expanded isothermally and reversibly to 15 bars in a steady-flow device. Heat is supplied to the steam from a reservoir at 550°C. Determine (*a*) the heat transfer and the work output, in kJ/kg, and (*b*) the total entropy change for the process in kJ/(kg)(°K). Is the process reversible, irreversible, or impossible? Sketch the process on a *Ts* diagram.

7-8M Steam at 15 bars and 320°C is contained in a closed system at constant pressure. During a process 100,000 N · m of work is performed on the steam by a paddle-wheel. At the same time heat in the amount of 12.70 kJ is transferred to the surroundings at 17°C. Determine (*a*) the entropy change of the steam, and (*b*) the entropy change for the overall process in kJ/(kg)(°K). Sketch the process on a *Ts* diagram.

7-9M A piston-cylinder device initially contains refrigerant 12 at 6.0 bars and 80°C. It is compressed quasistatically and at constant pressure, with a boundary-work input of 13.63 kJ/kg. Determine (*a*) the final specific volume, in cm^3/g, (*b*) the final specific entropy, in kJ/(kg)(°K), and (*c*) the heat transfer, in kJ/kg. (*d*) If the surrounding temperature is 20°C, determine the total entropy change for the overall process, in kJ/(kg)(°K). Sketch the process on a *Ts* diagram. Is the process reversible, irreversible, or impossible?

7-10M A piston-cylinder device contains 4.0 kg of steam at 30 bars and 360°C. During a certain quasistatic process, 1338 kJ of heat is removed from the steam at constant pressure to the surroundings at 18°C. Find (*a*) the entropy change of the steam, and (*b*) the total entropy change for the overall process, in kJ/°K. Sketch the process for the steam on a *Ts* diagram.

7-11M A rigid vessel contains 0.20 kg of steam at 7 bars and 280°C. Heat is transferred out of the system until the temperature drops to 200°C. The surrounding temperature is 17°C. Find (*a*) the entropy change of the steam, and (*b*) the total entropy change for the overall process, in kJ/°K. Sketch the process on a *Ts* diagram. Is the process reversible, irreversible, or impossible?

7-12M Refrigerant 12 flows at a low velocity through a well-insulated capillary tube. The initial and final conditions are (*a*) saturated liquid at 7 bars to 1.8 bars, (*b*) 8 bars, 40°C to 30°C, and (*c*) saturated vapor at 8 bars to 1 bar. (1) Determine the entropy change of the fluid, in kJ/(kg)(°K). (2) Does this value agree qualitatively with the second law? (3) What is the ratio of the final velocity to the initial velocity?

7-13M A steam turbine operates with an inlet condition of 100 bars and 520°C and an outlet state of 1 bar and 100°C. The device produces 2010 kW of power with a mass flow rate of 10,000 kg/h. The atmospheric temperature is 27°C. Determine the total entropy change for the overall process, in kJ/(kg)(°K).

7-14M Steam flows steadily through a nozzle from 30 bars, 320°C, and 20 m/s to 20 bars. The process is isothermal, and 40 kJ/kg of heat is transferred to the steam from a reservoir at 350°C. The surrounding temperature is 20°C. Determine (*a*) the nozzle exit velocity, in m/s, and (*b*) the total entropy change for the overall process, in kJ/(kg)(°K).

7-15M Steam enters a steady-flow compression process at 1.0 bar, 100°C, and exits at 10.0 bars,

200°C. The work input is measured to be 400 kJ/kg, and the kinetic- and potential-energy changes are negligible. The environment has a temperature of 27°C. Determine (*a*) the magnitude and direction of any heat transfer, in kJ/kg, (*b*) the entropy change of the fluid passing through the compressor, in kJ/(kg)(°K), and (*c*) the total entropy change for the overall process in kJ/(kg)(°K).

7-16M Saturated refrigerant-12 vapor at (*a*) 40°C, and (*b*) 9.0 bars is condensed in a steady-flow heat exchanger to saturated liquid by heat transfer to the surroundings at 20°C. Determine the net change in entropy for the fluid plus the surroundings, in kJ/(kg)(°K).

Second-law analysis of ideal-gas processes

7-17M A piston-cylinder device contains air at 350°K and 1 bar. The pressure is increased to 1.3 bars, but during the process heat transfer occurs so that the temperature remains constant.

(*a*) Determine the entropy change of the air, in kJ/(kg)(°K).

(*b*) If the temperature of the surroundings is 300°K, determine the entropy change of the surroundings, in kJ/(kg)(°K).

(*c*) Is the increase in entropy principle fulfilled?

7-18M A piston-cylinder device maintained at a constant pressure of 1 bar contains nitrogen gas. It is cooled from 150 to 40°C. Compute (*a*) the heat removed, in kJ/kg, (*b*) the work done, in kJ/kg, (*c*) the change in entropy of the nitrogen, in kJ/(kg)(°K), and (*d*) the total entropy change of the overall process if the temperature of the environment is 22°C, in kJ/(kg)(°K).

7-19M A rigid tank contains 50 g of carbon monoxide gas at 0.95 bar and 30°C. Heat is added until the pressure reaches 1.3 bars. Compute (*a*) the heat transferred, in kilojoules, (*b*) the change in entropy of CO, in kJ/K, and (*c*) the total entropy change for the overall process, in kJ/°K, if the heat is supplied from a reservoir at 450°K. (*d*) Is the answer to part *c* compatible with the second law?

7-20M Contained in a constant-pressure closed system is 0.5 kg of hydrogen gas at 6 bars and 17°C. Heat in the amount of 798 kJ is added to the gas from a reservoir at 450°K. Compute (*a*) the entropy change of the hydrogen, in kJ/°K, and (*b*) the total entropy change for the overall process, in kJ/°K. (*c*) Is the process reversible, irreversible, or impossible?

7-21M A rigid tank with a volume of 0.03 m^3 contains oxygen at an initial state of 87°C and 1.5 bar. During a process paddle-wheel work is carried out by applying a torque of 13 N · m for 25 revolutions, and a heat loss of 3.71 kJ occurs to the surroundings at 18°C. Determine (*a*) the entropy change of the oxygen, in kJ/°K, and (*b*) the total entropy change for the overall process, in kJ/°K. (*c*) Is the process reversible, irreversible, or impossible?

7-22M Air in the amount of 0.5 kg is compressed isothermally and quasistatically from 1 bar, 270°K, to a final state which requires 63.9 kJ of boundary-work input to the piston-cylinder device. Find (*a*) the final pressure, in bars, (*b*) the change in entropy of the air, in kJ/°K, and (*c*) the total entropy change of the overall process in kJ/°K if the temperature of the surroundings is 270°K. (*d*) Is the process reversible, irreversible, or impossible?

7-23M Air, in the amount of 0.5 kg, in a constant-volume system has an initial state of 0.10 m^3 and 27°C. Measurements during the process indicate that 9.50 kJ of heat were removed and paddle-wheel work was carried out by applying a torque of 15.7 N · m to the shaft for 400 revolutions. Find (*a*) the final temperature of the gas in the tank, in degrees Celsius, (*b*) the entropy change of the air, in kJ/°K, and (*c*) the total entropy change of the overall process, in kJ/°K, if the temperature of the environment to which heat is transferred is 22°C.

7-24M Carbon monoxide at 1 bar and 37°C is compressed in a closed system to 3 bars and 147°C. The required work input is 88.0 kJ/kg, and the surrounding temperature is 25°C. Determine (*a*) the entropy change of the CO in kJ/(kg)(°K), (*b*) the heat loss to the surroundings, in kJ/kg, and (*c*) the total entropy change for the CO and the environment in kJ/(kg)(°K).

7-25M Air initially at 2.3 bars and 62°C expands to (*a*) 1.4 bars and 22°C, and (*b*) 1.6 bars and 10°C. Can these changes of state be carried out adiabatically? Why?

7-26M Air enters a turbine at 6 bars and 277°C and leaves at 1 bar. The flow rate is 50 kg/min, and the power output is 180 kW. If the heat removed, which appears in the environment at 22°C, is 30 kJ/kg, find (*a*) the final temperature, in degrees Celsius, (*b*) the entropy change of the air, in kJ/(kg)(°K), and (*c*) the total entropy change for the overall process in kJ/(kg)(°K).

7-27M Air enters a steady-state compressor at 1 bar and 27°C at a rate of 1 kg/min, and leaves at 7 bars and 227°C. The power required to operate the compressor is 3.58 kW. Determine (*a*) the rate of heat transfer, in kJ/h, (*b*) the entropy change of the air, in kJ/(min)(°K), and (*c*) the entropy change of the environment which receives the heat transferred at 15°C, in kJ/(min)(°K). (*d*) Is the process reversible, irreversible, or impossible?

7-28M Argon gas enters a turbine at 0.35 MPa and 150°C, and leaves the device at 0.15 MPa. A heat loss of 3.7 kJ/kg occurs, and this energy appears in the surroundings at 22°C. The measured shaft-work output is 52.0 kJ/kg, and the kinetic-energy change is negligible. Find (*a*) the outlet temperature, in degrees Celsius, (*b*) the entropy change of the argon in kJ/(kg)(°K), and (*c*) the total entropy change for the overall process in kJ/(kg)(°K).

7-29M Nitrogen flows steadily through a nozzle from 0.5 MPa, 600°K, and 30 m/s to 0.3 MPa and 570°K. During the process, 30 kJ/kg of heat is transferred to the nitrogen from a reservoir at 630°K. Determine (*a*) the nozzle exit velocity, in m/s, and (*b*) the total entropy change for the overall process in kJ/(kg · mol)(°K).

7-30M Air passes steadily through a device from 1.3 bars and 27°C to 2.7 bars. During the process a work input of 64.4 kJ/kg is required, and a heat loss of 14.0 kJ/kg occurs with the energy appearing in the atmosphere at 27°C. Determine (*a*) the entropy change of the air, and (*b*) the entropy change of the atmosphere, both in kJ/(kg)(°K). (*c*) Is the process reversible, irreversible, or impossible?

7-31M Oxygen is throttled from 2 bars and 600°K to (*a*) 1.4 bars, and (*b*) 1.2 bars.

(1) Find the entropy change of the gas, in kJ/(kg)(°K).

(2) Is the process reversible, irreversible, or impossible?

Entropy changes for incompressible substances

7-32M Fifty kilograms of water at 1 bar and 20°C are mixed with 20 kg of water initially at 1 bar and 90°C. If the mixing process is adiabatic and at constant pressure, determine the total change in entropy for the 70 kg of water, in kJ/°K.

7-33M An 8-kg iron casting at 550°C is quenched in a tank containing 0.03 m^3 of water initially at 25°C. Assuming no heat loss to the surroundings and that the specific heat of the iron averages 0.45 kJ/(kg)(°C), determine the change in entropy of (*a*) the water, (*b*) the casting, and (*c*) the total process.

7-34M A 1-kg block of copper at 150°C and a 1-kg block of aluminum at 50°C initially are isolated from each other and from the local surroundings. They are then placed in thermal contact with each other, but are still isolated from the surroundings. Calculate ΔS for each block if they attain thermal equilibrium. The specific heat of copper averages 0.385 kJ/(kg)(°C), and for aluminum the value is 0.90 kJ/(kg)(°C). Report the answers, in kJ/°K.

7-35M A small solid object of constant heat capacity C equal to 0.70 kJ/°K is initially at 580°K. The object is thrown into a large lake which is at a temperature of 290°K. Calculate the total entropy change for this process, in kJ/°K. (*Note*: heat capacity C equals mass time specific heat, mc.)

7-36M A kilogram of ice at 0°C is mixed with 7 kg of water at 30°C. Compute the change in entropy of the composite system if the process is adiabatic and the pressure is maintained at 1 bar. The enthalpy of melting of ice is 335 kJ/kg. Report the answer, in kJ/°K.

7-37M Ten kilograms of liquid water at 20°C are mixed adiabatically at 1 bar pressure with 6 kg of water at 100°C.

(*a*) Compute the entropy change for each quantity of water, and the total change, in kJ/°K.

(*b*) Sketch the separate processes that each quantity of water undergoes on the same TS diagram.

7-38M A resistor of 20Ω, initially at 20°C, is maintained at a constant temperature while a current of 5 A is allowed to flow for 2 s.

(*a*) Determine the entropy change of the resistor and of the universe, in J/K.

(*b*) Now, insulate the resistor and carry out the same experiment.

For the latter case, determine the entropy change of the resistor and the universe. The mass of the resistor is 8.0 g and the specific heat of the resistor is 1.05 kJ/(kg)(°C). The environmental temperature is 20°C.

7-39M An electric resistor of 30Ω is maintained at a constant temperature of 17°C while a current of 6 A is maintained for 3 s.

(*a*) Determine the entropy change of the resistor and of the universe, in J/K.

(*b*) Now, insulate the resistor and carry out the same experiment.

For the latter case, determine the entropy change of the resistor and the universe. The mass of the resistor is 19.0 g and the specific heat of the resistor is 1.10 kJ/(kg)(°C). The environmental temperature is 17°C.

Second-law analysis of heat exchangers and mixing processes

7-40M Reconsider Prob. 5-41M. Determine (*a*) the entropy change of the air stream, (*b*) the entropy change of the refrigerant-12 stream, and (*c*) the total entropy change for the overall process, all in kJ/(°K)(s).

7-41M Reconsider Prob. 5-42M. Determine the entropy change in kJ/(°K)(min) for (*a*) the water stream, (*b*) the air stream, and (*c*) the overall process.

7-42M Reconsider Prob. 5-43M. Determine the entropy change in kJ/(°K)(min) for (*a*) the condensing steam, (*b*) the air stream, and (*c*) the overall process, if the exit air pressure is 1.10 bars.

7-43M Reconsider Prob. 5-45M. Determine the rate of entropy production for the process in kJ/(°K)(h) using the data of (*a*) 1750 kg/h, and (*b*) 1812 kg/h for the second source.

7-44M Reconsider Prob. 5-47M(*a*). Determine the rate of entropy production for the process in kJ/(°K)(min).

7-45M Superheated steam enters one section of an open feedwater heater at 5 bars and 240°C. Compressed liquid water at the same pressure and 35°C enters another inlet. The mixture of these two streams leaves at the same pressure as a saturated liquid. If the heater is adiabatic, calculate the change in entropy for the process per kilogram of mixture leaving the heater.

Steady-flow mechanical work

7-46M Carbon monoxide flows frictionlessly and isothermally through a 75-cm^2 constant-area duct. Upstream the pressure is 1.6 bars and the temperature is 100°C. After flowing against an impeller which produces shaft work, the gas reaches a downstream position where the pressure is 1 bar. If the mass flow rate is 1.8 kg/s, determine (*a*) the shaft work, and (*b*) the heat transfer, both in kJ/kg.

7-47M Air is compressed isothermally from 0.96 bar and 7°C to 4.8 bars. Flow through the compressor is steady at 0.95 kg/s. Kinetic and potential energies are negligible. Calculate (*a*) the power input, in kilowatts, and (*b*) the rate of heat removal, in kJ/s, if the process is frictionless.

7-48M A fluid flows frictionlessly through an open system which requires work input. The entrance conditions are: $P_1 = 3.4$ bars, $v_1 = 0.6$ m^3/kg, and $A_1 = 0.040$ m^2. At the exit $P_2 = 13.6$ bars and $A_2 = 0.0050$ m^2. The mass flow rate is 7 kg/s, and for the process, Pv is constant. Find the work required, in kJ/kg.

Isentropic processes of ideal gas

7-49M Air is contained in a piston-cylinder device initially at 157°C, 0.290 MPa, and 760 cm^3. The gas is expanded isentropically to 0.097 MPa. Determine (*a*) the final temperature, in degrees Celsius, (*b*) the mass of air within the device, in kg, (*c*) the final volume, in cm^3, and (*d*) the work output, in kJ. Use the air table for data.

7-50M Nitrogen at 3 bars and 400°K is allowed to expand adiabatically and reversibly to 1.7 bars in a closed system. Determine (*a*) the final temperature, in kelvins, and (*b*) the work output, in kJ/kg.

7-51M Air, in the amount of 0.5 kg, is compressed adiabatically and without friction from an initial state of 1.30 bars and 0.30 m^3 to a final volume of 0.060 m^3 in a piston-cylinder device. Determine (*a*) the work required, in kilojoules, and (*b*) the final temperature, in degrees Celsius.

7-52M A piston-cylinder device contains carbon dioxide at 1.05 bars, 310°K, and 1000 cm^3. The gas is compressed isentropically to 2.11 bars. Determine (*a*) the final temperature, in °K, and (*b*) the work required, in kJ. Use tabular data.

7-53M Air enters a nozzle at 7 bars and 57°C at a rate of 0.40 kg/s. The inlet velocity is negligible, and the exit pressure is 4 bars. If the process is isentropic, determine the exit area of the nozzle, in cm^2.

7-54M Argon, initially at 6.4 bars and 280°C, flows at a rate of 5 kg/s through an insulated, frictionless nozzle. The initial velocity is negligible, and the outlet pressure is 1.4 bars. Determine (*a*) the final temperature in degrees Celsius, (*b*) the final velocity, in m/s, and (*c*) the exit area, in square centimeters.

7-55M Air enters a diffuser at 0.70 bar and 7°C with a velocity of 300 m/s. The outlet temperature is 320°K, and the process is adiabatic and frictionless. Determine (*a*) the final velocity, in m/s, and (*b*) the final pressure, in bars.

7-56M Air enters a diffuser at 0.60 bar, −3°C, and 260 m/s. The air stream leaves the diffuser at a velocity of 130 m/s. For isentropic flow, find (*a*) the temperature at the outlet, in degrees Celsius, (*b*) the outlet pressure, in bars, and (*c*) the ratio of the outlet area to the inlet area.

7-57M Helium enters a turbine at 8.4 bars and 550°C and exhausts at 1.4 bars. Neglect kinetic- and potential-energy changes. Assuming isentropic flow, determine (*a*) the exhaust temperature, in degrees Celsius, and (*b*) the work output, in kJ/kg.

7-58M Carbon dioxide at 800°K and 200 bars enters a turbine with a velocity of 100 m/s through an inlet area of 10.0 cm^2. The gas expands isentropically to 500°K and leaves through an area of 30 cm^2. Find (*a*) the work output, in kJ/kg · mol, and (*b*) the mass flow rate, in kg · mol/s.

7-59M The inlet conditions for an air turbine are 6 bars and 327°C. The air expands isentropically to 1 bar. For a flow rate of 40 kg/min, compute the power output, in kilowatts.

7-60M Air at 20 m/s, 1.0 bar, and 27°C is taken into the compressor of a gas turbine of a large truck. The isentropic compressor delivers air at 8.0 bars and 60 m/s at a mass flow rate of 2.4 kg/s. Determine (*a*) the compressor exit temperature, in degrees Celsius, (*b*) the diameter of the circular inlet, in centimeters, and (*c*) the power required, in kilowatts. Use air table data.

7-61M Nitrogen gas flows through a compressor at a rate of 5 kg/s and undergoes an isentropic change from 1 bar and 17°C, to 2.7 bars. The measured kinetic-energy change is 5 kJ/kg. Compute the power input to the gas, in kilowatts.

7-62M Argon at a rate of 0.04 kg/s enters an adiabatic, frictionless compressor at 1.6 bars and 25°C. The exit pressure is 6.2 bars, and the kinetic- and potential-energy changes are negligible. Determine (*a*) the exit temperature, in kelvins, and (*b*) the power input, in kilowatts.

Isentropic processes with incompressible substances

7-63M Oil with a specific gravity of 0.85 is being pumped from a pressure of 0.70 bar to a pressure of 1.20 bars, and the outlet lies 2.0 m above the inlet. The fluid at 15°C flows at a rate of 0.10 m^3/s through an inlet cross-sectional area of 0.050 m^2, and the outlet area is 0.020 m^2. The local gravity is 9.8 m/s^2, and the flow is assumed to be adiabatic and frictionless. Determine the power input to the pump, in kilowatts.

7-64M Water at 20°C is pumped at 1.20 kg/s from the surface of an open tank into a constant-diameter piping system. At the discharge from the pipe, the velocity is 10 m/s and the gage pressure is 2.0 bars. The discharge is 15.0 m above the water surface in the open tank and g is 9.60 m/s^2. For isentropic flow, determine the power required for the pump, in kilowatts.

7-65M Water is pumped from an initial state of 2 bars, 15°C, and 2 m/s to a final state of 6.0 bars and 8 m/s. The pipe diameter at the exit is 2.0 cm, and the exit position lies 20.0 m vertically above the inlet position. If the pipe is insulated and the flow frictionless, determine (*a*) the mass flow rate, in kg/s, (*b*) the shaft work required, in kJ/kg, and (*c*) the power input, in kilowatts. Do not neglect any energy term if sufficient information is available. Local gravity is 9.80 m/s^2.

7-66M Water enters an inlet pipe to a water turbine at 1.0 bar and 20°C at an elevation 150 m above the pipe outlet. The inlet velocity is negligible. At the outlet pipe, following the turbine, the conditions are 1.0 bar and 9 m/s. If the flow is adiabatic and frictionless, determine the shaft-work output from the turbine, in kJ/kg. Local gravity is 9.65 m/s^2.

7-67M A hydraulic turbine is located 120 m below the surface of water behind a dam. Water is taken to the turbine through a 2.0-m-diameter pipe which begins just below the water level. At the outlet pipe of the same diameter which follows the turbine, the conditions are 1.0 bar, 15°C, and 12 m/s. If the flow is assumed to be isentropic and local gravity is 9.70 m/s^2, determine the power output of the turbine, in kilowatts.

7-68M Water enters a piping system at 20°C and 10 m/s. At a position downstream the conditions are 0.150 MPa and 20 m/s, and the elevation is 22.0 m above the inlet. The local gravity is 9.70 m/s^2. For adiabatic and frictionless flow, determine the required inlet pressure, in MPa.

7-69M Water enters a piping system at 3.50 bars, 15°C, and 7 m/s at a rate of 10.0 kg/s. Downstream at a given position the pressure is 1.80 bars, and the elevation is 12.0 m above the inlet. The local gravity is 9.75 m/s^2. Determine (*a*) the velocity at the downstream position, in m/s, and (*b*) the pipe diameter downstream, in centimeters, if the flow is assumed to be adiabatic and frictionless.

7-70M Water at 15°C flows over a dam and down into a vertical pipe with a 20-cm internal diameter and a 160-m length. Assume the flow is adiabatic and frictionless. Determine (*a*) the outlet velocity, in m/s, and (*b*) the mass flow rate, in kg/s, if the local gravity is 9.70 m/s^2.

Isentropic processes in superheat and saturation regions

7-71M Refrigerant 12 is contained initially in a piston-cylinder device at 1.0 bar and a quality of 95.0 percent. It is compressed isentropically to a final pressure of 9.0 bars. Determine (*a*) the final temperature, in degrees Celsius, and (*b*) the required work input, in kJ/kg. (*c*) Sketch the process on a *Ts* diagram.

7-72M Saturated water vapor at 7 bars, in the amount of 1 g, expands in a reversible, adiabatic process to a pressure of 1.5 bars in a closed system. Calculate the work done, in kilojoules. Sketch the process on a *Ts* diagram.

7-73M Steam is contained in a piston-cylinder device initially at 20 bars and 400°C. It expands isentropically to 2.0 bars. Determine the work output, in kJ/kg. Sketch the process on a *Ts* diagram.

7-74M Steam at 1.5 bars and 120°C expands isentropically through a nozzle to 1.0 bar.

(*a*) If the inlet velocity is negligible, what is the discharge velocity, in m/s?

(*b*) For a flow rate of 20 kg/min, what is the exit area of the nozzle, in square centimeters?

7-75M Refrigerant 12 enters a nozzle and expands from 6 bars and 120°C to a pressure of 1 bar. The nozzle is isentropic, and the inlet velocity may be neglected. Calculate the exit velocity, in m/s.

7-76M Refrigerant 12, at 0.6 bar and 20°C, enters a diffuser with a velocity of 200 m/s. The process is isentropic and the exit temperature is 50°C. Determine (*a*) the exit pressure, in bars, and (*b*) the exit velocity, in m/s.

7-77M Steam undergoes an isentropic turbine expansion from 160 bars and 640°C to 1.0 bar. Neglect the changes in kinetic and potential energies. If the mass flow rate is 20,000 kg/h, determine (*a*) the quality of the steam at the exhaust, and (*b*) the kilowatt output. Sketch the process on a *Ts* diagram.

7-78M Refrigerant 12 enters a steady-flow turbine at 10 bars and 120°C. It passes through isentropically and leaves at 1 bar. Neglecting kinetic- and potential-energy changes, determine the work output, in kJ/kg.

7-79M Refrigerant 12 is compressed from a saturated vapor at −5°C to a final pressure of 8 bars in a steady-flow process.

(*a*) Determine the lowest final temperature possible if the process is adiabatic.

(*b*) Determine the minimum shaft work required, in kJ/kg.

(*c*) If the actual final temperature is 11.3°C higher than the minimum temperature due to irreversibilities, what is the percent increase in shaft work required?

7-80M Dry, saturated steam at 1.5 bars is compressed in a reversible, adiabatic process to 7 bars.

(*a*) Determine the final temperature, in kelvins.

(*b*) Calculate the work required, in kJ/kg, if the process is steady-flow.

(*c*) If the mass flow rate is 10,000 kg/h, what power input is required, in kilowatts?

Adiabatic efficiencies

7-81M The nozzle of a turbojet engine receives air at 1.80 bars and 707°C, with a velocity of 70 m/s. The nozzle expands the gas adiabatically to a pressure of 0.70 bar. Determine the discharge velocity, in m/s, if the nozzle efficiency is (*a*) 90 percent, and (*b*) 93 percent. Use air table data, and plot the process on a Ts diagram.

7-82M Air enters a converging nozzle at 1.6 bars and 67°C. The final pressure is 1.0 bar, and the initial velocity is negligible. If the exhaust velocity is (*a*) 275 m/s, and (*b*) 283 m/s, determine the nozzle adiabatic efficiency.

7-83M Steam passes through a turbine from an initial state of 100 bars, 520°C, to a final pressure of 15 bars.

(*a*) If the flow is adiabatic and the measured outlet temperature is 280°C, determine the isentropic efficiency of the turbine.

(*b*) For a mass flow rate of 20,000 kg/h, determine the power output, in kilowatts.

7-84M Air expands in an adiabatic turbine from 3 bars, 117°C, and 70 m/s to a final pressure of 1 bar. The mass flow rate is 2.0 kg/s.

(*a*) Determine the maximum work output, in kJ/kg.

(*b*) If the actual outlet temperature is 30°C, determine the turbine isentropic efficiency.

(*c*) Determine the turbine inlet area, in square centimeters.

7-85M Steam at 20 bars, 440°C, and 80 m/s enters an adiabatic turbine and expands to 0.7 bar at a rate of 10,000 kg/h.

(*a*) Determine the turbine inlet area, in square centimeters.

(*b*) If the isentropic efficiency of the turbine is 80 percent, determine the steam outlet temperature, in degrees Celsius.

(*c*) Find the power output, in kilowatts.

7-86M Air is compressed adiabatically in steady flow from 1 bar and 17°C, to 6 bars. If the compressor efficiency is 82 percent, determine (*a*) the outlet temperature, in degrees Celsius, (*b*) the temperature rise, degrees Celsius, due to irreversibilities, and (*c*) the actual work input, in kJ/kg.

7-87M Air is compressed in a steady-state device from 1 bar and 27°C to 5 bars. If the outlet temperature is (*a*) 250°C, and (*b*) 240°C, determine the adiabatic efficiency of the compressor.

7-88M Refrigerant 12 is compressed from a saturated vapor at −10°C to a final pressure of 8 bars. If the process is adiabatic and the compressor efficiency is 78 percent, determine the outlet temperature, in degrees Celsius.

7-89M A small air turbine having an 80 percent isentropic efficiency is required to produce 100 kJ/kg of work. The turbine inlet temperature is 460°K and the turbine exhausts at 0.10 MPa. Determine (*a*) the required inlet pressure, in MPa, and (*b*) the exhaust temperature, in kelvins.

Incompressible, irreversible steady-flow processes

7-90M Water enters a hydraulic turbine at 6.0 bars and 15°C with a velocity of 1.4 m/s through an opening of 0.60 m^2. The exit conditions of the water are 1.0 bar and 4.8 m/s. If the adiabatic efficiency of the turbine is 88 percent, determine (*a*) the actual shaft-power output, in kW, and (*b*) the temperature change of the fluid, in degrees Celsius, for the adiabatic process.

7-91M Water enters a hydraulic turbine at 0.720 MPa, 20°C, and 4.2 m/s. The exit conditions are 0.098 MPa and 1.2 m/s. If the temperature of the fluid increases by 0.0120°C, determine (*a*) the actual shaft-work output, in kJ/kg, and (*b*) the adiabatic efficiency of the turbine.

7-92M Water enters a pump at 1.0 bar and 20°C with a velocity of 2.6 m/s through an opening of 22.0 cm^2. The exit conditions of the water are 7.0 bars and 7.8 m/s. If the required shaft-power input is 5.0 kW, determine (*a*) the adiabatic efficiency of the pump, and (*b*) the rise in temperature of the fluid for the adiabatic process.

7-93M A hydrocarbon fluid with a specific gravity of 0.82 enters a pump at 0.10 MPa and 25°C. Shaft work is done on the fluid in the amount of 2.40 kJ/kg. The specific heat of the fluid is 2.20 kJ/(kg)(°C). Find the pressure rise, in MPa, if (*a*) the process is isentropic, and (*b*) the fluid temperature rises 0.070°C during the process.

7-94M An oil with a specific gravity of 0.83 enters a nozzle at 3.2 bars, 20°C, and 0.60 m/s. The final velocity is 16.9 m/s. If the increase in internal energy is 0.017 kJ/kg for the process, determine (*a*) the exit pressure, in bars, and (*b*) the nozzle efficiency for the calculated exit pressure.

7-95M A fluid with a specific gravity of 0.86 enters a nozzle at 3.9 bars, 25°C, and 0.75 m/s. The exit conditions from the nozzle are 16.3 m/s and 2.66 bars. For the specified pressure drop, determine (*a*) the change in internal energy for the actual process, in kJ/kg, and (*b*) the nozzle efficiency.

7-96M Liquid water at 50 bars and 80°C is throttled to 25 bars. Estimate the change in temperature of the fluid, in degrees Celsius, if (*a*) the flow is incompressible, and (*b*) linear interpolation of compressed liquid data is used.

7-97M Liquid water at 75 bars and 80°C is throttled to 25 bars. Estimate the change in temperature of the fluid, in degrees Celsius, if (*a*) the flow is incompressible, and (*b*) linear interpolation of compressed liquid data is used.

7-98M Water enters a piping system at 10 m/s and 20°C. At a position downstream the velocity is 20 m/s and the elevation is 22 m above the inlet. The local gravity is 9.70 m/s^2, and the flow is adiabatic. Determine the percent increase in the pressure drop when the fluid experiences a 0.010°C temperature rise, compared to frictionless flow.

PROBLEMS (USCS)

Entropy tabulations for superheat and saturation states

7-1 Find the specific entropy, in kJ/(kg)(°K), for water substance in parts *a* through *j* listed in Prob. 4-3.

7-2 Find the specific entropy, in kJ/(kg)(°K), for water substance in parts *a* through *k* listed in Prob. 4-4.

7-3 Find the specific entropy, in kJ/(kg)(°K), for refrigerant 12 in parts *a* through *k* listed in Prob. 4-5.

7-4 Find the specific entropy, in kJ/(kg)(°K), for refrigerant 12 in parts *a* through *k* listed in Prob. 4-6.

Second-law analysis for superheat and saturation states

7-5 Refrigerant 12 at 40 psia and 140°F is compressed in a closed system to 200 psia. The process is isothermal and internally reversible, and the environmental temperature is 80°F. Determine (*a*) the heat transfer and the work done, in Btu/lb, and (*b*) the total entropy change for the overall process, in Btu/(lb)(°R). Sketch the process on a *Ts* diagram.

7-6 One-tenth pound of steam at 40 psia and 400°F is compressed isothermally to 200 psia in a piston-cylinder device. During this process 6450 ft · lb_f of work is done on the gas, and heat is lost to the environment which is at 80°F. Determine (*a*) the heat transfer, in Btu, (*b*) the entropy change of the steam, in Btu/°R, and (*c*) the total entropy change for the overall process, in Btu/°R. Is the process reversible, irreversible, or impossible?

7-7 Steam originally at 1000 psia, 1200°F, is expanded isothermally and reversibly to 200 psia in a steady-flow device. Heat is supplied to the steam from a reservoir at 1400°F. Determine (*a*) the heat transfer and the work output, in Btu/lb, and (*b*) the total entropy change for the process, in Btu/(lb)(°R). Is the process reversible, irreversible, or impossible? Sketch the process on a *Ts* diagram.

7-8 Refrigerant 12 at 120 psia and 100°F is contained in a closed system at constant pressure. During a process 29,280 ft · lb_f of work is performed on the steam by a paddle-wheel. At the same time heat in the amount of 20.80 Btu is transferred to the surroundings at 60°F. Determine (*a*) the entropy change of the steam, and (*b*) the entropy change for the overall process, in Btu/(lb)(°R). Sketch the process on a *Ts* diagram.

7-9 A piston-cylinder device initially contains refrigerant 12 at 50 psia and 120°F. It is compressed quasistatically and at constant pressure, with a boundary-work input of 2960 ft · lb_f/lb_m. Determine (*a*) the final specific volume, in ft³/lb, (*b*) the final specific entropy, in Btu/(lb)(°R), and (*c*) the heat transfer, in Btu/lb. (*d*) If the surrounding temperature is 70°F, determine the total entropy change for the overall process, in Btu/(lb)(°R). Sketch the process on a *Ts* diagram. Is the process reversible, irreversible, or impossible?

7-10 A piston-cylinder device contains 10.0 lb of steam at 400 psia and 700°F. During a quasistatic process 1570 Btu of heat is removed from the steam at constant pressure to the surroundings at 60°F. Find (*a*) the entropy change of the steam, and (*b*) the total entropy change for the overall process, in Btu/°R. Sketch the process for the steam on a *Ts* diagram.

7-11 A rigid vessel contains 0.50 lb of steam at 100 psia and 500°F. Heat is transferred out of the system until the temperature drops to 400°F. The surrounding temperature is 80°F. Find (*a*) the entropy change of the steam, and (*b*) the total entropy change for the overall process, in Btu/°R. Sketch the process on a *Ts* diagram. Is the process reversible, irreversible, or impossible?

7-12 Refrigerant 12 flows at a low velocity through a well-insulated capillary tube. The initial and final conditions are (*a*) saturated liquid at 100 psia to 20 psia, (*b*) 120 psia, 100°F to 80°F, and (*c*) saturated vapor at 120 psia to 15 psia. (1) Determine the entropy change of the fluid, in Btu/(lb)(°R). (2) Does this value agree qualitatively with the second law? (3) What is the ratio of the final velocity to the initial velocity?

7-13 A refrigerant-12 compressor operates with an inlet condition of saturated vapor at 10°F and an outlet state of 100 psia and 120°F. The device requires 1.72 hp of power input with a mass flow rate of 5 lb/min. The atmospheric temperature is 60°F. Determine the total entropy change for the overall process, in Btu/(lb)(°R).

7-14 Steam flows steadily through a nozzle from 500 psia, 700°F, and 20 ft/s to 200 psia. The process is isothermal and 40 Btu/lb of heat is transferred to the steam from a reservoir at 800°F. The surrounding temperature is 60°F. Determine (*a*) the nozzle exit velocity, in ft/s, and (*b*) the total entropy change for the overall process, in Btu/(lb)(°R).

7-15 Steam enters a steady-flow compression process at 14.7 psia, 250°F, and exits at 160 psia, 400°F. The work input is measured to be 200 Btu/lb, and the kinetic- and potential-energy changes are negligible. The environment has a temperature of 70°F. Determine (*a*) the magnitude and direction of any heat transfer, in Btu/lb, (*b*) the entropy change of the fluid passing through the compressor, in Btu/(lb)(°R), and (*c*) the total entropy change for the overall process, in Btu/(lb)(°R).

7-16 Saturated refrigerant-12 vapor at (*a*) 100°F, and (*b*) 160 psia is condensed in a steady-flow heat exchanger to saturated liquid by heat transfer to the surroundings at 60°F. Determine the net change in entropy for the fluid plus the surroundings, in Btu/(lb)(°R).

Second-law analysis of ideal-gas processes

7-17 A piston-cylinder device contains air at 640°R and 1 atm. The pressure is increased to 1.3 atm, but during the process heat transfer occurs so that the temperature remains constant.

(*a*) Determine the entropy change of the air in Btu/(lb)(°R).

(*b*) If the temperature of the surroundings is 500°R, determine the entropy change of the surroundings, in Btu/(lb)(°R).

(*c*) Is the increase in entropy principle fulfilled?

7-18 A piston-cylinder device maintained at a constant pressure of 1 atm contains nitrogen gas. It is cooled from 300 to 100°F. Compute (*a*) the heat removed, in Btu/lb, (*b*) the work done, in Btu/lb, (*c*) the change in entropy of the nitrogen, in Btu/(lb)(°R), and (*d*) the total entropy change of the overall process if the temperature of the environment is 70°F, in Btu/(lb)(°R).

7-19 A rigid tank contains 0.1 lb of carbon monoxide gas at 14.5 psia and 100°F. Heat is added until the pressure reaches 18.6 psia. Compute (*a*) the heat transferred, in Btu, (*b*) the change in entropy of CO, in Btu/°R, and (*c*) the total entropy change for the overall process, in Btu/°R if the heat is supplied from a reservoir at 360°F. (*d*) Is the answer to part *c* compatible with the second law?

7-20 Contained in a constant-pressure closed system is 0.5 lb of hydrogen gas at 100 psia and 40°F. Heat in the amount of 520 Btu is added to the gas from a reservoir at 550°F. Compute (*a*) the entropy change of the hydrogen, in Btu/°R, and (*b*) the total entropy change for the overall process, in Btu/°R. (*c*) Is the process reversible, irreversible, or impossible?

7-21 A rigid tank with a volume of 1.0 ft^3 contains oxygen at an initial state of 200°F and 1.5 atm. During a process paddle-wheel work is carried out by applying a torque of 9.55 $lb_f \cdot ft$ for 25 revolutions, and a heat loss of 3.52 Btu occurs to the surroundings at 65°F. Determine (*a*) the entropy change of the oxygen, in Btu/°R, and (*b*) the total entropy change for the overall process, in Btu/°R. (*c*) Is the process reversible, irreversible, or impossible?

7-22 Air in the amount of 1 lb is compressed isothermally and quasistatically from 15 psia, 500°R, to a final state which requires 55.1 Btu of boundary-work input to the piston-cylinder device. Find (*a*) the final pressure, in psia, (*b*) the change in entropy of the air, in Btu/°R, and (*c*) the total entropy change of the overall process, in Btu/°R, if the temperature of the surroundings is 40°F. (*d*) Is the process reversible, irreversible, or impossible?

7-23 Air, in the amount of 0.5 lb, in a constant-volume system has an initial state of 5.0 ft^3 and 80°F. Measurements during the process indicate that 4.50 Btu of heat were removed and paddle-wheel work was carried out by applying a torque of 23.9 $lb_f \cdot ft$ to the shaft for 50 revolutions. Find (*a*) the final temperature of the gas in the tank, in degrees Fahrenheit, (*b*) the entropy change of the air, in Btu/°R, and (*c*) the total entropy change of the overall process, in Btu/°R, if the temperature of the environment to which heat is transferred is 70°F.

7-24 Carbon monoxide at 1 atm and 100°F is compressed in a closed system to 3 atm and 300°F. The required work input is 38.0 Btu/lb, and the surrounding temperature is 80°F. Determine (*a*) the entropy change of the CO in Btu/(lb)(°R), (*b*) the heat loss to the surroundings, in Btu/lb, and (*c*) the total entropy change for the CO and the environment in Btu/(lb)(°R).

7-25 Air initially at 40 psia and 140°F expands to (*a*) 20 psia and 80°F, and (*b*) 25 psia and 40°F. Can these changes of state be carried out adiabatically? Why?

7-26 Air enters a turbine at 90 psia and 540°F and leaves at 15 psia. The flow rate is 110 lb/min, and the power output is 240 hp. If the heat removed, which appears in the environment at 70°F, is 18 Btu/lb, find (*a*) the final temperature in degrees Fahrenheit, (*b*) the entropy change of the air, in Btu/(lb)(°R), and (*c*) the total entropy change for the overall process, in Btu/(lb)(°R).

7-27 Air enters a steady-state compressor at 15 psia, 40°F, and 200 ft/s at a rate of 2 lb/s, and leaves at 30 psia, 160°F, and 400 ft/s. The power required to operate the compressor is 111 hp. Determine (*a*) the rate of heat transfer, in Btu/h, (*b*) the entropy change of the air, in Btu/(min)(°R), and (*c*) the entropy change of the environment which receives the heat transferred at 50°F, in Btu/(min)(°R). (*d*) Is the process reversible, irreversible, or impossible?

7-28 Argon gas enters a turbine at 50 psia and 300°F, and leaves the device at 20 psia. A heat loss of 2.0 Btu/lb occurs, and this energy appears in the surroundings at 60°F. The measured shaft-work output is 24.0 Btu/lb, and the kinetic-energy change is negligible. Find (*a*) the outlet temperature in degrees Fahrenheit, (*b*) the entropy change of the argon, in Btu/(lb)(°R), and (*c*) the total entropy change for the overall process, in Btu/(lb)(°R).

7-29 Nitrogen flows steadily through a nozzle from 75 psia, 620°F, and 60 ft/s to 45 psia and 560°F. During the process, 15 Btu/lb of heat is transferred to the nitrogen from a reservoir at 700°F. Determine (*a*) the nozzle exit velocity, in ft/s, and (*b*) the total entropy change for the overall process in Btu/(lb · mol)(°R).

7-30 Air passes steadily through a device from 20 psia and 80°F to 40 psia. During the process a work input of 24.2 Btu/lb is required, and a heat loss of 7.0 Btu/lb occurs with the energy appearing in the atmosphere at 80°F. Determine (*a*) the entropy change of the air, and (*b*) the entropy change of the atmosphere, both in Btu/(lb)(°R). (*c*) Is the process reversible, irreversible, or impossible?

7-31 Oxygen is throttled from 30 psia and 1000°R to (*a*) 20 psia, and (*b*) 15 psia.

(1) Find the entropy change of the gas in Btu/(lb)(°R).

(2) Is the process reversible, irreversible or impossible?

Entropy changes of incompressible substances

7-32 One hundred pounds of water at 1 atm and 60°F are mixed with 40 lb of water initially at 1 atm and 200°F. If the mixing process is adiabatic and at constant pressure, determine the total change in entropy for the 140 lb of water, in Btu/°R.

7-33 A 5-lb iron casting at 950°F is quenched in a tank containing 1 ft^3 of water initially at 75°F. Assuming no heat loss to the surroundings and that the specific heat of the iron casting averages 0.11 Btu/(lb)(°F), determine the change in entropy of (*a*) the water, (*b*) the casting, and (*c*) the universe.

7-34 A 1-lb block of copper at 250°F and a 1-lb block of aluminum at 120°F initially are isolated from each other and from the local surroundings. They are then placed in thermal contact with each other, but are still isolated from the surroundings. Calculate ΔS for each block if they attain thermal equilibrium. Specific-heat data for pure metals are given in Table A-19.

7-35 A small, solid object of constant heat capacity C equal to 12 Btu/°R is initially at 580°F. The object is then thrown into a large lake which is at a temperature of 60°F. Calculate the total entropy change for this process in Btu/°R. (*Note*: heat capacity C equals mass times specific heat, mc.)

7-36 A pound of ice at 32°F is mixed with 6.5 lb of water at 70°F. Compute the change in entropy of the composite system if the process is adiabatic and the pressure is maintained at 14.7 psia. (Enthalpy of melting of water is 144 Btu/lb.)

7-37 Twenty pounds of water at 180°F are mixed adiabatically at 1 atm pressure with 12 lb of water at 60°F.

(*a*) Compute the entropy change for each quantity of water and the total change in Btu/°R.

(*b*) Sketch the separate processes that each quantity of water undergoes on the same TS diagram.

7-38 A resistor of 20 Ω, initially at 80°F, is maintained at a constant temperature while a current of 5 A is allowed to flow for 2 s.

(*a*) Determine the entropy change for the resistor and of the universe, in Btu/°R.

(*b*) Now, insulate the resistor and carry out the same experiment.

For the latter case, determine the entropy change of the resistor and the universe. The mass of the resistor is 0.020 lb and the specific heat of the resistor is 0.55 Btu/(lb)(°F). The environmental temperature is 80°F.

7-39 An electric resistor of 30 Ω is maintained at a constant temperature of 70°F while a current of 6 A is maintained for 3 s.

(*a*) Determine the entropy change of the resistor and of the universe, in Btu/°R.

(*b*) Now, insulate the resistor and carry out the same experiment.

For the latter case, determine the entropy change of the resistor and the universe. The mass of the resistor is 0.032 lb and the specific heat of the resistor is 0.62 Btu/(lb)(°F). The environmental temperature is 70°F.

Second-law analysis of heat exchangers and mixing processes

7-40 Reconsider Prob. 5-41. Determine (*a*) the entropy change of the air stream, (*b*) the entropy change of the refrigerant-12 stream, and (*c*) the total entropy change for the overall process, all in Btu/(°R)(s).

7-41 Reconsider Prob. 5-42. Determine the entropy change in Btu/(°R)(min) for (*a*) the water stream, (*b*) the air stream, and (*c*) the overall process.

7-42 Reconsider Prob. 5-43. Determine the entropy change in Btu/(°R)(min) for (*a*) the condensing stream, (*b*) the air stream, and (*c*) the overall process, if the outlet air pressure is 1.10 atm.

7-43 Reconsider Prob. 5-45. Determine the rate of entropy production for the process in Btu/(°R)(h) using the data of (*a*) 4650 lb/h, and (*b*) 7890 lb/h for the second source.

7-44 Reconsider Prob. 5-47(*a*). Determine the rate of entropy production for the process in Btu/(°R)(min).

7-45 Superheated steam enters one section of an open feedwater heater at 60 psia and 300°F. Compressed liquid water at the same pressure and 100°F enters another inlet. The mixture of these two streams leaves at the same pressure as a saturated liquid. If the heater is adiabatic, calculate the change in entropy for the process, in Btu/°R, per pound of mixture leaving the heater.

Steady-flow mechanical work

7-46 Carbon monoxide flows frictionlessly and isothermally through a constant-area duct of 12.0 in^2. Upstream at position 1 the pressure is 20 psia and the temperature is 200°F. After flowing against an impeller which produces shaft work, the gas reaches a downstream position where the pressure is 15 psia. If the mass flow rate is 3.95 lb/s, determine (*a*) the shaft work, in $ft \cdot lb_f/lb_m$, and (*b*) the heat transfer, also in $ft \cdot lb_f/lb_m$.

7-47 Air is compressed isothermally from 14 psia, and 40°F, to 70 psia. Flow through the compressor is steady at 2.0 lb/s. Kinetic and potential energies are negligible. Assuming that air is an ideal gas, calculate the power input to the compressor and the heat removed from the air, both in Btu/s, if the process is frictionless.

7-48 A fluid flows frictionlessly through an open system which requires work input. The entrance conditions are: $P_1 = 50$ psia, $v_1 = 10$ ft^3/lb, and $A_1 = 0.25$ ft^2. At the exit $P_2 = 200$ psia and $A_2 = 0.20$ ft^2. The mass flow rate is 10 lb/s, and for the process, Pv is constant. Find (*a*) the work required, in Btu/lb, and (*b*) the horsepower required.

Isentropic processes of ideal gases

7-49 Air is contained in a piston-cylinder device initially at 300°F, 42.5 psia, and 46.3 in^3. The gas is expanded isentropically to 14.5 psia. Determine (*a*) the final temperature in degrees Fahrenheit, (*b*) the mass of air within the device, in lb, (*c*) the final volume, in in^3, and (*d*) the work output, in Btu. Use the air table for data.

7-50 Nitrogen at 45 psia and 240°F is allowed to expand adiabatically and reversibly to 25 psia in a closed system. Determine (*a*) the final temperature, in °R, and (*b*) the work output, in Btu/lb.

7-51 Air, in the amount of 0.5 lb, is compressed adiabatically and without friction from an initial state of 20 psia and 5.0 ft^3 to a final volume of 1.0 ft^3 in a piston-cylinder device. Determine (*a*) the work required, in Btu, and (*b*) the final temperature, in degrees Fahrenheit.

7-52 A piston-cylinder device contains air at 20 psia, 100°F, and 200 in^3. The gas is compressed isentropically at 58.6 psia. Determine (*a*) the final temperature, in degrees Fahrenheit, and (*b*) the work required, in Btu. Use tabular data.

7-53 Air enters a nozzle at 80 psia and 120°F at a rate of 0.40 lb/s. The inlet velocity is negligible, and the exit pressure is 50 psia. If the process is isentropic, determine the exit area of the nozzle, in in^2.

7-54 Argon, initially at 50 psia and 540°F, flows at a rate of 10 lb/s through an insulated, frictionless nozzle. The initial velocity is negligible, and the outlet pressure is 20 psia. Determine (*a*) the final temperature in degrees Fahrenheit, (*b*) the final velocity, in ft/s, and (*c*) the exit area, in square inches.

7-55 Air enters a diffuser at 10 psia and 40°F with a velocity of 900 ft/s. The outlet temperature is 104°F, and the process is adiabatic and frictionless. Determine (*a*) the final velocity, in ft/s, and (*b*) the final pressure, in psia.

7-56 Air enters a diffuser at 8 psia, 20°F, and 800 ft/s. The air stream leaves the diffuser at a velocity of 400 ft/s. For isentropic flow, find (*a*) the temperature at the outlet, in degrees Fahrenheit, (*b*) the outlet pressure, in psia, and (*c*) the ratio of the outlet area to the inlet area.

7-57 Helium enters a turbine at 100 psia and 1000°F and exhausts at 20 psia. Neglect kinetic- and potential-energy changes. Assuming isentropic flow, determine (*a*) the exhaust temperature, in degrees Fahrenheit, and (*b*) the work output, in Btu/lb.

7-58 Carbon dioxide at 1440°R and 200 atm enters a turbine with a velocity of 300 ft/s through an inlet area of 65.0 in^2. The gas expands isentropically to 900°R and leaves through an area of 195 in^2. Find (*a*) the work output, in Btu/lb · mol, and (*b*) the mass flow rate, in lb · mol/s.

7-59 The inlet conditions for an air turbine are 90 psia and 620°F. The air expands isentropically to 15 psia. For a flow rate of 100 lb/min, compute the power output, in horsepower, if the change in kinetic energy is small.

7-60 Air at 40 ft/s, 14.7 psia, and 80°F is taken into the compressor of a gas turbine of a large truck. The isentropic compressor delivers air at 112.5 psia and 120 ft/s at a mass flow rate of 6.8 lb/s. Determine (*a*) the compressor exit temperature, in degrees Fahrenheit, (*b*) the diameter of the circular inlet, in inches, and (*c*) the power required, in horsepower. Use air table data.

7-61 Nitrogen gas flows through a compressor at a rate of 10 lb/s and undergoes an isentropic change from 15 psia and 60°F, to 40 psia. The measured kinetic-energy change is 2.5 Btu/lb. Compute the horsepower input to the gas.

7-62 Argon at a rate of 0.10 lb/s enters an adiabatic, frictionless compressor at 20 psia and 80°F. The exit pressue is 100 psia, and the kinetic- and potential-energy changes are negligible. Determine (*a*) the exit temperature, in degrees Rankine, and (*b*) the power input, in horsepower.

Isentropic processes with incompressible substances

7-63 Oil with a specific gravity of 0.85 is being pumped from a pressure of 8 inHg vacuum to a pressure of 18.0 psig. The outlet lies 18.0 ft above the inlet. The fluid at 60°F flows at a rate of 3.0 ft^3/s through an inlet cross-sectional area of 0.50 ft^2, and the outlet area is 0.30 ft^2. The local gravity is 32.0 ft/s^2, and the flow is assumed to be adiabatic and frictionless. Determine the power input to the pump, in horsepower.

7-64 Water at 70°F is pumped at 2.50 lb/s from the surface of an open tank into a constant-diameter piping system. At the discharge from the pipe, the velocity is 30 ft/s and the gage pressure is 25.0 psig. The discharge is 48.0 ft above the water surface in the open tank and local gravity is 31.8 ft/s^2. For reversible, adiabatic flow, determine the power required for the pump, in kilowatts.

7-65 Water is pumped from an initial state of 20 psia, 60°F, and 4 ft/s to a final state of 80 psia and 20 ft/s. The pipe diameter at the exit is 3.0 in, and the exit position lies 90.0 ft vertically above the inlet position. If the pipe is insulated and the flow frictionless, determine (*a*) the mass flow rate, in lb/s, (*b*) the shaft work required ft · lb_f/lb_m, and (*c*) the power input, in horsepower. Do not neglect any energy term if sufficient information is available. Local gravity is 32.0 ft/s^2.

7-66 Water enters an inlet pipe to a water turbine at 1.0 atm and 70°F at an elevation 400 ft above the pipe outlet. The inlet velocity is negligible. At the outlet pipe, following the turbine, the conditions are 1.0 atm and 27 ft/s. If the flow is adiabatic and frictionless, determine the shaft-work output from the turbine , in ft · lb_f/lb_m. Local gravity is 31.6 ft/s^2.

7-67 A hydraulic turbine is located 360 ft below the surface of water behind a dam. Water is taken to the turbine through an 8.0-ft-diameter pipe which begins just below the water level. At the outlet pipe of the same diameter which follows the turbine, the conditions are 1.0 atm, 60°F, and 40 ft/s. If the flow is assumed to be isentropic and local gravity is 32.0 ft/s^2 determine the power output of the turbine, in horsepower.

7-68 Water enters a piping system at 70°F and 10 ft/s. At a position downstream the conditions are 20.0 psia and 20 ft/s, and the elevation is 40.0 ft above the inlet. The local gravity is 32.0 ft/s^2. For adiabatic and frictionless flow, determine the required inlet pressure, in psia.

7-69 Water enters a piping system at 50 psia, 60°F, and 15 ft/s at a rate of 20.0 lb/s. Downstream at a given position the pressure is 27.0 psia, and the elevation is 35.0 ft above the inlet. The local gravity is 31.8 ft/s^2. Determine (*a*) the velocity at the downstream position, in ft/s, and (*b*) the pipe diameter downstream, in inches, if the flow is assumed to be adiabatic and frictionless.

7-70 Water at 60°F flows over a dam and down into a vertical pipe with a 6-in internal diameter and a 500-ft length. Assume the flow is adiabatic and frictionless. Determine (*a*) the outlet velocity, in ft/s, and (*b*) the mass flow rate, in lb/s, if the local gravity is 31.8 ft/s^2.

Isentropic processes in superheat and saturation regions

7-71 Refrigerant 12 is contained initially in a piston-cylinder device at 20 psia and a quality of 98.8 percent. It is compressed isentropically to a final pressure of 180 psia. Determine (*a*) the final temperature, in degrees Fahrenheit, and (*b*) the required work input, in Btu/lb. (*c*) Sketch the process on a Ts diagram.

7-72 Saturated water vapor at 100 psia, in the amount of 0.1 lb, expands in a reversible, adiabatic process to a pressure of 20 psia in a closed system. Calculate the work done, in Btu. Sketch the process on a Ts diagram.

7-73 Steam is contained in a piston-cylinder device initially at 100 psia and 450°F, with a volume of 100 in^3. It expands isentropically to 14.7 psia. Determine the work output, in Btu. Sketch the process on a Ts diagram.

7-74 Steam at 60 psia and 350°F expands isentropically through a nozzle to 35 psia.

(*a*) If the inlet velocity is negligible, what is the discharge velocity, in ft/s?

(*b*) For a flow rate of 100 lb/min, what is the exit area of the nozzle, in square inches?

7-75 Refrigerant 12 enters a nozzle and expands from 90 psia and 240°F to a pressure of 40 psia. The nozzle is isentropic, and the inlet velocity may be neglected. Calculate the exit velocity, in ft/s.

7-76 Steam, at 40 psia and 300°F, enters a diffuser with a velocity of 1600 ft/s. The process is isentropic and the exit temperature is 400°F. Determine (*a*) the exit pressure, in psia, and (*b*) the exit velocity, in ft/s.

7-77 Steam undergoes an isentropic turbine expansion from 600 psia and 800°F to 15 psia. Neglect the changes in kinetic and potential energies. If the mass flow rate is 20,000 lb/h, determine (*a*) the quality of the steam at the exhaust, and (*b*) the horsepower output. Sketch the process on a Ts diagram.

7-78 Refrigerant 12 enters a steady-flow turbine at 90 psia and 240°F. It passes through isentropically and leaves at 30 psia. Neglecting kinetic- and potential-energy changes, determine the work output, in Btu/lb.

7-79 Refrigerant 12 is compressed from a saturated vapor at −20°F to a final pressure of 120 psia in a steady-flow process.

(*a*) Determine the lowest final temperature possible if the process is adiabatic.

(*b*) Determine the minimum shaft work required, in Btu/lb.

(*c*) If the actual final temperature is 23°F higher than the minimum temperature due to irreversibilities, what is the percent increase in shaft work required?

7-80 Saturated water vapor is compressed in a reversible, adiabatic process from 20 psia and 300°F, to 120 psia.

(*a*) Find the final temperature, in degrees Fahrenheit.

(*b*) If the system is steady-fow, calculate the work required, in Btu/lb.

(*c*) If the mass flow rate is 20,000 lb/h, compute the required power input, in horsepower.

Adiabatic efficiencies

7-81 The nozzle of a turbojet engine receives air at 27 psia and 1300°F, with a velocity of 200 ft/s. The nozzle expands the gas adiabatically to a pressure of 10.5 psia. Determine the discharge velocity, in ft/s, if the nozzle efficiency is (*a*) 90 percent, and (*b*) 93 percent. Use air table data, and plot the process on a Ts diagram.

7-82 Air enters a converging nozzle at 24.2 psia and 600°R. The final pressure is 14.7 psia, and the initial velocity is negligible. If the exhaust velocity is (*a*) 925 ft/s, and (*b*) 945 ft/s, determine the nozzle adiabatic efficiency.

7-83 Steam passes through a turbine from an initial state of 800 psia, 800°F, to a final pressure of 14.7 psia.

(*a*) If the flow is adiabatic and the measured outlet temperature is 250°F, determine the isentropic efficiency of the turbine.

(*b*) For a mass flow rate of 20,000 lb/h, determine the power output in horsepower.

7-84 Air expands in an adiabatic turbine from 45 psia, 240°F, and 200 ft/s to a final pressure of 15 psia. The mass flow rate is 4.0 lb/s.

(*a*) Determine the maximum work output, in Btu/lb.

(*b*) If the actual outlet temperature is 90°F, determine the turbine isentropic efficiency.

(*c*) Determine the turbine inlet area, in square inches.

7-85 Steam expands through a turbine from 2000 psia and 1000°F to 10 psia and 93 percent quality. Determine the turbine adiabatic efficiency.

7-86 Air is compressed adiabatically from 14.5 psia and 40°F to 98 psia. If the compressor efficiency is 82 percent, find the outlet temperature, in degrees Fahrenheit, and the work required, in Btu/lb, for a steady-flow process.

7-87 Air is compressed from 15 psia and 60°F to 90 psia and 480°F in an adiabatic, steady-flow process. Determine the adiabatic efficiency of the compressor.

7-88 Refrigerant 12 is compressed from a saturated vapor at −10°F to a final pressure of 100 psia. If the process is adiabatic and the compressor efficiency is 78.5 percent, determine the actual outlet temperature, in degrees Fahrenheit.

7-89 A small air turbine having an 80 percent isentropic efficiency is required to produce 50 Btu/lb of work. The turbine inlet temperature is 340°F and the turbine exhausts at 15 psia. Determine (*a*) the required inlet pressure, in psia, and (*b*) the exhaust temperature, in degrees Fahrenheit.

Incompressible, irreversible steady-flow processes

7-90 Water enters a hydraulic turbine at 175 psia and 50°F with a velocity of 4.4 ft/s through an opening of 6.2 ft^2. The exit conditions of the water are 15 psia and 4.8 ft/s. If the adiabatic efficiency of the turbine is 90 percent, determine (*a*) the actual shaft-power output, in horsepower, and (*b*) the temperature change of the fluid, in degrees Fahrenheit, for the adiabatic process.

7-91 Water enters a hydraulic turbine at 105 psia, 60°F, and 3.6 ft/s through a cross-sectional area of 5.4 ft^2. The exit conditions are 16 psia and 9.8 ft/s. If the actual shaft-work output is 415 hp, determine (*a*) the adiabatic efficiency of the turbine, and (*b*) the temperature rise of the fluid, in degrees Fahrenheit, for the adiabatic process.

7-92 Water enters a pump at 15 psia and 70°F with a velocity of 7.8 ft/s through an opening of 3.40 in^2. The exit conditions of the water are 105 psia and 24 ft/s. If the required shaft-power input is 6.7 hp, determine (*a*) the adiabatic efficiency of the pump, and (*b*) the rise in temperature of the fluid, in degrees Fahrenheit, for the adiabatic process.

7-93 A hydrocarbon fluid with a specific gravity of 0.82 enters a pump at 16 psia and 80°F. Shaft work is done on the fluid in the amount of 1.20 Btu/lb. The specific heat of the fluid is 0.53 Btu/(lb)(°F). Find the pressure rise, in psi, if (*a*) the process is isentropic, and (*b*) the fluid temperature rises 0.125°F during the process.

7-94 An oil with a specific gravity of 0.83 enters a nozzle at 48 psia, 65°F, and 2.0 ft/s. The final velocity is 55.0 ft/s. If the increase in internal energy is 0.0080 Btu/lb for the process, determine (*a*) the exit pressure, in psia, and (*b*) the nozzle efficiency for the calculated exit pressure.

7-95 Water enters a nozzle at 60 psia, 70°F, and 2.0 ft/s. The exit conditions from the nozzle are 45 ft/s and 45 psia. For the specified pressure drop, determine (*a*) the change in internal energy for the actual process, in Btu/lb, and (*b*) the nozzle efficiency for the adiabatic process.

7-96 Liquid water at 1000 psia and 150°F is throttled to 500 psia. Estimate the change in temperature of the fluid, in degrees Fahrenheit, if (*a*) the flow is incompressible, and (*b*) linear interpolation of compressed liquid data is used.

7-97 Liquid water at 1500 psia and 100°F is throttled to 500 psia. Estimate the change in temperature of the fluid, in degrees Fahrenheit, if (*a*) the flow is incompressible, and (*b*) linear interpolation of compressed liquid data is used.

7-98 Water enters a piping system at 10 ft/s and 70°F. At a position downstream the velocity is 20 ft/s and the elevation is 40 ft above the inlet. The local gravity is 32.0 ft/s^2, and the flow is adiabatic. Determine the percent increase in the pressure drop when the fluid experiences a 0.020°F temperature rise, compared to frictionless flow.

CHAPTER

EIGHT

CONCEPT OF AVAILABILITY

8-1 INTRODUCTION

One of the major goals of engineering design is the optimization of a process within given constraints. In the energy field this implies the optimal use of energy during transfer or transformation. As in other engineering areas, this requires that we have some way of measuring optimum performance. In Chap. 7 this was done by introducing the concept of equipment efficiencies, such as those for turbines, compressors, nozzles, etc. For these essentially adiabatic devices, the optimum performance is considered to be isentropic behavior. Efficiencies formulated on isentropic conditions are frequently called *first-law* efficiencies, and they are important in engineering analysis. However, such efficiencies have a serious drawback. The end state of the actual process and the isentropic process to which it is compared are not the same. Possibly a more fundamental approach would be to compare the actual process to the optimum process between the same end states. Such a comparison leads to the concept of *second-law* efficiencies.

The optimization of energy usage is based on the concept that energy has not only quantity, but also *quality*. Historically, when we speak of the quality of a given amount of energy, we mean the potential of that energy to produce useful work for us. If the work potential of a quantity of energy is reduced during a process, then we say say that the energy has been degraded. This occurs during irreversible processes. Thus the second law of thermodynamics is a law of degradation of energy. Whenever energy is transformed or transferred in real processes, its potential for producing useful work is reduced, forever. This chapter is devoted to methods of measuring optimum processes and degradation of energy during those processes.

8-2 CLOSED-SYSTEM AVAILABILITY

Consider a closed system X with equilibrium properties P, V, T, S, and U (and any other properties of interest) which is surrounded by a local environment or atmosphere of constant temperature T_0 and constant pressure P_0. Since the system and its environment are not in equilibrium (thermally or mechanically), these two regions may interact and work and heat interactions may occur. This situation is illustrated in Fig. 8-1. The work associated with the system is symbolized by δW_X. We now wish to develop a general expression for the optimum useful work (maximum work output or minimum work input) associated with the system-environment composite when the system passes between two given equilibrium sates. During this change of state heat may be exchanged only with the local atmosphere. (This restriction with regard to heat transfer will be removed later.) In addition, the system itself may be deformable. Thus boundary work may be associated with the change of state. However, we shall assume that the local environment is so large that its outer boundary is fixed.

The quantity δW_X is the total work associated with the system during the process. However, *useful work* is defined as the quantity of system work which does not include that portion done on or by the atmosphere. Since the atmosphere pressure P_0 is constant, then, for the system X,

$$\delta W_{\text{useful},\,X} = \delta W_x - (-P_0\, dV_X)$$

This quantity of useful work is not the optimum useful work between given end states. Two further restrictions are necessary. To optimize the work, the closed-system process must be totally reversible. Hence the system itself must undergo an internally reversible process. In addition, any heat transfer between the system and the environment must be reversible. To accomplish this, it is necessary that a reversible heat engine be inserted between the system and the environment, as shown in Fig. 8-1. This engine will produce an additional quantity of work, δW_E. For the overall process the optimum useful work, then, is

$$\delta W_{\text{useful, opt}} = \delta W_X - (-P_0\, dV_X) + \delta W_E \tag{a}$$

We now wish to transform Eq. (a) so that only thermodynamic properties will

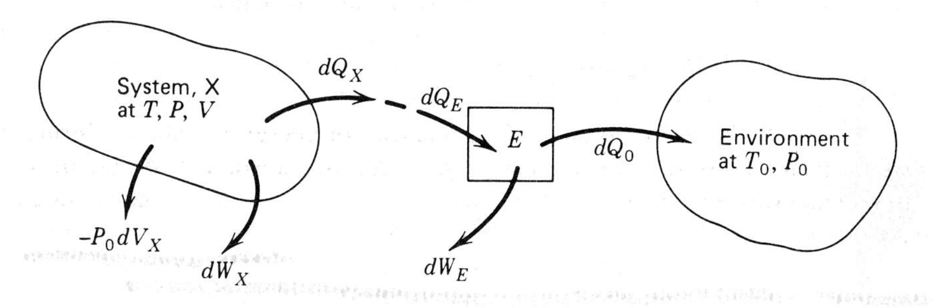

Figure 8-1 System-environment composite producing work.

appear on the right side of the equation. This requires an evaluation of the quantities δW_X and δW_E.

The quantity δW_X comes directly from the first law for the closed system, namely,

$$\delta W_{X;\,\mathrm{rev}} = dU - \delta Q_{X,\,\mathrm{rev}} \tag{b}$$

The engine work δW_E is determined from a first- and second-law analysis of the engine. First of all, from energy considerations for the cyclic device,

$$\delta W_E = -\delta Q_E - \delta Q_0 = \delta Q_X - \delta Q_0 \tag{c}$$

where it is recognized that δQ_X and δQ_E are of equal magnitude but of opposite sign when δQ_X is measured relative to the system and δQ_0 is relative to the engine. The second law for the engine states that

$$\frac{\delta Q_0}{T_0} = \frac{-\delta Q_E}{T} = \frac{\delta Q_X}{T} \tag{d}$$

Combining Eqs. (*c*) and (*d*), and recognizing that $\delta Q_X/T = dS_X$ by definition for a reversible process, we find that

$$\delta W_E = \delta Q_X - \delta Q_X \frac{T_0}{T} = \delta Q_X - T_0\, dS_X \tag{e}$$

Substitution of Eqs. (*b*) and (*e*) into Eq. (*a*) yields

$$\begin{aligned}\delta W_{\mathrm{useful,\,opt}} &= dU_X - \delta Q_{X,\,\mathrm{rev}} + P_0\, dV_X + (\delta Q_{X,\,\mathrm{rev}} - T_0\, dS_X)\\ &= dU_X + P_0\, dV_X - T_0\, dS_X \end{aligned} \tag{f}$$

Integration of this expression between states 1 and 2 of a closed system leads to the desired result, namely,

$$\begin{aligned}W_{\mathrm{useful,\,opt}} &= \Delta U + P_0\, \Delta V - T_0\, \Delta S\\ &= (U_2 - U_1) + P_0(V_2 - V_1) - T_0(S_2 - S_1)\end{aligned} \tag{8-1}$$

On a unit-mass basis this may be written as

$$\begin{aligned}w_{\mathrm{useful,\,opt}} &= \Delta u + P_0\, \Delta v - T_0\, \Delta s\\ &= (u_2 - u_1) + P_0(v_2 - v_1) - T_0(s_2 - s_1)\end{aligned} \tag{8-2}$$

These two equations allow evaluation of the optimum useful work since the overall process was assumed to be totally reversible. The equations predict the maximum useful work output or the minimum useful work input associated with the change of state of a simple, compressible closed system which exchanges heat solely with the local environment.

In addition to the general result given by Eq. (8-2), it is of interest to determine the maximum useful work when a closed system proceeds from a given state to a state of thermal and mechanical equilibrium with the environment. (Further chemical reactions also are assumed not to occur.) The final temperature and pressure of the system in this case would be that of the local environment, T_0 and P_0, respectively. When a system and its environment are in equilibrium with each other, the system is said to be in its *dead state*. The maximum useful work output that may be obtained from a system-atmosphere combination as the system proceeds from a given state to the dead state, while exchanging heat only with the atmosphere, is defined as the *availability* of the closed system in the initial state. When Eq. (8-2) is applied to a system which proceeds from a given state to the dead state, the maximum useful work is given by

$$w_{\text{useful, opt}} = (u_0 - u) + P_0(v_0 - v) - T_0(s_0 - s)$$

The numerical value obtained by using this equation will be negative, since work output is defined as negative in this text, by choice. However, in the technical literature the value of the availability of a closed system is usually considered to be positive. To maintain this latter convention, we shall define the *closed-system availability* per unit mass ϕ in the following manner.

$$\begin{aligned} \phi &= (u - u_0) + P_0(v - v_0) - T_0(s - s_0) \\ &= (u + P_0 v - T_0 s) - (u_0 + P_0 v_0 - T_0 s_0) \end{aligned} \tag{8-3}$$

where the symbols without a subscript refer to the actual state of the closed system. Although we speak of the availability of a system in a given state, note that the availability is a property which is a function of both the state of the system and the local environment. Its value depends upon T_0 and P_0 as well as the properties of the system. In addition, due to the manner in which ϕ is defined, the value of ϕ is always greater than zero for all states other than the dead state. When the system is not of unit mass, then the total availability Φ is given by $\Phi = m\phi$.

By applying Eq. (8-3) to the initial and final states 1 and 2 of a closed-system process, we find that

$$\Delta\phi = \phi_2 - \phi_1 = (u_2 - u_1) + P_0(v_2 - v_1) - T_0(s_2 - s_1) \tag{8-4}$$

When this result is inserted into Eq. (8-2) we obtain the general expression

$$w_{\text{useful, opt}} = \Delta\phi = \phi_2 - \phi_1 \tag{8-5}$$

For a work-producing process the value of $\Delta\phi$ will be negative, and it will measure the maximum useful work obtainable for a specified change of state. Although the concept of availability is introduced in terms of evaluating the maximum useful work output, it is useful in the analysis of any general process. For a work-consuming process the change in $\Delta\phi$ will provide a measure of the minimum useful work input required, and the value of $\Delta\phi$ will be positive. Even for processes where no useful work is present during the actual change of state, the value of $\Delta\phi$ is still a measure of the change in the work capability of the closed system. Hence the availability function is of considerable importance in the analysis of all types of processes of a closed stationary system. Finally, note that the availability can be increased only by doing work on the closed system or by transferring heat to it from another system at a temperature other than T_0.

Example 8-1M A tank of air at 4 bars and 600°K has a volume of 3 m^3. Heat is transferred from the air to the local environment until the temperature of the air in the tank is 300°K. The surrounding atmosphere is at 17°C and 1 bar. Calculate (*a*) the initial and final availability of the air, in kilojoules, and (*b*) the optimum useful work associated with the process, in kilojoules.

SOLUTION (*a*) Determination of the availability of a state requires information on *u*, *v*, and *s*. For the initial state

$$m = \frac{PV}{RT} = \frac{4(3)(29)}{0.08315(600)} = 6.98 \text{ kg}$$

$$v_0 = \frac{RT_0}{P_0} = \frac{0.08351(290)}{1(29)} = 0.832 \text{ m}^3/\text{kg}$$

$$v_1 = \frac{V}{m} = \frac{3}{6.98} = 0.430 \text{ m}^3/\text{kg}$$

$$s_1 - s_0 = c_p \ln \frac{T_1}{T_0} - R \ln \frac{P_1}{P_0} = 1.02 \ln \frac{600}{290} - \frac{8.314}{29} \ln \frac{4}{1}$$

$$= 0.742 - 0.397 = 0.345 \text{ kJ/(kg)(°K)}$$

Hence

$$\Phi_1 = m[(u_1 - u_0) + P_0(v_1 - v_0) - T_0(s_1 - s_0)]$$

$$= 6.98[0.740(600 - 290) + 1(0.430 - 0.832)(100) - 290(0.345)]$$

$$= 6.98(229.4 - 40.2 - 100.1) = 621.9 \text{ kJ}$$

For the final state, $v_2 = v_1$, since the tank is assumed rigid. Also, $P_2 = P_1(T_2/T_1) = 4(300/600) =$ 2 bars. In addition,

$$s_2 - s_0 = 1.01 \ln \tfrac{300}{290} - \tfrac{8.314}{29} \ln \tfrac{2}{1} = 0.0342 - 0.1987 = -0.165 \text{ kJ/(kg)(°K)}$$

Therefore the availability of the final state is

$$\Phi_2 = 6.98[0.718(300 - 290) + 1(0.430 - 0.832)(100) - 290(-0.165)]$$

$$= 6.98(7.18 - 40.2 + 47.85) = 103.5 \text{ kJ}$$

(*b*) The optimum useful work is given by Eq. (8-5), namely $w_{opt} = \phi_2 - \phi_1$. Accounting for the total mass of the system, we find that

$$W_{\text{useful, opt}} = 103.5 - 621.9 = -518.4 \text{ kJ}$$

The negative sign indicates that the maximum useful work output is 518.4 kJ.

Example 8-1 A tank of air at 50 psia and 660°F has a volume of 20 ft^3. Heat is transferred from the air to the local environment until the temperature of the air in the tank is 100°F. The surrounding atmosphere is at 60°F and 14.7 psia. Calculate (*a*) the initial and final availability of the air, in Btu, and (*b*) the optimum useful work associated with the process, in Btu.

SOLUTION Determination of the availability of a state requires information on *u*, *v*, and *s*. For the initial state

$$m = \frac{PV}{RT} = \frac{50(20)(29)}{10.73(1120)} = 2.41 \text{ lb}$$

$$v_0 = \frac{RT_0}{P_0} = \frac{10.73(520)}{29(14.7)} = 13.1 \text{ ft}^3/\text{lb}$$

$$v_1 = \frac{V}{m} = \frac{20}{2.41} = 8.30 \text{ ft}^3/\text{lb}$$

$$s_1 - s_0 = c_p \ln \frac{T_1}{T_0} - R \ln \frac{P_1}{P_0} = 0.246 \ln \frac{1120}{520} - \frac{1.986}{29} \ln \frac{50}{14.7}$$

$$= 0.1887 - 0.0838 = 0.105 \text{ Btu/(lb)(°R)}$$

Hence the availability of the initial state is

$$\Phi_1 = m[(u_1 - u_0) + P_0(v_1 - v_0) - T_0(s_1 - s_0)]$$

$$= 2.41\left[0.177(1120 - 520) + \frac{14.7(144)}{778}(8.30 - 13.1) - 520(0.105)\right]$$

$$= 2.41(106.2 - 13.1 - 54.6) = 92.8 \text{ Btu}$$

For the final state, $v_2 = v_1$, since the tank is assumed rigid. Also,

$P_2 = P_1(T_2/T_1) = 50(560/1120) = 25$ psia. In addition,

$$s_2 - s_0 = 0.240 \ln \frac{560}{520} - \frac{1.986}{29} \ln \frac{25}{14.7} = 0.0178 - 0.0364$$

$$= -0.0186 \text{ Btu/(lb)(°R)}$$

Therefore the availability of the final state is

$$\Phi_2 = 2.41\left[0.171(560 - 520) + \frac{14.7(144)}{778}(8.30 - 13.1) - 520(-0.0186)\right]$$

$$= 2.41(6.84 - 13.06 + 9.67) = 8.31 \text{ Btu}$$

(*b*) The optimum useful work is given by Eq. (8-5), namely $w_{\text{opt}} = \phi_2 - \phi_1$. On the basis of the total mass of the system,

$$W_{\text{useful, opt}} = 8.31 - 92.8 = -84.5 \text{ Btu}$$

The negative sign indicates that the maximum useful work output is 84.5 Btu.

In some availability analyses it is helpful to separarate the optimum work associated with the change of state of the system from that due to the heat engine work. Recall that when a system X undergoes a reversible process that

$$w_{\text{useful, opt}} = w_{\text{useful}, X} - w_{\text{eng}} = \Delta u + P_0\, \Delta v - T_0\, \Delta s$$

and from the first law for a closed system

$$q + w_{\text{useful}, X} - P_0\, \Delta v = \Delta u$$

Substitution of Δu from the energy equation into the expression for the availability change leads to

$$w_{\text{useful}, X} + w_{\text{eng}} = (q + w_{\text{useful}, X} - P_0\, \Delta v) + P_0\, \Delta v - T_0\, \Delta s$$

or

$$w_{\text{eng}} = q - T_0\, \Delta s \tag{8-6}$$

In Eq. (8-6) q is the heat transfer which occurs between the system and its local environment, and Δs is the specific entropy change of the substance within the closed system.

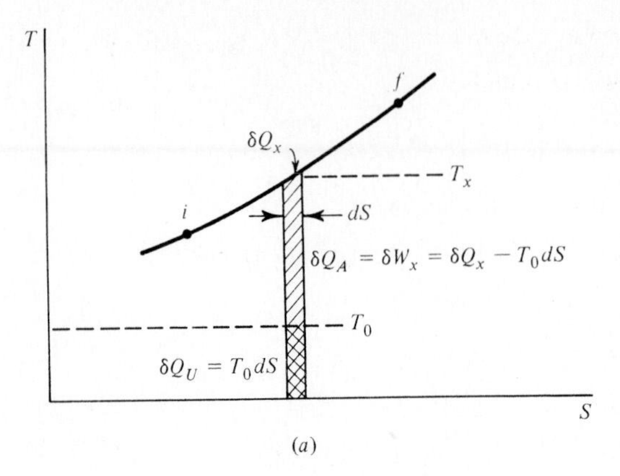

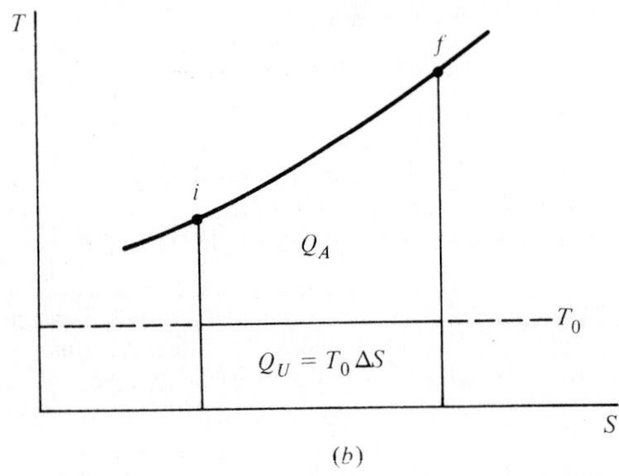

Figure 8-2 (*a*) Available and unavailable energy for a differential addition of heat to a heat engine. (*b*) Available and unavailable energy quantities represented on a TS diagram.

In the literature the quantities w_{eng} and $T_0\,\Delta s$ are sometimes referred to as the *available energy* q_A and the *unavailable energy* q_U. That is,

$$w_{\text{eng}} = q_A = q - q_U = q - T_0\,\Delta s \tag{8-7a}$$

for a system of unit mass, or, more generally,

$$W_{\text{eng}} = Q_A = Q - Q_U = Q - T_0\,\Delta S \tag{8-7b}$$

A graphical interpretation of these quantities is shown in Fig. 8-2. Figure 8-2*a* illustrates the available and unavailable energy for a differential removal of heat from a system, while Fig. 8-2*b* shows these same quantities as areas on a *TS* plot for a finite change of state. The unavailable energy Q_U is always determined by $T_0\,\Delta S$ and is represented by a rectangular area on a *TS* plot. The quantity Q is usually found from an energy balance on the closed system.

Example 8-2M Air in a closed system at 2 bars and 227°C is cooled in a quasistatic manner at constant pressure until its temperature reaches 27°C. How much of the heat removed during the process is available energy, in kJ/kg, if the atmospheric conditions are 7°C and 1 bar?

SOLUTION Figure 8-2*b* illustrates again the general problem, where the line passing through states *i* and *f* is now a constant-pressure line. The heat q removed from the air is found from the first law for the constant-pressure process. If the air is an ideal gas, then, from the air table

$$q = \Delta u - w = \Delta h = h_2 - h_1$$

$$= 300.2 - 503.0 = -202.8 \text{ kJ/kg}$$

The change in entropy of the air, since the pressure is constant, is

$$\Delta s = s_2^0 - s_1^0 - R \ln \frac{P_2}{P_1} = 1.70203 - 2.21952 = -0.5175 \text{ kJ/(kg)(°K)}$$

Hence,

$$q_U = T_0 \, \Delta s = 280(-0.5175) = -145 \text{ kJ/kg}$$

The available-energy part of the total heat supplied, then, is

$$q_A = q - q_U = -202.8 - (-145) = -57.8 \text{ kJ/kg}$$

Thus, of the 202.8 kJ/kg of heat removed from the air, 57.8 kJ/kg could theoretically be converted into work by use of a reversible heat engine. However, the available energy represents only 29 percent of the total heat added.

Example 8-2 Air in a closed system at 30 psia and 400°F is cooled in a quasistatic manner at constant pressure until its temperature reaches 80°F. How much of the heat removed during the process is available energy, in Btu/lb, if the atmospheric conditions are 40°F and 14.7 psia?

SOLUTION Figure 8-2*b* illustrates again the general problem, where the line passing through states *i* and *f* is now a constant-pressure line. The heat q removed from the air is found from the first law for the constant-pressure process. If the air is an ideal gas, then, from the air table,

$$q = \Delta u - w = \Delta h = h_2 - h_1$$

$$= 129.06 - 206.46 = -77.4 \text{ Btu/lb}$$

The change in entropy of the air, since the pressure is constant, is

$$\Delta s = s_2^0 - s_1^0 - R \ln P_2/P_1 = 0.60078 - 0.71323 = -0.11245 \text{ Btu/(lb)(°R)}$$

Hence,

$$q_U = T_0 \, \Delta s = 500(-0.11245) = -56.2 \text{ Btu/lb}$$

The available-energy part of the total heat supplied, then, is

$$q_A = q - q_U = -77.4 - (-56.2) = -21.2 \text{ Btu/lb}$$

Thus, of the 77.4 Btu/lb of heat removed from the air, only 21.2 Btu/lb could theoretically be converted into work by use of a reversible heat engine. This is 27 percent of the heat added.

One should be cautioned that some authors use the term available energy to represent the availability of a state. Consequently, the definitions of terms must be made clear in any article on second-law analysis.

8-3 AVAILABILITY IN STEADY-FLOW PROCESSES

Consider the control volume shown in Fig. 8-3, which is surrounded by a local environment at T_0 and P_0. During the passage of mass through the control

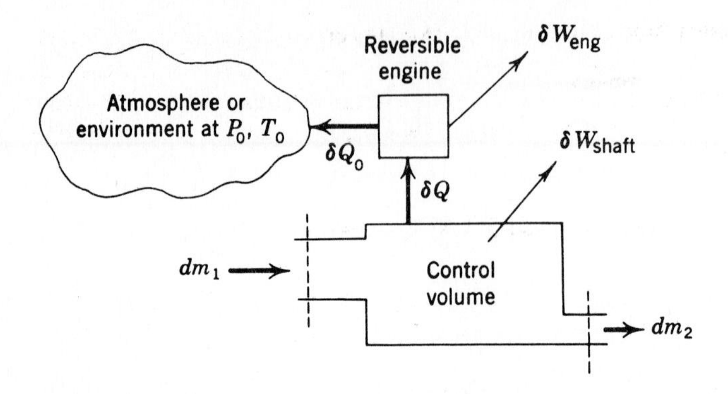

Figure 8-3 Control-volume-environment composite producing maximum work.

volume, a shaft-work interaction δW_{shaft} and a heat interaction δQ may occur. In order to optimize the total work associated with the change of state of the fluid (that is, maximize the work output or minimize the work input), the process within the control volume must be reversible. In addition, any heat exchange between the control volume and the environment must pass through a reversible heat engine, as shown in the figure. The heat rejected by the heat engine is δQ_0, and the reversible engine work is δW_{eng}. For the overall process the optimum steady-flow work is

$$\delta W_{opt,\,sf} = \delta W_{shaft,\,rev} + \delta W_{eng} \tag{a}$$

The terms on the right side of Eq. (*a*) can be expressed in terms of properties of the control volume or environment.

In the case of a steady-flow, steady-state process, the shaft work can be determined from the conservation of energy principle, namely,

$$\delta W_{shaft,\,rev} = (h + \mathrm{KE} + \mathrm{PE})_2\, dm - (h + \mathrm{KE} + \mathrm{PE})_1\, dm - \delta Q_{rev} \tag{b}$$

where the symbols 1 and 2 represent the inlet and exit states of the control volume, respectively. The engine work δW_{eng} is evaluated from both a first- and second-law analysis on the engine. First of all, from energy considerations on the cyclic device,

$$\delta W_{eng} = -\delta Q_{eng} - \delta Q_0 \tag{c}$$

where δQ_{eng} is measured relative to the engine. Since we measure δQ_{rev} relative to the control volume, then we recognize that δQ_{eng} and δQ_{rev} are of equal magnitude but of opposite sign. Hence Eq. (*c*) can be written as

$$\delta W_{eng} = \delta Q_{rev} - \delta Q_0 \tag{d}$$

The second-law statement for the control volume is expressed by Eq. (7-23), that is,

$$dS_{CV} + \sum_{out} (s\, dm) - \sum_{in} (s\, dm) - \sum \frac{\delta Q_i}{T_i} \geq 0 \tag{7-23}$$

Since the process is steady state, the quantity dS_{CV} is zero. Also, since the process is reversible, only the equality sign applies. (Recall that the summation sign on the quantity $\delta Q_i/T_i$ is necessary for the control volume, because the temperature varies along the control surface.) Therefore, for the case under consideration Eq. (7-23) reduces to

$$(s\,dm)_2 - (s\,dm)_1 - \sum \frac{\delta Q_i}{T_i} = 0 \tag{e}$$

Finally, for the series of Carnot heat engines which are receiving heat δQ_i at T_i and rejecting heat δQ_0 at T_0, we may write

$$\frac{\delta Q_0}{T_0} = \sum \frac{\delta Q_i}{T_i} \tag{f}$$

When Eqs. (*e*) and (*f*) are substituted into Eq. (*d*), we find that the engine work δW_{eng} can be expressed as

$$\delta W_{eng} = \delta Q_{rev} - T_0[(s\,dm)_2 - (s\,dm)_1] \tag{g}$$

Substitution of Eqs. (*b*) and (*g*) into Eq. (*a*) yields, after cancellation of terms and rearrangement,

$$\delta W_{opt,\,sf} = (h + KE + PE - T_0 s)_2\,dm - (h + KE + PE - T_0 s)_1\,dm \tag{h}$$

Integration of this expression over a time period during which unit mass enters and leaves the control volume leads to the following result.

$$\begin{aligned} w_{opt,\,sf} &= (h_2 + KE_2 + PE_2 - T_0 s_2) - (h_1 + KE_1 + PE_1 - T_0 s_1) \\ &= (h_2 - h_1) + (KE_2 - KE_1) + (PE_2 - PE_1) - T_0(s_2 - s_1) \end{aligned} \tag{8-8}$$

Equation (8-8) represents the optimum work associated with a control volume operating in steady state and steady flow with one inlet and one outlet, and exchanging heat solely with the atmosphere.

For a control volume with more than one inlet and one outlet it is necessary to put the above expression on a rate basis. This modification is

$$\dot{W}_{opt,\,sf} = \sum_{out} \dot{m}_e(h + KE + PE + T_0 s)_e - \sum_{in} \dot{m}_i(h + KE + PE + T_0 s)_i \tag{8-9}$$

where the subscripts e and i stand for the exit and inlet streams. Equation (8-9) predicts the maximum-power output or the minimum-power input for a steady-state, steady-flow control volume which exchanges heat only with the environment.

The *stream availability* of a fluid in steady-state flow is defined as the maximum-work output which can be obtained as the fluid is changed reversibly from the given state to a dead state while exchanging heat solely with the atmosphere. A dead state in this case implies not only thermal and mechanical equilibrium of the fluid with the atmosphere at T_0 and P_0, but also that the kinetic energy at the dead state is zero. Also, its potential energy must be at a minimum.

The stream availability is given by the symbol ψ for a unit mass and Ψ for the total mass, and $\Psi = m\psi$. The stream availability is measured by the value of the quantity $(h + \text{KE} + \text{PE} - T_0 s)$ at the given state relative to the dead state. That is,

$$\psi \equiv (h + \text{KE} + \text{PE} - T_0 s) - (h_0 + \text{PE}_0 - T_0 s_0) \tag{8-10}$$

Although we speak of the stream availability of a fluid in a certain state, the stream availability is a function of the state of the local atmosphere as well as the state of the fluid. The value of ψ can be greater than or less than zero for any state other than the dead state.

By applying Eq. (8-10) to the inlet state 1 and the outlet state 2 of a control volume, we find that

$$\begin{aligned} \Delta\psi = \psi_2 - \psi_1 &= (h_2 + \text{KE}_2 + \text{PE}_2 - T_0 s_2) - (h_1 + \text{KE}_1 + \text{PE}_1 - T_0 s_1) \\ &= (h_2 - h_1) + (\text{KE}_2 - \text{KE}_1) + (\text{PE}_2 - \text{PE}_1) - T_0(s_2 - s_1) \end{aligned} \tag{8-11}$$

When this result is substituted into Eq. (8-6) we obtain the general result,

$$w_{\text{opt, sf}} = \psi_2 - \psi_1 \tag{8-12}$$

Thus the change in the stream availability is a measure of the maximum work output (for a work producing device) or the minimum work input (for a work consuming process) for a control volume with one inlet and one outlet, and exchanging heat solely with the atmosphere. Even for processes with no shaft work present, the value of $\Delta\psi$ is still a measure of the change in the work capability of the control volume-atmosphere combination. Hence the stream availability is of considerable importance in the analysis of all types of steady-state, steady-flow processes. The stream availability ψ of a fluid can be increased only by doing work on the fluid or by transferring heat to it from another system at a temperature other than T_0.

In the case of a control volume with more than one inlet or outlet, we can write Eq. (8-12) on a rate basis in the following form.

$$\dot{W}_{\text{opt, sf}} = \sum_{\text{out}} \dot{m}_e \psi_e - \sum_{\text{in}} \dot{m}_i \psi_i \tag{8-13}$$

where the subscripts e and i again stand for the exit and inlet streams.

Example 8-3M An adiabatic turbine operates with air initially at 5 bars, 400°K, and 150 m/s. The exit conditions are 1 bar, 300°K, and 70 m/s. Determine the actual work output and compare this with the maximum work obtainable. Assume that c_p is constant at 1.01 kJ/(kg)(°K) and that the atmosphere is at 1 bar and 17°C.

SOLUTION The actual work output is found from the conservation of energy equation for a control volume around the turbine:

$$w_{act} = h_2 - h_1 + \frac{V_2^2 - V_1^2}{2}$$

$$= 1.01(300 - 400) + \frac{(70)^2 - (150)^2}{2(1000)}$$

$$= -101 - 8.8 = -109.8 \text{ kJ/kg}$$

The maximum work possible is calculated from Eq. (8-8):

$$w_{sf,\,max} = (h_2 - h_1) - T_0(s_2 - s_1) + \frac{V_2^2 - V_1^2}{2}$$

$$= w_{act} - T_0(s_2 - s_1)$$

$$= -109.8 - 290\left(1.01 \ln \frac{300}{400} - \frac{8.315}{29} \ln \frac{1}{5}\right)$$

$$= -109.8 - 49.6 = -159.4 \text{ kJ/kg}$$

The presence of irreversibilities within the turbine have led to a loss of 49.6 kJ/kg of work output, which is a sizable fraction of the maximum possible work output.

Example 8-3 An adiabatic turbine operates with air initially at 75 psia, 240°F, and 450 ft/s. The exit conditions are 15 psia, 80°F, and 200 ft/s. Determine the actual work output and compare this with the maximum useful work obtainable. Assume that c_p is constant at 0.240 Btu/(lb)(°F) and that the atmosphere is at 1 atm pressure and 70°F.

SOLUTION The actual work output is found from the conservation of energy equation for a suitable control volume around the turbine:

$$w_{act} = h_2 - h_1 + \frac{V_2^2 - V_1^2}{2}$$

$$= 0.24(80 - 240) + \frac{(200)^2 - (450)^2}{2(32.2)(788)}$$

$$= -38.4 - 3.3 = -41.7 \text{ Btu/lb}$$

The maximum work possible is calculated from Eq. (8-8):

$$w_{sf,\,max} = (h_2 - h_1) - T_0(s_2 - s_1) + \frac{V_2^2 - V_1^2}{2}$$

$$= w_{act} - T_0(s_2 - s_1)$$

$$= -41.7 - 530\left(0.24 \ln \frac{540}{700} - \frac{53.3}{778} \ln \frac{15}{75}\right)$$

$$= -41.7 - 25.3 = -67.0 \text{ Btu/lb}$$

The presence of irreversibilities within the turbine have led to a loss of 25.3 Btu/lb of work output, which is a sizable fraction of the maximum work possible from this steady-flow device.

Example 8-4M Nitrogen gas initially at 3.6 bars and 27°C is throttled through a well-insulated valve to a pressure of 1.1 bars. The atmospheric temperature is 15°C. Determine the maximum work output under steady-flow conditions.

SOLUTION Equation (8-8) is the basis for the calculation. Throttling is a process for which $\Delta h = 0$. Since Δh for an ideal gas equals $c_p \, \Delta T$, the temperature of the nitrogen remains constant at 27°C. Also, the kinetic-energy change is neglected. Hence the expression for the maximum work output becomes

$$w_{\text{sf, max}} = \Delta h + \Delta\text{KE} - T_0 \, \Delta s$$

$$= -T_0\left(c_p \ln \frac{T_2}{T_1} - R \ln \frac{P_2}{P_1}\right) = RT_0 \ln \frac{P_2}{P_1}$$

$$= \frac{8.314(288)}{29} \ln \frac{1.1}{3.6} = -97.9 \text{ kJ/kg}$$

For the actual process, of course, the work output is zero.

Example 8-4 Nitrogen gas initially at 50 psia and 100°F is throttled through a well-insulated valve to a pressure of 15 psia. The atmospheric temperature is 60°F. Determine the maximum work output under steady-flow conditions.

SOLUTION Equation (8-8) is the basis for the calculation. Throttling is a process for which $\Delta h = 0$. Since Δh for an ideal gas equals $c_p \, \Delta T$, the temperature of the nitrogen remains constant at 100°F. Also, kinetic-energy changes are neglected. Hence the expression for the maximum work output becomes

$$w_{\text{sf, max}} = \Delta h + \Delta\text{KE} - T_0 \, \Delta s$$

$$= -T_0\left(c_p \ln \frac{T_2}{T_1} - R \ln \frac{P_2}{P_1}\right) = RT_0 \ln \frac{P_2}{P_1}$$

$$= \frac{1.986(520)}{29} \ln \frac{15}{50} = -42.9 \text{ Btu/lb}$$

For the actual process, of course, the work output is zero.

The optimum work associated with a steady-flow process is the sum of the reversible work for the given end states of the fluid and the work from a heat engine if heat transfer occurs between the control volume and the environment. Similar to the development in Sec. 8-2, an equation for the engine work alone can be derived. Recall that the availability change is given by

$$w_{\text{opt, sf}} = w_{\text{CV}} + w_{\text{eng}} = \Delta h - T_0 \, \Delta s + \Delta\text{KE} + \Delta\text{PE}$$

In addition, the conservation of energy statement for a control volume with one inlet and one outlet is

$$q + w_{\text{CV}} = \Delta h + \Delta\text{KE} + \Delta\text{PE}$$

Elimination of Δh between these two equations leads to

$$w_{\text{CV}} + w_{\text{eng}} = (q + w_{\text{CV}} - \Delta\text{KE} - \Delta\text{PE}) - T_0 \, \Delta s + \Delta\text{KE} + \Delta\text{PE}$$

or

$$w_{\text{eng}} = q - T_0 \, \Delta s \tag{8-6}$$

This is the same result as developed for the change of state of a closed system. Again, we shall define the quantity w_{eng} as the *available energy* q_A of the process, and the quantity $T_0\,\Delta s$ as the unavailable energy q_U. Equation (8-7*a*) then is applicable again, that is,

$$w_{\text{eng}} = q_A = q - q_U = q - T_0\,\Delta s \tag{8-7a}$$

The graphical interpretation of these quantities is shown in Fig. 8-2.

8-4 EFFECT OF HEAT RESERVOIRS ON AVAILABILITY STUDIES

In the preceding two sections we have dealt with either closed systems or control volumes which exchange heat solely with the environment or local atmosphere. In more general circumstances the systems of interest may also exchange heat with some other reservoir at a temperature T_R. The preceding equations can be modified to fit this new situation. A thermal-energy (heat) reservoir is a closed system. Hence, as it exchanges heat δQ_R with the system of interest, the reservoir undergoes an availability change $d\Phi_R$. It is convenient to transform this quantity into a more familiar format.

On the basis of Eq. (8-4) the availability change of a reservoir for a differential change of state (and noting that the reservoir is not one of unit mass),

$$d\Phi_R = dU_R + P_0\,dV_R - T_0\,dS_R$$

In addition, recall that a heat reservoir undergoes only internally reversible changes of state and that no work is performed. As a result, $dV_R = 0$, $\delta Q_R = dU_R$, and $dS_R = \delta Q_R/T_R$. When these expressions are substituted into the above equation,

$$d\Phi_R = \delta Q_R - T_0\frac{\delta Q_R}{T_R} = \delta Q_R\left(1 - \frac{T_0}{T_R}\right) \tag{8-14}$$

where Q_R is measured with respect to the *reservoir*. Therefore, for the situation where heat may also be exchanged with a reservoir at T_R, we find that for a unit mass of a closed system, generally,

$$\begin{aligned} w_{\text{useful, opt}} &= \Delta u + P_0\,\Delta v - T_0\,\Delta s + q_R\left(1 - \frac{T_0}{T_R}\right) \\ &= \Delta\phi + q_R\left(1 - \frac{T_0}{T_R}\right) \end{aligned} \tag{8-15}$$

A similar equation can be written for a system not of unit mass.

Equation (8-14) is equally valid for a control volume which receives heat from or rejects heat to a reservoir at T_R. Consequently, we can write for a steady-state, steady-flow control volume with one inlet and one exit and exchanging heat also with a reservoir at T_R,

$$w_{\text{opt, sf}} = \Delta h + \Delta \text{KE} + \Delta \text{PE} - T_0\, \Delta s + q_R \left(1 - \frac{T_0}{T_R}\right)$$

$$= \Delta \psi + q_R \left(1 - \frac{T_0}{T_R}\right) \tag{8-16}$$

This equation can easily be extended to include control volumes with more than one inlet or outlet, in the light of Eqs. (8-9) and (8-13).

Example 8-5M A tank of air at 2 bars and 300°K has a volume of 3 m^3. Heat is transferred to the air from a reservoir at 1000°K until the temperature of the air in the tank is 600°K. The surrounding atmosphere is at 17°C and 1 bar. Determine the maximum useful work associated with the process, in kilojoules.

Solution The maximum useful work in this case is given by Eq. (8-15). The value of $\Delta\Phi$ for the system has already been calculated in Example 8-1M, except that the end states of the air in the tank have been reversed. Therefore, $\Delta\Phi$ in this case is 518.4 kJ. The value of Q_R is found from an energy balance on the air in the tank. Since the volume of the tank is constant,

$$Q = \Delta U - W = m(u_2 - u_1) = 6.98(0.740)(600 - 300) = 1550 \text{ kJ}$$

The value of Q_R with respect to the reservoir is the negative of Q, or -1550 kJ. Consequently, on the basis of Eq. (8-15) for the total mass,

$$W_{\text{useful, opt}} = \Delta\Phi + Q_R \left(1 - \frac{T_0}{T_R}\right) = 518.4 + (-1550)\left(1 - \frac{290}{1000}\right)$$

$$= 518 - 1100 = -528 \text{ kJ}$$

Thus a maximum useful work output of 582 kJ is associated with the stated process.

Example 8-5 A tank of air at 25 psia and 100°F has a volume of 20 ft^3. Heat is transferred from a reservoir at 1500°R until the temperature of the air in the tank is 660°F. The surrounding atmosphere is at 60°F and 14.7 psia. Determine the maximum useful work associated with the process, in Btu.

Solution The maximum useful work in this case is given by Eq. (8-15). The value of $\Delta\Phi$ for the system has already been calculated in Example 8-1, except that the end states of the air in the tank have been reversed. Therefore $\Delta\Phi$ in this case is 84.5 Btu. The value of Q_R is found from an energy balance on the air in the tank. Since the volume of the tank is constant,

$$Q = \Delta U - W = m(u_2 - u_1) = 2.41(0.177)(660 - 100) = 239 \text{ Btu}$$

The value of Q_R with respect to the reservoir is the negative of Q, or -239 Btu. Consequently, on the basis of Eq. (8-15) for the total mass,

$$W_{\text{useful, opt}} = \Delta\Phi + Q_R \left(1 - \frac{T_0}{T_R}\right) = 84.5 + (-239)\left(1 - \frac{520}{1500}\right)$$

$$= 84.5 - 156 = -71.5 \text{ Btu}$$

Thus a maximum useful work output of 71.5 Btu is associated with the stated process.

8-5 IRREVERSIBILITY FOR CLOSED AND OPEN SYSTEMS

In the analysis of either closed or open systems, the conservation of energy principle, in conjunction with auxiliary equations or data, is usually sufficient to ascertain the actual work delivered to or by a particular device. For engineering purposes it is important to have some standard to which actual performance can be compared. The concept of availability has been introduced for this purpose in the preceding sections. Maximum work output or minimum work input is associated with processes that are totally reversible. The presence of irreversibilities leads to a decrease in the amount of work which could be produced and delivered to devices external to the system and its local environment. This loss of work capability will be defined as the irreversibility I of any process, and will be taken to be a positive quantity. Hence the irreversibility is the difference between the actual work delivered and the maximum work obtainable for a given change of state for a system. Thus

$$I \equiv W_{\text{act}} - W_{\text{opt}} \tag{8-17}$$

In determining W_{opt} it is necessary that the system undergo a process between the same end states and exchange the same quantities of heat with various reservoirs. In addition, the system will probably also have to exchange heat with the environment. Equation (8-17) may also be written in terms of the irreversibility per unit mass i, where $i = I/m$.

In the analysis of the irreversibility associated with a closed-system process, it is convenient to express the work quantities in terms of useful work rather than total work. Note that the boundary work done on the environment is the same in any reversible process as in an actual process. If this latter quantity is subtracted from both W_{act} and W_{opt}, then

$$I \equiv W_{\text{useful, act}} - W_{\text{useful, opt}} \tag{8-18}$$

In conjunction with Eq. (8-15), the above equation may also be expressed as

$$I = W_{\text{useful, act}} - \Delta\Phi - Q_R\left(1 - \frac{T_0}{T_R}\right) \tag{8-19}$$

The quantity $W_{\text{useful, act}}$ normally would be found from an energy balance on the actual process.

The quantity I can be expressed in terms of the entropy function. For a closed system we may substitute Eq. (8-4) into Eq. (8-19). On a differential basis

$$\delta I = \delta W_{\text{useful, act}} - \delta W_{\text{useful, opt}}$$

$$= \delta W_{\text{total, act}} + P_0\,dV - (dU + P_0\,dV - T_0\,dS) - \delta Q_R\left(1 - \frac{T_0}{T_R}\right)$$

$$= \delta W_{\text{total, act}} - dU + T_0\,dS - \delta Q_R + \delta Q_R\,\frac{T_0}{T_R}$$

From the conservation of energy principle the quantity $\delta W_{\text{total, act}}$ may be replaced by $dU - \delta Q$, where δQ is the sum of all heat interactions. In this particular case, $-\delta Q$ is equal to $\delta Q_R + \delta Q_0$. Substitution of these quantities yields

$$\delta I = \delta Q_0 + T_0\, dS + \delta Q_R \frac{T_0}{T_R}$$

In addition, $\delta Q_0 = T_0\, dS_0$ and $\delta Q_R = T_R\, dS_R$, because any heat reservoir by concept is considered to be internally reversible. Thus the equation for the irreversibility δI of a process for a closed system becomes

$$\delta I = T_0\, dS_0 + T_0\, dS + T_0\, dS_R$$

$$= T_0\, (dS)_{\text{total}} \tag{8-20a}$$

or

$$I = T_0\, (\Delta S)_{\text{total}} \tag{8-20b}$$

The direct proportionality between the quantity I and the total entropy change associated with a process is not unexpected. The total increase in entropy of the parts of an isolated system is due solely to the presence of irreversibilities within the overall system.

An equation equivalent to Eq. (8-20) can also be developed for open systems in steady flow. On the basis of Eq. (8-16) we find that

$$\delta W_{\text{sf, opt}} = (h + \text{KE} - T_0\, s)_2\, dm - (h + \text{KE} - T_0\, s)_1\, dm + \delta Q_R \left(1 - \frac{T_0}{T_R}\right)$$

In addition, the steady-flow energy equation for a control volume is

$$\delta W_{\text{sf, act}} = -\delta Q + (h + \text{KE})_2\, dm - (h + \text{KE})_1\, dm$$

The potential-energy terms have been omitted from both of the preceding equations. As a result, the differential form of Eq. (8-17) becomes

$$\delta I_{\text{sf}} = -\delta Q + T_0(s_2 - s_1)\, dm - \delta Q_R + \delta Q_R \frac{T_0}{T_R}$$

However, $-\delta Q = \delta Q_0 + \delta Q_R$, $\delta Q_0 = T_0\, dS_0$, and $\delta Q_R = T_R\, dS_R$. With these substitutions the above equation becomes

$$\delta I_{\text{sf}} = T_0\, dS_0 + T_0(s_2 - s_1)\, dm + T_0\, dS_R$$

$$= T_0\, dS_{\text{isol}} \tag{8-21}$$

Per unit mass of fluid flowing through the control volume,

$$i_{\text{sf}} = T_0[\Delta s_0 + (s_2 - s_1) + \Delta s_R] \tag{8-22}$$

where Δs_0 and Δs_R are the entropy changes of the local surroundings and a heat reservoir at T_R, respectively. Thus the steady-flow analysis leads to an equation analogous to Eq. (8-20) for closed systems.

On the basis of the increase in entropy principle, which states that $dS_{\text{isol}} \geq 0$, we can conclude that

$$\delta I \geq 0 \tag{8-23}$$

or, for a finite change of state,

$$I \geq 0 \tag{8-24}$$

The equality sign in either of these two equations applies only in the case of a totally reversible process. For an irreversible process the irreversibility is a measure of the decrease in energy which can be converted into work.

Example 8-6M Calculate the irreversibility of the process considered previously in Example 8-5M, in kilojoules.

SOLUTION The irreversibility of the closed system is

$$I = W_{u,\,\text{act}} - W_{u,\,\text{max}}$$

In this particular problem the tank is rigid and the actual useful work is zero. The maximum useful work was previously found to be -582 kJ. Therefore the irreversibility of the process is simply 582 kJ. This value can also be calculated from the expression

$$I = T_0(\Delta S)_{\text{total}} = T_0(\Delta S_{\text{system}} + \Delta S_{\text{res}})$$

The entropy change of the system is given by

$$\Delta S_{\text{system}} = mc_v \ln \frac{T_0}{T_1} = 6.98(0.740) \ln \frac{600}{300} = 3.58 \text{ kJ/°K}$$

The entropy change of the reservoir is simply $Q_R/T_R = -1550/1000 = -1.55$ kJ/°K. Therefore, the total entropy change is $3.58 + (-1.55) = 2.03$ kJ/°K, and

$$I = 290(2.03) = 589 \text{ kJ}$$

This answer agrees with the preceding evaluation within the accuracy of the calculation.

Example 8-6 Calculate the irreversibility of the process considered previously in Example 8-5, in Btu.

SOLUTION The irreversibility of the closed system is

$$I = W_{u,\,\text{act}} - W_{u,\,\text{opt}}$$

In this particular problem the tank is rigid and the actual useful work is zero. The maximum useful work was previously found to be -71.5 Btu. Therefore the irreversibility of the process is simply 71.5 Btu. This value can also be calculated from the expression

$$I = T_0(\Delta S)_{\text{total}} = T_0(\Delta S_{\text{system}} + \Delta S_{\text{res}})$$

The entropy change of the system is given by

$$\Delta S_{\text{system}} = mc_v \ln \frac{T_2}{T_1} = 2.41(0.177) \ln \frac{1120}{560} = 0.296 \text{ Btu/°R}$$

The entropy change of the reservoir is simply $Q_R/T_R = -239/1500 = -0.159$ Btu/°R. Therefore, the total entropy change is $0.296 - 0.159 = 0.137$ Btu/°R, and

$$I = 520(0.137) = 71.2 \text{ Btu}$$

This answer agrees with the preceding evaluation within the accuracy of the calculation.

Example 8-7M and 8-7† Consider the expansion of nitrogen gas through a throttling valve under the conditions expressed in either Example 8-4M or 8-4. Determine the irreversibility of the process.

SOLUTION Per unit mass of nitrogen passing through the valve, $i = T_0 \Delta s_{total}$. Under steady-state conditions the only entropy change is that of the fluid as it passes through the valve. As a result, the irreversibility is equal in magnitude, but opposite in sign, to the steady-flow maximum work calculated in Examples 8-4M and 8-4. That is, $i = -w_{sf, opt}$. This result is not surprising, since the capability of work output associated with a finite pressure drop is completely lost when a fluid is throttled.

8-6 EFFECTIVENESS OF PROCESSES AND CYCLES

The concept of availability enables one to determine the optimum process between given end states in terms of work potential. For example, the change in availability predicts the maximum work output or minimum work input. When these optimum values are compared to the actual performance, then the irreversibility or degradation of work potential is found. This can be done for either a process or a cycle. One important criterion frequently used in analyzing work potential losses is called the *effectiveness* ϵ of the process or cycle. It is a measure of the conversion of energy to less useful forms. The law of degradation of energy states that work potential output will be less than work potential input. Hence one method found in the literature for defining the effectiveness is

$$\epsilon = \frac{\text{work potential output}}{\text{work potential input}} \tag{8-25}$$

The effectiveness is also defined in the following manner.

$$\epsilon = \frac{\text{increase in work potential}}{\text{decrease in work potential}} \tag{8-26}$$

Although these two definitions may appear to be equivalent, they can be interpreted quite differently. The possibility of different interpretations for ϵ is not of major concern. It is important, however, that a given interpretation be applied consistently when analyzing competitive processes.

In general, various increases and decreases in availability will occur in a system and the system's surroundings. However, the algebraic sum of the numerator and denominator in the above equations represents the loss in availability associated with the overall process as a result of irreversibilities. The evaluation of the effectiveness can best be illustrated by examples.

Example 8-8M Reconsider the performance of the air turbine discussed in Example 8-3M. Determine the effectiveness of the device.

† This example can be read with no concern for units.

SOLUTION In Example 8-3M we found the actual work output to be 109.8 kJ/kg, while the decrease in availability of the fluid was 159.4 kJ/kg. On the basis of Eq. (8-26), the effectiveness of the process is

$$\epsilon = \frac{\text{increase in work potential}}{\text{decrease in work potential}} = \frac{w_{\text{shaft}}}{\Delta\psi} = \frac{109.8}{159.4} = 0.69$$

Example 8-8 Reconsider the performance of the air turbine discussed in Example 8-3. Determine the effectiveness of the device.

SOLUTION In Example 8-3 we found the actual work output to be 41.7 Btu/lb, while the decrease in availability of the fluid was 67.0 Btu/lb. On the basis of Eq. (8-26), the effectiveness of the process is

$$\epsilon = \frac{\text{increase in work potential}}{\text{decrease in work potential}} = \frac{w_{\text{shaft}}}{\Delta\psi} = \frac{41.7}{67.0} = 0.62$$

PROBLEMS (METRIC)

Availability for closed systems

8-1M A tank with a volume of 0.30 m^3 contains air at (*a*) 6 bars and 300°K, and (*b*) 6 bars and 600°K. The surrounding atmosphere is at 0.95 bar and 300°K. Determine the availability of the air, in kilojoules.

8-2M The air in Prob. 8-1M undergoes a free expansion until its volume is (1) doubled, and (2) tripled. Determine the change in the closed-system availability for the process, in kilojoules, for an initial state of (*a*) 6 bars and 300°K, and (*b*) 6 bars and 600°K.

8-3M Determine the availability associated with 50 kg of water at 0°C and 0.95 bar if the surrounding atmosphere is at 0.95 bar and 20°C, in kilojoules.

8-4M Fifty kilograms of water at 0°C and 0.95 bar are allowed to mix in a closed system with 30 kg of water at 80°C and 0.95 bar. Determine the optimum useful work associated with the change of state of (*a*) the 50 kg of water, and (*b*) the 30 kg of water. (*c*) Find the change in availability for the overall adiabatic process, in kilojoules. $T_0 = 20°C$ and $P_0 = 1$ bar.

8-5M Determine the availability of steam in a closed system at 80 bars and 400°C, in kJ/kg. The environment is at 1 bar and 25°C.

8-6M The steam in Prob. 8-5M is cooled in a rigid tank until the pressure drops to 40 bars. Determine the optimum useful work associated with the change in state of the fluid, in kJ/kg.

8-7M Air initially at 1 bar and 27°C is contained in a well-insulated tank. An impeller inside the tank is turned by an external mechanism until the pressure is 1.2 bars. Determine (*a*) the actual work required, and (*b*) the optimum useful work associated with the change in state, both in kJ/kg. Let $T_0 = 27°C$ and $P_0 = 1$ bar.

8-8M A piston-cylinder device contains 0.40 kg of air at 0.10 MPa and 27°C. Determine the minimum work input, in kilojoules, required to compress the air to 0.40 MPa and 127°C. $T_0 = 20°C$ and $P_0 = 0.10$ MPa.

8-9M Refrigerant 12 is compressed from a saturated vapor at −10°C to a final pressure of 8 bars. The process is adiabatic and the compressor efficiency is 78 percent. Determine (*a*) the actual work required, in kJ/kg, and (*b*) the actual outlet temperature, in degrees Celsius. (*c*) Find the minimum work required, in kJ/kg, for the actual final state found in part *b*. The process occurs in a piston-cylinder device. $T_0 = 20°C$ and $P_0 = 1$ bar.

8-10M A storage battery is capable of delivering 1 kWh of energy. Determine the volume of air stored in a tank at 27°C and (*a*) 15 bars, and (*b*) 30 bars, that is needed to theoretically have the same work capability. The state of the environment is 27°C and 1 bar.

8-11M A tank of air at 12 bars and 227°C has a volume of 0.80 m^3. The air is cooled by heat transfer until the temperature is 27°C. The surroundings are at 1 bar and 27°C. Calculate the optimum useful work associated with the change in state of the fluid, in kilojoules.

8-12M Steam is contained in a piston-cylinder device. Before expansion the state is 10 bars, 280°C, and 0.010 m^3. After expansion the pressure and volume are 1.5 bars and 0.060 m^3. The heat transfer during the process is -0.80 kJ. The surroundings are at 20°C and 1 bar. Calculate (*a*) the actual work, in kilojoules, and (*b*) the optimum useful work for the process, in kilojoules.

8-13M A 5-kg block of aluminum at 300°C is brought into thermal contact with a 10-kg block of copper initially at -50°C. Contact is maintained until thermal equilibrium is reached. The process is adiabatic, and the specific heats of aluminum and copper may be taken to be 0.99 kJ/(kg)(°C) and 0.38 kJ/(kg)(°C), respectively. Determine the change in availability (*a*) of the aluminum block, (*b*) of the copper block, and (*c*) for the overall process, all in kilojoules. $T_0 = 27$°C.

8-14M Determine the availability of a unit mass of an ideal gas in a closed system at temperature T (different from T_0 of the surroundings), but at a pressure P which is the same as P_0 of the surroundings. By using your knowledge of property relations of ideal gases, express the answer in terms of T_0, P_0, T, and any required constants of the gas.

8-15M Determine the availability of a unit mass of an ideal gas in a closed system at temperature T_0 which is the same as that of the surroundings, but at a pressure P which is different from P_0 of the surroundings. By using your knowledge of property relations of ideal gases, express your answer in terms of T_0, P_0, P, and any required constants of the gas.

Available energy for closed systems

8-16M A constant-volume tank contains saturated water vapor at 90°C. Heat is added to the water until the pressure in the tank reaches (*a*) 1.5 bars, and (*b*) 1 bar. For a sink temperature of 10°C, compute how much of the heat added is available energy.

8-17M A kilogram of air at 1.8 bars is cooled at constant volume from 450 to 300°K. All the heat which leaves the system appears in the surroundings at (*a*) 12°C, and (*b*) 27°C. Determine how much of the heat removed is available energy, in kilojoules. Draw a Ts diagram for the process, and label the areas which represent available and unavailable energy.

8-18M Ten kilograms of air at 1 bar and 350°K are in a closed, rigid tank. Heat is transferred to the air from a thermal-energy reservoir that has a constant temperature of 500°K. The ambient temperature is 300°*K*.

(*a*) How much heat, in kilojoules, can be transferred to the air from the energy source?

(*b*) How much of the heat transferred to the air is available energy?

8-19M Air at 1 bar is cooled at constant pressure from 440 to 300°K, with all the heat given up by the system appearing in the environment, which has a temperature of 290°K. Determine how much of the heat removed is available energy for the closed system, in kJ/kg.

8-20M Steam at 3 bars and (*a*) 240°C, and (*b*) 280°C is allowed to cool at constant pressure in a closed system until it reaches thermal equilibrium with the surroundings at 20°C. Compute the loss in available energy due to the irreversible heat transfer, in kJ/kg.

8-21M Heat for a heat engine is available at a constant temperature of 1427°C, but actually is transferred to the working medium of the engine at 500°K. The lowest available environmental temperature is 300°K. What fraction of the heat supplied becomes unavailable because the heat transferred is received by the working fluid at 500°K rather than 1427°C.

Availability in open systems

8-22M Air enters a steady-flow turbine at 3 bars and 480°K and exhausts at 1 bar and 380°K. The process is adiabatic and the surroundings are at 1 bar and 20°C. Compute (*a*) the actual work output, and (*b*) the optimum shaft work output, in kJ/kg.

8-23M Steam enters a turbine at 40 bars and 400°C and expands to 1 bar and 100°C, in a steady-flow, adiabatic process. The ambient conditions are 1 bar and 27°C. Disregard changes in kinetic and potential energy. Find (*a*) the actual work delivered, and (*b*) the optimum shaft work, in kJ/kg.

8-24M Air enters a compressor in steady flow at 1.4 bars, 17°C, and 70 m/s. It leaves the adiabatic device at 4.2 bars, 147°C, and 110 m/s. Determine (*a*) the actual work input, and (*b*) the optimum work required, in kJ/kg, if $T_0 = 7°C$ and $P_0 = 1$ bar.

8-25M Refrigerant 12 is compressed in steady flow from a saturated vapor at −10°C to a final state of 8 bars and 50°C. For the adiabatic process determine (*a*) the actual work required, and (*b*) the minimum work required, in kJ/kg, if $T_0 = 20°C$ and $P_0 = 1$ bar.

8-26M Steam enters a turbine at 100 bars and 560°C at a rate of 50,000 kg/h. Partway through the turbine, 25 percent of the flow is bled off at 20 bars and 440°C. The rest of the steam leaves the turbine at 0.10 bar as a saturated vapor. Determine (*a*) the availability at the three states of interest, in kJ/kg, (*b*) the maximum power output possible, in kilowatts, and (*c*) the actual power output, in kilowatts, if the flow is adiabatic. The environment is at 1 bar and 20°C.

8-27M A hydrocarbon oil is to be cooled in a heat exchanger from 440 to 320°K by exchanging heat with water which enters the exchanger at 20°C at a rate of 3000 kg/h. The oil flows at a rate of 750 kg/h, and has an average specific heat of 2.30 kJ/(kg)(°C). Compute the change in flow availability, in kJ/h, for (*a*) the hydrocarbon oil stream, and (*b*) the water stream. (*c*) Find the loss in availability, in kJ/h, for the overall process if $T_0 = 17°C$.

8-28M Refrigerant 12 with a mass flow rate of 5 kg/min enters a condenser at 14 bars, 80°C, and leaves at a state of 52°C, 13.8 bars. Determine the loss in availability, in kJ/min, if the coolant in the condenser is (*a*) water which enters at 12°C and 7 bars and leaves at 24°C and 7 bars, and (*b*) air which experiences a 9°C temperature rise from 18°C at a constant pressure of 1.1 bars. $T_0 = 15°C$.

8-29M An open feedwater heater operates at 7 bars. Compressed liquid water at 35°C enters at one section, while superheated vapor enters at another section. The fluids mix and leave the heater as a saturated liquid. Determine the change in stream availability, in kJ/min, if the flow rate of compressed liquid is 1000 kg/min and the mass flow rate of superheated vapor is (*a*) 4520 kg/min, and (*b*) 4370 kg/min. $T_0 = 20°C$ and $P_0 = 1$ bar.

Heat reservoirs and availability changes

8-30M A piston-cylinder expands air from 6 bars, 77°C, and 0.060 m^3 to 3.5 bars and 0.150 m^3. During the process, 65 kJ of heat are added to the air from a source at 600°K. The atmosphere is at 1 bar and 300°K. Determine the optimum useful work associated with the process, in kilojoules.

8-31M Refrigerant 12 is contained in a rigid tank initially at 2 bars, a quality of 50.5 percent, and a volume of 0.10 m^3. Heat is added from a reservoir at 100°C until the pressure reaches 5 bars. Determine (*a*) the quantity of heat added, in kilojoules, and (*b*) the optimum useful work associated with the overall process, in kilojoules. $T_0 = 24°C$.

8-32M Carbon dioxide is contained in a frictionless piston-cylinder device at an initial condition of 2 bars and 17°C. Heat is added from a reservoir at 700°K to the 0.88 kg of gas until the volume is doubled. On the basis of data in Table A-9M determine (*a*) the quantity of heat supplied, in kilojoules, and (*b*) the optimum useful work associated with the overall process, in kilojoules, if the process is constant pressure. $T_0 = 290°K$ and $P_0 = 1$ bar.

Irreversibility in closed and open systems

8-33M Reconsider Prob. 8-7M. Determine the irreversibility of the process, in kJ/kg.

8-34M Reconsider Prob. 8-11M. Calculate the irreversibility of the process, in kilojoules.

8-35M Reconsider Prob. 8-12M. Determine the irreversibility of the process, in kilojoules.

8-36M Reconsider Prob. 8-30M. Determine the irreversibility of the process, in kilojoules.

8-37M Reconsider Prob. 8-22M. Calculate the irreversibility of the process, in kJ/kg.

8-38M Reconsider Prob. 8-23M. Determine the irreversibility of the process, in kJ/kg.

8-39M Reconsider Prob. 8-24M. Determine the irreversibility of the process, in kJ/kg.

8-40M Saturated liquid refrigerant 12 enters an expansion valve at 6 bars and leaves at 2 bars. Determine the irreversibility of the process, in kJ/kg, if (*a*) it is adiabatic, and (*b*) the fluid receives 4 kJ/kg of heat from the atmosphere. Atmospheric conditions are 1 bar and 27°C.

8-41M Saturated water vapor at 30 bars is throttled to 7 bars. The atmospheric temperature is 12°C. Calculate the irreversibility of the process, in kJ/kg.

8-42M Reconsider Prob. 8-27M. Determine the irreversibility of the overall process, in kJ/h.

8-43M Reconsider Prob. 8-28M. Determine the irreversibility of the overall process, in kJ/min.

8-44M One-fourth kilogram of an ideal gas initially at 1.4 bars and 25°C is in an insulated tank. An impeller within the tank is turned by an external motor until the pressure is 1.8 bars. Determine the irreversibility of the process if the gas is (*a*) nitrogen, (*b*) hydrogen, and (*c*) carbon dioxide. The atmosphere is at 0.96 bar and 22°C.

PROBLEMS (USCS)

Availability for closed systems

8-1 A tank with a volume of 10 ft^3 contains air at (*a*) 100 psia and 70°F, and (*b*) 100 psia and 300°F. The surrounding atmosphere is at 14.5 psia and 70°F. Determine the availability of the air, in Btu.

8-2 The air in Prob. 8-1 undergoes a free expansion until its volume is (1) doubled, and (2) tripled. Determine the change in the closed-system availability for the process, in Btu, for an initial state of (*a*) 100 psia and 70°F, and (*b*) 100 psia and 300°F.

8-3 Determine the availability associated with 50 lb of (*a*) ice, and (*b*) water at 32°F and 1 atm if the surrounding atmosphere is at 1 atm and 60°F, in Btu.

8-4 Fifty pounds of water at 40°F and 14.6 psia are allowed to mix in a closed system with 30 lb of water at 160°F and 14.6 psia. Determine the optimum useful work associated with the change of state of (*a*) the 50 lb of water, and (*b*) the 30 lb of water. (*c*) Find the change in availability for the overall adiabatic process, in Btu. $T_0 = 70$°F and $P_0 = 14.5$ psia.

8-5 Determine the availability of steam in a closed system at 1000 psia and 800°F, in Btu/lb. The environment is at 14.7 psia and 70°F.

8-6 The steam in Prob. 8-5 is cooled in a rigid tank until the pressure drops to 500 psia. Determine the optimum useful work associated with the change in state of the fluid, in Btu/lb.

8-7 Air initially at 15 psia and 90°F is contained in a well-insulated tank. An impeller inside the tank is turned by an external mechanism until the pressure is 18 psia. Determine (*a*) the actual work required, and (*b*) the optimum useful work associated with the change in state, both in Btu/lb. Let $T_0 = 90$°F and $P_0 = 15$ psia.

8-8 A piston-cylinder device contains 0.40 lb of air at 15 psia and 70°F. Determine the minimum work input, in Btu, required to compress the air to 60 psia and 250°F. $T_0 = 70$°F and $P_0 = 15$ psia.

8-9 Refrigerant 12 is compressed from a saturated vapor at -10°F to a final pressure of 100 psia. The process is adiabatic and the compressor efficiency is 78.5 percent. Determine (*a*) the actual work required, in Btu/lb, and (*b*) the actual outlet temperature, in degrees Fahrenheit. (*c*) Find the minimum work required, in Btu/lb, for the actual final state found in part *b*. The process occurs in a piston-cylinder device. $T_0 = 60$°F and $P_0 = 1$ atm.

8-10 A storage battery is capable of delivering 1 kWh of energy. Determine the volume of air stored in a tank at 80°F and (*a*) 500 psia, and (*b*) 750 psia, that is needed to theoretically have the same work capability. The state of the environment is 80°F and 1 atm.

8-11 A tank of air at 200 psia and 360°F has a volume of 30 ft^3. The air is cooled by heat transfer until the temperature is 80°F. The surroundings are at 1 atm and 80°F. Calculate the optimum useful work associated with the change in state of the fluid, in Btu.

8-12 Steam is contained in a piston-cylinder device. Before expansion the state is 160 psia, 500°F, and 0.10 ft^3. After expansion the pressure and volume are 20 psia and 0.60 ft^3. The heat transfer during the process is -0.80 Btu. The surroundings are at 80°F and 14.7 psia. Calculate (*a*) the actual work, in Btu, and (*b*) the optimum useful work for the process, in Btu.

8-13 A 5-lb block of aluminum at 250°F is brought into thermal contact with a 10-lb block of copper initially at 30°F. Contact is maintained until thermal equilibrium is reached. The process is adiabatic, and the average specific heats of aluminum and copper may be taken to be 0.225 Btu/(lb)(°F) and 0.092 Btu/(lb)(°F), respectively. Determine the change in availability (*a*) of the aluminum block, (*b*) the copper block, and (*c*) for the overall process, all in Btu. $T_0 = 80°F$.

8-14 Determine the availability of a unit mass of an ideal gas in a closed system at temperature T (different from T_0 of the surroundings), but at a pressure P which is the same as P_0 of the surroundings. By using your knowledge of property relations of ideal gases, express the answer in terms of T_0, P_0, T, and any required constants of the gas.

8-15 Determine the availability of a unit mass of an ideal gas in a closed system at temperature T_0 which is the same as that of the surroundings, but at a pressure P which is different from P_0 of the surroundings. By using your knowledge of property relations of ideal gases, express your answer in terms of T_0, P_0, P, and any required constants of the gas.

Available energy for closed systems

8-16 A constant-volume tank contains dry, saturated water vapor at 200°F. Heat is added to the water until the pressure in the tank reaches 20 psia. For a sink temperature of 50°F, compute how much of the heat added is available energy.

8-17 A pound of air at 25 psia is cooled at constant volume from 340 to 40°F. All the heat which leaves the system appears in the surroundings at 20°F. Determine how much of the heat removed is available energy, in Btu. Draw a Ts diagram for the process, and label the areas which represent available and unavailable energy.

8-18 Ten pounds of air at 15 psia and 600°R are in a closed, rigid tank. Heat is transferred to the air from an energy reservoir that has a constant temperature of 900°R. The ambient temperature is 500°R.

(*a*) How much heat can be transferred to this air from the energy source?

(*b*) How much of the heat transferred to the air is available energy?

8-19 Air at 15 psia is cooled at constant pressure from 340 to 80°F, with all the heat given up by the system appearing in the environment, which has a temperature of 60°F. Determine how much of the heat removed is available energy for the closed system, in Btu/lb.

8-20 Steam at 40 psia and (*a*) 500°F, and (*b*) 400°F is allowed to cool at constant pressure in a closed system until it reaches thermal equilibrium with the surroundings at 70°F. Compute the loss in available energy due to the irreversible heat transfer, in Btu/lb.

8-21 Heat for a heat engine is available at a constant temperature 2140°F, but actually is transferred to the working medium of the engine at 440°F. The lowest available environmental temperature is 80°F. What fraction of the heat supplied becomes unavailable because the heat transferred is received by the working fluid at 440°F rather than 2140°F?

Availability in open systems

8-22 Air enters a steady-flow turbine at 45 psia and 400°F and exhausts at 15 psia and 200°F. The process is adiabatic and the surroundings are at 14.7 psia and 70°F. Compute (*a*) the actual work output, and (*b*) the optimum shaft work output, in Btu/lb.

8-23 Steam enters a turbine at 400 psia and 700°F and expands to 14.7 psia and 250°F, in a steady-flow, adiabatic process. The ambient conditions are 14.7 psia and 80°F. Disregard changes in kinetic and potential energy. Find (*a*) the actual work delivered, and (*b*) the optimum shaft work, in Btu/lb.

8-24 Air enters a compressor in steady flow at 20 psia, 50°F, and 200 ft/s. It leaves the adiabatic device at 50 psia, 260°F, and 350 ft/s. Determine (*a*) the actual work input, and (*b*) the optimum work required, in Btu/lb, if $T_0 = 40°F$ and $P_0 = 1$ atm.

8-25 Refrigerant 12 is compressed in steady flow from a saturated vapor at −10°F to a final state of 100 psia and 120°F. For the adiabatic process determine (*a*) the actual work required, and (*b*) the minimum work required, in Btu/lb, if $T_0 = 60°F$ and $P_0 = 1$ atm.

8-26 Steam enters a turbine at 1000 psia and 1100°F at a rate of 100,000 lb/h. Partway through the turbine, 25 percent of the flow is bled off at 300 psia and 800°F. The rest of the steam leaves the turbine at 1 psia as saturated vapor. Determine (*a*) the availability at the three states of interest, in Btu/lb, (*b*) the maximum power output possible, in horsepower, and (*c*) the actual power output, in horsepower, if the flow is adiabatic. The environment is at 14.7 psia and 70°F.

8-27 A hydrocarbon oil is to be cooled in a heat exchanger from 260 to 120°F by changing heat with water which enters the exchanger at 50°F at a rate of 4000 lb/h. The oil flows at a rate of 1500 lb/h, and has an average specific heat of 0.55 Btu/(lb)(°F). Compute the change in flow availability, in Btu/h, for (*a*) the hydrocarbon oil stream, and (*b*) the water stream. (*c*) Find the loss in availability, in Btu/h, for the overall process if $T_0 = 60$°F.

8-28 Refrigerant 12 with a mass flow rate of 10 lb/min enters a condenser at 200 psia, 180°F, and leaves at a state of 120°F, 190 psia. Determine the loss in availability, in Btu/min, if the coolant in the condenser is (*a*) water which enters at 55°F and 90 psia and leaves at 75°F and 90 psia, and (*b*) air which experiences a 15°F temperature rise from 60°F at a constant pressure of 1.1 atm. $T_0 = 60$°F.

8-29 An open feedwater heater operates at 100 psia. Compressed liquid water at 100°F enters at one section, while superheated vapor enters at another section. The fluids mix and leave the heater as a saturated liquid. Determine the change in stream availability, in Btu/min, if the flow rate of compressed liquid is 1000 lb/min and the mass flow rate of superheated vapor is (*a*) 4470 lb/min, and (*b*) 4690 lb/min. $T_0 = 60$°F.

Heat reservoirs and availability changes

8-30 A piston-cylinder device expands air from 100 psia, 140°F, and 2.0 ft^3 to 60 psia and 5.0 ft^3. During the process, 75 Btu of heat are added to the air from a source at 1000°R. The atmosphere is at 1 atm and 500°R. Determine the optimum useful work associated with the process, in Btu.

8-31 Refrigerant 12 is contained in a rigid tank initially at 30 psia, a quality of 46.8 percent, and a volume of 3.0 ft^3. Heat is added from a reservoir at 300°F until the pressure reaches 80 psia. Determine (*a*) the quantity of heat added, in Btu, and (*b*) the optimum useful work associated with the overall process, in Btu. $T_0 = 70$°F.

8-32 Carbon dioxide is contained in a frictionless piston-cylinder device at an initial condition of 20 psia and 40°F. Heat is added from a reservoir at 740°F to the 0.88 lb of gas until the volume is doubled. On the basis of data in Table A-9 determine (*a*) the quantity of heat supplied, in Btu, and (*b*) the optimum useful work associated with the overall process, in Btu, if the process is constant pressure. $T_0 = 60$°F and $P_0 = 14.7$ psia.

Irreversibility in closed and open systems

8-33 Reconsider Prob. 8-7. Determine the irreversibility of the process, in Btu/lb.

8-34 Reconsider Prob. 8-11. Calculate the irreversibility of the process, in Btu.

8-35 Reconsider Prob. 8-12. Determine the irreversibility of the process, in Btu.

8-36 Reconsider Prob. 8-30. Determine the irreversibility of the process, in Btu.

8-37 Reconsider Prob. 8-22. Calculate the irreversibility of the process, in Btu/lb.

8-38 Reconsider Prob. 8-23. Determine the irreversibility of the process, in Btu/lb.

8-39 Reconsider Prob. 8-24. Determine the irreversibility of the process, in Btu/lb.

8-40 Saturated liquid refrigerant 12 enters an expansion valve at 100 psia and leaves at 30 psia. Determine the irreversibility of the process, in Btu/lb, if (*a*) it is adiabatic, and (*b*) the fluid receives 4 Btu/lb of heat from the atmosphere. Atmospheric conditions are 14.7 psia and 70°F.

8-41 Saturated water vapor at 400 psia is throttled to 100 psia. The atmospheric temperature is 50°F. Calculate the irreversibility of the process, in Btu/lb.

8-42 Reconsider Prob. 8-27. Determine the irreversibility of the overall process, in Btu/h.

8-43 Reconsider Prob. 8-28. Determine the irreversibility of the overall process, in Btu/min.

CHAPTER
NINE

A STATISTICAL VIEWPOINT OF ENTROPY AND THE SECOND LAW

In the last half of the nineteenth century, through the work of Maxwell, Clausius, Boltzmann, and others, the concept of the entropy function was developed in terms of microsopic parameters of a macroscopic system. The modern development of the subject, begun by Planck, Einstein, and others at the beginning of the twentieth century, requires the use of quantum mechanics and various techniques of statistical analysis. Since a statistical analysis of the behavior of a large number of particles is involved, this microscopic approach to thermodynamics is called statistical thermodynamics. It is based on its own set of postulates, which are independent of those of classical or macroscopic thermodynamics.

9-1 QUANTIZATION OF ENERGY

Before 1900, two major concepts regarding the nature of matter and energy had been established. One of these was the particle nature of matter. A quantity of mass was conceived of as being composed of atoms and molecules, and each particle in theory was a separate entity and identifiable. Also, the wave nature of electromagnetic radiation was well entrenched and in common usage. In the early 1900s, the concepts of energy and matter underwent a profound change. The work of Planck, Einstein, and others proposed that energy was particle-like. Light energy was considered to be absorbed or emitted in discrete quantities called quanta, or photons. In 1901, Planck derived an equation of the spectral distribution of blackbody radiation associated with electromagnetic radiation. He assumed that the energies corresponding to the various standing electromagnetic waves could take on only discrete values. These values were integral multiples of

$h\nu$, where ν is the frequency of the wave and h is constant known as Planck's constant. In 1905, Einstein employed the concept of light quanta introduced by Planck to explain the anomalous photoelectric effect. It had been discovered earlier that electrons were ejected from solids when radiation fell upon them. It was also known that the maximum energy of the ejected electrons was solely a function of the frequency of the incident radiation. Einstein assumed that radiation consisted of quanta of size $h\nu$, which could be absorbed by the electrons. A portion of this energy was expended as the electron escaped from the solid. Thus the photoelectric effect was explained by acknowledging the discrete nature of radiant energy.

Then, in the 1920s, de Broglie suggested that matter, especially electrons, might exhibit wave characteristics. The de Broglie equation relating the momentum p of a particle to its characteristic wavelength λ is $p = h/\lambda$. In 1927, the experimental work of Davisson and Germer showed that electrons are diffracted when they are passed through a crystal lattice. Hence these small particles exhibit a characteristic (diffraction) usually associated with wave phenomena. These and other experimental results have demonstrated the duality of nature. Both matter and energy, under various circumstances, exhibit the properties either of waves or of discrete quantities.

An example of the discreteness of energy values is provided by the emission of light from substances such as sodium vapor when they are heated in a gas flame. The measured wavelength of light emitted from a sodium source is 0.5890 μm. Electrons orbiting the nucleus have been thermally excited. When an excited electron "falls" or "decays" from the excited state to the normal or "ground" state of energy, the energy difference between these states of the electron is emitted as light. The radiant energy is always emitted at the same wavelength, and it is assumed to be emitted in the form of a single photon with an energy $h\nu$. This energy difference is given by the Einstein-Planck equation.

$$\epsilon_1 - \epsilon_0 = h\nu$$

where ϵ_1 = energy of excited state
ϵ_0 = ground state of energy
$h = 6.625 \times 10^{-34}\,\text{J} \cdot \text{s}$

For sodium emission at 0.5890 μm, this energy-level difference is 2.10 eV.

One of the important tasks of wave, or quantum, mechanics is to relate mathematically the wave nature of matter to the discrete energy values that are associated with individual particles or with large groups of particles. Schrödinger in the mid-1920s suggested an equation which represents the wave nature of matter in terms of a set of standing waves. The Schrödinger wave equation is taken as a basic postulate of quantum theory. Solution of this equation for the various energy modes leads to a set of quantum, or energy, levels for each mode. Recall that an energy mode is simply one of the physically distinguishable means by which a particle can have energy. The discrete or allowed energy states of particles are determined by (1) the nature of the particles, and (2) the circumstances of the system. By circumstances of the system we mean such constraints as the volume of the system or the strength of the applied magnetic field. For the present we shall simply accept the notion that energy is discrete, without any further examination of its quantitative aspects.

9-2 THE ENTROPY FUNCTION

A group of particles that constitutes an isolated thermodynamic system has an overall energy U. This energy is shared by the individual particles. However, each particle is not required to have the same energy ϵ. The results of quantum mechanics indicate that a whole series of energy values must be considered. Moreover, only certain discrete values of energy are allowed or need to be regarded. These discrete values are set by certain constraints on the system. For example, it is found that the translational (linear) kinetic energies of gas particles are inversely proportional to the volume to the two-thirds power. It is common practice to use the term energy level when referring to one of the allowed energy values. The series of allowed energy levels begins at some minimum value called the *ground level* (or zero level) of energy. The series then progresses through distinct steps to larger and larger energy values. Such a series of energy levels exists for each energy mode. For example, one would need to consider for a diatomic gas at low pressures at least those series of energy levels for the translational, rotational, and vibrational modes.

The term "energy level" allows one to present the concept of discrete energy values very simply. A typical diagram for energy levels is shown in Fig. 9-1. The symbols $\epsilon_0, \epsilon_1, \epsilon_2, \ldots$ represent the different energies per particle that a particle may have when occupying the ground level, the first level, etc. By occupying a level it is meant that a particle is considered to have the energy value associated with that particular level. In Fig. 9-1 the vertical distance of any line above the base line is proportional to the energy difference between that level and the ground level. The convenience of a diagram of this type as an aid in picturing the complex and chaotic nature of a thermodynamic system on a microscopic scale will soon become apparent. The actual values of the discrete energy levels for the possible energy modes are those given by solutions of the Schrödinger wave equation. The spacing between levels in Fig. 9-1 is shown to increase as the

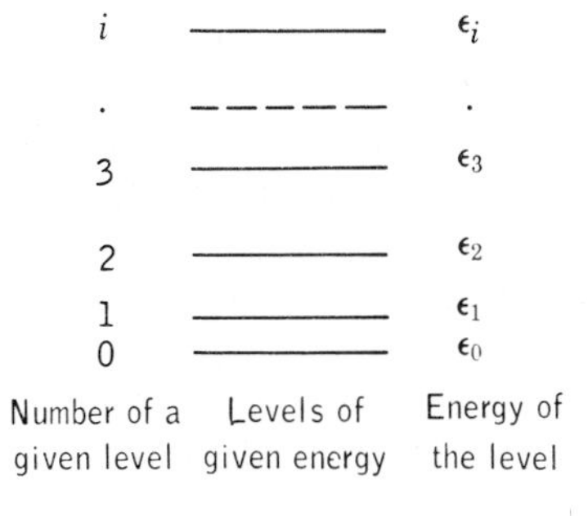

Number of a given level	Levels of given energy	Energy of the level
i	———	ϵ_i
.	– – – – –	.
3	———	ϵ_3
2	———	ϵ_2
1	———	ϵ_1
0	———	ϵ_0

Figure 9-1 A set of discrete energy levels.

energy ϵ_1 increases. This is in agreement with the analytical results obtained for an energy mode such as rotation. An energy mode such as vibration has equally spaced energy levels.

The microscopic picture of matter is one of an endless array of interactions or encounters between the various particles of the system. In the gas phase the particles move through the vessel undergoing encounters with other particles at the rate of, roughly, $10^{33}\ s^{-1}/cm^3$ under normal atmospheric pressure and temperature conditions. As a result of these encounters energy is transferred between the interacting bodies. The mechanism of the energy transfer is not presently of concern. What is of vital interest is that, at each new instant of time, the microscopic representation of any isolated system has changed. The exchange of energy by molecular encounters is restrained only by the conditions that the resultant energy values associated with the various energy modes of the particles afterward satisfy the values allowed by quantum restrictions and that the total energy be conserved. Thus any system, internally in equilibrium, has a microscopic structure which leads to definite macroscopic properties such as pressure, temperature, specific volume, specific internal energy, etc. However, a host of different descriptions on a particle basis will exist over a given time interval. A major task to be resolved is to ascertain what connection exists, if any, between the multitude of microscopic descriptions which are possible and the values we assign to properties based on macroscopic observations.

Consider an isolated system which contains n particles and has a total energy U. If the particles are considered independent of one another, then at any given time the energy of the ith molecule is given by ϵ_i. The total internal-energy requirements of the system must be fulfilled by the relationship

$$U = \sum_i n_i \epsilon_i \tag{9-1}$$

where n_i represents the number of molecules whose energy is ϵ_i. The values of ϵ_i are, of course, restricted to those values allowed by quantum theory. The summation is taken over all the levels which may be expected to contain one or more particles. (The largest value of ϵ_i in the summation obviously never can be greater than the value of U; that is, one particle has associated with it all the energy of the system.) It has been noted already that the energy ϵ_i possessed by a particle is relative to some ground-level energy ϵ_0. For simplicity in the following examples, the energy (quantum) levels will be regarded as equally spaced and the value of the ground-level energy will be taken as zero. First an isolated system will be chosen which contains four particles and a total energy of 6 units. Only one energy mode is considered, and the quantum levels for this energy mode are spaced 1 unit apart, beginning at zero. The following question naturally arises: In what manner may particles be distributed among the various energy levels and still satisfy the condition that the total energy be 6 units? Obviously, only the first seven allowed energy levels need be taken into account, since higher-valued ones could not possibly be occupied. A summary of the possible distributions of particles among energy levels for the chosen example is given in Table 9-1. The

Table 9-1 The nine possible macrostates of four particles having a total energy of 6 units, for equally spaced energy levels

Energy level	Distribution								
	A	*B*	*C*	*D*	*E*	*F*	*G*	*H*	*I*
ϵ_0	3	2	2	1	1	2			1
ϵ_1		1		2	1		3	2	
ϵ_2			1		1			2	3
ϵ_3					1	2	1		
ϵ_4			1	1					
ϵ_5		1							
ϵ_6	1								
W	4	12	12	12	24	6	4	6	4

nine possibilities appear in the vertical columns labeled *A* to *I*. The values in the horizontal rows give the number of particles of energy ϵ_1 for each distribution. The student may wish to verify that the vertical columns represent the only possible distributions.

We shall define a *macrostate* as any possible microscopic state of an assembly of particles described in terms of the number of particles in each energy level at a given instant of time. The description of a macrostate does not require enumeration of which particles have a given energy, but only an accounting of how many particles have a given energy. In Table 9-1 the nine distributions labeled *A* to *I* will now be identified as the nine possible macrostates for a system composed of four particles and having a total energy of 6 units. Each of these macrostates satisfies the macroscopic description of the state of the system.

Further analysis of Table 9-1 leads to the concept of a microstate if we assume that the particles are distinguishable. Consider any one of the nine macrostates, e.g., column *A*. Here three particles have a ground level of energy ϵ_0 and the remaining one is in the sixth quantum level of energy ϵ_6. Because of the random exchange of energy, the particle having 6 units of energy could be any one of the four particles in the system. Over a sufficient period of time macrostate *A* will occur a number of times. However, each of the four distinguishable particles will share in occupying the sixth level of energy for this particular macrostate. Thus the macrostate *A* really consists of four microstates all of which are equally likely to occur over a sufficiently long period of time. A *microstate*, then, is a description of which particles have certain energies. Of course, we cannot specify which microstate exists at a given instant of time, or even which macrostate. Microstates do exist, however, and they are continually changing with time. The other distributions, *B* to *I*, in Table 9-1 likewise may be formed in a number of ways, depending upon the placement of certain particles in various energy levels. The last horizontal row in the table, marked *W*, lists the number of microstates for each macrostate. We shall define the number of microstates *W* for a given macrostate as the *thermodynamic probability* of that macrostate. Unlike the usual mathematical probability which varies from zero to unity, the thermodynamic probability is never less than unity and is usually an extremely large number of thermodynamic systems. The value of W in the case of Table 9-1 is obtained from a permutation formula

$$W = \frac{n!}{n_1!n_2! \cdots n_k!}$$

This equation gives the number of arrangements (microstates) possible for n items where the identity of each item as residing in a particular group is recognized. However, the order of items within a group is immaterial. The symbols n_1, $n_2, \ldots$ represent the number of items in groups 1, 2, etc. As an example, four particles are placed in three groups containing 2, 1, and 1 particles, respectively. For this situation the above equation gives for the possible arrangements that

$$W = \frac{4!}{2!1!1!} = 12$$

Hence distributions B, C, and D all have 12 possible internal arrangements, or microstates. The other values of W in Table 9-1 are easily verified. It is seen from Table 9-1 that the total number of microstates is 84. Since all the microstates will be assumed to occur with equal likelihood, over a long period of time each microstate would be formed $\frac{1}{84}$ of the time.

In actual thermodynamic systems, the accounting of macrostates and microstates is even more complicated than the simple examples discussed above. First of all, quantum theory indicates that frequently one must consider energy states that are distinct, but of equal or nearly equal energy. When g_i quantum states of the same energy ϵ_i exist, the system of particles is said to be *degenerate*. We then speak of g_i degenerate states within an energy level. The values of g_i may be different for the various energy levels. The degeneracy of a system of particles does not alter the number of macrostates, but greatly increases the number of microstates. For example, in the case of distinguishable particles the number of microstates per energy level increases by the factor $g_i^{n_i}$. Therefore the total possible arrangements for given values of n_i and g_i for distinguishable particles becomes

$$W = n! \prod \frac{g_i^{n_i}}{n_i!}$$

A second important factor is the assumption of distinguishability of the particles. Distinguishability was assumed to provide a model that permitted an easy explanation of macrostates and microstates. Modern quantum statistics is based on indistinguishable particles. Therefore the equation above for W is incorrect for this latter model, and other equations for W must be developed. Nevertheless, regardless of the model, the concept of macrostates and microstates is a pertinent one for the development of a new themodynamic property.

Consider the following degenerate system of two indistinguishable particles which shares a total of 4 units of energy. The energy spacing again is 1 unit, the ground-state value is zero, and the degeneracies of the levels, starting from the ground level, are 2, 1, 2, 1, and 2. Figure 9-2 summarizes the macrostates and microstates for these data. The macrostates *a*, *b*, and *c* contain 4, 1, and 3 microstates, respectively. Note that interchange of the two particles does not constitute a new microstate, since the particles are indistinguishable. Each macrostate has a specified number of particles per energy level, but there is no further required specification as to which degenerate state within a level has particles associated with it.

For a system of *n* particles and total energy *U*, there is a finite fixed number of macrostates for each set of allowed energy levels. Since the number of macrostates is fixed in each case, the number of microstates is also set. The total number of microstates for a given thermodynamic state is represented by W_{tot}. This simply is the sum of the *W* values for each possible macrostate. In our first example W_{tot} would be 84, and in the second example the number would be 8. The number of microstates W_{tot} thus is a unique and identifiable number (at least in theory) for every equilibrium state. If W_{tot} has a certain fixed value for each equilibrium state, it partially fulfills the concept of a property in at least one respect. There is one and only one value for each equilibrium state. If a system is carried through a cyclic process, the value of W_{tot} will necessarily be the same for the initial and final states of the cycle.

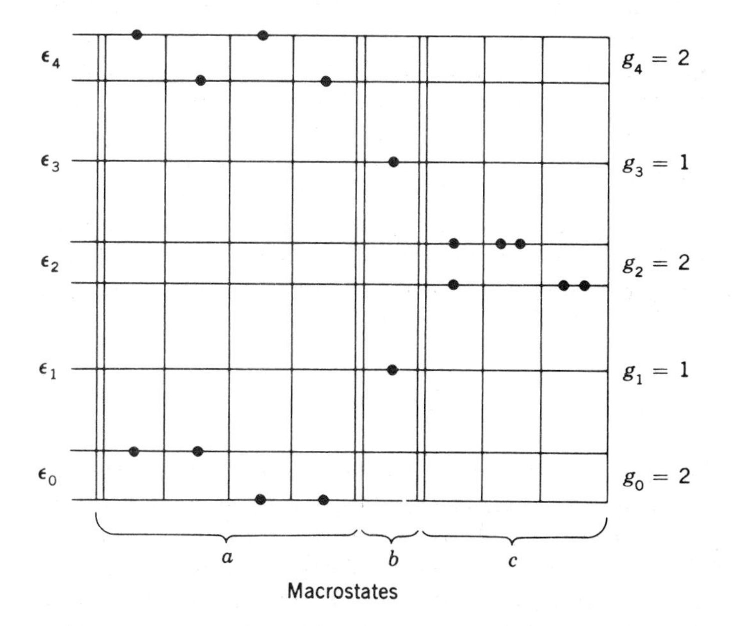

Figure 9-2 Enumeration of macrostates and microstates for a degenerate system of indistinguishable particles.

Although the value of W_{tot} is fixed for a given equilibrium state, it is not a useful measure of a thermodynamic property. If any new property is to be conveniently tabulated and correlated with other properties, its value should vary in direct proportion to its mass; i.e., it should be extensive in nature. The thermodynamic probability W is not an extensive variable. The theory of combinations and permutations states that the total arrangements of two independent groups taken together are the product of the arrangements of each individual group. In terms of thermodynamic probability, if system A can be arranged in 3 ways and system B in 2 ways, there are 6 possible arrangements that might occur when both systems are considered simultaneously. For two thermodynamic systems considered simultaneoulsy, $W_{AB} = W_A W_B$. Hence the property W is multiplicative in nature.

Although W itself is multiplicative, there exists a mathematical function of W such that one of its basic characteristics is its additivity. In fact, the only function of W which has this feature is a logarithmic function. We have already seen that for a composite system the value of W is

$$W_{AB} = W_A W_B$$

By taking the logarithm of each side, we find that

$$\log W_{AB} = \log W_A + \log W_B$$

Therefore we shall define a new thermodynamic property in terms of the thermodynamic probability of the system. This property is called the entropy of the system, and is designated by the symbol S. In general, S is related to W_{tot} by the definition,

$$S = k \ln W_{tot} \tag{9-2}$$

where k is a proportionality constant. It is necessary to choose the logarithm to the base e in order to make the numerical values of the entropy function agree with values calculated from macroscopic measurements. This equation is known as the Boltzmann-Planck equation. Its general validity has been well established. The constant k, which is called the Boltzmann constant, is important for two reasons: It determines both the magnitude and the units for the entropy function, since the $\ln W_{tot}$ term is a pure number. Quantitatively, it is given by

$$k = \frac{R_u}{N_A}$$

where R_u, again, is the universal gas constant, and N_A is Avogadro's number. In the metric system of units the value of k is 1.380×10^{-23} J/(°K)(molecule). This value is also listed in Table A-1M, for reference.

The development of this new thermodynamic function, which we have called the entropy of a system, may be made by two approaches on the microscopic level. The method of statistical mechanics based on the Boltzmann-Planck equation is well accepted in the modern literature. Recently, the work of Tribus, based on earlier papers by Shannon and Jaynes, has shown that an equivalent approach to the entropy function in thermodynamics is possible by extending the methods of information theory. This latter theory has been used primarily in the field of communications. In this text the statistical development of the entropy function will be based on the older of these methods, statistical mechanics. The reader may wish to refer to the work of Tribus for the parallel development of statistical concepts.

9-3 THE SECOND LAW OF THERMODYNAMICS

On the basis of microsopic concepts a new macroscopic property of matter, the entropy, has been introduced. It has been defined in terms of the number of microscopic arrangements, the thermodynamics probability W, of an isolated system of given energy U and volume V which contains n particles. There is a certain value of W_{tot} for each equilibrium state of a macroscopic system, and therefore a unique value of S associated with each equilibrium state. As any system undergoes a change of state, it generally is expected that W_{tot} will change and therefore that the system will have a new value for its entropy. There is also the possibility, however, that W_{tot} may not change in some cases. Hence the chance that the entropy will not be altered by some changes of state should not be ruled out. This would simply mean that the total number of arrangements did not change from one equilibrium state to another.

The reason for introducing this new macrosopic property is that it is intimately connected with a basic characteristic of nature. It is a matter of experience that there is a directional quality about physical phenomena. For example, if two masses at different temperature are brought into thermal contact (and they are insulated from the environment), it is observed that they always approach a common final temperature. The conservation of energy principle requires that the energy given up by one mass must equal that energy taken on by the other mass, but it does not require that the final temperatures be the same. In the study of chemical reactions it is found that many reactions do not proceed to completion under certain experimental conditions, such as a specified temperature and pressure of the system. For example, consider the reaction of CO and H_2O to form CO_2 and H_2 at 1500°K and 1 atm. The theoretical equation shows that $CO + H_2O \rightarrow CO_2 + H_2$. That is, for every mole of CO and H_2O introduced into a reaction vessel, theoretically 1 mole each of CO_2 and H_2 should be formed. However, the yield of these products is considerably less than this at the stated conditions. The conservation of energy principle places no restriction on the yield. As a third example, consider the mixing of two different gases by breaking a partition which initially separates them. After a sufficient period of time the gases are found to be uniformly mixed, within experimental accuracy. The spontaneous separation of the two gases, such that at a later time one gas is detected wholly on one side of the vessel and the other gas solely on the other side, is not observed. The example may appear trivial, for our experience tells us this separation process will not occur. However, it is important to be able to explain our experiences on the basis of basic laws. There certainly is no first-law restriction on the unmixing process.

As a final example, take the case of cyclic devices which are used to convert heat released from various sources into work output. Internal-combustion engines and nuclear steam-power stations, for example, fall into this general classification. Although great strides have been made in the design of such devices by engineers, the complete conversion of energy from fuels to work output has not been even closely approached for practical engines. (The conversion seldom approaches 40 percent.) Is this the fault of the designer or is there some theoretical reason for the observed limitations? The first law, on the other hand, permits the complete conversion of thermal energy into work by devices such as these. The above limited list of examples could be extended greatly, each example being illustrative of a process which satisfies the conservation of energy principle but whose final state is restricted. This restriction is either in a directional sense or a quantitative (extent of process) sense. Although more examples could be taken from widely diversified fields, it is found that a single postulate of thermodynamics is sufficient to explain all the observed results.

Before a postulate known as the second law of thermodynamics is presented, it may be helpful if we examine in a qualitative manner several examples of physical phenomena which illustrate the general principle. It must be kept in mind, however, that these examples are not proof of the validity of the statement to be made later. Laws are generalizations based on limited experimental evidence. Their validity is generally based on the ability of the laws, and corollaries of the laws, to successfully predict future behavior. The three examples below, then, are presented simply as common illustrations of behavior which lead to the general postulate. This postulate, the second law of thermodynamics, will be expressed in terms of the behavior of an isolated system. In each case we shall study an isolated system which changes state due to the removal of constraint. If the constraint is removed, various macroscopic (and microscopic) parameters change in value until a new equilibrium state is reached. We shall find that the final equilibrium state of the isolated system is intimately connected to the behavior of W_{tot} (and hence S) during the change of state.

First consider the free expansion of a gas from one-half of an insulated, rigid vessel into the other half of the vessel, which is initially evacuated. If the inside of the entire vessel walls is chosen as the boundary of the system, then the system is an isolated one, since both Q and W are zero and no mass crosses the boundaries. All changes of state occur within the vessel, and the environment is unaffected. On the basis of the conservation of energy principle, the energy of the gas does not change during the expansion process, (when the internal partition is removed). However the values of the permitted energy levels for the particles do change. As noted earlier, for a simple substance the energy levels for the translational kinetic energy of gases are inversely proportional to the volume to the two-thirds power. When the gas fills the entire vessel, the energy spacings $\Delta\epsilon$ are somewhat smaller than before. Thus we have the situation of a system of n particles with fixed energy U where the particles now redistribute themselves over allowed energy levels which are more closely spaced. It should be apparent that in this situation the value of W_{tot} must increase. That is, there are many more ways to distribute a given amount of energy among a given number of particles if the energy spacings decrease. For example, recall the data from Table 9-1, where four distinguishable particles having a total energy of 6 units were distributed over energy levels 1 unit apart. Nine macrostates resulted. If the energy levels become $\frac{1}{2}$ unit apart (due to a volume change), 30 macrostates would be possible. The important result is that the change in the total number of arrangements, and hence the entropy change, during a free expansion is always *positive*.

As a second example, consider the sliding of a block down an inclined plane. To eliminate the consideration of a change in the kinetic energy of the block as it slides down the plane, we shall assume that the solid-solid friction between block and plane is sufficient so that the block moves very slowly. After a finite time interval the block comes to rest at the bottom of the plane. To further simplify the analysis, we may assume that the heating effect at the solid-solid interface affects only the block and plane. That is, no energy in the form of heat is transferred to the environment. Our isolated system then is simply the block and plane. The conservation of energy principle for this isolated system reduces merely to $\Delta E = 0$, or $\Delta U + \Delta \text{PE} = 0$. The decrease in the potential energy of the block appears as an increase in the internal energy of both the block and the plane. Assuming that the particles of a solid act independently of one another and that the allowed energies of the particles are not altered by the process, the number of microscopic arrangements will increase for a solid as its energy increases. Recall again the data for Table 9-1. A system of four distinguishable particles with a total energy of 6 units has 9 macrostates and 84 microstates (for equally spaced energy levels). If the total energy is decreased to 4 units and all other data are the same, then the number of macrostates is reduced to 5, and the number of microstates is reduced from 84 to 35. A significant increase in microstates occurs, then, with an increase in energy. Therefore the value of W_{tot} for both the block and the plane increases during the frictional process. The end result is that the presence of solid-solid friction always increases the total entropy of those parts of an isolated system affected by the process.

As a final example, consider an isolated system consisting of two subsystems, X and Y. Initially they are isolated from each other by means of a rigid, thermally insulating partition. The partition acts as a constraint on the energy of each subsystem. If the total energy of the composite isolated system is $U_C = U_X + U_Y$, then the energy of X must remain constant, and likewise for subsystem Y. Associated with each subsystem are thermodynamic probabilities W_X and W_Y, and for the composite isolated system, $W_C = W_X W_Y$. The two subsystems are now brought into thermal contact through the rigid partition. Since the volume of each is fixed, no boundary work is possible. However, an energy exchange in the form of heat is permitted. Now the total energy U_C is shared by the two subsystems. On a microscopic basis there is no real limitation to the manner in which the energy is shared between X and Y. As the energy of each subsystem varies, the value for W for each subsystem changes.

Table 9-2 Microstates for two systems in thermal contact

U_x	U_Y	W_X	W_Y	W_C	P_i
0	8	1	45	45	0.015
1	7	4	36	144	0.048
2	6	10	28	280	0.093
3	5	20	21	420	0.140
4	4	35	15	525	0.175
5	3	56	10	560	0.186
6	2	84	6	504	0.168
7	1	120	3	360	0.120
8	0	165	1	165	0.055
				3003	1.000

As an illustrative example, let $\epsilon_0 = 0$ in both subsystems, with $\Delta\epsilon = 1$. Since the volumes of X and Y are constant, the energy-level values do not change for gases contained within the subsystems. Subsystem X will contain four distinguishable particles, subsystem Y will contain three distinguishable particles, and the total energy shared by the two subsystems will be 8 units. Table 9-2 summarizes the various possible microscopic descriptions, where each level is nondegenerate.

No matter what the original split in the total energy had been, the final value of 3003 for W_{tot} would have been much larger than the initial value once the constraint between the two subsystems was removed. Although the illustration involves distinguishable particles and small numbers, the same qualitative result would occur for indistinguishable particles and large degenerate systems. Apparently the removal of a thermal constraint, so that an isolated system changes to a new equilibrium state, always leads to an increase in W_{tot} and hence to an increase in the entropy of the isolated system.

(The preceding example simply illustrates the point that the removal of a thermal constraint brings about an increase in the thermodynamics probability of an isolated system. However, even though the total energy may now be divided between the two subsystems in nine different ways, we would not expect to find the energy of either subsystem to vary significantly at equilibrium if the subsystems are relatively large. That is, at macroscopic equilibrium all the thermodynamic properties, including each subsystem's energy, should be constant on the basis of experimental measurements. Therefore many of the possible splits in energy are never seen, even though they are theoretically possible. We shall use the data from Table 9-2 in a following section to illustrate another important point regarding the macroscopic equilibrium state.)

The preceding discussion can be summarized in the following way. If a constraint of some type is removed from an isolated system initially in equilibrium, the number of internal arrangements W_{tot} will either increase or remain the same. That is, $W_{\text{tot, final}} \geqslant W_{\text{tot, initial}}$. Although all our examples were ones during which W_t increased, the possibility that the thermodynamic probability of an isolated system will not change must be included. Experience simply dictates that the number of arrangements never decreases. Therefore we shall postulate the following statement as the *second law of thermodynamics.*

> The entropy of an isolated system always either increases or remains the same when the system changes from one equilibrium state to another.

This statement frequently is referred to as the principle of increase of entropy. The second law may be expressed mathematically for any differential change of state by

$$dS_{\text{isol system}} \geqslant 0 \tag{9-3a}$$

For a finite process it must be true that

$$\Delta S_{\text{isol system}} \geqslant 0 \tag{9-3b}$$

We must infer from these two equations not only that the overall change in entropy must be positive (or zero), but also that every differential step of the overall process must have an entropy change that is positive (or zero).

The general statement that the entropy of an isolated system always either increases or remains the same leads to the concept of reversible and irreversible processes. Consider first an isolated system which undergoes a change of state during which its entropy increases. This is permissible in the light of the second law. It is then proposed to reverse the direction of the original forward process in order to return the isolated system to its initial state. Since the entropy is a property of a system, the reverse process requires that the entropy of the isolated system decrease. But the second law states that the entropy cannot decrease if the system is isolated. Consequently the reverse process is impossible, and the original forward process is called *irreversible.* Any process is irreversible if the sum of the entropy changes for all subsystems involved in the process is positive.

There are some processes for which the entropy change of an isolated system will be zero. As a result, the entropy change would be zero for the reversed process. This latter reversed process does not violate the second law. Thus there are processes involving isolated systems which may proceed in either direction, since ΔS remains zero at all times. Such processes are called *reversible.* The entropy function then is a test for reversibility. Reversibility is an extremely important concept to the engineer, since the degree of irreversibility in a process determines, in part, its performance. The role of reversible and irreversible processes in engineering design will be amply illustrated throughout this text.

One of the most important features of the second law is its formulation in terms of reversible and irreversible processes. Another important feature of the second law is the recognition that it is a *nonconservation* law. Many of the basic laws studied in physics are conservation laws, and include the familiar ones involving mass, energy, momentum, and charge. A conservation principle is extremely useful in any field, since it provides a method of accounting for various changes which occur in a system. The second law, on the other hand, is the antithesis of a conservation principle. It states that in all those processes which might be termed irreversible, the entropy of the isolated system always increases. The decrease in entropy for any type of change of an isolated system is not possible. Hence entropy is not a conserved property. It is this characteristic that makes the second law a *directional* or *limiting* law. That is, it establishes a criterion for the direction toward or limit to which all real processes can proceed. If the entropy of an isolated system cannot decrease during any process, a large number of equilibrium states are excluded from the possibility of being the final equilibrium state. This may help to explain why certain phenomena are observed in nature, whereas other effects never occur. The second law of thermodynamics, expressed here in terms of an increase in entropy principle, has a broad application in providing guidelines to the understanding of natural phenomena and the design of engineering systems.

One final characteristic of the second law should be noted. Even though the entropy of an isolated system increases when a process is irreversible, it cannot increase indefinitely. The system must finally reach a new equilibrium state that is consistent with the constraints on the system. Therefore it may also be stated that the entropy of an isolated system always tends toward a maximum value, and when it reaches that maximum value, the system is characterized by a new set of equilibrium properties. The entropy function thus can be used as a criterion for equilibrium in an isolated system. The entropy increases as a process progresses, but in the limit as equilibrium is approached dS likewise approaches zero. Mathematically, $dS = 0$ is the criterion for equilibrium for an isolated system.

In summary, we have defined a new extensive property, the entropy, by the Boltzmann-Planck equation, $S = k \ln W_{\text{tot}}$. The constant k is Boltzmann's constant, and $k = R_u/N_A$, where N_A is Avogadro's number. The entropy is a measure of the number of microscopic arrangements in which a system could be found, consistent with the constraints on the system. The second law of thermodynamics then postulates that the entropy of an isolated system either increases or remains the same during a change of state; that is, $dS_{\text{isol}} \geqslant 0$. The inequality sign applies to that class of processes called irreversible, while the equality sign is valid for reversible processes. Thus the second law is a directional law, since it prohibits certain processes from ever occurring. An isolated system reaches a new equilibrium state, after the removal of a constraint, when the entropy reaches a maximum value. In a broad sense, then, the second law of thermodynamics is concerned with three subjects: (1) the direction of change when spontaneous processes occur, (2) the criteria for equilibrium in thermodynamic systems, and (3) the effect of irreversibilities on performance.

At this point a number of questions remain unanswered. What do we really imply in physical terms when we say a process is reversible or irreversible? If the entropy function is so important, how does one evaluate it by purely macroscopic techniques? How do we use the entropy function and the second law in the solution of scientific and engineering problems? Can the second law be applied in a modified form to closed systems, regardless of the environment? It is to these related questions that we must now turn.

9-4 REVERSIBLE AND IRREVERSIBLE PROCESSES

The reader at this point is directed to reread Sec. 6-5, and then follow this with Sec. 9-5.

9-5 MICROSCOPIC IMPLICATIONS OF WORK AND HEAT

Heat and work effects have been investigated extensively in a macroscopic sense in earlier chapters. In such considerations it was not necessary to acknowledge the existence of atomic particles or energy levels. However, it is important for us to examine these interactions on a microscopic basis, especially with regard to internally reversible processes. We shall find that the effects of heat and work interactions on changes at the microscopic level are quite different. In addition, the results of such an examination will be quite helpful in developing an expression for the entropy change of a closed system in terms of macroscopic variables.

In an earlier section it was noted that the discrete values of energy allowed for a given energy mode are set by the value of one of the generalized coordinates (displacements) of the system or by the nature of the substance. As an example, consider the translational kinetic energies of gas particles maintained in a box of volume V. The values of the ϵ_i's are found to be proportional to $1/V^{2/3}$. That is, the volume of the system determines the spacing of the energy levels for this particular form of energy. Other generalized displacements will determine the magnitude of the energy-level spacing for other energy modes. Consequently the energy-level spacing for a certain energy mode will change only if the corresponding generalized displacement is altered. Let us consider a simple system containing a monatomic gas. The volume is slowly reduced by a compression process. As the volume diminishes, the energy levels become farther apart. For each energy level, $\epsilon_i \rightarrow \epsilon_i + d\epsilon_i$, where $d\epsilon_i$ is positive for a compression process. Each level contains n_i particles. Therefore an increment of energy $n_i\, d\epsilon_i$ must be supplied at each level of energy. If the only energy added to all levels is due to boundary work, i.e., if the process is internally reversible, then

$$\delta W_{\text{int rev}} = \sum_i n_i\, d\epsilon_i \tag{9-4}$$

Internal reversibility is required, since any dissipative effects present (such as friction) would negate the statement that the generalized work interaction exactly equaled the change in the energy of the closed system. The above equation is not valid for nonquasistatic types of work interactions. For example, it has been pointed out that paddle-wheel effects, the flow of electric current through a resistor, and work done against frictional forces do not necessarily involve a change in a generalized displacement. Consequently these effects do not contribute to the change in the ϵ_i values of the particles. For simple systems the discrete energy values can be changed only by volume changes. It is also important to note that the n_i values do not change if the only interaction is the internally reversible change of a generalized displacement. The addition of energy to a closed system due to internally reversible work effects is not to be associated with a change in the particle distribution. The energy of each level changes with a constant population of particles for a given macrostate. Figure 9-3*a* illustrates a hypothetical change in volume of a system so that the energy spacing is doubled. Note that the n_i values are constant.

On a microscopic basis we have seen that $U = \sum_i n_i \epsilon_i$. Therefore, for a differential change of state,

$$dU = \sum_i n_i \, d\epsilon_i + \sum_i \epsilon_i \, dn_i$$

According to this equation any change in the energy U must be associated with two effects: (1) the population of each energy level is held fixed while the energy values themselves change, and (2) the values of the energy levels are held fixed while the distribution of particles among the levels changes. In addition, the conservation of energy principle for closed systems states that

$$dU = \delta Q + \delta W \tag{2-30}$$

For an internally reversible change of state, $W = \sum_i n_i \, d\epsilon_i$. Therefore it follows from the above equations that the macroscopic heat effects during an internally reversible process bring about a microscopic change in the state of the system given by

$$\delta Q_{\text{int rev}} = \sum_i \epsilon_i \, dn_i \tag{9-5}$$

Thus the population of each level changes by dn_i when a differential heating effect occurs. Since $\sum_i dn_i = 0$ for a closed system, some of the levels become more highly populated, whereas other levels decrease in their population. Figure 9-3*b* shows the hypotical case of heat addition in the absence of work interactions.

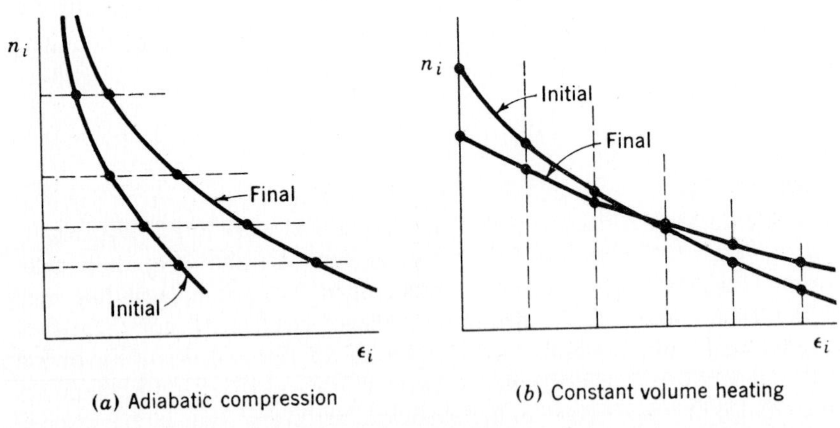

Figure 9-3 Shifting of ε_i or n_i by work or heat effects during an internally reversible process.

Equation (9-5) requires an internally reversible process because we placed this same restriction on Eq. (9-4).

The macroscopic conservation of energy principle for a closed system has been described in terms of microscopic changes within the system. However, such descriptions of δQ and δW are valid only if the system is free of irreversibilities. Note that the absence of irreversibilities refers only to the system. Hence the restriction on Eqs. (9-4) and (9-5) is internal reversibility, not total reversibility.

9-6 ENTROPY AND THE MOST PROBABLE MICROSTATE

The entropy function, as defined by the Boltzmann-Planck equation, is given by the relations $S = k \ln W_{tot}$. In practice, however, it is much more convenient to evaluate S by using a modified form of the Boltzmann-Planck equation. The validity of this modification can be justified by examining one further characteristic of systems on the microscopic level. Recall the examples presented in Sec. 9-2 which illustrated the concepts of macrostates and microstates. The first system considered had four distinguishable particles and a total energy of 6 units, which resulted in nine macrostates (see Table 9-1). With each macrostate is associated a certain number of microstates. All microstates are tacitly assumed to be equally likely in occurrence in a random molecular system. Macrostate E has 24 possible microstates out of the 84 total. This is 28 percent of the total number of arrangements. The macrostate which has the largest number of microstates is defined as the *most probable macrostate*. Hence macrostate E is the most probable macrostate for the given system. Associated with every thermodynamic system is a macrostate which can be called the most probable one.

The most probable macrostate defines the energy distribution which would be observed more frequently than any other distribution. For macroscopic thermodynamic systems it turns out that this distribution is overwhelmingly more probable than any other. Therefore it is the characteristics of the most probable macrostate that determine the equilibrium thermodynamic properties. The most probable distribution is, in a sense, the equilibrium distribution. Although there are innumerable macrostates possible for a given thermodynamic state, experience dictates that all of them can be ignored, except the most probable one, in computing thermodynamic properties.

The importance of the most probable macrostate in computing thermodynamic properties can be illustrated by the following example. Recall from Sec. 9-3 the situation where two subsystems X and Y were brought into thermal contact and allowed to share a total energy of 8 units. Table 9-2 summarized the total number of arrangements (3003 microstates) for each possible division of the energy between the two subsystems. It is seen that the most probable split of energy occurs when subsystem X has 5 units of energy. The probability of occurence p_i is 18.6 percent, since this description has 18.6 percent of the total number of microstates. The probability of occurrence of a situation where subsystem X has no energy is seen to be quite unlikely (1.5 percent), and this certainly fits our intuition. However, note that the probability of subsystem X having 4 or 6 units of energy is almost as probable as the case where X has 5 units, which is the most probable state. Thus fluctuations from the most probable state are quite probable, and this does not fit our expectation. In the final equilibrium state we do not anticipate any property of either subsystem, including the energy, to vary to any measurable extent. The reason this example contradicts our experience is that both subsystems are too small to be considered macroscopic systems.

A more meaningful analysis of the above situation is made when one studies two subsystems each of which has an extremely large number of particles and the total shared energy is large. When n is very large, it should be realized that W for a given subsystem is an extremely rapidly increasing function of its energy. In this case, since the total energy is shared, as subsystem X acquires an increasing amount of energy, the energy of Y decreases. Therefore, W_X increases extremely rapidly while W_Y decreases extremely rapidly as X acquires an increasing share of the total energy. Let us represent the thermodymamic probability of the composite system by W_C. Since $W_C = W_X W_Y$, the product of W_X and W_Y exhibits an extremely sharp maximum at some value of U_X. Therefore a plot of the probability of various microscopic descriptions versus U_X must also show the same sharp maximum. This state of maximum W_C is, of course, the most probable split in energy, and we might symbolize the energy of subsystem X in this state by $U_{X,\text{mp}}$. The general characteristics of a plot of p_i (the probability of a certain split in the shared energy) versus U_X is shown in Fig. 9-4. The region ΔU in the figure where p_i has a significant value is such that $\Delta U \ll U_{X,\text{mp}}$. We see that when two large systems are in thermal equilibrium, they do not exhibit a widely varying energy for either one, in a macroscopic sense. The probability that U_X will be appreciably different from $U_{X,\text{mp}}$ is negligible. On the microscopic level we see

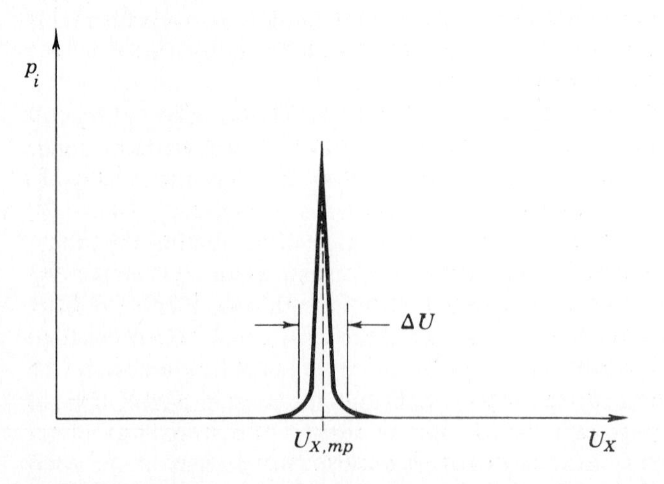

Figure 9-4 Probability of occurrence of subsystem X when its energy is U_X.

that wide variations are possible, but highly improbable. Hence our instruments never detect any fluctuations. The most probable arrangement and those only slightly different from it in their arrangement predominate over all other possibilities.

This preceding discussion was based on two systems which shared a common quantity of energy. The same general conclusions can be drawn for a single isolated system of energy U. Although the energy is fixed, there is a huge number of macrostates that might exist. However, for any large system of particles, only a small fraction of the possible macrostates account for a large fraction of the total number of microstates. In addition, the characteristics of the energy distributions of this small number of macrostates are not very different from the most probable distribution. That is, the probability of occurrence of a macrostate falls off extremely fast for very small changes in the n_i values associated with the most probable macrostate. For example, consider a degenerate system of distinguishable particles. Small changes δn_i are made at various energy levels, starting with the system in the most probable macrostate. The value of W_{mp} is altered to a new value $(W_{\text{mp}} + \delta W)$. A mathematical analysis of this situation leads to the following approximation:

$$\frac{W_{\text{mp}} + \delta W}{W_{\text{mp}}} \approx \exp\left[-\frac{1}{2}\sum_i \frac{(\delta n_i)^2}{n_i}\right]$$

Note that whether the δn_i values are positive or negative, the new macrostate always has fewer arrangements than the most probable one. Hence the equation is consistent with the definition of the most probable macrostate. In most systems n_i will usually be large, such as 10^{15} to 10^{20}. Therefore, even when δn_i is a fraction of a percent, the bracketed term is extremely large. This is due to the fact that the δn_i terms are all squared. Consequently small changes in δn_i lead to a drastic reduction in the number of microstates when starting from the most probable macrostate. Obviously only those macrostates which are nearly like the most probable one in their distribution of particles need to be considered. For this reason it is reasonable to base macroscopic properties solely on the particle distribution of the most probable macrostate.

If the most probable macrostate is a sufficient model for the prediction of thermodynamic properties, how is the concept of the entropy function affected, since it is based on W_{tot}, and not W_{mp}? Actually, since W_{mp} has a fixed value for any equilibrium state, it could serve as the basis for the definition of the entropy function, as well as W_{tot}. However, it is not necessary to modify our definition of S as given by the Boltzmann-Planck equation. As a result of the mathematics used when dealing with systems of a large number of particles, it turns out that for all practical purposes $\ln W_{\text{mp}} = \ln W_{\text{tot}}$. For evaluating the entropy, then, we use the most probable macrostate to find W, rather than account for all macrostates. The reasonableness of this approximation is illustrated by the following example. It is necessary to limit the example to a study of relatively small systems, but the extrapolation to larger systems is quite obvious.

Table 9-3 lists the arrangement of distinguishable particles among three

Table 9-3

n	W_{mp}	W_{tot}	W_{mp}/W_{tot}	$\log_{10} W_{mp}$	$\log_{10} W_{tot}$	Log ratio
62	1.4×10^{14}	4.25×10^{16}	3.3×10^{-3}	14.154	16.623	0.851
310	5.95×10^{75}	1.13×10^{80}	5.25×10^{-5}	75.783	80.053	0.948
1550	4.4×10^{386}	5.7×10^{395}	7.7×10^{-10}	386.642	395.757	0.978

equally spaced energy levels for systems x, y, and z. The specific macrostate shown in each case has been chosen to be the most probable one. The only difference among the systems is that the values of n and U have been varied. (Note that none of the systems is large enough to be considered macroscopic in size.) In each case the average energy per particle is exactly the same. For these three systems we wish to compare the values of W_{mp} to W_{tot} and $\ln W_{mp}$ to $\ln W_{tot}$. The results are summarized as shown. Note that the values of W_{mp} and W_{tot} are extremely large, even for these relatively small systems. As the number of particles increases, the ratio W_{mp}/W_{tot} becomes very small. More importantly, as the number of particles increases, it appears that $\log W_{mp}$ approaches $\log W_{tot}$. (*Note*: logarithms to the base 10 have been used in the table for convenience. The ratio of logarithms would be the same regardless of the base used.) If the example were extrapolated to a system of 10^{23} particles, the ratio of logarithms would be unity for all practical purposes. Although we have used an example involving a system of distinguishable particles with a limited size in all cases, the conclusion reached is applicable to real systems as well. That is, $\ln W_{mp} = \ln W_{tot}$ to any desired accuracy if the system is large. Hence the evaluation of the entropy function by use of the Boltzmann-Planck equation can be made in terms of a modified equation,

$$S = k \ln W_{mp} \tag{9-6}$$

In evaluating the entropy function, we need not bother with the details of the possible macrostates that could possibly exist for a particular equilibrium state. Attention is focused simply on the most probable macrostate.

In summary, for macroscopic thermodynamic systems, the most probable macrostate and those nearly identical to it predominate. Therefore the macroscopic properties of a system can be based entirely on the microscopic characteristics of the most probable macrostate. Calculation of properties based on this approach will agree with experimental measurements. (Exceptions to this rule must be permitted, since in a few cases fluctuations from the mean are significant.) It is of particular interest that the entropy function can be correctly evaluated in terms of the Boltzmann-Planck equation by employing the logarithm of the thermodynamic probability associated with the most probable macrostate.

9-7 ENTROPY CHANGES IN TERMS OF MACROSCOPIC VARIABLES

The entropy function S has been introduced and defined in terms of microscopic concepts. However, our day-to-day experiences, such as our observations and measurements, are macroscopic. Hence it is highly desirable to relate the entropy function to macroscopic phenomena. Recall that in the development of the conservation of energy principle we found it necessary to consider only changes in the internal energy and enthalpy functions. Absolute values of u and h were never required. We shall also find it necessary to consider only changes in the value of the entropy function. For an infinitesimal process the change in the entropy of a closed system is found by differentiating Eq. (9-6). Thus, in general,

$$dS = k\, d(\ln W_{\text{mp}}) \tag{9-7}$$

That is, the entropy change is determined by the change in the number of arrangements of the most probable distribution as the equilibrium state is altered. Our goal is to relate this change in $\ln W_{\text{mp}}$ to macroscopically measurable quantities.

We have already noted in preceding examples of isolated systems of given degeneracy that the thermodynamic probability W for any macrostate depends solely upon the n_i values associated with each energy level. Since the logarithm of W would also depend solely on the n_i values for that macrostate, we may denote functionally that

$$\ln W = f(n_1, n_2, n_3, \cdots, n_i, \cdots)$$

If a differential change of state occurred, the change in $\ln W$ for any macrostate would be given by the total differential,

$$d \ln W = \frac{\partial \ln W}{\partial n_1} dn_1 + \frac{\partial \ln W}{\partial n_2} dn_2 + \cdots = \sum_i \frac{\partial \ln W}{\partial n_i} dn_i$$

This equation is valid for any macrostate, including the most probable macrostate. Thus we may write specifically that

$$d \ln W_{\text{mp}} = \sum_i \frac{\partial \ln W_{\text{mp}}}{\partial n_i} dn_i \tag{9-8}$$

The change in the entropy function for an infinitesimal change of state between two equilibrium states then becomes, if we substitute Eq. (9-8) into Eq. (9-7),

$$dS = k \sum_i \frac{\partial \ln W_{\text{mp}}}{\partial n_i} dn_i \tag{9-9}$$

The evaluation of the quantity $\partial \ln W_{\text{mp}}/\partial n_i$ is fairly straightforward, although there is a considerable amount of unavoidable mathematics. Since the most probable macrostate is, by definition, the one with the largest number of arrangements, we may approach the problem as one requiring the maximization of a function through the techniques of the differential calculus. One additional complication must be considered, however. In addition to the general equation relating the dependent variable, $\ln W$, to the independent variables, the n_i's, there are two more equations relating the n_i's which must be satisfied during the maximization process. These are

$$\sum_i n_i = n \tag{9-10}$$

and
$$\sum_i \epsilon_i n_i = U \tag{9-1}$$

The problem we have, then, is one maximization in the presence of constraining equations.

The details of the maximization process are given in Appendix A-2. The major result is an equation relating $\partial \ln W_{mp}/\partial n_i$ to the value of the energy level where particles have been added or removed. For any energy level where dn_i particles are shifted when the system changes from one equilibrium state to another,

$$\frac{\partial \ln W_{mp}}{\partial n_i} = A + B\epsilon_i \tag{9-11}$$

The two quantities A and B represent two constants, as yet undetermined. This relationship is the one we sought in order to evaluate dS through the use of Eq. (9-9). If we combine Eqs. (9-9) and (9-11), then

$$\begin{aligned} dS &= k \sum_i \frac{\partial \ln W_{mp}}{\partial n_i} dn_i = k \sum_i (A + B\epsilon_i)\, dn_i \\ &= kA \sum_i dn_i + kB \sum_i \epsilon_i\, dn_i \end{aligned} \tag{9-12}$$

The terms A and B have been brought outside the summation signs, since they are constants. The first summation on the right side of Eq. (9-12) is zero, for the following reason. In a system of fixed number of particles,

$$\sum_i n_i = n \tag{9-10}$$

Differentiation of this equation yields

$$\sum_i dn_i = 0$$

Consequently Eq. (9-12) reduces to the form

$$dS = kB \sum_i \epsilon_i\, dn_i \tag{9-13}$$

Equation (9-13) represents the change in entropy due to a differential change of state between two equilibrium states, expressed in terms of the shifting of particles among the energy levels. The equation is completely general in this form. The change of state is not restricted to any particular type of process and may involve both heat and work interactions.

Equation (9-13) contains two quantities that must be replaced by their counterparts in macroscopic thermodynamics, namely, B and $\sum_i \epsilon_i\, dn_i$. The first of these is the constant which appears in the derivation of Eq. (9-11). This constant is evaluated in statistical thermodynamics by employing the constraining equations in a particular fashion. This type of evaluation is mathematically lengthy, and need not concern us here. Instead, a more qualitative and physical interpretation will be given.

Consider once more two simple, closed systems X and Y which are brought into thermal contact through a rigid wall. The volume of each system is constant, and the total energy of the composite system $X + Y$ is constant. The total entropy of the composite system is the sum of the entropies of the two individual systems. If the two systems are not intially in mutual equilibrium, a heat interaction will occur until a new equilibrium state is reached. On the basis of the second law of thermodynamics, the entropy of the composite system will increase until a maximum value is attained. As the maximum is reached, $dS = 0$. During the process the change in the entropy for either system can be expressed in the following way. For any simple system, there are two independent properties. In this case it is convenient to let these two properties be the energy and the volume. Hence, $S = S(U, V)$. For a differential change of state

$$dS = \left(\frac{\partial S}{\partial U}\right)_V dU + \left(\frac{\partial S}{\partial V}\right)_U dV$$

In our particular example the volume is being held fixed on each system, and therefore the second term on the right is zero for each system. Consequently the total entropy change for the composite system C may be written as

$$dS_C = dS_X + dS_Y = \left.\left(\frac{\partial S}{\partial U}\right)_V\right]_X dU_X + \left.\left(\frac{\partial S}{\partial U}\right)_V\right]_Y dU_Y \tag{9-14}$$

As the composite system approaches equilibrium $dS_C = 0$. The variation of the composite entropy with energy U_X is shown in Fig. 9-5. In addition, the total energy of the composite system is a constant given by $U_C = U_X + U_Y$. Thus $dU_X = -dU_Y$. These two facts in combination with Eq. (9-14) indicate that at the state of equilibrium

$$\left.\left(\frac{\partial S}{\partial U}\right)_V\right]_X = \left.\left(\frac{\partial S}{\partial U}\right)_V\right]_Y \tag{9-15}$$

The derivatives expressed in this equation are properties of the systems X and Y. However, the only property of each system that necessarily must be the same at the state of thermal equilibrium is the temperature. Therefore it is to be expected that the partial derivative $(\partial S/\partial U)_V$ is related in some way to the temperature of the system. The dimensions of the partial derivative give a clue to this relationship. From Eq. (9-6) it is seen that the dimensions of S are those of the Boltzmann constant k, which are energy/(absolute temperature). When divided by the dimension of U, which is energy, we see that the dimension of the partial derivative is the reciprocal of the absolute temperature T^{-1}. Hence it is appropriate to define a thermodynamic temperature which is the reciprocal of the derivative found in Eq. (9-15). Consequently,

$$T \equiv \frac{1}{(\partial S/\partial U)_V} \qquad \text{or} \qquad \left(\frac{\partial S}{\partial U}\right)_V = \frac{1}{T} \tag{9-16}$$

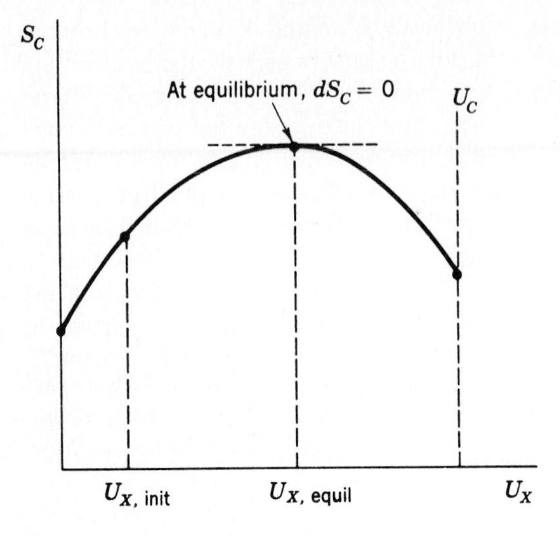

Figure 9-5 Variation of composite entropy with energy shared by systems X and Y.

It can be shown that this thermodynamic temperature is the same as that given by the ideal-gas temperature scale.

As a final step, the partial derivative in Eq. (9-16) can be related to the constant B. Recall that in general the change in U is given by $dU = \sum_i \epsilon_i \, dn_i + \sum_i n_i \, d\epsilon_i$. In our example involving systems X and Y the volume of each is constant. Hence the term $\sum_i n_i \, d\epsilon_i$ is zero in the expression of dU, since a change in ϵ depends upon a change in V. As a result we find that $dU_V = \sum_i \epsilon_i \, dn_i$. If we substitute this result into Eq. (9-13) for dS, then for this specific example

$$dS = kB \, dU \qquad \text{at constant volume}$$

or

$$\left(\frac{\partial S}{\partial U}\right)_V = kB \tag{9-17}$$

If Eqs. (9-16) and (9-17) are now equated, we find that

$$kB = \frac{1}{T} \tag{9-18}$$

Thus the constant B which appeared originally as one of the constants in Eq. (9-11) is inversely proportional to the thermodynamic temperature or ideal-gas temperature, measured either in degrees Rankine or degrees kelvin. Although derived for a special process, Eq. (9-18) is of general validity. Direct substitution of the above relation into Eq. (9-13) leads to the expression

$$dS = \frac{1}{T} \sum_i \epsilon_i \, dn_i \tag{9-19}$$

All that remains now is the replacement of the summation term by an equivalent macroscopic quantity.

In Sec. 9-5 it was demonstrated that a heat interaction affects the number of particles associated with each energy level if the process is internally reversible. That is,

$$\delta Q_{\text{int rev}} = \sum_i \epsilon_i \, dn_i \tag{9-5}$$

If this relation is substituted into Eq. (9-19), we obtain the result that

$$dS = \left(\frac{\delta Q}{T}\right)_{\text{int rev}} \tag{9-20}$$

If a process is carried out in an internally reversible manner, then only a knowledge of the transient heat effects and of the temperature of the system at each stage of the process is required in order to evaluate ΔS. The restriction of internal reversibility is inherent in the derivation and cannot be removed from this particular equation. Although internally reversible processes are limiting ones, Eq. (9-20) could be used to evaluate entropy changes from experimental measurements of heat transfer and temperature if irreversibilities within the system were negligible.

Integration of Eq. (9-20) leads to the relationship, on an intensive basis, that

$$s = \int \left(\frac{\delta q}{T}\right)_{\text{int rev}} + s_0 \tag{9-21}$$

This equation shows that integration of $\delta q/T$ gives the entropy only within an arbitrary constant. As long as we work with pure substances, or nonreacting mixtures, a knowledge of the constant s_0 is unimportant because only a change in the entropy is usually required. Therefore tables and graphs may be constructed for which the base, or reference, value of s is completely arbitrary. For example, in the saturation-temperature tables for water, A-12M and A-12, the specific entropy is chosen to be zero for saturated liquid water at the triple state (0.01°C or 32.02°F). Other reference states are chosen for other substances.

The dimensions of the entropy function are energy/(absolute temperature). The SI units for the specific entropy are kJ/(kg)(°K) or kJ/(kg · mol)(°K), while USCS units commonly are Btu/(lb)(°R) or Btu/(lb · mol)(°R). Other sets of units may also be found in the literature.

9-8 HEAT AND WORK RESERVOIRS

The reader at this point is directed to read Secs. 6-6 and 6-13, and then follow these sections with Sec. 9-9.

9-9 APPLICATION OF THE SECOND LAW TO CLOSED SYSTEMS

Our fundamental statement of the second law of thermodynamics has been made in terms of the behavior of isolated systems. In many cases, however, it is either desirable or convenient to focus our attention on the behavior of a closed system. The restatement of the second law in terms of a closed system may be made in the following manner.

Consider a closed system which, in general, may be affected by work and heat interactions. Any process it undergoes may be either reversible or irreversible. In order to focus attention on the effect of irreversibilities within the closed system, all changes external to the system which affect the state of the system will be made reversible. For example, any energy in the form of work supplied to the system must be from a reversible source, such as a flywheel, pulley-weight assembly, etc. In addition, any heat interaction must occur through an infinitesimal temperature difference and will be supplied from a heat reservoir. Figure 9-6 shows a closed system X at a temperature T undergoing reversible work and heat interactions. There is no restriction on the type of change within the system X.

The composite of the system and other systems with which it interacts is an isolated system to which we may apply the second law. The entropy change dS_C of the composite system is

$$dS_C = dS_X + dS_{\text{heat res}} + dS_{\text{work source}} \geqslant 0$$

Since the temperature difference between the heat reservoir and the closed system is an infinitesimal dT, the entropy change of the reservoir can be expressed in terms of system conditions. That is,

$$dS_{\text{res}} = -\frac{\delta Q}{T}$$

In addition, the entropy change of a reversible work source is zero, as noted in the preceding section. Therefore the inequality stated above reduces to

$$dS_X + \left(-\frac{\delta Q}{T}\right) \geqslant 0$$

or, upon rearrangement,

$$dS_{\text{closed}} \geqslant \left(\frac{\delta Q}{T}\right)_{\text{act}} \tag{9-22}$$

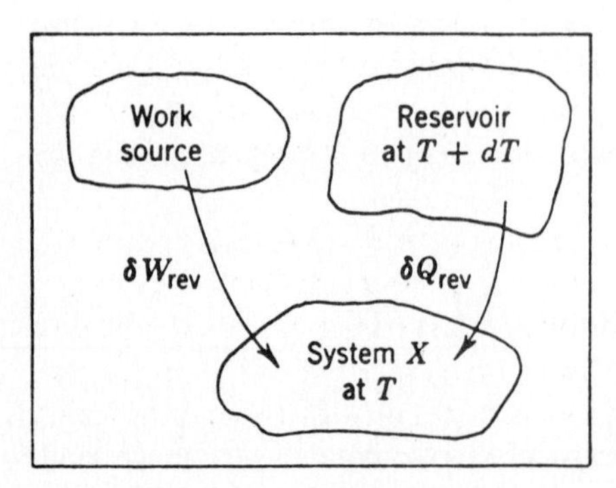

Figure 9-6 A closed system undergoing reversible interactions with its surroundings.

In Eq. (9-22) the quantity δQ represents the actual heat transfer to or from the system. The equality sign applies when the closed-system change is internally reversible, as we have noted previously. The inequality sign is valid if irreversibilities are present within the system.

Equation (9-22) leads to an important relation for any type of closed system. In the absence of heat transfer, the entropy change for any closed system must fulfill the condition that

$$dS_{\text{closed, adia}} \geqslant 0 \tag{9-23}$$

For a finite change of state,

$$\Delta S_{\text{adia}} \geqslant 0 \tag{9-24}$$

The entropy function always increases in the presence of internal irreversibilities for an adiabatic, closed system. In the limiting case of an internally reversible adiabatic process the entropy will remain constant.

Although Eqs. (9-23) and (9-24) are extremely useful in the analysis of engineering processes, there are a number of processes which are not adiabatic. In such situations recall that it is necessary to include in the second-law analysis every part of the surroundings which is affected by changes in the system. In effect, we must look at the total changes, and not just at the system change. For such isolated system changes we have already noted that

$$dS_{\text{isol system}} \geqslant 0 \tag{9-25}$$

For a finite change of state,

$$\Delta S_{\text{isol system}} \geqslant 0 \tag{9-26}$$

To emphasize the fact that for nonadiabatic systems we must take into account all parts of physical space that are being affected by a change in the state of a particular system, Eq. (9-26) may be written as

$$\Delta S_{\text{tot}} = \sum \Delta S_{\text{subsystems}} \geqslant 0 \tag{9-27}$$

That is, the algebraic sum of the entropy changes of *all* the subsystems participating in a process is zero or positive. If all the parts or subsystems of an overall system undergo reversible changes, and all interactions between these parts are reversible, then the total entropy of the overall (isolated) system will remain the same. If irreversibilities occur within any of the parts or are associated with any interaction between two or more of the subsystems, the entropy of the overall system must increase.

Equations (9-23) to (9-27) are the basic governing rules stemming from the second law of thermodynamics. Although other properties and further theorems may be developed which are more useful in certain types of engineering analyses, we begin with the entropy function and the "increase in entropy principles" stated above. Before proceeding, there are several important comments which must be made with respect to these principles.

1. The increase in entropy principles are directional statements. They limit the direction in which a process can proceed. A decrease in entropy is not possible for a closed system which is adiabatic, or for a composite of systems which interact among themselves.
2. The entropy function is a nonconserved property, and the increase in entropy principles are nonconservation laws. Only in the case of the reversible processes is entropy conserved. Reversible processes do not create entropy.
3. The second law states that all systems, left to themselves, are capable of reaching a state of equilibrium. The law now requires, in addition, that the entropy continually increase as that state of equilibrium is approached. Mathematically, the state of equilibrium must be attained when the entropy function reaches its maximum possible value, consistent with the constraints on the system. (Since a maximization process is involved, the techniques of the differential calculus are useful in evaluating the final equilibrium state.)
4. The increase in entropy principles are intimately connected with the concept of irreversibility. It is the presence of irreversibilities that leads to the increase in entropy. We shall find that the greater the magnitude of the irreversibilities, the greater the overall entropy change. Hence, the entropy function can be used quantitatively by the engineer to measure dissipative effects or losses associated with any type of process. It may also be used to establish performance criteria.

The last item above is extremely important with respect to the analysis of engineering systems. To conserve or make better use of energy supplies, it is necessary to have some measure of current or predicted performance. The entropy function, or other properties based on this function, is extremely valuable in providing guidelines in this direction.

9-10 HEAT ENGINE CONCEPTS

From an engineering viewpoint the interrelationships of heat and work as required by the first and second laws are extremely important in the analysis of work-producing devices. Thermal energy (heat) is available from a number of sources including fossil fuels, nuclear fission, and solar radiation. The conversion of heat into a more useful form of energy, work, is accomplished by a class of devices known as heat engines. A heat engine is a device which operates continuously or cyclically and produces work while exchanging heat across its boundaries. The restriction to continuous or cyclic operation implies that the substance within the device is returned to its initial state at regular intervals. For example, in a steam-power plant the steam is contained within a closed loop. It is heated, expanded, cooled, and compressed as it circulates around the loop in a continuous manner, returning to its initial state at the same point in the loop. One example of a cyclic heat engine is a piston-cylinder assembly. As the piston undergoes a cyclic change in position, the fluid contained within the cylinder undergoes heat and work interactions with its surroundings.

The second law places a severe restriction on the operation of a heat engine. A corollary to the second law states that

> It is impossible to operate in any manner a cyclic device that will produce work while exchanging heat with a single heat reservoir.

This corollary is known as the *Kelvin-Planck* statement of the second law. The proof of this statement follows directly from the increase in entropy principle for an isolated system. Three factors need to be checked in terms of entropy changes. Since the engine is cyclic in nature, it will undergo no net entropy change. The production of work in itself involves no creation of entropy if the energy equivalent to the work effect is stored in the surroundings in a reversible manner, as in a work reservoir. The third effect is the heat transfer from a heat reservoir to the cyclic heat engine. The entropy change of the reservoir is simply Q/T, and must be negative, since Q is negative with respect to the reservoir. Consequently the total entropy change of the cyclic device and its surroundings is negative. This is a violation of the second law as expressed by Eq. (9-3). Therefore the Kelvin-Planck statement is a direct consequence of the second-law statement given previously.

Machines which would operate in violation of the Kelvin-Planck statement are known as perpetual-motion machines of the second kind (PMM2). The existence of such machines would be a tremendous help to the world. Work could be derived from the enormous low-temperature heat sources, such as the oceans, in practically unlimited quantities. Unfortunately, such an operation is impossible. Note that neither the first nor second law restricts the inverse operation, that is, the complete conversion of work into heat.

On the basis of the Kelvin-Planck statement, all cyclic heat engines must reject a portion of the heat received to another heat reservoir. A schematic of a heat engine operating between reservoirs of temperatures T_H and T_L is shown in Fig. 9-7. Basically, the heat engine operates under these conditions:

1. Heat Q_H is supplied to the heat engine from a heat reservoir at a temperature T_H (high temperature).
2. The engine produces a net quantity of work W during the cyclic process.
3. Heat Q_L is rejected from the engine during the cycle to a heat reservoir at a temperature T_L (low temperature), where $T_L < T_H$.

For the heat engine as the closed system of interest, the conservation of energy principle is

$$\sum Q + \sum W = \Delta U$$

For a cyclic device, ΔU is zero. Thus for the engine shown in Fig. 9-7 we see

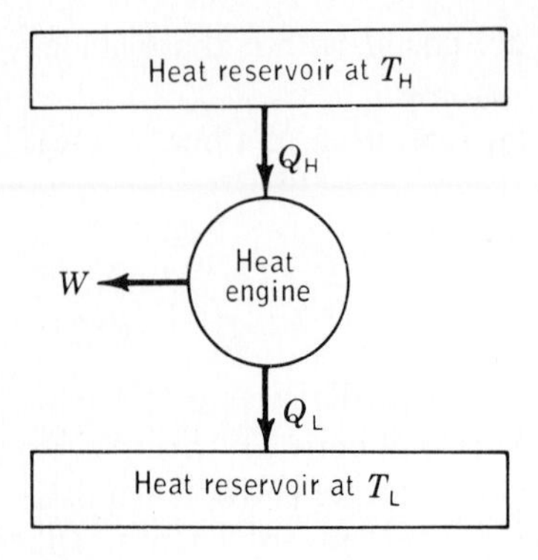

Figure 9-7 Schematic of a heat engine operating between two reservoirs at T_H and T_L.

that

$$|Q_H| - |Q_L| = W_{net}$$

where W_{net} is set as positive for heat engines. Since the heat added and the heat removed from the heat engine have different signs by convention, it is convenient to use absolute-value signs in the energy equation and insert a negative sign directly into the equation for the Q_L term. The object of a heat engine is to produce work from the energy added as heat. A measure of its performance is the ratio of work output to heat input. This ratio is defined as the thermal efficiency η_{th} of the engine. That is,

$$\eta_{th} = \frac{W_{net}}{Q_{in}} = \frac{W_{net}}{Q_H} \tag{9-28}$$

If we supply 100 units of energy to a heat engine and find that 70 units are rejected, then the net work output is 30 units and the thermal efficiency is 0.30 (or 30 percent).

We now wish to establish the maximum value of the thermal efficiency that any heat engine could achieve. If we apply the first law to a heat engine for a differential cycle, then

$$\delta Q_H - \delta Q_L - \delta W_{net} = 0 \tag{a}$$

Since δQ_L is out of the system, it is preceded by a negative sign in Eq. (*a*). When numerical values are substituted into this equation, only absolute values should be used, since the sign convention on the heat quantities has already been included. In addition, the total entropy change for the cyclic process is

$$dS_{tot} = dS_H + dS_L + dS_{engine} \tag{b}$$

The entropy change of the heat engine is zero because it is a cyclic device. The values of dS_H and dS_L for the heat reservoirs are $-\delta Q_H/T_H$ and $\delta Q_L/T_L$, respectively. The first term is negative since heat leaves the high-temperature reservoir. Consequently, Eq. (*b*) may be written as

$$dS_{tot} = -\frac{\delta Q_H}{T_H} + \frac{\delta Q_L}{T_L} \tag{c}$$

Substitution of the term δQ_L from Eq. (*a*) into Eq. (*c*) yields

$$dS_{tot} = -\frac{\delta Q_H}{T_H} + \frac{\delta Q_H}{T_L} - \frac{\delta W}{T_L} \tag{d}$$

Upon rearrangement, the thermal efficiency is found to be given by

$$\eta_{th} \equiv \frac{\delta W}{\delta Q_H} = \frac{T_H - T_L}{T_H} - \frac{T_L\, dS_{tot}}{\delta Q_H} \tag{9-29}$$

The first term on the right-hand side always lies between 0 and 1, and it is solely a function of the temperatures of the heat reservoirs. The second term on the right is always positive or zero, since the second law requires that $dS_{tot} \geqslant 0$. Hence the effect of the last term always is to reduce the thermal efficiency or in the limit to leave it unchanged for a reversible process. Equation (9-29) indicates that the maximum thermal efficiency is attained when dS_{tot} is zero. Under this condition Eq. (*c*) requires that $-\delta Q_H/T_H + \delta Q_L/T_L = 0$. For a finite totally reversible cycle this becomes

$$\frac{|Q_H|}{T_H} = \frac{|Q_L|}{T_L} \tag{9-30}$$

The maximum efficiency of a heat engine operating between two heat reservoirs is achieved when the heat quantities and reservoir temperatures satisfy the above relation. Under these circumstances the thermal efficiency is called the Carnot efficiency, and it is given simply by

$$\eta_{Carnot} = \frac{T_H - T_L}{T_H} \tag{9-31}$$

Equation (9-31) holds for any totally reversible heat engine operating between two reservoirs of fixed temperature. The temperature values must be expressed, of course, in either degrees kelvin or degrees Rankine.

The efficiency of any heat engine operating between two constant-temperature reservoirs is increased by either employing supply reservoirs of very high temperature or rejecting to heat reservoirs of extremely low temperature. The latter condition is difficult to accomplish since temperatures below atmospheric conditions must be artificially acquired. The use of heat from high-temperature sources is often restricted by the availability of materials that can withstand such extreme conditions. Much work is being carried on at the present time in the field of materials science to develop high-temperature-resistant solids, so that the theoretical thermal efficiencies of operating devices can be substantially increased. The maximum efficiency of unity is reached when T_H is infinite or T_L is absolute zero. It is important to recognize that, for reasonable values of T_H and T_L (for example, 1000 and 300°K, respectively, for continuously operating engines), the thermal efficiency never approached unity. When one considers that practical engines may have a heat supply the temperature of which varies considerably below the maximum value and that mechanical inefficiencies must also be included, it is not too surprising that modern heat engines have overall efficiencies of only 25 to 40 percent. It is essential to understand, however, that even under optimum theoretical conditions, a heat engine is relatively inefficient in converting heat into work by means of a cyclic process. This severe limitation has led to a renewed interest in work-producing devices which are not limited to the Carnot efficiency, such as fuel cells. This latter device has other inherent advantages, including its adaptability to unique environments and silent operation.

It has been demonstrated that the maximum efficiency of any heat engine operating between two fixed heat reservoirs occurs when the entropy change of the universe during the cyclic process is zero. From this we inferred earlier that the entire cycle must be carried out in a reversible manner. Several important consequences of this may be shown. These are usually known as the Carnot corollaries, or the *Carnot principle.* Normally, at least two parts, or statements, are included:

1. No heat engine is more efficient than a reversible engine which operates between the same heat reservoirs.
2. All reversible heat engines which operate between the same heat reservoirs will have the same thermal efficiency.

The first point has already been proved by use of Eq. (9-29). A more efficient engine would require that dS of the universe be negative, in violation of the second law. The point can also be shown in the following way. Equation (c) for a finite cycle becomes

$$\Delta S_{\text{tot}} = -\frac{Q_{\text{H}}}{T_{\text{H}}} + \frac{Q_{\text{L}}}{T_{\text{L}}} \geqslant 0$$

This can be rewritten in the form

$$\frac{Q_{\text{L}}}{Q_{\text{H}}} \geqslant \frac{T_{\text{L}}}{T_{\text{H}}}$$

where the inequality sign applies to an irreversible cycle. It is also generally true for an irreversible cycle and a reversible cycle that

$$\eta_{\text{irrev}} = 1 - \frac{Q_{\text{L}}}{Q_{\text{H}}}$$

and

$$\eta_{\text{Carnot}} = 1 - \frac{T_{\text{L}}}{T_{\text{H}}}$$

Combination of these two efficiency expressions leads to

$$\eta_{\text{irrev}} + \frac{Q_{\text{L}}}{Q_{\text{H}}} = \eta_{\text{Carnot}} + \frac{T_{\text{L}}}{T_{\text{H}}}$$

But it was noted above that for an irreversible cycle $Q_{\text{L}}/Q_{\text{H}} > T_{\text{L}}/T_{\text{H}}$. Therefore, as a general statement for heat engines

$$\eta_{\text{irrev}} < \eta_{\text{Carnot}} \tag{9-32}$$

The inequality in the second law requires that an irreversible engine be less efficient than a reversible engine when operating between the same heat reservoirs. The second part of Carnot's principle is self-apparent from Eq. (9-31).

One final point, which often is included in the Carnot principle, needs to be emphasized. The proof that a reversible engine has the maximum efficiency was accomplished without reference to the nature of the working medium or the operation of the engine itself. Whether the reversible engine operates on hydrogen gas, air, water, or mercury is of no consequence. Likewise, the sequence of processes which make up the cycle is not important.

(At this point, if the reader has used Chap. 9 as an introduction to the concept of entropy and the second law instead of Chap. 6, then one should follow Chap. 9 by Secs. 6-15 through 6-16 and Chap. 7.)

REFERENCES

Fast, J. D.: "Entropy," McGraw-Hill, New York, 1963.

Hatsopoulos, G. N., and J. H. Keenan: "General Principles of Thermodynamics," Wiley, New York, 1968.

Howerton, M. T.: "Engineering Thermodynamics," Van Nostrand, Princeton, N.J., 1962.

Lee, J. F., F. W. Sears, and D. L. Turcotte: "Statistical Thermodynamics," Addison-Wesley, Reading, Mass., 1963.

Reif, F.: "Fundamentals of Statistical and Thermal Physics," McGraw-Hill, New York, 1965.

Sonntag, R. E., and G. J. Van Wylen: "Fundamentals of Statistical Thermodynamics," Wiley, New York 1966.

Tribus, M.: "Thermostatics and Thermodynamics," Van Nostrand, Princeton, N.J., 1961.

PROBLEMS

9-1 What is the frequency of a photon of radiation which has an energy of 1 eV?

9-2 The frequency of a photon of radiation is $0.1 \times 10^{15}\ \text{s}^{-1}$. What is the energy associated with this photon in (*a*) joules, and (*b*) electronvolts?

Particle distribution

9-3 Determine the number of ways that seven distinguishable particles can be arranged into three groups so that the groups contain (*a*) 3, 2, and 2 particles, and (*b*) 4, 2, and 1 particle(s).

9-4 Determine the number of ways that eight distinguishable particles can be arranged into three groups so that the groups contain (*a*) 3, 3, and 2 particles, and (*b*) 4, 2, and 2 particles.

9-5 Four distinguishable balls are dropped at random into two boxes. After repeated tests, what fraction of the time will we expect to find that two balls are in each box if (*a*) the boxes are indistinguishable, and (*b*) the boxes are distinguishable?

9-6 Five distinguishable balls are dropped at random into three boxes. After repeated tests, what fraction of the time will we expect to find (*a*) that the distribution of balls is 3 : 1 : 1 if the boxes are indistinguishable, and (*b*) that 3 balls are in the first box, 1 ball is in the second box, and 1 ball is in the third box?

9-7 Consider the following two distributions A and B of particles among five energy levels:

Energy of level	A	B
0	12	13
1	6	6
2	3	1
3	2	2
4	1	2

Determine (*a*) the total number of particles in each distribution, (*b*) the total energy of each distribution, and (*c*) the ratio of the number of microstates of the most probable one to the other distribution.

9-8 An isolated system contains three particles with a total energy of 3 units. The energy levels for the energy mode under consideration are equally spaced one unit apart, and the ground level of energy is taken as zero.

(*a*) How many macrostates are possible for this system? How many microstates, if the particles are distinguishable? (Use a diagram similar to those used in the text to help develop your answers.)

(*b*) Is there a most probable macrostate in part (*a*)?

9-9 Consider the following three distributions, A, B, and C:

Energy of level	A	B	C
0	15	16	15
1	8	8	7
2	4	3	3
3	2	1	3
4	1	2	2

If the number of arrangements per distribution is given by the permutation formula $W = n!/\pi n_i!$, determine which distribution is most probable.

9-10 A system of three indistinguishable particles has a total energy of 3 units. The ground level is taken to be zero, and the spacing between levels is 1 unit. The degeneracy for the bottom four levels, starting from the ground level, is 1, 2, 2, and 2. Determine the number of macrostates and microstates.

9-11 Consider a system of three indistinguishable particles. The energies of each are restricted to values of 0, 1, 2, 3, and 4. Determine the number of macrostates and microstates if each of the energy levels has a degeneracy of unity. The total energy is 6.

9-12 Consider a system of three distinguishable particles with a total energy of 9 units. The degeneracy of each level is unity, and the particles are restricted to energy values of 0, 1, 2, 3, and 4. Determine the number of macrostates and microstates.

9-13 An isolated system contains three particles with a total energy of 5 units. The energy levels are equally spaced 1 unit apart and the ground level is taken as zero. How many macrostates and microstates are possible if the particles are distinguishable?

9-14 Consider a system of four distinguishable particles with a total energy of 6 units. The ground level is taken to be zero, the spacing between levels is 1 energy unit, and the levels are nondegenerate. Determine the number of macrostates and microstates.

9-15 Consider a system of ten distinguishable particles for which the energy levels are quantized according to the relation $\epsilon_i = i(i + 1)$, where $i = 0, 1, 2, \ldots$. The total energy is 18 units.

(*a*) How many macrostates are possible?

(*b*) How many microstates are possible?

(*c*) How many microstates are there for the most probable macrostate?

(*d*) How many particles are in the various energy levels, beginning at the ground level, for the most probable distribution?

9-16 Same as Prob. 9-15, except that there are 12 particles and the total energy is 24 units.

9-17 Consider a system of six distinguishable particles with a total energy of 12 units. The energy levels are quantized according to the relation $\epsilon_i = i + \frac{1}{2}$, where $i = 0, 1, 2, \ldots$.

(*a*) How many macrostates and microstates are there?

(*b*) Is there a most probable macrostate?

9-18 Same as Prob. 9-17, except that there are eight particles and the total energy is 14 units.

Effect of heat addition on particle distribution

9-19 Consider a system of three particles with a total energy of 4 units, initially. The ground level is taken to be zero, and the spacing between levels is 1 unit. A heat interaction occurs, and the total energy increases to 5 units. With the heat interaction, the spacing of energy levels remains the same.

(*a*) What is the number of macrostates initially and finally?

(*b*) If the particles are assumed to be distinguishable, what is the number of microstates initially and finally?

9-20 Same as Prob. 9-19, except that the total energy initially is 5 units and finally is 6 units.

Lagrangian technique

9-21 Find the values of x, y, and z which will make the function $R = x^2 + y^2 + z^2$ a minimum, with the restriction that $x + y + z = 9$ and $x + 2y + 3z = 20$.

9-22 Find the dimensions for a box with an open top, the volume of which is 500 in^2, such that the total surface area is a minimum.

9-23 Find the least and greatest distances of a point on the ellipse

$$\frac{x^2}{9} + \frac{y^2}{4} = 1$$

from the straight line $x + 2y = 8$.

9-24 Find the values of x, y, and z which make the function $F = x^2 + 4y^2 + 4z^2$ a minimum, subject to the restrictions that $x + y + z = 6$ and $x + 2y + 4z = 32$.

9-25 Determine the values of x, y, and z that make the function $u = xy^2z^3$ a minimum, subject to the relation $x + y + z = 36$.

9-26 Find the dimensions of an open box (no top) that has a total surface area of 108 in^2, such that the total volume is a maximum.

9-27 A system of 1000 distinguishable particles has three allowed energy levels which have values of 1, 2, and 3 energy units. By the method of Lagrangian multipliers determine the number of particles found in each of the three energy levels for the most probable macrostate if the total energy of the system is (*a*) 1600 units, (*b*) 2000 units, and (*c*) 2400 units.

9-28 A system of 10,000 distinguishable particles has three allowed energy levels. The energy associated with the levels is 0, 1, and 2 energy units. By the method of Lagrangian multipliers determine the number of particles found in each of the three levels for the most probable macrostate if the total energy of the system is (*a*) 7000 units, (*b*) 11,000 units, and (*c*) 14,000 units.

Carnot heat engine†

9-29 A totally reversible heat engine operates between temperatures of 427 and 17°C. Compute the ratio of the heat absorbed by the engine from the source to the work output.

9-30 A Carnot engine develops 1.5 kW of power output and rejects 7500 kJ/h to a sink at 20°C. Determine the source temperature, in degrees Celsius.

9-31 At what temperature is heat supplied to a Carnot engine that rejects 1000 kJ/min of heat at 7°C and produces (*a*) 40 kW of power, and (*b*) 50 kW of power?

9-32 A Carnot engine operates between 37 and 717°C. It is proposed to increase the temperature of the high-temperature source to 1027°C. A counterproposal is to lower the sink temperature to achieve the same new thermal efficiency as would have been achieved by raising the source temperature to 1027°C. What would be the new required sink temperature in degrees Celsius?

9-33 The efficiency of a Carnot engine discharging heat at 27°C is 30 percent. If the sink receives one million J/min, what is the power output of the engine, in kilowatts? What is the temperature of the high-temperature source?

9-34 In a Carnot engine it is found that an efficiency of 50 percent occurs, with 500 kJ/cycle taken from the source reservoir. Calculate the sink temperature, in degrees Celsius, and the heat rejected to the sink, in kJ/cycle, if the source temperature is 307°C.

9-35 A patent application claims that a heat engine which receives heat at 160°C and rejects to a sink at 5°C is capable of delivering 0.10 kWh of energy for every 1000 kJ received as heat by the engine. Is this a valid claim?

9-36 A Carnot engine with an efficiency of 40 percent receives 4000 kJ/h from a high-temperature source and rejects heat to a sink at 25°C. What is the power output, in kilowatts, and the temperature of the source, in degrees Celsius?

9-37 With reference to Prob. 9-30, compute the entropy changes of the heat source and heat sink, in kJ/(°K)(h).

9-38 With reference to Prob. 9-31, compute the entropy changes of the heat source and heat sink in kJ/(°K)(min) for part *a* and for part *b*.

9-39 With reference to Prob. 9-34, compute the entropy changes of the heat source and heat sink, in kJ/(°K)(cycle).

9-40 A reversible heat engine receives 1000 kJ of heat from a reservoir at 377°C and rejects heat to another reservoir at 27°C. Compute the entropy changes of the two reservoirs, in kJ/°K.

† Although the following problems are all expressed in metric units, the units are not essential in comprehending the principles exemplified.

CHAPTER

TEN

TRANSIENT FLOW ANALYSIS

10-1 INTRODUCTION

There are several prime reasons for studying nonsteady-flow problems. First of all, there are several interesting and useful applications of devices which operate under such conditions. For example, some wind tunnels operate on the discharge of a gas which has been stored under a high pressure in a large, rigid tank. These are known as blowdown wind tunnels. Compressed air stored in a cylinder might also be used to drive a gas turbine coupled with a generator as an auxiliary electric power source in the case of normal power failure. Second, the analysis of nonsteady-flow devices serves to strengthen one's understanding of and one's ability to use the general conservation principles related to mass and energy. The importance of the proper selection of a control volume often becomes apparent in such analyses. As has been pointed out earlier in this text, a control volume may change size and shape, as well as move in space relative to some reference point. In steady-state analyses we usually consider the control volume fixed in space, in size, and shape. In the following sections we wish to relinquish two of these constraints and allow the size and shape to vary, if need be. However, the control volume will still be considered to be fixed in location. Although the removal of this latter constraint leads to a highly profitable study with important engineering applications, it is more fitting to study such problems in fluid mechanics in conjunction with the conservation of momentum equation.

10-2 GENERAL CONSERVATION OF ENERGY PRINCIPLE FOR A CONTROL VOLUME

The conservation of energy principle may be applied to a control-volume analysis by properly extending the energy equation developed in Sec. 2-6 for a control mass. We shall consider the control volume shown enclosed within the dashed line in Fig. 10-1*a*. The control surface at sections 1 and 2 is open to the transfer of mass in or out of the control volume. At some initial time t we focus our attention on the control mass, which is the sum of the mass within the control volume at that instant and the mass m_A within the volume element marked A, which lies adjacent to section 1. At time $t + \Delta t$ this control mass has moved so that all the mass originally in region A is now just inside the control volume. In the same time interval, part of the control mass has been pushed out of the control volume into the region marked B, which is adjacent to section 2 (Fig. 10-1*b*). The mass and volume of regions A and B are taken to be small compared with those of the control volume. During the time interval Δt, heat and work interactions may have occurred to the control mass. These interactions are not shown in the figure. An energy balance on the control mass for the time interval is, simply

$$Q + W = \Delta E_{\text{CM}} = E_{\text{CM},\, t+\Delta t} - E_{\text{CM},\, t} \tag{10-1}$$

Since the masses associated with regions A and B are small compared with the control volume, the heat and work effects on the control mass are essentially the same as those which cross the boundary of the control volume at sections other than 1 and 2. Consequently, whether applied to the control volume or the control mass the values of Q and W are the same from either viewpoint. At this point, two items must be resolved in order to transform Eq. (10-1) into a relation directly applicable to the control volume itself. First, the right side of this equation is still in terms of the control mass. Second, the work term on the left needs to be split into several terms which represent distinct types of interactions.

The right-hand side of Eq. (10-1) may be expressed in terms of the control volume in the following manner: The energy E of the control mass at time t is the

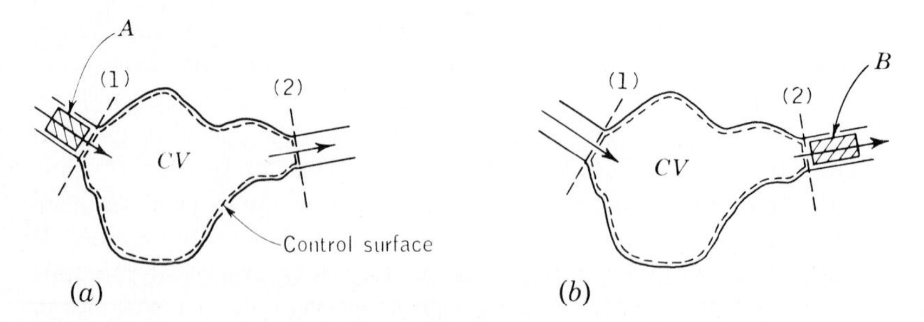

Figure 10-1 The development of conservation principles for a control-volume analysis. (*a*) Control mass at time t; (*b*) control mass at time $t + \Delta t$.

sum of the energies of the mass within the control volume at that instant plus the energy of the mass in region A. This may be expressed as

$$E_{\text{CM},\,t} = E_{\text{CV},\,t} + \Delta E_A \tag{a}$$

where the subscripts again indicate the physical significance of the terms and ΔE_A is the increment of energy associated with the mass which passes from region A into the control volume. In a similar fashion, the energy of the control mass at time $t + \Delta t$ would be

$$E_{\text{CM},\,t+\Delta t} = E_{\text{CV},\,t+\Delta t} + \Delta E_B \tag{b}$$

The quantity ΔE_B is the increment of energy associated with the mass which passes from the control volume into region B. Subtracting Eq. (*b*) from Eq. (*a*), we find that

$$E_{\text{CM},\,t+\Delta t} - E_{\text{CM},\,t} = (E_{\text{CV},\,t+\Delta t} - E_{\text{CV},\,t}) - \Delta E_A + \Delta E_B \tag{c}$$

The left side of Eq. (*c*) is equal to the right side of Eq. (10-1). Substitution of Eq. (*c*) into Eq. (10-1) yields

$$Q + W = E_{\text{CV},\,t+\Delta t} - E_{\text{CV},\,t} - \Delta E_A + \Delta E_B \tag{d}$$

The conservation of energy principle for a control volume is now placed on a rate basis by dividing Eq. (*d*) by the time interval Δt. Hence,

$$\frac{Q}{\Delta t} + \frac{W}{\Delta t} = \frac{E_{\text{CV},\,t+\Delta t} - E_{\text{CV},\,t}}{\Delta t} - \frac{\Delta E_A}{\Delta t} + \frac{\Delta E_B}{\Delta t} \tag{e}$$

The first term on the right represents the average rate of change of energy within the control volume during the time interval Δt. The last two terms are the average rates of energy transfer into and out of the control volume in the same time period. The limit is now taken as Δt approaches zero, so that each term may be expressed as a derivative. As a result

$$\dot{Q} + \dot{W} = \frac{dE_{\text{CV}}}{dt} - \frac{dE_A}{dt} + \frac{dE_B}{dt} \tag{10-2}$$

where the following definitions are introduced.

$$\lim_{\Delta t \to 0} \frac{Q}{\Delta t} = \frac{\delta Q}{dt} \equiv \dot{Q} \tag{10-3}$$

and

$$\lim_{\Delta t \to 0} \frac{W}{\Delta t} = \frac{\delta W}{dt} \equiv \dot{W} \tag{10-4}$$

$\dot{Q}$ is the instantaneous rate of heat transfer to the control volume, and $\dot{W}$ is the rate of work done on the system, or the power.

At this point it is desirable to change the last two quantities in Eq. (10-2) to terms identifiable with the mass crossing the control surface. Because the mass in region A would be differentially small for a time interval dt, this small mass which lies adjacent to section 1 would have the thermodynamic properties associated with that particular section if the condition of local equilibrium (see Sec. 5-1) were valid. Consequently, $E_A = e_1 m_1$, or $dE_A/dt = e_1(dm_1/dt)$. A similar expression holds for the transport of energy at section 2. Further, recall that dm/dt is defined as the mass flow rate $\dot{m}$. Then $dE_A/dt = e_{in}\dot{m}_{in}$ and $dE_B/dt = e_{out}\dot{m}_{out}$. Substitution of these two expressions into Eq. (10-2) leads to

$$\dot{Q} + \dot{W} = \frac{dE_{CV}}{dt} - e_{in}\dot{m}_{in} + e_{out}\dot{m}_{out} \tag{10-5}$$

The right-hand side of this equation is now expressed either in terms of changes in the control volume or in terms of effects at the control surface. The energy e represents all forms of conceivable energy that might be associated with a unit of mass.

Finally, the work term Eq. (10-5) must be separated into three parts. The first of these is the flow or displacement work associated with mass crossing a control surface. Recall from Sec. 5-3 that $\delta W_{flow} = Pv\dot{m}\,dt$, or $\dot{W}_{flow} = Pv\dot{m}$. Secondly, we must consider shaft work which is measured by devices external to the control volume, such as a dynamometer. Thirdly, other forms of work may be present in special situations. For example, if the boundary of the control volume is not rigid, boundary or $P\,dV$ work may occur. The second and third forms of work discussed above will be lumped together as $\dot{W}_{net}$ in the conservation of energy principle. On the basis of the preceding discussion, Eq. (10-5) may be written as

$$\dot{Q} + \dot{W}_{net} + (Pv)_{in}\dot{m}_{in} - (Pv)_{out}\dot{m}_{out} = \frac{dE_{CV}}{dt} - e_{in}\dot{m}_{in} + e_{out}\dot{m}_{out} \tag{10-6}$$

where the term $\dot{W}_{net}$ represents shaft work and any other forms of work associated with the control volume. It is important to remember that the product Pv is treated as an energy term only in those instances when fluid is crossing the control surface or boundary of an open system. Unlike the quantity $P\,\Delta v$, which results under special circumstances from the integration of $P\,dv$ for closed systems, the quantity $\Delta(Pv)$ found in open system equations is the change in a state function.

By collecting common terms in Eq. (10-6), the energy balance on a rate basis for a control volume with one inlet and one outlet is written as

$$\dot{Q} + \dot{W}_{net} + (e + Pv)_{in}\dot{m}_{in} - (e + Pv)_{out}\dot{m}_{out} = \frac{dE_{CV}}{dt} \tag{10-7}$$

All terms on the left represent instantaneous interactions which add or remove energy from the control volume, and the term on the right is the resultant rate of change of energy of the control volume due to these interactions. Since the instantaneous values on the left side of Eq. (10-7) generally may vary with time, this equation must be integrated over a time period t to give the total energy change of the control volume during this time interval. Thus

$$\int_0^t \dot{Q}\,dt + \int_0^t \dot{W}\,dt + \int_0^t (e + Pv)_{in}\dot{m}_{in}\,dt - \int_0^t (e + Pv)_{out}\dot{m}_{out}\,dt = \Delta E_{CV} \tag{10-8}$$

In order to carry out these integrations we will need to know how the various properties and rates vary with time. Another useful formulation of Eq. (10-7) is obtained by multiplying the equation by dt. The resulting relationship is valid for a differential change of state for a control volume over a period of time dt. Thus

$$\delta Q + \delta W_{\text{shaft}} + (e + Pv)_{\text{in}}\, dm_{\text{in}} - (e + Pv)_{\text{out}}\, dm_{\text{out}} = dE_{\text{CV}} \tag{10-9}$$

Since the properties at a boundary may vary with time as mass continues to flow in and out of the control volume, the total energy which crosses a boundary due to mass transfer must in general be found by integration of the $e + Pv$ term over the time interval of interest. Hence Eq. (10-9) may be written as

$$Q_{\text{net}} + W_{\text{shaft}} + \int (e + Pv)_{\text{in}}\, dm_{\text{in}} - \int (e + Pv)_{\text{out}}\, dm_{\text{out}} = \Delta E_{\text{CV}} \tag{10-10}$$

The integration indicated is in terms of the mass which crosses a boundary over a given time interval. Equations (10-7) to (10-10) represent in one form or another the general energy balance for a control-volume type of analysis. One important characteristic of these equations should be noted. The change of the energy within a control volume over an interval of time is due to two distinct effects. Heat and work interactions occur at boundaries not open to mass flow, and their values are found from measurements made external to the control surface. The remaining energy quantities are evaluated in terms of properties of the control volume at the control surface where mass is transferred.

In applying Eqs. (10-9) and (10-10) to transient flow problems, we shall restrict the types of energy associated with the mass within, entering, or leaving a control volume to internal, linear kinetic, and gravitational forms. That is,

$$e = u + \frac{V^2}{2} + zg \tag{10-11}$$

Thus we are neglecting any electric or magnetic field effects. The decision to include the last two terms on the right in any analysis will depend upon the particular circumstances in a given problem. Frequently these latter terms may be omitted, not because their change in value is zero, but because such changes are negligible when compared to other terms in the conservation of energy equation. Within this restriction we can rewrite Eq. (10-9) as

$$\delta Q + \delta W + \left(h + \frac{V^2}{2} + zg\right)_{\text{in}} dm_{\text{in}} - \left(h + \frac{V^2}{2} + zg\right)_{\text{out}} dm_{\text{out}} = dU_{\text{CV}} \tag{10-12}$$

In a similar manner Eq. (10-10) becomes

$$Q + W_{\text{net}} + \int \left(h + \frac{V^2}{2} + zg\right)_{\text{in}} dm_{\text{in}} - \int \left(h + \frac{V^2}{2} + zg\right)_{\text{out}} dm_{\text{out}} = \Delta U_{\text{CV}} \tag{10-13}$$

Although Eqs. (10-12) and (10-13) have been written for a control volume with one inlet and one outlet, it is apparent that a term

$$\left(h + \frac{V^2}{2} + zg\right) dm$$

must be included for every cross section of the control surface through which mass enters or leaves the control volume. Several examples will be presented to illustrate the use of the above equations. One point must be kept in mind. All energy values (and hence enthalpy values as well) must be measured from consistent reference states.

10-3 CHARGING AND DISCHARGING RIGID VESSELS

This section deals with two common unsteady-state problems. One of these is the charging of a tank with a fluid from a pressurized line. The other is the discharging of a fluid from a pressurized tank into a region of lower pressure.

a Charging Processes

A common and typical example of an unsteady-state flow problem is the charging of a rigid vessel by supplying a fluid from some outside high-pressure source. The rigid vessel may initially be evacuated, or may contain a finite amount of matter within it before the charging process begins. It is not necessary that the fluid used to charge the vessel be identical with the substance already in the vessel. The two fluids will be the same, though, in many practical problems. Generally the mass enters only at one section of the control surface, and no efflux of matter is permitted.

As more definite illustration of the problem, consider an evacuated, rigid tank connected through a closed valve to a high-pressure line. This situation is shown in Fig. 10-2*a*. A control volume has been selected, as indicated by the dashed line in the figure, which includes essentially the volume inside the rigid tank. We shall assume that the volume in the region of the valve and entrance

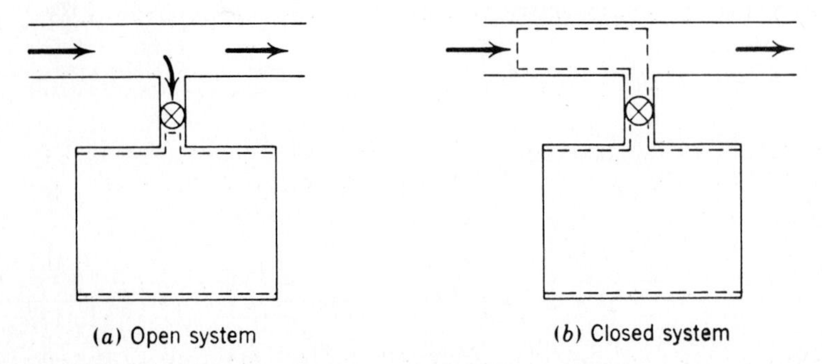

Figure 10-2 Two methods of analyzing the charging of a vessel.

pipe is negligible. Application of Eq. (10-13)) to this situation reveals that

$$Q + W_{\text{net}} + \int \left(h + \frac{V^2}{2} + zg \right) dm = \Delta U_{\text{CV}}$$

The net work term is zero in value on the basis of the statement of the process. If the velocity of the flow stream in the high-pressure line is relatively low (e.g., less than 200 ft/s) and the entrance pipe relatively short in the vertical direction, then the kinetic and potential energies of any unit mass on the line may be neglected. Thus the conservation of energy equation for the selected control volume reduces to

$$Q + \int h \, dm = \Delta U_{\text{CV}}$$

At this point one must make a decision about the quantity Q. In actual practice the value of the heat transfer could be estimated fairly accurately by independently evaluating the convective and radiative contributions to the overall heat loss or gain. However, if the tank were allowed to fill rapidly, the amount of heat transferred probably would be fairly small even if an appreciable temperature gradient existed between the contents of the tank and the environment. Therefore we shall analyze the process on the basis of negligible heat transfer. Realizing that the overall internal energy change of the control volume can be written as $\Delta(mu)$, we then note that the above energy equation becomes

$$\int h \, dm = m_f u_f - m_i u_i$$

where i and f represent the initial and final states within the control volume. The integration is carried out from zero mass to the final mass which enters from the line.

Frequently the integral of $h \, dm$ is difficult to evaluate because the enthalpy of the mass crossing the control surface varies with respect to time. However, in our particular case we can assume that the quantity of fluid bled from the line is small compared to the quantity flowing through the line. Consequently, the properties of the fluid at the entrance to the valve remain constant. If we designate the line conditions by the subscript L, then integration of the above equation leads to

$$h_L(m_L - 0) = m_f m_f - m_i u_i$$

where the terms on the right refer solely to the control volume. For this particular problem m_i is zero (the tank initially was evacuated), and m_f equals m_L. Hence

$$h_L m_L = m_L u_f$$

or

$$h_L = u_f$$

This relationship fixes the final state within the control volume. Under the conditions specified, we find that the enthalpy of the fluid in the high-pressure line is numerically equal to the internal energy of the fluid within the control volume at any instant. This latter statement assumes that the fluid equilibrates rapidly within the control volume, so that properties in general are defined and uniform in value at any instant.

The above result is completely general in the sense that the nature of the fluid has never been specified. Though equally applicable to any real gas, the equation above is of special interest when employed for the flow of an ideal gas into the tank. Since $h = c_p T$ and $u = c_v T$ for an ideal gas (if the values of the two specific heats are assumed constant over the required temperature range), then

$$c_p T_L = C_v T_f$$

or

$$T_f = k T_L$$

where k is the specific-heat ratio. This important result shows that T_f is always greater than T_L for flow of an ideal gas into a rigid tank under the following major idealizations.

1. Adiabatic, rigid control volume
2. Equilibrium within the tank at any instant
3. Negligible kinetic energy of the inflowing gas
4. Steady state for the inlet flow

Since k for many gases ranges from 1.25 to 1.40, the absolute temperature rise may be as much as 40 percent. For real gases, one probably will have to rely on tabular data in order to ascertain the final state of the fluid after it reaches a certain pressure. Such an analysis frequently is one involving iteration.

The problem of charging a tank from a steady-flow line is amenable to attack on the basis of a closed-system analysis as well. It is interesting to pursue this method of analysis, since it provides a striking contrast to the method previously used. The closed system selected for analysis is shown in Fig. 10-2*b*. Since we have chosen a closed system for analysis, all the mass that eventually enters the tank must initially be included within the system boundaries. Hence the boundaries extend beyond the valve and out into a portion of the line.

The conservation of energy equation in this case is

$$Q + W = \Delta E$$

Again we shall neglect kinetic-and potential-energy terms and assume adiabatic conditions. The energy equation becomes

$$W = \Delta U = m_f u_f - m_i u_i$$

Unlike the control-volume approach, however, neither the work term nor the initial mass term m_i is zero. The work done is the boundary work related to pushing the mass contained within the system, but outside the valve, into the tank. Recall that one idealization employed previously is that steady state exists in the line. Hence the pressure in the line in the vicinity of the valve is constant at some value P, and the boundary work is simply $-P\Delta V$. Also the initial mass in this closed system, m_i, is also that mass contained outside the valve initially, since the tank is evacuated at the start of the process. Consequently the energy equation becomes

$$-P(V_f - V_i) = m_f u_f - m_i u_i$$

or

$$-P(0 - m_L v_L) = m_f u_f - m_L u_L$$

Finally, combining the second term on the right with the term on the left, we find that

$$(P_L v_L + u_L) = m_f u_f$$

$$h_L m_L = m_f u_f$$

and

$$h_L = u_f$$

This result is identical to that obtained by a control-volume analysis. Both were based on comparable idealizations. However, it should be noted that the terms included in the two analyses are quite different. Whether transient flow problems are equally tractable to both control-volume and closed-system analyses depends largely on the particular circumstances. The closed-system approach is usually pertinent only to those situations where the position and state of the mass in transit is well defined.

b Discharging Processes

From a physical viewpoint the discharge of a fluid from a pressurized vessel is similar to the preceding case of the charging of a vessel. However, from a thermodynamic viewpoint there is little similarity in the relations which result from applying the conservation of energy principle to the two processes. We shall again neglect the kinetic and potential energies associated either with the control volume or the mass leaving the vessel through a suitable valve. Thus, $e = u$. The absence of shaft work again is noted. The presence or absence of appreciable heat-transfer effects again depends either upon the nature of the vessel's walls (adiabatic or diathermal), or upon the magnitude of the temperature difference between the system and the environment. In the absence of specific information on heat transfer, we shall assume that the effect is negligible, for one reason or another. If Eq. (10-12) is now applied to the control volume shown in Fig. 10-3*a* and the above comments are taken into consideration, we find that

$$(u + Pv)\, dm = dU_{CV}$$

or

$$h\, dm = d(mu)$$

This problem is unlike the charging of a vessel from a pressurized line in that the properties of the fluid crossing the control surface are continually changing due to the decrease in the mass within the control volume. For this reason we

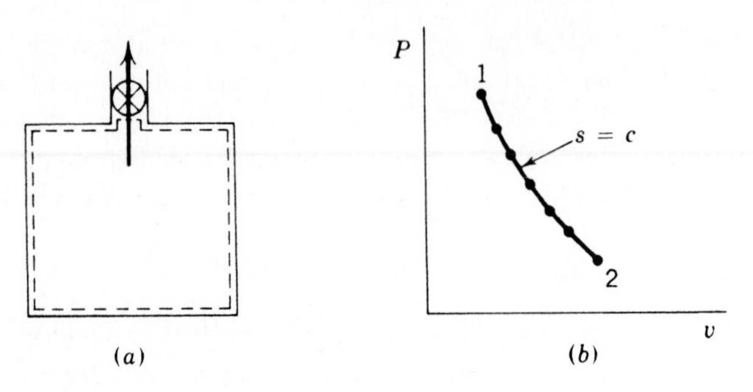

Figure 10-3 (*a*) Schematic diagram and (*b*) *Pv* process diagram for flow from a pressurized vessel.

have elected to write the conservation equation in its differential form. The equation can be rearranged into the following forms.

$$h\,dm = m\,du + u\,dm$$

$$(h - u)\,dm = m\,du$$

and

$$\frac{du}{h-u} = \frac{dm}{m}$$

It should be noted that the specification of property values, such as h and u, for the mass within the control volume requires that the fluid be essentially in equilibrium at all times. Hence the process of discharging the gas must be slow enough so that the process is quasistatic.

The equation presented above is completely general. That is, we have not specified the nature of the fluid leaving the pressurized tank. We may introduce two further relationships without losing this generality. First of all, from the definition of the enthalpy function, $h - u = Pv$. Second, since the volume of the control volume remains constant with time, then from the basic relationship $V = mv$ find that

$$dV = m\,dv + v\,dm = 0$$

Therefore

$$\frac{dm}{m} = -\frac{dv}{v}$$

Substitution of these two basic relations into the conservation of energy equation leads to

$$\frac{du}{Pv} = -\frac{dv}{v}$$

or

$$du + P\,dv = 0$$

This is a remarkable result, since for a closed system of fixed composition it is realized that

$$T\,ds = du + P\,dv$$

is a valid statement for a simple compressible substance, regardless of the nature of the process. Comparison of these last two equations forces one to conclude that, for the process,

$$ds = 0$$

Therefore, for any fluid which flows from a pressurized vessel under the stated assumptions or idealizations, the specific entropy within the control volume remains constant. The total entropy of the control volume decreases, since the amount of mass decreases. The overall process is highly irreversible, so that the net entropy change of the control volume plus its environment must be positive.

Figure 10-3*b* illustrates the process path for the discharging of a fluid from a pressurized tank. The path is for the fluid which remains inside the tank. Note that such a process must fulfill the following idealizations:

1. Adiabatic, rigid control volume
2. Equilibrium within the tank at any instant
3. Negligible kinetic- and potential-energy changes

The fact that the process is isentropic is not surprising, since it is adiabatic and quasistatic, and all internal dissipative effects and finite gradients have been ruled out. That is, the process within the tank is adiabatic and internally reversible.

As a specific example of such a process one might consider the flow of an ideal gas from a pressurized tank. The approximate validity of the ideal-gas equation will depend upon the pressure range chosen for the study. Since the process is isentropic, the isentropic relations presented in Sec. 7-9 are valid. For example, the relation $T(v)^{k-1}$ = constant is valid if one assumes constant specific heats. If one wishes to account for variable specific heats, a more accurate theoretical analysis can be made through the use of the p_r and v_r data in the abridged gas tables found in the Appendix.

A relationship for the mass remaining in the tank at any instant as a function of the properties of the mass within the control volume can also be obtained for an ideal gas. We have noted that

$$\frac{dm}{m} = -\frac{dv}{v}$$

and from the isentropic relations for an ideal gas

$$\frac{dT}{T} = (1 - k)\frac{dv}{v}$$

Therefore,

$$\frac{1}{k-1}\frac{dT}{T} = \frac{dm}{m}$$

Integration of the above equation yields

$$\frac{m_2}{m_1} = \left(\frac{T_2}{T_1}\right)^{1/(k-1)}$$

If the pressure ratio is known, the comparable equation is

$$\frac{m_2}{m_1} = \left(\frac{P_2}{P_1}\right)^{1/k}$$

A third equation can be derived for the mass as a function of the specific volume, or density.

Example 10-1M A tank with a volume of 1.5 m^3 is filled with air at a pressure of 7 bars and a temperature of 220°C. Determine the final temperature and the percent of the mass left in the tank if gas is permitted to leave the tank under adiabatic conditions until the pressure drops to 1 bar.

SOLUTION Under the conditions of the problem it is realistic to assume ideal-gas behavior. If the mass within the tank changes its state isentropically, then, since k is roughly equal to 1.4,

$$T_2 = T_1 \left(\frac{P_2}{P_1}\right)^{(k-1)/k} = 493\left(\frac{1}{7}\right)^{0.285} = 283\text{°K} = 10\text{°C}$$

The percent of the mass left in the tank can be found from the relation

$$\frac{m_2}{m_1} = \left(\frac{P_2}{P_1}\right)^{1/k} = \left(\frac{1}{7}\right)^{0.715} = 0.249$$

Therefore only 25 percent of the mass is left when a pressure of 1 bar is reached, and the temperature at that instant will be 10°C.

Example 10-1 A tank with a volume of 50 ft^3 is filled with air at a pressure of 100 psia and a temperature of 350°F. Determine the final temperature and the percent of the mass left in the tank for an adiabatic expansion down to the environment pressure of 15 psia.

SOLUTION For a maximum pressure of 100 psia it is realistic to assume ideal-gas behavior at the given temperature. If the mass within the tank changes its state isentropically, then, since k is roughly equal to 1.4,

$$\begin{aligned} T_2 &= T_1 \left(\frac{P_2}{P_1}\right)^{(k-1)/k} \\ &= 800\left(\frac{15}{100}\right)^{0.285} \\ &= 800(0.5825) \\ &= 466\text{°R} = 6\text{°F} \end{aligned}$$

The percent of the mass left in the tank can be found from the relation

$$\frac{m_2}{m_1} = \left(\frac{P_2}{P_1}\right)^{1/k} = \left(\frac{15}{100}\right)^{0.715} = 0.257$$

Therefore only 25 percent of the mass is left when pressure equilibrium with the environment is attained, and the temperature at that instant will be 6°F.

There are many other types of transient flow problems of theoretical as well as of practical interest. The solution of each type depends upon the restrictions and idealizations placed on the process. In every case it is best to begin with basic or fundamental equations, and proceed logically from that point. A discussion of a control-volume analysis on which boundary work is present might be helpful at this time.

10-4 TRANSIENT-SYSTEM ANALYSIS WITH BOUNDARY WORK

Consider the situation shown in Fig. 10-4. A weighted piston-cylinder assembly initially containing a known quantity of a gas at a known equilibrium state is connected through a valve to a high-pressure line. To simplify the problem, the fluid in the line is the same as that in the piston cylinder, although this is an unnecessary restriction. We are interested in determining the temperature and the quantity of mass within the cylinder after the volume has changed by a known amount.

The control volume under consideration is shown in Fig. 10-4. Since the upper boundary will move, the control volume changes size in this case, and boundary work occurs. It is necessary to place a restriction on the process if the boundary work is to be evaluated. Even though the pressure in the line may be substantially above the pressure within the cylinder (as maintained by the weighted piston), we must assume that the entering gas quickly equilibrates with the gas already present in the control volume so that a uniform pressure P is maintained on the piston by the gas. Then the work done by the gas on the surroundings is simply $-P\ \Delta V$. In the absence of sufficient information we shall assume that the process is adiabatic, which requires that the piston be frictionless. Although the center of mass of the control volume changes in elevation, this potential-energy effect can be neglected. The basic conservation of energy equation for the control volume in this case becomes

$$Q + W_{\text{shaft}} + W_{\text{boundary}} + \int (\mathrm{e} + Pv)\, dm = \Delta E_{CV}$$

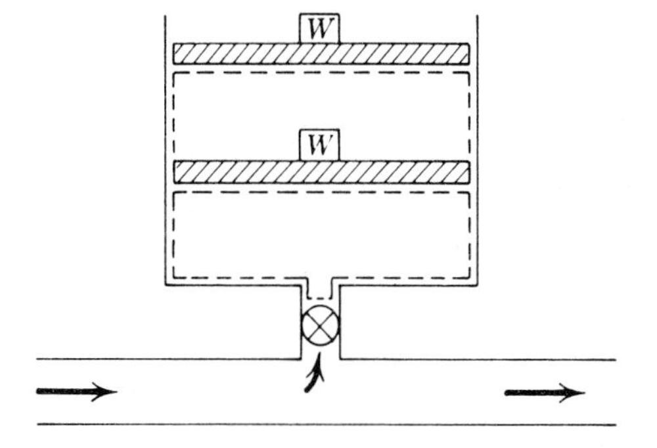

Figure 10-4 A control volume with a moving boundary.

This reduces to the relation

$$W_{\text{boundary}} + \int h\,dm = \Delta U_{CV}$$

In terms of the properties for the initial and final state we find that

$$-P(V_f - V_i) + \int h\,dm = m_f u_f - m_1 u_1$$

As a final step, since the flow through the line is steady, we can integrate the quantity $h\,dm$ directly and obtain the equation

$$-P(V_f - V_i) + h_L m_L = m_f u_f - m_1 u_i$$

The unknowns in this equation are m_L, m_f, and u_f. However, the mass quantities are related by

$$m_L = m_f - m_i$$

Thus we actually have one equation with two unknowns, m_f and u_f.

The second required relationship in this case is found from a knowledge of the final volume. Note that $V_f = m_f v_f$. Since V_f is given, the unknowns in this relationship are m_f and v_f. Hence another unknown, v_f, is introduced. However, u_f and v_f are not independent variables. On the basis of the state principle, for a given pressure within the control volume only u_f ir v_f can be varied independently, but not both. Thus a suitable equation of state which uniquely relates u_f and v_f for a given P is the desired third relation which permits the two equations already presented to be satisfied.

Example 10-2M A piston-cylinder assembly is attached through a valve to a source of air which is flowing in steady state through a pipe. Initially, inside the cylinder the volume is 0.01 m^3, the the temperature of the air inside is 40°C, and the pressure is 1 bar. The valve is then opened to the air-supply line which is maintained at 6 bars and 100°C. If the cylinder pressure remains constant, what will be the final equilibrium temperature, in degrees Celsius, and how much mass will have entered through the valve, in grams, when the cylinder volume reaches 0.02 m^3?

SOLUTION If the control volume is selected to lie just inside the solid boundaries of the piston-cylinder assembly (see Fig. 10-4) and the gas is assumed to behave ideally, then the conservation of energy equation developed above takes on the form

$$\begin{aligned} -P(V_f - V_i) + h_L(m_f - m_i) &= m_f c_v T_f - m_i c_v T_i \\ &= m_f c_v \frac{PV_f}{m_f R} - m_i c_v T_i \\ &= \frac{c_v P V_f}{R} - m_i c_v T_i \\ &= \frac{PV_f}{k-1} - m_i c_v T_i \end{aligned}$$

Hence by introducing the equation of state for an ideal gas, and the relation $c_v = R/(k-1)$, we find that the resulting equation contains only the single unknown m_f. Before substituting numerical values, it is convenient to calculate first the initial mass within the piston-cylinder assembly.

$$m_i = \frac{PV}{RT} = \frac{1 \text{ bar} \times 0.01 \text{ m}^3}{0.08314 \text{ bar} \cdot \text{m}^3/(\text{kg} \cdot \text{mol})(°\text{K}) \times 313°\text{K}} \times 29 \text{ kg/kg} \cdot \text{mol}$$

$$= 0.0111 \text{ kg} = 11.1 \text{ g}$$

Note also that $h_L = c_p T_L$, and c_p, c_v, and k values are in Table A-4M. Substitution of known values into the energy equation then yields (expressing each term in newton-meters or joules, and m_f in grams),

$$-1(0.02-0.01)10^5 + 1.01(m_f - 11.1)(373) = \frac{1(0.02)(10^5)}{0.4} - 11.1(0.72)(313)$$

$$-1000 + 377m_f - 4180 = 5000 - 2500$$

$$m_f = 7680/377 = 20.4 \text{ g}$$

Thus, the mass entering the assembly during the filling process is

$$m_L = m_f - m_i = 20.4 - 11.1 = 9.3 \text{ g}$$

The final equilibrium temperature is found from the ideal-gas relation:

$$T_f = \frac{PV_f}{m_f R} = \frac{1 \text{ bar} \times 0.02 \text{ m}^3}{20.4 \text{ g} \times 0.08314 \text{ bar} \cdot \text{m}^3/(\text{kg} \cdot \text{mol})(°\text{K})} \times 29 \text{ kg/kg} \cdot \text{mol} \times 1000 \text{ g/kg}$$

$$= 342°\text{K} = 69°\text{C}$$

Thus, the temperature of the resulting mixture in the assembly increases nearly 30°C above the initial temperature.

Example 10-2 A piston-cylinder assembly is attached through a valve to a source of air which is flowing in steady state through a pipe. Initially inside the cylinder the volume is 1.0 ft^3, the temperature of the air inside is 100°F, and the pressure is 14.8 psia. The valve is then opened to the air-supply line which is at 100 psia and 200°F. If the cylinder pressure remains constant, what will be the final equilibrium temperature, in degrees Fahrenheit, and how much mass will have entered through the valve, in pounds, when the cylinder volume reaches 2.0 ft^3?

Solution If the control volume is selected to lie just inside the solid boundaries of the piston-cylinder assembly (see Fig. 10-4) and the gas is assumed to behave ideally, then the conservation of energy equation developed above takes on the form

$$-P(V_f - V_i) + h_L(m_f - m_i) = m_f c_v T_f - m_i c_v T_i$$

$$= m_f C_v \frac{PV_f}{m_f R} - m_i c_v T_i$$

$$= \frac{c_v PV_f}{R} - m_i c_v T_i$$

$$= \frac{PV_f}{k-1} - m_i c_v T_i$$

Hence by introducing the equation of state for an ideal gas, and the relation $c_v = R(k-1)$, we find that the resulting equation contains only the single unknown m_f. Before substituting numerical values, it is convenient to calculate first the initial mass within the piston-cylinder assembly:

$$m_i = \frac{PV}{RT} = \frac{14.8(1)(29)}{10.73(560)} = 0.1715 \text{ lb}_m$$

Note that $h = c_p T_L$. Then the substitution of known values into the energy equation yields

$$\frac{-14.8(144)(2-1)}{778} + 0.24(m_f - 0.0715)(660) = \frac{14.8(144)(2.0)}{0.4(778)} - 0.0715(0.17)(560)$$

Then

$$-2.74 + 158.4m_f - 11.32 = 13.7 - 6.84$$

or

$$m_f = 0.132 \text{ lb}_m$$

Thus the mass entering the assembly during the filling process is

$$m_f - m_i = m_L = 0.132 - 0.0715$$

$$= 0.0606 \text{ lb}_m$$

The final equilibrium temperature is found from the ideal-gas relation:

$$T_f = \frac{PV_f}{m_f R} = \frac{14.0(2.0)(29)}{0.132(10.73)}$$

$$= 606°\text{R} = 146°\text{F}$$

If the fluid had been steam (or some other real gas), the solution to this problem would not have been quite as straightforward. In this latter case the property relations are usually not known in an explicit algebraic form. Consequently an iteration process often is required during the numerical evaluation step of the solution.

PROBLEMS (METRIC)

10-1M A pressurized tank contains 1.5 kg of air at 40°C and (*a*) 2 bars, and (*b*) 3 bars. Mass is allowed to flow from the tank until the pressure reaches 1 bar. However, during the process, heat is added to the air within the tank to keep it at constant temperature. Assuming constant specific heats, determine how much heat, in kilojoules, was added during the process.

10-2M An insulated tank with a volume of 0.50 m^3 contains air at 1.0 bar and 25°C. The tank is connected through a valve to a large compressed-air line, which carries air continuously at 7 bars and (*a*) 120°C, and (*b*) 160°C. If the valve is opened and air is allowed to flow into the tank until the pressure reaches 4 bars, (1) how much mass, in kilograms, has entered, and (2) what is the final temperature in the tank, in degrees Celsius?

10-3M A pressurized tank contains 1 kg of water vapor at 40 bars and 280°C. Mass is allowed to flow from the tank until the pressure reaches 7 bars. However, during the process, heat is added to the steam to keep it at constant temperature. How much heat, in kilojoules, was added during the process?

10-4M A tank with a volume of 3.0 m^3 contains steam at 20 bars and 280°C. The tank is heated until the temperature reaches 440°C. A relief valve is installed on the tank to keep the pressure constant during the process. Find (*a*) the amount of heat transferred, in kilojoules, and (*b*) the mass of steam that is bled from the tank, in kilograms.

10-5M A rigid insulated tank is initially evacuated. Atmospheric air at 0.1 MPa and 20°C is allowed to leak into the tank until the pressure reaches 0.1 MPa. What is the final temperature of the air within the tank, in degrees Celsius?

10-6M Consider the data of Prob. 10-5M, except that the tank initially contains air at 0.04 MPa and 20°C. Determine the final temperature, in degrees Celsius, in this case.

10-7M An ideal gas is contained in a rigid tank of volume V, initially at P_1 and T_1. Heat is supplied to the contents until the temperature reaches T_2. However, a relief valve allows gas to escape so that the pressure remains constant. Derive an expression for the heat transfer during the process in terms of the quantities T_1, T_2, P, V, c_p, and R.

10-8M Consider the situation in Prob. 10-7M, except that a valve holds the temperature constant while the pressure drops to P_2. Derive an expression for the heat transfer in terms of V, P_1, and P_2.

10-9M Two adiabatic tanks are interconnected through a valve. Tank A contains 0.25 m^3 of air at 40 bars and 90°C. Tank B contains 3.0 m^3 of air at 1 bar and 20°C. The valve is opened until the pressure in A drops to 15 bars. At this instant, determine (*a*) the temperature in tank A, (*b*) the temperature and pressure in tank B, (*c*) the mass remaining in tank A in kilograms, and (*d*) the total entropy change for the process, in kJ/°K. Assume constant specific heats.

10-10M Two adiabatic tanks are interconnected through a valve. Tank A contains 0.10 m^3 of nitrogen at 30 bars and 100°C. Tank B contains 2.5 m^3 of nitrogen at 2 bars and 30°C. The valve is opened until the pressure in A drops to 20 bars. At this instant, determine (*a*) the temperature in A, in degrees Celsius, (*b*) the temperature and pressure in tank B, (*c*) the mass remaining in tank A in kilograms, and (*d*) the total entropy change for the process, in kJ/°K. Assume constant specific heats.

10-11M A tank with a volume of 10 m^3 contains air initially at 6 bars and 40°C. Heat is added at a constant rate of 6 kJ/s while an automatic valve allows air to leave the tank at a constant rate of 0.03 kg/s. Determine the temperature of the air in the tank (*a*) 5 min, and (*b*) 7 min after the initial conditions.

10-12M With reference to Prob. 10-11M, how long will it take for the air in the tank to reach (*a*) 140°C, and (*b*) 180°C?

10-13M With reference to Prob. 10-11M, what is the pressure in the tank after 8 min, starting from the initial conditions?

10-14M A container of fixed volume V contains air at pressure P_1 and temperature T_a. It is surrounded by atmospheric air at a pressure P_a and temperature T_a. A valve is opened and atmospheric air is quickly admitted to the container until the pressure reaches the atmospheric value. At that instant the air in the container is at temperature T_2. Listing any necessary assumptions, derive an equation for the temperature ratio, $T_R = T_2/T_a$, in terms of the specific-heat ratio k and the pressure ratio, $P_R = P_1/P_a$.

10-15M A tank with a volume of 0.5 m^3 is half filled with liquid water and the remainder is filled with vapor. The pressure is (*a*) 20 bars, and (*b*) 30 bars. Heat is added until one-half of the liquid (by mass) is evaporated while an automatic valve lets saturated water vapor escape at such a rate that the pressure remains constant. Determine the amount of heat transfer, in kilojoules.

10-16M A pressure vessel with a volume of 0.5 m^3 contains saturated water at (*a*) 250°C, and (*b*) 300°C. The vessel initially contains 50 percent by volume of liquid. Liquid is slowly withdrawn from the bottom of the tank, and heat transfer takes place so that the contents are kept at constant temperature. How much heat must be added by the time half of the total mass has been removed?

10-17M A very large reservoir contains air at 12 bars and an unknown temperature T_a. Air flows from this reservoir into a small, insulated tank with a volume of 0.2 m^3. The small tank initially contains 0.2 kg of air at 27°C. Then a valve is opened to allow air to flow from the reservoir into the tank until the pressure in the tank becomes 3 bars. At this point the temperature in the tank is observed to be (*a*) 140°C, and (*b*) 180°C. What is the temperature of the air, T_a, in the reservoir? Assume constant specific heats.

10-18M A piston-cylinder assembly is attached through a valve to a source of constant-temperature and constant-pressure air. Initially inside the cylinder the volume is 0.1 m^3, the temperature of the air is 30°C, and the pressure is 1 bar. The valve is slowly opened to the air-supply line which is at 7 bars and 90°C. The piston moves out as air enters in order to maintain the cylinder pressure at the ambient condition of 1 bar. When the cylinder volume reaches 0.2 m^3, (*a*) what is the temperature inside the cylinder, in degrees Celsius, and (*b*) how much mass, in kilograms, has entered through the valve?

10-19M A tank with a volume of 0.5 m^3 contains carbon dioxide at 2 bars and 30°C. Nitrogen at a line pressure and temperature of 8 bars and 150°C, respectively, flows from a pipe into the tank until the pressure reaches 5 bars. If the entire process is adiabatic, determine the final temperature, in degrees Celsius, of the mixture within the tank.

10-20M A tank with a volume of 1.0 m^3 is half filled with liquid refrigerant 12 and the remainder is filled with vapor. The pressure is 8.0 bars. Heat is added until one-half of the liquid (by mass) is evaporated, while an automatic valve allows saturated vapor to escape at such a rate that the pressure remains constant within the tank. Determine the heat transfer required, in kilojoules.

PROBLEMS (USCS)

10-1 A pressurized tank contains 1.5 lb of air at 140°F and (*a*) 30 psia, and (*b*) 40 psia. Mass is allowed to flow from the tank until the pressure reaches 15 psia. However, during the process, heat is added to the air within the tank to keep it at constant tenperature. Assuming constant specific heats, determine how much heat, in Btu, was added during the process.

10-2 An insulated tank with a volume of 2.0 ft^3 contains air at 15 psia and 80°F. The tank is connected through a valve to a large compressed-air line, which carries air continuously at 100 psia and (*a*) 240°F, and (*b*) 300°F. If the valve is opened and air is allowed to flow into the tank until the pressure reaches 60 psia, (1) how much mass, in pounds, has entered, and (2) what is the final temperature in the tank, in degrees Fahrenheit?

10-3 A pressurized tank contains 1 lb of water vapor at 600 psia and 500°F. Mass is allowed to flow from the tank until the pressure reaches 100 psia. However, during the process, heat is added to the steam to keep it at constant temperature. How much heat, in Btu, was added during the process?

10-4 A tank with a volume of 90 ft^3 contains steam at 300 psia and 500°F. The tank is heated until the temperature reaches 800°F. A relief valve is installed on the tank to keep the pressure constant during the process. Find (*a*) the amount of heat transferred, in Btu, and (*b*) the mass of steam that is bled from the tank, in pounds.

10-5 A rigid insulated tank is initially evacuated. Atmospheric air at 1 atm and 70°F is allowed to leak into the tank until the pressure reaches 1 atm. What is the final temperature of the air within the tank, in degrees Fahrenheit?

10-6 Consider the data of Prob. 10-5, except that the tank initially contains air at 0.4 atm and 70°F. Determine the final temperature, in degrees Fahrenheit, in this case.

10-7 An ideal gas is contained in a rigid tank of volume V, initially at P_1 and T_1. Heat is supplied to the contents until the temperature reaches T_2. However, a relief valve allows gas to escape so that the pressure remains constant. Derive an expression for the heat transfer during the process in terms of the quantities T_1, T_2, P, V, c_p, and R.

10-8 Consider the situation in Prob. 10-7, except that a valve holds the temperature constant while the pressure drops to P_2. Derive an expression for the heat transfer in terms of V, P_1, and P_2.

10-9 Two adiabatic tanks are interconnected through a valve. Tank A contains 1.2 ft^3 of air at 500 psia and 200°F. Tank B contains 15.0 ft^3 of air at 20 psia and 100°F. The valve is opened until the pressure in A drops to 300 psia. At this instant, determine (*a*) the temperature in tank A, in degrees Fahrenheit, (*b*) the temperature and pressure in tank B, (*c*) the mass remaining in tank A, in pounds, and (*d*) the total entropy change for the process, in Btu/°R. Assume constant specific heats.

10-10 Two adiabatic tanks are interconnected through a valve. Tank A contains 3.0 ft^3 of nitrogen at 450 psia and 200°F. Tank B contains 40 ft^3 of nitrogen at 50 psia and 100°F. The valve is opened until the pressure in A drops to 300 psia. At this instant, determine (*a*) the temperature in A in degrees Fahrenheit, (*b*) the temperature and pressure in tank B, (*c*) the mass remaining in tank A in pounds, and (*d*) the total entropy change for the process, in Btu/°R. Assume constant specific heats.

10-11 A tank with a volume of 100 ft^3 contains air initially at 100 psia and 100°F. Heat is added at a constant rate of 6 Btu/s while an automatic valve allows air to leave the tank at a constant rate of 0.06 lb/s. Determine the temperature of the air, in degrees Fahrenheit, in the tank (*a*) 5 min, and (*b*) 7 min after the initial conditions.

10-12 With reference to Prob. 10-11, how long will it take for the air in the tank to reach (*a*) 350°F, and (*b*) 400°F, in minutes?

10-13 With reference to Prob. 10-11, what is the pressure in psia in the tank after 8 min, starting from the initial conditions?

10-14 A container of fixed volume V contains air at pressure P_1 and temperature T_a. It is surrounded by atmospheric air at a pressure P_a and temperature T_a. A valve is opened and atmospheric air is quickly admitted to the container until the pressure reaches the atmospheric value. At that instant the air in the container is at temperature T_2. Listing any necessary assumptions, derive an equation for the temperature ratio, $T_R = T_2/T_a$, in terms of the specific-heat ratio k and the pressure ratio, $P_R = P_1/P_a$.

10-15 A tank with a volume of 10 ft^3 is half filled with liquid water and the remainder is filled with vapor. The pressure is 500 psia. Heat is added until one-half of the liquid (by mass) is evaporated while an automatic valve lets saturated vapor escape at such a rate that the pressure remains constant. Determine the amount of heat transfer, in Btu.

10-16 A pressure vessel with a volume of 50 ft^3 contains saturated water at 600°F. The vessel initially contains 50 percent by volume liquid. Liquid is slowly withdrawn from the bottom of the tank, and heat transfer takes place so that the contents are kept at constant temperature. How much heat must be added by the time half of the total mass has been removed?

10-17 A very large reservoir contains air at 200 psia and an unknown temperature T_a. Air flows from this reservoir into a small insulated tank with a volume of 2 ft^3. The small tank initially contains 0.1 lb of air at 80°F. Then a valve is opened to allow air to flow from the reservoir into the tank until the pressure in the tank becomes 50 psia. At this point the temperature in the tank is observed to be 340°F. What is the temperature of the air, T_a, in the reservoir? Assume constant specific heats.

10-18 A piston-cylinder assembly is attached through a valve to a source of constant-temperature and constant-pressure air. Initially inside the cylinder the volume is 1 ft^3, the temperature of the air is 100°F, and the pressure is 14.7 psia. The valve is slowly opened to the air-supply line which is at 100 psia and 200°F. The piston moves out as air enters in order to maintain the cylinder pressure at the ambient condition of 14.7 psia. When the cylinder volume reaches 2 ft^3, (*a*) what is the temperature inside the cylinder, in degrees Fahrenheit and (*b*) how much mass, in pounds, has entered through the valve?

CHAPTER

ELEVEN

NONREACTIVE IDEAL-GAS MIXTURES

11-1 INTRODUCTION

The basic thermodynamic laws introduced so far are of general validity. In the application of these laws to closed and open systems, however, we have dealt primarily with systems containing a single chemical species. Analytical expressions, tables, and graphs have been presented which relate such properties as P, v, T, u, h, s, c_v, and c_p for systems of a single component. In view of the fact that many engineering applications involve multicomponent systems, it is vital to have some understanding of methods of evaluating the properties of such systems.

A complete description of a multicomponent system requires a specification not only of two properties such as the pressure and the temperature of the mixture, but also of the composition. Thus properties such as u, h, v, and s of a mixture are different for every different composition. Note, however, that the properties of individual components are readily available. Hence one method of evaluating mixture properties is to devise rules for averaging the properties of the individual pure components, so that the resulting value is representative of the overall composition. This approach is used to model the behavior of ideal-gas mixtures, and a few other classes of mixtures.

11-2 MASS ANALYSIS OF GAS MIXTURES

In the study of thermodynamic systems there are two general bases for measuring the mass within the system. One of these involves the use of units like kilograms or pounds. The other basis of mass is the mole unit, i.e., kilogram-moles or pound-moles. Thus, in dealing with gas mixtures, we may analyze systems from either viewpoint. When the analysis of a gas mixture is made on the basis of mass or weight, it is called a *gravimetric* analysis. For a nonreacting gas mixture it is apparent that the total mass of the mixture m_m is the sum of the masses of each of the k components. That is,

$$m_m = m_1 + m_2 + m_3 + \cdots + m_k = \sum_{i=1}^{k} m_i \tag{11-1}$$

The mass fraction mf_i of the ith component is defined as

$$mf_i \equiv \frac{m_i}{m_m} \tag{11-2}$$

The sum of the mass fractions of all the components in a mixture is unity. If an analysis of a gas mixture is based on the number of moles of each component present, the analysis is termed a *molar* analysis. The total number of moles N_m for a mixture is given by

$$N_m = N_1 + N_2 + N_3 + \cdots + N_k = \sum_{i=1}^{k} N_i \tag{11-3}$$

and the mole fraction of any component y_i is defined as

$$y_i \equiv \frac{N_i}{N_m} \tag{11-4}$$

The sum of the mole fractions of all the components in a mixture also is unity. From the definition of the molar mass (or molecular weight) M_i, the mass of a component is related to the number of moles of that component by

$$m_i = N_i M_i \tag{11-5}$$

If Eq. (11-5) is substituted into Eq. (11-1) for each of the components, then

$$m_m = N_1 M_1 + N_2 M_2 + N_3 M_3 + \cdots + N_k M_k = N_m M_m \tag{11-6}$$

where M_m is an average, or apparent, molar mass, or molecular weight, for the mixture. The solution of Eq. (11-6) for M_m yields, in terms of y_i,

$$M_m = \sum_{i=1}^{k} y_i M_i \tag{11-7}$$

The average molar mass of a gas mixture, then, is the sum over all the components of the mole fraction times the molar mass.

As an example of this latter relationship, the average molar mass of atmospheric air can be approximated in the following way. If we neglect the presence of other gas components, air is composed of roughly 78.1 percent N_2, 21.0 percent O_2, and 0.9 percent argon. Substitution of these values into Eq. (11-7) yields

$$M_{\text{air}} = 0.781(28.0) + 0.210(32.0) + 0.009(40) = 28.95$$

When other gases are included in the calculation for the average composition of air, the commonly quoted value of 28.97 results.

11-3 *PvT* Relationships for Mixtures of Ideal Gases

The *PvT* relationship for a mixture of gases is usually based on two models. The first of these is known as *Dalton's law* of additive pressures. This rule states that the total pressure exerted by a mixture of gases is the sum of the component pressures p_i, each measured alone at the temperature and volume of the mixture. Hence Dalton's rule can be written in the form

$$P = p_1 + p_2 + p_3 + \cdots + p_k = \sum_i p_i \tag{11-8}$$

where p_i is the component pressure of the ith component. A physical representation of the additive-pressure rule is shown in Fig. 11-1 for the case of two gases A and B. It would be expected that ideal gases fulfill Dalton's rule exactly, since the concept of an ideal gas implies that intermolecular forces are negligible, and thus the gases act independently of one another.

The pressure exerted by an ideal gas in a gas mixture, in view of Dalton's rule, can be expressed as

$$p_i = \frac{N_i R_u T}{V} \tag{11-9}$$

where T and V are the absolute temperature and volume of the mixture, respectively. If the component pressures of each chemical species are now substituted into Eq. (11-8), we find that

$$P = \frac{N_1 R_u T}{V} + \frac{N_2 R_u T}{V} + \cdots + \frac{N_i R_u T}{V}$$

$$= (N_1 + N_2 + \cdots + N_i)\frac{R_u T}{V} = \frac{N_m R_u T}{V} \tag{11-10}$$

Thus the gas mixture also follows the ideal-gas equation, as would be expected. Equation (11-10) has been confirmed experimentally for gases at relatively low pressures. Dalton's rule may also be applied to mixtures of real gases as an approximation technique. However, the results will not necessarily agree with experiment, due to the influence of intermolecular forces between real-gas particles.

Recall from Eq. (11-6) that N_m equals m_m/M_m. Making this substitution into

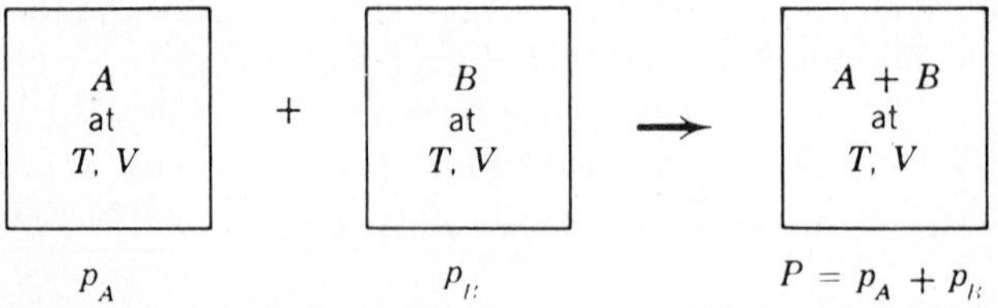

A at T, V + B at T, V → $A + B$ at T, V

p_A p_B $P = p_A + p_B$

Figure 11-1 Schematic representation of Dalton's law of additive pressures.

the above equation, we obtain the following relation:

$$P = \frac{m_m}{M_m}\frac{R_u T}{V} = \frac{m_m T}{V}\frac{R_u}{M_m} = \frac{m_m R_m T}{V} \tag{11-11}$$

where R_m is an apparent gas constant for the gas mixture on a mass basis and is defined as

$$R_m = \frac{R_u}{M_m} \tag{11-12}$$

The value of M_m is obtained from Eq. (11-7). Equation (11-11) is the ideal-gas relation for a gas mixture on a mass basis, rather than the molar basis given by Eq. (11-10).

A relationship between the component pressure p_i of an ideal-gas mixture and its mole fraction y_i is found by dividing Eq. (11-9) by Eq. (11-10). This leads to

$$\frac{p_i}{P} = \frac{N_i R_u T/V}{N_m R_u T/V} = \frac{N_i}{N_m} = y_i$$

or

$$p_i = y_i P \tag{11-13}$$

In the thermodynamic literature the product $y_i P$ is frequently defined as the *partial* pressure p'_i of a gas.

Another model or description of gas mixtures is that based on the *Amagat-Leduc law* of additive volumes. This law states that the total volume of a mixture of gases is the sum of the volumes that each gas would occupy if measured individually at the pressure and the temperature of the mixture. This may be expressed by the relation

$$V = V_1 + V_2 + V_3 + \cdots + V_k = \sum_{i=1}^{k} V_i \tag{11-14}$$

where V_i is the volume of the ith component measured at the pressure and the temperature of the mixtures. This relation may be applied to both ideal-gas and real-gas mixtures, although in the latter case it is only approximately valid. When Amagat's law is applied to ideal gases in a mixture, the ideal-gas equation for each component becomes $PV_i = N_i RT$. Substitution of this equation into the Amagat-Leduc model results in an expected form, namely,

$$\begin{aligned} V &= \frac{N_1 R_u T}{P} + \frac{N_2 R_u T}{P} + \cdots + \frac{N_k R_u T}{P} \\ &= (N_1 + N_2 + \cdots + N_k)\frac{R_u T}{P} \\ &= \frac{N_m R_u T}{P} \end{aligned} \tag{11-15}$$

Now if the equation for V_i is divided by Eq. (11-15), we find that

$$\frac{V_i}{V} = \frac{N_i R_u T/P}{N_m R_u T/P} = \frac{N_i}{N_m} = y_i \tag{11-16}$$

A physical interpretation of the component volume V_i can be obtained for ideal-gas mixtures in the following way: Consider a mixture of two ideal gases in a system of volume V with a total pressure P and a temperature T. Hypothetically, the gases could be separated so that one species occupied a certain part of the volume, and the other species filled the remaining volume. The temperature and pressure of each separate gas would still be identical. The actual volume filled by a component volumes is the total volume of the closed system. A schematic representation of the additive-volume rule is shown in Fig. 11-2 for the case of two gases.

On the basis of Eqs. (11-13) and (11-16) it is evident that

$$\frac{p_i}{P} = y_i = \frac{N_i}{N_m} = \frac{V_i}{V} \tag{11-17}$$

Hence, for ideal-gas mixtures, the mole fraction, the volume fraction, and the ratio of component pressure to total pressure are all equal for a given gas. The direct relationship between mole fractions and volume fractions enables one to convert between volumetric analyses and mass analyses. The next two examples are illustrations of such conversion calculations. Note that a tabular type of calculation is quite convenient, where the calculations proceed from left to right.

Example 11-1M A gaseous fuel is composed of 20 percent CH_4, 40 percent C_2H_6, and 40 percent C_3H_8, where all percentages are by volume. Determine the gravimetric analysis of the fuel, the apparent molecular weight of the mixture, and the apparent gas constant for the mixture.

SOLUTION It is convenient to select 100 moles as a basis for the subsequent calculation. Since the volume fraction is also the mole fraction, if it is assumed that the gases are ideal, then the percentages given here are also the number of moles of each component per 100 mol of mixture. These values are listed in the second column of the accompanying table. The third column lists the molecular weight, or molar mass, of each. Recall that the mass of a component is given by $N_i M_i$ [Eq. (11-5)]. Consequently, the mass of each component per 100 mol of mixture is found by multiplying the value in column 2 by the value in column 3 for each species. The result is

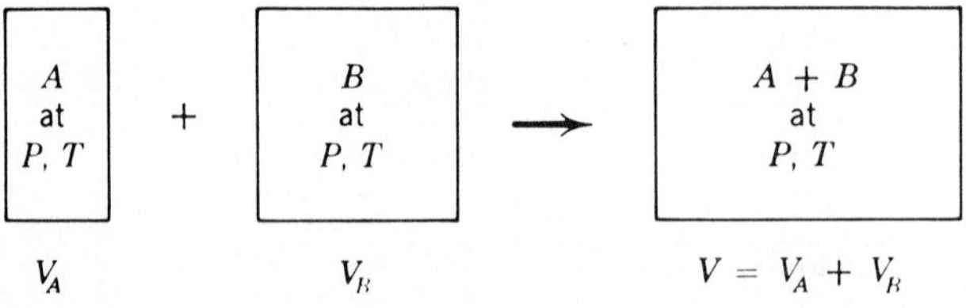

Figure 11-2 Schematic representation of Amagat's law of additive volumes.

shown in column 4. The number at the bottom is the summation for the masses. The gravimetric analysis, then, is found simply by dividing the mass of each component by the total mass, as given by column 5. It should be noted that the gravimetric analysis shown by column 5 is significantly different from the volumetric, or molar, analysis of the fuel mixture. The sum of column 4 is the mass of mixture per 100 mol of mixture. Therefore the mass per mole must be $\frac{3280}{100} = 32.80$. But this is the definition of the apparent molar mass of a mixture. The gas constant on a mass basis may now be found.

Metric:

$$R_m = \frac{R_u}{M_m} = \frac{0.08314}{32.80} = 2.54 \times 10^{-3} \text{ bar} \cdot \text{m}^3/(\text{kg})(°\text{K})$$

USCS:

$$R_m = \frac{R_u}{M_m} = \frac{0.730}{32.80} = 0.0223 \text{ atm} \cdot \text{ft}^3/(\text{lb})(°\text{R})$$

Other values of R_m is metric and USCS units may be evaluated by using other values of R_u found in Table A-1.

(1) Component	(2) Moles per 100 moles of mixture	(3) Molar-mass	(4) Mass per 100 moles of mixture	(5) Mass analysis, %
CH_4	20	16	320	9.76
C_2H_6	40	30	1200	36.59
C_3H_8	40	44	1760	53.65
Total			3280	100.00

Example 11-1 See Example 11-1M for the appropriate calculations.

Example 11-2M An ideal-gas mixture consists of 10 percent hydrogen, 48 percent oxygen, and 42 percent carbon monoxide by mass. Determine the volumetric analysis, in percent, and the apparent molar mass.

Solution In the table below the first three columns list the chemical species, the mass fraction, and the molar mass. Column 4 is the moles of each per unit mass of mixture, found by dividing column 2 by column 3. The mole fraction of each constituent in column 5 is then determined by dividing each value in column 4 by the total in column 4. Since the volume fraction is equal to the mole fraction, the volumetric analysis in column 6 is based directly on the values in column 5.

(1) Component	(2) Mass fraction	(3) Molar mass	(4) Moles per unit mass of mixture	(5) Mole fraction	(6) Volume analysis, %
H_2	0.10	2	0.050	0.6250	62.50
O_2	0.48	32	0.015	0.1875	18.75
CO	0.42	28	0.015	0.1875	18.75
Total			0.080	1.0000	100.00

Note that hydrogen, which is present in the smallest amount in the gravimetric analysis, has the largest percentage on a volumetric basis. The apparent molar mass is obtained from the last value in column 4.

$$M_m = \frac{1}{\text{moles/unit mass of mixture}} = \frac{1}{0.080} = 12.5$$

Example 11-2 See Example 11-2M as an appropriate problem.

11-4 PROPERTIES OF MIXTURE OF IDEAL GASES

In a system of ideal gases the temperature T applies to all the gases within the system when occupying a volume V at a total pressure P. The component pressures are given, of course, by the quantity $y_i P$. Other thermodynamic properties of the individual gases and of the mixture may be obtained by the application of the *Gibbs-Dalton law*, which is simply a generalization of Dalton's rule of additive pressures. It states that in a mixture of ideal gases each component of the mixture acts as if it were alone in the system at the volume V and the temperature T of the mixture. Consequently all extensive properties of the multicomponent mixture could be found by summing the contributions made by each gas component.

On the basis of the Gibbs-Dalton law the total internal energy of the mixture U_m is given by

$$U_m = U_1 + U_2 + U_3 + \cdots + U_k = \sum_{i=1}^{k} U_i \tag{11-18}$$

One way the total internal energy of each component may be expressed is by $N_i u_i$, where u_i is the specific internal energy on a mole basis. As a result we may write Eq. (11-8) in the following manner.

$$U_m = N_m u_m = N_1 u_1 + N_2 u_2 + N_3 u_3 + \cdots + N_k u_k \tag{11-19}$$

In the energy analysis of closed systems it is the change in internal energy that is needed. On the basis of Eq. (11-19),

$$\Delta U_m = N_m \Delta u_m = N_1 \Delta u_1 + N_2 \Delta u_2 + \cdots + N_k \Delta u_k = \sum_{i=1}^{k} N_i \Delta u_i \tag{11-20a}$$

If this equation is divided by N_m, we obtain the specific internal-energy change Δu_m of the mixture. That is,

$$\Delta u_m = y_1 \Delta u_1 + y_2 \Delta u_2 + \cdots + y_k \Delta u_k = \sum_{i=1}^{k} y_i \Delta u_i \tag{11-20b}$$

The u_i data needed for these equations is obtained from ideal-gas property tables for the individual gases.

The enthalpy of a mixture of ideal gases will simply be the sum of the enthalpies of the individual components. On a mole basis

$$H_m = N_m h_m = H_1 + H_2 + H_3 + \cdots + H_k$$
$$= N_1 h_1 + N_2 h_2 + N_3 h_3 + \cdots + N_k h_k \tag{11-21}$$

where h_m is the specific enthalpy of the mixture. For energy analyses we are interested in the enthalpy change. Similar to the equations above for the internal-energy change of ideal-gas mixtures, we may write

$$\Delta H_m = N_m \Delta h_m = N_1 \Delta h_1 + N_2 \Delta h_2 + \cdots + N_k h_k = \sum_{i=1}^{k} N_i \Delta h_i \tag{11-22a}$$

and
$$\Delta h_m = y_1 \Delta h_1 + y_2 \Delta h_2 + \cdots + y_k \Delta h_k = \sum_{i=1}^{k} y_i \Delta h_i \tag{11-22b}$$

These equations for the enthalpy change could be applied to a mixture contained within a closed system at constant pressure. This same format would apply to a steady-state control-volume analysis with one inlet and one exit stream. In either case the h_i data would be obtained from ideal-gas property tables.

The changes in internal energy and enthalpy of an ideal gas can also be evaluated from specific-heat data. Recall that

$$\Delta u_i = c_{v,i}\,\Delta T \qquad \text{and} \qquad \Delta h_i = c_{p,i}\,\Delta T$$

where $c_{v,i}$ and $c_{p,i}$ are usually taken to be constant or the arithmetic average values in the given temperature range. When specific-heat data are employed, Eq. (11-20*b*) becomes

$$\Delta u_m = \Delta T \sum_{i=1}^{k} y_i c_{v,i} = c_{v,m}\,\Delta T \tag{11-23}$$

and Eq. (11-22*b*) becomes

$$\Delta h_m = \Delta T \sum_{i=1}^{k} y_i c_{p,i} = c_{p,m}\,\Delta T \tag{11-24}$$

where $c_{v,i}$ and $c_{p,i}$ in these equations must be expressed on a mole basis, since mole fractions are involved. The quantities $c_{v,m}$ and $c_{p,m}$ are the composite values of c_v and c_p for the mixture. Frequently, the specific-heat data are given on a mass basis. In such cases, mass fractions must be used when evaluating $c_{v,m}$ and $c_{p,m}$. That is, on a mass basis

$$c_{v,m} = \sum_{i=1}^{k} mf_i c_{v,i} \tag{11-25a}$$

$$c_{p,m} = \sum_{i=1}^{k} mf_i c_{p,i} \tag{11-25b}$$

Thus in general we can evaluate Δu_m and Δh_m either on a mole basis or a mass basis, depending on whether the u_i, h_i, $c_{v,i}$, and $c_{p,i}$ data are on a mole or mass basis.

Example 11-3M Consider the mixture of three ideal gases used in Example 11-2M at an initial state of 300°K and 3 bars. The temperature is raised to 500°K at constant pressure. Determine the specific enthalpy change of the mixture, in kJ/kg and kJ/kg · mol, by using (*a*) c_p data and (*b*) tabular h data.

SOLUTION A summary of the necessary data is shown in the accompanying table. The gravimetric and volumetric analyses are taken from Example 11-2M, and the values of c_p and h for each gas are taken from Tables A-4M, A-7M, A-8M, and A-11M.

Component	Gravimetric analysis, %	Volumetrical analysis, %	c_p, kJ/(kg)(°K) 300°K	c_p, kJ/(kg)(°K) 500°K	h, kJ/kg · mol 300°K	h, kJ/kg · mol 500°K
H_2	10	62.50	14.307	14.513	8522	14,350
O_2	48	18.75	0.918	0.972	8736	14,770
CO	42	18.75	1.040	1.063	8723	14,600

(*a*) The specific enthalpy change of the mixture is first found by employing the format of Eq. (11-24). However, the c_p data above are on a mass basis. Therefore we must use mass fractions for the mixture, rather than mole fractions, as shown in Eq. (11-25*b*). In addition, the arithmetic average of the c_p data between 300 and 500°K is used.

$$\Delta h_m = [0.10(14.420) + 0.48(0.945) + 0.42(1.051)]\,(500 - 300)$$
$$= 467.4\,\text{kJ/kg}$$

To convert this value to a mole basis, we find from Example 11-2M that the molar mass of the mixture M_m is 12.5. Therefore

$$\Delta h_m = 467.4\,\text{kJ/kg} \times 12.5\,\text{kg/kg}\cdot\text{mol} = 5843\,\text{kJ/kg}\cdot\text{mole}$$

(*b*) Since the enthalpy data are reported on a mole basis, the mixture enthalpy is found in this case by using Eq. (11-22*b*). That is,

$$\Delta h_m = 0.6250(14{,}350 - 8522) + 0.1875(14{,}770 - 8736)$$
$$+ 0.1875(14{,}600 - 8723)$$
$$= 5876\,\text{kJ/kg}\cdot\text{mol}$$

This answer differs from that found in part *a* by about one-half percent, and is due to using average specific heats in part *a*. The answer in part *b* could be converted to a mass basis by dividing the answer by M_m.

Example 11-3 Consider the mixture of three ideal gases used in Example 11-2 at an initial state of 100°F and 14.5 psia. The temperature is raised to 500°F at constant pressure. Determine the specific enthalpy change of the mixture, , in Btu/lb and Btu/lb · mol, by using (*a*) c_p data and (*b*) tabular h data.

SOLUTION A summary of the necessary data is shown in the accompanying table. The gravimetric and volumetric analyses are taken from Example 11-2, and the values of c_p and h for each gas are taken from Tables A-4, A-7, A-8, and A-11.

Component	Gravimetric analysis, %	Volumetric analysis, %	c_p, Btu/(lb)(°F)		h, Btu/lb · mol	
			100°F	500°F	100°F	500°F
H_2	10	62.50	3.426	3.469	3799	6585
O_2	48	18.75	0.220	0.235	3887	6786
CO	42	18.75	0.249	0.256	3890	6705

(*a*) The specific enthalpy change of the mixture is first found by employing the format of Eq. (11-24). However, the c_p data above are on a mass basis. Therefore we must use mass fractions in the calculation, rather than mole fractions, as shown in Eq. (11-25*b*). In addition, the arithmetic average of the c_p data between 100 and 500°F will be used. Hence

$$\Delta h_m = [0.10(3.448) + 0.48(0.2275) + 0.42(0.2525)]\ (500 - 100)$$

$$= 224.0 \text{ Btu/lb mixture}$$

To convert this value to a mole basis, we use the value of 12.5 for the molar mass M_m found from Example 11-2. Therefore

$$\Delta h_m = 224.0 \text{ Btu/lb} \times 12.5 \text{ lb/lb} \cdot \text{mol} = 2800 \text{ Btu/lb} \cdot \text{mol}$$

(*b*) Since the enthalpy data are reported on a mole basis, the mixture enthalpy is found in this case by using Eq. (11-22*b*). That is,

$$\Delta h_m = 0.6250(6585 - 3799) + 0.1875(6786 - 3887) + 0.1875(6705 - 3890)$$

$$= 2813 \text{ Btu/lb} \cdot \text{mol}$$

The answers from parts *a* and *b* are in good agreement, due to the small variation of c_p over the given temperature range for each gas. Finally, the answer in part *b* could be converted to a mass basis by dividing the answer by the molar mass M_m.

The entropy of a mixture of ideal gases can also be determined on the basis of the Gibbs-Dalton rule. Since each gas behaves as if it alone occupied the volume V of the system at the mixture temperature T, we may write

$$S_m = S_1(T, V) + S_2(T, V) + \cdots + S_k(T, V) \tag{11-26}$$

The total change of entropy for a mixture of gases is determined from the sum of the entropy changes of the individual constituents. That is,

$$\Delta S_m = N_1\,\Delta s_1 + N_2\,\Delta s_2 + \cdots + N_k\,\Delta s_k \tag{11-27a}$$

or

$$\Delta S_m = m_1\,\Delta s_1 + m_2\,\Delta s_2 + \cdots + m_k\,\Delta s_k \tag{11-27b}$$

In Eq. (11-27*a*) the specific entropy change Δs_i must be expressed on a mole basis, while in Eq. (11-27*b*) it must be expressed on a mass basis.

The entropy change of an ideal gas is expressed usually as a function of temperature and pressure, rather than T and V of the system. Recall that

$$ds = c_p \frac{dT}{T} - R \frac{dp}{p} \tag{7-15}$$

The symbol p has been used for the pressure in this case, since, at the temperature and the volume of the mixture, the pressure of any gas is measured by its partial, or component, pressure and not the total pressure P of the system. The change in entropy of an ideal gas which is part of a mixture, in terms of its component or partial pressure p_i, is given by

$$\Delta s_i = c_{p,i} \ln \frac{T_2}{T_1} - R_i \ln \frac{p_{i2}}{p_{i1}} \tag{11-28}$$

Recall that in the gas tables the integral of $c_p\, dT/T$ is given by $s_2^0 - s_1^0$. Consequently, the entropy change of the ith component in a mixture of ideal gases may also be found from

$$\Delta s_i = s_{i2}^0 - s_{i1}^0 - R_i \ln \frac{p_{i2}}{p_{i1}} \tag{11-29}$$

if the gas tables are available for the component. The use of the entropy function in calculations for processes involving ideal-gas mixtures will now be illustrated by several examples.

Example 11-4M A mixture of ideal gases consisting of 0.20 kg of nitrogen and 0.30 kg of carbon dioxide is compressed from 2 bars and 300°K to 6 bars adiabatically and reversibly in a closed system. Accounting for variable specific heats, determine (*a*) the final temperature, (*b*) the work required, in kilojoules, and (*c*) the entropy change of the nitrogen and the carbon dioxide, in kJ/°K. Finally, (*d*) determine the final temperature using constant specific-heat data.

SOLUTION (*a*) The process is isentropic. Therefore the entropy change of the mixture is zero. The final temperature can be found by employing Eqs. (11-27*a*) and (11-29) and tabular data from the Appendix. We find that

$$\Delta S_m = 0 = \frac{0.20}{28}\left[(s_{N_2}^0 - 191.682) - 8.314 \ln \tfrac{6}{2}\right]$$

$$+ \frac{0.30}{44}\left[(s_{CO_2}^0 - 213.915) - 8.314 \ln \tfrac{6}{2}\right]$$

Note that for each gas the term $\ln (p_{i2}/p_{i1})$ has been replaced by $\ln (P_2/P_1)$. This is valid because the mole fraction of a given gas remains constant during the process. Rearranging,

$$0.00714\, s_{N_2}^0 + 0.00682\, s_{CO_2}^0 = 2.955 \text{ kJ/°K}$$

The solution to this equation is found by iteration. A final temperature is assumed, and the s^0 values for the two gases are found in their respective tables. If these two values of s^0 do not satisfy the equation, another temperature must be assumed, until the correct final temperature is determined. In this case

At 390°K: $\sum N_i s_i^0 = 0.00714(199.331) + 0.00682(224.182) = 2.952 \text{ kJ/°K}$

At 400°K: $\sum N_i s_i^0 = 0.00714(200.071) + 0.00682(225.225) = 2.965 \text{ kJ/°K}$

By interpolation, the final temperature is close to 392°K.

(*b*) The work required for the adiabatic-compression process of a closed system is simply

$$W = U_2 - U_1 = \sum N_i\ \Delta u_i$$

Interpolating again for the final u_i data, we find that

$$W = 0.00714(8146 - 6229) + 0.00682(9784 - 6939) = 33.1 \text{ kJ}$$

(*c*) The entropy changes of the individual gases in the mixture are found by applying Eq. (11-29). Thus, after interpolating for the final s_i^0 values,

$$\Delta S_{N_2} = 0.00714[(199.479 - 191.682) - 8.314 \ln \tfrac{6}{2}] = 0.0095 \text{ kJ/°K}$$

$$S_{CO_2} = 0.00682[(224.391 - 213.915) - 8.314 \ln \tfrac{6}{2}] = -0.0092 \text{ kJ/°K}$$

Since the overall process is isentropic, the sum of ΔS for the two gases should be zero. The slight difference here is due to round-off error. Note, however, that each gas undergoes a definite entropy change of its own. For a mixture of two gases, this change is always equal in magnitude but opposite in sign.

(*d*) The final temperature can be estimated from the relationship, $T_2/T_1 = (P_2/P_1)^{(k-1)/k}$. The specific-heat ratio must be the average for the mixture, that is, $k_m = c_{p,m}/c_{v,m}$. Equation (11-25) is used to evaluate $c_{p,m}$ and $c_{v,m}$. Since T_2 is unknown, we either must use the specific-heat data at the original temperature or guess the final temperature and use the average specific-heat data in the temperature range. If we assume that T_2 is around 400°K, and then use average values at 350°K, then

$$c_{v,m} = \frac{0.20(0.744) + 0.30(0.706)}{0.20 + 0.30} = 0.721 \text{ kJ/(kg)(°K)}$$

$$c_{p,m} = \frac{0.20(1.041) + 0.30(0.895)}{0.20 + 0.30} = 0.953 \text{ kJ/(kg)(°K)}$$

Therefore the specific-heat ratio k_m for the mixture is 0.953/0.721 = 1.32. The final temperature T_2, then is

$$T_2 = T_1\left(\frac{P_2}{P_1}\right)^{(k-1)/k} = 300\left(\frac{6}{2}\right)^{0.242} = 391\text{°K}$$

Due to the relatively small temperature change, this method essentially agrees with the answer obtained in part *a*.

Example 11-4 A mixture of ideal gases consisting of 0.2 lb of nitrogen and 0.30 lb of carbon dioxide is compressed from 14.5 psia and 60°F to 58 psia adiabatically and reversibly in a closed system. Accounting for variable specific heats, determine (*a*) the final temperature, (*b*) the work required, in Btu, and (*c*) the entropy change of the nitrogen and the carbon dioxide, in Btu/°R. Finally, (*d*) determine the final temperature using constant specific-heat data.

SOLUTION (*a*) The process is isentropic. Therefore the entropy change of the mixture is zero. The final temperature can be found by employing Eqs. (11-27*a*) and (11-29) and tabular data from the Appendix. We find that

$$\Delta S_m = 0 = \frac{0.20}{28}\left[(s^0_{N_2} - 45.519) - 1.986 \ln \frac{58}{14.5}\right]$$

$$+ \frac{0.30}{44}\left[(s^0_{CO_2} - 50.750) - 1.986 \ln \frac{58}{14.5}\right]$$

Note that for each gas the term $\ln (p_{i2}/p_{i1})$ has been replaced by $\ln (P_2/P_1)$. This is valid because the mole fraction of a given gas remains constant during the process. Rearranging,

$$0.00714\, s^0_{N_2} + 0.00682\, s^0_{CO_2} = 0.7096 \text{ Btu/°R}$$

The solution to this equation is found by iteration. A final temperature is assumed, and the s^0 values for the two gases are found in their respective tables. If these two values of s^0 do not satisfy the equation, another temperature must be assumed, until the correct final temperature is determined. In this case

At 720°R: $\quad \sum N_i s_i^0 = 0.00714(47.785) + 0.00682(53.780) = 0.7080 \text{ Btu/°R}$

At 740°R: $\sum N_i s_i^0 = 0.00714(47.977) + 0.00682(54.051) = 0.7112$ Btu/°R

By interpolation, the final temperature is close to 730°R

(*b*) The work required for the adiabatic-compression process of a closed system is simply

$$W = U_2 - U_1 = \sum N_i \, \Delta u_i$$

Interpolating again for the final u_i data, we find that

$$W = 0.00714(3625 - 2579) + 0.00682(4398 - 2848) = 24.2 \text{ Btu}$$

(*c*) The entropy changes of the individual gases in the mixture are found by applying Eq. (11-29). Thus, after interpolating for the final s_i^0 values,

$$\Delta S_{N_2} = 0.00714\left[(47.881 - 45.519) - 1.986 \ln \frac{58}{14.5}\right] = 0.00279 \text{ Btu/°R}$$

$$\Delta S_{CO_2} = 0.00682\left[(53.916 - 50.750) - 1.986 \ln \frac{58}{14.5}\right] = -0.00281 \text{ Btu/°R}$$

Since the overall process is isentropic, the sum of ΔS for the two gases should be zero. The slight difference here is due to round-off error. Note, however, that each gas undergoes a definite entropy change of its own. For a mixture of two gases, this change is always equal in magnitude but opposite in sign.

(*d*) The final temperature can be estimated from the relationship, $T_2/T_1 = (P_2/P_1)^{(k-1)/k}$. The specific-heat ratio must be the average for the mixture, that is, $k_m = c_{p,m}/c_{v,m}$. Equation (11-25) is used to evaluate $c_{p,m}$ and $c_{v,m}$. Since T_2 is unknown, we either must use the specific-heat data at the original temperature or guess the final temperature and use the average values of the specific heats in that temperature range. If we assume that T_2 is around 730°R, and then use average values around 625°R or 165°F, then

$$c_{v,m} = \frac{0.20(0.178) + 0.30(0.168)}{0.20 + 0.30} = 0.172 \text{ Btu/(lb)(°F)}$$

$$c_{p,m} = \frac{0.20(0.249) + 0.30(0.213)}{0.20 + 0.30} = 0.227 \text{ Btu/(lb)(°F)}$$

Therefore the specific-heat ratio k_m for the mixture is 0.227/0.172 = 1.32. The final temperature T_2, then, is estimated to be

$$T_2 = T_1 \left(\frac{P_2}{P_1}\right)^{(k-1)/k} = 520\left(\frac{58}{14.5}\right)^{0.242} = 727\text{°R}$$

Due to the relatively small temperature change, this method essentially agrees with the answer obtained in part *a*.

11-5 MIXING PROCESSES INVOLVING IDEAL GASES

When an ideal-gas mixture undergoes a process change without a change in composition, the component pressure of each species remains the same. Consequently, the value of p_{i2}/p_{i1} is the same as P_2/P_1 for the total mixture. However, when two or more pure gases are mixed, or two gas mixtures come into contact, the component or partial pressure may change significantly. This change must be accounted for, especially when evaluating the entropy change of the individual gases.

Consider a rigid tank divided into k compartments by partitions. Different ideal gases fill each of the k compartments, and the total pressure and temperature of each pure gas initially are the same. The tank is insulated, and then the partitions are pulled from the tank. Eventually, each gas spreads into the total volume of the tank, and a new equilibrium state is attained. In the absence of heat and work intereactions, the energy equation for the closed system reduces to $\Delta U = 0$. In terms of the individual components this takes the form

$$U_{\text{init}} = U_{\text{final}}$$

or

$$\sum N_{i1} u_{i1} = \sum N_{i2} u_{i2} \tag{11-30}$$

where u_{i1} and u_{i2} are the initial and final molar internal energies of the ith component, and they are solely a function of the temperature for ideal gases. The conservation of energy principle requires that

$$U_2 = \sum N_{i2} u_{i2} = \text{const}$$

The values of N_i are constant, since chemical reactions are not under consideration. In addition, it must be noted that u_i for any component increases with increasing temperature. If the temperature increases upon mixing, each of the u_{i2} values would be larger. Such an increase in the u_{i2} values would increase the overall energy U_2. However, the conservation of energy principle requires that U_2 be constant. Therefore, the final temperature must be equal to the initial value for the case under consideration.

The effect of mixing of two gases initially at the same pressure and temperature on the final pressure may be found by using the ideal-gas equation. For any component i we may write

$$\frac{p_{i2} V_{i2}}{p_{i1} V_{i1}} = \frac{N_i R_u T}{N_i R_u T}$$

or

$$\frac{p_{i2}}{p_{i1}} = \frac{V_{i1}}{V_{i2}} = \frac{V_{i1}}{V} = y_i$$

where V is the total volume of the mixture. Hence $p_{i2} = y_i p_{i1} = y_i P_{\text{init}}$. If we sum over all components,

$$P_{\text{final}} = \sum p_{i2} = \sum y_i P_{\text{init}} = P_{\text{init}}$$

Therefore the pressure also does not change upon mixing in this special case.

If the initial temperatures before mixing are not the same, then the first law may be used to determine the final temperature. In general, either of the two following forms could be used for a closed-system analysis.

$$\Delta U_m = 0 = \sum_{i=1}^{k} N_i \, \Delta u_i \tag{11-20a}$$

or

$$\Delta U_m = 0 = \sum_{i=1}^{k} m_i c_{v,i} \, \Delta T = \sum_{i=1}^{k} N_i c_{v,i} \, \Delta T \tag{11-31}$$

In the first equation tabular u_i data on a mole basis would be appropriate, while in the second equation specific-heat data on a mass or molar basis would be appropriate. When the components initially are at different pressures and temperatures, the final mixture pressure is determined directly from the ideal-gas equation, after an energy balance has been made to find the final temperature.

The mixing of two or more gases is highly irreversible, and under adiabatic conditions a positive entropy change would be expected. The entropy change of each component is given by Eq. (11-29) multiplied by the moles of that species.

$$\Delta S_i = N_i\left(s_{i2}^0 - s_{i1}^0 - R_u \ln \frac{p_{i2}}{p_{i1}}\right)$$

If we again consider the special case of mixing different gases initially at the same temperature and pressure, then $T_2 = T_1$. Consequently the first two terms on the right cancel. Also, $p_{i1} = P$ and $p_{i2} = y_i P$, since the initial and final total pressures are equal. Therefore for the ith species,

$$\Delta S_i = -N_i R_u \ln \frac{y_i P}{P} = -N_i R_u \ln y_i \tag{11-32}$$

The sum of the ΔS_i terms for the mixture is the total entropy change for the adiabatic mixing process.

$$\Delta S_m = -R_u \sum N_i \ln y_i \tag{11-33}$$

Since the y_i values are always less than unity, the entropy change given by Eq. (11-33) is always positive. In addition, the total entropy change is independent of the composition of the gases, and depends soley on the number of moles of each gas involved in the mixing process. Finally, the entropy change for the mixing of gases developed above is valid only if all the gases are distinguishable from one another. The entropy change is zero when the same gas is mixed at constant pressure and temperature.

Example 11-5M A rigid, insulated tank is divided into two compartments by a partition. Initially 0.02 kg · mol of nitrogen fills one compartment at 2 bars and 100°C. The other compartment contains 0.03 kg · mol of carbon dioxide at 1 bar and 20°C. The partition is removed and the gases allowed to mix. Determine the temperature and pressure of the mixture and the entropy change for the mixing process.

SOLUTION For the low pressures involved the gases are assumed to behave as ideal gases. Since both Q and W are zero, then the first law for a closed system dictates that $\Delta U = 0$, or U_{init} equals U_{final}. On the basis of Eq. (11-31)

$$[Nc_v(T_2 - T_1)]_{N_2} + [Nc_v(T_2 - T_1)]_{CO_2} = 0$$

The final temperature T_2 is the same for each gas, and lies between 20 and 100°C. Based on Table A-4M we shall let c_v for nitrogen and carbon dioxide be constant and have values of 0.744 and 0.680 kJ/(kg)(°K), respectively. These are 20.8 and 29.9 kJ/(kg · mol)(°K) on a molar basis. Substitution of numerical values into the above equation yields

$$0.02(20.8)(T_2 - 100) + 0.03(29.9)(T_2 - 20) = 0$$

Therefore

$$T_2 = 45.3°\text{C}$$

The final pressure is determined from the ideal-gas relation, $PV = NR_uT$. The total volume is the sum of the volumes of the original compartments, which may also be determined from the same equation. Hence

$$V_{N_2} = \frac{NR_uT}{P} = \frac{0.02(0.08314)(373)}{2} = 0.310\,\text{m}^3$$

$$V_{CO_2} = \frac{0.03(0.08314)(293)}{1} = 0.731\ \text{m}^3$$

$$V_{tot} = 0.310 + 0.731 = 1.041\ \text{m}^3$$

As a result,

$$P_2 = \frac{N_m R_u T_2}{V_m} = \frac{0.05(0.08314)(318)}{1.041} = 1.27\ \text{bars}$$

The entropy change is the sum of the entropy changes for the individual components. If we use Eq. (11-28), we will need c_p data. Recall that for ideal gases $c_p = c_v + R$. Hence for N_2 and CO_2 the c_p values are 29.1 and 38.2 kJ/(kg · mol)(°K), respectively. Keep in mind that each gas exerts only its component pressure at the total volume and temperature of the system. For the two gases we find that

$$\Delta S_{N_2} = 0.02\left[29.1\ \ln\frac{318}{373} - 8.314\ \ln\frac{0.4(1.27)}{2}\right]$$

$$= 0.02(-4.64 + 11.40) = 0.135\ \text{kJ/°K}$$

$$\Delta S_{CO_2} = 0.03\left[38.2\ \ln\frac{318}{293} - 8.314\ \ln\frac{0.6(1.27)}{1}\right]$$

$$= 0.03(3.13 + 2.26) = 0.162\ \text{kJ/°K}$$

Therefore, for the entire system,

$$\Delta S_{tot} = 0.135 + 0.162 = 0.297\ \text{kJ/°K}$$

The entropy change for each gas could also be calculated from Eq. (11-29), which employs the s^0 values from the gas tables A-6M and A-9M. For example, the entropy change for nitrogen would be

$$\Delta S_{N_2} = 0.2\left[193.377 - 198.027 - 8.314\ \ln\frac{0.4(1.27)}{2}\right]$$

$$= 0.2(-4.65 + 11.40) = 0.135\ \text{kJ/°K}$$

This agrees with the previously calculated value, because of the small temperature range involved.

Example 11-5 A rigid, insulated tank is divided into two compartments by a partition. Initially 0.2 mol of nitrogen is introduced into one compartment at a pressure of 30 psia and a temperature of 140°F. At the same time 0.3 mol of carbon dioxide is introduced into the other compartment at 15 psia and 60°F. The partition is then removed and the gases are allowed to mix. Determine the temperature and pressure of the mixture at equilibrium and the entropy change for the mixing process. The constant-volume molar specific heats for nitrogen and carbon dioxide will be assumed to be constant for the process, and are 5.0 Btu/(mol)(°R) and 6.9 Btu/(mol)(°R), respectively.

SOLUTION For the low pressures involved, the gases behave essentially as ideal gases. We shall assume that the mixture obeys the Gibbs-Dalton law. The internal energy of ideal gases is solely a function of temperature. Thus the final equilibrium temperature depends upon the internal energy of the mixture. There are no work or heat interactions for the process if the system is chosen to include the entire tank. Therefore the internal energy of the system does not change. In terms of the mixing process, this means that U_{init} equals U_{final}, or $\Delta U = 0$. Consequently, on the basis of Eq. (11-31),

$$[Nc_v(T_2 - T_1)]_{N_2} + [Nc_v(T_2 - T_1)]_{CO_2} = 0$$

The final temperature T_2 is the same for each gas at equilibrium. Substitution of the proper numerical values into the equation yields

$$0.2(5.0)(T_2 - 140) + 0.3(6.9)(T_2 - 60) = 0$$

Therefore

$$T_2 = 86°\text{F}$$

The final pressure may now be determined from the ideal-gas relation, $PV = NR_uT$. The total volume of the mixture is the sum of the volumes of the original compartments, which may also be determined from the same equation. Hence

$$V_{N_2} = \frac{NR_uT}{P} = \frac{0.2(10.73)(600)}{30} = 42.9 \text{ ft}^3$$

$$V_{CO_2} = \frac{0.3(10.73)(520)}{15} = 111.6 \text{ ft}^3$$

$$V_{\text{tot}} = 42.9 + 111.6 = 154.5 \text{ ft}^3$$

$$P_2 = \frac{N_m R_u T_2}{V_2} = \frac{(0.2 + 0.3)(10.73)(460 + 86)}{154.5} = 19.0 \text{ psia}$$

On the basis of the Gibbs-Dalton law, the entropy change is the sum of the entropy changes for the individual components. We may apply Eq. (11-28) to each constituent, keeping in mind that a gas exerts only its component pressure at the volume and temperature of the system. The c_p data required for Eq. (11-28) may be calculated from the ideal-gas relation, $c_p = c_v + R$. Using the c_v data given in the statement of the problem, we find that c_p values for nitrogen and carbon dioxide are approximately 7.0 and 8.9 Btu/(lb · mol)(°R), respectively. On this basis

$$\Delta S_{N_2} = N\left(c_p \ln \frac{T_2}{T_1} - R_u \ln \frac{p_2}{p_1}\right)$$

$$= 0.2\left[7.0 \ln \frac{546}{600} - 1.986 \ln \frac{0.4(19.0)}{30}\right]$$

$$= 0.2(-0.659 + 2.74) = 0.416 \text{ Btu/°R}$$

$$\Delta S_{CO_2} = 0.3\left[8.9 \ln \frac{546}{520} - 1.986 \ln \frac{0.6(19.0)}{15}\right]$$

$$= 0.3(0.434 + 0.554) = 0.988 \text{ Btu/°R}$$

Therefore, for the entire system,

$$\Delta S_{\text{tot}} = 0.416 + 0.988 = 1.404 \text{ Btu/°R}$$

The entropy change for each gas could also be calculated from Eq. (11-29), which employs the s^0 values from gas tables A-6 and A-9. For example, the entropy change for nitrogen on this basis would be

$$\Delta S_{N_2} = 0.2\left[45.858 - 46.514 - 1.986 \ln \frac{0.4(19.0)}{30}\right]$$

$$= 0.2(-0.656 + 2.74) = 0.416 \text{ Btu/°R}$$

This agrees with the previously calculated value, because of the small temperature range involved.

Mixing of ideal gases also occurs in steady-flow, open systems. The calculations are essentially the same as for the closed-system analysis shown above, except a different energy equation must be used. In addition, there is no general way to evaluate the final pressure unless some additional data on flow conditions entering and leaving the control volume are given.

11-6 PROPERTIES OF A MIXTURE OF AN IDEAL GAS AND A VAPOR

Although the relationships developed in the preceding sections for the properties of ideal-gas mixtures are of general usefulness, there is one additional complication that must be recognized when dealing with gas mixtures. There is always the possibility that one or more of the gases may exist in a state that is close to a saturation state for the given component. We have seen that each gas exerts a pressure which is equal to its component pressure. But the component pressure can never be greater than the saturation pressure for that component at the mixture temperature. Any attempt to increase the component pressure beyond the saturation pressure results in partial condensing of the vapor. For example, consider increasing the total pressure of an ideal-gas mixture at constant temperature. Since the mole fractions of every component gas are fixed (at least temporarily), the component pressure of each increases in direct proportion to the increase in the total pressure. However, if the component pressure of any constituent eventually exceeds its saturation pressure for that temperature, the gas will begin to condense out as the pressure is increased further. This process is illustrated on a Pv diagram in Fig. 11-3*a*. The gas which condenses under this circumstance is usually spoken of as a vapor. Consequently, the ideal-gas mixtures are referred to as gas-vapor mixtures.

A similar situation also occurs when a gas-vapor mixture is cooled at constant pressure In this case the partial pressure of the vapor remains constant (up to the point of condensation), but the temperature eventually is lowered sufficiently to equal the saturation temperature for the given partial pressure. As the temperature is lowered still further, the saturation pressure corresponding to the temperature becomes less than the actual partial pressure, and hence some of the vapor must condense. The effect of lowering the temperature of a mixture containing condensable vapor is shown on a Ts diagram in Fig. 11-3*b*, where state 2 is the state where condensation begins. A well-known example of a gas mixture

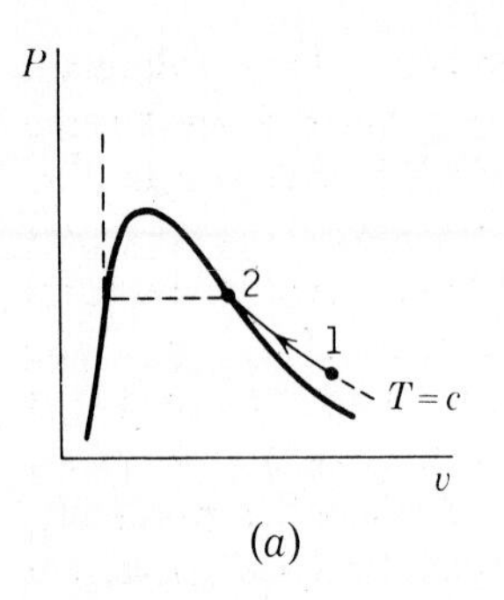

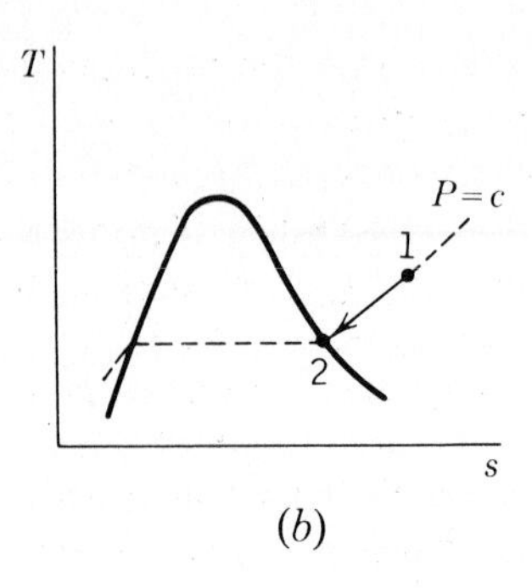

Figure 11-3 The effect of raising the total pressure or lowering the temperature of a gas mixture containing a vapor.

containing a condensable vapor is the air in the atmosphere. The condensation of water from the air as the temperature is lowered, forming dew, is a common experience. Although there are a number of systems of engineering interest which involve the use of gas-vapor mixtures, we shall direct our attention in this chapter to air–water-vapor mixtures. Such a mixture is frequently referred to as atmospheric air. The study of the basic properties of such a mixture is quite important, since it is realized that the composition of atmospheric air plays a decisive role in the proper functioning of the human body.

The temperature of the mixture as measured by a conventional thermometer is called the *dry-bulb* temperature, such as T_1 in Fig. 11-3. In Fig. 11-3*b* a process was illustrated for which the temperature of a gas-vapor mixture was lowered and the total pressure of the mixture remained constant. Until the saturation state of the vapor is reached, the partial pressures of the constituents also remain constant. The temperature at which the mixture becomes saturated, or condensation begins, when a mixture of dry air and water vapor is cooled at constant pressure from an unsaturated state, is called the *dew-point* temperature. Hence the temperature at state 2 on the Ts diagram is the dew point of any mixture for which the partial pressure of the water vapor is represented by the constant-pressure line. The dew-point temperature is the saturation temperature of water which corresponds to the partial pressure of the water vapor actually in the atmospheric air. A dry-air–water-vapor mixture which is saturated with water is frequently called saturated air.

For air–water-vapor mixtures which are not saturated, we need to be able to denote the quantity of water vapor present at a given state of the mixture. This is done conventionally in two ways: by the relative humidity and by the specific humidity (or humidity ratio). The *relative humidity* ϕ is defined as the ratio of the partial pressure of the vapor in a mixture to the saturation pressure of the vapor at the same temperature and pressure of the mixture. If p_v represents the actual vapor pressure and p_g represents the saturation pressure at the same temperature, then

$$\phi \equiv \frac{p_v}{p_g} \tag{11-34}$$

The pressures used to define the relative humidity are shown in the Ts diagram of Fig. 11-4, which is an extension of the data shown in Fig. 11-3*b*. State 1 is the initial state of the water vapor in the mixture, and its vapor pressure at this state is p_1. If this same vapor were present in saturated air at the same temperature, its pressure would necessarily have to be that given at state 3, which is the saturation pressure p_g for that temperature. In terms of the figure, $\phi = p_1/p_3$. The relative humidity ϕ is always less than or equal to unity. It should be noted that since the relative humidity is defined solely in terms of the vapor in the mixture, it is independent of the state of the air in the mixture. Based on the assumption that both the dry air and the water vapor in the mixture behave as ideal gases, the equation for the relative humidity can be expressed in terms of specific volumes (or densities) as well as partial pressures. That is,

$$\phi = \frac{p_v}{p_g} = \frac{RT_v/v_v}{RT_g/v_g} = \frac{v_g}{v_v} = \frac{\rho_v}{\rho_g} \tag{11-35}$$

since the temperatures and the gas constants in the relation are equal.

The *humidity ratio* (or specific humidity) ω describes the quantity of water vapor in a mixture relative to the amount of dry air present. It is formally defined as the ratio of the mass of water vapor present, m_v, to the mass of dry air, m_a. These masses could be on a kilogram or a pound basis. The humidity ratio is not a measure of the mass fraction of the water vapor in the mixture, as should be carefully noted. In equation form, the humidity ratio is

$$\omega \equiv \frac{m_v}{m_a} \tag{11-36}$$

Recall that the specific volume is defined as $v = V/m$. Since both the water vapor and the dry air occupy the same total volume, the mass ratio can be replaced by the ratio of specific volumes. Hence

$$\omega = \frac{V/v_v}{V/v_a} = \frac{v_a}{v_v} = \frac{\rho_v}{\rho_a} \tag{11-37}$$

Going one step further, the humidity ratio can also be expressed in terms of the partial pressures of the two constituent gases. The specific volume of an ideal gas

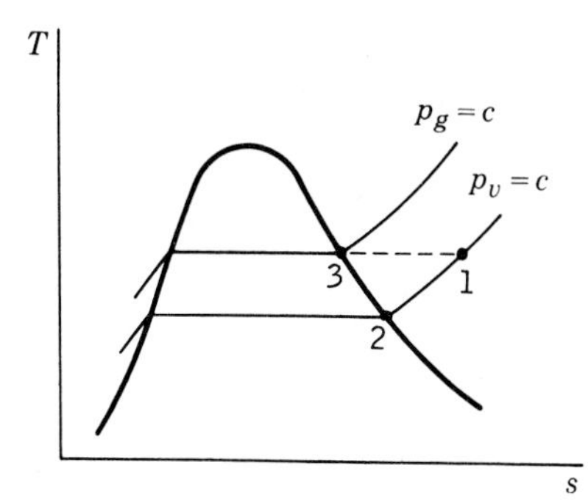

Figure 11-4 Ts diagram for water vapor in atmospheric air.

in a gas mixture is given by $v_i = RT/p_i = R_u T/p_i M_i$, where p_i is the partial pressure and M_i is the molar mass of the component. By substituting this equation into Eq. (11-37), we find that

$$\omega = \frac{R_u T/p_a M_a}{R_u T/p_v M_v} = \frac{M_v p_v}{M_a p_a}$$

The ratio of molar masses for water to air is 0.622, and $p_a = P - p_v$; thus

$$\omega = 0.622 \frac{p_v}{P_a} = 0.622 \frac{p_v}{P - p_v} \tag{11-38}$$

Finally, we are able to relate the humidity ratio and the relative humidity of a mixture by combining Eqs. (11-34) and (11-38). This yields

$$\phi = \frac{p_v}{p_g} = \frac{p_a \omega}{0.622 p_g} \tag{11-39}$$

Any equipment used to alter the state of atmospheric air mixtures is usually operated as a steady-state flow device. Thus, energy balances on such equipment involve the use of the enthalpy of the mixture. The enthalpy of a dry-air–water-vapor mixture is the sum of the enthalpies of the individual components. That is,

$$H_{\text{mix}} = H_{\text{dry air}} + H_{\text{water vapor}}$$
$$= m_a h_a + m_v h_v$$

On the basis of a unit mass of dry air, the specific enthalpy becomes

$$h_{\text{mix}} = h_a + \omega h_v$$

where h_a is the enthalpy of the dry-air component and h_v is the enthalpy of the water vapor. In atmospheric air problems, due to the low vapor pressure of water, h_v may be evaluated by using h_g for a saturated vapor at the given dry-bulb temperature. In addition, due to the small temperature range of interest (typically -10 to 40°C or 20 to 110°F), the constant-pressure specific heat of dry air is essentially a constant. Consequently, the specific enthalpy of the mixture can be represented by

$$h_{\text{mix}} = c_{p,a} T + \omega h_g \tag{11-40}$$

on a dry-air basis. In the temperature range of interest the specific heat of dry air will be taken to be 1.005 kJ/(kg) (°K) or 0.240 Btu/(lb) (°F).

It is helpful in some situations to express h_g in Eq. (11-40) as a function of temperatures. Between the freezing point of water (0°C) and a temperature around 40°C, Eq. (11-40) can be written in the following form with reasonable accuracy for SI units. In units of kJ/kg,

$$h_{\text{mix}} = 1.005T + \omega(2501.7 + 1.82T) \tag{11-41a}$$

where T is the dry-bulb temperature in degrees Celsius, and the datum state for dry air has been selected as 0°C. In USCS units the equation has the form

$$h_{\text{mix}} = 0.240T + \omega(1061 + 0.444T) \tag{11-41b}$$

where in this case T is the dry-bulb temperature in degrees Fahrenheit, and the datum for dry air is 0°F. Equation (11-41*b*), in units of Btu/lb, retains good accuracy between 32 and 100°F.

Finally, it is convenient to have values of the specific volume of the gas-vapor mixture on the basis of a unit mass of dry air. Consider a sample of atmospheric air with a volume V such that it contains a unit mass of dry air, plus water vapor. The volume occupied by the mixture is the same as the volume occupied by the dry-air component. Consequently,

$$v_{\text{mix}} = \frac{V_{\text{mix}}}{m_a} = \frac{V_{\text{a}}}{m_a} = v_a = \frac{R_a T}{p_a} = \frac{R_a T}{P - p_v} \tag{11-42}$$

where p_a is the component pressure and R_a is the specific gas constant for dry air. The evaluation of the above quantities is illustrated in the following example.

Example 11-6M An air–water-vapor mixture at 25°C and 1 bar has a relative humidity of 50 percent. Compute (*a*) the humidity ratio, (*b*) the dew point, (*c*) the enthalpy, in kJ/kg dry air (where $h = 0$ at 0°C), and (*d*) the specific volume, in m^3/kg dry air.

SOLUTION (*a*) The humidity ratio is evaluated from Eq. (11-38). From steam table A-12M it is found that the saturation pressure p_g at 25°C is 0.0317 bar. Therefore the actual vapor pressure p_v is $\phi p_g = 0.50(0.0317) = 0.0159$ bar. Then

$$\omega_1 = \frac{0.622(0.0159)}{1.00 - 0.0159} = 0.01005 \text{ kg water/kg dry air}$$

(*b*) The dew point by definition is the temperature at which the actual vapor pressure becomes the saturation pressure. From Table A-12M the vapor pressure at 10°C is 0.01228 bar and at 15°C it is 0.01705 bar. By linear interpolation, the dew point is roughly 13.8°C.

(*c*) The enthalpy is determined by employing Eq. (11-40). The value of h_g at 25°C found in the saturation steam table is 2547.2 kJ/kg. Thus

$$h_{\text{mix}} = 1.005(25) + 0.01005(2547.2) = 50.72 \text{ kJ/kg dry air}$$

This value can be compared with the approximate value obtained by Eq. (11-41*a*). This latter equation yields

$$h_{\text{mix}} = 1.005(25) + 0.01005[2501.7 + 1.82(25)] = 50.72 \text{ kJ/kg dry air}$$

At this particular state the approximation equation gives the same value.

(*d*) The specific volume of the mixture on a dry-air basis is found from Eq. (11-42).

$$v = \frac{R_a T}{p_a} = \frac{0.08314}{29} \times \frac{298}{1.0 - 0.0159} = 0.868 \text{ m}^3\text{/kg dry air}$$

The specific volume of an atmospheric air mixture typically runs between 0.80 and 0.95 m^3/kg.

Example 11-6 An air–water-vapor mixture at 75°F and 14.7 psia has a relative humidity of 50 percent. Compute (*a*) the humidity ratio, (*b*) the dew point in degrees Fahrenheit, (*c*) the enthalpy, in Btu/lb dry air (where $h = 0$ at 0°F for the dry-air component), and (*d*) the specific volume, in ft^3/lb dry air.

SOLUTION (*a*) The humidity ratio is evaluated from Eq. (11-38). From steam table A-12 it is found that the saturation pressure p_g at 75°F is 0.4300 psia. Therefore the actual vapor pressure

p_v is $\phi P_g = 0.50(0.4300) = 0.2150$ psia. Then

$$\omega_1 = \frac{0.622(0.2150)}{14.7 - 0.2150} = 0.00925 \text{ lb water/lb dry air}$$

Since the humidity ratio is usually a very small number, it is frequently tabulated or referred to on the basis of grains of water per pound of dry air. There are 7000 gr in a pound by definition; hence, for the initial state,

$$\omega_1 = 0.00925(7000) = 64.75 \text{ gr water/lb dry air}$$

(*b*) The dew point by definition is the temperature at which the actual vapor pressure becomes the saturation pressure. From Table A-3 the vapor pressure at 55°F is 0.2141 psia. Consequently, the dew-point temperature of the initial mixture is close to 55°F, since the actual vapor pressure is 0.2150 psia.

(*c*) The enthalpy is determined by employing Eq. (11-40). The value of h_g at 75°F, as found in the saturation steam table, is 1094.2 Btu/lb. Thus

$$h_{\text{mix}} = 0.240(75) + 0.00925(1094.2) = 28.12 \text{ Btu/lb dry air}$$

This value can be compared with the approximate value obtained from Eq. (11-41*b*). This latter equation yields

$$h_{\text{mix}} = 0.240(75) + 0.00925[1061 + 0.444(75)] = 28.12 \text{ Btu/lb dry air}$$

At this particular state the approximation equation gives the same value.

(*d*) The specific volume of the mixture on a dry-air basis is found from Eq. (11-42).

$$v = \frac{R_a T}{p_a} = \frac{10.73}{29} \times \frac{535}{14.7 - 0.2150} = 13.67 \text{ ft}^3\text{/lb dry air}$$

The specific volume of an atmospheric air mixture typically falls in the range of 13 to 14.5ft^3/lb dry air.

11-7 THE ADIABATIC-SATURATION AND WET-BULB TEMPERATURES

In the preceding section the definitions of relative humidity and humidity ratio were introduced as useful quantities in the analysis of problems dealing with the conditioning of water-vapor–dry-air mixtures. Such quantities are not, however, measured directly. In the preceding examples we have presupposed a knowledge of their numerical values. We must therefore investigate methods for evaluating either the relative humidity or the humidity ratio from other easily measured thermodynamic parameters.

One method of evaluating the humidity ratio of a mixture is based on the technique of adiabatic saturation. The experimental apparatus is sketched in Fig. 11-5*a*, and the *Ts* diagram for the process is shown in Fig. 11-5*b*.

Referring to Fig. 11-5*a*, we see that an unsaturated mixture of gases enters a steady-flow channel with a dry-bulb temperature T_1 and a relative humidity which is less than 100 percent. If the channel is long enough, the mixture will pick up additional moisture as it passes over the liquid water in the bottom of the channel, and it will leave the device as a saturated mixture at some temperature

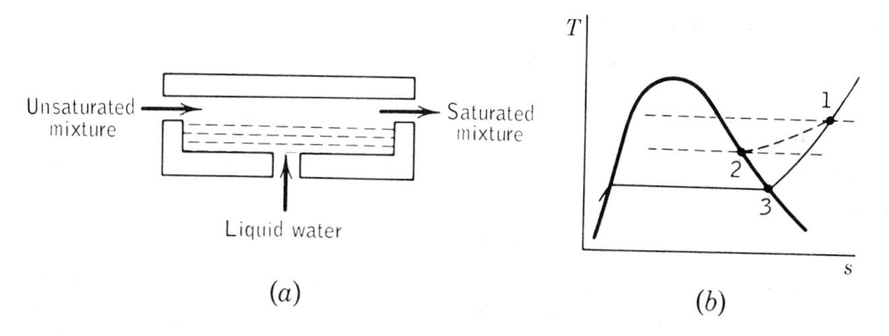

Figure 11-5 Physical description of the adiabatic-saturation process and its representation on a *Ts* diagram.

T_2. The device is normally insulated, so that the final temperature achieved when completely saturated is known as the adiabatic-saturation temperature. The adiabatic-saturation temperature is always less than the dry-bulb temperature T_1 since the evaporation of water into the mixture requires energy, which comes from both the air mixture passing through and the liquid water in the channel. Consequently, the gas mixture is cooled as it becomes saturated. A steady flow of liquid water is added at temperature T_2 to make up for the evaporation of water into the gas stream. The process itself for the mixture is shown by the dashed line from state 1 to state 2 in Fig. 11-5*b*. The temperature at state 3 is the dew-point temperature for the mixture, which is lower than both the initial dry-bulb temperature and the adiabatic-saturation temperature.

In most cases the kinetic- and potential-energy changes of the air stream are negligible; hence the conservation of energy principle for the steady-flow process under adiabatic conditions reduces simply to $H_{in} = H_{out}$. The enthalpy brought into the control volume which lies inside the channel consists of two parts, that due to the entering gas stream and that due to the liquid makeup water. The exit enthalpy is associated solely with the mixture stream leaving the control volume. On the basis of a unit mass of dry air entering and leaving the control volume in a given time interval for the adiabatic-saturation process, the energy balance becomes

$$h_{a1} + \omega_1 h_{v1} + (\omega_2 - \omega_1) h_{f2} = h_{a2} + \omega_2 h_{v2}$$

where h_f is the enthalpy of saturated liquid water. On the basis of Eq. (11-40) this equation becomes

$$c_{p,a} T_{a1} + \omega_1 h_{v1} + (\omega_2 - \omega_1) h_{f2} = c_{p,a} T_{a2} + \omega_2 h_{v2}$$

Noting that $h_{v2} - h_{f2} = h_{fg2}$ and rearranging the equation, we obtain the following expression for ω_1.

$$\omega_1 = \frac{c_{p,a}(T_2 - T_1) + \omega_2 h_{fg2}}{h_{v1} - h_{f2}} \tag{11-43}$$

Every quantity of the right-hand side of the equation is known, once the temperatures T_1 and T_2 are measured. The humidity ratio ω_2 at T_2 is known, since this state is a saturation state. Thus the humidity ratio (and hence the relative humidity) of an unsaturated gas-vapor mixture could be ascertained by measuring the inlet and outlet temperatures and the total pressure of the mixture which undergoes a process of adiabatic saturation.

Although the technique of adiabatic saturation leads to the desired results, it is difficult in practice to attain a saturated state by this method without employing an extremely long flow channel, which is impractical. In lieu of this, a temperature equivalent to the adiabatic-saturation temperature for water-vapor–air mixtures, known as the *wet-bulb* temperature, is used. This temperature is easily measured by the following techniques. The bulb of an ordinary thermometer is covered with a wick which has been moistened with water. The unsaturated atmospheric air of undetermined humidity ratio is then passed over the wetted wick until dynamic equilibrium is attained and the temperature of the wick (and hence the thermometer) reaches a stable value. It is found that, for air–water-vapor mixtures at normal temperatures and pressures, the wet-bulb temperature, the determination of which relies on heat- and mass-transfer rates, is very close in value to the adiabatic-saturation temperature. Thus the temperature T_2 used in Eq. (11-43) to obtain the initial humidity ratio is normally the wet-bulb temperature, and this leads to an answer of sufficient accuracy. This method does not usually lead to correct answers if the pressure of air–water-vapor mixtures is quite different from the usual atmospheric pressures, or for any other gas-vapor mixtures.

Example 11-7M A sample of atmospheric air at 1 bar has a dry-bulb temperature of 24°C and a wet-bulb temperature of 16°C. Determine (*a*) the humidity ratio, (*b*) the relative humidity, and (*c*) the enthalpy of the mixture per kilogram of dry air.

SOLUTION (*a*) The humidity ratio, or specific humidity, is calculated from Eq. (11-43). The value of ω_2 in this equation is determined from Eq. (11-38), where in this special case p_v at state 2 is the saturation pressure at 16°C. Therefore,

$$\omega_2 = \frac{0.622(0.01818)}{1.0 - 0.01818} = 0.0115 \text{ kg water/kg dry air}$$

The enthalpies of the water in the liquid and vapor phases are found in Table A-12M. Hence, using Eq. (11-43),

$$\omega_1 = \frac{1.0(16 - 24) + 0.0115(2463.6)}{2545.4 - 67.2} = 0.00820 \text{ kg water/kg dry air}$$

(*b*) The relative humidity is now computed from Eqs. (11-38) and (11-39). First,

$$\omega_1 = 0.00820 = \frac{0.622p_v}{1.0 - p_v}$$

The solution of this equation yields a value of $p_v = 0.0130$ bar. Thus the relative humidity, by Eq. (11-39), is

$$\phi = p_v/p_g = 0.0130/0.02985 = 0.436 \qquad \text{or } 43.6\%$$

(*c*) The enthalpy of the mixture per unit mass of dry air is

$$h_m = h_a + \omega h_v$$

where h_v again is the enthalpy of saturated vapor, and h_a of dry air is given by $c_p T$. We obtain, then,

$$h_1 = 1.005(24) + 0.00820(2545.4) = 44.99 \text{ kJ/kg dry air}$$

It is important to note the arbitrariness of this enthalpy value. The enthalpy of air is chosen to be zero at 0°C, and the enthalpy of the water vapor, on the basis of the steam tables, is zero as a saturated liquid at the triple state of 0.01°C. If differences in enthalpies are to be calculated, the enthalpy at each state must be based on the same reference values.

Example 11-7 It is found from measurements that a certain sample of atmospheric air at 14.7 psia has a dry-bulb temperature of 74°F and a wet-bulb temperature of 59°F. Determine (*a*) the humidity ratio, (*b*) the relative humidity, and (*c*) the enthalpy of the mixture per pound of dry air.

SOLUTION (*a*) The humidity ratio, or specific humidity, is found from Eq. (11-43). The value of ω_2 in this equation is determined from Eq. (11-38), where in this special case p_v at state 2 is the saturation pressure at 59°F. Consequently,

$$\omega_2 = \frac{0.622(0.247)}{14.7 - 0.247} = 0.0106 \text{ lb water/lb dry air}$$

The enthalpies of the water in the liquid and vapor phases are found from the steam tables in the appropriate places. Therefore

$$\omega_1 = \frac{0.24(59 - 74) + 0.0106(1060.5)}{1094.1 - 27.1}$$

$$= 0.00716 \text{ lb water/lb dry air} = 50.1 \text{ gr water/lb dry air}$$

(*b*) The relative humidity is now computed from Eqs. (11-38) and (11-39). First,

$$\omega_1 = 0.00716 = \frac{0.622 p_v}{14.7 - p_v}$$

The solution of this equation yields a value of p_v equal to 0.167 psia. Then the relative humidity $\phi = p_v/p_g = 0.167/0.415 = 0.402$, or roughly 40 percent.

(*c*) The enthalpy of the mixture is given by $h_a + \omega h_v$, where h_v again is the enthalpy of saturated water vapor at the given temperature, and the enthalpy of dry air is given by $c_p T$. One obtains, then,

$$h_1 = 0.24(74) + 0.00716(1094.1) = 25.60 \text{ Btu/lb dry air}$$

It is important to note the complete arbitrariness of this enthalpy value. The enthalpy of the air was chosen to be zero at 0°F, and the enthalpy of the water vapor, on the basis of the steam tables, is zero as a liquid at the triple state of 32.02°F. If differences in enthalpies are to be calculated for engineering purposes, the enthalpy at each state must be based on the same reference values.

With the equations and definitions introduced in the preceding material, we are now in a position to analyze engineering processes involving air–water-vapor mixtures. No further concepts or definitions are necessary other than the first law of thermodynamics. To make such analyses easier, though, it is convenient to plot the important parameters of such mixtures on a diagram known as a psychrometric chart. A brief discussion of this chart is presented in the following section.

11-8 THE PSYCHROMETRIC CHART

To facilitate computations of process changes for dry-air–water-vapor mixtures, it is convenient to plot some of the important parameters on a diagram known as a psychrometric chart. The evaluation of the humidity ratio (specific humidity) requires a specification of the total pressure; consequently, this chart is usually based on a total pressure of one standard atmosphere. Normally, the humidity ratio appears on the ordinate and the dry-bulb temperature on the abscissa. Both of these appear as linear scales. Equation (11-38) shows that the humidity ratio is directly a function of the vapor pressure of the water, at a given total pressure. Therefore the vapor pressure is also frequently plotted on the ordinate, in psia, but this scale will not be linear. We have also seen from Eq. (11-41) that the humidity ratio is determined from a knowledge of the wet-bulb and dry-bulb temperatures for the mixture. Hence the main parameter on the psychrometric chart which relates the humidity ratio and the dry-bulb temperature is the wet-bulb temperature. Lines of wet-bulb temperatures run from the upper left to the lower right of the chart. An outline of a psychrometric chart is shown in Fig. 11-6. The wet-bulb temperatures begin at a 100 percent saturation line, and at this saturation line the wet-bulb and dry-bulb temperatures have the same value. Inasmuch as the relative humidity is directly related to the humidity ratio, lines of

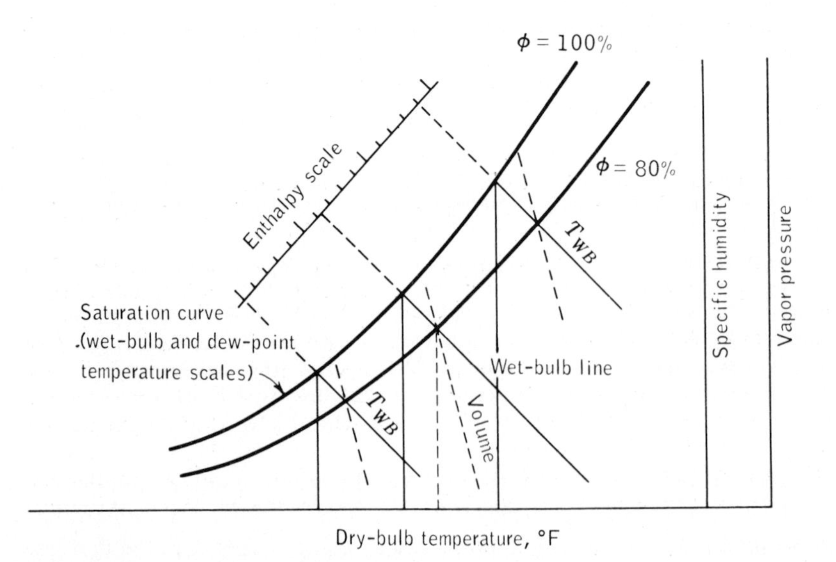

Figure 11-6 Outline of a psychrometric chart.

relative humidity are also conveniently plotted on this diagram. The saturation line is, of course, a line of 100 percent relative humidity. Other relative-humidity lines follow the same general shape to the right of the saturation line.

In addition to these parameters, it is also useful to have lines of constant enthalpy and constant specific volume on the psychrometric chart. It will be noted that lines of constant enthalpy (Btu per pound of dry air) lie approximately parallel to lines of constant wet-bulb temperatures, although in reality they are not exactly parallel. During a process of adiabatic saturation the wet-bulb temperature remains the same and the total enthalpy of the stream remains very nearly constant, except for the slight enthalpy change due to the addition of water to the air stream. As a first approximation, the wet-bulb temperatures and the enthalpies are frequently plotted as parallel lines, if great accuracy is not desired. Other charts take this slight deviation into account by providing separate scales for the two properties. Lines of constant enthalpy appear as straight lines on the chart. In SI units the enthalpy values tabulated on the chart are based on a zero value for the dry air and for the water vapor at 0°C. On the chart using USCS units the enthalpy values are based on a zero value for the dry air at 0°F and a zero value for the water vapor at 32°F. Finally, lines of constant specific volume also appear. In SI units the values are reported in cubic meters per kilogram of dry air, while in engineering units the values are in terms of cubic feet of mixture per pound of dry air. Correction tables are frequently attached to the psychrometric chart, which allow the user to correct for different total pressures. These corrections are normally small, and will be neglected here. The psychrometric chart provides a convenient and fast method for ascertaining the properties of dry-air–water-vapor mixtures. A psychrometric chart in SI and USCS units is given in Figs. A-20M and A-20 in the Appendix.

Example 11-8M The dry-bulb and wet-bulb temperatures of atmospheric air at a total pressure of 1 bar are 23°C and 16°C, respectively. From the psychrometric chart, Fig. A-20M, determine (*a*) the humidity ratio, (*b*) the relative humidity, (*c*) the vapor pressure, in bars, (*d*) the dew point, (*e*) the enthalpy, and (*f*) the specific volume, in m^3/kg.

Solution The properties are found from the chart by first finding the point at the intersection of the vertical dry-bulb temperature line and the sloping wet-bulb-temperature line.

(*a*) The humidity ratio is read from the ordinate at the right to be about 0.0087 kg water/kg dry air.

(*b*) It is found that the 50 percent relative humidity line runs through the point selected.

(*c*) The vapor pressure is also read from the ordinate scale at the right, and is, roughly, 0.0140 bar.

(*d*) The dew point is the temperature at which condensation would just begin if the mixture were cooled at constant pressure. Since the vapor pressure and the humidity ratio remain constant until condensation begins, the dew-point temperature is found by moving horizontally to the left from the initial state until the saturation line is reached. The temperature at this point is close to 12°C, which is the dew point.

(*e*) The enthalpy is found by following the wet-bulb line (which is also an enthalpy line) from the initial state up to the enthalpy scale. The value read is approximately 45.3 kJ/kg of dry air.

(*f*) At the initial state a constant-specific-volume line of 0.86 m^3/kg passes just to the left of the point. The actual value is close to 0.861 m^3/kg of dry air.

Example 11-8 The dry-bulb and wet-bulb temperatures of atmospheric air at a total pressure of 14.7 psia are 74°F and 59°F, respectively. From the psychrometric chart, Fig. A-20, determine (*a*) the humidity ratio, (*b*) the relative humidity, (*c*) the vapor pressure, in psia, (*d*) the dew point, (*e*) the enthalpy, and (*f*) the specific volume per pound of dry air.

SOLUTION The properties are found from the chart by first finding the point at the intersection of the vertical dry-bulb temperature line and the sloping wet-bulb-temperature line.

(*a*) The humidity ratio is read from the ordinate to be slightly more than 50 gr/lb of dry air.

(*b*) It is found that the 40 percent relative humidity line runs through the point selected.

(*c*) The vapor pressure is also read from the ordinate, and is, roughly, 0.167 psia.

(*d*) The dew point is the temperature at which condensation would just begin if the mixture were cooled at constant pressure. Since the vapor pressure and the humidity ratio remain constant until condensation begins, the dew point is found by moving horizontally to the left from the initial state until the saturation line is reached. The dry-bulb temperature at the intersection of the constant-humidity ratio line and the saturation line is found to be 48°F, which is the dew-point temperature.

(*e*) The enthalpy of the mixture is found by following the wet-bulb line (which is also an enthalpy line) from the initial state up to the enthalpy scale in the upper left of the diagram. The enthalpy is read to be approximately 25.9 Btu/lb dry air.

(*f*) At the initial state a constant-specific-volume line equal to 13.6 ft^3/lb dry air passes through the point on the diagram.

11-9 AIR-CONDITIONING PROCESS

A person generally feels more comfortable when the air within a building is maintained in a fairly limited range of temperatures and relative humidities. However, due to mass- and heat-transfer between the inside of the building and the local environment, and due to internal effects such as cooking, baking, and clothes washing in the home, the temperature and the relative humidity frequently reach undesirable levels. To achieve values of T and ϕ within the desired ranges (the comfort zone), it is usually necessary to alter the state of the air. As a result, equipment must be designed to raise or lower the temperature and the relative humidity, individually or simultaneously. In addition to altering the state of a specific air stream by heating, cooling, humidifying, or dehumidifying, a change in state can also be attained by mixing directly the internal or building air with another air stream from, for example, outside the building. Thus there are a number of basic processes to be considered with respect to the conditioning of atmospheric air.

The basic relations available for the evaluation of such processes are three in number. An energy balance on the flow stream(s) may be necessary, as well as mass balances on the water vapor and dry air. In addition, of course, property data for dry air and water substance must be known. The psychrometric chart will be used to describe qualitatively various process designs. The chart is extremely helpful as an aid in visualizing changes of state brought about by process equipment, as well as for estimating property values.

1 Dehumidification with Heating

A fairly common condition within industrial and residential buildings, especially in the summer, is the tendency toward high temperatures and high relative humidities. The discomfort of the body in this situation is well known. A major method lowering both T and ϕ simultaneously is illustrated in Fig. 11-7*a*. The air to be treated is passed through a flow channel which contains cooling coils. The fluid inside the coils might be, for example, relatively cold water or a refrigerant which has been cooled in a vapor-compression refrigeration cycle. The initial

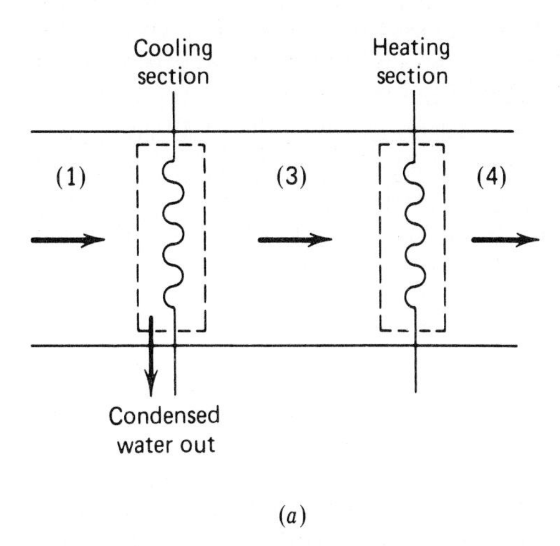

(*a*)

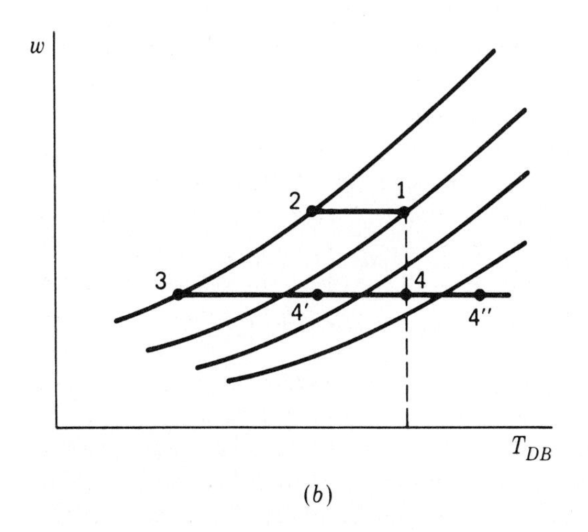

(*b*)

Figure 11-7 Dehumidification process with heating. (*a*) Equipment; (*b*) psychrometric chart process diagram.

state of the air stream is shown as state 1 on the sketch of a psychrometric chart in Fig. 11-7*b*. As the air passes through the cooling coil its temperature decreases and its relative humidity increases, at constant specific humidity. If the air remains in contact with the cooling coil sufficiently, the air stream will reach its dew point, indicated by state 2 in Fig. 11-7*b*. Further cooling requires the air to remain saturated, and its state follows the 100 percent relative humidity line to the left toward state 3. During this latter process water condenses out from the air, and its specific humidity is lowered. Hence, by sufficient contact with the coil, both the temperature and water content of the air are lowered. In many cases this conditioned-air stream flows directly back and mixes with the air in the building.

However, in some cases the conditioned air may be at too low a temperature. This is overcome by then passing the air stream leaving the cooling coil section through a heating section. By proper choice of the temperature of the fluid within the heating coil, the temperature of the air stream leaving the overall equipment may be adjusted to the desired value. Three possibilities are shown as states 4, 4′, and 4″ in Fig. 11-7*b*. By proper adjustment of the amount of cooling (which controls the position of state 3) and the amount of heating, a suitable state 4 may be attained. The following examples illustrate the overall process.

Example 11-9 Outside atmospheric air at 32°C and 70 percent relative humidity is to be conditioned so it enters a home at 22°C and 45 percent relative humidity. First the air passes over cooling coils. The air is cooled below its dew point, and water condenses from the air stream until the desired humidity ratio is reached. The air then passes over a heating coil until the temperature reaches 22°C. Determine (*a*) the amount of water removed, in kg/kg dry air, (*b*) the heat removed by the cooling system, in kJ/kg dry air, and (*c*) the quantity of heat added in the final section, in kJ/kg dry air.

SOLUTION The heat-transfer quantities are given by $q = h_{\text{out}} - h_{\text{in}}$ and the water removed is measured by $\Delta\omega$ for process 2-3 on Fig. 11-7*b*. Therefore the h and ω values are the important properties to evaluate for the overall process. The basic expressions are

$$h_m = c_p T_{\text{air}} + \omega h_g$$

$$\omega = \frac{0.622 p_v}{P - p_v}$$

From Table A-12M the vapor pressures at 32 and 22°C are 0.04759 and 0.02645 bar, respectively. We assume $P = 1$ bar, and recall that c_p for air is 1.005 kJ/(kg)(°C). Therefore

$$\omega_1 = \omega_2 = \frac{0.622(0.70)(0.04759)}{1.00 - 0.70(0.04759)} = 0.0214 \text{ kg water/kg dry air}$$

$$\omega_3 = \omega_4 = \frac{0.622(0.45)(0.02645)}{1.00 - 0.45(0.02645)} = 0.00749 \text{ kg water/kg dry air}$$

At states 1 and 4 the enthalpies are

$$h_1 = 1.005(32) + 0.0214(2559.9) = 86.94 \text{ kJ/kg dry air}$$

$$h_4 = 1.005(22) + 0.00749(2541.7) = 41.15 \text{ kJ/kg dry air}$$

To find h_2 and h_3, we need information on T_2 and T_3; T_2 is the dew-point temperature and is found from the fact that $p_{v2} = p_g$ at $T_2 = 0.7(0.04759) = 0.0333$ bar. From Table A-12M a pressure of 0.0333 bar falls between saturation temperatures of 25 and 26°C. By linear interpolation, $T_2 = 25.8$°C. Thus

$$h_2 = 1.005(25.8) + 0.0214(2548.7) = 80.47 \text{ kJ/kg dry air}$$

Finally, $p_{v3} = p_g$ at $T_3 = 0.45(0.02645) = 0.01190$ bar. This pressure falls between 8 and 10°C in Table A-12M. By linear interpolation, $T_3 = 9.5$°C. Hence

$$h_3 = 1.005(9.5) + 0.00749(2518.9) = 28.42 \text{ kJ/kg dry air}$$

(*a*) The quantity of water removed is given by the difference in the humidity ratios between states 2 and 3. Consequently,

$$\Delta\omega = 0.00749 - 0.0214 = -0.01391 \text{ kg water/kg dry air}$$

The negative sign indicates water was removed from the flow.

(*b*) The heat removed in the cooling-coil section is found from an energy balance on the section. Neglecting kinetic-energy changes, the steady-flow equation per kilogram of dry air is

$$\begin{aligned} q &= (1)h_3 - (\omega_3 - \omega_1)h_{f3} - (1)h_1 \\ &= 1(28.37) - (-0.01391)(39.9) - 1(86.78) \\ &= 28.37 + 0.56 - 86.78 = -57.85 \text{ kJ/kg dry air} \end{aligned}$$

Note that the energy removed by the condensed-liquid stream is extremely small, and might be neglected as a first approximation.

(*c*) The heat added in the final section is simply the enthalpy change of the air stream, if it is assumed that the process is adiabatic. Hence

$$q_{\text{in}} = h_4 - h_3 = 41.15 - 28.42 = 12.73 \text{ kJ/kg dry air}$$

The data for state 2 are not necessary for the solution, but were listed merely to indicate the dew-point properties of the mixture. The values calculated for the enthalpies and the humidity ratios at the various states should now be checked by means of the psychrometric chart in the Appendix.

Example 11-9 Outside atmospheric air at 88°F and 60 percent relative humidity is to be conditioned so that it enters a home at 72°F and 40 percent relative humidity. First the air passes over cooling coils. The air is cooled below its dew point, and water condenses from the air stream until the desired humidity ratio is reached. The air then passes over a heating coil until the temperature reaches 72°F. Determine (*a*) the amount of water removed per pound of dry air, (*b*) the heat removed by the cooling system, in Btu/lb dry air, and (*c*) the quantity of heat added in the heating section, in Btu/lb dry air.

SOLUTION The heat-transfer quantities are given by $q = h_{\text{out}} - h_{\text{in}}$ and water removed is measured by $\Delta\omega$ for process 2-3 in Fig. 11-7*b*. Therefore the h and ω valves are the important properties for evaluating the overall process. The basic expressions are

$$h_m = c_p T_{\text{air}} + \omega h_g$$

$$\omega = \frac{0.622 p_v}{(P - p_v)}$$

From Table A-12 the vapor pressures at 88 and 72°F are 0.6562 and 0.3887 psia, respectively.

We shall assume that $P = 14.7$ psia, and recall that c_p for air is 0.240 Btu/(lb)(°F). Therefore

$$\omega_1 = \omega_2 = \frac{0.622(0.60)(0.6562)}{14.7 - 0.60(0.6562)} = 0.0171 \text{ lb water/lb dry air}$$

$$= 120 \text{ gr water/lb dry air}$$

$$\omega_3 = \omega_4 = \frac{0.622(0.40)(0.3887)}{14.7 - 0.40(0.3887)} = 0.00665 \text{ lb water/lb dry air}$$

$$= 46.6 \text{ gr water/lb dry air}$$

At states 1 and 4 the enthalpies are

$$h_1 = 0.24(88) + 0.0171(1099.9) = 39.9 \text{ Btu/lb dry air}$$

$$h_4 = 0.24(72) + 0.00665(1092.9) = 24.6 \text{ Btu/lb dry air}$$

To find h_2 and h_3 we need information on T_2 and T_3; T_2 is the dew-point temperature for state 1 and is found from the fact that $p_{v2} = p_g$ at $T_2 = 0.6(0.6562) = 0.3937$ psia. From Table A-12 a pressure of 0.3937 psia falls between saturation temperatures of 72 and 74°F. By linear interpolation, $T_2 = 72.4$°F. Thus

$$h_2 = 0.24(72.4) + 0.0171(1093.1) = 36.1 \text{ Btu/lb dry air}$$

Finally, $p_{v3} = p_g$ at $T_3 = 0.4(0.3887) = 0.1555$ psia. This pressure falls between 46 and 48°F in Table A-12. By linear interpolation, $T_3 = 46.4$°F. Hence

$$h_3 = 0.24(46.4) + 0.00665(1081.7) = 18.3 \text{ Btu/lb dry air}$$

(*a*) The quantity of water removed is given by the difference in the humidity ratios between states 2 and 3. Consequently

$$\Delta\omega = 46.6 - 120.0 = -73.4 \text{ gr water/lb dry air}$$

$$= -0.01045 \text{ lb water/lb dry air}$$

The negative sign indicates water was removed from the flow stream.

(*b*) The heat removed in the cooling-coil section is found from an energy balance on the section. Neglecting kinetic-energy changes, the steady-flow equation per pound of dry air is

$$q = (1)h_3 - (\omega_3 - \omega_1)h_{f3} - (1)h_1$$

$$= 1(18.3) - (-0.01045)(14.0) - 1(39.9)$$

$$= 18.3 + 0.1 - 39.9 = -21.5 \text{ Btu/lb dry air}$$

Note that the energy removed by the condensed-liquid stream is extremely small, and might be neglected as a first approximation.

(*c*) The heat added in the final section is simply the enthalpy change of the air stream between states 3 and 4, if it is assumed that the process is adiabatic. Hence

$$q_{\text{in}} = h_4 - h_3 = 24.6 - 18.3 = 6.3 \text{ Btu/lb dry air}$$

The data for state 2 are not necessary, but were listed merely to indicate the dew-point properties of the initial air. The values calculated for the enthalpies and the humidity ratios at the various states should now be checked by means of the psychrometric chart in the Appendix.

2 Evaporative cooling

In desert climates the air in the atmosphere is frequently hot and dry (very low relative humidity). Rather than pass the air through a refrigerated cooling section, which is costly, it is possible to take advantage of the low humidity to achieve cooling. This is accomplished by passing the air stream through a water-spray section, as shown in Fig. 11-8*a*. (The equivalent effect may be carried out by passing the air through a filter bed of some type, through which water is allowed to trickle. This provides reasonably good air–liquid-stream contact.) Due to the low relative humidity, part of the liquid-water stream evaporates. The energy for the evaporation process comes from the air stream, so that it is cooled. The overall effect is a cooling and a humidification of the air stream. Since the air is so dry to begin with, the additional moisture added to the air does not particu-

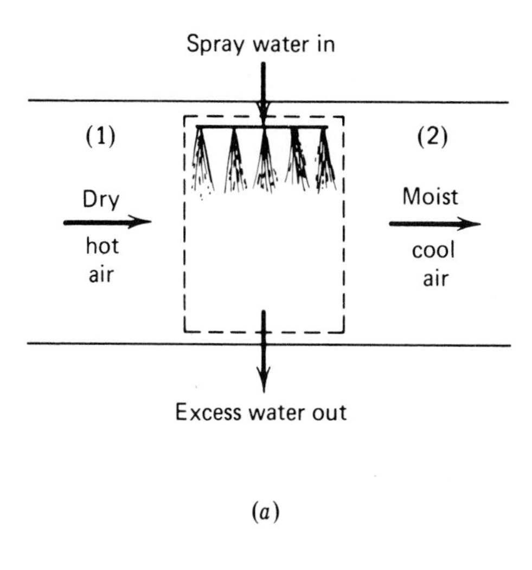

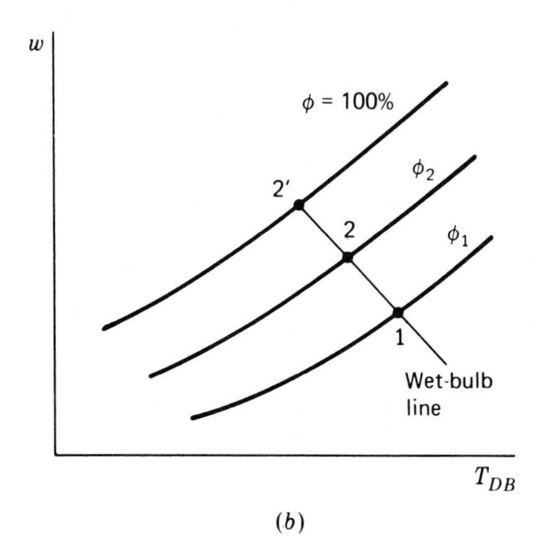

Figure 11-8 Evaporative cooling. (*a*) Process equipment; (*b*) process on a psychrometric chart.

larly make the living conditions uncomfortable. This process is essentially equivalent to the adiabatic-saturation process discussed in Sec. 11-6. Consequently the path of the process follows a constant-wet-bulb line on a psychrometric chart, as shown in Fig. 11-8*b*. There is a minimum temperature which can be achieved by such a process, which is the saturation state 2′ on the figure. The examples below illustrate the analysis of such a process.

Example 11-10M Desert air at 36°C, 1 bar, and 10 percent relative humidity passes through an evaporative cooler. Water is added at 20°C. (*a*) If the final air temperature is 20°C, how much water is added per kilogram of dry air? (*b*) What is the final relative humidity? (*c*) What is the minimum temperature that could be achieved by this process?

SOLUTION (*a*) The amount of water added is $\Delta\omega$. For the initial state, using Eq. (11-38) and Table A-12M,

$$\omega_1 = \frac{0.622(0.10)(0.05947)}{1.00 - 0.10(0.05947)} = 0.00372 \text{ kg water/kg dry air}$$

An energy balance on the process shown in Fig. 11-8*a* leads to

$$h_1 + \Delta\omega h_{f2} = h_2$$

$$c_p T_1 + \omega_1 h_{g1} + (\omega_2 - \omega_1)h_{f2} = c_p T_2 + \omega_2 h_{g2}$$

$$1.005(36) + 0.00372(2567.1) + (\omega_2 - 0.00372)(83.96) = 1.005(20) + \omega_2(2538.1)$$

$$\omega_2 = \frac{36.2 + 9.55 - 0.31 - 20.1}{2538.1 - 83.96} = 0.0103 \text{ kg water/kg dry air}$$

If the liquid water contribution had been neglected, the final humidity ratio would calculate to be 0.0101 kg water/kg dry air. This contribution usually is small. Therefore the amount of water added is

$$\omega_2 - \omega_1 = 0.0103 - 0.00372 = 0.00658 \text{ kg water/kg dry air}$$

(*b*) To determine the relative humidity, we must first find the vapor pressure by applying Eq. (11-38).

$$p_v = \frac{\omega P}{\omega + 0.622} = \frac{0.0103(1)}{0.0103 + 0.622} = 0.0163 \text{ bar}$$

At 20°C the saturation vapor pressure is 0.02339 bar. Hence

$$\phi_2 = \frac{p_v}{p_g} = \frac{0.0163}{0.02339} = 0.697 \qquad \text{or } 69.7\%$$

The values calculated above may be checked on a psychrometric chart.

(*c*) The minimum temperature is the adiabatic-saturation value, as represented by T_2 in Eq. (11-43). Unfortunately, this equation cannot be solved directly for T_2. A trial-and-error solution is necessary. Based on known data, Eq. (11-43) becomes

$$0.00372 = \frac{1.005(T_2 - 36) + \omega_2 h_{fg2}}{2567.1 - h_{f2}}$$

In addition, the equation for ω_2 at the final saturation state is

$$\omega_2 = \frac{0.622 p_{g2}}{1.0 - p_{g2}}$$

To solve these two equations, we assume first that $T_2 = 16°C$. Then

$$\omega_2 = \frac{0.622(0.01818)}{1.0 - 0.01818} = 0.0115 \text{ kg water/kg dry air}$$

Finally, the right-hand side of Eq. (11-43) becomes

$$\frac{1.005(16 - 36) + 0.0115(2463.6)}{2567.1 - 67.2} = 0.00330$$

When 17°C is used an an estimate, the right-hand side of Eq. (11-43) is equal to 0.00451. Therefore the minimum temperature lies between 16 and 17°C, and this state is denoted by state 2′ on Fig. 11-8*b*. This minimum temperature can also be found on a psychrometric chart by following a constant wet-bulb line from the initial state to a state of 100 percent relative humidity.

Example 11-10 Desert air at 98°F, 14.7 psia, and 10 percent relative humidity passes through an evaporative cooler. Water is added at 70°F.

(*a*) If the final air temperature is 70°F, how much water is added per pound of dry air? (*b*) What is the final relative humidity? (*c*) What is the minimum temperature that could be achieved by this process?

SOLUTION (*a*) The amount of water added is $\Delta\omega$. For the initial state, using Eq. (11-38) and Table A-12,

$$\omega_1 = \frac{0.622(0.10)(0.8945)}{1.00 - 0.10(0.8945)} = 0.00381 \text{ lb water/lb dry air}$$

An energy balance on the process shown in Fig. 11-8a leads to

$$h_1 + \Delta\omega h_{f2} = h_2$$

$$c_p T_1 + \omega_1 h_{g1} + (\omega_2 - \omega_1)h_{f2} = c_p T_2 + \omega_2 h_{g2}$$

$$0.240(98) + 0.00381(1104.2) + (\omega_2 - 0.00381)(38.09) = 0.240(70) + \omega_2(1092.0)$$

$$\omega_2 = \frac{23.5\&4.2 - 0.15 - 16.8}{1092.0 - 38.09} = 0.0102 \text{ lb water/lb dry air}$$

If the liquid water contribution had been neglected, the final humidity ratio would calculate to be 0.00998 lb water/lb dry air. This contribution usually is small. Threfore the amount of water added is

$$\omega_2 - \omega_1 = 0.0102 - 0.00381 = 0.00639 \text{ lb water/lb dry air}$$

(*b*) To determine the relative humidity, we must first find the vapor pressure by applying Eq. (11-38).

$$P_v = \frac{\omega P}{\omega + 0.622} = \frac{0.0102(14.7)}{0.0102 + 0.622} = 0.237 \text{ psia}$$

At 70°F the saturation vapor pressure is 0.3632 psia. Hence

$$\phi_2 = \frac{p_v}{p_g} = \frac{0.237}{0.3632} = 0.653 \qquad \text{or } 65.3\%$$

The values calculated above may be checked on a psychrometric chart.

(*c*) The minimum temperature is the adiabatic-saturation value, which is T_2 in Eq. (11-43). Unfortunately, this equation cannot be solved directly for T_2. A trial-and-error solution is necessary. Based on known data, Eq. (11-43) becomes

$$0.00381 = \frac{0.24(T_2 - 98) + \omega_2 h_{fg2}}{1104.2 - h_{f2}}$$

In addition, the equation for ω_2 at the final saturation state is

$$\omega_2 = \frac{0.622 p_{g2}}{14.7 - p_{g2}}$$

To solve these two equations, we shall first let $T_2 = 62°\text{F}$. Hence

$$\omega_2 = \frac{0.622(0.2751)}{14.7 - 0.2751} = 0.0119 \text{ lb water/lb dry air}$$

Then the right-hand side of Eq. (11-43) above becomes

$$\frac{0.24(62 - 98) + 0.0119(1058.5)}{1104.2 - 30.1} = 0.00364$$

This result is very close to the desired value of 0.00381. Hence 62°F is the minimum possible temperature and this state is denoted by state 2′ on Fig. 11-8*b*. This minimum temperature value can also be found on a psychrometric chart by following a constant wet-bulb line from the initial state to a state of 100 percent relative humidity.

3 Heating with Humidification

In winter or at high altitudes the air in the atmosphere frequently is dry (low relative humidity) and cold. Thus the engineering problem is one of increasing both the water content and the temperature of any inlet air into a building. One method of achieving humidification with heating is shown in Fig. 11-9*a*. The air stream passes first over a heating coil, and then through a spray section. Process 1-2 is well defined for the heating section, but process 2-3, shown in Fig. 11-9*b*, has a number of possible end states. State 3 is a function of the temperature of the water which enters the air stream. Usually the water will be nearly the same temperature as the air stream. Thus process 2-3 will be evaporative cooling, as described in the preceding section. As another possibility, steam may be introduced in place of liquid water. This results in humidification with additional heating, as shown by state 3′ in Fig. 11-9*b*. The processes could be reversed, of course, with water introduction followed by heating. Heating with humidification is illustrated by the following examples.

Example 11-11M An air stream at 8°C and 30 percent relative humidity is first heated to 32°C and then passed through an evaporative cooler until the temperature reaches 26°C. Determine (*a*) the heat added, in kJ/kg dry air, and (*b*) the final relative humidity. Total pressure is 1 bar.

SOLUTION (*a*) The initial enthalpy and humidity ratio are evaluated as follows from Eqs. (11-38) and (11-40):

$$\omega_1 = \frac{0.622(0.30)(0.01072)}{1.00 - 0.30(0.01072)} = 0.00201 \text{ kg water/kg dry air}$$

$$h_1 = 1.005(8) + 0.00201(2516.1) = 13.10 \text{ kJ/kg dry air}$$

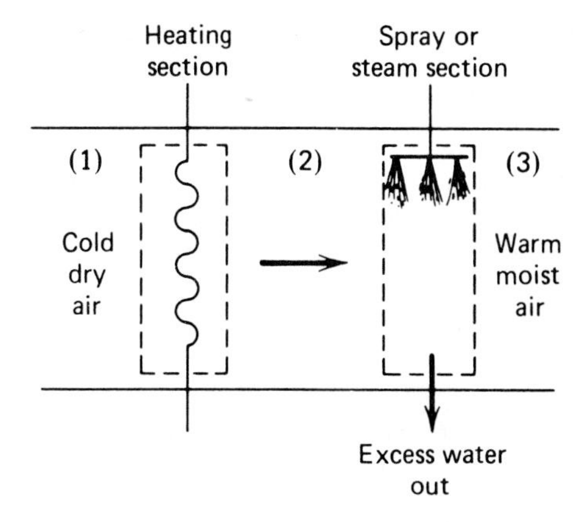

(a)

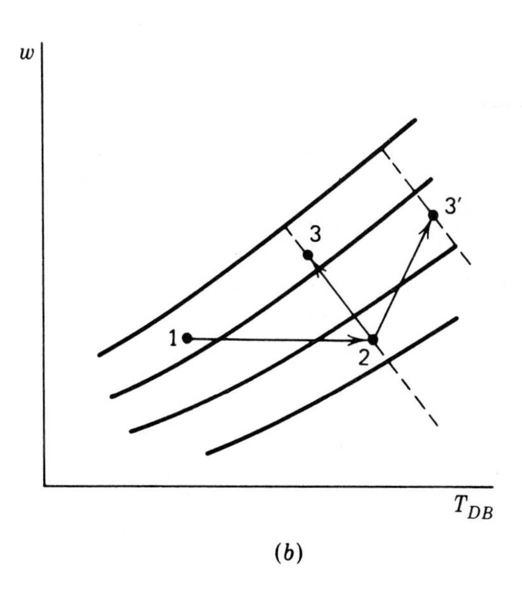

(b)

Figure 11-9 Heating with humidification. (a) Schematic of the process equipment; (b) process diagram on a psychrometric chart.

Since the humidity ratio is constant during heating.

$$h_2 = 1.005(32) + 0.00201(2559.9) = 37.30 \text{ kJ/kg dry air}$$

Therefore the heat added is

$$q = h_2 - h_1 = 37.30 - 13.10 = 24.2 \text{ kJ/kg dry air}$$

(b) The humidity ratio at state 3 is determined by equating the initial and final enthalpies for the spray process. Thus, when the liquid enthalpy is neglected,

$$c_p T_2 + \omega_2 h_{g2} = c_p T_3 + \omega_3 h_{g3}$$

$$1.005(32) + 0.00201(2559.9) = 1.005(26) + \omega_3(2549.0)$$

$$\omega_3 = \frac{32.1 + 5.15 - 26.1}{2549.0} = 0.00437 \text{ kg water/kg dry air}$$

This value of ω_3 is used to find the vapor pressure at state 3 through the use of Eq. (11-38).

$$0.00437 = \frac{0.622 p_v}{1 - p_v} \qquad p_v = 0.00700 \text{ bar}$$

At 26°C the vapor pressure from Table A-12M is 0.03363 bar. Hence the relative humdity at state 3 is

$$\phi_3 = \frac{p_v}{p_g} = \frac{0.00700}{0.03363} = 0.208 \qquad \text{or } 20.8\%$$

Example 11-11 An air stream at 50°F and 40 percent relative humidity is first heated to 96°F and then passed through an evaporative cooler until the temperature reaches 70°F. Determine (*a*) the heat added, in Btu/lb dry air, and (*b*) the final relative humidity. The total pressure is 14.7 psia.

Solution (*a*) The initial enthalpy and humidity ratio are evaluated as follows from Eqs. (11-38) and (11-40):

$$\omega_1 = \frac{0.622(0.40)(0.178)}{14.7 - 0.40(0.178)} = 0.00303 \text{ lb water/lb dry air}$$

$$h_1 = 0.24(50) + 0.00303(1083.3) = 15.28 \text{ Btu/lb dry air}$$

Since the humidity ratio is constant during heating,

$$h_2 = 0.24(96) + 0.00303(1103.3) = 26.2 \text{ Btu/lb dry air}$$

Therefore the heat added is

$$q = h_2 - h_1 = 26.2 - 15.3 = 10.9 \text{ Btu/lb dry air}$$

(*b*) The humidity ratio at state 3 is determined by equating the initial and final enthalpies for the spray process. Thus, when the liquid enthalpy is neglected,

$$c_p T_2 + \omega_2 h_{g2} = c_p T_3 + \omega_3 h_{g3}$$

$$0.24(96) + 0.00303(1103.3) = 0.24(70) + \omega_3(1092.0)$$

$$\omega_3 = \frac{26.2 - 16.8}{1092} = 0.00855 \text{ lb water/lb dry air}$$

This value of ω_3 is next used to find the vapor pressure at state 3 through the use of Eq. (11-38).

$$0.00855 = \frac{0.622 \phi p_g}{14.7 - \phi p_g} = \frac{0.622(0.363)\phi}{14.7 - 0.363\phi}$$

$$\phi = 0.55 \qquad \text{or } 55\%$$

The values calculated above can be checked on the psychrometric chart.

4 Adiabatic Mixing of Two Streams

Another important application in air conditioning is the mixing of two dry air–water-vapor streams, as shown in Fig. 11-10*a*. Three basic relations can be written for the overall control volume on a rate basis:

1. mass balance on dry air

$$\dot{m}_{a1} + \dot{m}_{a2} = \dot{m}_{a3} \tag{11-44a}$$

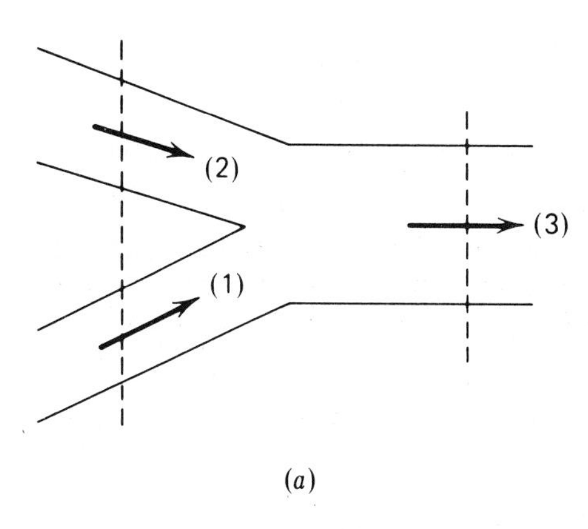

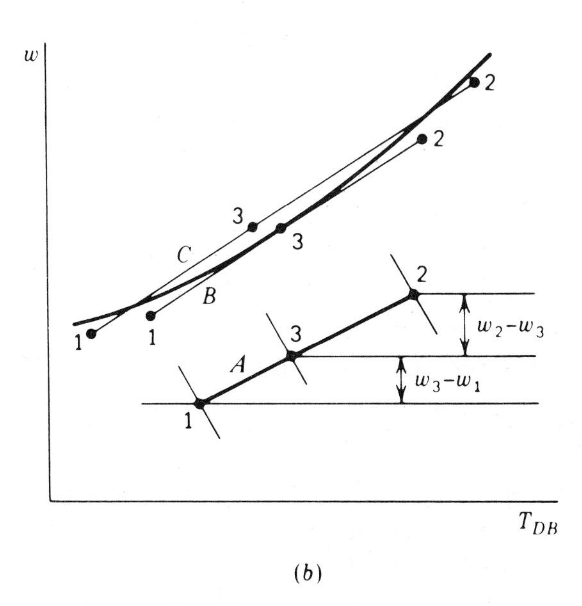

Figure 11-10 Adiabatic mixing of two air streams. (*a*) Schematic of the mixing process; (*b*) process on a psychrometric chart.

2. mass balance on water vapor

$$\dot{m}_{a1}\omega_1 + \dot{m}_{a2}\,\omega_2 = \dot{m}_{a3}\,\omega_3 \tag{11-44b}$$

3 energy balance for adiabatic mixing

$$\dot{m}_{a1}(h_{a1} + \omega_1\,h_{g1}) + \dot{m}_{a2}(h_{a2} + \omega_2\,h_{g2}) = \dot{m}_{a3}(h_{a3} + \omega_3\,h_{g3}) \tag{11-44c}$$

When properties of two of the flow streams are known, these three equations are sufficient to evaluate the properties of the third stream.

When the properties of the two inlet streams are known, the three equations above may be used to evaluate ω_3 and h_3 of the exit stream. If Eqs. (11-44*b*) and (11-44*c*) are divided by $\dot{m}_{a3}$, then

$$\frac{\dot{m}_{a1}}{\dot{m}_{a3}}\,\omega_1 + \frac{\dot{m}_{a2}}{\dot{m}_{a3}}\,\omega_2 = \omega_3 \tag{11-45}$$

and

$$\frac{\dot{m}_{a1}}{\dot{m}_{a3}}\,h_1 + \frac{\dot{m}_{a2}}{\dot{m}_{a3}}\,h_2 = h_3 \tag{11-46}$$

The ratios $\dot{m}_{a1}/\dot{m}_{a3}$ and $\dot{m}_{a2}/\dot{m}_{a3}$ represent the fractions of the total flow which enter the mixing process at states 1 and 2. Thus knowledge of the inlet states and the inlet mass flow rates of dry air is sufficient to determine the exit humidity ratio and enthalpy. These latter two properties fix all the other properties of the exit stream, such as its dry-bulb and wet-bulb temperatures and its relative humidity.

The mixing process also has an interesting interpretation on a psychrometric chart. For this pupose it is useful to combine Eqs. (11-44*a*) through (11-44*c*) to form two additional expressions. When Eqs. (11-44*a*) and (11-44*b*) are combined so that $\dot{m}_{a3}$ is eliminated, then

$$\frac{\dot{m}_{a1}}{\dot{m}_{a2}} = \frac{\omega_2 - \omega_3}{\omega_3 - \omega_1} \tag{11-47}$$

If Eq. (11-44*a*) is substituted into Eq. (11-44*c*), the result is

$$\frac{\dot{m}_{a1}}{\dot{m}_{a2}} = \frac{h_2 - h_3}{h_3 - h_1} \tag{11-48}$$

Both Eqs. (11-47) and (11-48) have a geometric interpretation with respect to the psychrometric chart. For example, note process line *A* on Fig. 11-10*b*. State 3 will have a humidity ratio which lies between that of states 1 and 2. Equation (11-47) dictates that the vertical distances between states 2 and 3 and between states 3 and 1 on the psychrometric chart are in proportion to the ratio of the mass flow rates of dry air for streams 1 and 2. A similar analysis may be made for Eq. (11-48) in terms of the constant-enthalpy lines on the chart.

Figure 11-10*b* illustrates three possible situations with respect to state 3. For process line *A*, states 1 and 2 are situated so that state 3 must lie below the 100 percent relative humidity line. In this case $\phi_3 < 1$, regardless of the mass ratio $\dot{m}_{a1}/\dot{m}_{a2}$. Process line *B* indicates that state 3 may be saturated ($\phi = 1$). For given positions of states 1 and 2, this requires a definite value for $\dot{m}_{a1}/\dot{m}_{a2}$. Finally, if states 1 and 2 lie close to or on the 100 percent relative humidity line, then state 3 may lie to the left of the saturation line. In this case, water will condense during the mixing process, and frequently will remain suspended as foglike droplets in the exit flow stream. Generally, this would be an undesirable condition if the flow stream goes directly into a home or business area. The following examples illustrate the adiabatic mixing process of two flow streams.

Example 11-12M An air stream (1) enters an adiabatic mixing chamber at a rate of 150 m^3/min at 10°C and $\phi = 0.80$. It is mixed with another stream (2) at 32°C and $\phi = 0.60$ at a rate of 100 m^3/min. Determine the final temperature and relative humidity of the exit stream, if the total pressure is 1 bar.

SOLUTION Equations (11-47) and 11-48) require a knowledge of the mass flow rates of dry air. The volume rates given are for the total flow, including water vapor. However, $\dot{m}_a$ can be found by dividing the volume rate by the specific volume of the air. It was noted in Sec. 11-5 that $v = R_a T/p_a$. Thus, for state 1

$$p_{v1} = \phi_1 p_g = 0.8(0.01228) = 0.0098 \text{ bar}$$

$$p_{a1} = P - p_{v1} = 1.0 - 0.0098 = 0.9902 \text{ bar}$$

$$v_1 = \frac{0.08314(283)}{29(0.9902)} = 0.819 \text{ m}^3\text{/kg dry air}$$

$$\dot{m}_{a1} = 150/0.819 = 183 \text{ kg dry air/min}$$

$$\omega_1 = \frac{0.622(0.0098)}{0.9902} = 0.00616 \text{ kg water/kg dry air}$$

$$h_1 = 1.005(10) + 0.00616(2519.8) = 25.5 \text{ kJ/kg dry air}$$

Similarly for state 2,

$$p_{v2} = \phi_2 p_g = 0.6(0.04759) = 0.0286 \text{ bar}$$

$$p_{a2} = P - p_{v2} = 1.0 - 0.0286 = 0.9714 \text{ bar}$$

$$v_2 = \frac{0.08314(305)}{29(0.9714)} = 0.900 \text{ m}^3\text{/kg dry air}$$

$$\dot{m}_{a2} = 100/0.900 = 111 \text{ kg dry air/min}$$

$$\omega_2 = \frac{0.622(0.0286)}{0.9714} = 0.0183 \text{ kg water/kg dry air}$$

$$h_2 = 1.005(32) + 0.0183(2559.9) = 79.0 \text{ kJ/kg dry air}$$

From these data we can evaluate ω_3 and h_3 from Eqs. (11-47) and (11-48).

$$\frac{183}{111} = \frac{\omega_3 - 0.0183}{0.00616 - \omega_3}$$

$$\omega_3 = 0.0108 \text{ kg water/kg dry air}$$

$$\frac{183}{111} = \frac{h_3 - 79.0}{25.5 - h_3}$$

$$h_3 = 45.7 \text{ kJ/kg dry air}$$

The temperature T_3 is found from the value of h_3, since

$$h_3 = 45.7 = 1.005(T_3) + 0.0108\,(h_{v3})$$

One method of solving this equation is by trial and error. A temperature is guessed, the value of h_g at this temperature is found from the steam tables, and T_3 and h_{g3} are substituted into the equation. If the equation is not satisfied, another temperature must be tried. A simpler method is to employ the approximation equation for h given by Eq. (11-41*a*). Thus,

$$h_3 = 45.7 = 1.005T_3 + 0.0108(2501.7 + 1.82T_3)$$

$$T_3 = \frac{18.68}{1.025} = 18.2°\text{C}$$

Although an approximation technique, the answer is probably within 0.1°C of that found from tabular h data. To determine the relative humidity, we must first find p_v from the value of ω_3.

$$\omega_3 = 0.0108 = \frac{0.622p_v}{1 - p_v} \qquad p_v = 0.01707 \text{ bar}$$

Since the saturation pressure at 18.2°C is around 0.0209 bar, then

$$\phi = p_v/p_g = \frac{0.01707}{0.0209} = 0.816 \qquad \text{or } 81.6\%$$

The initial values in this problem can be checked against values from the psychrometric chart. The final state can be located on the chart by using the inlet mass flow rates in terms of a "lever rule" on the straight line connecting the initial states.

Example 11-12 An air stream (1) enters an adiabatic mixing chamber at a rate of 1500 ft^3/min at 50°F and $\phi = 0.80$. It is mixed with another stream (2) at 90°F and $\phi = 0.60$ entering at a rate of 1000 ft^3/min. Determine the final temperature and relative humidity of the exit stream, if the total pressure is 14.7 psia.

SOLUTION Equations (11-47) and (11-48) require a knowledge of the mass flow rates of dry air. The volume rates are given for the total flow, including water vapor. However, $\dot{m}_a$ can be found by dividing the volume rate by the specific volume of the air stream. It was noted in Sec. 11-5 that $v = R_a T/p_a$. Thus, for state 1

$$p_{v1} = \phi_1 p_g = 0.80(0.1780) = 0.1424 \text{ psia}$$

$$p_{a1} = P - p_{v1} = 14.7 - 0.14 = 14.56 \text{ psia}$$

$$v_1 = \frac{10.73(510)}{29(14.56)} = 12.96 \text{ ft}^3\text{/lb dry air}$$

$$\dot{m}_{a1} = \frac{1500}{12.96} = 115.7 \text{ lb dry air/min}$$

$$\omega_1 = \frac{0.622(0.1424)}{14.56} = 0.00608 \text{ lb water/lb dry air}$$

$$h_1 = 0.240(50) + 0.00608(1083.3) = 18.59 \text{ Btu/lb dry air}$$

Similarly for state 2,

$$p_{v2} = \phi_2 p_g = 0.60(0.6988) = 0.4193 \text{ psia}$$

$$p_{a2} = P - p_{v2} = 14.7 - 0.42 = 14.28 \text{ psia}$$

$$v_2 = \frac{10.73(550)}{29(14.28)} = 14.25 \text{ ft}^3/\text{lb dry air}$$

$$\dot{m}_{a2} = \frac{1000}{14.25} = 70.2 \text{ lb dry air/min}$$

$$\omega_2 = \frac{0.622(0.4193)}{14.28} = 0.01826 \text{ lb water/lb dry air}$$

$$h_2 = 0.240(90) + 0.01826(1100.7) = 41.70 \text{ Btu/lb dry air}$$

From these data we can evaluate ω_3 and h_3 by employing Eqs. (11-45) and (11-46). Noting that $\dot{m}_{a3} = 185.9$ lb/min,

$$\omega_3 = \frac{115.7(0.00608)}{185.9} + \frac{70.2(0.01826)}{185.9} = 0.0107 \text{ lb water/lb dry air}$$

$$h_3 = \frac{115.7(18.59)}{185.9} + \frac{70.2(41.70)}{185.9} = 27.3 \text{ Btu/lb dry air}$$

These two values fix state 3. This state must also lie on a line connecting states 1 and 2 on the psychrometric chart. Using a lever rule, we can estimate that $T_3 = 65°\text{F}$ and $\phi_3 = 82$ percent. Another method of finding T_3 is to use Eq. (11-41*b*). Substituting in the known values for state 3,

$$h_3 = 27.3 = 0.240T_3 + 0.0107(1061 + 0.444T_3)$$

$$T_3 = \frac{15.95}{0.2448} = 65.2°\text{F}$$

Although an approximation technique, the answer based on Eq. (11-41*b*) is probably within 0.1°F of that using steam table data. Finally, to determine the relative humidity, we must first find p_{v3} from ω_3.

$$\omega_3 = 0.0107 = \frac{0.622p_v}{14.7 - p_v} \qquad p_v = 0.249 \text{ psia}$$

Since the saturation pressure at 65.2 psia is around 0.308 psia, then

$$\phi_3 = \frac{p_v}{p_g} = \frac{0.249}{0.308} = 0.808 \qquad \text{or } 80.8\%$$

The calculated initial values in this problem can be checked against values on the psychrometric chart.

5 Wet Cooling Tower

At fossil- and nuclear-fueled power plants a considerable portion of the energy released by the fuel must be rejected to the environment. Cooling water from natural sources such as rivers and lakes is commonly used to carry off some or all of the rejected energy. Due to environmental concern, there is a limitation on the temperature of the cooling-water effluent from the plant which is discharged back to the natural source. (The resulting temperature rise is known as thermal pollution.) Since the air of the atmosphere is so large in size, it can also be used as a sink for the energy rejected to the environment. The engineering problem is the transfer of this energy from the cooling water to the atmospheric air.

One method of transferring energy from any water stream to the atmosphere is through the use of a wet cooling tower. A schematic of the device is shown in Fig. 11-11. The warm cooling water is sprayed from the top of the tower and falls downward under gravitational force. Atmospheric air is inducted at the bottom of the tower by a fan and flows upward and in counterflow to the falling water droplets. The water and air streams are thus brought into intimate contact, and a small fraction of the water stream evaporates into the air stream. The evaporation process requires energy, and this energy transfer results in a cooling of the remaining water stream. The cooled water stream is then returned to the power plant to pick up additional waste energy. Since a portion of the cooling water is evaporated into the air stream, an equivalent amount of water must be added as makeup somewhere in the flow cycle.

To make an energy analysis of a wet cooling tower, a control volume is drawn around the entire tower, as shown by the dashed line in Fig. 11-11. The

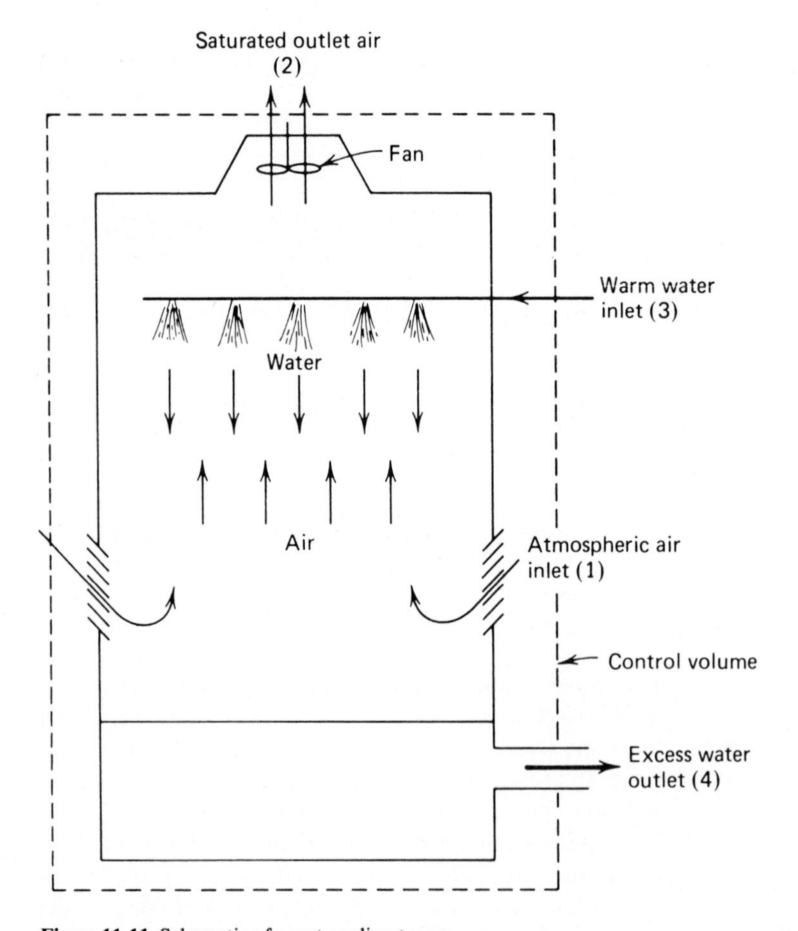

Figure 11-11 Schematic of a wet cooling tower.

process is assumed to be adiabatic, the fan work is neglected, and changes in kinetic and potential energy are negligible. Consequently, the basic energy balance given by Eq. (5-16) reduces to

$$\sum (\dot{m}h)_{out} - \sum (\dot{m}h)_{in} = 0$$

In terms of Fig. 11-11 this may be written as

$$\dot{m}_{a1} h_{m1} + \dot{m}_{w3} h_{w3} = \dot{m}_{a2} h_{m2} + \dot{m}_{w4} h_{w4} \tag{11-49}$$

or

$$m_{a1}(h_{a1} + \omega_1 h_{v1}) + m_{w3} h_{w3} = m_{a2}(h_{a2} + \omega_2 h_{v2}) + m_{w4} h_{w4} \tag{11-50}$$

where h_m is the mixture enthalpy per unit mass of dry air. This value of h_m can be read from the psychrometric chart or calculated by means of Eq. (11-40). The water stream enthalpy h_w may be evaluated by h_f at the given temperature, and note that $\dot{m}_{a1} = \dot{m}_{a2} = \dot{m}_a$. As a result, Eq. (11-50) can be written in the form

$$\dot{m}_a[c_{p,a}(T_1 - T_2) + \omega_1 h_{g1} - \omega_2 h_{g2}] = \dot{m}_{w4} h_{f4} - \dot{m}_{w3} h_{f3} \tag{11-51}$$

In addition to the energy balance, it is necessary to relate $\dot{m}_{w3}$ and $\dot{m}_{w4}$ by means of a mass balance on the water passing through the control volume. This mass balance leads to

$$\dot{m}_{w4} = \dot{m}_{w3} - \dot{m}_a(\omega_2 - \omega_1) \tag{11-52}$$

In the analysis of wet cooling towers the value of $\dot{m}_{w3}$ is known, as well as the temperature drop needed for the water stream. For assumed inlet and outlet air conditions, the above equations can be used to calculate the mass flow rate of dry air required. The inlet volume flow rate then can be found when the inlet specific volume of the atmospheric air is known.

Example 11-13M Water enters a cooling tower at 40°C and leaves at 25°C. The tower receives atmospheric air at 20°C and 40 percent relative humidity. The air leaves the tower at 35°C and 95 percent relative humidity. Find the mass flow rate of dry air, kg/min, passing through the tower if the water rate at the inlet is 12,000 kg/min.

SOLUTION The main data needed, in addition to steam-table information, are the humidity ratios for the inlet and exit air streams. These are obtained as follows:

$$\omega_1 = \frac{0.622(0.40)(0.02339)}{1.0 - 0.40(0.02339)} = 0.00603 \text{ kg water/kg dry air}$$

$$\omega_2 = \frac{0.622(0.95)(0.05628)}{1.0 - 0.95(0.05628)} = 0.0351 \text{ kg water/kg dry air}$$

Other data include:

$$h_{g1} = 2538.1 \text{ kJ/kg} \qquad h_{f3} = 167.6 \text{ kJ/kg}$$

$$h_{g2} = 2565.3 \text{ kJ/kg} \qquad h_{f4} = 104.9 \text{ kJ/kg}$$

With these data, and recalling that $c_{p,a} = 1.005$ kJ/(kg) (°C), we find that Eq. (11-51) yields

$$\dot{m}_a[1.005(20 - 35) + 0.00603(2538.1) - 0.0351(2565.3)] = \dot{m}_{w4}(104.9) - 12{,}000(167.6)$$

In addition, Eq. (11-52) yields

$$\dot{m}_{w4} = 12{,}000 - \dot{m}_a(0.0351 - 0.00603)$$

When $\dot{m}_{w4}$ from the last equation is substituted into the first equation, then

$$\dot{m}_a(-15.08 + 15.30 - 90.04) = -7.524 \times 10^6 - 3.05\dot{m}_a$$

$$\dot{m}_a = 8{,}670 \text{ kg dry air/min}$$

If the air enters at 1 bar, the specific volume would be 0.848 m^3/kg, and the required inlet volume rate would be 7350 m^3/min. This problem could also be solved by using Eq. (11-49) in conjunction with data from the psychrometric chart and the steam table.

Example 11-13 Water enters a cooling tower at 100°F and leaves at 80°F. The tower receives atmospheric air at 70°F, 14.7 psia, and 40 percent relative humidity, and the air leaves at 95°F and 95 percent relative humidity. Find the volume flow rate of atmospheric air, in ft^3/min, required at the inlet if the water rate at the inlet is 25,000 lb/min.

SOLUTION The main data needed, in addition to steam-table information, are the humidity ratios and enthalpies for the inlet and outlet air streams. These values can be calculated from Eqs. (11-38) and (11-40), and checked against data from the psychrometric chart. The calculated data are

$$\omega_1 = \frac{0.622(0.40)(0.3632)}{14.7 - 0.40(0.3632)} = 0.00621 \text{ lb water/lb dry air}$$

$$\omega_2 = \frac{0.622(0.95)(0.816)}{14.7 - 0.95(0.816)} = 0.0346 \text{ lb water/lb dry air}$$

$$h_1 = 0.240(70) + 0.00621(1092.0) = 23.58 \text{ Btu/lb dry air}$$

$$h_2 = 0.240(95) + 0.0346(1102.9) = 60.96 \text{ Btu/lb dry air}$$

(The values of ω_2 and h_2 cannot be checked on the psychrometric chart, because the exit state falls off the chart.) Employing Eq. (11-49), and noting that h_f is 48.1 Btu/lb at 80°F and 68.1 Btu/lb at 100°F,

$$\dot{m}_a(23.58) + 25{,}000(68.1) = \dot{m}_a(60.96) + \dot{m}_{w4}(48.1)$$

In addition, Eq. (11-52) yields

$$\dot{m}_{w4} = 25{,}000 - \dot{m}_a(0.0346 - 0.00621)$$

By eliminating $\dot{m}_{w4}$ from the pair of equations, we find that

$$\dot{m}_a(23.58) + 1.70 \times 10^6 = 60.96\dot{m}_a + 1.20 \times 10^6 - 1.36\dot{m}_a$$

$$\dot{m}_a = 13{,}880 \text{ lb dry air/min}$$

Since

$$v = R_a \frac{T}{p_a} = \frac{10.73(530)}{14.7 - 0.145(29)} = 13.47 \text{ ft}^3\text{/lb dry air}$$

then $$\text{Volume rate} = v_1\dot{m}_a = 13.47(13{,}880) = 187{,}000 \text{ ft}^3\text{/min}$$

Although the volume rate through a cooling tower may be large, the velocity is small because these devices have a large cross-sectional area.

PROBLEMS (METRIC)

Mass, volume, pressure, temperature relations

11-1M A rigid tank contains 0.5 kg of oxygen at 3 bars and 40°C. Nitrogen is added to the tank until the pressure reaches (*a*) 4 bars and (*b*) 6 bars at the same temperature. Determine the mass of nitrogen required, in kilograms.

11-2M A rigid tank contains 0.2 kg of hydrogen at 200 kPa and 27°C. Nitrogen is added to the tank until the pressure reaches (*a*) 300 kPa and (*b*) 500 kPa at the same temperature. Determine the mass of nitrogen required, in kilograms.

11-3M A gas mixture has the following volumetric analysis: N_2, 60 percent; CO_2, 33 percent; O_2, 7 percent.

(*a*) Determine the gravimetric analysis.

(*b*) What is the mass, in kilograms, of 20 m^3 of the gas mixture at 1 bar and 27°C?

11-4M The volumetric analysis of a mixture of ideal gases is N_2, 70 percent; CO_2, 20 percent, O_2, 10 percent.

(*a*) Determine the gravimetric analysis.

(*b*) Compute the component pressure of CO_2 if the mixture is at 1 bar and 60°C.

(*c*) How many kilograms of the mixture would be contained in 15 m^3 at 1 bar and 60°C?

11-5M A mixture of ideal gases has the following volumetric analysis: CO_2, 50 percent; N_2, 40 percent; H_2O, 10 percent.

(*a*) Calculate the gravimetric analysis of the mixture and the apparent molar mass.

(*b*) A tank of 0.0224 m^3 capacity contains 0.80 kg of the mixture at 7°C. Determine the pressure in the tank, in bars.

11-6M A mixture at 300°K and 180 kPa has the following volumetric analysis: O_2, 60 percent; CO_2, 40 percent. Compute (*a*) the mass analysis, (*b*) the partial pressure of O_2, in kilopascals, and (*c*) the apparent molar mass of the mixture, in kg/kg · mol.

11-7M A gas mixture has the following gravimetric analysis: O_2, 32 percent; CO, 56 percent; He, 12 percent. Determine (*a*) the volumetric analysis, and (*b*) the apparent gas constant, in kJ/(kg)(°K).

11-8M A gas mixture contains 0.28 kg of CO, 0.16 kg of O_2, and 0.66 kg of CO_2 at 1.4 bars and 17°C. Compute (*a*) the partial pressure of each component, in millibars, and (*b*) the volume occupied by each component measured, in cubic meters, at its own component pressure and mixture temperature.

11-9M A 0.1 m^3 tank contains 0.7 kg of N_2 and 1.1 kg of CO_2 at 27°C. Compute (*a*) the component pressure of N_2, in kilopascals, (*b*) the component volume of each species, in cubic meters, (*c*) the total pressure of the mixture, in kilopascals, and (*d*) the gas constant of the mixture, in kJ/(kg)(°K).

11-10M An ideal-gas mixture is made up of 20 percent O_2 and 80 percent H_2 by weight. The total pressure is 3 bars and the temperature is 27°C. Determine (*a*) the partial pressure of hydrogen, in bars, (*b*) the apparent molar mass, in kg/kg · mol, and (*c*) the gas constant, in kJ/(kg)(°K).

11-11M In what molar ratio should helium and krypton be mixed so that the gas mixture may have (*a*) the same average molar mass as pure argon, and (*b*) the same average gas constant as pure argon?

11-12M In what molar ratio should helium and carbon dioxide be mixed so that the gas mixture may have (*a*) the same average molar mass as pure diatomic oxygen, and (*b*) the same average gas constant as pure diatomic oxygen?

Energy analysis of ideal-gas mixtures

11-13M A gas sample at 77°C and 100 kPa has the following volumetric analysis: H_2, 4 percent; CO, 12 percent; CO_2, 35 percent; and N_2, 49 percent. The mixture is heated in a steady-flow process until the temperature reaches 227°C. Determine (*a*) the gravimetric analysis, (*b*) the heat transfer based on tabular ideal-gas tables, and (*c*) the heat transfer based on average specific-heat data, in kJ/kg.

11-14M A rigid tank contains 0.90 kg of H_2O, 4.0 kg of O_2, and 7.0 kg of CO at an initial state of 77°C and 200 kPa. The mixture is heated until the pressure is 280 kPa. Determine the quantity of heat required, in kilojoules, based on (*a*) tabular ideal-gas data and (*b*) average specific-heat data.

11-15M For the mixture composition described in Prob. 11-4M, determine the heat transfer, in kJ/kg · mol, based on (*a*) tabular ideal-gas data and (*b*) average specific-heat data if the mixture is cooled in a rigid tank from 227 to 27°C.

11-16M The mixture described in Prob. 11-8M is now heated at constant pressure in a piston-cylinder device from the given initial state of 1.4 bars and 17°C to a final temperature of 277°C. Determine the heat required, in kilojoules, by employing (*a*) ideal-gas tabular data and (*b*) average specific-heat data.

11-17M The mixture described in Prob. 11-3M is cooled as it passes through a steady-state heat exchanger from 550 to 325°K. Compute the heat transferred, in kJ/kg · mole and kJ/kg, using tabular ideal-gas data.

11-18M A piston-cylinder assembly contains 0.06 kg · mol of N_2 and 0.04 kg · mol of O_2. The mixture is compressed adiabatically from 300°K and 100 kPa to 500°K. What is the required work input, in kilojoules?

11-19M Carbon dioxide (CO_2) in the amount of 1 g · mol and initially at 2 bars and 27°C is mixed adiabatically with 2 g · mol of O_2 initially at 5 bars and 152°C. During the constant-volume mixing process electric energy equivalent to 0.67 kJ/g · mol of mixture is added. Determine (*a*) the final temperature of the mixture, in degrees Celsius, if tabular data are used, and (*b*) the final pressure, in bars.

11-20M Carbon dioxide (CO_2), in the amount of 1 kg · mol and initially at 2 bars and 327°C, is mixed adiabatically in a closed system at constant pressure with 2 kg · mol of N_2, initially at 2 bars and 27°C. During the process 1000 kJ are added in the form of electric work. Determine the final temperature of the mixture, in degrees Celsius.

11-21M Nitrogen in the amount of 2 kg and initially in a rigid tank at 700 kPa and 177°C, is mixed with 1 kg of oxygen, initially in another tank at 300 kPa and 27°C, by opening a valve connecting the two tanks. If the final equilibrium temperature is 102°C, determine (*a*) the heat transfer, in kilojoules, and (*b*) the final pressure, in kilopascals.

11-22M A rigid tank contains 0.2 kg of nitrogen and 0.1 kg of carbon dixide at 2 bars (200 kPa) and 37°C. During a process, 4.90 kJ of heat are added, and a current of 4.5 A passes through a resistor within the tank for a period of 1.50 min. If the final temperature of the gas mixture is 147°C, determine the constant voltage applied across the resistor. Use tabular data from the ideal-gas tables.

11-23M An ideal-gas mixture consists of 56 percent CO, 32 percent O_2, and 12 percent He by weight at 3.4 bars and 327°C. It expands adiabatically through a steady-flow turbine until the exit temperature is 77°C. Determine (*a*) the shaft-work output, in kJ/kg · mol, and (*b*) the power, in kilowatts, for 2.0 m^3/min.

11-24M An equimolar mixture of two ideal gases expands through an abiabatic nozzle from an initial state of 400°K and 100 m/s to a final state of 340°K. Determine the exit velocity, in m/s, if the mixture is (*a*) He and CO_2, and (*b*) He and CO.

11-25M An ideal-gas mixture has the following volumetric analysis: CO, 33.3 percent; CO_2, 50.0 percent; O_2, 16.7 percent. It enters a steady-state compressor at 37°C and 60 m/s and leaves the device at 237°C and 100 m/s.

(*a*) Determine the shaft-work input required, in kJ/kg, if a heat loss of 2.0 kJ/kg occurs during the process.

(*b*) If the volume flow rate at the entrance is 4.0 m^3/min and the pressure is 110 kPa, determine the power input, in kilowatts.

11-26M An equimolar mixture of two gases flows steadily through a heat exchanger. The initial and final temperature of the mixture are 300 and 1000°K, respectively. If the total flow rate is 0.010

kg · mol/s, what is the rate of heat transfer to the gas, in kJ/s, for a mixture of (*a*) N_2 and CO_2, (*b*) CO and CO_2, and (*c*) N_2 and Ar?

11-27M An equimolar mixture of two gases flows steadily through a diffuser. The initial state is 300°K and 200 m/s, and the final velocity is 30 m/s. Using specific-heat data, estimate the final temperature, in degrees kelvin, for a mixture of (*a*) N_2 and CO_2 and (*b*) CO and Ar, if the process is adiabatic.

Isentropic processes of ideal-gas mixtures

11-28M A mixture of three gases with component pressures in the ratio 1 : 2 : 5 is expanded isentropically from 600°K and 1000 kPa to 100 kPa in a piston-cylinder device. Using s^0 data, determine how much work is done, in kJ/kg · mol of mixture, for mixtures of (*a*) H_2O, CO, and N_2, and (*b*) CO, CO_2, and N_2.

11-29M A gas mixture consisting of N_2, CO_2, and H_2O in a molar ratio of 4 : 1 : 1 enters a turbine at 1000°K. The gases expand isentropically through (*a*) an 8 : 1 and (*b*) a 6 : 1 pressure ratio. Using s^0 data, compute the work output, in kJ/kg · mol, of the mixture.

11-30M Reconsider the mixture composition data given in Prob. 11-28M for a piston-cylinder device. Now, however, consider a compression process from 1 bar and 300°K to 4 bars. Using s^0 data, determine (1) the final temperature, in degrees kelvin, and (2) the work required, in kJ/kg · mol of mixture, for mixtures (*a*) and (*b*).

11-31M Reconsider the data of Prob. 11-29M, except the process is isentropic compression from an initial temperature of 300°K. Using s^0 data, determine the work required, in kJ/kg · mol of mixture, for (*a*) an 8 : 1 and (*b*) a 6 : 1 pressure ratio.

Entropy changes of ideal-gas mixtures

11-32M Carbon monoxide and argon in separate streams enter an adiabatic mixing chamber in a 2 : 1 mass ratio. At the inlet the carbon monoxide is at 120 kPa and 300°K, and the argon is at 120 kPa and 450°K. The mixture leaves at 110 kPa. Determine (*a*) the final temperature of the mixture, in degrees kelvin, and (*b*) the entropy change of the carbon monoxide, in kJ/(kg · mol)(°K).

11-33M Hydrogen and nitrogen in separate streams enter an adiabatic mixing chamber in a 3 : 1 molar ratio. At the inlet the hydrogen is at 2 bars and 77°C, and the nitrogen is at 2 bars and 277°C. The mixture leaves at 1.9 bars.

(*a*) Determine the final temperature of the mixture, in degrees kelvin.

(*b*) Compute the entropy change of the hydrogen gas, in kJ/(kg · mol)(°K).

11-34M An insulated 0.3-m^3 tank is divided into two sections by a partition. One section is 0.2 m^3 in volume and initially contains hydrogen gas at 2 bars and 127°C. The remaining section initially holds nitrogen gas at 4 bars and 27°C. The adiabatic partition is then removed, and the gases are allowed to mix. Determine (*a*) the temperature of the equilibrium mixture, in degrees kelvin, (*b*) the pressure of the mixture, and (*c*) the entropy change for each component and the total value, in kJ/°K.

11-35M A system consists of two tanks interconnected with a pipe and valve. One tank with a volume of 0.83 m^3 contains 2 kg of Ar at 1.5 bars and 27°C. The other tank has a volume of 0.33 m^3 and holds 1.6 kg of O_2 at 5 bars and 127°C. The valve is opened and the gases mix. The atmospheric temperature is 17°C; the final temperature of the mixture is 77°C. Determine (*a*) the heat transfer, in kilojoules, (*b*) the final pressure, in bars, (*c*) the change in the entropy of the Ar and the O_2, in kJ/°K, and (*d*) the entropy change of the atmosphere, in kJ/°K.

11-36M Consider Prob. 11-35M, except the final temperature is now 177°C. To accomplish this, heat is transferred to the mixture from a heat reservoir at 600°K. The initial data remain the same. Answer the same questions.

11-37M An insulated 0.06-m^3 tank is divided into two sections by a partition. One section is 0.02 m^3 in volume and initially contains 0.070 kg of carbon monoxide at 267°C. The remaining section initially holds 0.010 kg of helium at 17°C. The adiabatic partition is then removed, and the gases are allowed to mix. Determine (*a*) the temperature of the equilibrium mixture, in degrees kelvin, (*b*) the pressure of the final mixture, in kPa, and (*c*) the entropy change of the carbon monoxide, in kJ/°K.

11-38M At one inlet to a control volume 33 kg/min of CO_2 and 9 kg/min of H_2O enter as a mixture at 2.0 bars and 440°K. At another inlet 8 kg/min of O_2 and 14 kg/min of N_2 enter as a mixture at 2.0 bars and 340°K. The mixture of four gases leaves the control volume at 1.9 bars and 400°K. The shaft work is zero. Use the ideal-gas tables to determine (*a*) the magnitude and direction of any heat transfer, in kJ/h, (*b*) the rate of entropy change of the CO_2 component, in kJ/(h)(°K), and (*c*) the cross-sectional area at the control surface where CO_2 and H_2O enter, in cm^2, if the gas velocity at that surface is 8 m/s.

Properties of air–water-vapor mixtures

11-39M A tank contains 15 g of dry air and 50 g of saturated water vapor at 70°C. Determine (*a*) the volume of the tank, in liters, and (*b*) the total pressure, in millibars.

11-40M A tank contains 7 g of dry air and 70 g of dry, saturated steam at a temperature of 80°C. Determine (*a*) the volume of the tank, in liters, and (*b*) the total pressure, in millibars.

11-41M If the partial pressure of water vapor in atmospheric air at 1 bar is 30 mbar (3 kPa) at 30°C, determine (*a*) the relative humidity, (*b*) the approximate dew-point temperature, (*c*) the humidity ratio, (*d*) the enthalpy, in kJ/kg (based on $h = 0$ at 0°C for both dry air and water), and (*e*) the specific volume of the mixture. Use steam-table data where necessary.

11-42M The partial pressure of water vapor in atmospheric air is 2.5 kPa (25 mbar) at 25°C. Determine (*a*) the relative humidity, (*b*) the dew-point temperature, (*c*) the specific humidity, in g/kg, (*d*) the enthalpy, in kJ/kg dry air (based on $h = 0$ at 0°C for both dry air and water), and (*e*) the specific volume, in m^3/kg dry air, for 98 kPa (980 mbar) total pressure. Use the steam tables where necessary.

11-43M An atmospheric air mixture at 960 mbar (96 kPa) contains 2 percent water by volume. If the dry-bulb temperature is 25°C, determine (*a*) the relative humidity, (*b*) the dew-point temperature, (*c*) the humidity ratio, in g/kg, and (*d*) the enthalpy, in kJ/kg dry air (based on $h = 0$ at 0°C for both dry air and water). Use steam-table data.

11-44M Atmospheric air with a relative humidity of 50 percent is held at 35°C and 970 mbar. Determine (*a*) the specific humidity, in g/kg, (*b*) the dew-point temperature, and (*c*) the enthalpy, in kJ/kg dry air, where $h = 0$ at 0°C. Use steam-table data.

11-45M Atmospheric air at a barometric pressure of 96 kPa (960 mbar) and a dry-bulb temperature of 20°C has a relative humidity of 40 percent. Determine, through the use of the steam tables, (*a*) the specific humidity, in g/kg, (*b*) the dew-point temperature, in degrees Celsius, and (*c*) the enthalpy, in kJ/kg (based on $h = 0$ at 0°C).

11-46M Condensation on cold-water pipes often occurs in warm, humid rooms. The outside of a pipe may reach a minimum of 10°C, and the room temperature is kept at 25°C.

(*a*) What is the maximum limit of relative humidity in the room to avoid condensation at normal atmospheric pressure?

(*b*) How high must the minimum temperature be raised if condensation is avoided and the room relative humidity is 54 percent at 25°C?

11-47M On a cold winter's day the inside surface of a wall in a home is found to be 15°C and the air within the room is 22.5°C.

(*a*) What is the maximum relative humidity that the air can have without the occurrence of condensation of water on the wall?

(*b*) If added insulation in the wall raises the maximum permissible relative humidity to 75 percent, what is the new, permissible inside-wall temperature, in degrees Celsius?

11-48M Arrange the temperature values (*a*) 19, 35, and 8°C, (*b*) 4, 10, and 16°C, and (*c*) 6, 15, and 26°C in the following order: dry-bulb, dew-point, and wet-bulb temperature.

11-49M Atmospheric air at 1 bar has a dry-bulb temperature of 25°C and a wet-bulb temperature of

20°C. Determine, through the use of steam-table data and Eq. (11-43), (*a*) the humidity ratio, in g/kg, (*b*) the relative humidity, and (*c*) the enthalpy, in kJ/kg (based on $h = 0$ at 0°C).

11-50M Atmospheric air at 970 mbar has a dry-bulb temperature of 30°C and a wet-bulb temperature of 20°C. Determine, through the use of steam-table data and Eq. (11-43), (*a*) the humidity ratio, in g/kg, (*b*) the relative humidity, and (*c*) enthalpy, in kJ/kg (based on $h = 0$ at 0°C).

11-51M Same as Prob. 11-49M, except that the dry-bulb and wet-bulb temperatures are 27 and 21°C, respectively.

Property evaluation from a psychrometric chart

11-52M Reconsider Prob. 11-49M. Estimate the answers by using the psychrometric chart.

11-53M Reconsider Prob. 11-50M. Estimate the answers by using the psychrometric chart.

11-54M Reconsider Prob. 11-51M. Estimate the answers by using the psychrometric chart.

11-55M The dry-bulb and wet-bulb temperatures for atmospheric air are 30 and 23°C, respectively. Determine (*a*) the partial pressure of the water vapor, (*b*) the humidity ratio, (*c*) the specific volume per mass of dry air, (*d*) the relative humidity, and (*e*) the enthalpy, in kJ/kg, of dry air. Employ the psychrometric chart, and assume the total pressure is 1 bar.

11-56M Atmospheric air has a dry-bulb temperature of 32°C and a wet-bulb temperature of 26°C. From the psychrometric chart, determine (*a*) the humidity ratio or specific humidity, (*b*) the dew point, (*c*) the enthalpy, (*d*) the specific volume, and (*e*) the relative humidity.

11-57M Atmospheric air at a pressure of 1 bar and 28°C has a specific humidity of 0.0090 kg/kg. Determine, from the psychrometric chart, (*a*) the relative humidity, (*b*) the vapor pressure, (*c*) the enthalpy, and (*d*) the specific volume, and (*e*) the dew-point temperature.

11-58M Atmospheric air at 1 bar has dry-bulb and wet-bulb temperatures of 25 and 18°C, respectively. From the psychrometric chart, determine (*a*) the humidity ratio, (*b*) the dew point, (*c*) the specific volume, (*d*) the relative humidity, and (*e*) the enthalpy.

Heating, cooling, and dehumidification

11-59M Air which originally exists at 36°C, 1 bar, and a relative humidity of 40 percent is cooled at constant pressure to 24°C. Determine (*a*) the relative humidity, (*b*) the humidity ratio, and (*c*) the dew-point temperature at the final state, employing the psychrometric chart.

11-60M Atmospheric air with dry-bulb and wet-bulb temperatures of 30 and 20°C, respectively, is cooled to 18°C at a constant pressure of 1 bar. If the mass flow rate through the cooling unit is 500 kg dry air/h, calculate the heat removed, in kJ/h. Check calculated property data against the psychrometric chart.

11-61M Atmospheric air at 98 kPa (0.98 bar) and 26°C dry-bulb temperature, with a relative humidity of 70 percent, is cooled to 10°C.

(*a*) Determine the grams of water vapor condensed per kilogram of dry air.

(*b*) How much heat is removed, in kJ/kg of dry air? Calculate the required property data, and check values against the psychrometric chart.

11-62M Atmospheric air with dry-bulb and wet-bulb temperatures of 28 and 20°C, respectively, is cooled to 17°C at a constant pressure of 995 mbar (99.5 kPa). The volume flow rate entering the cooling unit is 100 m^3/h.

(*a*) Determine the mass flow rate, in kg/h, of dry air.

(*b*) Determine the heat removed, in kJ/h. Calculate the required property data, and check values against the psychrometric chart.

11-63M Atmospheric air at 28°C and 70 percent relative humidity flows over a cooling coil at a rate of 500 m^3/min. The liquid which condenses is removed from the system at 15°C. Subsequent heating of the air results in a final temperature of 30°C and a relative humidity of 30 percent. Determine (*a*) the heat removed in the cooling section, in kJ/min, (*b*) the heat added in the heating section, in kJ/min, and (*c*) the amount of vapor condensed, in kg/min. Calculate required property data, and check these values against the psychrometric chart.

11-64M A 3-m^3 storage tank initially contains air at 5 bars and 150°C with a relative humidity of 10 percent. The air then is cooled back to the ambient temperature of 17°C. Determine (*a*) the dew point of the mixture, (*b*) the temperature at which condensation begins, (*c*) the amount of water condensed, and (*d*) the heat transferred from the tank, in kilojoules.

11-65M Atmospheric air at 1.01 bar, 30°C, and 60 percent relative humidity flows over a set of cooling coils at an inlet flow rate of 1500 m^3/min. The liquid which condenses is removed from the system at 17°C. Subsequent heating of the air results in a final state of 25°C and 60 percent relative humidity. Determine (*a*) the heat removed in the cooling section, in kJ/min, (*b*) the heat added in the heating section, in kJ/min, and (*c*) the amount of vapor condensed, in kg/min.

11-66M Ambient air from outside a store is to be conditioned from a temperature of 29°C and 80 percent relative humidity to a final state of 21°C and 40 percent relative humidity. If the pressure remains constant at 1 bar, calculate (*a*) the amount of water removed, (*b*) the heat removed in the cooling process, and (*c*) the heat added in the final step of the process, in kJ/kg dry air.

Evaporative cooling

11-67M Atmospheric air at 34°C and 20 percent relative humidity passes through an evaporative cooler until the final temperature is (*a*) 24°C, (*b*) 21°C, and (*c*) 19°C. Determine how much water is added to the air and what the final relative humidity is in each case. Calculate the required property data, and check these values against the psychrometric chart.

11-68M Atmospheric air at 28°C and 10 percent relative humidity passes through an evaporative cooler until the final relative humidity is (*a*) 40 percent, (*b*) 60 percent, and (*c*) 80 percent. For each case find the amount of water added to the air and the final dry-bulb temperature, by employing the psychrometric chart.

11-69M Atmospheric air at 36°C and 10 percent relative humidity passes through an evaporative cooler.

(*a*) On the basis of the psychrometric chart, what is the minimum temperature that could be reached by this process?

(*b*) If the final dew-point temperature is (1) 12°C, and (2) 15°C, estimate from the psychrometric chart how much water is added in the cooler and what is the final temperature.

Heating and humidification

11-70M An atmospheric air stream at 10°C and 40 percent relative humidity is first heated to 33°C and then passed through an evaporative cooler until the temperature reaches (*a*) 26°C and (*b*) 20°C. Calculate the required property data, using steam-table data, and determine (1) the heat added, (2) the water added, amd (3) the final relative humidity. Check calculations by employing the psychrometric chart.

11-71M Atmospheric air at 12°C and 30 percent relative humidity is first heated to 35°C and then passed through an evaporative cooler until the temperature reaches (*a*) 26°C, and (*b*) 20°C. Determine the heat added in the first section and the water added in the second section. Calculate the required property data, using steam-table data, and check the results against the psychrometric chart.

11-72M An air stream at 34°C with a wet-bulb temperature of 15°C passes through an evaporative cooler until the temperature is (*a*) 21°C, and (*b*) 17°C. It is then heated to 30°C. Determine the amount of water added in the first section and the heat added in the second section of the equipment. Calculate the required property data.

Adiabatic mixing

11-73M Two streams of atmospheric air at 1 bar are mixed together in an adiabatic, steady-flow process. One air stream, at 36°C and 40 percent relative humidity, enters at a rate of 5 kg dry air/min, while the second stream, at 5°C and 100 percent relative humidity, enters at (*a*) 10 kg dry air/min, and (*b*) 15 kg dry air/min. Determine (1) the exit humidity ratio, (2) the exit stream enthalpy, and (3) the exit dry-bulb temperature, in degrees Celsius. Compute the required property data, and check these values against the psychrometric chart.

11-74M One atmospheric air stream, at 15°C and 30 percent relative humidity, is mixed adiabatically with another air stream, at 30°C and 70 percent relative humidity, in a steady-flow process. If the mass flow rate of the dry air in the hot stream is (*a*) twice, and (*b*) three times that of the colder stream, determine the final mixture dry-bulb temperature and its relative humidity. Compute the required property data, using steam table data where necessary, and compare values with the chart, if P is 1 bar.

11-75M In an air-conditioning process 50 m^3/min of outside air at 29°C and 80 percent relative humidity are mixed adiabatically with 64 m^3/min of inside air initially at 20°C and 30 percent relative humidity. Determine for the resultant mixture (*a*) the dry-bulb temperature, (*b*) the wet-bulb temperature, (*c*) the humidity ratio, and (*d*) the relative humidity. Compute initial property values that are required, using steam table data where necessary, and compare values with the psychrometric chart if the pressure is 1 bar.

11-76M An atmospheric air mixture at 29°C dry bulb and 21°C wet bulb mixes adiabatically with another air stream originally at 14°C dry bulb and 12°C wet bulb. The total pressure is 1 bar. The ratio of the volume flow rate of the cold stream to that of the hot stream is 2.0. Determine for the resultant mixture (*a*) the dry-bulb temperature, (*b*) the humidity ratio, and (*c*) the relative humidity. Compute the required inlet property data, and compare against the psychrometric chart.

Wet cooling tower

11-77M Water enters a cooling tower at 36 m^3/min and is cooled from 30 to 20°C. The entering atmospheric air at 1 bar has dry-bulb and wet-bulb temperatures of 21 and 15°C, respectively. The air is saturated when it leaves the tower. Determine (*a*) the volume rate of inlet air required, in m^3/min, and (*b*) the rate of water evaporated, in kg/h. The exit air temperature is 28°C.

11-78M Water enters a cooling tower at 40°C and leaves at 22°C. The tower receives 10,000 m^3/min of atmospheric air at 1 bar, 20°C, and 40 percent relative humidity. The air exit condition is 32°C and 90 percent relative humidity. Determine (*a*) the mass flow rate of dry air passing through the tower, in kg/min, (*b*) the mass flow rate of entering water, in kg/min, and (*c*) the amount of water evaporated, in kg/min.

11-79M It is desired to cool 1000 kg/min of water from 38 to 25°C. The cooling tower receives 800 m^3/min of air at 1 bar with dry-bulb and wet-bulb temperatures of 29 and 21°C, respectively. If the evaporation rate from the water stream is 1200 kg/h, determine the temperature of the exit air stream.

11-80M It is desired to cool water from 40 to 26°C. The cooling tower receives 800 m^3/min of atmospheric air at 1 bar with dry-bulb and wet-bulb temperatures of 29 and 21°C. The outlet water rate is 1250 kg/min and the exit air relative humidity is 95 percent. Determine (*a*) the outlet temperature of the air, in °C, and (*b*) the evaporation rate of the water, in kg/min.

PROBLEMS (USCS)

Mass, volume, pressure, temperature relations

11-1 A rigid tank contains 0.8 lb of oxygen at 30 psia and 75°F. Nitrogen is added to the tank until the pressure reaches (*a*) 40 psia and (*b*) 60 psia at the same temperature. Determine the mass of nitrogen required, in pounds.

11-2 A rigid tank contains 0.2 lb of hydrogen at 20 psia and 60°F. Nitrogen is added to the tank until the pressure reaches (*a*) 30 psia and (*b*) 40 psia at the same temperature. Determine the mass of nitrogen required, in pounds.

11-3 A gas mixture has the following volumetric analysis: N_2, 60 percent; CO_2, 33 percent; O_2, 7 percent.

(*a*) Determine the gravimetric analysis.

(*b*) What is the mass, in pounds, of 1000 ft^3 of the gas mixture at 30 psia and 80°F?

11-4 The volumetric analysis of a mixture of ideal gases is N_2, 70 percent; CO_2, 20 percent, O_2, 10 percent.

(*a*) Determine the gravimetric analysis.

(*b*) Compute the component pressure of CO_2 if the mixture is at 15 psia and 140°F.

(*c*) How many pounds of the mixture would be contained in 450 ft^2 at 15 psia and 140°F?

11-5 A mixture of ideal gases has the following volumetric analysis: CO_2, 50 percent; N_2, 40 percent; H_2O, 10 percent.

(*a*) Calculate the gravimetric analysis of the mixture and the apparent molar mass.

(*b*) A tank of 3.59 ft^3 capacity contains 0.32 lb of the mixture at 32°F. Determine the pressure in the tank, in atm.

11-6 A mixture at 50°F and 35 psia has the following volumetric analysis; O_2, 60 percent; CO_2, 40 percent. Compute (*a*) the mass analysis, (*b*) the partial pressure of O_2 in psia, and (*c*) the apparent molar mass of the mixture, in lb/lb · mol.

11-7 A gas mixture has the following gravimetric analysis; O_2, 32 percent; CO, 56 percent; He, 12 percent. Determine (*a*) the volumetric analysis, and (*b*) the apparent gas constant, in Btu/(lb)(°R).

11-8 An ideal-gas mixture contains 0.28 lb of CO, 0.16 lb of O_2, and 0.66 lb of CO_2 at 20 psia and 90°F. Compute (*a*) the partial pressure of each component, in psia, and (*b*) the volume occupied by each component measured, in cubic feet, at its own component pressure and mixture temperature.

11-9 A 10-ft^3 tank contains 0.7 lb of N_2 and 1.1 lb of CO_2 at 90°F. Compute (*a*) the component pressure of N_2, in psia, (*b*) the component volume of each species, in cubic feet, (*c*) the total pressure of the mixture, in psia, and (*d*) the gas constant of the mixture, in Btu/(lb)(°R).

11-10 An ideal-gas mixture is made up of 20 percent O_2 and 80 percent H_2 by weight. The total pressure is 50 psia and the temperature is 50°F. Determine (*a*) the partial pressure of hydrogen, in psia, (*b*) the apparent molar mass, in lb/lb · mol, and (*c*) the gas constant, in Btu/(lb)(°R).

11-11 In what molar ratio should helium and krypton be mixed so that the gas mixture may have (*a*) the same average molar mass as pure argon, and (*b*) the same average gas constant as pure argon?

11-12 In what molar ratio should helium and carbon dioxide be mixed so that the gas mixture may have (*a*) the same average molar mass as pure diatomic oxygen, and (*b*) the same average gas constant as pure diatomic oxygen?

Energy analysis of ideal-gas mixtures

11-13 A gas sample at 100°F and 14.5 psia has the following volumetric analysis: H_2, 4 percent; CO, 12 percent; CO_2, 35 percent; N_2, 49 percent. The mixture is heated in a steady-flow process until the temperature reaches 300°F. Determine (*a*) the gravimetric analysis, (*b*) the heat transfer based on tabular ideal-gas tables, and (*c*) the heat transfer based on average specific-heat data, in Btu/lb.

11-14 A rigid tank contains 0.90 lb of H_2O, 4.0 lb of O_2, and 7.0 lb of CO at an initial state of 80°F and 15 psia. The mixture is heated until the pressure is 20 psia. Determine the quantity of heat required, in Btu, based on (*a*) tabular ideal-gas data and (*b*) average specific-heat data.

11-15 For the mixture composition described in Prob. 11-4, determine the heat transfer, in Btu/lb · mol, based on (*a*) tabular ideal-gas data and (*b*) average specific-heat data if the mixture is cooled in a rigid tank from 500 to 100°F.

11-16 The mixture described in Prob. 11-8 is now heated at constant pressure in a piston-cylinder device from the given initial state of 20 psia and 90°F to a final temperature of 440°F. Determine the heat required, in Btu, by employing (*a*) ideal-gas tabular data and (*b*) average specific-heat data.

11-17 The mixture described in Prob. 11-3 is cooled as it passes through a steady-state heat exchanger from 540 to 140°F. Compute the heat transferred, in Btu/lb · mol and Btu/lb, using tabular ideal-gas data.

11-18 A piston-cylinder assembly contains 0.06 lb · mol of N_2 and 0.04 lb · mol of O_2. The mixture is compressed adiabatically from 600°R and 1 atm to 900°R. What is the required work input, in Btu?

11-19 Carbon dioxide (CO_2) in the amount of 1 lb · mol and initially at 30 psia and 100°F is mixed adiabatically with 2 lb · mol of O_2 initially at 80 psia and 300°F. During the constant-volume mixing process electric energy equivalent to 200 Btu/lb · mol of mixture is added. Determine (*a*) the final temperature of the mixture, in degrees Fahrenheit, if tabular data are used, and (*b*) the final pressure, in psia.

11-20 Carbon dioxide (CO_2) in the amount of 1 lb · mol and initially at 30 psia and 200°F, is mixed adiabatically in a closed system at constant pressure with 2 lb · mol of N_2, initially at 30 psia and 80°F. During the process 400 Btu are added in the form of electric work. Determine the final temperature of the mixture, in degrees Fahrenheit.

11-21 Nitrogen, in the amount of 2 lb and initially in a rigid tank at 100 psia and 140°F, is mixed with 1 lb of oxygen, initially in another tank at 50 psia and 60°F, by opening a valve connecting the two tanks. If the final equilibrium temperature is 95°F, determine (*a*) the heat transfer, in Btu, and (*b*) the final pressure, in psia.

11-22 A rigid tank contains 1.0 lb of nitrogen and 1.0 lb of carbon dioxide at 30 psia and 100°F. During a process, 47.0 Btu of heat are added and a current of 4.5 A passes through a resistor within the tank for a period of 15.0 min. If the final temperature of the gas mixture is 300°F, determine the constant voltage applied across the resistor. Use tabular data from the ideal-gas tables.

11-23 An ideal-gas mixture consists of 56 percent CO, 32 percent O_2, and 12 percent He by weight at 50 psia and 620°F. It expands adiabatically through a steady-flow turbine until the exit temperature is 160°F. Determine (*a*) the shaft-work output, in Btu/lb · mol, and (*b*) the power, in kilowatts, for 50 ft^3/min at the inlet.

11-24 An equimolar mixture of two ideal gases expands through an adiabatic nozzle from an initial state of 260°F and 100 ft/s to a final state of 180°F. Determine the exit velocity, in ft/s, if the mixture is (*a*) He and CO_2, and (*b*) He and CO.

11-25 An ideal-gas mixture has the following volumetric analysis: CO, 33.3 percent; CO_2, 50.0 percent; O_2, 16.7 percent. It enters a steady-state compressor at 100°F and 150 ft/s and leaves the device at 440°F and 300 ft/s.

(*a*) Determine the shaft-work input required, in Btu/lb, if a heat loss of 2.0 Btu/lb occurs during the process.

(*b*) If the volume flow rate at the entrance is 100 ft^3/min and the pressure is 15.0 psia, determine the power input, in horsepower.

11-26 An equimolar mixture of two gases flows steadily through a heat exchanger. The initial and final temperatures of the mixture are 500 and 1500°R, respectively. If the total flow rate is 0.050 lb · mol/s, what is the rate of heat transfer to the gas, in Btu/s, for a mixture of (*a*) H_2 and CO_2, (*b*) CO and CO_2, and (*c*) N_2 and Ar?

11-27 An equimolar mixture of two gases flows steadily through a diffuser. The initial state is 540°R and 600 ft/s, and the final velocity is 100 ft/s. Using specific-heat data, estimate the final temperature, in degrees Rankine, for a mixture of (*a*) N_2 and CO_2 and (*b*) CO and Ar, if the process is adiabatic.

Isentropic processes of ideal-gas mixtures

11-28 A mixture of three gases with component pressures in the ratio 1 : 2 : 5 is expanded isentropically from 2000°R and 10 atm to 1 atm in a piston-cylinder device. Using s^0 data, determine how much work is done, in Btu/lb · mol of mixtures of (*a*) H_2O, CO, and N_2, and (*b*) CO, CO_2, and N_2.

11-29 A gas mixture consisting of N_2, CO_2, and H_2O in a molar ratio of 4 : 1 : 1 enters a turbine at 2500°R. The gases expand isentropically through (*a*) an 8 : 1, and (*b*) a 6 : 1 pressure ratio. Using s^0 data, compute the work output, in Btu/lb · mol, of the mixture.

11-30 Reconsider the mixture composition data given in Problem 11-28 for a piston-cylinder device. Now, however, consider a compression process from 1 atm and 80°F to 4 atm. Using s^0 data, determine (1) the final temperature, in degrees Fahrenheit, and (2) the work required, in Btu/lb · mol of mixture, for mixtures *a* and *b*.

11-31 Reconsider the data of Prob. 11-29, except the process is an isentropic compression from an initial temperature of 540°R. Using s^0 data, determine the work required, in Btu/lb · mol of mixture, for (*a*) an 8 : 1 and (*b*) a 6 : 1 pressure ratio.

Entropy changes of ideal-gas mixtures

11-32 Carbon monoxide and argon in separate streams enter an adiabatic mixing chamber in a 2 : 1 mass ratio. At the inlet the carbon monoxide is at 18 psia and 540°R, and the argon is at 18 psia and 840°R. The mixture leaves at 16 psia. Determine (*a*) the final temperature of the mixture, in degrees Rankine, and (*b*) the entropy change of the carbon monoxide, in Btu/(lb · mol)(°R).

11-33 Hydrogen and nitrogen in separate streams enter an adiabatic mixing chamber in a 3 : 1 molar ratio. At the inlet the hydrogen is at 25 psia, 140°F, and the nitrogen is at 25 psia, 540°F. The mixture leaves the chamber at 24 psia.

(*a*) Determine the final temperature of the mixture, in degrees Rankine.

(*b*) Compute the entropy change of the hydrogen gas, in Btu/(lb · mol)(°R).

11-34 An insulated 3-ft^3 tank is divided into two sections by a partition. One section is 2 ft^3 in volume and initially contains hydrogen gas at 30 psia and 110°F. The remaining section initially holds nitrogen gas at 50 psia and 50°F. The adiabatic partition is then removed, and the gases are allowed to mix. Determine (*a*) the temperature of the equilibrium mixture, (*b*) the pressure of the mixture, and (*c*) the entropy change for the process, in Btu/°R.

11-35 A system consists of two tanks interconnected with a pipe and valve. One tank with a volume of 37.0 ft^3 contains 5.0 lb of argon at 20 psia and 90°F. The other tank has a volume of 8.0 ft^3 and holds 4.0 lb of oxygen at 100 psia and 140°F. The valve is opened, and the gases are allowed to mix. The atmospheric temperature is 40°F, and during the mixing process 31 Btu of heat leave the uninsulated system. Determine (*a*) the final temperature, (*b*) the final mixture pressure, (*c*) the apparent molecular weight of the mixture, (*b*) the change in entropy of the argon, in Btu/°R, and (*e*) the entropy change of the atmosphere, in Btu/°R.

11-36 Consider Prob. 11-35, except that 49.0 Btu of heat is transferred to the mixture from a heat reservoir at 540°F. The initial data remain the same. Answer the same questions, except part *e* is for the reservoir.

11-37 An insulated 1.5-ft^3 tank is divided into two sections by a partition. One section is 0.5 ft^3 in volume and initially contains 0.10 lb of carbon monoxide at 500°F. The remaining section initially holds 0.020 lb of helium at 60°F. The adiabatic partition is then removed, and the gases are allowed to mix. Determine (*a*) the temperature of the equilibrium mixture, in degrees Fahrenheit, (*b*) the pressure of the final mixture, in psia, and (*c*) the entropy change of the carbon monoxide, in Btu/°R.

11-38 At one inlet to a control volume, 33 lb/min of CO_2 and 9 lb/min of H_2O enter as a mixture at 30 psia and 300°F. At another inlet, 8 lb/min of O_2 and 14 lb/min of N_2 enter as a mixture at 30 psia and 200°F. The mixture of four gases leaves the control volume at 28 psia and 260°F. The shaft work is zero. Use the ideal-gas tables to determine (*a*) the magnitude and direction of any heat transfer, in Btu/h, (*b*) the rate of entropy change of the CO_2 component, in Btu/(h)(°R), and (*c*) the cross-sectional area at the control surface where CO_2 and H_2O enter, in ft^2, if the gas velocity at that surface is 30 ft/s.

Properties of air–water-vapor mixtures

11-39 A tank contains 0.018 lb of air and 9.12 lb of saturated water vapor at 150°F. Determine (*a*) the volume of the tank, in cubic feet, and (*b*) the total pressure, in psia.

11-40 A tank contains 0.015 lb of air and 0.15 lb of dry, saturated steam at a temperature of 180°F. Determine (*a*) the volume of the tank, and (*b*) the total pressure in the tank.

11-41 If the partial pressure of water vapor in atmospheric air at 14.7 psia is 0.400 psia at 90°F, determine (*a*) the relative humidity, (*b*) the approximate dew-point temperature, (*c*) the humidity ratio, (*d*) the enthalpy, in Btu/lb, based on $h = 0$ at 0°F for dry air, and (*e*) the specific volume of the mixture. Use steam-table data where necessary.

11-42 The partial pressure of water vapor in atmospheric air is 0.30 psia at 80°F. Determine (*a*) the relative humidity, (*b*) the dew-point temperature, (*c*) the specific humidity, in lb/lb, (*d*) the enthalpy, in Btu/lb dry air (based on $h = 0$ at 0°F for dry air, and (*e*) the specific volume, in ft^3/lb dry air, for 14.6 psia total pressure. Use the steam tables where necessary.

11-43 An atmospheric air mixture at 14.5 psia contains 2 percent water by volume. If the dry-bulb temperature is 74°F, determine (*a*) the relative humidity, (*b*) the dew-point temperature, (*c*) the humidity ratio, in gr/lb, and (*d*) the enthalpy, in Btu/lb dry air (based on $h = 0$ at 0°F for dry air). Use steam-table data.

11-44 Atmospheric air with a relative humidity of 50 percent is held at 94°F and 14.30 psia. Determine (*a*) the specific humidity, in gr/lb, (*b*) the dew-point temperture, and (*c*) the enthalpy, in Btu/lb dry air, where $h = 0$ at 0°F. Use the steam-table data.

11-45 Atmospheric air at a barometric pressure of 14.5 psia and a dry-bulb temperature of 70°F has a relative humidity of 40 percent. Determine, through the use of the steam tables, (*a*) the specific humidity, in gr/lb, (*b*) the dew-point temperature, in degrees Fahrenheit, and (*c*) the enthalpy, in Btu/lb (based on $h = 0$ at 0°F).

11-46 Condensation on cold-water pipes often occurs in warm humid rooms. If the water temperature in the pipes may reach a minimum of 44°F and the room temperature is kept at 74°F, what is the maximum limit of relative humidity in the room to avoid condensation at normal atmospheric pressure?

11-47 On a cold winter day the inside surface of a wall in a home is found to be 58°F. If the air within the room is at 72°F, what is the maximum relative humidity that the air can have without the occurrence of condensation of water on the wall?

11-48 Arrange the temperature values (*a*) 66, 97, and 46°F (*b*) 50, 60, and 74°F, and (*c*) 60, 70, and 90°F in the following order: dry-bulb, dew-point, and wet-bulb temperature.

11-49 Atmospheric air at 14.7 psia has a dry-bulb temperature of 78°F and a wet-bulb temperature of 68°F. Determine, through the use of steam-table data and Eq. (11-43), (*a*) the humidity ratio, in lb/lb, (*b*) the relative humidity, and (*c*) the enthalpy, in Btu/lb.

11-50 Atmospheric air at 14.6 psia has a dry-bulb temperature of 86°F and a wet-bulb temperature of 68°F. Determine, through the use of the steam-table data and Eq. (11-43), (*a*) the humidity ratio, in lb/lb, (*b*) the relative humidity, and (*c*) the enthalpy, in Btu/lb.

11-51 Same as Prob. 11-49, except that the dry-bulb and wet-bulb temperatures are 82°F and 72°F, respectively.

Property evaluation from a psychrometric chart

11-52 Reconsider Prob. 11-49. Estimate the answers by using the psychrometric chart.

11-53 Reconsider Prob. 11-50. Estimate the answers by using the psychrometric chart.

11-54 Reconsider Prob. 11-51. Estimate the answers by using the psychrometric chart.

11-55 The dry-bulb and wet-bulb temperatures for atmospheric air are 85 and 74°F, respectively. Determine (*a*) the partial pressure of the water vapor, (*b*) the humidity ratio (*c*) the specific volume per pound of dry air, (*d*) the relative humidity, and (*e*) the enthalpy, in Btu/lb of dry air. Employ the psychrometric chart, and assume the total pressure is 14.696 psia.

11-56 Atmospheric air has a dry-bulb temperature of 90°F and a wet-bulb temperature of 84.5°F. From the psychrometric chart determine (*a*) the specific humidity, (*b*) the dew point, (*c*) the enthalpy, (*d*) the specific volume, and (*e*) the relative humidity.

11-57 Atmospheric air at a pressure of 14.7 psia and 90°F has a specific humidity of 0.0095 lb water vapor/lb dry air. Determine from the psychrometric chart (*a*) the relative humidity, (*b*) the vapor pressure, (*c*) the total enthalpy, (*d*) the specific volume, and (*e*) the dew-point temperature.

11-58 Atmospheric air at 14.7 psia has dry-bulb and wet-bulb temperatures of 78 and 64°F, respectively. From the psychrometric chart, determine (*a*) the humidity ratio, (*b*) the dew point, (*c*) the specific volume, (*d*) the relative humidity, and (*e*) the enthalpy.

Heating, cooling, and dehumidification

11-59 Air which originally exists at 100°F, 14.7 psia, and a relative humidity of 40 percent is cooled at constant pressure to 70°F. Determine (*a*) the relative humidity, (*b*) the humidity ratio, and (*c*) the dew-point temperature at the final state, employing the psychrometric chart.

11-60 Atmospheric air with dry-bulb and wet-bulb temperatures of 90 and 70°F, respectively, is cooled to 66°F at a constant pressure of 14.7 psia. If the mass flow rate through the cooling unit is 500 lb dry air/h, calculate the heat removed, in Btu/h. Check calculated property data against the psychrometric chart.

11-61 Atmospheric air at 14.6 psia and 80°F dry-bulb temperature, and a relative humidity of 70 percent, is cooled to 50°F.

(*a*) Determine the pounds of water vapor condensed per pound of dry air.

(*b*) How much heat is removed, in Btu/lb of dry air? Calculate the required property data, and check values against the psychrometric chart.

11-62 Atmospheric air with dry-bulb and wet-bulb temperatures of 82 and 68°F, respectively, is cooled to 61°F at a constant pressure of 14.7 psia. The volume flow rate entering the cooling unit is 100 ft^3/min.

(*a*) Determine the mass flow rate, in lb/min of dry air.

(*b*) Determine the heat removed, in Btu/min. Calculate the required property data, and check values against the psychrometric chart.

11-63 Atmospheric air at 86°F and 60 percent relative humidity flows over a cooling coil at a rate of 2000 ft^3/min. The liquid which condenses is removed from the system at 60°F. Subsequent heating of the air results in a final temperature of 75°F and a relative humidity of 50 percent. Determine (*a*) the heat removed in the cooling section, in Btu/h, (*b*) the heat added in the heating section, in Btu/h, and (*c*) the amount of vapor condensed, in lb/h. Calculate required property data, and check these values against the psychrometric chart.

11-64 A 100-ft^3 storage tank initially contains air at 80 psia and 300°F and a relative humidity of 10 percent. The air then cools back to the ambient temperature of 60°F. Determine (*a*) the dew point of the mixture, (*b*) the temperature at which condensation begins, (*c*) the amount of water condensed per pound of dry air, and (*d*) the heat transferred to or from the tank during the process, in Btu/lb dry air.

11-65 Atmospheric air at 84°F and 70 percent relative humidity and a rate of 15,000 ft^3/min flows over a set of cooling coils. The liquid which condenses is removed from the system at 50°F. Subsequent heating of the air results in a final temperature of 75°F and a relative humidity of 40 percent. Determine the heat removed in the cooling section and the heat added in the heating section, in Btu/h.

11-66 Ambient air from outside a store is to be conditioned from a temperature of 84°F and 80 percent relative humidity to a final state of 70°F and 40 percent relative humidity. On the assumption that the process occurs at a constant pressure of 14.7 psia, calculate (*a*) the amount of water removed per pound of dry air, (*b*) the heat removed in the cooling process, and (*c*) the heat added in the final step of the process.

Evaporative cooling

11-67 Atmospheric air at 95°F and 20 percent relative humidity passes through an evaporative cooler until the final temperature is (*a*) 68°F, (*b*) 75°F, and (*c*) 80°F. Determine how much water is added to the air and what the final relative humidity is in each case. Calculate the required property data, and check these values against the psychrometric chart.

11-68 Atmospheric air at 95°F and 10 percent relative humidity passes through an evaporative cooler until the final relative humidity is (*a*) 40 percent, (*b*) 60 percent, and (*c*) 80 percent. For each case find the amount of water added to the air and the final dry-bulb temperature, by employing the psychrometric chart.

11-69 Atmospheric air at 90°F and 10 percent relative humidity passes through an evaporative cooler.

(*a*) On the basis of the psychrometric chart, what is the minimum temperature that could be reached by this process?

(*b*) If the final dew-point temperature is (1) 50°F, and (2) 55°F, estimate from the psychrometric chart how much water is added in the cooler and what is the final temperature.

Heating and humidification

11-70 An atmospheric air stream at 50°F and 40 percent relative humidity is first heated to 90°F and then passed through an evaporative cooler until the temperature reaches (*a*) 80°F, and (*b*) 70°F. Calculate the required property data, using steam-table data, and determine (1) the heat added, (2) the water added, and (3) the final relative humidity. Check calculations by employing the psychrometric chart.

11-71 Atmospheric air at 55°F and 30 percent relative humidity is first heated to 94°F and then passed through an evaporative cooler until the temperature reaches (*a*) 80°F, and (*b*) 68°F. Determine the heat added in the first section and the water added in the second section. Calculate the required property data, using steam-table data, and check the results against the psychrometric chart.

11-72 An air stream at 90°F with a wet-bulb temperature of 60°F passes through an evaporative cooler until the temperature is (*a*) 70°F, and (*b*) 62°F. It is then heated to 80°F. Determine the amount of water added in the first section and the heat added in the second section of the equipment. Calculate the required property data.

Adiabatic mixing

11-73 Two streams of atmospheric air at 14.7 psia are mixed together in an adiabatic, steady-flow process. One air stream, at 100°F and 40 percent relative humidity, enters at a rate of 5 lb dry air/min, while the second stream, at 40°F and 100 percent relative humidity, enters at (*a*) 10 lb dry air/min, and (*b*) 15 dry air/min. Determine (1) the exit humidity ratio, (2) the exit stream enthalpy, and (3) the exit dry-bulb temperature, in degrees Fahrenheit. Compute the required property data, and check these values against the psychrometric chart.

11-74 One atmospheric air stream, at 60°F and 30 percent relative humidity, is mixed adiabatically with another air stream, at 90°F and 80 percent relative humidity, in a steady-flow process. If the mass flow rate of the dry air in the hot stream is (*a*) twice, and (*b*) three times that of the colder stream, determine the final mixture dry-bulb temperature and its relative humidity. Compute the required property data, using stream-table data where necessary, and compare values with the chart, if P is 14.7 psia.

11-75 In an air-conditioning process 500 ft^3/min of outside air at 86°F and 80 percent relative humidity are mixed adiabatically with 640 ft^3/min of inside air initially at 70°F and 30 percent relative humidity. Determine for the resultant mixture (*a*) the dry-bulb temperature, (*b*) the humidity ratio, and (*c*) the relative humidity. Compute initial property values that are required, using steam-table data where necessary, and compare values with the psychrometric chart if the pressure is 14.7 psia.

11-76 An atmospheric air mixture at 90°F dry bulb and 80°F wet bulb mixes adiabatically with another air stream originally at 60°F dry bulb and 50°F wet bulb. The total pressure is 14.7 psia. The ratio of the volume flow rate of the cold stream to that of the hot stream is 2.0. Determine for the resultant mixture (*a*) the dry-bulb temperature, (*b*) the humidity ratio, and (*c*) the relative humidity. Compute the required inlet property data, and compare against the psychrometric chart.

Wet cooling tower

11-77 Water enters a cooling tower at 9000 gal/min and is cooled from 86 to 68°F. The entering atmospheric air at 14.7 psia has dry-bulb and wet-bulb temperatures of 70 and 60°F, respectively. The air is saturated when it leaves the tower. Determine (*a*) the volume rate of inlet air required, in ft^3/min, and (*b*) the rate of water evaporated, in lb/h. The exit air temperature is 82°F.

11-78 Water enters a cooling tower at 105°F and leaves at 72°F. The tower receives 300,000 ft^3/min of atmospheric air at 14.7 psia, 68°F, and 40 percent relative humidity. The air exit condition is 90°F and 90 percent relative humidity. Determine (*a*) the mass flow rate of dry air passing through the tower, in lb/min, (*b*) the mass flow rate of entering water, in lb/min, and (*c*) the amount of water evaporated, in lb/min.

11-79 It is desired to cool 3000 lb/min of water from 100 to 77°F. The cooling tower receives 24,000 ft^3/min of air at 14.7 psia with dry-bulb and wet-bulb temperatures of 85 and 70°F, respectively. If the evaporation rate from the water stream is 3600 lb/h, determine the temperature of the exit air stream.

11-80 It is desired to cool water from 100 to 80°F. The cooling tower receives 24,000 ft^3/min of atmospheric air at 14.7 psia with dry-bulb and wet-bulb temperatures of 85 and 70°F. The outlet water rate is 2800 lb/min and the exit air relative humidity is 95 percent. Determine (*a*) the outlet temperature of the air, in °F, and (*b*) the evaporation rate of the water, in lb/min.

CHAPTER

TWELVE

PvT BEHAVIOR OF REAL GASES AND REAL-GAS MIXTURES

In Chap. 1 some qualitative characteristics of the PvT behavior of real gases were presented. Quantitative data for real gases were presented in Chap. 4 in the form of tables like those which appear in the Appendix. However, in many cases it would be convenient to have analytical expressions for the relationships among these three properties. There is no set method for doing this; consequently, we shall look at a few of the many correlations available for the equations of state of real gases. The accuracy of these PvT relations varies with the type of gas and the range of the properties under consideration. The best methods have not necessarily been chosen for analysis in this chapter, since the purpose here is simply to provide some insight into typical correlations. When the occasion arises, a search of the literature will reveal numerous equations of state which might profitably be used for particular investigations.

12-1 THE VIRIAL EQUATION OF STATE

For simple, compressible systems we have postulated that two independent properties are required in order to fix the equilibrium state. If Pv is chosen as the dependent thermodynamic property, it may be considered to be a function of two independent properties such as pressure and temperature, or volume and temperature. The data of Fig. 12-1 illustrate a functional relationship of the type

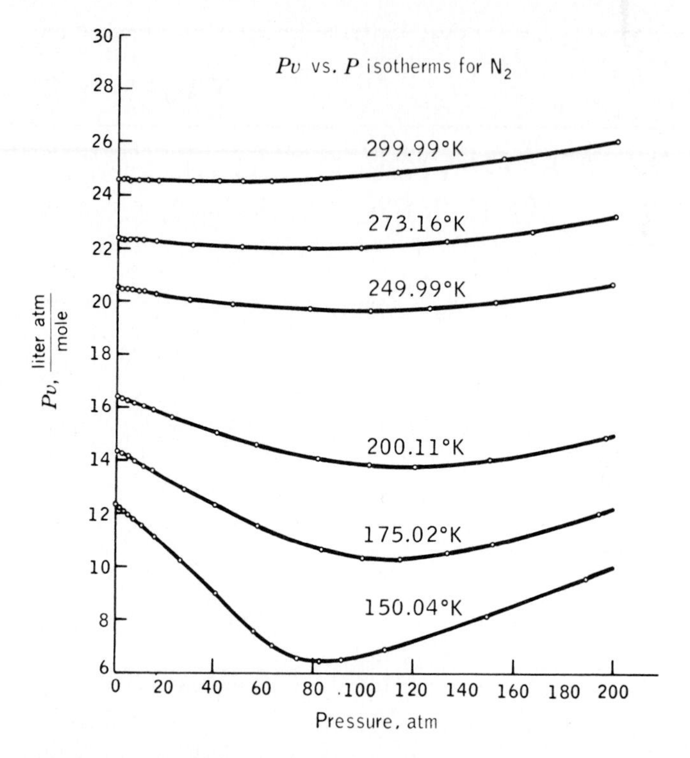

Figure 12-1 Variation of Pv of nitrogen with pressure at constant temperature. (*From M. W. Zemansky, "Heat and Thermodynamics," 4th ed., McGraw-Hill, New York, 1957.*)

$Pv = f(P, T)$. This functional relationship may be expressed to any desired accuracy by an infinite-series expansion of the type

$$Pv = a + bP + cP^2 + dP^3 + \cdots \tag{12-1}$$

If the independent variables were selected as v and T, the following form of an infinite series would be applicable:

$$Pv = a\left(1 + \frac{b'}{v} + \frac{c'}{v^2} + \frac{d'}{v^3} + \cdots\right) \tag{12-2}$$

Equations of this type, which relate P, v, and T, are known as *virial equations* of state. The coefficients a, b, c, etc., are called the first, second, third, etc., virial coefficients for the respective equations. These coefficients are solely a function of the temperature of the system. They are of particular interest since they can be given a physical significance on a molecular scale. As a result, they can be determined, at least in theory, from statistical mechanics, as well as from direct macroscopic measurements illustrated in Fig. 12-1. The virial coefficients are also, of course, a function of the substance of interest. Since all PvT equations of state should reduce to $Pv = R_u T$ as the pressure approaches zero, it is seen that the constant a, common to both equations, is equal to $R_u T$.

As an example, a virial equation for nitrogen gas at 0°C and for pressure up to 200 atm has been developed as follows:

$$Pv = 22{,}414.6 - 10.281P + 0.065189P^2 + 5.1955 \times 10^{-7} P^4 - 1.3156 \times 10^{-11} P^6 + 1.009 \times 10^{-16} P^8 \quad (12\text{-}3)$$

The volume in this equation is expressed in $cm^3/g \cdot mol$, and the pressure is in atmospheres.

When the pressure is relatively low, it is seen from Eq. (12-3) that the latter terms for Pv contributed little compared with the first several terms. Thus at low pressures a virial equation of the type

$$Pv = RT + bP \quad (12\text{-}4)$$

is often of sufficient accuracy, where b is the second virial coefficient. This coefficient is negative at low temperatures, but increases with increasing temperature and eventually is positive. The temperature for which the second virial coefficient is zero is known as the Boyle temperature. An empirical equation for the second virial coefficient b of nitrogen has the form

$$b = 39.5 - \frac{1.00 \times 10^4}{T} - \frac{1.084 \times 10^6}{T^2} \quad (12\text{-}5)$$

where b is in $cm^3/g \cdot mol$, and T is in degrees kelvin. We shall find in the following chapter that equations similar to (12-4) and (12-5) are useful not only in correlating PvT data, but also in evaluating other properties of matter.

12-2 THE VAN DER WAALS EQUATION OF STATE

In 1873 van der Waals proposed an equation of state which was an attempt to correct the ideal-gas equation so that it would be applicable to real gases. On the basis of simple kinetic theory particles are assumed to be point masses, and there are no intermolecular forces between particles. However, as the pressure increases on a gaseous system, the volume occupied by the particles may become a significant part of the total volume. In addition, the intermolecular attractive forces become important under this condition. To account for the volume occupied by the particles, van der Waals proposed that the specific volume in the ideal-gas equation of state be replaced by the term $v - b$. At the same time the ideal pressure was to be replaced by the term $P + a/v^2$. The constant b is the covolume of the particles, and the constant a is a measure of the attractive forces. Thus the van der Waals equation is

$$\left(P + \frac{a}{v^2}\right)(v - b) = RT \quad (12\text{-}6)$$

Both a and b have units which must be consistent with those employed for P, v, and T. Note that, as the pressure approaches zero and the specific volume approaches infinity, the correction terms are negligible and the equation reduces to $Pv = RT$. Empirically, it is observed that, as the pressure increases, the a/v^2 term usually becomes important sooner than the b correction factor.

The van der Waals equation is moderately successful, but it has a fundamental weakness in that the constants a and b in actuality vary with temperature. Hence their values should be determined empirically for particular regions of pressure and temperature. One specific method of evaluating the two constants is based on the experimental fact that the critical isotherm on a Pv diagram has a zero slope at the critical state for all gases. In addition, this isotherm goes through a point of inflection at the critical state. This phenomenon is illustrated in Fig. 12-2, which shows some typical isotherms based on the van der Waals equation of state. The curve marked T_c is the critical isotherm, and it is fairly representative of actual experimental data. On the basis that the critical isotherm goes through a point of inflection at the critical state, we may write the following equations which are valid at the critical state:

$$\left(\frac{\partial P}{\partial v}\right)_{T_c} = 0 \qquad \text{and} \qquad \left(\frac{\partial^2 P}{\partial v^2}\right)_{T_c} = 0$$

These equations permit the evaluation of the constants in any two-constant equation of state. For example, the van der Waals constants a and b may now be determined by this method. It will be found that, on the basis of Eq. (12-6),

$$\left(\frac{\partial P}{\partial v}\right)_{T_c} = \frac{-RT_c}{(v_c - b)^2} + \frac{2a}{v_c^3} = 0$$

and

$$\left(\frac{\partial^2 P}{\partial v^2}\right)_{T_c} = \frac{2RT_c}{(v_c - b)^3} - \frac{6a}{v_c^4} = 0$$

The subscript c indicates that the property must be evaluated at the critical state. By combining these two expressions with the van der Waals equation of state it is

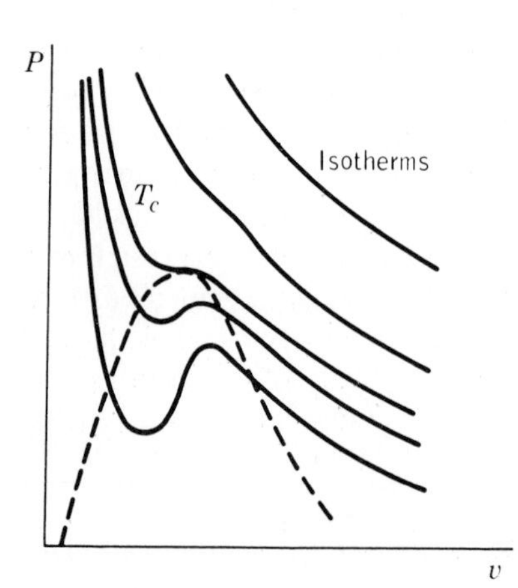

Figure 12-2 Isotherms for a van der Waals gas on a Pv diagram.

found that

$$a = \frac{27}{64}\frac{R^2 T_c^2}{P_c} \qquad b = \frac{RT_c}{8P_c} \qquad \frac{P_c v_c}{RT_c} = \frac{3}{8}$$

The above equations for a and b allow the determination of these constants from P_c and T_c, which are obtained experimentally. However, since it is observed that the values of a and b actually need to be varied in order to relate P, v, and T accurately, the use of critical data to evaluate the two constants must lead to an equation of state of limited accuracy. Another indication of this is the fact that the quantity $P_c v_c/RT_c$ is 0.375 according to the van der Waals equation. Experimentally, this is found to lie between 0.20 and 0.30 for the majority of gases. Thus the van der Waals equation is an approximation which is sometimes in serious error. Nevertheless, it is of historical interest as one of the first attempts to correct the ideal-gas equation so that real-gas behavior might be predicted. Tables A-3M and A-3 list some typical values for the van der Waals constants based on critical data.

12-3 OTHER EQUATIONS OF STATE

Two other two-constant equations of state are those of Berthelot and Dieterici. These have the form

Berthelot:
$$P = \frac{RT}{v - b} - \frac{a}{T_v^2} \tag{12-7}$$

Dieterici:
$$P = \frac{RT}{v - b} e^{-a/RTv} \tag{12-8}$$

The Berthelot equation is quite similar to the van der Waals equation, except that the term a/v^2 also contains the temperature in the denominator. As a result, the constants a and b in this equation are not the same as those for the van der Waals equation of state, when based on critical data. The Dieterici equation was developed primarily to give better agreement with the quantity $P_c v_c/RT_c$ determined experimentally. It was noted earlier that this quantity based on the van der Waals equation was considerably in error. Like the van der Waals equation, these equations are also of limited accuracy.

Although many other equations of state could be presented, of varying complexity, the Redlich-Kwong equation is of considerable interest. This equation of state contains only two constants, and is empirical in nature. Nevertheless, it appears to have considerable accuracy over a wide range of *PvT* conditions. On the basis of theoretical and practical considerations, Redlich and Kwong proposed in 1949 the following relationship:

$$P = \frac{RT}{v - b} - \frac{a}{T^{1/2}v(v + b)} \tag{12-9}$$

The constants a and b again can be evaluated from critical data. It is found that $a = 0.4275R^2 T_c^{2.5}/P_c$, and $b = 0.0867RT_c/P_c$. One of the primary considerations in the development of this equation was that, at high pressures, the volume of all gases tends to approach the limiting value of $0.26v_c$, and this value is essentially independent of the temperature. Consequently, b also equals $0.26v_c$. As a result, it is found that the equation gives quite good results at high pressures. The equation also appears to be fairly accurate for temperatures above the critical value. When T is less than T_c, the equation is found to deviate from experimental data as the temperature is lowered. However, in several instances it has been shown that the Redlich-Kwong equation is as accurate as the eight-constant Benedict-Webb-Rubin equation of state. Consequently, the Redlich-Kwong equation of state has the advantages of simplicity combined with a fair degree of accuracy, especially with respect to other two-constant equations of state.

Since the Redlich-Kwong equation is a cubic equation in v, a rearrangement of the equation leads to a more convenient method for finding v when P and T are known. By multiplying the equation by v/RT, we find that the Redlich-Kwong equation can be written as

$$\frac{Pv}{RT} = Z_{\mathbf{RK}} = \frac{v}{v-b} - \frac{a}{(v+b)RT^{1.5}}$$

$$= \frac{1}{1-k} - \left(\frac{a}{bRT^{1.5}}\right)\left(\frac{k}{1+k}\right) \tag{12-10a}$$

where k is defined as b/v and $Z_{\mathbf{RK}}$ is a compressibility factor for a Redlich-Kwong gas. (See Sec. 4-6 for an introductory discussion of the compressibility factor Z.) Moreover it has been noted above that a and b are functions of the critical temperature and pressure. As a result, $a/bRT^{1.5} = 4.934(T_c/T)^{1.5}$, and

$$Z_{\mathbf{RK}} = \frac{1}{1-k} - \frac{4.934}{(T/T_c)^{1.5}}\left(\frac{k}{1+k}\right) \tag{12-10b}$$

Also

$$k = \frac{b}{v} = \frac{bP}{ZRT} = \frac{0.0867P_R}{ZT_R} \tag{12-10c}$$

where $P_R = P/P_c$ and $T_R = T/T_c$. When P and T are given and P_c and T_c are known, Eqs. (12-10b) and (12-10c) contain two unknowns, namely, Z and k. The two equations may be solved by iteration by hand or computer techniques. For example, as a first guess let $Z = 1$. Then use this value in Eq. (12-10c) to find a value of k. Use of this first estimate of k in Eq. (12-10b) yields a new approximation of Z. This in turn is used again in Eq. (12-10c) to find a new value of k. This procedure is continued until the iteration produces changes in Z and k which are smaller than some prescribed small values.

Example 12-1M Estimate the pressure which would be exerted by 3.7 kg of CO in a 0.030-m^3 container at 215°K, employing (*a*) the ideal-gas equation, (*b*) the van der Waals equation and (*c*) the Redlich-Kwong equation of state.

SOLUTION The specific volume of the gas is $0.030/3.7 = 0.0081$ m^3/kg $= 0.227$ m^3/kg mole, and R_u is 0.08314 bar · m^3/(kg · mol)(°K).

(*a*) The pressure is computed directly from the ideal-gas equation as

$$P = \frac{RT}{v} = \frac{0.08314(215)}{28(0.00811)} = 78.7 \text{ bars}$$

(*b*) The constants for the van der Waals equation are found in Table A-3M. For CO we find that $a = 1.463$ bar · m^6/(kg · mol)^2 and $b = 0.0394$ m^3/kg · mol. Substitution of these values in the equation of state leads to

$$\left(P + \frac{a}{v^2}\right)(v - b) = RT$$

$$\left[P + \frac{1.463}{(0.227)^2}\right][0.00811(28) - 0.0394] = 0.08314(215)$$

The solution to this equation gives the pressure to be 66.9 bars.

(*c*) The specific volume is 0.227 m^3/kg · mol and the temperature is 215°K. The constants a and b are found in Table A-21M to be 17.26 bar(m^6)(°K)$^{1/2}$/(kg · mol)^2 and 0.02743 m^3/kg · mol, respectively. Making the proper substitutions, we find that

$$P = \frac{RT}{v - b} - \frac{a}{T^{1/2}v(v + b)} = \frac{0.08314(215)}{0.227 - 0.0274} - \frac{17.26}{(215)^{1/2}(0.227)(0.227 + 0.0274)} = 69.2 \text{ bars}$$

Example 12-1 Estimate the pressure which would be exerted by 8.2 lb of CO in a 1-ft^3 container at -78°F, employing (*a*) the ideal-gas equation, (b) the van der Waals equation of state, and (*c*) the Redlich-Kwong equation.

SOLUTION The temperature of -78°F is equal to 382°R. The specific volume of the gas is $1/8.2 = 0.122$ ft^3/lb.

(*a*) The pressure is computed directly from the ideal-gas equation as

$$P = \frac{RT}{v} = \frac{10.73(382)}{28(0.122)} = 1200 \text{ psia} = 81.7 \text{ atm}$$

(*b*) The constants for the van der Waals equation are found in Table A-3. For CO we find that $a = 372$ atm · ft^6/(lb · mol)^2 and $b = 0.63$ ft^3/lb · mol. In this system of units the value of R_u will be 0.730 atm · ft^3/(lb · mol)(°R). In addition, the specific volume to be used is $0.122(28) = 3.42$ ft^3/lb · mol. Substitution of these values in the proper equation leads to

$$\left(P + \frac{a}{v^2}\right)(v - b) = RT$$

$$\left[P + \frac{372}{(3.42)^2}\right](3.42 - 0.63) = 0.73(382)$$

The solution of this equation gives the pressure to be 68.2 atm.

(*c*) The specific volume is 3.42 ft^3/lb · mol and the temperature is 382°R. The constants a and b are found in Table A-21 to be 5870 (atm · ft^6)(°R)$^{1/2}$(lb · mol)^2 and 0.4395 ft^3/lb · mol, respectively.

Making the proper substitutions, we find that

$$P = \frac{RT}{v - b} - \frac{a}{T^{1/2}v(v + b)} = \frac{0.73(382)}{3.42 - 0.440} - \frac{5870}{(382)^{1/2}(3.42)(3.42 + 0.440)} = 70.8 \text{ atm}$$

One of the inherent limitations in two-constant equations of state is the lack of accuracy over a wide range of conditions. To achieve higher accuracy requires equations of state with a considerably larger number of terms, or constants. An example of a more complex equation which has been successful in predicting the PvT behavior especially of hydrocarbons is the Benedict-Webb-Rubin equation. This equation is

$$P = \frac{RT}{v} + \left(B_0 RT - A_0 - \frac{C_0}{T_2}\right)\frac{1}{v^2} + \frac{bRT - a}{v^3} + \frac{a\alpha}{v^6} + \frac{c}{v^3T^2}\left(1 + \frac{\gamma}{v^2}\right)\exp\left(-\frac{\gamma}{v^2}\right) \tag{12-11}$$

This equation has eight adjustable constants for a given substance. Values of these constants for five common substances are found in Tables A-21M and A-21. The units on these constants must be consistent with those used for P, v, R, and T. Another example of a complex equation of state is that suggested by Martin and Hou. It fits polar and nonpolar substances with densities up to 50 percent greater than the critical density. The original form of this equation in 1955 had nine adjustable constants. Since that time (1959), the equation has been modified to include eleven constants.

Example 12-2M Estimate the pressure of CO for the conditions expressed in Example 12-1M, on the basis of the Benedict-Webb-Rubin equation of state.

SOLUTION The values of the constants for the B-W-R equation for CO are listed in Table A-21M. The specific volume is 0.227 L/g · mol, and $R_u = 0.08315$ bar (L)/(g · mol)(°K). Making the proper substitutions, we find that

$$P = \frac{0.08314(215)}{0.227} + \left[0.05454(0.08314)(215) - 1.3587 - \frac{8.673 \times 10^3}{(215)^2}\right]\frac{1}{(0.227)^2}$$

$$+ \frac{0.002632(0.08314)(215) - 0.0371}{(0.227)^3} + \frac{0.0371(1.35 \times 10^{-4})}{(0.227)^6}$$

$$+ \frac{1.054 \times 10^3}{(0.227)^3(215)^2}\left[1 + \frac{0.0060}{(0.227)^2}\right]\exp\left[-0.0060/(0.227)^2\right]$$

$$= 78.75 - 11.09 + 0.86 + 0.04 + 1.94 = 70.50 \text{ bars}$$

NBS Technical Note 202 reports a pressure of 70.91 bars for the given specific volume and temperature. Thus the B-W-R equation is in error less than 0.6 percent. In comparison, the results of Example 12-1M indicate that the van der Waals equation is 5 percent too low, while the Redlich-Kwong equation is 1.8 percent too low.

Example 12-2 Estimate the pressure of CO for the conditions expressed in Example 12-1, on the basis of the Benedict-Webb-Rubin equation of state.

SOLUTION The values of the constants for the B-W-R equation for CO are listed in Table A-21.

The specific volume is 3.42 $ft^3/lb \cdot mol$ and T is 382°R. Making the proper substitutions, we find that

$$P = \frac{0.730(382)}{3.42} + \left[0.8740(0.73)(382) - 344.5 - \frac{7.124 \times 10^6}{(382)^2}\right]\frac{1}{(3.42)^2} + \frac{0.676(0.73)(382) - 150.73}{(3.42)^3}$$

$$+ \frac{150.73(0.5556)}{(3.42)^6} + \frac{1.387 \times 10^7}{(3.42)^3(382)^2}\left[1 + \frac{1.541}{(3.42)^2}\right] \exp\left[-1.541/(3.42)^2\right]$$

$$= 81.54 - 12.79 + 0.94 + 0.05 + 2.36 = 72.1 \text{ atm}$$

Because of the complexity of the B-W-R equation, the above answer is probably a good approximation to the true value. In comparison, the results of Example 12-1 indicate that the van der Waals equation is 5.4 percent too low, while the Redlich-Kwong equation is 1.8 percent too low.

12-4 GENERALIZED COMPRESSIBILITY CHARTS

If we are to represent the PvT behavior of real gases accurately, we have seen that we require fairly complicated empirical expressions. Each of these equations of state requires its own set of experimentally determined constants. A general method of correlating PvT data, which has a reasonable degree of accuracy combined with an inherent simplicity, is based on the van der Waals principle of corresponding states. This principle was first introduced in Sec. 4-6. The reader should review that section at this point.

12-5 REAL-GAS MIXTURES

In Chap. 11 it was pointed out that the evaluation of the properties of mixtures is a difficult one, since any one property depends not only upon two independent properties such as pressure and temperature, but also upon a specification of the composition of the mixture such as the mole fractions of each component. Hence the tabulation of the properties of mixtures is not too fruitful because of the tremendous quantity of data required. As in the approach for ideal gases presented earlier, one solution is to employ the properties of the individual pure constituents in some sort of mixture rule. The two most common rules are those of additive pressures and additive volumes introduced in the preceding chapter. For example, the total pressure of a mixture of gases might be computed by Dalton's law of additive pressures as the sum of the pressures exerted by the individual constituents. That is,

$$P_m = p_1 + p_2 + p_3 + \cdots]_{T,V} \tag{12-12}$$

where the pressures of the components 1, 2, 3, etc., are computed from more exact methods for each individual gas. Each component pressure is evaluated at the volume and the temperature of the mixture as noted in Eq. (12-12). The real error in this approach is that each gas is assumed to act as if it alone occupied the entire volume and its behavior were not affected by the presence of other chemical species. As a consequence, it might be expected that the law of additive pressures will be more valid at relatively low pressures or densities, and considerable error might occur at higher pressures.

If each gas in a gas mixture were assumed to follow the van der Waals equation, the equation for component i would be

$$p_i]_{T,V} = \frac{R_u T}{v_i - b_i} - \frac{a_i}{v_i^2}$$

where v_i is the molar volume of the component. However, the molar volume of any species is related by definition to the molar volume v_m of the mixture by $v_m = y_i v_i$. The substitution of this relation into the above equations yields

$$p_i]_{T,V} = \frac{y_i R_u T}{v_m - y_i b_i} - \frac{y_i^2 a_i}{v_m^2} \tag{12-13}$$

If this general form of the equation is substituted into the additive-pressure law, we find upon rearrangement that

$$P_m = R_u T\left(\frac{y_1}{v_m - y_1 b_1} + \frac{y_2}{v_m - y_2 b_2} + \cdots\right) - \frac{1}{v_m^2}(a_1 y_1^2 + a_2 y_2^2 + \cdots) \tag{12-14}$$

For any mixture of given temperature, volume, and composition, the right-hand side of this equation can be computed directly, assuming that the van der Waals constant for each constituent has already been determined.

Another approach to the use of the additive-pressure law would be to assume that the compressibility factors for the individual components could be used. If an equation of the type $p_i V = Z_i N_i R_u T$ is substituted in the additive-pressure law for each component, the result is

$$P_m = \frac{R_u T}{V}(N_1 Z_1 + N_2 Z_2 + \cdots)$$

However, if an average compressibility factor Z_m for the mixture is defined in terms of the equation $P_m V = Z_m N_m R_u T$, then we see that

$$Z_m N_m = Z_1 N_1 + Z_2 N_2 + \cdots$$

or

$$Z_{m,V,T} = y_1 Z_1 + y_2 Z_2 + \cdots \tag{12-15}$$

where y again is the mole fraction for that species. The subscripts V and T on Z_m emphasize that the compressibility factors for each component are to be evaluated at the volume and temperature of the mixture. In general it can be stated that the rule of additive pressures in the form of Eq. (12-15) tends to give values of Z_m which are greater than the experimental value at low pressures or densities and to give values which are too low at high pressures. To overcome this difficulty, a method, frequently called the Bartlett rule of additive pressures, may be employed. By this method the same form of Eq. (12-15) is used, but the individual compressibility factors are evaluated, using the temperature and the molar volume of the mixture rather than the molar volume of the component. This approach tends to give better results in the low-pressure region. Other equations of state can be used, of course, in conjunction with the law of additive pressures.

Amagat's rule of additive volumes may also be applied to gas mixtures as an approximation technique. In this case

$$V_m = V_1 + V_2 + V_3 + \cdots]_{P,\,T} \tag{12-16}$$

where the individual volumes are computed from appropriate equations of state based on the pressure and the temperature of the mixture. If the compressibility-factor methods is employed, then $P_m V_i = Z_i N_i R_u T$. The use of this expression in Eq. (12-16) leads to the following relationship for the overall mixture compressibility factor Z_m:

$$Z_{m,\,P,\,T} = y_1 Z_1 + y_2 Z_2 + \cdots \tag{12-17}$$

The subscripts P and T emphasize that the individual Z factors are to be evaluated at the pressure and the temperature of the mixture. Although Eqs. (12-15) and (12-17) are identical in form for the additive-pressure and additive-volume rules, the method of evaluation is quite different. As a result, the two methods give different values for a given problem, when applied to real gases. Since Amagat's rule takes into account the total pressure of the system, it effectively is accounting for the influence of intermolecular forces between different chemical species, as well as between like molecules. It is a matter of experience that the rule of additive volumes for real-gas mixtures leads to Z_m values which are too low in the low-pressure region. In general, it appears that Amagat's rule is probably superior to the additive-pressure rule except at relatively low pressures.

For preliminary engineering estimates another mixture rule known as Kay's rule is satisfactory to within, roughly, 10 percent. In this method, pseudocritical temperature and pressure for mixtures are defined in the following manner:

$$T'_c = y_1 T_{c1} + y_2 T_{c2} + y_3 T_{c3} + \cdots \tag{12-18a}$$

and

$$P'_c = y_1 P_{c1} + y_2 P_{c2} + Y_3 P_{c3} + \cdots \tag{12-18b}$$

This technique is quite useful since it requires merely a knowledge of the critical temperatures and pressures of the constituent gases. Other types of correlations for real-gas mixtures are available in the literature.

Example 12-3 A system contains a mixture of hydrogen and nitrogen which is 75 mole percent hydrogen and 25 mole percent nitrogen at 77°F(25°C). When the specific volume is 1.355 $ft^3/lb \cdot mol$ ($0.0845\ m^3/kg \cdot mol$), the pressure experimentally is found to be 400 atm. Estimate what the pressure would be on the basis of (*a*) ideal-gas behavior, (*b*) additive-pressure rule and van der Waals gases, (*c*) additive-volume rule and van der Waals gases, (*d*) additive-pressure rule and compressibility factors, (*e*) additive-volume rule and compressibility factors, (*f*) the Bartlett rule of additive pressures, and (*g*) pseudocritical temperature and pressure approach.

SOLUTION The critical data and the van der Waals constants for hydrogen and nitrogen are as follows:

$$T_{c,H_2} = 59.8 + 14.4 = 74.2°\text{R} \qquad a_{H_2} = 62.80 \text{ atm} \cdot \text{mol}^2/\text{ft}^6$$

$$P_{c,H_2} = 12.8 + 8.0 = 20.8 \text{ atm} \qquad b_{H_2} = 0.426 \text{ ft}^3/\text{lb} \cdot \text{mol}$$

$$T_{c,N_2} = 227°\text{R} \qquad a_{N_2} = 346 \text{ atm} \cdot \text{mol}^2/\text{ft}^6$$

$$P_{c,N_2} = 33.5 \text{ atm} \qquad b_{N_2} = 0.618 \text{ ft}^3/\text{lb} \cdot \text{mol}$$

For use with the generalized compressibility charts the critical temperature and pressure of hydrogen have been corrected by the factors of 14.4 and 8.0, as indicated in the main text, for greater accuracy. The value of the universal gas constant R_u to use with the above set of units is 0.73 atm · ft³/(lb · mol)(°R).

(*a*) The pressure based on the ideal-gas equation is

$$P = \frac{NRT}{V} = \frac{0.73(537)}{1.355} = 290 \text{ atm}$$

This value is in error by over 25 percent.

(*b*) For the additive-pressure rule with the van der Waals equation, we may employ Eq. (12-14). Thus

$$P = 0.73(537)\left[\frac{0.75}{1.355 - 0.75(0.426)} + \frac{0.25}{1.355 - 0.25(0.618)}\right]$$

$$- \frac{1}{(1.355)^2}[62.80(0.75)^2 + 346(0.25)^2]$$

$$= 392(0.9335) - 0.545(56.9) = 335 \text{ atm}$$

The result is better than the ideal-gas relation, but still quite low compared with the experimental value.

(*c*) According to the additive-volume rule, the total volume is the sum of the volumes of the components, measured at the pressure and the temperature of the mixture. The pressure is unknown; so we write the following three equations:

$$1.355 = 0.75v_{H_2} + 0.25v_{N_2}$$

$$P_{H_2} = \frac{RT}{v - b} - \frac{a}{v^2} = \frac{0.73(537)}{v - 0.426} - \frac{62.80}{v^2} = \frac{392}{v - 0.426} - \frac{62.80}{v^2}$$

$$P_{N_2} = \frac{392}{v - 0.618} - \frac{346}{v^2}$$

For a satisfactory solution the pressures of the two gases must be equal. Employing an iteration process, we assume that v of the hydrogen is 1.38 ft³/lb · mol. Then, from the first equation above,

$$v_{N_2} = \frac{1.355 - 0.75(1.38)}{0.25} = 1.28$$

Substituting these two values into the last two equations, we find that

$$P_{H_2} = \frac{392}{1.38 - 0.426} - \frac{62.80}{(1.38)^2} = 378 \text{ atm}$$

$$P_{N_2} = \frac{392}{1.28 - 0.618} - \frac{346}{(1.28)^2} = 381 \text{ atm}$$

These two answers are sufficiently close together not to warrant another trial. This solution is, roughly, 5 percent lower than the measured value.

(*d*) The additive-pressure rule and the compressibility chart can be combined into a very rapid method of approximation if pseudoreduced-volume lines are available. Recall that $v'_R = vP_c/RT_c$, where v is the specific volume an individual gas would have if it filled the entire system. On this basis the pseudoreduced volumes and the reduced temperatures for the two gases become

$$v'_{R,\,H_2} = \frac{1.355(20.8)(2120)}{0.75(74.2)(1545)} = 0.71 \qquad T_{R,\,H_2} = \frac{537}{74.2} = 7.25$$

$$v'_{R,\,N_2} = \frac{1.355(33.5)(2120)}{0.25(227)(1545)} = 1.10 \qquad T_{R,\,N_2} = \frac{537}{227} = 2.37$$

From Figs. A-25M and A-26M the Z factors for hydrogen and nitrogen are found to be 1.06 and 0.99, respectively. Hence

$$Z_m = 0.75(1.06) + 0.25(0.99) = 1.05$$

and

$$P = Z_m P_{\text{ideal}} = 1.05(290) = 305 \text{ atm}$$

(*e*) According to the additive-volume rule, each gas exists at the pressure and the temperature of the mixture. Since the pressure is unknown, it must be assumed under an iteration process. If we first assume that the system pressure is 350 atm, then

$$P_{R,\,H_2} = \frac{350}{20.8} = 16.8 \qquad T_{R,\,H_2} = 7.25$$

$$P_{R,\,N_2} = \frac{350}{33.5} = 10.45 \qquad T_{R,\,N_2} = 2.37$$

The individual compressibility factors based on these reduced properties are 1.20 and 1.21 for hydrogen and nitrogen, respectively. Therefore

$$Z_m = 0.75(1.20) + 0.25(1.21) = 1.20$$

and

$$P = Z_m P_{\text{ideal}} = 1.20(290) = 352 \text{ atm}$$

This value agrees substantially with the assumed value; so no further trial is necessary. In this case the additive-volume rule was a better approximation than the additive-pressure rule, although both give too low a value.

(*f*) The Bartlett rule of additive pressures is based on using the molar specific volume of the mixture for the specific volume of any component. If the value of 1.355 is used in the pseudo-reduced-volume definition, we find that, from the calculation made in part *d*,

$$v'_{R,\,H_2} = 0.75(0.71) = 0.53 \qquad T_{R,\,H_2} = 7.25$$

$$v'_{R,\,N_2} = 0.25(1.10) = 0.28 \qquad T_{R,\,N_2} = 2.37$$

The compressibility factors for hydrogen and nitrogen under these conditions are approximately 1.25 and 1.17, respectively. Thus

$$Z_m = 0.75(1.25) + 0.25(1.17) = 1.23$$

and

$$P = Z_m P_{\text{ideal}} = 1.23(290) = 360 \text{ atm}$$

Note that the answer is much improved over that obtained by the unmodified rule of additive pressures.

(*g*) If, finally, the definitions of the pseudocritical temperature and pressure defined above are employed, then

$$T'_c = y_1 T_{c1} + y_2 T_{c2} = 0.75(74.2) + 0.25(227) = 112.4°\text{R}$$

$$P'_c = y_1 P_{c1} + y_2 P_{c2} = 0.75(20.8) + 0.25(33.5) = 24.0 \text{ atm}$$

Consequently,

$$v'_R = \frac{vP'_c}{RT'_c} = \frac{1.355(24.0)}{0.73(112.4)} = 0.397$$

$$T'_R = \frac{T}{T'_c} = \frac{537}{112.4} = 4.79$$

From Fig. A-26M, the Z factor is 1.31, and the corresponding pressure of the system is $P = 1.31(290) = 379$ atm. All the methods employed above gave answers which were too low, although some were much better than others.

REFERENCES

Benedict, O., G. B. Webb, and L. C. Rubin: *J. Chem. Phys.*, **8**(4): 334–345 (1940).
Lewis, G. N., and M. Randall: "Thermodynamics," 2d ed., McGraw-Hill, New York, 1961.
Martin, J. J., and Y. C. Hou: *A.I. Ch. E. J.*, **1**(2): 142–151 (1955).
Martin, J. J., R. M. Kapoor and N. deNevers: *A.I. Ch. E. J.*, **5**(2): 159–164 (1959).
Obert, E. F., "Concepts of Thermodynamics," McGraw-Hill, New York, 1960.
Otto, J., A. Michels, and H. Wouters: *Physik. Z.*, **35**(3): 97–100 (1934).
Redlich, O., and J. N. S. Kwong: *Chem. Rev.*, **44**(1): 233–244 (1949).

PROBLEMS (METRIC)

12-1M Determine the expressions for the constants a and b in the Berthelot equation of state in terms of the critical-state values of P, v, and T.

12-2M Determine the equations for a and b in the Dieterici equation of state in terms of the critical-state values of P, v, and T.

12-3M Determine equations for a and b in the Redlich-Kwong equation of state in terms of the critical values of P, v, and T.

12-4M Steam at 100 bars and 360°C is a real gas. Determine the specific volume, in cm^3/g, by using (*a*) the ideal gas equation, (*b*) the van der Waals equation, (*c*) the Redlich-Kwong equation, (*d*) the compressibility factor, and (*e*) the tabulated value.

12-5M Carbon dioxide at 313°K and 73 bars has an observed specific volume of 200 $\text{cm}^3/\text{g} \cdot \text{mol}$. Calculate the value based on (*a*) the ideal-gas equation, (*b*) the Redlich-Kwong equation, (*c*) the van der Waals equation, and (*d*) the compressibility factor.

12-6M Compute the specific volume of steam at 140 bars and 400°C by means of (*a*) the ideal-gas equation, (*b*) the van der Waals equation, (*c*) the Redlich-Kwong equation, (*d*) the compressibility factor, and (*e*) the tabulated value.

12-7M The specific volume of refrigerant 12 at 60°C is 19.41 cm^3/g. Determine the pressure of the fluid by means of (*a*) the ideal-gas equation, (*b*) the van der Waals equation, (*c*) the Redlich-Kwong equation, (*d*) the compressibility factor, and (*e*) the tabulated value.

12-8M Same as Prob. 12-6M, except find the pressure if the given state is 360°C and 30.9 cm^3/g.

12-9M Same as Prob. 12-7M, except that the state is 80°C and 12.0 cm^3/g.

12-10M Same as Prob. 12-6M, except find the temperature for the given state of 100 bars and 23.3 cm^3/g.

12-11M Same as Prob. 12-7M, except find the temperature for the given state of 9 bars and 22.0 cm^3/g.

12-12M The specific volume of saturated CO_2 vapor at 25°C is 4.19 cm^3/g. Estimate the pressure, in bars, on the basis of (*a*) the ideal-gas equation, (*b*) the Redlich-Kwong equation, (*c*) the Benedict-Webb-Rubin equation, (*d*) the van der Waals equation, and (*e*) the compressibility factor. The tabulated value is 64.0 bars.

12-13M Determine the pressure of N_2 at -123°C if the specific volume is 2.39 cm^3/g, employing (*a*) the van der Waals equation, (*b*) the Redlich-Kwong equation, (*c*) the Benedict-Webb-Rubin equation, and (*d*) the principle of corresponding states. (*e*) Compare to tabulated value, in bars.

12-14M Methane gas is maintained at 23 bars and -82°C. Compute the specific volume, in m^3/kg, on the basis of (*a*) the Redlich-Kwong equation, (*b*) the van der Waals equation, and (*c*) the principle of corresponding states.

12-15M Calculate the pressure of methane (CH_4) at 12°C and 0.0187 m^3/kg on the basis of (*a*) the principle of corresponding states, (*b*) the Redlich-Kwong equation, and (*c*) the Benedict-Webb-Rubin equation.

12-16M Calculate the pressure required to compress 100 L of nitrogen at 745 mmHg and 23°C to 1.21 L at -73°C, using Z data. Compare to tabular data, in bars.

12-17M On the basis of the Redlich-Kwong equation calculate the specific volume, in $cm^3/g \cdot mol$, at (*a*) 350°K and 101 bars, and (*b*) 400°K and 101 bars, for carbon dioxide.

12-18M The constants for the virial equation of state in the form $Pv = A[1 + (b'/v) + (c'/v^2) + \cdots]$ have been determined experimentally for nitrogen at -100°C. The values are: $A = 14.39$, $b' = -0.05185$ $m^3/kg \cdot mol$, $c = 0.002125$ $m^6/(kg \cdot mol)^2$. Also, P is in bars, T is in degrees kelvins, and v is in $m^3/kg \cdot mol$

(*a*) Determine the compressibility factor at 68 bars and -100°C from the above equation.

(*b*) Compare the above result to that obtained from a generalized compressibility chart.

12-19M Calculate the specific volume of N_2, in cm^3/g, at 200°K and 150 bars, using the compressibility-factor method. What would be the percent error in assuming nitrogen to be an ideal gas under these conditions? Compare to tabulated value.

12-20M Diatomic oxygen is at 100 bars and -73°C. Calculate how many kilograms can be stored in a tank with a volume of 3 m^3 if we use (*a*) the ideal-gas equation, and (*b*) the corresponding-states concept.

12-21M Estimate the temperature of 1 kg of *n*-octane (C_8H_{18}) at 27 bars if the volume is 0.0125 m^3.

12-22M On the basis of the Z factor estimate the kilograms of propane gas in a 0.3-m^3 cylinder if the pressure is 200 bars and the temperature is (*a*) 207°C, and (*b*) 283°C.

12-23M Nitrogen gas at a pressure of 100 bars and -70°C is contained in a tank of0.25 m^3. Heat is added until the temperature is 37°C. Determine approximately, through the use of the Z factor, (*a*) the specific volume of the gas, in $m^3/kg \cdot mol$, and (*b*) the final pressure, in bars.

12-24M Ethane gas is being transported through a pipeline at a pressure of 95 bars and 55°C. Using the generalized chart, determine the increase, in percent, in the velocity of the ethane that is required to deliver the same mass flow rate if the line pressure remains the same but the temperature increases to 116°C.

12-25M Diatomic oxygen at a pressure of 101 bars and -27°C is contained in a tank of 0.20 m^3. Heat is transferred until the temperature is 5°C. Estimate (*a*) the specific volume of the gas, in $m^3/kg \cdot mol$, and (*b*) the final pressure, in bars, using the compressibility chart.

12-26M Calculate the specific volume, in $m^3/kg \cdot mol$, of nitrogen at 102 bars and -45°C by (*a*) the ideal-gas equation, (*b*) the compressibility chart, and (*c*) the virial equation. The virial coefficients for nitrogen at this temperature are $b = -2.34 \times 10^{-2}$, $c = 3.61 \times 10^{-5}$, and $d = 5.18 \times 10^{-7}$, where P is in bars, T is in degrees kelvin, and v is in $m^3/kg \cdot mol$.

12-27M Determine the specific volume of nitrogen at 200 bars and 0°C by means of Eq. (12-3). Compare with the answer found by using Eqs. (12-4) and (12-5). Check against tabulated value, in $cm^3/g \cdot mol$.

12-28M Same as Prob. 12-27M, except the pressure is 150 bars.

12-29M On the basis of the generalized chart, determine the kilograms of ethylene (C_2H_4) in a 0.5-m^3 tank if the temperature is 38°C and the pressure is 82 bars. If the temperature is increased to 123°C, find the final pressure, in bars.

12-30M Determine the specific volume of nitrogen at 150°K and 64 bars by means of Eqs. (12-4) and (12-5), and compare with the result based on the generalized compressibility chart.

12-31M *NBS Technical Note* 202 (1963) lists the specific volume of carbon monoxide at 200°K and 60 atm to be 0.2356 $m^3/kg \cdot mol$. Estimate the value of P at the given v and T by means of (*a*) the ideal-gas equation, (*b*) the van der Waals equation, (*c*) the Redlich-Kwong equation, (*d*) the Benedict-Webb-Rubin equation, and (*e*) the generalized compressibility chart.

12-32M A mixture consists of 1 kg of CO_2 and 1 kg of water vapor at 20 bars and 200°C. Calculate (*a*) the partial pressure of each component, (*b*) the component pressure based on the additive-pressure rule and CO_2 an ideal gas, (*c*) the component pressure based on the additive-pressure rule if CO_2 is a real gas, and (*d*) the volume of the mixture.

12-33M On the basis of compressibility factors, determine the isothermal work necessary to compress carbon dioxide from 70 to 250 bars at a temperature of 92°C in a steady-flow process.

12-34M Same as Prob. 12-33M, except that the CO_2 is at 153°C.

Real-gas mixtures

12-35M Assume air is a mixture of two gases, and the composition is 78 mole percent nitrogen and 22 mole percent oxygen. What is the pressure of the mixture if 14 m^3 of it at 20°C and 1 bar is compressed to 0.028 m^3 and 37°C? Assume the following: (*a*) ideal-gas law; (*b*) van der Waals gases, using the additive-pressure law; (*c*) van der Waals gases, using the additive-volume law; and (*d*) reduced-coordinates and additive-volume law.

12-36M Two gases A and B, each in the amount of 1 $kg \cdot mol$, exist in a mixture at 260°K and a total pressure of 100 bars. Find the total volume, in cubic meters, using (*a*) the additive-pressure rule, and (*b*) the additive-volume rule. The critical pressures and temperatures of A and B are: $P_{cA} = 3$ bars, $P_{cB} = 4$ bars, $T_{cA} = T_{cB} = 200°K$.

12-37M Same as Prob. 12-36M, except that the mixture temperature is 240°K and the total pressure is 90 bars.

12-38M One wishes to prepare a mixture of 60 mole percent acetylene (C_2H_2) and 40 mole percent CO_2 at 47°C and 100 bars, in a 1-m^3 tank. The tank initially contains acetylene at 47°C and pressure P_1. Carbon dioxide is then bled into the tank from a line containing CO_2 at 47°C and 100 bars until the tank pressure reaches 100 bars. What is the value of P_1 such that when the tank pressure reaches 100 bars the composition within the tank will be 60 mole percent acetylene. Assume the validity of Kay's rule for the mixture.

12-39M Calculate the pressure exerted by a mixture of 0.5 $kg \cdot mol$ of methane (CH_4) and 0.5 $kg \cdot mol$ of propane (C_3H_8) for a temperature of 90°C and a volume of 0.48 m^3. Use (*a*) van der Waals equation and the additive-pressure law, (*b*) the compressibility chart and additive volumes, and (*c*) the compressibility chart and Kay's rule. Compare with the ideal-gas value. The observed value is 50.6 bars.

PROBLEMS (USCS)

12-1 Show that, for an adiabatic reversible expansion of a van der Waals gas, $T(v - b)^{k-1} =$ constant.

12-2 Determine the equations for a and b in the Dieterici equation of state in terms of the critical-state values of P, v, and T. Then compute the value of Z_c for such a gas, and compare with experimental data for real gases.

12-3 Steam at 1600 psia and 740°F is a real gas. Determine the specific volume by using (*a*) the compressibility factor, (*b*) the ideal-gas equation, (*c*) the van der Waals equation, and (*d*) the tabulated values.

12-4 Carbon dioxide at 563°R and 72 atm has an observed specific volume of 3.21 $ft^3/lb \cdot mol$. Calculate the value based on (*a*) the principle of corresponding states, (*b*) the ideal-gas equation, and (*c*) van der Waals equation.

12-5 Compute the specific volume of steam at 2000 psia and 800°F by means of (*a*) the ideal-gas equation and (*b*) the compressibility chart. (*c*) Compare with tabulated data.

12-6 The specific volume of nitrogen at −110°F is 0.115 ft^3/lb. Determine the pressure of the substance by means of (*a*) the ideal-gas equation, (*b*) the compressibility factor, (*c*) the van der Waals equation, (*d*) the Redlich-Kwong equation, and (*e*) the tabulated value.

12-7 Determine the pressure of nitrogen at −160°F if the specific volume is 0.03990 ft^3/lb, employing (*a*) the van der Waals equation, (*b*) the Redlich-Kwong equation, (*c*) the principle of corresponding states, and (*d*) the superheat table.

12-8 Compute the pressure of methane (CH_4) at a temperature of 200°F and a specific volume of 0.235 ft^3/lb by using (*a*) the compressibility factor, (*b*) the van der Waals equation, and (*c*) the Redlich-Kwong equation.

12-9 Methane gas is maintained at 335 psia and −115°F. Compute the specific volume of the gas, in ft^3/lb, on the basis of the principle of corresponding states.

12-10 Calculate the volume of 1 lb of methane (CH_4) at 40°F and 1000 psia, using compressibility data.

12-11 Calculate the pressure required to compress 10 ft^3 of nitrogen at 14.4 psia and 73°F to 0.1 ft^3 at −167°F, employing Z data.

12-12 (*a*) Evaluate the constants a and b in the Redlich-Kwong equation for carbon dioxide in units compatible with P in atmospheres, T in degrees Rankine, and v in $ft^3/lb \cdot mol$.

(*b*) Calculate v for 100 atm and 630°R.

(*c*) Calculate v for 100 atm and 1260°R.

(*d*) Compare the results with data for v from the *National Bureau of Standards Circular* 564.

12-13 The constants for the virial equation of state in the form $Pv = A[1 + (b'/v) + (c'/v^2) + \cdots]$ have been determined experimentally for nitrogen at −148°F. The values are: $A = 0.634$, $b' = -2.3146 \times 10^{-3}$, and $c' = 4.235 \times 10^{-6}$.

(*a*) Determine the compressibility factor at 67 atm and −148°F from the above equation.

(*b*) Compare the above result to that obtained from a generalized compressibility chart.

12-14 (*a*) Calculate the specific volume, in ft^3/lb, for N_2 at 1130°R and 735 psia, using the compressibility factor.

(*b*) What would be the percentage error in assuming nitrogen to be an ideal gas under these conditions?

12-15 Oxygen (O_2) is at 100 atm and −100°F. Calculate how many pounds can be stored in a tank with a volume of 100 ft^3 if we (*a*) use the ideal-gas equation and (*b*) use the principle of corresponding states.

12-16 Estimate the temperature of 1.0 lb of *n*-octane (C_8H_{18}) at 27 atm if the volume is 0.20 ft^3.

12-17 Determine the pounds of propane gas in a 10-ft^3 cylinder if the pressure is 2900 psia and the temperature is (*a*) 405°F and (*b*) 540°F.

12-18 Nitrogen gas at a pressure of 1500 psia and −97°F is contained in a tank of 10 ft^3. Heat is transferred until the temperature is 100°F. Determine approximately (*a*) the specific volume of the gas, in $ft^3/lb \cdot mol$, and (*b*) the final pressure, in psia, using the compressibility chart.

12-19 Ethane gas is being transported through a pipeline at a pressure of 1400 psia and 130°F. Using the generalized compressibility chart, determine, in percent, the increase in velocity of the ethane required to deliver the same mass flow rate if the line pressure remains the same but the temperature increases to 240°F.

12-20 Calculate the specific volume, in $ft^3/lb \cdot mol$, of nitrogen at 1500 psia and −50°F by (*a*) the ideal-gas law, (*b*) the compressibility-factor method, and (*c*) the virial equation. (The virial coefficients for nitrogen at this temperature are $b = -3.75 \times 10^{-1}$, $c = 5.86 \times 10^{-4}$, and $d = 8.53 \times 10^{-6}$, where P is in atmospheres and T is in degrees Rankine.)

12-21 Determine the specific volume of nitrogen at 200 atm and 0°C by means of Eq. (12-3). Compare with the answer found by using Eqs. (12-4) and (12-5).

12-22 On the basis of the generalized compressibility chart, determine the pounds of ethylene (C_2H_4) in a 10-ft tank if the temperature is 73°F and the pressure is 1225 psia. If the temperature is increased to 252°F, find the final pressure, in psia.

12-23 Determine the specific volume of nitrogen at 150°K and 60 atm by means of Eqs. (12-4) and (12-5), and compare with the result based on the generalized charts, Figs. A-25M and A-26M.

12-24 Air is a mixture of two real gases. What is the pressure of the mixture if 500 ft^3 of it at 70°F are compressed to 1.0 ft^3? The original pressure was 1 atm, and the final temperature is 100°F. Assume the following: (*a*) ideal gas; (*b*) van der Waals gases, using the additive-pressure law; (*c*) van der Waals gases, using additive-volume law; and (*d*) reduced coordinates and additive-volume law.

12-25 A mixture consists of 1.0 lb. of CO_2 (an ideal gas) and 1.0 lb of water vapor (a real gas) at 300 psia and 400°F. Calculate (*a*) the partial pressure of each component, (*b*) the component pressure based on the additive-pressure rule, (*c*) the component pressure based on the additive-pressure rule if CO_2 is a real gas, and (*d*) the volume of the mixture.

12-26 On the basis of compressibility factors, determine the isothermal work necessary to compress carbon dioxide from 1000 to 4000 psia at a temperature of 197°F in a steady-flow process,

12-27 Two gases A and B, each in the amount of 1 $lb \cdot mol$, exist in a mixture at 250°R and a total pressure of 100 atm. Find the total volume, in cubic feet, using (*a*) the additive-pressure rule; and (*b*) the additive-volume rule. The critical pressures and temperatures of A and B are: $P_{cA} = 3$ atm, $P_{cB} = 4$ atm, $T_{cA} = T_{cB} = 100$°R.

12-28 One wishes to prepare a mixture of 60 mole percent acetylene (C_2H_2) and 40 percent CO_2 at 120°F, 1500 psia, in a 1.0 ft^3 tank. The tank initially contains acetylene at 120°F and pressure P_1. Carbon dioxide is then bled into the tank from a line containing CO_2 at 120°F and 1500 psia until the tank pressure reaches 1500 psia. What is the value of P_1 such that when the tank pressure reaches 1500 psia the composition within the tank will be 60 percent acetylene? Assume the validity of Kay's rule for the mixture.

12-29 On the basis of Eqs. (12-4) and (12-5) determine the specific volume of nitrogen, in ft^3/lb, at 1000 psia and 300°R, and compare the result to the experimental value of 0.0828 ft^3/lb. What percent error is made if one assumes ideal-gas behavior at this state?

CHAPTER

THIRTEEN

GENERALIZED THERMODYNAMIC RELATIONSHIPS

In the solution of engineering and scientific problems it is essential to be able to determine the values of thermodynamic properties. Since classical thermodynamics is an experimental science, a great deal of empirical data has been obtained over the past decades. The student must recognize, however, that only a relatively few properties can be evaluated by direct experimentation. Of these, the correlation among pressure, specific volume, and temperature and the relationship between the specific heats and temperature at low pressures are most easily measured. The evaluation of such properties as the internal energy, the enthalpy, and the entropy is made on the basis of calculations involving the directly measurable data mentioned above. Consequently, one of the major tasks of thermodynamics is to provide basic equations which enable one to evaluate properties such as u, h, and s, and others, from measurable-property data.

A second point to recognize is that, in many cases, insufficient data are available to calculate properties, even if the basic mathematical relations have been developed. Hence approximation techniques are needed, since in numerous situations either insufficient time or money may require that alternative methods of evaluating properties be available. A case in point is the use of the generalized compressibility factor Z, discussed in Chaps. 4 and 12, which provides a means of correlating PvT data in the absence of sufficient direct experimental information. This particular method will be extended to other properties, in this chapter. Before investigating some general thermodynamic relations for the qualitative and quantitative insight they provide, a brief review of some rules of partial differential calculus is in order.

13-1 FUNDAMENTALS OF PARTIAL DERIVATIVES

Many of the expressions to be developed in this chapter will involve a dependent variable expressed as a function of two independent variables, since we are primarily interested in simple systems. With this in mind, we shall first consider three thermodynamic variables represented by x, y, and z. Their functional relationship may be expressed either in the form $f(x, y, z) = 0$ or in the form $x = x(y, z)$. In the latter case we could also write that $y = y(x, z)$ or that $z = z(x, y)$. The differential of the dependent variable x is given by the equation

$$dx = \left(\frac{\partial x}{\partial y}\right)_z dy + \left(\frac{\partial x}{\partial z}\right)_y dz \tag{a}$$

Similar expressions for dy and dz may also be written, depending upon which variable is selected as the dependent variable.

The reason for not considering a partial derivative as a fraction is easily seen by recalling that the equilibrium states of a simple system may be represented by a three-dimensional surface. Such a surface is shown in Fig. 13-1 for a single-phase region, where the thermodynamic properties are symbolized by x, y, and z. As an example, this surface could represent the superheat region of a PvT diagram similar to that shown in Fig. 4-1. Sections of the surface have been cut away around the equilibrium state D, so that the curvature of the surface is more clearly seen. Consider a plane of constant z which intersects the surface. The curve of intersection is labeled with the points C, D, and E on the diagram. The quantity $(\partial x/\partial y)_z$ is the slope of the surface at any state along this curve of intersection. In particular, at state D a tangent has been drawn to the curve which is the line AB. The value of the slope of this line is also the value of $(\partial x/\partial y)_z$ at state D. This partial derivative exists, of course, when any other plane of constant z intersects the equilibrium surface, as long as the surface is continuous in the z direction at the state of interest. It is apparent that similar interpretations of the quantities $(\partial x/\partial z)_y$ and $(\partial y/\partial z)_x$ can be made when planes of constant y and x, respectively, intersect the surface of equilibrium states. The parameters x, y, and z represent any combination of three intrinsic properties of a thermodynamic system.

A mathematical test is available from partial differential calculus to determine whether or not the total derivative of a function is an exact differential. This test is a necessary and sufficient condition for the total derivative to be exact. Since differential equations relating thermodynamic properties are exact, the test for exactness provides a means of developing additional relations among thermostatic properties. This method is especially useful when a property is presented as a function of two independent variables, as in the case of a simple system. Recalling Eq. (a), if $x = x(y, z)$, then

$$dx = \left(\frac{\partial x}{\partial y}\right)_z dy + \left(\frac{\partial x}{\partial z}\right)_y dz$$

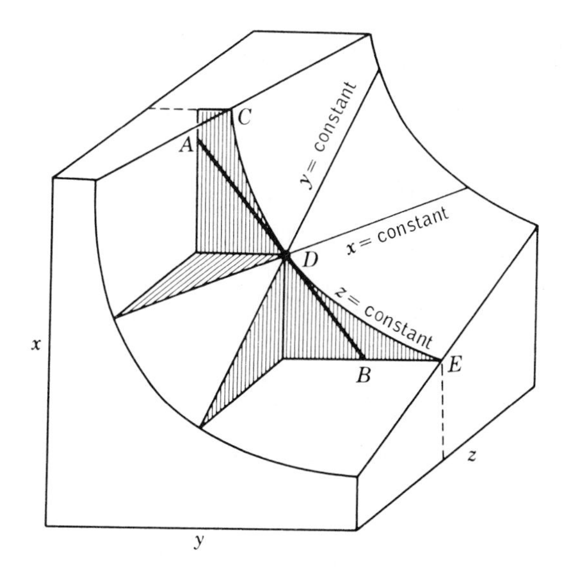

Figure 13-1 A representation of a partial derivative for a thermodynamic surface.

By denoting the coefficient of dy by M and the coefficient of dz by N, the above equation becomes

$$dx = M\,dy + N\,dz \tag{b}$$

Partial differentiation of M and N with respect to z and y, respectively, leads to

$$\frac{\partial M}{\partial z} = \frac{\partial^2 x}{\partial y\,\partial z} \quad \text{and} \quad \frac{\partial N}{\partial y} = \frac{\partial^2 x}{\partial z\,\partial y}$$

If these partial derivatives exist, it is known from calculus that the order of differentiation is immaterial, so that

$$\frac{\partial M}{\partial z} = \frac{\partial N}{\partial y} \tag{13-1a}$$

When Eq. (13-1*a*) is satisfied for any function x, then dx is an exact differential. Equation (13-1*a*) is known as the test for exactness. Its usefulness in terms of thermostatic properties is demonstrated in the next section.

Another expression from partial differential calculus which is useful in relating thermodynamic variables is the *cyclic* relation. This relation, which is not derived in the text, is

$$\left(\frac{\partial w}{\partial x}\right)_y \left(\frac{\partial x}{\partial y}\right)_w \left(\frac{\partial y}{\partial w}\right)_x = -1 \tag{13-1b}$$

Equation (13-1*b*) is important in interrelating thermodynamic variables. It also illustrates the fallacy of handling partial derivatives as fractions. When the variables held constant on the partial derivatives are different, the quantities within the partials cannot be canceled.

13-2 SOME FUNDAMENTAL RELATIONS FOR SIMPLE COMPRESSIBLE SYSTEMS

In Chap. 7 several imporant equations were developed for the entropy change of simple, compressible systems of fixed chemical composition by combining the first and second laws of thermodynamics. The first law in this particular case is $\delta q + \delta w = du$. For an internally reversible process, the heat and work interactions may be represented by $T\,ds$ and $-P\,dv$. Substitution of these latter two expressions into the first law yields

$$T\,ds - P\,dv = du \tag{13-2}$$

Since $h = u + Pv$, then $du = dh - P\,dv - v\,dP$. Replacement of du in Eq. (13-2) by this relationship leads to the second $T\,ds$ equation, namely

$$T\,ds + v\,dP = dh \tag{13-3}$$

Two additional equations of interest may be formed by defining two other properties of matter. The Helmholtz function a is defined by the equation

$$a \equiv u - Ts \tag{13-4}$$

Hence,

$$da = du - T\,ds - s\,dT \tag{13-5}$$

If Eq. (13-2) is substituted into Eq. (13-5), then

$$da = -P\,dv - s\,dT \tag{13-6}$$

The Gibbs function g is defined by the equation

$$g \equiv h - Ts \tag{13-7}$$

Therefore,

$$dg = dh - T\,ds - s\,dT \tag{13-8}$$

Substitution of dh from Eq. (13-3) into Eq. (13-8) yields

$$dg = v\,dP - s\,dT \tag{13-9}$$

In order to summarize these four important relationships among properties of simple systems, they are collected together as a set and presented below.

$$du = T\,ds - P\,dv \tag{13-2}$$

$$dh = T\,ds + v\,dP \tag{13-3}$$

$$da = -P\,dv - s\,dT \tag{13-6}$$

$$dg = v\,dP - s\,dT \tag{13-9}$$

Note that the variables on the right-hand side of these equations include only P, v, s, and T. These equations are sometimes referred to as the Gibbsian equations.

They relate the change in various properties of a simple compressible system during a differential change between equilibrium states.

One of the most important sets of thermodynamic relations which arise from the Gibbsian, or $T\,ds$, equations above is obtained by applying the test for exactness to the expressions for du, dh, da, and dg. Since these quantities are exact differentials, the test for exactness must lead to an equality. Application of Eq. (13-1a) to the four equations above yields the following set of relations among partial derivatives:

$$\left(\frac{\partial T}{dv}\right)_s = -\left(\frac{\partial P}{\partial s}\right)_v \tag{13-10}$$

$$\left(\frac{\partial T}{\partial P}\right)_s = \left(\frac{\partial v}{\partial s}\right)_P \tag{13-11}$$

$$\left(\frac{\partial P}{\partial T}\right)_v = \left(\frac{\partial s}{\partial v}\right)_T \tag{13-12}$$

$$\left(\frac{\partial v}{\partial T}\right)_P = -\left(\frac{\partial s}{\partial P}\right)_T \tag{13-13}$$

This set of equations is referred to as the Maxwell relations. Their importance is not apparent at this point, but a simple example may help illustrate their usefulness. Consider a system at a given equilibrium state. For a differential change of state it is desirable to know the rate of change of entropy of the system as the volume is altered isothermally. This could apply, for example, to a gas contained within a piston-cylinder assembly which is expanded isothermally. It is not possible to measure $(\partial s/\partial v)_T$ directly for the process, since entropy variations cannot be evaluated directly. However, Eq. (13-12) states that it is necessary only to measure the rate of change of pressure with temperature at constant volume, since $(\partial P/\partial T)_v = (\partial s/\partial v)_T$. From an experimental viewpoint, it is relatively easy to measure pressure and temperature variations.

13-3 GENERALIZED RELATIONS FOR CHANGES IN ENTROPY, INTERNAL ENERGY, AND ENTHALPY FOR SIMPLE, COMPRESSIBLE SUBSTANCES

One of the most important functions of thermodynamics is to provide fundamental equations for the evaluation of properties or the change in properties under the most general considerations. These equations, for example, should be independent of the type or the phase of the substance under consideration. Once these "generalized" equations have been developed, it will then be necessary to provide experimental information on specific substances if further numerical evaluation is to be profitable. It would be desirable to have the experimental information in an analytical form, but in many cases it may be necessary to rely on tabular or graphical forms of equations of state. In any event, our present goal is the derivation of these generalized equations. Again we shall restrict ourselves to simple, compressible thermodynamic systems. In this way the number of independent variables will be restricted to two. Of primary interest in this section will be equations for the changes in entropy, internal energy, and enthalpy. Relationships for the specific heats will appear in the following section.

It is most convenient to begin with a derivation of an equation for the change in entropy of a simple, homogeneous system. Once this is accomplished, equations for du and dh may then be obtained directly by employing the $T\,ds$ equations already developed. It is desirable to derive generalized relations in terms of easily measured independent variables. For this reason we shall select the independent properties to be any pair of the group (P, v, T). Thus three equations for ds can be determined. For practical reasons, we are usually concerned only with the two pairs of variables (T, v) and (T, P). If the entropy is chosen to be a function of T and v, we may write that the total derivative of s is given by

$$ds = \left(\frac{\partial s}{\partial T}\right)_v dT + \left(\frac{\partial s}{\partial v}\right)_T dv \tag{13-14}$$

It is now desirable to attempt to express ds solely in terms of measurable quantities. This requires replacing the partial derivatives in Eq. (13-14) with other terms which include only the variables P, v, T and the specific heats c_v and c_p.

The first partial derivative on the right is related to c_v and T. If $u = u(T, v)$, then

$$du = \left(\frac{\partial u}{\partial T}\right)_v dT + \left(\frac{\partial u}{\partial v}\right)_T dv$$

$$= c_v\, dT + \left(\frac{\partial u}{\partial v}\right)_T dv \tag{13-15}$$

When du in the above expression is replaced by Eq. (13-2), then, upon rearrangement

$$ds = \frac{c_v\, dT}{T} + \frac{1}{T}\left[\left(\frac{\partial u}{\partial v}\right)_T + P\right] dv \tag{13-16}$$

Equations (13-14) and (13-16) are both valid expressions for ds. Hence, the coefficients in front of the dT terms on the right must be equal, and

$$\left(\frac{\partial s}{\partial T}\right)_v = \frac{c_v}{T} \tag{13-17}$$

The second partial derivative is contained in a Maxwell relation:

$$\left(\frac{\partial s}{\partial v}\right)_T = \left(\frac{\partial P}{\partial T}\right)_v \tag{13-12}$$

By making the proper substitutions into Eq. (13-14), we find that

$$ds = \frac{c_v\, dT}{T} + \left(\frac{\partial P}{\partial T}\right)_v dv \qquad (13\text{-}18)$$

This is the result we sought, since the right-hand side of the equation is now expressed solely in terms of measurable quantities. An equivalent equation for ds in terms of the variables T and P may be obtained by starting with the relation

$$ds = \left(\frac{\partial s}{\partial T}\right)_P dT + \left(\frac{\partial s}{\partial P}\right)_T dP \qquad (13\text{-}19)$$

Again, the second partial derivative may be replaced by a Maxwell relation. In addition, it can be shown in a manner similar to the development of Eq. (13-17) that

$$\left(\frac{\partial s}{\partial T}\right)_P = \frac{c_p}{T} \qquad (13\text{-}20)$$

As a result, the substitution of Eqs. (13-13) and (13-20) into Eq. (13-19) yields the desired result,

$$ds = \frac{c_p\, dT}{T} - \left(\frac{\partial v}{\partial T}\right)_P dP \qquad (13\text{-}21)$$

Integration of Eqs. (13-18) and (13-21) between the same two equilibrium states should lead to the same value for Δs, since the change in the value of a property is independent of the method employed to calculate it. These equations are generalized relations, since they are not restricted to any particular substance or to any particular phase of a substance. They are restricted, however, to simple compressible systems.

A generalized equation for the sensible internal-energy change in now possible by recalling the first $T\, ds$ equation developed previously.

$$du = T\, ds - P\, dv \qquad (13\text{-}2)$$

The quantity ds is eliminated from this equation by substituting Eq. (13-18) for it. After separation of variables it turns out that

$$du = c_v\, dT + \left[T\left(\frac{\partial P}{\partial T}\right)_v - P\right] dv \qquad (13\text{-}22)$$

The generalized equation for the change in enthalpy is found by employing the second $T\, ds$ equation, namely,

$$dh = T\, ds + v\, dP \qquad (13\text{-}3)$$

Substitution of Eq. (13-21) for ds and subsequent rearrangement leads to

$$dh = c_p\, dT + \left[v - T\left(\frac{\partial v}{\partial T}\right)_P\right] dP \qquad (13\text{-}23)$$

Integration of Eqs. (13-18), (13-21), (13-22), and (13-23) for actual systems requires experimental knowledge of the PvT behavior of the substance in the region of interest, plus experimental information on the relationship between the specific heats and temperature.

The change in enthalpy, for example, of a simple, compressible substance is found by integration of Eq. (13-23). As a result

$$h_2 - h_1 = \int_1^2 c_p\, dT + \int_1^2 \left[v - T\left(\frac{\partial v}{\partial T}\right)_P\right] dP \qquad (13\text{-}24)$$

In order to integrate the first term, information is required on the variation of c_p with temperature at a fixed pressure. (Frequently this pressure is chosen to be essentially zero, so that ideal-gas specific-heat data, $c_{p,0}$, are used.) The integration of the second term requires knowledge of the PvT behavior of the substance for the range of pressure desired at a given temperature. Due to the format of the coefficient of dP in Eq. (13-24), it is helpful if the PvT equation of state is explicit in v. On the other hand, an equation explicit in P would be advantageous when Eq. (13-22) for du is integrated. Figure 13-2 shows one possible path of integration between two real-gas states, 1 and 2, on PT and Ts diagrams. For this particular path the first term of Eq. (13-24) is integrated at zero pressure between states x and y. The second term in Eq. (13-24) must be integrated twice in this case. One integration is at constant temperature T_1 between states 1 and x, and the other integration is at T_2 between states y and 2.

Examples 13-1M and 13-1 In Chap. 3 it was pointed out that the internal energy of gases at low pressures could be approximated quite successfully by a relation of the type $du = c_v\, dT$. This assumption that the internal energy was solely a function of temperature was based on Joule's law, which stated that $(\partial u/\partial v)_T$ at low pressures was experimentally determined to be zero. Demonstrate that Joule's law holds exactly for ideal gases.

SOLUTION In order to evaluate Joule's law in terms of an ideal gas, $(\partial u/\partial v)_T$ must be expressed in terms of P, v, and T. This is easily done by writing the total derivative of u as a function of T and v and then comparing this equation with Eq. (13-22). These two equations are

$$du = \left(\frac{\partial u}{\partial T}\right)_v dT + \left(\frac{\partial u}{\partial v}\right)_T dv$$

and

$$du = c_v\, dT + \left[T\left(\frac{\partial P}{\partial T}\right)_v - P\right] dv \qquad (13\text{-}22)$$

Since the latter expression is of general validity, it is apparent on comparison of these two

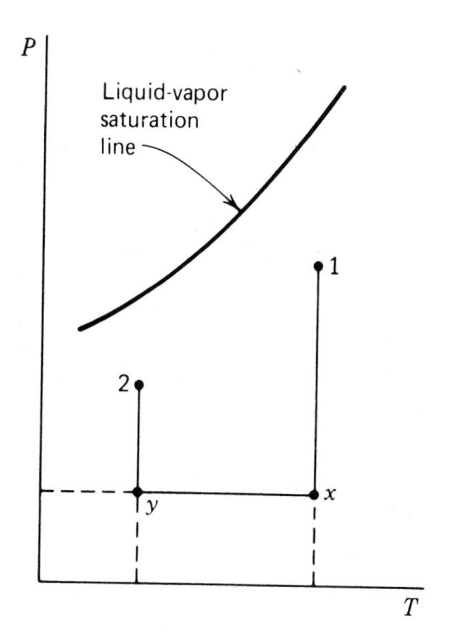

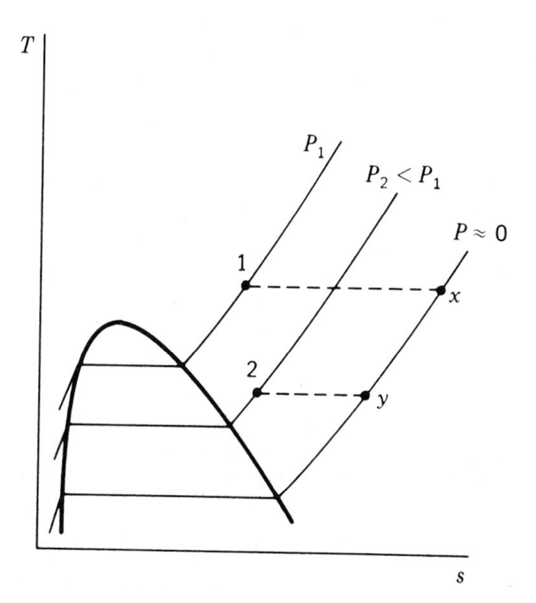

Figure 13-2 Possible path of integration between two real-gas states.

equations that

$$\left(\frac{\partial u}{\partial v}\right)_T = T\left(\frac{\partial P}{\partial T}\right)_v - P$$

This relation is valid for any simple, compressible system. We shall now apply the ideal-gas relationship to it, namely, $Pv = RT$. It is easily shown that, for an ideal gas, $(\partial P/\partial T)_v = R/v$.

Substituting this into the equation for $(\partial u/\partial v)_T$, we find that, for an ideal gas,

$$\left(\frac{\partial u}{\partial v}\right)_T = T\left(\frac{R}{v}\right) - P = P - P = 0$$

Thus an ideal gas is always a substance such that the internal energy is truly only a function of temperature, as given by $du = c_v\, dT$. Since many gases at low pressures approximately follow the ideal-gas law, their internal energies also are only a function of temperature, to a high degree of accuracy. If a similar treatment is carried out for the enthalpy function, it may be shown that $(\partial h/\partial P)_T$ is always zero for ideal gases. This proves that, for hypothetical ideal gases, the enthalpy change is always given by $dh = c_p\, dT$.

Examples 13-2M and 13-2 Find the change in enthalpy and entropy for a real gas along an isothermal path between pressures P_1 and P_2. Assume for the range of pressures involved that the PvT behavior of the gas is represented with reasonable accuracy by the relation $v = RT/P + b - a/RT$, where a and b are constants.

SOLUTION The change in enthalpy is given by the second term of Eq. (13-24), namely,

$$(h_2 - h_1)_T = \int_1^2 \left[v - T\left(\frac{\partial v}{\partial T}\right)_P\right] dP$$

The equation of state is explicit in v. Hence $(\partial v/\partial T)_P = R/P + a/RT^2$, and

$$v - T\left(\frac{\partial v}{\partial T}\right)_P = \frac{RT}{P} + b - \frac{a}{RT} - T\left(\frac{R}{P} + \frac{a}{RT^2}\right) = b - \frac{2a}{RT}$$

Therefore,

$$(h_2 - h_1)_T = \left(b - \frac{2a}{RT}\right)(P_2 - P_1)$$

To evaluate the entropy change we choose Eq. (13-21), rather than Eq. (13-18), because it requires an equation of state explicit in v. Hence

$$(s_2 - s_1)_T = -\int_1^2 \left(\frac{\partial v}{\partial T}\right)_P dP = -\int_1^2 \left(\frac{R}{P} + \frac{a}{RT^2}\right) dP$$

$$= -R \ln \frac{P_2}{P_1} - \frac{a(P_2 - P_1)}{RT^2}$$

13-4 GENERALIZED RELATIONS FOR c_p AND c_v

Two expressions for the specific heats have already been developed in this chapter:

$$c_v = T\left(\frac{\partial s}{\partial T}\right)_v \tag{13-17}$$

$$c_p = T\left(\frac{\partial s}{\partial T}\right)_P \tag{13-20}$$

The latter two expressions are generalized relations for c_v and c_p, and they may be used in any single-phase region where PvT data are available. An alternative method of evaluating specific-heat data is based on the experimental fact that it is relatively easier to measure specific-heat data at low pressures than at elevated pressures. For example, we have already seen from Chap. 3 that much is known about the specific heats of common gases as a function of temperature at low pressures. Such data were called "zero-pressure" specific heats. Consequently, we are primarily concerned with determining in what manner specific-heat values vary with increasing pressure (or decreasing specific volume) at constant temperature. Such an evaluation must again be based solely on the use of measured PvT data in the desired range of equilibrium states. In a mathematical sense what we seek are expressions for the two terms $(\partial c_v/\partial v)_T$ and $(\partial c_p/\partial P)_T$. Generalized equations for these two expressions are obtained by starting with Eqs. (13-18) and (13-21); that is,

$$ds = \frac{c_v\, dT}{T} + \left(\frac{\partial P}{\partial T}\right)_v dv \tag{13-18}$$

and

$$ds = \frac{c_p\, dT}{T} - \left(\frac{\partial v}{\partial T}\right)_P dP \tag{13-21}$$

Since these equations are expressions for exact differentials, the test for exactness may be employed. When applied to Eq. (13-18), the following relation results:

$$\left(\frac{\partial c_v}{\partial v}\right)_T = T\left(\frac{\partial^2 P}{\partial T^2}\right)_v \tag{13-25}$$

This is the desired relation. If we start with Eq. (13-21), it can be shown by an analogous procedure that

$$\left(\frac{\partial c_p}{\partial P}\right)_T = -T\left(\frac{\partial^2 v}{\partial T^2}\right)_P \tag{13-26}$$

In order to obtain the value of c_p, for example, at an elevated pressure, Eq. (13-26) is integrated from zero pressure to the desired value. Hence

$$c_p - c_{p,0} = -T \int_0^P \left(\frac{\partial^2 v}{\partial T^2}\right)_P dP \tag{13-27}$$

where $c_{p,0}$ again is the zero-pressure, or ideal-gas, specific heat. The integration of the right-hand side requires a knowledge of the PvT behavior of the substance in either tabular or analytical form.

Another thermodynamic relation of interest is the difference between the constant-pressure and constant-volume specific heats, that is, $c_p - c_v$. One reason for this interest is that c_p values usually are easier to measure than c_v values. Hence c_v values could in theory be evaluated solely from c_p and PvT data. Since the change in any property value is not dependent upon the method of evaluation, we may equate the two equations for ds previously presented. Equating

Eqs. (13-18) and (13-21), we find

$$\frac{c_v\,dT}{T} + \left(\frac{\partial P}{\partial T}\right)_v dv = \frac{c_p\,dT}{T} - \left(\frac{\partial v}{\partial T}\right)_P$$

or
$$\frac{c_p - c_v}{T}\,dT = \left(\frac{\partial P}{\partial T}\right)_v dv + \left(\frac{\partial v}{\partial T}\right)_P dP$$

Differentiation with respect to pressure at constant volume yields

$$\frac{c_p - c_v}{T}\left(\frac{\partial T}{\partial P}\right)_v = \left(\frac{\partial v}{\partial T}\right)_P$$

or upon rearrangement,

$$c_p - c_v = T\left(\frac{\partial v}{\partial T}\right)_P \left(\frac{\partial P}{\partial T}\right)_v \tag{13-28}$$

An equivalent form for $c_p - c_v$ can be obtained by replacing $(\partial P/\partial T)_v$ in terms of the cyclic rule $(\partial P/\partial T)_v = -(\partial v/\partial T)_P(\partial P/\partial v)_T$. Use of this expression in Eq. (13-28) leads to the relation

$$c_p - c_v = -T\left(\frac{\partial v}{\partial T}\right)_P^2 \left(\frac{\partial P}{\partial v}\right)_T \tag{13-29}$$

A number of important qualitative results stem from Eq. (13-29).

First of all, on the basis of experimental data it is known that $(\partial P/\partial v)_T$ is always negative for all substances in all phases. Since the first partial in Eq. (13-29) is a squared term, then $c_p - c_v$ must always be positive, or zero. This quantity becomes zero on two occasions. The first of these is, apparently, when T is absolute zero on the thermodynamic scale, if the remaining terms remain finite at this state. Consequently, the specific heats at constant pressure and constant volume are identical at the absolute zero of temperature. The specific heats will also be equal if the value of $(\partial v/\partial T)_P$ is ever zero. This occurs, for example, in the case of liquid water at, roughly, 4°C, where the fluid is at its state of maximum density. It should also be noted that even at temperatures above zero, the difference in specific heats will generally be small for liquids and solids. This is true since the value of $(\partial v/\partial T)_P$ is very small for most equilibrium states, which can be verified by referring to a PvT surface for a substance, such as was shown in Fig. 4-1. Hence one frequently speaks of the specific heat of a liquid or a solid without specifying the type, since the c_v and c_p values are not significantly different in many cases. The data quoted are usually c_p values. The fact that $c_p \geq c_v$ also leads to the generalization that constant-volume lines always have steeper slopes than constant-pressure lines at the same point on a Ts diagram.

Example 13-3M and 13-3 Determine the isothermal change in c_p with pressure for the same gas studied in Example 13-2.

SOLUTION The isothermal change in c_p with pressure is given by Eq. (13-26), namely

$(\partial c_p/\partial P)_T = -T(\partial^2 v/\partial T^2)_P$. In Example 13-2 the equation of state was $v = RT/P + b - a/RT$, and the first derivative was found to be

$$\left(\frac{\partial v}{\partial T}\right)_P = \frac{R}{P} + \frac{a}{RT^2}$$

The second partial derivative, then, is

$$\left(\frac{\partial^2 v}{\partial T^2}\right)_P = -\frac{2a}{RT^3}$$

Consequently,

$$(c_{P2} - c_{p1})_T = -\int_1^2 T\left(-\frac{2a}{RT^3}\right) dP = \frac{2a(P_2 - P_1)}{RT^2}$$

The property variation of solids and liquids is often expressed in terms of the volumetric expansion coefficient β and the isothermal coefficient of compressibility K_T. These two quantities are defined as

$$\beta = \frac{1}{v}\left(\frac{\partial v}{\partial T}\right)_P \tag{13-30}$$

and

$$K_T = -\frac{1}{v}\left(\frac{\partial v}{\partial P}\right)_T \tag{13-31}$$

Substitution of these two equations in Eq. (13-29) leads to

$$c_p - c_v = \frac{vT\beta^2}{K_T} \tag{13-32}$$

The use of β and K_T is quite helpful in many calculations since their values may often be assumed to be constant during a given process. The slow variation of these two properties with temperature is illustrated by the data of Table 13-1, which shows data for solid copper.

Table 13-1 β and K_T for copper as a function of temperature

T, °K	$\beta \times 10^6$, °K^{-1}	$K_T \times 10^7$, cm^2/N
100	31.5	0.721
150	41.0	0.733
200	45.6	0.748
250	48.0	0.762
300	49.2	0.776
500	54.2	0.837
800	60.7	0.922

Examples 13-4M and 13-4 At 500°K the values of v, β, and K_T for solid copper are 7.115 $cm^3/g \cdot mol$, $54.2 \times 10^{-6}\ °K^{-1}$, and $0.837 \times 10^{-7}\ cm^2/N$, respectively.

(*a*) Determine the value of $c_p - c_v$ in J/g · mol)(°C).

(*b*) If the value of c_p is 26.15 J/(g · mol)(°C) at this temperature, what percent error would be made in c_v if we assume that $c_p = c_v$?

SOLUTION (*a*) The difference between c_p and c_v is obtained directly by substituting the appropriate values into Eq. (13-32). Thus,

$$c_p - c_v = \frac{vT\beta^2}{K_T} = \frac{7.115(500)(54.2 \times 10^{-6})^2}{0.837 \times 10^{-7}} N \cdot \text{cm}/(\text{g} \cdot \text{mol})(°\text{K}) \times \frac{\text{m}}{10^2\ \text{cm}}$$

$$= 1.249\ \text{J}/(\text{g} \cdot \text{mol})(°\text{K})$$

(*b*) If $c_p = 26.15$ J/(g · mol)(°K), then the actual value of c_v is 24.90 J/(g · mol)(°K). Therefore the percent error in assuming that c_v equals c_p is 1.249/24.90 = 0.050, or 5 percent. As a result, one must be careful not to assume that c_p and c_v are equal for solid materials if the temperature is sufficiently high.

13-5 VAPOR PRESSURE AND THE CLAPEYRON EQUATION

The vapor pressures of all liquids vary with saturation temperature in essentially the same manner. The dependency of the saturation pressure on the temperature will now be developed from theoretical considerations. The generalized relationship that results is also valid for solid-gas and solid-liquid phase changes. We shall begin by calculating the entropy change of a simple substance during a phase change. This entropy change in terms of the variables v and T has already been presented in the form of Eq. (13-14), namely,

$$ds = \left(\frac{\partial s}{\partial v}\right)_T dv + \left(\frac{\partial s}{\partial T}\right)_v dT \qquad (13\text{-}14)$$

However, for any process involving a change in phase, we realize that the temperature is constant during the phase change. Therefore the above equation reduces to

$$ds = \left(\frac{\partial s}{\partial v}\right)_T dv$$

The quantity $(\partial s/\partial v)_T$ can be replaced by the Maxwell relation given by Eq. (13-12), which shows that $(\partial s/\partial v)_T = (\partial P/\partial T)_v$. Hence

$$ds = \left(\frac{\partial P}{\partial T}\right)_v dv$$

The term $(\partial P/\partial T)_v$ is the slope of the saturation curve at a given saturation state, and this quantity is independent of the volume during a change of phase. Consequently, the partial derivative may be written as a total derivative, dP/dT, and it may be moved outside of the integral sign during the integration of the above equation. Integration leads to

$$s_2 - s_1 = \frac{dP}{dT}(v_2 - v_1)$$

or

$$\frac{dP}{dT} = \frac{s_2 - s_1}{v_2 - v_1} \tag{13-33}$$

where the subscripts 1 and 2 represent the saturation phases for the process. For example, they may represent the saturated-vapor and saturated-liquid phases during a vaporization process.

The entropy change during a phase change may be evaluated from the first and second laws. From the second law $ds = \delta q/T$, and for a constant-pressure process (such as a phase change) the first law for a closed system is $\delta q = dh$. Thus $ds = dh/T$ and $s_2 - s_1 = (h_2 - h_1)/T$. Equation (13-33) then becomes

$$\frac{dP}{dT} = \frac{h_2 - h_1}{T(v_2 - v_1)} = \frac{\Delta h}{T\,\Delta v} \tag{13-34}$$

Equation (13-34) is called the Clapeyron equation. It is generally valid for any phase change which occurs at constant pressure and temperature. For a liquid-vapor phase change this equation might be written as

$$\frac{dP}{dT} = \frac{h_{fg}}{Tv_{fg}}$$

where the subscripts follow nomenclature introduced in Chap. 4. In general, Δh and Δv are the enthalpy and volume changes between any two saturation states at the same pressure and temperature. Note that the Clapeyron equation permits the evaluation of enthalpy changes for phase changes from a knowledge of only PvT data.

For liquid-vapor and solid-vapor phase changes, Eq. (13-34) can be further modified by introducing several approximations. For purposes of discussion we shall consider only the first of these, but the results are equally applicable to solid-vapor phase changes. For liquid-vapor phase changes at relatively low pressures, the value of v_g is many times the size of v_f. Thus a good approximation is to replace v_{fg} by v_g in the above equation. Also, at these low pressures, the PvT relation for the vapor closely follows that for an ideal gas; that is, $v_g = RT/P$. By making these two successive approximations in Eq. (13-34), we find that

$$\frac{dP}{dT} = \frac{Ph_{fg}}{RT^2}$$

or

$$\frac{dP}{P} = \frac{h_{fg}\,dT}{RT^2} \tag{13-35}$$

Equation (13-35) is frequently called the Clapeyron-Clausius equation. Integration of this equation depends upon the variation of h_{fg} with temperature. If a small variation of pressure (or temperature) is chosen so that the change in h_{fg} over the interval of integration is small, then integration yields

$$\ln P = -\frac{h_{fg}}{R}\left(\frac{1}{T}\right) + C \tag{13-36}$$

where C is a constant of integration. This equation indicates that the vapor pressure of a liquid is very closely an exponential function of the saturation temperature. The general form of the equation is also valid for saturation data below the triple state in the sublimation region.

It should be kept in mind that Eq. (13-36) is only an approximation. A more accurate analytical expression for the variation of the saturation pressure with respect to temperature requires that additional terms be added. For example, a better approximation might be given by an equation of the form

$$\ln P_{\text{sat}} = A + \frac{B}{T} + C \ln T + DT + ET^2 + \cdots \tag{13-37}$$

The constants A, B, C, etc., are adjusted to obtain the best fit with the experimental data. Nevertheless, the simple exponential form given by Eq. (13-36) is fairly accurate in many cases.

An equation like Eq. (13-37) is important for the following reason. If the equation represents a precise fit to experimental data, then the derivative of the equation will give an accurate value of dP/dT. Substitution of this value of dP/dT into the Clapeyron equation, Eq. (13-34), will lead to an accurate evaluation of Δh for a liquid-gas or a solid-gas phase transformation.

The relationship given by Eq. (13-34) is quite useful to demonstrate one other point: The slope of a saturation line on a PT diagram apparently depends upon the signs of Δh and Δv. In most cases, when heat is added to a closed system to bring about a phase change, the volume also increases. Hence dP/dT is usually positive. However, in the case of the melting of water and a few other substances, the volume decreases. The slope of the melting curve on a phase diagram for these few substances must then be negative. This was pointed out in Chap. 4 during the discussion of phase diagrams, but now the Clapeyron equation theoretically substantiates what is empirically observed. The freezing temperature of any substance which expands on freezing is lowered when the pressure is increased.

Example 13-5M Estimate the enthalpy of vaporization of water at 200°C using PvT data from Table A-12M.

SOLUTION On the basis of Eq. (13-34) it is seen that

$$\Delta h = T\,\Delta v\,\frac{dP}{dT}$$

The value of dP/dT may be approximated to a high degree by $\Delta P/\Delta T$ in the region of interest. The saturation data at a 10° temperature interval on either side of 200°C are as follows: At

190°C the saturation pressure is 12.54 bars, and at 210°C the pressure is 19.06 bars. Hence $\Delta P/\Delta T$ equals (19.06 − 12.54)(210 − 190), or 0.326 bars/°C. The change in volume v_{fg} at 200°C is given as 127.3 cm^3/g. Substitution of these values into the above equation yields

$$\Delta h = \frac{473(127.3)(0.326)}{10} = 1963 \text{ kJ/kg}$$

where the factor of 10 converts units of cm^3 · bar to N · m. Table A-12M lists the value of h_{fg} as 1941 kJ/kg. An error of roughly 1 percent results from this approximation technique.

Example 13-5 Estimate the enthalpy of vaporization of water at 400°F using PvT data from Table A-12.

SOLUTION On the basis of Eq. (13-34) it is seen that

$$\Delta h = T \, \Delta v \frac{dP}{dT}$$

The value of dP/dT may be approximated to a high degree by $\Delta P/\Delta T$ in the region of interest. The saturation data at a 10° temperature interval on either side of 400°F are as follows: At 390°F the saturation pressure is 220.20 psia, and at 410°F the pressure is 276.50 psia. Hence $\Delta P/\Delta T$ equals (276.50 − 220.20)/(410 − 390), or 2.819 psi/°F. The change in volume v_{fg} at 400°F is given as 1.8474 ft^3/lb. Substitution of these values into the above equation yields

$$\Delta h = \frac{860(1.8474)(2.819)(144)}{778} = 827 \text{ Btu/lb}$$

Table A-12 lists the value as 826.8 Btu/lb, so good agreement is obtained by the calculation above.

Example 13-6M The saturation pressure and the enthalpy of vaporization of refrigerant 12 are 20°C are found to be 5.673 bars and 140.9 J/g, respectively. Without any additional experimental data, estimate the saturation pressure at 0°C.

SOLUTION Definite integration of Eq. (13-35) leads to

$$\ln \frac{P_2}{P_1} = \frac{h_{fg}(T_2 - T_1)}{RT_1T_2}$$

Although it would be better to use the average value of h_{fg} between the temperatures of interest, we must use the value at T_1, for lack of information. Substituting in the proper values, we find that

$$\ln \frac{P_2}{P_1} = \frac{140.9(121)(0 - 20)}{8.314(293)(273)} = -0.513$$

Therefore $P_2/P_1 = 0.599$, and $p_2 = 0.599(5.673) = 3.398$ bars. The tabulated value at 0°C is 3.086 bars. Although the estimate is in error by 10 percent, it is a fair approximation in the absence of further experimental data.

Example 13-6 The saturation pressure and the enthalpy of vaporization for refrigerant 12 at 70°F are measured and found to be 84.89 psia and 60.31 Btu/lb, respectively. Without any additional experimental data, estimate the saturation pressure at 10°F.

SOLUTION Definite integration of Eq. (13-35) leads to

$$\ln \frac{P_2}{P_1} = \frac{h_{fg}(T_2 - T_1)}{RT_1T_2}$$

Although it would be better to use the average value of h_{fg} between the temperatures of interest, we must use the value at T_1, for lack of more information. Substituting in the proper values, we find that

$$\ln \frac{P_2}{P_1} = \frac{60.31(121)(470 - 530)}{1.986(470)(530)} = -0.885$$

Therefore $P_2/P_1 = 0.413$, and $P_2 = 0.413(84.89) = 35.0$ psia. The tabulated value is, roughly, 29.3 psia at 10°F. Although the estimate is in error by some 20 percent, it is seen that a fair approximation is possible without further experimental data.

13-6 THE JOULE-THOMSON COEFFICIENT

Consider the flow of a fluid down a duct which contains a restriction to the flow. This restriction might be some type of porous plug, such as steel wool or cotton. The effect of this porous plug is a significant pressure drop across the restriction. A schematic of the equipment is shown in Fig. 13-3*a*. The work effects within the control surfaces 1 and 2 are zero, and the flow passage is heavily insulated: consequently, the heat effect is negligible. The flow of the fluid is adjusted to steady-state conditions. In addition, the changes in the kinetic and potential energies of the flow stream across the porous plug may also be made negligible. The conservation of energy principle for a control volume around the restriction is

$$q + w = \Delta h + \Delta KE + \Delta PE$$

In light of the above statements, this equation reduces merely to

$$h_1 = h_2$$

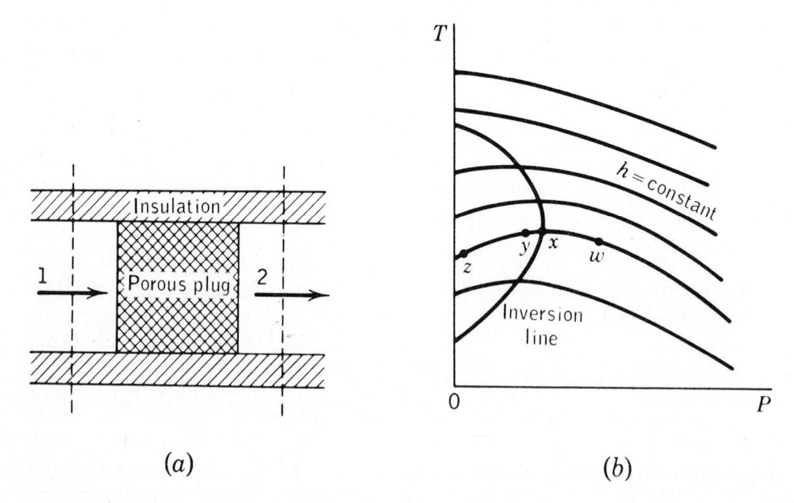

Figure 13-3 Equipment for Joule-Thomson experiment and a plot of data resulting from typical throttling measurements.

We have noted previously that a process for which the inlet and outlet enthalpies are the same is called throttling. The effect of throttling has a number of important scientific and engineering applications. Based on some original experiments by Joule and Thomson, the outcome of allowing a fluid to flow through a restriction from a higher to a lower pressure such as throttling is frequently called the Joule-Thomson effect. The importance of the Joule-Thomson effect is severalfold. First of all, other thermostatic properties may be related or evaluated from measurements of the Joule-Thomson effect. These include, for example, specific volumes, specific heats, and enthalpies. Second, we shall find that under certain conditions the result of throttling is a reduction in the temperature of the fluid. Thus low temperatures may be achieved with a device which has no moving parts. In fact, under proper conditions, it is possible that one or more of the components of a gas flow stream might pass into the liquid phase during the throttling process. Such liquefaction might provide a means of separating components of a gas mixture.

It is an experimental fact that throttling of a fluid leads to a final temperature which may be higher or lower than the initial value, depending upon the values of P_1, T_1, and P_2. A mathematical measure of this effect is given by Joule-Thomson coefficient μ_{JT}, which is defined as

$$\mu_{JT} = \left(\frac{\partial T}{\partial P}\right)_h \tag{13-38}$$

The Joule-Thomson coefficient may be readily determined at various states by plotting experimental data in terms of a family of constant-enthalpy lines on a TP diagram. To obtain this plot the values of P_1 and T_1 upstream from the restriction are held fixed, and the pressure P_2 downstream is varied experimentally. For each setting of P_2 the downstream temperature T_2 is measured. Under throttling conditions the state for each measurement made downstream has the same enthalpy as the initial state upstream. After making a sufficient number of measurements downstream for a given state upstream, a line of constant enthalpy can be drawn on the TP diagram. Then either the initial pressure or temperature is altered, and the measurement procedure is repeated for this new value of the enthalpy. In this manner a whole family of constant-enthalpy lines on a TP plot may be obtained. A typical result is shown in Fig. 13-3*b*. The slope of a constant-enthalpy line at any state is a measure of the Joule-Thomson coefficient at that state, i.e., a measure of $(\partial T/\partial P)_h$.

Figure 13-3*b* shows that a number of the constant-enthalpy lines have a state of maximum temperature. The line shown in the figure which passes through these states of maximum temperature is called the inversion line, and the value of the temperature for that state is the inversion temperature. A pressure line will cut the inversion curve at two different states; hence one speaks of the upper and lower inversion temperatures for a given pressure. This line has an important physical significance. To the right of the inversion line on a TP plot the Joule-Thomson coefficient is negative. That is, in this particular region the temperature will increase as the pressure decreases through the throttling device. A heating effect occurs. On the other hand, to the left of the inversion curve the Joule-Thomson coefficient is positive, which means that cooling will occur for expansions in this region. Hence, on the throttling of a fluid, the final temperature after a porous plug may be greater than, equal to, or less than the initial temperature, depending upon the final pressure for any given set of initial conditions. For example, in Fig. 13-3*b* a typical initial state might be point w. Expansion to the inversion curve (point x) results in heating of the fluid. If further expansion to point y is permitted, some cooling will occur, but this is not sufficient to lower the temperature back to that of the initial state. However, if expansion to point z is possible, enough cooling will occur to bring the final temperature to a lower value than that for the initial state.

It should also be noted that, for some initial states, a cooling process is impossible. The upper part of the inversion curve passes through zero pressure at some finite temperature for all substances. Consequently, many enthalpy lines at high temperatures never pass through the inversion line, as seen in Fig. 13-3*b*. For these enthalpy lines the Joule-Thomson coefficient is always negative throughout the range of pressures. Examples of this are hydrogen and helium, which have negative coefficients at ordinary temperatures and low pressures. Hence, for these two gases, the temperature must be artificially lowered considerably before throttling can be employed for an additional cooling effect. For most substances, however, at ordinary temperatures the Joule-Thomson coefficient is negative at high pressures, and it becomes positive at low pressures. For a given pressure drop it is seen that the maximum cooling effect is attained only if the initial state lies on the inversion line. If the initial state lies to the right of the inversion curve, part of the expansion results in heating, which counters the desired effect.

Another useful diagram is the Joule-Thomson coefficient plotted against temperature for various pressures. This is obtained by measuring the slopes of the constant-enthalpy lines at various states from a diagram such as Fig. 13-3*b*. Data for the Joule-Thomson coefficient of nitrogen determined from experiment work are shown in Fig. 13-4. Note that the coefficient is zero at approximately -135 and $+220°C$ for a pressure of 200 atm. These temperatures correspond to the lower and upper inversion temperatures at this pressure. The following example makes use of this diagram to estimate the cooling effect that can be achieved by the throttling of nitrogen from a high pressure to atmospheric conditions.

Example 13-5M and 13-5 Estimate the final temperature which could be obtained by throttling nitrogen gas from $-50°C$ and 100 atm pressure to 1 atm pressure.

Solution In order to estimate the final temperature, one needs to know the average Joule-Thomson coefficient in the given pressure range. Two properties are required to evaluate a third property at a given equilibrium state, but the final temperature is unknown. To evaluate the coefficient at the final state we shall first assume that there is no temperature change on throttling. From Fig. 13-4 it is estimated that the average coefficient between 100 atm and 1 atm at $-50°C$ is 0.31°C/atm. On this basis a drop of 99 atm in pressure would give a temperature drop of 99(0.31), or 31°C. Thus a better approximation to the final temperature at 1 atm is $-81°C$. Now a new average coefficient may be estimated. At 100 atm, $-50°C$, and 1 atm,

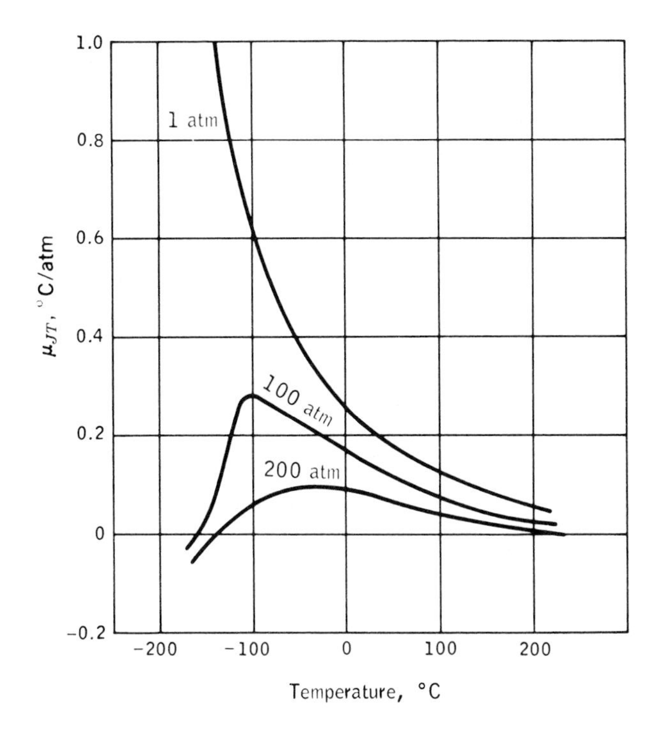

Figure 13-4 The Joule-Thomson coefficient for nitrogen. [*Based on the data of Roebuck and Osterling, Phys. Rev.,* **48**(*5*) *: 450–457* (*1935*).]

−81°C, the coefficients are, roughly, 0.23 and 0.51°C/atm, respectively. The average then is 0.37°C/atm, and the corresponding temperature drop is 37°C. Another trial, based on the new final-temperature estimate of −87°C, leads to a final temperature of −89°C. Further refinement is unnecessary on the basis of the available data. Hence it is estimated that the throttling of nitrogen gas from 100 atm and −50°C to 1 atm will result, approximately, in a temperature drop of 39°C, or 70°F.

It is useful to examine the Joule-Thomson coefficient in terms of a generalized equation, that is, its relation to the variables P, v, and T and the specific heats. This is easily obtained by recalling the generalized equation for the enthalpy, namely,

$$dh = c_p\,dT + \left[v - T\left(\frac{\partial v}{\partial T}\right)_P\right] dP \qquad (13\text{-}23)$$

By differentiating this equation with respect to the pressure at constant enthalpy, one obtains the following result:

$$\left(\frac{\partial T}{\partial P}\right)_h = \frac{1}{c_p}\left[T\left(\frac{\partial v}{\partial T}\right)_p - v\right] = \mu_{JT} \qquad (13\text{-}39)$$

Thus the Joule-Thomson coefficient may be calculated from a knowledge of the PvT relationship of the fluid and the specific heat at constant pressure for that state. In practice, one could use the Joule-Thomson coefficient, which is easily measured, to evaluate specific-heat data at elevated pressures. If this generalized relationship is applied to an ideal gas, an interesting result occurs. We find that, since $(\partial v/\partial T)_P$ for an ideal gas is simply R/P,

$$\mu_{\mathrm{JT,\,ideal\ gas}} = \frac{1}{c_p}\left(\frac{RT}{P} - v\right) = 0$$

Hence an ideal gas undergoes no change in temperature upon throttling. This is not surprising since it has already been pointed out in Chap. 4 that the enthalpy of an ideal gas is solely a function of its temperature. If the initial and final enthalpies of a fluid are equal for a throttling process by definition, an ideal gas under this condition would also have the same initial and final temperatures. No actual gas is an ideal gas; however, many gases at low pressures approximate this condition. Consequently, the temperature change upon throttling of real gases at low pressures is often quite small.

13-7 GENERALIZED THERMODYNAMIC CHARTS

The principle of corresponding states discussed in Sec. 4-6 is extremely useful in predicting property values other than P, v, and T. These three values were previously correlated through the compressibility factor Z and the reduced properties P_R, v'_R, and T_R. The compressibility factor and reduced coordinates may be used to evaluate such properties as the enthalpy, the entropy, and the specific heat at constant pressure for gases at elevated pressures. The usefulness of such a method will be that only the critical pressure and temperature are required for any given substance. The correlations for these properties again will be presented in graphical form. The method of evaluation involves the generalized equations previously developed in this chapter.

a Generalized Enthalpy Chart

Recall from Sec. 13-3 that the enthalpy of simple homogeneous substances may be evaluated from the generalized equation

$$dh = c_p\,dT + \left[v - T\left(\frac{\partial v}{\partial T}\right)_P\right] dP \tag{13-23}$$

The change in the enthalpy of a gas with temperature is fairly easily computed, since it requires a knowledge only of the variation of c_p with temperature at the desired pressure. Hence the first term on the right of the above equation is not too difficult to evaluate in many cases. However, the variation of h with pressure is not so straightforward because it requires a knowledge of the PvT behavior of each substance of interest. Since detailed data for many compounds will be lacking, a more general method must be employed.

At constant temperature it is noted that the enthalpy change is given by

$$dh = \left[v - T\left(\frac{\partial v}{\partial T}\right)_P\right] dP$$

If the compressibility relation $Pv = ZRT$ is used, one finds that

$$dh_T = \left[\frac{ZRT}{P} - \frac{ZRT}{P} - \frac{RT^2}{P}\left(\frac{\partial Z}{\partial T}\right)_P\right] dP = -\frac{RT^2}{P}\left(\frac{\partial Z}{\partial T}\right)_P dP$$

Before integrating this expression it must be transformed into reduced coordinates so that the result will be of general validity. By definition, $T = T_c T_R$ and $P = P_c P_R$. Hence

$$dT = T_c\, dT_R \qquad \text{and} \qquad dP = P_c\, dP_R$$

Substitution of these expressions into the equation for dh_T yields

$$dh_T = -\frac{RT_c^2 T_R^2}{P_c P_R}\left(\frac{dZ}{T_c\, \partial T_R}\right)_{P_R} P_c\, dP_R = -RT_c T_R^2\left(\frac{\partial Z}{\partial T_R}\right)_{P_R} d \ln P_R$$

Upon integrating at constant temperature we obtain the expression

$$\frac{\Delta h_T}{R_u T_c} = -\int_i^f T_R^2\left(\frac{\partial Z}{\partial T_R}\right)_{P_R} d \ln P_R \tag{13-40}$$

where the subscripts i and f signify the initial and final limits of integration for the reduced pressure. For convenience, the enthalpy should be evaluated from the ideal-gas to a real-gas state at the same temperature. The lower limit on the right, then, is zero pressure, for which state P_R is likewise zero. The enthalpy of an ideal gas will be indicated by an asterisk, that is, h^*. The upper limit is the actual real-gas enthalpy h at some elevated pressure P. Hence

$$\frac{h^* - h}{R_u T_c} = \int_0^P T_R^2\left(\frac{\partial Z}{\partial T_R}\right)_{P_R} d \ln P_R \tag{13-41}$$

The value of the integral is obtained by graphical integration, employing data from the generalized compressibility chart. The result of integration leads to values of $(h^* - h)/R_u T_c$ as a function of P_R and T_R. A plot of these data is called a generalized enthalpy chart, and a typical chart is shown in Fig. A-27M. An example of the use of this chart is given in the following.

Examples 13-7M and 13-7 Methane gas (CH_4) is cooled in a constant-pressure process from T_1 to T_2. Calculate the heat transfer per unit mass of methane (*a*) using the generalized enthalpy chart, Fig. A-26, and (*b*) assuming ideal-gas behavior.

SOLUTION A sketch of the process from state 1 to state 2 on a Ts diagram is shown in Fig. 13-5. The actual path follows the constant-pressure line. The first law for a closed system at constant

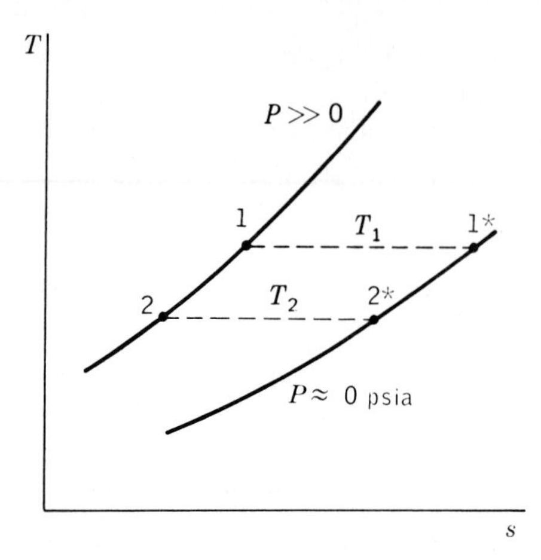

Figure 13-5 Process diagram for Examples 13-7M and 13-7.

pressure reduces to $q = \Delta h$. However, the enthalpy change $h_2 - h_1$ cannot be obtained directly from the generalized enthalpy chart, since the chart always gives the enthalpy difference between an actual state and an ideal-gas state at the same temperature. Hence the sketch also shows a line of approximately zero pressure, which is the ideal-gas state. The generalized enthalpy chart, Fig. A-27M, gives the enthalpy change between such states at 1 and 1*, or 2 and 2*. Since the enthalpy is a state function, we may calculate the change of this property by the path 1-1*-2*-2 as well as by the direct path 1-2. Consequently,

$$q = \Delta h = h_2 - h_1 = (h_1^* - h_1) + (h_2^* - h_1^*) - (h_2^* - h_2)$$

The first and last terms on the right are obtained from Fig. A-27M, the generalized enthalpy chart.

Metric analysis: (*a*) The constant-pressure process occurs at 65 bars, and the initial and final temperatures are 70 and $-6°C$, respectively. The value of c_p at zero is given by $c_{p,0} = 18.9 + 0.0555T$, where T is in degrees kelvin and the specific heat is in J/(g · mol)(°K). From the statement of the problem,

$$P_{R1} = 65/46.4 = 1.40 \qquad P_{R2} = 1.40$$

$$T_{R1} = 343/191 = 1.80 \qquad T_{R2} = 267/191 = 1.40$$

At these values we find from the enthalpy chart that $(h^* - h)/R_u T_c$, at states 1 and 2, has values of 0.38 and 0.80, respectively. Hence

$$h_1^* - h_1 = 0.38(8.314)(191) = 603 \text{ J/g} \cdot \text{mol}$$

$$h_2^* - h_2 = 0.80(8.314)(191) = 1270 \text{ J/g} \cdot \text{mol}$$

The value of $h_2^* - h_1^*$ is found from the integration of the zero-pressure specific-heat data; that is

$$h_2^* - h_1^* = \int_{343}^{267} (18.9 + 0.0555T)\, dT = -1436 - 1286 = -2722 \text{ J/g} \cdot \text{mol}$$

Substitution of these values into the first law yields

$$q = 603 + (-2722) - 1270 = -3389 \text{ J/g} \cdot \text{mol} = -212 \text{ J/g}$$

(*b*) If the effect of pressure is neglected, the change in enthalpy for this process is the same as for an ideal gas; that is

$$q = h_2^* - h_1^*$$

This value has already been calculated and found to be -2722 J/g · mol, which is equivalent to -170 J/g. Thus a 20 percent error will result if the effect of pressure on the enthalpy is neglected.

USCS analysis: (*a*) The constant-pressure process occurs at 55 atm, and the initial and final temperatures are 160 and 20°F, respectively. The value of c_p at zero pressure is given by $c_{p,0} = 4.52 + 0.00737T$, where T is in degrees Rankine and the specific heat is in Btu/(lb · mol)(°R). From the statement of the problem,

$$P_{R1} = 55/45.8 = 1.20 \qquad P_{R2} = 1.20$$

$$T_{R1} = 620/344 = 1.80 \qquad T_{R2} = 480/344 = 1.40$$

At these values we find from the enthalpy chart that $(h^* - h)/R_u T_c$, at states 1 and 2, has values of 0.32 and 0.67, respectively. Hence

$$h_1^* - h_1 = 0.32(1.986)(344) = 219 \text{ Btu/lb} \cdot \text{mol}$$

$$h_2^* - h_2 = 0.67(1.986)(344) = 458 \text{ Btu/lb} \cdot \text{mol}$$

The value of $h_2^* - h_1^*$ is found from the integration of the zero-pressure specific-heat data; that is

$$h_2^* - h_1^* = \int_{620}^{480} (4.52 + 0.00737T)\, dT = -633 - 567 = -1200 \text{ Btu/lb} \cdot \text{mol}$$

Substitution of these values into the first law yields

$$q = 219 + (-1200) - 458 = -1439 \text{ Btu/lb} \cdot \text{mol} = -89.9 \text{ Btu/lb}$$

(*b*) If the effect of pressure is neglected, the change in enthalpy for this process is the same as for an ideal gas; that is

$$q = h_2^* - h_1^*$$

This value has already been calculated and found to be -1200 Btu/lb · mol, which is equivalent to -75.0 Btu/lb. Thus a 17 percent error will result if the effect of pressure on the enthalpy is neglected.

b Generalized Entropy Chart

For a number of different processes, including those with heat transfer, it is important to have a generalized entropy chart as well as the enthalpy chart. The entropy chart is based on the generalized equation for the entropy change of a simple substance, namely,

$$ds = \frac{c_p\, dT}{T} - \left(\frac{\partial v}{\partial T}\right)_P dP \tag{13-21}$$

As in the situation for the enthalpy function, we note that the first term on the right requires only specific-heat data for a substance at the required pressure. The term in Eq. (13-21) that is difficult to evaluate in many cases is the second one, simply because sufficient PvT data may not be available for the substance of interest. Therefore a generalized approach is necessary in many cases. Following the procedure for the enthalpy function, we shall integrate Eq. (13-21) from essentially zero pressure to the desired pressure at constant temperature. This is denoted mathematically by

$$(s_P - s_0^*)_T = -\int_0^P \left(\frac{\partial v}{\partial T}\right)_P dP \tag{13-42}$$

Normally, the next procedure would be to insert the definition of the compressibility factor and the reduced pressure and temperature into this expression. However, Eq. (13-42) is not directly useful since the entropy at the ideal-gas state of zero pressure is infinite in value. This dilemma is circumvented in the following way: We shall apply Eq. (13-21) to an isothermal change between zero pressure and a given finite pressure P, but we shall assume that the gas behaves as an ideal gas at all times. Hence

$$(s_P^* - s_0^*)_T = -\int_0^P \left(\frac{\partial v}{\partial T}\right)_P dP = -R\int_0^P \frac{dP}{P} \tag{13-43}$$

The state represented by s_P^* is a fictional state since an ideal gas, for which the equation of state $Pv = RT$ is valid, exists only at zero pressure. However, we may still assign values to the state even if it is nonexistent. If Eq. (13-42) is now subtracted from Eq. (13-43), we obtain

$$(s_P^* - s_P)_T = -\int_0^P \left[\frac{R}{P} - \left(\frac{\partial v}{\partial T}\right)_P\right] dP \tag{13-44}$$

From the definition of the compressibility factor it is found that

$$\left(\frac{\partial v}{\partial T}\right)_P = \frac{RZ}{P} + \frac{RT}{P}\left(\frac{\partial Z}{\partial T}\right)_P$$

The use of this equation permits Eq. (13-44) to be written as

$$(s_P^* - s_P)_T = -R\int_0^P \left[\frac{1-Z}{P_R} - \frac{T}{P}\left(\frac{\partial Z}{\partial T}\right)_P\right] dP$$

This latter result can now be expressed in terms of reduced properties,

$$(s_P^* - s_P)_T = -R\int_0^{P_R} \frac{1-Z}{P_R}\, dP_R + RT_R\int_0^{P_R} \left(\frac{\partial Z}{\partial T_R}\right)_{P_R} \frac{dP_R}{P_R}$$

By comparing the last term of this equation with Eq. (13-41) one finds that this term can be written as a function of $h^* - h$. The final result is

$$\frac{(s_P^* - s_P)_T}{R_u} = \frac{h^* - h}{R_u T_R T_c} - \int_0^{P_R} (1 - Z)\, \frac{dP_R}{P_R} \tag{13-45}$$

The value of the first term on the right is available from a generalized enthalpy chart. The last term on the right must be evaluated by graphical integration of compressibility data. Equation (13-45) permits evaluation of the departure of the entropy value from the ideal-gas value at the same pressure and temperature. A graphical presentation of $(s_P^* - s_P)_T/R_u$ versus the reduced pressure and temperature is given in Fig. A-28M in the form of a generalized entropy chart. Among other applications, a generalized entropy chart is useful for isentropic processes of real gases.

Examples 13-8M and 13-8 Carbon dioxide is compressed reversibly and adiabatically from known values of P_1 and T_1 to P_2 in a steady-flow process. Determine the final temperature with the aid of the generalized entropy chart, if the gas at P_1 is ideal.

SOLUTION The process is reversible and adiabatic, hence isentropic. Thus we seek the value of T_2 for which $s_2 = s_1$. The calculation will require two steps if we make use of the generalized entropy chart. First we shall calculate the change in entropy from the initial to the final state as if the gas were ideal. Then we shall evaluate the entropy change from the ideal-gas to a real-gas state at the final pressure and temperature. The sum of these two changes must be zero if the process is isentropic. Thus

$$s_1 - s_2 = s_1^* - s_2 = (s_1^* - s_2^*) + (s_2^* - s_2) = 0$$

State 1 equals 1* since the initial state is an ideal-gas state. The quantity $s_2^* - s_2$ is found from Fig. A-28M in terms of P_{R2} and T_{R2}. The term $s_1^* - s_2^*$ may be found from the ideal-gas relation

$$s_2^* - s_1^* = \int_1^2 c_p \, d \ln T - R \ln \frac{P_2}{P_1} = s_2^0 - s_1^0 - R \ln \frac{P_2}{P_1}$$

where the s^0 values are found in the ideal-gas table. Therefore the total entropy change becomes

$$s_1 - s_2 = s_1^0 - s_2^0 + R \ln \frac{P_2}{P_1} + (s_2^* - s_2)_{\text{chart}} = 0$$

The values of s_2^0 and the chart correction are unknown, however, since the final temperature is not known. Hence the solution becomes one of iteration. A trial temperature is chosen repeatedly until the above equation is satisfied.

Metric analysis: The initial pressure and temperature are 1 bar and 220°K, and the final pressure is 40 bars. The following data are available:

$$P_{R2} = 40/73.9 = 0.541 \qquad s_1^0 = 202.966$$

$$R \ln \frac{P_2}{P_1} = 8.314 \ln 40 = 30.67 \text{ kJ/(kg} \cdot \text{mol)(°K)}$$

As a first trial we shall assume that T_2 is 500°K. On this basis,

$$T_{R2} = 500/304 = 1.64 \qquad (s_2^* - s_2)/R_u = 0.073 \qquad s_2^0 = 234.814$$

Substitution of all the data into the equation for $s_1 - s_2$ yields

$$s_1 - s_2 = 202.97 - 234.81 + 30.67 + 0.073(8.314) = -0.56 \text{ kJ/(kg} \cdot \text{mol)(°K)}$$

The assumption of T_2 equal to 500°K gives a total entropy change close to zero. If a further assumption of $T_2 = 490$°K is made, we find that

$$s_1 - s_2 = 202.97 - 233.92 + 30.67 + 0.078(8.314) = 0.37 \text{ kJ/(kg} \cdot \text{mol)(°K)}$$

The first approximation leads to a negative Δs value, while the second evaluation yields a positive Δs. Hence the best approximation to the final temperature lies between 490 and 500°K, and is probably around 494°K.

USCS analysis: The initial pressure and temperature are 1 atm and 400°R, and the final pressure is 40 atm. The following data are available:

$$P_{R2} = \frac{40}{72.9} = 0.548 \qquad s_1^0 = 48.555 \text{ Btu/(lb} \cdot \text{mol)(°R)}$$

$$R \ln \frac{P_2}{P_1} = 1.986 \ln 40 = 7.33 \text{ Btu/(lb} \cdot \text{mol)(°R)}$$

As a first trial we shall assume that T_2 is 900°R. On this basis,

$$T_{R2} = 900/548 = 1.64 \qquad (s_2^* - s_2)/R_u = 0.073 \qquad s_2^0 = 56.070$$

Substitution of all the data into the equation for $s_1 - s_2$ yields

$$s_1 - s_2 = 48.555 - 56.070 + 7.33 + 0.15 = -0.040 \text{ Btu/(lb} \cdot \text{mol)(°R)}$$

The assumption of this temperature gives a total of change in entropy close to zero. If a further assumption of $T_2 = 880$°R is made, we find that

$$s_1 - s_2 = 48.555 - 55.831 + 7.33 + 0.16 = +0.213 \text{ Btu/(lb} \cdot \text{mol)(°R)}$$

Consequently, the best approximation to the final temperature is probably around 897°R. As a comparison we can calculate the final temperature based solely on ideal-gas behavior from the gas table A-9. At 400°R the relative pressure p_{r1} is 0.04153. Hence $p_{r2} = 40(0.04153) = 1.661$, which is the relative pressure for a final temperature of about 884°R. The use of the generalized entropy chart predicts a final temperature about 13° higher than this. The entropy correction for real-gas behavior is especially significant at low values of the reduced temperature.

Although only two generalized charts have been presented here, it should be apparent that any number of these might be devised, once a generalized equation is available for a property in terms of the variables P and T. For example, a generalized chart which permits the estimation of c_p values at high pressures is available. Chemists and chemical engineers find a generalized chart for the property called fugacity to be quite useful. In the absence of an abundance of PvT data for a substance, generalized charts are powerful tools for predicting the properties of a fluid—gas, or liquid.

REFERENCES

Hall, N. A., and W. E. Ibele: "Engineering Thermodynamics," Prentice-Hall, Englewood Cliffs, N.J., 1960.

Zemansky, M. W.: "Heat and Thermodynamics," 5th ed., McGraw-Hill, New York, 1968.

PROBLEMS (METRIC)

Generalized relations

13-1M Prove that the constant-pressure lines in the wet region of an hs diagram are straight and not parallel and that the slope of a constant-pressure line in the superheat region increases with temperature.

13-2M From any of the four Maxwell relations, derive the other three, making use of Eq. (13-1*b*).

13-3M Derive expressions for (*a*) $(\partial u/\partial P)_T$, and (*b*) $(\partial u/\partial v)_T$ that involve only P, T, and v.

13-4M Derive the relation $c_p = T(\partial s/\partial T)_P$.

13-5M From Eqs. (13-17) and (13-20), show that the slope of a constant-volume line is greater than that for a constant-pressure line through the same state point in the gas region of a *Ts* diagram.

13-6M Derive the expression $c_p = T(\partial P/\partial T)_s(\partial v/\partial T)_P$.

13-7M Derive the expression $c_v = -T(\partial v/\partial T)_s(\partial P/\partial T)_v$.

13-8M Approximate the value of c_p for steam at (*a*) 60 bars and 400°C, and (*b*) 120 bars and 480°C, by employing the equation derived in Problem 13-6M. Compare with the value obtained by approximating the derivative, $c_p = (\partial h/\partial T)_P$.

13-9M Approximate the value of c_p for refrigerant 12 at (*a*) 6 bars and 50°C, and (*b*) 12 bars and 80°C, by employing the equation derived in Prob. 13-6M. Compare with the value obtained by approximating the derivative, $c_p = (\partial h/\partial T)_P$.

13-10M Show that the slope of a constant-pressure line in the vapor region of a *Ts* diagram normally increases with increasing temperature.

13-11M Derive the relation $(\partial^2 g/\partial T^2)_P = -c_p/T$.

13-12M Derive the relation $(\partial^2 a/\partial T^2)_v = -c_v/T$.

13-13M What can be concluded qualitatively about the change in the enthalpy of a fluid during an isentropic compression?

13-14M Prove that, for any homogeneous system, $c_p/c_v = (\partial P/\partial v)_s(\partial P/\partial v)_T$.

13-15M Check the validity of Eq. (13-13) by finding approximate values of the derivatives on both sides of the equation for steam at (*a*) 15 bars and 360°C, and (*b*) 80 bars and 400°C.

13-16M Check the validity of Eq. (13-13) by finding approximate values of the derivatives on both sides of the equation for refrigerant 12 at (*a*) 7 bars and 50°C, and (*b*) 12 bars and 80°C.

13-17M On the basis of the van der Waals equation of state and the generalized relations:

(*a*) Show that $(h_2 - h_1)_T = (P_2 v_2 - P_1 v_1) + a(1/v_1 - 1/v_2)$.

(*b*) Show that $(s_2 - s_1)_T = R \ln [(v_2 - b)/(v_1 - b)]$.

(*c*) Evaluate $(h_2 - h_1)_T$ in kJ/kg · mol for oxygen at 27°C when it is compressed from 1 bar to 100 bars. Compare with the ideal-gas solution to the problem.

13-18M Develop an expression for the isothermal change in internal energy for a substance which follows the Redlich-Kwong equation of state. The result should be given in terms of T, v, a, b, and a constant.

13-19M On the basis of the equation derived in Prob. 13-18M, evaluate with the data in Table A-21M the internal-energy change of steam as it is compressed isothermally at 360°C from 23.31 cm^3/g to 11.05 cm^3/g. Check the answer against tabular data for steam.

13-20M Same as Prob. 13-19M, except that we wish to compress refrigerant 12 isothermally at 80°C from 27.48 cm^3/g to 11.98 cm^3/g.

13-21M Derive expressions for the isothermal change (*a*) of enthalpy, and (*b*) of entropy for a substance which follows the Berthelot equation of state.

13-22M A pressure-enthalpy diagram is frequently used in the refrigeration industry. Determine an equation for the slope of an isentropic line on this diagram in terms of PvT data only.

13-23M Use Eqs. (13-27), (12-4) and (12-5) to predict the c_p value of nitrogen, in kJ/(kg · mol)(°C), at (*a*) 200°K and 40 bars, and (*b*) 190°K and 70 bars.

13-24M A gas has a compressibility factor Z given by $Z = 1 + aP/T^2$. Derive an expression for the change in enthalpy between two states at the same temperature but different pressures.

13-25M Derive an expression for the change in enthalpy in terms of temperature and pressure changes from a state T_1 and P_1 to a state of higher values of T_2 and P_2 for a gas whose equation of state is $Pv/RT = 1 + AP/T$ and whose specific heat at a pressure P_0 is given by $c_{p,0} = 1 + BT$; A and B are constants, and P_0 is less than P_1 and P_2.

Use of β and K_T

13-26M Show that the change in volume of a substance can be related to β and K_T such that $dv/v = \beta\, dT - K_T\, dP$.

13-27M On the basis of the result shown in Prob. 13-26M, estimate the percent change in the volume of copper when it changes state from 200°K and 1 bar to 300°K and 1000 bars.

13-28M A 0.1-kg mass of copper is heated from 250 to 500°K and is compressed from 1 bar to 500 bars. Estimate the change in volume in cubic centimeters if the value of v at 500°K is 7.115 cm^3/g · mol, on the basis of Prob. 13-26M.

13-29M Evaluate (*a*) the coefficient β, and (*b*) the coefficient K_T symbolically for a van der Waals gas.

13-30M At 800°K the value of v is 7.215 cm^3/g · mol.

(*a*) Determine the value of $c_p - c_v$ for solid copper, in kJ/(kg · mol)(°K).

(*b*) If c_p is 28.48 kJ/(kg · mol)(°K), what percent error would be made in c_v if we assumed that $c_p = c_v$ at this temperature?

13-31M At 300°K the value of v for solid copper is 7.062 cm^3/g · mol. Determine the value of $c_p - c_v$, in kJ/(kg · mol)(°K).

13-32M At 0°C the values of v, β, and K_T for liquid mercury are 14.67 cm^3/g · mol, 174×10^{-6} °K^{-1}, and 3.79×10^{-7} cm^2/N, respectively.

(*a*) Determine the value of $c_p - c_v$, in kJ/kg · mol)(°K).

(*b*) If c_p is 28.0 kJ/(kg · mol)(°K), what percent error would be made in c_v if we assumed that $c_p = c_v$?

Phase changes

13-33M The vapor pressure of carbon tetrachloride at several temperatures is as follows:

T, °C	25	35	45	55
P, mbar	151.7	232.5	345.1	498.0

Plot ln P versus $1/T$, and from the slope evaluate the mean enthalpy of vaporization, in kJ/kg.

13-34M Estimate the enthalpy of vaporization of water at (*a*) 100°C, (*b*) 150°C, and (*c*) 200°C by employing (1) the Clapeyron equation, and (2) the Clapeyron-Clausius equation. Compare with data from Table A-12M.

13-35M Estimate from tabular data the enthalpy of vaporization of refrigerant 12 at (*a*) 8°C, (*b*) 24°C, and (*c*) 44°C by using (1) the Clapeyron equation, and (2) the Clapeyron-Clausius equation. Compare result with data from Table A-16M.

13-36M The specific volumes of liquid water and ice at 0°C are 1.0002 and 1.0911 cm^3/g, respectively. Estimate the melting-point temperature of ice at (*a*) 250 bars, (*b*) 500 bars, and (*c*) 750 bars, if the enthalpy of fusion is 333.4 kJ/kg.

13-37M The vapor pressure of water can be represented with reasonable accuracy between 5 and 50 bars by the equation, $\ln P = (-4692/T) + 0.0124 \ln T + 12.58$, with P in bars and T in degrees kelvin. Compute the enthalpy of vaporization at (*a*) 160°C, (*b*) 200°C, and (*c*) 250°C, using values of the specific volume from tables.

13-38M The triple state of carbon dioxide is −56.6°C and 5.178 bars. Predict the slopes of the three saturation lines in the vicinity of the triple state on a PT diagram on the basis of the following triple-state data.

Phase	Solid	Liquid	Vapor
h, J/g	181.3	380.2	728.5
v, cm^3/g	0.661	0.849	72.22

13-39M Helium 4 boils at 4.22°K at 1 bar, and the enthalpy of vaporization is 83.3 J/g · mol. By producing a vacuum over the liquid phase, the fluid boils at a lower temperature. Estimate what pressure, in millibars, is necessary to produce a temperature of (*a*) 2°K, (*b*) 1°K, and (*c*) 0.5°K.

13-40 At the triple state of water the pressure and temperature are 6.12 mbar and 0.010°C. The enthalpy of melting is 333.4 kJ/kg, and the specific volumes of the liquid and solid phases are 1.0002 and 1.0911 cm^3/g, respectively. A person skates on ice at 30°F on blades with a contact area of 0.32 cm^2. What must the weight and the mass, in N and kg, respectively, of a person be to just melt the ice beneath the blades.

13-41M Consider the thermodynamic data given in Prob. 13-40M. A person with a mass of 80 kg is ice skating on blades which have a total area of 0.25 cm^2 in contact with the ice. The temperature of the ice is (*a*) −2°C, and (*b*) −3°C. Will the ice melt under the blades?

Joule-Thomson coefficient

13-42M Determine the Joules-Thomson coefficient for water at (*a*) 30 bars and 320°C, (*b*) 60 bars and 320°C, and (*c*) 100 bars and 400°C, in °C/bar.

13-43M A gas obeys the relation $P(v - b) = RT$, where b is a positive constant. Determine the Joule-Thomson coefficient of the gas. Could this gas be cooled effectively by throttling?

13-44M Calculate the Joule-Thomson coefficient for a Dieterici gas, and prove that the inversion temperature for such a gas is $2a(v - b)/Rbv$.

13-45M By employing the generalized equation for the Joule-Thomson coefficient and the tabular data for refrigerant 12, estimate the coefficient, in °C/bar, at (*a*) 10 bars, and 80°C, (*b*) 12 bars and 80°C, and (*c*) 14 bars and 80°C.

13-46M The Joule-Thomson coefficient for nitrogen at 40 bars and −73°C is approximately 0.4°C/bars on the basis of Fig. 13-4. Determine the value of c_p, in J/(g · mol)(°C), if Eqs. (12-4) and (12-5) represent the PvT behavior of nitrogen at this state.

13-47M An acceptable equation of state for helium gas is given by $Pv = RT - aP/T + bP$, where $a = 386.7$°K · cm^3/g · mol and $b = 15.29$ cm^3/g · mol.

(*a*) Compute the Joule-Thomson coefficient at 150 and 15°K.

(*b*) Find the inversion temperature, in degrees kelvin.

(*c*) Estimate the temperature reached in an ideal throttling process from 25 bars and 15°K to 1 bar.

13-48M By the use of Eq. (12-5), employ a graphical technique to determine roughly the inversion temperature for nitrogen, in degrees kelvin.

13-49M On the basis of Eq. (12-5) determine the Joule-Thomson coefficient for nitrogen at 1 bar and (*a*) 300°K, (*b*) 500°K, and (*c*) 700°K.

Generalized charts

13-50M Calculate the value, in kJ/kg, of $(h^* - h)$ at 40 bars and 320°C for steam by use of (*a*) the generalized chart, and (*b*) steam-table data.

13-51M Same as Prob. 13-50M, except for a state of 80 bars and 360°C.

13-52M Calculate the value, in kJ/kg, of $(h^* - h)$ at 16 bars and 80°C for refrigerant 12 by use of (*a*) the generalized chart, and (*b*) tabular data.

13-53M Refrigerant 12 is throttled from 16 bars and 100°C to 4.0 bars. Determine the final temperature based on a generalized chart. Compare the result with that obtained from tabulated data. The zero-pressure specific heat in this temperature range may be taken as 0.65 J/(g)(°C).

13-54M Same as Prob. 13-53M, except that the fluid is throttled from 16 bars and 80°C, to 4 bars.

13-55M By means of a generalized chart, estimate the change in enthalpy, in kJ/kg · mol, accompanying the isothermal expansion of ethane from 20 bars, 30°C, to 5 bars.

13-56M Methane gas is cooled in a constant-pressure process at 55 bars from 100 to 30°C. The equation for $c_{p,0}$ is given in Table A-4M. Calculate the heat transferred, in kJ/kg.

13-57M Ethane is compressed from 30°C and 15 bars to 60°C and 98 bars in a steady-flow process.

The molar specific heat in kJ/(kg · mol)(°C) is given by $c_{p,0} = 16.8 + 0.123T$, where T is in degrees kelvin. Using generalized charts, determine the enthalpy change, in kJ/kg · mol, if (*a*) the gas is an ideal gas, and (*b*) the gas is a real gas.

13-58M Nitrogen gas is compressed isothermally from the ideal-gas state to 50 bars and 250°K. Determine the value of $h^* - h$ by employing (*a*) Eq. (12-5) and the generalized equation, and (*b*) the generalized chart.

13-59M Methane gas (CH_4) is compressed in a steady-flow process from 13.9 bars and 71°C to 186 bars and 300°C. Using generalized charts, determine (*a*) the change in enthalpy, in kJ/kg · mol, and (*b*) the change in entropy, in kJ/(kg · mol)(°K).

13-60M Nitrogen at 100 bars and 200°K is contained in a tank of 0.3 m^3. Heat is transferred to the nitrogen until the temperature is 350°K. Determine the heat transferred and the final pressure (*a*) using generalized charts, and (*b*) assuming ideal-gas behavior.

13-61M Calculate (*a*) the work of compression, and (*b*) the heat transfer, in kJ/kg, when ethane is compressed reversibly and isothermally from 5 to 98 bars at 30°C in a steady-flow process.

13-62M Carbon dioxide is compressed in a reversible, adiabatic, steady-flow process from 5 bars, 30°C, to 40 bars. Determine the work of compression, in kJ/kg.

13-63M Ethylene at 67°C and 1 bar is compressed isothermally in a reversible, nonflow process to 255 bars. On the basis of generalized charts, determine (*a*) the change in entropy, in kJ/(kg · mol)(°K), (*b*) the change in internal energy, in kJ/kg · mol, (*c*) the heat transferred, and (*d*) the work required, in kJ/kg · mol.

13-64M Ethylene is being expanded in a gas turbine from 180°C and 307 bars to 67°C and 61 bars. The intake capacity is 1 m^3/min. A mean c_p value of 50.0 kJ/(kg · mol)(°K) may be assumed at zero pressure. Estimate the shaft-output power, in kilowatts, for the adiabatic process.

13-65M Steam at 280 bars and 520°C is expanded adiabatically and reversibly in a steady-flow process to 100 bars. Determine the final temperature, in degrees Celsius, and the change in enthalpy using (*a*) generalized charts, and (*b*) tabular data.

13-66M Propane (C_3H_8) enters a pipe at 34.2 bars, 100°C, and a velocity of 40 m/s. The gas flows adiabatically through the pipe until the pressure reaches 10.7 bars. If the zero-pressure value of c_p is relatively constant at 1.76 kJ/(kg)(°C), estimate the exit temperature, in degrees Celsius, and the exit velocity, in m/s.

13-67M Oxygen initially at 40 bars and −27°C is compressed adiabatically in a steady-flow process to a final pressure of 200 bars. Determine the minimum work of compression if (*a*) the generalized charts are used, and (*b*) the gas is assumed to be ideal.

13-68M Carbon dioxide is compressed isentropically in steady flow from 4 bars and 0°C to 60 bars. Determine the work of compression, in kJ/kg · mol, if based on (*a*) a generalized chart, and (*b*) an ideal gas.

PROBLEMS (USCS)

Generalized relations

13-1 Prove that the constant-pressure lines in the wet region of an *hs* diagram are straight and not parallel and that the slope of a constant-pressure line in the superheat region increases with temperature.

13-2 From any of the four Maxwell relations, derive the other three, making use of Eq. (13-1*b*).

13-3 Derive expressions for (*a*) $(\partial u/\partial P)_T$ and (*b*) $(\partial u/\partial v)_T$ that involve only the pressure, temperature, and specific volume.

13-4 Derive the relation $c_p = T(\partial s/\partial T)_P$.

13-5 From Eqs. (13-17) and (13-20), show that the slope of a constant-volume line is greater than that for a constant-pressure line through the same state point on a *Ts* diagram.

13-6 Derive the expression $c_p = T(\partial P/\partial T)_s(\partial v/\partial T)_p$.

13-7 Derive the expression $c_v = -T(\partial v/\partial T)_s(\partial P/\partial T)_v$.

13-8 Approximate the value of c_p for steam at 800 psia and 650°F by employing the equation derived in Prob. 13-6. Compare with the value obtained directly from the definition of this property.

13-9 Approximate the value of c_v for steam at 600 psia and 600°F by employing the equation derived in Prob. 13-7.

13-10 Show that the slope of a constant-pressure line in the vapor region of a Ts diagram normally increases with increasing temperature.

13-11 Derive the relation $(\partial^2 g/\partial T^2)_P = -c_p/T$.

13-12 Derive the relation $(\partial^2 a/\partial T^2)_v = -c_v/T$.

13-13 What can be concluded qualitatively about the change in the enthalpy of a fluid during an isentropic compression?

13-14 Prove that, for any homogeneous system, $c_p/c_v = (\partial P/\partial v)_s(\partial P/\partial v)_T$.

13-15 Find the approximate values of the derivatives on both sides of Eq. (13-13) for steam at 200 psia and 500°F in the same set of units.

13-16 Develop the general relations $(\partial h/\partial v)_T = T(\partial P/\partial T)_v + v(\partial P/\partial v)_T$, and evaluate $(\partial h/\partial v)_T$ for an ideal gas.

13-17 On the basis of the van der Waals equation of state and the generalized relations:

(a) Show that $(h_2 - h_1)_T = (P_2 v_2 - P_1 v_1) + a(1/v_1 - 1/v_2)$.

(b) Show that $(s_2 - s_1)_T = R \ln [(v_2 - b)/(v_1 - b)]$.

(c) Evaluate $(h_2 - h_1)_T$, in Btu/lb · mol, for oxygen at 80°F which is compressed from 1 to 100 atm. Compare with the ideal-gas solution.

13-18 Derive expressions for the isothermal change of enthalpy and entropy for a substance which follows the Berthelot equation of state.

13-19 A pressure-enthalpy diagram is frequently employed in the refrigeration industry. Determine an equation for the slope of an isentropic line on this diagram in terms of PvT data only.

13-20 A gas has a compressibility factor Z given by $Z = 1 + aP/T^2$. Derive an expression for the change in enthalpy between two states at the same temperature but different pressures.

13-21 Derive an expression for the change in enthalpy in terms of temperature and pressure changes from a state T_1 and P_1 to a state of higher values T_2 and P_2 for a gas whose equation of state is $Pv/RT = 1 + AP/T$ and whose specific heat at a pressure P_0 is given by $c_{p,0} = 1 + BT$; A and B are constants, and P_0 is less than P_1 and P_2.

13-22 A gas, for which the equation of state is $P(v - b) = RT$, undergoes a change of state from 20 psia and 500°R to 800 psia and 700°R. Compute the enthalpy change if $c_{p,0} = 0.24 + 1.3 \times 10^{-4}T$ and $b = 0.0039$, with T in degrees Rankine and P in psia.

13-23 At 900°R the values of v, β, and K_T for solid copper at 0.114 ft³/lb · mol, 30.1×10^{-6} °R^{-1}, and 400.3×10^{-12} ft²/lb$_f$, respectively.

(a) Determine the value of $c_p - c_v$, in Btu/(lb · mol)(°F).

(b) If c_p is 6.25 Btu/(lb · mol)(°F), what percent error would be made in c_v if we assumed that $c_p = c_v$?

13-24 At 32°F and 14,225 psia the values of v, β, and K_T for liquid mercury are 0.235 ft³/lb · mol, 96.7×10^{-6} °R^{-1}, and 1686×10^{-12} ft²/lb$_f$, respectfully.

(a) Determine the value of $c_p - c_v$, in Btu/(lb · mol)(°F).

(b) If c_p is 6.69 Btu/(lb · mol)(°F), what percent error would be made in c_v if we assumed that $c_p = c_v$?

Phase changes

13-25 The vapor pressure of carbon tetrachloride at several temperatures is as follows:

T, °F	77	95	113	131
P, mmHg	113.8	174.4	258.9	373.6

Plot log P versus $1/T$, and from the slope evaluate the mean enthalpy of vaporization in the given temperature range, in Btu/lb.

13-26 Find the enthalpy of vaporization of water at 400°F by employing (*a*) the Clapeyron equation, and (*b*) the Clapeyron-Clausius equation. Compare with tabulated data.

13-27 From tabular data determine the enthalpy of vaporization of refrigerant 12 at 70°F using (*a*) the Clapeyron equation, and (*b*) the Clapeyron-Clausius equation.

13-28 The effect of pressure on the melting point of sodium has been measured by Bridgman [*Phys. Rev.*, **3**:127 (1914)]. Some of the data include:

P, atm	T, °F	$\Delta v \times 10^5$, ft^3/lb
5810	288.5	30.06
7740	310.6	27.46
9680	332.0	24.97

Determine the approximate enthalpy of fusion at 310.6°F, and compare with the tabulated value of 51.6 Btu/lb.

13-29 The specific volumes of water and ice at 32°F are 0.01602 and 0.01747 ft^3/lb, respectively. Estimate the melting-point temperature of ice at 500 atm if the enthalpy of fusion at 32°F is 143.32 Btu/lb.

13-30 The vapor pressure of water can be represented by the equation, $\ln P = (-8445/T) + 0.0124 \ln T + 15.252$, with P in psia and T in degrees Rankine. Compute the enthalpy of vaporization at 400°F using values of the specific volumes from tables.

13-31 The enthalpy of fusion of water is practically constant at 143.8 Btu/lb. Calculate the freezing-point temperature at a pressure of (*a*) 10,000 psia and (*b*) 20,000 psia. See Prob. 13-29 for data.

Joule-Thomson coefficient

13-32 Using data from Table A-14 and noting that $c_p = 0.48$ Btu/(lb)(°F), estimate the Joule-Thomson coefficient for steam at 100 psia and 700°F.

13-33 Estimate the Joule-Thomson effect on the temperature of nitrogen which is throttled from 200 atm and 25°C to 1 atm.

13-34 Determine the Joule-Thomson coefficient for water at 500 psia and 550°F, in °F/psi.

13-35 Determine the change in the Gibbs function for the vaporization of refrigerant 12 at 10°F.

13-36 A gas obeys the relation $P(v - b) = RT$, where b is a positive constant. Determine the Joule-Thomson coefficient of the gas. Could this gas be cooled effectively by throttling?

13-37 By employing the generalized equation for the Joule-Thomson coefficient and the tabular data for refrigerant 12 in Table A-18, estimate the coefficient, in °F/atm, at (*a*) 300 psia and 200°F, and (*b*) 400 psia and 240°F.

13-38 The Joule-Thomson coefficient for nitrogen at 40 atm and −100°F (−73°C) is approximately 0.4°C/atm on the basis of Fig. 13-4. Determine the value of c_p, in Btu/(lb · mol)(°F), if Eqs. (12-4) and (12-5) represent the PvT behavior of nitrogen at this state.

13-39 Estimate the enthalpy deviation of carbon monoxide at 1500 psia and −100°F, in Btu/lb · mol.

13-40 An acceptable equation of state for helium gas is given by $Pv = RT - aP/T + bP$, where $a = 11.17°R\ ft^3/lb \cdot mol$ and $b = 0.245\ ft^3/lb \cdot mol$.

(*a*) Compute the Joule-Thomson coefficient for helium at 270° and 27°R.

(*b*) Find the inversion temperature, in degrees Rankine.

(*c*) Estimate the temperature reached in an ideal throttling process from 24 atm and 27°R to 1 atm.

13-41 By the use of Eq. (12-5), employ a graphical technique to determine approximately the inversion temperature for nitrogen, in degrees kelvin.

13-42 On the basis of Eq. (12-5), determine the Joule-Thomson coefficient of nitrogen at 1 atm and (*a*) 300°K, (*b*) 500°K, and (*c*) 700°K.

Generalized charts

13-43 Carbon dioxide is compressed in a reversible, adiabatic steady-flow process from 5 atm and 460°R, to 40 atm. Determine the work of compression, in Btu/lb, if based on (*a*) a generalized chart, and (*b*) an ideal gas.

13-44 Refrigerant 12 is throttled from 600 psia and 280°F to 100 psia. Determine the final temperature based on generalized charts. The zero-pressure specific heat in this temperature range may be taken as 0.155 Btu/(lb)(°F). Compare with the result obtained from tabulated values of properties.

13-45 By means of a generalized chart, estimate the change in enthalpy, in Btu, accompanying the isothermal expansion of 1 lb · mol of ethane from 300 atm and 90°F, to 30 atm.

13-46 Nitrogen at a pressure of 1500 psia and −100°F is contained in a tank of 10 ft^3. Heat is transferred to the nitrogen until the temperature is 200°F. Determine the heat transfer and the final pressure of the nitrogen (*a*) using compressibility data, and (*b*) assuming ideal-gas behavior.

13-47 Methane gas is cooled in a constant-pressure process at 800 psia from 200 to 100°F. The molar specific heat at zero pressure is given by $c_{p,0} = 4.52 + 0.00737T$, where T is in degrees Rankine. Calculate the heat transferred per pound.

13-48 Ethane is compressed from 90°F and 15 atm to 145°F and 96 atm in a steady-flow process. The molar specific heat at zero pressure is given by $c_{p,0} = 4.01 + 0.01636T$, where T is in degrees Rankine. Determine the enthalpy change of the gas, in Btu/lb · mol (*a*) if the gas is an ideal gas, and (*b*) if the gas is a real gas, using generalized charts.

13-49 Calculate (*a*) the work of compression and (*b*) the heat transfer per pound when ethane is compressed reversibly and isothermally from 70 to 1420 psia at 90°F in a steady-flow process.

13-50 Nitrogen gas is compressed isothermally from the ideal-gas state to 1000 psia and 450°R. Determine the value of $h^* - h$ by (*a*) employing Eq. (12-5) and the generalized equation, (*b*) employing the generalized chart, and (*c*) using tabular data.

13-51 Calculate the value of $(h^* - h)_T$, in Btu/lb, at 200 psia and 140°F for refrigerant 12 by use of (*a*) the generalized chart and (*b*) the data of Table A-18.

13-52 Calculate the value of $(h^* - h)_T$ at 500 psia and 600°F for steam, in Btu/lb, by use of (*a*) the generalized chart and (*b*) the data of Table A-14 or the Mollier diagram.

13-53 Nitrogen gas is compressed isothermally at −100°F from 15 psia to 300 psia. Determine the enthalpy change, in Btu/lb, by employing the generalized equation in conjunction with Eqs. (12-4) and (12-5).

13-54 The equation of state of a gas is $v = RT/P - aT/(T - b)$, where a and b are positive constants. T is greater than b.

(*a*) Derive an expression for $h_2 - h_1$ between two states at the same temperature but at different pressures.

(*b*) If this gas flows through a throttling valve, does the temperature decrease, increase, or remain unchanged?

(*c*) Is c_p a function of P for this gas?

13-55 Listed below are some data for the values of b and $T(db/dT)$ for air as a function of temperature, on the basis of the virial equation, $Pv = RT + bP$. Compute the entropy change, in Btu/(lb · mol)(°R), for an isothermal compression at (*a*) 360°R and (*b*) 450°R from 1 to 50 atm.

T, °R	b, ft³/lb · mol	$T(db/dT)$, ft³/lb · mol
360	−0.613	1.57
450	−0.310	1.17

13-56 Ethylene at 127°F and 14.7 psia is compressed isothermally in a reversible, nonflow process to 4080 psia. Using the principle of corresponding states, determine (*a*) the change in the specific entropy, (*b*) the change in the specific internal energy, (*c*) the heat added or removed, and (*d*) the work done per pound.

13-57 Propane is being pumped a distance of 250 m. At a booster pump a gas is compressed from 73°F and 185 psia to 340°F and 1235 psia. The mean-zero-pressure specific heat c_p is 0.44 Btu/(lb)(°F). It is desired to deliver 100 ft³/min at the compressor discharge. Determine the horsepower required for the compression process, if adiabatic.

13-58 Methane gas initially at 90°F and 400 psia is compressed isothermally and reversibly in a steady-flow device to a pressure of 2700 psia with a flow rate of 800,000 lb/h. Determine the following four quantities for the condition of (*a*) a gas which follows the principle of corresponding states, and (*b*) a gas which behaves as an ideal gas at all pressures: (1) the entropy change, in Btu/(lb · mol)(°R), (2) the heat transferred, in Btu/lb · mol, (3) the work required, in Btu/lb · mol, and (4) the required area of the compressor outlet, in square inches, if the outlet velocity is 100 ft/s.

13-59 Steam at 4500 psia and 1000°F is expanded adiabatically and reversibly in a steady-flow process to 1600 psia. Determine the final temperature and the change in enthalpy employing (*a*) the principle of corresponding states, and (*b*) the experimental data for steam.

13-60 Ethylene (C_2H_4) at 100°F and 1 atm is compressed isothermally in a reversible, nonflow process to 101 atm. Using the principle of corresponding states, determine (*a*) the change in specific internal energy, (*b*) the change in entropy per mole, and (*c*) the work done, in Btu/lb · mol.

13-61 Methane gas (CH_4) is compressed in a steady-flow process from 168 psia and 160°F to 2860 psia and 570°F. Using generalized charts, determine (*a*) the change in enthalpy, in Btu/lb · mol, (*b*) the change in internal energy, in Btu/lb · mol, and (*c*) the change in entropy, in Btu/(lb · mol)(°R).

CHAPTER

FOURTEEN

COMBUSTION AND THERMOCHEMISTRY

In the preceding chapters attention was focused on the thermodynamic analysis of nonreacting systems. There are many practical examples, however, of engineering systems in which chemical reactions play a major role. It is our purpose in this chapter to apply the basic concepts of the first law of thermodynamics to reacting systems.

14-1 STOICHIOMETRY OF REACTIONS

Although chemical reactions could involve many different types of reactants, our interest will be directed mainly toward combustion reactions. Combustion normally involves the reaction of fuels containing primarily carbon and hydrogen with oxygen or air to form carbon dioxide (CO_2), carbon monoxide (CO), and water (H_2O) as the primary products. Other possible products will be discussed later, when the effects of dissociation are introduced.

One basic consideration in the analysis of combustion processes is the *theoretical or stoichiometric* reaction for a given fuel. A theoretical reaction requires the complete combustion of carbon, hydrogen, and any other combustible elements in the fuel. For example, all the carbon present is assumed to be burned to carbon dioxide and all the hydrogen is converted into water. In addition, no oxygen is present in the products of combustion. Hence the complete combustion of methane (CH_4) with oxygen is written as

$$CH_4 + 2O_2 \rightarrow CO_2 + 2H_2O$$

The theoretical oxygen requirement for a given fuel is the minimum oxygen required for the complete combustion. For the combustion of methane, two moles of O_2 are required per mole of fuel. The balanced chemical equation for the complete combustion of a fuel, such as methane, is called the *stoichiometric* equation, and in this case, theoretically, no oxygen will appear in the products of combustion. In general, one might consider the chemical reaction

$$\nu_A A + \nu_B B + \cdots \rightarrow \nu_L L + \nu_M M + \cdots \tag{14-1}$$

where the uppercase letters A, B, etc., represent the chemical species of the reactants and L, M, etc., represent the chemical species of the products. The ν_i terms are known as the stoichiometric coefficients of the various species. In the combustion of methane, $\nu_A = 1$, $\nu_B = 2$, $\nu_L = 1$, and $\nu_M = 2$. In general, any number of reactants or products might need to be considered.

The combustion of fuels in industrial practice is normally accomplished by employing air as the oxidizer. The major components of air are considered to be approximately 21 percent oxygen by volume, 78 percent nitrogen, and 1 percent argon. Small amounts of carbon dioxide and other gases are present, of course. It is convenient to assume that air is composed of 21 percent oxygen and 79 percent nitrogen by volume. Hence there is 21 moles of oxygen to every 79 moles of nitrogen in our assumed composition of atmospheric air. Therefore we may write

$$1 \text{ mol } O_2 + 3.76 \text{ mol } N_2 = 4.76 \text{ mol air}$$

or

$$1 \text{ lb } O_2 + 3.31 \text{ lb } N_2 = 4.31 \text{ lb air}$$

or

$$1 \text{ kg } O_2 + 3.31 \text{ kg } N_2 = 4.31 \text{ kg air}$$

The average molecular weight of air is 28.97, which we shall round off to 29.0 in most calculations.

When air is used as the oxidizer, we speak of the *theoretical air* requirements of a fuel. In the case of the oxidation of methane, the theoretical or stoichiometric reaction with air is written in the form

$$CH_4 + 2O_2 + 2(3.76)N_2 \rightarrow CO_2 + 2H_2O + 7.52N_2$$

The theoretical or chemically correct combustion of propane (C_3H_8) with air is given by the chemical equation

$$C_3H_8 + 5O_2 + 5(3.76)N_2 \rightarrow 3CO_2 + 4H_2O + 18.80N_2$$

In each case no oxygen appears in the products of combustion, and the nitrogen has been assumed to undergo no chemical change.

For the complete combustion of carbon and hydrogen to CO_2 and H_2O we may use the term theoretical, stoichiometric, or chemically correct oxygen, or air, requirements. When this quantity is not used in a process, we speak of the *percent theoretical* oxygen, or air, actually used. The stoichiometric quantity is the 100 percent theoretical requirement. When a deficiency is used, the percent theoretical is somewhere between 0 and 100 percent, and an excess of oxygen (or air) means that some value greater than the 100 percent theoretical value was employed. Thus 200 percent theoretical air means that twice as much air is supplied as is necessary for complete combustion. In such a case oxygen will necessarily appear in the product gases. Other terms in frequent usage are the *percent excess* and the *percent deficiency* of oxygen (or air). As examples, 150 percent theoretical air is equivalent to 50 percent excess air, and 80 percent theoretical air is a 20 percent deficiency of air. When 150 percent theoretical air or 50 percent excess air is supplied, the combustion reaction for propane becomes

$$C_3H_8 + 7.5O_2 + 28.20N_2 \rightarrow 3CO_2 + 4H_2O + 2.5O_2 + 28.20N_2$$

This is a theoretical reaction, since in practice other products such as carbon monoxide may appear in small quantities. In writing chemical equations like the one above, no information about the products of the actual process in necessary.

In addition to the preceding nomenclature, the relationship between the fuel and air supplied to a combustion process is frequently given in terms of the *air-fuel* or *fuel-air* ratio. The air-fuel ratio (AF) is defined as the mass of air supplied per unit mass of fuel supplied. The fuel-air ratio (FA) is the reciprocal of the above definition. For the 100 percent theoretical combustion of propane, for example, the chemical equation shows that 23.80 mol of air (5 mol of O_2 + 18.80 mol of N_2) are required per mole of fuel. Consequently, the air-fuel ratio for theoretical combustion of this fuel is

$$\mathrm{AF} = \frac{23.80 \text{ kg}\cdot\text{mol air}}{\text{kg}\cdot\text{mol fuel}} \times \frac{29 \text{ kg air}}{\text{kg}\cdot\text{mol air}} \times \frac{\text{kg}\cdot\text{mol fuel}}{44 \text{ kg fuel}} = 15.7 \frac{\text{kg air}}{\text{kg fuel}}$$

Since the air-fuel ratio is expressed as mass of air per mass of fuel, its value is the same whether expressed as kilograms per kilogram as above, or as pounds per pound. The fuel-air ratio for the same combustion process would be 0.0637, in units of kg fuel/kg air or lb fuel/lb air. Many hydrocarbon fuels from oil or natural gas require an air-fuel ratio in the neighborhood of 15 to 16 for stoichiometric combustion.

Finally, the relationship between the amounts of fuel and air supplied to a combustion process is also given by the *equivalence* ratio ϕ. By definition,

$$\phi = \frac{\mathrm{FA}_{\text{actual}}}{\mathrm{FA}_{\text{stoich}}}$$

where FA in the numerator represents the fuel-air ratio used under actual combustion conditions and FA in the denominator is the stoichiometric or chemically correct value. The value of ϕ is less than 1 when an excess of oxidant (such as air or oxygen) is used. This is also called a lean mixture. A rich mixture is one where ϕ is greater than unity, and the fuel is in excess of the stoichiometric requirement. The term equivalence ratio is used frequently with respect to spark ignition and compression ignition engine operation, and to gas turbine analysis.

Example 14-1M A gaseous fuel contains the following components on a volumetric or mole basis: hydrogen, 2 percent; methane, 64 percent; and ethane, 34 percent. Calculate (*a*) the air-fuel ratio required, kg air/kg fuel, (*b*) the equivalence ratio used, and (*c*) the volume of air required per kilogram and per kilogram mole of fuel, if 20 percent excess air is used and the air conditions are 27°C and 0.98 bar (98 kPa).

SOLUTION (*a*) The first step is to write the chemical reactions for the combustion of the fuel per mole of fuel. In terms of the theoretical or chemically correct oxygen requirements,

$$0.02\ H_2 + 0.01\ O_2 \rightarrow 0.02\ H_2O$$

$$0.64\ CH_4 + 1.28\ O_2 \rightarrow 0.64\ CO_2 + 1.28\ H_2O$$

$$0.34\ C_2H_6 + 1.19\ O_2 \rightarrow 0.68\ CO_2 + 1.02\ H_2O$$

As a result we find that 2.48 moles of oxygen per mole of fuel are required for complete combustion. The theoretical air-fuel ratio is

$$AF = \frac{2.48(4.76)(29)}{0.02(2) + 0.64(16) + 0.34(30)} = \frac{342.3}{20.48} = 16.7$$

For 20 percent excess air the required value is

$$AF = 16.7(1.20) = 20.0 \text{ kg air/kg fuel}$$

(*b*) The equivalence ratio is defined as FA_{act}/FA_{stoich}. If this is written in terms of the AF ratio, then, since 20 percent excess air is used,

$$\phi = \frac{AF_{stoich}}{AF_{act}} = \frac{1}{1.2} = 0.83$$

(*c*) Under the stated conditions, the air is assumed to behave as an ideal gas. The volume occupied by each kilogram of air is found from the ideal-gas equation.

$$v = \frac{RT}{P} = 0.08314 \text{ bar m}^3/(\text{kg}\cdot\text{mole})(°\text{K}) \times \frac{300°\text{K}}{0.98 \text{ bar}} \times \frac{\text{kg}\cdot\text{mol}}{29 \text{ kg}} = 0.878 \text{ m}^3/\text{kg}$$

In part *a* the air-fuel ratio was found to be 20.0 for the 20 percent excess air supplied. Therefore the volume of air required per kilogram of fuel is

$$V = 0.878 \text{ m}^3/\text{kg air} \times 20.0 \text{ kg air/kg fuel} = 17.6 \text{ m}^3/\text{kg fuel}$$

Finally, in part *a* it was determined that there are 20.48 kg fuel per kilogram mole of fuel. Therefore the volume of air per kilogram mole of fuel is

$$V = 17.6 \text{ m}^3/\text{kg fuel} \times 20.48 \text{ kg fuel/kg}\cdot\text{mol fuel} = 360 \text{ m}^3/\text{kg}\cdot\text{mol fuel}$$

Example 14-1 A gaseous fuel contains the following components on a volumetric or mole basis: hydrogen, 2 percent; methane, 64 percent; and ethane, 34 percent. Calculate (*a*) the air-fuel ratio required, lb air/lb fuel, (*b*) the equivalence ratio used, and (*c*) the volume of air required per pound and per pound mole of fuel, if 20 percent excess air is used and the air conditions are 80°F and 14.5 psia.

SOLUTION (*a*) The first step is to write the chemical reactions for the combustion of the fuel per mole of fuel. In terms of the theoretical or chemically correct oxygen requirements,

$$0.02\ H_2 + 0.01\ O_2 \rightarrow 0.02\ H_2O$$

$$0.64\ CH_2 + 1.28\ O_2 \rightarrow 0.64\ CO_2 + 1.28\ H_2O$$

$$0.34\ C_2H_6 + 1.19\ O_2 \rightarrow 0.68\ CO_2 + 1.02\ H_2O$$

As a result we find that 2.48 moles of oxygen are required per mole of fuel for complete combustion. The theoretical air-fuel ratio is

$$AF = \frac{2.48(4.76)(29)}{0.02(2) + 0.64(16) + 0.34(30)} = \frac{342.3}{20.48} = 16.7$$

For 20 percent excess air the required value is

$$AF = 16.7(1.20) = 20.0 \text{ kg air/kg fuel}$$

(*b*) The equivalence ratio is defined as FA_{act}/FA_{stoich}. If this is written in terms of the AF ratio, then, since 20 percent excess air is used,

$$\phi = \frac{AF_{stoich}}{AF_{act}} = \frac{1}{1.2} = 0.83$$

(*c*) Under the stated conditions, the air is assumed to behave as an ideal gas. The volume occupied by each pound of air is found from the ideal-gas equation.

$$v = \frac{RT}{P} = 10.73 \text{ (psia)(ft}^3\text{)/(lb} \cdot \text{mol)(}^\circ\text{R)} \times \frac{540^\circ\text{R}}{14.5 \text{ psia}} \times \frac{\text{lb} \cdot \text{mol}}{29 \text{ lb}} = 13.78 \text{ ft}^3\text{/lb}$$

In part *a* the air-fuel ratio was found to be 20.0 for the 20 percent excess air supplied. Therefore the volume of air required per pound of fuel is

$$V = 13.78 \text{ ft}^3\text{/lb air} \times 20.0 \text{ lb air/lb fuel} = 276 \text{ ft}^3\text{/lb fuel}$$

Finally, in part *a* it was determined that there are 20.48 lb fuel per pound mole of fuel. Therefore the volume of air per pound mole of fuel is

$$V = 276 \text{ ft}^3\text{/lb fuel} \times 20.48 \text{ lb fuel/lb} \cdot \text{mol fuel} = 5650 \text{ ft}^3\text{/lb} \cdot \text{mol fuel}$$

In the incomplete combustion of carbon in a fuel, the carbon reacts according to the reaction $C + \frac{1}{2}O_2 \rightarrow CO$. Since oxygen has a greater affinity for combining with hydrogen than it does with carbon, all the hydrogen in a fuel normally is converted to water. If there is insufficient oxygen to assure complete combustion, it is always the carbon which is not completely reacted. In actual practice, there is usually CO in the products, even though an excess of oxygen was supplied. This may be attributed either to incomplete mixing during the process or to insufficient time for complete combustion. In addition, the nitrogen in the air and in the fuel partially reacts with oxygen to form oxides of nitrogen. While the quantity formed is small (usually less than 2000 parts per million), oxides of nitrogen are recognized as air pollutants.

The following example involves a combustion process carried out with a deficiency of air. The example illustrates the use of a basic tool in combustion analysis—the principle of the conservation of mass of each of the elements present in the overall reaction. By making appropriate mass balances on each element, we can systematically determine additional information on the initial reactant state or the state of the products. This technique will be used in subsequent examples as well.

Example 14-2M Propane gas is allowed to react with 80 percent theoretical air. Determine the theoretical equation for the reaction.

SOLUTION With a deficiency of air it is assumed that all the hydrogen is converted into water, but the carbon is converted into both CO and CO_2. The equation for 100 percent theoretical air has been shown to be

$$C_3H_8 + 5O_2 + 18.80N_2 \rightarrow 3CO_2 + 4H_2O + 18.80N_2$$

For 80 percent theoretical air we may write, in general, that

$$C_3H_8 + 4O_2 + 15.04N_2 \rightarrow aCO + bCO_2 + 4H_2O + 15.04N_2$$

The problem is to predict theoretically the quantities of CO and CO_2 in the products, as given by the unknowns a and b. One of the basic premises for reacting systems of this sort is that all atomic species are conserved. Therefore we can write mass balances for each of the atomic species present in a reaction. Balances on hydrogen and nitrogen are not informative in this particular case. However, a carbon and an oxygen balance lead to the following equations:

C balance: $$3 = a + b$$

O balance: $$8 = a + 2b + 4$$

The solution to these two equations is that a equals 2 and b equals 1. Hence the correct theoretical equation for the combustion of propane with 80 percent theoretical air is

$$C_3H_8 + 4O_2 + 15.04N_2 \rightarrow 2CO + 1CO_2 + 4H_2O + 15.04N_2$$

The values of a and b would vary, depending upon the percent theoretical air used. This equation would not be applicable to the actual combustion process, since there may be O_2 in the products, as well as other species in small amounts. However, the chemical equation can be used as a first approximation of what should be expected in the way of products for this particular reaction under the given conditions.

Example 14-2 Read Example 14-2M above.

Up to this point we have not designated the phases of the reactants and the products for a reaction, since we have been interested only in reviewing the chemistry of combustion processes and in introducing new terminology pertinent to the subject. Nevertheless, acknowledgment of the fact that different phases might be present during the reaction is important for several reasons. One of these is the problem of the dew point of the product gases. In the combustion of hydrocarbon fuels, one of the main products is water. In a mixture of ideal product gases, this water vapor has a certain partial pressure. If the partial pressure ever becomes greater than the saturation pressure of water at a given temperature, some of the water will condense out as the temperature is lowered further. The presence of liquid-water droplets in the combustion gases may lead to corrosion problems, for one thing. Consequently, it is useful to be able to predict the dew point of a given product gas. This requires a knowledge of the partial pressure of the water vapor in a gas, which in turn is a function of the mole fraction of the water vapor. The example below illustrates a typical calculation for the dew point of a combustion gas, when dry air is used.

Example 14-3M Gaseous propane is burned with 150 percent theoretical air at a pressure of 970 mbar (97 kPa). If the entering air is dry, determine (a) the mole analysis of the product gas assuming complete combustion, (b) the dew point of the gas mixture, in degrees Celsius, and (c) the percent of the H_2O formed that is condensed if the product gases are cooled to 20°C.

SOLUTION (a) The chemical equation for the reaction of propane (C_3H_8) with 150 percent theoretical air has been given previously as

$$C_3H_8 + 7.5O_2 + 28.20N_2 \rightarrow 3CO_2 + 4H_2O + 2.5O_2 + 28.20N_2$$

The total moles of products is 37.7; therefore the molar analysis of the products is simply

$$\text{Mole fraction } CO_2 = \frac{3}{37.7} = 0.0796 \text{ (7.96 percent)}$$

$$\text{Mole fraction } H_2O = \frac{4}{37.7} = 0.1061 \text{ (10.61 percent)}$$

$$\text{Mole fraction } O_2 = \frac{2.5}{37.7} = 0.0663 \text{ (6.63 percent)}$$

$$\text{Mole fraction } N_2 = \frac{28.20}{37.7} = 0.7480 \text{ (74.80 percent)}$$

(*b*) The partial pressure of the water vapor in the product gas is 0.1061(970) = 103 mbar = 0.103 bar. From Table A-12M the saturation temperature corresponding to this pressure is, roughly, 46°C. When the gas mixture is cooled to 46°C at constant pressure, the dew point is reached. For temperatures below this value water will condense out of the product gas.

(*c*) Since the dew point of the mixture is 46°C, water will condense out as the gases are cooled to 20°C. At 20°C, the gas mixture will be saturated with water, and the partial pressure of the water will be the saturation pressure at 20°C of 0.02339 bar. Assuming an ideal-gas mixture, this partial pressure of water must equal its mole fraction in the gas phase times the total pressure, that is, $p_i = y_i P = (N_i/N_m)P$. The moles of dry products is (3 + 2.5 + 28.2), or 33.7. Representing the unknown moles of water still in the vapor phase by the symbol W gives

$$p_g = 23.39 = \frac{W(970)}{33.7 + W} \qquad \text{or} \qquad W = 0.83 \text{ mol}$$

Because the moles of water formed is 4, the percent water vapor which condenses is

$$\text{Percent condensed} = \frac{4 - 0.83}{4}(100) = 79 \text{ percent}$$

Thus roughly 20 percent of the water formed remains in the gas phase as a saturated vapor.

Example 14-3 Gaseous propane is burned with 150 percent theoretical air at a pressure of 14.5 psia. If the entering air is dry, determine (*a*) the mole analysis of the product gas assuming complete combustion, (*b*) the dew point of the product gas mixture, in degrees Fahrenheit, and (*c*) the percent of the H_2O formed that is condensed if the product gases are cooled to 70°F.

SOLUTION (*a*) The evaluation of the mole analysis is found in the solution of part *a* of Example 14-3M.

(*b*) The mole fraction of water vapor is found to be 0.1061; therefore, the partial pressure of the water vapor is 0.1061(14.5) = 1.54 psia. From Table A-12 the saturation temperature corresponding to this pressure is, roughly, 117°F. When the gas mixture is cooled to 117°F at constant pressure, the dew point of the gas mixture is reached.

(*c*) Since the dew point of the mixture is roughly 117°F, water will condense out as the gases are cooled to 70°F. At 70°F, the gas mixture will be saturated with water vapor, and the partial pressure of the water will be the saturation pressure of 0.3632 psia at 70°F. Assuming an ideal-gas mixture, this partial pressure must equal its mole fraction in the gas phase times the total pressure, that is, $p_i = y_i P = (N_i/N_m)P$. The moles of dry products is (3 + 2.5 + 28.2), or 33.7. Representing the unknown moles of water still in the vapor phase by the symbol W gives

$$p_g = 0.3632 = \frac{W(14.7)}{33.7 + W} \qquad \text{or} \qquad W = 0.85 \text{ mol}$$

Because the moles of water formed in the combustion process is 4, the percent water vapor which condenses is

$$\text{Percent condensed} = \frac{4 - 0.85}{4}(100) = 79 \text{ percent}$$

Thus roughly 20 percent of the water formed remains in the gas phase as a saturated vapor.

14-2 ACTUAL COMBUSTION PROCESSES

In the preceding section the discussion and examples were based on the premise that complete information was available on the reactants entering into a combustion process. In addition, it was necessary to assume that, in the presence of excess air, all carbon within a fuel would be converted completely into carbon dioxide. However, it is a common experience based on measurements of product gases that carbon monoxide is frequently present in significant quantities, even if an excess of air has been used. As a further point there are a number of applications for which actual measurement of the air-fuel ratio is difficult to ascertain. The flow of fuel, whether solid, liquid, or gas, into the reaction chamber is normally fairly well known; however, the airflow might be hard to measure accurately. To overcome these problems, an analysis of the gaseous products may be made with a good degree of accuracy. From this analysis a significant amount of information on the overall combustion process is learned.

There are numerous experimental methods that can be used to determine the concentration of various components in the actual gaseous products of combustion. The need for more accurate techniques increased with the passage of strict air-pollution emission standards by the state and federal governments, beginning in the late 1960s. Today there is a large number of manufacturers of suitable equipment. The analysis of combustion gases is usually reported on either a "dry" or "wet" basis. On a dry basis the percent water vapor in the gas stream is not reported. The well-established Orsat analyzer is a typical piece of equipment which reports the overall analysis on a dry basis. Although the mole fraction of water vapor in the original gas sample is not reported by such a measurement, this does not limit the usefulness of the technique.

Typical calculations based on representative "dry" analyses are given below. The general methodology is based on the use of mass balances on the atomic species present in the reacting mixture. This approach was introduced earlier, and illustrated in Example 14-2. We shall now make extensive use of the fact that atomic species are conserved in chemical reactions when nuclear effects are absent.

Examples 14-4M and 14-4 Propane is reacted with air in such a ratio that an analysis of the products of combustion gives CO_2, 11.5 percent; O_2, 2.7 percent; and CO, 0.7 percent. What is the percent theoretical air used during the test?

SOLUTION The proper basis for beginning the analysis is 100 mol of dry product gases. From the analysis above the moles of N_2 in the dry products must be 85.1 mol. With this in mind, a chemical equation for the reaction might be

$$xC_3H_8 + aO_2 + 3.76aN_2 \rightarrow 11.5CO_2 + 0.7CO + 2.7O_2 + 85.1N_2 + bH_2O$$

A balance on each element enables us to evaluate each of the unknown coefficients. For nitrogen,

N_2 balance: $3.76a = 85.1$

$$a = 22.65$$

A carbon balance leads directly to the quantity of fuel used.

C balance: $3x = 11.5 + 0.7 = 12.2$

$$x = 4.07$$

The only unknown at this point is the value of b, the moles of water in the products. However, two mass balances have not been used, namely, those for oxygen and hydrogen. The additional mass balance provides in this case, where the composition of the fuel is known, an independent check on the preceding calculations. From the oxygen balance,

O balance: $2(22.65) = 2(11.5) + 0.7 + 2(2.7) + b$

$$b = 16.2$$

Then, employing a hydrogen balance, we check the computation.

H balance: $8x = 2b$

$$8(4.07) = 2(16.2)$$

$$32.56 = 32.4$$

This check is fairly good, considering the limited significant figures used for the calculations. The actual reaction, then, is represented by the chemical equation

$$4.07C_3H_8 + 22.65O_2 + 85.1N_2 \rightarrow 11.5CO_2 + 0.7CO + 2.7O_2 + 85.1N_2 + 16.2H_2O$$

By dividing through by 4.07, the equation could be placed on the basis of 1 mol of fuel. The stoichiometric equation for propane has been shown to be

$$C_3H_8 + 5O_2 + 18.80N_2 \rightarrow 3CO_2 + 4H_2O + 18.80N_2$$

The moles of oxygen used theoretically per mole of fuel is 5. For the actual combustion, this ratio is $22.65/4.07 = 5.57$. Therefore the percent theoretical air (or oxygen) used in the actual combustion process was

$$\text{Percent theoretical air} = \frac{5.57(100)}{5.0} = 111 \text{ percent}$$

or the percent excess air was 11 percent.

Examples 14-5M and 14-5 An unknown hydrocarbon fuel C_xH_y was allowed to react with air. An Orsat analysis was made of a representative sample of the product gases, with the following result: CO_2, 12.1 percent; O_2, 3.8 percent; and CO, 0.9 percent. Determine (*a*) the chemical equation for the actual reaction, (*b*) the composition of the fuel, (*c*) the air-fuel ratio used during the test, and (*d*) the excess or deficiency of air used.

SOLUTION The percentages of the three gases in the Orsat analysis added up to 16.8 percent. If it is assumed that the remaining gas in the sample is nitrogen, the volumetric percent of nitrogen must be 83.2 percent. With this information we may now write a general chemical equation for the reaction in the form

$$C_xH_y + aO_2 + 3.76aN_2 \rightarrow 12.1CO_2 + 3.8O_2 + 0.9CO + 83.2N_2 + bH_2O$$

The equation actually reverses the technique used in the preceding section. Basically, we ask ourselves what initial composition of fuel and what air-fuel ratio would be required to produce certain known percentages of product gases. The calculation is based on 100 mol of dry product gases as the starting point. The unknown values of x and y determine the fuel composition, and the value of a will establish the air-fuel ratio. In addition, the value of b, the moles of water vapor formed, will need to be found if the dew point is desired. A nitrogen balance determines the value of a.

N_2 balance:

$$3.76a = 83.2$$

$$a = 22.1$$

An oxygen balance enables one to determine the moles of water formed during the reaction.

O balance:

$$2(22.1) = 2(12.1) + 2(3.8) + 0.9 + b$$

$$b = 11.5$$

Now carbon and hydrogen balances may be used to find the values of x and y.

C balance:

$$x = 12.1 + 0.9 = 13.0$$

H balance:

$$y = 2b = 2(11.5) = 23.0$$

With a knowledge of the values of the initially unknown quantities, the chemical equation may now be written as

$$C_{13}H_{23} + 22.1O_2 + 83.2N_2 \rightarrow 12.1CO_2 + 3.8O_2 + 0.9CO + 83.2N_2 + 11.5H_2O$$

The fact that the values of x and y came out to be whole numbers has no significance, since 3-figure accuracy is involved. Usually, the values of x and y for similar problems of this type will not be whole numbers. Also, the formula $C_{13}H_{23}$ should not be thought of as belonging to a single chemical species, but as the average formula for a fuel which probably contains a large number of different compounds.

From the balanced chemical equation above, the air-fuel ratio can be computed. Since 105.3 mol of air were used per mole of fuel, then,

$$\text{AF} = \frac{105.3 \text{ mol air}}{1 \text{ mol fuel}} \times \frac{29 \text{ lb air}}{1 \text{ mol air}} \times \frac{\text{mol fuel}}{179 \text{ lb fuel}} = 17.1 \text{ lb air/lb fuel (or kg air/kg fuel)}$$

Finally, the excess or deficiency of air can be found by first computing the theoretical AF value for this fuel. The chemical equation for combustion of $C_{13}H_{23}$ with the theoretical-air requirements is

$$C_{13}H_{23} + 18.75O_2 + 3.76(18.75)N_2 \rightarrow 13CO_2 + 11.5H_2O + 18.75(3.76)N_2$$

Since 18.75 mol of O_2 are required and 22.1 mol of O_2 were actually used, the percent excess air used during the test is

$$\text{Percent excess} = \frac{22.1 - 18.75}{18.75}(100) = 18 \text{ percent}$$

The preceding discussion has dealt with the chemistry of reactive systems in terms of theoretical and actual considerations. Terminology used in the engineering studies of reactive systems has been reviewed. The presentation has been based primarily on the principle of conservation of atomic species during chemical reactions. We are now in a position to consider reactive systems in the light of the conservation of energy principle.

14-3 STEADY-FLOW ENERGY ANALYSIS OF REACTING MIXTURES

The application of the conservation of energy principle to reacting systems is merely an extension of the ideas presented earlier in the text for systems of fixed composition. Engineering applications of reacting systems are generally directed toward steady-state, steady-flow processes. In the absence of magnetic and electric fields, the conservation of energy principle reduces to the familiar equation

$$Q + W_{\text{shaft}} = \Delta H + \Delta \text{KE} + \Delta \text{PE} \tag{14-2}$$

For a chemically reactive system the ΔH term in this latter expression may be written as

$$\begin{aligned} \Delta H &= H_{\text{prod}} - H_{\text{reac}} \\ &= \sum_i (N_i h_i)_{\text{prod}} - \sum_i (N_i h_i)_{\text{reac}} \end{aligned} \tag{14-3}$$

where h_i is the molar enthalpy of any product or reactant at the temperature and pressure of the reaction, and N_i is the moles of any product or reactant.

The evaluation of the h_i quantities in Eq. (14-3) introduces a difficulty unique to reactive systems. In Chaps. 3 and 4, methods for evaluating the enthalpy of an ideal or real gas were introduced. A common way of presenting such data is in tabular format. The Appendix contains a number of tables of thermodynamic data for ideal and real gases. Recall, however, that the values of h in these tables are dependent upon the choice of a reference state. The ideal-gas data, for example, are based arbitrarily on a zero value of the enthalpy at absolute zero of temperature. The steam data, on the other hand, are based on a zero reference value for the saturated liquid at the triple state. In the literature we may find tables of these same substances with reference states different from those discussed above, since it is merely an arbitrary choice of the author of a given table. Since reference states are completely arbitrary, different values of ΔH in Eq. (14-3) will result when tables based on different reference states are employed. We must seek another technique for evaluating the enthalpy function which avoids this difficulty.

The tables of h data discussed above still are useful if we correctly account for another major contribution to the enthalpy of any substance when considering chemical reactions. Energy is associated with forces which bind atoms together into a specific compound. As a result, each reactant and product of a chemical reaction has a definite "bond" energy at a given temperature and pressure. When reactants form products, certain bonds are disrupted and others are formed. The sum of the bond energies of the reactants may be quite different from that of the products of a reaction. This difference in bond energy must be strictly accounted for in an energy balance on the reacting system. In general, the energy associated with the products may be more or less than the energy of the reactants, depending upon (1) the chemical nature of the reactants and products, and (2) the state of the reactants and products.

A consistent accounting of this change in "chemical" energy is accomplished by introducing the concept of the enthalpy of formation Δh_f of a pure substance. The *enthalpy of formation* is defined as the enthalpy change that occurs when a chemical compound is formed isothermally from its elements. In a steady-flow process, in the absence of shaft work, this enthalpy change would be equal to the heat released or absorbed during the chemical change. For example, if $H_2(g)$ and $O_2(g)$ are reacted in a steady-flow system at 25°C, it would be necessary to remove 241,810 kJ/kg · mol of gaseous water formed in order to keep the temperature constant. Thus the enthalpy of formation of gaseous water at 25°C is −241,810 kJ/kg · mol. On the basis of Eq. (14-3) this formation process is denoted symbolically by

$$\Delta h_f = h_{\text{compound}} - \sum_i (v_i h_i)_{\text{elements}} \tag{14-4}$$

where v_i again is the stoichiometric coefficient for a given element. If we rearrange this equation, then

$$h_{\text{compound}} = \Delta h_f + \sum_i (v_i h_i)_{\text{elements}} \tag{14-5}$$

Hence the concept of the enthalpy of formation enables us to evaluate the enthalpy of any specific compound in terms of the quantities on the right-hand side of Eq. (14-5).

The actual numerical evaluation of Eq. (14-5) requires the following four considerations:

1. The enthalpy of formation Δh_f is determined by laboratory measurement or by the methods of statistical thermodynamics, which employ spectroscopic data for the species of interest. Values of the enthalpy of formation are available for hundreds of pure substances. The enthalpy of formation for a number of common substances is given in Tables A-22M and A-22 in the Appendix. These data are reported for a state of 25°C (298.15°K) or 77°F (536.7°R) and 1 atm. This is a *standard reference* state for thermochemical calculations. Properties at 1 atm are symbolized by the superscript "°." Hence the values of the enthalpy of formation in Tables A-22M and A-22 are listed as Δh_f°.
2. The sign convention for Δh_f° is the same as that for Q and W. When energy released as a compound is formed from its stable elements, the value of Δh_f° is negative.
3. The term "elements" must be strictly defined. The reactants in this case are taken arbitrarily as the *stable* form of the elements at the specified state. For example, the stable forms of elements such as hydrogen, nitrogen, and oxygen

at 1 atm and room temperature are $H_2(g)$, $N_2(g)$, and $O_2(g)$. On the other hand, the stable form of carbon under these same conditions is solid graphite, C(s), and not diamond.

4. The enthalpy of formation of any *stable* element at any temperature, by concept, is zero in value. For example, the enthalpy of formation of the stable form of oxygen, O_2, is listed as zero in Tables A-22M and A-22. However, note in these tables that the unstable form of oxygen (the monatomic species, O) has a nonzero value for its enthalpy of formation.
5. For thermochemical calculations the sensible enthalpy of any species will be set, by convention, to be zero at the standard reference temperature of 298°K (537°R). (This is different from the method used for the ideal gas tables in the appendices, where the enthalpy of any species is zero at the absolute zero of temperature.) Because the enthalpy of formation of stable elements is also zero at the standard reference temperature, the overall enthalpy (the sum of the chemical and sensible contributions) of all *stable* elements at 1 atm and 25°C (77°F) is assigned a value of zero. This choice establishes a consistent and convenient basis for measuring energy transformations in chemically reactive systems.

If we now apply the considerations discussed above to Eq. (14-5), then for a formation reaction at 25°C (77°F) and 1 atm the equation becomes

$$h_{\text{compound}} \text{ (at 25°C and 1 atm)} = \Delta h^\circ_{f,298} = \Delta h^\circ_{f,537} \tag{14-6}$$

Hence the enthalpy of any compound at 298°K (537°R) and 1 atm is equal to its enthalpy of formation at that state, while the enthalpy of stable elements at this same state is zero. Note that the tables in the Appendix present Δh°_f data for a particular phase of a substance. The difference between the enthalpy of formation for the gas phase and the liquid phase can be approximated quite closely by the enthalpy of vaporization h_{fg} at that temperature. Thus, to convert data one applies the relation

$$\Delta h^\circ_f \text{(gas phase)} = \Delta h^\circ_f \text{(liquid phase)} + h_{fg} \tag{14-7}$$

(Note that when a substance vaporizes at the standard reference temperature of 298°K, the saturation pressure would not normally be the standard pressure of 1 atm. However, the additional corrections in the vapor and liquid phase enthalpies for this pressure difference are very small and can be neglected.)

The final step is to evaluate the enthalpy at a specified temperature and pressure which is different from the standard reference state. To accomplish this, we must add to the value given by Eq. (14-6) the change in enthalpy between the reference state of 25°C and 1 atm and the specified state. This contribution is simply that found from conventional ideal-gas tables and steam tables, for example. If we restrict ourselves to reacting ideal-gas mixtures, then the enthalpy of each gas is independent of pressure. In this case the total enthalpy $h_{i,T}$ of an ideal gas at temperature T is given by

$$h_{i,T} = \Delta h^\circ_{f,298} + (h_T - h_{298})_i \tag{14-8}$$

where h_T is the enthalpy at the specified temperature T and h_{298} is the enthalpy at the reference temperature of 298°K or 537°R. If tabular data are not available to evaluate the last term in Eq. (14-8), the enthalpy change $h_T - h_{298}$ must be computed from the integral of $c_p\, dT$. Thus the enthalpy of any compound is composed of two parts, that associated with its formation from elements at a given temperature and pressure and that associated with a change of state at constant composition. It is convenient to call the second term on the right of Eq. (14-8) the *sensible* enthalpy change of a substance.

If we now combine Eqs. (14-2), (14-3), and 14-8), the steady-state, steady-flow energy balance for processes which include chemical reactions becomes

$$Q + W_{\text{shaft}} = \sum_{\text{prod}} N_i(\Delta h_f^\circ + h_T - h_{298} + \text{KE})_i - \sum_{\text{reac}} N_i(\Delta h_f^\circ + h_T - h_{298} + \text{KE})_i \tag{14-9}$$

The potential energy change of the flow stream has been neglected in the above expression. In many combustion processes there is no shaft work for the selected control volume and the change in kinetic energy frequently is negligible. Under these additional constraints the energy balance reduces to

$$Q = \sum_{\text{prod}} N_i(\Delta h_{f,298}^\circ + h_T - h_{298})_i - \sum_{\text{reac}} N_i(\Delta h_{f,298}^\circ + h_T - h_{298})_i \tag{14-10}$$

The value of Q, of course, may be positive or negative. Chemical reactions which liberate energy in the form of heat are called exothermic, and those which absorb energy are known as endothermic.

Example 14-6M In Section 14-1 it was shown that the combustion of methane with the stoichiometric amount of oxygen is given by the chemical equation

$$CH_4(g) + 2O_2(g) \longrightarrow CO_2(g) + 2H_2O(g)$$

where the symbol (g) after the chemical formula indicates the gas phase. Determine the heat released or absorbed if this reaction occurs at 1 atm and 25°C.

SOLUTION Equation (14-10) is applicable to this problem, in conjunction with data from Table A-22M. In this particular case the sensible enthalpy change $h_T - h_{298}$ is zero for reactants and products, since both the initial and final temperatures represent the standard reference temperature. Substitution of appropriate data yields

$$\begin{aligned} Q &= \sum_{\text{prod}} N_i\, \Delta h_{f,i}^\circ - \sum_{\text{reac}} N_i\, \Delta h_{f,i}^\circ \\ &= 1(-393{,}520) + 2(-241{,}820) - 1(-74{,}870) - 2(0) \\ &= -802{,}290 \text{ kJ/kg} \cdot \text{mol } CH_4 \end{aligned}$$

It should be noted that the above answer is independent of the amount of oxidant supplied to the reaction. If air is supplied, even in an excess amount, the oxygen and nitrogen enter and leave at the standard reference temperature. Hence the value of $h_T - h_{298}$ is zero for these elements, regardless of the quantity of each.

Example 14-6 In Sec. 14-1 it was shown that the combustion of methane with the stoichiometric amount of oxygen is given by the chemical equation

$$CH_4(g) + 2O_2(g) \longrightarrow CO_2(g) + 2H_2O(g)$$

where the symbol (g) after the chemical formula indicates the gas phase. Determine the heat released or absorbed if this reaction occurs at 1 atm and 77°F.

SOLUTION Equation (14-10) is applicable to this problem, in conjunction with data from Table A-22. In this particular case the sensible enthalpy change $h_T - h_{537}$ is zero for reactants and products, since both the initial and final temperatures represent the standard reference temperature. Substitution of appropriate data yields

$$Q = \sum_{\text{prod}} N_i \, \Delta h^\circ_{f,i} - \sum_{\text{reac}} N_i \, \Delta h^\circ_{f,i}$$

$$= 1(-169{,}300) + \; + 2(-104{,}040) - 1(-32{,}210) - 2(0)$$

$$= -345{,}170 \text{ Btu/lb} \cdot \text{mol } CH_4$$

It should be noted that the above answer in independent of the amount of oxidant supplied to the reaction. If air is supplied, even in an excess amount, the oxygen and nitrogen enter and leave at the standard reference temperature. Hence the value of $h_T - h_{537}$ is zero for these elements, regardless of the quantity of each.

Example 14-7M Methane gas initially at 400°K is burned with 50 percent excess air which enters the combustion chamber at 500°K. The reaction, which occurs at 1 atm, goes to completion and the temperature of the product gases is 1800°K. Determine the heat transfer to or from the combustion chamber, in kJ/kg · mol of fuel.

SOLUTION The chemical equation for the complete combustion of methane with 50 percent excess air is

$$CH_4(g) + 3O_2(g) + 3(3.76)N_2(g) \longrightarrow CO_2(g) + 2H_2O(g) + 11.28N_2(g) + O_2(g)$$

The state of the water in the products is indicated as a gas, since the final temperature is well above the dew point. Also, since the partial pressure of the water vapor is only 132 mbar, the vapor may be assumed to be an ideal gas along with the other product gases. The enthalpy data for all the gases except methane are obtained from Tables A-6M through A-10M, and the enthalpy of formation data are listed in Table A-22M. No tabular data are available in this text that permit the calculation of the enthalpy change for methane from 298 to 400°K. This quantity must be determined by integrating c_p data. An equation for c_p of methane, found in Table A-4M in the Appendix, is

$$\frac{c_p}{R_u} = 3.826 - 3.979 \times 10^{-3}T + 24.558 \times 10^{-6}T^2 - 22.733 \times 10^{-9}T^3 + 6.963 \times 10^{-12}T^4$$

where T is expressed in degrees kelvin. The appropriate value of R_u is 8.314 kJ/(kg · mol)(°K). The enthalpy change between 298 and 400°K for methane is found from the integral of $c_p \, dT$. Using the above equation for c_p, and omitting the integration steps themselves, we find that

$$\Delta h(CH_4) = \int_{298}^{400} c_p \, dT = 8.314(390.25 - 141.64 + 307.27 - 100.67 + 10.99)$$

$$= 3876 \text{ kJ/kg} \cdot \text{mol } CH_4$$

Substitution of the above value and data from tables in the Appendix into Eq. (14-10) yields

$$Q = 1(-393{,}520 + 88{,}806 - 9364) + 2(-241{,}820 + 72{,}513 - 9904) + 11.28(0 + 57{,}651 - 8669)$$

$$+ 1(0 + 60{,}371 - 8682) - 1(-74{,}870 + 3876) - 3(0 + 14{,}770 - 8682)$$

$$- 11.28(0 + 14{,}581 - 8669) = -82{,}260 \text{ kJ/kg} \cdot \text{mol } CH_4$$

In Example 14-6M for the stoichiometric combustion of methane gas at 25°C it was determined that the heat released is 802,290 kJ/kg · mol. When excess air is present and the products are heated to 1800°K, as in this example, roughly 90 percent of the energy released at 25°C is used to heat the products to 1800°K.

Example 14-7 Methane gas initially at 720°R is burned with 50 percent excess air which enters the combustion chamber at 900°R. The reaction, which occurs at 1 atm, goes to completion and the temperature of the product gases is 3300°R. Determine the heat transfer to or from the combustion chamber, in Btu/lb · mol of fuel.

SOLUTION The chemical equation for the complete combustion of methane with 50 percent excess air is

$$CH_4(g) + 3O_2(g) + 3(3.76)N_2(g) \longrightarrow CO_2(g) + 2H_2O(g) + 11.28N_2(g) + O_2(g)$$

The state of the water in the products is indicated as a gas, since the final temperature is well above the dew point. Also, since the partial pressure of the water vapor is only 1.92 psia, the water vapor may be assumed to be an ideal gas along with the other product gases. The enthalpy data for all the gases except methane are obtained from Tables A-6 through A-10, and the enthalpy of formation data are listed in Table A-22. No tabular data are available in this text that permit the calculation of the enthalpy change for methane from 537 to 700°R. This quantity must be determined by integrating $c_p\, dT$. The equation for methane, found in Table A-4 in the Appendix, is

$$c_p/R_u = 3.826 - 2.211 \times 10^{-3}T + 7.580 \times 10^{-6}T^2$$
$$- 3.898 \times 10^{-9}T^3 + 0.663 \times 10^{-12}T^4$$

where T is expressed in degrees Rankine. The appropriate value of R_u to use is 1.986 Btu/(lb · mol)(°R). Using the above equation for c_p, and omitting the integration procedure itself, we find that

$$\Delta h(CH_4) = \int_{537}^{720} c_p\, dT = 1.986(700.16 - 254.30 + 551.81 - 180.85 + 19.74)$$
$$= 1661 \text{ Btu/lb} \cdot \text{mol } CH_4$$

Substitution of the above value and data from tables in the Appendix into Eq. (14-10) yields

$$Q = 1(-169{,}300 + 39{,}087 - 4028) + 2(-104{,}040 + 31{,}918 - 4258)$$
$$+ 11.28(0 + 25{,}306 - 3730) + 1(0 + 26{,}412 - 3725) - 1(-32{,}210 + 1661)$$
$$- 3(0 + 6338 - 3725) - 11.28(0 + 6268 - 3730) = -26{,}850 \text{ Btu/lb} \cdot \text{mol } CH_4$$

In Example 14-6 for the stoichiometric combustion of methane gas at 537°R the heat released was found to be 345,170 Btu/lb · mol of methane. When excess air is present and the products are heated to 3300°R, as in this example, more than 90 percent of the energy released at 537°R is used to heat the products to 3300°R.

Example 14-8M Liquid butane (C_4H_{10}) at 25°C is sprayed into a combustion chamber. In addition, 400 percent theoretical air is supplied at an inlet temperature of 600°K. The gaseous products of combustion leave the chamber at 1100°K. If complete combustion is assumed, because of the large excess of air, find how much heat, per kilogram-mole of fuel, must be removed from the gases in the combustion chamber.

SOLUTION For the fuel under consideration the stoichiometric chemical equation is

$$C_4H_{10}(l) + 6.5O_2(g) + 24.2N_2(g) \longrightarrow 4CO_2(g) + 5H_2O(g) + 24.4N_2(g)$$

For 400 percent theoretical air the equation becomes

$$C_4H_{10}(l) + 26.0O_2(g) + 97.6N_2(g) \longrightarrow 4CO_2(g) + 5H_2O(g) + 19.5O_2(g) + 97.6N_2(g)$$

The heat transfer is found through use of Eq. (14-10), namely,

$$Q = \sum_{\text{prod}} N_i(\Delta h^\circ_{f,298} + h_T - h_{298})_i - \sum_{\text{reac}} N_i(\Delta h^\circ_{f,298} + h_T - h_{298})_i$$

The tabulated value of $\Delta h^\circ_{f,298}$ for *n*-butane in Table A-22M is listed for the gas phase. This value must be corrected by h_{fg}, as shown in Eq. (14-7), when substituted into the above energy balance. The other required enthalpy values are obtained from Tables A-6M through A-10M in the Appendix. Use of these data and the corrected enthalpy of formation of *n*-butane yields

$$\begin{aligned} Q &= 4(-393{,}350 + 48{,}258 - 9364) + 5(-241{,}810 + 40{,}071 - 9904) \\ &\quad + 19.5(0 + 34{,}899 - 8682) + 97.6(0 + 33{,}426 - 8669) \\ &\quad - 1(-126{,}150 - 21{,}060) - 26.0(0 + 17{,}929 - 8682) \\ &\quad - 97.6(0 + 17{,}563 - 8669) \\ &= -510{,}470 \text{ kJ/kg} \cdot \text{mol } C_4H_{10} \end{aligned}$$

Example 14-8 Liquid butane (C_4H_{10}) at 77°F is sprayed into a combustion chamber which is operating at a constant pressure of 5 atm. In addition, 400 percent theoretical air is supplied by an air compressor which has compressed the air to 5 atm and 620°F. The gaseous products of combustion leave the combustion chamber and enter a turbine which has a maximum permissible temperature of 1520°F. If complete combustion is assumed, because of the large excess of air, how much heat per mole of fuel must be removed from the gases in the combustion chamber to assure no damage to the turbine?

SOLUTION For the particular fuel under consideration the stoichiometric equation is

$$C_4H_{10}(l) + 6.5O_2(g) + 24.4N_2(g) \longrightarrow 4CO_2(g) + 5H_2O(g) + 24.4N_2(g)$$

For 400 percent theoretical air (or 300 percent excess air) the equation becomes

$$C_4H_{10}(l) + 26.0O_2(g) + 97.6N_2(g) \longrightarrow 4CO_2(g) + 5H_2O(g) + 19.5O_2(g) + 97.6N_2(g)$$

The heat transfer is found through use of Eq. (14-10), namely,

$$Q = \sum_{\text{prod}} N_i(\Delta h^\circ_{f,537} + h_T - h_{537})_i - \sum_{\text{reac}} N_i(\Delta h^\circ_{f,537} + h_T - h_{537})_i$$

The tabulated value of $\Delta h^\circ_{f,537}$ for *n*-butane in Table A-22 is listed for the gas phase. This value must be corrected by h_{fg}, as shown in Eq. (14-7), when substituted into the above energy balance. The other required enthalpy values are obtained from Tables A-6 through A-10 in the Appendix. Use of these data and the corrected enthalpy of formation of *n*-butane yields

$$\begin{aligned} Q &= 4(-169{,}290 + 20{,}753 - 4028) + 5(-104{,}040 + 17{,}235 - 4258) \\ &\quad + 19.5(0 + 14{,}995 - 3725) + 97.6(0 + 14{,}375 - 3730) \\ &\quad - 1(-54{,}270 - 9060) - 26.9(0 + 7697 - 3725) - 97.6(0 + 7551 - 3730) \\ &= -219{,}730 \text{ Btu/lb} \cdot \text{mol } C_4H_{10} \end{aligned}$$

14-4 ADIABATIC-COMBUSTION TEMPERATURE

In the absence of work effects and any appreciable kinetic-energy change of the flow stream, the energy released by a chemical reaction in a steady-flow reactor appears in two forms: heat loss to the surroundings and a temperature rise of the product gases. The smaller the heat loss, the larger the temperature rise becomes. In the limit of adiabatic operation of the reactor, the maximum temperature rise will occur. In a number of engineering applications of reacting systems, such as rocket-propulsion and gas-turbine cycles, it is desirable to be able to predict the maximum attainable temperature by the product gases. This maximum temperature is referred to as the *adiabatic-flame* or *adiabatic-combustion* temperature of the reacting mixture. On the basis of Eq. (14-10) the energy balance for a reacting mixture in steady flow and under adiabatic conditions becomes

$$\sum_{\text{prod}} N_i(\Delta h^\circ_{f,\,298} + h_T - h_{298})_i = \sum_{\text{reac}} N_i(\Delta h^\circ_{f,\,298} + h_T - h_{298})_i \qquad (14\text{-}11)$$

Since the initial temperature and composition of the reactants normally is known, the right-hand side of Eq. (14-11) can be evaluated directly. To obtain the maximum temperature rise of the products, a reaction must go to completion. Hence the N_i values of the products are also known from the chemistry of the reaction. In addition, the values of Δh°_f and h_{298} for each of the products is obtainable from tables of thermochemical data. Therefore, the only unknowns in Eq. (14-11) are the h_T values for each of the product gases at the unknown adiabatic-combustion temperature. Since the h_T values are tabulated against temperature in the Appendix, the solution to Eq. (14-11) is one of iteration. That is, a temperature must be guessed, and then the enthalpy values of the product gases at that temperature are found from their respective tables. If the guessed temperature is correct, then the numerical values of the left-hand and right-hand sides of Eq. (14-11) are equal. If not, then another estimate of the final temperature must be tried.

An estimate of the maximum combustion temperature made on this basis normally is conservative. That is, the calculated value frequently will be several hundred degrees higher than a measured value. In actual practice the calculated combustion temperature is not attained because of several effects. First of all, combustion is seldom complete. In addition, heat losses may be minimized but not eliminated. Last, some of the products of combustion will dissociate into other chemical species as a result of the high temperatures present. These dissociation reactions normally are endothermic, and consume some of the energy released by the overall reaction. The next two examples illustrate adiabatic-combustion-temperature calculations.

Example 14-9M Liquid butane (C_4H_{10}) at 25°C and 400 percent theoretical air at 600°K react at constant pressure in a steady-flow process. Determine the adiabatic-combustion temperature in degrees kelvin, assuming complete combustion.

SOLUTION The equation for the reaction under consideration is the same as in Example 14-8M.

$$C_4H_{10}(l) + 26O_2(g) + 97.6N_2(g) \longrightarrow 4CO_2(g) + 5H_2O(g) + 19.5O_2(g) + 97.6N_2(g)$$

Equation (14-11) is applicable to this problem, namely,

$$\sum_{\text{reac}} N_i(\Delta h^\circ_{f,\,298} + h_T - h_{298})_i = \sum_{\text{prod}} N_i(\Delta h^\circ_{f,\,298} + h_T - h_{298})_i$$

As noted in the discussion which precedes this example, all the values of the terms in this equation are known, except the h_T values for the products. These values are unknown because the final temperature of the products is not known. Substitution of the known values from tables in the Appendix yields

$$1(-126{,}150 - 21{,}060) + 26(0 + 17{,}929 - 8682) + 97.6(0 + 17{,}563 - 8669)$$
$$= 4(-393{,}520 + h_{T,\,CO_2} - 9364) + 5(-241{,}820 + h_{T,\,H_2O} - 9904)$$
$$+ 19.5(0 + h_{T,\,O_2} - 8682) + 97.6(0 + h_{T,\,N_2} - 8669)$$

In the above equation the value 21,060 in the first term is the enthalpy of vaporization of *n*-butane. Solving for the unknown enthalpy values at the adiabatic-combustion temperature, we find that

$$4h_{T,\,CO_2} + 5h_{T,\,H_2O} + 19.5h_{T,\,O_2} + 97.6h_{T,\,N_2} = 4{,}846{,}800 \text{ kJ}$$

We are now involved in an iteration, or trial-and-error, type of solution. The value of the adiabatic-flame temperature is guessed, which enables us to look up the corresponding h_T values for each of the products. These values permit the evaluation of the left-hand side of the above equation. Several estimates may be required before the approximate answer is obtained.

A reasonably fast convergence on the correct temperature is achieved in the following manner. The chemical equation for the overall reaction shows that most of the combustion product is nitrogen (roughly 77 percent on a mole basis). Hence as a first approximation, we shall assume the product gas is entirely nitrogen. In this case the energy balance reduces to

$$126.1h_{T,\,N_2} = 4{,}846{,}800 \text{ kJ}$$

or

$$h_{T,\,N_2} = 38{,}400 \text{ kJ/kg}\cdot\text{mol}$$

From Table A-6M it is found that this enthalpy value corresponds roughly to a temperature of 1240°K. This value will be found to be too high. The actual product gases contain CO_2, H_2O, and O_2 as well as N_2. The triatomic gases, CO_2 and H_2O, have a larger average specific heat c_p than that of nitrogen in the given temperature range. As a result, they absorb more energy per mole for a given temperature change than does nitrogen. The effect of the presence of these triatomic gases, then, is to lower the final temperature from that predicted solely on the basis of nitrogen. The table below shows a summary of the iteration process in the vicinity of the final answer.

	1200°K	1220°K
$4h_T(CO_2)$	215,400	219,900
$5h_T(H_2O)$	221,900	226,300
$19.5h_T(O_2)$	749,700	763,700
$97.6h_T(N_2)$	3,589,400	3,655,300
$\sum h_T(\text{prod})$	4,776,400	4,865,200

The sum of the enthalpy values for the products of combustion should be 4,846,800 kJ. The data in the table above indicate that the adiabatic-combustion temperature is close to 1215°K.

Example 14-9 Liquid butane (C_4H_{10}) at 77°F and 400 percent theoretical air at 1080°R react at constant pressure in a steady-flow process. Determine the adiabatic-combustion temperature, in degrees Rankine, assuming complete combustion.

SOLUTION The equation for the reaction under consideration is the same as in Example 14-8.

$$C_4H_{10}(l) + 26O_2(g) + 97.6N_2(g) \longrightarrow 4CO_2(g) + 5H_2O(g) + 19.5O_2(g) + 97.6N_2(g)$$

Equation (14-11) is applicable to this problem, namely,

$$\sum_{\text{reac}} N_i(\Delta h^\circ_{f,537} + h_T - h_{537})_i = \sum_{\text{prod}} N_i(\Delta h^\circ_{f,537} + h_T - h_{537})_i$$

As noted in the discussion which precedes this example, all the values of the terms in this equation are known, except the h_T values for the products. These values are unknown because the final temperature of the products is not known. Substitution of the known values from tables in the Appendix yields

$$1(-54{,}270 - 9060) + 26(0 + 7697 - 3725) + 97.6(0 + 7551 - 3730)$$
$$= 4(-167{,}300 + h_{T,CO_2} - 4028) + 5(-104{,}040 + h_{T,H_2O} - 4258)$$
$$+ 19.5(0 + h_{T,O_2} - 3725) + 97.6(0 + h_{T,N_2} - 3730)$$

In the above equation the value 9060 in the first term is the enthalpy of vaporization of n-butane. Solving for the unknown enthalpy values at the adiabatic-combustion temperature, we find that

$$4h_{T,CO_2} + 5h_{T,H_2O} + 19.5h_{T,O_2} + 97.6h_{T,N_2} = 2{,}084{,}400 \text{ Btu}$$

We are now involved in an iteration, or trial-and-error, type of solution. The value of the adiabatic-flame temperature is guessed, which enables us to look up the corresponding h_T values for each of the products. These values permit the evaluation of the left-hand side of the above equation. Several estimates may be required before the approximate answer is obtained.

A reasonably fast convergence on the correct temperature is achieved in the following manner. The chemical equation for the overall reaction shows that most of the combustion product is nitrogen (roughly 77 percent on a mole basis). Hence, as a first approximation, we shall assume the product gas is entirely nitrogen. In this case the energy balance reduces to

$$126.1h_{T,N_2} = 2{,}084{,}400 \text{ Btu}$$

or

$$h_{T,N_2} = 16{,}530 \text{ Btu/lb} \cdot \text{mol}$$

From Table A-6 it is found that this enthalpy value corresponds roughly to a temperature of 2260°R. This value will be found to be too high. The actual product gases contain CO_2, H_2O, and O_2 as well as N_2. The triatomic gases, CO_2 and H_2O, have a larger average specific heat c_p than that of nitrogen in the given temperature range. As a result, they absorb more energy per mole for a given temperature change than does nitrogen. The effect of the presence of these triatomic gases, then, is to lower the final temperature from that predicted solely on the basis of nitrogen. The table below shows a summary of the iteration process in the vicinity of the final answer.

	2180°R	2220°R
$4h_T(CO_2)$	93,700	95,900
$5h_T(H_2O)$	96,500	98,600
$19.5h_T(O_2)$	325,500	332,200
$97.6h_T(N_2)$	1,559,500	1,591,100
$\sum h_T(\text{prod})$	2,075,200	2,117,800

The sum of the enthalpy values for the products of combustion should be 2,084,400 Btu. The data in the table above indicate that the adiabatic-combustion temperature is close to 2190°R.

Example 14-10M Determine the adiabatic-flame temperature for the reaction of liquid butane (C_4H_{10}) with the theoretical air requirements at 1 atm if all the reactants enter the combustion chamber at 25°C in steady flow. Assume complete combustion.

SOLUTION The chemical equation for the reaction is

$$C_4H_{10}(l) + 6.5O_2(g) + 24.4N_2(g) \longrightarrow 4CO_2(g) + 5H_2O(g) + 24.4N_2(g)$$

The energy balance for the steady-flow process again reduces to

$$\sum_{\text{reac}} N_i(\Delta h^\circ_{f,298} + h_T - h_{298})_i = \sum_{\text{prod}} N_i(\Delta h^\circ_{f,298} + h_T - h_{298})_i$$

The only reactant terms in this case are Δh°_f and the enthalpy of vaporization for butane. Substitution of tabular data yields

$$1(-126{,}150 - 21{,}060) = 4(-393{,}520 + h_{T,\,CO_2} - 9364) + 5(-241{,}820 + h_{T,\,H_2O} - 9904) + 24.4(h_{T,\,N_2} - 8669)$$

Therefore,

$$4h_{T,\,CO_2} + 5h_{T,\,H_2O} + 24.4h_{T,\,N_2} = 2{,}934{,}500 \text{ kJ}$$

To shorten the iteration process, we again assume that all the products are nitrogen. This leads to the approximation,

$$h_{T,\,N_2} = \frac{2{,}934{,}500}{33.4} = 87{,}860 \text{ kJ/kg} \cdot \text{mol}$$

From the nitrogen table, A-6M, the temperature corresponding to this enthalpy is between 2600 and 2650°K. The actual temperature must lie below this value. A summary of several trial calculations in the vicinity of the final answer is shown below.

	2350°K	2400°K
$4h_T(CO_2)$	488,400	500,600
$5h_T(H_2O)$	504,200	517,500
$24.4h_T(N_2)$	1,890,900	1,935,400
$\sum h_T(\text{prod})$	2,883,500	2,935,500

Since the summation of enthalpies for the products should equal 2,934,500 kJ, the preceding table indicates that the adiabatic-flame temperature is very nearly 2400°K.

It is interesting to compare the above result with that obtained in Example 14-9M. Even though the air in that case is preheated to 600°K, the adiabatic-combustion temperature is only around 1220°K when 400 percent theoretical air is used. This is some 1200 degrees less than the value for stoichiometric combustion. In the original example the excess nitrogen and oxygen molecules are a large sink for the energy released by the reaction. The presence of this excess gas greatly reduces the combustion temperature. It should also be pointed out that temperatures in the range of 2300 to 2500°K, as found in this example, are fairly typical for the adiabatic combustion of many hydrocarbons with the theoretical or stoichiometric quantity of air.

Example 14-10 Determine the adiabatic-flame temperature for the reaction of liquid butane (C_4H_{10}) with the theoretical air requirements at 1 atm if all the reactants enter the combustion chamber at 77°F in steady flow. Assume complete combustion.

SOLUTION The chemical equation for the reaction is

$$C_4H_{10}(l) + 6.5O_2(g) + 24.4N_2(g) \longrightarrow 4CO_2(g) + 5H_2O(g) + 24.4N_2(g)$$

The energy balance for the steady-flow process again reduces to

$$\sum_{\text{reac}} N_i(\Delta h^\circ_{f,537} + h_T - h_{537})_i = \sum_{\text{prod}} N_i(\Delta h^\circ_{f,537} + h_T - h_{537})_i$$

The only reactant terms in this case are Δh_f° and the enthalpy of vaporization for butane. Substitution of tabular data yields

$$1(-54{,}270 - 9060) = 4(-169{,}300 + h_{T,\,CO_2} - 4028) + 5(-104{,}040 + h_{T,\,H_2O} - 4258)$$
$$+ 24.4(h_{T,\,N_2} - 3730)$$

Therefore

$$4h_{T,\,CO_2} + 5h_{T,\,H_2O} + 24.4h_{T,\,N_2} = 1{,}262{,}500 \text{ Btu}$$

To shorten the iteration process, we again assume that all the products are nitrogen. This leads to the approximation,

$$h_{T,\,N_2} = \frac{1{,}262{,}500}{33.4} = 37{,}800 \text{ Btu/lb}\cdot\text{mol}$$

From the nitrogen table, A-6, the temperature corresponding to this enthalpy is around 4740°R. The actual temperature must lie below this value. A summary of several trial calculations in the vicinity of the final answer is shown below.

	4300°R	4340°R
$4h_T(CO_2)$	214,500	216,900
$5h_T(H_2O)$	221,400	223,900
$24.4h_T(N_2)$	828,100	836,700
$\sum h_T$(prod)	1,264,000	1,277,500

Since the summation of enthalpies for the products should equal 1,262,500 Btu, the table above indicates that the adiabatic-flame temperature is very nearly 4300°R.

It is interesting to compare the above result with that obtained in Example 14-9. Even though the air in that case is preheated to 1080°R, the adiabatic-combustion temperature is only around 2180°R when 400 percent theoretical air is used. This is some 2100 degrees less than the value for stoichiometric combustion. In the original example the excess nitrogen and oxygen molecules are a large sink for the energy released by the reaction. The presence of this excess gas greatly reduces the combustion temperature. It should be pointed out that temperatures in the range of 4200 to 4500°R, as found in this example, are fairly typical for the adiabatic combustion of many hydrocarbons with the theoretical or stoichiometric quantity of air.

14-5 THERMOCHEMICAL ANALYSIS OF CONSTANT-VOLUME SYSTEMS

If a chemical reaction occurs in a closed system maintained at constant volume, then the basic energy balance is

$$Q = \Delta U = U_{\text{prod}} - U_{\text{reac}} = \sum_{\text{prod}} N_i u_i - \sum_{\text{reac}} N_i u_i \tag{14-12}$$

The specific internal energy u_i of any component is related to its specific enthalpy h_i by recalling that $u = h - Pv$, by definition. Using this definition and Eq. (14-7) for h_i, we find that

$$u_{i,\,T} = h_{i,\,T} - (Pv)_{i,\,T} = [\Delta h_{f,\,298} + h_T - h_{298} - (Pv)_T]_i \tag{14-13}$$

Furthermore, the sensible enthalpy change given by $h_T - h_{298}$ can be replaced by the definition relating h to u. This leads to a second format for u_i, namely,

$$u_{i,T} = [\Delta h_{f,298} + u_T - u_{298} + (Pv)_T - (Pv)_{298} - (Pv)_T]_i$$
$$= [\Delta h_{f,298} + u_T - u_{298} - (Pv)_{298}]_i \qquad (14\text{-}14)$$

The substitution of Eqs. (14-13) and (14-14) into Eq. (14-12) leads to two possible equations for the heat transfer under constant volume conditions. These are

$$Q = \sum_{\text{prod}} N_i \, \Delta h_{f,298,i} - \sum_{\text{reac}} N_i \, \Delta h_{f,298,i} + \sum_{\text{prod}} N_i \, (h_T - h_{298}) - \sum_{\text{reac}} N_i (h_T - h_{298})_i - \sum_{\text{prod}} N_i (Pv)_{i,T} + \sum_{\text{reac}} N_i (Pv)_{i,T}$$

and

$$Q = \sum_{\text{prod}} N_i \, \Delta h_{f,298,i} - \sum_{\text{reac}} N_i \, \Delta h_{f,298,i} + \sum_{\text{prod}} N_i (u_T - u_{298})_i - \sum_{\text{reac}} N_i (u_T - u_{298})_i - \sum_{\text{prod}} N_i (Pv)_{i,298} + \sum_{\text{reac}} N_i (Pv)_{i,298}$$

Furthermore, the volume of solid and liquid components in a chemical reaction is usually negligible in comparison to the volume occupied by the gaseous components. Hence in the summations above which involve Pv terms, it is necessary to account for only the gaseous species. In addition, if we assume that all the gaseous components are ideal gases, then Pv can be replaced by RT. Consequently the conservation of energy principle for the constant volume process can be written in the following two forms.

$$Q = \sum_{\text{prod}} N_i (\Delta h_{f,298} + h_T - h_{298})_i - \sum_{\text{reac}} N_i (\Delta h_{f,298} + h_T - h_{298})_i - \sum_{\text{prod}} N_i RT + \sum_{\text{reac}} N_i RT \qquad (14\text{-}15)$$

and

$$Q = \sum_{\text{prod}} N_i (\Delta h_{f,298} + u_T - u_{298})_i - \sum_{\text{reac}} N_i (\Delta h_{f,298} + u_T - u_{298})_i - \Delta N R T_{298} \qquad (14\text{-}16)$$

where ΔN in the last equation represents the moles of gaseous products minus the moles of gaseous reactants. In general, ΔN may be positive, negative, or zero. The above two equations are equally valid in USCS units, if data at 298°K are replaced by values at 537°R. The only major difference between these two equations is that one requires sensible enthalpy data, while the other makes use of sensible internal energy data. The following examples illustrate the use of both sets of data for ideal gases.

Example 14-11M Carbon monoxide with 200 percent theoretical air at 25°C reacts completely in a constant-volume reaction chamber. After a given time interval the gas temperature is measured to be 1200°K. Determine the heat loss from the chamber, in kJ/kg · mol of carbon monoxide, based on (*a*) Eq. (14-15), and (*b*) Eq. (14-16).

SOLUTION The overall chemical reaction is given by

$$CO(g) + O_2(g) + 3.76N_2(g) \longrightarrow CO_2(g) + \frac{1}{2}O_2(g) + 3.76N_2(g)$$

(*a*) The heat loss will first be determined by means of Eq. (14-15). Substituting in the required data from the Appendix, we find that

$$\begin{aligned} Q &= 1(-393{,}520 + 53{,}848 - 9364) + \tfrac{1}{2}(0 + 38{,}447 - 8682) \\ &\quad + 3.76(0 + 36{,}777 - 8669) - 1(-110{,}530 + 0) - 1(0) - 3.76(0) \\ &\quad - 5.26(8.314)(1200) - 5.76(8.314)(298) \\ &= -156{,}140 \text{ kJ/kg} \cdot \text{mol CO} \end{aligned}$$

(*b*) The heat loss is now found from Eq. (14-16).

$$\begin{aligned} Q &= 1(-393{,}520 + 43{,}871 - 6885) + \tfrac{1}{2}(0 + 28{,}469 - 6203) \\ &\quad + 3.75(0 + 26{,}799 - 6190) - 1(-110{,}530 + 0) - 1(0) - 3.76(0) \\ &\quad - (-0.5)(8.314)(298) \\ &= -156{,}110 \text{ kJ/kg} \cdot \text{mol CO} \end{aligned}$$

The answers differ slightly due to round-off error.

Example 14-11 Carbon monoxide with 200 percent theoretical air at 77°F reacts completely in a constant-volume reaction chamber. After a given time interval the gas temperature is measured to be 2100°R. Determine the heat loss from the chamber, in Btu/lb · mol of carbon monoxide, based on (*a*) Eq. (14-15), and (*b*) Eq. (14-16).

SOLUTION The overall chemical reaction is given by

$$CO(g) + O_2(g) + 3.76N_2(g) \rightarrow CO_2(g) + \tfrac{1}{2}O_2(g) + 3.76N_2(g)$$

(*a*) The heat loss will first be determined by means of Eq. (14-15). Substituting in the required data from the Appendix, we find that

$$\begin{aligned} Q &= 1(-169{,}290 + 22{,}353 - 4028) + \tfrac{1}{2}(0 + 16{,}011 - 3725) \\ &\quad + 3.76(0 + 15{,}334 - 3730) - 1(-47{,}540 + 0) - 1(0) - 3.76(0) \\ &\quad - 5.26(1.986)(2100) - 5.76(1.986)(537) \\ &= -69{,}450 \text{ Btu/lb} \cdot \text{mol CO} \end{aligned}$$

(*b*) The heat loss is now found from Eq. (14-16).

$$\begin{aligned} Q &= 1(-169{,}290 + 18{,}182 - 2964) + \tfrac{1}{2}(0 + 11{,}841 - 2659) \\ &\quad + 3.75(0 + 11{,}164 - 2663) - 1(-47{,}540 + 0) - 1(0) - 3.76(0) \\ &\quad - (-0.5)(1.986)(537) \\ &= -69{,}450 \text{ Btu/lb} \cdot \text{mol CO} \end{aligned}$$

The answers are in excellent agreement.

When chemical reactions occur umder constant-volume conditions, not only elevated temperatures but also higher pressures usually result. It is of practical interest to the engineer to estimate the maximum possible pressures that might develop. This would occur theoretically under adiabatic conditions, with the simultaneous achievement of the adiabatic-flame temperature. From a more practical viewpoint the heat release would have to be nearly instantaneous and conditions of internal equilibrium also would have to be met. For calculation purposes these criteria will be assumed to be valid. This will lead to an upper limit, which is of real interest and importance. Dissociation effects at these high temperatures will be neglected at this time.

In general, two methods of calculation present themselves. One requires a knowledge of the initial volume; the other requires information on the initial bomb pressure. In order to simplify both methods, we shall assume that the gases involved will behave ideally. If one applies the ideal-gas relation to the resultant gases after a chemical reaction at constant volume, the expression for the final pressure is

$$P_f = \frac{N_f R_u T_f}{V} \tag{14-17}$$

where T_f = adiabatic-flame temperature
N_f = total number of gaseous moles of products
V = bomb volume

In this case V will be known through measurement, and T_f may be calculated by methods previously outlined. The value of N_f may be determined from the basic chemical equation representing the reaction.

Example 14-12M Liquid benzene (C_6H_6), in the amount of 2.5 g, is placed in a 0.030-m^3 constant-volume bomb. The initial temperature of the fuel and oxidizer is 25°C. If the fuel is burned with 20 percent excess air and the reaction goes to completion, determine the maximum pressure in the bomb, in bars.

SOLUTION The chemical equation for the reaction is

$$C_6H_6(l) + 9O_2(g) + 33.8N_2(g) \longrightarrow 6CO_2(g) + 3H_2O(g) + 1.5O_2(g) + 33.8N_2(g)$$

Before determining the maximum pressure, we must calculate the maximum final temperature expected. In this case we shall set Eq. (14-16) equal to zero, and solve by iteration for the unknown temperature. Note that Eq. (14-15) could be used in a similar fashion. Before substituting the appropriate data into the equation, one correction must be made. Table A-22M lists the enthalpy of formation for benzene in the gaseous state, but it enters the reaction in the liquid state. The enthalpy of vaporization is given in the same table as 33,830 kJ/kg · mol. The enthalpy of formation must be decreased by this amount in the energy balance. Other appropriate data from the Appendix are now substituted into Eq. (14-16).

$$\begin{aligned} 0 = {} & 6(-393{,}520 + u_{T,CO_2} - 6885) + 3(-241{,}810 + u_{T,H_2O} - 7425) \\ & + 1.5(0 + u_{T,O_2} - 6203) + 33.8(0 + u_{T,N_2} - 6190) \\ & - 1(82{,}930 - 33{,}830 + 0) - 9(0) - 33.8(0) - 1.5(8.314)(298) \end{aligned}$$

Solving for the unknown quantities, we write the energy balance as

$$6u_{T,CO_2} + 3u_{T,H_2O} + 1.5u_{T,O_2} + 33.8u_{T,N_2} = 3{,}421{,}500 \text{ kJ}$$

A temperature must now be found which satisfies the above equation. Results of the iteration process are summarized below.

	2650°K	2700°K
$3u(H_2O)$	284,870	291,800
$6u(CO_2)$	711,000	727,030
$1.5u(O_2)$	106,320	108,650
$33.8u(N_2)$	2,246,180	2,294,340
$\sum u(\text{prod})$	3,348,370	3,421,830

Results of the table indicate that the maximum combustion temperature is essentially 2700°K. Finally, there remains the determination of N_f, the number of kilogram-moles of gaseous products at equilibrium. From the chemical equation it is noted that 44.3 mol of gaseous products are formed for each mole of benzene. The initial quantity of 2.5 g of benzene is equivalent to 3.21×10^{-5} kg · mol. The number of kilogram · moles of gaseous products then is $44.3(3.21 \times 10^{-5}) = 0.00142$ kg · mol. Therefore the maximum combustion pressure is

$$P_f = \frac{N_f R_u T_f}{V} = \frac{0.00142(0.08314)(2700)}{0.030} = 10.6 \text{ bars}$$

Example 14-12 Liquid benzene (C_6H_6), in the amount of 2.5 g, is placed in a 1.0-ft^3 constant-volume bomb. The initial temperature of the fuel and oxidizer is 77°F. If the fuel is burned with 20 percent excess air and the reaction goes to completion, determine the maximum pressure in the bomb, in psia.

SOLUTION The chemical equation for the reaction is

$$C_6H_6(l) + 9O_2(g) + 33.8N_2(g) \rightarrow 6CO_2(g) + 3H_2O(g) + 1.5O_2(g) + 33.8N_2(g)$$

Before determining the maximum pressure, we must calculate the maximum final temperature expected. In this case we shall set Eq. (14-16) equal to zero, and solve by iteration for the unknown temperature. Note that Eq. (14-15) could be used in a similar fashion. Before substituting the appropriate data into the equation, one correction must be made. Table A-22 lists the enthalpy of formation for benzene in the gaseous state, but it enters the reaction in the liquid state. The enthalpy of vaporization is given in the same table as 14,552 Btu/lb · mol. The enthalpy of formation must be decreased by this amount in the energy balance. Other appropriate data from the Appendix are now substituted into Eq. (14-16).

$$\begin{aligned} 0 = {} & 6(-169{,}290 + u_{T,\text{CO}_2} - 2964) + 3(-104{,}040 + u_{T,\text{H}_2\text{O}} - 3192) \\ & + 1.5(0 + u_{T,\text{O}_2} - 2659) + 33.8(0 + u_{T,\text{N}_2} - 2663) \\ & - 1(35{,}680 - 14{,}552 + 0) - 9(0) - 33.8(0) - 1.5(1.986)(537) \end{aligned}$$

Solving for the unknown quantities, we write the energy balance as

$$6u_{T,\text{CO}_2} + 3u_{T,\text{H}_2\text{O}} + 1.5u_{T,\text{O}_2} + 33.8u_{T,\text{N}_2} = 1{,}471{,}950 \text{ Btu}$$

A temperature must now be found which satisfies the above equation. Results of the iteration process are shown in the following table.

	4800°R	4900°R
$3u(H_2O)$	123,500	127,000
$6u(CO_2)$	309,000	317,000
$33.8u(N_2)$	974,000	996,000
$1.5u(O_2)$	46,000	47,100
$\sum u(\text{prod})$	1,452,500	1,487,100

From the summations found in the table, it is seen that the maximum combustion temperature is around 4850°R.

Finally, there remains the determination of N_f, the number of pound · moles of gaseous products at equilibrium. From the chemical equation it is noted that 44.3 moles of gaseous products are formed for each mole of benzene in the reaction. The initial quantity of 2.5 g of benzene is equivalent to 7.06×10^{-5} lb · mol. Therefore the explosion pressure becomes

$$P_f = \frac{N_f R_u T_f}{V} = \frac{(7.06 \times 10^{-5})(44.3)(1545)(4850)}{1} = 23{,}400 \text{ psfa} = 162.5 \text{ psia}$$

A second method of determining explosion pressures is based on a knowledge of the initial pressure in the reaction vessel. By applying the ideal-gas relation to conditions both before and after the reaction takes place, the following relation results:

$$P_f = \frac{P_i N_f T_f}{N_i T_i} \tag{14-18}$$

where the subscripts f and i indicate the final and initial conditions, respectively.

Example 14-13M Consider again the combustion of 2.5 g of liquid benzene with 20 percent excess air, both initially at 25°C. In this case, however, the volume of the bomb will be such that the initial pressure P_i is 1.0 bar. Determine the maximum bomb pressure under this new condition.

Solution The following data are known from Example 14-12M: $N_i/N_f = 42.8/44.3$, $T_i =$ 298°K, $T_f = 2700$°K. In addition, $P_i = 1.0$ bar. Therefore

$$P_f = \frac{P_i N_f T_f}{N_i T_i} = \frac{1.0(44.3)(2700)}{42.8(298)} = 9.38 \text{ bars}$$

Example 14-13 Consider again the combustion of 2.5 g of liquid benzene with 20 percent excess air, both initially at 77°F. In this case the volume of the bomb will be sufficient so that the initial pressure P_i is 14.7 psia. Now determine the explosion pressure under this new condition.

Solution The following data are known from Example 14-12: $P_i = 14.7$ psia, $N_i = 42.8(7.06 \times 10^{-5})$ lb · mol, $N_f = 44.3(7.06 \times 10^{-5})$ lb · mol, $T_i = 537$°R, $T_f = 4850$°R. Therefore

$$P_f = \frac{14.7(44.3)(4850)}{42.8(537)} = 137.5 \text{ psia}$$

14-6 ENTHALPY OF REACTION AND HEATING VALUES

In the preceding sections in this chapter the energy analysis of reacting substances in constant-volume or steady-flow systems was presented. In either type of system the only thermodynamic information required was the enthalpy of formation data and sensible enthalpy or internal energy data for each of the substances. There are practical situations, however, where the enthalpy of formation will not be known for the fuel. For example, a fuel oil may contain dozens of individual compounds, and the exact mole analysis of the fuel may not be known. Similarly, solid fuels such as coal have a variable composition, depending upon which mine supplied the fuel. In addition, coal is not considered to be a mixture of various compounds, but rather a material which is represented by some overall chemical composition. This composition is usually expressed in terms of a gravimetric analysis known as the *ultimate* analysis. Such an analysis reports the carbon, hydrogen, oxygen, nitrogen, and sulfur content of a particular sample. In addition, the moisture (water) and ash content are also noted for coal. In Table 14-1 the ultimate analysis of some typical coals in the United States is listed.

Due to the wide range of chemical composition that fuel oil and coal generally have, the enthalpy of formation is not a useful concept in such cases. A different approach must be used when an energy analysis is made for the combustion of these fuels. To replace enthalpy of formation data, experiments are carried out with the fuels to determine the enthalpy of reaction or the heating value of the fuel. Solid and liquid fuels are commonly tested in a constant-volume system known as the bomb calorimeter, while gaseous fuels are tested in a steady-flow system known as a steady-flow calorimeter. In either case, a fuel is burned completely in such a manner that the products are returned to the same temperature as the initial reactants. The required heat removal per unit mass of fuel is measured accurately. This heat quantity on a mole basis for the steady-flow calorimeter is called the *enthalpy of reaction* $\Delta h_{R,T}$. It is commonly reported at the standard reference state of 298°K (537°R) and 1 atm, and thus is symbolized by $\Delta h^\circ_{R,298}$ or $\Delta h^\circ_{R,537}$. If we apply Eq. (14-10) to the process, it is seen that for a reaction at 298°K (537°R)

$$Q = \Delta h^\circ_{R,298} = \sum_{\text{prod}} N_i \Delta h^\circ_{f,298,i} - \sum_{\text{reac}} N_i \Delta h^\circ_{f,298,i} \tag{14-19}$$

As a result, by measuring the overall enthalpy change $\Delta h_{R,298}$ at the given reference temperature experimentally, we bypass the problem of not knowing the value of $\Delta h^\circ_{f,298}$ for the fuel.

If Eq. (14-19) is now substituted back into Eq. (14-10) for the general situation where both reactants and products could be at temperatures other than 298°K, then

$$Q = \Delta h^\circ_{R,298} + \sum_{\text{prod}} N_i(h_T - h_{298})_i - \sum_{\text{reac}} N_i(h_T - h_{298})_i \tag{14-20}$$

Values at 298°K are replaced by values at 537°R when USCS units are employed. Similar to the use of Eq. (14-10), Eq. (14-20) can be used to find heat quantities or

Table 14-1 Ultimate analysis of representative coals in the United States on a moisture-and-ash basis, percent by mass

Type (Rank)	State	Ash	S	H	C	N	O	H_2O	Heating value†	
									Btu/lb	kJ/kg
Anthracite	Pa.	10.7	0.5	1.7	81.6	0.6	0.8	4.1	12,590	29,290
Bituminous	W.Va.	5.1	0.7	5.2	78.8	1.6	5.8	2.8	13,980	32,520
	Ill.	8.1	0.9	4.0	68.5	1.1	7.6	9.8	12,015	27,950
	Ind.	5.6	1.1	4.4	66.0	1.5	7.9	13.5	11,788	27,420
Subbituminous	Wyo.	6.2	0.4	4.4	60.6	1.0	15.4	12.0	10,640	24,750
Lignite	S.Dak.	8.3	2.2	2.2	38.0	0.5	9.6	39.2	6,307	14,670

† Heating value listed is the higher heating value at 25°C (77°F).

to calculate adiabatic-combustion temperatures in those situations where enthalpy of reaction data are available for a fuel, rather than enthalpy of formation data.

In this chapter enthalpy of reaction data will be used primarily in the energy analysis of reacting systems when enthalpy of formation data are not directly available. However, this parameter is important in several other areas. In Chap. 15 the equilibrium composition of a reacting ideal-gas mixture will be related to the equilibrium constant K_p. The enthalpy of reaction may be related theoretically to K_p. Secondly, in Chap. 19, Δh_R is used to measure the energy-conversion efficiency of fuel cells and batteries. Hence knowledge of the value of Δh_R is required in situations other than energy analyses. Note that Eq. (14-19) can be used to evaluate enthalpy of reaction data when enthalpy of formation data are available for all the reactants and products. Otherwise it must be measured directly, as described above.

The enthalpy of reaction at a reference temperature of 298°K (537°R) is determined by means of Eq. (14-19). By combining Eqs. (14-19) and (14-20) we obtain an equation for the enthalpy of reaction at any temperature T, namely,

$$\Delta h_{R,T} = \sum_{\text{prod}} N_i(\Delta h_{f,298} + h_T - h_{298})_i - \sum_{\text{reac}} N_i(\Delta h_{f,298} + h_T - h_{298})_i \quad (14\text{-}21)$$

where h_T is measured at the same temperature for both the reactant and product terms. (The subscript 298 again becomes 537 when the temperature is measured in degrees Rankine.)

Although we have specifically been discussing the enthalpy of reaction in terms of fuel-oxidant systems, the term is a general one. That is, it is a measure of the isothermal enthalpy change for any type of reaction. However, there are many engineering applications of chemical reactions which involve the combustion of hydrocarbon fuels with oxygen or air. The enthalpy of reaction for a combustion process is frequently called the *enthalpy of combustion* Δh_c. The enthalpy of combustion of a fuel is also referred to as the *heating value* of the fuel.

The heating value is defined as energy released from the combustion process, and is always positive in value. Therefore the enthalpy of combustion Δh_c and the heating value for a given fuel are opposite in sign. Another difference is that Δh_c is usually quoted on a mole basis, while heating values are nearly always listed on a mass (kilogram or pound) basis.

Two different heating values are quoted in the literature. The value commonly cited is the *higher heating value*, which is based on liquid water in the products of combustion. When water in the product gases is in the vapor state, the energy released is called the *lower heating value*. This value is of particular importance in engineering calculations, because combustion gases normally leave equipment before the dew point of the gases is reached. When a higher heating value q_H is listed at 25°C or 77°F, the lower heating value q_L may be calculated by

$$q_L = q_H - 2442 \frac{m_w}{m_f} \qquad \text{kJ/kg fuel} \tag{14-22a}$$

or

$$q_L = q_H - 1050 \frac{m_w}{m_f} \qquad \text{Btu/lb fuel} \tag{14-22b}$$

where m_w/m_f is the mass of water formed per mass of fuel. The numerical constants in the above equations are simply the enthalpy of vaporization values at the given temperature. The examples below illustrate the difference between lower and higher heating values of a given fuel. The notation of whether $H_2O(l)$ or $H_2O(g)$ appears in a chemical equation is quite important from an energy standpoint.

Example 14-14M In Example 14-6M the enthalpy of reaction (or combustion) of gaseous methane (CH_4) at 25°C was found to be −802,290 kJ/kg · mol when water appears in the products as a gas. Determine the lower and higher heating values for the fuel at the same temperature.

SOLUTION The lower heating value of methane gas is found by dividing the given enthalpy of combustion by the molar mass of methane, and changing the sign on the answer. Thus

$$q_L(CH_4) = \frac{802{,}290}{16.04} = 50{,}020 \text{ kJ/kg fuel}$$

The higher heating value is found by using Eq. (14-22*a*) and the fact that two moles of water are formed per mol methane. Hence,

$$q_H(CH_4) = 50{,}020 + 2442\left(\frac{2}{1}\right)\left(\frac{18.02}{16.04}\right) = 55{,}510 \text{ kJ/kg fuel}$$

In this case the higher heating value is approximately 11 percent higher than the lower heating value at 25°C.

Example 14-14 In Example 14-6 the enthalpy of reaction (or combustion) of gaseous methane (CH_4) at 77°F was found to be −345,170 Btu/lb · mol when water appears as a gas in the products. Determine the lower and higher heating values for the fuel at the same temperature.

SOLUTION The lower heating value of methane gas is found by dividing the given enthalpy of combustion by the molar mass of methane, and changing the sign on the answer. Thus

$$q_L(CH_4) = \frac{345{,}170}{16.04} = 21{,}520 \text{ Btu/lb fuel}$$

The higher heating value is found by using Eq. (14-22*b*) and the fact that two moles of water are formed per mole of methane. Hence,

$$q_H(CH_4) = 21{,}520 + 1050\left(\frac{2}{1}\right)\left(\frac{18.02}{16.04}\right) = 23{,}800 \text{ Btu/lb fuel}$$

In this case the higher heating value is approximately 11 percent higher than the lower heating value at 77°F.

Example 14-15M Consider an Indiana coal with the ultimate analysis given in Table 14-1.

(*a*) Determine the reaction equation for complete combustion with 20 percent excess air and the air-fuel ratio employed.

(*b*) If the reactants enter a steady-flow combustor at 25°C and the products are cooled to 500°K, determine the heat transfer, in kJ/kg fuel, on a dry fuel basis.

Solution (*a*) The basis for the calculation is 1 kg of moist fuel. To write a suitable reaction equation, the gravimetric analysis of the fuel must be converted to a mole basis, by using the appropriate molar masses of the atomic reactants. For example, the moles of carbon per kilogram of fuel = 0.66/12 = 0.055 kg · mol. Note that any oxygen in the fuel reduces the oxygen that needs to be supplied in the air. Also it is assumed that any nitrogen in the fuel ends up as gaseous N_2. On this basis the equation for the chemically correct amount of air for complete combustion is found to be

$$0.055\ C + 0.044\ H + 0.00034\ S + 0.00107\ N + 0.00494\ O + 0.0639\ O_2 + 0.2402\ N_2 \longrightarrow$$
$$0.055\ CO_2 + 0.022\ H_2O + 0.00034\ SO_2 + 0.2407\ N_2$$

The reaction equation for 20 percent excess air would be

$$0.055\ C + 0.044\ H + 0.00034\ S + 0.00107\ N + 0.00494\ O + 0.0767\ O_2 + 0.2883\ N_2 \longrightarrow$$
$$0.055\ CO_2 + 0.022\ H_2O + 0.00034\ SO_2 + 0.2888\ N_2 + 0.0128\ O_2$$

The air-fuel ratio for the 20 percent excess air case is

$$AF = \frac{0.0767(4.76)(29)}{1.0} = 10.59 \text{ kg air/kg fuel}$$

(*b*) Because the mole fraction of SO_2 in the product gas is less than 0.1 percent, its contribution to the energy analysis will be neglected. The heating value of 27,420 kJ/kg in Table 14-1 for the coal is the higher heating value. Because the products of combustion are cooled only to 500°K, we need the lower heating value, in kJ/kg. In part *a* we found that 0.022 mol of water, or 0.396 kg of water, is formed per kilogram of fuel. Thus Eq. (14-22*a*) shows that

$$q_L = q_H - 2442\frac{m_w}{m_f} = 27{,}420 - 2442\frac{0.396}{1} = 26{,}450 \text{ kJ/kg}$$

The energy analysis is based on Eq. (14-20), where the reactant term is zero in this case.

Therefore,

$$Q = \Delta h^\circ_{R,\,298} + \sum_{\text{prod}} N_i(h_T - h_{298})_i - \sum_{\text{reac}} N_i(h_T - h_{298})_i$$

$$= -26{,}450 + 0.055(17{,}678 - 9364) + 0.022(16{,}828 - 9904)$$

$$+ 0.2888(14{,}581 - 8669) + 0.0128(14{,}770 - 8682)$$

$$= -24{,}055 \text{ kJ/kg fuel}$$

Note that the heating value has to have its sign changed when substituted into the energy equation. This calculation indicates that roughly 9 percent of the energy released by the fuel is used to heat the products to 500°K. The remaining energy appears as a heat loss from the system.

Example 14-15 Consider an Indiana coal with the ultimate analysis given in Table 14-1.

(*a*) Determine the reaction equation for complete combustion with 20 percent excess air and the air-fuel ratio employed.

(*b*) If the reactants enter a steady-flow combustor at 77°F and the products are cooled to 800°R, determine the heat transfer, in Btu/lb fuel, on a dry fuel basis.

SOLUTION (*a*) The basis for the calculation will be 1 lb of moist fuel. Regardless of the system of units employed, the determination of the reaction equation is the same. The method of solution is given in part *a* of Prob. 14-15M. The air-fuel ratio for 20 percent excess air is the same as determined in Prob. 14-15M, since it is a mass ratio. Hence,

$$\text{AF (20 percent excess air)} = 10.59 \text{ lb air/lb fuel}$$

(*b*) Because the mole fraction of SO_2 in the product gas is less than 0.1 percent, its contribution to the energy analysis will be neglected. The heating value of 11,788 Btu/lb in Table 14-1 for the coal is the higher heating value. Because the products of combustion are cooled only to 800°R, we need the lower heating value. In part *a* we found that 0.022 mol of water, or 0.396 lb of water, are formed per pound of fuel. Thus Eq. (14-22*b*) shows that

$$q_L = q_H - 1050\,\frac{m_w}{m_f} = 11{,}788 - 1050\,\frac{0.396}{1} = 11{,}370 \text{ kJ/kg fuel}$$

The energy analysis is based on Eq. (14-20), where the reactant term is zero in this case. Therefore,

$$Q = \Delta h^\circ_{R,\,537} + \sum_{\text{prod}} N_i(h_T - h_{537})_i - \sum_{\text{reac}} N_i(h_T - h_{537})_i$$

$$= -11{,}370 + 0.055(6553 - 4028) + 0.022(6397 - 4258)$$

$$+ 0.2888(5564 - 3730) + 0.0128(5602 - 3725)$$

$$= -10{,}630 \text{ Btu/lb fuel}$$

Note that the heating value has to have its sign changed when substituted into the energy equation. This calculation indicates that roughly 7 percent of the energy released by the fuel is used to heat the products to 800°R. The remaining energy appears as a heat loss from the system.

14-7 ENTROPY CHANGES FOR REACTING MIXTURES

In the discussion up to this point we have relied solely upon the conservation of energy principle, in conjunction with the conservation of atomic species, in analyzing reactive systems. As in the case of systems of fixed compositions, the second law and the evaluation of the change of the entropy for processes is of particular value in the study of chemical reactions. In the application of the first law, we found it necessary to assign arbitrary reference values to the enthalpy of stable elements at 77°F and 1 atm pressure, in order to establish meaningful enthalpy values for pertinent chemical species. Only by considering both sensible- and chemical-enthalpy changes could the change in the enthalpy of a reacting system be obtained in a consistent and straightforward manner. The same problem of a reference state occurs in evaluating the entropy change of a process for a chemical reaction, since to date we have established equations only for the change in entropy of pure substances or ideal-gas mixtures of fixed composition. In this case, however, a more fundamental approach is available for establishing consistent entropy values of substances, based on what is known as the third law of thermodynamics.

The *third law* of thermodynamics, on the basis of experimental data, states that the entropy of a pure crystalline substance may be taken to be zero at the absolute zero thermodynamic temperature, that is, 0°K or 0°R. This law was fairly well accepted by around 1920, following the work of Boltzmann, Planck, Nernst, Einstein, Lewis, and others in the two or three decades preceding this time. It may also be shown that the entropy of crystalline substances at absolute zero temperature is not a function of pressure; that is $(\partial s/\partial P)_{T=0} = 0$. At temperatures above absolute zero, however, the entropy of a substance is a function of the pressure. Because of this the tabulated values of the absolute entropy, relative to a value of zero at a temperature of absolute zero, are usually given at the standard reference pressure of 1 atm. Similar to the enthalpy function, this reference state is denoted by the superscript "0" on $s_{i,T}$, that is, by $s^0_{i,T}$. (The absolute entropy of a substance can be evaluated by two distinct methods. One of these requires specific-heat and latent-heat data, while the other is statistical in origin and requires molecular information.)

The absolute entropy values of a large number of substances at 1 atm and a wide range of temperatures are available in the literature. In this text the ideal-gas tables, A-5M through A-11M and A-5 through A-11, list values of $s^0_{i,T}$ along with the values of u and h at selected temperatures.

The evaluation of entropy changes for reactive mixtures requires a means of determining the entropy of a substance, either pure or in a mixture, at pressures other than that used in reference tables. In order to simplify the presentation, the following discussion will be restricted to ideal gases. Recall from Chap. 11 that, based on the Gibbs-Dalton rule, each constituent of a mixture of ideal gases behaves as if it existed alone at the temperature and volume of the mixture. Thus, each ideal gas exerts a pressure equal to its component or partial pressure, p_i. The entropy of a pure ideal gas at a pressure other than that found in a table is determined by employing a basic equation derived in Chap. 7, namely,

$$ds = c_p \frac{dT}{T} - R \frac{dP}{P} \tag{7-9}$$

This equation is integrated at constant temperature from a reference state where the gas is pure and at a reference pressure P_0 to a mixture state where its pressure is its partial pressure p_i

$$s_{i,T,P} - s^0_{i,T} = -R \ln \frac{p_i}{P_0}$$

Because the standard reference pressure is usually 1 atm, the expression for the absolute entropy of an ideal gas at any other pressure and the given temperature is found to be

$$s_{i,T,P} = s^0_{i,T} - R \ln p_i \tag{14-23}$$

where p_i must be measured in atmospheres. For a mixture of ideal gases the total entropy is

$$S_{m,T,P} = \sum_i N_i s_i(T, p_i) \tag{14-24}$$

and for a chemical reaction the change in entropy is given by

$$\Delta S = \sum_i (N_i s_i)_{\text{prod}} - \sum_i (N_i s_i)_{\text{reac}} \tag{14-25}$$

where N_i is the number of moles of the ith component in the mixture, and s_i is the absolute entropy of the pure substances at the required temperature and partial pressure.

Example 14-16M Carbon monoxide and oxygen in a stoichiometric ratio at 25°C and 1 atm react to form carbon dioxide at the same pressure and temperature. If the reaction goes to completion, evaluate the entropy change, in kJ/°K per kilogram · mole of CO.

SOLUTION The basic reaction is $CO + \frac{1}{2}O_2 \rightarrow CO_2$ in the gas phase. The overall entropy change is given by Eq. (14-25), namely,

$$\Delta S = \sum_i (N_i s_i)_{\text{prod}} - \sum_i (N_i s_i)_{\text{reac}}$$

The absolute entropies of the three reacting components are calculated below on the basis of Eq. (14-23). The values of s_i^0 at 25°C are from Tables A-7M, A-8M, and A-9M. In units of kJ/(kg · mol)(°K),

$$s(CO_2) = 213.685 - 8.314 \ln 1.0 = 213.685$$

$$s(CO) = 197.543 - 8.314 \ln \tfrac{2}{3} = 200.914$$

$$s(O_2) = 205.033 - 8.314 \ln \tfrac{1}{3} = 214.168$$

Therefore the total change in entropy for the complete reaction is

$$\Delta S = 213.685 - 200.914 - \tfrac{1}{2}(214.168) = -94.463 \text{ kJ/°K}$$

The negative change in entropy for the gas-phase reaction is not a violation of the second law. To keep the mixture at 25°C the energy released by the reaction appears as a heat loss to the surroundings. This heat quantity is found by applying Eq. (14-10). Using the appropriate data, we find that

$$Q = \sum_i (N_i \, \Delta h_{f,i})_{\text{prod}} - \sum_i (N_i \, \Delta h_{f,i})_{\text{reac}}$$

$$= 1(-393{,}520) - 1(-110{,}530) - \tfrac{1}{2}(0) = -282{,}990 \text{ kJ/kg} \cdot \text{mol CO}$$

Now, if the surroundings are at 25°C, they undergo an entropy change of

$$\Delta S_{surr} = \frac{Q}{T} = -\frac{-282{,}900}{298} = 949.6 \text{ kJ/°K}$$

Hence the increase in entropy of the surroundings overshadows the decrease in entropy of the chemical reaction.

Example 14-16 Carbon monoxide and oxygen in a stoichiometric ratio at 77°F and 1 atm react to form carbon dioxide at the same pressure and temperature. If the reaction goes to completion, evaluate the entropy change, in Btu/°R per pound-mole of CO.

SOLUTION The basic reaction is $CO + \frac{1}{2}O_2 \rightarrow CO_2$ in the gas phase. The overall entropy change is given by Eq. (14-25), namely,

$$\Delta S = \sum_i (N_i s_i)_{prod} - \sum_i (N_i s_i)_{reac}$$

The absolute entropies of the three reacting components are calculated below on the basis of Eq. (14-23). The values of s_i^0 at 77°F are from Tables A-7, A-8, and A-9. In units of Btu/(lb · mol) (°R),

$$s(CO_2) = 51.032 - 1.986 \ln 1.0 = 51.032$$

$$s(CO) = 47.272 - 1.986 \ln \tfrac{2}{3} = 48.077$$

$$s(O_2) = 48.982 - 1.986 \ln \tfrac{1}{3} = 51.164$$

Therefore the total change in entropy for the complete reaction is

$$\Delta S = 51.032 - 48.077 - \tfrac{1}{2}(51.164) = -22.627 \text{ Btu/°R}$$

The negative change in entropy for the gas-phase reaction is not a violation of the second law. To keep the mixture at 77°F the energy released by the reaction appears as heat loss to the surroundings. This heat quantity is found by applying Eq. (14-10). Using the appropriate data, we find that

$$Q = \sum_i (N_i h_{f,i})_{prod} - \sum_i (N_i h_{f,i})_{reac}$$

$$= 1(-169{,}290) - 1(-47{,}540) - \tfrac{1}{2}(0) = -121{,}750 \text{ Btu/lb · mol CO}$$

Now, if the surroundings are at 77°F, they undergo an entropy change of

$$\Delta S_{surr} = \frac{Q}{T} = -\frac{-121{,}750}{537} = 226.7 \text{ Btu/°R}$$

Hence the increase in entropy of the surroundings overshadows the decrease in entropy of the chemical reaction.

Example 14-17M A 1 : 1 molar ratio of gaseous carbon monoxide and water vapor enters a steady-flow, adiabatic reactor at 400°K and 1 atm. If it is assumed that the pressure is constant and that the reaction goes to completion according to the following equation,

$$CO(g) + H_2O(g) \longrightarrow CO_2(g) + H_2(g)$$

determine the overall entropy change, in kJ/°K per kilogram · mole of CO.

SOLUTION To evaluate the entropy change, we must first determine the final temperature. By neglecting kinetic- and potential-energy changes, the steady-flow energy balance reduces to

$$0 = \sum_{prod} N_i(\Delta h_f^\circ + h_T - h_{298})_i - \sum_{reac} N_i(\Delta h_f^\circ + h_T - h_{298})_i$$

Using Δh_f° data from Table A-22M and sensible-enthalpy data for ideal gases from Tables A-8M through A-11M, we find that

$$0 = 1(-393{,}520 - h_{T,\,CO_2} - 9364) + 1(0 + h_{T,\,H_2} - 8468) - 1(-110{,}530 + 11{,}644 - 8669)$$
$$-1(-241{,}820 + 13{,}356 - 9904)$$

Upon rearrangement the equation becomes

$$h_{T,\,CO_2} + h_{T,\,H_2} = 65{,}430 \text{ kJ}$$

From Tables A-9 and A-11M at 920°K, it is found that

$$h_{T,\,CO_2} + h_{T,\,H_2} = 38{,}467 + 26{,}747 = 65{,}210 \text{ kJ}$$

Similarly, at 960°K, it is found that

$$h_{T,\,CO_2} + h_{T,\,H_2} = 40{,}607 + 27{,}947 = 68{,}555 \text{ kJ}$$

Therefore the final temperature of the products is very close to 920°K. The entropy change is found by evaluating $s_{i,\,T}^0 - R \ln p_i$ for each chemical species in the reaction. Since the mole fraction of each reactant and each product is $\frac{1}{2}$ and the total pressure is constant, the partial pressure in each constituent is the same. Hence the value of $R \ln p_i$ is the same for each, and the terms will cancel out when the change in entropy is calculated. As a consequence, Eq. (14-25) for this problem reduces to

$$\Delta S = (s_{CO_2} + s_{H_2})_{920°K} - (s_{CO} + s_{H_2O})_{400°K}$$
$$= 264.728 + 163.607 - 206.125 - 198.673$$
$$= 23.537 \text{ kJ/°K}$$

The positive answer is in accord with the second law.

Example 14-17 A 1 : 1 molar ratio of gaseous carbon monoxide and water vapor enters a steady-flow, adiabatic reactor at 260°F and 1 atm. If it is assumed that the pressure is constant and that the reaction goes to completion according to the following equation,

$$CO(g) + H_2O(g) \longrightarrow CO_2(g) + H_2(g)$$

determine the overall entropy change, in Btu/°R per pound-mole of CO.

SOLUTION In order to evaluate the entropy change, we must first determine the final temperature. By neglecting kinetic- and potential-energy changes, the steady-flow energy balance reduces to

$$0 = \sum_{\text{prod}} N_i(\Delta h_f^\circ + h_T - h_{537})_i - \sum_{\text{reac}} N_i(\Delta h_f^\circ + h_T - h_{537})_i$$

Using Δh_f° data from Table A-22 and sensible-enthalpy data for ideal gases from Tables A-8 through A-11, we find that

$$0 = 1(-169{,}300 + h_{T,\,CO_2} - 4028) + 1(0 + h_{T,\,H_2} - 3640) - 1(-47{,}540 + 5006 - 3725)$$
$$-1(-104{,}040 + 5{,}739 - 4258)$$

Upon rearrangement the equation becomes

$$h_{T,\,CO_2} + h_{T,\,H_2} = 28{,}150 \text{ Btu}$$

From Tables A-9 and A-11 at 1600°R, it is found that

$$h_{T,\,CO_2} + h_{T,\,H_2} = 15{,}829 + 11{,}093 = 26{,}920 \text{ Btu}$$

Similarly, at 1700°R, it is found that

$$h_{T,\,CO_2} + h_{T,\,H_2} = 17{,}101 + 11{,}807 = 28{,}910 \text{ Btu}$$

Therefore, the final temperature of the products is very close to 1660°R. The entropy change is found by evaluating $s^0_{i,\,T} - R \ln p_i$ for each chemical species in the reaction. Since the mole fraction of each reactant and each product is $\frac{1}{2}$ and the total pressure is constant, the partial pressure in each constituent is the same. Hence the value of $R \ln p_i$ is the same for each, and the terms will cancel out when the change in entropy is calculated. As a consequence, Eq. (14-25) for this problem reduces to

$$\begin{aligned}\Delta S &= (s_{CO_2} + s_{H_2})_{1660^\circ R} - (s_{CO} + s_{H_2O})_{720^\circ R} \\ &= 63.250 + 39.090 - 49.317 - 47.450 \\ &= 5.573 \text{ Btu/}^\circ\text{R}\end{aligned}$$

The positive answer is in accord with the second law.

In the two preceding examples it is noted that the entropy change is positive for this adiabatic reaction, in accordance with the second-law requirement. However, this does not necessarily mean that the reaction as proposed is possible. If the entropy of the system attains a maximum value at some other concentration of the four chemical species present, the reaction will stop at that point. That is, whatever composition of reactants and products leads to the highest entropy value will be the composition at the equilibrium state. In order to determine the equilibrium composition of a reacting mixture of CO and H_2O initially at 260°F (400°K) in an adiabatic-reaction vessel at 1 atm pressure, we need to evaluate the total absolute entropy of the system for different degrees or stages of reaction. Recall that the stoichiometric equation is

$$CO(g) + H_2O(g) \longrightarrow CO_2(g) + H_2(g)$$

We shall evaluate the total absolute entropy of the reactive gases under those conditions for which 0, 0.2, 0.4, 0.6, 0.7, 0.8, 0.9, and 1.0 mol of CO_2 are formed under adiabatic conditions. Only one actual calculation will be shown, since the same method is used throughout. For example, if 0.8 mol of CO_2 is formed from the initial 1 : 1 molar mixture of CO and H_2O, the chemical reaction proceeds accordingly as

$$CO(g) + H_2O(g) \longrightarrow 0.8CO_2(g) + 0.8H_2(g) + 0.2CO(g) + 0.2H_2O(g)$$

As in Examples 14-17M and 14-17, one must employ the first law to determine the final temperature that would be achieved for this reaction, which only goes to 80 percent completion. By equating the enthalpy of the products with that of the reactants, it is found by an iteration process that the final temperature of the four product gases is 834°K, or roughly 1040°F. This is approximately 30°C or 60°F cooler than for complete combustion, but this is to be expected, since the total enthalpy of reaction is not released. The absolute entropy of each gas may now be found from

$$S_i = N_i(s^0_{i,\,T} - R \ln p_i)$$

Recall that $s^0_{i,T}$ is the value of the absolute entropy at 1 atm and the desired temperature. The values of $s^0_{i,T}$ used in the calculation below are based on data from the ideal-gas tables in the Appendix. The following calculations are in terms of metric data, with the USCS results shown in parentheses. Hence

$$S_{CO_2} = 0.8(259.560 - 8.314 \ln 0.4) = 213.743 \text{ kJ/°K } (51.035 \text{ Btu/°R})$$

$$S_{H_2} = 0.8(160.673 - 8.314 \ln 0.4) = 134.634 \text{ kJ/°K } (32.154 \text{ Btu/°R})$$

$$S_{CO} = 0.2(228.493 - 8.314 \ln 0.1) = 49.528 \text{ kJ/°K } (11.846 \text{ Btu/°R})$$

$$S_{H_2O} = 0.2(225.311 - 8.314 \ln 0.1) = 48.891 \text{ kJ/°K } (11.676 \text{ Btu/°R})$$

$$S_{tot} = 446.796 \text{ kJ/°K } (106.711 \text{ Btu/°R})$$

Thus the total entropy, when the reaction goes to 80 percent completion, is 446.796 kJ/°K or 106.711 Btu/°R per mole of carbon monoxide. The total entropy for the system for no reaction and for complete reaction can be found from data used in Examples 14-17M and 14-17. For metric data

$$S_{init} = 206.125 + 198.673 - 2(8.314)(\ln 0.5) = 416.325 \text{ kJ/°K}$$

$$S_{complete} = 264.728 + 163.607 - 2(8.314)(\ln 0.5) = 439.862 \text{ kJ/°K}$$

For the same calculation in USCS units

$$S_{init} = 49.317 + 47.450 - 2(1.986)(\ln 0.5) = 99.520 \text{ Btu/°R}$$

$$S_{complete} = 63.250 + 39.090 - 2(1.986)(\ln 0.5) = 105.093 \text{ Btu/°R}$$

For these data, in either metric or USCS, it is seen that the entropy for 80 percent completion is greater than that for 100 percent completion, and both values are greater than the initial value. This indicates that the entropy function is a maximum before the reaction reaches completion. To ascertain the actual state of equilibrium, one needs to carry out similar calculations at other degrees of completion. A summary of these computations appears in the table below. The system temperature at each degree of completion is also indicated. The data from this table are plotted in Fig. 14-1. It is found that the state of maximum entropy occurs when approximately 70 percent of the CO and H_2O have reacted. This, then, is the state at equilibrium.

Moles CO_2 formed	Metric		USCS	
	S_{tot}, kJ/°K	T_{final}, °K	S_{tot}, Btu/°R	T_{final}, °R
0	416.325	400	99.520	720
0.2	433.870	520	103.759	940
0.4	442.525	630	105.828	1140
0.6	447.218	740	106.815	1330
0.7	447.752	790	106.935	1420
0.8	446.796	834	106.711	1500
1.0	439.862	920	105.093	1660

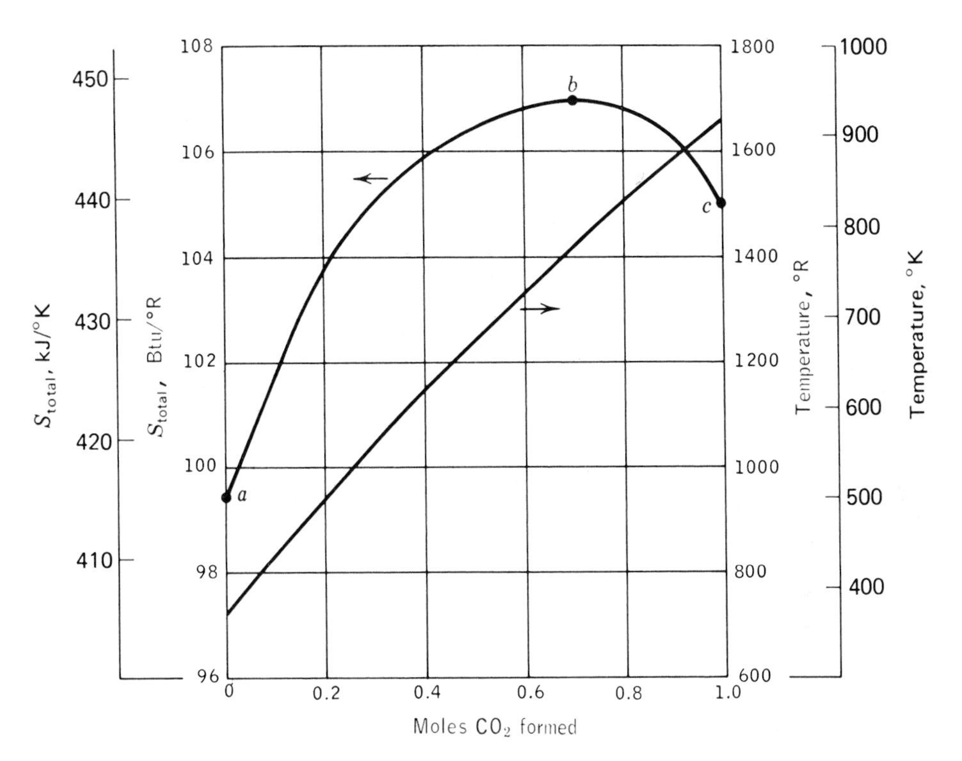

Figure 14-1 The total entropy and the combustion temperature for the adiabatic reaction of an equimolar mixture of CO and H_2O initially at 400°K (720°R) as a function of the extent of the reaction, for the chemical reaction $CO(g) + H_2O(g) \rightleftharpoons CO_2(g) + H_2(g)$.

The foregoing discussion illustrates the fact that a second-law analysis is frequently as important as a first-law analysis when dealing with reactive systems. In the absence of the second law it is possible to arrive at conclusions which are incorrect. In the next chapter we shall extend our analysis of reactive systems in the light of the second law. We shall find a more direct method of ascertaining the equilibrium state of a chemical reaction.

14-8 AVAILABILITY ANALYSIS OF REACTING SYSTEMS

Thermodynamics, through the use of the second law, provides a means of measuring the quality of energy as well as its quantity. Toward this goal the concepts of reversible work, availability, and irreversibility were introduced in Chap. 8. In this section we shall apply these concepts to chemically reacting systems. This is of practical importance. For example, consider a heat engine. The chemical energy released by a combustion reaction is used to heat a gas to a high temperature. This high-temperature gas then is a source of thermal energy for the heat engine. The performance of the heat engine is characterized by its thermal efficiency, the ratio of net work output to heat input. Rarely does the actual thermal efficiency exceed 40 percent. However, a more fundamental question exists. Regardless of the system used to produce the work, what is the maximum work output associated with a given chemical reaction for specified end states of the reaction?

If we restrict ourselves to steady-state, steady-flow processes, the optimum work associated with a chemical reaction is given by the change in the stream availability for the specified end states and the state of the surroundings or environment. In the absence of significant changes in kinetic and potential energy,

$$W_{\text{opt, sf}} = \Psi_f - \Psi_i = H_f - H_i - T_0(S_f - S_i)$$

where f and i represent the final and initial states. For a mixture of ideal gases this becomes

$$\begin{aligned} W_{\text{opt, sf}} &= \sum_{\text{prod}} N_i(h_i - T_0 s_i) - \sum_{\text{reac}} N_i(h_i - T_0 s_i) \\ &= \sum_{\text{prod}} N_i[(\Delta h_{f,\,298} + h_T - h_{298})_i - T_0(s^0_{i,\,T} - R \ln p_i)] \\ &\quad - \sum_{\text{reac}} N_i[(\Delta h_{f,\,298} + h_T - h_{298})_i - T_0(s^0_{i,\,T} - R \ln p_i)] \end{aligned} \tag{14-26}$$

Values at 298°K are replaced by values at 537°R when USCS units are used. Equation (14-26) is a general solution. One special case of interest is the situation where the reactants and products are in thermal equilibrium with the surroundings at T_0. For such a case the values of h_i, T_0, and s_i are all measured at the same temperature. Therefore

$$h_i - T_0 s_i = h_i - T s_i = g_i$$

As a result, each term on the right side of Eq. (14-26) can be replaced by the specific Gibbs function g_i. Hence, when the reactants start out at T_0 and the products are brought back to T_0,

$$W_{\text{opt, sf}} = \Delta\Psi = \sum_{\text{prod}} N_i g_i - \sum_{\text{reac}} N_i g_i \tag{14-27}$$

The Gibbs function g_i of a substance is found in the same manner as the enthalpy h_i, that is,

$$g_i = (\Delta g_{f,\,298} + g_T - g_{298})_i = (\Delta g_{f,\,537} + g_T - g_{537})_i \tag{14-28}$$

The Gibbs function of formation, Δg_f, at the standard reference state of 298°K (537°R) and 1 atm is listed for a number of substances in Tables A-22M and A-22. Note that the Gibbs function of formation of stable elements, like the enthalpy of formation, is equal to zero by convention. When Eq. (14-28) is substituted into Eq. (14-27) for the situation where both reactants and products are at T_0, then

$$W_{\text{opt, sf}} = \sum_{\text{prod}} N_i \, \Delta g_{f,i,298} - \sum_{\text{reac}} N_i \, \Delta g_{f,i,298} \tag{14-29}$$

In this special case only information of the Gibbs function of formation is required.

Finally, the irreversibility of a process is still defined, as in Chap. 8, by the two relations,

$$I = W_{\text{act}} - W_{\text{opt}} = T_0 \, \Delta S_{\text{tot}} \tag{14-30}$$

where ΔS is the total entropy change of all subsystems involved in the overall process. The following two examples illustrate the calculation procedures. Generally, the reactants and products of a combustion reaction are in a mixture state. However, in some cases the reactants enter and the products leave the system essentially in the pure state. These two possibilities are noted in the examples.

Example 14-18M Acetylene gas (C_2H_2) and 80 percent excess air enter a steady-flow combustion chamber at 25°C. Assume that the reaction is complete, and that the products leave at 25°C. As a further simplification for this example, we shall assume that each species in the reaction either enters or leaves the reaction chamber separately at 1 atm. In addition, for comparative purposes with later examples, the water in the products is assumed to be in a hypothetical vapor state. Determine (*a*) the heat released by the reaction, (*b*) the maximum work output, and (*c*) the irreversibility of the reaction, all in kJ/kg mol of fuel, if the gases are ideal gases.

Solution The equation for the gas phase chemical reaction is

$$C_2H_2 + 4.50\ O_2 + 16.92\ N_2 \longrightarrow 2\ CO_2 + H_2O + 2.00\ O_2 + 16.92\ N_2$$

(*a*) The heat released is found from Eq. (14-10). Employing data from Table A-22M, we find that

$$Q = 2(-393{,}520) + 1(-241{,}810) + 2(0) + 16.92(0) - 1(226{,}730) - 4.50(0) - 16.92(0)$$
$$= -1{,}255{,}580 \text{ kJ/kg} \cdot \text{mol } C_2H_2$$

(*b*) The maximum work is provided in this case either by Eq. (14-26) or Eq. (14-29). If we first use Eq. (14-29) and data from Table A-22M, then

$$W_{\text{opt, sf}} = 2(-394{,}360) + 1(-228{,}590) - 1(209{,}170)$$
$$= -1{,}226{,}480 \text{ kJ/kg} \cdot \text{mol } C_2H_2$$

This calculation indicates that the combustion at 25°C of acetylene gas with 80 percent excess air has the theoretical capacity of producing 1,226,480 kJ of work output per mole of fuel for the stated conditions.

An alternative method of evaluating the maximum or reversible work associated with the combustion process is to employ Eq. (14-26), namely,

$$W_{\text{opt, sf}} = \Delta H - T_0 \, \Delta S$$

The enthalpy change for the reaction has already been calculated, since this quantity is simply the heat transfer found in part *a*. Hence the only additional information required is the entropy change for the reaction. Using the absolute entropy values found in Table A-22M, we find that

$$\Delta S = 2(213.64) + 1(188.72) + 2.00(205.03) + 16.92(191.50)$$

$$- 1(200.85) - 4.50(205.03) - 16.92(191.50)$$

$$= -97.425 \text{ kJ/°K per mole } C_2H_2$$

The use of the ΔH and ΔS values leads to

$$W_{\text{opt, sf}} = -1{,}255{,}580 - 298.15(-97.425) = -1{,}226{,}530 \text{ kJ/kg} \cdot \text{mol } C_2H_2$$

The slight difference in the two answers is due to round-off error.

(*c*) The irreversibility is most easily determined by the difference between the actual work and the optimum or reversible work for the given end states. The actual process involves no work because the total energy released by the reaction appears as a heat loss. Therefore the irreversibility of the process is

$$I = W_{\text{act}} - W_{\text{opt}} = 0 - (-1{,}226{,}480) = 1{,}226{,}480 \text{ kJ/kg} \cdot \text{mol } C_2H_2$$

Thus the opportunity to produce a large amount of work is totally lost.

Example 14-18 Acetylene gas (C_2H_2) and 80 percent excess air enter a steady-flow combustion chamber at 77°F. Assume that the reaction is complete, and that the products leave at 77°F. As a further simplification for this example, we shall assume that each species in the reaction either enters or leaves the reaction chamber separately at 1 atm. In addition, for comparative purposes with later examples, the water in the products is assumed to be in a hypothetical vapor state. Determine (*a*) the heat released by the reaction, (*b*) the maximum work output, and (*c*) the irreversibility of the reaction, all in Btu/lb · mol of fuel, if the gases are ideal gases.

SOLUTION The equation for the gas phase chemical reaction is

$$C_2H_2 + 4.50\ O_2 + 16.92\ N_2 \rightarrow 2\ CO_2 + H_2O + 2.00\ O_2 + 16.92\ N_2$$

(*a*) The heat released is found from Eq. (14-10). Employing data from Table A-22, we find that

$$Q = 2(-169{,}290) + 1(-104{,}040) + 2(0) + 16.92(0) - 1(97{,}540) - 4.50(0) - 16.92(0)$$

$$= -540{,}160 \text{ Btu/lb} \cdot \text{mol } C_2H_2$$

(*b*) The maximum work is provided in this case either by Eq. (14-26) or Eq. (14-29). If we first use Eq. (14-29) and data from Table A-22, then

$$W_{\text{opt, sf}} = 2(-169{,}680) + 1(-98{,}350) - 1(87{,}990)$$

$$= -525{,}700 \text{ Btu/lb} \cdot \text{mol } C_2H_2$$

This calculation indicates that the combustion at 77°F of acetylene gas with 80 percent excess air has the theoretical capacity of producing 540,160 Btu of work output per mole of fuel for the stated initial and final states.

An alternative method of evaluating the maximum or reversible work associated with the combustion process is to employ Eq. (14-26), namely,

$$W_{\text{opt, sf}} = \Delta H - T_0\ \Delta S$$

The enthalpy change for the reaction has already been calculated, since this quantity is simply the heat transfer found in part *a*. Hence the only additional information required is the entropy change for the reaction. Using the absolute entropy values found in Table A-22, we find that

$$\Delta S = 2(51.07) + 1(45.11) + 2.00(49.00) + 16.92(45.77)$$
$$- 1(48.00) - 4.50(49.00) - 16.92(45.77)$$
$$= -23.25 \text{ Btu/°R per mole } C_2H_2$$

The use of the ΔH and ΔS values leads to

$$W_{\text{opt, sf}} = -540{,}160 - 536.67(-23.25) = -527{,}680 \text{ Btu/lb} \cdot \text{mol } C_2H_2$$

The slight difference in the two answers is due to round-off error.

(*c*) The irreversibility is most easily determined by the difference between the actual work and the optimum or reversible work for the given end states. The actual process involves no work, because the total energy released by the reaction appears as a heat loss. Therefore the irreversibility of the process is

$$I = W_{\text{act}} - W_{\text{opt}} = 0 - (-525{,}700) = 525{,}700 \text{ Btu/lb} \cdot \text{mol } C_2H_2$$

Thus the opportunity to produce a large amount of work is totally lost.

Example 14-19M Reconsider Example 14-18M in the following manner. Rather than having each reactant enter and each product leave separately at 1 atm, the reactants now enter as a mixture at 1 atm and the products leave as a mixture at 1 atm. Determine parts *a* and *b* of the example.

SOLUTION (*a*) The heat released is still 1,255,580 kJ/kg · mol of C_2H_2, because the enthalpies of ideal gases are not a function of pressure.

(*b*) The value of the maximum work is altered from the original value, because the entropies of ideal gases are a function of the component or partial pressures. The correction for each gas is given by $N_i(-R \ln p_i)$, and this correction applies whether Eq. (14-26) or Eq. (14-29) is used. On the basis of the overall chemical reaction

$$C_2H_2 + 4.50\ O_2 + 16.92\ N_2 \rightarrow 2\ CO_2 + H_2O + 2.00\ O_2 + 16.92\ N_2$$

The total correction to the ΔS term is evaluated in the following manner.

$$\sum_i N_i \ln p_i = 2 \ln \frac{2}{21.92} + \ln \frac{1}{21.92} + 2.00 \ln \frac{2}{21.92}$$
$$+ 16.92 \ln \frac{16.92}{21.92} - \ln \frac{1}{22.42}$$
$$- 4.50 \ln \frac{4.50}{22.42} - 16.92 \ln \frac{16.92}{22.42}$$
$$= -1.946$$

$$\Delta S_{\text{corr}} = -R_u \sum_i N_i \ln p_i = -8.314(-1.946) = 16.18 \text{ kJ/°K}$$

When this result is used to correct the value of $-1{,}226{,}480$ kJ/°K found in Example 14-18M, then

$$W_{\text{opt, sf}} = -1{,}226{,}480 - 298.15(16.18) = -1{,}231{,}300 \text{ kJ/°K}$$

In this case the maximum work output is little affected by whether the reactants and products are separate or in a mixture at 1 atm.

Example 14-19 Reconsider Example 14-18 in the following manner. Rather than having each reactant enter and each product leave separately at 1 atm, the reactants now enter as a mixture at 1 atm and the products leave as a mixture at 1 atm. Determine parts *a* and *b* of the example.

SOLUTION (*a*) The heat released is still 540,160 Btu/lb · mol of C_2H_2, because the enthalpies of ideal gases are not a function of pressure.

(*b*) The value of the maximum work is altered from the original value, because the entropies of ideal gases are a function of the component or partial pressures. The correction for each gas is given by $N_i(-R \ln p_i)$, and this correction applies whether Eq. (14-26) or Eq. (14-29) is used. On the basis of the overall chemical reaction

$$C_2H_2 + 4.50\ O_2 + 16.92\ N_2 \rightarrow 2\ CO_2 + H_2O + 2.00\ O_2 + 16.92\ N_2$$

The total correction to the ΔS term is evaluated in the following manner.

$$\begin{aligned}\sum_i N_i \ln p_i &= 2 \ln \frac{2}{21.92} + \ln \frac{1}{21.92} + 2.00 \ln \frac{2}{21.92} \\ &\quad + 16.92 \ln \frac{16.92}{21.92} - \ln \frac{1}{22.42} \\ &\quad - 4.50 \ln \frac{4.50}{22.42} - 16.92 \ln \frac{16.92}{22.42} \\ &= -1.946\end{aligned}$$

$$\Delta S_{\text{corr}} = -R_u \sum_i N_i \ln p_i = -1.986(-1.946) = 3.865\ \text{Btu/°R}$$

When this result is used to correct the value of $-525{,}700$ Btu/°R found in Example 14-18, then

$$W_{\text{opt, sf}} = -525{,}700 - 536.67(3.865) = -527{,}770\ \text{Btu/°R}$$

In this case the maximum work output is little affected by whether the reactants and products are separate or in a mixture at 1 atm.

The preceding examples illustrate that the availability change associated with combustion reactions is quite large. The engineer must develop methods of utilizing this vast work potential. A major and conventional way is to carry out the combustion process adiabatically. The hot product gases are then used as the heat source for some type of heat engine. By comparing the work output of the engine to the optimum work output of the chemical reaction, we have some measure of the effectiveness of the process. There is, however, an even more basic question. Combustion processes by nature are highly irreversible. We would expect that the availability of the product gases, even at the high adiabatic-combustion temperature, have a lower availability than the initial reactants. What is the typical loss in availability, then, of a combustion process, even before the energy is used in a heat engine? The following example attempts to provide an answer.

Example 14-20M Reconsider Example 14-18M for the combustion of acetylene gas (C_2H_2) with 80 percent excess air, both initially at 25°C. The steady-flow process is now carried out adiabatically. Determine (*a*) the adiabatic-combustion temperature, (*b*) the optimum work, in kJ/kg · mol of fuel, and (*c*) the irreversibility of the process.

Solution (*a*) To find the adiabatic-combustion temperature, one equates the enthalpy of the reactants to that of the products. Since we have already determined the heat release at 25°C in Example 14-18M, we can also find the final temperature by equating Q to the sensible enthalpy change of the product gases. That is

$$Q = \sum_{\text{prod}} N_i \, \Delta h_i = \sum_{\text{prod}} N_i(h_T - h_{298})_i$$

$$1{,}255{,}580 = 2(h_{CO_2} - 9364) + 1(h_{H_2O} - 9904) + 2.00(h_{O_2} - 8682) + 16.92(h_{N_2} - 8669)$$

Upon rearrangement,

$$2h_{CO_2} + h_{H_2O} + 2.0h_{O_2} + 16.92h_{N_2} = 1{,}448{,}260 \text{ kJ}$$

The correct temperature is found by iteration. At 1920°K as a trial,

$$\sum_i N_i h_i = 2(95{,}995) + 78{,}527 + 2(64{,}868) + 16.92(61{,}936) = 1{,}448{,}210 \text{ kJ}$$

Thus the adiabatic-combustion temperature is essentially 1920°K.

(*b*) The optimum work is the change in availability for the adiabatic process. Rather than calculate this directly, a more convenient method may be chosen. The optimum work is a state function, because it depends solely on the state of the system-environment composite. Therefore, the optimum work associated with the adiabatic combustion process, plus that associated with the cooling of these gases back to 25°C, must be the same as the optimum work associated with the isothermal conversion of reactants to products at 25°C. This latter value was found to be −1,226,480 kJ/kg · mol in Example 14-18M. The optimum work associated with the cooling of the gases from the flame temperature to 25°C is simply

$$\begin{aligned} W_{\text{opt}} &= \Psi_0 - \Psi_T = \sum_{\text{prod}} N_i[(h_0 - h_T) - T_0(s_0 - s_T)] \\ &= 2(9364 - 95{,}995) + 1(9904 - 78{,}527) + 2(8682 - 64{,}868) \\ &\quad + 16.92(8669 - 61{,}936) - 298.15[2(213.685 - 306.751) \\ &\quad + 1(188.720 - 262.497) + 2(205.033 - 267.115) \\ &\quad + 16.92(191.502 - 250.502)] \\ &= -1{,}255{,}580 + 412{,}150 = -843{,}380 \text{ kJ/kg} \cdot \text{mol } C_2H_2 \end{aligned}$$

Hence the optimum work for the adiabatic-combustion process is

$$\begin{aligned} W_{\text{opt}} &= W_{\text{opt}}(\text{isothermal, 25°C}) - W_{\text{opt}}(\text{cooling}) \\ &= -1{,}226{,}480 - (-843{,}380) = -383{,}100 \text{ kJ/kg} \cdot \text{mol } C_2H_2 \end{aligned}$$

It is important to note that this value applies only to the situation where the products are separate and each is at 1 atm at the adiabatic-combustion temperature. In practice the product gases are all in a mixture at 1 atm, and this should be taken into account. Example 14-19M has shown, however, that the contributions of the (R ln p_i) terms are fairly small. Hence the value we have obtained for the optimum work is reasonably close to that which we would find for a mixture of product gases.

(*c*) Similar to Example 14-18M, the irreversibility is simply the optimum work calculated, because the actual work output is zero. That is,

$$I = W_{\text{act}} - W_{\text{opt}} = 0 - (-383{,}100) = 383{,}100 \text{ kJ}$$

An alternative method is to evaluate $T_0\,\Delta S$ for the adiabatic process.

$$\Delta S = \sum_{\text{prod}} N_i s_i - \sum_{\text{reac}} N_i s_i$$

$$= 2(306.751) + 1(262.497) + 2(267.115) + 16.92(250.502)$$

$$- 1(200.85) - 4.50(205.03) - 16.92(191.50)$$

$$= 1285.06 \text{ kJ/°K}$$

As a result, the irreversibility is given by

$$I = T_0\,\Delta S = 298.15(1285.06) = 383{,}140 \text{ kJ/kg} \cdot \text{mol } C_2H_2$$

The answers by the two methods are in good agreement.

Example 14-20 Reconsider Example 14-18 for the combustion of acetylene gas (C_2H_2) with 80 percent excess air, both initially at 77°F. The steady-flow process is now carried out adiabatically. Determine (*a*) the adiabatic-combustion temperature, (*b*) the optimum work, in Btu/lb · mol of fuel, and (*c*) the irreversibility of the process.

SOLUTION (*a*) To find the adiabatic-combustion temperature, one equates the enthalpy of the reactants to that of the products. Since we have already determined the heat release at 77°F in Example 14-18, we can also find the final temperature by equating Q to the sensible enthalpy change of the product gases. That is,

$$Q = \sum_{\text{prod}} N_i\,\Delta h_i = \sum_{\text{prod}} N_i (h_T - h_{537})_i$$

$$540{,}760 = 2(h_{CO_2} - 4028) + 1(h_{H_2O} - 4258) + 2.00(h_{O_2} - 3725)$$

$$+ 16.92(h_{N_2} - 3729)$$

Upon rearrangement,

$$2h_{CO_2} + h_{H_2O} + 2.0h_{O_2} + 16.92h_{N_2} = 623{,}020 \text{ Btu}$$

The correct temperature is found by iteration. At 3460°R as a trial,

$$\sum_i N_i h_i = 2(41{,}388) + 33{,}839 + 2(27{,}914) + 16.92(26{,}673)$$

$$= 623{,}750 \text{ Btu}$$

Thus the adiabatic-combustion temperature is essentially 3460°R.

(*b*) The optimum work is the change in availability for the adiabatic process. Rather than calculate this directly, a more convenient method may be chosen. The optimum work is a state function, because it depends solely on the state of the system-environment composite. Therefore, the optimum work associated with the adiabatic combustion process, plus that associated with the cooling of these gases back to 77°F, must be the same as the optimum work associated with the isothermal conversion of reactants to products at 77°F. This latter value was found to be −525,700 Btu/lb · mol in Example 14-18. The optimum work associated with the cooling of the gases from the flame temperature to 77°F is simply

$$W_{\text{opt}} = \Psi_0 - \Psi_T = \sum_{\text{prod}} N_i[(h_0 - h_T) - T_0(s_0 - s_T)]$$

$$= 2(4028 - 41{,}388) + 1(4258 - 33{,}839) + 2(3725 - 27{,}914)$$

$$+ 16.92(3730 - 26{,}673) - 536.67[2(51.032 - 73.297)$$

$$+ 1(45.079 - 62.738) + 2(48.982 - 63.811)$$

$$+ 16.92(45.743 - 59.846)]$$

$$= -540{,}870 + 177{,}350 = -363{,}520 \text{ Btu/lb} \cdot \text{mol } C_2H_2$$

Hence the optimum work for the adiabatic-combustion process is

$$W_{opt} = W_{opt}(\text{isothermal, } 25°\text{C}) - W_{opt}(\text{cooling})$$

$$= -525{,}700 - (-363{,}520) = -162{,}180 \text{ Btu/lb} \cdot \text{mol } C_2H_2$$

It is important to note that this value applies only to the situation where the products are separate and each is at 1 atm at the adiabatic-combustion temperature. In practice the product gases are all in a mixture at 1 atm, and this should be taken into account. Example 14-19 has shown, however, that the contributions of the (R ln p_i) terms are fairly small. Hence the value we have obtained for the optimum work is reasonably close to that which we would find for a mixture of product gases.

(*c*) Similar to Example 14-18, the irreversibility is simply the optimum work calculated, because the actual work output is zero. That is,

$$I = W_{act} - W_{opt} = 0 - (-162{,}180) = 162{,}180 \text{ Btu}$$

An alternative method is to evaluate $T_0\ \Delta S$ for the adiabatic process.

$$\Delta S = \sum_{prod} N_i s_i - \sum_{reac} N_i s_i$$

$$= 2(73.297) + 1(62.738) + 2(63{,}811) + 16.92(59.846)$$

$$- 1(48.00) - 4.50(48.982) - 16.92(45.743)$$

$$= 307.16 \text{ Btu/°R}$$

As a result the irreversibility is given by

$$I = T_0\ \Delta S = 536.67(307.16) = 164{,}800 \text{ Btu/lb} \cdot \text{mol } C_2H_2$$

The answers by the two methods are in reasonably good agreement.

Examples 14-20M and 14-20 illustrate a significant point with respect to adiabatic-combustion processes. The primary objective of creating a high-temperature gas (1920°K or 3460°R in this case) for use as a thermal energy source for a heat engine is achieved. However, the theoretical work potential of the hot gas is only 69 percent of the work potential of the reactants, for the specified initial and final states. That is, 31 percent of the work potential of the reactants is already lost, due to the irreversibility of the adiabatic-combustion process. If the heat engine process were totally reversible, only 69 percent of the work potential of the reactants would be delivered. Since actual heat engines rarely reach 40 percent actual thermal efficiency, the actual work output will be less than 28 percent ($0.69 \times 0.40 \times 100$) of the theoretical capability of the reactants in this given case. When the air supplied is closer to the stoichiometric requirements, the work potential of the hot gas will be somewhat higher than 69 percent. If the excess air used with a hydrocarbon fuel is several hundred percent, the work potential of the product gases could approach only 50 percent of the original reactants. The message is clear. Irreversibilities associated with the adiabatic combustion of hydrocarbon fuels lead to a significant reduction in the work potential, even though the quantity of energy is still the same. Irreversibilities associated with the heat engine process will further reduce this process.

To make full use of the work potential of a chemical reaction, the irreversibilities which degrade the process must be reduced. As an ultimate goal, the chemical reaction should be reversible. The galvanic or reversible chemical cell discussed briefly in Sec. 2-9 is a device for converting chemical energy into electrical energy by means of a controlled chemical reaction. Rather than the unconstrained electron exchange between reacting species which occurs in a combustion reaction, in the electrolytic cell we have a controlled electron exchange. When the electric potential between anode and cathode approaches a small value, some reactions take place essentially in a reversible manner. Theoretically, the electrical work output should approach that predicted by the change in availability of the chemical reaction.

In practice this direct conversion of chemical to electric energy occurs in a device known as a fuel cell. The first practical fuel cells used hydrogen and oxygen as the fuel and oxidizer. Such fuel cells have provided electric power on manned space missions. Research and development currently is aimed at the use of gaseous hydrocarbons and air, and at an improvement in the conversion efficiency. A more detailed discussion of fuel cells appears in Sec. 19-1.

PROBLEMS (METRIC)

Fuel-air combustion analysis

14-1M Ethane (C_2H_6) is burned with dry air in a mass ratio of 1 : 17. Compute (*a*) the percent excess air used, (*b*) the equivalence ratio, (*c*) the percent CO_2 by volume in the product gases, (*d*) the dew-point temperature, in degrees Celsius, and (*e*) the percent of the water vapor condensed if the product gases are cooled to 20°C, assuming complete combustion and a total pressure of 1.0 bar.

14-2M Same as Prob. 14-1M, except that the fuel is ethylene (C_2H_4).

14-3M Same as Prob. 14-1M, except that the fuel is acetylene (C_2H_2).

14-4M Same as Prob. 14-1M, except that the fuel is propylene (C_3H_6).

14-5M One mole of ethane (C_2H_6) is burned with 20 percent excess dry air. Determine (*a*) the air-fuel ratio used, (*b*) the equivalence ratio, (*c*) the mole percent of N_2 in the product gas, (*d*) the dew-point temperature in degrees Celsius, and (*e*) the percent of the water vapor condensed if the product gases are cooled to 25°C, assuming complete combustion and a total pressure of 1.03 bars (103 kPa).

14-6M Same as Prob. 14-5M, except that the fuel is ethylene (C_2H_4).

14-7M Same as Prob. 14-5M, except that the fuel is acetylene (C_2H_2).

14-8M Same as Prob. 14-5M, except that the fuel is propylene (C_3H_6).

14-9M Determine the volume flow rate in m^3/min of air at 1 bar and 27°C, required to burn 1 kg/min of fuel for the fuel-air mixtures specified in (*a*) Prob. 14-1M, (*b*) Prob. 14-2M, (*c*) Prob. 14-3M, (*d*) Prob. 14-4M, (*e*) Prob. 14-5M, (*f*) Prob. 14-6M, (*g*) Prob. 14-7M, and (*h*) Prob. 14-8M.

14-10M If the actual air supplied has a humidity ratio (specific humidity) of 15 g H_2O/kg dry air, determine the dew-point temperature, in degrees Celsius, for the products of the reaction specified in (*a*) Prob. 14-1M, (*b*) Prob. 14-2M, (*c*) Prob. 14-3M, (*d*) Prob. 14-4M, (*e*) Prob. 14-5M, (*f*) Prob. 14-6M, (*g*) Prob. 14-7M, and (*h*) Prob. 14-8M.

14-11M A fuel gas with a volumetric composition of: CH_4, 60 percent; C_2H_6, 30 percent; N_2,

10 percent; is burned to completion using 20 percent excess of dry air. Determine (*a*) the mole percent of the products on a dry basis, (*b*) the air-fuel ratio used, (*c*) the dew-point temperature, in degrees Celsius, if the total pressure is 1.0 bar, and (*d*) the mole percent of CO_2 in the product gas on a wet basis if the gas is cooled to 20°C.

14-12M A gaseous fuel having a volumetric analysis of 65 percent CH_4, 25 percent C_2H_6, 5 percent CO_2, and 5 percent N_2 is burned with 30 percent excess of dry air. Determine (*a*) the air-fuel ratio used, (*b*) the mole percent CO_2 in the total products, (*c*) the dew-point temperature if the pressure is 1.0 bar, and (*d*) the mole percent N_2 in the total gaseous products if the gas is cooled to 18°C.

14-13M A gaseous fuel having a volumetric analysis of 70 percent CH_4, 20 percent C_2H_6, and 10 percent C_3H_8 is burned with 20 percent excess of dry air. Determine (*a*) the air-fuel ratio used, (*b*) the mole percent N_2 in the total gaseous products, (*c*) the dew-point temperature if the pressure is 1.05 bars (105 kPa), and (*d*) the mole percent N_2 in the gaseous products if the gas is cooled to 20°C.

14-14M A gaseous fuel which is 60 mole percent CH_4 and 40 mole percent C_2H_4 is burned with 10 percent excess dry air. Determine (*a*) the air-fuel ratio used, (*b*) the mole percent of CO_2 in the total products, (*c*) the dew-point temperature if the pressure is 1.05 bars (105 kPa), and (*d*) the mole percent CO_2 in the gaseous products if the mixture is cooled to 25°C.

14-15M Ethylene gas (C_2H_4) is burned with 10 percent excess air. Due to incomplete mixing, only 96 percent of the carbon in the fuel is converted to CO_2, the rest appearing as CO. Determine (*a*) the air-fuel ratio used, (*b*) the mole fraction of CO_2 in the total products, and (*c*) the dew point, in degrees Celsius, if the pressure is 104 kPa (1.04 bars).

14-16M Liquid benzene (C_6H_6) is burned with (*a*) 120 percent and (*b*) 140 percent theoretical-air requirements. Compute (1) the air-fuel ratio used, (2) the mole fraction of N_2 in the total products, and (3) the dew point, in degrees Celsius, if the pressure is 1.10 bars (110 kPa).

Dew-point temperature evaluation

14-17M Pentane gas (C_5H_{12}) is burned to completion with the stoichiometric amount of air at a pressure of 1.03 bars.

(*a*) If the air supplied is dry, find (1) the dew point, in degrees Celsius, and (2) the number of moles of water per mole of fuel burned which would condense if the product gases were cooled to 22°C.

(*b*) If the air supplied has a specific humidity of 14 g H_2O/kg dry air, find the answers to the same questions asked in part *a*.

14-18M Methane gas (CH_4) is burned with 50 percent excess air at 0.95 bars.

(*a*) If the air supplied is dry, find (1) the dew point, in degrees Celsius, and (2) the number of moles of water per mole of fuel which would condense if the product gases were cooled to 18°C.

(*b*) If the air supplied has a specific humidity of 16 g H_2O/kg dry air, find the answers to the same questions asked in part *a*.

14-19M Ethyl alcohol (C_2H_5OH) is burned with (*a*) 25 percent excess air, and (*b*) 50 percent excess air at 1.0 bar (100 kPa). Determine the dew point of the product gases, in degrees Celsius, if the air supplied has a specific humidity (humidity ratio) of 15 g H_2O/kg dry air.

Analysis of product gases

14-20M Octane (C_8H_{18}) is burned with dry air, and a volumetric analysis on a dry basis of the products reveals: 10.39 percent CO_2, 4.45 percent O_2, and 1.48 percent CO. Compute (*a*) the actual air-fuel ratio used, and (*b*) the percent theoretical air used.

14-21M The combustion of methane gas (CH_4) with dry air leads to the following volumetric analysis during a test: 9.7 percent CO_2, 0.5 percent CO, and 3.0 percent O_2. Determine (*a*) the moles of air used per mole of fuel, (*b*) the percent theoretical air used, and (*c*) the air-fuel ratio.

14-22M Solid carbon is burned with dry air in a combustion test. A volumetric analysis reveals that the products of combustion include 3.5 percent CO, 13.8 percent CO_2, and 5.2 percent O_2 on a dry basis. Determine (*a*) the air-fuel ratio used, and (*b*) the percent theoretical air used.

14-23M A fuel gas with a volumetric analysis of 60 percent CO, 20 percent H_2, and 20 percent N_2 is burned with dry air. A volumetric analysis of the product gases gives 20.0 percent CO_2, 5.0 percent CO, and 2.8 percent O_2 on a dry basis. Calculate (*a*) the percent theoretical air used, and (*b*) the air-fuel ratio used.

14-24M A gaseous fuel is composed of 20 percent CH_4, 40 percent C_2H_6, and 40 percent C_2H_8, where all percentages are by volume. The volumetric analysis of the dry combustion gases gives 11.4 percent CO_2, 1.7 percent O_2, and 1.2 percent CO. Determine (*a*) the theoretically correct air-fuel ratio required, and (*b*) the percent excess air used.

14-25M A gas has the following volumetric analysis by percent: CH_4, 80.62; C_2H_6, 5.41; C_3H_8, 1.87; C_4H_{10}, 1.60; N_2, 10.50. A volumetric analysis of the products of combustion shows 7.8 percent CO_2, 7.0 percent O_2, 0.2 percent CO. Calculate the actual fuel-air ratio.

14-26M Ethylene (C_2H_4) is burned with 33 percent excess air. An analysis of the products of combustion on a dry basis reveals 6.06 percent O_2 by volume. The remaining data for the product analysis are missing. What percent of the carbon in the fuel was converted to CO instead of CO_2?

14-27M Propylene (C_3H_6) is burned with 20 percent excess air. An analysis of the products of combustion on a dry basis reveals 4.31 percent O_2 by volume. What percent of the carbon in the fuel was converted to CO instead of CO_2?

14-28M The volumetric analysis of the dry products of combustion of a hydrocarbon fuel described by the general formula C_xH_y is: CO_2, 13.6 percent; O_2, 0.4 percent; CO, 0.8 percent; CH_4, 0.4 percent; and N_2, 84.8 percent. There are 13.6 moles of CO_2 formed per mole of fuel. Determine (*a*) the values of *x* and *y*, and (*b*) the air-fuel ratio used.

14-29M The volumetric analysis of the dry products of combustion of a hydrocarbon described by the general formula C_xH_y is: CO_2, 12.37 percent; CO, 0.87 percent; O_2, 2.47 percent; and N_2, 84.29 percent. There are 1.42 moles of O_2 in the products of combustion per mole of fuel. Determine (*a*) the values of *x* and *y*, and (*b*) the air-fuel ratio used.

Energy analysis with heat effects

14-30M Liquid octane ($n - C_8H_{18}$) at 25°C and air at 500°K enter a combustion chamber in steady flow at 1-bar pressure. Determine the amount of heat transfer that occurs, in kJ/kg · mol of fuel, if the amount of air used and the temperature of the product gases are (*a*) 150 percent theoretical and 1000°K, (*b*) 200 percent theoretical and 800°K, and (*c*) 400 percent theoretical and 1000°K.

14-31M Ethane gas (C_2H_6) and 25°C and air at 227°C enter a combustion chamber in steady flow at 1 bar. The products of combustion leave at 1100°K. If the percent excess air used is (*a*) 25 percent, and (*b*) 15 percent, and combustion is complete, determine the heat loss, in kJ/kg · mol of fuel.

14-32M Propane (C_3H_8) is burned in a steady-flow system from an original temperature of 25°C. Determine the heat transfer required, in kJ/kg · mol of fuel, if the amount of air used and the temperature of the product gases are (*a*) 100 percent excess and 600°K, and (*b*) 50 percent excess and 800°K.

14-33M Carbon monoxide is burned with air at an initial state of 25°C. The final temperature of the products of combustion is (*a*) 1100°K, and (*b*) 1000°K, and during the process, 38,000 kJ of heat per kilogram-mole of CO are removed. Determine the percent excess air used if combustion is complete.

14-34M Hydrogen gas is burned with air with each at an initial temperature of 400°K. The final temperature of the products is (*a*) 800°K, and (*b*) 1000°K. During the process, 80,000 kJ of heat are removed per kilogram-mole of H_2. Determine the percent excess air used if combustion is complete.

14-35M Hydrogen gas is burned with (*a*) stoichiometric air, and (*b*) 100 percent excess air at 25°C and 1 bar. During the process, a heat loss of 35,000 kJ/kg · mol of fuel occurs. Find the final gas temperature, in degrees kelvin.

14-36M Consider the combustion reaction discussed in Prob. 14-15M. The fuel enters a steady-state combustion chamber at 25°C, while the air enters at 400°K. If the final temperature of the products of combustion is 1500°K, determine the magnitude and direction of any heat transfer, in kJ/kg · mol of fuel.

14-37M Propane gas (C_3H_8) with 20 percent excess air enters a combustion chamber in steady flow at 1 bar and 25°C. The carbon in the fuel becomes either CO or CO_2. Determine the heat transfer, in kJ/kg · mol of fuel, if the percent carbon converted to CO_2 and the temperature of the product gases are (*a*) 94 percent and 900°K, and (*b*) 96 percent and 700°K.

Adiabatic-combustion processes

14-38M If ethylene (C_2H_4) at 25°C is burned with 200 percent excess air supplied at 400°K and the reaction occurs at constant pressure in the gas phase, estimate the maximum combustion temperature, in degrees kelvin.

14-39M Hydrogen gas at 25°C is reacted with (*a*) 350 percent, (*b*) 400 percent, and (*c*) 500 percent theoretical oxygen requirements which enter at 500°K. Estimate the maximum combustion temperature, in degrees kelvin.

14-40M Propane gas (C_3H_8) is burned at constant pressure with (*a*) 20 percent, and (*b*) 40 percent excess air starting at 25°C. Determine the maximum combustion temperature, in degrees kelvin.

14-41M Determine the maximum theoretical combustion temperature for the reaction of ethane with (*a*) 30 percent, and (*b*) 50 percent excess air in a steady-flow process. The reactants enter at 25°C and the reaction goes to completion.

14-42M Determine the maximum temperature under adiabatic conditions when methane (CH_4) is burned with (*a*) stoichiometric air, and (*b*) 20 percent excess air, both entering at 25°C. Assume that 10 percent of the carbon is burned only to CO.

14-43M Carbon monoxide undergoes adiabatic combustion with the stoichiometric amount of air. Neglecting dissociation of products, determine the maximum combustion temperature, in °K, if the initial reactants are (*a*) at 25°C, and (*b*) 1000°K.

14-44M A gas turbine combustion chamber is supplied with air at 400°K and liquid octane at 25°C. The products of combustion leave at (*a*) 1400°K, and (*b*) 1600°K. Calculate the air-fuel ratio used if the flow is steady, the combustion is complete, and the heat loss is negligible.

14-45M Determine the adiabatic-combustion temperature for the reaction of liquid methanol (CH_3OH) with the theoretical air requirements at 1 bar. The methanol enters at 25°C and the air enters at (*a*) 25°C, and (*b*) 400°K. Assume complete combustion and steady flow.

Constant-volume combustion

14-46M Determine the adiabatic-combustion temperature, in degrees kelvin, and the explosion pressure, in bars, for the constant-volume combustion of CO with 50 percent excess air. Initial conditions are 1 bar and 27°C.

14-47M An equimolar mixture of hydrogen and carbon monoxide, together with the theoretical amount of air for complete combustion, is ignited in a constant-volume bomb. The initial conditions are 3 bars and 25°C. Estimate the maximum temperature and pressure that would be attained, assuming complete combustion.

14-48M Two cubic centimeters of liquid benzene (C_6H_6) are placed in a 28.3-L constant-volume bomb at 25°C. If the fuel is burned with the stoichiometric amount of air initially at 1 bar, determine the explosion pressure, in bars.

14-49M Same as Prob. 14-48M, except the fuel is liquid methyl alcohol (CH_3OH).

Use of heating values

14-50M Consider a Pennsylvania coal with the ultimate analysis given in Table 14-1.

(*a*) Determine the moles of CO_2, N_2, and O_2 present in the products of combustion per kilogram of fuel when 20 percent excess air is used, and the air-fuel ratio used.

(*b*) Neglecting the effect of SO_2, determine the heat transfer, in kJ/kg fuel, if the reactants enter a steady-flow combustor at 25°C and the product gases leave at 460°K.

14-51M Consider the Illinois coal with the ultimate analysis given in Table 14-1. Answer the same questions presented in Prob. 14-50M.

14-52M Consider the Wyoming coal with the ultimate analysis given in Table 14-1. Answer the same questions presented in Prob. 14-50M.

14-53M Consider the South Dakota coal with the ultimate analysis given in Table 14-1. Answer the same questions presented in Prob. 14-50M.

Entropy changes for reacting mixtures

14-54M Methane gas (CH_4) is burned with the stoichiometric amount of dry air, both initially at 25°C, in a steady-flow process. The fuel and the air each enter separately at 1 atm, combustion is complete, and the environmental temperature is 25°C.

(*a*) Determine the entropy change for the reaction and the total entropy change for the overall process, in kJ/°K per kilogram-mole of fuel, if water appears as a liquid in the product mixture which is at 25°C and 1 atm.

(*b*) Determine the entropy change for the reaction and the total entropy change for the overall process, in kJ/°K per kilogram-mole of fuel, if water appears as a vapor in the product mixture at 1 atm and 25°C.

14-55M Same as Prob. 14-54M, except that the fuel is gaseous ethane (C_2H_6).

14-56M Same as Prob. 14-54M, except that the fuel is gaseous propane (C_3H_8).

14-57M Same as Prob. 14-54M, except that the fuel is liquid methyl alcohol (CH_3OH).

14-58M Consider the process described in Prob. 14-40M, where (*a*) is for 20 percent excess air, and (*b*) is for 40 percent excess air. Evaluate the entropy change for the reaction, in kJ/°K per kilogram-mole of fuel, if the fuel and air initially are separate and each is at 1 atm.

14-59M Consider the process described in Prob. 14-41M, where (*a*) is for 30 percent excess air, and (*b*) is for 50 percent excess air. Evaluate the entropy change for the reaction, in kJ/°K per kilogram-mole of fuel, if the fuel and air initially are separate and each is at 1 atm.

14-60M Consider the process described in Prob. 14-43M. Evaluate the entropy change for the reaction, in kJ/°K per kilogram-mole of fuel, if the fuel and air initially are separate and each is at 1 atm.

14-61M Consider the process described in Prob. 14-45M, part *a*. Evaluate the entropy change for the reaction, in kJ/°K per kilogram-mole of fuel, if the fuel and air initially are separate and each is at 1 atm.

Availability of reacting systems

14-62M Propane gas (C_3H_8) is burned in steady flow at 1 atm with (*a*) 20 percent, and (*b*) 40 percent excess air starting at 25°C. Assume the reaction is complete and that the products leave at the environmental temperature of 25°C. The fuel and the oxidant each enter separately at 1 atm, the products leave as a mixture at 1 atm, and the water is in the liquid state in the products. Determine (1) the heat released by the reaction, (2) the optimum work for the process, and (3) the irreversibility of the reaction, all in kJ/kg · mol of fuel.

14-63M Reconsider Prob. 14-62M. The steady-flow process is now carried out adiabatically, as described in Prob. 14-40M. For (*a*) 20 percent, and (*b*) 40 percent excess air at 25°C initially, determine (1) the availability of the hot product gases, (2) the optimum work for the adiabatic process, (3) the value of $T_0 \, \Delta S$ for the process, all in kJ/kg · mol, and (4) the ratio of the availability of the hot gases to that of the original reactants.

14-64M Ethane gas (C_2H_6) reacts with (*a*) 30 percent, and (*b*) 50 percent excess air in a steady-flow process starting a 25°C. The reaction is complete, and the products leave as a mixture at 1 atm at the environmental temperature of 25°C. The fuel and air enter separately at 1 atm, and water in the products is in the liquid state. Determine (*a*) the heat released by the reaction, (*b*) the optimum work, and (*c*) the irreversibility of the process, all in kJ/kg · mol of fuel.

14-65M Reconsider Prob. 14-64M. The steady-flow process is now carried out adiabatically, as described in Prob. 14-41M. The fuel and air still enter separately at 1 atm and 25°C, and the products are a mixture at 1 atm. For (*a*) 30 percent, and (*b*) 50 percent excess air, determine (1) the availability of the hot product gases, (2) the optimum work for the adiabatic process, (3) the value of $T_0 \Delta S$ for the process, all three answers in kJ/kg · mol of fuel, and (4) the ratio of the availability of the product gases to that of the original reactants.

14-66M Carbon monoxide (CO) reacts with the stoichiometric quantity of air in a steady-flow process starting at 25°C. The reaction is complete, and the products leave as a mixture at 1 atm at an environmental temperature of 25°C. The carbon monoxide and air enter separately at 1 atm. Determine (*a*) the heat released by the reaction, (*b*) the optimum work, and (*c*) the irreversibility of the process, all in kJ/kg · mol of fuel.

14-67M Reconsider Prob. 14-66M. The steady-flow process is now carried out adiabatically, as described in Prob. 14-43M(*a*). The fuel and air still enter separately at 1 atm and 25°C, and the products are a mixture at 1 atm. For carbon monoxide and stoichiometric air, determine (1) the availability of the hot product gases, (2) the optimum work for the adiabatic process, (3) the value of $T_0 \Delta S$ for the process, all three answers in kJ/kg · mol of fuel, and (4) the ratio of the availability of the hot product gases to that of the original reactants.

14-68M Liquid methyl alcohol (CH_3OH) is reacted with the stoichiometric amount of air in a steady-flow process starting at 25°C. The reaction is complete, and the products leave as a mixture at 1 atm and the environment temperature of 25°C. The fuel and air enter separately at 1 atm, and the water produced is assumed to be a liquid. Determine (*a*) the heat released by the reaction, (*b*) the maximum work output, and (*c*) the irreversibility of the reaction, all in kJ/kg · mol of fuel.

14-69M Reconsider Prob. 14-68M. The steady-flow process is now carried out adiabatically, as described in Prob. 14-45M(*a*). The fuel and air still enter separately at 1 atm and 25°C, and the products are a mixture at 1 atm. For methyl alcohol and stoichiometric air, determine (*a*) the availability of the hot product gases, (*b*) the optimum work for the adiabatic process, (*c*) the value of $T_0 \Delta S$ for the process, all three answers in kJ/kg · mol of fuel, and (*d*) the ratio of the availability of the hot product gases to that of the original reactants.

PROBLEMS (USCS)

Fuel-air combustion analysis

14-1 Ethane (C_2H_6) is burned with dry air in a mass ratio of 1 : 17. Compute (*a*) the percent excess air used, (*b*) the equivalence ratio, (*c*) the percent CO_2 by volume in the product gases, (*d*) the dew-point temperature, in degrees Fahrenheit, and (*e*) the percent of the water vapor condensed if the product gases are cooled to 70°F, assuming complete combustion and a total pressure of 14.5 psia.

14-2 Same as Prob. 14-1, except that the fuel is ethylene (C_2H_4).

14-3 Same as Prob. 14-1, except that the fuel is acetylene (C_2H_2).

14-4 Same as Prob. 14-1, except that the fuel is propylene (C_3H_6).

14-5 One mole of ethane (C_2H_6) is burned with 20 percent excess dry air. Determine (*a*) the air-fuel ratio used, (*b*) the equivalence ratio, (*c*) the mole percent of N_2 in the product gas, (*d*) the dew-point temperature, in degrees Fahrenheit, and (*e*) the percent of the water vapor condensed if the product gases are cooled to 80°F, assuming complete combustion and a total pressure of 14.8 psia.

14-6 Same as Prob. 14-5, except that the fuel is ethylene (C_2H_4).

14-7 Same as Prob. 14-5, except that the fuel is acetylene (C_2H_2).

14-8 Same as Prob. 14-5, except that the fuel is propylene (C_3H_6).

14-9 Determine the volume flow rate, in ft^3/min of air, at 14.7 psia and 80°F, required to burn 1 lb/min of fuel for the fuel-air mixtures specified in (*a*) Prob. 14-1 (*b*) Prob. 14-2, (*c*) Prob. 14-3, (*d*) Prob. 14-4, (*e*) Prob. 14-5, (*f*) Prob. 14-6, (*g*) Prob. 14-7, and (*h*) Prob. 14-8.

14-10 If the actual air supplied has a humidity ratio (specific humidity) of 0.015 lb H_2O/lb dry air, determine the dew-point temperature, in degrees Fahrenheit, for the products of the reaction specified in (*a*) Prob. 14-1, (*b*) Prob. 14-2, (*c*) Prob. 14-3, (*d*) Prob. 14-4, (*e*) Prob. 14-5, (*f*) Prob. 14-6, (*g*) Prob. 14-7, and (*h*) Prob. 14-8.

14-11 A fuel gas with a volumetric composition of: CH_4, 60 percent; C_2H_6, 30 percent; N_2, 10 percent, is burned to completion using 20 percent excess of dry air. Determine (*a*) the mole percent of the products on a dry basis, (*b*) the air-fuel ratio used, (*c*) the dew-point temperature, in degrees Fahrenheit, if the total pressure is 14.7 psia, and (*d*) the mole percent of CO_2 in the product gas on a wet basis if the gas is cooled to 70°F.

14-12 A gaseous fuel having a volumetric analysis of 65 percent CH_4, 25 percent C_2H_6, 5 percent CO_2, and 5 percent N_2 is burned with 30 percent excess of dry air. Determine (*a*) the air-fuel ratio used, (*b*) the mole percent CO_2 in the total products, (*c*) the dew-point temperature if the pressure is 14.7 psia, and (*d*) the mole percent N_2 in the total gaseous products if the gas is cooled to 65°F.

14-13 A gaseous fuel having a volumetric analysis of 70 percent CH_4, 20 percent C_2H_6, and 10 percent C_3H_8 is burned with 20 percent excess of dry air. Determine (*a*) the air-fuel ratio used, (*b*) the mole percent N_2 in the total gaseous products, (*c*) the dew-point temperature if the pressure is 14.8 psia, and (*d*) the mole percent N_2 in the gaseous products if the gas is cooled to 70°F.

14-14 A gaseous fuel which is 60 mole percent CH_4 and 40 mole percent C_2H_4 is burned with 10 percent excess dry air. Determine (*a*) the air-fuel ratio used, (*b*) the mole percent of CO_2 in the total products, (*c*) the dew-point temperature, in degrees Fahrenheit, if the pressure is 14.9 psia, and (*d*) the mole percent CO_2 in the gaseous products if the mixture is cooled to 80°F.

14-15 Ethylene gas (C_2H_4) is burned with 10 percent excess air. Due to incomplete mixing only 96 percent of the carbon in the fuel is converted to CO_2, the rest appearing as CO. Determine (*a*) the air-fuel ratio used, (*b*) the mole fraction of CO_2 in the total products, and (*c*) the dew point, in degrees Fahrenheit, if the pressure is 14.8 psia.

14-16 Liquid benezene (C_6H_6) is burned with (*a*) 120 percent and (*b*) 140 percent theoretical-air requirements. Compute (1) the air-fuel ratio used, (2) the mole fraction of N_2 in the total products, and (3) the dew point, in degrees Fahrenheit, if the pressure is 15.0 psia.

Dew-point temperature evaluation

14-17 Pentane gas (C_5H_{12}) is burned to completion with the stoichiometric amount of air at a pressure of 14.8 psia.

(*a*) If the air supplied is dry, find (1) the dew point, in degrees Fahrenheit, and (2) the number of moles of water per mole of fuel burned which would condense if the product gases were cooled to 75°F.

(*b*) If the air supplied has a specific humidity of 0.014 lb H_2O/lb dry air, find the answers to the same questions asked in part *a*.

14-18 Methane gas (CH_4) is burned with 50 percent excess air at 14.4 psia.

(*a*) If the air supplied is dry, find (1) the dew point, in degrees Fahrenheit, and (2) the number of moles of water per mole of fuel which would condense if the product gases were cooled to 66°F.

(*b*) If the air supplied has a specific humidity 0.016 lb H_2O/lb dry air, find the answers to the same questions asked in part *a*.

14-19 Ethyl alcohol (C_2H_5OH) is burned with (*a*) 25 percent excess air, and (*b*) 50 percent excess air at 14.7 psia. Determine the dew point of the product gases, in degrees Fahrenheit, if the air supplied has a specific humidity (humidity ratio) of 0.015 lb H_2O/lb dry air.

Analysis of product gases

14-20 through 14-29 Use Probs. 14-20M through 14-29M above.

Energy analysis with heat effects

14-30 Liquid octane (n-C_8H_{18}) at 77°F and air at 900°R enter a combustion chamber in steady flow at 14.7 psia pressure. Determine the amount of heat transfer that occurs, in Btu/lb · mol of fuel, if the amount of air used and the temperature of the product gases are (*a*) 150 percent theoretical and 1800°R, (*b*) 200 percent theoretical and 1600°R, and (*c*) 400 percent theoretical and 1800°R.

14-31 Ethane gas (C_2H_6) at 77°F and air at 540°F enter a combustion chamber in steady flow at 14.7 psia. The products of combustion leave at 2000°R. If the percent excess air used is (*a*) 25 percent, and (*b*) 15 percent, and combustion is complete, determine the heat loss, in Btu/lb · mol of fuel.

14-32 Propane (C_3H_8) is burned in a steady-flow system from an original temperature of 77°F. Determine the heat transfer required, in Btu/lb · mol of fuel, if the amount of air used and the temperature of the product gases are (*a*) 100 percent excess and 620°F, and (*b*) 50 percent excess and 940°F.

14-33 Carbon monoxide is burned with air at an initial state of 77°F. The final temperature of the products of combustion is (*a*) 2000°R, and (*b*) 1800°R, and during the process 17,000 Btu of heat per pound-mole of CO are removed. Determine the percent excess air used if combustion is complete.

14-34 Hydrogen gas is burned with air with each at an initial temperature of 720°R. The final temperature of the products is (*a*) 1500°R, and (*b*) 1800°R. During the process, 70,000 Btu of heat are removed per pound-mole of H_2. Determine the percent excess air used if combustion is complete.

14-35 Hydrogen gas is burned with (*a*) stoichiometric air, and (*b*) 100 percent excess air at 77°F and 1 atm. During the process, a heat loss of 17,000 Btu/lb · mol of fuel occurs. Find the final gas temperature, in degrees Rankine.

14-36 Consider the combustion reaction discussed in Prob. 14-15. The fuel enters a steady-state combustion chamber at 77°F, while the air enters at 720°R. If the final temperature of the products of combustion is 2700°R, determine the magnitude and direction of any heat transfer, in Btu/lb · mol of fuel.

14-37 Propane gas (C_3H_8) with 20 percent excess air enters a combustion chamber in steady flow at 1 atm and 77°F. The carbon in the fuel becomes either CO or CO_2. Determine the heat transfer, in Btu/lb · mol of fuel, if the percent carbon converted to CO_2 and the temperature of the product gases are (*a*) 94 percent and 1600°R and (*b*) 96 percent and 1300°R.

Adiabatic-combustion processes

14-38 If ethylene (C_2H_4) at 77°F is burned with 200 percent excess air supplied at 260°F and the gaseous reaction occurs in steady flow, estimate the maximum combustion temperature, in degrees Rankine.

14-39 Hydrogen gas at 77°F is reacted with (*a*) 500 percent, (*b*) 600 percent, and (*c*) 700 percent theoretical oxygen requirements which enter at 1000°R. For complete combustion, estimate the maximum combustion temperature, in degrees Rankine.

14-40 Propane gas (C_3H_8) is burned at a constant pressure of 1 atm with 20 percent excess air starting at 77°F. Assuming no dissociation, determine the maximum adiabatic temperature.

14-41 Determine the maximum theoretical combustion temperature for the reaction of ethane with 50 percent excess air in a steady-flow process. The ethane and the air both enter the combustor at 1 atm and 77°F, and it is assumed that the reaction goes to completion.

14-42 Determine the maximum temperature under adiabatic conditions when methane (CH_4) is burned with (*a*) stoichiometric air, and (*b*) 20 percent excess air, both entering at 77°F. Assume that 10 percent of the carbon is burned only to CO, the rest to CO_2.

14-43 Carbon monoxide undergoes adiabatic combustion with the stoichiometric amount of air. Neglecting the dissociation of products, determine the maximum combustion temperature, in degrees Rankine, if the initial reactants are at (*a*) 77°F, and (*b*) 1000°R.

14-44 A gas turbine combustion chamber is supplied with air at 440°F and liquid octane at 77°F. The products of combustion leave at (*a*) 2060°F, and (*b*) 2240°F. Calculate the air-fuel ratio used if the flow is steady, the combustion is complete, and the heat loss is negligible.

14-45 Determine the adiabatic-combustion temperature for the reaction of liquid methanol

(CH_3OH) with the theoretical air requirements at 1 atm. The methanol enters at 77°F and the air enters at (*a*) 77°F, and (*b*) 340°F. Assume complete combustion and steady flow.

Constant-volume combustion

14-46 Determine the adiabatic-flame temperature and the explosion pressure for the constant-volume combustion of CO with 50 percent excess air. The initial conditions are 1 atm and 80°F.

14-47 An equimolar mixture of hydrogen and carbon monoxide, together with the theoretical amount of air for complete combustion, is ignited in a constant-volume bomb. The initial pressure and temperature are 3 atm and 77°F, respectively. Estimate the maximum temperature and pressure that would be attained, assuming complete combustion and ideal-gas behavior.

14-48 Two cubic centimeters of liquid benzene (C_6H_6) are placed in a 1.0-ft^3 constant-volume bomb. The initial temperature of the reactants is 77°F. If the fuel is burned with the stoichiometric amount of air, determine the explosion pressure in psia.

14-49 Two cubic centimeters of liquid methyl alcohol (CH_3OH) are placed in a 1.0-ft^3 constant-volume bomb. The initial reactant temperature is 77°F. If the fuel is burned with the stoichiometric amount of air, determine the explosion pressure, in psia.

Use of heating values

14-50 Consider a Pennsylvania coal with the ultimate analysis given in Table 14-1.

(*a*) Determine the moles of CO_2, N_2, and O_2 present in the products of combustion per pound of fuel when 20 percent excess air is used, and the air-fuel ratio used.

(*b*) Neglecting the effect of SO_2, determine the heat transfer, in Btu/lb fuel, if the reactants enter a steady-flow combustor at 77°F and the product gases leave at 340°F.

14-51 Consider the Illinois coal with the ultimate analysis given in Table 14-1. Answer the same questions presented in Prob. 14-50.

14-52 Consider the Wyoming coal with the ultimate analysis given in Table 14-1. Answer the same questions presented in Prob. 14-50.

14-53 Consider the South Dakota coal with the ultimate analysis given in Table 14-1. Answer the same questions presented in Prob. 14-50.

Entropy changes for reacting mixtures

14-54 Methane gas (CH_4) is burned with the stoichiometric amount of dry air, both initially at 77°F, in a steady-flow process. The environmental temperature is 60°F, combustion is complete, and both fuel and air enter separately at 1 atm.

(*a*) Determine the entropy change for the reaction and the total entropy change for the overall process, in Btu/°R per pound-mole of fuel, if water appears as a liquid in the products which are at 77°F.

(*b*) Determine the entropy change for the reaction and the total entropy change for the overall process, in Btu/°R per pound-mole of fuel, if water appears as a vapor in the products which are at 77°F.

14-55 Same as Prob. 14-54, except that the fuel is gaseous ethane (C_2H_6).

14-56 Same as Prob. 14-54, except that the fuel is gaseous propane (C_3H_8).

14-57 Same as Prob. 14-54, except that the fuel is liquid methyl alcohol (CH_3OH).

14-58 Consider the process described in Prob. 14-40. Evaluate the entropy change for the reaction, in Btu/°R per pound-mole of fuel, if the fuel and air initially are separate and each is at 1 atm.

14-59 Consider the process described in Prob. 14-41. Evaluate the entropy change for the reaction, in Btu/°R per pound-mole of fuel, if the fuel and air initially are separate and each is at 1 atm.

14-60 Consider the process described in Prob. 14-43(*a*). Evaluate the entropy change for the reaction, in Btu/°R per pound-mole of fuel, if the fuel and air initially are separate and each is at 1 atm.

14-61 Consider the process described in Prob. 14-45(*a*). Evaluate the entropy change for the reaction, in Btu/°R per pound-mole of fuel, if the fuel and air initially are separate and each is at 1 atm.

Availability of reacting systems

14-62 Propane gas (C_3H_8) is burned in steady flow at 1 atm with 20 percent excess air starting at 77°F. Assume the reaction is complete and that the products leave at the environmental temperature of 77°F. The fuel and the oxidant each enter separately at 1 atm, the products leave as a mixture at 1 atm, and the water is in the liquid state in the products. Determine (*a*) the heat released by the reaction, (*b*) the optimum work for the process, and (*c*) the irreversibility of the reaction, all in Btu/lb · mol of fuel.

14-63 Reconsider Prob. 14-62. The steady-flow process is now carried out adiabatically, as described in Prob. 14-40. For 20 percent excess air at 77°F initially, determine (*a*) the availability of the hot product gases, (*b*) the optimum work for the adiabatic process, (*c*) the value of $T_0\ \Delta S$ for the process, all three answers in Btu/lb · mol of fuel, and (*d*) the ratio of the availability of the hot gases to that of the original reactants.

14-64 Ethane gas (C_2H_6) is reacted with 50 percent excess air in a steady-flow process starting at 77°F. The reaction is complete, and the products leave as a mixture at 1 atm at the environmental temperature of 77°F. The fuel and air enter separately at 1 atm, and water in the products is in the liquid state. Determine (*a*) the heat released by the reaction, (*b*) the optimum work, and (*c*) the irreversibility of the process, all in Btu/lb · mol of fuel.

14-65 Reconsider Prob. 14-64. The steady-flow process is now carried out adiabatically, as described in Prob. 14-41. The fuel and air still enter separately at 1 atm and 77°F, and the products are a mixture at 1 atm. For 50 percent excess air, determine (*a*) the availability of the hot product gases, (*b*) the optimum work for the adiabatic process, (*c*) the value of $T_0\ \Delta S$ for the process, all three answers in Btu/lb · mol of fuel, and (*d*) the ratio of the availability of the product gases to that of the original reactants.

14-66 Carbon monoxide (CO) is reacted with the stoichiometric quantity of air in a steady-flow process starting at 77°F. The reaction is complete, and the products leave as a mixture at 1 atm and the environment temperature of 77°F. The carbon monoxide and air enter separately at 1 atm. Determine (*a*) the heat released by the reaction, (*b*) the optimum work, and (*c*) the irreversibility of the process, all in Btu/lb · mol of fuel.

14-67 Reconsider Prob. 14-66. The steady-flow process is now carried out adiabatically, as described in Problem 14-43(*a*). The fuel and air still enter separately at 1 atm and 77°F, and the products are a mixture at 1 atm. For carbon monoxide and stoichiometric air, determine (1) the availability of the hot product gases, (2) the optimum work for the adiabatic process, (3) the value of $T_0\ \Delta S$ for the process, all three answers in Btu/lb · mol of fuel, and (4) the ratio of the availability of the hot product gases to that of the original reactants.

14-68 Liquid methyl alcohol (CH_3OH) is reacted with the stoichiometric amount of air in a steady-flow process starting at 77°F. The reaction is complete, and the products leave as a mixture at 1 atm at an environmental temperature of 77°F. The fuel and air enter separately at 1 atm, and the water produced is assumed to be liquid. Determine (*a*) the heat released by the reaction, (*b*) the maximum work output, and (*c*) the irreversibility of the reaction, all in Btu/lb · mol of fuel.

14-69 Reconsider Prob. 14-68. The steady-flow process is now carried out adiabatically, as described in Prob. 14-45(*a*). The fuel and air still enter separately at 1 atm and 77°F, and the products are a mixture at 1 atm. For methyl alcohol and stoichiometric air, determine (*a*) the availability of the hot product gases, (*b*) the optimum work for the adiabatic process, (*c*) the value of $T_0\ \Delta S$ for the process, all three answers in Btu/lb · mol of fuel, and (*d*) the ratio of the availability of the hot product gases to that of the original reactants.

CHAPTER

FIFTEEN

CHEMICAL EQUILIBRIUM

15-1 INTRODUCTION

In conjunction with the third law of thermodynamics, which establishes a base, or reference state, for the evaluation of absolute entropy values, the entropy change for any theoretical chemical reaction may be computed. This type of calculation for reactive ideal-gas mixtures was illustrated at the end of the preceding chapter. In general, this entropy change could be either positive or negative, depending upon the initial and final states of the system. However, if a reaction occurs in an isolated or an adiabatic, closed system, the second law of thermodynamics requires that the value of the entropy of the system must increase for all spontaneous, or irreversible, processes. In view of this requirement, the entropy of the system must reach a maximum value at the final equilibrium state, consistent with the external constraints on the system. In Sec. 14-7 the variation of the entropy for the reactive system, $CO + H_2O \rightarrow CO_2 + H_2$, was illustrated. For any change of state of an adiabatic, closed system, the entropy can never decrease in value. Consequently, the process on a macroscopic basis will stop at state *b*, where the numbers of moles of CO_2 formed is considerably less than unity. Hence the degree to which any reaction reaches completion, as given by the theoretical equation, depends upon the composition of the system which leads to a maximum value of the entropy under these conditions.

Although the entropy function is quite useful for the analysis of reactive systems under adiabatic conditions, it is impractical for the majority of systems of engineering interest. More generally, we wish to determine the extent of a chemical reaction for a given temperature and pressure. What is a suitable criterion for chemical equilibrium in this case? Any such criterion must be established in terms of the properties of the system alone, since information on the surroundings generally will not be known. In this chapter we shall first investigate the use of the second law in establishing a suitable criterion for determining the state of equilibrium of a system at a specified temperature and pressure. Such a criterion will then be employed to ascertain the equilibrium composition of a reactive mixture. Although we shall restrict ourselves to ideal-gas mixtures, the major points presented are of general usefulness.

15-2 EQUILIBRIUM CRITERIA

For an isolated system the second law of thermodynamics requires that

$$(dS)_{\text{isol}} \geq 0$$

Any simple, compressible system which is isolated is one of constant internal energy, since both heat and work effects are prohibited. It is also one of constant volume, since work is zero, which negates expansion or compression processes. Hence it is quite common to find the above equation written in the following notation:

$$(dS)_{U,V,N} > 0 \qquad \text{spontaneous process} \tag{15-1a}$$

$$(dS)_{U,V,N} = 0 \qquad \text{reversible process} \tag{15-1b}$$

$$(dS)_{U,V,N} < 0 \qquad \text{unnatural process} \tag{15-1c}$$

The subscript N is employed to indicate that the system is one of fixed mass. The set of equations above simply states that, for a spontaneous, irreversible process of a system of fixed mass and constant internal energy and volume, the change in entropy must be positive. In the limit of a reversible change, the variation of S is zero. Under these same conditions a process involving a negative change in S is deemed unnatural, or impossible. We may conclude that when the entropy reaches a maximum value such that, for a further infinitesimal change, the value of dS is zero, then the system is in thermodynamic equilibrium. No further change will occur in the system unless some further interaction between the system and the surroundings is permitted.

As discussed in the preceding section, from a practical standpoint it is important to develop an equilibrium criterion for a system maintained at a certain pressure and temperature. This requires that we make use of the Gibbs function G, which is defined as $G = H - TS$. An equivalent statement is $G = U + PV - TS$. A differential change in G is given by

$$dG = dU + P\,dV + V\,dP - T\,dS - S\,dT \tag{15-2}$$

At constant temperature and pressure this reduces to

$$dG_{T,P} = dU + P\,dV - T\,dS \tag{15-3}$$

However, the first and second laws for a fixed mass system indicate that

$$dU = \delta Q + \delta W_{\text{tot}} \tag{15-4}$$

and

$$\delta Q \leq T\, dS \tag{15-5}$$

If Eq. (15-4) is first used to eliminate dU in Eq. (15-3), and then Eq. (15-5) is used to eliminate δQ from that resulting expression, then we shall find that

$$dG_{T,P} \leq \delta W_{\text{tot}} - \delta W_{P\,dV} = \delta W_{\text{net}} \tag{15-6}$$

In Eq. (15-6), δW_{net} represents the sum of all work interactions except those associated with a volume change for the system.

Now consider a closed system for which all forms of work interactions are absent except boundary work. In such a case the net-work term in Eq. (15-6) reduces to zero, and the equation takes on a new form and meaning, namely,

$$dG_{T,P} \leq 0 \tag{15-7}$$

This equation simply states that the Gibbs function always decreases for a spontaneous, isothermal, isobaric change of a closed system in the absence of all work effects except boundary work. As a process approaches equilibrium, the Gibbs function attains a minimum value, and in this limiting case of equilibrium, dG is zero. For comparison with the previous set of equations (15-1), one may write

$$dG_{T,P,N} < 0 \qquad \text{spontaneous process} \tag{15-8a}$$

$$dG_{T,P,N} = 0 \qquad \text{reversible process} \tag{15-8b}$$

$$dG_{T,P,N} > 0 \qquad \text{unnatural process} \tag{15-8c}$$

This set of equations would be directly applicable, for example, to a chemically reactive system at a given pressure and temperature.

A graphical interpretation of Eq. (15-7) may be obtained in a manner similar to that used in Sec. 14-7 for the entropy function. Consider the gas-phase reaction $CO + H_2O \rightarrow CO_2 + H_2$. The temperature will be chosen to be 1000°K, and the pressure is low enough so that the gases are essentially ideal. On the basis of 1 mole each of the initial reactants, the reaction might proceed until 1 mole each of CO_2 and H_2 is formed. As CO and H_2O are consumed, the composition of the system continually changes. Consequently, the Gibbs function for the total system also changes. Employing the concepts of total-enthalpy and absolute-entropy values for each component introduced in Chap. 14, one may calculate the value of G_{tot} at the given temperature and pressure for various compositions of the system. These compositions depend, of course, upon the extent of the reaction. The values of G_{tot} (in kJ/kg · mol) for various values of the number of moles of CO_2 formed are summarized for a temperature of 1000°K:

Moles CO_2	0	0.2	0.4	0.5	0.6	0.7	0.8	1.0
G_{tot}	−783.2	−792.2	−795.6	−796.3	−796.3	−795.5	−793.8	−786.3

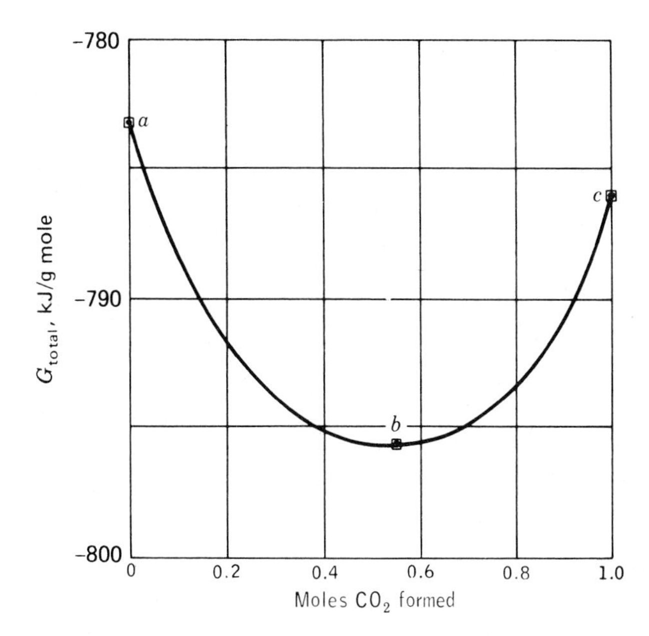

Figure 15-1 The total Gibbs function for the reaction of an equimolar mixture of CO and H_2O at 1000°K as a function of the extent of the reaction, for the chemical reaction $CO(g) + H_2O(g) \leftrightharpoons CO_2(g) + H_2(g)$.

These data are plotted in Fig. 15-1. When an equimolar mixture of CO and H_2O is allowed to react, the Gibbs function of the system decreases until approximately 0.55 mol of CO_2 is formed, for a temperature of 1000°K. A further change in composition represented by the process from state *b* to state *c* is impossible, since this requires that the Gibbs function increase for a process at constant temperature and pressure. Although state *c* has a lower value of *G* than state *a*, it is not the minimum value based on the composition of the original system. Consequently, only state *b* will eventually prevail. It should also be noted that if an equimolar mixture of CO_2 and H_2 is heated to 1000°K, it will react, and the quantity of CO_2 will decrease to 0.55 mol for each initial mole in the mixture. Hence the reaction from state *c* to state *b* is also possible, since this is not a violation of the criterion established by Eq. (15-7). It must be kept in mind that the application of the Gibbs function as a criterion for equilibrium is equivalent to the usual second-law statement in terms of the entropy function. The use of the Gibbs function, however, has the advantage that easily controlled properties such as the temperature and the pressure are involved in its application to reactive systems.

15-3 EQUILIBRIUM AND THE CHEMICAL POTENTIAL

The equilibrium criterion established in the preceding section, $dG_{T,P} = 0$, is valid for any type of homogeneous or heterogeneous system. We wish to use this criterion to find the equilibrium composition of a reacting mixture. To do this, we need to develop a general expression for dG in terms of the moles of reactants and products present at any time. This expression is then set equal to zero, in accordance with the above criterion. Equating the derivative dG to zero is equivalent to setting the total Gibbs function of the mixture at its minimum value. The solution to the resulting equation leads to an evaluation of the equilibrium composition.

On the basis of the state postulate for a simple, compressible substance, the Gibbs function is a function of two independent variables, such as T and P, for a system of fixed composition. For a system of variable composition, however, the Gibbs function is also a function of the number of moles of each of the substances present at a given time. This is expressed mathematically by

$$G = G(T, P, N_1, N_2, \ldots, N_i) \tag{15-9}$$

where the quantities N_i represent the number of moles of each chemical species within the system at some instant. A change of state may be brought about by altering the temperature or the pressure or the composition. The overall change in G is given by the total differential of G, that is,

$$dG = \left(\frac{\partial G}{\partial P}\right)_{T,\,N_i} dP + \left(\frac{\partial G}{\partial T}\right)_{P,\,N_i} dT + \sum_i \left(\frac{\partial G}{\partial N_i}\right)_{P,\,T,\,N_j} dN_i \tag{15-10}$$

In Eq. (15-10) the subscript N_i on the first two partial derivatives implies that the moles of every species are held fixed during the change in either P or T. In the summation the subscript N_j on the partial indicates that the moles of every component are held constant except one, along with fixed values of P and T.

We wish to apply Eq. (15-10) to a system at a fixed temperature and pressure. Consequently, the first two terms on the right side of the equation are zero in value, because $dT = dP = 0$. In addition, the partial derivative which appears in the summation in Eq. (15-10) will be defined as the *chemical potential* μ_i. Hence

$$\mu_i \equiv \left(\frac{\partial G}{\partial N_i}\right)_{P,\,T,\,N_j} \tag{15-11}$$

On this basis Eq. (15-10) can now be written as

$$dG_{T,\,P} = \sum_i \mu_i \, dN_i \tag{15-12}$$

This equation is a fundamental relationship which relates changes in the Gibbs function to changes in composition of any phase of a simple, compressible system at constant temperature and pressure.

At this point we are specifically interested in obtaining an expression for the change in the Gibbs function of a reactive system when infinitesimal amounts of reactants are converted into products. Consider a generalized chemical reaction represented by the equation

$$\nu_A A + \nu_B B \rightleftarrows \nu_E E + \nu_F F \tag{15-13}$$

The ν symbols represent the stoichiometric coefficients for the balanced chemical equation and the uppercase letters stand for the chemical species involved in the reaction. Although the reaction we have chosen has two reactants and two products, it must be realized that the format of the resulting equations will be valid for any number of reactants and products. The values of ν_A, ν_B, ν_E, etc., are not related in any way to the actual numbers of moles N_i of each component actually placed in the reaction vessel, or the actual numbers of moles of products produced by the reaction. The values of the stoichiometric coefficients ν_i are always known, once the reaction is specified.

Consider now that a reactive mixture having a reaction represented by Eq. (15-13) undergoes an infinitesimally small change of composition. If the temperature and pressure remain the same, then the Gibbs-function change for the overall reaction, in terms of Eq. (15-12), is

$$\begin{aligned} dG_{T,P} &= \sum_i \mu_i \, dN_i \\ &= \mu_A \, dN_A + \mu_B \, dN_B + \mu_E \, dN_E + \mu_F \, dN_F \end{aligned} \tag{15-14}$$

The dN_i terms may be positive or negative, depending upon whether the reaction is proceeding from left to right or right to left. The dN_i terms in Eq. (15-14) are not independent of each other, since the change in the moles of reactants and products is always in proportion to the stoichiometric coefficients. Thus we may write for the forward direction of the reaction

$$dN_A = -k\nu_A \qquad dN_B = -k\nu_B \qquad dN_E = k\nu_E \qquad dN_F = k\nu_F$$

where k is a proportionality constant and is an extremely small number. The negative signs are necessary for the two reactant terms since the stoichiometric coefficients such as ν_A and ν_B are always considered to be positive in value. Substitution of the above expressions for dN_i into Eq. (15-14) yields an equation for the Gibbs-function change in terms of the stoichiometric coefficients. The result is

$$dG_{T,P} = (-\mu_A \nu_A - \mu_B \nu_B + \mu_E \nu_E + \mu_F \nu_F)k$$

However, if such an infinitesimal reaction occurred essentially at the point of chemical equilibrium, the net change in the Gibbs function would be zero. This is the particular case in which we are interested. Therefore, for $dG_{T,P} = 0$ at equilibrium, the criterion for equilibrium is

$$\nu_E \mu_E + \nu_F \mu_F - \nu_A \mu_A - \nu_B \mu_B = 0 \tag{15-15}$$

The proportionality constant k has dropped out since it appears in each term of the equation as a multiplying factor. When Eq. (15-15) is satisfied, the Gibbs function of a system at a given temperature and pressure will be a minimum, in line with the general criterion set up earlier.

Equation (15-15) is known as the *equation of reaction equilibrium*, and it relates intensive properties of the reactants and products. The equation is valid for any chemical reaction, regardless of the phases of the reacting species. To establish the equilibrium composition of a reacting mixture we must first determine expressions for the chemical potential of a given component as a function of the temperature, pressure, and composition. In the next section this is done for a mixture of ideal gases.

15-4 THE CHEMICAL POTENTIAL OF AND IDEAL GAS

According to the Gibbs-Dalton rule introduced in Chap. 11, an ideal gas in a mixture of gases behaves as if it alone occupies the volume of the system at the given temperature. Under this circumstance the gas exerts a pressure equal to its component or partial pressure p_i. Therefore the chemical potential of the ideal gas in the mixture is determined from a knowledge of its temperature and component pressure p_i. The Gibbs-Dalton rule also provides us with the method of evaluating the chemical potential of the ideal gas under the above conditions. Because the gases do behave independently, then at the temperature and total pressure of the ideal gas mixture

$$G_{\text{mix}} = \sum_i N_i g_i = N_1 g_1 + N_2 g_2 + \cdots + N_i g_i \tag{15-16}$$

where each g_i term is measured at the temperature and component pressure of that species. If we take the partial derivative of G with respect to N_i, holding temperature, pressure, and the remaining N's constant, then

$$\left(\frac{\partial G}{\partial N_i}\right)_{T,\,P,\,N_j} = g_{i,\,\text{pure},\,p_i}$$

Since the term on the left is the definition of the chemical potential of the ith species, then for an ideal gas in an ideal-gas mixture

$$\mu_{i,\,\text{mix},\,P} = \mu_{i,\,\text{pure},\,p_i} = g_{i,\,\text{pure},\,p_i} \tag{15-17}$$

Hence the chemical potential of an ideal gas in a mixture may be evaluated in terms of the specific Gibbs function of the pure component at a pressure p_i and temperature T. The specific Gibbs function of a substance at temperature T is, by definition,

$$g_{i,\,T} = h_{i,\,T} - Ts_{i,\,T} \tag{a}$$

The enthalpy and the entropy of an ideal gas must be evaluated in terms of the standard reference state discussed in Chap. 14. Recall that the standard state of an ideal gas is taken as 1 atm, and that it is symbolized by the superscript $\circ$. Since the enthalpy of an ideal gas is not a function of pressure, we may write

$$h_{i,\,T} = h^\circ_{i,\,T} \tag{b}$$

where $h^\circ_{i,\,T}$ accounts for the enthalpy of formation as well as the enthalpy difference between 298°K (537°R) and the specified temperature T. Also, the absolute entropy at temperature T and pressure p_i is given by Eq. (14-20), namely,

$$s_{i,\,T} = s^\circ_{i,\,T} - R \ln p_i \tag{c}$$

The chemical potential of a component in an ideal-gas mixture is then found by combining Eqs. (*a*), (*b*), and (*c*) with Eq. (15-17). The result is

$$\mu_{i,T} = g_{i,T} = h_{i,T}^{\circ} - Ts_{i,T}^{\circ} + RT \ln p_i$$

With reference to Eq. (*a*), it is noted that the first two terms on the right-hand side of the equation above is the standard-state Gibbs function, $g_{i,T}^{\circ}$. Therefore the chemical potential of an ideal gas at temperature T and pressure p_i is given by

$$\mu_{i,\,\text{ideal}} = g_i^{\circ} + RT \ln p_i \qquad (15\text{-}18)$$

where the quantity p_i must be measured in atmospheres. This relation, in conjunction with the equation of reaction equilibrium, enables us to determine the equilibrium composition of a reacting ideal-gas mixture.

15-5 THE EQUILIBRIUM CONSTANT K_p

The equilibrium composition of a reacting ideal-gas mixture is determined by means of (1) the equation of reaction equilibrium and (2) the equation for the chemical potential of an ideal gas. In review, these equations are

$$\nu_E \mu_E + \nu_F \mu_F - \nu_A \mu_A - \nu_B \mu_B = 0 \qquad (15\text{-}15)$$

and

$$\mu_i = g_i^{\circ} + RT \ln p_i \qquad (15\text{-}18)$$

where A and B represent reactants and E and F are products. By using Eq. (15-18) to replace the μ_i terms in Eq. (15-15), we find that

$$\nu_E(g_E^{\circ} + RT \ln p_E) + \nu_F(g_F^{\circ} + RT \ln p_F) - \nu_A(g_A^{\circ} + RT \ln p_A) - \nu_B(g_B^{\circ} + RT \ln p_B) = 0 \qquad (15\text{-}19)$$

At this point it is convenient to rearrange Eq. (15-19) and collect common terms. This leads to

$$(\nu_E g_E^{\circ} + \nu_F g_F^{\circ} - \nu_A g_A^{\circ} - \nu_B g_B^{\circ}) + \nu_E RT \ln p_E + \nu_F RT \ln p_F - \nu_A RT \ln p_A - \nu_B RT \ln p_B = 0 \qquad (15\text{-}20)$$

The quantity in the parentheses is called the standard-state Gibbs-function change for a reaction and is given the symbol ΔG_T°. From the definition of the Gibbs function, it is easily seen that

$$\Delta G_T^{\circ} = \Delta H_T^{\circ} - T\,\Delta S_T^{\circ} \qquad (15\text{-}21)$$

Recall that the standard state for ideal gases is defined as 1 atm pressure. Equation (15-21) then indicates that the standard-state Gibbs-function change can be evaluated from a knowledge of the enthalpy of reaction and the entropy change for the stoichiometric reaction occurring at 1 atm pressure and temperature T. Methods for computing these quantities were introduced in the preceding chapter. Consequently, the value of ΔG_T° is known once the stoichiometric chemical equation and the temperature are specified.

We shall now substitute this definition of ΔG_T° into Eq. (15-20). At the same time the terms containing logarithms can be combined into a single term. Thus we find that

$$-\Delta G_T^\circ = \nu_E RT \ln p_E + \nu_F RT \ln p_F - \nu_A RT \ln p_A - \nu_B RT \ln p_B$$

$$= RT[\ln (p_E)^{\nu E} + \ln (p_F)^{\nu F} - \ln (p_A)^{\nu A} - \ln (p_B)^{\nu B}]$$

$$= RT \ln \frac{(p_E)^{\nu E}(p_F)^{\nu F}}{(p_A)^{\nu A}(p_B)^{\nu B}} \tag{15-22}$$

where the p_i values are the actual component pressures of the reacting gases at equilibrium. The exponents of the component pressures are the stoichiometric coefficients based on the balanced theoretical chemical equation. One final change in the form of the relationship is now necessary. The left-hand side of expression (15-22) is solely a function of temperature for a given reaction. Hence its value could be tabulated in standard tables of thermodynamic data. It is more convenient, however, to define first an *equilbrium constant* K_p for ideal-gas reactions by the relation

$$K_p \equiv \frac{(p_E)^{\nu E}(p_F)^{\nu F}}{(p_A)^{\nu A}(p_B)^{\nu B}} \tag{15-23}$$

In general, of course, the number of terms in the numerator and denominator depends upon the number of product and reactant species in the theoretical equation. Based on this definition for an ideal-gas reaction, Eq. (15-22) becomes

$$\Delta G_T^\circ = -RT \ln K_p$$

or

$$K_p = e^{-\Delta G_T^\circ/RT} \tag{15-24}$$

As a result of the introduction of the term K_p, it is more common to see tabulations of K_p or $\log_{10} K_p$ in thermodynamic tables than ΔG_T° itself. Since ΔG_T° is solely a function of temperature, the values of K_p or $\log_{10} K_p$ are formally tabulated against the temperature. The use of the base 10 on the logarithm, rather than the base e, is one of convenience. A tabulation of $\log_{10} K_p$ values over a range of temperatures of some common ideal-gas reactions is presented in Table A-29M. For the purposes of this text we shall assume that the values of K_p for any reaction of interest are always available.

Although the definition and the subsequent evaluation of K_p values appear straightforward, the following items need to be emphasized.

1. As noted above, the value of K_p is independent of the pressure during the actual reaction. The reason for this is that the standard-state Gibbs-function change ΔG_T° is defined as the value in the standard state of 1 atm pressure, regardless of the actual conditions.
2. The equilibrium constant based on component pressure K_p is defined with

products in the numerator of the definition and reactants in the denominator. This is fairly standard procedure, but there are authors who invert the definition. Some care must be taken in this respect when obtaining K_p values from tables which are unfamiliar.

3. The value of K_p is a function of the method of writing a chemical equation for particular reactants and products. For example, consider the following three reactions at a given temperature:

$$CO + \tfrac{1}{2}O_2 \longrightarrow CO_2 \qquad (a)$$

$$2CO + O_2 \longrightarrow 2CO_2 \qquad (b)$$

$$CO_2 \longrightarrow CO + \tfrac{1}{2}O_2 \qquad (c)$$

The standard state Gibbs-function change for reaction *b* is twice that for reaction *a*. Consequently, doubling the stoichiometric equation in effect will square the value of K_p, that is, $(K_p)_b = (K_p)_a^2$. On the other hand, reversal of the direction of a chemical equation reverses the sign on ΔG_T°. Thus the K_p value for reaction *c* is the reciprocal of the value for reaction *a*. These results stem from the exponential relationship between K_p and ΔG_T°. One can conclude from these results that tabulated values of K_p as a function of temperature are meaningless unless the chemical equation to which they refer is also cited.

4. The difference in $dG_{T,P}$ and ΔG_T° should be kept in mind. The former term is the criterion for equilibrium at any given temperature and pressure. It must be zero at equilibrium under any conditions. The second quantity is the standard-state Gibbs-function change at the required temperature, but at 1 atm pressure. Its value is usually finite (positive or negative) and is zero at only one particular temperature for each possible reaction. When ΔG_T° is zero, it merely means that K_p is unity. The value of ΔG_T° (and hence K_p) enables one to calculate the equilibrium composition, as we shall see shortly.

15-6 CALCULATION OF K_p VALUES

As demonstrated in the preceding section, the equilibrium constant K_p is directly related to the standard-state Gibbs-function change, ΔG_T°. For a generalized chemical reaction represented by the equation

$$\nu_A A + \nu_B B \leftrightharpoons \nu_E E + \nu_F F \qquad (15\text{-}13)$$

the standard state Gibbs-function change is determined by

$$\Delta G_T^\circ = \nu_E g_E^\circ + \nu_F g_F^\circ - \nu_A g_A^\circ - \nu_B g_B^\circ$$

This quantity appears as the first part of Eq. (15-20). Any term g_i° is found by evaluating the quantity $h_i^\circ - Ts_i^\circ$. By employing Eq. (14-7) for the enthalpy h_i° of any reacting species, then the standard-state Gibbs-function for any species is

$$g_i^\circ = (\Delta h_f^\circ + h_T - h_{298} - Ts_T^\circ)_i$$

The standard-state Gibbs-function change for any reaction, then, is given by

$$\Delta G_T^\circ = \sum_{\text{prod}} v_i(\Delta h_f^\circ + h_T - h_{298} - Ts_T^\circ)_i$$
$$- \sum_{\text{reac}} v_i(\Delta h_{f,\,298}^\circ + h_T - h_{298} - Ts_T^\circ)_i \qquad (15\text{-}25)$$

For evaluation purposes, Table A-22M provides $\Delta h_{f,\,298}^\circ$ data, and Tables A-6M through A-11M are sources of data for the values of h and s at the required temperature. Once ΔG_T° has been determined from Eq. (15-25), then Eq. (15-24) is used to evaluate the equilibrium constant K_p. That is,

$$K_p = \exp(-\Delta G_T^\circ/RT) \qquad (15\text{-}24)$$

The evaluation technique is illustrated in the following example.

Example 15-1M Evaluate the equilibrium constant K_p at 1000°K for the gas-phase reaction, $CO + H_2O \rightleftharpoons CO_2 + H_2$.

SOLUTION First the standard-state Gibbs-function change will be evaluated for the reaction, making use of Eq. (15-24). In this case

$$\Delta G_T^\circ = 1[-393{,}520 + 42{,}769 - 9364 - 1000(269.215)]$$
$$+ 1[0 + 29{,}154 - 8468 - 1000(166.114)]$$
$$- 1[-110{,}530 + 30{,}355 - 8669 - 1000(234.421)]$$
$$- 1[-241{,}810 + 35{,}882 - 9904 - 1000(232.597)]$$
$$= -3064 \text{ kJ/kg} \cdot \text{mol CO}$$

This value is now substituted into Eq. (15-24). At 1000°K

$$K_p = \exp \frac{3064}{(8.314)(1000)} = \exp(0.369) = 1.45$$

This value is in substantial agreement with the value found in Table A-29M.

15-7 CALCULATION OF EQUILIBRIUM COMPOSITIONS

In theory, the standard-state Gibbs-function change ΔG_T° can be computed for any desired reaction. Therefore the value of K_p for any reaction involving ideal gases is also known, once the temperature is known. Moreover, the equilibrium constant K_p is related to the component pressures of the gases at chemical equilibrium by Eq. (15-23), namely,

$$K_p = \frac{(p_E)^{v_E}(p_F)^{v_F}\cdots}{(p_A)^{v_A}(p_B)^{v_B}\cdots} \qquad (15\text{-}23)$$

where the dots indicate that, in general, any number of products and reactants may be involved. The equilibrium composition of the reacting species is uniquely related by Eq. (15-23). This relationship is more meaningful and useful to us if it is written in terms of the number of moles of each constituent present at equilibrium, rather than the partial pressures. Since the component pressure of any ideal gas is defined by

$$p_i = y_i P = \frac{N_i}{N_m} P$$

the expression for K_p may be modified to the following form for a reaction with two reactants and two products:

$$K_p = \frac{(N_E)^{\nu_E}(N_F)^{\nu_F}}{(N_A)^{\nu_A}(N_B)^{\nu_B}} \left(\frac{P}{N_m}\right)^{\Delta\nu} \tag{15-26}$$

In Eq. (15-26) N_m is the sum of the total number of moles of mixture present in the reaction vessel at equilibrium, P is the total system pressure, and $\Delta\nu$ by definition is the sum of the stoichiometric coefficients of the products minus the sum of the stoichiometric coefficients of the reactants all from the balanced chemical equation. In Eq. (15-26), $\Delta\nu$ equals $\nu_E + \nu_F - \nu_A - \nu_B$.

At a given temperature and pressure for a reaction, the only unknowns in Eq. (15-26) are the values of N_i, that is, the number of moles of each reacting chemical species present at chemical equilibrium. The value of N_m is related, of course, to the N_i values by

$$N_m = N_A + N_B + \cdots + N_E + N_F + \cdots + N_{\text{inerts}} \tag{15-27}$$

It is important not to omit the last term in this expression. The presence of inert gases certainly affects the component pressure of each reacting species. If we again restrict ourselves for the moment to a reaction with two reactants and two products, Eqs. (15-26) and (15-27) constitute two equations with five unknowns. The remaining equations necessary for the solution are based on the principle of the conservation of atomic species previously used, for example, in determining equations for chemical reactions based on product gas analyses, in Chap. 14. In practice it will be found that there are sufficient mass balances to make a solution possible. The method employed in obtaining the equilibrium composition of a reacting mixture of ideal gases is illustrated below by several pertinent examples.

Examples 15-2M and 15-2 Carbon monoxide (CO) and oxygen (O_2) in equimolar proportions are allowed to attain equilibrium at 1 atm and 3000°K (5400°R). Determine the composition of the equilibrium mixture.

SOLUTION The main reaction to consider is

$$CO(g) + \tfrac{1}{2}\,O_2(g) \rightleftharpoons CO_2(g)$$

Another possible reaction might be the further dissociation of O_2 into atomic oxygen (O). It is a matter of experience that appreciable dissociation of O_2 requires temperatures much higher than 3000°K. Consequently, we shall assume that the only chemical species present at equilibrium are CO, O_2, and CO_2. The equilibrium constant K_p at this temperature is listed as 3.06 in

Table A-29M. Therefore

$$K_p = \frac{(N_{CO_2})^1}{(N_{O_2})^{1/2}(N_{CO})^1}\left(\frac{P}{N_m}\right)^{1-1-1/2}$$

or

$$3.06 = \frac{N_{CO_2}}{(N_{O_2})^{1/2}N_{CO}}\left(\frac{1}{N_m}\right)^{-1/2}$$

The actual chemical reaction itself, which does not go to completion, may be written as

$$1CO + 1O_2 \longrightarrow xCO + yO_2 + zCO_2$$

where x, y, and z represent the numbers of moles of CO, O_2, and CO_2 present in the mixture at equilibrium. Since $N_m = x + y + z$, the expression for K_p becomes, upon rearrangement,

$$3.06 = \frac{z(x + y + z)^{1/2}}{x(y)^{1/2}}$$

In addition to this relationship, which contains three unknowns, two balances may be made on the carbon and oxygen atoms. Hence

C balance: $\quad 1 = x + z$

O balance: $\quad 3 = x + 2y + 2z$

Solving for y and z, we find that

$$z = 1 - x$$

and

$$y = \tfrac{1}{2}(3 - x - 2z) = \tfrac{1}{2}(1 + x)$$

Also,

$$N_m = x + y + z = x + (1 - x) + \tfrac{1}{2}(1 + x) = \tfrac{1}{2}(3 + x)$$

We have chosen to evaluate the numbers of moles of O_2 and CO_2 in terms of CO. This is an arbitrary choice, and either y or z could have been selected as the remaining unknown. Substitution of the equations for y, z, and N_m into the expression for K_p yields

$$3.06 = \frac{(1 - x)[(3 + x)/2]^{1/2}}{x[(1 + x)/2]^{1/2}} = \frac{(1 - x)(3 + x)^{1/2}}{x(1 + x)^{1/2}}$$

This equation for x may be solved by iteration, synthetic division, Newton's method, or any other suitable technique. A value of 0.34 satisfies the equation within the desired accuracy. Therefore the correct chemical equation for the reaction becomes

$$1CO + 1O_2 \longrightarrow 0.34CO + 0.67O_2 + 0.66CO_2$$

This compares with the theoretical reaction for complete combustion as

$$1CO + 1O_2 \longrightarrow CO_2 + \tfrac{1}{2}O_2$$

For the particular reactant mixture and final pressure and temperature, the CO_2 formed is, roughly, two-thirds that expected for complete combustion.

Examples 15-3M and 15-3 An equimolar mixture of carbon monoxide and oxygen is allowed to attain equilibrium at 3000°K and 5 atm pressure. Determine the composition of the equilibrium mixture.

SOLUTION On the basis of the analysis made in the preceding example, we may write the following set of equations:

$$1CO + 1O_2 \longrightarrow xCO + yO_2 + zCO_2$$

$$z = 1 - x$$

$$y = \tfrac{1}{2}(1 + x)$$

$$N_m = \tfrac{1}{2}(3 + x)$$

$$K_p = \frac{z(5)^{-1/2}}{x(y)^{1/2}(x + y + z)^{-1/2}} = 3.06$$

or, upon rearrangement,

$$3.06(5)^{1/2} = \frac{(1 - x)(3 + x)^{1/2}}{x(1 + x)^{1/2}}$$

The method of solution is exactly the same as in the preceding example. However, we see that the change in pressure does affect the numerical answer. The right-hand side of the equation is the same as before, but the numerical value of the left-hand side has increased by the square root of the pressure, expressed in atmospheres. The value of x which satisfies the above relation is approximately 0.193 mol. Thus 0.807 mol of CO_2 is formed at equilibrium. When the pressure is 1 atm, the CO_2 formed is 0.66 mol per mole of initial CO. Apparently, the effect of pressure on the equilibrium composition is quite pronounced for this reaction.

As a generalization, any time the exponent of the pressure term in the K_p expression is negative (i.e., the sum $\nu_E + \nu_F + \nu_A - \nu_B$ is negative), an increase in the pressure will always increase the number of moles of products formed at a given temperature, with a concomitant decrease in the number of moles of reactants present at equilibrium. The opposite result is true when the exponent on the pressure is positive. When the exponent is zero, the pressure has no effect on the equilibrium composition.

Examples 15-4M and 15-4 In order to determine the effect of the presence of inert gases on the equilibrium composition, compute the mixture composition at 3000°K (5400°R) and 1 atm pressure for a mixture initially composed of 1 mol of carbon monoxide and 4.76 mol of air.

SOLUTION The approach is basically the same as in the two preceding examples, except that the equation for the total number of moles N_m must be revised. The chemical equation for the ideal-gas reaction now becomes

$$1CO + 1O_2 + 3.76N_2 \longrightarrow xCO + yO_2 + zCO_2 + 3.76N_2$$

Again, from the carbon and oxygen balances, $z = 1 - x$ and $y = \frac{1}{2}(1 + x)$. However, the total number of moles of mixture at equilibrium is now

$$N_m = x + y + z + 3.76 = \tfrac{1}{2}(10.52 + x)$$

The expression for K_p is then

$$3.06 = \frac{(1 - x)(1)^{-1/2}}{x[(1 + x)/2]^{1/2}[(10.52 + x)/2]^{-1/2}} = \frac{(1 - x)(10.52 + x)^{1/2}}{x(1 + x)^{1/2}}$$

A suitable solution for x from this equation is 0.47 mol of CO at the equilibrium state of 3000°K and 1 atm pressure. Compared with the original example, we find that the presence of inert nitrogen has decreased the CO_2 formed from 0.66 to 0.53 mol for the same pressure and temperature. It is now apparent that the pressure of the system and the presence of inert gases must be taken into account when evaluating the equilibrium composition of a reacting ideal-gas mixture at a given temperature.

The preceding examples have illustrated the general method for evaluating the equilibrium composition of an ideal-gas mixture when the final pressure and temperature are known. In order to make any computations for a reacting mixture, the identity of the chemical species expected in the equilibrium mixture must be assumed. This is necessary, because the presence of various species determines which equilibrium reactions must be considered. To date we have examined mixtures for which only one equilibrium reaction was significant. Other reactions undoubtedly would occur, such as the dissociation of O_2 into atomic oxygen (O) or N_2 into atomic nitrogen (N), but the extent of these reactions was assumed to be negligible. Experience enables one to predict which reactions can be safely neglected and which reactions should be taken into account. The magnitude of the equilibrium constant K_p is often a good clue to the importance of a reaction. As a rule of thumb, when K_p is less than 0.001 (or $\log_{10} K_p$ is less than -3.0), the extent of the reaction is usually not significant. When K_p is greater than 1000 (or $\log_{10} K_p$ is greater than $+3.0$), the reaction probably proceeds very close to completion. For example, consider the reaction $CO + \frac{1}{2}O_2 \leftrightharpoons CO_2$. The equilibrium constant K_p at 3600°R (2000°K) is, roughly, 730. For an equimolar ratio of CO and O_2 at 1 atm pressure, the CO present at equilibrium at this temperature is approximately 0.0024 mol per initial mole of CO. The reaction essentially goes to completion. Consequently, for temperatures below, roughly 2200°K or 4000°R, the dissociation of CO_2 into CO and O_2 may usually be neglected, except at pressures very much below atmospheric. At higher temperatures the effects of dissociation must be considered, as shown by the examples previously given, since the value of K_p falls rapidly with increasing temperatures above 2200°K or 4000°R. (See K_p data in Table A-29M.) We shall prove later that the value of K_p always decreases with increasing temperatures for all exothermic reactions.

15-8 IONIZATION REACTIONS

The concept of the dissociation of a compound into two or more smaller particles can be extended to ionization effects. Elements, for example, will ionize into a positively charged ion and an electron. At elevated temperatures diatomic nitrogen dissociates into its monatomic form according to the equation

$$N_2 \leftrightharpoons 2N$$

As the gas is heated to higher temperatures the following reaction also occur:

$$N \leftrightharpoons N^+ + e^-$$

It is reasonable to assume in many situations that the positive ions and the electrons behave as ideal-gas particles. Consequently the K_p expression [Eq.

(15-26)] is equally valid for a mixture of neutral particles, ions, and electrons at a given temperature and pressure. (In the presence of electric fields the temperature of the electrons may not necessarily be the same as the temperature of the ions and neutral particles. In the presence of moderate fields, one can assume that the temperature of all particles is the same.) In general, the degree of ionization increases as the temperature is raised and the pressure is lowered.

Examples 15-5M and 15-5 A hypothetical monatomic gas species A ionizes according to the relation $A \leftrightarrows A^+ + e^-$. If at some temperature T the value of K_p is 0.1 and the pressure is 0.1 atm, determine the percent ionization,

SOLUTION The K_p expression for the reaction is

$$K_p = \frac{N_{A^+}N_{e^-}}{N_A}\frac{P}{N_m}$$

where $N_m = N_{A^+} + N_{e^-} + N_A$. In addition, since charge must be conserved, $N_{A^+} = N_{e^-}$. Also, every ion must come originally from a neutral particle. Therefore, on the basis of 1 initial mole of A, $1 - N_A = N_{A^+}$. If these latter three equalities are substituted into the K_p expression, we find that

$$K_p = 0.1 = \frac{(1 - N_A)(1 - N_A)}{N_A}\frac{0.1}{2 - N_A}$$

or

$$N_A(2 - N_A) = (1 - N_A)^2$$

The solution to this equation is

$$N_A = 0.293$$

Therefore the degree of ionization is close to 70 percent.

15-9 FIRST-LAW ANALYSIS OF REACTING IDEAL-GAS MIXTURES

In Chap. 14 the conservation of energy principle was applied to chemical reactions which were assumed to proceed to completion. Two types of processes were analyzed. In the first of these, the initial and final states of the reactants and products, respectively, were known. On the basis of this information, the heat interactions could be evaluated. In the second case, the process was assumed to be adiabatic. From a knowledge of the initial state of the reactants, the maximum theoretical reaction temperature (the adiabatic-flame temperature) was computed. These two types of calculations may now be repeated in the light of the second-law restrictions on chemical reactions. We realize now that in actual combustion processes two things may occur. Either the principal reaction itself may not go to completion, or products of the principal reaction may dissociate into other chemical species not initially present. Hence computations for chemical reactions should take into account these two factors if it is thought that they might be important. If sufficient heat is removed, for example, during a combustion process, it is quite possible that neither effect need be considered. The two examples that follow illustrate the use of both the first and second laws in the analysis of chemical reactions of ideal-gas mixtures.

Examples 15-6M and 15-6 An equimolar mixture of carbon monoxide and water vapor enters a steady-flow device at 400°K or 260°F and 1 atm pressure. The water-gas reaction occurs, and the final products leave at 1500°K or 2240°F. Compute the quantity and direction of the heat transferred per mole of initial CO in the reactants.

SOLUTION The theoretical water-gas reaction is given by

$$CO(g) + H_2O(g) \leftrightharpoons CO_2(g) + H_2(g)$$

In lieu of the expected equilibrium reaction among the constituents, the actual reaction must be written as

$$CO + H_2O \longrightarrow wCO + xCO_2 + yH_2O + zH_2$$

The use of mass balances leads to relationships among w, x, y, and z:

O balance: $2 = w + 2x + y$

C balance: $1 = w + x$

H balance: $2 = 2y + 2z$

A further relationship among the four variables is provided by the equilibrium-constant expression for the water-gas reaction:

$$K_p = \frac{(N_{CO_2})(N_{H_2})}{(N_{CO})(N_{H_2O})}\left(\frac{P}{N_m}\right)^0$$

At 1500°K, the values of K_p for this reaction are listed in Table A-29M as 0.390. Therefore, we may write

$$0.390 = \frac{xz}{wy}$$

These four equations can now be solved simultaneously. The solution in terms of the actual chemical equation is

$$1CO + 1H_2O \longrightarrow 0.615CO + 0.385CO_2 + 0.615H_2O + 0.385H_2$$

Thus the reaction goes to only 38.5 percent of completion. With this information the value of the heat interaction may now be calculated either in metric or USCS units.

(*a*) METRIC SOLUTION The conservation of energy principle for this steady-flow process is

$$Q = \sum_{\text{prod}} N_i h_i - \sum_{\text{reac}} N_i h_i$$

$$= \sum_{\text{prod}} N_i(\Delta h_f^\circ + h_T - h_{298})_i - \sum_{\text{reac}} N_i(\Delta h_f^\circ + h_T - h_{298})_i$$

where the enthalpy of formation of each constituent is evaluated at 298°K. Substituting Δh_f data from Table A-22M and h data from Tables A-8M to A-11M, we find that

$$Q = 0.615(-110{,}530 + 47{,}517 - 8669) + 0.385(-393{,}520 + 71{,}078 - 9364)$$
$$+ 0.615(-241{,}810 + 57{,}999 - 9904) + 0.385(0 + 44{,}738 - 8468)$$
$$- 1(-110{,}530 + 11{,}644 - 8669) - 1(-241{,}810 + 13{,}356 - 9904)$$
$$= 69{,}910 \text{ kJ/kg} \cdot \text{mol CO}$$

The positive value of Q indicates that 69,910 kJ of heat would have to be added per initial kilogram-mole of CO to permit the products to reach 1500°K. At this temperature, dissociation of CO_2 and H_2O according to the chemical equations, $CO_2 \rightleftharpoons CO + \frac{1}{2}O_2$ and $H_2O \rightleftharpoons H_2 + \frac{1}{2}O_2$, may be neglected.

(*b*) USCS SOLUTION The conservation of energy principle for this steady-flow process is

$$Q = \sum_{\text{prod}} N_i h_i - \sum_{\text{reac}} N_i h_i$$

$$= \sum_{\text{prod}} N_i(\Delta h_f^\circ + h_T - h_{537})_i - \sum_{\text{reac}} N_i(\Delta h_f^\circ + h_T - h_{537})_i$$

where the enthalpy of formation of each constituent is evaluated at 537°R. Substituting Δh_f data from Table A-22 and h data from Tables A-8 to A-11, we find that

$$Q = 0.615(-47{,}540 + 20{,}434 - 3725) + 0.385(-169{,}300 + 30{,}581 - 4028)$$

$$+ 0.615(-104{,}040 + 24{,}957 - 4258) + 0.385(0 + 19{,}238 - 3640)$$

$$- 1(-47{,}540 + 5006 - 3725) - 1(-104{,}040 + 5739 - 4258)$$

$$= 29{,}700 \text{ Btu/lb} \cdot \text{mol CO}$$

The positive value of Q indicates that 29,700 Btu of heat would have to be added per initial pound-mole of CO in order for the products to reach 2700°R. At this temperature, dissociation of CO_2 and H_2O according to the chemical equations, $CO_2 \rightleftharpoons CO + \frac{1}{2}O_2$ and $H_2O \rightleftharpoons H_2 + \frac{1}{2}O_2$, may be neglected.

Examples 15-7M and 15-7 One mole of carbon monoxide and 220 percent theoretical oxygen requirements, both initially at 25°C or 77°F, undergo a reaction in a steady-flow process at 1 atm pressure. Neglecting dissociation of O_2, determine the final equilibrium composition and the final temperature if the process is adiabatic.

SOLUTION The theoretical equation for complete combustion would be

$$CO + 2.2(0.5)O_2 \longrightarrow CO_2 + 0.6O_2$$

In this particular case, however, when the effects of chemical equilibrium are to be considered, we must write

$$CO + 1.1O_2 \longrightarrow xCO_2 + yCO + zO_2$$

The mass balances on the equation are

O balance: $\quad 1 + 2.2 = 2x + y + 2z$

C balance: $\quad 1 = x + y$

The expression for K_p for the reaction $CO + \frac{1}{2}O_2 \rightleftharpoons CO_2$ is

$$K_p = \frac{N_{CO_2}}{(N_{CO})(N_{O_2})^{1/2}}\left(\frac{P}{N_m}\right)^{-1/2} = \frac{x(x+y+z)^{1/2}}{y(z)^{1/2}}$$

At this point we have three equations and four unknowns. The unknowns are x, y, z, and K_p. The value of K_p, of course, is solely a function of the temperature. The additional required equation is the conservation of energy principle, which in this case reduces to

$$H_{\text{reac}} = H_{\text{prod}}$$

The energy balance may be evaluated in either metric (SI) or USCS units.

(*a*) METRIC SOLUTION The above energy balance can be written more explicitly as

$$\sum_{\text{reac}} N_i(\Delta h_f^\circ + h_T - h_{298})_i = \sum_{\text{prod}} N_i(\Delta h_f^\circ + h_T - h_{298})_i$$

In terms of the above reaction, this can be written as

$$1(-110{,}530 + 0) + 1.1(0) = x(-393{,}520 + h_{T,\,CO_2} - 9364) + y(-110{,}530 + h_{T,\,CO} - 8669) + z(0 + h_{T,\,O_2} - 8682)$$

where the enthalpy terms are to be evaluated at the final but as yet unknown temperature. This equation introduces no more unknowns; hence we now have a sufficient number of equations for a solution. The method of solution, however, requires an iteration, or trial-and-error, technique. Before starting the iteration process, the mass balances and the K_p expression may be combined into the form

$$K_p = \frac{x(2.1 - 0.5x)^{1/2}}{(1 - x)(1.1 - 0.5x)^{1/2}}$$

At the same time a combination of the mass balances and the energy equation leads to

$$-110{,}530 = x(-402{,}884 + h_{T,\,CO_2}) + (1 - x)(-119{,}199 + h_{T,\,CO}) + (1.1 - 0.5x)(h_{T,\,O_2} - 8682)$$

Thus we have two equations with two unknowns, primarily x and T.

As a first approximation we shall let T equal 2900°K. The sensible-enthalpy values in the energy balance may now be obtained from gas tables A-7M to A-9M, and the linear energy equation can be solved for x. This yields $x = 0.707$. This value of x is now substituted into the expression for K_p. Thus

$$K_p = \frac{0.707(2.1 - 0.354)^{1/2}}{0.293(1.1 - 0.354)^{1/2}} = 3.69$$

At 2900°K, however, the equilibrium constant is found to be 4.7 by interpolation in Table A-29M. As a second guess, we shall let T equal 3000°K. The energy balance then gives $x = 0.739$. Substitution of this quantity into the relation for K_p yields a value of 4.36. At 3000°K the value of K_p given in Table A-29M is 3.06. Since this latter value is smaller than the value calculated from the value of x, we have now guessed too high a value for the temperature. The correct answer must lie between 2900 and 3000°K. We need not refine the method further, since the general approach has been sufficiently demonstrated.

(*b*) USCS SOLUTION The energy balance can be written more explicitly as

$$\sum_{\text{reac}} N_i(\Delta h_f^\circ + h_T - h_{537})_i = \sum_{\text{prod}} N_i(\Delta h_f^\circ + h_T - h_{537})_i$$

In terms of the above reaction, this can be written as

$$1(-47{,}540 + 0) + 1.1(0) = x(-169{,}300 + h_{T,\,CO_2} - 4028) + y(-47{,}540 + h_{T,\,CO} - 3725) + z(0 + h_{T,\,O_2} - 3725)$$

where the enthalpy terms are to be evaluated at the final but as yet unknown temperature. This equation introduces no more unknowns; hence we now have a sufficient number of equations for a solution. The method of solution, however, requires an iteration, or trial-and-error, technique. Before starting the iteration process, the mass balances and the K_p expression may be combined into the form

$$K_p = \frac{x(2.1 - 0.5x)^{1/2}}{(1 - x)(1.1 - 0.5x)^{1/2}}$$

At the same time a combination of the mass balances and the energy equations leads to

$$-47{,}540 = x(-173{,}328 + h_{T,\,CO_2}) + (1 - x)(-51{,}265 + h_{T,\,CO}) + (1.1 - 0.5x)(h_{T,\,O_2} - 3725)$$

Thus we have two equations with two unknowns, primarily x and T.

As a first approximation we shall let T equal 5220°R (2900°K). The sensible-enthalpy values in the energy balance may now be obtained by linear interpolation from gas tables A-7 to A-9 and the energy equation can be solved for x. This yields $x = 0.704$. This value of x is now substituted into the K_p expression. Thus,

$$K_p = \frac{0.704(2.1 - 0.352)^{1/2}}{0.296(1.1 - 0.352)^{1/2}} = 3.64$$

At 2900°K, however, the equilibrium constant is found to be 4.7 by interpolation in Table A-29. As a second guess we might try T equal to 5400°R. The energy equation then gives $x = 0.739$. Substitution of this value into the relation for K_p yields a K_p value of 4.36. At 5400°R (or 3000°K) the value of K_p is given in the tables as 3.06. Since this tabulated value is smaller than the value calculated from the value of x, we have now guessed too high a value of the temperature. The correct answer must lie between 5220 and 5400°R, and closer to 5300°R. We need not refine the method further, since the general approach has been sufficiently demonstrated.

The foregoing example is fairly simple because only one equilibrium reaction is involved. In actual practice there may be a number of equilibrium reactions involved for a given set of initial reactions. Nevertheless, the method of solution is exactly the same, whether the process is adiabatic or nonadiabatic. Mass balances based on the conservation of atomic species, the conservation of energy principle, and the second law in the form of K_p expressions must be employed. Care must be taken to assure that all the K_p expressions are independent of each other. In addition, it is important to keep in mind that our study of chemical reactions involving the use of K_p requires ideal-gas behavior.

15-10 SIMULTANEOUS REACTIONS

In the preceding discussion we have focused our attention on determining the equilibrium state for a single chemical reaction at a given temperature and pressure. To attain equilibrium, the total Gibbs function of all the reacting species must be minimized. It is quite common, however, for two or more reactions to occur simultaneously in a reacting mixture. In addition, some of the reacting species will appear in several of the competing chemical reactions. As might be anticipated, the correct analysis of this more complex process requires the use of the equation of reaction equilibrium for every independent reaction which occurs in the mixture.

As a generalization, consider a system for which there are R *independent* reactions. The solution to this problem of simultaneous reactions requires that R equations of the form

$$\sum_{\text{prod}} \nu_i \mu_i - \sum_{\text{reac}} \nu_i \mu_i = 0$$

be written. By substituting the chemical potential of an ideal gas for μ_i in these R

equations, there will arise R different expressions for K_p. Each K_p is evaluated from the general reationship

$$K_p = \exp\left(\frac{-\Delta G_T^\circ}{RT}\right)$$

where a different ΔG_T° value exists for each reaction at the given temperature. In addition, atomic balances are written in terms of the initial state of the reacting mixture. The K_p expressions plus the atomic balances will provide sufficient information to determine the equilibrium composition of the ideal-gas mixture.

As an example, consider the adiabatic combustion of a hydrocarbon fuel C_xH_y with a small excess of air. In order to calculate the adiabatic-combustion temperature, a number of competing reactions may have to be considered in addition to the main combustion reaction. Among the possible reactions at a high temperature would be

$$CO_2 \leftrightharpoons CO + \tfrac{1}{2}O_2 \qquad H_2O \leftrightharpoons H_2 + \tfrac{1}{2}O_2$$

$$O_2 \leftrightharpoons 2O \qquad H_2O \leftrightharpoons H + OH$$

$$N_2 \leftrightharpoons 2N \qquad N_2 + O_2 \leftrightharpoons 2NO$$

$$H_2 \leftrightharpoons 2H \qquad 2NO + O_2 \leftrightharpoons 2NO_2$$

If sulfur were present in the fuel, other reactions may be important. It is apparent that calculations involving a number of reactions are difficult by hand. However, this type of problem is easily solved by use of a digital computer.

It has been stressed that the R chemical reactions must be independent, in order to achieve a valid solution. As an example, consider the situation where the reacting species include CO, CO_2, H_2, H_2O, and O_2. In this case we might write

$$CO + \tfrac{1}{2}O_2 \leftrightharpoons CO_2 \tag{a}$$

$$CO + H_2O \leftrightharpoons CO_2 + H_2 \tag{b}$$

A third reaction we also might have written is

$$H_2 + \tfrac{1}{2}O_2 \leftrightharpoons H_2O \tag{c}$$

These three reactions are not independent, however, since Eq. (*c*) can be formed by subtracting Eq. (*b*), algebraically, from Eq. (*a*). Hence, only two K_p expressions should be written in this case, although which two is a matter of choice.

Examples 15-8M and 15-8 One mole of CO and one mole of H_2O are heated to 2500°K(4500°R) and 1 atm. Determine the equilibrium composition, if it is assumed that only CO, CO_2, H_2, H_2O, and O_2 are present and that all these gases are ideal gases.

SOLUTION The theoretical equations we shall choose are

$$CO + \tfrac{1}{2}O_2 \leftrightharpoons CO_2 \tag{a}$$

and

$$CO + H_2O \leftrightharpoons CO_2 + H_2 \tag{b}$$

From Table A-29M we find the K_p values for these reactions as written are 27.5 and 0.164, respectively. Thus

$$K_{p,a} = 27.4 = \frac{N_{CO_2}}{N_{CO}(N_{O_2})^{1/2}}\left(\frac{P}{N_T}\right)^{-1/2}$$

and
$$K_{p,b} = 0.164 = \frac{(N_{CO_2})(N_{H_2})}{(N_{CO})(N_{H_2O})}\left(\frac{P}{N_T}\right)^0$$

where
$$N_T = N_{CO} + N_{CO_2} + N_{O_2} + N_{H_2} + N_{H_2O}.$$

The overall chemical equation relating the reactants and the products is

$$1CO + 1H_2O \longrightarrow N_{CO} + N_{CO_2} + N_{O_2} + N_{H_2} + N_{H_2O}$$

This relationship can be used to write mass balances on the atomic species present in the overall reaction. These are

C balance: $1 = N_{CO} + N_{CO_2}$

O balance: $2 = N_{CO} + 2N_{CO_2} + N_{H_2O} + 2N_{O_2}$

H_2 balance: $1 = N_{H_2} + N_{H_2O}$

The two equilibrium-constant expressions and the above three species balances constitute five equations with five unknowns. Hence a unique numerical solution exists. First, we solve the three atomic balances for three of the N_i quantities in terms of the remaining two independent N_i values. As an arbitrary choice, let N_{CO_2} and H_{H_2O} be the two independent variables. Then the expressions for the three dependent variables, and N_T, are

$$N_{CO} = 1 - N_{CO_2} \qquad N_{H_2} = 1 - N_{H_2O}$$

$$N_{O_2} = \tfrac{1}{2}(1 - N_{CO_2} - N_{H_2O}) \qquad N_T = \tfrac{1}{2}(5 - N_{CO_2} - N_{H_2O})$$

Substitution of these four relations into the K_p relations yields

$$27.4 = \frac{N_{CO_2}(5 - N_{CO_2} - N_{H_2O})^{1/2}}{(1 - N_{CO_2})(1 - N_{CO_2} - N_{H_2O})^{1/2}}$$

and
$$0.164 = \frac{N_{CO_2}(1 - N_{H_2O})}{(1 - N_{CO_2})(N_{H_2O})}$$

By suitable numerical techniques the simultaneous solution of the two preceding equations may be carried out. The table below summarizes the results.

Species	Moles initially	Moles finally
CO	1	0.71
H_2O	1	0.71
CO_2	0	0.29
H_2	0	0.29
O_2	0	0.005

From these data it is seen that nearly 30 percent of both CO and H_2O have reacted at the state of equilibrium.

15-11 A RELATIONSHIP BETWEEN K_p AND THE ENTHALPY OF REACTION

In a preceding section the equilibrium constant K_p was related to the standard-state Gibbs-function change ΔG_T°. Since this latter quantity is a function of the enthalpy of reaction and the entropy of reaction at 1 atm pressure and a given temperature, its value theoretically can be determined for any desired reaction from basic thermodynamic data. Consequently, the values of K_p are also known as a function of temperature. It is informative to derive a general expression for the variation of $\ln K_p$ with temperature.

From Eq. (15-24) it is seen that $-R \ln K_p = \Delta G_T^\circ/T$. If the relation $\Delta G_T = \Delta H_T - T\,\Delta S_T$ is substituted and the resulting expression differentiated with respect to temperature, we find that

$$-R\,\frac{d \ln K_p}{dT} = -\frac{\Delta H_T}{T^2} + \frac{d(\Delta H_T)}{T\,dT} - \frac{d(\Delta S_T)}{dT}$$

However, if one applies the basic equation $T\,dS = dH - V\,dP$ to a chemical reaction at constant pressure, then $T\,d(\Delta S)/dT = d(\Delta H)/dT$. Hence the second and third terms on the right of the above equation cancel, and it reduces to

$$\frac{d \ln K_p}{dT} = \frac{\Delta H_T^\circ}{RT^2} = \frac{\Delta h_R^\circ}{RT^2} \tag{15-28}$$

The expression for the change in $\ln K_p$ with temperature may be written also as

$$\frac{d \ln K_p}{d(1/T)} = \frac{-\Delta h_R^\circ}{R} \tag{15-29}$$

Either of these equations is referred to as the van't Hoff isobar equation.

In order to integrate these equations, the functional relationship between Δh_R° and T must be known. The enthalpy of reaction for many reactions is nearly independent of temperature. Consequently, it is often possible to assume that Δh_R° is constant over the range of temperatures of interest. If some average or initial value for the enthalpy of reaction is chosen, integration of Eq. (15-29) leads directly to

$$\ln \frac{K_{p2}}{K_{p1}} = -\frac{\Delta h_R^\circ}{R}\left(\frac{1}{T_2} - \frac{1}{T_1}\right) \tag{15-30}$$

This approximation is generally quite good for small temperature intervals.

Equation (15-30) leads to an interesting qualitative result. If a reaction is exothermic (heat released), Δh_R° is negative by convention. In addition, when $T_2 > T_1$ for such a reaction, the right-hand side of Eq. (15-30) is negative. Hence K_{p2} must be less than K_{p1}. Thus, for exothermic reactions, the equilibrium constant K_p decreases with increasing temperature. As K_p decreases, the tendency for the reaction to proceed to completion is reduced. Under adiabatic conditions, then, the energy released by an exothermic reaction is diminished since the reaction does not proceed to completion.

Examples 15-9M and 15-9 The equilibrium constant K_p for the gas-phase reaction $CO + \frac{1}{2}O_2 \leftrightharpoons CO_2$ is found to be 3.055 at 3000°K (5400°R). Estimate the value at 2000°K, based on the van't Hoff isobar equation.

SOLUTION The Δh_R° values at 2000 and 3000°K are −277,950, and −272,690 kJ/kg · mol, respectively. Since the change in Δh_R° is relatively small, Eq. (15-30) should lead to a fairly good evaluation of K_p at the lower temperature. The average value of Δh_R° is −275,320 kJ/kg · mol. Therefore

$$\ln \frac{K_{p2}}{K_{p1}} = \frac{275{,}320}{8.314}\left(\frac{1}{2000} - \frac{1}{3000}\right) = 33{,}115(1.667 \times 10^{-4}) = 5.52$$

Hence, $K_{p2}/K_{p1} = 252$, or $K_{p2} = 770$. The tabulated value found in Table A-33 is 766. The error is 0.5 percent, but this is for a temperature difference of 1000°K. The calculation does indicate that reasonable accuracy is obtained if Δh_R° is relatively constant.

Examples 15-10M and 15-10 At high temperature the potassium atom is ionized according to the equation $K \leftrightharpoons K^+ + e^-$. The values of equilibrium constant K_p for this gas-phase reaction at 3000 and 3500°K are 8.33×10^{-6} and 1.33×10^{-4}, respectively. Estimate the average enthalpy of reaction in the given temperature range, in joules per gram-mole and electronvolts per molecule.

SOLUTION The average enthalpy of reaction can be determined for the van't Hoff isobar equation. Use of Eq. (15-29) yields

$$\ln \frac{1.33 \times 10^{-4}}{8.33 \times 10^{-6}} = -\frac{\Delta h_R^\circ}{8.314}\left(\frac{1}{3500} - \frac{1}{3000}\right)$$

or

$$\Delta h_R^\circ = \frac{-2.77(8.314)}{-4.76 \times 10^{-5}} = +483{,}700 \text{ J/g} \cdot \text{mol}$$

The conversion of this quantity into electronvolts per molecule is carried out as follows:

$$\Delta h_R^\circ = 483{,}700 \text{ J/g} \cdot \text{mol} \times \frac{\text{g} \cdot \text{mol}}{6.023 \times 10^{23} \text{ molecule}} \times \frac{\text{eV}}{1.06 \times 10^{-19} \text{ J}}$$

$$= +7.58 \text{ eV/molecule}$$

This value of the enthalpy of reaction is also known as the *ionization potential* of the potassium atom. Note that the reaction is highly endothermic.

PROBLEMS†

Calculation of K_p data

15-1M Consider the reaction $H_2 + \frac{1}{2}O_2 \rightarrow H_2O$ at (*a*) 298°K, (*b*) 1600°K, and (*c*) 2000°K.

(1) Calculate the value of g° for each gas in the reaction, in kJ/kg · mol.

(2) Evaluate Δg_T° for the stoichiometric reaction, in kilojoules.

(3) On the basis of Eq. (15-24), calculate $\ln K_p$.

(4) Compare this value of $\ln K_p$ to that given in Table A-29M, after first converting the answer in part 3 to the base 10 logarithm.

15-2M Same as Prob. 15-1M, except the reaction is $CO + \frac{1}{2}O_2 \rightarrow CO_2$.

† All problems for this chapter are metric.

15-3M Same as Prob. 15-1M, except the reaction is $O_2 \rightarrow 2O$, and the temperatures are (*a*) 298°K, (*b*) 2200°K, and (*c*) 2600°K.

15-4M Same as Prob. 15-1M, except the reaction is $OH + \frac{1}{2}H_2 \rightarrow H_2O$, and the temperatures are (*a*) 298°K, (*b*) 2200°K, and (*c*) 2600°K.

15-5M A gas mixture at 10 atm consists of 0.1 mol of CO, 0.6 mol of CO_2, 0.3 mol of O_2, and 2 mol of N_2. The mixture is in equilibrium for the reaction $CO + \frac{1}{2}O_2 \rightleftharpoons CO_2$. Determine (*a*) the value of K_p, and (*b*) the mixture temperature, in degrees kelvin, using data from Table A-29M.

15-6M Consider the reaction $O_2 + N_2 \rightleftharpoons 2NO$. Measurement of an equilibrium mixture indicates the presence of 0.942 mol of O_2, 2.942 mol of N_2, and 0.116 mol of NO. Determine (*a*) the value of K_p, and (*b*) the temperature of the mixture, in degrees kelvin, using data from Table A-29M.

15-7M Consider the gas-phase reaction $CO + H_2O \rightleftharpoons CO_2 + H_2$. At 2 atm tests indicate an equilibrium composition of 1.56 mols of CO_2, 1.44 mols of CO, 2.44 mols of H_2O, and 0.60 mols of H_2. Determine (*a*) the value of K_p, and (*b*) the temperature of the mixture, using data from Table A-29M.

15-8M Consider the chemical reaction

$$CO(g) + 3H_2(g) \rightleftharpoons CH_4(g) + H_2O(g)$$

Initially, a constant-pressure reaction vessel is filled with 2 mol of CO, 5 mol of H_2, and 2 mol of nitrogen (an inert gas) at temperature T. For the reaction given above, it is found that, at chemical equilibrium, the products include 0.5 mol CO at 9 atm total pressure and temperature T. Determine the equilibrium constant K_p for the reaction.

15-9M At what temperature is (*a*) 10 percent, and (*b*) 7 percent of CO_2 dissociated at 2 atm pressure, using Table A-29M?

15-10M One mole of water vapor is heated at 2 bars until it is (*a*) 10 percent, and (*b*) 7 percent dissociated. Using Table A-29M, determine the final mixture temperature if the sole products of dissociation are H_2 and O_2.

Equilibrium-composition calculations

15-11M At what temperature, in degrees kelvin, will CO be (*a*) 10 percent, and (*b*) 7 percent of the total moles of products if CO is burned with the stoichiometric amount of O_2 at 2 atm pressure?

15-12M What temperature is necessary to dissociate diatomic oxygen to a state where the monatomic species comprises 20 percent of the total moles present at equilibrium for a pressure of (*a*) 0.30, and (*b*) 0.20 atm?

15-13M The equilibrium constant K_p for the reaction $I_2 \rightleftharpoons 2I$ at 1500°K is 1.22.

(*a*) If 1 mol of diatomic iodine is heated to 1500°K at 0.5 atm, determine the number of moles of I_2 at equilibrium.

(*b*) Determine the moles of I_2 at equilibrium at 1500°K and 0.5 atm if the initial reactants include 1 mole of I_2 and 1 mole of argon.

(*c*) Determine the moles of I_2 at equilibrium at 1500°K and 0.3 atm if the initial reactants include 1 mole of I_2 and 2 moles of argon.

15-14M The standard-state Gibbs-function change for the dissociation of diatomic fluorine (F_2) into the monatomic species at 12°K is 12,550 kJ/kg · mol.

(*a*) Compute the percent dissociation of pure F_2 at this temperature and (1) 1 atm and (2) 0.1 atm.

(*b*) If the initial reactants include 1 mol of F_2 and 1 mol of N_2 (an inert gas) at 1200°K, determine the moles of F_2 present at (1) 1 atm and (2) 0.1 atm.

15-15M At 300°K the equilibrium constant for the reaction $N_2O_4 \rightleftharpoons 2NO_2$ is $K_p = 0.18$. What system pressure will be required for (*a*) 20 percent and (*b*) 15 percent of the initially pure N_2O_4 to dissociate at this temperature?

15-16M Consider the dissociation reaction $NO \rightleftharpoons \frac{1}{2}O_2 + \frac{1}{2}N_2$ at 2500°K.

(*a*) Determine the percent dissociation for initially pure NO at 2 atm pressure.

(*b*) Determine the moles of NO present at 1 atm if the initial reactants include 2 mol of NO and 1 mol of He.

(*c*) Repeat part *b* if the initial reactants are 2 mol of NO and 1 mol of O_2.

15-17M One mole of water at 3000°K dissociates according to the reaction $H_2O \rightleftharpoons H_2 + \frac{1}{2}O_2$.

(*a*) Determine the percent dissociation at this temperature and 1 atm pressure.

(*b*) Repeat the calculation for the same temperature and pressure if the initial reactants include 1 mole of water and 1 mole of nitrogen.

(*c*) Repeat the calculation for the same temperature and reactants as part *a*, but with the pressure 0.5 atm.

15-18M The equilibrium constant K_p for the reaction $\frac{3}{2}H_2 + \frac{1}{2}N_2 \rightleftharpoons NH_3$ is 0.0068 at 450°C. Determine the equilibrium composition at 450°C if 3 mol of H_2 and 1 mol of N_2 are mixed at a pressure of (*a*) 20 atm and (*b*) 50 atm.

15-19M A gas mixture enters a reaction vessel with the following composition by volume: SO_2, 7.8 percent; O_2, 10.8 percent; N_2, 81.4 percent. The pressure and temperature are 1 atm and 500°C. If the K_p value is 56, then determine the number of moles of SO_3 per mole of reactant gas at equilibrium for the overall reaction $SO_2 + \frac{1}{2}O_2 \rightleftharpoons SO_3$.

15-20M A mixture consisting of 1 mol of H_2, 0.7 mol of O_2, and 1 mol of N_2 react at 3000°K and (*a*) 0.5 atm, and (*b*) 0.3 atm. For the reaction $H_2 + \frac{1}{2}O_2 \rightleftharpoons H_2O$, determine the number of moles of H_2O present at equilibrium.

15-21M At 1000°K the value of K_p for the reaction $NO + \frac{1}{2}O_2 \rightleftharpoons NO_2$ is 0.110.

(*a*) Determine the moles of NO present at 1000°K and 1.1 atm if the initial reactant is 1 mol of pure NO_2.

(*b*) Determine the moles of NO present if the initial reactants include 1 mol of NO and 2 mol of NO_2 at 1000°K and 1.1 atm.

(*c*) Repeat part *b* if the pressure changes to 0.5 atm.

15-22M One mole of CO_2 is mixed with 1 mol of H_2. Determine the equilibrium composition for the reaction $CO_2 + H_2 \rightleftharpoons CO + H_2O$ if the temperature is (*a*) 1000°K, and (*b*) 2000°K.

15-23M Repeat Prob. 15-22M if the initial composition is 1 mol of CO_2 and 2 mol of H_2.

15-24M Reconsider Prob. 15-22M if the initial reactants include 2 mol of CO_2, 1 mol of CO, and 2 mol of H_2O for (*a*) 1000°K, and (*b*) 2000°K.

Ionization reactions

15-25M The equilibrium constant for the ionization reaction for argon ($Ar \rightleftharpoons Ar^+ + e^-$) at 10,000°K is 0.00042. Determine the mole fraction of ionized argon atoms at (*a*) 0.01 atm and (*b*) 0.05 atm.

15-26M Cesium vapor ionizes at elevated temperatures according to the reaction $Cs \rightleftharpoons Cs^+ + e^-$. Determine the percent ionization at (*a*) 1400°K, (*b*) 1800°K, and (*c*) 2000°K, if the corresponding $\log_{10} K_p$ values are −1.01, 0.609, and 1.19, respectively, and the pressure is 1 atm.

15-27M For the ionization reaction $Na \rightleftharpoons Na^+ + e^-$ the value of $\log_{10} K_p$ at 2000°K is −0.175. Determine the percent ionization at this temperature at (*a*) 1 atm and (*b*) 0.5 atm.

15-28M The ionization equilibrium constant for the reaction $N \rightleftharpoons N^+ + e^-$ is (*a*) 6.26×10^{-4} at 10,000°K, (*b*) 1.51×10^{-2} at 12,000°K, and (*c*) 0.151 at 14,000°K. What system pressure is necessary for 5 percent of N to ionize, assuming this is the sole species present initially.

Energy analysis of equilibrium mixtures

15-29M Diatomic oxygen (O_2) is heated at a constant pressure of 0.1 atm from 298°K to (*a*) 3000°K, and (*b*) 2800°K. Determine the heat transferred to or from the reaction vessel, in kJ/kg · mol.

15-30M Diatomic hydrogen (H_2) is heated at constant pressure from 25°C to 2800°K at (*a*) 0.1 atm, and (*b*) 0.2 atm. Determine the heat transferred to or from the reaction vessel, in kJ/kg · mol.

15-31M One mole of carbon monoxide and 80 percent stoichiometric air initially at 25°C are allowed to react adiabatically at a pressure of 1 atm. Compute the maximum combustion temperature with dissociation.

15-32M Carbon monoxide is reacted with a 10 percent excess of the stoichiometric amount of air, both initially at 500°K.

(*a*) Determine the adiabatic-flame temperature without dissociation.

(*b*) Determine the final equilibrium-flame temperature and the percent dissociation for the same initial conditions if the pressure is (1) 2 atm, (2) 5 atm, and (3) 10 atm.

15-33M Carbon monoxide is burned at constant volume, using the theoretical amount of air for combustion to CO_2. The initial reactant pressure and temperature at 1 atm and 500°K, respectively. Determine the maximum combustion temperature if dissociation is considered.

15-34M One mole of carbon monoxide is burned with 20 percent excess air at a pressure of 0.5 atm. If the original temperature of the reactants is 25°C, determine the adiabatic-combustion temperature with dissociation.

15-35M Reconsider Prob. 14-39M. For the same input data, determine the adiabatic-combustion temperature, in degrees kelvins, for (*a*) 350 percent, and (*b*) 400 percent theoretical oxygen which enters at 500°K, accounting for the dissociation of water into hydrogen and oxygen at 1 atm.

15-36M Solid carbon at 25°C and the stoichiometric quantity of air at 600°K are burned adiabatically at a constant pressure of 1 atm. Determine the moles of CO_2 formed per mole of carbon, assuming that only CO_2, CO, O_2, and N_2 are present in the products.

15-37M Carbon monoxide is burned with 120 percent of the theoretical air requirement, both being supplied at 25°C and burned at a pressure of 0.1 atm. It is desired to supply the products of the reaction at 2500°K, assuming that only CO_2, CO, O_2, and N_2 are present in the products.

(*a*) Determine the equilibrium composition at 2500°K.

(*b*) Determine the heat which must be added or removed from the product gases (per mole of initial CO) in order to achieve the required final temperature.

(*c*) Assume that CO initially is at 25°C, and the products again are at 2500°K and 0.1 atm. To what temperature must the air be preheated before the reaction so that the reaction itself is adiabatic, in degrees kelvin?

15-38M Carbon monoxide and water vapor in a 1 : 1 molar ratio enter a reaction vessel maintained at 1 atm at 500°K. The mixture is heated at 1500°K. Determine the magnitude and direction of any heat transfer, in kJ/mol of initial CO, if the only gases present are CO, H_2O, CO_2, and H_2.

Simultaneous reactions

15-39M A mixture of 1 mol of CO_2, 1 mol of H_2, and $\frac{1}{2}$ mol of O_2 is raised to 3000°K at 1 atm. Assume that the mixture consists of CO_2, H_2, O_2, CO, and H_2O. Determine the equilibrium composition, assuming H, O, and OH are absent.

15-40M The reaction of 1 mol of CO_2, 1 mol of H_2, and 0.5 mol of O_2 yields an equilibrium mixture of 0.5 mol of CO_2, 0.11 mol of H_2, 0.305 mol of O_2, 0.5 mol of CO, and 0.89 mol of H_2O. Determine the temperature, in degrees kelvin, at which this equilibrium condition exists, if the pressure is 30.3 atm.

15-41M One mole of H_2O vapor is heated to (*a*) 2800°K, and (*b*) 3000°K. Determine the equilibrium composition assuming that H_2O, H_2, O_2, and OH are present.

Relationship between K_p and Δh_R°

15-42M On the basis of the van't Hoff isobar equation, (1) estimate at (*a*) 1700°K, (*b*) 2000°K, and (*c*) 3000°K, the enthalpy of formation of O_2 from monatomic oxygen (O), in kJ/kg · mol of O_2. (2) For a comparison, calculate the Δh_R° value calculated from $\Delta h_{f,298}^\circ$ and sensible enthalpy data.

15-43M From the equilibrium-constant data for the reaction $H_2 \rightleftharpoons 2H$ (1) estimate the energy required to break H_2 down into its monatomic species at (*a*) 1500°K, (*b*) 2000°K, and (*c*) 2500°K and 1 atm in kJ/kg · mol of H_2. (2) For a comparison, calculate the Δh_R° value at the given temperature by using $\Delta h^\circ_{f,\,298}$ and by noting that the sensible enthalpy change of monatomic hydrogen for parts *a*, *b*, and *c* is 24,980, 35,375, and 45,770 kJ/kg, respectively.

15-44M Calculate Δh_R at (*a*) 1800°K, (*b*) 2200°K, and (*c*) 2500°K for the reaction $CO + \frac{1}{2}O_2 \leftrightarrows CO_2$ by using equilibrium-constant data. For comparison purposes, calculate the Δh°_R value at the given temperature by using $\Delta h^\circ_{f,\,298}$ and sensible enthalpy data.

15-45M On the basis of data given in Prob. 15-28M, estimate the enthalpy of ionization for the reaction $N \leftrightarrows N^+ + e^-$ at 12,000°K, in kJ/kg · mol and eV/molecule of N.

15-46M On the basis of the answer to Prob. 15-42M, will the amount of O_2 formed from the reaction $2O \leftrightarrows O_2$ increase or decrease with an increase in temperature at constant pressure?

15-47M For the reaction $N_2O_4 \leftrightarrows 2NO_2$, the heat of reaction at 300°K is 57,930 kJ/kg · mol. Will the amount of N_2O_4 in the equilibrium mixture increase or decrease with (*a*) an increase in temperature at constant pressure and (*b*) an increase in pressure at constant temperature?

15-48M For the dissociation of SO_3 according to the equation $SO_3 \leftrightarrows SO_2 + \frac{1}{2}O_2$, the equilibrium constant is given by the approximate relation, $\ln K_p = -11{,}400/T + 10.75$, where T is in degrees kelvin.

(*a*) Estimate the enthalpy of reaction, kJ/kg · mol.

(*b*) If 0.01 mol of SO_3 is sealed in a rigid, but uninsulated, vessel at 1 atm and 300°K, determine the temperature to which the system must be heated to permit 5 percent of the SO_3 to dissociate.

(*c*) For the final conditions of part *b*, determine the final vessel pressure in atm.

15-49M The equilibrium-constant data for the reaction $Cs \leftrightarrows Cs^+ + e^-$ are

T, °K	1400	1600	1800	2000
$\log_{10} K_p$	−1.010	−0.108	0.609	1.194

Estimate the enthalpy of ionization for the above reaction at (*a*) 1600°K, and (*b*) 1800°K, in kJ/kg · mol and eV/molecule.

15-50M The equilibrium-constant data for the reaction $Na \rightleftharpoons Na^+ + e^-$ are

T, °K	1600	1800	2000	2200
$\log_{10} K_p$	−1.819	−0.913	−0.175	0.438

Estimate the enthalpy of ionization for the above reaction at (*a*) 1800°K, and (*b*) 2000°K, in kJ/kg · mol and eV/molecule.

CHAPTER

SIXTEEN

GAS POWER CYCLES

An important use of thermodynamics is made in the study of cyclic devices for power production. In this chapter we shall restrict ourselves to devices which employ a gas as the working fluid. Modern automotive, truck, and gas-turbine engines are examples of the extremely fruitful application of thermodynamic analysis. In the next two chapters the study will be directed toward power and refrigeration processes which involve the presence of two phases during the course of the process.

16-1 THE AIR-STANDARD CYCLE

Because of the complexities of the actual processes, it is profitable in the initial study of gas power and refrigeration cycles to examine the general characteristics of each cycle without going into a detailed analysis. The advantage of a simple model is that the main parameters which govern the cycle are made more apparent. By stripping the actual process of all its complications and retaining only a bare minimum of detail, the engineer is able to examine the influence of major operating variables on the performance of the device. However, it must be kept in mind that any numerical values calculated from such models may not be strictly representative of the actual process. Thus modeling is an important tool in engineering analysis, but at times it is highly qualitative.

Gas cycles are those in which the working fluid remains a gas throughout the cycle. In actual gas power cycles the fluid consists mainly of air, plus the products of combustion such as carbon dioxide and water vapor. Since the gas is predominantly air, especially in gas-turbine cycles, it is convenient to examine gas power cycles in terms of an air-standard cycle. An *air-standard* cycle is an idealized cycle based on the following approximations: (1) the working fluid is taken to be solely air throughout the cycle, and the air behaves like an ideal gas; (2) any combustion process which might appear in actual practice is replaced by a heat-addition process from an external source; and (3) a heat-rejection process to the surroundings is used to restore the fluid to its initial state and complete the cycle.

In applying the air-standard cycle restrictions to various processes, it is sometimes customary to place additional constraints on the property values of air. For example, in the *cold*-air-standard cycle the specific-heat quantities c_v and c_p and the specific-heat ratio k are assumed to have constant values, and these are measured at room temperature. This approach is frequently used, but the numerical results may be considerably different from those obtained by accounting for variable specific heats. This is due to the wide temperature variation in most gas power cycles, which severely alters the values of c_v and c_p throughout the cycle. In actual practice, of course, it would be desirable to employ information on the actual gases which result from the combustion of hydrocarbon fuels mixed with air.

16-2 THE AIR-STANDARD CARNOT CYCLE

In Chap. 6 it is pointed out the the maximum thermal efficiency of any heat engine operating between two fixed temperature levels is the Carnot efficiency. The relation for the Carnot efficiency is

$$\eta_{\text{Carnot}} = 1 - \frac{T_L}{T_H} \tag{6-4}$$

where T_L and T_H represent the sink temperature and the source temperature, respectively. A Carnot heat engine, as described in Sec. 6-14, undergoes a cyclic process composed of two isothermal, reversible processes and two adiabatic, reversible processes. The PV and TS diagrams for air undergoing a Carnot cycle are shown in Fig. 16-1. This cycle may take place in a closed system, such as a reciprocating piston-cylinder device, or in a steady-flow device. The equipment required for a steady-flow Carnot cycle using air as the working fluid is shown in Fig. 16-2. Steady-flow turbines are required for the isothermal, reversible and adiabatic, reversible expansion processes (steps 1-2 and 2-3). Similarly, steady-flow compressors are needed for the isothermal, reversible and adiabatic, reversible compression process (steps 3-4 and 4-1). Heat addition Q_H occurs to the fluid in the isothermal turbine, while heat removal Q_L from the air occurs in the isothermal compressor.

To approach the Carnot efficiency given by Eq. (6-4), an actual engine must be relatively free of dissipative effects, such as friction. In addition, the fluid temperature should be constant during the heat-addition and heat-removal processes. In practice these restrictions, among others, are impossible to meet. The achievement of an engine which approximates the Carnot cycle is not practical. Hence the efficiency of an actual engine is always considerably less than that for a

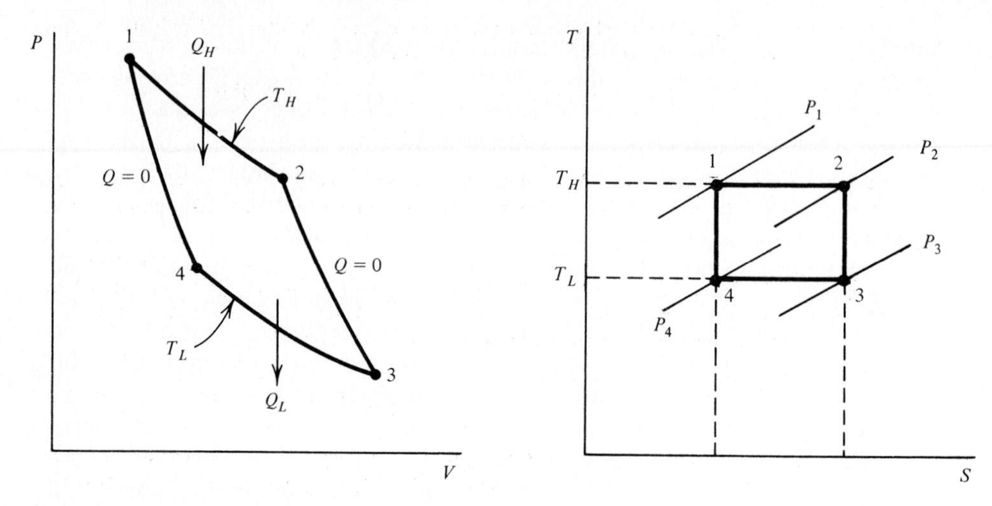

Figure 16-1 The *PV* and *TS* diagrams for a Carnot heat-engine cycle.

Carnot engine operating between the same maximum and minimum temperatures. Nevertheless the performance of a Carnot heat engine operating on an air-standard cycle is an imporatnt standard against which actual engines may be compared.

Example 16-1M The pressure, volume, and temperature in a Carnot heat engine using air as the working medium are, at the beginning of the isothermal expansion, 5 bars, 500 cm^3, and 260°C, respectively. During the isothermal expansion, 0.30 kJ of heat is added and the maximum volume for the cycle is 3300 cm^3. Determine (*a*) the volume after isothermal expansion, (*b*) the sink temperature, in degrees Celsius, (*c*) the volume after isothermal compression, in cm^3, (*d*) the heat rejected per cycle, and (*e*) the thermal efficiency of the cycle.

SOLUTION (*a*) The final volume after isothermal expansion is set by the amount of heat added during the process. Because the process is isothermal and internally reversible, $Q_H = T_H \Delta S_H$.

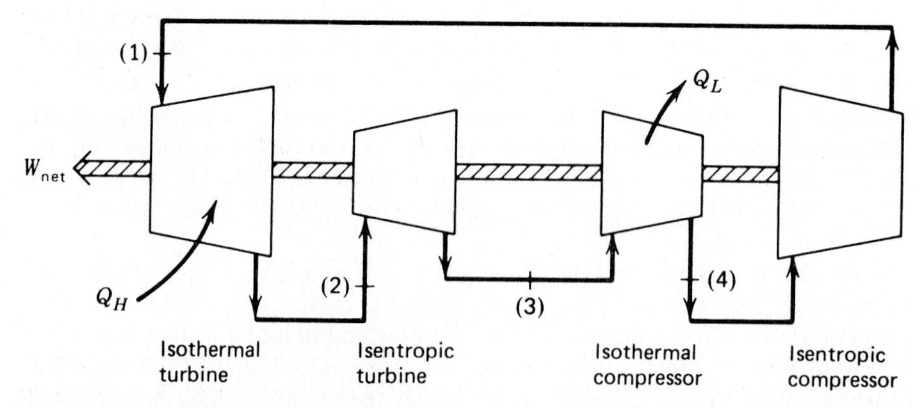

Figure 16-2 Steady-flow Carnot heat engine.

The entropy change of an ideal gas at constant temperature is simply $mR \ln V_2/V_1$. Hence, $Q_H = mRT_H \ln V_2/V_1 = P_1V_1 \ln V_2/V_1$. Substitution of data leads to

$$0.30 \text{ kJ} = 5 \text{ bar} \times 550 \text{ cm}^3 \times 10^{-4} \text{ kJ/bar} \cdot \text{cm}^3 \times \ln V_2/V_1$$

$$V_2 = 2.977V_1 = 2.977(550) = 1637 \text{ cm}^3$$

Another way to evaluate V_2/V_1 is to note that for an isothermal process of an ideal gas $Q = W = \int P\, dV$. By equating Q_H to the integral of $P\, dV$ for the isothermal expansion, the same algebraic expression results.

(*b*) The sink temperature $T_L(=T_3)$ is determined by examining process 2-3, the isentropic expansion. For this process T_2 $(=T_H)$, V_2, and V_3 are known. In Sec. 7-9 it was shown for isentropic processes involving ideal gases that $V_3/V_2 = (T_2/T_3)^{1/(1-k)}$ if the specific heats are constant. Using the data for this problem and assuming $k = 1.40$, we find that

$$\frac{3300}{1637} = \left(\frac{533}{T_3}\right)^{1/(1-1.4)} \qquad \text{or} \qquad T_3 = \frac{533}{1.326} = 402°\text{K} = 129°\text{C}$$

(*c*) The volume after isothermal compression is found by employing the isentropic relation used in part *b*: $V_4/V_1 = (T_1/T_4)^{1/(1-k)}$. However, $T_1/T_4 = T_2/T_3$. Therefore, for a Carnot cycle, $V_4/V_1 = V_3/V_2$. As a result,

$$V_4 = V_3 \frac{V_1}{V_2} = \frac{3300}{2.977} = 1100 \text{ cm}^3$$

(*d*) The heat rejected is determined from the Carnot cycle relation, $Q_H/Q_L = T_H/T_L$.

$$Q_L = Q_H \frac{T_L}{T_H} = 0.30 \frac{402}{533} = 0.226 \text{ kJ}$$

Note that parts *c* and *d* may be interchanged. Once Q_L is determined from a knowledge of the two temperatures, then the method of part *a* can be employed to find V_4.

(*e*) The thermal efficiency for the cycle is simply

$$\eta_{\text{Carnot}} = 1 - \frac{T_L}{T_H} = 1 - \frac{402}{533} = 0.246 \qquad \text{(or 24.6 percent)}$$

The efficiency is relatively low, because of the low temperature of supply.

Example 16-1 The pressure, volume, and temperature in a Carnot heat engine using air as the working medium are, at the beginning of the isothermal expansion, 75 psia, 0.20 ft^3, and 500°F, respectively. During the isothermal expansion 3.0 Btu of heat are added, and the maximum volume for the cycle is 1.20 ft^3. Determine (*a*) the volume after isothermal expansion, in ft^3, (*b*) the sink temperature, in degrees Fahrenheit, (*c*) the volume after isothermal compression, in ft^3, (*d*) the heat rejected per cycle, and (*e*) the thermal efficiency of the cycle. The system is closed.

SOLUTION (*a*) The final volume after isothermal expansion is set by the amount of heat added during the process. Because the process is isothermal and internally reversible, $Q_H = T_H\, \Delta S_H$. The entropy change of an ideal gas at constant temperature is simply $mR \ln V_2/V_1$. Hence, $Q_H = mRT_H \ln V_2/V_1 = P_1V_1 \ln V_2/V_1$. Substitution of data leads to

$$3.0 \text{ Btu} = (75 \times 144) \text{ lb}_f/\text{ft}^2 \times 0.20 \text{ ft}^3 \times \frac{\text{Btu}}{778 \text{ ft} \cdot \text{lb}_f} \times \ln \frac{V_2}{V_1}$$

$$V_2 = 2.946V_1 = 2.946(0.20) = 0.589 \text{ ft}^3$$

Another way to evaluate V_2/V_1 is to note that for an isothermal process of an ideal gas $Q = W = \int P\, dV$. By equating Q_H to the integral of $P\, dV$ for the isothermal expansion, the same algebraic expression results.

(*b*) The sink temperature T_L $(=T_3)$ is determined by examining process 2-3, the isentropic expansion. For this process T_2 $(=T_H)$, V_2, and V_3 are known. In Sec. 7-9 it was shown for isentropic processes involving ideal gases that $V_3/V_2 = (T_2/T_3)^{1/(1-k)}$ if the specific heats are constant. Using the data for this problem and assuming $k = 1.40$, we find that

$$\frac{1.20}{0.589} = \frac{960^{1/(1-1.4)}}{T_3} \qquad \text{or} \qquad T_3 = \frac{960}{1.329} = 722°\text{R} = 262°\text{F}$$

(*c*) The volume after isothermal compression is found by employing the isentropic relation used in part *b*: $V_4/V_1 = (T_1/T_4)^{1/(1-k)}$. However, $T_1/T_4 = T_2/T_3$. Therefore, for a Carnot cycle, $V_4/V_1 = V_3/V_2$. As a result,

$$V_4 = V_3 \frac{V_1}{V_2} = \frac{1.20}{2.946} = 0.407 \text{ ft}^3$$

(*d*) The heat rejected is determined from the Carnot cycle relation, $Q_H/Q_L = T_H/T_L$.

$$Q_L = Q_H \frac{T_H}{T_L} = 3.0 \frac{722}{960} = 2.256 \text{ Btu}$$

Note that parts *c* and *d* may be interchanged. Once Q_L is determined from a knowledge of the two temperatures, then the method of part *a* can be employed to find V_4.

(*e*) The thermal efficiency for the cycle is simply

$$\eta_{\text{Carnot}} = 1 - \frac{T_L}{T_H} = 1 - \frac{722}{960} = 0.248 \qquad \text{(or 24.8 percent)}$$

The efficiency is relatively low, because the supply temperature T_H is low.

16-3 SOME INTRODUCTORY NOMENCLATURE FOR RECIPROCATING DEVICES

A number of applications make use of a piston-cylinder arrangement in which the piston is observed to undergo cycles, or revolutions. The *bore* of the piston is its diameter, and the distance the piston moves in one direction is known as the *stroke*. When the piston has moved to a position such that a minimum volume of fluid is left in the cylinder, the piston is said to be at top dead center (TDC). This minimum volume is called the *clearance volume*. When the piston has moved the distance of the stroke so that the fluid now occupies the maximum volume, the piston is in the bottom dead center (BDC) position. The volume displaced by the piston as it moves the distance of the stroke between TDC and BDC is the displacement volume. The clearance volume is frequently cited in terms of the *percent clearance*, which is the percent of the piston displacement equal to the clearance volume. The *compression ratio r* of a reciprocating device is defined as the volume of the fluid at BDC divided by the volume of the fluid at TDC; that is,

$$r = \frac{V_{\text{BDC}}}{V_{\text{TDC}}} = \frac{\text{clearance volume} + \text{displacement volume}}{\text{clearance volume}} \tag{16-1}$$

The compression ratio is always expressed in terms of a volume ratio.

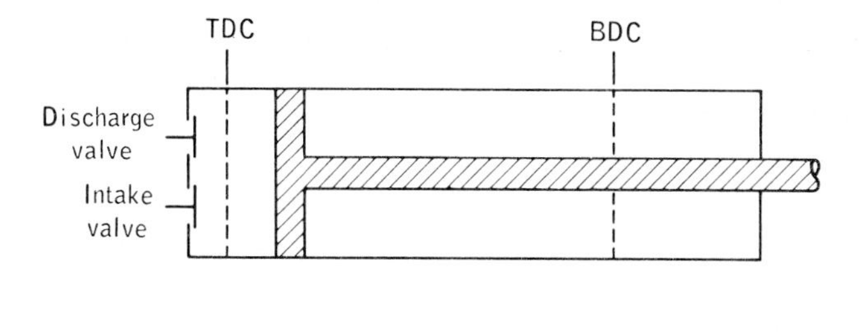

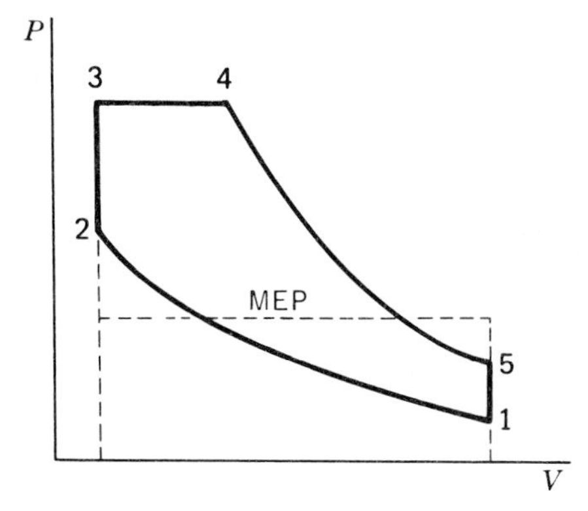

Figure 16-3 Interpretation of mean effective pressure on a *PV* diagram.

The *mean effective pressure* (MEP) is a useful parameter in the study of reciprocating devices that are used for power production. It is defined as an average pressure which, if it acted on the piston during the entire power or outward stroke, would produce the same work output as the net work output for the actual cyclic process. It is seen that the work per cycle is given by

$$\begin{aligned} W_{\text{cycle}} &= (\text{MEP})(\text{piston area})(\text{stroke}) \\ &= (\text{MEP})(\text{displacement volume}) \end{aligned} \tag{16-2}$$

For reciprocating engines of comparable size, a larger mean effective pressure is an indication of better performance in terms of power produced at the same rated speed. For the purpose of illustration, consider the hypothetical clockwise cycle 1-2-3-4-5-1 in Fig. 16-3. The net work produced is represented by the enclosed area on the diagram. The mean effective pressure of the cycle is shown by the horizontal line, and the area under this line equals the enclosed area of the actual cycle.

16-4 THE AIR-STANDARD OTTO CYCLE

The four-stroke spark-ignition engine is an important component in the operation of a technology to meet the modern needs of society. Although it has undergone some modifications in order to meet pollution standards for mobile equipment, this engine will undoubtedly continue to play a significant role as a device for producing relatively small quantities of power. A typical *PV* diagram for such an engine at wide-open throttle is shown in Fig. 16-4. The series of events includes the intake stroke *ab*, the compression stroke *bc*, the expansion or power stroke *cd*, and finally the exhaust stroke *da*. The intake and exhaust strokes occur essentially at atmospheric pressure. The process lines *ab* and *da* do not lie on top of each other, but since Fig. 16-4 is drawn to scale it is difficult to show the separation between the intake- and exhaust-process lines, except near the BDC position. Normally the point of ignition occurs on the compression stroke before the TDC position, since flame propagation across the combustion chamber takes a finite time. For a given engine, the point of ignition can be altered until the setting for maximum power is determined. Note also that the exhaust valve is opened before the piston reaches BDC. This allows the pressure of the exhaust gases to nearly reach atmospheric pressure before the exhaust stroke begins.

As discussed in Sec. 16-1, the initial step in analyzing the performance of a four-stroke reciprocating spark-ignition engine is to prepare a simple model of the overall process. While such a model is of value only in a qualitative sense, it does provide some information on the influence of major operating variables on performance.

A theoretical cycle of interest in analyzing the behavior of reciprocating spark-ignition engines is the Otto cycle. A four-stroke Otto cycle is composed of four internally reversible processes, plus an intake and an exhaust portion of the cycle. Both *PV* and *TS* diagrams for the theoretical cycle are shown in Fig. 16-5. Consider a piston-cylinder assembly containing air and the piston situated at the bottom dead center position. This is shown as point 1 on the diagrams. As the piston moves to the top dead center position, compression of the air occurs adiabatically. Since the processes are reversible, the compression process is isentropic, ending at state 2. Heat is then added to the air instantaneously, so that

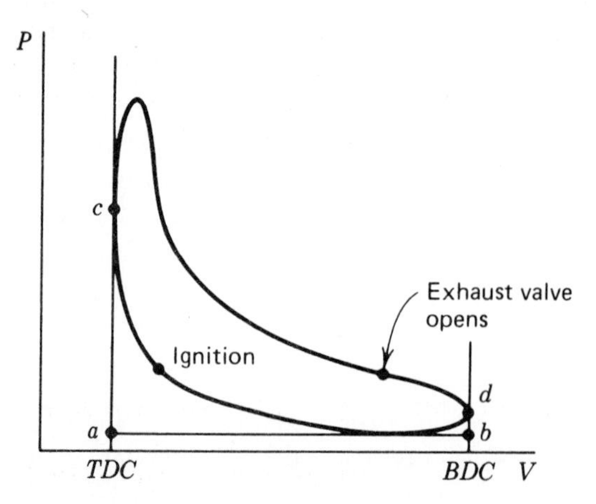

Figure 16-4 Typical *PV* diagram for wide-open throttle for a 4-stroke spark-ignition engine.

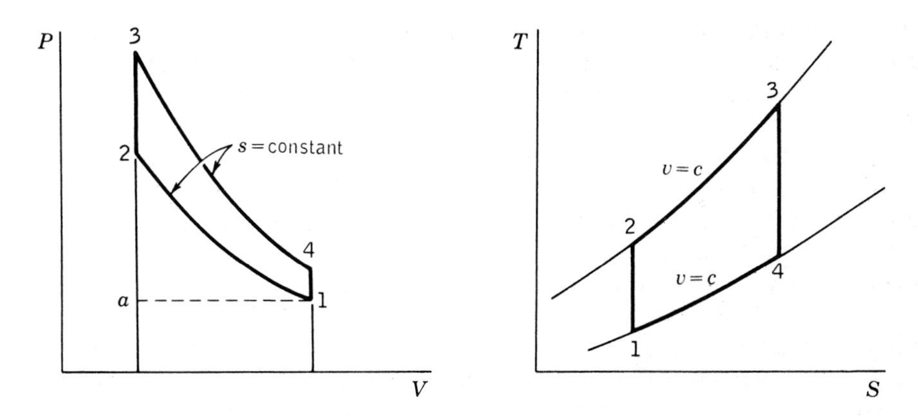

Figure 16-5 PV and TS diagrams for an air-standard Otto cycle.

both the pressure and the temperature rise to high values during a constant-volume process 2-3. As the piston now moves toward the BDC position once more, the expansion is carried out adiabatically and internally reversible, i.e., isentropically, to state 4. Now with the piston in its BDC position, heat is rejected at constant volume until the initial state is achieved.

At this point the fluid, theoretically, could begin to go through another cycle. In order to make the cycle somewhat more realistic, the following sequence might be considered before resuming the cyclic pattern: In actual practice, the gases would contain the products of hydrocarbon combustion, so that an exhaust stroke would be necessary. Consequently, the exhaust valve now opens, and the piston moves from BDC to TDC, expelling the gases to the ambient surroundings. Then the exhaust valve closes and the intake valve opens, while the piston returns to the BDC position. During this suction stroke the cylinder is filled with fresh air for the next cycle. In the air-standard cycle this recharging of the cylinder is not necessary, since the same fluid continually undergoes the cyclic variations. Note that the work required to push the charge from the cylinder is the same in magnitude but opposite in sign to that required to suck in the new charge. Hence these two portions of the theoretical cycle do not affect the net work done by the cycle, and it is only the cycle 1-2-3-4-1 which is important in the thermodynamic analysis. In review, the theoretical Otto cycle is composed of the following internally reversible processes:

1. Adiabatic compression, 1-2
2. Constant-volume heat addition, 2-3
3. Adiabatic expansion, 3-4
4. Constant-volume heat rejection, 4-1

In addition, one may consider an exhaust stroke, 1-*a*, and an intake stroke, *a*-1, for completeness, although this is not necessary.

Since the air acts as a closed system, the conservation of energy principle, when applied to the various processes, leads to the following equations: For the adiabatic compression and expansion processes, since $q = 0$,

$$w = \Delta u$$

For the constant-volume heat input and rejection processes, since $w = 0$,

$$q = \Delta u$$

At this point it is informative to analyze the Otto cycle on the basis of the cold air-standard cycle, since it provides some insight as to the important parameters that determine the thermal efficiency of the cycle. Employing the cold-air-standard cycle, we find that

$$q_{\text{in}} = q_{23} = u_3 - u_2 = c_v(T_3 - T_2)$$

and

$$q_{\text{out}} = -q_{41} = u_4 - u_1 = c_v(T_4 - T_1)$$

Since the net work is the sum of q_{23} and q_{41}, the thermal efficiency is given by

$$\eta_{\text{th}} = \frac{w_{\text{net}}}{q_{\text{in}}} = \frac{c_v(T_3 - T_2) - c_v(T_4 - T_1)}{c_v(T_3 - T_2)}$$

$$= 1 - \frac{T_4 - T_1}{T_3 - T_2} = 1 - \left(\frac{T_1}{T_2}\right)\frac{T_4/T_1 - 1}{T_3/T_2 - 1}$$

Note that $V_2 = V_3$ and $V_1 = V_4$. Since the isentropic relations show that

$$\frac{T_2}{T_1} = \left(\frac{V_1}{V_2}\right)^{k-1} \quad \text{and} \quad \frac{T_3}{T_4} = \left(\frac{V_4}{V_3}\right)^{k-1} = \left(\frac{V_1}{V_2}\right)^{k-1}$$

then $T_2/T_1 = T_3/T_4$, or $T_4/T_1 = T_3/T_2$. When this result is substituted into the equation for the thermal efficiency of an air-standard Otto cycle, it is seen that

$$\eta_{\text{th, Otto}} = 1 - \frac{T_1}{T_2} = 1 - \left(\frac{V_2}{V_1}\right)^{k-1} = 1 - \frac{1}{r^{k-1}} \tag{16-3}$$

where r is the *compression ratio* for the theoretical cycle. Equation (16-3) indicates that the major parameters governing the thermal efficiency of an Otto cycle are the compression ratio and the specific-heat ratio. The influence of these factors on the thermal efficiency is demonstrated in Fig. 16-6. For a given specific-heat ratio, the value of the thermal efficiency increases with increasing compression ratio. It should be noted, however, that the curves flatten out at compression ratios above 10 or so. Hence the advantage of operating at high compression ratios lessens rapidly. From a practical viewpoint, the compression ratio is limited by the occurrence of preignition or engine knock when the compression ratio rises much above 10, for common hydrocarbon fuels. Figure 16-6 also shows that the thermal efficiency increases with increasing specific-heat ratio. This means that higher values of the thermal efficiency are obtained when simple monatomic or diatomic molecules are present in the working fluid. Unfortunately, the presence of carbon dioxide and water vapor and other heavier molecules in actual practice makes the attainment of high specific-heat-ratio values a practical impossibility.

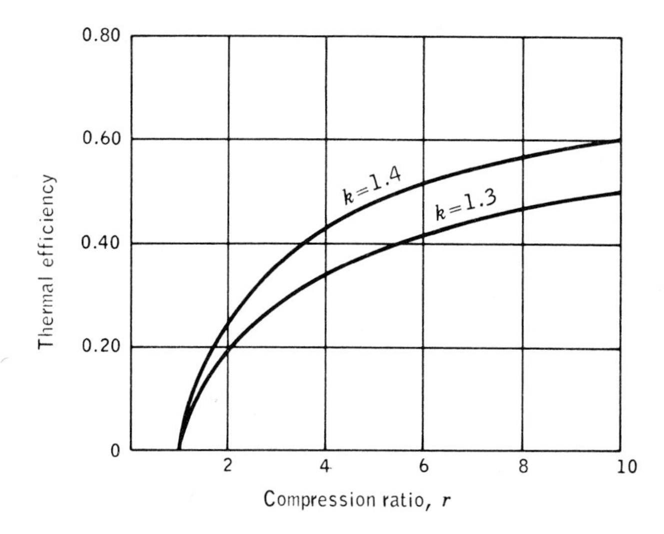

Figure 16-6 Thermal efficiency of the air-standard Otto cycle as a function of compression ratio and specific-heat ratio.

To account for variable specific-heat values, the thermal efficiency must be determined by the relation

$$\eta_{th} = 1 - \frac{u_4 - u_1}{u_3 - u_2} \qquad \text{(Otto cycle)} \qquad (16\text{-}4)$$

In addition, the u values must be read from the air table in the Appendix at the corresponding temperatures around the cycle. The temperatures at states 2 and 4 are found from the isentropic relations

$$v_{r2} = v_{r1}\left(\frac{V_2}{V_1}\right) = \frac{v_{r1}}{r} \qquad \text{and} \qquad v_{r4} = v_{r3}\left(\frac{V_4}{V_3}\right) = rv_{r3}$$

Recall that the v_r data are solely a function of temperature.

Example 16-2M The initial conditions for an air-standard Otto cycle operating with a compression ratio of 8 : 1 are 0.95 bar and 17°C. At the beginning of the compression stroke the cylinder volume is 2.20 L, and 3.60 kJ of heat are added during the constant-volume heating process. Calculate the pressure and temperature at the end of each process of the cycle, and determine the thermal efficiency and mean effective pressure of the cycle.

SOLUTION By denoting the states in the same manner as in Fig. 16-5, we let $P_1 = 0.95$ bar and $T_1 = 290°\text{K}$. From the ideal-gas equation of state,

$$v_1 = \frac{RT}{P} = \frac{0.08314(290)}{29(0.95)} = 0.875 \text{ m}^3/\text{kg}$$

The state after isentropic compression is determined by use of relative volume v_r data, which account for variable specific heats. On the basis of data from Table A-5M,

$$v_{r2} = v_{r1}\frac{v_2}{v_1} = \frac{676.1}{8.0} = 84.5$$

Linear interpolation in the table shows that T_{2s} is 652°K, $p_{r2} = 22.17$, and $u_{2s} = 475.1$ kJ/kg. One method of evaluating P_{2s} is

$$P_{2s} = P_1 \frac{p_{r2}}{p_{r1}} = 0.95 \frac{22.17}{1.2311} = 17.1 \text{ bars}$$

It might be noted that P_{2s} can also be calculated from the relation $P_2 = P_1(V_1/V_2)(T_2/T_1)$. To determine state 3, the amount of heat added per kilogram of air must be found.

$$q = \frac{Q}{m} = \frac{Qv_1}{V_1} = \frac{3.60 \text{ kJ}(0.875 \text{ m}^3/\text{kg})}{2.2 \times 10^{-3} \text{ m}^3} = 1432 \text{ kJ/kg}$$

Since $q_{in} = u_3 - u_2$, the internal energy at state 3 is found by

$$u_3 = u_2 + q = 475.1 + 1432 = 1907.1 \text{ kJ/kg}$$

Again, by linear interpolation in the air table, we find that $T_3 = 2235$°K, $p_{r3} = 3369$, and $v_{r3} = 1.907$. The pressure after constant-volume heating is simply $P_3 = P_2(T_3/T_2) = 17.1(2235/652) = 58.6$ bars. State 4 is found again from isentropic relations.

$$v_{r4} = v_{r3} \frac{v_4}{v_3} = 1.907(8) = 15.26$$

Linear interpolation in the air table reveals that $T_{4s} = 1180$°K, $p_{r4} = 222.2$, and $u_{4s} = 915.6$ kJ/kg. Thus $P_4 = P_3(p_{r4}/p_{r3}) = 58.6(222.2/3369) = 3.9$ bars.

The thermal efficiency is found from the heat supplied and rejected. The heat rejected is

$$q_{out} = u_4 - u_1 = 915.6 - 206.9 = 708.7 \text{ kJ/kg}$$

Consequently,

$$\eta_{th} = 1 - \frac{q_{out}}{q_{in}} = 1 - \frac{708.7}{1432} = 0.505 \qquad \text{(or 50.5 percent)}$$

Equation (16-3) was not used to calculate the thermal efficiency, because it is based on constant specific-heat data. Finally, the mean effective pressure of the cycle is found by applying Eq. (16-2). First, the net work $w_{net} = q_{in} - q_{out} = 1432 - 709 = 723$ kJ/kg. Then

$$\text{MEP} = \frac{w_{net}}{v_1 - v_2} = \frac{723 \text{ kJ/kg}}{0.875 - 0.109 \text{ m}^3/\text{kg}} \times \frac{\text{bar m}^2}{10^5 \text{ N}} \times 10^3 \text{ g/kg} = 9.44 \text{ bars}$$

where v_2 is one-eighth of v_1.

Example 16-2 The initial conditions for an air-standard Otto cycle operating with a compression ratio of 8 : 1 are 14.4 psia and 60°F. At the beginning of the compression stroke the total cylinder volume is 180 in^3, and 4.80 Btu of heat are added during the constant-volume heating process. Calculate the pressure and temperature at the end of each process of the cycle, and determine the thermal efficiency and mean effective pressure of the cycle.

Solution By denoting the states in the same manner as in Fig. 16-5, we let $P_1 = 14.4$ psia and $T_1 = 520$°R. From the ideal-gas equation of state,

$$v_1 = \frac{RT}{P} = \frac{10.73(520)}{29(14.4)} = 13.33 \text{ ft}^3/\text{lb}$$

The state after isentropic compression is determined by use of relative volume v_r data, which account for variable specific heats. On the basis of data from Table A-5 for air,

$$v_{r2} = v_{r1} \frac{v_2}{v_1} = \frac{158.58}{8.0} = 19.82$$

Linear interpolation in the table shows that $T_{2s} = 1170°R$, $p_{r2} = 21.93$, and $u_{2s} = 203.6$ Btu/lb. One method of evaluating P_{2s} is

$$P_{2s} = P_1 \frac{p_{r2}}{p_{r1}} = 14.4 \frac{21.93}{1.2147} = 260 \text{ psia}$$

It might be noted that P_{2s} can also be found from the relation $P_2 = P_1(V_1/V_2)(T_2/T_1)$. To determine state 3, the amount of heat added per pound of air must be found. This requires use of the initial cylinder volume.

$$q = \frac{Q}{m} = \frac{Qv_1}{V_1} = \frac{4.80 \text{ Btu } (13.33 \text{ ft}^3/\text{lb})}{(180/1728) \text{ ft}^3} = 614.2 \text{ Btu/lb}$$

Since $q_{in} = u_3 - u_2$, the internal energy at state 3 is found by

$$u_3 = u_2 + q = 203.6 + 614.2 = 817.8 \text{ Btu/lb}$$

Again, by linear interpolation in Table A-5, we find that $T_3 = 4016°R$, $p_{r3} = 3339$, and $v_{r3} = 0.4458$. The pressure after constant-volume heating is simply $P_3 = P_2(T_3/T_2) = 260(4016/1170) = 892$ psia. State 4 is found again from isentropic relations.

$$v_{r4} = v_{r3} \frac{v_4}{v_3} = 0.4458(8) = 3.566$$

Linear interpolation in the air table reveals that $T_{4s} = 2120°R$, $p_{r4} = 220.5$, and $u_{4s} = 392.8$ Btu/lb. Thus $P_4 = P_3(p_{r4}/p_{r3}) = 892(220.5/3339) = 58.9$ psia. The thermal efficiency is found from the net work output and the heat supplied. The heat rejected is

$$q_{out} = u_4 - u_1 = 392.8 - 88.6 = 304.2 \text{ Btu/lb}$$

The net work, then, is $q_{in} - q_{out} = 614.2 - 304.2 = 310.0$ Btu/lb. Consequently,

$$n_{th} = \frac{w_{net}}{q_{in}} = \frac{310.0}{614.2} = 0.505 \qquad \text{(or 50.5 percent)}$$

Finally, the mean effective pressure of the cycle is found from Eq. (16-2).

$$\text{MEP} = \frac{w_{net}}{v_1 - v_1} = \frac{310.0(778)}{13.33 - 1.67} = 20{,}685 \text{ psf} = 143.6 \text{ psi}$$

where v_2 is one-eighth of v_1.

16-5 THE AIR-STARDARD DIESEL CYCLE AND THE DUAL CYCLE

In a spark-ignition engine the fuel is ignited by energy supplied from an external source. An alternative method for initiating the combustion process in a reciprocating engine is to raise the fuel-air mixture above its autoignition temperature. An engine built on this principle is called a compression-ignition (CI) engine. By using compression ratios in the range of 14 : 1 to 24 : 1 and using diesel fuel instead of gasoline, the temperature of the air within the cylinder will exceed the ignition temperature at the end of the compression stroke. If the fuel were premixed with the air, as in a spark-ignition engine, combustion would begin throughout the mixture when the ignition temperature is reached. As a result we would have no control on the timing of the combustion process. To overcome this difficulty, the fuel is injected into the cyclinder in a separate operation. Injection begins when the piston is near the TDC position. Thus the CI engine differs from the spark-ignition (SI) engine primarily in the method of achieving combustion and in the adjustment of the timing of the combustion process. The rest of the four-stroke CI cycle is similar to that of the SI cycle discussed in Sec. 16-4.

A typical *PV* diagram for a compression-ignition engine at rated load is shown in Fig. 16-7. Note that the modern CI engine, frequently called a diesel engine, has a *PV* diagram very similar to that of a spark-ignition engine at full load. In the earlier history of this engine the combustion part of the cycle was somewhat flatter, so that the initial section of the expansion process was closer to a constant-pressure process. As a result the compression-ignition engine was modeled early in its history by a theoretical cycle known as a Diesel cycle. The modern compression-ignition engine is better modeled by the Otto cycle discussed in Sec. 16-4 or the dual cycle discussed later in this section. Since the theoretical Diesel cycle is of limited use, only its basic characteristics will be outlined.

The theoretical Diesel cycle for a reciprocating engine is shown in Fig. 16-8 on both *PV* and *Ts* diagrams. This cycle, like the Otto cycle, is composed of four internally reversible processes. The only difference between the two cycles is that a Diesel cycle models the combustion as occurring at constant pressure, while the Otto cycle assumes constant-volume heat addition. A useful analysis of the Diesel cycle is made possible by an air-standard cycle based on constant specific heats. Under this circumstance the heat input and output for the cycle are given by

$$q_{\text{in}} = c_p(T_3 - T_2) \qquad \text{and} \qquad q_{\text{out}} = c_v(T_4 - T_1)$$

Consequently,

$$\eta_{\text{th, Diesel}} = \frac{q_{\text{in}} - q_{\text{out}}}{q_{\text{in}}} = \frac{c_p(T_3 - T_2) - c_v(T_4 - T_1)}{c_p(T_3 - T_2)}$$

$$= 1 - \frac{T_4 - T_1}{k(T_3 - T_2)}$$

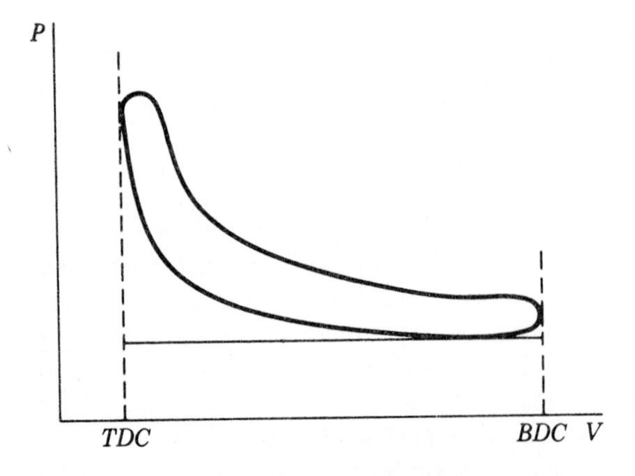

Figure 16-7 Typical *PV* diagram for a CI engine at rated load.

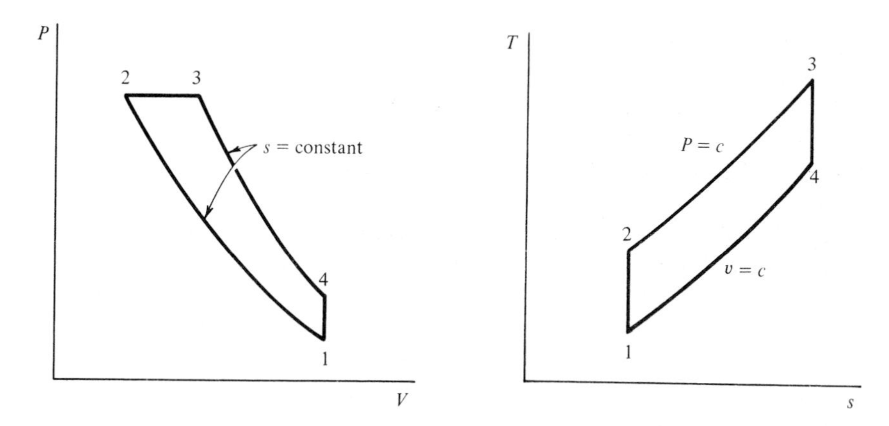

Figure 16-8 *PV* and *TS* diagrams for the air-standard Diesel cycle.

The above equation is more informative if one introduces the concept of the *cutoff ratio* r_c, which is defined as V_3/V_2. By recalling that the compression ratio r is defined as V_1/V_2, it can be shown that the preceding equation containing temperatures can be expressed in terms of volumes in the following manner:

$$\eta_{\text{th, Diesel}} = 1 - \frac{1}{r^{k-1}}\left[\frac{r_c^k - 1}{k(r_c - 1)}\right] \tag{16-5}$$

This equation indicates that the theoretical Diesel cycle is primarily a function of the compression ratio r, the cutoff ratio r_c, and the specific-heat ratio k.

It may be demonstrated that the term in the brackets in Eq. (16-5) is always equal to or greater than unity. Hence the thermal efficiency of a Diesel cycle is always less than that of an Otto cycle for the same compression ratio when r_c is greater than unity. In fact, an increase in the cutoff ratio has a drastic effect on the thermal efficiency of an air-standard Diesel cycle. This effect is shown in Fig. 16-9 for one particular compression ratio. Hence manufacturers of modern compression-ignition engines strive to design their engines so that the performance more nearly equals that of an Otto cycle.

In the case where variable specific-heat values are to be considered, the equation for the thermal efficiency of the Diesel cycle becomes

$$\eta_{\text{th, Diesel}} = 1 - \frac{u_4 - u_1}{h_3 - h_2} \tag{16-6}$$

To use Eq. (16-6) it is necessary to evaluate the u and h data from the air table in the Appendix. In this case the temperatures at states 2 and 4 are found from the isentropic relations

$$v_{r2} = v_{r1}\frac{V_2}{V_1} = \frac{v_{r1}}{r} \qquad \text{and} \qquad v_{r4} = v_{r3}\frac{V_4}{V_3} = \frac{rv_{r3}}{r_c} \tag{16-7}$$

Recall that the v_r data in the air table are solely a function of temperature.

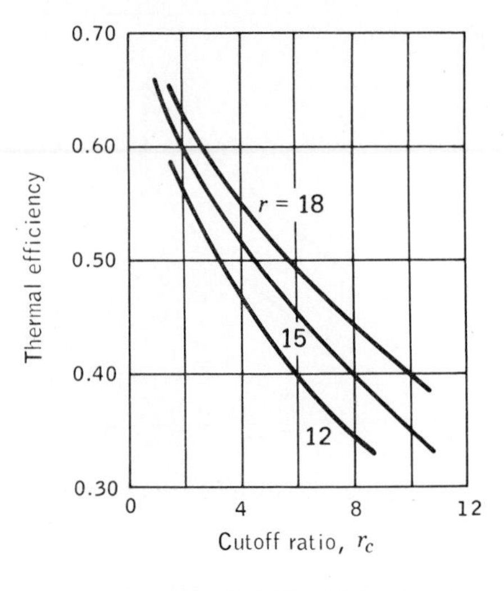

Figure 16-9 The thermal efficiency versus cutoff ratio and compression ratio (for $k = 1.4$) for an air-standard Diesel cycle.

A theoretical cycle which comes closer than the Diesel cycle to matching the actual performance of modern compression-ignition engines is the *dual* cycle. As shown in Fig. 16-10, a short heat-addition process 2-x at constant volume is followed by a second heat-addition process x-3 at constant pressure. The other three parts of the cycle are similar to those found in the Otto and Diesel cycles. The dual cycle is also called a mixed, or limited-pressure, cycle. Note that the use of a two-step heat-addition process allows the theoretical dual cycle to model fairly closely the upper left-hand portion of the actual performance curve shown in Fig. 16-7 for a compression-ignition engine.

The thermal efficiency of the air-standard dual cycle is a function of the heat-input and heat-output quantities. The heat added during the constant-

$$(T_x - T_2) + k(T_3 - T_x)$$

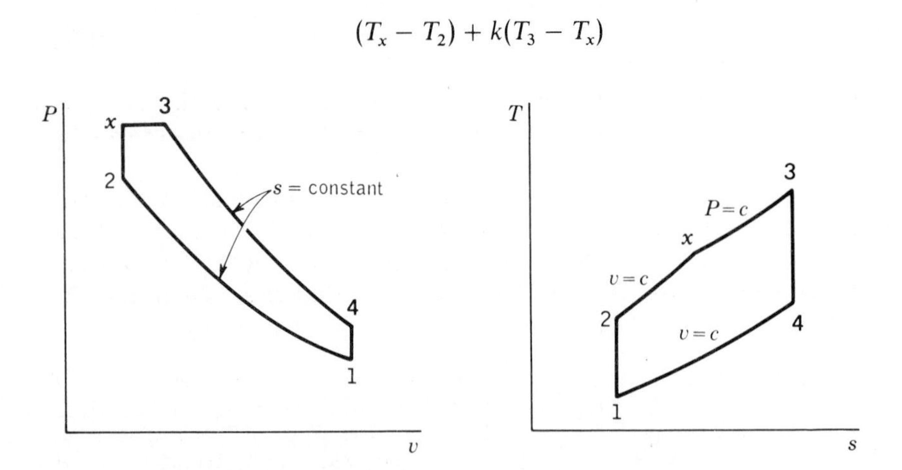

Figure 16-10 PV and TS diagrams for the air-standard dual cycle.

volume process 2-x is

$$q_{\text{in}} = c_v(T_x - T_2)$$

if we assume that the specific heats are constant. For the constant-pressure path (x–3) the heat addition is

$$q_{\text{in}} = c_p(T_3 - T_x)$$

Therefore the total heat input for the dual cycle is

$$q_{\text{in, tot}} = c_v(T_x - T_2) + c_p(T_3 - T_x)$$

The heat rejected along the constant-volume path is

$$q_{\text{out}} = c_v(T_4 - T_1)$$

Consequently, the thermal efficiency of an air-standard dual cycle for constant-specific-heat data is

$$\eta_{\text{th, dual}} = \frac{q_{\text{in}} - q_{\text{out}}}{q_{\text{in}}} = 1 - \frac{c_v(T_4 - T_1)}{c_v(T_x - T_2) + c_p(T_3 - T_x)}$$

$$= 1 - \frac{T_4 - T_1}{(T_x - T_2) + k(T_3 - T_x)} \tag{16-8a}$$

where k is the specific-heat ratio. Recall that the compression ratio $r = V_1/V_2$ and the cutoff ratio $r_c = V_3/V_x = V_3/V_2$. If, in addition, we define the pressure ratio $r_{p,\,v}$ during the constant-volume combustion process as

$$r_{p,\,v} = \frac{P_x}{P_2} = \frac{P_3}{P_2}$$

then the thermal efficiency as evaluated by Eq. (16-8a) may be expressed solely as a function of r, r_c, $r_{p,\,v}$, and k. The result is

$$\eta_{\text{th, dual}} = 1 - \frac{1}{r^{k-1}}\left[\frac{r_{p,\,v}r_c^k - 1}{kr_{p,\,v}(r_c - 1) + r_{p,\,v} - 1}\right] \tag{16-8b}$$

When $r_{p,\,v}$ is unity, Eq. (16-8b) reduces to the Diesel efficiency equation given by Eq. (16-5). Also Eq. (16-8b) reduces to Eq. (16-3) for the Otto cycle when r_c is unity.

On the basis of the same heat input and compression ratio, the thermal efficiency of the three theoretical cycles decreases in the following order: Otto cycle, dual cycle, Diesel cycle. This is a major reason that modern CI engines are designed to operate close to the Otto- or dual-cycle models, rather than the Diesel cycle.

Example 16-3M The intake conditions for an air-standard dual cycle operating with a compression ratio of 15 : 1 are 0.95 bar and 17°C. The pressure ratio during constant-volume heating is 1.5 : 1 and the volume ratio during the constant-pressure part of the heating process is 1.8 : 1. Calculate (a) the temperatures and pressures around the cycle, (b) the heat input and heat rejection, and (c) the thermal efficiency, using air-table data.

SOLUTION (*a*) We shall use the notation of Fig. 16-10. The pressure and temperature after isentropic compression are found from v_r data.

$$v_{r2} = v_{r1}\frac{v_2}{v_1} = \frac{676.1}{15} = 45.07$$

By linear interpolation in the table, $T_{2s} = 819°\text{K}$, $u_{2s} = 607.4$ kJ/kg, and $p_{r2} = 52.15$. Using the last value, we find that $P_2 = P_1(p_{r2}/p_{r1}) = 0.95(52.15/1.2311) = 40.24$ bars. The pressure after constant-volume heating is simply

$$P_x = 1.5P_2 = 1.5(40.24) = 60.4 \text{ bars}$$

On the basis of the ideal-gas equation,

$$T_x = T_2\frac{P_x}{P_2} = 819(1.5) = 1229°\text{K}$$

The pressure at state 3 is the same as that at state x, namely, 60.4 bars. Also at state 3

$$T_3 = T_x\frac{V_3}{V_x} = 1229(1.8) = 2212°\text{K}$$

State 4 is determined from isentropic relations. First, however, we need to determine V_3/V_4. This ratio is: $V_3/V_4 = V_3/V_1 = (V_3/V_2)(V_2/V_1) = r_c/r$. Therefore, since $v_{r3} = 1.98$ at 2212°K,

$$v_{r4} = v_{r3}\frac{V_4}{V_3} = 1.98\frac{15}{1.8} = 16.5$$

By linear interpolation in Table A-5M, $T_{4s} = 1150°\text{K}$, $p_{r4} = 200.2$, and $u_{4s} = 889.1$ kJ/kg. Finally, based on $p_{r3} = 3216$ at 2212°K,

$$P_4 = P_3\frac{p_{r4}}{p_{r3}} = 60.4\frac{200.2}{3216} = 3.76 \text{ bars}$$

(*b*) The heat supplied to the cycle is the sum of terms for processes 2-x and x-3. Hence

$$q_{\text{in}} = q_{2x} + q_{x3} = (u_x - u_2) + (h_3 - h_x)$$
$$= (959.1 - 607.4) + (2518.4 - 1311.9) = 1558.2 \text{ kJ/kg}$$

The heat rejected from the cycle is

$$q_{\text{out}} = q_{41} = u_1 - u_4 = 206.9 - 889.1 = -682.2 \text{ kJ/kg}$$

(*c*) Finally, the thermal efficiency for this dual cycle is

$$\eta_{\text{th}} = 1 - \frac{q_{\text{out}}}{q_{\text{in}}} = 1 - \frac{682.2}{1558.2} = 0.562 \qquad \text{(or 56.2 percent)}$$

If $k = 1.4$ for a cold-air cycle, Eq. (16-8*b*) would predict a thermal efficiency of 62.5 percent.

Example 16-3 The intake conditions for an air-standard dual cycle operating with a compression ratio of 15 : 1 are 14.4 psia and 60°F. The pressure ratio during constant-volume heating is 1.5 : 1 and the volume ratio during the constant-pressure part of the heating process is 1.8 : 1. Calculate (*a*) the temperatures and pressures around the cycle, (*b*) the heat input and heat rejection, and (*c*) the thermal efficiency, using air-table data.

SOLUTION (*a*) We shall use the notation of Fig. 16-10. The pressure and temperature after isentropic compression are found from v_r data. At 520°R, $v_{r1} = 158.58$, so that

$$v_{r2} = v_{r1}\frac{V_2}{V_1} = \frac{158.58}{15} = 10.57$$

By linear interpolation in the table, $T_{2s} = 1469°\text{R}$, $p_{r2} = 51.57$, and $u_{2s} = 260.3$ Btu/lb. Using the p_r data, we find that $P_2 = P_1(p_{r2}/p_{r1}) = 14.4(51.57/1.2147) = 611$ psia. The pressure after constant-volume heating is simply

$$P_x = 1.5P_2 = 1.5(611) = 917 \text{ psia}$$

On the basis of the ideal-gas equation, at constant volume,

$$T_x = T_2 \frac{P_x}{P_2} = 1469(1.5) = 2204°\text{R}$$

At this temperature $u_x = 410.6$ Btu/lb and $h_x = 561.7$ Btu/lb. The pressure at state 3 is the same as that at state x, namely, 917 psia. Also at state 3

$$T_3 = T_x \frac{V_3}{V_x} = 2204(1.8) = 3967°\text{R}$$

At this temperature $h_3 = 1078.3$ Btu/lb and $p_{r3} = 3163$. State 4 is determined from isentropic relations. First, however, we need to determine V_3/V_4. This ratio is $V_3/V_4 = V_3/V_1 = (V_3/V_2)(V_2/V_1) = r_c/r$. Therefore, since $v_{r3} = 0.4648$ at 3967°R,

$$v_{r4} = v_{r3} \frac{V_4}{V_3} = 0.4648 \frac{15}{1.8} = 3.87$$

By linear interpolation in Table A-5, $T_{4s} = 2064°\text{R}$, $p_{r4} = 197.8$, and $u_{4s} = 381.0$ Btu/lb. Finally,

$$P_4 = P_3 \frac{P_{r4}}{P_{r3}} = 917 \frac{197.8}{3163} = 57.3 \text{ psia}$$

(*b*) The heat supplied to the cycle is the sum of terms for processes 2-x and x-3. Hence

$$q_{\text{in}} = q_{2x} + q_{x3} = (u_x - u_2) + (h_3 - h_x)$$
$$= (410.6 - 260.3) + (1078.3 - 561.7) = 666.9 \text{ Btu/lb}$$

The heat rejected from the cycle is

$$q_{\text{out}} = q_{41} = u_1 - u_4 = 88.6 - 381.0 = -292 \text{ Btu/lb}$$

(*c*) Finally, the thermal efficiency for the dual cycle is

$$\eta_{\text{th}} = 1 - \frac{q_{\text{out}}}{q_{\text{in}}} = 1 - \frac{292.4}{666.9} = 0.562 \qquad \text{(or 56.2 percent)}$$

If $k = 1.4$ for a cold-air cycle, Eq. (16-8*b*) would predict a thermal efficiency of 62.5 percent.

16-6 THE AIR-STANDARD BRAYTON CYCLE

In a simple gas-turbine power cycle, separate equipment is used for the various processes of the cycle. Initially, air is compressed adiabatically in a rotating axial or centrifugal compressor. At the end of this process the air enters a combustion chamber where fuel is injected and burned at essentially constant pressure. The products of combustion are then expanded through a turbine until they reach the ambient pressure of the surroundings. A cycle composed of these three steps is called an open cycle, because the cycle is not actually completed. Actual gas-turbine cycles are open cycles, since new air must continually be introduced into the compressor. If one wishes to examine a closed cycle, the products of combustion which have expanded through the turbine must be sent through a heat exchanger, where heat is rejected from the gas until the initial temperature is attained. The open and closed gas-turbine cycles are shown in Fig. 16-11. In the analysis of gas-turbine cycles it is useful in the beginning to employ an air-standard cycle. An air-standard gas-turbine cycle with isentropic compression and expansion is called a Brayton cycle. In the Brayton cycle it is necessary to replace the actual combustion process by a heat-addition process. The use of air as the sole working medium throughout the cycle is a fairly good model since a relatively high air-fuel ratio of, roughly, at least 50 : 1 on a mass basis is quite common in actual operation with conventional hydrocarbon fuels.

In the Brayton cycle the compression and expansion processes are assumed to be isentropic, and the heat-addition and heat-removal processes occur at constant pressure. Both Pv and Ts diagrams for this ideal cycle are shown in Fig. 16-12. For processes 1-2 and 3-4, which are isentropic, the conservation of energy principle reduces to

$$w = \Delta h + \Delta \text{KE}$$

A steady-flow energy balance for the heat exchangers used in processes 2-3 and 4-1 is of the form

$$q = \Delta h + \Delta \text{KE}$$

In the absence of appreciable kinetic-energy changes, the thermal efficiency of the ideal Brayton cycle is given by

$$\eta_{\text{th, Brayton}} = 1 - \frac{q_{\text{out}}}{q_{\text{in}}} = 1 - \frac{h_{4s} - h_1}{h_3 - h_{2s}} \tag{16-9}$$

For a cold-air-standard Brayton cycle with constant specific-heat values, Eq.

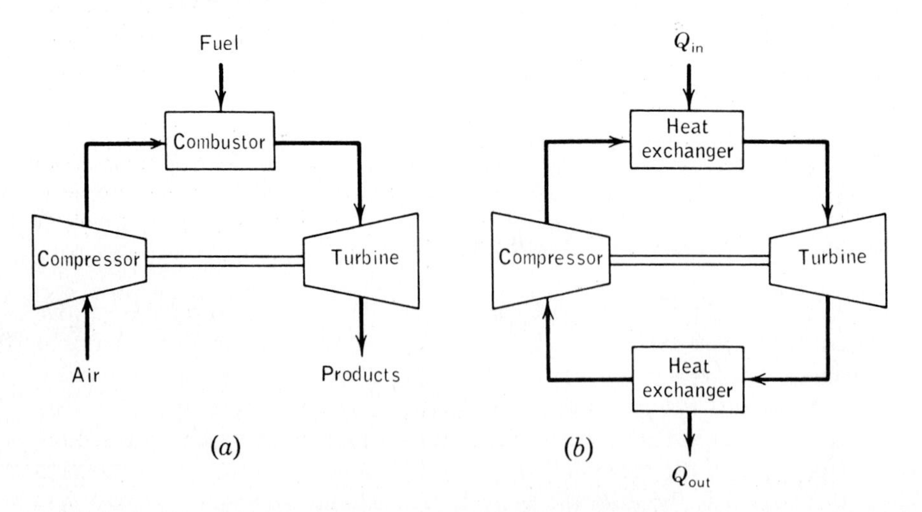

Figure 16-11 Gas turbine operating on the (*a*) open and (*b*) closed Brayton cycle.

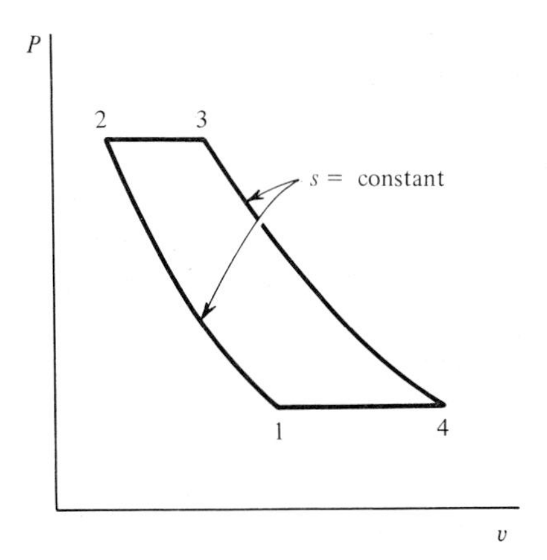

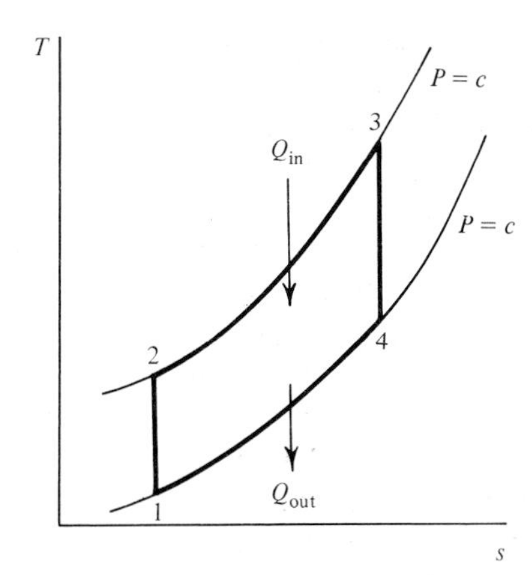

Figure 16-12 Typical PV and Ts diagrams of the air-standard Brayton cycle.

(16-9) becomes

$$\eta_{\text{th, Brayton}} = 1 - \frac{T_{4s} - T_1}{T_3 - T_{2s}} = 1 - \frac{1}{r^{k-1}} = 1 - \frac{1}{r_p^{(k-1)/k}} \tag{16-10}$$

where r is defined as v_1/v_2 and r_p as P_2/P_1. Thus the thermal efficiency of an air-standard Brayton cycle is primarily a function of the overall pressure ratio.

The use of constant-specific-heat data, which led to Eq. (16-10), is quite useful in the initial modeling of a gas-turbine power cycle. However, to obtain reasonable values for the heat and work terms in a cycle analysis, it is necessary to account for the variation of c_p with temperature. This requires the use of the air tables in the Appendix. For the isentropic compression and expansion processes in the Brayton cycle, the final state can be found by employing either s^0 or p_r data. For a review of the use of the entropy function s^0 and the relative pressure function p_r, see Sec. 7-9.

Example 16-4M An ideal air-standard Brayton cycle operates with air entering the compressor at 0.95 bar and 22°C. The pressure ratio r_p is 6 : 1, and the air leaves the combustion chamber at 1100°K. Compute the compressor-work input, the turbine-work output, and the thermal efficiency of the cycle, based on the data of Table A-5M.

Solution The compressor outlet condition can be determined from isentropic relationships. For the given inlet state, it is found from Table A-5M that $h_1 = 295.17$ kJ/kg and $p_{r1} = 1.3068$. The relative pressure at the compressor outlet is

$$p_{r2} = p_{r1}\frac{P_2}{P_1} = 1.3068(6) = 7.841$$

The enthalpy and temperature at state 2 are then approximated by interpolation of the p_{r2} data in Table A-5M. We find that $T_{2s} = 490$°K and $h_{2s} = 492.7$ kJ/kg. In addition, since T_3 is given as 1100°K, the value of $h_3 = 1161.1$ kJ/kg and the relative pressure $p_{r3} = 167.1$. Hence, for state 4 we find that

$$p_{r4} = p_{r3}\frac{P_4}{P_3} = 167.1(\tfrac{1}{6}) = 27.85$$

Again interpolating in Table A-5M, we estimate that $T_{4s} = 694°\text{K}$ and $h_{4s} = 706.5$ kJ/kg. On the basis of these data, the following quantities are determined:

$$w_{C,\,\text{in}} = h_{2s} - h_1 = 492.7 - 295.2 = 197.5 \text{ kJ/kg}$$

$$w_{T,\,\text{out}} = h_3 - h_{4s} = 1161.1 - 706.5 = 454.6 \text{ kJ/kg}$$

$$q_{\text{in}} = h_3 - h_{2s} = 1161.1 - 492.7 = 668.4 \text{ kJ/kg}$$

$$\eta_{\text{th}} = \frac{w_T - w_C}{q_{\text{in}}} = \frac{454.6 - 197.5}{668.4} = \frac{257.1}{668.4} = 0.385 \qquad \text{(or 38.5 percent)}$$

Note that states 2s and 4s can be determined from s^0 data. At state 1 the value of s^0 is 1.68515 and at state 3 the value is 3.07732. Hence,

$$s_2^0 = s_1^0 + R \ln r_p = 1.68515 + \frac{8.314}{29} \ln 6 = 2.1988 \text{ kJ/(kg)(°K)}$$

and

$$s_4^0 = s_3^0 + R \ln \frac{1}{r_p} = 3.07732 + \frac{8.314}{29} \ln \frac{1}{6} = 2.5636 \text{ kJ/(kg)(°K)}$$

From Table A-5M it is found that T_{2s} is 490°K and T_{4s} is 694°K. These temperatures are in excellent agreement with the values obtained above by using p_r data.

Example 16-4 An ideal Brayton cycle operates with air entering the compressor at 14.5 psia and 80°F. The pressure ratio is 6 : 1, and the air leaves the combustion chamber at 1540°F. Compute the compressor-work input, the turbine-work output, and the thermal efficiency of the cycle, based on an air-standard cycle and employing Table A-15.

SOLUTION The compressor outlet conditions can be determined from isentropic relationships. For the given inlet state, it is found that $h_1 = 129.06$ Btu/lb and $p_{r1} = 1.386$. The relative pressure at the compressor outlet, then, is

$$p_{r2} = p_{r1}\frac{P_2}{P_1} = 1.386(6) = 8.316$$

The enthalpy and temperature at state 2 can be approximated by interpolation in the abridged tables in the Appendix. If the unabridged gas tables are used, one finds that $h_{2s} = 215.52$ Btu/lb and $T_{2s} = 897°\text{R}$. In addition, since T_3 is given as 1540°F (2000°R), the value of $h_3 = 504.71$ Btu/lb, and the relative pressure $p_{r3} = 174.00$. Hence

$$p_{r4} = p_{r3}\frac{P_4}{P_3} = 174.00(\tfrac{1}{6}) = 29.00$$

Again employing the unabridged tables for accuracy, one finds that $h_{4s} = 307.25$ Btu/lb and $T_{4s} = 1262°\text{R}$ (802°F). On the basis of these data, the following quantities are determined:

$$w_{C,\,\text{in}} = h_{2s} - h_1 = 215.52 - 129.06 = 86.46 \text{ Btu/lb}$$

$$w_{T,\,\text{out}} = h_3 - h_{4s} = 504.71 - 307.25 = 197.46 \text{ Btu/lb}$$

$$q_{\text{in}} = h_3 - h_{2s} = 504.71 - 215.52 = 289.19 \text{ Btu/lb}$$

$$\eta_{\text{th}} = \frac{w_T - w_C}{q_{\text{in}}} = \frac{197.46 - 86.46}{289.19} = \frac{111.0}{289.19} = 0.384 \qquad \text{(or 38.4 percent)}$$

Note that states 2s and 4s can be determined from s^0 data. At state 1 the value of s^0 is 0.60078 and at state 3 the value is 0.93205. Hence,

$$s_2^0 = s_1^0 + R \ln r_p = 0.60078 + \frac{1.986}{29} \ln 6 = 0.7235 \text{ Btu/(lb)(°R)}$$

and
$$s_4^0 = s_3^0 + R \ln \frac{1}{r_p} = 0.93205 + \frac{1.986}{29} \ln \frac{1}{6} = 0.8093 \text{ Btu/(lb)(°R)}$$

From Table A-5 it is found that T_{2s} is 897°R and T_{4s} is 1262°R. These temperatures are in excellent agreement with the values obtained above by using p_r data.

The data of the preceding example illustrate an important feature of the gas-turbine power cycle. In the ideal Brayton cycle analyzed above, the ratio of the compressor work to the turbine work is roughly 0.44, or 44 percent. This quantity, w_C/w_T, is known as the *back work ratio* of a power cycle. In practice this ratio typically ranges from 40 to 80 percent. As the next several sections will show, the effect of irreversibilities in the compressor and turbine is to greatly increase the back work ratio. Obviously, the ratio cannot exceed unity, for in that limiting case the net work output becomes zero. In the case of vapor power cycles, discussed in Chap. 17, it is found that the back work ratio of pump work to turbine work is several percent or less.

Figure 16-13 illustrates the effect of the pressure ratio r_p and the combustor-outlet temperature T_3 on the Brayton cycle. Equation (16-10) shows that the thermal efficiency increases with an an increasing pressure ratio, but as r_p increases, it is found that T_3 increases for the same heat rejection. For example, increasing the compressor-outlet pressure from state 2 to state 5 on the figure requires that the combustor-outlet temperature increase from state 3 to state 6. The heat rejections for the two cycles 1-2-3-4-1 and 1-5-6-4-1 are exactly the same, as given by the area under the curve 4-1. Increasing T_3 to T_6 may be disadvantageous, since the temperature may now exceed the maximum allowable

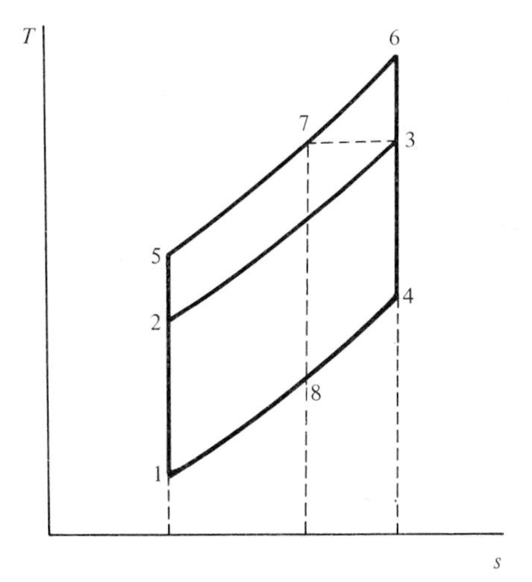

Figure 16-13 Illustration of the effect of the pressure ratio and the combustor-outlet temperature on the thermal efficiency of the Brayton cycle.

at the inlet to the turbine. To overcome this difficulty, it might be proposed that r_p be increased but that the combustor-outlet temperature be restricted to a value not exceeding T_3. Such a cycle would be represented by 1-5-7-8-1. The efficiency of this cycle would be greater than that for the basic cycle 1-2-3-4-1, but the figure easily demonstrates that the net work output would be much less. To achieve the same power output as the basic cycle, a larger mass flow rate would be required, which in turn means a larger physical system would be required. This may be undesirable. In the limit, as r_p increases and the combustor-outlet temperature is restricted to a value equivalent to T_3, the net work approaches zero. It may also be shown that as r_p decreases for a fixed value of T_3, the net work decreases. Consequently, there must be a value of T_2 (and hence P_2) which leads to a maximum net work output. This is found by differentiating the equation for the net work with respect to T_2, for fixed values of T_1 and T_3. It is found that the maximum net work for the Brayton cycle operating on a cold-air-standard cycle under these restrictions occurs when $T_2 = (T_1 T_3)^{1/2}$. The solid curve in Fig. 16-14 shows the variation of thermal efficiency with pressure ratio for an ideal air-standard gas-turbine cycle. The data are based on constant-specific-heat values and a turbine-inlet to compressor-inlet temperature ratio of 4 : 1. The net work of the cycle is shown as the dashed line. Plotted as a dimensionless parameter $w_{net}/c_p T_1$, the net work maximizes at a pressure ratio of 12 : 1 and then falls off very slowly at higher pressure ratios.

Compared with the Otto cycle, the Brayton cycle operates over a wider range of volume but a smaller range of pressures and temperatures. This feature makes the Brayton cycle unadaptable for use with reciprocating machinery. Instead, one makes use of turbomachinery, which comprises steady-flow devices. The Brayton cycle is a standard for the practical design of modern gas-turbine power plants. It should be kept in mind that irreversibilities occur in the compressor and turbine

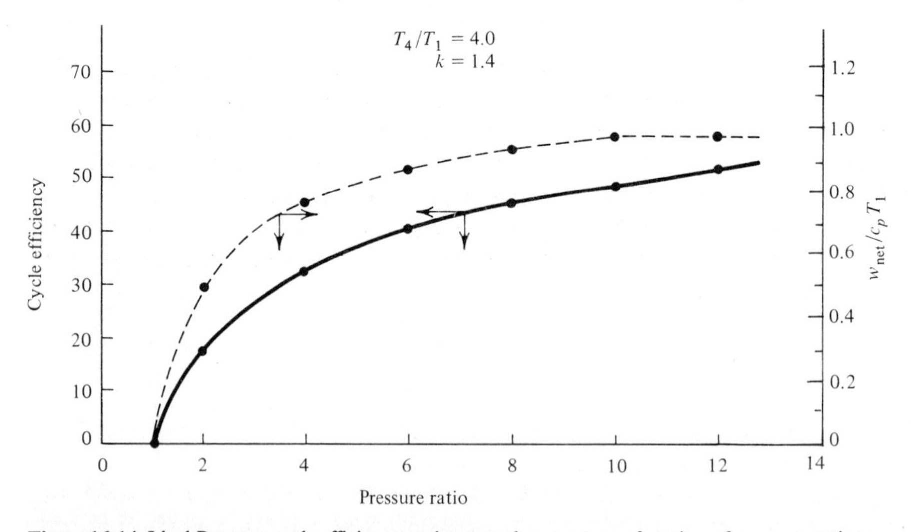

Figure 16-14 Ideal Brayton-cycle efficiency and net work output as a function of pressure ratio.

in actual operation. Hence the compressor- and turbine-work quantities are not equal to the isentropic enthalpy change across each device. The effect of these irreversibilities is to decrease the heat input in the combustor and increase the heat output in the heat-rejection part of the cycle. The overall result of this is a decrease in the thermal efficiency of the cycle. Other differences appear between the actual and the theoretical performance of the cycle. In actuality, P_3 does not equal P_2, and P_4 does not equal P_1, because of pressure drops in the combustor and the heat exchanger. In the cycle analysis presented here, these pressure drops are ignored. Heat losses from the compressor and turbine may also be present.

The effect of irreversibilities in the compressor and turbine is frequently accounted for by the use of a term called the adiabatic efficiency of these devices. Without actual test data on a particular model, the adiabatic efficiency can be predicted fairly well from experience with similar devices. The adiabatic efficiency permits one to estimate the actual work quantities of turbines and compressors, as will be seen in the next section. It will be found that the thermal efficiency of a gas-turbine power cycle is quite sensitive to the efficiencies of the turbine and the compressor.

16-7 THE ADIABATIC EFFICIENCY OF WORK DEVICES

For turbines and compressors which are close to being adiabatic in practice, the isentropic work associated with these devices is a standard, or model, to which all actual equipment may be compared. The actual performance of work-producing or work-absorbing devices which are essentially adiabatic is described by an adiabatic efficiency. An *adiabatic turbine efficiency* η_T is defined by

$$\eta_T \equiv \frac{\text{actual work output}}{\text{isentropic work output}} = \frac{W_a}{W_s} \tag{16-11}$$

This quantity is frequently termed, simply, the turbine efficiency. Figure 16-15 illustrates reversible and irreversible adiabatic processes through a turbine. It is noted that the specific entropy always increases for the actual adiabatic processes (with end states $2a$, $2a'$, and $2a''$), as dictated by the second law. If the changes in kinetic and potential energies are neglected, the work terms may be evaluated as a function of the enthalpy changes for a steady-flow process. Use of the notation from Fig. 16-15 allows the adiabatic efficiency of a work-producing device such as a turbine to be written as

$$\eta_T = \frac{h_1 - h_{2a}}{h_1 - h_{2s}} \tag{16-12}$$

where the subscript a represents the actual outlet condition and the subscript s represents the isentropic outlet state.

To assess the effect of irreversibilities on performance, consider an adiabatic turbine with an ideal gas as the working medium. For constant specific heats the

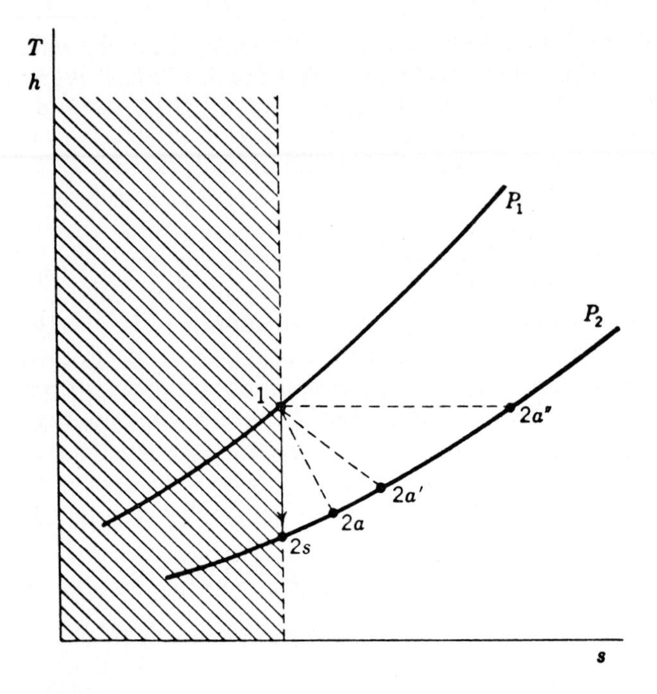

Figure 16-15 Illustration of the effect of irreversibilities on adiabatic turbine performance on a *Ts* diagram.

conservation-of-energy equation reduces to

$$w_{\text{shaft}} = c_p(T_2 - T_1)$$

if we neglect the kinetic-energy change. The equation for the entropy change can be written as

$$\Delta s + R \ln \frac{P_2}{P_1} = c_p \ln \frac{T_2}{T_1}$$

The fixed conditions for a turbine include P_1, T_1, and P_2. When Δs is zero (isentropic), T_2 is fixed, which in turn fixes the shaft-work output. The isentropic process from state 1 to state 2*s* is shown in Fig. 16-15. Note that the ordinate on the diagram is either *T* or *h*. This is possible since *h* is proportional to *T* for an ideal gas. Thus we are plotting a *Ts* and an *hs* diagram simultaneously. Recall that $w_{\text{shaft}} = \Delta h$. Therefore the vertical distance between any two states on the diagram represents the shaft work on an *hs* diagram.

For the internally irreversible process, Δs is positive. From the preceding equation for Δs it is seen that as Δs increases, T_2 must increase for a given T_1, P_1, and P_2. Therefore the final state 2*a* for an irreversible process lies up the P_2 line at a higher *T* and higher *s*. This also means that state 2*a* has a greater enthalpy than state 2*s*. Consequently the shaft-work output is less. State 2*a*′ represents the final state for a more irreversible process than that represented by $2a$. In the limit, state $2a''$ would be the final state, and for the given pressure drop Δh would be zero. The shaft work also would be zero. In this extreme case the turbine acts like a throttling valve. This would be a highly undesirable method of operation.

For devices which require work input, the *adiabatic compressor efficiency* is defined as

$$\eta_C \equiv \frac{\text{isentropic work input}}{\text{actual work input}} = \frac{W_s}{W_a} \tag{16-13}$$

It is convenient to speak simply of the compressor efficiency. Figure 16-16 illustrates reversible and irreversible flow through an adiabatic compressor. The isentropic process, 1-2s, again is shown on a composite Ts and hs diagram. The vertical distance on the hs plot represents the shaft-work input when the kinetic-energy change is negligible. In this latter case Eq. (16-13) may be written as

$$\eta_C = \frac{h_{2s} - h_1}{h_{2a} - h_1} \tag{16-14}$$

For the irreversible situation we shall again examine the entropy change equation, namely,

$$c_p \ln \frac{T_2}{T_1} = R \ln \frac{P_2}{P_1} + \Delta s$$

This equation shows that T_2 increases as the entropy increases due to irreversibilities, for fixed values of P_1, T_1, and P_2. States $2a$ and $2a'$ show this effect in Fig. 16-16. The major consequence, however, is the large change in shaft work required. Based on $w_{\text{shaft}} = \Delta h$, we see that there are sizable increases in shaft work as the degree of irreversibility increases. More shaft work is required to attain the same required pressure. The isentropic process is an obvious standard of performance.

Slightly different definitions for these efficiencies are employed sometimes; so the student should not accept them as having universal applicability. For instance, the kinetic-energy change may be included in either the numerator or denominator, or both, of the efficiency expression. This is a matter of choice. In the absence of kinetic-energy effects it is seen that the efficiencies defined above are ratios of vertical distances taken from the Mollier diagram. Efficiencies of 80 to 85 percent are not uncommon for modern turbines and compressors. The effect of the irreversibilities is to require a larger compressor-work input, which is available from a smaller turbine output. Hence the compressor in actual practice may consume 40 to 70 percent of the turbine output.

Example 16-5M An air-standard gas-turbine cycle operates under the conditions used in Example 16-4M, except that the compressor and turbine are only 82 and 85 percent efficient, respectively. Compute the compressor- and turbine-work quantities and the thermal efficiency for this cycle, again using the data of Table A-5M for air.

SOLUTION The data for the compressor- and turbine-inlet states remain the same as before. The

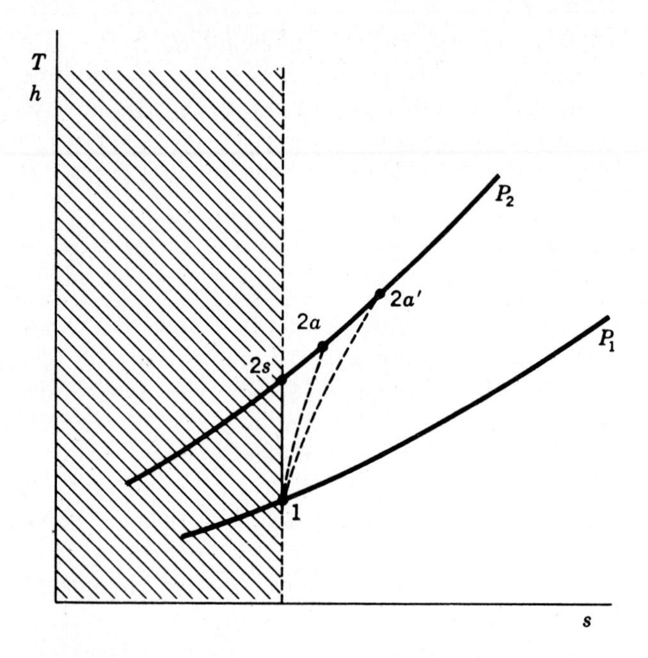

Figure 16-16 Illustration of the effect of irreversibilities on adiabatic compressor performance on a Ts diagram.

isentropic values previously calculated for the outlet states of these two devices are employed to find the actual states, in conjunction with the definitions of compressor and turbine efficiencies. For the compressor, in the absence of kinetic-energy effects

$$\Delta h_a = \frac{\Delta h_s}{\eta_C} = \frac{197.5}{0.82} = 240.9 \text{ kJ/kg}$$

Since $\Delta h_a = h_{2a} - h_1$, the actual compressor-outlet enthalpy is

$$h_{2a} = 240.9 + 295.2 = 536.1 \text{ kJ/kg}$$

From Table A-5M it is seen that this corresponds to an outlet temperature of 532°K. Consequently, the irreversibilities within the compressor have resulted in a temperature rise of 42°K over that for the isentropic compression. For the turbine,

$$\Delta h_a = \eta_T \, \Delta h_s = 0.85(-454.6) = -386.4 \text{ kJ/kg}$$

Since $\Delta h_a = h_{4a} - h_3$, the actual turbine-outlet enthalpy is

$$h_{4a} = 1161.1 - 386.4 = 774.7 \text{ kJ/kg}$$

This value of h corresponds to a temperature of 757°K, which is 63°K above that for the isentropic expansion. The thermal efficiency becomes

$$\eta_{th} = \frac{w_{T,a} - w_{C,a}}{h_3 - h_{2a}} = \frac{386.4 - 240.9}{1161.1 - 536.1} = 0.233 \qquad \text{(or 23.3 percent)}$$

Therefore the presence of irreversibilities within the compressor and turbine has led to a decrease in the thermal efficiency from 38.5 to 23.3 percent, which is quite significant. In addition, irreversibilities have increased the ratio of compressor work to turbine work (back work ratio) to 0.62 from the value of 0.44 for isentropic flow. Relatively small changes in adiabatic efficiencies greatly influence the net work output and the back work ratio of the device.

Example 16-5 A gas-turbine cycle operates under the conditions used in Example 16-4, except that the compressor and turbine are only 82 and 85 percent efficient, respectively. Compute the compressor- and turbine-work quantities and the thermal efficiency for this cycle, again using Table A-5, the table of properties for air.

SOLUTION The data for the compressor- and turbine-inlet states remain the same as before. The isentropic values for the outlet states of these two devices are employed to find the actual states, in conjunction with the definitions of compressor and turbine efficiencies. For the compressor, in the absence of kinetic-energy effects,

$$\Delta h_a = \frac{\Delta h_s}{\eta_C} = \frac{86.46}{0.82} = 105.4 \text{ Btu/lb}$$

Since $\Delta h_a = h_{2a} - h_1$, the actual compressor-outlet enthalpy is

$$h_{2a} = 105.4 + 129.06 = 234.5 \text{ Btu/lb}$$

From Table A-5 it is seen that this corresponds to an outlet temperature of 974°R. Consequently, the irreversibilities within the compressor have resulted in a temperature rise of 77°F over that for the isentropic compression. For the turbine,

$$\Delta h_a = \eta_T \, \Delta h_s = 0.85(-197.46) = -167.8 \text{ Btu/lb}$$

Since $\Delta h_a = h_{4a} - h_3$, the actual turbine-outlet enthalpy is

$$h_{4a} = 504.71 - 167.8 = 336.9 \text{ Btu/lb}$$

This value corresponds to a temperature which is 115°F above that for the isentropic expansion. The thermal efficiency becomes

$$\eta_{\text{th}} = \frac{167.8 - 105.4}{504.7 - 234.5} = \frac{62.4}{270.2} = 0.232 \qquad \text{(or 23.2 percent)}$$

Therefore the presence of fluid friction within the compressor and turbine has led to a decrease in the thermal efficiency from 38.4 to 23.2 percent, which is quite significant. In addition, irreversibilities have increased the ratio of compressor work to turbine work to 0.63 from the value of 0.44 for isentropic flow. Small changes in adiabatic efficiencies greatly influence the net work output of the device.

The general effect of irreversibilities in the compressor and turbine is more clearly shown in Fig. 16-17. This figure is based on a cold-air-standard cycle with $k = 1.40$, and the compressor and turbine efficiencies are 0.82 and 0.85, respectively. Figure 16-17*a* shows the thermal efficiency of the irreversible cycle as a function of pressure ratio and selected ratios of the turbine-inlet temperature to the compressor-inlet temperature, T_3/T_1. Unlike the curve in Fig. 16-14 for the ideal Brayton cycle, these efficiency curves exhibit maximum points. Note also that the thermal efficiency maximizes at higher values of the pressure ratio as the temperature ratio T_3/T_1 increases. Figure 16-17*b* illustrates the effect of pressure ratio and temperature ratio on the net work output. Similar to Fig. 16-14, the net work is presented as a dimensionless parameter, w_{net}/c_pT_1. Note again that the net work output maximizes at larger values of r_p as the temperature ratio increases. For comparative purposes, the net work curve for an ideal cycle with

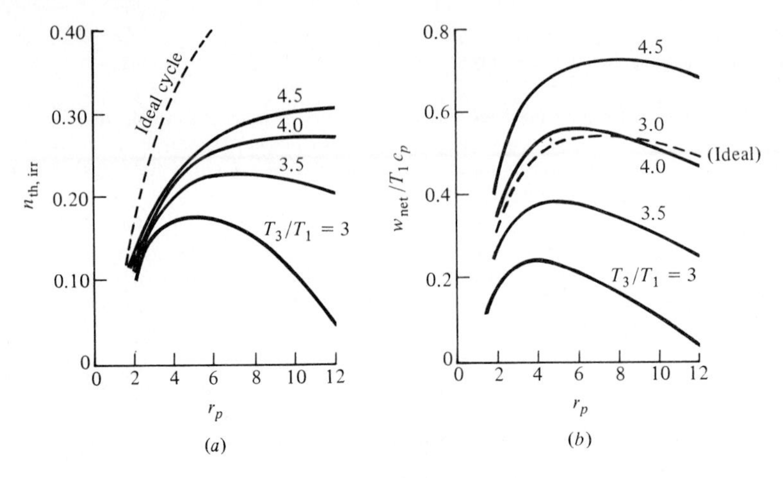

Figure 16-17 The effect of compressor and turbine irreversibilities on (*a*) the thermal efficiency, and (*b*) the net work output of a gas-turbine power cycle as a function of pressure ratio and the ratio of turbine-inlet temperature to compressor-inlet temperature. (Data based on $k = 1.40$, $\eta_C = 0.82$, and $\eta_T = 0.85$.)

T_3/T_1 of 3.0 is also shown. In this case two effects are clear when the two curves for T_3/T_1 of 3.0 are compared in Fig. 16-17*b*. The net work maximizes at a lower pressure ratio when irreversibilities are taken into account. Secondly, the net work is drastically reduced over the full range of pressure ratios, even though the compressor and turbine efficiencies chosen are relatively high.

16-8 THE REGENERATIVE GAS-TURBINE CYCLE

The basic gas-turbine cycle can be modified in several important ways to increase its overall efficiency. One of these ways is based on the concept of regeneration. In the preceding examples for the Brayton cycle, it is noted that the gases leaving the turbine are at a relatively high temperature. In many cases the outlet-turbine temperature is higher than the outlet-compressor temperature. It is possible, then, to reduce the amount of fuel injected into the combustor by preheating the air leaving the compressor with energy taken from the turbine exhaust gases. The exchange of heat between the two flow streams takes place in a heat exchanger usually called a regenerator. A flow diagram for the regenerative gas-turbine cycle is shown in Fig. 16-18*a*. In the ideal situation it is assumed that the flow through the regenerator occurs at constant pressure. If an internally reversible heat exchanger is assumed, the heat transfer between the two flow streams may be represented as an area on a *Ts* diagram. Since the heat flow from the turbine-exhaust flow stream must equal the heat input into the compressor-outlet stream, the two crosshatched areas in Fig. 16-18*b* must be equal in magnitude; that is, $Q_{2x} = -Q_{45}$. Note that T_4 is considerably higher than T_2.

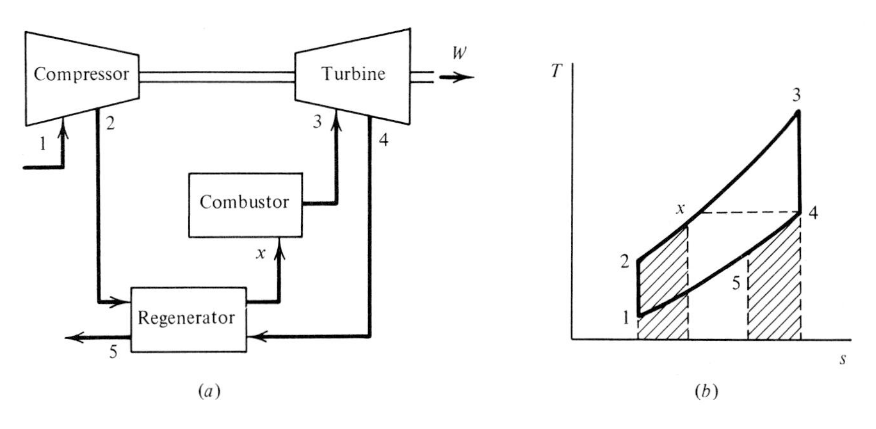

Figure 16-18 Flow diagram and Ts diagram for a regenerative gas-turbine cycle.

If the regenerator were ideal in its operation, it would be possible to preheat the compressor-outlet stream up to the temperature of the turbine-outlet stream. In this situation, state x in Fig. 16-18b would lie horizontally across from state 4. This is impractical, though, because a very large surface area is required for heat transfer as the temperature difference between the two flow streams approaches zero. As a measure of the approach to this limiting condition, the regenerator effectiveness, η_{eff}, is defined as

$$\eta_{\text{eff}} \equiv \frac{\text{actual heat transfer}}{\text{maximum possible heat transfer}} = \frac{h_x - h_2}{h_4 - h_2} \tag{16-15}$$

where the temperature corresponding to h_x is somewhat lower than the temperature corresponding to h_4. In the presence of a regenerator $q_{\text{in}} = h_3 - h_x$, and $q_{\text{out}} = h_5 - h_1$. Consequently, the thermal efficiency becomes

$$\eta_{\text{th, regen}} = 1 - \frac{h_5 - h_1}{h_3 - h_x} \tag{16-16}$$

If the gases behave ideally and the specific heats can be considered constant, it may be shown that the thermal efficiency reduces to (for ideal regeneration)

$$\eta_{\text{th, regen}} = 1 - \frac{T_1}{T_3} r_p^{(k-1)/k} \tag{16-17}$$

and at the same time the effectiveness becomes

$$\eta_{\text{eff}} = \frac{T_x - T_2}{T_4 - T_2} \tag{16-18}$$

Thus the thermal efficiency for a regenerative-type gas-turbine cycle is a function not only of the pressure ratio but also of the ratio of the minimum to maximum temperatures occurring in the cycle.

Figure 16-19 illustrates the variation of thermal efficiency with pressure ratio

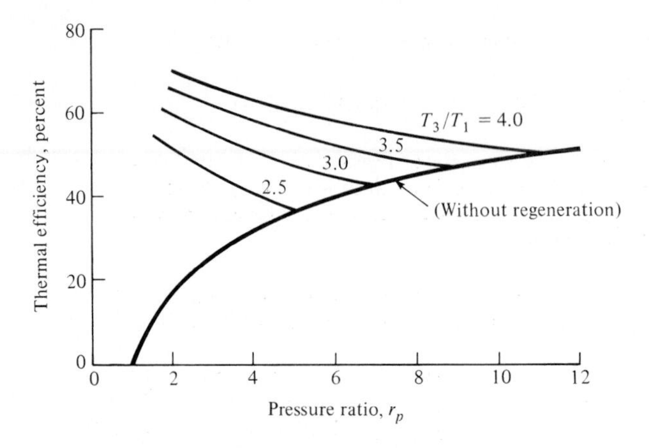

Figure 16-19 Gas-turbine-cycle thermal efficiency as a function of pressure ratio, for $k = 1.40$, with and without ideal regeneration.

and temperature ratio (T_3/T_1) for an ideal regenerative cycle on a cold air standard. The efficiency curve for the ideal Brayton cycle is also shown for comparison. Unlike the Brayton cycle, the thermal efficiency of the regenerative cycle decreases with increasing pressure ratio for a fixed value of T_3/T_1. This trend is predicted by Eq. (16-17). To increase the thermal efficiency, the value of h_x, and hence T_x, should be as high as possible, because the heat input is reduced but the net work remains the same. The usual value of the effectiveness is somewhat less than 0.7 for gas-turbine stationary power-plant applications. To increase it beyond this value usually leads to equipment costs which negate any advantage of the higher thermal efficiency. Also, a higher effectiveness requires a greater heat-transfer area. However, this leads to an increased pressure drop through the regenerator, which is a loss in terms of cycle efficiency. Pressure drops due to the regenerator are an important factor when deciding whether a regenerator should be added.

Example 16-6M Employing the data used in Example 16-4M for an ideal air-standard Brayton cycle, determine the effect on the thermal efficiency of the cycle if an ideal regenerator is inserted into the cycle.

SOLUTION The data for the enthalpies at the inlet and outlet of the compressor and turbine remain the same. Employing the symbols shown in Fig. 16-18, we note that $h_1 = 295.2$, $h_{2s} = 492.7$, $h_3 = 1161.1$, and $h_{4s} = 706.5$, all values in kJ/kg. For an ideal regenerator the enthalpy of the fluid leaving the regenerator h_x must be equal to h_{4s}. The heat input saved because of the regenerator is

$$q_{\text{saved}} = h_x - h_{2s} = h_{4s} - h_2 = 706.5 - 492.7 = 213.8 \text{ kJ/kg}$$

The required heat input now becomes

$$q_{\text{in, regen}} = h_4 - h_3 = h_4 - h_5 = 1161.1 - 706.5 = 454.6 \text{ kJ/kg}$$

The percent saved over the case without regeneration is

$$\%q_{\text{saved}} = \frac{213.8}{668.4} 100 = 32.0 \text{ percent}$$

Since the turbine and compressor work are exactly the same as in Example 16-3M,

$$\eta_{th} = \frac{257.1}{454.6} = 0.565 \qquad \text{(or 56.5 percent)}$$

This compares with an efficiency of 38.5 percent without regeneration. Thus the effect of regeneration on the thermal efficiency is considerable, although in actual cases the effect would not be quite as large, since ideal regeneration is not possible. In addition, internal irreversibilities and heat losses will reduce the efficiency further.

Example 16-6 Employing the data used in Example 16-4 for an ideal Brayton cycle, determine the effect on the thermal efficiency of the cycle if an ideal regenerator is inserted into the cycle.

SOLUTION The data for the enthalpies at the inlet and outlet of the compressor and turbine remain the same. Employing the symbols shown in Fig. 16-18, we note that $h_1 = 129.06$, $h_{2s} = 215.52$, $h_3 = 504.71$, and $h_{4s} = 307.25$, all values in Btu/lb. For an ideal regenerator the enthalpy of the fluid leaving the regenerator h_x must be equal to h_{4s}. The heat input saved because of the regenerator is

$$q_{saved} = h_x - h_{2s} = h_{4s} - h_{2s} = 307.25 - 215.52 = 91.73 \text{ Btu/lb}$$

The required heat input now becomes

$$q_{in,\,regen} = h_4 - h_3 = h_4 - h_5 = 504.71 - 307.25 = 197.46 \text{ Btu/lb}$$

The percent heat saved over the case without regeneration is

$$\% q_{saved} = \frac{91.73}{289.19} 100 = 31.7 \text{ percent}$$

Since the turbine and compressor work are exactly the same as before,

$$\eta_{th} = \frac{111.0}{197.46} = 0.562 \qquad \text{(or 56.2 percent)}$$

This compares with an efficiency of 38.4 percent without regeneration. It is seen that the effect of regeneration on the thermal efficiency is considerable, although in actual cases the effect would not be quite as large since ideal regeneration is not possible. In addition, internal irreversibilities and heat losses will reduce the efficiency further.

Example 16-7M Employing the data of Example 16-5M, determine the effect on the thermal efficiency if a regenerator of 70 percent effectiveness is inserted into the cycle.

SOLUTION In Example 16-5M the compressor and turbine efficiencies are 82 and 85 percent, respectively. The data for the enthalpies at the inlet and outlet of the compressor and turbine remain the same. Using the symbols shown in Fig. 16-20, we note that $h_1 = 295.2$, $h_{2a} = 536.1$, $h_3 = 1161.1$, and $h_{4a} = 774.7$, all values in kJ/kg. Use of Eq. (16-15) for the effectiveness enables us to determine the heat added between states 2 and x.

$$0.70 = \frac{h_x - h_2}{h_4 - h_2} = \frac{h_x - h_2}{774.7 - 536.1}$$

or $$h_x - h_2 = 0.7(238.6) = 167.0 \text{ kJ/kg}$$

This quantity represents the heat input saved by regeneration. Without regeneration the heat

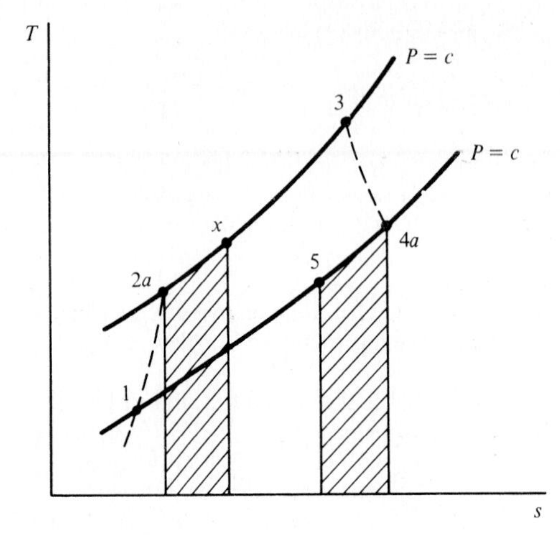

Figure 16-20 The Ts diagram for the regenerative gas-turbine cycle in Examples 16-7M and 16-7.

input would need to be

$$q_{\text{without regen}} = h_3 - h_{2a} = 1161.1 - 536.1 = 625.0 \text{ kJ/kg}$$

The percent heat input saved over the case without regeneration is

$$\%q_{\text{saved}} = \frac{167.0}{625.0} 100 = 26.7 \text{ percent}$$

Since the turbine and compressor work are exactly the same as in Example 16-5M,

$$\eta_{\text{th}} = \frac{386.4 - 240.9}{625.0 - 167.0} = \frac{145.5}{458.0} = 0.318 \qquad \text{(or 31.8 percent)}$$

This compares to an efficiency of 23.3 percent without regeneration. It is also a more realistic value than the 56.5 percent value found in Example 16-6M, because reasonable turbine and compressor efficiencies and regenerator effectiveness have been assumed.

Example 16-7M Employing the data of Example 16-5, determine the effect on the thermal efficiency if a regenerator of 70 percent effectiveness is inserted into the cycle.

SOLUTION The data for the enthalpies at the inlet and outlet of the compressor and turbine remain the same. Employing the symbols shown in Fig. 16-20, we note that $h_1 = 129.1$, $h_{2a} = 234.5$, $h_3 = 504.7$, and $h_{4a} = 336.9$, all values in Btu/lb. Use of Eq. (16-15) for the effectiveness enables us to determine the heat added between states 2 and x:

$$0.70 = \frac{h_x - h_2}{336.9 - 234.5}$$

or

$$h_x - h_2 = 0.7(102.4) = 71.1 \text{ Btu/lb}$$

This quantity represents the heat input saved by regeneration. Without regeneration the heat input would need to be

$$q_{\text{without regen}} = h_3 - h_{2a} = 504.7 - 234.5 = 270.2 \text{ Btu/lb}$$

The percent heat input saved over the case without regeneration is

$$\%q_{\text{saved}} = \frac{71.7}{270.2} 100 = 26.5 \text{ percent}$$

Since the turbine and compressor work are exactly the same as in Example 16-5,

$$\eta_{\text{th}} = \frac{62.4}{270.2 - 71.7} = \frac{62.4}{198.5} = 0.315 \qquad \text{(or 31.5 percent)}$$

This compares to an efficiency of 23.2 percent without regeneration. It is also a more realistic value than the 56.2 percent value found in Example 16-6, since realistic turbine and compressor efficiencies and regenerator effectiveness have been assumed.

16-9 THE POLYTROPIC PROCESS

When an ideal gas is expanded or compressed in an internally reversible, adiabatic manner, the relationship between P and v is given by the isentropic equation

$$Pv^k = \text{constant} \tag{7-35b}$$

This process equation is a reasonably good model when the heat transfer during an actual process is quite small, compared to the work interaction. Even if the heat transfer is not negligible during a process, the format of Eq. (7-35*b*) can be retained by writing the equation in the form

$$Pv^n = \text{constant} \tag{16-19}$$

A process which follows this relation between equilibrium states is called a *polytropic* process. The equation is restricted to a quasistatic process for a simple, compressible gaseous system. If the relation $Pv = RT$ is valid, then it is easily shown that two other equations are equally valid for a polytropic process. Analogous to equations developed for adiabatic, reversible processes, we find that for polytropic processes,

$$T(v)^{n-1} = \text{constant} \tag{16-20}$$

and

$$TP^{-(n-1)/n} = \text{constant} \tag{16-21}$$

The simplicity of the polytropic equations is a distinct advantage when deriving mathematical relationships for various thermodynamic functions such as δw, δq, and du, if the validity of the model is justified.

One of the notable features of the polytropic process is that, when n takes on certain specific values, the process may be identified with a more familiar change of state. As noted above, when a process is adiabatic and internally reversible, $n = k$. The other three familiar cases are: (1) when $n = 1$, the process is isothermal, (2) when $n = 0$, the process is constant pressure, and (3) when $n = \infty$, the process is constant volume. Figure 16-21 illustrates the polytropic process for the

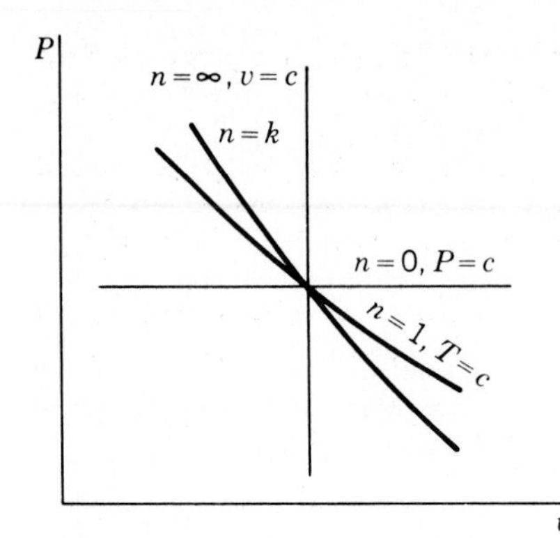

Figure 16-21 Polytropic processes.

specific values of n discussed above. In some instances the polytropic expression $Pv^n = c$ is a good approximation, even though $Pv \neq RT$. Under such circumstances it must be recognized that Eqs. (16-20) and (16-21) will no longer be valid.

Example 16-8M Air is compressed polytropically along a path for which $n = 1.30$ in a closed system. The initial temperature and pressure are 17°C and 1 bar, respectively, and the final pressure is 5 bars. Find (*a*) the final temperature, in degrees Celsius, (*b*) the work done on the gas, in kJ/kg, and (*c*) the heat transferred, in kJ/kg.

SOLUTION Air is essentially an ideal gas in the range of pressures and temperatures involved. Consequently, we may use Eq. (16-21) to determine the final temperature.

(*a*)
$$T_2 = T_1\left(\frac{P_2}{P_1}\right)^{(n-1)/n} = 290\left(\frac{5}{1}\right)^{(1.30-1)/1.30}$$

$$= 290(1.45) = 420°\text{K} = 147°\text{C}$$

(*b*) The first-law statement for a closed system is $q + w = \Delta u$. Both q and w are unknown. Since we have no equations for calculating q independently, we must first evaluate w by means of the integral of $-P\,dv$. Substitution of the process equation $Pv^n = c$ and subsequently integration leads to the following equation for the work associated with polytropic processes within a closed system:

$$w = -\int P\,dv = -\int cv^{-n}\,dv = \frac{c(v_2)^{1-n} - c(v_1)^{1-n}}{n-1}$$

$$= \frac{P_2v_2 - P_1v_1}{n-1} = \frac{R(T_2 - T_1)}{n-1}$$

Employing the proper values, we find that

$$w = \frac{8.314(420 - 290)}{29(1.3 - 1)} = 124.2 \text{ kJ/kg}$$

(*c*) The heat transferred is determined by the conservation-of-energy statement for the closed system, $q = \Delta u - w$. The specific internal-energy change for an ideal gas is simply $c_v\Delta T$, and for the temperature range involved the average value of c_v is close to 0.723 kJ/(kg)(°K). Hence the heat transferred is

$$q = 0.723(420 - 290) - 124.2 = -30.2 \text{ kJ/kg}$$

The negative sign indicates that heat is removed from the gas during the compression process. The ratio of q to w for this specific polytropic process is 30.2/124.1 = 0.24.

Example 16-8 Air is compressed polytropically along a path for which $n = 1.3$ in a closed system. The initial temperature is 40°F, the initial pressure is 15 psia, and the final pressure is 75 psia. Find (*a*) the final temperature, (*b*) the work done on the gas, and (*c*) the heat transferred.

SOLUTION Air will follow closely the ideal-gas equation of state in the range of pressures and temperatures involved. Consequently, we may employ Eq. (16-21) to determine the final temperature.

(*a*)

$$T_2 = T_1\left(\frac{P_2}{P_1}\right)^{(n-1)/n} = 500\left(\frac{75}{15}\right)^{(1.3-1)/1.3}$$

$$= 500(1.45) = 725°\text{R} = 265°\text{F}$$

(*b*) The work done on the gas normally might be computed from the first-law statement for a closed system, namely, $q + w = \Delta u$. However, in the absence of a means of calculating the value of q independently, we must turn to the analytical expression for boundary work and determine the value of w independent of other quantities. The boundary work is given by the inegral of $-P\,dv$, and the relationship between P and v is simply $Pv^n = c$. Subsequent substitution and definite integration leads to the following equation for the work associated with polytropic processes:

$$w = -\int P\,dv = -\int cv^{-n}\,dv = \frac{c(v_2)^{1-n} - c(v_1)^{1-n}}{n-1}$$

$$= \frac{P_2 v_2 - P_1 v_1}{n-1} = \frac{R(T_2 - T_1)}{n-1}$$

Employing the proper numbers, we find that

$$w = \frac{1545(265 - 40)}{29(1.3 - 1)} = +40{,}000 \text{ ft} \cdot \text{lb}_f/\text{lb}_m$$

$$= +51.5 \text{ Btu/lb}_m$$

(*c*) The heat transferred is determined by the conservation-of-energy statement for the closed system, $q = \Delta u - w$. The specific internal-energy change for an ideal gas is simply $c_v\,\Delta T$, and in the temperature range involved the average value of c_v is, roughly, 0.172 Btu/(lb$_m$)(°F). Hence the heat transferred per pound of gas is

$$q = 0.172(265 - 40) - 51.5 = -12.8 \text{ Btu/lb}$$

The negative sign indicates that heat was removed from the gas during the compression.

16-10 STEADY-FLOW ANALYSIS OF COMPRESSORS

The steady-flow equation for the conservation of energy with regard to a compressor reduces to

$$w = h_2 - h_1 + \text{KE}_2 + \text{KE}_1 - q$$

Often the kinetic-energy difference is small; hence

$$w = (h_2 - h_1) - q$$

For ideal gases the enthalpy values can be found in the ideal-gas tables or calculated from the integral of $c_p\,dT$. For real gases the enthalpy can be obtained from tables or calculated from generalized equations introduced in Chap. 13. In analyzing any compression process it is convenient to consider three particular cases. We shall restrict ourselves in the following discussion to ideal gases, and kinetic-energy changes are neglected.

1 Adiabatic Compression with Constant Specific Heats

For an adiabatic process the conservation-of-energy principle reduces to

$$w = (h_2 - h_1) = c_p(T_2 - T_1)$$

If, in addition, the process is reversible, then $T_2/T_1 = (P_2/P_1)^{(k-1)/k}$. It may be shown that, under these circumstances, the steady-flow mechanical work becomes

$$w_{sf} = \frac{kRT_1[1 - (P_2/P_1)^{(k-1)/k}]}{1 - k} \tag{16-22}$$

2 Isothermal Compression

In the case of an ideal gas, the enthalpy change for an isothermal process is zero. Hence the steady-state energy balance now reduces to $w = -q$. An analytical expression for the steady-flow mechanical work of a frictionless process can be obtained from the integral of $v\,dP$; thus

$$w_{sf} = \int v\,dP = \int \frac{RT}{P}\,dP = RT \ln \frac{P_2}{P_1} = RT \ln \frac{v_1}{v_2}$$

3 Polytropic Compression

In Sec. 16-9 it was shown that, under certain circumstances, a quasistatic process with heat transfer may follow a path represented by the equation $Pv^n =$ constant. In this situation it is found that the steady-flow work is given by

$$w_{sf} = \int v\,dP = \frac{n(P_2 v_2 - P_1 v_1)}{n - 1} = \frac{nR(T_2 - T_1)}{n - 1} \tag{16-23}$$

In the above equation the pressures and temperatures at inlet and outlet are related by Eq. (16-21) for a polytropic process.

It is helpful to compare the frictionless steady-flow mechanical work for the three types of processes just described. In the absence of significant kinetic-energy changes, the work of compression for a steady-flow process is evaluated by the integral of $v\,dP$. A graphical representation of such a process is shown in Fig. 16-22 for a fixed pressure ratio P_2/P_1. Since the area to the left of the process curve is a measure of the steady-flow work, it is seen that the maximum friction-

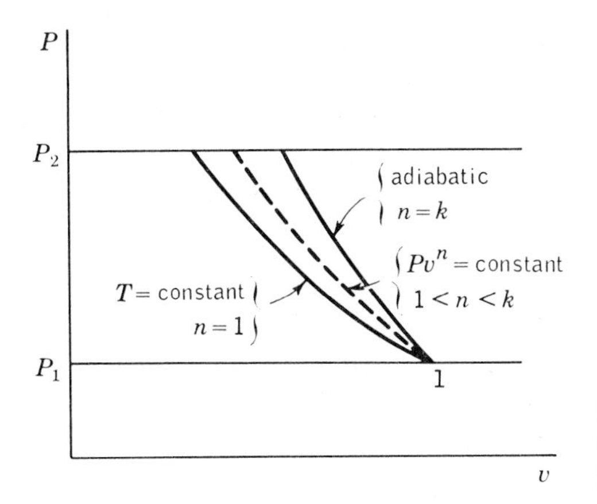

Figure 16-22 The effect of cooling on the required work of compression for a steady-flow process between fixed pressure limits.

less work is required for an adiabatic compression, where $n = k$. As the value of n is decreased by removal of heat during the process, the work requirements also decrease. When sufficient heat is removed during a process that n is unity, the process is isothermal. Figure 16-19 illustrates the simple fact that the removal of heat during a compression process is advantageous with regard to the work-input requirements. This result is equally true for both reciprocating and rotating (axial and centrifugal) compressors. Consequently, it might be practical to use cooling jackets around the casing of a compressor in order to lower the value of the polytropic coefficient during the compression process. For air the specific-heat ratio at normal temperatures is 1.4, but by external cooling the value of n for the process may be reduced to the neighborhood of 1.35. The practice of cooling a fluid during compression depends upon the use to which the fluid will be placed. If a gas at high pressure is the primary purpose of a compression process, cooling is beneficial. If a fluid of high enthalpy is desired at a high pressure, adiabatic operation is more desirable.

4 Multistage Compression with Intercooling

Although the practice of cooling the gas passing through a compressor has its advantages, in many cases it is either not possible or practical to have much heat transfer through the compressor casing. To achieve the benefits of cooling, another physical arrangement is frequently used, called multistage compression with intercooling. This is especially effective when a comparatively large pressure change is desired. In multistage compression, the fluid is first compressed to some intermediate pressure which lies between P_1 and P_2. The fluid then is passed through an intercooler (a heat exchanger), where it is cooled down to a lower temperature at essentially constant pressure. In some instances this lower temperature could be the initial temperature T_1. Then it passes through another stage of the compressor, where its pressure is further raised. This would be followed by another intercooler process, and then another staging of the compressor, until the final high pressure is achieved. The overall result is, of course, a lowering of the net work input required for a given pressure ratio.

The effect of intercooling on a two-state compressor is demonstrated in Fig. 16-23. In this special case it has been assumed that the intercooler cools the fluid down to the initial temperature before it enters the second stage. The compression processes themselves between pressures P_1 to P_x and P_x to P_2 are of the type Pv^n = constant, where the value of n depends upon the amount of heat transfer that occurs during compression. The crosshatched area on the Pv diagram represents the work input that is saved by the process of intercooling. Diagrams similar to the Pv and Ts diagrams shown in Fig. 16-23 could be drawn for compressors with a large number of stages. The compression processes on the Ts diagram have been sketched for a polytropic process, where n lies between 1 and k.

It is of theoretical and practical interest to determine the conditions under which the work input would be minimized, that is, the crosshatched area in Fig. 16-23a would be maximized. For a two-stage compressor, both the intermediate pressure and temperature can be varied, which leads to an indefinitely large number of ways of carrying out the compression process. In order to simplify the calculation, we shall assume that it is possible to design an *ideal* intercooler that will reduce the temperature of the fluid to the initial temperature of the system before it enters the second state. By using Eq. (16-23), the total work of compression for the two stages, neglecting kinetic-energy changes, is

$$w = \frac{nRT_1\left[\left(\frac{P_x}{P_1}\right)^{(n-1)/n} - 1\right]}{n-1} + \frac{nRT_1\left[\left(\frac{P_2}{P_x}\right)^{(n-1)/n} - 1\right]}{n-1} \tag{16-24}$$

In order to find the minimum total work of compression, the above equation for

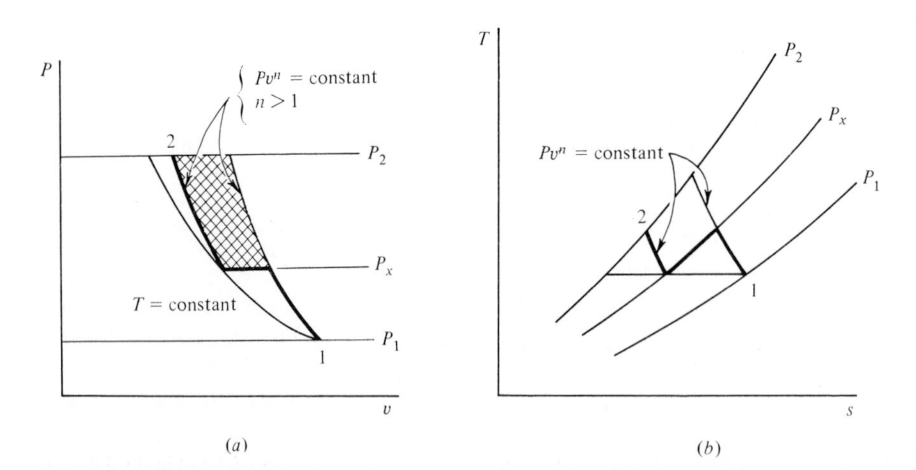

Figure 16-23 Two-stage compression with intercooling to the initial temperature.

w is differentiated with respect to the variable pressure P_x and the resulting equation is set equal to zero. It is found that the minimum work input occurs when the pressures are related by the expression

$$P_x = (P_1 P_2)^{1/2} \qquad \text{or} \qquad \frac{P_x}{P_1} = \frac{P_2}{P_x} \tag{16-25}$$

If this result is substituted into Eq. (16-24), it is noted that the work done by each stage is equal; that is, $w_{1x} = w_{x2}$.

16-11 GAS-TURBINE CYCLES WITH INTERCOOLING AND REHEATING

Another method of increasing the overall efficiency of a gas-turbine cycle is to decrease the work input to the compression process and/or increase the work output of the turbine. The effect of either of these procedures is to increase the net work output. In Sec. 16-10, it was pointed out that the work of a compressor operating between fixed pressure levels can be reduced by intercooling between stages of the compressor. Figure 16-24 shows an equipment schematic and a Ts diagram for a cycle with intercooling between a two-stage compressor. Note that the final temperature T_2 is less than the temperature which would occur without intercooling. As a consequence of this low temperature, a gas-turbine cycle with intercooling is especially adaptable to regeneration, as shown in the figure. In fact, intercooling is promising only if a regenerator is employed at the same time; otherwise a considerable amount of heat must be supplied to the cycle at a relatively low temperature. It must also be kept in mind that a considerably larger regenerator will be required than in the case of no intercooling.

In conjunction with intercooling, it is often found effective to use turbine staging as well. Instead of expanding directly through a single turbine, the gases

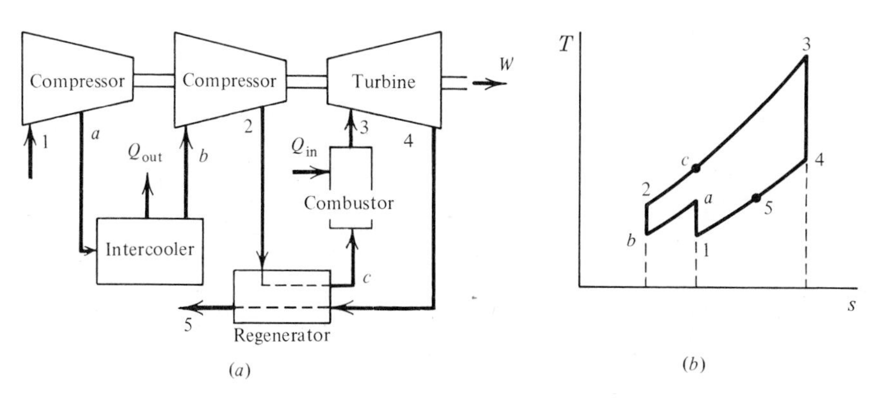

Figure 16-24 (*a*) Schematic equipment diagram; (*b*) Ts diagram for a regenerative gas-turbine cycle with intercooling.

are allowed to expand only partially before they are returned to another combustion chamber, where heat ideally is added at constant pressure until the limiting temperature is reached. Then, in the case of a two-stage turbine, further expansion occurs until the ambient pressure is attained. Figure 16-25 illustrates the process path for a cycle with reheating on a *Ts* diagram. The use of reheat increases the turbine-work output without changing either the compressor work or the maximum limiting temperature. It is noted that the use of reheat requires a substantial increase in the heat input to the cycle. However, the final turbine-exhaust temperature T_6 shown in the figure is somewhat above the outlet-turbine temperature without reheat, T_7. Consequently, reheating is quite effective when used in conjunction with regeneration, since the quantity of heat exchanged in the regenerator can be greatly increased.

Under the condition of ideal reheating (T_3 and T_5 are equal in Fig. 16-25), the total work output from the two stages depends upon the pressure ratio across each stage. Analogous to two-stage compression with ideal intercooling, the maximum work output is found by differentiation of a general equation for the work of the two stages. Equation (16-24) again applies, except T_1 in this equation becomes T_3 and the pressure ratios in the equation now represent those across the two turbine stages. Since the derivation is the same, the answer for maximum work output is the same as minimum work input for a two-stage compressor. That is, the intermediate pressure is the geometric mean between the inlet pressure and the outlet pressure, or the pressure ratio across each stage is the square root of the overall pressure ratio. In terms of Fig. 16-25, the maximum work output would occur when $P_4 = P_5 = (P_3 P_6)^{1/2}$ for two-stage expansion with ideal reheating. As a result, the work output of each stage will be equal for the case of ideal reheating. An equipment schematic and a *Ts* diagram for an ideal-gas-turbine cycle employing intercooling, reheat, and regeneration is shown in Fig. 16-26.

From the foregoing discussion it should be apparent that, from practical considerations, a gas-turbine power cycle is improved most when a combination of intercooling and reheating is employed with regeneration. Under any circumstances, the effects of irreversibilities in the turbine and compressor, or pressure

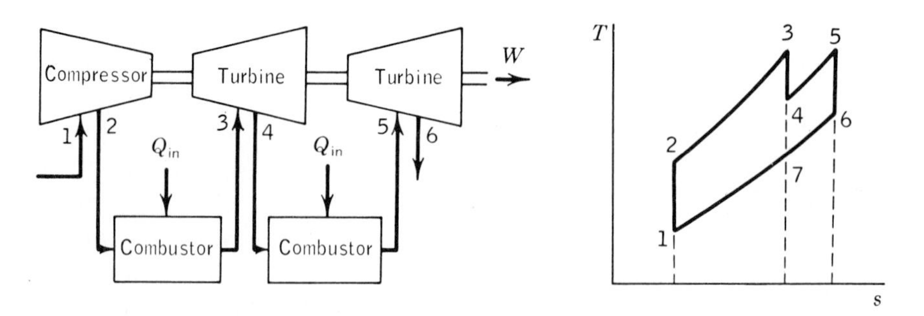

Figure 16-25 The air-standard gas-turbine cycle with reheat.

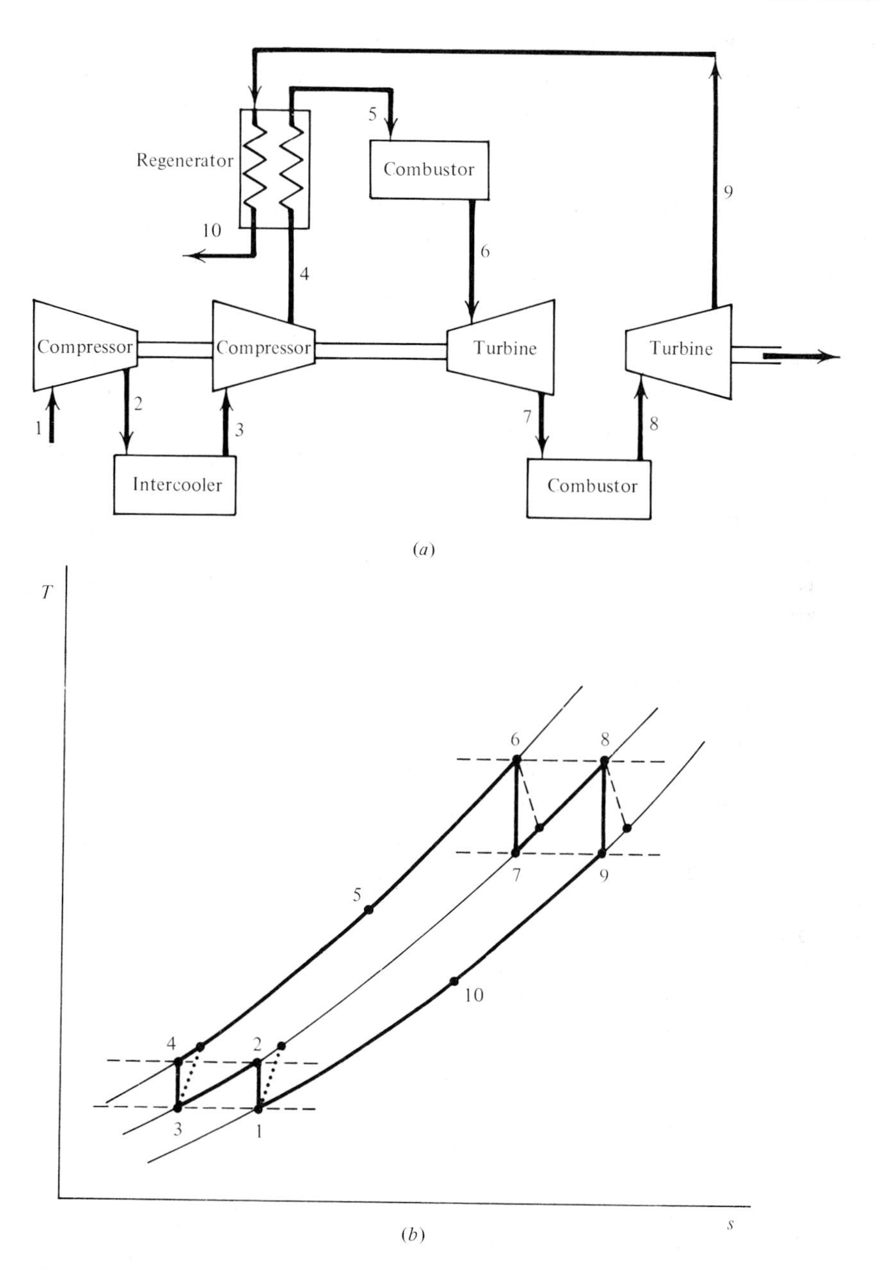

Figure 16-26 (*a*) Schematic equipment diagram; (*b*) *Ts* process diagram for a gas-turbine power cycle with intercooling, reheat, and regeneration.

losses in the combustor, etc., must be considered in predicting the actual performance of a gas-turbine cycle. One should not conclude, however, that intercooling and reheating without regeneration improve the thermal efficiency. In fact, the effect without regeneration is always to decrease the thermal efficiency. The reason is that intercooling and reheating used alone decrease the average temperature of the heat supply and increase the average temperature of heat rejection. This argument can be seen qualitatively from a Ts diagram such as Fig. 16-26*b*. The major purpose in employing intercooling and reheating is to increase the effective use of a regenerator.

Example 16-9M Ideal intercooling and reheating are added to the cycle proposed in Example 16-7M. In addition, the pressure ratios across the two-stage compressor and two-stage turbine are adjusted to provide the minimum work input, and the maximum work output, respectively. Determine the effect of this additional equipment on the thermal efficiency.

Solution In order to minimize the compressor work and maximize the turbine work, the pressure ratio across each stage must be the same. This requires the pressure ratio across each stage to be the square root of the overall pressure ratio. Since this latter value is 6, the ratio for each stage (compressor or turbine) is 2.45. At the inlet to the compressor's first stage $h_1 =$ 295.2 kJ/kg and $p_{r1} = 1.3068$. The relative pressure at the outlet of this stage for isentropic flow would be

$$p_{r2} = p_{r1}\frac{P_2}{P_1} = 1.3068(2.45) = 3.202$$

Interpolation in the abridged air table A-5M indicates that $h_{2s} = 381.8$ kJ/kg and $T_{2s} = 381°\text{K}$. The actual compressor-enthalpy rise is

$$\Delta h_{C,a} = \frac{\Delta h_{C,s}}{\eta_C} = \frac{381.8 - 295.2}{0.82} = 105.6 \text{ kJ/kg}$$

The actual compressor work is twice this value, that is,

$$w_C = 2(105.6) = 211.2 \text{ kJ/kg}$$

The actual outlet enthalpy at state 4 (see dotted lines in Fig. 16-26*b*) for the compressor is

$$h_{4a} = h_3 + \Delta h_{C,a} = h_1 + \Delta h_{C,a}$$
$$= 295.2 + 105.6 = 400.8 \text{ kJ/kg}$$

For the first stage of the turbine, $h_6 = 1161.1$ kJ/kg, $p_{r6} = 167.1$, and

$$p_{r7} = p_{r6}\frac{P_7}{P_6} = 167.1\frac{1}{2.45} = 68.2$$

From the abridged air table $h_{7s} = 907.6$ kJ/kg and $T_{7s} = 877°\text{K}$. The actual turbine-enthalpy drop is

$$\Delta h_{T,a} = \Delta h_{T,s}\eta_T = (116.1 - 907.6)(0.85)$$
$$= 253.5(0.85) = 215.5 \text{ kJ/kg}$$

The total turbine output is

$$w_T = 2(215.5) = 431.0 \text{ kJ/kg}$$

The actual enthalpy at state 9 for the turbine outlet is

$$h_{9a} = h_8 - \Delta h_{T,a} = h_6 - \Delta h_{T,a}$$
$$= 1161.1 - 215.5 = 945.6 \text{ kJ/kg}$$

For the regenerator, $\eta_{\text{eff}} = (h_5 - h_4)/(h_9 - h_4)$. Hence

$$0.70 = \frac{h_5 - h_4}{945.6 - 400.8}$$

or

$$h_5 - h_4 = q_{\text{saved}} = 0.7(544.8) = 381.4 \text{ kJ/kg}$$

The heat input required in the combustor becomes

$$q_{\text{comb}} = (h_6 - h_4) - (h_5 - h_4)$$

$$= (1161.1 - 400.8) - 381.4 = 378.9 \text{ kJ/kg}$$

The total heat input, combustor plus reheat, is

$$q_{\text{in}} = q_{\text{comb}} + (h_8 - h_{7a}) = q_{\text{comb}} + (h_6 - h_{7a})$$

$$= q_{\text{comb}} + w_{T,a} = 378.9 + 215.5 = 594.4 \text{ kJ/kg}$$

The thermal efficiency then is

$$\eta_{\text{th}} = \frac{w_T - w_C}{q_{\text{in}}} = \frac{431.0 - 211.2}{594.4} = 0.370 \qquad \text{(or 37.0 percent)}$$

In comparison to Example 16-7M, the addition of intercooling and reheating to the cycle with regeneration increases the cycle thermal efficiency from 31.8 percent to 37.0 percent.

Example 16-9 Ideal intercooling and reheating are added to the cycle proposed in Example 16-7. In addition, the pressure ratios across the two-stage compressor and two-stage turbine are adjusted to provide the minimum work input, and maximum work output, respectively. Determine the effect on the thermal efficiency.

SOLUTION In order to minimize the compressor work and maximize the turbine work, the pressure ratio across each stage must be the same. This requires the pressure ratio across each stage to be the square root of the overall pressure ratio. Since this latter value is 6, the ratio for each stage (compressor or turbine) is 2.45. At the inlet to the compressor's first stage $h_1 =$ 129.1 Btu/lb and $p_{r1} = 1.386$. The relative pressure at the outlet of this stage for isentropic flow would be

$$p_{r2} = p_{r1} \frac{P_2}{P_1} = 1.386(2.45) = 3.40$$

Interpolation in the abridged tables indicates that $h_{2s} = 167$ Btu/lb and $T_{2s} = 697°\text{R}$. The actual compressor-enthalpy rise is

$$\Delta h_{C,a} = \frac{\Delta h_s}{\eta_C} = \frac{167 - 129}{0.82} = 46.4 \text{ Btu/lb}$$

The total compressor work is twice this value, that is,

$$w_C = 2(46.4) = 92.8 \text{ Btu/lb}$$

The actual outlet enthalpy at state 4 (see Fig. 16-26*b*) for the compressor is

$$h_{4a} = h_3 + \Delta h_{C,a} = h_1 + \Delta h_{C,a}$$

$$= 129.1 + 46.4 = 175.5 \text{ Btu/lb}$$

For the first stage of the turbine, $h_6 = 504.7$ Btu/lb, $p_{r6} = 174.00$, and

$$p_{r7} = p_{r6} \frac{P_7}{P_6} = 174.0 \frac{1}{2.45} = 71.0$$

From the abridged air tables $h_{7s} = 394.7$ and $T_{7s} = 1596°R$. The actual turbine-enthalpy drop is

$$\Delta h_{T,a} = \Delta h_{T,s}\eta_T = (504.7 - 394.7)(0.85)$$
$$= 110(0.85) = 93.5 \text{ Btu/lb}$$

The total turbine output is

$$w_T = 2(93.5) = 187.0 \text{ Btu/lb}$$

The actual enthalpy at state 9 for the turbine outlet is

$$h_{9a} = h_8 - \Delta h_{T,a} = h_6 - \Delta h_{T,a}$$
$$= 504.7 - 93.5 = 411.2 \text{ Btu/lb}$$

For the regenerator, $\eta_{eff} = (h_5 - h_4)/(h_9 - h_4)$. Hence,

$$0.70 = \frac{h_5 - h_4}{411.2 - 175.5}$$

or
$$h_5 - h_4 = q_{saved} = 0.7(235.7) = 165.0 \text{ Btu/lb}$$

The heat input required in the combustor becomes

$$q_{comb} = (h_6 - h_4) - (h_5 - h_4)$$
$$= (504.7 - 175.5) - (165.0)$$
$$= 164.2 \text{ Btu/lb}$$

The total heat input, combustor plus reheat, is

$$q = q_{comb} + (h_8 - h_{7a}) = q_{comb} + (h_6 - h_{7a})$$
$$= q_{comb} + w_{T,a} = 164.2 + 93.5 = 257.7 \text{ Btu/lb}$$

The thermal efficiency then is

$$\eta_{th} = \frac{w_T - w_C}{q_{in}} = \frac{187.0 - 92.8}{257.7} = 0.365 \qquad \text{(or 36.5 percent)}$$

In comparison to Example 16-7 the addition of intercooling and reheating to a cycle with regeneration increases the cycle efficiency from 31.5 percent to 36.5 percent.

16-12 DIFFUSER AND NOZZLE PERFORMANCE

Before proceeding to the discussion of another application of the gas-turbine power cycle—aircraft propulsion—it is appropriate to analyze briefly diffuser and nozzle performance under reversible and irreversible operation. A diffuser decelerates flow and increases the pressure. If the heat transfer is negligible, the steady-flow energy equation reduces to $\Delta h + \Delta KE = 0$. To simplify the analysis in this section, a cold-air-standard basis will be used. In this case the energy equation simplifies to

$$c_p \, \Delta T = -\Delta KE \qquad \text{or} \qquad c_p(T_2 - T_1) = -\frac{V_2^2 - V_1^2}{2} \tag{a}$$

where 1 and 2 represent the inlet and outlet states. If the final velocity is small compared to the initial value, then V_1 fixes the kinetic-energy change, which in turn fixes ΔT. Hence, the temperature change is independent of whether irreversibilities are present or not. Likewise, T_2 is not dependent on irreversibilities for a given value of T_1. The property that is affected by irreversibilities is the pressure. This is clearly seen from a second-law analysis of the flow through the diffuser. Recall for an ideal gas with constant specific heat that

$$\Delta s = c_p \ln \frac{T_2}{T_1} - R \ln \frac{P_2}{P_1} \tag{b}$$

We have shown for the diffuser that the temperature ratio T_2/T_1 is fixed. A comparison of a reversible and an irreversible process through the diffuser in terms of the entropy function leads to

$$\Delta s + R \ln \frac{P_{2a}}{P_1} = R \ln \frac{P_{2s}}{P_1} \tag{c}$$

where the subscripts s and a again represent the isentropic and actual (irreversible) cases. On the basis of Eq. (b), P_{2s} is fixed by the values of P_1, T_1, and T_2. Thus the right side of Eq. (c) is fixed by the kinetic-energy change. Therefore as Δs increases due to increasing irreversibilities within the diffuser, the above equation shows that P_{2a} decreases for a given ΔKE. This effect is illustrated on a Ts diagram in Fig. 16-27. Note that higher values of s_2 require lower values of P_{2a} for a fixed value of T_2.

One measure of the performance of a diffuser in terms of pressure data is the

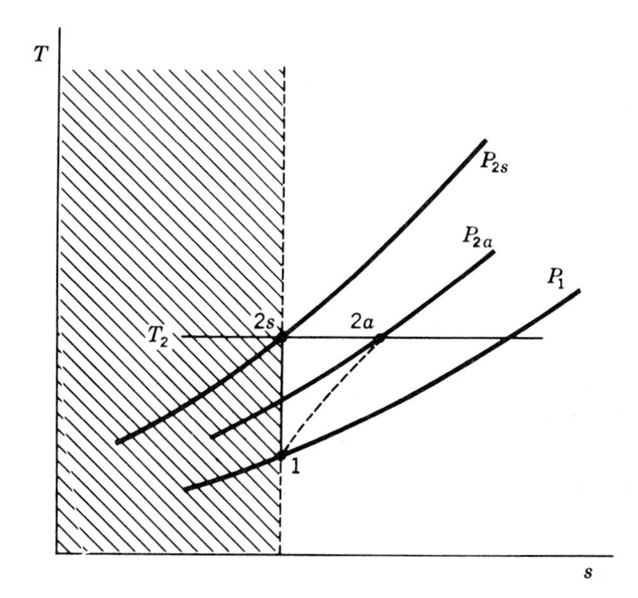

Figure 16-27 Effect of irreversibilities on diffuser-outlet pressure.

pressure coefficient K_P. For a given value of ΔKE for a process,

$$K_P = \frac{\text{actual pressure rise}}{\text{isentropic pressure rise}} = \frac{P_{2a} - P_1}{P_{2s} - P_1} \tag{16-26}$$

(Note that K_P is sometimes defined in a slightly different way. The definition may require P_{2s} to be the pressure reached isentropically when the final kinetic energy is zero. This limiting pressure, which lies vertically above P_{2s} in Fig. 16-27 if the final kinetic energy is not zero, is known as the stagnation pressure for the ideal diffuser.)

In the analysis of nozzles we normally neglect heat transfer and potential-energy changes. Thus the conservation-of-energy statement for a nozzle is the same as for a diffuser,

$$\Delta \text{KE} = -\Delta h = c_p(T_1 - T_2)$$

The entropy change for an ideal gas is still governed by Eq. (*b*). Unlike the diffuser analysis, however, the independent variables for nozzle flow are T_1, P_1, and P_2. The final pressure is fixed in the operation of a nozzle, and V_2 becomes an unknown. For isentropic flow Eq. (*b*) dictates that T_2 be fixed by the three variables. Since T_{2s} is fixed, the energy equation above dictates that ΔKE be fixed. Hence, the acceleration of a gas through a nozzle is determined solely by the three variables if the flow is isentropic. The isentropic path is shown in Fig. 16-28.

The effect of internal irreversibilities on nozzle flow is most clearly seen by rearranging the equation for Δs into the form

$$c_p \ln \frac{T_2}{T_1} = R \ln \frac{P_2}{P_1} + \Delta s \tag{d}$$

Keep in mind that T_1, P_1, and P_2 are fixed for a given mode of operation. As Δs increases with increasing irreversibilities, T_{2a} likewise increases. This is shown in Fig. 16-28. That is, $T_{2a} > T_{2s}$. Since $\Delta \text{KE} = c_p(T_1 - T_2)$, we see that isentropic expansion gives a larger ΔT, and thus a larger ΔKE. An isentropic expansion through a nozzle is more effective in accelerating the flow than an irreversible expansion.

An appropriate measure of the performance of a nozzle is the adiabatic efficiency η_N of the device. This was defined in Sec. 7-10 as

$$\eta_N = \frac{\text{KE}_{2a}}{\text{KE}_{2s}} = \frac{V_{2a}^2/2}{V_{2s}^2/2} \tag{7-46}$$

If the inlet velocity is negligible compared to V_{2a} and V_{2s} at the exit, then Eq. (7-46) may be expressed also by

$$\eta_N = \frac{(h_1 - h_2)_a}{(h_1 - h_2)_s} \tag{7-47}$$

Figure 16-28 is plotted as an *hs* diagram as well as a *Ts* diagram for an ideal gas. The efficiency cited by Eq. (7-47) is a ratio of the vertical distances 1-2*a* and 1-2*s* on the figure.

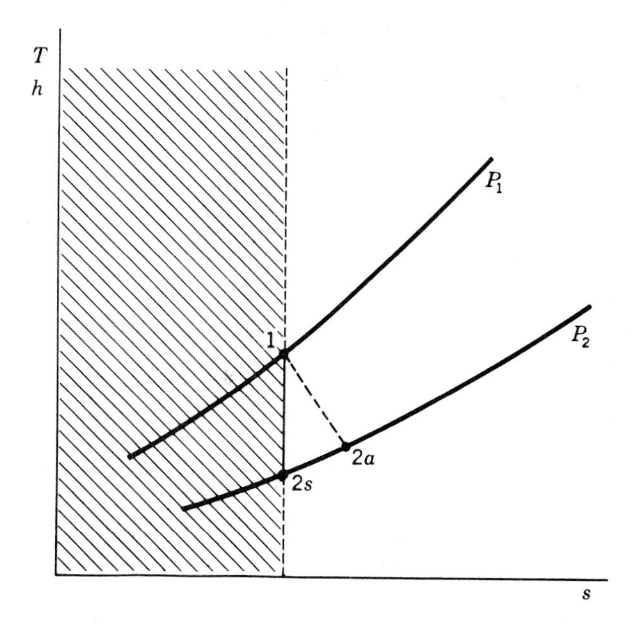

Figure 16-28 Effect of irreversibilities on nozzle performance.

16-13 AIRCRAFT GAS TURBINES

One of the most effective adaptations of the gas-turbine power cycle has been for the propulsion of aircraft. This is due to the favorable power to weight and power to volume ratios for a gas-turbine unit. The stationary gas-turbine cycle studied earlier and the jet engine cycle do have some major differences, however. One of these is in the operation of the compressor and turbine. In a turboject propulsion unit there is no work output required from the turbine beyond that needed to drive the compressor and auxiliary equipment. (For a turboprop engine, the turbine also drives the propeller.) Hence, the turbine drives the compressor, and there is no net output shaft. This is seen in Fig. 16-29, which is a schematic of a turbojet engine. The center section of the engine contains the three major components of a gas-turbine unit—compressor, combustor, and turbine. Since the work requirement is less, the gas does not expand back to ambient pressure in the turbine. The final expansion occurs in the nozzle which follows the turbine. Here the fluid is accelerated to a relatively high velocity. The pressure ratio P_4/P_5 in the nozzle may by two or more. A third difference in operation is the placement of a diffuser in front of the compressor. Its purpose is to slow down the entering fluid and increase the pressure. A small pressure rise of a few decibars (or psi) accompanies the decrease in kinetic energy. This pressure rise is referred to as the ram effect.

The solid lines in Fig. 16-30 show the general thermodynamic characteristics of an ideal turbojet engine on a Ts diagram. Process y-1 shows a pressure rise in the diffuser due to a decrease in kinetic energy. The following three processes are

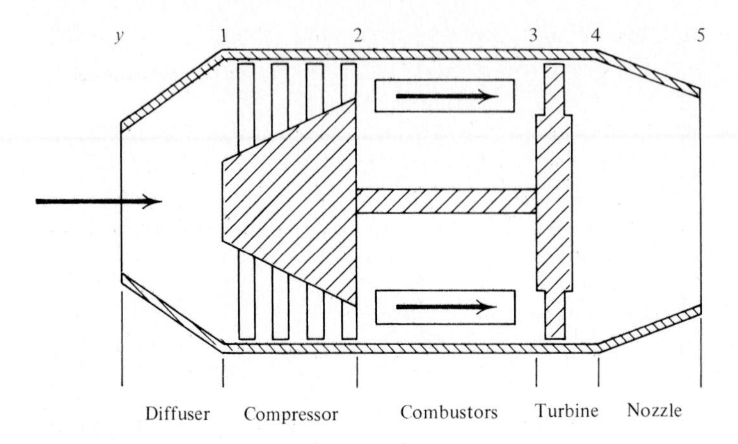

Figure 16-29 Turbojet engine schematic.

typical of a gas-turbine cycle: process 1-2 is isentropic compression, process 2-3 is constant-pressure heat addition, and process 3-4 is isentropic expansion. Finally, process 4-5 shows isentropic expansion through the nozzle, where the pressure decrease is accompanied by a significant increase in kinetic energy. In actual practice irreversibilities occur in the engine. The compressor, turbine, diffuser, and nozzle are not isentropic. In addition, a pressure drop occurs in the com-

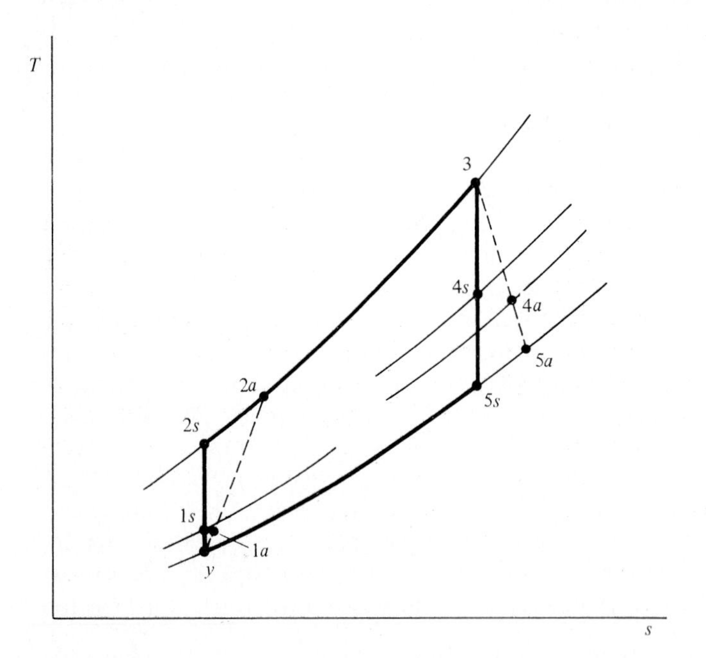

Figure 16-30 Ts diagram for a turbojet engine showing ideal performance (solid lines) and irreversible performance (dashed lines).

bustor. These irreversibilities modify the basic cycle shown in Fig. 16-30. The dashed lines in the figure show the approximate location of states in the actual cycle.

Compared to stationary gas-turbine power plants, turbojet engines operate at higher pressure ratios and higher inlet-turbine temperatures. Pressure ratios from 10 : 1 to 25 : 1 are common. Inlet-turbine temperatures are in the range of 1000-1400°K or 1800-2500°F. The following discussion covers the basic energy analysis of a turbojet cycle as well as the basic parameters used to measure the performance of the cycle.

a Energy Analysis

The energy analysis of a turbojet engine is based on a knowledge of the following data: the air-inlet temperature T_y, pressure P_y, and velocity V_y, the pressure ratio across the compressor r_p, and the limiting turbine-inlet temperature T_3. Subscripts for the preceding symbols are based on the notation of Fig. 16-29. In addition, for irreversible flow, it is necessary to have information of the diffuser pressure coefficient K_P, the compressor efficiency η_C, the turbine efficiency η_T, and the nozzle efficiency η_N. As a model an air-standard cycle will be used, and the variation in specific heat with temperature will be accounted for by using air-table data. As a further model, the exit velocity from the diffuser is assumed to be very small in comparison to the inlet value, and the fluid expands through the nozzle to the ambient pressure.

The actual numerical analysis of a gas-turbine cycle using tabular data follows the general pattern illustrated in preceding examples in this chapter. With the inclusion of the diffuser and nozzle in the cycle, however, three comments may be pertinent. First, the diffuser analysis begins with an energy balance. This leads to a value of h_1, and hence T_1, p_{r1}, and s_1^0 at the diffuser outlet. The outlet pressure P_1 is then found by either $P_1 = P_y(p_{r1}/p_{ry})$ or $s_1^0 - s_y^0 = -R \ln P_1/P_y$. If irreversible, the pressure coefficient K_P is then used to determine the actual pressure rise across the diffuser. Secondly, the turbine analysis is altered from previous examples, because the pressure ratio P_3/P_4 across it is unknown. However, for the overall cycle $w_{T,a} = -w_{C,a} = h_{4a} - h_3$ in the irreversible case. From this relation h_{4a} is found, and the corresponding value of T_{4a}. At the same time, $w_{T,s} = w_{T,a}/\eta_T = h_{4s} - h_3$. The only unknown in this expression is h_{4s}, and its value leads to values of T_{4s}, p_{r4}, and s_4^0. Since T_3 (the turbine-inlet temperature) is input data, p_{r3} and s_3^0 are known. Therefore the turbine-outlet pressure P_4 can now be found by the use of p_r or s^0 data for states 3 and 4, similar to the diffuser calculation outlined above. Finally, because the nozzle-outlet pressure is assumed to be the ambient value, the value of p_{r5} or s_5^0 can be determined for the nozzle outlet. This in turn fixes T_{5s} and h_{5s}. Since the enthalpy change across the nozzle under isentropic conditions is now known, the kinetic-energy change is now known from the basic relation $\Delta \text{KE}_s = -\Delta h_s$. The nozzle efficiency may then be used to evaluate the actual kinetic-energy increase.

Thus the method of determining values in the aircraft gas-turbine cycle does not follow the same pattern as that for stationary gas-turbine units. The following examples illustrate the numerical calculations outlined above for a reversible gas-turbine cycle.

Example 16-10M A turbojet aircraft flies at 260 m/s at an altitude of 5000 m, where the atmospheric pressure is 0.60 bar and the temperature is 250°K. The compressor pressure ratio is 8 : 1, and the maximum temperature at the turbine inlet is 1300°K. Assuming ideal performance of the various components of the engine, determine the compressor-work input, the pressures and temperatures throughout the cycle, and the exit-jet velocity.

SOLUTION The values calculated will correspond to the states shown in Fig. 16-30. For the diffuser process, 1-y, we shall assume that the outlet velocity is negligible compared to the inlet value. At 250°K the enthalpy is 250.1 kJ/kg and $p_{r1} = 0.7329$ (Table A-5M). For isentropic flow through the diffuser, we find that

$$h_y + \frac{V_y^2}{2} = h_1 + \frac{V_1^2}{2}$$

$$250.1 + \frac{260^2}{2(1000)} = h_1 + 0$$

$$h_1 = 250.1 + 33.8 = 283.9 \text{ kJ/kg}$$

From the air table, $T_1 = 284°\text{K}$ and $p_{r1} = 1.141$. Hence

$$P_1 = P_y \frac{p_{r1}}{p_{ry}} = 0.60 \frac{1.141}{0.7329} = 0.934 \text{ bar}$$

Thus there is a 0.33-bar pressure rise in the diffuser. For the compressor, with $r_p = 8$, $P_2 = 8(0.934) = 7.47$ bars. Therefore

$$p_{r2} = p_{r1} \frac{P_2}{P_1} = 1.141(8) = 9.13$$

From the air table A-5M, $T_{2s} = 511°\text{K}$, $h_{2s} = 514.9$ kJ/kg. The isentropic compressor work is

$$w_{C,s} = h_{2s} - h_1 = 514.9 - 283.9 = 231.0 \text{ kJ/kg}$$

At the turbine inlet, where $T_3 = 1300°\text{K}$, we find that $h_3 = 1396.0$ kJ/kg and $p_{r3} = 330.9$. Since the turbine work must equal the compressor work,

$$h_{4s} = h_3 - w_C = 1396.0 - 231.0 = 1165.0 \text{ kJ/kg}$$

For this value of the enthalpy, Table A-5M shows that $T_{4s} = 1102°\text{K}$ and $p_{r4} = 169.3$. Hence

$$P_4 = P_3 \frac{p_{r4}}{p_{r3}} = 7.47 \frac{169.3}{330.9} = 3.82 \text{ bars}$$

Note that the turbine-outlet pressure is considerably above the atmospheric value, and hence a reasonably large pressure drop still exits across the nozzle. The nozzle-outlet state is found from

$$p_{r5} = p_{r4} \frac{P_5}{P_4} = 169.3 \frac{0.60}{3.82} = 26.6$$

At this condition, $T_{5s} = 685°\text{K}$ and $h_{5s} = 697.4$ kJ/kg. An energy balance on the nozzle yields, neglecting the inlet velocity,

$$h_4 + \frac{V_4^2}{2} = h_5 + \frac{V_5^2}{2}$$

$$1165.0 + 0 = 697.4 + \frac{V_5^2}{2(1000)}$$

$$V_5^2 = 2(1000)(1165.0 - 697.4) = 935{,}200$$

$$V_5 = 967 \text{ m/s}$$

This velocity is supersonic, with a Mach number around 1.8. A more meaningful calculation would include compressor, turbine, and nozzle efficiencies, and a diffuser pressure coefficient.

Example 16-10 A turbojet aircraft flies at 575 mi/h at an altitude of 16,500 ft, where $P_a =$ 8.7 psia and $T_a = 460°\text{R}$. The compressor pressure ratio is 8, and the maximum temperature in the combustion chamber is 1840°F. Assuming ideal performance of the various components of the engine, determine the compressor-work input, the pressures and temperatures throughout the cycle, and the exit-jet velocity.

SOLUTION The values calculated will correspond to the states shown on Fig. 16-30. For the diffuser process, 1-y, we shall assume that the outlet velocity is negligible compared to the inlet value. At 460°R the enthalpy is 109.9 Btu and $p_{ry} = 0.7913$ (Table A-5). Assuming isentropic flow through the diffuser, we find that

$$h_y + \frac{V_y^2}{2} = h_1 + \frac{V_1^2}{2}$$

$$109.9 + \frac{843^2}{2(32.2)(778)} = h_1 + 0$$

$$h_1 = 109.9 + 14.2 = 124.1 \text{ Btu/lb}$$

From the air table, $T_1 = 519°\text{R}$ and $p_{r1} = 1.21$. Hence

$$P_1 = P_y \frac{p_{r1}}{p_{ry}} = 8.7 \frac{1.21}{0.7913} = 13.3 \text{ psia}$$

There is a 4.6-psi pressure rise in the diffuser. For the compressor, with $r_p = 8$, $P_2 = 8(13.3) =$ 106.4 psia.

$$p_{r2} = p_{r1} \frac{P_2}{P_1} = 1.21(8) = 9.68$$

From the air table, $T_2 = 936°\text{R}$, $h_2 = 225.1$ Btu/lb. The isentropic compressor work is

$$W_{C,s} = 225.1 - 124.1 = 101.0 \text{ Btu/lb}$$

At the turbine inlet, where $T_3 = 2300°\text{R}$, we find that $h_3 = 588.8$ Btu/lb and $p_{r3} = 308.1$. Since the turbine work must equal the compressor work,

$$h_{4s} = h_3 - w_C = 588.8 - 101.0 = 487.8 \text{ Btu/lb}$$

Table A-5 shows that $T_{4s} = 1939°\text{R}$ and $p_{r4} = 153.6$. Hence

$$P_4 = P_3 \frac{p_{r4}}{P_{r3}} = 106.4 \frac{153.6}{308.1} = 53.0 \text{ psia}$$

Note that a considerable drop still exists across the nozzle. The nozzle-outlet state is found from

$$p_{r5} = p_{r4} \frac{P_5}{P_4} = 153.6 \frac{8.7}{53.0} = 25.2$$

At this condition, $T_5 = 1216°\text{R}$ and $h_5 = 295.3$ Btu/lb. An energy balance on the nozzle yields, neglecting the inlet velocity,

$$h_4 + \frac{V_4^2}{2} = h_5 + \frac{V_5^2}{2}$$

$$487.9 + 0 = 295.3 + \frac{V_5^2}{2(32.2)(778)}$$

$$V_5^2 = 2(32.2)(778)(487.8 - 295.3) = 9.62 \times 10^6$$

$$V_5 = 3100 \text{ ft/s}$$

This velocity is supersonic, with a Mach number around 1.8. A more meaningful calculation would include compressor, turbine, and nozzle efficiencies, and a diffuser pressure coefficient.

b Performance Parameters for a Jet Engine Cycle

In the analysis of stationary gas-turbine cycles we examined the effect of pressure ratio r_p on the net work output and the thermal efficiency. For an aircraft gas-turbine cycle we need to examine the effect of inlet and exit air velocity (relative to the aircraft) on thrust and efficiency. The following discussion is based on an air-standard cycle, so that the effect of fuel addition on mass flow rates is neglected.

To an observer on the ground a jet aircraft does work in overcoming the resistance to motion known as fluid drag. In addition, an energy analysis on the overall engine reveals enthalpy and kinetic-energy terms at the inlet and exit of the device. Finally, a heat term must be included to account for energy released by the combustion of fuel. For the control volume surrounding the engine in flight we may write

$$h_{in} + \text{KE}_{in} + q_{in} = h_{out} + \text{KE}_{out} + w_{out} \tag{a}$$

In terms of the notation of Fig. 16-29, $h_{in} = h_y$ and $h_{out} = h_5$. The kinetic-energy terms are based on the absolute velocities noted by the observer on the ground. For still air $\text{KE}_{in} = 0$, and KE_{out} is based on the relative velocity between the aircraft speed and the jet exhaust velocity. In still air the aircraft velocity is V_y, so that the relative exit velocity seen by the observer is $V_5 - V_y$. Hence Eq. (*a*) becomes

$$h_y + 0 + q_{in} = h_5 + \frac{(V_5 - V_y)^2}{2} + w \tag{b}$$

It is now desired to express this equation for w solely in terms of the inlet and exit velocities V_y and V_5.

The enthalpy terms and the heat term in Eq. (*b*) can be replaced by making another energy analysis on the control volume surrounding the engine, except that the observer now is on the engine. In this case the steady-flow energy equation becomes

$$h_{in} + \text{KE}_{in} + q_{in} = h_{out} + \text{KE}_{out}$$

or

$$h_y + \frac{V_y^2}{2} + q_{in} = h_5 + \frac{V_5^2}{2} \tag{c}$$

If Eq. (*c*) is solved for the quantity q_{in}, and that result is substituted into Eq. (*b*), then

$$w = \frac{V_5^2}{2} - \frac{V_y^2}{2} - \frac{(V_5 - V_y)^2}{2} = V_y(V_5 - V_y) \tag{d}$$

This equation is important in determining the propulsive efficiency of the engine, as well as the thrust.

The thrust developed by the engine is found from the relations for the power developed, namely,

$$\dot{W} = \dot{m}_a w = F V_y \tag{e}$$

where F is the thrust or force which acts to overcome fluid drag. When Eq. (*d*) is substituted into Eq. (*e*) we find that

$$F = \dot{m}_a(V_5 - V_y) \tag{16-27}$$

where V_5 and V_y are measured relative to the engine.

The propulsive efficiency η_P is a measure of how well energy available in the cycle is converted to work to overcome drag forces. It is defined as the ratio of the work done in flight divided by the sum of this work quantity and the kinetic energy of the exhaust stream relative to ground. Thus, in terms of quantities in Eqs. (*b*) and (*d*),

$$\eta_P = \frac{w}{w + \text{KE}_{out}} = \frac{V_y(V_5 - V_y)}{V_y(V_5 - V_y) + \dfrac{(V_5 - V_y)^2}{2}} = \frac{2}{1 + \dfrac{V_5}{V_y}} \tag{16-28}$$

It should be noted that when the plane speed is zero ($V_y = 0$), the propulsive efficiency is zero and the thrust is a maximum for given values of mass flow rate and exhaust velocity. When the plane speed equals the exhaust velocity, the propulsive efficiency is unity (or 100 percent) and the thrust is zero.

Finally, the condition for maximum work or power can be determined by setting the first derivative of w with respect to V_y equal to zero. The result based on Eq. (*d*) is

$$V_y = \frac{V_5}{2} \qquad \text{(maximum power)} \tag{16-29}$$

Hence the maximum power is developed when the plane speed is one-half of the exhaust velocity. In terms of Eq. (16-28), this condition is not particularly good with respect to propulsive efficiency.

16-14 CLOSED-LOOP GAS-TURBINE CYCLES

The modern gas-turbine engine has two major applications: (1) as a power source for the production of electric power when coupled with an electric generator and (2) as a power source for the propulsion of aircraft. The device is also applicable as a power source for ground-level propulsion equipment, such as passenger cars, trucks, and trains. The thermodynamic fundamentals of the basic air-breathing gas-turbine cycle have been discussed in Secs. 16-6 through 16-8. The Brayton cycle may also be used as a power system for orbiting vehicles in the space environment outside our atmosphere. In this case the space power system is being used to generate electric power aboard the spacecraft, and not to propel the vehicle. The major features of the cycle described previously remain the same for space applications. Typical turbine expansion ratios are around 2 : 1 or less, and cycle temperature ratios (minimum to maximum temperature) are around 0.25 : 0.30. However three major modifications must be made to permit operation in a space environment. These modifications are:

1. The device must operate as a closed cycle, since the working fluid must be conserved for reuse. Air is not available as in terrestrial applications of open gas-turbine cycles.
2. The heat source must be of a different nature than the conventional hydrocarbons which undergo combustion directly in the cycle. This is necessary since the cycle loop is closed, and hence accumulation of products of combustion cannot be permitted.
3. The temperature of the gas prior to reentering the compressor in the cycle must be reduced to its initial value. In an open cycle relatively hot gases are simply thrown out into the atmosphere, even if a regenerator (recuperator) has been employed to preheat the gases between the compressor and the heat source.

The first modification to a closed cycle increases the physical complexity of the system. However, the closed-loop operation is advantageous, since it permits the use of working fluids which have properties more desirable than those of air. The use of inert gases, for example, reduces the problem of corrosion. Fundamental studies of turbomachinery indicate that the number of turbine blade rows decreases rapidly as the molecular weight increases. An application requiring 10 stages using helium (molar mass of 4) requires two stages with neon (molar mass of 20) and one stage with argon (molar mass of 40). For gases of even higher molecular weight, a single stage could be used at lower speeds than for argon, and hence at lower stresses. The turbomachinery diameter also increases as the molecular weight increases. For the relatively low-power turbomachinery used in space applications, this also is an advantage. For power levels in the range of 10-50 kW, gases under study have molecular weights in the range of 40-80. However, high-molecular-weight gases do have disadvantages. A major one is discussed when the third modification of the basic system is presented.

The energy sources of principal interest are solar radiation and nuclear reactors, although radioisotopes are also a possibility. Nuclear reactors are compact and require no special orientation with respect to the space environment. However, they are hazardous. A solar heat source is not hazardous, but it does require special means of orientation with respect to the sun. In addition, a relatively large solar concentrator (mirror) with good collection efficiency is required. For example, for a 10-kW Brayton power cycle with a maximum temperature of 2000°R, a 25-ft concentrator may be required. Whichever energy source is used, the heat generated must be transferred to the working fluid by means of a heat exchanger. This heat exchanger is shown as the "energy-source heat exchanger" in Fig. 16-31. The regenerator is, of course, a second heat exchanger in the loop. Regenerators working with argon have been built with an effectiveness of 90 percent and a 2 percent pressure drop.

The third modification for a Brayton space power system involves the "heat-sink heat exchanger" shown in Fig. 16-31. After passing through the regenerator (recuperator), the working fluid is cooled further by passing through this third heat exchanger. The energy discharged from the working fluid to a secondary liquid in the radiator loop is then radiated into space. The radiator size is significantly reduced by increasing the turbine-inlet temperature. This third heat-exchange process reduces the temperature of the primary working fluid to the required value at the compressor inlet. A disadvantage of the liquid loop is the added complexity. Passage of the inert gas (primary fluid) directly through the radiator, without a liquid loop, could be considered. However, for a given radiator size the extra gas-pressure loss in the radiator is found to be greater than the equivalent power required to pump the liquid.

It has been mentioned that high-molecular-weight gases (molecular weight greater than air) have properties favorable to turbomachinery design. Unfortunately, the thermal conductivity of a gas falls rapidly with increasing molecular

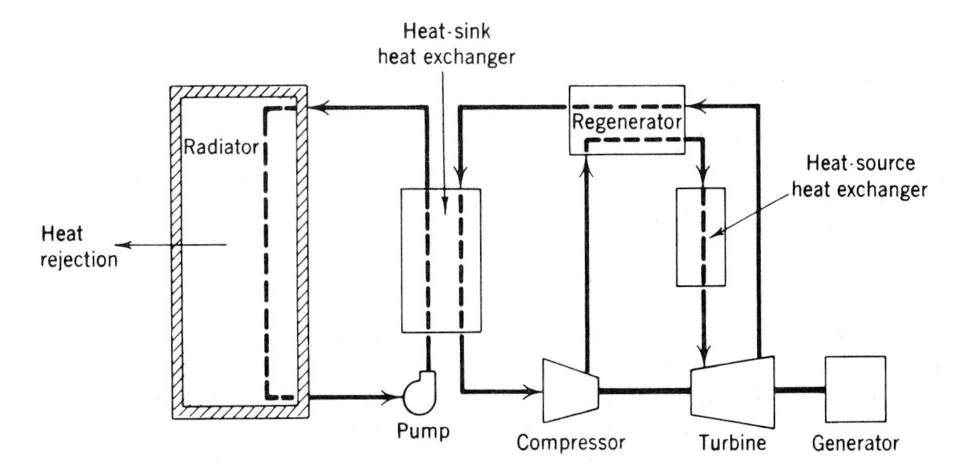

Figure 16-31 Gas-turbine-cycle space-power system.

weight. The conductivity of krypton (molecular weight of 83) is roughly $\frac{1}{15}$ of that of helium (molecular weight of 4). This implies a severe penalty in heat-exchanger size. Recall from Fig. 16-31 that the working fluid passes through three different heat exchangers. A heat exchanger designed for krypton would be about 10 times the size of one designed for helium. The increased weight and volume suggest that the choice of a working fluid is a compromise between turbomachinery needs and heat-exchanger requirements.

An interesting engineering-design feature is that the benefits of high molecular weight to turbomachinery design can be met in a subtle way without sacrificing heat-exchanger size. This is done by using a mixture of inert gases, rather than a pure gas. For example, helium and krypton could be mixed to give a gas which has the same molecular weight as argon. However, the average thermal conductivity of the gas mixture is greatly increased by the presence of helium. The result is a reduction in heat-exchanger size of about one-half of that for pure argon.

In summary, the Brayton cycle for space-power applications appears quite promising for several reasons:

1. It has a relatively high cycle efficiency (which reduces the required size of a solar collector or space radiator).
2. It can cover a range of power levels for a given system.
3. It can be used with a number of different energy sources.
4. It can operate over extended periods of time.

16-15 THE ERICSSON AND STIRLING CYCLES

In Sec. 16-11 it was demonstrated that the composite effect of intercooling, reheat, and regeneration is an increase in the thermal efficiency of a gas-turbine power cycle. It is interesting to examine the situation where the number of stages of both intercooling and reheat beome indefinitely large. In this situation the isentropic compression and expansion processes become isothermal ones. That is, for an indefinitely large number of stages the cycle can be represented by two constant-temperature processes and two constant-pressure processes with regeneration. Such a cycle is called an Ericsson cycle.

Figure 16-32 shows the *PV* and *TS* diagrams for the cycle and also a schematic diagram of an Ericsson engine operating as a steady-flow device. The fluid expands isothermally through the turbine from state 1 to 2. Work is produced and heat is absorbed reversibly from a reservoir at T_H. The fluid is then cooled at constant pressure in a regenerator (heat exchanger). From state 3 to state 4 the fluid is compressed isothermally. This requires work input and reversible heat rejection to a reservior at a temperature T_L. Finally, the fluid is heated at constant pressure to the initial state by passing it in a countercurrent direction through the regenerator and picking up the heat rejected to the regenerator during the constant-pressure cooling process. In this manner the temperature

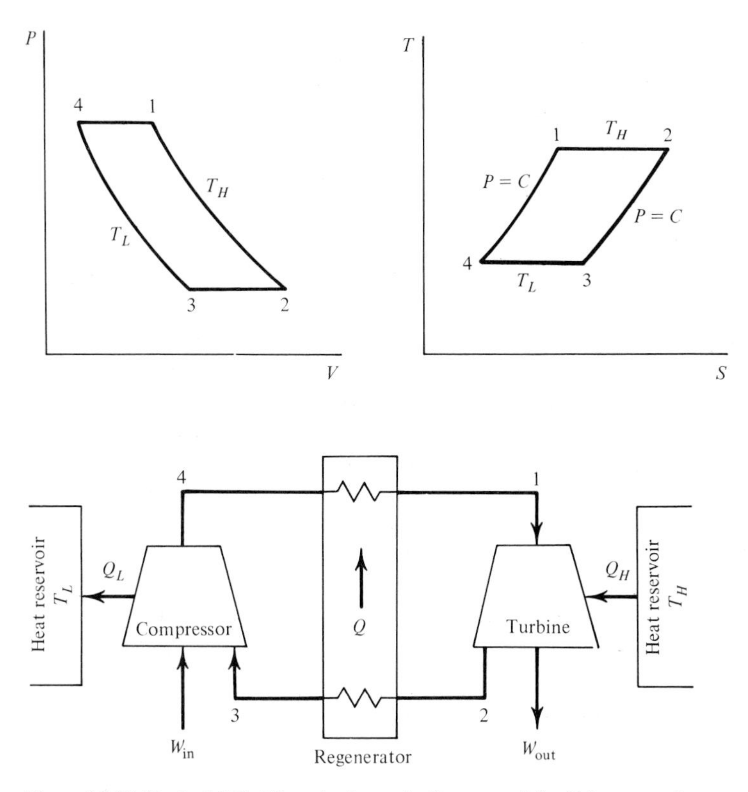

Figure 16-32 Typical *PV*, *TS*, and schematic diagrams of the Ericsson cycle.

difference across the heat exchanger is always infinitesimal throughout the length of the exchanger, as required for reversible heat transfer. The regenerator again acts as an energy-storage unit within the system. For any complete cycle the net energy storage is zero. This requires that the area under curve 4-1 on the *TS* diagram in Fig. 16-32 equal the area under curve 2-3.

Since the only external heat transfer is between heat reservoirs, and all processes are described as reversible, the thermal efficiency of the Ericsson cycle equals that of the Carnot cycle, given by $1 - T_L/T_H$. The original engine constructed by Ericsson (1803–1889) was a nonflow device, but the thermodynamic cycle is the same as for the steady-flow device described above. Such a design is impractical, however, in terms of intercooling and reheating, because the cost and size requirements are prohibitive.

Although the Ericsson cycle is impractical, the cycle does demonstrate how a regenerator might be placed in a cycle in order to increase the thermal efficiency. Another theoretical cycle of more practical importance which incorporates a regenerator in the basic scheme is the Stirling cycle. Proposed by Robert Stirling (1790–1878), the cycle is composed of two isothermal, reversible processes and two constant-volume, reversible processes. The *PV* and *TS* diagrams for the cycle are shown in Fig. 16-33.

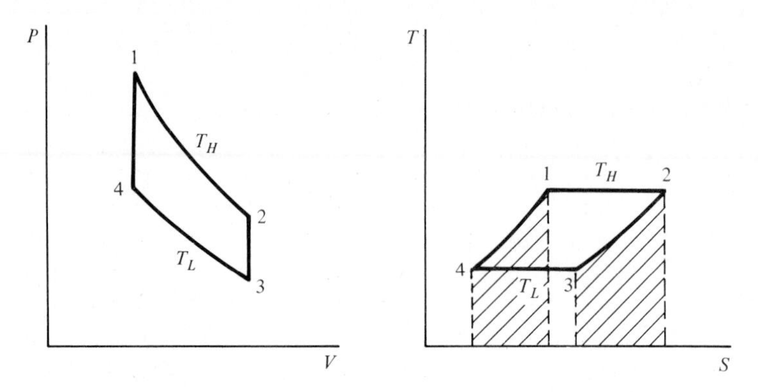

Figure 16-33 PV and TS diagrams for the Stirling cycle.

From an initial state 1 the gas is expanded isothermally to state 2 and heat is added reversibly from a reservoir at T_H. From state 2 to state 3 heat is removed at constant volume until the temperature of the fluid reaches T_L. The volume is then reduced to its original value isothermally, and heat is removed reversibly to a second reservoir at T_L. Finally, heat is added at constant volume from state 4 to state 1. The cycle would operate between two fixed-temperature reservoirs if the heat quantities for the processes 2-3 and 4-1 could be kept within the system. Application of an energy balance on the closed system for these two processes shows that the two heat quantities are equal in magnitude. This is shown by the cross-hatched areas under process curves 4-1 and 2-3 on the TS plot. What is needed is simply a means of storing the heat given up by process 2-3 and then supplying this same energy to the working medium during the process 4-1. This requirement of energy storage within the system necessitates the use of a regenerator. Thus the only heat effect external to the system during each Stirling cycle is heat exchange between two fixed-temperature reservoirs. As a result, the thermal efficiency of a Stirling cycle will equal that of a Carnot cycle operating between the same temperatures.

For many years the Stirling cycle has been of only theoretical interest. However, since the early 1950s considerable work has been carried out toward the development of a practical engine working on the Stirling cycle. Although it matches the Carnot cycle in thermal efficiency, it is difficult to build a Stirling engine without introducing some inherent disavantages. For example, it operates at very high pressures, and the most suitable working fluids are helium and hydrogen. Its weight-to-horsepower ratio is not too favorable, except possibly for large vehicles such as trucks and buses. The high temperature in the cycle also leads to problems, since the pistons are not lubricated to avoid fouling of the regenerator. However one major advantage of the Stirling engine is its excellent emission quality. This engine is an "external-combustion engine," as opposed to the common internal-combustion type for automotive use. Hence the combustion process is much more complete than in an internal-combustion type in terms of carbon monoxide, hydrocarbon, and nitrogen oxide content in the exhaust. Other advantages of Stirling engines are their relatively silent operation, reliability and long life, and multifuel capability.

In 1976, the Ford Motor Company installed an experimental 170-hp Stirling engine in a 1975 Ford Torino automobile. The four-cylinder engine had a 0.86-L (52.5-in^3) displacement and operated with hydrogen as the working fluid. The maximum working pressure in the cycle was around 18 MPa (2600 psi) and the heater head temperature was around 760°C (1400°F). Research and development work on a practical Stirling engine continues throughout the world. The Stirling cycle has some inherent advantages when operated in the reverse direction so that it produces a refrigeration effect. Stirling refrigeration devices are particularly effective in achieving temperatures in the range of −100 to −200°C (−150 to −330°F). A discussion of the Stirling cycle as a refrigeration technique appears in Sec. 18-7.

REFERENCES

Burghardt, M. D.: "Engineering Thermodynamics with Applications," Harper & Row, New York, 1978.

Obert, E. F., and R. A. Gaggioli: "Thermodynamics," McGraw-Hill, New York, 1963.

Reynolds, W. C., and H. C. Perkins: "Engineering Thermodynamics," 2nd ed., McGraw-Hill, New York, 1977.

Van Wylen, G. J., and R. E. Sonntag: "Fundamentals of Classical Thermodynamics," 2nd ed., Wiley, 1973.

PROBLEMS (METRIC)

Air-standard cycles

16-1M An air-standard cycle is proposed which begins at an initial state of 27°C, 1 bar, and 0.860 m^3/kg. Process 1-2 is constant-volume heating to 2 bars; process 2-3 is constant-pressure heating to 1200°K and 1.72 m^3/kg; process 3-4 is isentropic expansion to 1 bar; process 4-1 is constant-pressure cooling to the initial state.

(*a*) Sketch Pv and Ts diagrams for the cycle.

(*b*) Determine the total amount of heat added, in kJ/kg.

(*c*) Determine the net work output, in kJ/kg.

(*d*) Compute the thermal efficiency.

(*e*) What is the Carnot efficiency for a heat engine with the same maximum and minimum temperatures as the above cycle? Use u and h data from the air table.

16-2M An air-standard cycle is proposed which begins at an initial state of 47°C, 1.2 bars, and 0.765 m^3/kg. Process 1-2 is constant-volume heating to 2.7 bars; process 2-3 is constant-pressure heating to 1.53 m^3/kg; process 3-4 is isentropic expansion to 1.2 bars; process 4-1 is constant-pressure cooling to the initial state.

(*a*) Sketch Pv and Ts diagrams for the cycle.

(*b*) Determine the total amount of heat added, in kJ/kg.

(*c*) Find the net work output, in kJ/kg.

(*d*) Compute the thermal efficiency.

(*e*) What is the Carnot efficiency for a heat engine with the same maximum and minimum temperatures as the above cycle? Use u and h data from the air table.

16-3M A proposed air-standard cycle begins at 1 bar, 27°C, and 0.860 m^3/kg. Process 1-2 is isentropic compression to 2.75 bar; process 2-3 is constant-pressure heating to 1200°K and 1.25 m^3/kg; process 3-4 is constant-volume cooling to 1 bar; process 4-1 is constant-pressure cooling to the initial state.

(*a*) Sketch *Pv* and *Ts* diagrams of the cycle.

(*b*) Determine the amount of heat added and the net work output, in kJ/kg.

(*c*) Compute the thermal efficiency.

(*d*) What is the Carnot efficiency for a heat engine with the same maximum and minimum temperatures as the above cycle? Use air-table data.

16-4M A proposed air-standard cycle begins at 1.1 bars, 37°C, and 0.808 m^3/kg. Process 1-2 is isentropic compression to 3.2 bars; process 2-3 is constant-pressure heating to 1050°K and 0.940 m^3/kg; process 3-4 is constant-volume cooling to 1.1 bars; process 4-1 is constant-pressure cooling to the initial state.

(*a*) Sketch *Pv* and *Ts* diagrams of the cycle.

(*b*) Determine the amount of heat added and the net work output, in kJ/kg.

(*c*) Compute the thermal efficiency.

(*d*) What is the Carnot efficiency for a heat engine with the same maximum and minimum temperatures as the above cycle? Use air-table data.

Air-standard Carnot cycles

16-5M An air-standard Carnot cycle rejects 100 kJ/kg to a sink at 300°K. The minimum and maximum pressures in the closed cycle are 0.10 and 17.4 MPa. On the basis of the air table, determine (*a*) the pressure after isothermal compression, (*b*) the temperature of the heat supply reservoir, in degrees kelvin, (*c*) the specific volume after isothermal compression and after isentropic compression, both in m^3/kg, and (*d*) the thermal efficiency.

16-6M An air-standard Carnot cycle for a closed system is supplied with 200 kJ/kg of heat from a source at 1000°K. The minimum and maximum pressures in the cycle are 1 and 69.3 bars. On the basis of the air table, determine (*a*) the pressure after isothermal heat addition, (*b*) the temperature of heat rejection, in degrees kelvin, (*c*) the specific volume, in m^3/kg, after isothermal heat addition and after isentropic expansion, and (*d*) the thermal efficiency.

16-7M An air-standard Carnot cycle operates in a piston-cylinder device between temperatures of 300 and 900°K. The minimum pressure in the cycle is 0.10 MPa. Determine (1) the maximum pressure in the cycle, in MPa, and (2) the specific volume, in m^3/kg, after isentropic compression, if the heat rejection is (*a*) 60 kJ/kg, and (*b*) 40 kJ/kg.

16-8M An air-standard Carnot cycle operates between temperatures of 300 and 1100°K. The minimum pressure in the cycle is 1 bar. Determine (1) the maximum pressure in the cycle, and (2) the pressure after isothermal compression in the closed system, in bars, if the amount of heat supplied is (*a*) 120 kJ/kg, and (*b*) 150 kJ/kg.

16-9M A Carnot heat engine which produces 10 kJ of work for one cycle has a thermal efficiency of 50 percent. The working fluid is 0.50 kg of air, and the pressure and volume at the beginning of the isothermal expansion are 7 bars and 0.119 m^3, respectively. Find (*a*) the maximum and minimum temperatures for the cycle, in degrees kelvin, (*b*) the heat and work for each of the four processes, in kJ/cycle, and (*c*) the volume at the end of the isothermal expansion process and at the end of the isentropic expansion process, in m^3.

Compression ratio and mean effective pressure

16-10M For the data in Prob. 16-5M, determine the compression ratio and the mean effective pressure for a reciprocating device.

16-11M For the data in Prob. 16-6M, determine the compression ratio and the mean effective pressure for a reciprocating device.

16-12M For the data in Prob. 16-7M, determine the compression ratio and the mean effective pressure for a reciprocating device if the heat rejection is (*a*) 60 kJ/kg, and (*b*) 40 kJ/kg.

16-13M For the data in Prob. 16-8M, determine the compression ratio and the mean effective pressure for a reciprocating device if the amount of heat supplied is (*a*) 120 kJ/kg, and (*b*) 150 kJ/kg.

Otto cycle

16-14M The compression ratio of an Otto cycle is 8 : 1. Before the compression stroke of the cycle begins, the pressure is 0.98 bar and the temperature is 27°C. The amount of heat added to the air per cycle is 1332 kJ/kg. On the basis of the air-standard cycle and employing air table A-5M, determine (*a*) the pressure and temperature at the end of each process of the cycle, (*b*) the theoretical thermal efficiency, and (*c*) the mean effective pressure for the cycle, in bars.

16-15M Same as Prob. 16-14M, except that the amount of heat added is 1188 kJ/kg.

16-16M The air at the beginning of the compression stroke of an air-standard Otto cycle is at 0.095 MPa and 22°C, and the cylinder volume is 2800 cm^3. The compression ratio is 9, and 4.30 kJ are added during the heat-addition process. On the basis of Table A-5M, determine (*a*) the temperature and pressure after the heat-addition and expansion processes, (*b*) the thermal efficiency, and (*c*) the mean effective pressure, in MPa.

16-17M Same as Prob. 16-16M, except that the amount of heat added is 3.54 kJ.

16-18M An air-standard Otto cycle operates with a compression ratio of 8.55, and at the beginning of compression the air is at 0.98 bar and 32°C. During the heat-addition process the pressure is tripled. On the basis of Table A-5M, determine (*a*) the temperatures around the cycle, in degrees kelvin, (*b*) the thermal efficiency of the cycle, and (*c*) the thermal efficiency of a Carnot engine operating between the same overall temperature limits.

16-19M Consider an air-standard Otto cycle that has a compression ratio of 8.3 and a heat addition of 1310 kJ/kg. If the pressure and temperature at the beginning of the compression process are 0.095 MPa and 7°C, determine on the basis of Table A-5M (*a*) the maximum pressure and temperature for the cycle, (*b*) the net work output, in kJ/kg, (*c*) the thermal efficiency, and (*d*) the mean effective pressure, in MPa.

16-20M Same as Prob. 16-19M, except that the amount of heat added is 1213 kJ/kg.

16-21M Reconsider the Otto cycle in Prob. 16-14M. Determine (*a*) the closed-system availability of the air at the end of isentropic expansion, in kJ/kg, if $T_0 = 27°\text{C}$ and $P_0 = 0.98$ bar, and (*b*) the ratio of this availability quantity to the net work output of the cycle.

16-22M Reconsider the Otto cycle in Prob. 16-16M. Determine (*a*) the closed-system availability of the air at the end of isentropic expansion, in kJ/kg, if $T_0 = 22°\text{C}$ and $P_0 = 0.095$ MPa, and (*b*) the ratio of this availability quantity to the net work output of the cycle.

16-23M Reconsider the Otto cycle in Prob. 16-19M. Determine (*a*) the closed-system availability of the air at the end of isentropic expansion, in kJ/kg, if $T_0 = 7°\text{C}$ and $P_0 = 0.095$ MPa, and (*b*) the ratio of this availability quantity to the net work output of the cycle.

Diesel cycle

16-24M An air-standard Diesel cycle operates with a compression ratio of 16.7 and a cutoff ratio of 2. At the beginning of compression the air temperature and pressure are 37°C and 0.10 MPa, respectively. Determine (*a*) the maximum temperature in the cycle, in degrees kelvin, (*b*) the pressure after isentropic expansion, in MPa, and (*c*) the heat input per cycle, in kJ/kg. Use data from Table A-5M.

16-25M An engine operates on the theoretical Diesel cycle with a compression ratio of 16 : 1, and fuel is injected for 10 percent of the stroke. The pressure and temperature of the air entering the cylinder are 0.98 bar and 17°C, respectively. Determine (*a*) the cutoff ratio, (*b*) the temperature, in degrees kelvin, at the end of the compression process and at the end of the heat-addition process, (*c*) the pressure after isentropic expansion, in bars, and (*d*) the heat input, in kJ/kg. Use Table A-5M.

16-26M The intake conditions for an air-standard Diesel cycle operating with a compression ratio of 15 : 1 are 0.95 bar and 17°C. At the beginning of the compression stroke the cylinder volume is 380 L, and 7.5 kJ of heat are added to the gas during the constant-pressure heating process.

(*a*) Calculate the pressure and temperature at the end of each process of the cycle.
(*b*) Determine the thermal efficiency and the mean effective pressure of the cycle.

16-27M A Diesel air-standard cycle has a compression ratio of 15 : 1. The pressure and temperature at the beginning of compression are 1 bar and 17°C, respectively. If the maximum temperature of the cycle is 2250°K, determine (*a*) the cutoff ratio, (*b*) the thermal efficiency, and (*c*) the mean effective pressure, in bars.

16-28M An air-standard Diesel cycle is supplied with 1490 kJ/kg of heat per cycle. The pressure and temperature at the beginning of compression are 0.095 MPa and 27°C, respectively, and the pressure after compression is 4.33 MPa. Determine (*a*) the compression ratio, (*b*) the maximum temperature in the cycle, in degrees kelvin, (*c*) the cutoff ratio, and (*d*) the pressure after isentropic expansion, in MPa. Use Table A-5M for data.

16-29M Reconsider the Diesel cycle in Prob. 16-24M. Determine (*a*) the closed system availability of the air at the end of isentropic expansion, in kJ/kg, if $T_0 = 17°C$ and $P_0 = 0.10$ MPa, and (*b*) the ratio of this availability quantity to the net work output of the cycle.

16-30M Reconsider the Diesel cycle of Prob. 16-26M. Determine (*a*) the closed system availability of the air at the end of isentropic expansion, in kJ/kg, if $T_0 = 17°C$ and $P_0 = 0.95$ bar, and (*b*) the ratio of this availability quantity to the net work output of the cycle.

Dual cycle

16-31M An air-standard dual cycle operates with a compression ratio of 15 : 1. At the beginning of compression the conditions are 17°C, 0.95 bar, and 3.80 L. The amount of heat added is 6.3 kJ, of which one-third is added at constant volume and the remainder at constant pressure. Determine (*a*) the pressure after the constant-volume heat-addition process, (*b*) the temperature before and after the constant-pressure heat-addition process, in degrees Kelvin, (*c*) the temperature after isentropic expansion, and (*d*) the thermal efficiency.

16-32M An air-standard dual cycle operates with a compression ratio of 14 : 1. At the beginning of the isentropic compression the conditions are 27°C and 0.96 bar. The total heat addition is 1480 kJ/kg, of which one-fourth is added at constant volume and the remainder at constant pressure. Determine (*a*) the temperatures at the end of each process around the cycle, in degrees kelvin, (*b*) the thermal efficiency, and (*c*) the mean effective pressure.

16-33M Same as Prob. 16-31M, except that the heat addition is 6.0 kJ, of which 30 percent is added at constant volume.

16-34M Same as Prob. 16-32M, except that the heat addition is 1530 kJ/kg, of which one-third is added at constant volume.

Ideal open gas-turbine cycle

16-35M A gas-turbine power plant operates on an air-standard cycle between pressure limits of 0.10 MPa and 0.60 MPa. The inlet-air temperature is 22°C, and the air enters the turbine at 747°C. Using the data of Table A-5M, determine (*a*) the net work output and the heat input, in kJ/kg, and (*b*) the thermal efficiency if the cycle is ideal.

16-36M A gas-turbine power plant operates on an air-standard cycle between pressure limits of 1 and 6.4 bars. The inlet-air temperature is 22°C, and the temperature limitation on the turbine is 807°C. Calculate (*a*) the net work output, in kJ/kg, and (*b*) the thermal efficiency if the cycle is ideal. Use the air table for data.

16-37M A gas-turbine power plant operates on an air-standard cycle between pressure limits of 1 bar and 6 bars. The inlet-air temperature to the compressor is 17°C, and the inlet-air temperature to the turbine is 1080°K. Calculate (*a*) the net work output, in kJ/kg, (*b*) the heat input, and (*c*) the thermal efficiency if the cycle is ideal. Use the air table for data.

16-38M A stationary gas-turbine power plant has maximum and minimum cycle temperatures of 827 and 27°C, and a pressure ratio of 5.2 : 1. Find (*a*) the ratio of compressor work to turbine work, (*b*) the thermal efficiency, and (*c*) the mass flow rate of air required, in kg/min, for a net power output of 1000 kW. Use air-table data for the ideal cycle analysis.

16-39M The pressure ratio of an air-standard Brayton cycle is 4.5, and the inlet conditions to the compressor are 0.10 MPa and 27°C. The turbine is limited to a temperature of 827°C, and the mass flow rate is 4 kg/s. Determine (*a*) the compressor and turbine work, in kJ/kg, (*b*) the thermal efficiency, (*c*) the net power output, in kilowatts, and (*d*) the volume flow rate at the compressor inlet, in m^3/min. (*e*) If the heat addition is accomplished by the complete combustion of a fuel with a heating value of 42,900 kJ/kg, estimate the fuel-air ratio used in the combustor, in kg/kg. Use the air table for all property data.

16-40M The pressure ratio of an air-standard Brayton cycle is 6.0 : 1, and the inlet conditions are 1.0 bar and 17°C. The turbine is limited to a temperature of 1000°K, and the mass flow rate is 3.5 kg/s. Determine (*a*) the compressor and turbine work, in kJ/kg, (*b*) the thermal efficiency, (*c*) the net power output, in kilowatts, and (*d*) the volume flow rate at the compressor inlet, in m^3/min. (*e*) If the heat addition is accomplished by the complete combustion of a fuel with a heating value of 44,000 kJ/kg, estimate the fuel-air ratio used in the combustor, in kg/kg, using air data.

16-41M If the sink temperature T_0 is the same as the compressor-inlet temperature, then determine the steady-flow availability of the turbine exhaust stream, in kJ/kg, for the Brayton cycle described in (*a*) Prob. 16-35M, (*b*) Prob. 16-36M, (*c*) Prob. 16-37M, and (*d*) Prob. 16-38M. Also, calculate the percent increase in net work output if this availability could be completely converted into work output.

16-42M Prove that the maximum net work for a simple Brayton cycle, for fixed compressor- and turbine-inlet temperatures, occurs when $T_2 = (T_1 T_3)^{1/2}$ if the specific heats are constant.

16-43M Show that the pressure ratio which yields the largest net work for a simple Brayton cycle with fixed compressor- and turbine-inlet temperatures is given by $P_2/P_1 = (T_3/T_1)^n$, where $n = k/[2(k-1)]$ and k is a constant.

Nonideal simple gas-turbine power cycle

16-44M Using the data of Prob. 16-35M, calculate the heat input and the thermal efficiency if the compressor and turbine adiabatic efficiencies are 84 and 87 percent, respectively.

16-45M Using the data of Prob. 16-36M, calculate the required quantities if the adiabatic efficiencies of the compressor and turbine are 82 and 85 percent, respectively.

16-46M Using the data of Prob. 16-37M, calculate the required quantities if the compressor and turbine adiabatic efficiencies are 82 and 85 percent, respectively.

16-47M Using the data of Prob. 16-38M, calculate the required quantities if the adiabatic efficiencies of the compressor and turbine are 81 and 86 percent, respectively.

16-48M Using the data of Prob. 16-39M, calculate the required quantities if the compressor and turbine adiabatic efficiencies are 83 and 86 percent, respectively.

16-49M Using the data of Prob. 16-40M, calculate the required quantities if the compressor and turbine adiabatic efficiencies are 84 and 88 percent, respectively.

16-50M A gas-turbine power cycle operates with a pressure ratio of 12 : 1. The compressor and turbine adiabatic efficiencies are 85 and 90 percent, respectively. The compressor-inlet temperature is 22°C, and the turbine-inlet temperature is 1027°C. For a mass flow rate of 1 kg/s, find the power generated by the cycle. Now, double the pressure ratio and find the power output in this latter case, in kilowatts.

16-51M Air enters a gas-turbine power plant at 17°C and 1 bar. The compressor and turbine both operate with a pressure ratio of 8, and the adiabatic efficiencies of each are 80 and 85 percent, respectively. Gases enter the combustor at 887°C, and the mass flow rate is 10 kg/s. Determine what percent of the total output of the turbine may be used to drive devices other than the compressor, on the basis of data in Table A-5M.

16-52M If the sink temperature T_0 is the same as the compressor-inlet temperature, then determine the steady-flow availability of the turbine exhaust stream, in kJ/kg, for the irreversible cycle described in (*a*) Prob. 16-44M, (*b*) Prob. 16-45M, (*c*) Prob. 16-46M, and (*d*) Prob. 16-47M. Also calculate the percent increase in net work output if this availability could be completely converted into work output.

Regenerative gas-turbine cycle

16-53M Using the data of Prob. 16-44M, compute (1) the thermal efficiency, and (2) the percent fuel saved if a regenerator is installed with an effectiveness of (*a*) 80 percent, and (*b*) 65 percent.

16-54M Using the data of Prob. 16-45M, compute (1) the thermal efficiency, and (2) the percent fuel saved if a regenerator were installed with an effectiveness of (*a*) 50 percent, and (*b*) 70 percent.

16-55M Using the data of Prob. 16-46M, determine (1) the thermal efficiency, and (2) the percent fuel saved if a regenerator is installed with an effectiveness of (*a*) 60 percent, and (*b*) 70 percent.

16-56M Using the data of Prob. 16-47M, determine (1) the thermal efficiency, and (2) the percent fuel saved if a regenerator is added with an effectiveness of (*a*) 70 percent, and (*b*) 50 percent.

16-57M In an air-standard gas-turbine cycle with regeneration, the compressor is driven directly by the turbine. The following enthalpy data were taken on a test of the gas turbine with a pressure ratio of 5.41 : 1. Determine (*a*) the thermal efficiency of the actual cycle, (*b*) the effectiveness of the regenerator, (*c*) the adiabatic efficiency of the compressor, and (*d*) the adiabatic efficiency of the turbine. Data are in kJ/kg.

System	Entering	Leaving
Compressor	290.2	505.0
Regenerator	505.0	629.4
Combustor	629.4	1046.0
Turbine	1046.0	713.7
Regenerator	713.7	590.1

16-58M An automobile manufacturer is contemplating the use of a regenerative-type gas turbine for a power source of a new model. Fresh air enters the compressor at 22°C and 1 bar, with a flow rate of 1 kg/s, and is compressed at a ratio of 4 : 1. Air from the compressor enters the regenerator, where it is circulated until its temperature is approximately 537°C. It then passes to the combustion chamber, where it is heated to 927°C. After expanding through the two-stage turbine, the hot gases pass through the other half of the regenerator and then are exhausted to the atmosphere. Assuming ideal compression and expansion and negligible pressure losses through the burner and regenerator, determine (*a*) the thermal efficiency of the unit, (*b*) the net power output, in kilowatts, (*c*) the regenerator effectiveness, and (*d*) the exhaust stream temperature, in degrees Celsius.

16-59M If the sink temperature T_0 is the same as the compressor-inlet temperature, then determine the steady-flow availability of the stream leaving the regenerator and entering the atmosphere, in kJ/kg, for the irreversible regenerative cycle described in (*a*) Prob. 16-53M, (*b*) Prob. 16-54M, (*c*) Prob. 16-55M, and (*d*) Prob. 16-56M.

Polytropic process

16-60M Air is compressed reversibly in a piston-cylinder device from 0.10 MPa, 7°C, to a pressure of 0.50 MPa. Calculate (1) the work required, and (2) the heat transferred, in kJ/kg, if the process is (*a*) polytropic with $n = 1.30$, (*b*) adiabatic, and (*c*) isothermal.

16-61M Work Prob. 16-60M for the same data, except the initial and final pressures are 1 and 6 bars.

16-62M A steady-flow air compressor operates between an inlet condition of 1 bar and 17°C, and an outlet pressure of 5 bars. Determine (1) the work input, and (2) the heat transferred, in kJ/kg, for the following reversible processes: (*a*) isothermal, (*b*) $n = 1.30$, and (*c*) adiabatic.

16-63M Work Prob. 16-62M, except change the inlet conditions to 0.10 MPa and 37°C and the outlet condition to 0.60 MPa.

16-64M Find the power requirements, in kilowatts, if 7 m^3/min of air at 0.95 bar and 12°C are compressed irreversibly and polytropically ($n = 1.30$) to 7.60 bars. Cooling water runs through the compressor jacket at a rate of 14 kg/min and has a temperature rise of 6.0°C. The air velocity at the inlet is small, but it is 160 m/s at the outlet.

16-65M Find the power requirements, in kilowatts, if 0.50 m^3/min of air at 0.11 MPa and 17°C are compressed irreversibly and polytropically ($n = 1.32$) to 0.68 MPa. Cooling water runs through the compressor jacket at a rate of 5.0 kg/min and has a temperature rise of 5.1°C. The air velocity at the inlet is small, but it is 130 m/s at the outlet.

Compressor staging with intercooling

16-66M The inlet conditions of a two-stage steady-flow compressor are 0.95 bar and 27°C. The outlet pressure is 9.5 bars, and the pressure ratio across each stage is the same. If an intercooler cools the air to the initial temperature and the stages are isentropic, (*a*) determine the total work input, in kJ/kg, and (*b*) compare the result of part *a* with the work required for a single-stage compressor. Use Table A-5M for data.

16-67M Same as Prob. 16-66M, except that the inlet conditions are 1.05 bars and 37°C, and the outlet pressure is 6.3 bars.

16-68M The inlet conditions of a two-stage steady-flow compressor are 0.10 mPa and 27°C. The outlet pressure is 0.80 MPa, the stages are isentropic, and the intercooler cools the air to the initial temperature. Determine the total work input, in kJ/kg, if (*a*) the pressure ratio across each stage is the same, and (*b*) the pressure ratio across the first stage is twice that across the second stage.

16-69M Same as Prob. 16-68M, except that the inlet conditions are 1 bar and 17°C, and the outlet pressure is 4.5 bars.

16-70M Consider a three-stage steady-flow compressor with intercoolers which cool the ideal gas to the initial inlet temperature. The outlet pressures of the first and second stages of the isentropic compressor will be denoted by P_a and P_b, respectively. Show that the minimum work of compression is attained when $P_a = (P_1^2 P_2)^{1/3}$ and $P_b = (P_1 P_2^2)^{1/3}$.

16-71M A three-stage steady-flow compressor has the same inlet and outlet conditions as given in Prob. 16-66M. Based on the results of Prob. 16-70M, determine the minimum work of compression, in kJ/kg. Use Table A-5M for data.

16-72M A three-stage steady-flow compressor has the same inlet and outlet conditions as given in Prob. 16-67M. Based on the results of Prob. 16-70M, determine the minimum work of compression, in kJ/kg. Use Table A-5M for data.

Gas-turbine cycle with intercooling, reheating, and regeneration

16-73M Reconsider Prob. 16-53M. Determine the net work output and the thermal efficiency if two-stage compression and expansion are used with a regenerator of (*a*) 80 percent, and (*b*) 65 percent effectiveness. The compressor and turbine efficiencies are given in Prob. 16-44M. The intercooling and reheating are ideal.

16-74M Reconsider Prob. 16-56M. Determine the net work output and the thermal efficiency if two-stage compression and expansion are used with a regenerator effectiveness of (*a*) 70 percent, and (*b*) 50 percent. The compressor and turbine efficiencies are given in Prob. 16-47M. The intercooling and reheating are ideal.

16-75M Reconsider Prob. 16-54M. Determine the net work output and the thermal efficiency if two-stage compression and expansion are used with a regenerator effectiveness of (*a*) 50 percent, and (*b*) 70 percent. The compressor and turbine efficiencies are given in Prob. 16-45M. The intercooling and reheating are ideal.

16-76M Reconsider Prob. 16-55M. Determine the net work output and the thermal efficiency if two-stage compression and expansion are used with a regenerator effectiveness of (*a*) 60 percent, and (*b*) 70 percent. The compressor and turbine efficiencies are given in Prob. 16-46M. The intercooling and reheating are ideal.

16-77M A gas-turbine cycle operates with two stages of compression and two stages of expansion. The pressure ratio across each stage is 2. The inlet temperature is 22°C to each stage of compression and 827°C to each stage of expansion. The compressor and turbine efficiencies are 78 and 84 percent, respectively, and the regenerator has an effectiveness of 70 percent. Determine (*a*) the compressor and turbine work, in kJ/kg, (*b*) the thermal efficiency, and (*c*) the temperature of the air stream leaving the regenerator and entering the atmosphere, in degrees Celsius. Use air-table data.

16-78M Reconsider Prob. 16-77M, and calculate the same required quantities, except that the compressor and turbine efficiencies are now 81 and 86 percent, respectively, and the regenerator effectiveness is now 75 percent.

16-79M A gas-turbine cycle operates with two-stage compression and expansion. The pressure ratio across each stage of compression is 2.0, and the compressor efficiency is 81 percent. The compressor-inlet temperature is 22°C, but intercooling reduces the temperature only to 37°C before the air enters the second stage. The inlet temperature to each stage of expansion is 827°C, but a pressure drop between the compressor and turbine reduces the expansion pressure ratio of 1.9 : 1 across each turbine stage, which has an adiabatic efficiency of 86 percent. The regenerator effectiveness is 75 percent. Determine (*a*) the compressor work, (*b*) the turbine work, (*c*) the heat removed in the intercooler, (*d*) the thermal efficiency, (*e*) the temperature of the air stream leaving the regenerator and entering the combustor, and (*f*) the stream availability of the fluid leaving the regenerator and entering the environment, which has a temperature of 22°C.

16-80M A gas-turbine power plant employs two-stage compression and expansion, with intercooling, reheating, and regeneration. The temperature at the outlet of the compressor second stage is 390°K, and the combustor-inlet temperature is 750°K. The turbine-inlet temperature is limited to 1180°K.

(*a*) Determine the gross maximum work output from the two-stage turbine if the overall pressure ratio is 6 : 1.

(*b*) Determine the regenerator effectiveness.

Turbojet cycles

16-81M Rework Example 16-10M in the text under the following conditions: (1) the actual pressure rise in the diffuser is 92 percent of theoretical, (2) the compressor efficiency is 82 percent, (3) the turbine efficiency is 86 percent and (4) the nozzle efficiency is 95 percent. Determine (*a*) the pressure and temperature throughout the cycle, (*b*) the compressor work, and (*c*) the exit-jet velocity, in m/s.

16-82M The airspeed of a turbojet aircraft is 300 m/s in still air at 0.25 bar and 220°K. The compressor pressure ratio is 9, and the maximum temperature in the cycle is 1320°K. Assume ideal performance of the various components. Determine (*a*) the temperatures and pressures throughout the cycle, (*b*) the compressor work required, in kJ/kg, and (*c*) the exit-jet velocity, in m/s. Use Table A-5M.

16-83M Rework Prob. 16-82M under the following conditions: (1) the actual pressure rise in the diffuser is 90 percent of the isentropic value, (2) the compressor and turbine adiabatic efficiencies are 83 and 87 percent, respectively, and (3) the nozzle efficiency is 96 percent. Determine the quantities specified in Prob. 16-82M.

16-84M The airspeed of a jet aircraft is 280 m/s in still air at 0.050 MPa and 250°K. The pressure ratio across the compressor is 11, and the maximum cycle temperature is 1360°K. Assume ideal performance of the various components. Determine (*a*) the compressor work, in kJ/kg, (*b*) the pressure at the turbine outlet, in MPa, and (*c*) the exit-jet velocity in m/s. Use Table A-5M for data.

16-85M Rework Prob. 16-84M under the following conditions: (1) the actual pressure rise in the diffuser is 92 percent of the isentropic value, (2) the compressor and turbine efficiencies are 84 and 87 percent, respectively, and (3) the nozzle efficiency is 94 percent.

16-86M Reconsider Example 16-10M. Determine (*a*) the total thrust, in newtons, (b) the specific work required, in kJ/kg, and (*c*) the propulsive efficiency, if the mass flow rate of air is 60 kg/s.

16-87M Reconsider Prob. 16-81M. Determine (*a*) the total thrust, in newtons, (*b*) the specific work required, in kJ/kg, and (*c*) the propulsive efficiency, if the mass flow rate is 50 kg/s.

16-88M Reconsider Prob. 16-83M. Determine (*a*) the total thrust, in newtons, (*b*) the specific work required, in kJ/kg, and (*c*) the propulsive efficiency, if the mass flow rate is 75 kg/s.

16-89M Reconsider Prob. 16-85M. Determine (*a*) the total thrust, in newtons, (*b*) the specific work required, in kJ/kg, and (*c*) the propulsive efficiency, if the mass flow rate is 55 kg/s.

Closed gas-turbine cycles

16-90M A closed gas-turbine power plant operates with helium as the working fluid. The compressor and turbine efficiencies are each 85 percent. The compressor-inlet state is 5 bars and 47°C, while the turbine-inlet state is 12 bars and 980°K. If the mass flow rate is 5 kg/s, determine (*a*) the net power output, in kilowatts, and (*b*) the rate of heat removed from the cycle, in kJ/min.

16-91M A closed-cycle gas-turbine power plant operates with argon as the working fluid. The compressor and turbine efficiencies are 0.83 and 0.86, respectively. Compressor-inlet conditions are 8 bars and 340°K, while the turbine-inlet state is 15 bars and 1000°K. If the mass flow rate is 7 kg/s, find (*a*) the net power output, in kilowatts, and (*b*) the rate of heat removed from the cycle, in kJ/min.

16-92M A closed-cycle gas-turbine power plant operates with argon as the working fluid. The compressor and turbine efficiencies are 84 and 86 percent, respectively. The compressor-inlet conditions are 6 bars and 37°C, while the turbine-inlet conditions are 21 bars, and 652°C. If the net power output is 3500 kW, determine (*a*) the required mass flow rate, in kg/s, (*b*) the gross kilowatt output of the turbine, and (*c*) the rate of heat removed from the cycle and radiated to outer space, in kJ/min.

Ericsson cycle

16-93M An Ericsson cycle operates on air, and the minimum volume of the cycle is 0.01 m^3. The maximum pressure is 6 bars, and at the end of constant-pressure expansion the volume is 0.02 m^3. Draw a *PV* diagram, and compute, for each 40 kJ of heat added, the amount of heat that must be rejected.

16-94M Same as Prob. 16-93M, except that the minimum volume is 5 L, the maximum pressure is 5 bars, and the volume at the end of constant-pressure expansion is 12 L. The heat added is 20 kJ.

Stirling cycle

16-95M A Stirling cycle operates with air, and at the beginning of isothermal expansion the state is 447°C and 8 bars. The minimum pressure in the cycle is 2 bar, and at the end of isothermal compression the volume is 60 percent of the maximum volume. Determine (*a*) the thermal efficiency of the cycle, and (*b*) the mean effective pressure.

16-96M A Stirling cycle operates with air, and at the beginning of isothermal compression the state is 77°C and 2 bars. The maximum pressure is 6 bars, and during isothermal expansion the volume increases by 40 percent. Determine (*a*) the thermal efficiency, and (*b*) the mean effective pressure of the cycle.

16-97M A Stirling cycle operates with air, and at the beginning of the isothermal expansion the pressure is 5 bars and the temperature is 257°C. The thermal efficiency is 44 percent, and at the end of the isothermal compression the volume is two-thirds the maximum volume. Determine the mean effective pressure of the cycle.

PROBLEMS (USCS)

Air standard cycles

16-1 An air-standard cycle is proposed which begins at an initial state of 70°F, 14.7 psia, and 13.35 ft^3/lb. Process 1-2 is constant-volume heating to 29.4 psia; process 2-3 is constant-pressure heating to

1660°F and 26.7 ft^3/lb; process 3-4 is isentropic expansion to 14.7 psia; process 4-1 is constant-pressure cooling to the initial state.

(*a*) Sketch *Pv* and *Ts* diagrams for the cycle.
(*b*) Determine the total heat added, in Btu/lb.
(*c*) Determine the net work output, in Btu/lb.
(*d*) Compute the thermal efficiency.
(*e*) What is the Carnot efficiency for a heat engine with the same maximum and minimum temperature as the above cycle? Use *u* and *h* data from the air table.

16-2 An air-standard cycle is proposed which begins at an initial state of 100°F, 16 psia, and 12.95 ft^3/lb. Process 1-2 is constant-volume heating to 42 psia; process 2-3 is constant-pressure heating to 2340°F and 25.90 ft^3/lb; process 3-4 is isentropic expansion to 16 psia; process 4-1 is constant-pressure cooling to the initial state.

(*a*) Sketch *Pv* and *Ts* diagrams for the cycle.
(*b*) Determine the total heat added, in Btu/lb.
(*c*) Find the net work output, in Btu/lb.
(*d*) Compute the thermal efficiency.
(*e*) What is the Carnot efficiency for a heat engine with the same maximum and minimum temperatures as the above cycle? Use *u* and *h* data from the air table.

16-3 A proposed air-standard cycle begins at 15 psia, 80°F, and 13.32 ft^3/lb. Process 1-2 is isentropic compression to 41.2 psia; process 2-3 is constant-pressure heating to 1700°F and 19.41 ft^3/lb; process 3-4 is constant-volume cooling to 15 psia; process 4-1 is constant-pressure cooling to the initial state.

(*a*) Sketch *Pv* and *Ts* diagrams of the cycle.
(*b*) Determine the heat added and the net work output, in Btu/lb.
(*c*) Compute the thermal efficiency.
(*d*) What is the Carnot efficiency for a heat engine with the same maximum and minimum temperatures as the above cycle? Use air-table data.

16-4 A proposed air-standard cycle begins at 16 psia, 100°F, and 12.95 ft^3/lb. Process 1-2 is isentropic compression to 35.0 psia; process 2-3 is constant-pressure heating to 1290°F and 18.5 ft^3/lb; process 3-4 is constant-volume cooling to 16 psia; process 4-1 is constant-pressure cooling to the initial state.

(*a*) Sketch *Pv* and *Ts* diagrams of the cycle.
(*b*) Determine the heat added and the net work output, in Btu/lb.
(*c*) Compute the thermal efficiency.
(*d*) What is the Carnot efficiency for a heat engine with the same maximum and minimum temperatures as the above cycle? Use air-table data.

Air-standard Carnot cycles

16-5 An air-standard Carnot cycle rejects 50 Btu/lb to a sink at 500°R. The minimum and maximum pressures in the closed cycle are 1 and 292 atm. On the basis of the air table, determine (*a*) the pressure after isothermal compression, in atm, (*b*) the temperature of the heat supply reservoir, in degrees Rankine, (*c*) the specific volume after isothermal compression and after isentropic compression, both in ft^3/lb, and (*d*) the thermal efficiency.

16-6 An air-standard Carnot cycle for a closed system is supplied with 100 Btu/lb of heat from a source at 1200°R. The minimum and maximum pressures in the cycle are 1 and 88 atm. On the basis of the air table, determine (*a*) the pressure after isothermal heat addition, (*b*) the temperature of heat rejection, in degrees Rankine, (*c*) the specific volume, in ft^3/lb, after isothermal heat addition and after isentropic expansion, and (*d*) the thermal efficiency.

16-7 An air-standard Carnot cycle operates in a piston-cylinder device between temperatures of 500 and 1500°R. The minimum pressure in the cycle is 1 atm. Determine (1) the maximum pressure in the cycle, in atm, and (2) the specific volume, in ft^3/lb, after isentropic compression, if the heat rejection is (*a*) 30 Btu/lb, and (*b*) 20 Btu/lb.

16-8 An air-standard Carnot cycle operates between temperatures of 540 and 2000°R. The minimum pressure in the cycle is 1 atm. Determine (1) the maximum pressure in the cycle, and (2) the pressure after isothermal compression in the closed system, in atm, if the amount of heat supplied is (*a*) 80 Btu/lb, and (*b*) 70 Btu/lb.

16-9 A Carnot heat engine which produces 10 Btu of work for one cycle has a thermal efficiency of 50 percent. The working fluid is 1.0 lb of air, and the pressure and volume at the beginning of the isothermal expansion are 100 psia and 4.0 ft^3, respectively. Find (*a*) the maximum and minimum temperatures for the cycle, in degrees Rankine, (*b*) the heat and work for each of the four processes, in Btu/cycle, and (*c*) the volume at the end of the isothermal expansion process and at the end of the isentropic expansion process, in ft^3.

Compression ratio and mean effective pressure

16-10 For the data in Prob. 16-5, determine the compression ratio and the mean effective pressure for a reciprocating device.

16-11 For the data in Prob. 16-6, determine the compression ratio and the mean effective pressure for a reciprocating device.

16-12 For the data in Prob. 16-7, determine the compression ratio and the mean effective pressure for a reciprocating device if the heat rejection is (*a*) 30 Btu/lb, and (*b*) 20 Btu/lb.

16-13 For the data in Prob. 16-8, determine the compression ratio and the mean effective pressure for a reciprocating device if the heat supplied is (*a*) 80 Btu/lb, and (*b*) 70 Btu/lb.

Otto cycle

16-14 The compression ratio of an Otto cycle is 8 : 1. Before the compression stroke of the cycle begins, the pressure is 14.5 psia and the temperature is 80°F. The heat added to the air per cycle is 840 Btu/lb of air. On the basis of the air-standard cycle, and employing air table A-5, determine (*a*) the pressure and temperature at the end of each process of the cycle, (*b*) the theoretical thermal efficiency, and (*c*) the mean effective pressure for the cycle, in lb/in^2.

16-15 Same as Prob. 16-14, except that the amount of heat added is 792 Btu/lb.

16-16 The air at the beginning of the compression stroke of an air-standard Otto cycle is at 14 psia and 80°F, and the cylinder volume is 0.20 ft^3. The compression ratio is 9, and 8.80 Btu are added during the heat-addition process. On the basis of Table A-5, determine (*a*) the temperature and pressure after the heat-addition and expansion processes, (*b*) the thermal efficiency, and (*c*) the mean effective pressure, in psi.

16-17 Same as Prob. 16-16, except that the amount of heat added is 8.30 Btu.

16-18 An air-standard Otto cycle operates with a compression ratio of 8.50, and at the beginning of compression the air is at 14.5 psia and 90°F. During the heat-addition process the pressure is tripled. On the basis if Table A-5, determine (*a*) the temperatures around the cycle, in degrees Rankine, (*b*) the thermal efficiency of the cycle, and (*c*) the thermal efficiency of a Carnot engine operating between the same overall temperature limits.

16-19 Consider an air-standard Otto cycle that has a compression ratio of 9.0 and a heat addition of 870 Btu/lb. If the pressure and temperature at the beginning of the compression process are 14.0 psia and 40°F, determine on the basis of Table A-5 (*a*) the maximum pressure and temperature for the cycle, (*b*) the net work output, in Btu/lb, (*c*) the thermal efficiency, and (*d*) the mean effective pressure, in psi.

16-20 Same as Prob. 16-19, except that the amount of heat added is 750 Btu/lb.

16-21 Reconsider the Otto cycle in Prob. 16-14. Determine (*a*) the closed-system availability of the air at the end of isentropic expansion, in Btu/lb, if $T_0 = 80°F$ and $P_0 = 14.5$ psia, and (*b*) the ratio of this availability quantity to the net work output of the cycle.

16-22 Reconsider the Otto cycle in Prob. 16-16. Determine (*a*) the closed-system availability of the air at the end of isentropic expansion, in Btu/lb, if $T_0 = 80°F$ and $P_0 = 14.0$ psia, and (*b*) the ratio of this availability quantity to the net work output of the cycle.

16-23 Reconsider the Otto cycle in Prob. 16-19. Determine (*a*) the closed-system availability of the air at the end of isentropic expansion, in Btu/lb, if $T_0 = 40°\text{F}$ and $P_0 = 14.0$ psia, and (*b*) the ratio of this availability quantity to the net work output of the cycle.

Diesel cycle

16-24 An air-standard Diesel cycle operates with a compression ratio of 14.8 and a cutoff ratio of 2. At the beginning of compression the air temperature and pressure are 100°F and 14.5 psia, respectively. Determine (*a*) the maximum temperature in the cycle, in degrees Rankine, (*b*) the pressure after isentropic expansion, in psia, and (*c*) the heat input per cycle, in Btu/lb. Use data from Table A-5.

16-25 An engine operates on the theoretical Diesel cycle with a compression ratio of 15.3 : 1, and fuel is injected for 10 percent of the stroke. The pressure and temperature of the air entering the cylinder are 14.0 psia and 60°F, respectively. Determine (*a*) the cutoff ratio, (*b*) the temperature, in degrees Rankine, at the end of the compression process and at the end of the heat-addition process, (*c*) the pressure after isentropic expansion, in psia, and (*d*) the heat input, in Btu/lb. Use Table A-5.

16-26 The intake conditions for a Diesel cycle operating with a compression ratio of 15 are 14.4 psia and 60°F. The cutoff ratio for the cycle is (*a*) 2.84, and (*b*) 2.0 Determine (1) the pressure and temperature at the end of each process of the cycle, and (2) the thermal efficiency and the mean effective pressure.

16-27 A Diesel cycle operating on an air-standard cycle has a compression ratio of 15. The pressure and temperature at the beginning of compression are 15 psia and 60°F. If the maximum temperature of the cycle is 4200°R, determine (*a*) the thermal efficiency, and (*b*) the mean effective pressure.

16-28 An air-standard Diesel cycle is supplied with 700 Btu/lb of heat per cycle. The pressure and temperature at the beginning of compression are 14.0 psia and 80°F, respectively, and the pressure after compression is 540 psia. Determine (*a*) the compression ratio, (*b*) the maximum temperature in the cycle, in degrees Rankine, (*c*) the cutoff ratio, and (*d*) the pressure after isentropic expansion, in psia. Use Table A-5 for data.

16-29 Reconsider the Diesel cycle in Prob. 16-24. Determine (*a*) the closed system availability of the air at the end of isentropic expansion, in Btu/lb, if $T_0 = 60°\text{F}$ and $P_0 = 14.5$ psia, and (*b*) the ratio of this availability quantity to the net work output of the cycle.

16-30 Reconsider the Diesel cycle of Prob. 16-26. Determine (1) the closed system availability of the air at the end of isentropic expansion, in Btu/lb, if $T_0 = 60°\text{F}$ and $P_0 = 14.4$ psia, and (2) the ratio of this availability quantity to the net work output of the cycle, for parts (*a*) and (*b*).

Dual cycle

16-31 An air-standard dual cycle operates with a compression ratio of 15 : 1. At the beginning of compression the conditions are 60°F, 14.6 psia, and 230 in^3. The amount of heat added is 7.5 Btu, of which one-third is added at constant volume and the remainder at constant pressure. Determine (*a*) the pressure after the constant-volume heat-addition process, (*b*) the temperature before and after the constant-pressure heat-addition process, (*c*) the temperature after isentropic expansion, and (*d*) the thermal efficiency.

16-32 An air-standard dual cycle operates with a compression ratio of 14 : 1. At the beginning of isentropic compression the conditions are 80°F and 14.5 psia. The total heat addition is 800 Btu/lb, of which one-fourth is added at constant volume and the remainder at constant pressure. Determine (*a*) the temperatures at the end of each process around the cycle, in degrees Rankine, (*b*) the thermal efficiency, and (*c*) the mean effective pressure.

16-33 Same as Prob. 16-31, except that the heat addition is 6.9 Btu, of which 30 percent is added at constant volume.

16-34 Same as Prob. 16-32 except that the heat addition is 900 Btu/lb, of which one-third is added at constant volume.

Ideal open gas-turbine cycle

16-35 A gas-turbine power plant operates on an air-standard cycle between pressure limits of 14.5 psia and 87.0 psia. The inlet-air temperature is 80°F, and the air enters the turbine at 1290°F. Using the data of Table A-5M, determine (*a*) the net work output and the heat input, in Btu/lb, and (*b*) the thermal efficiency if the cycle is ideal.

16-36 A gas-turbine power plant operates on an air-standard cycle between pressure limits of 1 and 4.0 atm. The inlet-air temperature is 60°F, and the temperature limitation on the turbine is 1540°F. Calculate (*a*) the net work output, in Btu/lb, and (*b*) the thermal efficiency if the cycle is ideal. Use the air table for data.

16-37 A gas-turbine power plant operates on an air-standard cycle between pressure limits of 14.7 psia and 94 psia. The inlet-air temperature to the compressor is 60°F, and the inlet-air temperature to the turbine is 1480°F. Calculate (*a*) the net work output, in Btu/lb, (*b*) the heat input, and (*c*) the thermal efficiency if the cycle is ideal. Use the air table for data.

16-38 A stationary gas-turbine power plant has maximum and minimum cycle temperatures of 1540 and 80°F, and a pressure ratio of 5.0 : 1. Find (*a*) the ratio of compressor work to turbine work, (*b*) the thermal efficiency, and (*c*) the mass flow rate of air required, in lb/min, for a net power output of 1000 hp. Use air-table data for the ideal cycle analysis.

16-39 The pressure ratio of an air-standard Brayton cycle is 6.0, and the inlet conditions to the compressor are 15 psia and 40°F. The turbine is limited to a temperature of 1440°F, and the mass flow rate is 10 lb/s. Determine (*a*) the compressor and turbine work, in Btu/lb, (*b*) the thermal efficiency, (*c*) the net power output, in horsepower, and (*d*) the volume flow rate at the compressor inlet, in ft^3/min. (*e*) If the heat addition is accomplished by the complete combustion of a fuel with a heating value of 18,500 Btu/lb, estimate the fuel-air ratio used in the combustor, in lb/lb. Use the air table for all property data.

16-40 The pressure ratio of an air-standard Brayton cycle is 5.4 : 1, and the inlet conditions are 14.6 psia and 60°F. The turbine is limited to a temperature of 1490°F, and the mass flow rate is 7.0 lb/s. Determine (*a*) the compressor and turbine work, in Btu/lb, (b) the thermal efficiency, (*c*) the net power output, in horsepower, and (*d*) the volume flow rate at the compressor inlet, in ft^3/min. (*e*) If the heat addition is accomplished by the complete combustion of a fuel with a heating value of 19,000 Btu/lb, estimate the fuel-air ratio used in the combustor, in lb/lb, using air data.

16-41 If the sink temperature T_0 is the same as the compressor-inlet temperature, then determine the steady-flow availability of the turbine exhaust stream, in Btu/lb, for the Brayton cycle described in (*a*) Prob. 16-35, (*b*) Prob. 16-36, (*c*) Prob. 16-37, and (*d*) Prob. 16-38. Also calculate the percent increase in net work output if this availability could be completely converted into work output.

16-42 Prove that the maximum net work for a simple Brayton cycle, for fixed compressor- and turbine-inlet temperatures, occurs when $T_2 = (T_1 T_3)^{1/2}$.

16-43 Show that the pressure ratio which yields the largest net work for a simple Brayton cycle with fixed compressor- and turbine-inlet temperatures is given by $P_2/P_1 = (T_3/T_1)^{k/[2(k-1)]}$.

Nonideal simple gas-turbine power cycle

16-44 Using the data of Prob. 16-35, calculate the heat input and the thermal efficiency if the compressor and turbine adiabatic efficiencies are 83 and 86 percent, respectively.

16-45 Using the data of Prob. 16-36, calculate the required quantities if the adiabatic efficiencies of the compressor and turbine are 78 and 84 percent, respectively.

16-46 Using the data of Prob. 16-37, calculate the required quantities if the compressor and turbine adiabatic efficiencies are 82 and 85 percent, respectively.

16-47 Using the data of Prob. 16-38, calculate the required quantities if the adiabatic efficiencies of the compressor and turbine are 81 and 86 percent, respectively.

16-48 Using the data of Prob. 16-39, calculate the required quantities if the compressor and turbine adiabatic efficiencies are 80 and 85 percent, respectively.

16-49 Using the data of Prob. 16-40, calculate the required quantities if the compressor and turbine adiabatic efficiencies are 84 and 88 percent, respectively.

16-50 A gas-turbine power cycle operates with a pressure ratio of 12 : 1. The compressor and turbine efficiencies are 85 and 90 percent, respectively. The compressor-inlet temperature is 60°F, and the turbine-inlet temperature is 2200°F. For a mass flow rate of 1 lb/s, find the power generated by the cycle. Now double the pressure ratio. What is the power output in this case?

16-51 Air enters a gas-turbine power plant at 60°F and 14.5 psia. The compressor and turbine operate with a pressure ratio of 8, and because of fluid friction, the efficiency of each is 80 and 85 percent, respectively. Gases leave the combustor at 1540°F, and the mass flow rate of inlet air is 20 lb/s. Part of the turbine output is used to drive the compressor. Determine what percent of the total output of the turbine may be used to drive devices other than the compressor, on the basis of air-table calculations.

16-52 If the sink temperature T_0 is the same as the compressor-inlet temperature, then determine the steady-flow availability of the turbine exhaust stream, in kJ/kg, for the irreversible cycle described in (*a*) Prob. 16-44, (*b*) Prob. 16-45, (*c*) Prob. 16-46, and (*d*) Prob. 16-47. Also calculate the percent increase in net work output if this availability could be completely converted into work output.

Regenerative gas-turbine cycle

16-53 Using the data of Prob. 16-44, compute (1) the thermal efficiency, and (2) the percent fuel saved if a regenerator is installed with an effectiveness of (*a*) 80 percent, and (*b*) 65 percent.

16-54 Using the data of Prob. 16-45, compute (1) the thermal efficiency, and (2) the percent fuel saved if a regenerator were installed with an effectiveness of (*a*) 80 percent, and (*b*) 60 percent.

16-55 Using the data of Prob. 16-46, determine (1) the thermal efficiency, and (2) the percent fuel saved if a regenerator is installed with an effectiveness of 70 percent.

16-56 Using the data of Prob. 16-47, determine (1) the thermal efficiency, and (2) the percent fuel saved if a regenerator is added with an effectiveness of (*a*) 70 percent, and (*b*) 60 percent.

16-57 In an air-standard gas-turbine cycle with regeneration, the compressor is driven directly by the turbine. The following enthalpy data were taken on a test of a gas turbine with a pressure ratio of 5.41 : 1. Determine (*a*) the thermal efficiency of the actual cycle, (*b*) the regenerator effectiveness, (*c*) the compressor efficiency, and (*d*) the turbine efficiency. Data are in Btu/lb.

System	Entering	Leaving
Compressor	124.3	216.3
Regenerator	216.3	266.4
Combustor	266.4	449.7
Turbine	449.7	306.7
Regenerator	306.7	256.6

16-58 An automotive manufacturer is contemplating the use of a regenerative-type gas turbine for the power source of a new model. Fresh air enters the turbine compressor at 60°F and 1 atm, with a flow rate of 2.2 lb/s, and is compressed at a ratio of 4 : 1. Air from the compressor enters the regenerator, where it is circulated until its temperature is approximately 1000°F. It then passes to the combustion chamber, where it is heated to 1700°F. After expanding through the two-stage turbine, the hot gases pass through the other half of the regenerator and are then exhausted to the atmosphere. Assuming ideal compression and expansion and negligible pressure losses through the burner and regenerator, determine (*a*) the thermal efficiency of the unit (*b*) the power output, in horsepower, (*c*) the regenerator effectiveness and (*d*) the exhaust stream temperature, in degrees Fahrenheit.

16-59 If the sink temperature T_0 is the same as the compressor-inlet temperature, then determine the steady-flow availability of the stream leaving the regenerator and entering the atmosphere, in Btu/lb, for the irreversible regenerative cycle described in (*a*) Prob. 16-53, (*b*) Prob. 16-54, (*c*) Prob. 16-55, and (*d*) Prob. 16-56.

Polytropic process

16-60 Air is compressed in a piston-cylinder device from 14 psia, 40°F, to a pressure of 70 psia. Calculate (1) the work required, and (2) the heat transferred, in Btu/lb, if the process is (*a*) polytropic with $n = 1.30$, (*b*) isentropic, and (*c*) isothermal.

16-61 Work Prob. 16-60 for the same data, except the final pressure is 84 psia. List the answers in $\text{ft} \cdot \text{lb}_f/\text{lb}_m$.

16-62 A steady-flow air compressor operates between an inlet condition of 15 psia and 60°F, and an outlet pressure of 75 psia. Determine (1) the work input, and (2) the heat transferred, in Btu/lb, for the following reversible processes: (*a*) isothermal, (*b*) $n = 1.30$, and (*c*) adiabatic.

16-63 Work Prob. 16-62, except change the inlet conditions to 15 psia and 100°F and the outlet condition to 90 psia.

16-64 Find the power requirements, in horsepower, if 220 ft^3/min of air at 14 psia and 50°F are compressed polytropically ($n = 1.30$) to 112 psia. Cooling water runs through the compressor jacket at a rate of 28.5 lb/min and has a temperature rise of 10.0°F. The air velocity at the inlet is small, but it is 500 ft/s at the outlet. The process is irreversible.

16-65 Find the power requirements, in kilowatts, if 180 ft^3/min of air at 16.0 psia and 60°F are compressed polytropically ($n = 1.32$) to 100 psia. Cooling water runs through the compressor jacket at a rate of 11.0 lb/min and has a temperature rise of 9.2°F. The air velocity at the inlet is small, but it is 420 ft/s at the outlet. The process is irreversible.

Compressor staging with intercooling

16-66 The inlet conditions of a two-stage steady-flow compressor are 14 psia and 80°F. The outlet pressure is 140 psia, and the pressure ratio across each stage is the same. If an intercooler cools the air to the initial temperature and the stages are isentropic, (*a*) determine the total work input, in Btu/lb, and (*b*) compare the result of *a* with the work required for a single-stage compressor. Use Table A-5 for data.

16-67 Same as Prob. 16-66, except that the inlet conditions are 16 psia and 100°F, and the outlet pressure is 96 psia.

16-68 The inlet conditions of a two-stage steady-flow compressor are 15 psia and 80°F. The outlet pressure is 120 psia, the stages are isentropic, and the intercooler cools the air to the initial temperature. Determine the total work input, in Btu/lb, if (*a*) the pressure ratio across each stage is the same, and (*b*) the pressure ratio across the first stage is twice that across the second stage.

16-69 Same as Prob. 16-68, except that the inlet conditions are 14 psia and 60°F, and the outlet pressure is 63 psia.

16-70 Consider a three-stage steady-flow compressor with intercoolers which cool the ideal gas to the initial inlet temperature. The outlet pressures of the first and second stages of the isentropic compressor will be denoted by P_a and P_b, respectively. Show that the minimum work of compression is attained when $P_a = (P_1^2 P_2)^{1/3}$ and $P_b = (P_1 P_2^2)^{1/3}$.

16-71 A three-stage steady-flow compressor has the same inlet and outlet conditions as given in Prob. 16-66. Based on the results of Prob. 16-70, determine the minimum work of compression, in Btu/lb. Use Table A-5 for data.

16-72 A three-stage steady-flow compressor has the same inlet and outlet conditions as given in Prob. 16-67. Based on the results of Prob. 16-70, determine the minimum work of compression, in Btu/lb. Use Table A-5 for data.

Gas-turbine cycle with intercooling, reheating, and regeneration

16-73 Reconsider Prob. 16-53. Determine the net work output and the thermal efficiency if two-stage compression and expansion are used with a regenerator effectiveness of (*a*) 80 percent, and (*b*) 65 percent. The compressor and turbine efficiencies are given in Prob. 16-44. The intercooling and reheating are ideal.

16-74 Reconsider Prob. 16-54. Determine the net work output and the thermal efficiency if two-stage compression and expansion are used with a regenerator effectiveness of (*a*) 60 percent, and (*b*) 80 percent. The compressor and turbine efficiencies are given in Prob. 16-45. The intercooling and reheating are ideal.

16-75 Reconsider the data of Prob. 16-55. Determine the net work output and the thermal efficiency if two-stage compression and expansion are used with a regenerator of 70 percent effectiveness. The compressor and turbine efficiencies are given in Prob. 16-46. The intercooling and reheating are ideal.

16-76 Reconsider Prob. 16-56. Determine the net work output and the thermal efficiency if two-stage compression and expansion are used with a regenerator effectiveness of 70 percent. The compressor and turbine efficiencies are given in Prob. 16-47. The intercooling and reheating are ideal.

16-77 A gas-turbine cycle operates with two stages of compression and two stages of expansion. The pressure ratio across each stage is 2. The inlet temperature is 60°F to each stage of compression and 1540°F to each stage of expansion. The compressor and turbine efficiencies are 78 and 84 percent, respectively, and the regenerator has an effectiveness of 70 percent. Determine (*a*) the compressor and turbine work, in Btu/lb, (*b*) the thermal efficiency, and (*c*) the temperature of the air stream leaving the regenerator and entering the atmosphere, in degrees Fahrenheit. Use air-table data.

16-78 Reconsider Prob. 16-77, and calculate the same required quantities, except that the compressor and turbine efficiencies are now 81 and 86 percent, respectively, and the regenerator effectiveness is now 75 percent.

16-79 A gas-turbine cycle operates with two-stage compression and expansion. The pressure ratio across each stage of compression is 2.0, and the compressor efficiency is 81 percent. The compressor-inlet temperature is 60°F, but intercooling reduces the temperature only to 100°F before the air enters the second stage. The inlet temperature to each stage of expansion is 1540°F, but a pressure drop between the compressor and turbine reduces the expansion pressure ratio to 1.9 : 1 across each turbine stage, which has an adiabatic efficiency of 86 percent. The regenerator effectiveness is 75 percent. Determine (*a*) the compressor work, (*b*) the turbine work, (*c*) the heat removed in the intercooler, in Btu/lb, (*d*) the thermal efficiency, (*e*) the temperature of the air stream leaving the regenerator and entering the combustor, in degrees Fahrenheit, and (*f*) the stream availability of the fluid leaving the regenerator and entering the environment, which has a temperature of 60°F, in Btu/lb.

16-80 A gas-turbine power plant employs two-stage compression and expansion, with intercooling, reheating, and regeneration. The temperature at the outlet of the compressor second stage is 700°R, and the combustor-inlet temperature is 1360°R. The turbine-inlet temperature is limited to 2100°R.

(*a*) Determine the total maximum work output from the two-stage turbine if the overall pressure ratio is 6 : 1.

(*b*) Determine the regenerator effectiveness.

Turbojet cycles

16-81 Rework Example 16-10 in the text under the following conditions: (1) the actual pressure rise in the diffuser is 92 percent of theoretical, (2) the compressor efficiency is 82 percent, (3) the turbine efficiency is 86 percent, and (4) the nozzle efficiency is 95 percent. Determine (*a*) the pressures and temperatures throughout the cycle, (*b*) the compressor work input, and (*c*) the exit-jet velocity, in ft/s.

16-82 The airspeed of a turbojet aircraft is 900 ft/s in still air at 4.0 psia and −60°F. The compressor pressure ratio is 9, and the maximum temperautre in the cycle is 1740°F. Assume ideal performance of the various components. Determine (*a*) the temperatures and pressures throughout the cycle, (*b*) the compressor work required, in Btu/lb, and (*c*) the exit-jet velocity, in ft/s. Use Table A-5.

16-83 Rework Prob. 16-82 under the following conditions: (1) the actual pressure rise in the diffuser is 90 percent of the isentropic value, (2) the compressor and turbine adiabatic efficiencies are 83 and 87 percent, respectively, and (3) the nozzle efficiency is 96 percent. Determine the quantities specified in Prob. 16-82.

16-84 The airspeed of a jet aircraft is 800 ft/s in still air at 8.0 psia and 0°F. The pressure ratio across the compressor is 11, and the maximum cycle temperature is 1840°F. Assume ideal performance of the various components. Determine (*a*) the compressor work, in Btu/lb, (*b*) the pressure at the turbine outlet, in psia, and (*c*) the exit-jet velocity, in ft/s. Use Table A-5 for data.

16-85 Rework Prob. 16-84 under the following conditions: (1) the actual pressure rise in the diffuser is 92 percent of the isentropic value, (2) the compressor and turbine efficiencies are 84 and 87 percent, respectively, and (3) the nozzle efficiency is 94 percent.

16-86 Reconsider Example 16-10. Determine (*a*) the total thrust, in lb_f, (*b*) the specific work required, in Btu/lb, and (*c*) the propulsive efficiency, if the mass flow rate of air is 120 lb/s.

16-87 Reconsider Prob. 16-81. Determine (*a*) the total thrust, in lb_f, (*b*) the specific work required, in Btu/lb, and (*c*) the propulsive efficiency, if the mass flow rate is 100 lb/s.

16-88 Reconsider Prob. 16-83. Determine (*a*) the total thrust, in lb_f, (*b*) the specific work required, in Btu/lb, and (*c*) the propulsive efficiency, if the mass flow rate is 150 lb/s.

16-89 Reconsider Prob. 16-85. Determine (*a*) the total thrust, in lb_f, (*b*) the specific work required, in Btu/lb, and (*c*) the propulsive efficiency, if the mass flow rate is 110 lb/s.

Closed gas-turbine cycles

16-90 A closed-cycle gas-turbine power plant operates with helium as the working fluid. The compressor and turbine efficiencies are each 85 percent. The compressor-inlet state is 80 psia and 120°F, while the turbine-inlet state is 180 psia and 1300°F. If the mass flow rate is 10 lb/s, determine (*a*) the net power output, in horsepower, and (*b*) the rate of heat removed from the cycle, in Btu/min.

16-91 A closed-cycle gas-turbine power plant operates with argon as the working fluid. The compressor and turbine efficiencies are 83 and 86 percent, respectively. Compressor-inlet conditions are 120 psia and 150°F, while the turbine-inlet state is 250 psia and 1400°F. If the mass flow rate is 15 lb/s, determine (*a*) the net power output, in horsepower, and (*b*) the rate of heat removed from the cycle, in Btu/min.

16-92 A closed-cycle gas-turbine power plant operates with argon as the working fluid. The compressor and turbine efficiencies are 0.86 and 0.84, respectively. The compressor-inlet conditions are 100 psia and 100°F, while the turbine-inlet conditions are 350 psia and 1220°F. If the net power output is 2500 hp, determine (*a*) the required mass flow rate, in lb/s, (*b*) the gross output of the turbine, in horsepower, and (*c*) the rate of heat removed from the cycle and radiated to outer space, in Btu/min.

Ericsson cycle

16-93 An Ericsson cycle operates on air, and the minimum volume of the cycle is 1 ft^3. The maximum pressure is 100 psia, and at the end of the constant-pressure expansion process the volume is 2 ft^3. Draw a *PV* diagram, and compute, for each 100 Btu of heat added, the amount of heat that must be rejected.

16-94 An Ericsson cycle operates on air, and the minimum volume of the cycle is 0.20 ft^3. The maximum pressure is 120 psia, and the volume at the end of constant-pressure expansion is 0.44 ft^3. Draw a *PV* diagram, and compute for each 24 Btu of heat added, the amount of heat that must be rejected.

Stirling cycle

16-95 A Stirling cycle operates with air, and at the beginning of isothermal expansion the state is 800°F and 80 psia. The minimum pressure in the cycle is 20 psia, and at the end of isothermal compression the volume is 60 percent of the maximum volume. Determine (*a*) the thermal efficiency, and (*b*) the mean effective pressure of the cycle.

16-96 A Stirling cycle operates with air, and at the beginning of isothermal compression the state is 240°F and 25 psia. The maximum pressure is 75 psia, and during isothermal expansion the volume increases by 40 percent. Determine (*a*) the thermal efficiency, and (*b*) the mean effective pressure of the cycle.

16-97 A Stirling cycle operates with air, and at the beginning of the isothermal expansion the pressure is 60 psia and the temperature is 500°F. The thermal efficiency is 43.7 percent, and at the end of the isothermal compression the volume is two-thirds the maximum volume. Assuming constant specific heats of air, determine the mean effective pressure of the cycle.

CHAPTER

SEVENTEEN

VAPOR POWER CYCLES

One of the basic industries of interest to the engineer is electric power generation. A major portion of the electricity generated commercially is through the use of steam power plants. These plants operate essentially on the same basic cycle, whether the input energy is from the combustion of fossil fuels or from a fission process in a nuclear reactor. The steam cycle differs from the gas cycles presented in the preceding chapter in that both the vapor and liquid phases appear during the process. To a large degree the following discussion deals with idealized cycles. More comprehensive analyses of vapor power cycles may be obtained from standard textbooks on the subject.

17-1 THE CARNOT VAPOR CYCLE

The thermal efficiency of a power cycle is maximized if all the heat supplied from an energy source occurs at the maximum possible temperature and all the energy rejected to a sink occurs at the lowest possible temperature. For a reversible cycle operating under these conditions the thermal efficiency is the Carnot efficiency given by $(T_H - T_L)/T_H$. One theoretical cycle which fulfills these conditions is the Carnot heat engine cycle introduced in Sec. 6-14. In review, a Carnot cycle is composed of two reversible, isothermal processes and two reversible, adiabatic processes (or isentropic processes). If the working fluid is to appear in both the liquid and vapor phases during various parts of the cycle, then the *Ts* diagram for the cycle would be similar to that shown in Fig. 17-1*a*. A schematic of the equipment needed for such a cycle is shown in Fig. 17-1*b*.

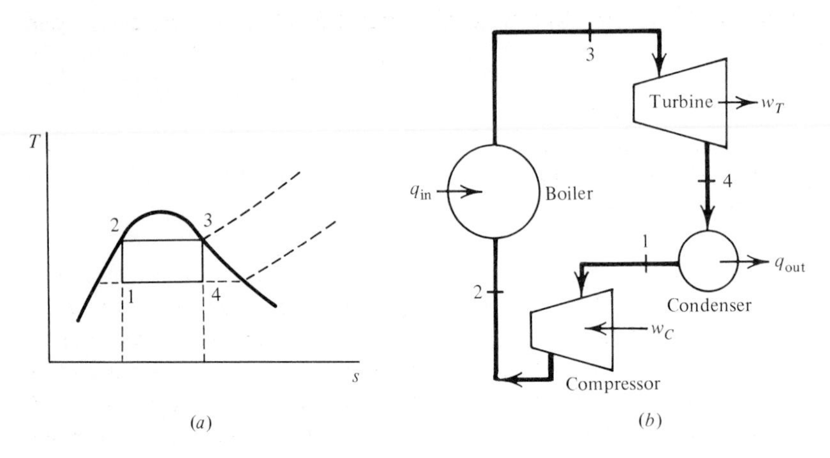

Figure 17-1 *Ts* diagram and equipment schematic of a Carnot vapor power cycle.

For the purpose of this discussion, assume water is the working fluid. Wet steam at state 1 is compressed isentropically to saturated liquid at state 2. At this elevated pressure, energy is added at constant pressure until the water is completely evaporated to saturated vapor at state 3. It is then allowed to expand isentropically through a turbine to state 4. The wet steam leaving the steam turbine is then partially condensed at constant pressure back to state 1. The thermal efficiency of the cycle is, of course, the highest for any engine operating between temperatures T_1 and T_2, and is given by $(T_2 - T_1)/T_2$. The Carnot cycle is impractical to use, however, with fluids which undergo phase changes. For example, it is difficult, first of all, to compress a two-phase mixture isentropically, as required by process 1-2. Second, the condensing process 4-1 would have to be controlled very accurately to end up with the desired quality of state 1. Third, the efficiency of a Carnot cycle is greatly affected by the temperature T_2 at which energy is added. For steam, the critical temperature is only 705°F. Therefore, if the cycle is to be operated within the wet region, the maximum possible temperature is severely limited.

17-2 THE RANKINE CYCLE

The objections listed in the preceding section for the basic Carnot vapor can be eliminated by a slight modification of the model. Instead of condensing to a low-quality vapor (state 1 in Fig. 17-1*a*), the condensation process is carried out so that the wet steam leaving the turbine is condensed to saturated liquid at the turbine-outlet pressure. The compression process is now handled by a liquid pump, which isentropically compresses the liquid leaving the condenser to the pressure desired in the heat-addition process. This model for a steam power cycle is called the Rankine cycle. The basic cycle is presented schematically and on a

Ts diagram in Fig. 17-2. The ideal cycle of a simple Rankine steam power cycle then consists of:

1. Isentropic compression in a pump
2. Constant-pressure heat addition in a boiler
3. Isentropic expansion in a turbine
4. Constant-pressure heat removal in a condenser

The heat added in process 2-3 may come from the combustion of conventional fuels, from a solar source, or from a nuclear reactor. If the changes in potential and kinetic energies may be neglected, the heat transfer to the fluid in the boiler is represented on the *Ts* diagram by the area enclosed by states 2-2′-3-*b*-*a*-2. The area enclosed by states 1-4-*b*-*a*-1 then represents the heat removed from the fluid in the condenser. The first law for an open cyclic process indicates that the net heat effect equals the net work effect. Hence the net work is represented by the difference in the areas for the heat input and heat rejection, i.e., area 1-2-2′-3-4-1. The thermal efficiency for the cycle is again defined as W_{net}/Q_{in}.

Expressions for the work and heat interactions in the ideal cycle are found by applying the steady-flow energy equation to each separate piece of equipment. If we may neglect kinetic-and potential energy changes, the basic equation for each process reduces to $q + w = h_{out} - h_{in}$. The isentropic pump work is given by

$$w_{in,\ pump} = h_2 - h_1$$

However, the isentropic pump work may also be computed from the steady-flow mechanical-work equation

$$w = \int v\, dP \tag{7-28}$$

Since the change in the specific volume of liquid water from the saturation state to pressures in the compressed-liquid state normally encountered in steam power

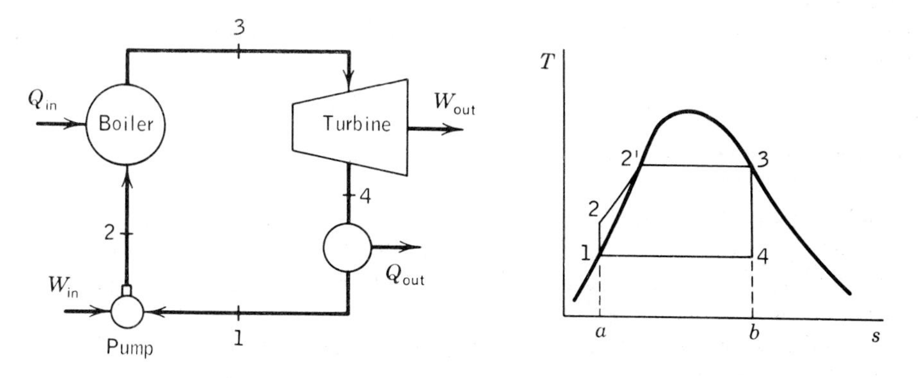

Figure 17-2 An equipment schematic and *Ts* diagram for a simple steam power plant operating on the Rankine cycle.

plants is less than 1 percent, the fluid in the pump may be considered incompressible. Consequently, the pump work is frequently determined within the desired accuracy from the relation

$$w_{\text{in, pump}} = v_f(P_2 - P_1) \qquad s_1 = s_2$$

where v_f is the saturated-liquid specific volume at state 1. Note in Fig. 17-2 that the length of line 1-2 is greatly exaggerated. The temperature rise due to isentropic compression is actually quite small.

The heat input, the isentropic work output from the turbine, and the heat rejection in the condenser, all on a unit-mass basis, are

$$q_{\text{in, boiler}} = h_3 - h_2 \qquad P_3 = P_2$$

$$w_{\text{out, turbine}} = h_3 - h_4 \qquad s_3 = s_4$$

$$q_{\text{out, condenser}} = h_4 - h_1 \qquad P_4 = P_1$$

The thermal efficiency of an ideal Rankine cycle then may be written as

$$\eta_{\text{th}} = \frac{w_{\text{T}} - w_{\text{P}}}{q_{\text{in}}} = \frac{(h_3 - h_4) - v_{f,1}(P_2 - P_1)}{h_3 - h_2}$$

Typical examples are given below.

Example 17-1M Compute the thermal efficiency of an ideal Rankine cycle for which steam leaves the boiler as saturated vapor at 30 bars and is condensed at 1.0 bar.

SOLUTION The evaluation of the thermal efficiency requires calculating the turbine and pump work and the heat input. Using the notation employed in Fig. 17-2, we find the following data from the steam tables:

$$h_3 = h_g \text{ at 30 bars} = 2804.2 \text{ kJ/kg}$$

$$h_1 = h_f \text{ at 1.0 bar} = 417.5 \text{ kJ/kg}$$

The pump work is

$$w_{\text{in, pump}} = w_P = 104 \text{ cm}^3/\text{g} \times 29.0 \text{ bars} \times 10^5 \text{ N/bar}\cdot\text{m}^2 \times \frac{\text{m}^3}{10^6 \text{ cm}^3}$$

$$= 3.0 \text{ kJ/kg}$$

Therefore $\qquad h_2 = h_1 + w_\text{P} = 417.5 + 3.0 = 420.5 \text{ kJ/kg}$

The enthalpy at state 4 is the value for a pressure of 1.0 bar and for $s_4 = s_3 = s_g$ at 30 bars. Since $s_3 = 6.1869$ kJ/(kg)(°K), $6.1869 = (s_f + xs_{fg})_{\text{at 1.0 bar}} = 1.3026 + x(7.3594 - 1.3026)$

$$x = \frac{4.8843}{6.0568} = 0.806 \qquad \text{(or 80.6 percent)}$$

Then the turbine-outlet enthalpy can be found.

$$h_4 = h_f + xh_{fg} = 417.5 + 0.806(2258.0) = 2238.4 \text{ kJ/kg}$$

Consequently,

$$w_{\text{T, out}} = h_3 - h_4 = 2804.2 - 2238.4 = 565.8 \text{ kJ/kg}$$

$$q_{\text{in}} = h_3 - h_2 = 2804.2 - 420.0 = 2384.2 \text{ kJ/kg}$$

The thermal efficiency is

$$\eta_{th} = \frac{565.8 - 3.0}{2384.2} = 0.236 \qquad \text{(or 23.6 percent)}$$

Note that some of the heat addition occurs during process 2-2′ in Fig. 17-2, for which the temperature is lower than the maximum temperature of 233.9°C (the saturation temperature at 30 bars). Since the average temperature during the overall heat-addition process has been lowered, the thermal efficiency calculated above should be less than that of a Carnot cycle operating between the maximum and minimum temperatures of the actual Rankine cycle. This is confirmed by noting that

$$\eta_{Carnot} = 1 - \frac{T_L}{T_H} = 1 - \frac{99.6 + 273.1}{233.9 + 273.1} = 0.265 \qquad \text{(or 26.5 percent)}$$

Example 17-1 Compute the thermal efficiency of an ideal Rankine cycle for which steam leaves the boiler as saturated vapor at 400 psia and is condensed at 14.7 psia.

SOLUTION The evaluation of the thermal efficiency requires calculating the turbine and pump work and the heat input. Using the notation employed in Fig. 17-2, we find the following data from the steam tables:

$$h_3 = h_g \text{ at 400 psia} = 1205.5 \text{ Btu/lb}$$

$$h_1 = h_f \text{ at 14.7 psia} = 180.2 \text{ Btu/lb}$$

The pump work is

$$w_{in,\,pump} = w_P = \frac{0.01614(400 - 14.7)(144)}{778} = 1.2 \text{ Btu/lb}$$

Therefore $$h_2 = h_1 + w_P = 180.2 + 1.2 = 181.4 \text{ Btu/lb}$$

The enthalpy at state 4 is the value for a pressure of 1 psia and for $s_4 = s_3 = s_g$ at 400 psia. Since $s_3 = 1.4856$ Btu/(lb)(°R),

$$1.4856 = (s_f + xs_{fg})_{\text{at 14.7 psia}} = 0.3121 + x(1.4446)$$

or $$x = \frac{1.1735}{1.4446} = 0.812 \qquad \text{(or 81.2 percent)}$$

Then $$h_4 = 180.2 + 0.812(970.4) = 968.2 \text{ Btu/lb}$$

Consequently,

$$w_{T,\,out} = h_3 - h_4 = 1205.5 - 968.2 = 237.3 \text{ Btu/lb}$$

$$q_{in} = h_3 - h_2 = 1205.5 - 181.4 = 1024.1 \text{ Btu/lb}$$

The thermal efficiency is

$$\eta_{th} = \frac{237.3 - 1.2}{1024.1} = 0.231 \qquad \text{(or 23.1 percent)}$$

Note that some of the heat addition occurs during process 2-2′ in Fig. 17-2, for which the temperature is lower than the maximum temperature of 444.7°F (the saturation temperature at 400 psia). Since the average temperature during the overall heating process has been lowered, the thermal efficiency calculated above should be less than that of a Carnot cycle operating between the maximum and minimum temperatures of the actual Rankine cycle. This is confirmed by noting that

$$\eta_{Carnot} = 1 - \frac{T_L}{T_H} = 1 - \frac{212 + 460}{445 + 460} = 0.257 \qquad \text{(or 25.7 percent)}$$

It is interesting to note from the preceding examples that the pump work is a very small fraction of the turbine work. In the analysis of the gas-turbine cycle introduced in Chap. 16, it is found that the compressor consumes a large fraction of the turbine output. The ratio of work input to work output is known as the back work ratio, and for the ideal Rankine cycle analyzed above this ratio is less than 0.01. This major difference between a Rankine vapor power cycle and a Brayton gas-turbine power cycle is due to the difference between the work required to compress a liquid and the work needed to compress a gas. Since the work required to compress a fluid reversibly over a given increment of pressure is directly proportional to the specific volume of the fluid [Eq. (7-28)], it takes considerably less work to compress a liquid than to compress a gas for the same pressure change.

The efficiency of the simple Rankine cycle discussed above may be increased, on the basis of the theoretical Carnot cycle, either by decreasing the temperature at which heat is rejected or by increasing the average temperature at which heat is added. The first of these effects is accomplished by lowering the exhaust pressure, which in turn lowers the value of T_4 or (T_1) shown in Fig. 17-2. This point is illustrated by Fig. 17-3. The crosshatched area enclosed by the states 1-2-2″-1″-4″-4-1 is a measure of the decrease in heat rejected when the exhaust pressure is lowered from P_4'' to P_4. This area is also a measure of the increase in net work output. The increase in heat input is measured by the total area under the curve 2-2″. Hence, as a result of the lower exhaust pressure, both the quantity of heat input required and the net work output increase. The overall effect of these two changes is an increase in thermal efficiency. The following examples illustrate this.

Example 17-2M Compute the thermal efficiency of an ideal Rankine cycle for which the steam leaves the boiler as saturated vapor at 30 bars and is condensed to 0.10 bar.

SOLUTION The evaluation of the thermal efficiency is carried out in a manner similar to that shown in Example 17-1M. The notation is the same as that used in Fig. 17-3, where the important states are 1, 2, 3, and 4. A summary of the calculations follows:

$$h_3 = h_g \text{ at 30 bars} = 2804.2 \text{ kJ/kg}$$

$$h_1 = h_f \text{ at 0.10 bar} = 191.8 \text{ kJ/kg}$$

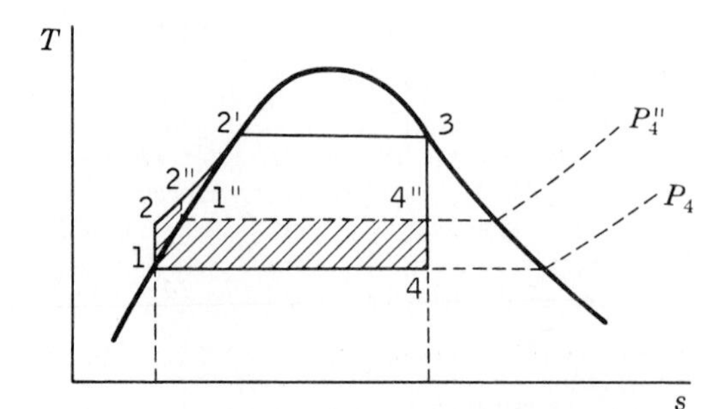

Figure 17-3 Effect of turbine-exhaust pressure on the efficiency of an ideal Rankine cycle.

$$w_{\text{in, pump}} = v_f(P_2 - P_1) = 1.01\,\frac{29.9}{10} = 3.0\ \text{kJ/kg}$$

$$h_2 = h_1 + w_P = 191.8 + 3.0 = 194.8\ \text{kJ/kg}$$

$$s_3 = 6.1869 = (s_f + xs_{fg})_{\text{at 0.1 bar}} = 0.6493 + x(8.1502 - 0.6493)$$

$$x = \frac{5.5376}{7.5009} = 0.738 \qquad \text{(or 73.8 percent)}$$

$$h_4 = h_f + xh_{fg} = 191.8 + 0.738(2392.8) = 1958.3\ \text{kJ/kg}$$

$$w_{\text{T, out}} = h_3 - h_4 = 2804.2 - 1958.3 = 845.9\ \text{kJ/kg}$$

$$q_{\text{in}} = h_3 - h_2 = 2804.2 - 194.8 = 2609.4\ \text{kJ/kg}$$

$$\eta_{\text{th}} = \frac{845.9 - 3.0}{2609.4} = 0.323 \qquad \text{(or 32.3 percent)}$$

In comparison to Example 17-1M, the thermal efficiency has increased from 23.6 to 32.3 percent, and the turbine-work output has increased from 565.8 kJ/kg to 845.9 kJ/kg. This is accomplished by lowering the exhaust pressure from 1 to 0.1 bar, which is equivalent to lowering the exhaust temperature from 100 to 46°C. The Carnot efficiency based on the maximum and minimum temperatures of this modified cycle will be found to be 37.1 percent.

Example 17-2 Compute the thermal efficiency of an ideal Rankine cycle for which steam leaves the boiler as saturated vapor at 400 psia and is condensed at 1 psia.

SOLUTION The evaluation of the thermal efficiency requires calcualting the turbine and pump work and the heat input. Using the notation seen in Fig. 17-3, we find the following data:

$$h_3 = h_g \text{ at 400 psia} = 1205.5\ \text{Btu/lb}$$

$$h_1 = h_f \text{ at 1 psia} = 69.7$$

The pump work is

$$w_{\text{in, pump}} = w_P = \frac{0.01614(400 - 1)(144)}{778} = 1.2\ \text{Btu/lb}$$

Therefore

$$h_2 = h_1 + w_P = 69.7 + 1.2 = 70.9\ \text{Btu/lb}$$

The enthalpy at state 4 is the value for a pressure of 1 psia and for $s_4 = s_3 = s_g$ at 400 psia. Since $s_3 = 1.4856$ Btu/(lb)(°R),

$$1.4856 = (s_f + xs_{fg})_{\text{at 1 psia}} = 0.1327 + x(1.8453)$$

or

$$x = \frac{1.3529}{1.8453} = 0.733 \qquad \text{(or 73.3 percent)}$$

Then

$$h_4 = 69.70 + 0.733(1036.0) = 829.3\ \text{Btu/lb}$$

Consequently,

$$w_{\text{T, out}} = h_3 - h_4 = 1205.5 - 829.3 = 376.2\ \text{Btu/lb}$$

$$q_{\text{in}} = h_3 - h_2 = 1205.5 - 70.85 = 1134.6\ \text{Btu/lb}$$

The thermal efficiency is

$$\eta_{th} = \frac{376.3 - 1.2}{1134.6} = 0.33 \qquad \text{(or 33 percent)}$$

In comparison to Example 17-1, the thermal efficiency has increased from 23.1 to 33 percent, and the turbine-work ouptput has increased from 237.3 to 376.2 Btu/lb. This is accomplished by lowering the exhaust pressure from 14.7 to 1 psia, which is equivalent to lowering the exhaust temperature from 212 to 102°F. The Carnot efficiency based on the maximum and minimum temperatures of this modified cycle will be found to be 37.9 percent.

The preceding examples illustrate the large effect of the exhaust pressure on the work output and thermal efficiency of a simple Rankine cycle. There is a limit, however, to the minimum pressure in the condenser. Heat is transferred from the condensing steam to cooling water or atmospheric air. The temperature of the cooling water or air normally would vary only over a very narrow range. Typically this might be 15 to 30°C, or 60 to 90°F. A temperature differential of 10 to 15°C, or 15 to 25°F, must exist across the heat-transfer surface in order to maintain adequate heat-transfer rates. Hence the minimum condensing temperature for the steam would range from 25 to 45°C or 75 to 115°F. From the saturated-steam tables, the saturation pressures corresponding to these ranges of temperature are roughly from 0.03 to 0.10 bar or 0.5 to 1.5 psia. Thus, fairly low pressures, well below atmospheric values, are possible in the condensers of modern steam power plants. However, this pressure varies over a fairly small range.

Although the effect of lowering the exhaust pressure is advantageous from the standpoint of increasing the thermal efficiency, it has the great disadvantage of increasing the moisture content (decreasing the quality) of the fluid leaving the turbine. This increased moisture content throughout the turbine decreases the efficiency of an actual turbine. In addition, the impingement of liquid droplets on the turbine blades may lead to a serious erosion problem. In practice, it is desirable to keep the moisture content less than about 10 percent at the low-pressure end of the turbine. It was also mentioned previously that increasing the average temperature at which heat is supplied would increase the efficiency of the Rankine cycle. Both the increase of the efficiency of the cycle by raising the temperature of the fluid entering the turbine and the removal of the moisture problem in the turbine can be met by the addition of a superheater to the simple Rankine cycle already presented. The process of superheating leads to a higher temperature at the turbine inlet without increasing the maximum pressure in the cycle. After the saturated vapor leaves the boiler, the fluid passes through another heat-input section, where the temperature is increased theoretically at constant pressure. The steam leaves the superheater at a temperature which is usually restricted only by metallurgical effects. Temperatures in the range of 540 to 600°C, or 1000 to 1100°F, are generally permissible. A Ts diagram and an equipment schematic for a Rankine cycle with superheating are shown in Fig. 17-4. The shaded area on the Ts diagram represents the additional net work output, and the area beneath the curve 3-3′ represents the additional heat added in the

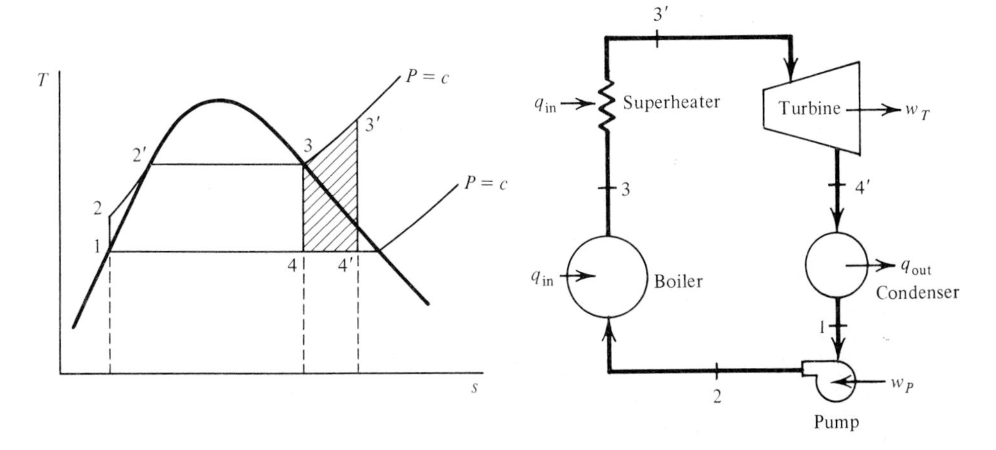

Figure 17-4 *Ts* diagram and equipment schematic for a Rankine cycle with superheating.

superheater section. Note that the average temperature at which heat is added during process 3-3′ is greater than that during the heat-addition process in the boiler section, whereas the temperature of rejection is still the same. On the basis of a Carnot-engine analysis, the efficiency of the cycle would be expected to have increased. The other important point to note is that the quality at state 4′ is considerably higher than that at state 4. Hence the moisture problem in the turbine has been alleviated.

Example 17-3M Determine the thermal efficiency and the change in the steam quality leaving the turbine of an ideal Rankine cycle with superheating, for which the turbine-inlet conditions are 30 bars and 500°C and the condenser pressure is 0.1 bar.

SOLUTION This example contains the same pressure conditions as Example 17-2M. However, the steam entering the turbine has been superheated from the saturation state of 234 to 500°C. The pump work will remain the same, i.e., 3.0 kJ/kg, but all other heat and work quantities are altered. The turbine-inlet enthalpy is now 3456.5 kJ/kg, and the entropy at this state is 7.2338 kJ/(kg)(°K). Hence, using h_2 from the preceding example, we find that

$$q_{in} = h'_3 - h_2 = 3456.5 - 194.8 = 3261.7 \text{ kJ/kg}$$

where the notation of Fig. 17-4 is being used. Calculation of the turbine work requires that the final isentropic state 4′ be found first. Since $s'_3 = s'_4 = 7.2338$ kJ/(kg)(°K),

$$7.2338 = (s_f + xs_{fg})_{\text{at 0.1 bar}} = 0.6493 + x(7.5009)$$

or

$$x = \frac{6.5845}{7.5009} = 0.878 \qquad \text{(or 87.8 percent)}$$

$$h'_4 = (h_f + xh_{fg})_{\text{at 0.1 bar}} = 191.8 + 0.878(2392.8) = 2292.7 \text{ kJ/kg}$$

Therefore

$$w_{T,\,in} = h'_3 - h'_4 = 3456.5 - 2292.7 = 1163.8 \text{ kJ/kg}$$

The thermal efficiency becomes

$$\eta_{th} = \frac{1163.8 - 3.0}{3261.7} = 0.356 \qquad \text{(or 35.6 percent)}$$

The result of superheating is an increase in the thermal efficiency from 31.4 to 35.6 percent, and an increase in the quality of the steam leaving the turbine from 74.8 to 87.8 percent. Although this latter value of the quality is still objectionable, it must be remembered that the above calculations are based on isentropic flow through the turbine. The presence of irreversibilities in the flow, although decreasing the work output of the turbine, will be found to increase the quality at the outlet. As another point, the Carnot efficiency based on the maximum and minimum temperatures of the superheat cycle (500 and 45.8°C) is now 58.7 percent. This is considerably above the 35.6 percent for the actual ideal cycle. Because heat is added to the fluid at 500°C only at the very end of the heat-addition process, the average temperature of heat addition is well below 500°C. Consequently, the cycle efficiency begins to deviate considerably from that predicted for the Carnot cycle.

Example 17.3 Determine the thermal efficiency of an ideal Rankine cycle with superheating, for which the turbine-inlet conditions are 400 psia and 900°F and the condenser pressure is 1 psia.

SOLUTION This example contains the same pressure conditions as Example 17-2. However, the steam entering the turbine has been superheated from the saturation state of 444.6 to 900°F. The pump work will remain the same, i.e., 1.2 Btu/lb, but all other heat and work quantities are altered. The turbine-inlet enthalpy is now 1470.1 Btu/lb, and the entropy in this state is 1.7252 Btu/(lb)(°R). Hence, using h_2 from the preceding example, we find that

$$q_{\text{in, tot}} = h'_3 - h_2 = 1470.1 - 70.9 = 1399.2 \text{ Btu/lb}$$

Calculation of the turbine work requires that the final isentropic state at 4′ be found first. Since $s'_3 = s'_4 = 1.7252$ Btu/(lb)(°R),

$$1.7252 = (s_f + xs_{fg})_{\text{at 1 psia}} = 0.1327 + x(1.8453)$$

or

$$x = 0.863 \qquad \text{(or 86.3 percent)}$$

$$h'_4 = (h_f + xh_{fg})_{\text{at 1 psia}} = 69.7 + 0.863(1036.0) = 964.0 \text{ Btu/lb}$$

Therefore

$$w_{\text{T, in}} = h'_3 - h'_4 = 1470.1 - 964.0 = 506.1 \text{ Btu/lb}$$

The thermal efficiency becomes

$$\eta_{\text{th}} = \frac{506.1 - 1.2}{1399.2} = 0.361 \qquad \text{(or 36.1 percent)}$$

The result of superheating is an increase in the thermal efficiency from 33 to 36 percent and an increase in the quality of steam leaving the turbine from 73.3 to 86.3 percent. Although this latter quality is still objectionable, it must be remembered that the calculations above are based on isentropic flow through the turbine. The presence of irreversibilities in the flow, although decreasing the work output of the turbine, will be found to increase the outlet enthalpy even further, thus eliminating the moisture problems to a large extent. As another point, the Carnot efficiency based on the maximum and minimum temperatures of the superheat cycle (900 and 101.7°F) is now 58.7 percent. This is considerably above the 36.1 percent for the actual ideal cycle. Because heat is added to the fluid at 900°F only at the very end of the heat-addition process, the average temperature of heat addition is well below 900°F. Consequently, the cycle efficiency begins to deviate considerably from that predicted for the Carnot cycle.

17-3 THE REHEAT CYCLE

In the ideal Rankine cycle the efficiency may be increased by the use of a superheater section. The process of superheating in general raises the average temperature at which heat is supplied to the cycle, thus raising the theoretical efficiency. An equivalent gain in the average temperature during the heat-input process may be accomplished by raising the maximum pressure of the cycle, i.e., the boiler pressure. This may result in a higher initial cost of the steam generator (boiler plus superheater), because of the higher pressure that must be contained, but over a period of years the higher efficiency of the overall unit may more than compensate for this factor. However, for a given maximum temperature in the steam generator, an increase in the generator pressure results in a decrease in the quality of the steam leaving the turbine. To avoid the erosion problem and still take advantage of the higher temperatures made available to increasing the boiler pressure, the reheat cycle has been developed.

In the reheat cycle the steam is not allowed to expand completely to the condenser pressure in a single stage. After partial expansion the steam is withdrawn from the turbine and reheated at constant pressure. Then it is returned to the turbine for further expansion to the exhaust pressure. The turbine may be considered to consist of two stages, a high-pressure one and a low-pressure one. Figure 17-5 illustrates the reheat cycle on a *Ts* diagram and the equipment schematic. The position of state 4 after the first stage of expansion is usually close to the saturation line. The temperature upon reheating to state 5 in Fig. 17-5 is usually equal to or slightly less than the inlet temperature to the first stage of the turbine.

Extreme care must be used in selecting the path 4-5 for reheating, because the average temperature for the reheat process may turn out to be less than the average temperature of the heat-addition process 2-3. Thus reheating does not necessarily increase the thermal efficiency of the basic Rankine cycle. Correct use of reheating will, however, remove the objectionable feature of high moisture content at the turbine exhaust as well as increase the thermal efficiency. In addition, there is a reheat pressure which will maximize the thermal efficiency for the given values of P_3, T_3, T_5, and P_6 in Fig. 17-5. For conventional values of these parameters the maximum efficiency of an ideal reheat cycle typically occurs

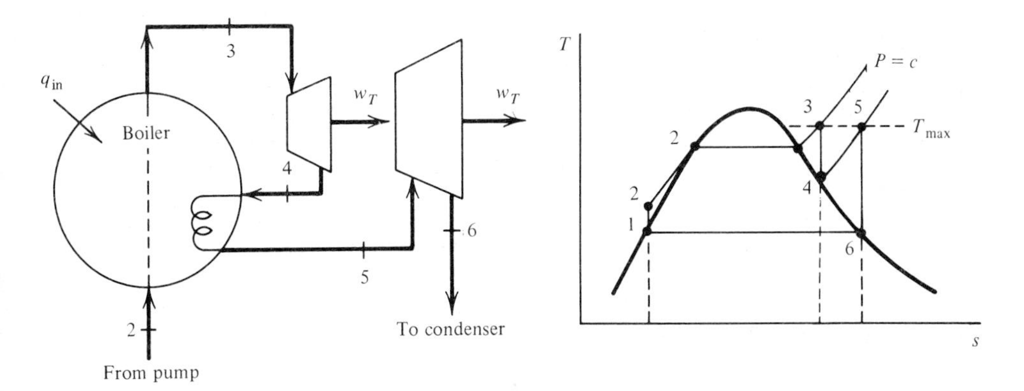

Figure 17-5 Equipment schematic and *Ts* diagram for an ideal reheat vapor cycle.

when the ratio P_4/P_3 is in the range of 0.15 to 0.35. In computing the thermal efficiency of a reheat cycle, one must remember to account for the work output of both stages of the turbine, as well as the heat inputs to both the boiler-superheater section and the reheat section. In terms of the notation of Fig. 17-5, the thermal efficiency is

$$\eta_{th} = \frac{(h_3 - h_4) + (h_5 - h_6) - w_P}{(h_3 - h_2) + (h_5 - h_4)} \qquad \text{(reheat cycle)} \qquad (17\text{-}1)$$

The pump work is found from $v\ \Delta P$.

Example 17-4M In a steam power plant utilizing the reheat cycle, the turbine-inlet condition is 30 bars and 500°C. After expansion to 5.0 bars, the steam is reheated to 500°C and then expanded to the condenser pressure of 0.1 bar. Compute the cycle efficiency and the state of the steam at the outlet of the turbine.

SOLUTION The problem is based on the data of Example 17-3M, except for the addition of the reheat section. From the preceding example, the following data are still valid. The subscripts on the property values are consistent with the notation in Fig. 17-5.

$$h_1 = 191.8 \text{ kJ/kg}$$

$$h_2 = 194.8 \text{ kJ/kg}$$

$$h_3 = 3456.5 \text{ kJ/kg}$$

$$w_P = h_2 - h_1 = 3.0 \text{ kJ/kg}$$

$$q_{\text{boiler-superheater}} = h_3 - h_2 = 3261.7 \text{ kJ/kg}$$

$$s_3 = 7.2338 \text{ kJ/(kg)(°K)}$$

From the last value we can estimate the enthalpy at state 4, where the pressure is 5 bars. The entropy of saturated vapor at 5 bars is 6.8213 kJ/(kg)(°K). Therefore state 4 is still in the superheat region. From superheat table A-14M it is found by linear interpolation that T_4 is close to 241°C and that h_4 is 2941.6 kJ/kg. State 5 is at 5 bars and 500°C, for which $h_5 = 3483.9$ kJ/kg and $s_5 = 8.0873$ kJ/(kg)(°K). Finally, state 6 is the result of isentropic expansion from state 5 to 0.1 bar. From the saturation table at 0.1 bar

$$8.0873 = (s_f + xs_{fg})_{\text{at 0.1 bar}} = 0.6493 + x(8.5009)$$

or

$$x = \frac{8.0873 - 0.6493}{7.5009} = 0.992 \qquad \text{(or 99.2 percent)}$$

Hence

$$h_6 = h_f + xh_{fg} = 191.8 + 0.992(2392.8) = 2565.5 \text{ kJ/kg}$$

Consequently,

$$w_{\text{T, high pressure}} = h_3 - h_4 = 3456.5 - 2941.6 = 514.9 \text{ kJ/kg}$$

$$w_{\text{T, low pressure}} = h_5 - h_6 = 3483.9 - 2565.5 = 918.4 \text{ kJ/kg}$$

$$q_{\text{reheat}} = h_5 - h_4 = 3483.9 - 2941.6 = 542.3 \text{ kJ/kg}$$

The thermal efficiency, then, is

$$\eta_{th} = \frac{514.9 + 918.4 - 3.0}{3261.7 + 542.3} = 0.376 \qquad \text{(or 37.6 percent)}$$

Compared with Example 17-3M, without reheat, the thermal efficiency has increased only from 35.6 to 37.6 percent. However, the outlet quality of the steam at the turbine exhaust has increased from 87.8 to 99.2 percent.

Example 17-4 In a steam power plant utilizing the reheat cycle, the turbine-inlet condition is 400 psia and 900°F. After expansion to 60 psia, the steam is reheated to 900°F, and then expanded to the condenser pressure of 1 psia. Compute the cycle efficiency and the net work output per pound of steam.

SOLUTION The problem is based on the data of Example 17-3, except for the addition of the reheat section. From the preceding example, the following data are still valid. The subscripts on the property values are consistent with the notation in Fig. 17-5.

$$h_1 = 69.7 \text{ Btu/lb}$$

$$h_2 = 70.9 \text{ Btu/lb}$$

$$h_3 = 1470.1 \text{ Btu/lb}$$

$$w_P = h_2 - h_1 = 1.2 \text{ Btu/lb}$$

$$q_{\text{boiler-superheater}} = h_3 - h_2 = 1399.2 \text{ Btu/lb}$$

$$s_3 = 1.7252 \text{ Btu/(lb)(°R)}$$

From the last value we can estimate the enthalpy at state 4, where the pressure is 60 psia. The entropy of saturated vapor at 60 psia is 1.6443 Btu/(lb)(°R). Therefore state 4 is still in the superheat region. From Table A-14 it is found that T_4 is close to 422°F and that h_4 is 1244.2 Btu/lb. State 5 is at 60 psia and 900°F, for which $h_5 = 1481.8$ Btu/lb and $s_5 = 1.9408$ Btu/(lb)(°R). Finally, state 6 is the result of isentropic expansion from state 5 to 1 psia. From the saturation tables at 1 psia,

$$1.9408 = s_f + xs_{fg} = 0.1327 + x(1.8453)$$

or

$$x = 0.980 \qquad \text{(or 98 percent)}$$

Hence

$$h_6 = h_f + xh_{fg} = 69.7 + 0.98(1036.0) = 1085.0 \text{ Btu/lb}$$

Consequently,

$$w_{\text{T, high pressure}} = h_3 - h_4 = 1470.1 - 1244.2 = 225.9 \text{ Btu/lb}$$

$$w_{\text{T, low pressure}} = h_5 - h_6 = 1481.8 - 1085.0 = 396.8 \text{ Btu/lb}$$

$$q_{\text{reheat}} = h_5 - h_4 = 1481.1 - 1244.2 = 237.6 \text{ Btu/lb}$$

The thermal efficiency, then is

$$\eta_{\text{th}} = \frac{225.9 + 396.8 - 1.2}{1399.2 + 237.6} = 0.379 \qquad \text{(or 37.9 percent)}$$

Compared to Example 17-3, without reheat, the efficiency has increased only from 36.1 to 37.9 percent. However, the outlet quality of the steam at the turbine exhaust has increased from 86.3 to 98 percent.

17-4 THE REGENERATIVE CYCLE

Referring to Fig. 17-4 for the simple Rankine cycle with superheating reveals a serious disadvantage of the basic cycle. For the portion of the heat-addition process 2-2′, the average temperature is much below the temperature of the vaporization and superheating process 2′-3-3′. From the viewpoint of the second law, the cycle efficiency is greatly reduced as a result of this relatively low-temperature heat-addition process. If the average temperature for this portion of the heat-addition process could be raised, the efficiency of the cycle would more nearly approach that of the Carnot cycle. One practical method of accomplishing this is by the use of a regeneration process internal to the overall cycle.

The ideal regeneration vapor power cycle, shown in Fig. 17-6, is accomplished in the following way: Part of the superheated steam which enters the turbine at state 3 is bled or extracted from the turbine at state 4, which is an intermediate state in the turbine expansion process. The extracted steam is directed into a heat exchanger known as a feedwater heater. The portion of the steam which is not extracted expands completely to the condenser pressure (state 5), and it is then condensed to saturated liquid at state 6. A pump then increases the pressure of the liquid leaving the condenser isentropically to the same pressure as that of the extracted steam. The compressed liquid at state 7 then enters the feedwater heater, where it mixes directly with the flow stream extracted from the turbine. Because of this direct-mixing process, the feedwater heater in Fig. 17-6 is called an *open* or *direct-contact* type of heater. In the ideal situation the mass flow rates for the two streams entering the heater are adjusted so that the state of the mixture leaving the heater is a saturated liquid at the heater pressure (state 1). A second pump then isentropically raises the pressure of the liquid to state 2, which corresponds to the steam-generator pressure.

a Open Feedwater Heater

For the ideal, internally reversible cycle the areas beneath the curves on the Ts diagram in Fig. 17-6 represent heat-addition or heat-rejection quantities. One

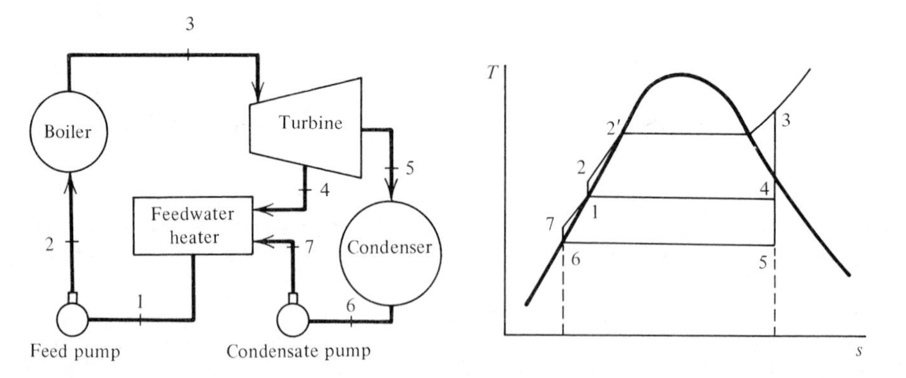

Figure 17-6 Equipment schematic and Ts diagram for an ideal regenerative vapor cycle with one open feedwater heater.

must keep in mind, however, that the mass flow rates through certain parts of the cycle are not equal to the total mass flow rate through the steam generator. For example, the area under line 5-6 is a measure of the heat rejected in the condenser, but only for each pound which passes through the condenser. Before presenting a numerical example for an ideal regenerative cycle, it is convenient to employ the conservation of mass and energy principles to a control volume which lies just inside the open feedwater heater. On the basis of Fig. 17-6 we may write

$$\dot{m} = \dot{m}_4 + \dot{m}_7$$

and

$$\dot{m}_1 h_1 = \dot{m}_4 h_4 + \dot{m}_7 h_7$$

Combining the two equations so that $\dot{m}_7$ is eliminated, one observes that

$$\dot{m}_1 h_1 = \dot{m}_4 h_4 + (\dot{m}_1 - \dot{m}_4) h_7$$

or

$$h_1 = \frac{\dot{m}_4}{\dot{m}_1} h_4 + \left(1 - \frac{\dot{m}_4}{\dot{m}_1}\right) h_7$$

If the fraction of the steam extracted from the turbine $\dot{m}_4/\dot{m}_1$ at state 4 is represented by y_4, then

$$1(h_1) = y_4 h_4 + (1 - y_4) h_7 \qquad \text{or} \qquad y_4 = \frac{h_1 - h_7}{h_4 - h_7} \qquad \text{(open heater)} \tag{17-2}$$

Since the enthalpies of the flow streams entering and leaving the open feedwater heater are known for the ideal cycle, the above relation permits the evaluation of the quantity of steam which must be extracted from the turbine at state 4 to ensure that the flow stream leaving the heater will be a saturated liquid.

It is important to note that both the work output of the turbine and the work input to the (condensate) pump between states 6 and 7 in Fig. 17-6 are affected by the fraction of the steam bled from the turbine at state 4. On the basis of a unit mass passing through the boiler-superheater section, the total turbine work output is

$$w_{T,\,\text{out}} = 1(h_3 - h_4) + (1 - y_4)(h_4 - h_5) \tag{17-3}$$

On the same basis the condensate pump work under isentropic conditions is

$$w_{P,\,6\text{-}7} = v(P_7 - P_6)(1 - y_4) \tag{17-4}$$

The (feed) pump work input between states 1 and 2 is still given by $v\,\Delta P$, because the total flow passes through this pump.

Example 17-5M An ideal regenerative steam-power cycle operates so that steam enters the turbine at 30 bars and 500°C and exhausts at 0.1 bar. A single, open feedwater heater is employed which operates at 5 bars. Compute the thermal efficiency of the cycle.

SOLUTION The Ts and flow diagrams for the cycle are given in Fig. 17-6. From the preceding examples the following data are still valid:

$$P_3 = 30 \text{ bars}, \ T_3 = 500°\text{C}, \ h_3 = 3456.5 \text{ kJ/kg}$$

$$P_4 = 5 \text{ bars}, \ s_4 = s_3, \ h_4 = 2941.6 \text{ kJ/kg}$$

$$P_5 = 0.1 \text{ bar}, \ s_5 = s_3, \ h_5 = 2292.7 \text{ kJ/kg}$$

$$P_6 = 0.1 \text{ bar, saturated liquid}, \ h_6 = 191.8 \text{ kJ/kg}$$

In addition, state 1 is a saturated-liquid state at 5 bars; thus $h_1 = h_f$ at 5 bars = 640.2 kJ/kg. The enthalpies at states 2 and 7 can be found by adding the pump work (w_P) to the enthalpies at states 1 and 6, respectively. The pump work is approximated by the value of $v \, \Delta P$. Hence

$$w_{P,\,1\text{-}2} = 1.09(25)\frac{10^5}{10^6} = 2.7 \text{ kJ/kg}$$

$$h_2 = h_1 + w_{P,\,1\text{-}2} = 640.2 + 2.7 = 642.9 \text{ kJ/kg}$$

$$w_{P,\,6\text{-}7} = 1.01(4.9)\frac{10^5}{10^6} = 0.5 \text{ kJ/kg}$$

$$h_7 = h_6 + w_{P,\,6\text{-}7} = 191.8 + 0.5 = 192.3 \text{ kJ/kg}$$

In order to determine the turbine work output, it is necessary to find, first, the fraction of the steam which is extracted at 5 bars. Substitution of the enthalpy values into Eq. (17-2) yields

$$y_4 = \frac{h_1 - h_7}{h_4 - h_7} = \frac{640.2 - 192.3}{2941.6 - 192.3} = 0.163$$

Therefore 0.163 kg of steam is extracted for every kilogram which enters the turbine. The work output of the turbine becomes

$$w_{T,\,\text{out}} = h_3 - h_4 + (1 - y_4)(h_4 - h_5)$$

$$= (3456.5 - 2941.6) + 0.837(2941.6 - 2292.7)$$

$$= 514.9 + 543.1 = 1058.0 \text{ kJ/kg}$$

On the basis of a unit mass passing through the boiler, the total pump work is

$$w_{P,\,\text{tot}} = 2.7 + 0.837(0.5) = 3.1 \text{ kJ/kg}$$

and the heat input is

$$q_{\text{in}} = h_3 - h_2 = 3456.5 - 642.9 = 2813.6 \text{ kJ/kg}$$

Consequently, the thermal efficiency for the cycle is

$$\eta_{\text{th}} = \frac{w_{\text{net}}}{q_{\text{in}}} = \frac{1058.0 - 3.1}{2813.6} = 0.375 \qquad \text{(or 37.5 percent)}$$

In Example 17-3 the conditions given were the same as for the above cycle, except that no regeneration was used. In that case the thermal efficiency was 35.6 percent. Thus the inclusion of one open feedwater heater in the ideal cycle has increased the cycle efficiency by some 5.3 percent. Note, however, that the work output of the turbine has been lowered by nearly 10 percent as compared to the case without regeneration.

Example 17-5 An ideal regenerative steam power cycle operates so that steam enters the turbine at 400 psia and 900°F and exhausts at 1 psia. A single, open feedwater heater is employed which operates at 60 psia. Compute the thermal efficiency of the cycle.

SOLUTION The Ts and flow diagrams for the cycle are given in Fig. 17-6. From the preceding examples the following data are still valid:

$$P_3 = 400 \text{ psia}, \ T_3 = 900°\text{F}, \ h_3 = 1470.1 \text{ Btu/lb}$$

$$P_4 = 60 \text{ psia}, \ s_4 = s_3, \ h_4 = 1244.2 \text{ Btu/lb}$$

$$P_5 = 1 \text{ psia}, \ s_5 = s_3, \ h_5 = 964.0 \text{ Btu/lb}$$

$$P_6 = 1 \text{ psia, saturated liquid}, \ h_6 = 69.7 \text{ Btu/lb}$$

In addition, state 1 is a saturated-liquid state at 60 psia; thus $h_1 = 262.2$ Btu/lb. The enthalpies at states 2 and 7 can be found by adding the pump work to the enthalpies at states 1 and 6, respectively. The pump work (w_P) can be approximated again by the value of $v \, \Delta P$. Hence

$$w_{P,\,1\text{-}2} = \frac{0.01727(400 - 60)(144)}{778} = 1.1 \text{ Btu/lb}$$

$$h_2 = h_1 + w_{P,\,1\text{-}2} = 262.2 + 1.1 = 263.3 \text{ Btu/lb}$$

$$w_{P,\,6\text{-}7} = \frac{0.01614(60 - 1)(144)}{778} = 0.2 \text{ Btu/lb}$$

$$h_7 = h_6 + w_{P,\,6\text{-}7} = 69.7 + 0.2 = 69.9 \text{ Btu/lb}$$

In order to determine the turbine work output, it is necessary to find, first, the fraction of the steam which is extracted at 60 psia. Substitution of the enthalpy values into Eq. (17-2) yields

$$h_1 = yh_4 + (1 - y)h_7$$

$$262.2 = y(1244.2) + (1 - y)(69.9)$$

$$y = 0.164$$

Therefore 0.164 lb of steam is extracted for every pound which enters the turbine. The work output of the turbine becomes

$$w_{T,\,\text{out}} = h_3 - h_4 + (1 - y)(h_4 - h_3)$$

$$= (1470.1 - 1244.2) + 0.836(1244.2 - 964.0)$$

$$= 225.9 + 234.2 = 460.1 \text{ Btu/lb}$$

The total pump work is

$$w_{P,\,\text{tot}} = 1.1 + 0.2(0.836) = 1.3 \text{ Btu/lb}$$

and the heat input is

$$q_{\text{in}} = h_3 - h_2 = 1470.1 - 263.3 = 1206.8 \text{ Btu/lb}$$

Consequently, the thermal efficiency for the cycle is

$$\eta_{\text{th}} = \frac{w_{\text{net}}}{q_{\text{in}}} = \frac{460.1 - 1.3}{1206.8} = 0.381 \qquad \text{(or 38.1 percent)}$$

In Example 17-3 the conditions given were the same as for the above cycle, except that no regeneration was used. In that case the thermal efficiency was 36.1 percent. Thus the inclusion of one open feedwater heater in the ideal cycle has increased the cycle efficiency by some 5.5 percent. In modern high-pressure steam power plants several open feedwater heaters operating at different pressures may be employed.

b Closed Feedwater Heater

A modification of the regenerative cycle described above occurs when *closed* feedwater heaters are used to preheat the water being returned from the condenser to the boiler. As the term might imply, in a closed feedwater heater the two entering flow streams do not mix directly. The feedwater leaving the condenser flows inside tubes that pass through the heater. The steam extracted from the turbine enters the heater and condenses on the outside of the tubes which carry the feedwater. A schematic diagram of a closed feedwater heater is shown in Fig. 17-7. Two alternatives are shown for removal of the condensate. A pump may be used to return the condensate directly back into the feedwater line (line *a*), or the condensate may be collected in a trap which permits only liquid to flow to another region of lower pressure (line *b*). This lower-pressure region may be the condenser itself or another feedwater heater. On the basis of Fig. 17-7 the steady-flow energy balance on the closed feedwater heater may be written as $(\dot{m}\,\Delta h)_{\text{extracted}} = (\dot{m}\,\Delta h)_{\text{feedwater}}$. Note that the $\dot{m}$ values in this equation are not the same.

In the ideal closed heater the extracted steam condenses and leaves the feedwater heater as a saturated liquid at the turbine extraction pressure. The feedwater from the condenser is assumed to leave as a compressed liquid at the same temperature as the condensed extracted steam in the ideal case. One advantage of a closed feedwater heater is that the pressures of the extracted steam and of the feedwater can be significantly different. Due to their type of construction, closed heaters usually operate at higher pressures than the open type. However, since open feedwater heaters operate at fairly low pressures, they are less expensive. Another advantage of the open heater is that it brings the feedwater up to its saturation temperature at the heater pressure.

c Optimum Extraction Pressure

When one considers the open-heater regenerative cycle shown in Fig. 17-6, the question arises as to the appropriate extraction pressure P_4 to be used. The extraction pressure must lie between the boiler-superheater pressure P_3 and the condenser pressure P_5. If the steam is extracted at the limiting conditions of either before the turbine inlet at P_3 or after the turbine at P_5, then it will be

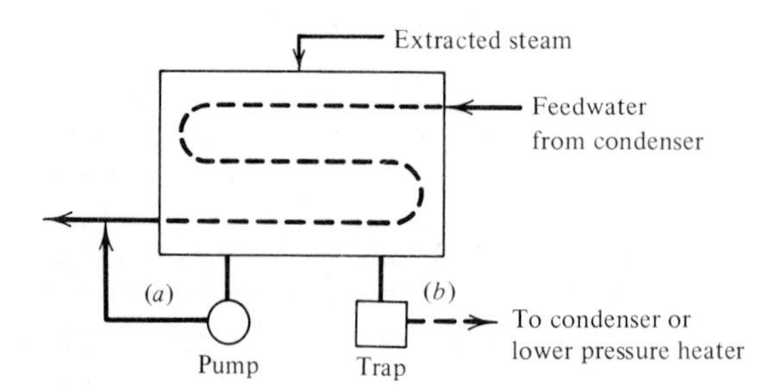

Figure 17-7 Schematic of a closed feedwater heater, with alternatives for condensate removal.

found that the thermal efficiency is unaffected by the presence of the heater. Since regeneration does increase the thermal efficiency, there must exist an optimum pressure which will maximize the thermal efficiency for given values of P_3, T_3, and P_5. For the case of one open heater it may be shown that the feedwater exit temperature at state 1 should be close to halfway between the saturation temperature in the boiler and the saturation temperature in the condenser.

Example 17-6M Reconsider the ideal regenerative cycle in Example 17-5M. Estimate the optimum extraction pressure for the cycle.

SOLUTION The optimum extraction pressure will be based on the temperature which is halfway between the boiler and condenser saturation temperatures. At 30 and 0.10 bars the saturation temperatures are 234 and 46°C, respectively. The average of these values is 140°C. The saturation pressure corresponding to this temperature is 3.6 bars, and this pressure would be a good estimate of the extraction pressure to use to optimize the thermal efficiency.

Example 17-6 Reconsider the ideal regenerative cycle in Example 17-5. Estimate the optimum extraction pressure for the cycle.

SOLUTION The optimum extraction pressure will be based on the temperature which is halfway between the boiler and condenser saturation temperatures. At 400 and 1 psia the saturation temperatures are 445 and 102°F, respectively. The average of these values is 274°F, and the saturation pressure corresponding to this temperature is roughly 45 psia. This pressure would be a good estimate of the extraction pressure to use to optimize the thermal efficiency.

d Multiple Use of Feedwater Heaters

Modern high-pressure steam power plants usually employ several open and closed feedwater heaters in a given cycle. The maximum number is around 6 to 8. Although the thermal efficiency does increase as the number of heaters is increased, so does the capital costs. Therefore, beyond a certain point the use of additional heaters cannot be economically justified. Methods of optimizing a regenerative cycle with two or more heaters with respect to using the proper extraction pressures are well known but beyond the scope of this text. The examples below illustrate a cycle with one closed and one open heater. The method of analyzing a closed heater is examined in detail. A schematic of the equipment and a *Ts* diagram for the cycle are shown in Fig. 17-8. Note that the process also includes reheat after the first extraction point. Recall that reheat helps prevent excessive moisture content in the turbine exhaust stream.

Example 17-7M An ideal, regenerative steam power cycle operates so that steam enters the turbine at 30 bars and 500°C and exhausts at 0.1 bar. Steam is bled from the turbine at 10 bars for a closed feedwater heater and at 5 bars for an open heater. Condensate from the closed heater is pumped to 30 bars and joins the feedwater stream after it leaves the closed heater. Compute the thermal efficiency of the cycle if the steam is reheated to 500°C at the 10-bar extraction point.

SOLUTION The *Ts* and flow diagrams for the cycle are shown in Fig. 17-8. The following data are still valid from Example 17-5M, except the numbering system has been altered. Based on the

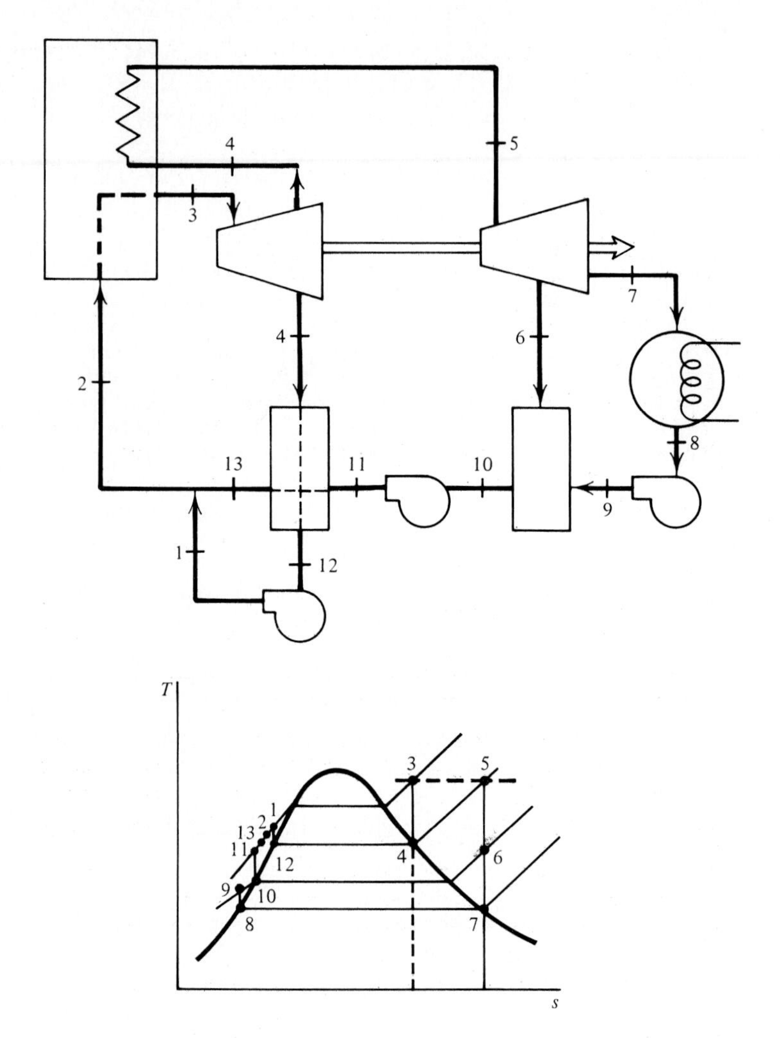

Figure 17-8 Schematic of equipment and Ts diagram for a regenerative-reheat cycle with one closed and one open feedwater heater.

notation of Fig. 17-8,

$$P_3 = 30 \text{ bars}, \; T_3 = 500°\text{C}, \; h_3 = 3456.5 \text{ kJ/kg}$$

$$P_8 = 0.1 \text{ bar, saturated liquid}, \; h_8 = 191.8 \text{ kJ/kg}$$

$$w_{P,\,8\text{-}9} = 1.01(4.9)\frac{10^5}{10^6} = 0.5 \text{ kJ/kg}$$

$$h_9 = h_8 + w_{P,\,8\text{-}9} = 191.8 + 0.5 = 192.3 \text{ kJ/kg}$$

$$P_{10} = 5 \text{ bars, saturated liquid}, \; h_{10} = 640.2 \text{ kJ/kg}$$

$$w_{P,\,10\text{-}11} = 1.09(25)\frac{10^5}{10^6} = 2.7 \text{ kJ/kg}$$

$$h_{11} = h_{10} + w_{P,\,10\text{-}11} = 640.2 + 2.7 = 642.9 \text{ kJ/kg}$$

In addition, the enthalpy values for states 4, 5, 6, 7, 12, 13, and 1 may be found from the tables or by direct calculations.

$$P_4 = 10 \text{ bars},\ s_4 = s_3 = 7.2338 \text{ kJ/(kg)(°K)},\ h_4 = 3116.9 \text{ kJ/kg}$$

$$P_5 = 10 \text{ bars},\ T_5 = 500\text{°C},\ h_5 = 3478.5 \text{ kJ/kg}$$

$$P_6 = 5 \text{ bars},\ s_6 = s_5 = 7.7622 \text{ kJ/(kg)(°K)},\ h_6 = 3251.3 \text{ kJ/kg}$$

$$P_7 = 0.1 \text{ bar},\ s_7 = s_5,\ h_7 = 2460.9 \text{ kJ/kg}$$

$$P_{12} = 10 \text{ bars, saturated liquid},\ h_{12} = 762.8 \text{ kJ/kg}$$

$$P_{13} = 30 \text{ bars},\ T_{13} = T_{12} = 179.9\text{°C, compressed liquid},\ h_{13} = 765.0 \text{ kJ/kg}$$

$$w_{P,\,12\text{-}1} = 1.13(20)\frac{10^5}{10^6} = 2.3 \text{ kJ/kg}$$

$$h_1 = h_{12} + w_{P,\,12\text{-}1} = 762.8 + 2.3 = 765.1 \text{ kJ/kg}$$

At this point every enthalpy value is known except that for state 2. Generally this is found from energy and mass balances around the point where streams 1 and 13 enter and stream 2 leaves. That is,

$$\dot{m}_1 h_1 + \dot{m}_{13} h_{13} = \dot{m}_2 h_2 \qquad \text{and} \qquad \dot{m}_1 + \dot{m}_{13} = \dot{m}_2$$

However, the split in total flow between streams 1 and 13 has not yet been determined. Once this split is found, then state 2 may be evaluated. In this particular case a numerical solution is available without determining the fraction bled at state 4. This occurs because the enthalpies at states 1 and 13 are essentially identical, that is, 765.0 and 765.1 kJ/kg. Thus, regardless of the split in flow, h_2 after mixing will also be 765.0 kJ/kg.

The fractional flow extracted to the two heaters is important, however, for the evaluation of the turbine and pump work and the heat input. The fraction of flow through the boiler-superheater section which is extracted at state 4 is found from energy and mass balances around the closed heater.

$$\dot{m}_1(h_4 - h_{12}) = \dot{m}_{13}(h_{13} - h_{11}) \qquad \text{and} \qquad \dot{m}_1 + \dot{m}_{13} = \dot{m}_3$$

If we let $\dot{m}_4/\dot{m}_3 = \dot{m}_1/\dot{m}_3 = y_4$ and $\dot{m}_{13}/\dot{m}_3 = 1 - y_4$, then

$$y_4(h_4 - h_{12}) = (1 - y_4)(h_{13} - h_{11})$$

$$y_4(3116.9 - 762.8) = (1 - y_4)(765.0 - 642.9)$$

$$y_4 = 0.0493$$

Hence 4.93 percent of the total flow around the cycle is bled to the closed heater. An energy balance around the open heater, similar to Eq. (17-2), will lead to the fraction of the flow bled to the open heater.

$$\dot{m}_6 h_6 + \dot{m}_9 h_9 = \dot{m}_{10} h_{10} \qquad \text{and} \qquad \dot{m}_6 + \dot{m}_9 = \dot{m}_{10}$$

If we let $\dot{m}_6/\dot{m}_{10} = z_6$ and $\dot{m}_9/\dot{m}_{10} = (1 - z_6)$, then

$$z_6(3251.3) + (1 - z_6)(192.3) = 1(640.2)$$

$$z_6 = 0.1464$$

Thus 14.64 percent of the flow reheated and passing through the second stage of the turbine is extracted to the open heater. Note that the symbol z_6 has been used rather than y_6, because we wish to let y_6 represent the fraction of the *total flow* bled to the open heater. It can easily be shown that $y_6 = 0.1392$ on the basis of the values of y_4 and z_6.

Finally we may evaluate the turbine and pump work and the heat input.

$$w_{T,\,out} = 1(h_3 - h_4) + (1 - y_4)(h_5 - h_6) + (1 - y_4 - y_6)(h_6 - h_7)$$

$$= (3456.5 - 3116.9) + 0.9507(3478.5 - 3251.3) + 0.8115(3251.3 - 2460.9)$$

$$= 339.6 + 216.0 + 641.4 = 1197.0 \text{ kJ/kg}$$

$$w_P = w_{P,\,8\text{-}9}(1 - y_4 - y_6) + w_{P,\,10\text{-}11}(1 - y_4) + w_{P,\,12\text{-}1}y_4$$

$$= 0.5(0.8115) + 2.7(0.9507) + 2.3(0.0493) = 3.1 \text{ kJ/kg}$$

$$q_{in} = 1(h_3 - h_2) + (1 - y_4)(h_5 - h_4)$$

$$= (3456.5 - 765.0) + 0.9507(3478.5 - 3116.9) = 3035.3 \text{ kJ/kg}$$

As a result of the above calculations the thermal efficiency becomes

$$\eta_{th} = \frac{w_{T,\,out} - w_P}{q_{in}} = \frac{1197.0 - 3.1}{3035.3} = 0.393 \qquad \text{(or 39.3 percent)}$$

In Example 17-5M the efficiency was found to be 37.5 percent for a single open heater without reheat. The increase to 39.3 percent in this example is due both to a second heater in the cycle and to the reheat section.

Example 17-7 An ideal, regenerative steam power cycle operates so that steam enters the turbine at 400 psia and 900°F and exhausts at 1 psia. Steam is bled from the turbine at 120 psia for a closed feedwater heater and at 60 psia for an open heater. Condensate from the closed heater is pumped to 400 psia and joins the feedwater stream after it leaves the closed heater. Compute the thermal efficiency of the cycle if the steam is reheated to 900°F at the 120 psia extraction point.

SOLUTION The Ts and flow diagrams for the cycle are shown in Fig. 17-8. The following data are still valid from Example 17-5, except that the numbering system has been altered. Based on the notation of Fig. 17-8,

$$P_3 = 400 \text{ psia}, \; T_3 = 900°\text{F}, \; h_3 = 1470.1 \text{ Btu/lb}$$

$$P_8 = 1 \text{ psia, saturated liquid}, \; h_8 = 69.7 \text{ Btu/lb}$$

$$w_{P,\,8\text{-}9} = \frac{0.01614(60 - 1)(144)}{778} = 0.2 \text{ Btu/lb}$$

$$h_9 = h_8 + w_{P,\,8\text{-}9} = 69.7 + 0.2 = 69.9 \text{ Btu/lb}$$

$$P_{10} = 60 \text{ psia, saturated liquid}, \; h_{10} = 262.2 \text{ Btu/lb}$$

$$w_{P,\,10\text{-}11} = \frac{0.01727(400 - 60)(144)}{778} = 1.1 \text{ Btu/lb}$$

$$h_{11} = h_{10} + w_{P,\,10\text{-}11} = 262.2 + 1.1 = 263.3 \text{ Btu/lb}$$

In addition, the enthalpy values for states 4, 5, 6, 7, 12, 13, and 1 may be found from the tables or by direct calculations.

$$P_4 = 120 \text{ psia}, \; s_4 = s_3 = 1.7252 \text{ Btu/(lb)(°R)}, \; h_4 = 1315.8 \text{ Btu/lb}$$

$P_5 = 120$ psia, $T_5 = 900°F$, $h_5 = 1479.8$ Btu/lb

$P_6 = 60$ psia, $s_6 = s_5 = 1.8633$ Btu/(lb)(°R), $h_6 = 1384.3$ Btu/lb

$P_7 = 1$ psia, $s_7 = s_5$, $h_7 = 1041.3$ Btu/lb

$P_{12} = 120$ psia, saturated liquid, $h_{12} = 312.7$ Btu/lb

$P_{13} = 400$ psia, $T_{13} = T_{12} = 341°F$, compressed liquid, $h_{13} = 313.2$ Btu/lb

$$w_{P,\,12\text{-}1} = \frac{0.01789(400 - 120)(144)}{778} = 0.9 \text{ Btu/lb}$$

$$h_1 = h_{12} + w_{P,\,12\text{-}1} = 312.7 + 0.9 = 313.6 \text{ Btu/lb}$$

The enthalpy values are now known except for state 2. Generally this is found from energy and mass balances around the mixing point where streams 1 and 13 enter and stream 2 leaves. That is,

$$\dot{m}_1 h_1 + \dot{m}_{13} h_{13} = \dot{m}_2 h_2 \qquad \text{and} \qquad \dot{m}_1 + \dot{m}_{13} = \dot{m}_2$$

However, the split in total flow between streams 1 and 13 has not yet been determined. Once this split is found, then h_2 may be evaluated. In this particular example a numerical solution is available without determining first the fraction bled at state 4. This occurs because the enthalpies at states 1 and 13 are essentially identical, that is, 313.2 and 313.6 Btu/lb. Thus, regardless of the split in flow, h_2 after mixing will be around 313.3 Btu/lb.

The fractional flow extracted to the two heaters is important, however, for the evaluation of the turbine and pump work and the heat input. The fraction of the flow through the boiler-superheater section which is extracted at state 4 is found from energy and mass balances around the closed heater.

$$\dot{m}_1(h_4 - h_{12}) = \dot{m}_{13}(h_{13} - h_{11}) \qquad \text{and} \qquad \dot{m}_1 + \dot{m}_{13} = \dot{m}_3$$

If we let $\dot{m}_4/\dot{m}_3 = \dot{m}_1/\dot{m}_3 = y_4$ and $\dot{m}_{13}/\dot{m}_3 = 1 - y_4$, then

$$y_4(h_4 - h_{12}) = (1 - y_4)(h_{13} - h_{11})$$

$$y_4(1315.8 - 312.7) = (1 - y_4)(313.2 - 263.3)$$

$$y_4 = 0.0474$$

Hence 4.74 percent of the total flow around the cycle is bled to the closed heater. An energy balance around the open heater, similar to Eq. (17-2), will lead to the fraction of the flow bled to the open heater.

$$\dot{m}_6 h_6 + \dot{m}_9 h_9 = \dot{m}_{10} h_{10} \qquad \text{and} \qquad \dot{m}_6 + \dot{m}_9 = \dot{m}_{10}$$

If we let $\dot{m}_6/\dot{m}_{10} = z_6$ and $\dot{m}_9/\dot{m}_{10} = (1 - z_6)$, then

$$z_6(1384.3) + (1 - z_6)(69.9) = 1(262.2)$$

$$z_6 = 0.1463$$

Thus 14.63 percent of the flow reheated and returned to the second stage of the turbine is extracted to the open heater. Note that the symbol z_6 has been used rather than y_6, because we wish to let y_6 represent the fraction of the *total flow* bled to the open heater. It can easily be shown that $y_6 = 0.1394$ on the basis of the values of y_4 and z_6.

Finally we may evaluate the turbine and pump work and the heat input.

$$w_{T,\,\text{out}} = 1(h_3 - h_4) + (1 - y_4)(h_5 - h_6) + (1 - y_4 - y_6)(h_6 - h_7)$$

$$= (1470.1 - 1315.8) + 0.9526(1479.8 - 1384.3) + 0.8132(1384.3 - 1041.3)$$

$$= 154.3 + 91.0 + 278.9 = 524.2 \text{ Btu/lb}$$

$$w_{P,\,tot} = w_{P,\,8\text{-}9}(1 - y_4 - y_6) + w_{P,\,10\text{-}11}(1 - y_4) + w_{P,\,12\text{-}1}y_4$$

$$= 0.2(0.8132) + 1.1(0.953) + 0.9(0.0474) = 1.3 \text{ Btu/lb}$$

$$q_{in} = 1(h_3 - h_2) + (1 - y_4)(h_5 - h_4)$$

$$= (1470.1 - 313.3) + 0.9526(1479.8 - 1315.8) = 1313.0 \text{ Btu/lb}$$

As a result of the above calculations the thermal efficiency becomes

$$\eta_{th} = \frac{w_{T,\,out} - w_{P,\,tot}}{q_{in}} = \frac{524.2 - 1.3}{1313.0} = 0.398 \qquad \text{(or 39.8 percent)}$$

In Example 17-5 the efficiency was found to be 38.1 percent for a single open heater without reheat. The increase to 39.8 percent in this example is due both to a second heater in the cycle and to the reheat section.

17-5 EFFECT OF IRREVERSIBILITIES ON VAPOR-POWER-CYCLE PERFORMANCE

It must be kept in mind that the various cycles which have been presented earlier in this chapter are ideal cycles, and irreversibilities have not been considered. In all cases there are frictional losses in the piping which lead to pressure drops. Heat-transfer losses occur throughout the equipment. Irreversibilities caused by the flow of the fluid are especially important in the turbine and the pump. If heat losses are negligible in a turbine or pump, the adiabatic efficiency of these pieces of equipment must be considered to bring the ideal analysis more in line with actual performance. The concept of an adiabatic efficiency for various flow devices was introduced in Sec. 7-10. In review, for turbines and pumps

$$\eta_T = \frac{w_{act}}{w_{isen}} \qquad \text{and} \qquad \eta_P = \frac{w_{isen}}{w_{act}}$$

Fig. 17-9 shows the change in the position of the exit states for the turbine and pump when irreversibilities are present.

Example 17-8M The turbine-inlet conditions in a simple, steam power cycle are 30 bars and 500°C, and the turbine-exhaust condition is 0.1 bar. The turbine operates adiabatically with an efficiency of 82 percent, and the efficiency of the pump is 78 percent. Compute the thermal efficiency of the cycle.

SOLUTION Losses other than those in the turbine and pump will be neglected. From Example 17-3M, it is known that the inlet enthalpy h_3 and entropy s_3 to the turbine are 3456.5 kJ/kg and 7.2338 kJ/(kg)(°K), respectively. Also, isentropic expansion to 0.1 bar gives a turbine-outlet enthalpy h_{4s} of 2292.7 kJ/kg. The actual work output is the isentropic enthalpy change times the turbine-adiabatic efficiency. Hence

$$w_{T,\,act} = 0.82(3456.5 - 2292.7) = 954.3 \text{ kJ/kg}$$

$$= h_3 - h_{4a}$$

Since h_3 is 3456.5 kJ/kg, $h_{4a} = 3456.5 - 954.3 = 2502.2$ kJ/kg. Since h_g at 0.1 bar is 2584.7 kJ/kg,

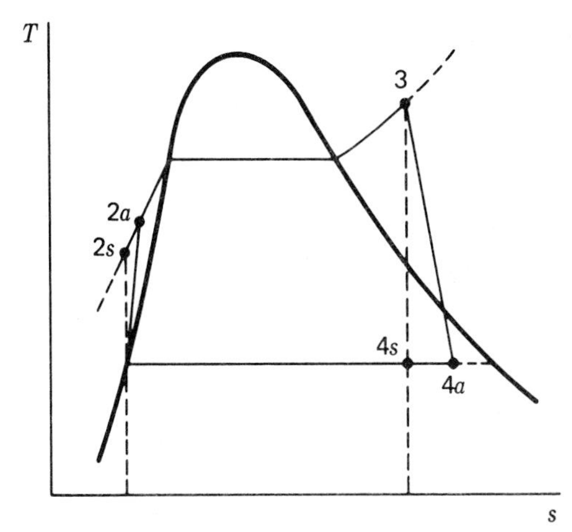

Figure 17-9 *Ts* diagram of a simple steam power cycle with irreversible turbine and pump performance.

the steam leaving the turbine is in the wet region. The exit quality is, roughly, 96 percent. The pump work for an isentropic process was previously determined to be 3.0 kJ/kg. For the irreversible process this becomes

$$w_{P,\,act} = \frac{3.0}{0.78} = 3.8 \text{ kJ/kg}$$

Therefore the enthalpy of the water leaving the pump is

$$h_2 = h_1 + w_P = 191.8 + 3.8 = 195.6 \text{ kJ/kg}$$

The thermal efficiency then becomes

$$\eta_{th} = \frac{w_T - w_P}{h_3 - h_2} = \frac{954.3 - 3.8}{3456.5 - 195.6} = 0.291 \qquad \text{(or 29.1 percent)}$$

Thus irreversibilities lead to a considerable decrease from the ideal Rankine-cycle efficiency of 35.6 percent obtained in Example 17-3M.

Example 17-8 The turbine-inlet conditions in a simple, steam power cycle are 400 psia and 900°F, and the exhaust condition is 1 psia. The turbine operates adiabatically with an efficiency of 82 percent, and the efficiency of the pump is 78 percent. Compute the thermal efficiency of the cycle.

SOLUTION Losses other than those in the turbine and pump will be neglected. From Example 17-3 it is known that the inlet enthalpy h_3 and entropy s_3 to the turbine are 1470.1 Btu/lb and 1.7252 Btu/(lb)(°R), respectively. Also, isentropic expansion to 1 psia gives the turbine-outlet condition as 964.0 Btu/lb for the enthalpy h_{4s}. The actual work output is the isentropic enthalpy change times the adiabatic efficiency. Hence

$$w_{T,\,act} = 0.82(1470.1 - 964.0) = 415.0 \text{ Btu/lb}$$

$$= h_3 - h_{4a}$$

Since h_3 is 1470.1 Btu/lb, $h_{4a} = 1470.1 - 415.0 = 1055.1$ Btu/lb. This corresponds to an exit steam quality of, roughly, 95 percent. The pump work for an isentropic process was previously determined to be 1.2 Btu/lb. For the irreversible process this becomes

$$w_{P,\,act} = \frac{1.2}{0.78} = 1.5 \text{ Btu/lb}$$

Therefore the enthalpy of the steam leaving the pump is

$$h_2 = h_1 + w_P = 69.7 + 1.5 = 71.2 \text{ Btu/lb}$$

The thermal efficiency then becomes

$$\eta_{\text{th}} = \frac{w_T - w_P}{h_3 - h_2} = \frac{415.0 - 1.5}{1470.1 - 71.2} = 0.296 \qquad \text{(or 29.6 percent)}$$

This is a considerable decrease from the ideal Rankine-cycle efficiency of 36.1 percent obtained in Example 17-3.

17-6 SUPERCRITICAL RANKINE CYCLE

In the preceding discussions of the vapor power cycle, the final pressure at the outlet of the pump preceding the heat-addition section of the cycle was always chosen to be below the critical pressure of the fluid. That is, the maximum pressure was less than 221 bars, or 3200 psia, when the working fluid was water. A number of modern steam power plants operate on a cycle for which the turbine-inlet pressure is supercritical. A typical value might be 250 to 325 bars, or 3500 to 5000 psia.

A Rankine cycle with a supercritical turbine-inlet pressure is shown in Fig. 17-10*a*. Note that during the heat-addition process (4-1) a phase change does not occur. The pressurized fluid enters the heat-exchanger tubes in the furnace and gradually expands in volume with no bubbling occurring as it passes through the tubes. Inlet-turbine temperatures may approach 620°C or 1150°F. For large power units working under these conditions, two reheat sections are often employed, and as many as five to eight feedwater heaters are commonly required.

One of the major thermodynamic considerations in the analysis of power cycles is the average temperature level of the fluid during heat addition. For

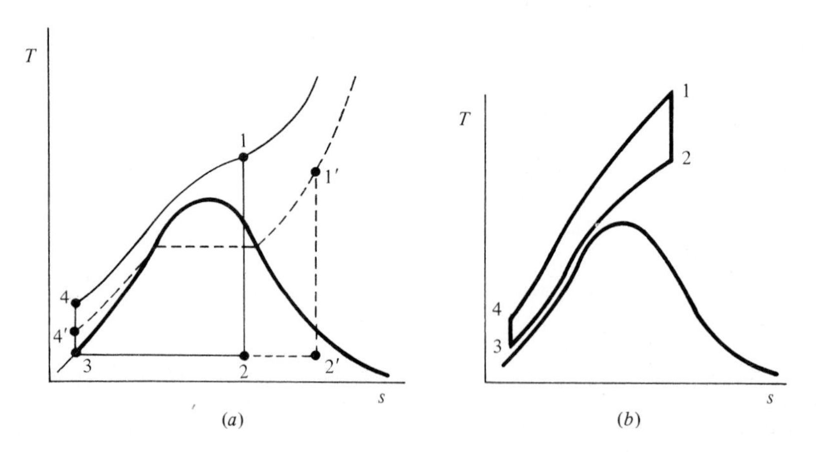

Figure 17-10 Supercritical Rankine cycle with heat rejection (*a*) in the wet region, and (*b*) at a supercritical pressure.

comparative purposes a subcritical cycle is also sketched on Fig. 17-10*a* as a dashed line. During the heat-addition process 4′-1′ for the subcritical cycle, a considerable portion of the heating occurs at a constant temperature which is less than the critical temperature of the fluid. For the supercritical cycle the temperature continually increases during the heat-addition process 4-1. For the same limiting temperature at the turbine inlet for both types of cycles, the average temperature level during heat addition for the supercritical cycle will be greater. Theoretically, this should lead to a higher thermal efficiency, all other factors being the same. The actual thermal efficiency of supercritical units is around 40 percent. The theoretical Rankine cycle does not require expansion to a low pressure followed by condensation of the working fluid. Another possibility is shown in Fig. 17-10*b*. In this figure the upper and lower pressures are each supercritical, but the lowest temperature represents a state which is in the subcooled liquid region.

In a totally supercritical power cycle such as described in Fig. 17-10*b* it is not necessary for the working fluid to be water. Carbon dioxide, with a critical pressure of 73.9 bars (1072 psia) and a critical temperature of 304°K (88°F), has been proposed. Conservative calculations for a totally supercritical power cycle indicate that competitive performance with present equipment will be attained only if a turbine-inlet temperature above 650°C (1200°F) is used. One advantage of a totally supercritical cycle is that it can be designed to operate similar to a Brayton gas turbine cycle with regeneration (see Sec. 16-8). The *Ts* diagram for such a cycle is shown in Fig. 17-11, with a subcritical Rankine cycle shown as a dashed line for comparison. The regenerative heating occurs during process 2-*x*, and this energy is supplied by the internal cooling process 4-*y*. The effect of regeneration is to raise the level of the average temperature during heat addition.

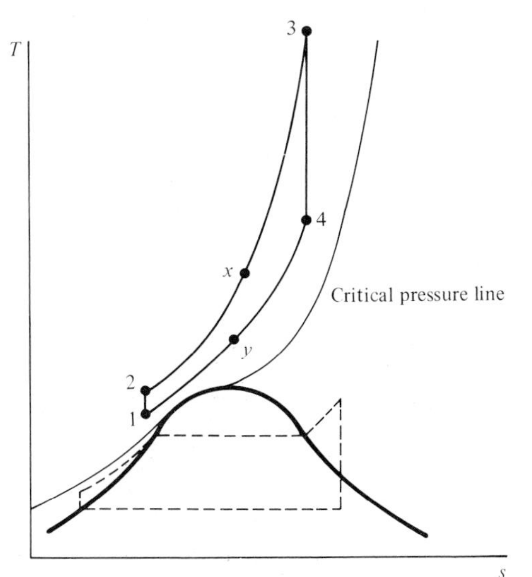

Figure 17-11 *Ts* diagram for a totally supercritical vapor cycle with regeneration.

Note that with regeneration heat is added from an external source only during process x-3 in Fig. 17-11. A major disadvantage of the totally supercritical cycle is the high pressures involved for those substances, such as CO_2, which are attractive working fluids.

17-7 HIGH-TEMPERATURE AND BINARY VAPOR CYCLES

On the basis of the Carnot efficiency equation for the conversion of thermal to mechanical energy, it is apparent that heat addition to the working fluid should occur at the highest possible temperature. It must be kept in mind, however, that the use of higher temperatures is restricted to the availability of materials which can withstand such conditions. Materials technology has limited the steam power cycle to around 560°C (1050°F) and the industrial gas-turbine cycle to around 1000 to 1200°C (1800 to 2200°F). As technology develops materials to withstand higher temperatures, it may become possible to operate a Rankine-type cycle at temperatures now reserved for the gas-turbine cycle. In such a case it would be highly desirable to use some other fluid than water. As noted in the discussion in the preceding section, the great disadvantage of water is its critical temperature. When a boiler is used, the heat is added to the fluid at a constant and relatively low temperature. This low-temperature heat addition could be partially overcome by employing a substance with a much higher critical temperature.

It has been found that the properties of alkali metals are fairly suitable for high-temperature Rankine cycles. Most promising are potassium, sodium, and mixtures of these two metals. In addition to the appropriateness of their thermodynamic properties, such as vapor pressure and heat capacity, these substances have reasonably high heat-transfer coefficients. As a result, any heat-exchanger size is significantly reduced. Such a cycle has been proposed for the generation of power aboard a spacecraft.

The maximum temperature for a Rankine cycle operating with potassium, for example, may be as high as 1200°C or 2200°F. The degree of superheating at the turbine inlet typically might be 30 to 80°C, or 50 to 150°F. The vapor pressure of potassium at 1150°C is roughly 13.5 bars, which is equivalent to 200 psia at 2100°F. The condensing temperature of potassium may be as low as 600°C, with a vapor pressure of 0.17 bar. This is equivalent to 1100°F, with a vapor pressure of 2.5 psia. These temperature limits give a turbine expansion ratio of 80, which is relatively high when only two stages of expansion are used. Hence the expansion ratio places a practical lower limit on the cycle-temperature ratio (minimum to maximum temperature). Typical space power Rankine cycles are constrained to operate with cycle-temperature ratios on the order of 0.65 to 0.75. Consequently the cycle efficiency is relatively low, probably 25 percent or less.

The low cycle-temperature ratio does have one advantage in terms of space applications. Since the condensing temperature is high, the temperature of the fluid in the heat-injection loop is also high. A high radiator rejection temperature leads to a small radiator size, since the radiant heat transfer per unit area varies as the fourth power of the temperature. Tabular data for potassium in the saturation and superheat regions are presented in Tables A-24M and A-24.

The use of alkali metals in a high-temperature Rankine cycle for space applications has led to a renewed interest in binary vapor power cycles. A binary cycle is one in which the heat removed during the heat rejection process of one power cycle is used as the heat input for another power cycle. It was noted above that the condensing temperature of a potassium cycle may be around 600°C or 1100°F. Heat removed at this temperature may be used to supply a Rankine cycle operating on steam and rejecting at atmospheric temperature. Because one cycle operates at temperatures above that in the other cycle, the high-temperature cycle is frequently called a *topping* cycle. An equipment schematic and a Ts diagram for a binary cycle using potassium and water are shown in Fig. 17-12. Only a portion of the saturation curve for potassium is shown on the Ts diagram, and the processes are idealized. In fact, the pump work is neglected on the potassium cycle. In practice, irreversible performance must be considered, as well as the possible presence of a reheat section and feedwater heaters for the steam cycle. Also, heat addition in the steam cycle might be supercritical. Note from Fig. 17-12 that a finite difference must exist between the potassium condensing temperature and the boiler-superheat temperatures of the steam cycle.

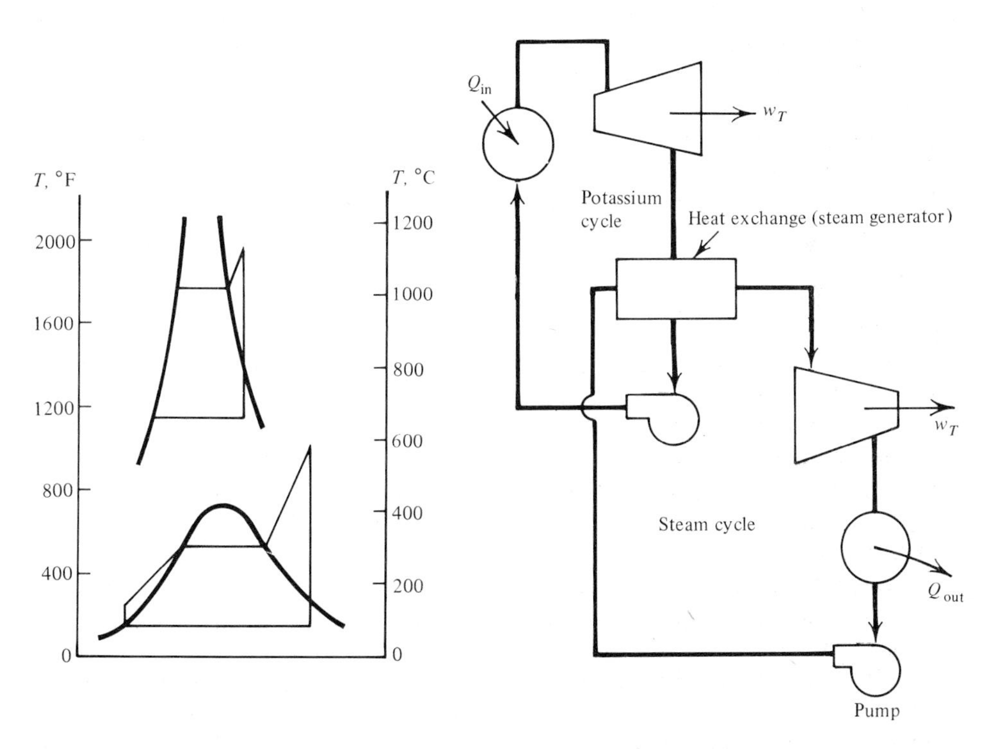

Figure 17-12 Equipment schematic and Ts diagram for a binary vapor power cycle involving steam and potassium.

For reasonable temperature values and turbine and pump efficiencies the predicted thermal efficiency of a potassium cycle runs around 20 to 30 percent. When a potassium cycle is used as a topping cycle for a steam cycle, the thermal efficiency of the overall binary cycle may approach 50 to 60 percent. A disadvantage of the binary cycle is a high capital cost. However, this becomes less important as fuel costs rise.

REFERENCES

Burghardt, M. D.: "Engineering Thermodynamics with Applications," Harper & Row, New York, 1978.

"Steam, Its Generation and Uses," The Babcock and Wilcox Company, New York, 1972.

PROBLEMS (METRIC)

Carnot vapor cycle

17-1M A Carnot cycle operates with steam as the working medium. At the end of adiabatic compression the pressure is 1.5 MPa and the quality is 20 percent. Heat is added during the isothermal expansion until the steam becomes a saturated vapor. The fluid then expands adiabatically until the pressure is 0.10 MPa. Determine (*a*) the quality at the end of adiabatic expansion, (*b*) the thermal efficiency, (*c*) the heat input, in kJ/kg, and (*d*) the net work output, in kJ/kg.

17-2M A Carnot engine operates with steam as the working medium. At the end of adiabatic compression the pressure is 10 bars and the quality is 10 percent. At the end of isothermal heat addition the steam is a saturated vapor, and during isothermal rejection the pressure is 0.3 bar. Determine (*a*) the quality at the end of adiabatic expansion, (*b*) the thermal efficiency, (*c*) the heat rejected, in kJ/kg, and (*d*) the net work output per cycle, in kJ/kg.

17-3M A Carnot engine contains 0.10 kg of water. At the beginning of the heat-addition process the fluid is a saturated liquid, and at the end of this process it is a saturated vapor. Heat addition occurs at 120 bars and heat rejection is at 0.3 bar. Determine (*a*) the quality at the end of adiabatic expansion and at the end of isothermal heat rejection, (*b*) the thermal efficiency, (*c*) the heat added per cycle, in kilojoules, and (*d*) the net work output per cycle.

The Rankine cycle

17-4M A Rankine cycle has steam entering the turbine at 80 bars and 440°C. If the turbine output is 10,000 kW, determine (1) the quality at the turbine outlet, (2) the thermal efficiency, and (3) the mass flow rate of steam for condenser pressures of (*a*) 0.08 bar and (*b*) 0.04 bar.

17-5M A Rankine cycle has an exhaust pressure of 0.08 bar and a turbine-inlet pressure of 60 bars. Determine (1) the moisture content at the turbine outlet, (2) the thermal efficiency, and (3) the mass flow rate of steam for a net power output of 10 MW for turbine-inlet temperatures of (*a*) 540°C and (*b*) 440°C.

17-6M A Rankine cycle has an exhaust pressure of 0.08 bar and a turbine-inlet temperature of 520°C. Determine (1) the moisture content at the turbine outlet and (2) the thermal efficiency of the cycle for a turbine-inlet pressure of (*a*) 120 bars and (*b*) 80 bars.

17-7M An ideal Rankine cycle generates steam at 140 bars and 560°C, and condenses steam at 0.06 bar. The net power output is 20 MW. Determine (*a*) the heat input, in kJ/kg, (*b*) the thermal efficiency, (*c*) the mass flow rate of steam, in kg/h, and (*d*) the mass flow rate of cooling water required in the condenser if the water experiences a 10°C temperature rise.

17-8M An ideal Rankine steam power cycle which produces 125 MW of turbine power has the following operating conditions: turbine inlet, 80 bars and 560°C; condenser pressure, 0.06 bar. Determine (*a*) the heat input, in kJ/kg, (*b*) the thermal efficiency, (*c*) the mass flow rate of steam, in kg/h, and (*d*) the mass flow rate of cooling water required in the condenser if the water experiences an 8°C temperature rise.

17-9M A steam power unit operating on the Rankine cycle has a mass flow rate of 23,800 kg/h through it. Water enters the boiler at 30 bars, and it leaves the condenser at 0.10 bar and 45.8°C. Cooling water in the condenser is circulated at a rate of 1.31×10^6 kg/h, and it experiences a temperature rise of 8.50°C. Determine (*a*) the enthalpy and entropy at the turbine outlet, (*b*) the enthalpy at the turbine inlet, (*c*) the heat added, in kJ/kg, (*d*) the thermal efficiency, and (*e*) the power output of the turbine, in kilowatts.

17-10M A Rankine power cycle uses solar energy for the heat input and refrigerant 12 as the working fluid. Fluid enters the pump as a saturated liquid at 7 bars, and is pumped to 14 bars. The turbine-inlet temperature is 140°C and the mass flow rate is 1200 kg/h. Determine (*a*) the net work output, in kJ/kg, (*b*) the thermal efficiency, and (*c*) the area, in square meters, of the solar collector needed if the collector picks up 630 J/(m^2)(s).

17-11M A Rankine power cycle uses solar energy for the heat input and refrigerant 12 as the working fluid. The fluid enters the pump as a saturated liquid at 9 bars, and is pumped to 16 bars. The turbine-inlet temperature is (*a*) 120°C, and (*b*) 160°C and the mass flow rate is 1000 kg/h. Determine (1) the net work output, in kJ/kg, (2) the thermal efficiency, and (3) the area, in square meters, of the solar collector needed if the collector picks up 650 J/(m^2)(s).

Reheat cycles

17-12M Modify the data of Prob. 17-4M in the following manner. The steam in the turbine is expanded to 15 bars, reheated to 440°C, and then expanded to (*a*) 0.08 bar, and (*b*) 0.04 bar. Determine (1) the quality at the condenser outlet, and (2) the thermal efficiency.

17-13M Modify the data of Prob. 17-5M in the following manner. (*a*) Steam at 60 bars and 540°C is expanded to 5 bars, reheated to 500°C, and expanded to 0.08 bar. (*b*) Steam at 60 bars and 440°C is expanded to 10 bars, reheated to 440°C, and expanded to 0.08 bar. Determine (1) the quality at the condenser inlet, and (2) the thermal efficiency.

17-14M Modify the data of Prob. 17-6M in the following manner. (*a*) Steam at 120 bars and 520°C is expanded to 10 bars, reheated to 500°C, and expanded to 0.08 bar. (*b*) Steam at 80 bars and 520°C is expanded to 7 bars, reheated to 500°C, and expanded to 0.08 bar. Calculate (1) the quality at the condenser inlet, and (2) the thermal efficiency.

17-15M The data of Prob. 17-7M are modified in the following manner. Reheat occurs at 10 bars to a temperature of 540°C. Answer the same questions.

17-16M The data of Prob. 17-8M are modified in the following manner. After expansion to 5 bars, the fluid is reheated to 500°C. Answer the same questions.

17-17M An ideal reheat cycle operates with turbine-inlet conditions of 160 bars and 560°C, a condenser pressure of 0.06 bar, and reheat to 560°C. Determine the thermal efficiency of the cycle for reheat pressures of (*a*) 20 bars, (*b*) 30 bars, (*c*) 40 bars, (*d*) 60 bars, and (*e*) 80 bars. Plot the efficiency of the reheat cycle versus reheat pressure.

Regenerative cycles

17-18M Modify the data of Prob. 17-4M as follows. The steam is expanded to 15 bars, where a portion is bled to a single, open feedwater heater operating at this same pressure. Answer the same questions for a condenser pressure of (*a*) 0.08 bar, and (*b*) 0.04 bar.

17-19M The data of Prob. 17-5M are modified as follows. (*a*) Steam at 60 bars and 540°C is expanded to 5 bars, where a portion is bled to a single open heater and the remainder is expanded to 0.08 bar. (*b*) Steam at 60 bars and 440°C is expanded to 10 bars, where a portion is bled to a single open heater and the remainder is expanded to 0.08 bar. Determine (1) the quality at the condenser inlet, and (2) the thermal efficiency.

17-20M Modify the data of Prob. 17-6M as follows. (*a*) Steam at 120 bars and 520°C is expanded to 10 bars, where a portion is bled to a single, open heater and the remainder is expanded to 0.08 bar. (*b*) Steam at 80 bars and 520°C is expanded to 7 bars, where a portion is bled to a single, open feedwater heater and the remainder is expanded to 0.08 bar. Compute (1) the quality at the turbine outlet, and (2) the thermal efficiency.

17-21M The data of Prob. 17-7M are modified in the following manner. The steam is expanded to 10 bars, where a portion is bled to (*a*) a single, open feedwater heater, and (*b*) a single, closed feedwater heater followed by a pump. Determine (1) the fraction of the total flow bled to the heater, (2) the mass flow rate of steam required, and (3) the thermal efficiency.

17-22M The data of Prob. 17-8M are modified as follows. The steam is expanded to 5 bars, where a portion is bled to (*a*) a single, open feedwater heater, and (*b*) a single, closed feedwater heater followed by a pump. Determine (1) the fraction of the total flow bled to the heater, (2) the mass flow rate of steam required, (3) the thermal efficiency.

17-23M The boiler-superheater of an ideal regenerative steam cycle produces steam at 120 bars and 600°C. A closed feedwater heater receives steam from the turbine at 30 bars and an open feedwater heater operates at 10 bars. The condenser operates at 0.08 bars, and the liquid condensate from the closed heater is throttled back into the open heater. There is a pump after the condenser and after the open heater. Determine (*a*) the fraction of the total flow which goes to the closed heater and to the open heater, (*b*) the work output of the turbine and the total pump work, in kJ/kg of total flow, and (*c*) the thermal efficiency.

17-24M Steam enters the turbine of an ideal regenerative cycle at 40 bars and 500°C. Steam is extracted at 7 and 3 bars and introduced into two open feedwater heaters that are in series. Appropriate pumps are employed after the condenser, which operates at 0.06 bar, and after each heater. Determine (*a*) the fraction of the total flow which goes to the 7 bars heater and the 3 bars heater, (*b*) the work output of the turbine and the total pump work, in kJ/kg of total flow, and (*c*) the thermal efficiency.

17-25M Steam is generated at 140 bars and 520°C. The steam is expanded through the first stage of a turbine to 40 bars. Part of the exhaust stream from this turbine is supplied to a closed feedwater heater, and the remainder is reheated to 520°C. The reheated steam is then expanded in a second stage of the turbine to an exhaust pressure of 0.08 bar. Some steam is bled from the second turbine at 7 bars for use in an open feedwater heater. The steam bled from the first turbine condenses as it passes through the closed heater and is throttled back into the open heater at 7 bars. The saturated liquid leaving the open heater first passes through a pump to 140 bars and then through the other side of the closed heater to the boiler. If we assume an ideal cycle and neglect pump work, determine (*a*) the percent of the steam entering the first turbine which goes to the closed heater, (*b*) the percent of the steam entering the second turbine which is bled to the open heater, and (*c*) the thermal efficiency.

Reheat-regenerative cycles

17-26M Consider the following modification of Probs. 17-12M and 17-18M. The steam in the turbine is expanded to 15 bars, where a portion is bled to an open feedwater heater operating at this pressure. The remaining portion is reheated to 440°C and then expanded to (*a*) 0.08 bar, and (*b*) 0.04 bar. Determine the thermal efficiency.

17-27M Consider the following modification of Probs. 17-13M and 17-19M. (*a*) Steam at 60 bars and 540°C is expanded to 5 bars, where a portion is bled to an open feedwater heater. The remaining portion is reheated to 500°C and then expanded to 0.08 bar. (*b*) Steam at 60 bars and 440°C is expanded to 10 bars, where a portion is bled to an open feedwater heater. The remaining portion is reheated to 440°C and then expanded to 0.08 bar. Find the thermal efficiency.

17-28M Consider the following modifications of Probs. 17-14M and 17-20M. (*a*) Steam at 120 bars and 520°C is expanded to 10 bars, where a portion is bled to a single, open feedwater heater. The remaining portion is reheated to 500°C and expanded to 0.08 bar. (*b*) Steam at 80 bars and 520°C is expanded to 7 bars, where a portion is bled to a single, open feedwater heater. The remaining portion is reheated to 500°C and expanded to 0.08 bar. Determine the thermal efficiency.

17-29M Consider the following modification of Prob. 17-23M. The portion of the flow which does not go to the closed heater is reheated to 540°C before it enters the second stage of the turbine. Determine (*a*) the fraction of the total flow which goes to the open heater, (*b*) the heat input, in kJ/kg, and (*c*) the thermal efficiency.

17-30M Consider the following modification of Prob. 17-24M. The portion of the flow which is not extracted to the open heater at 7 bars is reheated to 500°C before it enters the second stage of the turbine. Find (*a*) the fraction of the total flow which goes to the open heater at 3 bars, (*b*) the heat input, in kJ/kg, and (*c*) the thermal efficiency.

17-31M An ideal, reheat-regenerative cycle operates with conditions at the turbine inlet of 140 bars and 600°C, and reheat is at 7 bars to 500°C. A closed feedwater heater operates at 15 bars, and the drain from the closed heater is trapped back into an open feedwater heater which operates at 3 bars. The condenser pressure is 0.06 bar. Determine (*a*) the thermal efficiency of the cycle, and (*b*) the mass flow rate through the steam generator required for a turbine output of 100,000 kW.

Effect of irreversibilities on vapor cycle performance

17-32M The adiabatic efficiency of the turbine is 85 percent and that of the pump is 70 percent in a simple steam power cycle. On this basis determine the thermal efficiency for the data given in (*a*) Prob. 17-4M(*a*), (*b*) Prob. 17-4M(*b*), (*c*) Prob. 17-5M(*a*), (*d*) Prob. 17-5M(*b*), (*e*) Prob. 17-6M(*a*), (*f*) Prob. 17-6M(*b*), (*g*) Prob. 17-7M, and (*h*) Prob. 17-8M.

17-33M The adiabatic efficiency of the turbine is 85 percent and that of the pump is 70 percent in a reheat cycle. On this basis determine the thermal efficiency for the data given in (*a*) Prob. 17-12M(*a*), (*b*) Prob. 17-12M(*b*), (*c*) Prob. 17-13M(*a*), (*d*) Prob. 17-13M(*a*), (*e*) Prob. 17-14M(*a*), (*f*) Prob. 17-14M(*b*), (*g*) Prob. 17-15M, and (*h*) Prob. 17-16M.

17-34M The adiabatic efficiencies of the turbine and pump in a regenerative cycle are 85 and 70 percent, respectively. On this basis determine the thermal efficiency of the cycle for the data given in (*a*) Prob. 17-18M(*a*), (*b*) Prob. 17-18M(*b*), (*c*) Prob. 17-19M(*a*), (*d*) Prob. 17-19M(*b*), (*e*) Prob. 17-20M(*a*), (*f*) Prob. 17-20M(*b*), (*g*) Prob. 17-21M(*a*), (*h*) Prob. 17-22M(*a*), and (*i*) Prob. 17-23M.

17-35M The adiabatic efficiencies of the turbine and pump in a reheat-regenerative cycle are 85 and 70 percent, respectively. On this basis determine the thermal efficiency for the data given in (*a*) Prob. 17-26M(*a*), (*b*) Prob. 17-26M(*b*), (*c*) Prob. 17-27M(*a*), (*d*) Prob. 17-27M(*b*), (*e*) Prob. 17-28M(*a*) (*f*) Prob. 17-28M(*b*), (*g*) Prob. 17-29M, and (*h*) Prob. 17-30M.

17-36M A simple steam power cycle uses solar energy for the heat input. Water in the cycle enters the pump as a saturated liquid at 40°C, and is pumped to 2 bars. It then evaporates in the boiler at this pressure, and enters the turbine as saturated vapor. At the turbine exhaust the conditions are 40°C and 8 percent moisture. The mass flow rate is 1500 kg/h, and the pump is rated at 0.3 kW. Determine (*a*) the net work output in kJ/kg, (*b*) the energy-conversion efficiency, and (*c*) the area, in square meters of the solar collector needed if the collector picks up 700 J/(m^2)(*s*).

Supercritical steam-power cycle

17-37M A supercritical steam power cycle operates with reheat and regeneration. The steam enters the turbine at 240 bars and 600°C and expands to 0.06 bar. Steam leaves the first stage at 30 bars, where part of it enters a closed heater, while the rest is reheated to 540°C. Both sections of the turbine have an adiabatic efficiency of 88 percent. There is a condensate pump between the condenser and the heater. Another pump for the extracted steam that is condensed lies between the heater and the condensate line from the heater. The total feedwater enters the steam generator at 280 bars. Determine (*a*) the enthalpies at the 10 states around the cycle, (*b*) the fraction of the total mass flow rate which is extracted to the heater, and (*c*) the thermal efficiency of the cycle.

17-38M In a supercritical steam power cycle steam expands ideally from 320 bars and 600°C to 240 bars. It is then cooled at constant pressure to 400°C, followed by isentropic compression to 320 bars. Heat is then added until the temperature reaches 600°C. Determine (*a*) the amount of heat added in kJ/kg, (*b*) the amount of heat rejected, and (*c*) the thermal efficiency.

17-39M A supercritical steam power plant operates with reheat and regeneration. The steam enters the turbine at 240 bars and 560°C and expands to 0.08 bar. Steam leaves the first stage at 40 bars, where part of it enters a closed heater, while the rest is reheated to 540°C. Both sections of the turbine have an adiabatic efficiency of 87 percent. There is a condensate pump between the main condenser and the closed heater. Another pump for the extracted steam that is condensed lies between the heater outlet and the condenser-outlet line from the heater. Determine (*a*) the four enthalpies at the inlets and outlets of the closed heater, (*b*) the enthalpy at the condenser inlet and the enthalpy at the inlet to the heat-addition section, (*c*) the fraction of the total mass flow which is extracted to the heater, and (*d*) the thermal efficiency of the cycle.

High-temperature and binary vapor cycles

17-40M A Rankine cycle operating on potassium as the working fluid has a turbine-inlet state of 1325°K and 6 atm, and the temperature of the condensing fluid is 950°K.

(*a*) Determine the thermal efficiency for an ideal cycle.

(*b*) If the turbine and pump efficiencies are 85 and 60 percent, respectively, determine the thermal efficiency.

17-41M A Rankine cycle using potassium as the working fluid has a turbine-inlet state of 1450°K and 10 atm, and the condenser operates at 900°K.

(*a*) Determine the quality at the exit of the turbine and the thermal efficiency for an ideal cycle.

(*b*) Repeat part *a* if the turbine and pump efficiencies are 82 and 50 percent, respectively.

17-42M A potassium Rankine cycle operates with a turbine-inlet state of 1200°K and 3 atm, and the condenser temperature is 900°K.

(*a*) Determine the thermal efficiency for an ideal cycle.

(*b*) Repeat part *a* if the turbine and pump efficiencies are 86 and 65 percent, respectively.

17-43M A binary vapor power cycle operates with the potassium cycle described in Prob. 17-40M(*b*). The heat rejected from this cycle enters the boiler-superheater section of a reheat-regenerative steam cycle. This cycle is described by Prob. 17-35M(*c*). Assume that the heat for the reheat section of the steam cycle is provided from another source, and not from the potassium cycle. Determine (*a*) the ratio of the mass flow rate of potassium to that of steam, and (*b*) the overall thermal efficiency.

17-44M A binary vapor power cycle operates with the potassium cycle described in Prob. 17-41M(*b*). The heat rejected from this cycle enters the boiler-superheater section of a reheat-regenerative steam cycle. This cycle is described by Prob. 17-35M(*e*). Assume that the heat for the reheat section of the steam cycle is provided from another source, and not from the potassium cycle. Determine (*a*) the ratio of the mass flow rate of potassium to that of steam, and (*b*) the overall thermal efficiency.

17-45M A binary vapor power cycle operates with the potassium cycle described in Prob. 17-42M(*b*). The heat rejected from this cycle enters the boiler-superheater section of a reheat-regenerative steam cycle. This cycle is described by Prob. 17-35M(*a*). Assume that the heat for the reheat section of the steam cycle is provided from another source, and not from the potassium cycle. Determine (*a*) the ratio of the mass flow rate of potassium to that of steam, and (*b*) the overall thermal efficiency.

PROBLEMS (USCCS)

Carnot vapor cycle

17-1 A Carnot cycle operates with steam as the working medium. At the end of adiabatic compression the pressure is 250 psia and the quality is 20 percent. Heat is added during the isothermal expansion until the steam becomes a saturated vapor. The fluid then expands adiabatically until the pressure is 15 psia. Determine (*a*) the quality at the end of adiabatic expansion, (*b*) the thermal efficiency, (*c*) the heat input, in Btu/lb, and (*d*) the net work output in Btu/lb.

17-2 A Carnot engine operates with steam as the working medium. At the end of adiabatic compression the pressure is 200 psia and the quality is 10 percent. At the end of isothermal heat addition the steam is a saturated vapor, and during isothermal heat rejection the pressure is 10 psia. Determine (*a*) the quality at the end of adiabatic expansion, (*b*) the thermal efficiency, (*c*) the heat rejected, in Btu/lb, and (*d*) the net work output per cycle, in Btu/lb.

17-3 A Carnot engine contains 0.10 lb of water. At the beginning of the heat-addition process the fluid is a saturated liquid, and at the end of this process it is a saturated vapor. Heat addition occurs at 640°F and heat rejection is at 40°F. Determine (*a*) the quality at the end of adiabatic expansion and at the end of isothermal heat rejection, (*b*) the thermal efficiency, (*c*) the heat added per cycle, in Btu, and (*d*) the net work output per cycle.

The Rankine cycle

17-4 A Rankine cycle has steam entering the turbine at 550 psia and 800°F. If the turbine output is 10,000 kW, determine (1) the quality at the turbine outlet, (2) the thermal efficiency, and (3) the mass flow rate of steam, in lb/h, for condenser pressures of (*a*) 1 psia and (*b*) 0.60 psia.

17-5 A Rankine cycle has an exhaust pressure of 1 psia and a turbine-inlet pressure of 800 psia. Determine (1) the moisture content at the turbine outlet, (2) the thermal efficiency, and (3) the mass flow rate of steam, in lb/h, for a net power output of 10 MW and for turbine-inlet temperatures of (*a*) 1000°F and (*b*) 800°F.

17-6 A Rankine cycle has an exhaust pressure of 1 psia and a turbine-inlet temperature of 1000°F. Determine (1) the moisture content at the turbine outlet and (2) the thermal efficiency of the cycle for a turbine-inlet pressure of (*a*) 1600 psia and (*b*) 1200 psia.

17-7 An ideal Rankine cycle generates steam at 2000 psia and 1000°F, and condenses steam at 0.80 psia. The net power output is 100 MW. Determine (*a*) the heat input, in Btu/lb, (*b*) the thermal efficiency, (*c*) the mass flow rate of steam, in lb/h, and (*d*) the mass flow rate of cooling water required in the condenser if the water experiences a 14°F temperature rise.

17-8 An ideal Rankine steam power cycle which produces 125 MW of turbine power has the following operating conditions: turbine inlet, 1000 psia and 1000°F; condenser pressure, 1.0 psia. Determine (*a*) the heat input, in Btu/lb, (*b*) the thermal efficiency, (*c*) the mass flow rate of steam, in lb/h, and (*d*) the mass flow rate of cooling water required in the condenser, in lb/h, if the water experiences an 12°F temperature rise.

17-9 A steam power unit operating on the Rankine cycle has a mass flow rate of 52,400 lb/h through it. Water enters the boiler at 500 psia, and it leaves the condenser at 1.2 psia. Cooling water in the condenser is circulated at a rate of 3.20×10^6 lb/h, and it experiences a temperature rise of 14.0°F. Determine (*a*) the enthalpy and entropy at the turbine outlet, (*b*) the enthalpy at the turbine inlet, (*c*) the heat added in Btu/lb, (*d*) the thermal efficiency, and (*e*) the power output of the turbine, in kilowatts.

17-10 A Rankine power cycle uses solar energy for the heat input and refrigerant 12 as the working fluid. Fluid enters the pump as a saturated liquid at 100 psia, and is pumped to 300 psia. The turbine-inlet temperature is 240°F and the mass flow rate is 1200 lb/h. Determine (*a*) the net work output, in Btu/lb, (*b*) the thermal efficiency, and (*c*) the area, in square feet, of the solar collector needed if the collector picks up 200 Btu/(ft^2)(h).

17-11 A Rankine power cycle uses solar energy for the heat input and refrigerant 12 as the working fluid. The fluid enters the pump as a saturated liquid at 120 psia, and is pumped to 400 psia. The turbine-inlet temperature is (*a*) 220°F, and (*b*) 260°F and the mass flow rate is 1000 lb/h. Determine (1) the net work output, in Btu/lb, (2) the thermal efficiency, and (3) the area, in square feet, of the solar collector needed if the collector picks up 220 Btu/(ft^2)(h).

Reheat cycles

17-12 Modify the data of Prob. 17-4 in the following manner. The steam in the turbine is expanded to 100 psia, reheated to 800°F, and then expanded to (*a*) 1 psia, and (*b*) 0.60 psia. Determine (1) the quality at the condenser inlet, and (2) the thermal efficiency.

17-13 Modify the data in Prob. 17-5 in the following manner. (*a*) Steam at 800 psia and 1000°F is expanded to 60 psia, reheated to 1000°F, and expanded to 1 psia. (*b*) Steam at 800 psia and 800°F is expanded to 120 psia, reheated to 800°F, and expanded to 1 psia. Determine (1) the quality at the condenser inlet, and (2) the thermal efficiency.

17-14 Modify the data in Prob. 17-6 in the following manner. (*a*) Steam at 1600 psia and 1000°F is expanded to 140 psia, reheated to 1000°F, and expanded to 1 psia. (*b*) Steam at 1200 psia and 1000°F is expanded to 80 psia, reheated to 1000°F, and expanded to 1 psia. Calculate (1) the quality at the condenser inlet, and (2) the thermal efficiency.

17-15 The data of Prob. 17-7 are modified in the following manner. Reheat occurs at 160 psia to a temperature of 900°F. Answer the same questions.

17-16 The data of Prob. 17-8 are modified in the following manner. After expansion to 60 psia, the fluid is reheated to 900°F. Answer the same questions.

17-17 An ideal reheat cycle operates with turbine-inlet conditions of 2500 psia and 1000°F, a condenser pressure of 0.80 psia, and reheat to 1000°F. Determine the thermal efficiency of the cycle for reheat pressures of (*a*) 200 psia, (*b*) 450 psia, (*c*) 600 psia, (*d*) 800 psia, and (*e*) 1200 psia. Plot the efficiency of the reheat cycle versus reheat pressure.

Regenerative cycles

17-18 Modify the data of Prob. 17-4 as follows. The steam is expanded to 100 psia, where a portion is bled to a single, open feedwater heater operating at this same pressure. Answer the same questions for condenser pressures of (*a*) 1 psia, and (*b*) 0.60 psia.

17-19 The data of Prob. 17-5 are modified as follows. (*a*) Steam at 800 psia and 1000°F is expanded to 60 psia, where a portion is bled to a single open heater and the remainder is expanded to 1 psia. (*b*) Steam at 800 psia and 800°F is expanded to 120 psia, where a portion is bled to a single open heater and the remainder is expanded to 1 psia. Determine (1) the quality at the condenser inlet, and (2) the thermal efficiency.

17-20 Modify the data of Prob. 17-6 as follows. (*a*) Steam at 1600 psia and 1000°F is expanded to 140 psia, where a portion is bled to a single, open heater and the remainder is expanded to 1 psia. (*b*) Steam at 1200 psia and 1000°F is expanded to 80 psia, where a portion is bled to a single, open feedwater heater and the remainder is expanded to 1 psia. Compute (1) the quality at the turbine outlet, and (2) the thermal efficiency.

17-21 The data of Prob. 17-7 are modified in the following manner. The steam is expanded to 160 psia, where a portion is bled to (*a*) a single, open feedwater heater, and (*b*) a single, closed feedwater heater followed by a pump. Determine (1) the fraction of the total flow bled to the heater, (2) the mass flow rate of steam required, and (3) the thermal efficiency.

17-22 The data of Prob. 17-8 are modified as follows. The steam is expanded to 60 psia, where a portion is bled to (*a*) a single, open feedwater heater, and (*b*) a single, closed feedwater heater followed by a pump. Determine (1) the fraction of the total flow bled to the heater, (2) the mass flow rate of steam required, and (3) the thermal efficiency.

17-23 The boiler-superheater of an ideal regenerative steam cycle produces steam at 1800 psia and 1000°F. A closed feedwater heater receives steam from the turbine at 400 psia and an open feedwater heater operates at 140 psia. The condenser operates at 0.80 psia, and the liquid condensate from the closed heater is throttled back into the open heater. There is a pump after the condenser and after the open heater. Determine (*a*) the fraction of the total flow which goes to the closed heater and to the open heater, (*b*) the work output of the turbine and the total pump work, in Btu/lb of total flow, and (*c*) the thermal efficiency.

17-24 Steam enters the turbine of an ideal regenerative cycle at 500 psia and 900°F. Steam is extracted at 100 and 40 psia and introduced into two open feedwater heaters that are in series. Appropriate pumps are employed after the condenser, which operates at 1 psia, and after each heater. Determine (*a*) the fraction of the total flow which goes to the 100 psia heater and the 40 psia heater, (*b*) the work output of the turbine and the total pump work, in Btu/lb of total flow, and (*c*) the thermal efficiency.

17-25 Steam is generated at 2000 psia and 1000°F. The steam is expanded through the first stage of a turbine to 600 psia. Part of the exhaust stream from this turbine is supplied to a closed-feedwater heater, and the remainder is reheated to 1000°F. The reheated steam is then expanded in a second turbine to an exhaust pressure of 1 psia. Some steam is bled from the second turbine at 100 psia for use in an open feedwater heater. The steam bled from the first turbine condenses as it passes through the closed heater and is throttled back into the open heater at 100 psia. The saturated liquid leaving the open heater first passes through a pump to 2000 psia and then through the other side of the closed heater to the boiler-superheater. If we assume an ideal cycle and neglect pump work, determine (*a*) the percent of the steam entering the first turbine which goes to the closed heater, (*b*) the percent of the steam entering the second turbine which is bled to the open heater, and (*c*) the thermal efficiency.

Reheat-regenerative cycles

17-26 Consider the following modification of Probs. 17-12 and 17-18. The steam in the turbine is expanded to 100 psia, where a portion is bled to an open feedwater heater operating at this pressure. The remaining portion is reheated to 800°F and then expanded to (*a*) 1 psia, and (*b*) 0.60 psia. Determine the thermal efficiency.

17-27 Consider the following modification of Prob. 17-13 and 17-19. (*a*) Steam at 800 psia and 1000°F is expanded to 60 psia, where a portion is bled to an open feedwater heater. The remaining portion is reheated to 1000°F and then expanded to 1 psia. (*b*) Steam at 800 psia and 800°F is expanded to 120 psia, where a portion is bled to an open feedwater heater. The remaining portion is reheated to 800°F and then expanded to 1 psia. Find the thermal efficiency.

17-28 Consider the following modifications of Probs. 17-14 and 17-20. (*a*) Steam at 1600 psia and 1000°F is expanded to 140 psia, where a portion is bled to a single, open feedwater heater. The remaining portion is reheated to 1000°F and expanded to 1 psia. (*b*) Steam at 1200 psia and 1000°F is expanded to 80 psia, where a portion is bled to a single, open feedwater heater. The remaining portion is reheated to 1000°F and expanded to 1 psia. Determine the thermal efficiency.

17-29 Consider the following modification of Prob. 17-23. The portion of the flow which does not go to the closed heater at 400 psia is reheated to 1000°F before it enters the second state of the turbine. Determine (*a*) the fraction of the total flow which goes to the open heater, (*b*) the heat input, in Btu/lb, and (*c*) the thermal efficiency.

17-30 Consider the following modification of Prob. 17-24. The portion of the flow which is not extracted to the open heater at 100 psia is reheated to 900°F before it enters the second stage of the turbine. Find (*a*) the fraction of the total flow which goes to the open heater at 40 psia, (*b*) the heat input, in Btu/lb, and (*c*) the thermal efficiency.

17-31 An ideal reheat-regenerative cycle operates with throttle conditions at the turbine inlet of 2000 psia and 1100°F, and reheat is at 100 psia to 1000°F. A closed feedwater heater operates at 200 psia, and the drain from the closed heater is trapped back into an open feedwater heater which operates at 30 psia. The condenser pressure is 1 psia. Determine (*a*) the efficiency of the cycle, and (*b*) the required mass flow rate through the steam generator for a turbine output equivalent to 100,000 kW.

Effect of irreversibilities on vapor cycle performance

17-32 The adiabatic efficiency of the turbine is 85 percent and that of the pump is 70 percent in a simple steam power cycle. One this basis determine the thermal efficiency for the data given in (*a*) Prob. 17-4*a*, (*b*) Prob. 17-4*b*, (*c*) Prob. 17-5*a*, (*d*) Prob. 17-5*b*, (*e*) Prob. 17-6*a*, (*f*) Prob. 17-6*b*, (*g*) Prob. 17-7, and (*h*) Prob. 17-8.

17-33 The adiabatic efficiency of the turbine is 85 percent and that of the pump is 70 percent in a reheat cycle. On this basis determine the thermal efficiency for the data given in (*a*) Prob. 17-12*a*, (*b*) Prob. 17-12*b*, (*c*) Prob. 17-13*a*, (*d*) Prob. 17-13*b*, (*e*) Prob. 17-14*a*, (*f*) Prob. 17-14*b*, (*g*) Prob. 17-15, and (*h*) Prob. 17-16.

17-34 The adiabatic efficiencies of the turbine and pump in a regenerative cycle are 85 and 70 percent, respectively. On this basis determine the thermal efficiency of the cycle for the data given in (*a*) Prob. 17-18*a*, (*b*) Prob. 17-18*b*, (*c*) Prob. 17-19*a*, (*d*) Prob. 17-19*b*, (*e*) Prob. 17-20*a*, (*f*) Prob. 17-20*b*, (*g*) Prob. 17-21(*a*), (*h*) Prob. 17-22(*a*), and (*i*) Prob. 17-23.

17-35 The adiabatic efficiencies of the turbine and pump in a reheat-regenerative cycle are 85 and 70 percent, respectively. On this basis determine the thermal efficiency for the data given in (*a*) Prob. 17-26*a*, (*b*) Prob. 17-26*b*, (*c*) Prob. 17-27*a*, (*d*) Prob. 17-27*b*, (*e*) Prob. 17-28*a*, (*f*) Prob. 17-28*b*, (*g*) Prob. 17-29, and (*h*) Prob. 17-30.

17-36 A simple steam power cycle uses solar energy for the heat input. Water in the cycle enters the pump as a saturated liquid at 120°F, and is pumped to 30 psia. It then evaporates in the boiler at this pressure, and enters the turbine as saturated vapor. At the turbine exhaust the conditions are 120°F and 6 percent moisture. The flow rate is 300 lb/h, and the pump is rated at $\frac{1}{2}$ hp. Determine (*a*) the net work output, (*b*) the energy-conversion efficiency, and (*c*) the area, in square feet, of solar collector needed if the collectors pick up 225 Btu/(ft^2)(h).

Supercritical steam power cycle

17-37 A supercritical steam power cycle operates with regeneration and reheat. The steam enters the turbine at 3500 pisa, 1200°F, and expands to 1 psia. Steam leaves the first stage at 200 psia, where part of it enters a closed heater, while the rest is reheated to 1000°F. Both sections of the turbine have an abiabatic efficiency of 89 percent. There is a condensate pump between the condenser and the heater. Another pump for the extracted steam that is condensed lies between the heater and the condensate-outlet line from the heater. The total feedwater enters the steam generator at 4000 psia. Determine (*a*) the enthalpies at the 10 states around the cycle, (*b*) the fraction of the total mass flow rate which is extracted to the heater, and (*c*) the thermal efficiency of the cycle.

17-38 In a supercritical steam power cycle steam expands ideally from 4800 psia and 1100°F to 3500 psia. It is then cooled at constant pressure to 650°F, followed by compression to 4800 psia. Heat is then added until the temperature reaches 1100°F. Determine (*a*) the amount of heat added, in Btu/lb, (*b*) the amount of heat rejected, and (*c*) the thermal efficiency.

17-39 A supercritical steam power plant operates with reheat and regeneration. The steam enters the turbine at 4400 psia, 1100°F, and expands to 1 psia. Steam leaves the first stage at 400 psia, where part of it enters a closed heater, while the rest is reheated to 1000°F. Both sections of the turbine have an adiabatic efficiency of 88 percent. There is a condensate pump between the main condenser and the heater. Another pump for the extracted steam that is condensed lies between the heater and the condensate-outlet line from the heater. Determine (*a*) the four enthalpies at the inlets and outlets of the closed heater, (*b*) the enthalpy at the condenser inlet and the enthalpy at the inlet to the heat-addition section, (*c*) the fraction of the total mass flow which is extracted to the heater, and (*d*) the thermal efficiency of the cycle.

High-temperature and binary vapor cycles

17-40 A Rankine cycle with potassium as the working fluid has a turbine-inlet state of 2400°R and 90 psia, and the temperature of the condensing fluid is 1600°R.

(*a*) Determine the thermal efficiency for an ideal cycle.

(*b*) Determine the thermal efficiency if the turbine- and pump-adiabatic efficiencies are 85 and 60 percent, respectively.

17-41 A Rankine cycle using potassium as the working fluid has a turbine-inlet state of 2600°R and 150 psia, and the condenser operates at 1700°R.

(*a*) Determine the quality at the turbine exit and the thermal efficiency for an ideal cycle.

(*b*) Repeat part *a* if the turbine and pump efficiencies are 84 and 50 percent, respectively.

17-42 A Potassium-Rankine cycle operates with a turbine-inlet state of 2200°R and 50 psia, and the condenser temperature is 1600°R.

(*a*) Determine the thermal efficiency for an ideal cycle.

(*b*) Repeat part *a* if the turbine and pump efficiencies are 86 and 65 percent, respectively.

17-43 A binary vapor power cycle operates with the potassium cycle described in Prob. 17-40(*b*). The heat rejected from this cycle enters the boiler-superheater section of a reheat-regenerative steam cycle. This cycle is described by Prob. 17-35(*c*). Assume that the heat for the reheat section of the steam cycle is provided from another source, and not from the potassium cycle. Determine (*a*) the ratio of the mass flow rate of potassium to that of steam, and (*b*) the overall thermal efficiency.

17-44 A binary vapor power cycle operates with the potassium cycle described in Prob. 17-41(*b*). The heat rejected from this cycle enters the boiler-superheater section of a reheat-regenerative steam cycle. This cycle is described by Prob. 17-35(*e*). Assume that the heat for the reheat section of the steam cycle is provided from another source, and not from the potassium cycle. Determine (*a*) the ratio of the mass flow rate of potassium to that of steam, and (*b*) the overall thermal efficiency.

17-45 A binary vapor power cycle operates with the potassium cycle described in Prob. 17-42(*b*). The heat rejected from this cycle enters the boiler-superheater section of a reheat-regenerative steam cycle. This cycle is described by Prob. 17-35(*a*). Assume that the heat for the reheat section of the steam cycle is provided from another source, and not from the potassium cycle. Determine (*a*) the ratio of the mass flow rate of potassium to that of steam, and (*b*) the overall thermal efficiency.

CHAPTER

EIGHTEEN

REFRIGERATION SYSTEMS

A refrigeration system is used to maintain a region of space at a temperature which is below that of the environment. The working fluid employed in the system may remain in a single phase (gas refrigeration), or may appear in two phases (vapor-compression refrigeration). It is common to associate refrigeration with the preservation of foods and the air conditioning of buildings. However, there are many other needs for refrigeration techniques. The use of liquid fuels for rocket propulsion, liquefied oxygen for steel making, liquid nitrogen for low-temperature (cyrogenic) research and surgery techniques, and liquefied natural gas for shipment between continents are a few examples where refrigeration is essential. The heat pump, which is capable of providing both a cooling and a heating effect with the same equipment, continues its popular use in residential and commercial buildings. In this chapter we shall examine some of the basic thermodynamic cycles used for the maintenance of low temperatures.

18-1 THE REVERSED CARNOT CYCLE

In the study of cyclic devices which operate for the purpose of removing heat continuously from a low-temperature source, it is useful to recall the features of the reversed Carnot cycle introduced in Chap. 6. A diagram of a reversed Carnot-cycle engine operating as a heat pump or refrigerator is shown in Fig. 18-1*a*. A quantity of heat Q_L is transferred reversibly from a low-temperature source of

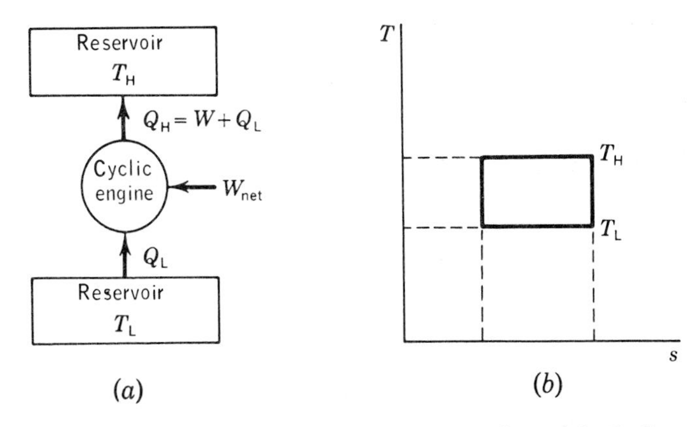

Figure 18-1 Schematic of a reversed Carnot heat engine and the Ts diagram for the cycle.

temperature T_L to the reversed heat engine. The reversed heat engine operates through a cycle during which net work W is added to the engine and a quantity of heat Q_H is transferred reversibly to a higher-temperature sink of temperature T_H. On the basis of the first law for a closed cyclic process, $Q_L + W = Q_H$. From the second law for a totally reversible process it is found that $T_H/T_L = Q_H/Q_L$. The reversed Carnot heat engine is useful as a standard of comparison because it requires the minimum work input for a given refrigeration effect between two given bodies of fixed temperature.

In the place of the thermal efficiency, which is used as a criterion in the analysis of heat engines, the energy-efficiency standard for refrigeration processes is the coefficient of performance. A performance standard is commonly defined as the ratio of the desired output to the costly input. The objective of a refrigerator is to remove heat from a low-temperature region in order to maintain the temperature at a desired value. Hence the coefficient of performance (COP) for a refrigerator is defined as

$$\mathrm{COP}_{\mathrm{refrig}} = \frac{Q_L}{W_{\mathrm{in}}} \tag{18-1}$$

Recall that the areas under the T_H and T_L lines in Fig. 18-1b represent Q_H and Q_L, respectively. Therefore, for a Carnot refrigerator,

$$\mathrm{COP}_{\mathrm{refrig,\,Carnot}} = \frac{T_L}{T_H - T_L} \tag{18-2}$$

It is important to note that the value of the COP can exceed unity, and, in fact, does so for a well-designed unit. Note also that the main variable which controls the COP of a Carnot refrigerator is the temperature difference $T_H - T_L$. For a Carnot heat engine, performance is improved by increasing T_H and decreasing T_L. The reverse is true for the Carnot refrigerator, in that T_H should be as low as possible and T_L should be as high as possible. However, T_H cannot be less than the temperature of the environment to which heat is rejected, and T_L cannot be greater than the temperature of the cold region from which heat is removed.

18-2 THE VAPOR COMPRESSION-REFRIGERATION CYCLE

Although the reversed Carnot cycle is a standard with which all actual cycles may be compared, it is not a practical device for refrigeration purposes. It would be highly desirable, however, to approximate the constant-temperature heat-addition and -rejection processes in order to achieve the highest coefficient of performance possible. This is accomplished in a large degree by operating a refrigeration device on a vapor-compression cycle. The schematic of the equipment for the cycle, along with *Ts* and *Ph* diagrams of the ideal cycle, is shown in Fig. 18-2. Saturated vapor at state 1 is compressed isentropically to a superheated vapor at state 2. The refrigerant vapor then enters a condenser, where heat is removed at constant presssure until the fluid becomes a saturated liquid at state 3. In order to return the fluid to a lower pressure, it is expanded adiabatically through a valve or a capillary tube to state 4. Process 3-4 is a throttling process, and $h_3 = h_4$. At state 4 the refrigerant is a low-quality wet mixture. Finally, it passes through the evaporator at constant pressure. Heat enters the evaporator

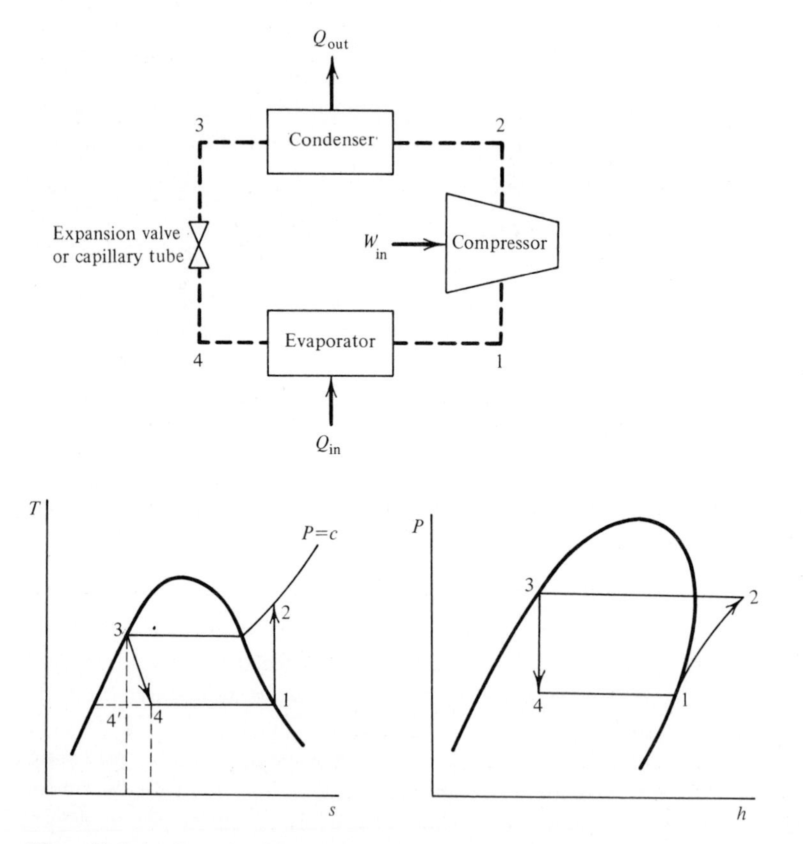

Figure 18-2 A schematic of the equipment and *Ts* and *Ph* diagrams for a vapor-compression refrigeration cycle.

from the cold-temperature source and vaporizes the fluid to the saturated-vapor state. Thus the cycle is completed. Note that all of process 4-1 and a large portion of process 2-3 occur at constant temperature. Unlike many other ideal cycles, the vapor-compression cycle modeled in Fig. 18-2 contains an irreversible process within it, the throttling process. All other portions of the cycle are assumed to be reversible.

The entire cycle could be made internally reversible if the throttling process 3-4 on the *Ts* diagram of Fig. 18-2 were replaced by the isentropic expansion process 3-4′ shown on the diagram. Theoretically, the work output of the expander could be used to help drive the compressor. Additionally, the refrigeration effect per unit mass of refrigerant is increased, because q_L now would occur from state 4′ to 1, rather than from 4 to 1. In other words, when throttling is employed the refrigeration effect is decreased by an amount equal to the area under the line 4′-4 in Fig. 18-2. Both the effect of a decreased amount of net work input and of an increased amount of refrigeration would increase the COP when a work expander is used, in comparison to throttling. However, in practice, a throttling or free-expansion process is used. First of all, the work output of an expander would be small because the fluid is mainly a liquid with a small specific volume. Secondly, a throttling device is much less expensive than a work expander and it is nearly maintenance free.

The rating of refrigeration systems is frequently given on the basis of the tons of refrigeration provided by the unit when operating at design conditions. A *ton of refrigeration* is defined as a heat-removal rate from the cold region (or the heat absorption rate by the fluid passing through the evaporator) of 211 kJ/min or 200 Btu/min. Another quantity frequently cited for a refrigeration device is the volume flow rate of the refrigerant at the compressor inlet. This is the effective displacement of the compressor.

Example 18-1M An ideal vapor-compression-refrigeration cycle with refrigerant 12 as the working fluid operates with an evaporator temperature of $-20°C$ and a condenser pressure of 9.0 bars. The mass flow rate of the refrigerant is 3 kg/min through the cycle. Compute the coefficient of performance, the tons of refrigeration, and the coefficient of performance of a Carnot reversed heat engine operating under the same maximum and minimum temperatures as the actual cycle.

SOLUTION Table A-16M to A-18M provide the following data for the states illustrated in Fig. 18-2:

$$h_1 = h_{g,\,\mathrm{at}\,-20°C} = 178.74\ \mathrm{kJ/kg} \qquad s_1 = s_g = 0.7087\ \mathrm{kJ/(kg)(°K)}$$

In addition, the fluid pressure in the evaporator is roughly 1.5 bars. At 9 bars the entropy of saturated vapor is 0.6832 kJ/(kg)(°K). Hence isentropic compression to state 2 leads to a superheated vapor. Interpolation in the superheat table at 9 bars indicates that T_2 is, roughly, 43°C and $h_{2s} = 208.2$ kJ/kg. At the end of the condensation process

$$h_3 = h_{f,\,\mathrm{at\ 9\ bars}} = 71.93\ \mathrm{kJ/kg} \qquad T_3 = T_{\mathrm{sat}} = 37.37°C$$

The enthalpy at state 3 is also the enthalpy at state 4, after the fluid passes through the throttling process. Consequently,

$$\text{COP} = \frac{h_1 - h_4}{h_2 - h_1} = \frac{178.74 - 71.93}{208.2 - 178.74} = 3.63$$

$$\text{Tons of refrigeration} = \frac{3(178.74 - 71.93)}{211} = 1.52$$

$$\text{COP}_{\text{Carnot}} = \frac{T_1}{T_2 - T_1} = \frac{253}{63} = 4.02$$

Example 18-1 An ideal vapor-compression-refrigeration cycle with refrigerant 12 as the working fluid operates with an evaporator temperature of 0°F and a condenser pressure of 100 psia. The mass flow rate of the refrigerant is 6 lb/min. Compute the coefficient of performance, the tons of refrigeration, the horsepower input per ton of refrigeration, and the coefficient of performance of a reversed Carnot heat engine operating under the same maximum and minimum temperatures.

SOLUTION Tables A-16 to A-18 provide the following data for the states illustrated in Fig. 18-2:

$$h_1 = h_{g,\text{ at } 0^\circ\text{F}} = 77.27 \text{ Btu/lb} \qquad s_1 = s_g = 0.1689 \text{ Btu/(lb)(°R)}$$

In addition, the fluid pressure in the evaporator is roughly 24 psia. At 100 psia the entropy of saturated vapor is 0.1639 Btu/(lb)(°R). Hence isentropic compression to state 2 leads to a slightly superheated vapor. Interpolation in the superheat table at 100 psia indicates that T_{2s} is, roughly, 96.6°F and $h_{2s} = 88.09$ Btu/lb. At the end of the condensation process

$$h_3 = h_{f,\text{ at } 100 \text{ psia}} = 26.54 \text{ Btu/lb} \qquad T_3 = T_{\text{sat}} = 80.76^\circ\text{F}$$

The enthalpy at state 3 is also the enthalpy at state 4, after the fluid passes through the throttling process. Consequently,

$$\text{COP} = \frac{h_1 - h_4}{h_2 - h_1} = \frac{77.27 - 26.54}{88.09 - 77.27} = 4.69$$

$$\text{Tons of refrigeration} = \frac{6(77.27 - 26.54)}{200} = 1.52$$

$$\text{hp/ton} = \frac{h_2 - h_1}{42.4} \frac{200}{h_1 - h_4} = \frac{4.715}{\text{COP}} = 1.01$$

$$\text{COP}_{\text{Carnot}} = \frac{T_1}{T_2 - T_1} = \frac{460}{96.6} = 4.76$$

In actual operation, a refrigeration cycle differs from the ideal cycle in a number of ways. The presence of fluid friction results in pressure drops throughout the cycle and in nonisentropic flow through the compressor. In addition, the presence of heat transfer must also be taken into account. Since it is not possible to control exactly the state of the fluid leaving the evaporator, it usually leaves as a superheated vapor, instead of as the saturated vapor found in the ideal cycle. Irreversibilities in the flow through the compressor lead to an increase in the entropy of the fluid during the process and a concomitant increase in the final temperature over that found in the ideal case. If heat losses from the compressor are great enough, however, the actual entropy of the fluid at the compressor exit can be less than that at the inlet. Even if pressure losses in the condenser are small, the fluid will probably leave the condenser as a subcooled liquid rather than as the saturated-liquid state assumed in the ideal cycle. This is a beneficial effect, since the lower enthalpy which results from the subcooling effect permits a larger quantity of heat to be absorbed by the fluid during the evaporation process.

The evaluation of certain parameters of interest in refrigeration cycles has been based on the saturation temperatures of the refrigerant in the evaporator and condenser. Nevertheless, the operating temperatures in an actual cycle are really established by the temperature desired to be maintained in the cold region and the temperature of the cooling water or air that is available for use in the condenser. Figure 18-3 illustrates this point. In order to obtain sufficient heat-transfer rates, the temperature difference between two fluids must be at least on the order of 10°C or 20°F. In the evaporator, heat is transferred from a cold region to the refrigerant, which undergoes a phase change at constant temperature. If the temperature of the cold region (T_{cr} on the figure) is to be −18°C (0°F), for example, the refrigerant would have to be maintained at a saturation temperature corresponding to, say, −25°C (−15°F) for effective heat transfer. At the same time the refrigerant is being condensed in the condenser by the transfer of heat to a coolant flow stream external to the cycle. Cooling water and atmospheric air are two typical coolants which might be passed over the condenser tubes. Since these two substances usually would be available at temperatures from 15 to 30°C or 60 to 90°F, roughly (T_{cw} in the figure), the saturation temperature of the refrigerant in the condenser must be above this range.

Example 18-2M Reconsider the ideal vapor-compression-refrigeration cycle introduced in Example 18-1M. The cold region is to be maintained at −20°C. To allow for a finite temperature difference between the cold region and the evaporator fluid, the evaporator will be set to operate at −30°C. The air which passes over the condenser coils in the preceding example is at the saturation temperature corresponding to 9.0 bars, or 37.37°C. Again, to allow for a finite temperature difference, the condenser pressure will be raised to 12.0 bars, for which the saturation temperature is 49.31°C. Determine the coefficient of performance of the cycle under the new conditions, and compare with the result of Example 18-1.

SOLUTION Table A-16M to A-18M provide data for the problem. At the evaporator outlet

$$h_1 = h_{g,\text{ at }-30^\circ\text{C}} = 174.20 \text{ kJ/kg} \qquad s_1 = s_g = 0.7170 \text{ kJ/(kg)(°K)}$$

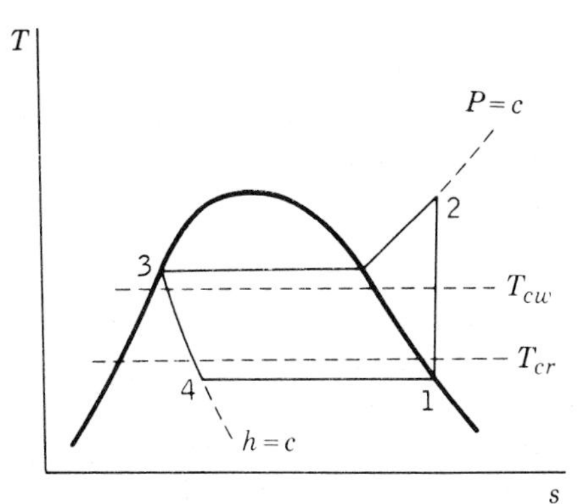

Figure 18-3 Effect of irreversible heat transfer on the performance of a vapor-compression refrigeration cycle.

State 2 at the compressor outlet is found from knowledge of P_2 and the fact that $s_2 = s_1$. By interpolation at 12 bars in the superheat table, we find that T_{2s} is, roughly, 64.6°C and $h_{2s} = 218.56$ kJ/kg. At the end of the condensation process

$$h_3 = h_{f,\,\text{at 12 bars}} = 84.21 \text{ kJ/kg}$$

The enthalpy at state 3 is also the enthalpy at state 4, after the fluid passes through the throttling process. Consequently,

$$\text{COP} = \frac{h_1 - h_4}{h_2 - h_1} = \frac{174.20 - 84.21}{218.56 - 174.20} = 2.03$$

By allowing for a 10°C temperature differential at the evaporator end of the cycle and an 11.9°C difference at the condenser, the COP has decreased from 3.63 to 2.03. Although these calculations were based on an idealized cycle, the trend in the actual case would be similar. The coefficient of performance is severely affected by the requirement of finite temperature differences across the condenser and evaporator heat-transfer surfaces.

Example 18-2 Reconsider the ideal vapor-compression-refrigeration cycle introduced in Example 18-1. The cold region is to be maintained at 0°F. To allow for a finite temperature difference between the cold region and the evaporator fluid, the evaporator will be set to operate at −20°F. The air which passes over the condenser coil in the preceding example is at the saturation temperature corresponding to 100 psia, or 80.76°F. Again, to allow for a finite temperature difference, the condenser pressure will be raised to 120 psia, for which the saturation temperature is 93.29°F. Determine the coefficient of performance for the cycle under the new conditions, and compare with the result of Example 18-1.

SOLUTION Tables A-16 to A-18 provide data for the problem. At the evaporator outlet

$$h_1 = h_{g,\,\text{at} -20^\circ\text{F}} = 75.11 \text{ Btu/lb} \qquad s_1 = s_g = 0.1710 \text{ Btu/(lb)(°R)}$$

State 2 at the compressor outlet is found from knowledge of P_2 and the fact that $s_2 = s_1$. By interpolation at 120 psia in the superheat table, we find that T_{2s} is, roughly, 117.4°F and $h_{2s} = 90.78$ Btu/lb. At the end of the condensation process

$$h_3 = h_{f,\,\text{at 120 psia}} = 29.49 \text{ Btu/lb}$$

The enthalpy at state 3 is also the enthalpy at state 4, after the fluid passes through the throttling process. Consequently,

$$\text{COP} = \frac{h_1 - h_4}{h_2 - h_1} = \frac{75.11 - 29.49}{90.78 - 75.11} = 2.91$$

By allowing for a 20°F temperature differential at the evaporator end of the cycle and a 12.5°F difference at the condenser, the COP has decreased from 4.69 to 2.91. Although these calculations were based on an idealized cycle, the trend in the actual case would be similar. The coefficient of performance is severely affected by the requirement of finite temperature differences across the condenser and evaporator heat-transfer surfaces.

In the vapor-refrigeration cycle the two desired saturation temperatures for the evaporation and condensation processes determine the operating pressures of the cycle for a given refrigerant. Consequently, the choice of a refrigerant partially depends upon the saturation pressure-temperature relationship in the range of interest. Normally, the minimum pressure in the cycle should be above 1 atm to avoid leakage into the equipment, but maximum pressures over 150 to 200 psia (10 to 15 bars) would be undesirable. In addition, the fluid needs to be nontoxic, stable, and of low cost and to have a relatively high enthalpy of vaporization. These restrictions and others limit the number of compounds which are suitable for refrigerants. In fact, because of the range of applicability of refrigeration cycles, no one fluid is suitable for all cases. Even with a proper choice of the refrigerant, a number of modifications may be made on the basic cycle to improve the coefficient of performance. Such changes are discussed in standard textbooks and handbooks on refrigeration.

Example 18-3M Reconsider the vapor-compression cycle analyzed in Example 18-2M. The evaporator operates at $-30°C$, and the condenser pressure is 12.0 bars, for which the saturation temperature is 49.31°C. To further modify the cycle, we shall consider that the fluid leaves the evaporator superheated by 10°C, and that it leaves the condenser subcooled by 1.31°C. In addition, the compressor is essentially adiabatic, with an efficiency of 75 percent. Calculate the coefficient of performance.

SOLUTION The saturation pressure at $-30°C$ is very close to 1.0 bar. Thus the fluid leaving the evaporator is at 1 bar and $-20°C$. From the superheat table the enthalpy h_1 and the entropy s_1 are found to be 179.99 kJ/kg and 0.7406 kJ/(kg)(°K), respectively. The isentropic enthalpy h_{2s} at the compressor outlet (12 bars) is found from the superheat table to be 227.86 kJ/kg at roughly 76.5°C. Consequently, the actual outlet enthalpy is

$$h_{2a} = h_1 + \frac{w_s}{\eta_C} = 179.99 + \frac{228.86 - 179.99}{0.75} = 243.82 \text{ kJ/kg}$$

Since the fluid at the condenser outlet is subcooled, its enthalpy at state 3 is approximated by the saturated-liquid enthalpy at $T_3 = 49.31 - 1.31 = 48.0°C$. Hence $h_3 = 82.83$ kJ/kg, and this value is also h_4. Therefore,

$$\text{COP} = \frac{h_1 - h_4}{h_2 - h_1} = \frac{179.99 - 82.83}{243.82 - 179.99} = 1.52$$

The effects of superheating at the evaporator outlet, subcooling at the condenser outlet, and irreversibilities in the compressor lower the COP found in Example 18-2M from 2.03 to 1.52.

Example 18-3 Reconsider the vapor-compression cycle analyzed in Example 18-2. The evaporator operates at $-20°F$, and the condenser pressure is 120 psia, for which the saturation temperature is 93.29°F. To further modify the cycle, we shall consider that the fluid leaves the evaporator superheated by 10°F, and that it leaves the condenser subcooled by 3.29°F. In addition, the compressor is essentially adiabatic, with an efficiency of 75 percent. Calculate the coefficient of performance.

SOLUTION The saturation pressure at $-20°F$ is 15.267 psia. Thus the fluid leaving the evaporator is at 15.267 psia and $-10°F$. To approximate the data at state 1, we shall use the pressure of 15.0 psia in the superheat table. By interpolation between the saturation temperature of $-20.75°F$ and 0°F, we find that the enthalpy h_1 and the entropy s_1 are 76.52 Btu/lb and 0.1744 Btu/(lb)(°R), respectively. The isentropic enthalpy h_{s2} at the compressor outlet (120 psia) is found from the superheat table to be 92.75 Btu/lb at roughly 124°F. Consequently, the actual outlet enthalpy is

$$h_{2a} = h_1 + \frac{w_s}{\eta_C} = 76.52 + \frac{92.75 - 76.52}{0.75} = 98.16 \text{ Btu/lb}$$

Since the fluid at the condenser outlet is subcooled, its enthalpy at state 3 is approximated by the saturated-liquid enthalpy at $T_3 = 93.29 - 3.29 = 90°F$. Hence $h_3 = 28.71$ Btu/lb, and this value is also h_4. Therefore

$$\text{COP} = \frac{h_1 - h_4}{h_2 - h_1} = \frac{76.52 - 28.71}{98.16 - 76.52} = 2.21$$

The effects of superheating at the evaporator outlet, subcooling at the condenser outlet, and irreversibilities in the compressor lower the COP found in Example 18-2 from 2.91 to 2.21.

18-3 THE HEAT PUMP

A refrigerator removes heat Q_L from a cold region and rejects heat Q_H to the environment. Its major purpose is heat removal from the cold region. However, the same basic cycle could have as a major purpose the supply of heat Q_H to a living space, such as a home or commercial building. In this latter case the heat removed comes from the cooler environment. In fact, the modern heat pump combines both heating and cooling of a region of space in the same unit. When cooling is required, the heat-pump system operates like an air conditioner. In this refrigeration mode, heat Q_L is removed from a living space and heat Q_H is rejected outside the building to the environment. In the air-conditioning mode the COP is given by Eq. (18-1). In terms of the ideal refrigeration cycle shown in Fig. 18-2, the COP for cooling is

$$\text{COP}_{\text{heat pump, cooling}} = \frac{Q_L}{W_{\text{in}}} = \frac{h_1 - h_4}{h_2 - h_1} \tag{18-3}$$

In the heating mode, the heat-pump system in the wintertime removes heat Q_L from the environment and rejects heat Q_H to the living space. On the basis of desired output and costly input, the coefficient of performance for a heat pump in the heating mode is defined as

$$\text{COP}_{\text{heat pump, heating}} = \frac{Q_H}{W_{\text{in}}} = \frac{h_2 - h_3}{h_2 - h_1} \tag{18-4}$$

Because the areas under the T_H and T_L lines on the Ts diagram in Fig. 18-1*b* represent Q_H and Q_L, respectively, the COP of a Carnot heat pump operating in the heating mode is

$$\text{COP}_{\text{heat pump, heating, Carnot}} = \frac{T_H}{T_H - T_L} \tag{18-5}$$

Thus the performance of a heat pump in the heating mode is improved by decreasing the temperature difference $T_H - T_L$, similar to that of a refrigerator or air conditioner.

One method of decreasing $T_H - T_L$ is to increase T_L through the use of a solar collector. Such units are called solar-assisted heat pumps. Solar energy is collected by a fluid circulating through solar panels. The relatively hot fluid is stored in a large, insulated tank. By proper regulation of the flow, the temperature of the fluid within the tank may rise to a range of 20 to 30°C, or 70 to 90°F,

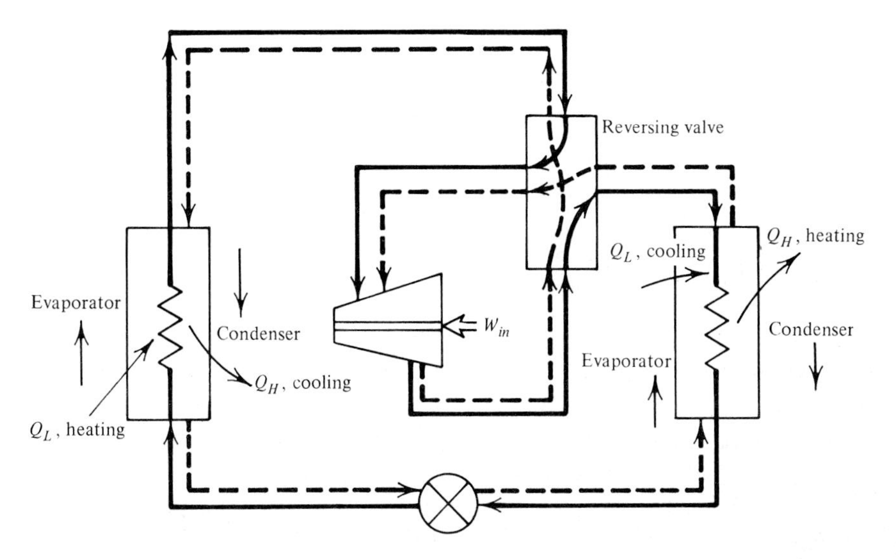

Figure 18-4 Equipment flow diagram for a heat pump in a heating mode (solid line) and in a cooling mode (dashed line).

for example, even though the outside temperature may be below the freezing point of water. The evaporator of the heat pump is placed internally within the tank, so that the fluid within the heat-pump system evaporates at a higher temperature (and pressure) than normal. Although more equipment is required, the solar-assisted heat pump will have a much higher coefficient of performance when compared to operation of the evaporator in the cold environment. For the same heating load, the electrical work input will be reduced sizably when solar energy assists the operation.

When a heat pump is used to air condition a building, the evaporator is within the building and the condenser is outside the building. However, in a heating mode, the evaporator is outside the building and the condenser is inside. It would be impractical to have two sets of equipment, so each heat exchanger (one inside and one outside the building) must act as both a condenser and an evaporator, depending upon the mode of operation. One method of accomplishing this is to add a reversing valve in the cycle, in addition to the compressor and throttling device. A schematic of this design is shown in Fig. 18-4. The solid flow line is the flow direction for the heating mode of operation, and the dashed line is for the cooling mode. Note that the direction of flow through the compressor is always the same, regardless of the mode of operation.

18-5 GAS-REFRIGERATION CYCLES

The adiabatic expansion of gases can be used to produce a refrigeration effect. In its simplest form this is accomplished by reversing the Brayton cycle used as a

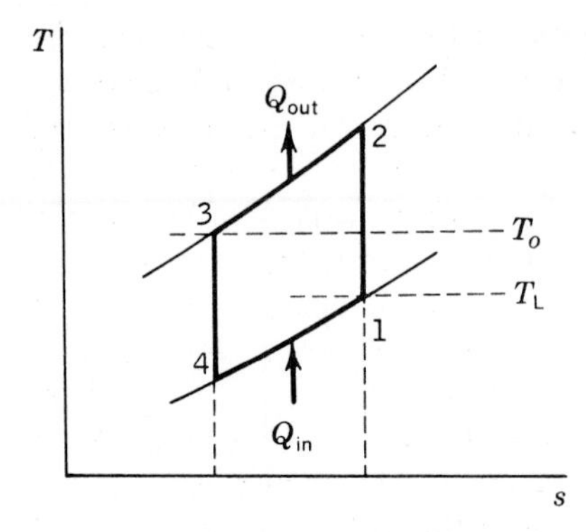

Figure 18-5 Gas-refrigeration cycle using a reversed Brayton cycle.

standard for gas power cycles. Figure 18-5 is a Ts diagram for the ideal reversed Brayton cycle. The gas is first compressed isentropically from state 1 to state 2, which is at a relatively high temperature compared with the ambient temperature T_0. It then passes through a heat exchanger, where in the limit it can be cooled to the ambient temperature. (In practice, state 3 may be 10° or so above the ambient condition.) The gas then enters an adiabatic-expansion device, such as a turbine, where work is extracted and the gas is cooled to a temperature T_4, which is considerably below the temperature T_L of the cold region. Consequently, heat may be transferred from the cold region to the gas through a heat exchanger, and ideally the gas will be heated to the temperature T_L, which is also T_1 in the cycle. The fluid in the ideal-gas-refrigeration cycle is assumed to pass through the cycle in an internally reversible manner. For steady-flow operation, and in the absence of kinetic- and potential-energy changes, the heat and work interactions for each piece of equipment may be evaluated in terms of the enthalpy change during the process. On the Ts diagram in Fig. 18-5, the area under the curve 4-1 represents the heat removed from the cold region, the area enclosed by the cycle 1-2-3-4-1 represents the net work input, and the ratio of these areas is a measure of the COP of the device. In this case

$$\text{COP} = \frac{q_{\text{refrig}}}{w_{\text{net, in}}} = \frac{q_{\text{in}}}{w_{\text{C}} - w_{\text{T}}} = \frac{h_1 - h_4}{(h_2 - h_1) - (h_3 - h_4)} \tag{18-6}$$

The reversed Brayton cycle has the disadvantage of large temperature variations of the fluid during the heat-addition and heat-removal processes. Hence, the coefficient of performance of this cycle is considerably below that for a reversed Carnot engine operating between the same minimum temperatures, T_0 and T_L.

Modifications of the reversed Brayton cycle lead to some useful applications of gas-refrigeration cycles. For example, in Fig. 18-5, it is noted that the temperature of the fluid after picking up heat from the cold region is below that of state 3, where the fluid enters the expansion engine. If the gas at state 1 could be used to cool the gas further, below the temperature of state 3, subsequent expansion would lead to a temperature below that of state 4. In this manner extremely low temperatures might be achieved. Physically, this is accomplished by inserting a heat exchanger internally into the cycle, as shown in Fig. 18-6. Heat transfer external to the cycle results in the usual temperature drop from state 2 to state 3 in the figure. The additional regenerator, however, permits the gas to be cooled further to state 4 in the ideal case. The gas, after expansion, receives heat from the cold region from state 5 to state 6, and receives a further quantity of heat in the heat exchanger from state 6 to state 1. Such use of heat exchangers internal to the cycle is important in processes for the liquefaction of gases. The gas refrigeration cycle in the form of an open cycle is employed for the purpose of aircraft cooling, since it has a definite weight advantage over vapor-compression refrigeration. There are a number of modifications of the technique, depending, for example, on the type of aircraft. In general, however, air is compressed, cooled by rejecting heat to the ambient atmosphere, and then expanded through a turbine. The cool air leaving the turbine goes directly into the cabin of the aircraft.

Example 18-4M A reversed Brayton cycle is proposed to operate between $T_0 = 27°C$ and $T_L = -8°C$ (see Fig. 18-5 for notation). The compression and expansion ratios are 3 : 1. Determine the coefficient of performance.

Solution From the air table A-5M for $T_1 = -8°C$, $h_1 = 265.1$ kJ/kg and $p_{r1} = 0.900$. State 2 is found by noting that

$$p_{r2} = p_{r1} \frac{P_2}{P_1} = 0.900(3) = 2.70$$

For this p_r value, the air table indicates that $T_{2s} = 363°K$ (90°C) and $h_{2s} = 363.6$ kJ/kg. A similar calculation can be carried out for process 3-4. At 27°C, $h_3 = 300.2$ kJ/kg and $p_{r3} = 1.386$. Consequently, $p_{r4} = 1.386/3 = 0.462$, $T_{4s} = 219°K$ (−54°C), and $h_{4s} = 219.0$ kJ/kg. The compressor work, the turbine work, and the heat removed from the cold region are found by

$$w_{C,\,in} = h_{2s} - h_1 = 363.6 - 265.1 = 98.5 \text{ kJ/kg}$$

$$w_{T,\,out} = h_3 - h_{4s} = 300.2 - 219.0 = 81.2 \text{ kJ/kg}$$

$$q_{in} = h_1 - h_{4s} = 265.1 - 219.0 = 46.1 \text{ kJ/kg}$$

Therefore the coefficient of performance is

$$\text{COP} = \frac{q_{in}}{w_C - w_T} = \frac{46.1}{98.5 - 81.2} = 2.66$$

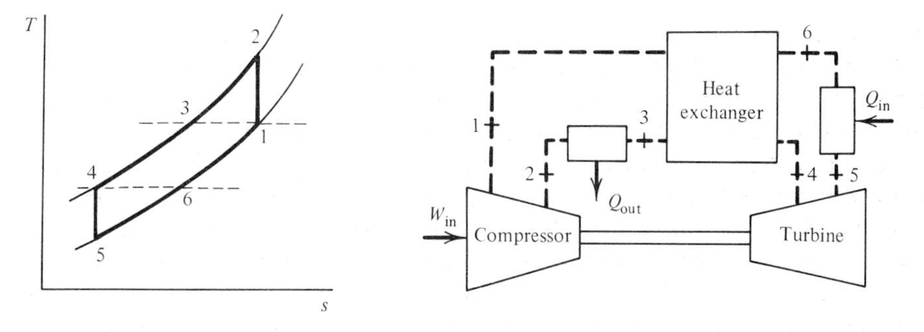

Figure 18-6 Gas-refrigeration cycle using a heat exchanger (regenerator) internal to the cycle.

This coefficient of performance is somewhat high, since we have not accounted for the irreversibilities in the turbine and compressor, among other things.

Example 18-4 A reversed Brayton cycle is proposed to operate between $T_0 = 80°\text{F}$ and $T_L = 20°\text{F}$ (see Fig. 18-5 for notation). The compression and expansion ratios are 3 : 1. Determine the coefficient of performance.

SOLUTION From the air table A-5 for $T_1 = 20°\text{F}$, $h_1 = 114.7$ Btu/lb and $p_r = 0.9182$. State 2 is found by noting that

$$p_{r2} = p_{r1}\frac{P_2}{P_1} = 0.9182(3) = 2.755$$

For this p_r value, $T_{2s} = 655°\text{R}$ (195°F) and $h_{2s} = 157.0$ Btu/lb. A similar calculation can be carried out for process 3-4. At 80°F, $h_3 = 129.1$ Btu/lb and $p_{r3} = 1.386$. Consequently $p_{r4} = 1.386/3 = 0.462$, $T_{4s} = 395°\text{R}$ (−65°F), and $h_{4s} = 94.3$ Btu/lb. The compressor work, the turbine work, and the heat removed from the cold region are found by

$$w_{\text{C, in}} = h_2 - h_1 = 157.0 - 114.7 = 42.3 \text{ Btu/lb}$$

$$w_{\text{T, out}} = h_3 - h_4 = 129.1 - 94.3 = 34.8 \text{ Btu/lb}$$

$$q_{\text{in}} = h_1 - h_4 = 144.7 - 94.3 = 20.4 \text{ Btu/lb}$$

Therefore the coefficient of performance is

$$\text{COP} = \frac{q_{\text{in}}}{w_C - w_T} = \frac{20.4}{42.3 - 34.8} = 2.72$$

This coefficient of performance will be lowered when we account for the irreversibilities in the turbine and compressor, among other things.

18-5 LIQUEFACTION AND SOLIDIFICATION OF GASES

In modern technology the preparation of liquids at temperatures below −75°C or −100°F is quite important. In the study of the properties and the behavior of substances at low temperatures, the materials are placed in baths composed of liquefied gases. Mixtures of gases may be separated by liquefaction techniques. For example, liquid oxygen and nitrogen are separated from air in this manner, and in the same manner rarefied gases such as helium may be obtained. Liquefied gases are also successful rocket propellants. Liquid helium and hydrogen are especially useful in research studies in the temperature range from 2 to 30°K, such as those of superconductivity and superfluidity. The following discussion provides a brief introduction to the thermodynamic cycles employed in the liquefaction and solidification of gases.

In general, two criteria must be met if a process is to alter a gas into the liquid or solid phase successfully. First, the gas must be cooled below its critical temperature, since only below the critical temperature can one make a distinction between the liquid and gas phases. The critical temperatures of a number of substances are given in Table A-3. For common monatomic and diatomic gases, the critical temperature is considerably below atmospheric values. Hence these substances exist in the liquid phase only at extremely low temperatures. The second criterion involves the Joule-Thompson coefficient and the maximum inversion temperature of a substance. Liquefaction processes rely on throttling to achieve a portion of the required cooling effect. The change in temperature that occurs when the pressure is decreased during throttling is measured in terms of $(\partial T/\partial P)_h$. This latter quantity is known as the Joule-Thompson coefficient μ_{JT} (see Sec. 13-6). When μ_{JT} is positive, a cooling effect will occur, because the temperature decreases with a decrease in pressure. For a given substance at a given pressure, there is a temperature where $\mu_{JT} = 0$ and above which the value of μ_{JT} is negative. This temperature is known as the inversion temperature for that pressure. Every gas has a maximum inversion temperature, and for most gases this is well above room temperature, except for hydrogen, helium, and neon. Hence, considerable precooling of these three gases is necessary before one can take advantage of the Joule-Thompson effect.

A simple cycle for liquefying gases is illustrated in Fig. 18-7. Makeup gas enters the steady-flow system at state 6 and after mixing, the gas is compressed to an elevated pressure and temperature (state 1). Note that the compression is shown not to be isentropic on the *Ts* diagram. In fact, in practice, multistaging with intercooling is used, so that the compression process is closer to isothermal than isentropic. Before undergoing a throttling process, the gas is cooled to state 2 by passing it through an efficient counterflow heat exchanger. (The use of the aftercooler shown in the figure is discussed in a subsequent paragraph of this section.) The gas is now throttled to state *a*, which must lie within the wet region (or solid-gas region) if a phase change is to occur. The liquid is collected in the

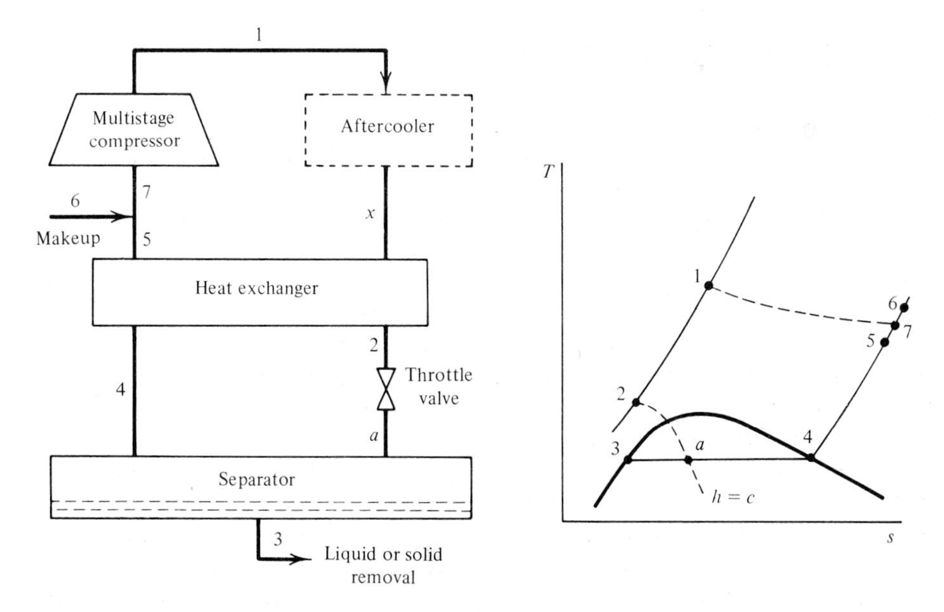

Figure 18-7 Equipment schematic and *Ts* diagram of a basic cycle for liquefying gases.

bottom of the separation chamber as a saturated liquid at state 3, and the remaining gas is passed back through the heat exchanger. Since the gas leaving the separator as saturated vapor at state 4 may be considerably colder than the gas at state 1, it may be used effectively to cool the gas stream passing from the compressor to the throttling valve. The gas now at state 5 is mixed with makeup gas, and the cycle is repeated. The process is traced on the *Ts* diagram in Fig. 18-7.

Consider a control volume drawn around the heat exchanger, throttling valve, and separator shown in Fig. 18-7. It is assumed that these three pieces of equipment are adiabatic and the changes in kinetic and potential energies are negligible. The steady-state energy equation for this control volume consequently reduces to $\Delta H = 0$. Let a unit mass enter the control volume and let y equal the fraction of liquid or solid formed in the separator per unit mass entering. Therefore, on the basis of symbols shown in Fig. 18-7,

$$yh_3 + (1 - y)h_5 = 1(h_1)$$

Solving for y, we find that

$$y = \frac{h_5 - h_1}{h_5 - h_3} \tag{18-7}$$

The value of h_3 is set by the pressure in the separator, P_3. At the same time h_5 is set by the design of the heat exchanger. The enthalpy h_1 is fixed by T_1 and P_1; however, T_1 is set by the design of the separator and heat exchanger. Consequently, the compressor-outlet pressure P_1 is the main variable controlling the fraction of the gas which is liquefied or solidified. It is noted that y is made larger by decreasing the value of h_1. It may be shown that the maximum degree of liquefaction occurs when the inital value of P_1 for a given temperature of the gas entering the heat exchanger is the inversion pressure at that temperature.

An interesting application of the apparatus described in Fig. 18-7 is in the production of solid carbon dioxide, or dry ice. A *Ts* diagram for carbon dioxide appears as Fig. A-24. Figure 18-8 is a representation of the approximate path of

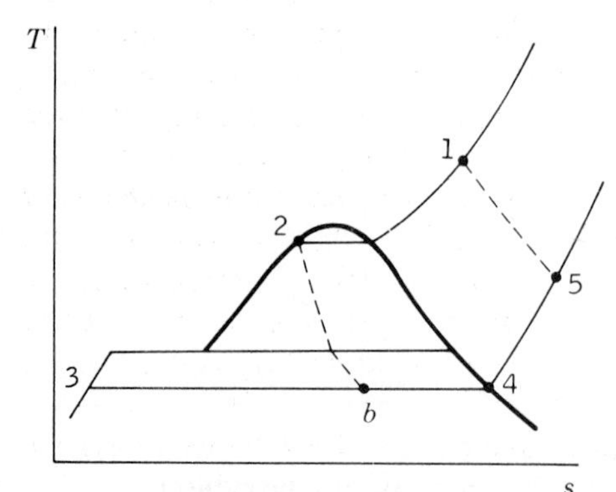

Figure 18-8 *Ts* diagram of the cycle for the production of solid carbon dioxide.

the process on a *Ts* diagram. Carbon dioxide at atmospheric conditions (state 5) is compressed by multistaging to a high pressure (roughly, 70 bars or 1000 psia), represented by state 1. It then passes through the heat exchanger, where it is cooled and condensed to a saturated liquid at state 2. The corresponding temperature in this liquid state is room temperature. The liquid is then expanded through a throttling valve to atmospheric pressure in the separator. The resulting mixture is solid and gas at state *b*. The corresponding saturation states are states 3 and 4 on the figure. Solid is formed because the triple state of carbon dioxide is, roughly, 5 atm and $-57°C$ or $-70°F$. Consequently, liquid carbon dioxide cannot exist at a pressure of 1 atm. The sublimation temperature of carbon dioxide at atmospheric pressure is approximately $-80°C$ or $-110°F$. Solid carbon dioxide will be formed by cooling to this temperature by throttling. The solid phase is pressed into bars, and the gas phase at state 4 and the required makeup gas return to the compressor at state 5.

A number of modifications of the basic cycle lead to improved operation and efficiency. One of these is the use of an aftercooler or precooler, which appears between the compressor and the heat exchanger. This is indicated by the dashed box in Fig. 18-7. In practice, a separate refrigerating system is used to precool the gas as it passes through the aftercooler. Any auxiliary refrigeration device such as this is inherently more efficient in cooling the gas, compared with a throttling process, because the refrigeration device can be made more reversible than a throttling process. The inherent irreversibilities of a throttling process can be partially overcome by another method of modifying the basic liquefaction cycle discussed above. In addition to cooling by the throttling process, a gas can be cooled by expansion through a device such as a turbine. This was discussed in Sec. 18-4 with respect to gas-refrigeration cycles. If the throttling device is simply replaced by an expander, operating difficulties will ensue, because of the complications introduced by two-phase flow through a turbine, which is highly undesirable. This is overcome in the Claude system for liquefying gases by employing both a throttling valve and an expansion engine. Figure 18-9 illustrates the process schematically and on a *Ts* diagram. Multistage compression with intercooling changes the gas from state 9 to state 1. An auxiliary refrigerating system cools the gas further to state 2, at which point the flow stream splits. A small portion continues through another heat exchanger, and then through the throttling valve into the separator. The other portion passes through an expansion engine, where it is cooled, and it then rejoins the flow stream coming from the separator. After passing through additional heat exchangers, the main flow returns to the compressor and begins the cycle again. By placing the expander early in the cycle, a cooling effect is achieved, but the fluid, even after expansion, still lies outside the wet region. Thus two-phase flow in the expansion device is avoided. The work output of the expander can be used as work input to the compressor section of the cycle. The overall result is a smaller expenditure of net work input for a given amount of liquefaction.

Other modifications for systems which liquefy gases at low temperatures are possible. These are discussed fully in texts and handbooks on the subject.

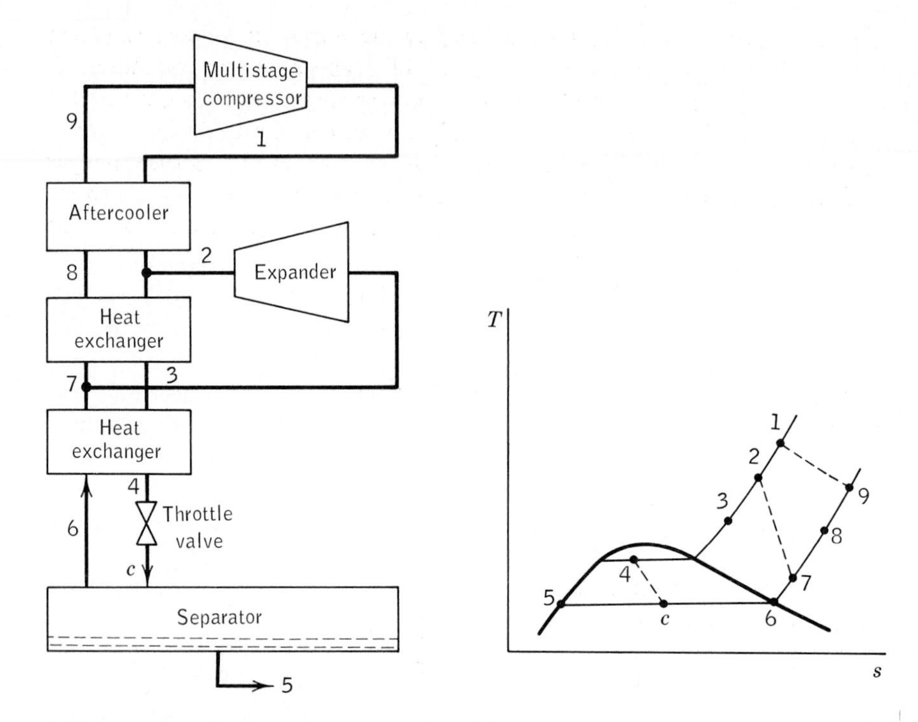

Figure 18-9 Schematic and *Ts* diagram for gas-liquefaction process involving a turbine expander.

18-6 CASCADE AND MULTISTAGED VAPOR-COMPRESSION SYSTEMS

This section is devoted to two variations of the basic vapor-compression refrigeration cycle. The first of these is the cascade cycle, which permits the use of a vapor-compression cycle when the temperature difference between the evaporator and condenser is quite large. The second variation involves the use of multistaged compression with intercooling, which reduces the required compressor-work input.

a Cascade Cycle

The discussion in the preceding section was directed toward methods of attaining extremely low (cyrogenic) temperatures by a combination of vapor compression and throttling. Such methods are valuable and indispensable for the liquefaction and solidification of gases. There are industrial applications, however, where only moderately low temperatures are required and less complex systems are needed. This is especially true when temperatures in the range of -25 to $-75°C$ (-10 to $-100°F$) are required. Unfortunately, a single vapor-compression cycle usually cannot be used to achieve these moderately low temperatures. The temperature difference between the condenser and evaporator is now quite large. Consequently, the vapor pressure-saturation temperature variation for any single refrigerant would not fit the desired values for the evaporator and condenser. To overcome this difficulty, and still rely on vapor compression, a cascade system may be employed. A *cascade cycle* is simply an arrangement of simple vapor-compression cycles in series, such that the condenser of a lower temperature cycle provides the heat input to the evaporator of a higher temperature cycle, as shown in Fig. 18-10*a*. Although only two units are shown, the use of three or four units

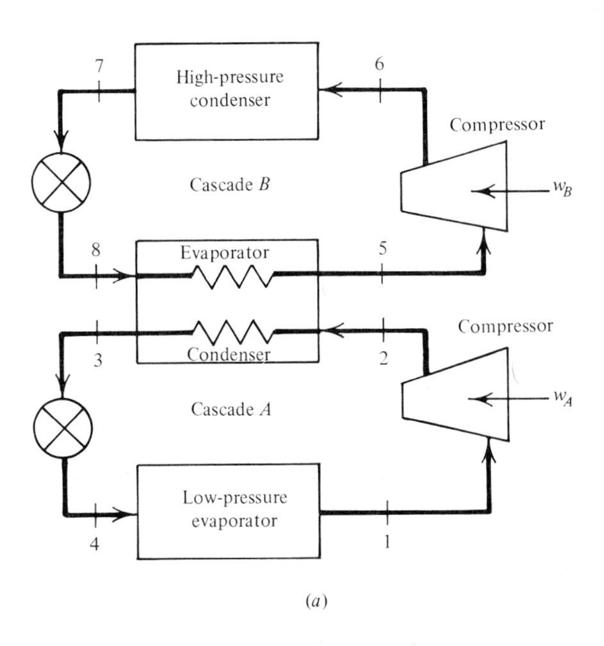

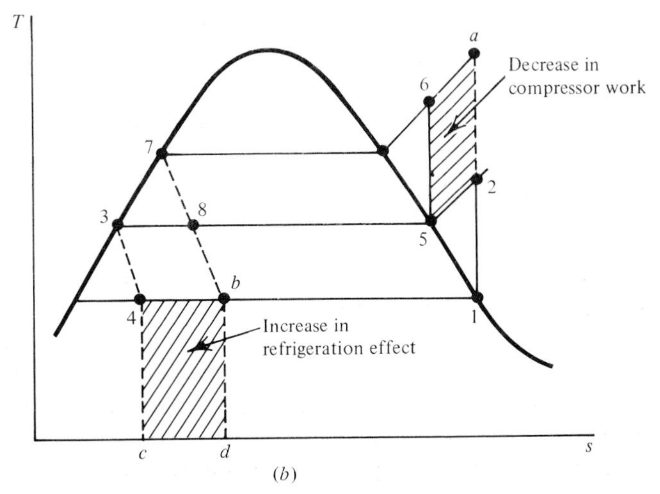

Figure 18-10 Equipment schematic and Ts diagram for a cascade refrigeration cycle.

in series is practical, if needed. Normally a different refrigerant would be used in each separate cycle, in order to match the desired ranges of temperature and pressure. When choosing the two refrigerants in Fig. 18-10*a*, for example, it is important that the triple-state temperature of the fluid in cycle *B* be lower than the critical temperature of the fluid in cycle *A*.

Figure 18-10*b* shows a *Ts* diagram for an ideal double cascade system employing the same refrigerant in each loop. (If two different refrigerants are used in a cascade system, then two separate *Ts* diagrams must be used.) Although not the normal practice, as noted above, the use of the same refrigerant in each loop does provide a basis for discussing the virtues of a cascade system. The positions of cycle *A* (1-2-3-4) and cycle *B* (5-6-7-8) are clearly shown in the figure. The mass flow rates of refrigerant in the two cycles are usually not the same, whether the refrigerant is the same or is different. The mass flow rate $\dot{m}_A$ is fixed by the tons of refrigeration required in the evaporator of cycle *A*. In addition, the rate of heat transfer from the condenser of cycle *A* must equal the rate of heat transfer to the fluid in the evaporator of cycle *B*, if the overall heat exchanger is well insulated. An energy balance on the condenser-evaporator heat exchanger reveals that

$$\dot{m}_A(h_2 - h_3) = \dot{m}_B(h_5 - h_8) \qquad \text{(cascade system)} \qquad (18\text{-}8)$$

Thus the ratio of the mass flow rates through each cycle is fixed by the enthalpy changes of each fluid as it passes through the heat exchanger.

If a single refrigeration cycle could be used for the overall temperature range, then this would be represented by the cycle 1-*a*-7-*b*-1 on Fig. 18-10*b*. Two significant effects are apparent from the *Ts* diagram. First, for the single cycle the compressor work is increased by the area 2-*a*-6-5, in comparison to the cascade system. Secondly, there is a decrease in the refrigeration capacity when a single unit is used, for the same mass flow rate through the low-temperature evaporator. This loss is represented by the area 4-*b*-*d*-*c* on the *Ts* diagram. These two effects would lead to a higher COP for the cascade system in comparison to the single unit.

b Multistage Vapor Compression

Another modification of the vapor-compression-refrigeration cycle involves multistage compression with intercooling in order to decrease the work input. For gas power cycles (see Sec. 16-11), the heat removed from the intercooler is usually transferred to the environment. In a refrigeration cycle the sink for the energy can be the circulating refrigerant itself, because in many sections of the cycle the temperature of the refrigerant is below the environmental temperature. Hence, the intercooler heat exchanger becomes a regenerative heat exchanger, since the heat transfer now occurs internal to the system. One scheme for two-stage compression with regenerative intercooling is shown in Fig. 18-11*a*. The liquid leaving the condenser is throttled (process 5-6) into a flash chamber maintained at a pressure between the evaporator and condenser pressures. All the vapor separated from the liquid in the flash chamber is transferred to a mixing chamber,

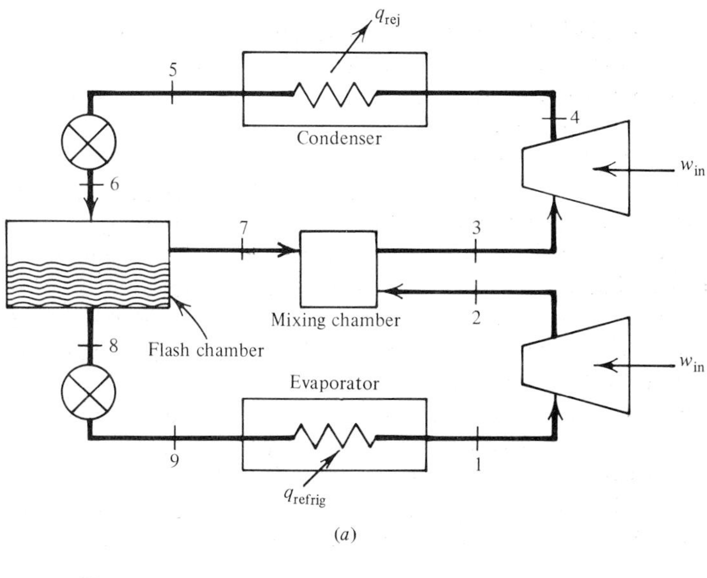

(a)

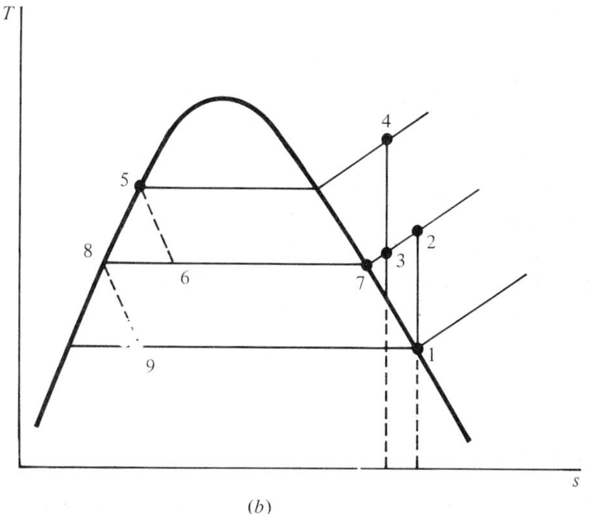

(b)

Figure 18-11 Equipment schematic and Ts diagram for a two-stage vapor-compression refrigeration cycle with regenerative intercooling.

where it mixes intimately with the vapor leaving the low-pressure compressor at state 2. The mixing chamber acts like a regenerative intercooler in that it cools the vapor leaving the low-pressure compressor before the total mixture enters the high-pressure stage of the compressor at state 3. The saturated liquid from the flash chamber is throttled to the evaporator pressure at state 9.

The two-stage compression process with regenerative intercooling is shown on the Ts diagram in Fig. 18-11b. Isentropic compression has been assumed.

Although the same refrigerant circulates through both loops of the overall system, the mass flow rates through the loops are not the same. For the purpose of analysis of the system it is convenient to assume that unit mass circulates in one of the loops, but the choice is arbitrary. For the purpose of this discussion let us assume that unit mass passes through states 3-4-5-6 in the high-pressure loop. The fraction of vapor formed in the flash chamber is the quality x of the fluid at state 6 in Fig. 18-11*b*, and this is the fraction of the flow through the condenser which passes through the mixing chamber from the flash chamber. The fraction of liquid formed is $(1 - x)$, and this is the fraction of the total flow which passes through the evaporator. The value of the enthalpy at state 3 may be ascertained from an energy balance on the mixing chamber under adiabatic conditions. Neglecting kinetic-energy effects, we find that

$$xh_7 + (1 - x)h_2 = 1(h_3) \tag{18-9}$$

where h_3 is the only unknown. The refrigeration effect per unit mass through the condenser is

$$q_{\text{refrig}} = (1 - x)(h_1 - h_9) \tag{18-10}$$

The total work input to the compressor per unit mass passing through the condenser is the sum of the two terms for the two stages, namely,

$$w_{\text{comp}} = (1 - x)(h_2 - h_1) + 1(h_4 - h_3) \tag{18-11}$$

The COP of the two-stage vapor-compression cycle with regenerative intercooling is still defined as $q_{\text{refrig}}/w_{\text{comp}}$.

18-7 STIRLING REFRIGERATION CYCLE

The attainment of extremely low temperatures (less than 200°K or −100°F) is usually achieved by three well-established methods.

1. Vaporization of a liquid.
2. Joule-Thomson effect by isenthalpic expansion
3. Adiabatic expansion in an engine with production of work

Since the early 1950s considerable work has been done on developing a practical refrigeration device based on the Stirling cycle. (See Sec. 16-15 for a discussion of the Stirling heat engine.) Devices arising from this work have proved useful in the temperature range of 100 to 200°K. Other methods for maintaining temperatures in this range are not too plentiful. As noted in Sec. 16-15, the Stirling cycle is composed to two constant-temperature processes and two constant-volume processes. Figure 18-12 is a repeat of Fig. 16-33 and it shows the *PV* and *TS* diagrams for the cycle. If the cycle is reversible, then the heat quantities for the cycle are represented by areas on the *TS* diagram. In the presence of an ideal regenerator in the cycle the heat quantities Q_{14} and Q_{32}, which are equal in

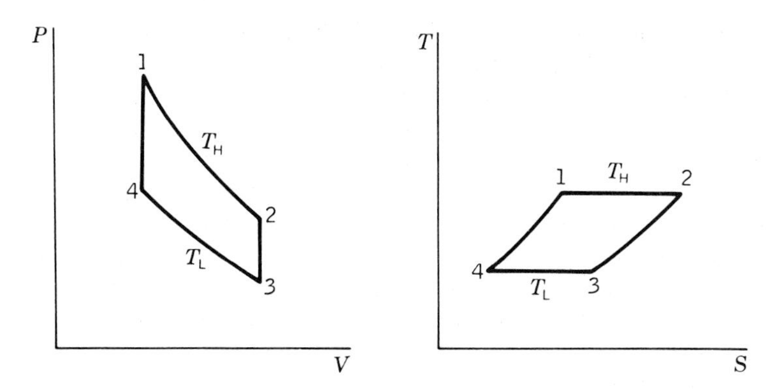

Figure 18-12 *PV* and *TS* diagrams of a Stirling cycle for a gaseous substance.

magnitude but opposite in sign, are exchanged between fluid streams within the device. Hence the only external heat transfer occurs in processes 4-3 and 2-1 at constant temperatures T_L and T_H. Consequently the coefficient of performance of the Stirling refrigerator theoretically equals that of the Carnot refrigerator, namely, $T_L/(T_H - T_L)$. The COP will be quite small if T_L is small, since $T_H - T_L$ will also be quite large. For example, if T_L is 100°K and T_H is 300°K, a value for the COP of 0.5 results. Thus considerable work must be performed per unit of heat removal, compared to a household refrigerator or air-conditioning unit.

An ideal reciprocating Stirling refrigeration unit is shown in Fig. 18-13. A cold temperature is produced by a reversible expansion of a gas in region *E*, while the gas is heated by compression in region *D*. Particles of the gas oscillate between the two spaces which are connected by a regenerator *F*. The regenerator must be composed of a material with a high heat capacity. Cylinder *A* encloses

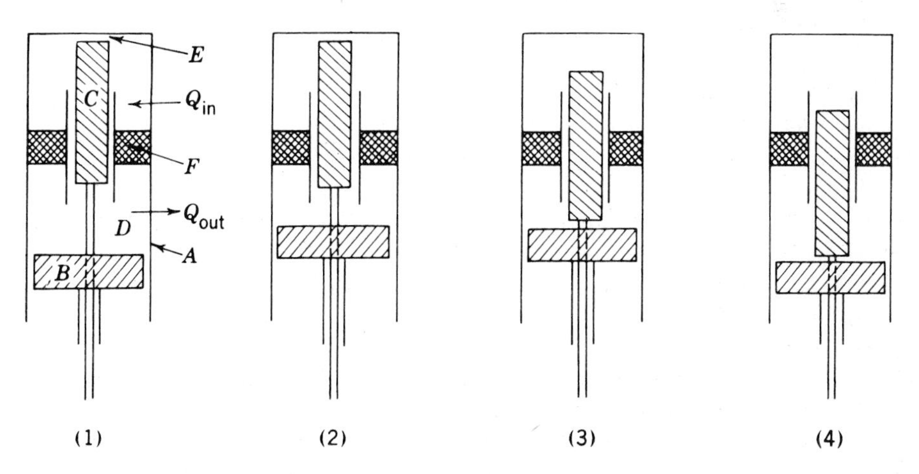

Figure 18-13 Relative positions of regular piston and displacer piston in an ideal reciprocating Stirling refrigeration cycle.

the regular piston B and a displacer piston C. The shaft of the displacer piston passes through piston B. Consider the piston initially in position 1 shown in Fig. 18-13. Four distinct ideal processes take place in the cycle.

1. The piston moves upward, compressing the gas in region D. To keep the process isothermal at temperature T_H, heat is removed through the cylinder walls during the compression process. This is equivalent to process 2-1 in Fig. 18-13.
2. The displacer piston now moves downward, forcing part of the gas through the regenerator to space E. The gas is cooled as it passes through the regenerator, the energy being stored in the regenerator material. This creates a temperature gradient in the regenerator, the temperature increasing from region E to region D. Since piston B does not move, this transfer of part of the gas from E to D, with energy storage in E, occurs at constant volume. This corresponds to process 1-4.
3. Both pistons now move downward, with the resulting expansion of the gas in region E. The expansion process would tend to cool the gas, but heat added from an outside source keeps the temperature at T_L. This heat removal at very low T_L values constitutes the refrigeration effect. For example, a gas circulated around the outside of region E could be liquefied. This process is equivalent to process 4-3 in Fig. 18-13.
4. Finally, the displacer piston moves upward to its initial position, forcing gas from region D to region E. As the gas passes through the regenerator from the cold to the hot side, it is reheated. This internal heat-addition process corresponds to process 3-2. Note that the temperature gradient in the regenerator makes the internal heat transfer between the gas and the refrigerator material reversible.

Several discrepancies between theory and practice should be pointed out. First, the entire gas within the system does not follow the process path shown in Fig. 18-12. A considerable fraction of the gas remains either in region D or region E. Hence the TS and PV diagrams do not have their usual clear-cut representation of the process. The system at any instant is nonhomogeneous, and therefore cannot be represented by a single state. Different gas particles describe different cycles. Secondly, the simple piston displacements described above would be difficult to match, since they occur discontinuously. In an actual device the reciprocating motions of the pistons would be harmonic. To match more nearly the theoretical cycle, the displacer and regular pistons are placed out of phase by some angle ϕ (on the same drive shaft), so that the compression region D lags in phase with respect to the expansion space E. This compromise tends to blur even further the four steps of the theoretical cycle.

As noted earlier, the working fluid for a practical Stirling refrigeration unit is either helium or hydrogen. Maximum and minimum pressures may range around 35 and 15 bars (or 500 and 200 psia), respectively, with an engine speed of

1500 rpm. A number of applications include:

1. Means for cooling electronic equipment and superconducting research magnets
2. Freeze-drying of materials
3. A precooler for production of liquid hydrogen and helium
4. An air liquefier
5. Gas separation, e.g., as a liquid nitrogen generator from air

18-8 ABSORPTION REFRIGERATION

In any refrigeration process the energy removed from the cold region eventually must be rejected to another region which is at a considerably higher temperature. This second region is usually the surrounding environment. In order to carry out the heat-rejection process, the temperature of the fluid within the refrigeration cycle must be raised to a value above that of the environment. In a vapor-compression-refrigeration cycle, as discussed in Sec. 18-2, the temperature of the vapor leaving the evaporator is raised by a compression process. The work input required in an ideal steady-flow compression process is given by the integral of $v\,dP$. The pressure limits on the integral are set by the saturation temperatures required in the evaporator and condenser of the cycle. Once the pressure range is determined for a given refrigerant, the main variable which controls the amount of work input is the specific volume of the fluid. In a vapor-compression refrigeration cycle the value of v is relatively large, since the fluid is in the superheat region throughout the compression process. Therefore, the work input is also relatively large. One method of overcoming this disadvantage is to design a refrigeration cycle in which the fluid is a liquid during the compression process. Then the work input will be significantly smaller.

The technique of absorption refrigeration is based on this approach. To accomplish this, however, the overall cycle becomes physically more complex. In addition, a two-component mixture, such as ammonia and water or lithium bromide and water, must be used as the circulating fluid in part of the cycle, rather than the single component used in a vapor-compression cycle. Two-component fluids have an important characteristic which must be recognized. When two phases are present at equilibrium, the composition of a given component is not the same in the two phases. The vapor phase will contain more of that component which is more volatile at the given temperature. For example, consider an ammonia-water mixture. At 43°C (110°F) the saturation pressure of ammonia is 17 bars (247 psia), while that of water is 0.09 bar (1.3 psia). Therefore ammonia has a much greater tendency to vaporize at a given temperature than does water. Hence for an ammonia-water solution, the vapor phase contains much more ammonia (is richer in ammonia) than the liquid phase in equilibrium with it. This fact is extremely important when making mass and energy balances on equipment used in absorption refrigeration.

A schematic diagram of a simple absorption-refrigeration cycle is shown in Fig. 18-14. A condenser, throttle valve, and evaporator are shown on the left-hand side of the diagram. These three pieces of equipment also are used in a conventional vapor-compression cycle, as shown earlier in Fig. 17-13. The compressor in that cycle, however, is now replaced by four pieces of equipment. These are an absorber, a pump, a generator, and a valve. For the purpose of discussion, we shall consider ammonia and water as the two components in the cycle. Essentially pure ammonia vapor leaves the evaporator at state 1 and enters the absorber. The absorbing medium is a weak solution (low ammonia concentration) of ammonia and water which continually enters at state 5. The process of absorption releases energy, hence cooling water must be circulated through the absorber in order to keep the solution at a constant temperature. The temperature of the absorbing fluid must be kept at as low a temperature as possible, since the amount of pure refrigerant (ammonia) which can be absorbed decreases as the temperature increases. However, the absorber must operate at 10 to 20° above the cooling water temperature to allow for adequate heat-transfer rates. The liquid which leaves the absorber at state 2 is a rich or strong solution (high ammonia concentration). This binary liquid mixture is now compressed by a pump to state 3, which is at the desired condenser pressure.

The temperature rise of the binary mixture due to the pump work usually is quite small. Thus the strong solution is subcooled liquid as it enters the generator shown in Fig. 18-14. Heat Q_G must now be added to the solution in the generator to warm the incoming liquid to the saturation temperature and to drive out of solution some of the ammonia. This nearly-pure ammonia passes to the con-

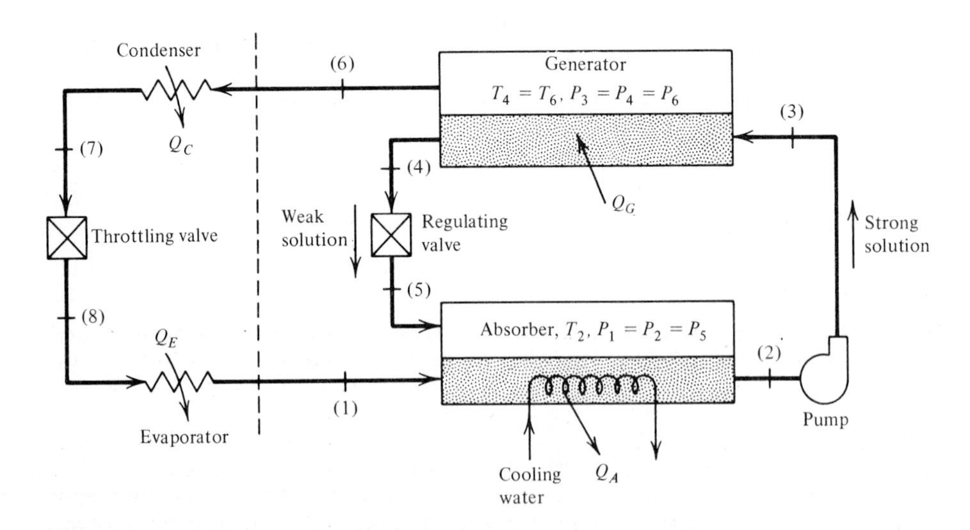

Figure 18-14 Schematic diagram of a simple absorption-refrigeration cycle.

denser at state 6, and eventually returns to the absorber at state 1. The weak solution left in the generator (state 4) now flows through a regulating valve, which drops the pressure of the solution to that in the absorber. It then mixes with the solution remaining in the absorber, and cold vapor coming from the evaporator is added to the overall liquid solution. The strong solution is cooled, as before, and the cycle is repeated. Hence the ammonia-water solution which cycles through the absorber, pump, generator, and valve merely serves as a transport medium for the ammonia refrigerant. Note that any absorption-refrigeration device requires an external heat source for the generation of refrigerant vapor. Thus absorption refrigeration is especially attractive if a low-temperature (100 to 200°C, 200 to 400°F) source of thermal energy is readily available.

In actual practice, absorption-refrigeration units have at least two modifications not shown in Fig. 18-14. First, the cold, strong solution at state 3 needs warming before it enters the generator, and the warm, weak solution at state 4 needs cooling before it enters the absorber. Consequently, a heat exchanger is placed between the absorber and generator, which permits heat transfer from the weak solution to the strong solution. Second, a major requirement is that the ammonia from the generator should be essentially free of water as it passes through the condenser–throttling-valve–evaporator loop. Any traces of water will freeeze in the expansion valve and evaporator. Hence the vapor leaving the generator passes through a device called a rectifier before it enters the condenser. The rectifier separates any remaining water vapor from the vapor stream leaving the generator, and returns the water to the generator.

REFERENCES

American Society of Heating, Refrigerating, and Air Conditioning Engineers, "Fundamentals Handbook," New York, 1981.

American Society of Heating, Refrigerating, and Air Conditioning Engineers, "Systems Handbook," New York, 1980.

PROBLEMS (METRIC)

Carnot refrigeration cycles

18-1M Refrigerant 12 is used as the working fluid in a refrigerator which follows a reversed Carnot cycle and operates between 2 and 8 bars. In the condenser the refrigerant changes from saturated vapor to saturated liquid. Determine (*a*) the coefficient of performance, (*b*) the quality of the fluid leaving the evaporator, and (*c*) the work input to the compressor, in kJ/kg.

18-2M A refrigerator operates on a reversed Carnot cycle between an evaporator temperature of 4°C and a condenser temperature of 28°C. The working fluid is refrigerant 12 which changes from saturated vapor to saturated liquid in the condenser. Determine (*a*) the coefficient of performance, (*b*) the quality of the fluid leaving the expansion process, and (*c*) the work input to the compressor in kJ/kg.

18-3M Same as Prob. 18-2M, except that the evaporator and condenser temperatures are −15 and 32°C, respectively.

Ideal vapor-compression refrigeration cycle

18-4M Saturated refrigerant 12 vapor enters the compressor of an ideal vapor-compression-refrigeration cycle at 1.4 bars; saturated liquid enters the expansion valve at 7 bars. Determine (*a*) the temperature of the fluid leaving the compressor, (*b*) the coefficient of performance, (*c*) the percent increase in refrigeration capacity if a work expander were used instead of a throttling device, (*d*) the percent decrease in net work input required if a work expander were used instead of a throttling device, and (*e*) the power input to the compressor for a 5-ton unit, in kilowatts.

18-5M An ideal vapor-compression-refrigeration cycle uses refrigerant 12 and operates between 1.8 and 8 bars. Entering the compressor, the fluid is a saturated vapor. Determine (*a*) the temperature of the fluid leaving the isentropic compressor, in degrees Celsius, (*b*) the coefficient of performance, (*c*) the percent increase in refrigeration capacity if a work expander were used instead of a free-expansion device, (*d*) the effective displacement of the compressor, in L/min, for a nominal 7-ton refrigeration plant, and (*e*) the power input to the compressor, in kilowatts, for a 7-ton unit.

18-6M The pressures in the evaporator and condenser of a 5-ton refrigeration plant operating on refrigerant 12 are 0.20 and 0.70 MPa, respectively. For the ideal cycle the fluid enters the compressor as a saturated vapor, and no subcooling occurs in the condenser. Determine (*a*) the temperature of the fluid leaving the isentropic compressor, in degrees Celsius, (*b*) the coefficient of performance, (*c*) the effective displacement, in L/min, and (*d*) the power input, in kilowatts.

Vapor-compression refrigeration with superheating and subcooling

18-7M Reconsider Prob. 18-4M under the following conditions. The refrigerant leaves the evaporator at 1.4 bars and −10°C, and leaves the condenser at 7 bars and 20°C. Determine, relative to the cycle in Prob. 18-4M, (*a*) the percent change in compressor-work input, and (b) the percent change in refrigeration capacity, for the same mass flow rate.

18-8M Reconsider Prob. 18-5M under the following conditions. The refrigerant leaves the evaporator at 1.8 bars and −10°C, and enters the throttling device at 8 bars and 24°C. Determine, relative to Prob. 18-5M, (*a*) the percent change in compressor work input, and (*b*) the percent change in refrigeration capacity, for the same mass flow rate.

18-9M Reconsider Prob. 18-6M under the following conditions. The refrigerant leaves the evaporator at 0.20 MPa and 0°C, and enters the throttling device at 0.70 MPa and 24°C. Determine, relative to Prob. 18-6M, (*a*) the percent change in compressor work input, and (*b*) the percent change in refrigeration capacity, for the same mass flow rate.

Vapor-compression refrigeration with irreversible heat transfer

18-10M Reconsider Prob. 18-4M under the following conditions. To allow for a finite temperature difference between the cold region and the evaporator fluid, the evaporator will be set to operate at 1.0 bar. Likewise, the pressure of the fluid in the condenser will be raised to 8 bars. Other information remains the same. Determine, relative to Prob. 18-4M, the percent change in (*a*) the compressor work, (*b*) the refrigeration capacity, and (*c*) the coefficient of performance.

18-11M Reconsider Prob. 18-5M under the following conditions. To allow for a finite temperature difference between the cold region and the evaporator fluid, the evaporator pressure will be set at 1.4 bars. Likewise, the pressure of the fluid in the condenser will be raised to 10 bars. Other information remains the same. Determine, relative to Prob. 18-5M, the percent change in (*a*) the compressor work, (*b*) the refrigeration capacity, and (*c*) the coefficient of performance.

18-12M Reconsider Prob. 18-6M under the following conditions. To allow for a finite temperature difference between the cold region and the evaporator fluid, the evaporator pressure will be set at 0.16 MPa. Likewise, the pressure of the fluid in the condenser will be raised to 0.90 MPa. Other information remains the same. Determine, relative to Prob. 18-6M, the percent change in (*a*) the compressor work, (*b*) the refrigeration capacity, and (*c*) the coefficient of performance.

Nonideal vapor-compression cycles

18-13M Refrigerant 12 leaves the evaporator of a vapor-compression-refrigeration plant at 1.4 bars and −20°C, and is compressed to 7 bars and 50°C. The temperature of the fluid leaving the condenser is 24°C. Determine (*a*) the coefficient of performance, (*b*) the effective displacement of the compressor in L/min per ton of refrigeration, and (*c*) the compressor efficiency.

18-14M In a vapor-compression-refrigeration cycle, refrigerant 12 leaves the evaporator as saturated vapor. The evaporator and condenser pressures are 1.4 and 10 bars, respectively. The fluid entering the condenser is at 80°C, and the refrigeration capacity is 10 tons. Determine (*a*) the mass flow rate required, in kg/min, (*b*) the power input, in kilowatts, and (*c*) the compressor adiabatic efficiency.

18-15M In a vapor-compression-refrigeration cycle which circulates refrigerant 12 at a rate of 6 kg/min, the refrigerant enters the compressor superheated 1.9°C at 1.4 bars and leaves at 7 bars. The fluid leaves the condenser as a saturated liquid. The compressor adiabatic efficiency is 66.7 percent. Determine (*a*) the compressor-outlet temperature, in degrees Celsius, (*b*) the coefficient of performance, and (*c*) the tons of refrigeration produced by the cycle.

Heat-pump cycles

18-16M An ideal vapor-compression heat pump cycle operates between an evaporator temperature of 0°C and a condenser pressure of 8.0 bars. Refrigerant 12 leaves the evaporator as saturated vapor and enters the expansion valve as saturated liquid. If the heat pump supplies 1000 kJ/min to a high-temperature region, determine (*a*) the temperature at the exit of the isentropic compressor, in degrees Celsius, (*b*) the coefficient of performance, (*c*) the effective displacement of the compressor, in L/min, (*d*) the power input to the compressor, in kilowatts, and (*e*) the power input required if electric resistance heating is used, in kilowatts.

18-17M An actual vapor-compression heat pump opererates with refrigerant 12. Underground water pumped into the building at 8°C will be the source of heat, which is used to maintain the inside air supply at 30°C. The evaporator is designed to operate at a temperature 8 degrees below the cold-water-inlet temperature, and the condenser operates at 9.0 bars. The fluid leaving the evaporator is saturated vapor, and it is subcooled 1.37°C as it leaves the condenser. The compressor has an adiabatic efficiency of 72 percent. Determine (*a*) the pressure in the evaporator, in bars, (*b*) the minimum temperature difference in the condenser between the refrigerant and the heated air supply, (*c*) the temperature at the compressor outlet, in degrees Celsius, (*d*) the COP of the device, and (*e*) the volume flow rate of supply water required, in m^3/min, if the supply water has a temperature change of 2.0°C as it passes through the evaporator, and the heat supplied to the air is at a rate of 1200 kJ/min.

18-18M In a heat-pump cycle using refrigerant 12 the evaporator and condenser pressures are 2.4 and 9 bars, respectively. Vapor enters the compressor at 0°C and leaves at 60°C. Liquid leaves the condenser at 36°C. If the cycle is to supply 80,000 kJ/h to a building, calculate (*a*) the refrigerant flow rate, in kg/min, (*b*) the power input to the compressor, in kilowatts, (*c*) the compressor adiabatic efficiency, (*d*) the coefficient of performance, and (*e*) the power input, in kilowatts, if electric resistance heating is employed instead of the heat pump.

18-19M A building requires 200,000 kJ/h of heat to maintain the interior air supply at 35°C when the outside temperature is −10°C. The heat will be supplied by a heat-pump system using refrigerant 12. The evaporator operates at a temperature 10° below the outside air temperature and the condenser operates at 10 bars. The compressor has an adiabatic efficiency of 75 percent. The fluid leaving the evaporator is a saturated vapor, and it is a saturated liquid leaving the condenser. Determine (*a*) the pressure in the evaporator, in bars, (*b*) the minimum temperature difference in the condenser between the refrigerant and the heated air supply, (*c*) the temperature at the compressor outlet, in degrees Celsius, (*d*) the quality of the fluid leaving the throttle valve, and (*e*) the percent increase in input power if direct electric resistance heating were used instead of the heat pump.

18-20M The price of electricity is $0.06/kWh and fuel oil costs $1.40/gal and the heating value of oil is 130,000 kJ/gal. Determine the heating cost in $/day for (*a*) Prob. 18-18M, and (*b*) Prob. 18-19M if the energy source is (1) a heat pump, (2) direct electric resistance heating, and (3) an oil burner which operates with 70 percent efficiency.

Gas-refrigeration cycles

18-21M A reversed Brayton cycle with a pressure ratio of 3 is used to produce a refrigeration effect. The compressor-inlet temperature is 27°C and the turbine-inlet temperature is 7°C. If the compressor and turbine perform ideally, determine (*a*) the maximum and minimum temperatures in the cycle, in degrees Celsius, and (*b*) the coefficient of performance. If the compressor and turbine adiabatic efficiencies are 83 and 88 percent, respectively, determine (*c*) the maximum and minimum temperatures in the cycle, in degrees Celsius, and (*d*) the coefficient of performance.

18-22M The compressor of an air-standard Brayton refrigeration cycle operates between 0.10 MPa, 280°K, and 0.50 MPa. The turbine-inlet temperature is 360°K. If the compressor and turbine perform ideally, determine (*a*) the maximum and minimum temperatures in the cycle, in degrees Celsius, and (*b*) the coefficient of performance. If the compressor and turbine adiabatic efficiencies are 84 and 87 percent, respectively, determine (*c*) the maximum and minimum temperatures in the cycle, in degrees Celsius, (*d*) the modified coefficient of performance, and (*e*) the mass flow of air required, in kg/min, to remove 211 kJ/min of heat from the cold region.

18-23M A reversed Brayton cycle operates with air entering the compressor at 1 bar and 7°C and leaving at 4 bars. Air enters the turbine at 37°C. If the compressor and turbine perform ideally, determine (*a*) the minimum temperature in the cycle, in degrees Celsius (*b*) the coefficient of performance, and (*c*) the mass flow rate of air required, in kg/min, to remove 211 kJ/min of heat from the cold region. Now, if the compressor and turbine efficiencies are 84 and 88 percent, respectively, determine (*d*) the minimum temperature, (*e*) the coefficient of performance, and (*f*) the mass flow rate for 211 kJ/min of cooling.

18-24M An ideal reversed Brayton cycle operates with a pressure ratio of 3. Air enters the compressor at 7°C, and enters the turbine normally at 27°C. The cycle is now modified by considering a regenerator in the cycle similar to that shown in Fig. 18-6. The heat exchanger allows the air to be precooled down to the compressor-inlet temperature before it enters the turbine. Determine (*a*) the old and new minimum cycle temperature, in degrees Celsius, and (*b*) the old and new COP values.

18-25M Using the data of Prob. 18-23M, consider a regenerator placed in the cycle similar to that shown in Fig. 18-6. Assume that the air entering the turbine is precooled by the regenerator to the compressor-inlet temperature. For the ideal cycle, determine (*a*) the minimum temperature in the cycle, in degrees Celsius, and (*b*) the coefficient of performance.

18-26M A regenerative-type gas refrigerator similar to that shown in Fig. 18-6 operates with helium between pressure limits of 1.0 and 6.0 bars. Compression and expansion are isentropic, and pressure drops are negligible. The compressor-inlet temperature (state 1) and regenerator-inlet temperature (state 3) are both 20°C, and the refrigerator-outlet temperature is −50°C. Determine (*a*) the minimum temperature achieved, in degrees Celsius, (*b*) the amount of refrigeration possible, in kJ/kg, and (*c*) the net amount of work required, in kJ/kg.

18-27M Air at 1 bar and −13°C enters an adiabatic compressor, which discharges the fluid at 5.0 bars to a heat exchanger cooled by the ambient surroundings. The compressed air leaves the heat exchanger at 4.9 bars and 42°C and enters an adiabatic turbine which expands the fluid to 1.1 bars. The expanded and cold air now passes through a second heat exchanger where it picks up heat from a low-temperature region and exits at 1 bar and −13°C. The cycle is then repeated. The turbine and compressor are both 80 percent efficient, and the turbine helps drive the compressor. On the basis of air-table data, determine (*a*) the net work input required per unit of heat picked up from the low-temperature region, and (*b*) the heat removed to the ambient surroundings per unit of heat picked up from the low-temperature region. (*c*) Sketch a Ts diagram of the process.

Liquefaction and solidification

18-28M Consider the basic process for liquefying gases shown in Fig. 18-7, but ignore the compressor and aftercooler. Refrigerant 12 enters the heat exchanger at 16 bars and 80°C and is cooled by saturated vapor which is withdrawn from the heavily insulated separator. The high-pressure cooled gas leaves the heat exchanger and passes through a throttling valve into the separator. The separator contains liquid and vapor in equilibrium at 1.4 bars. The vapor leaving at low pressure from the heat exchanger has a temperature of 60°C. As a first approximation neglect all pressure losses except that across the throttling valve. Determine (*a*) the fraction of the inlet high-pressure gas which is liquefied, and (*b*) the temperature (if superheated) or the quality (if saturated) of the flow stream entering the valve.

18-29M Work Prob. 18-28M under the following set of conditions. Refrigerant 12 enters the heat exchanger at 14 bars and 60°C. The separator operates at 1.8 bars, and the vapor leaving the heat exchanger at low pressure has a temperature of 50°C.

Cascade and multistage vapor-compression cycles

18-30M A cascade refrigeration system employs refrigerant 12 in both of the closed loops. The low-temperature evaporator is designed to operate at −40°C, while the fluid condenses at 3.2 bars. The high-temperature cycle has an evaporator temperature of −10°C and a condenser pressure of 8.0 bars. The refrigeration capacity required for the low-temperature evaporator is 5 tons, and the compressors are isentropic. Determine (*a*) the mass flow rates, in kg/min, in both loops, (*b*) the power input to both compressors, in kilowatts, (*c*) the COP of the cascade system, (*d*) the percent change in compressor work required if a single vapor-compression cycle is used between −40°C and 8.0 bars, for the same mass flow rate through the low-temperature evaporator, and (*e*) the percent change in refrigeration capacity achieved if a single vapor-compression cycle is used with the same mass flow rate through the low-temperature evaporator.

18-31M Reconsider Prob. 18-30M, except the compressor efficiencies are both 75 percent. Answer the same questions.

18-32M A two-stage refrigeration system with regenerative intercooling operates with refrigerant 12 and has pressures of 2.0, 4.0, and 10.0 bars in the evaporator, flash chamber and mixing chamber, and condenser, respectively. Assume isentropic compression and isenthalpic throttling, and no other losses. If the refrigeration load is 5 tons, determine (*a*) the mass flow rate through the evaporator, in kg/min, (*b*) the power input to the low-pressure compressor, in kilowatts, (*c*) the mass flow rate leaving the flash chamber and entering the mixing chamber, in kg/min, (*d*) the power input to the high-pressure compressor, in kilowatts, and (*e*) the COP of the cycle.

18-33M Reconsider Prob. 18-32M. Repeat the required calculations if the adiabatic efficiency for both compressors is 77 percent. Other losses such as pressure drops are neglected.

18-34M Reconsider Prob. 18-32M. Repeat the required calculations if the three pressures cited are changed to 1.6, 2.8, and 8.0 bars. All other conditions remain the same.

18-35M Reconsider Prob. 18-32M. Repeat the required calculations if the three pressures cited are changed to 1.6, 2.8, and 8.0, and the adiabatic efficiency for both compressors is 75 percent.

Stirling refrigeration cycle

18-36M A Stirling cycle is used for refrigeration purposes. Air at 230°K and 10 bars is expanded isothermally to 1 bar. It is then heated to 315°K at constant volume. Compression at a constant temperature of 315°K follows, and the cycle is completed by constant-volume heat removal. Compute the coefficient of performance for the cycle and the quantity q_L, in kJ/kg.

18-37M Air enters a Stirling cycle used for refrigeration purposes at 240°K and 8 bars and expands isothermally to 1.2 bars. It is then heated to 320°K at constant volume. Compression at a constant temperature of 320°K follows, and the cycle is completed by constant-volume heat removal. Compute the coefficient of performance for the cycle and the quantity q_L, in kJ/kg.

18-38M Work Prob. 18-36M if the working fluid is carbon dioxide and it is assumed to behave as an ideal gas.

18-39M Work Prob. 18-37M if the working fluid is argon.

PROBLEMS (USCS)

Carnot reversed cycles

18-1 Refrigerant 12 is used as the working fluid in a refrigerator which follows a reversed Carnot cycle and operates between 40 and 160 psia. In the condenser the refrigerant changes from saturated vapor to saturated liquid. Determine (*a*) the coefficient of performance, (*b*) the quality of the fluid leaving the evaporator, and (*c*) the work input to the compressor, in Btu/lb.

18-2 A refrigerator operates on a reversed Carnot cycle between an evaporator temperature of 30°F and a condenser temperature of 90°F. The working fluid is refrigerant 12 which changes from saturated vapor to saturated liquid in the condenser. Determine (*a*) the coefficient of performance, (*b*) the quality of the fluid leaving the expansion process, and (*c*) the work input to the compressor, in Btu/lb.

18-3 Same as Prob. 18-2, except that the evaporator and condenser temperatures are 0 and 100°F, respectively.

Ideal vapor-compression-refrigeration cycle

18-4 Saturated refrigerant 12 vapor enters the compressor of an ideal vapor-compression-refrigeration cycle at 20 psia; saturated liquid enters the expansion valve at 160 psia. Determine (*a*) the temperature of the fluid leaving the compressor, (*b*) the coefficient of performance, (*c*) the percent increase in refrigeration capacity if a work expander were used instead of a throttling device, and (*d*) the percent decrease in net work input required if a work expander were used instead of a throttling device, and (*e*) the power input to the compressor for a 5-ton unit, in horsepower.

18-5 An ideal vapor-compression-refrigeration cycle uses refrigerant 12 and operates between 40 and 160 psia. Entering the compressor, the fluid is a saturated vapor. Determine (*a*) the temperature of the fluid leaving the isentropic compressor, in degrees Fahrenheit, (*b*) the coefficient of performance, (*c*) the percent increase in refrigeration capacity if a work expander were used instead of a free expansion device, (*d*) the effective displacement of the compressor, in ft^3/min, for a nominal 7-ton refrigeration plant, and (*e*) the power input to the compressor, in horsepower, for a 7-ton unit.

18-6 The pressures in the evaporator and condenser of a 5-ton refrigeration plant operating on refrigerant 12 are 30 and 100 psia, respectively. For the ideal cycle the fluid enters the compressor as a saturated vapor, and no subcooling occurs in the condenser. Determine (*a*) the temperature of the fluid leaving the isentropic compressor, in degrees Fahrenheit, (*b*) the coefficient of performance, (*c*) the effective displacement, in ft^3/min, and (*d*) the power input, in horsepower.

Vapor-compression refrigeration with superheating and subcooling

18-7 Reconsider Prob. 18-4 under the following conditions. The refrigerant leaves the evaporator at 20 psia and 0°F, and leaves the condenser at 160 psia and 110°F. Determine, relative to the cycle in Prob. 18-4, (*a*) the percent change in compressor-work input, and (*b*) the percent change in refrigeration capacity, for the same mass flow rate.

18-8 Reconsider Prob. 18-5 under the following conditions. The refrigerant leaves the evaporator at 40 psia and 30°F, and enters the throttling device at 160 psia and 110°F. Determine, relative to Prob. 18-5, (*a*) the percent change in compressor-work input, and (*b*) the percent change in refrigeration capacity, for the same mass flow rate.

18-9 Reconsider Prob. 18-6 under the following conditions. The refrigerant leaves the evaporator at 30 psia and 20°F, and enters the throttling device at 100 psia and 80°F. Determine, relative to Prob. 18-6, (*a*) the percent change in compressor-work input, and (*b*) the percent change in refrigeration capacity, for the same mass flow rate.

Vapor-compression refrigeration with irreversible heat transfer

18-10 Reconsider Prob. 18-4 under the following conditions. To allow for a finite temperature difference between the cold region and the evaporator fluid, the evaporator will be set to operate at 15 psia. Likewise, the pressure of the fluid in the condenser will be raised to 200 psia. Other information remains the same. Determine, relative to Prob. 18-4, the percent change in (*a*) the compressor work, (*b*) the refrigeration capacity, and (*c*) the coefficient of performance.

18-11 Reconsider Prob. 18-5 under the following conditions. To allow for a finite temperature difference between the cold region and the evaporator fluid, the evaporator pressure will be set at 30 psia. Likewise, the pressure of the fluid in the condenser will be raised to 200 psia. Other information remains the same. Determine, relative to Prob. 18-5, the percent change in (*a*) the compressor work, (*b*) the refrigeration capacity, and (*c*) the coefficient of performance.

18-12 Reconsider Prob. 18-6 under the following conditions. To allow for a finite temperature difference between the cold region and the evaporator fluid, the evaporator pressure will be set at 20 psia. Likewise, the pressure of the fluid in the condenser will be raised to 120 psia. Other information remains the same. Determine, relative to Prob. 18-6, the percent change in (*a*) the compressor work, (*b*) the refrigeration capacity, and (*c*) the coefficient of performance.

Nonideal vapor-compression cycles

18-13 Refrigerant 12 leaves the evaporator of a vapor-compression-refrigeration cycle at 20 psia and 0°F, and is compressed to 100 psia and 120°F. The temperature leaving the compressor is 80°F. Determine (*a*) the coefficient of performance, (*b*) the effective displacement of the compressor, in ft^3/min per ton of refrigeration, and (*c*) the compressor adiabatic efficiency.

18-14 In a vapor-compression-refrigeration cycle, refrigerant 12 leaves the evaporator as saturated vapor. The evaporator and condenser pressures are 20 and 180 psia, respectively. The fluid entering the condenser is at 200°F, and the refrigeration capacity is 10 tons. Determine (*a*) the mass flow rate, in lb/min, (*b*) the power input, in horsepower, and (*c*) the compressor adiabatic efficiency.

18-15 In a vapor-compression refrigeration cycle, refrigerant 12 enters the compressor superheated by 8.1°F at 20 psia, and leaves at 100 psia. The compressor adiabatic efficiency is 63.1 percent, and the fluid leaves the condenser as a saturated liquid at a rate of 12.0 lb/min. Determine (*a*) the compressor-outlet temperature, in degrees Fahrenheit, (*b*) the coefficient of performance, and (*c*) the tons of refrigeration produced by the cycle.

Heat-pump cycles

18-16 An actual vapor-compression heat pump operates between an evaporating temperature of 40°F and a condensing pressure of 120 psia. Refrigerant 12 leaves the evaporator as saturated vapor, and enters the expansion valve at 95°F. The heat pump supplies 1000 Btu/min to an air stream at 80°F, and the compressor efficiency is 80 percent. Determine (*a*) the temperature at the exit of the compressor, in degrees Fahrenheit, (*b*) the coefficient of performance, (*c*) the minimum temperature difference in the condenser between the refrigerant and the heated air supply, (*d*) the effective displacement of the compressor, in ft^3/min, (*e*) the power input to the compressor, in kilowatts, and (*f*) the power input required by electric resistance heating, in kilowatts.

18-17 An ideal vapor-compression heat-pump cycle operates with refrigerant 12. The fluid leaves the evaporator as a saturated vapor at 30°F, and enters the expansion valve as a saturated liquid at 140 psia. If the heat pump supplies 1200 Btu/min to the high-temperature region, determine (*a*) the temperature of the fluid at the compressor outlet, in degrees Fahrenheit, (*b*) the coefficient of performance, (*c*) the percent decrease in net work input if a work expander were used instead of a free-expansion device, (*d*) the effective displacement of the compressor, in ft^3/min, (*e*) the power input to the compressor, in horsepower and kilowatts, and (*f*) the power input required, in kilowatts, if the energy source is direct electric-resistance heating.

18-18 In a heat-pump cycle using refrigerant 12 the evaporator and condenser pressures are 40 and 180 psia, respectively. Vapor enters the compressor at 30°F and leaves at 160°F. Liquid leaves the condenser at 120°F. If the cycle is to supply 80,000 Btu/h to a building, calculate (*a*) the refrigerant flow rate, in lb/min, (*b*) the power input to the compressor, in kilowatts and horsepower, (*c*) the compressor adiabatic efficiency, (*d*) the coefficient of performance, and (*e*) the power input, in kilowatts, if electric-resistance heating is employed instead of the heat pump.

18-19 A building requires 200,000 Btu/lb of heat to maintain the interior air supply at 95°F when the outside temperature is 10°F. The heat will be supplied by a heat-pump system using refrigerant 12.

The evaporator operates at a temperature 20° below the outside air temperature and the condenser operates at 160 psia. The compressor has an adiabatic efficiency of 75 percent. The fluid leaving the evaporator is a saturated vapor, and it is a saturated liquid leaving the condenser. Determine (*a*) the pressure in the evaporator, in bars, (*b*) the minimum temperature difference in the condenser between the refrigerant and the heated air supply, (*c*) the temperature at the compressor outlet, in degrees Fahrenheit, (*d*) the quality of the fluid leaving the throttle valve, and (*e*) the percent increase in input power if direct electric-resistance heating were used instead of the heat pump.

18-20 The price of electricity is $0.06 kWh and fuel oil, with a heating value of 125,000 Btu/gal, is priced at $1.40/gal. Determine the heating cost in $/day for (*a*) Prob. 18-18, and (*b*) Prob. 18-19 if the energy source is (1) a heat pump, (2) direct electric-resistance heating, and (3) an oil burner with 70 percent combustion efficiency.

Gas-refrigeration cycles

18-21 A reversed Brayton cycle with a pressure ratio of 3 is used to produce a refrigeration effect. The compressor-inlet temperature is 80°F and the turbine-inlet temperature is 40°F. If the compressor and turbine perform ideally, determine (*a*) the maximum and minimum temperatures in the cycle, in degrees Fahrenheit, and (*b*) the coefficient of performance. If the compressor and turbine adiabatic efficiencies are 83 and 88 percent, respectively, determine (*c*) the maximum and minimum temperatures in the cycle, in degrees Fahrenheit, and (*d*) the coefficient of performance.

18-22 The compressor of an air-standard Brayton refrigeration cycle operates between 14 psia, 40°F, and 63 psia. The turbine-inlet temperature is 190°F. If the compressor and turbine perform ideally, determine (*a*) the maximum and minimum temperatures in the cycle, in degrees Fahrenheit, and (*b*) the coefficient of performance. If the compressor and turbine adiabatic efficiencies are 84 and 87 percent, respectively, determine (*c*) the maximum and minimum temperatures in the cycle, in degrees Fahrenheit, (*d*) the coefficient of performance, and (*e*) the mass flow rate of air required, in lb/min, to remove 200 Btu/min of heat from the cold region.

18-23 A reversed Brayton cycle operates with air entering the compressor at 14.5 psia and 80°F and leaving at 58 psia. Air enters the turbine at 140°F. If the compressor and turbine perform ideally, determine (*a*) the minimum temperature in the cycle, in degrees Fahrenheit, (*b*) the coefficient of performance, and (*c*) the mass flow rate of air required, in lb/min, to remove 200 Btu/min of heat from the cold region. Now, if the compressor and turbine efficiencies are 84 and 88 percent, respectively, determine (*d*) the minimum temperature, (*e*) the coefficient of performance, and (*f*) the mass flow rate for 200 Btu/min of cooling.

18-24 An ideal, reversed Brayton cycle operates with a pressure ratio of 3. Air enters the compressor at 45°F, and normally would enter the turbine at 80°F. However, the cycle is now modified by considering a regenerator in the cycle similar to that shown in Fig. 18-6. The heat exchanger allows the air to be precooled down to the compressor-inlet temperature. Determine (*a*) the original and new minimum cycle temperatures, in degrees Fahrenheit, and (*b*) the original and new COP values.

18-25 Using the data of Prob. 18-23, consider a regenerator placed in the cycle similar to that shown in Fig. 18-6. Assume that the air entering the turbine is precooled to the compressor-inlet temperature. For an ideal cycle, determine (*a*) the minimum temperature in the cycle, in degrees Fahrenheit, and (*b*) the coefficient of performance.

18-26 A regenerative-type gas refrigerator similar to that shown in Fig. 18-6 operates with helium between pressure limits of 1.0 and 6.0 atm. Compression and expansion are isentropic, and pressure drops are negligible. The compressor-inlet temperature (state 1) and regenerator-inlet temperature (state 3) are both 80°F, and the refrigerator-outlet temperature is −60°F. Determine (*a*) the minimum temperature achieved, in degrees Fahrenheit, (*b*) the amount of refrigeration possible, in Btu/lb, and (*c*) the net amount of work required, in Btu/lb.

18-27 Air at 1 atm and 10°F enters an adiabatic compressor, which discharges the fluid at 5.0 atm to a heat exchanger cooled by the ambient surroundings. The compressed air leaves the heat exchanger at 4.9 atm and 110°F and enters an adiabatic turbine which expands the fluid to 1.1 atm. The expanded and cold air now passes through a second heat exchanger where it picks up heat from a low-temperature region and exits at 1 atm and 10°F. The cycle is then repeated. The turbine and compressor are both 80 percent efficient, and the turbine helps drive the compressor. On the basis of air-table data, determine (*a*) the net work input required per unit of heat picked up from the low-temperature region, and (*b*) the heat removed to the ambient surroundings per unit of heat picked up from the low-temperature region. (*c*) Sketch a *Ts* diagram of the process.

Liquefaction and solidification

18-28 Consider the basic process for liquefying gases as shown in Fig. 18-7, but ignore the compressor and aftercooler. Refrigerant 12 enters the heat exchanger at 300 psia and 180°F and is cooled by saturated vapor which is withdrawn from the heavily insulated separator. The high-pressure cooled gas leaves the heat exchanger and passes through a throttling valve into the separator. The separator contains liquid and vapor refrigerant in equilibrium at 20 psia. The vapor leaving at low pressure from the heat exchanger has a temperature of 160°F. As a first approximation neglect all pressure losses except that across the throttling valve. Determine (*a*) the fraction of the inlet high-pressure gas which is liquefied, and (*b*) the temperature (if superheated) or the quality (if saturated) of the flow stream entering the valve.

18-29 Work Prob. 18-28 under the following set of conditions. Refrigerant 12 enters the heat exchanger at 200 psia and 160°F. The separator operates at 20 psia and the vapor leaving the heat exchanger from the separator has a temperature of 150°F.

18-30 Consider the equipment diagram shown in Fig. 18-7 and the *Ts* diagram for the solidification of CO_2 shown in Fig. 18-8. Omitting the aftercooler, state 1 has a pressure of 1000 psia and a temperature of 90°F. After passing through the heat exchanger, the fluid is throttled to 50 psia. Employing the data from Fig. A-26, determine (*a*) the percent solid in the separation process, and (*b*) the temperature at state 5, in degrees Fahrenheit. The fluid at state 2 is 50 percent liquid at 1000 psia.

18-31 Solid carbon dioxide is to be produced by throttling the fluid under the following steady-flow conditions:

(*a*) The gas is compressed to 600 psia and then precooled to 100°F. A pressure drop of 10 psi is incurred in the precooler. The fluid is passed through a heat exchanger and then throttled to 20 psia into a separator. The gas phase in the separator is passed back through the heat exchanger and discharged at 15 psia and 80°F. The solid CO_2 is removed at 20 psia. Determine (1) the mass fraction of solid CO_2 produced, and (2) the temperature and condition of the CO_2 entering the throttle device.

(*b*) Now the same process is carried out, except that the fluid is precooled another 60° down to 40°F before it enters the heat exchanger. The outlet temperature at 15 psia is also 60° lower than before. Repeat parts (1) and (2).

Cascade and multistage vapor-compression cycles

18-32 A cascade refrigeration system employs refrigerant 12 in both of the closed loops. The low-temperature evaporator is designed to operate at −40°F, while the fluid condenses at 40 psia. The high-temperature cycle has an evaporator temperature of 10°F and a condenser pressure of 140 psia. The refrigeration capacity required for the low-temperature evaporator is 5 tons, and the compressors are isentropic. Determine (*a*) the mass flow rates, in lb/min, in both loops, (*b*) the power input to both compressors, in horsepower, (*c*) the COP of the cascade system, (*d*) the percent change in compressor work required if a single vapor-compression cycle is used between −40°F and 140 psia, for the same mass flow rate through the low-temperature evaporator, and (*e*) the percent change in refrigeration capacity achieved if a single vapor-compression cycle is used with the same mass flow rate through the low-temperature evaporator.

18-33 Reconsider Prob. 18-32. Repeat the required calculations if the adiabatic efficiency for both compressors is 75 percent. Other losses such as pressure drops are neglected.

18-34 A two-stage refrigeration system with regenerative intercooling operates with refrigerant 12 and has pressures of 30, 60, and 140 psia in the evaporator, flash chamber and mixing chamber, and condenser, respectively. Assume isentropic compression and isenthalpic throttling, and no other losses. If the refrigeration load is 5 tons, determine (*a*) the mass flow rate through the evaporator, in lb/min, (*b*) the power input to the low-pressure compressor, in horsepower, (*c*) the mass flow rate leaving the flash chamber and entering the mixing chamber, in lb/min, (*d*) the power input to the high-pressure compressor, in horsepower, and (*e*) the COP of the cycle.

18-35 Reconsider Prob. 18-34. Repeat the required calculations if the adiabatic efficiency for both compressors is 77 percent. Other losses such as pressure drops are neglected.

18-36 Reconsider Prob. 18-34. Repeat the required calculations if the three pressures cited are changed to 20, 40, and 120 psia, and the refrigeration load is 6 tons.

18-37 Reconsider Prob. 18-34. Repeat the required calculations if the three pressures cited are changed to 20, 40, and 120 psia, the adiabatic efficiency for both compressors is 75 percent, and the refrigeration load is 6 tons.

Stirling refrigeration cycle

18-38 Consider the use of a Stirling cycle for refrigeration purposes. Air at −40°F and 10 atm is expanded isothermally to 1 atm. It is then heated to 100°F at constant volume. Compression at a constant temperature of 100°F follows, and the cycle is completed by constant-volume heat removal. Compute the coefficient of performance, the quantity q_L in Btu/lb, and P_{max}.

18-39 A Stirling cycle is used for refrigeration purposes. Air at −20°F and 8 atm is expanded isothermally to 1.2 atm. It is then heated to 120°F at constant volume. Compression at a constant temperature of 120°F follows, and the cycle is completed by constant-volume heat removal. Compute the coefficient of performance for the cycle, the quantity q_L in Btu/lb, and P_{max}.

CHAPTER

NINETEEN

ADVANCED AND INNOVATIVE ENERGY SYSTEMS

19-1 BATTERIES AND FUEL CELLS

There are a number of conventional devices which convert thermal energy into some form of work. Sources of thermal energy include solar radiation, solid, liquid, and gaseous fuels, and nuclear reactions. The work produced commonly is in the form of a rotating shaft or electrical work. The maximum work obtainable from a cyclic device which receives heat from a high-temperature source and rejects to the environment is limited to the Carnot efficiency. If the theoretical thermal efficiency is to be relatively large, the temperature of the source must be fairly high, e.g., upwards to 1500 to 1800°K or 3000 to 3500°R. Because of the presence of irreversibilities in actual practice, modern heat engines seldom achieve thermal efficiencies greater than about 40 percent. A considerable fraction of the energy supplied to the cyclic device is discharged as heat to the local surroundings. It would be highly desirable if other methods of energy conversion were available which did not rely on the conversion of heat into work, since such methods would not be limited to the Carnot efficiency.

One well-known device which bypasses the heat-work energy-conversion step is the conventional battery. By means of a controlled chemical reaction, a battery converts energy stored in the chemical bonds of the reactants into electrical work. Usually a small quantity of heat is also discharged by the battery, but the operation is nearly isothermal. A fuel cell is a form of a battery with several important changes. The electrode materials are not consumed during its operation, but remain invariant. As a result, the fuel and oxidizer must be supplied continuously from an outside source. In addition, a means of eliminating the products of the reaction must be provided. The conventional battery is a closed system, whereas the fuel cell operates as an open system.

The electrical work output of a battery or fuel cell can be ascertained from a thermodynamic analysis of either a closed or open system. Consider the battery shown in Fig. 19-1*a*. We shall assume that heat transfer to or from the battery results in an isothermal process. In addition to electrical work, boundary work in the amount $P\,dV$ may also be present, since the system is maintained at a constant pressure. An energy balance on the closed system is of the form

$$\delta q + \delta w_{\text{elec}} + (-P\,dV) = du$$

or

$$\delta w_{\text{elec}} = du + P\,dv - \delta q = dh - \delta q$$

In addition, the maximum electrical work output occurs when the process is internally reversible. Under this constraint, $\delta q = T\,ds$. Replacing δq in the energy balance by this expression, we find that

$$\delta w_{\text{elec}} = dh - T\,ds$$

However, since $g = h - Ts$, then under isothermal conditions we see that $dg_T = dh - T\,ds$. Therefore, for the closed system described in Fig. 19-1*a*,

$$\delta w_{\text{elec}} = dg_T \tag{19-1}$$

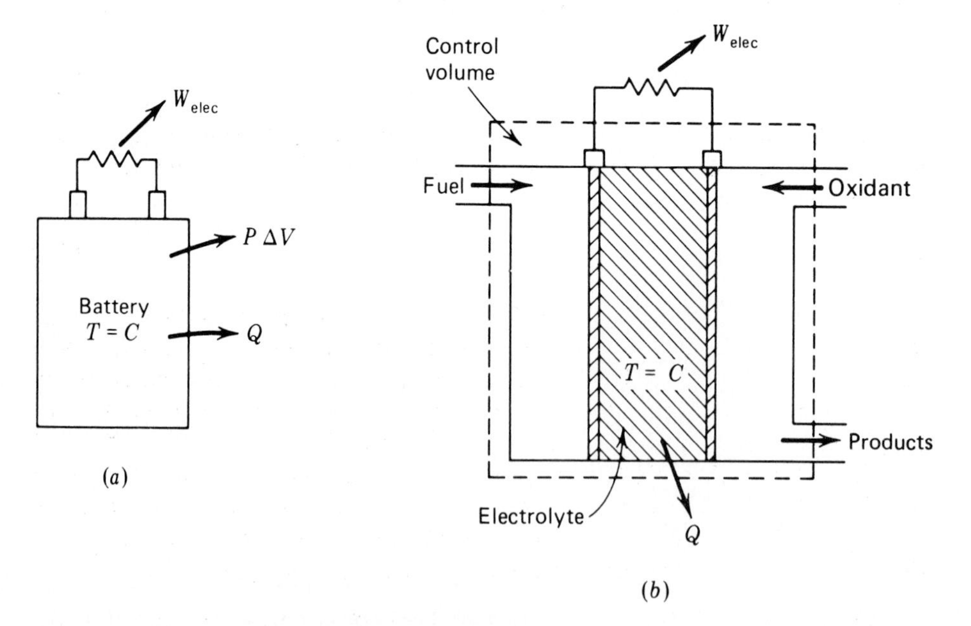

Figure 19-1 Schematic of (*a*) a battery, and (*b*) a fuel cell.

or, for a finite change of state,

$$w_{\text{elec}} = \Delta g_T \qquad \text{(closed, isothermal)} \tag{19-2}$$

Therefore the electrical work output of a battery under isothermal, internally reversible conditions is measured by the change in the Gibbs function.

A schematic of a steady-state fuel cell is shown in Fig. 19-1*b*. Again we shall assume isothermal operation. An energy balance on the control volume around the fuel cell shows that

$$\delta q + \delta w_{\text{elec}} = dh + d\text{KE} + d\text{PE}$$

In the operation of most fuel cells the changes in kinetic and potential energy may be neglected. In addition, for a unit mass passing through the control volume, $\delta q = T\,ds$. As a result, we obtain the identical result that was developed above for a closed system such as a battery, namely,

$$w_{\text{elec}} = \Delta g_T \qquad \text{(steady flow, isothermal)} \tag{19-2}$$

Hence the maximum electrical work output of a battery or fuel cell operating under isothermal conditions is measured by the decrease in the Gibbs function for the overall process.

There are several different efficiencies frequently defined for the operation of batteries and fuel cells. One possible standard of performance is the ratio of the maximum useful-work output to the energy input. We have seen above that the maximum useful-work output is given by Δg_R. The energy input is the enthalpy of reaction Δh_R released by the overall chemical reaction. The ideal efficiency or effectiveness of a battery or fuel cell, therefore, may be defined by the relation

$$\eta_i = \frac{\Delta g_R}{\Delta h_R} \tag{19-3}$$

where both Δg_R and Δh_R normally have negative values. For isothermal operation it is also true that $\Delta g = \Delta h - T\,\Delta s$. Therefore, the ideal efficiency frequently is seen in the form

$$\eta_i = 1 - \frac{T\,\Delta s_R}{\Delta h_R} \tag{19-4}$$

For isothermal, reversible processes the quantity $T\,\Delta s$ represents the heat transfer to or from the system. When the reversible heat transfer is out of the system (Δs is negative), then the ideal efficiency is less than unity. An ideal efficiency greater than unity occurs when Δs is positive. This is possible, and simply indicates that heat must be added to keep the process isothermal.

The evaluation of the ideal efficiency requires information on the enthalpy of reaction, Δh_R, the Gibbs function change for the reaction, Δg_R, and the entropy change for the reaction, Δs_R, at the specified temperature. Recall from Chap. 14 that the values of Δh_R and Δs_R for a system of reacting ideal gases are

$$\Delta h_{R,T} = \sum_{\text{prod}} \nu_i(\Delta h^0_{f,298} - h_T - h_{298})_i - \sum_{\text{reac}} \nu_i(\Delta h^0_{f,298} + h_T - h_{298})_i \tag{14-21}$$

and $$\Delta s_{R,T} = \sum_{\text{prod}} \nu_i (s^0_{i,T} - R \ln p_i) - \sum_{\text{reac}} \nu_i (s^0_{i,T} - R \ln p_i) \tag{14-25}$$

where the subscript T in this case represents the same temperature for the reactants and products, since the process is isothermal. For reactions which occur at the standard reference temperature of 298°K, the equation for the enthalpy of reaction reduces to the form

$$\Delta h_R = \sum_i (\nu_i \,\Delta h_{f,i})_{\text{prod}} - \sum_i (\nu_i \,\Delta h_{f,i})_{\text{reac}} \tag{14-19}$$

A similar equation holds for the Gibbs function. That is,

$$\Delta g_R = \sum_i (\nu_i \,\Delta g_{f,i})_{\text{prod}} - \sum_i (\nu_i \,\Delta g_{f,i})_{\text{reac}} \tag{19-5}$$

Typical values of the Gibbs function of formation of compounds are found in Table A-22M for conditions of 1 atm and 25°C (77°F).

In addition to the ideal efficiency of a battery or fuel cell, another important parameter is the ideal open circuit voltage developed by the cell. We have already seen that the Gibbs-function change for the reversible case is a measure of the electrical work produced by the cell. The electrical work is the product of the amount of charge Q_e that passes from the cell per mole of reacting fuel and the ideal electrostatic potential V_i developed. That is, $w_{\text{elec}} = -Q_e V_i$. The quantity of charge Q_e is equal to the number of moles of electrons j produced by the cell reaction per mole of reacting fuel multiplied by the number of coulombs (C) per mole of electrons, $\mathcal{F}$. Hence $Q_e = j\mathcal{F}$. Therefore

$$w_{\text{elec}} = -j\mathcal{F}\, V_i = \Delta g_R$$

or $$V_i = \frac{-\Delta g_R}{j\mathcal{F}} \tag{19-6}$$

The quantity Δg_R is negative for battery and fuel cell reactions, so that the ideal voltage V_i is a positive value.

The quantity of charge $\mathcal{F}$ is called a faraday. Its value is

$$\mathcal{F} = \frac{6.023 \times 10^{23} \text{ electrons}}{\text{g} \cdot \text{mol electrons}} \times \frac{1.602 \times 10^{-19} \text{ C}}{\text{electron}}$$

$$= 96{,}487 \text{ C/g} \cdot \text{mol electrons}$$

If this value is multiplied by the identity, 1 J = 1V · C, then

$$\mathcal{F} = 96{,}487 \text{ kJ/(V)(kg} \cdot \text{mol electrons)} \tag{19-7}$$

Therefore Eq. (14-24) becomes

$$V_i = \frac{-\Delta g_R}{96{,}487 j} \tag{19-8}$$

where Δg_R is expressed in kilojoules per kilogram mole and V_i is in volts.

The method of evaluation of the quantity j in Eq. (19-8) is made more clear by examining two specific examples. Figure 19-2a is a schematic of the well-

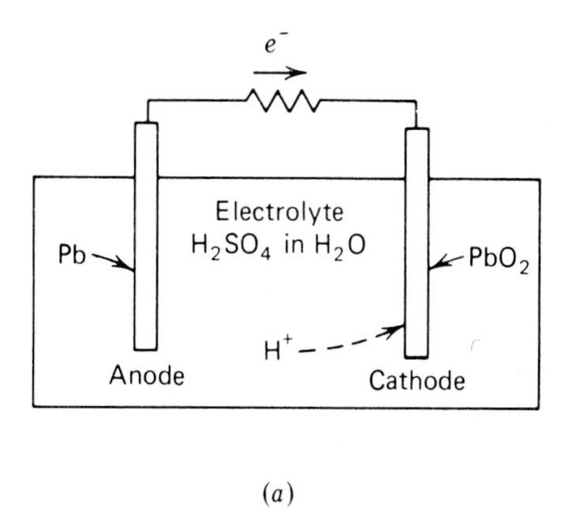

(a)

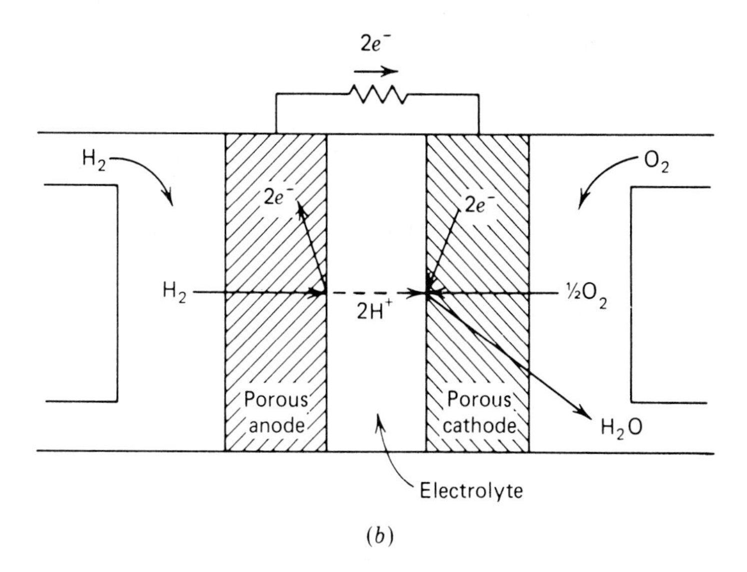

(b)

Figure 19-2 Schematic of (a) a lead-acid battery, and (b) a hydrogen-oxygen fuel cell.

known lead-acid storage battery. This cell consists of two electrodes: one of metallic lead and the other of lead oxide, PbO_2. The electrode plates are immersed in a water solution of sulfuric acid. In the solution, called the *electrolyte*, the sulfuric acid dissociates into hydrogen and sulfate ions. The overall reaction during discharge of the cell is

$$Pb + PbO_2 + 2H_2SO_4 \rightarrow 2PbSO_4 + 2H_2O \tag{19-9}$$

Thus, as the cell is discharged, both of the plates are converted to lead sulfate and the sulfuric acid is consumed while excess water is produced. It is now convenient to express the overall chemical reaction in terms of the individual reactions which occur at each electrode. The reactions that occur at the anode and cathode are called the *half-cell* reactions. For the lead-acid battery the anode half-cell reaction is

$$Pb + SO_4^{--} \rightarrow PbSO_4 + 2e^-$$

The electrons released at the anode now pass through an external circuit in order to complete the reaction. By placing an external load in the circuit, useful work can be produced by the process, even though the reaction is isothermal. Since chemical energy is converted directly to electric energy, there is no Carnot limitation of the conversion process. At the cathode the reaction is

$$PbO_2 + SO_4^{--} + 4H^+ + 2e^- \rightarrow PbSO_4 + 2H_2O$$

This reaction requires the electrons that were released at the anode, and which pass through the external circuit. In this particular case it is seen that j has a value of 2. That is, two electrons are released when a molecule of reactant at the anode is consumed.

A special comment should be made with respect to the lead-acid storage battery. The values of Δg_R and Δh_R are dependent upon the molality of the solution. This is shown by the data at the bottom of Table A-22M. Since water is a net product of the reaction in a lead-acid battery, the solution becomes more dilute as the battery is discharged. Hence the values of the maximum work, the ideal efficiency, and the ideal voltage change as the cell is discharged.

A fuel cell operating on hydrogen and oxygen has already been used on space missions to the moon. The overall reaction at room temperature is

$$H_2(g) + \tfrac{1}{2}O_2(g) \rightarrow H_2O(l) \tag{19-10}$$

A schematic of the cell operating with an acidic electrolyte is shown in Fig. 19-2*b*. The half-cell reactions are:

Anode: $$H_2 \rightarrow 2e^- + 2H^+$$

Cathode: $$2H^+ + 2e^- + \tfrac{1}{2}O_2 \rightarrow H_2O(l)$$

Thus, during the process hydrogen ions migrate through the electrolyte from the anode to the cathode, while electrons pass through the external circuit. In this case j is again 2, since two electrons are liberated at the anode for every molecule of hydrogen consumed.

Example 19-1M Consider the hydrogen-oxygen fuel cell shown in Fig. 19-2*b*, which contains an acidic electrolyte. For a temperature of 25°C, and assuming that the water formed as a product at the cathode is all in the liquid state, determine (*a*) the maximum work output, in kJ/kg of fuel, (*b*) the ideal efficiency, and (*c*) the ideal voltage.

SOLUTION (*a*) In this particular fuel cell, all reactants and products may be assumed to be pure, and the operating pressure will be taken to be 1 atm. Consequently, the maximum work output as given by the Gibbs-function change for the reaction is determined from the standard state value Δg^0_{298}. Using Eq. (19-5) and the data from Table A-22M, we find that

$$w_{max} = \Delta g^0_{298} = 1(-237{,}180) - 1(0) - \tfrac{1}{2}(0) = -237{,}180 \text{ kJ/kg} \cdot \text{mol}$$

(*b*) The ideal efficiency requires the evaluation of the enthalpy change for the reaction, which in this case of pure substances each at 1 atm is given by Δh_{298}^0. Hence from Eq. (14-19) and data from Table A-22M,

$$\Delta h_{298}^0 = 1(-285{,}830) - 1(0) - \tfrac{1}{2}(0) = -285{,}830 \text{ kJ/kg} \cdot \text{mol}$$

On the basis of Eq. (19-3), then, the ideal efficiency becomes

$$\eta_i = \frac{\Delta g_R}{\Delta h_R} = \frac{-237{,}180}{-285{,}830} = 0.830 \qquad \text{(or 83.0 percent)}$$

Note that the energy output Δg_R is less than the energy input Δh_R in this particular case. The difference between these values, $\Delta h_R - \Delta g_R$, is a measure of the heat transfer necessary to the environment if this cell is to be maintained at constant temperature.

(*c*) The ideal voltage of the cell is determined by using Eq. (19-8). In this case the value of j is 2, as shown by the anode or cathode half-cell reaction discussed above. Therefore,

$$V_i = \frac{-\Delta g_R}{96{,}487j} = \frac{-(-237{,}180)}{96{,}487(2)} = 1.23 \text{ V}$$

When other substances are used in batteries and fuel cells, they frequently exhibit a similar value for the ideal voltage. Values may typically range from 1 to 2 V per cell. Higher overall voltages are obtained by placing several cells in series.

The number of moles of electrons, j, produced by the cell reaction per mole of reacting fuel can be ascertained in the following manner. For fuel cells with any hydrocarbon C_xH_y and oxygen O_2, the half-cell reactions must be considered. At the anode the reactants are C_xH_y and water from the electrolyte, and the products are gaseous carbon dioxide, hydrogen ions, and electrons. For example,

$$\text{Anode:} \qquad C_xH_y + (2x)H_2O \rightarrow xCO_2 + (4x + y)H^+ + (4x + y)e^-$$

At the cathode the electrons and hydrogen ions react with oxygen introduced to the cell to form water. In this case,

$$\text{Cathode:} \quad \left(x + \frac{y}{4}\right)O_2 + (4x + y)H^+ + (4x + y)e^- \rightarrow \left(2x + \frac{y}{2}\right)H_2O$$

The overall chemical reaction may be written as

$$C_xH_y + \left(x + \frac{y}{4}\right)O_2 \rightarrow xCO_2 + \frac{y}{2}H_2O$$

which is simply the sum of the reactions at the anode and cathode. For any hydrocarbon which does not contain oxygen in its formula, the number of electrons, j, released per molecule of fuel C_xH_y is $4x + y$.

One important consideration in the analysis of any fuel cell is the effect of temperature on performance. First of all, the ideal efficiency may be affected by temperature. This depends upon the variation of Δg_R and Δh_R with temperature. These properties are fairly sensitive to temperature for the hydrogen-oxygen cell, for example. The influence of temperature on the rate of reaction is also important. Cells employing carbon or hydrocarbons as the fuel frequently must be operated at elevated temperatures in order to achieve sufficient rates.

Finally, we need to consider the factors which influence the performance of actual fuel cells. In the actual operation there are a number of irreversible effects within the cell which greatly reduce the terminal voltage of the cell. The losses within a fuel cell are generally spoken of as overvoltages or polarization effects. These frequently are grouped into three classes: (1) resistance or ohmic polarization, (2) activation or chemical polarization, and (3) concentration polarization. The magnitude of each of these effects is a function of the current density J_i. The first of these is caused by the internal resistance of the cell, and it includes losses in the electrolyte as well as in the electrodes. The activation polarization arises from chemical changes occurring at the surface of the electrodes, as well as adsorption and desorption effects on the surface. The concentration polarization is caused by the concentration gradients set up in the electrolyte and in the gas streams in the vicinity of the electrodes.

The net result of the three polarization effects or overvoltages is a reduction in the terminal voltage at a given current density. Figure 19-3 presents a typical diagram for the cell voltage versus the current density. Current densities typically range from 0 to 1 A/cm^2. The rapid drop at low current is due to the activation or chemical polarization. In the middle range neither the activation or concentration effects are changing appreciably, and the curve varies linearly due to the ohmic drop. Finally, at high current densities the concentration polarization drops the cell output sharply.

The fuel cell played an important role in the 1960s as a power source on manned space flights. In the future it is expected that the fuel cell system will evolve into an important contributor to the production of power on the scale of a large commercial power plant. Through the use of first-generation fuel-cell technology, a demonstration test of the magnitude of 5 MW is scheduled in New York City before the mid-1980s. This first large-scale test program uses hydrogen and oxygen with a phosphoric acid electrolyte. The operating temperature of this type cell must be held between 150 to 200°C. Stacks of individual cells (each of which produce between 0.6 and 1.0 V direct current) are capable of generating

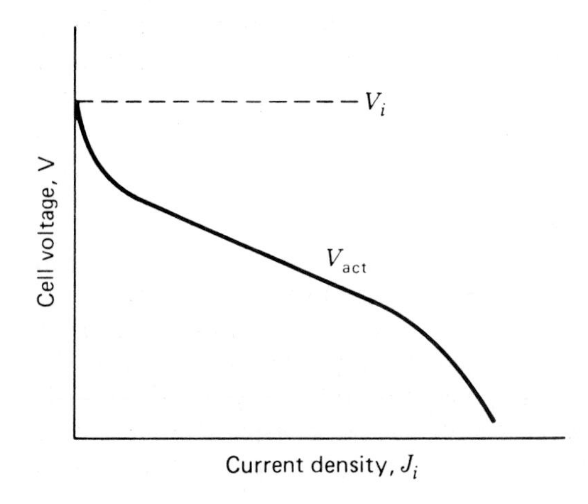

Figure 19-3 A typical voltage-current density diagram for a fuel cell.

250 to 500 V direct current at power ratings of 200 to 500 kW. Megawatts of power at 2000 to 3000 V direct current can be produced when such stacks are connected in series and parallel arrangements. A more advanced, second-generation fuel cell would use molten carbonate technology with operating temperatures in the 600 to 700°C range. This latter technology is estimated to be five years behind that of the phosphoric acid fuel cell.

The large-scale fuel-cell power plant has a number of distinct advantages over conventional fossil-fuel plants. A fuel cell operates as a direct-energy converter, and not as a heat engine. Therefore it is not affected by the Carnot efficiency limitation of the second law. Thus it has the potential for greater energy conversion efficiencies than conventional heat engine devices. First-generation fuel cells would have efficiencies around 40 percent, while second-generation cell efficiency is expected to approach 45 to 50 percent. This compares to 30 to 40 percent for fossil-fuel plants currently in use. In addition, the efficiency of a fuel cell system is directly related to that of the individual cell. Hence plant size will not affect efficiency. In comparison, the thermal efficiency of fossil-fuel plants decrease with decreasing size. For a given size, a conventional plant will be much less efficient at part load. On the other hand, the efficiency of fuel-cell power plants will remain roughly the same when operating anywhere between 25 to 100 percent of rated load. Another major advantage of the fuel-cell power plant is its ability to respond quickly to load variations. This response can be less than a minute. Large conventional power plants have very large response times. Therefore, large load variations are handled by some other means.

The fuel-cell power plant has several other desirable characteristics. For example, sulfur oxide and nitrogen oxide emissions from the plant would be extremely small, because no direct combustion process occurs. In addition, since the device is not a heat engine, large quantities of waste heat energy do not have to be rejected to the environment. Thus large quantities of cooling water are not required, and "thermal pollution" problems are nonexistent. If a given type of fuel cell requires thermodynamically that some waste heat be rejected, this thermal energy can be used to produce hot water for commercial purposes. This use of thermal energy in a power plant is called cogeneration, and is discussed further in Sec. 19-3. Although many technological problems remain before the realization of large-scale production of power by fuel cell systems, the decade of the 1980s may prove to be an important era for commercial fuel-cell development.

19-2 THE COMBINED CYCLE

The thermal efficiency of the gas-turbine power cycle discussed in Chap. 16 and of the vapor power cycle discussed in Chap. 17 is typically less than 40 percent in practice. Although techniques such as reheating and regeneration do improve the cycle performance, the rejected or waste energy is still a large fraction of the energy input in either case. One possible way of achieving further improvement is by an arrangement called a combined or coupled cycle. A *combined power cycle* is one based on the coupling of two different power cycles such that the waste heat from one cycle is used partially or totally as the heat source for the other cycle. This basic concept was introduced in Sec. 17-7 under the title of binary vapor cycles. In this latter case a metallic fluid such as potassium is employed in a Rankine *topping* cycle. That is, a potassium Rankine cycle operates with the temperature level of the waste heat higher than that of the maximum temperature of steam in a conventional Rankine power cycle. Consequently the potassium cycle is able to reject waste heat to the boiler-superheater of the steam cycle, as illustrated earlier in Fig. 17-12. One important combined cycle under active development involves the use of a gas-turbine (Brayton) topping cycle with a steam-turbine (Rankine) cycle.

In a gas-turbine cycle the exhaust stream leaving the turbine is relatively hot. As discussed in Sec. 16-8, one method of utilizing this high-temperature energy is through the concept of regeneration. A portion of the energy is returned to the air as it passes from the compressor outlet to the combustor inlet. An alternative possibility is shown in Fig. 19-4. In place of regeneration, the gas-turbine exhaust stream is used as the energy source in the boiler of a conventional steam power cycle. Although this is not illustrated, the steam cycle would probably employ feedwater heaters. However, reheat in the turbine section of the steam cycle would not normally be possible. The energy in the hot gas-turbine exhaust is

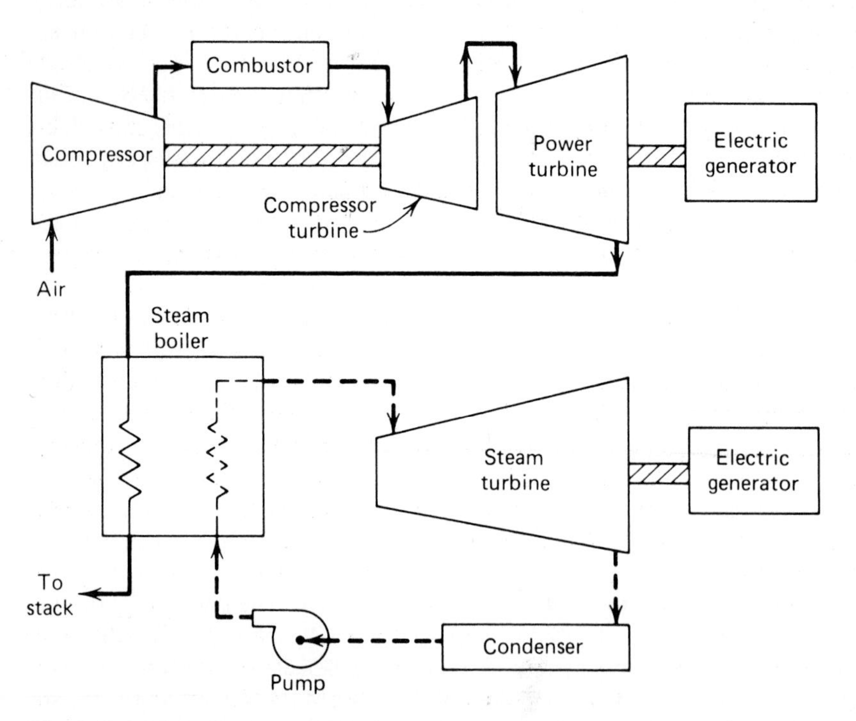

Figure 19-4 Schematic of a combined cycle composed of a basic Rankine steam cycle and a gas-turbine topping cycle.

needed primarily for the boiling and superheating of the steam. The thermodynamic analysis of this combined cycle would follow the procedures established in sections on the Brayton and Rankine cycles.

The practical development of the combined gas-steam-turbine cycle was delayed until modern technology provided the means of building gas-turbine power plants which operate at relatively high pressure ratios and with turbine-inlet temperatures which exceed 1350°K (2400°R). Pressure ratios of 10 : 1 to 13 : 1 are typical for this temperature. When turbine-inlet temperatures around 1600°K (2900°R) are possible on commercially available units, pressure ratios in the vicinity of 20 : 1 may be used. Commercial gas-steam-turbine power plants are available on the market. As gas-turbine technology improves, the advantages of this combined cycle will improve in relation to conventional steam power plants. This cycle appears to be extremely useful in conjunction with coal gasification. Not only is cycle efficiency improved by the combined cycle itself, but also coal gasification as a source of fuel for the gas-turbine combustor offers the additional advantage of removing potential air pollutants before they enter the combustion process. An additional discussion of the combined cycle and coal gasification is presented in the latter part of Sec. 19-3 on cogeneration systems.

There are some power cycles, such as a diesel engine, for which the temperature of rejection may be a few hundred degrees above that of the environment. In such cases another power cycle could operate between the low exhaust temperature of the main cycle and the atmospheric temperature. These low-temperature power cycles used in conjunction with another power cycle are called *bottoming* cycles. Only a very few fluids have the reasonably high critical temperature and large enthalpy of vaporization required to be a successful working medium for a bottoming cycle. Fluorocarbons are only fair in meeting these constraints, and usually are restricted in their use to temperatures below 150°C (300°F) or slightly higher. Some relatively low-molecular-weight hydrocarbons, such as isobutane and ammonia, have potential as working fluids at low temperatures.

The overall thermal efficiency η_0 of a combined cycle and the ratio of the work outputs of the two cycles can be determined in terms of the individual thermal efficiencies of the separate cycles. Consider the schematic shown in Fig. 19-5, where the equipment diagram of Fig. 19-4 has been reduced to a simple layout which shows heat and work exchanges for the combined cycle. Q_H and Q_L represent for each cycle the heat supplied and rejected, respectively. Heat engine 1 represents a gas-turbine cycle, for example. The heat Q_{L1} rejected from this cycle is used as the heat supply Q_{H2} to heat engine 2, which could represent the steam-vapor cycle shown in Fig. 19-4. To make the derivation more general, an additional heat source Q_E is shown which supplies heat to engine 2. Q_E might represent either of the two following cases. In practice, the gases leaving the gas-turbine cycle might pass through an afterburner before entering the heat exchanger or steam boiler-superheater. Fuel is burned in the afterburner and raises the temperature of the gas-turbine exhaust gases. Another possibility is heat addition to the steam after it leaves the heat exchanger and before it enters

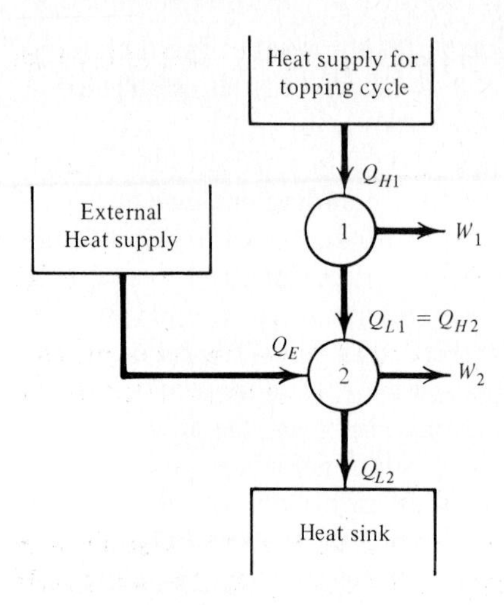

Figure 19-5 Schematic showing heat and work effects for a combined cycle with possible external heat supply.

the turbine. Such heat addition would increase the degree of superheating of the steam before it enters the turbine. Either case can be represented by an external heat addition Q_E.

The thermal efficiencies of the individual cycles are defined in the usual manner, that is, work delivered divided by heat supplied. Therefore,

$$\eta_1 = \frac{W_1}{Q_{H1}} = 1 - \frac{Q_{L1}}{Q_{H1}} \quad \text{and} \quad \eta_2 = \frac{W_2}{Q_{H2} + Q_E} = \frac{W_2}{Q_{L1} + Q_E} \tag{19-11}$$

Note that the equation for η_2 requires that all the heat rejected by engine 1 appear as supply heat to engine 2. In practice this would not be true, because it is unlikely that the heat exchanger can remove all the waste heat in the gas-turbine exhaust stream. However, the following equations are based on this assumption. In addition to the efficiency equations, it must be stated how much of the total heat added to engine 2 comes from an external source. This will be done by defining a parameter σ such that

$$\sigma \equiv \frac{\text{heat externally supplied to engine 2}}{\text{heat internally exchanged between cycles}} = \frac{Q_E}{Q_{H2}} = \frac{Q_E}{Q_{L1}} \tag{19-12}$$

These three definitions allow us to predict the overall combined cycle thermal efficiency and the work ratio.

The overall thermal efficiency η_0 of the combined cycle is

$$\eta_0 = \frac{\text{net work output}}{\text{heat supplied}} = \frac{W_1 + W_2}{Q_{H1} + Q_E} \tag{19-13}$$

and the work ratio β of heat engine 2 to heat engine 1 is simply

$$\beta \equiv \frac{W_2}{W_1} \tag{19-14}$$

When Eqs. (19-11) and (19-12) are substituted into Eqs. (19-13) and (19-14) we find that

$$\eta_0 = \frac{\eta_1 + \eta_2(\sigma + 1)(1 - \eta_1)}{1 + \sigma(1 - \eta_1)} \tag{19-15}$$

and

$$\beta = \frac{(\eta_2 + \sigma)(1 - \eta_1)}{\eta_1} \tag{19-16}$$

There are two situations of combined cycle operation of immediate interest. One is the degenerate case where $\sigma = \infty$. In this case Eqs. (19-15) and (19-16) reduce to

$$\eta_0 = \eta_2 \qquad \text{and} \qquad \beta = \infty$$

This result is not surprising. When σ approaches ∞, the external heat addition is very large compared to the heat added to engine 1. Therefore, the combined cycle behaves as if only heat engine 2 were in operation, and the overall efficiency is the same as that for the low-temperature cycle. The other situation, which is of practical importance, is the case where $\sigma = 0$. When heat engine 2 receives no heat from an external source, then

$$\eta_0 = \eta_1 + \eta_2 - \eta_1\eta_2 \qquad \text{and} \qquad \beta = \frac{\eta_2(1 - \eta_1)}{\eta_1} \tag{19-17}$$

where η_1 and η_2 refer to the high-temperature and low-temperature cycles, respectively. Keep in mind that the above equations are valid only if $Q_{H2} = Q_{L1}$.

19-3 COGENERATION SYSTEMS

The various descriptions of gas power cycles in Chap. 16 and vapor power cycles in Chap. 17 deal with systems for which the sole purpose is power production. It must be recognized that there are industrial and commercial situations where thermal energy is also required. For example, a power plant on a college campus may provide steam for heating buildings as well as for electric power. Consider also the following five major energy-intensive industries: chemicals, oil refining, steel making, food processing, and pulp and paper production. Large plants in these basic industries require steam for the operation of various processes in addition to their electrical needs. Although the multiple use of steam generation has been in practice for a number of decades, there will probably be an increased interest in such systems in the future. This is due to the decreasing availability and increasing cost of fossil fuels.

For a large power plant which produces only electricity, the thermal efficiency typically ranges from 0.30 to 0.40, or 30 to 40 percent. For energy conservation it is important that a larger fraction of the energy from basic sources be utilized. One method of achieving this is to integrate the use of steam for heating or industrial purposes with the general production of electric power.

This technique of sequential production of energy (usually electrical and thermal) from a single energy source is called *cogeneration*. The performance of a cogeneration system is frequently measured in terms of its effectiveness ϵ, which is defined as

$$\epsilon = \frac{\text{electric energy delivered} + \text{thermal energy delivered}}{\text{combustion heat absorbed}} \tag{19-18}$$

Compared to the thermal efficiency range quoted earlier, with cogeneration it is estimated that an effectiveness as high as 55 to 70 percent is possible. A schematic of one possible cogeneration system, as might be used on a college campus, is shown in Fig. 19-6. Steam is bled from an intermediate point in the turbine to provide energy for the thermal load. In practice the heating load might be integrated into a cycle which includes reheat and regeneration.

It should be noted that the thermal energy from a steam power plant can be delivered two ways—as steam bled from the turbine as noted above, or as waste heat removed from the turbine exhaust stream. In this latter case the energy removed from the exhaust stream is used to heat water, which is circulated separately. Which technique is used depends upon the end use of the thermal energy. Process heat supplied by steam in the temperature range of 150 to 200°C, or 300 to 400°F, is a common industrial need. Extraction of steam commonly occurs at pressures of 5 to 7 bars (75 to 100 psia). For conventional turbine inlet and exit pressures the steam will have produced over 60 percent of the total work possible before extraction occurs. The steam which is not extracted, of course, continues to produce work as it expands to the condenser pressure. Although the

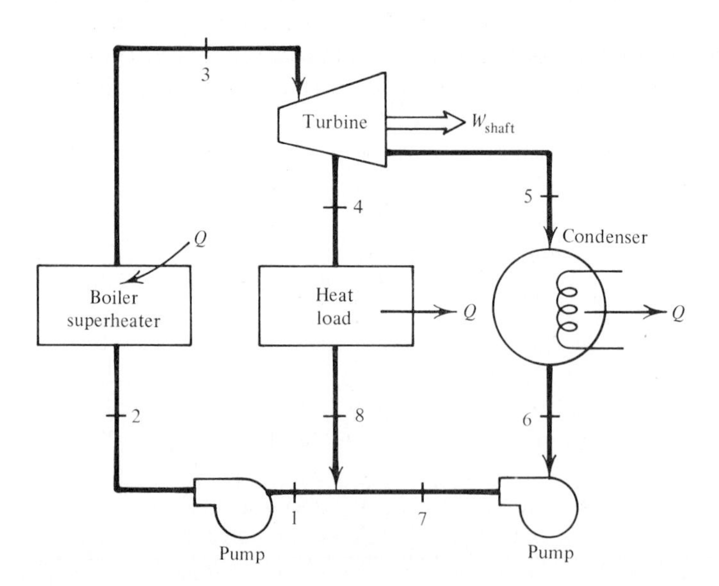

Figure 19-6 Schematic of a cogeneration plant employing steam bled from the turbine for the thermal load.

total work output is decreased for a given mass flow rate, a larger fraction of the total heat input is used for a useful purpose when cogeneration of this type is employed. Hot water for residential heating obtained by removing heat from the turbine exhaust stream would be supplied at temperatures considerably below that for extracted steam.

Cogeneration is especially attractive when a power plant can be integrated into a community so that it provides also for residential and commerical heating. The distribution of thermal energy from a central source to industrial, commercial, and residential customers for space heating, domestic hot water, and process needs is called *district heating*. Steam district heating grew in usage in the United States from its introduction in the late 1800s until around 1940. Its growth was curtailed around that time by the availability of inexpensive fuel oil and natural gas. By the early 1980s steam district heating again became a viable technology, due to the rapid rise in oil and gas prices in the 1970s. The state of Minnesota, for example, is actively planning for hot water district heating in St. Paul and other cities. Although cogeneration is an attractive approach to energy conservation, a plant operating in this mode must be carefully designed. Such a plant must be able to follow the power and heat loads on a local basis, and possibly integrate its electric load with the large electrical utilities which produce the base electrical needs of a state or region. This situation is fairly complex and requires careful evaluation.

An alternative to cogeneration based on a steam power cycle is one based on gas-turbine operation. To avoid the use of a "high-quality" energy source such as oil or natural gas, one can consider a coal-fired gas-turbine cogeneration system. In this case the heat supplied to the cycle might be provided by an atmospheric fluidized-bed combustion process. A schematic of one possible arrangement for this situation is shown in Fig. 19-7. Such equipment would probably operate with pressure ratios from 4 to 8. Studies indicate that cogeneration based on a closed-

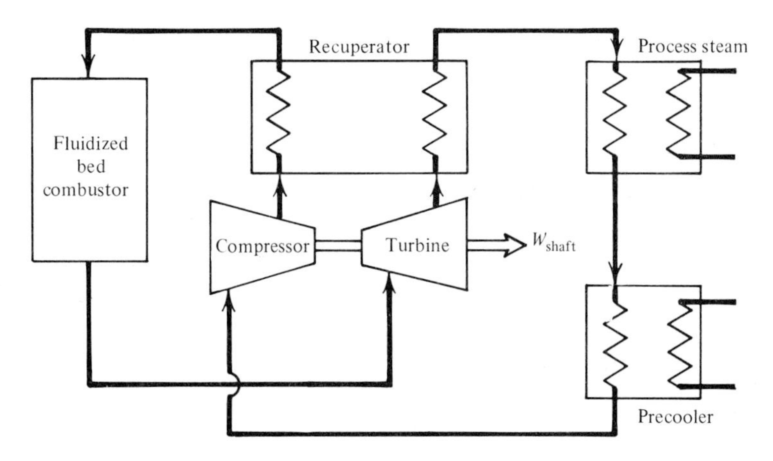

Figure 19-7 Schematic of a gas-turbine cogeneration plant based on atmospheric fluidized-bed combustion of coal.

cycle Brayton cycle would have excellent load-following performance, in terms of independently varying the required electric power and process heat loads.

A more advanced approach is to incorporate into an overall system the concepts of coal gasification, a combined cycle, and cogeneration. The basic elements of this system are shown in the schematic in Fig. 19-8. Coal gasification in this case provides a gaseous fuel for both the combustor of the gas-turbine cycle and the boiler-superheater (waste heat reboiler) of the steam-turbine cycle. The reboiler in this case can be designed to provide not only superheated steam for the steam cycle, but also process steam or hot water. Additional process steam can be bled from some intermediate point in the turbine. Preliminary design calculations for plants with a thermal-to-electric ratio between 1.5 and 2.5 indicate an overall efficiency or effectiveness of 60 to 70 percent. Advantages of coal gasification and the combined cycle include high efficiency, relatively low cost, low emissions to the environment, and low water usage. One of the first large-scale plants of this type will be a 100-MW unit built in southern California. Around 1000 tons (907 metric tons) per day of bituminous coal will be gasified, and the resulting medium-Btu synthetic gas will have a heating value of roughly 80,600 kJ/SCM (270 Btu/SCF). The turbine-inlet temperature will be 1365°K (2000°F), and the expected heat rate is roughly 11,500 kJ/kWh (10,900 Btu/kWh). Total cost is estimated to be 275 million dollars.

In addition to the systems discussed above, cogeneration is also feasible in conjunction with fuel cell operation. Research and development organizations are

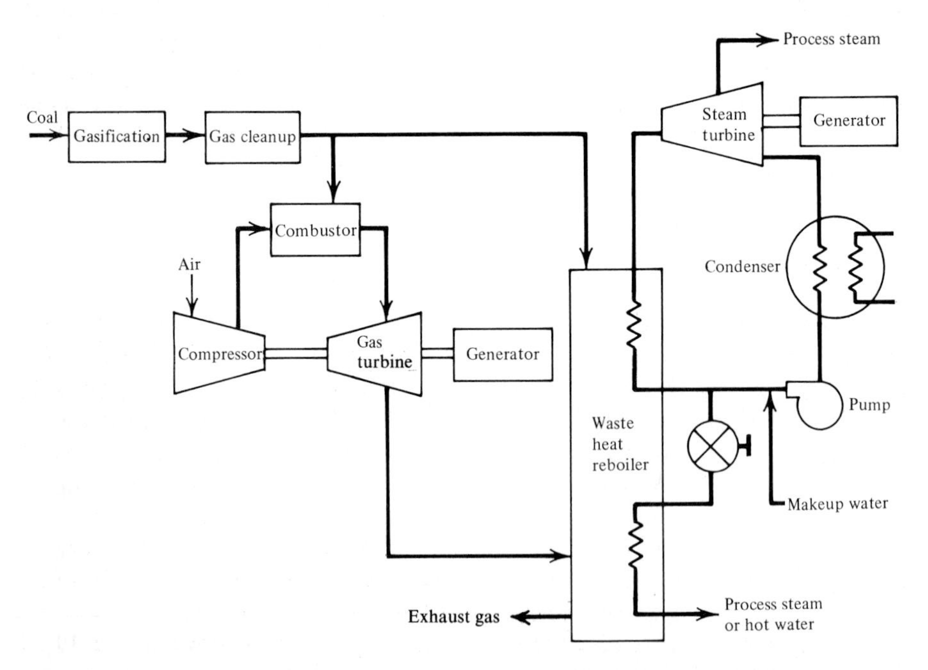

Figure 19-8 Schematic of a cogeneration plant based on coal gasification and a combined cycle.

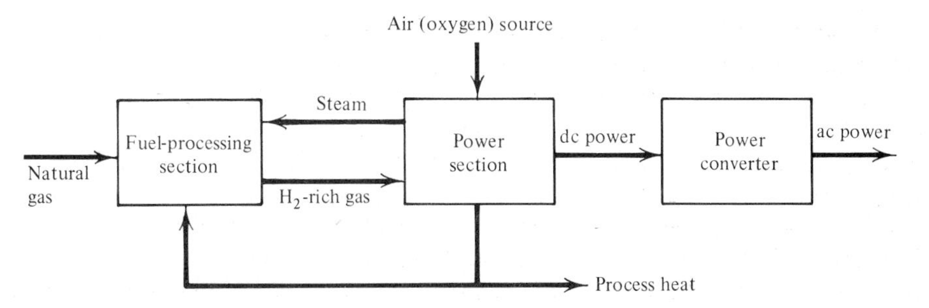

Figure 19-9 Schematic of a fuel-cell system employed for cogeneration needs.

investigating the potential of phosphoric acid and molten carbonate fuel cells for industrial cogeneration applications. One of the distinct advantages of fuel-cell energy systems is their excellent emission characteristics in terms of environmental air quality. A functional diagram of a fuel-cell power plant is shown in Fig. 19-9. Current fuel cells generate electricity with an efficiency around 40 percent over a wide operating range. It is anticipated that the thermal contribution to the efficiency could be an additional 30 to 40 percent. Thus the overall efficiency for energy usage could approach 70 to 80 percent.

The phosphoric acid fuel cell is restricted to producing primarily low-grade thermal energy in the form of hot water in the temperature range of 70 to 80°C (160 to 180°F). A small quantity of thermal energy might be available at temperatures approaching 135°C (275°F). For a 40-kW power plant the by-product thermal energy may be as high as 44 kW. The major advantage of a molten carbonate system is the availability of high-grade thermal energy, due to the high operating temperatures of the cell itself (around 700°C or 1300°F). The availability of high-grade thermal energy makes the molten carbonate fuel cell especially attractive for industrial cogeneration applications.

19-4 MAGNETOHYDRODYNAMICS

In very simple terms, the conventional electric generator works on the basic principle that the motion of a conducting medium, such as a wire, at a right angle to a magnetic field H will induce an electric field E at right angles to both the direction of motion and of the magnetic field. The current generated by the electric field may be passed through an external load and produce a useful effect. This process in practice requires two pieces of rotating machinery to accomplish the desired result. Work developed from the expansion of a fluid through a gas or steam turbine is used to drive the electric generator. This same effect can be accomplished without rotating parts. *Magnetohydrodynamics* or MHD is a process for producing electricity by passing an ionized gas (the electric conductor) through a channel across which exists a magnetic field M, as shown in Fig. 19-10. An electric field E is generated across the bottom to top plates of the channel

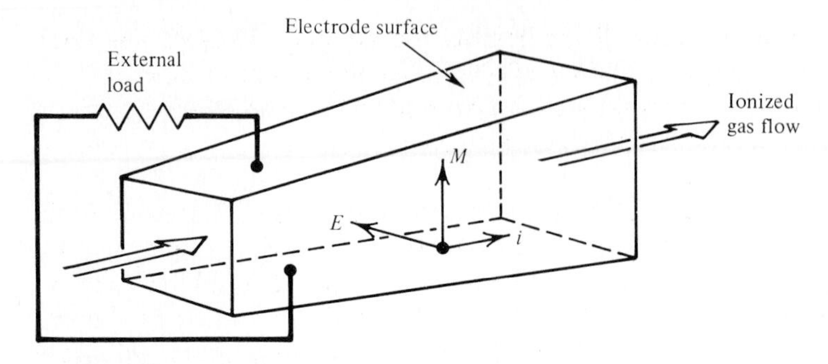

Figure 19-10 Sketch of a MHD channel, showing the directions of the gas flow and of the electric and magnetic fields.

shown in the figure, which act as electrode walls. When the electrode plates are connected externally, the voltage developed causes a current to pass through an external load. The electric energy expended comes at the expense of a portion of the energy of the ionized gas.

There are a number of scientific and engineering problems associated with the operation of an MHD facility for commercial power production. One of these is the formation of the ionized gas itself, frequently called a *plasma*. One method of ionizing a gas is simply by thermal means, that is, heating the gas to a high enough temperature to free electrons from the substance. In an MHD system the hot gas may be produced by the combustion of a fossil fuel. To conserve high-quality energy it would be better to use coal rather than natural gas or oil. However, the thermal ionization of gases such as oxygen and nitrogen is not significant at the combustion temperatures provided by the combustion of fossil fuels with air. To overcome this problem, the combustion gas may be seeded with a small amount of a substance which will ionize sufficiently at available temperatures. One substance of interest is cesium, which will provide adequate ionization (production of free electrons) at conventional combustion temperatures of 1900 to 2200°K or 3000 to 3500°F. A major disadvantage of cesium, however, is its cost. As a result, cesium generally is considered a fruitful seeding material when the cycle is closed so that the gases are not discharged into the environment. If a closed system were to be used, it appears that helium or argon may be more attractive working fluids than air.

In the development of a practical MHD power plant, an open MHD cycle probably is more attractive as a first-generation facility. In this case a much lower priced seeding material such as potassium may be used. The major disadvantage of potassium is that it must be heated to a considerably higher temperature than cesium to obtain the same degree of ionization. (See Sec. 15-8 for ionization calculations.) In fact, the desired temperature range is around 2300 to 2800°K, or 3600 to 4600°F. This temperature range, unfortunately, is above the adiabatic combustion temperature of common fossil fuels, as noted earlier. However, the difference is not all that great. To overcome this difficulty the air may be preheated, or oxygen enrichment of the combustion air may be used. The lower temperature limit of roughly 2300°K (3600°F) noted above is the minimum outlet value desired as the gases pass through the main MHD channel. Below this value the ionization of potassium is insufficient to provide a suitable conducting medium.

Figure 19-10 illustrates the general layout for the MHD channel where the energy conversion process occurs. The overall system is much more complicated than this, as shown by the equipment diagram in Fig. 19-11. In this illustration, coal and oxygen-enriched air are fired in a combustor. The result is a combustion gas at a temperature level around 2500°K (4600°F). In this section the gas is seeded with potassium carbonate. Following the combustor is a nozzle which accelerates the flow as it enters the MHD channel. The temperature of the gas decreases as it passes through the channel due to the removal of the electric energy. However, as noted earlier, the gases leave the channel at a temperature around 2300°K or 3600°F. Such gases still contain a large amount of high-quality energy. A considerable fraction of this energy may be recovered in a steam-generator section. This steam passes through a turbine which drives a conventional electric generator. Thus additional electric energy is produced, and the MHD system becomes another type of combined cycle, as discussed in Sec. 19-2. Present designs indicate that the MHD channel and the steam turbine-generator set would each produce about the same amount of power for the combined cycle. Additional energy might be recovered in an air preheater which follows the steam generator unit, as shown in Fig. 19-11. Finally, a unit for the recovery of the ionizing seed and for pollution control completes the flow pattern before the gases are discharged through a stack to the atmosphere.

Due to the possible use of coal as the primary fuel for future MHD energy converters, the emission of SO_2 must be controlled. One method of control is internal chemical removal. If a properly chosen potassium salt, such as potassium carbonate, is used for the seeding material, the sulfur dioxide produced from the sulfur-bearing coal will combine chemically in the gas and form potassium sul-

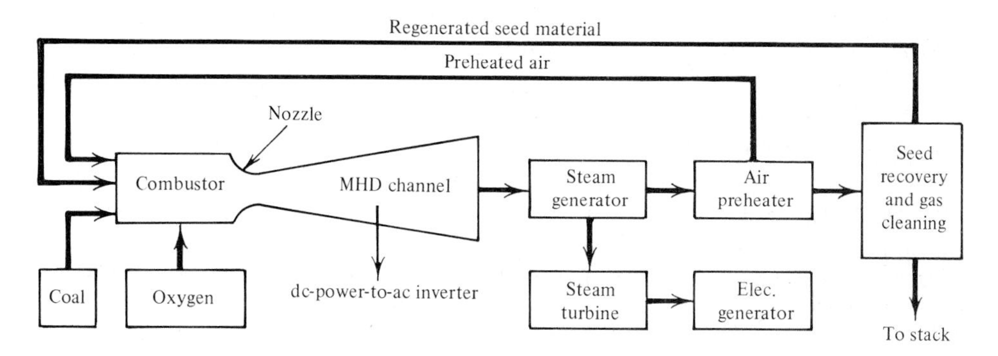

Figure 19-11 Equipment schematic for an open-cycle MHD power plant which includes a steam power section.

fate. However, it may be desirable to recover the potassium carbonate seed from the potassium sulfate product, and this type of process has not been developed to date. As an alternative, especially for the early commercial plants, conventional SO_2 scrubbing techniques used on present-day coal-fired boilers could be used.

Besides the SO_2 removal problem, there are other basic areas of concern in the development of a practical MHD converter. The ash contained within coal will produce a slag which may be carried though the system described in Fig. 19-11. The presence of this material could adversely affect, for example, the heat exchange process in the steam generator or the air preheater. A design which would remove a large fraction of the liquid slag while the material is still in the combustion section would greatly alleviate this problem. Another area of concern is the production of high levels of nitrogen oxide in the process due to high combustion temperatures. Some type of control measure will be needed in order for the system to meet the EPA emission standard for this pollutant. In addition, it should be noted that superconducting magnets must be used to achieve the desired magnetic field strength. Thus a cryogenic (very low temperature) support system is required for the magnet.

Although large-scale operation of a commercial MHD power plant is not expected until around the year 2000, research and development work is proceeding at a rapid pace throughout the world. A 20-MW pilot unit operating on natural gas is currently in use in the USSR. The United States has plans for a 40- to 50-MW component development and integration facility at a new plant in Butte, Montana, based on coal firing. After reliable MHD components have been developed, a 250- to 500-MW test facility is expected to be the next step in this country around 1990. Due to the high temperatures associated with MHD, the coupling of MHD with nuclear fission and fusion reactors as the energy source appears attractive as a long-range goal.

19-5 GEOTHERMAL ENERGY CONVERSION

Geothermal energy is the natural energy stored in the earth at depths close enough to the surface to be tapped and used either for electric generation or for thermal energy usage. This type of energy is stored in the form of dry steam, hot water, hot water with dissolved methane gas, and dry hot rock. The first use of geothermal energy in the United States began in 1960, when dry steam from The Geysers, an area north of San Francisco, was used to produce 11 MW of commercial electric power. The steam conditions at The Geysers are around 0.76 MPa (110 psi) and 180°C (350°F). By 1980, geothermal plants in this area produced about 800 MW of power, and by the end of the 1980s this capacity may nearly double. Unfortunately, natural dry steam occurs in very few places in the United States, and much of this is within national park boundaries. It is estimated that dry steam resources constitute only about one-half percent of the total U.S. geothermal resources.

Geothermal hot water, however, underlies a considerable portion of the continental United States. Twenty-four states have known hot water resources with temperatures above 90°C (195°F) and located no deeper than 3 km (2 mi) below the surface. The presence of underground hot water resources in the western half of the country is well established. It is significant that 60°C (140°F) water was discovered in 1979 in Maryland in the Atlantic Coastal Plain. Hence geothermal sources span the continent. The recoverable energy from hot water reservoirs has been estimated to be as high as 200 to 400 quadrillion (10^{15}) kJ (or Btu).

When water is above 150°C (300°F), it is said to be of electrical grade. All hot water of this type lies in the western states. For those water sources which have a temperature above this lower limit, there are two distinct technologies used to produce electric power. When the hydrothermal fluid is above 210°C (410°F), the fluid can be throttled to a lower pressure so that part of the liquid will flash or vaporize into steam. The dry steam produced is then passed through a steam turbine of the same type used in the dry-steam geothermal plants discussed above, such as at The Geysers and other sites around the world. Energy conversion efficiency based on this method is around 15 percent. When the fluid temperature is below 210°C (410°F), then the available temperature difference between turbine inlet and outlet is too small for economic operation by the above method. Instead, in this latter case of moderate temperatures one must rely on the operation of a *binary* cycle. Binary in this case indicates that two separate fluids are involved in the overall energy conversion system. However, only one of these fluids is responsible for the actual power production cycle. A demonstration plant of this type is planned for operation in California's Imperial County, near Mexico. The plant would operate with 180°C (355°F) water of low salinity and would produce 45 MW of electric power.

In a binary cycle the hydrothermal fluid transfers thermal energy across a heat exchanger surface to a second fluid that vaporizes at a much lower temperature than water for a given pressure. This vaporized secondary fluid then passes through a turbine that drives an electric generator, as shown in Fig. 19-12. The fluid then passes through a condenser and feedwater pump as in any conventional Rankine vapor power cycle. Both fluids in this system pass through closed loops. That is, the geothermal water is returned to the hot water reservoir through injection wells after it passes through the heat exchanger. In the Imperial County project, water would be returned at roughly 70°C (160°F). It is significant to note that in order to produce the same amount of power as a direct-flash cycle, the binary cycle requires only about two-thirds the water flow rate. Binary cycles are designed to use a low molecular weight hydrocarbon (three to six carbon atoms per molecule) as the secondary working fluid in the Rankine cycle. Propane and isobutane are typical examples. One engineering problem in the design of geothermal binary cycles is the scale-up of turbines to the large size required when operating with hydrocarbon fluids. Scaling and corrosion are two other problem areas in working with geothermal fluids, especially with respect to the heat exchanger operation. Another major problem area is the heat rejection

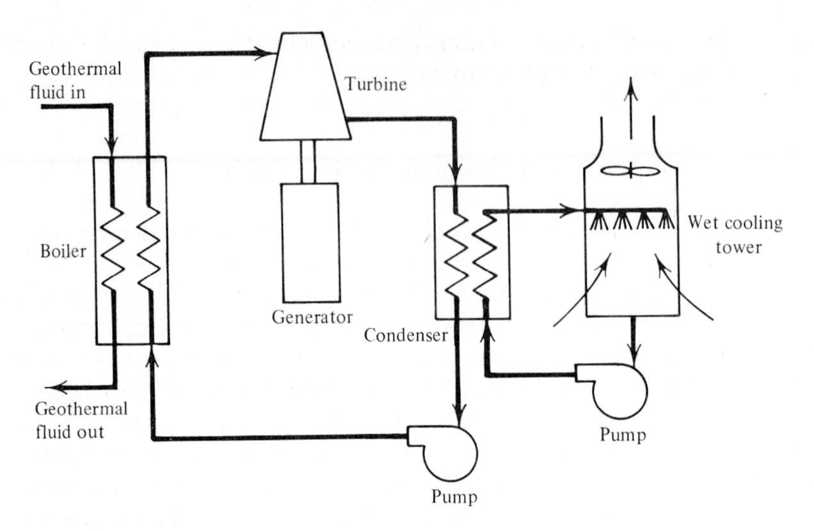

Figure 19-12 Schematic of a binary geothermal power cycle.

process. Because the thermal efficiency of geothermal plants is so low, a much higher fraction of the input energy must be rejected, in comparison to conventional power plants based on fossil fuels. In direct-flash power plants, the turbine condensate passes through an evaporative cooling tower and then back through the condenser. Binary cycles, as shown in Fig. 19-12, may have to use conventional wet cooling towers with water as the cooling medium for the condenser.

A third source of geothermal energy is called geopressured water, and it accounts for roughly 15 percent of the accessible energy in geothermal form in the United States. Most of this energy lies beneath the Gulf Coast. Water and organic material were deposited in past geological times beneath shale caps, and this organic material has been transformed into natural gas, or methane. These geopressure sources exhibit temperatures around 100 to 180°C (220 to 350°F) at depths of 2 to 6 km (1 to 4 mi). Interestingly, the pressures in these deposits are far greater than those based solely on hydrostatic effects at these depths, since they run as high as 70 MPa or 10,000 psi. Although research is being done in this area, the use of geopressured energy on a large scale is not expected in the near future.

Finally, one must consider petrothermal or magma energy. This is the thermal energy found in vast regions of molten rock, or magma, or the hot dry rock that is formed when magma is cooled. Magma itself can produce thermal energy at temperatures around 1000 to 1100°C (1800 to 2000°F). Principal magma-type locations are in the western United States. As might be expected, there currently are no known methods or materials for drilling into magma. For hot rock regions, it is expected that water would be injected into a region which has been fractured, and after heating the warm water would be pumped to the surface through another well in the vicinity. Current research deals with drilling methods, fracture techniques, and water circulation, among others. Although petrothermal energy accounts for over 80 percent of accessible geothermal energy in the United States, its commercial development lies in the future.

19-6 OCEAN THERMAL ENERGY CONVERSION

Heat engines are cyclic devices which receive heat from a high-temperature source, produce a net work output, and reject heat to a low-temperature sink. The larger the temperature difference between heat source and sink, the greater the theoretical energy conversion efficiency. When fuels (such as fossil sources) become more costly and less available, efforts are made to increase the thermal efficiency by operating at a higher temperature of supply. Other methods include the use of combined or cogeneration cycles (as discussed in preceding sections) which transform a larger fraction of the energy input into useful output. Nevertheless, if an energy source were relatively free in cost and of nearly unlimited availability, then heat engines operating between heat reservoirs with a very small difference in temperature might be attractive, even though the thermal efficiency would be extremely low. It is this latter situation that is encouraging research and development in ocean thermal energy conversion, or OTEC.

An ocean thermal energy conversion device is simply a heat engine designed to operate between the relatively warm temperature at the surface of an ocean and the cooler water temperature found deep below the surface. This approach was first suggested by the French physicist d'Arsonval in 1881. In practice, for economical operation the temperature difference required is around 20°C or 36°F. At the 600- to 900-m (2000- to 3000-ft) level below the ocean surface the temperature is around 5°C or 40°F. To achieve the desired temperature difference quoted above, one must seek out geographically those regions of the ocean's sun-warmed surface where the temperature averages 25 to 26°C or 78 to 80°F, at least. As might be anticipated, such regions exist only at latitudes close to the equator. The greatest temperature differences are found in the western Pacific Ocean. However, regions east and west of Central America are also satisfactory, and areas off the coast of southern United States and east of Florida are marginally good.

An OTEC system is simply a Rankine vapor power cycle operating under some rather special conditions. Because the temperatures in the evaporator and condenser are low, a working fluid must be chosen so that its vapor pressure is fairly high for these temperatures. For example, propane (C_3H_8) has a vapor pressure of about 0.55 MPa (80 psia) at 5°C (40°F), and the value is around 0.95 MPa (140 psia) at 25°C (78°F). Similarly, ammonia (NH_3) at these cited temperatures has vapor pressures of 0.48 MPa (70 psia) and 1.03 MPa (150 psia), respectively. Other fluids are possible choices. Warm water is drawn from near the surface of the ocean into a heat exchanger, or evaporator, where the liquid working fluid of the Rankine cycle is evaporated (see Fig. 19-13). The vapor then passes through a turbine, which powers an electric generator, and then passes through the condenser. Cold water pumped up from the depths of the ocean

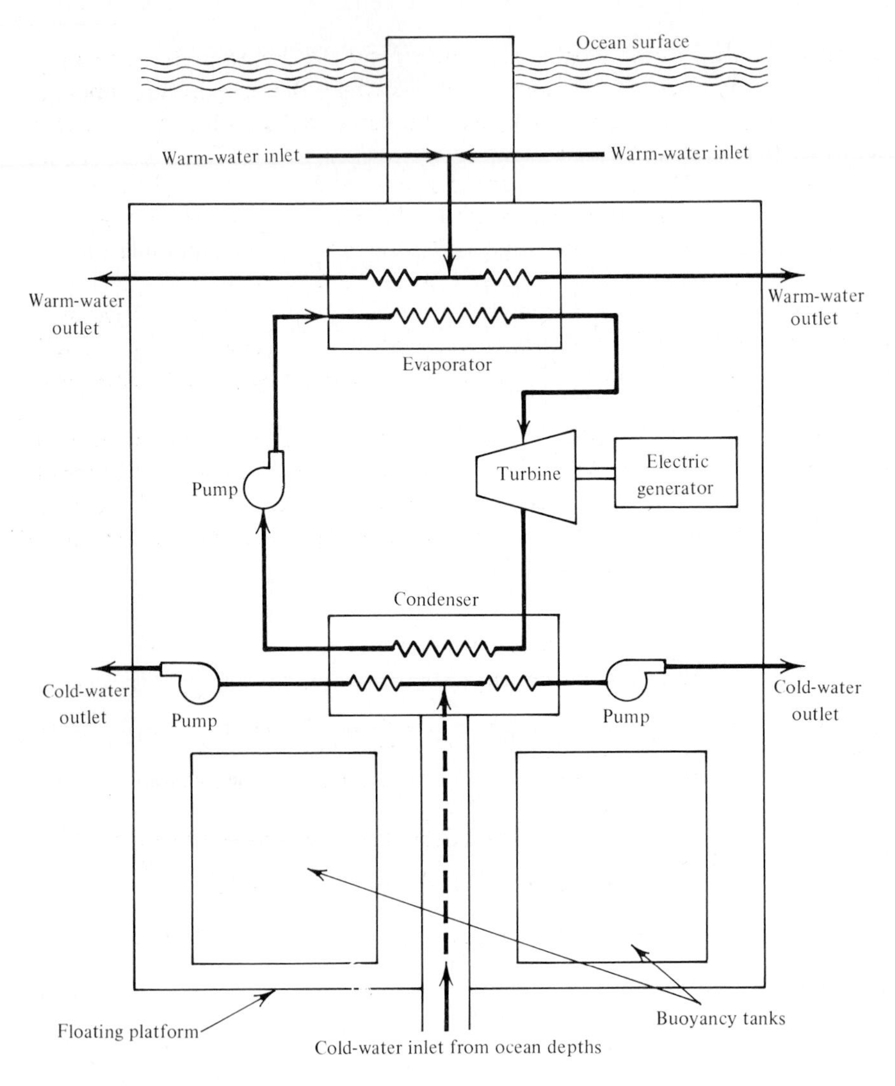

Figure 19-13 One possible equipment layout for a floating platform containing an ocean thermal energy conversion (OTEC) system.

through a long pipe is used to cool the working medium, which condenses back to a liquid. The liquid is then pumped to the evaporator for another loop around the cycle. In addition to the vapor pressure requirements, another special condition of an OTEC plant is the obvious enormous dimensions of the circulating systems for the warm and cold water streams. Note also that the cold water is pumped from depths of 600 to 900 m or 2000 to 3000 ft.

When OTEC units are used for the generation of electricity, riser cables will carry electricity from the floating platform down 4000 to 6000 ft to the ocean bottom where a fixed cable will carry electricity to shore. These riser cables must withstand the severe forces created by ocean currents, waves, and the floating platform itself. Unfortunately, in some locations where the temperature differential is favorable for electric generation, the distance is too great for the economic transport of power to shore. However, in such cases the OTEC plants might function as factories. By using raw materials brought from shore and the power generated on the platform, products such as aluminum and methanol might be produced, among others. Another important product could be ammonia for the fertilizer industry. The electricity generated could be used to separate hydrogen from seawater by hydrolysis. A combination of hydrogen with nitrogen separated from air leads to ammonia. Another possible product is liquid hydrogen itself, which could be brought ashore and used in fuel cells for large-scale peak production of electricity.

Although there are a number of technical problems to be resolved, such as biofouling of heat exchanger surfaces by marine organisms, there are a number of test facilities in the United States and elsewhere in the world which are actively pursuing the development of ocean thermal energy conversion as a viable technology by the end of the twentieth century.

REFERENCES

Levi, E.: MHD's Target: Payoff by 2000, *IEEE Spectrum*, May 1978, pp. 46–51. "MHD: Direct Channel from Heat to Electricity," *EPRI Journal*, **5** (3): 21–25 (1981).

Proceedings of the 16th Intersociety Energy Conversion Engineering Conference, *ASME*, New York, 1981.

Seventh International Conference on Magnetohydrodynamics Electric Power Generation, MIT, Cambridge, Mass., 1980.

Tapping the Main Stream of Geothermal Energy, *EPRI Journal*, **5** (4): 6–15 (1981).

PROBLEMS (METRIC)

Batteries and fuel cells

19-1M A fuel cell operates on an equimolar mixture of CO and H_2 as the fuel. The fuel and the stoichiometric amount of oxygen enter at the anode and cathode, respectively, at the environmental conditions of 25°C and 1 atm. Assume that all the water formed at the cathode leaves as a liquid, and that the CO_2 formed at the anode leaves at 25°C and 1 atm as a pure gas. Determine (*a*) the maximum work output for complete reaction in a steady-flow process, in kJ/kg · mol of mixture, (*b*) the ideal efficiency of the cell, in percent, and (*c*) the ideal voltage.

19-2M Methyl alcohol and stoichiometric oxygen in the gas phase enter a fuel cell each at 1 atm and 25°C, and form complete products also at 25°C. All the water formed at the cathode leaves as a liquid, and the CO_2 formed at the anode leaves at 1 atm as a pure gas. Determine (*a*) the maximum work output for the steady-flow process, in kJ/kg · mol of fuel, (*b*) the ideal efficiency, and (*c*) the ideal voltage.

19-3M Same as Prob. 19-2M, except that the fuel is ethane, C_2H_6.

19-4M Same as Prob. 19-2M, except that the fuel is ethene, C_2H_4.

19-5M Same as Prob. 19-2M, except that the fuel is methane, CH_4.

19-6M Same as Prob. 19-2M, except that the fuel is hydrogen, H_2, and the reactants enter and the products leave at (*a*) 400°K, (*b*) 600°K, and (*c*) 800°K. Note that water is now a gaseous product at 1 atm.

19-7M A lead-acid battery operates at 25°C. Determine (1) the ideal efficiency and (2) the ideal voltage if the electrolyte is (*a*) pure H_2SO_4, and (*b*) a 1.0 molal solution of H_2SO_4.

19-8M An ordinary dry cell involves Zn at the anode and MnO_2 at the cathode. At the anode, Zn is converted to ZnO by the migration of OH ions from the cathode, with the simultaneous formation of water. At the cathode, MnO_2 is converted to Mn_2O_3 by the reaction with water and the simultaneous formation of OH ions. Determine, for a temperature of 25°C, (*a*) the maximum work output, in kJ/kg · mol of zinc, (*b*) the ideal efficiency, (*c*) the moles of electrons released per mole of zinc, (*d*) the ideal or open-circuit voltage of the cell, and (*e*) the capacity of the battery expressed as ampere-hours per gram of battery. Assume that the initial reactants are present in their stoichiometric amounts.

19-9M A battery involves Zn at the anode and silver oxide (AgO) at the cathode. At the anode, Zn is converted to ZnO by the migration of OH ions from the cathode, with the simultaneous formation of water. At the cathode AgO is converted to Ag_2O by the reaction with water and the simultaneous formation of OH ions. Determine the same quantities requested in Prob. 19-8M. The data for AgO include: $\Delta h_f = -11{,}600$ kJ/kg · mol, $\Delta g_f = 14{,}500$ kJ/kg · mol, and $s^0 = 57.8$ kJ(kg · mol)(°K) at 25°C.

19-10M A battery involves zinc (Zn) at the anode and mecuric oxide (HgO) at the cathode. At the anode, Zn is converted to ZnO by the migration of OH ions from the cathode, with the simultaneous formation of water. At the cathode HgO reacts with water to form Hg and OH ions. Determine the same quantities requested in Prob. 19-8M.

Combined gas-steam turbine cycle

19-11M A gas-turbine power plant operates on a pressure ratio of 12 : 1 with compressor- and turbine-inlet temperatures of 300 and 1400°K, respectively. The adiabatic efficiencies of the compressor and turbine are 85 and 87 percent, respectively. The turbine exhaust, used as the energy source for a steam cycle, leaves the boiler at 480°K. The inlet conditions to the 85 percent efficient turbine in the steam cycle are 140 bars and 520°C. The condenser pressure is 0.10 bar and the pump is 75 percent efficient. Determine (*a*) the required heat input, in kJ/kg of air, (*b*) the mass-flow-rate ratio of air to steam, (*c*) the net work output of the gas turbine cycle, in kJ/kg of air, (*d*) the net work output of the steam cycle, in kJ/kg of steam, and (*e*) the overall thermal efficiency of the combined cycle.

19-12M Consider the data of Prob. 19-11M. In addition, an open feedwater heater is operated in the steam cycle with steam bled from the turbine at 7 bars. For this modified cycle, determine the quantities that were required in Prob. 19-11M.

19-13M A gas-turbine power plant operates on a pressure ratio of 13 : 1 with compressor- and turbine-inlet temperatures of 290 and 1440°K, respectively. The adiabatic efficiencies of the compressor and turbine are 84 and 88 percent, respectively. The turbine exhaust, used as the energy source for a steam cycle, leaves the boiler at 500°K. The inlet conditions to the 86 percent efficient turbine in the steam cycle are 160 bars and 560°C. The condenser pressure is 0.08 bar and the pump is 70 percent efficient. Determine (*a*) the required heat input, in kJ/kg of air, (*b*) the mass-flow-rate ratio of air to steam, (*c*) the net work output of the gas-turbine cycle, in kJ/kg of air, (*d*) the net work output of the steam cycle, in kJ/kg of steam, and (*e*) the overall thermal efficiency of the combined cycle.

19-14M Consider the data of Prob. 19-13M. In addition, an open feedwater heater is operated in the steam cycle with steam bled to the heater from the turbine at 10 bars. For this modified cycle, determine the quantities that were required in Prob. 19-13M.

19-15M A gas-turbine power plant operates on a 10 : 1 pressure ratio with compressor- and turbine-inlet temperatures of 300 and 1340°K, respectively. The adiabatic efficiencies of the compressor and turbine are 84 and 87 percent, respectively. The turbine exhaust, used as the energy source for a steam cycle, leaves the boiler at 460°K. The inlet conditions to the 86 percent efficient turbine in the steam cycle are 100 bars and 480°C. The condenser pressure is 0.08 bar and the pump is 70 percent efficient. Determine (*a*) the required heat input, in kJ/kg of air, (*b*) the mass-flow-rate ratio of air to steam, (*c*) the net work output of the gas-turbine cycle, in kJ/kg of air, (*d*) the net work output of the steam cycle, in kJ/kg of steam, and (*e*) the overall thermal efficiency of the combined cycle.

19-16M Determine the quantities that were required in Prob. 19-15M if in addition, an open feed-water heater is added to the steam cycle with steam bled from the turbine to the heater at 7 bars.

19-17M A combined cycle is composed of a gas-turbine cycle with an efficiency of 28 percent and a steam-turbine cycle with a thermal efficiency of 36 percent. In addition, the heat added to the steam cycle from an external source is (*a*) 20 percent, and (*b*) 30 percent of the heat rejected from the gas-turbine cycle. If all the heat rejected from the gas-turbine cycle appears as heat supply to the steam cycle, determine (1) the overall cycle thermal efficiency, and (2) the ratio of the work output of the steam cycle to the gas cycle.

19-18M A combined cycle is composed of a gas-turbine cycle with a thermal efficiency of 30 percent and a steam-turbine cycle with an efficiency of 38 percent. In addition, the heat added to the steam cycle from an external source is (*a*) 12 percent, and (*b*) 18 percent of the heat rejected from the gas-turbine cycle. If all the heat rejected from the gas cycle appears as heat supply to the steam cycle, determine (1) the overall thermal efficiency, and (2) the ratio of the work output of the steam cycle to the gas cycle.

Steam cycles for heating and power

19-19M Steam expands through a turbine in a cogeneration power cycle from 60 bars and 440°C to 0.08 bar. The turbine has an efficiency of 85 percent and delivers 50 MW of power. Steam is extracted at 5 bars for a heating load of 1×10^6 kJ/min. The steam leaves the heating load as a saturated liquid at 4 bars and mixes with the steam leaving the condensate pump at 4 bars (see Fig. 19-6). The pump efficiencies are both 70 percent. Determine (*a*) the mass flow rates, in kg/min, through the heating load and through the condenser, (*b*) the heat input in the boiler-superheater section, (*c*) the ratio of the heating load plus the turbine output to the heat input, and (*d*) the ratio of the turbine output to the heat input if the cycle were operated without the heating load, but at the same total mass flow rate.

19-20M A turbine in a cogeneration system expands steam from 100 bars and 480°C to 0.06 bar with an efficiency of 84 percent. The total mass flow rate through the boiler-superheater is 0.30×10^6 kg/h, and 25 percent of the flow is extracted at 7 bars for a heating load (see Fig. 19-6). Assume the fluid leaves the heating load at 5 bars and 150°C and then mixes with the water leaving the condensate pump at 5 bars. The pump efficiencies are both 65 percent. Determine (*a*) the heating load, in kJ/min, (*b*) the turbine output in megawatts, (*c*) the ratio of the heating load plus the turbine output to the heat input, and (*d*) the ratio of the turbine output to the heat input if the cycle were operated without the heating load, but at the same total mass flow rate.

19-21M Steam expands through a turbine in a cogeneration cycle from 12 MPa and 440°C to 0.008 MPa. The turbine has an efficiency of 87 percent and delivers 60 MW of power. Steam leaves the first stage of the turbine at 0.70 MPa, where a portion is extracted for a heating load of 2×10^6 kJ/min. The remaining portion of the steam is reheated to 400°C before expansion to the condenser pressure. The fluid leaves the heating load as a saturated liquid at 0.50 MPa and then mixes with the water leaving the condensate pump at 0.50 MPa (see Fig. 19-6). The efficiency of both pumps is 60 percent. Determine (*a*) the mass flow rates, in kg/min, through the heating load and through the condenser, (*b*) the heat input to the boiler-superheater section, (*c*) the ratio of the heating load plus the turbine output to the heat input, and (*d*) the ratio of the turbine output to the heat input if the cycle were operated without the heating load, but at the same total mass flow rate.

19-22M A steam plant is used for the supply of both heat and power. Steam enters the turbine at 30 bars and 320°C and expands to 0.06 bar. After expansion to 2 bars, steam is removed for heating purposes, but 2000 kg/h must always go through the low-pressure end of the turbine to keep the blades from overheating. The adiabatic efficiency of the turbine is 80 percent and its output is 3000 kW. Determine the maximum number of kilograms per hour of steam which can be supplied for heating.

Ocean thermal energy conversion

19-23M An OTEC system operates in a region of the ocean where the surface temperature is 27°C and the cold water temperature is 7°C. Assume both water streams experience a 3°C temperature change in the evaporator and condenser, and that the minimum temperature between either water stream and the refrigerant 12 used as the working fluid is 2°C. The power requirement for the device is 10 MW. Determine (*a*) the refrigerant-12 flow rate, in kg/min, (*b*) the evaporator heat transfer rate, in kJ/min, (*c*) the condenser heat transfer rate, in kJ/min, (*d*) the refrigerant-12 pump power rating, in kilowatts, and (*e*) the required power for the cold water pump which draws water from a depth of 800 m, in kilowatts. Assume all pumps and the turbine are isentropic in operation.

PROBLEMS (USCS)

Combined gas-stream-turbine cycle

19-1 A gas-turbine power plant operates on a pressure ratio of 12 : 1 with compressor- and turbine-inlet temperatures of 80 and 2040°F, respectively. The adiabatic efficiencies of the compressor and turbine are 85 and 87 percent, respectively. The turbine exhaust, used as the energy source for a steam cycle, leaves the boiler at 400°F. The inlet conditions to the 85 percent efficient turbine in the steam cycle are 2000 psia and 900°F. The condenser pressure is 1 psia and the pump is 75 percent efficient. Determine (*a*) the required heat input, in Btu/lb air, (*b*) the mass-flow-rate ratio of air to steam, (*c*) the net work output of the gas-turbine cycle, in Btu/lb of air, (*d*) the net work output of the steam cycle, in Btu/lb of steam, and (*e*) the overall thermal efficiency of the combined cycle.

19-2 Consider the data of Prob. 19-1. In addition, an open feedwater heater is operated in the steam cycle with steam bled from the turbine at 100 psia. For this modified cycle determine the quantities that were required in Prob. 19-1.

19-3 A gas-turbine power plant operates on a pressure ratio of 13 : 1 with compressor- and turbine-inlet temperatures of 60 and 2140°F, respectively. The adiabatic efficiencies of the compressor and turbine are 84 and 88 percent, respectively. The turbine exhaust, used as the energy source for a steam cycle, leaves the boiler at 440°F. The inlet conditions to the 86 percent efficient turbine in the steam cycle are 2500 psia and 1000°F. The condenser pressure is 0.8 psia and the pump is 70 percent efficient. Determine (*a*) the required heat input, in Btu/lb of air, (*b*) the mass-flow-rate ratio of air to steam, (*c*) the net work output of the gas-turbine cycle, in Btu/lb of air, (*d*) the net work output of the steam cycle, in Btu/lb of steam, and (*e*) the overall thermal efficiency of the combined cycle.

19-4 Consider the data of Prob. 19-3. In addition, an open feedwater heater is operated in the steam cycle with steam bled from the turbine at 150 psia. For this modified cycle determine the quantities that were required in Prob. 19-3.

19-5 A gas-turbine power plant operates on a 10 : 1 pressure ratio with compressor- and turbine-inlet temperatures of 80 and 1940°F, respectively. The adiabatic efficiencies of the compressor and turbine are 84 and 87 percent, respectively. The turbine exhaust, used as the energy source for a steam cycle, leaves the boiler at 380°F. The inlet conditions to the 86 percent efficient turbine in the steam cycle are 1600 psia and 900°F. The condenser pressure is 1 psia and the pump is 70 percent efficient. Determine (*a*) the required heat input, in Btu/lb of air, (*b*) the mass-flow-rate ratio of air to steam, (*c*) the net work output of the gas-turbine cycle, in Btu/lb of air, (*d*) the net work output of the steam cycle, in Btu/lb of steam, and (*e*) the overall thermal efficiency of the combined cycle.

19-6 Determine the quantities that were required in Prob. 19-5 if, in addition, an open feedwater heater is added with steam bled from the turbine to the heater at 80 psia.

19-7 A combined cycle is composed of a gas-turbine cycle with a thermal efficiency of 26 percent and a steam-turbine cycle with an efficiency of 34 percent. In addition, the heat added to the steam cycle from an external source is (*a*) 15 percent, and (*b*) 22 percent of the heat rejected from the gas-turbine cycle. If all the heat rejected from the gas cycle appears as heat supply to the steam cycle, determine (1) the overall thermal efficiency of the combined cycle, and (2) the ratio of the work output of the steam cycle to that of the gas cycle.

19-8 A combined cycle is composed of a gas-turbine cycle with a thermal efficiency of 32 percent and a steam-turbine cycle with an efficiency of 36 percent. In addition, the heat added to the steam cycle from an external source is (*a*) 8 percent, and (*b*) 16 percent of the heat rejected from the gas-turbine cycle. If all the heat rejected from the gas cycle appears as heat supplied to the steam cycle, determine (1) the overall thermal efficiency of the combined cycle, and (2) the ratio of the work output of the steam cycle to that of the gas cycle.

Steam cycles for heating and power

19-9 Steam expands through a turbine in a cogeneration power cycle from 1000 psia and 800°F to 0.80 psia. The turbine has an efficiency of 85 percent and delivers 50 MW of power. Steam is extracted at 80 psia for a heating load of 1×10^6 Btu/min. The steam leaves the heating load as a saturated liquid at 60 psia and mixes with the steam leaving the condensate pump at 60 psia (see Fig. 19-6). The pump efficiencies are both 70 percent. Determine (*a*) the mass flow rates, in lb/min, through the heating load and through the condenser, (*b*) the heat input in the boiler-superheater section, in Btu/lb min, (*c*) the ratio of the heating load plus the turbine output to the heat input, and (*d*) the ratio of the turbine output to the heat input if the cycle were operated without the heating load, but at the same total mass flow rate.

19-10 A turbine in a cogeneration system expands steam from 1600 psia and 900°F to 0.80 psia with an efficiency of 84 percent. The total mass flow rate through the boiler-superheater is 0.30×10^6 lb/h, and 25 percent of the flow is extracted at 100 psia for a heating load (see Fig. 19-6). Assume the fluid leaves the heating load at 80 psia and 310°F and then mixes with the water leaving the condensate pump at 80 psia. The pump efficiencies are both 65 percent. Determine (*a*) the heating load, in Btu/min, (*b*) the turbine output, in megawatts, (*c*) the ratio of the heating load plus the turbine output to the heat input, and (*d*) the ratio of the turbine output to the heat input if the cycle were operated without the heating load, but at the same total mass flow rate.

19-11 Steam expands through a turbine in a cogeneration cycle from 1800 psia and 800°F to 1 psia. The turbine has an efficiency of 87 percent and delivers 60 MW of power. Steam leaves the first stage of the turbine at 100 psia, where a portion is extracted for a heating load of 2×10^6 Btu/min. The remaining portion of the steam is reheated to 700°F before expansion to the condenser pressure. The fluid leaves the heating load as a saturated liquid at 60 psia and then mixes with the water leaving the condensate pump at 60 psia (see Fig. 19-6). The efficiency of both pumps is 60 percent. Determine (*a*) the mass flow rates, in lb/min, through the heating load and through the condenser, (*b*) the heat input to the boiler-superheater section, in Btu/min, (*c*) the ratio of the heating load plus the turbine output to the heat input, and (*d*) the ratio of the turbine output to the heat input if the cycle were operated without the heating load, but at the same total mass flow rate.

19-12 A steam plant is used for supplying both heat and power. Steam enters the turbine at 250 psia and 450°F, and expands to 1 psia. After expansion to 40 psia, steam is removed for heating purposes, but 3000 lb/h must always go through the low-pressure end of the turbine to keep the blades from overheating. The efficiency of the turbine is 80 percent and its output is 3000 kW. Determine the maximum number of pounds per hour of steam which can be supplied for heating.

Ocean thermal energy conversion

19-13 An OTEC system operates in a region of the ocean where the surface temperature is 80°F and the cold water temperature is 40°F. Assume both water streams experience a 5°F temperature change in the evaporator and condenser, and that the minimum temperature between either water stream and the refrigerant 12 used as the working fluid is 5°F. The power requirement from the turbine is 10,000 hp. Determine (*a*) the refrigerant-12 flow rate, in lb/min, (*b*) the evaporator heat transfer rate, in Btu/min, (*c*) the condenser heat transfer rate, in Btu/min, (*d*) the pump power rating in the power cycle, in horsepower, and (*e*) the required power for the cold water pump which draws water from a depth of 2000 ft, in horsepower. Assume all pumps and the turbine are isentropic.

APPENDIX

A-1

THE METHODS OF LAGRANGIAN UNDETERMINED MULTIPLIERS

For a function represented by $F(x, y, z)$, the calculus requires that, at an extremum point such as a maximum or minimum, the function satisfy the relation

$$dF = \left(\frac{\partial F}{\partial x}\right)_{y,z} dx + \left(\frac{\partial F}{\partial y}\right)_{x,z} dy + \left(\frac{\partial F}{\partial z}\right)_{x,y} dz = 0 \tag{A-1}$$

When x, y, and z are independent variables, the coefficients of dx, dy, and dz are required to be zero at the extremum of the function in order for Eq. (A-1) to be zero. Hence at a maximum or minimum point,

$$\left(\frac{\partial F}{\partial x}\right)_{y,z} = 0 \qquad \left(\frac{\partial F}{\partial y}\right)_{x,z} = 0 \qquad \left(\frac{\partial F}{\partial z}\right)_{x,z} = 0 \tag{A-2}$$

The partial derivatives in Eqs. (A-2) are a function of x, y, and z, in general. This set of equations gives us the required number of equations to solve for the values of the independent variables at the extremum condition.

In some cases the approach to finding a maximum or minimum point is complicated somewhat since there may be one or more constraints on the possible values of x, y, and z. That is, the values of the independent variables at the extremum must satisfy one or more relations of the form $G_n(x, y, z) = c$, where c is a constant and n is the number of constraining equations. The number of constraining equations must be less than the number of independent variables; otherwise an extremum point cannot be found for the function.

In this new situation a conditional maximum or minimum exists at some state (x_0, y_0, z_0). Equation (A-1) is still valid, but we realize that dx, dy, and dz are no longer independent of one another, since relations of the type $G_n(x, y, z) = c$ must also be fulfilled. Since these functional relationships equal different constants, the total derivatives of the constraining equations must be zero. Hence we may write for one such equation

$$dG = \left(\frac{\partial G}{\partial x}\right)_{y,z} dx + \left(\frac{\partial G}{\partial y}\right)_{x,z} dy + \left(\frac{\partial G}{\partial z}\right)_{x,y} dz = 0 \tag{A-3}$$

This shows that dx, dy, and dz are interrelated by Eq. (A-3) as well as Eq. (A-1). One straightforward method of solving for such a conditional extremum is as follows: For the case of one constraining equation the relationship $G(x, y, z) = c$ may be used to eliminate one of the variables from the function $F(x, y, z)$. This function can be maximized with respect to the remaining two variables, which are now independent of each other. By setting the coefficients of the two remaining variables equal to zero, the values of x_0, y_0, and z_0 at the extremum condition are determined.

A more general method of solution for the foregoing type of problem was developed by Lagrange. It is termed the method of undetermined multipliers. In general, we wish to formulate an equation like (A-1) in which the variables dx, dy, and dz are independent of one another. If any additional constraints are present, this normally will not be true. Once the variables are made independent, however, their coefficients may then be set equal to zero. This is accomplished in the following manner.

At the conditional extremum both dF and dG equal zero. Then it is also true that $A\,dG = 0$, where A is some arbitrary undetermined constant multiplier. Thus we may write

$$dF - A\,dG = 0 \tag{A-4}$$

The quantities dF and dG are now replaced by Eqs. (A-1) and (A-3), and common terms are collected. The result is

$$\left[\left(\frac{\partial F}{\partial x}\right) - A\left(\frac{\partial G}{\partial x}\right)\right]dx + \left[\left(\frac{\partial F}{\partial y}\right) - A\left(\frac{\partial G}{\partial y}\right)\right]dy + \left[\left(\frac{\partial F}{\partial z}\right) - A\left(\frac{\partial G}{\partial z}\right)\right]dz = 0 \tag{A-5}$$

The subscripts on the partial derivatives have been omitted to simplify the equation. In expression (A-5) still only two of the variations dx, dy, and dz are independent. However, one term in Eq. (A-5) can be eliminated by the proper choice of the constant A, since A is any arbitrary value. One proper choice of A is that

$$A = \frac{(\partial F/\partial z)_{x_0,\,y_0,\,z_0}}{(\partial G/\partial z)_{x_0,\,y_0,\,z_0}} \tag{A-6}$$

The subscripts x_0, y_0, and z_0 again simply denote that the partials must be evaluated at the maximum point. Since the location of the extremum is yet unknown, the value of A likewise is not yet known. With the assignment to A of the value given by Eq. (A-6), the final term in Eq. (A-5) drops out. At this point the remaining variables dx and dy can be varied independently. Thus we are required to set the coefficients of dx and dy equal to zero in order to satisfy Eq. (A-5), since dx and dy can take on nonzero values. Hence, as a solution,

$$\left(\frac{\partial F}{\partial x}\right)_{y,z} - A\left(\frac{\partial G}{\partial x}\right)_{y,z} = 0 \quad \text{and} \quad \left(\frac{\partial F}{\partial y}\right)_{x,z} - A\left(\frac{\partial G}{\partial y}\right)_{x,z} = 0 \qquad \text{(A-7)}$$

The two expressions in Eq. (A-7) constitute in this case two equations with three unknowns, namely, x_0, y_0, and A. Equation (A-6) gives a third equation, but at the same time introduces a fourth unknown, z_0. The additional relationship we need is given by the constraining equation itself, $G(x_0, y_0, z_0) = c$. Such an expression introduces no other unknown quantities than those already sought.

In actual practice, one follows a less formal procedure once the mathematical justification is recognized. An equation like Eq. (A-5) is set up which contains m differential terms with coefficients. All the m coefficients are set equal to zero. In addition, there are n constraining equations. These $m + n$ equations are solved for the $m + n$ unknowns, where m is the number of independent variables and n is the number of Lagrangian multipliers. The following example illustrates the general method.

It is desired to find the minimum distance from the line $ax + by + c = 0$ to the origin. It should be clear from geometric considerations that such a minimum exists. If a circle of radius r is drawn with the origin at its center, the distance we seek is the radius of a circle which is just tangent to the line. In such a case the radius will be perpendicular to the line and hence will represent the minimum distance. Thus we must determine the minimum radius of a circle such that a single point on it also lies on the line $ax + by + c = 0$. Let the function to be minimized be given by

$$F(x, y) = x^2 + y^2 = r^2$$

subject to the constraint

$$G(x, y) = ax + by + c = 0$$

By differentiating each of these equations we obtain

$$dF = 2x\,dx + 2y\,dy \qquad \text{and} \qquad dG = a\,dx + b\,dy$$

The expression for dG is now multiplied by the multiplier A, and the result is subtracted from dF. In order to be a minimum radius, the final expression must equal zero; hence

$$2x\,dx + 2y\,dy - Aa\,dx - Ab\,dy = 0$$

Separation of variables yields

$$(2x - Aa)\,dx + (2y - Ab)\,dy = 0$$

By proper selection of the value of A, the coefficient of dy becomes zero. Consequently, x becomes an independent variable, and the coefficient of dx may be set equal to zero. Hence $A = 2x/a$. By substituting this result into the coefficient of dy, which was arbitrarily made equal to zero, we obtain $b = 2y/A = 2ya/2x = ya/x$. This relation, in conjunction with the constraining equation $ax + by + c = 0$, enables one to determine x and y in terms of the constants a, b, and c, and hence to evaluate the radius r. It may easily be verified that the simultaneous solution of these two equations leads to

$$x = \frac{-ac}{a^2 + b^2} \qquad y = \frac{-bc}{a^2 + b^2} \qquad r = \frac{c}{(a^2 + b^2)^{1/2}}$$

This may be confirmed in an actual example by considering the line passing through the points (2, 0) and (0, 2) on cartesian rectangular coordinates and showing mathematically and graphically that $r = 2^{1/2}$.

APPENDIX

A-2

DERIVATION OF EQ. (9-11)

In Sec. 9-7 it is desired to develop an expression for $(\partial \ln W_{mp})/\partial n_i$. Mathematically, this requires the maximization of the function, ln W, in conjunction with certain constraining equations. This type of problem is usually solved in the following manner. First, the constraining equations are substituted into the general equation, thus eliminating several of the independent variables. For example, if there are two constraining equations, one can use these equations to eliminate two of the independent variables in the general equation. Following this substitution, the general equation is differentiated with respect to the remaining independent variables, and the total differential is set equal to zero. Then by equating to zero the coefficients in front of the derivative of each independent variable, one determines the values of the independent variables which maximize the function. Our approach will be basically the same as this, with one slight modification. Instead of substituting the constraining equations and then differentiating the general equation, we shall first differentiate both the general equation and the constraining equations and then substitute the differentiated forms of the general equation. Using this combination of equations, we then set the coefficients in front of the remaining independent variables equal to zero. The end result is exactly the same as the method above, since the order of the mathematical steps is immaterial.

First we differentiate ln W with respect to n_i for each energy level, and then set the resulting total differential equal to zero. Therefore,

$$d \ln W = \sum_i \frac{\partial \ln W}{\partial n_i} dn_i = 0 \tag{A-8}$$

where the summation is over all the energy levels, $i = 1, 2, 3$, etc. Since we are maximizing the function, $\ln W$, the partial derivatives in the above equation refer solely to the most probable macrostate. In order to simplify the notation, the subscript "mp" has not been placed on the partial derivatives to indicate this restriction. However, this restriction must be kept in mind. Now we introduce the constraining equations. In past discussions we have required that the total energy U and the total number of particles n be conserved when considering macrostates. These constraining equations are given by

$$n = \sum_i n_i \tag{A-9}$$

and

$$U = \sum_i n_i \epsilon_i \tag{A-10}$$

If these equations are differentiated, holding the ϵ_i values constant during the differentiation, we find that

$$dn_1 + dn_2 + \cdots + dn_a + dn_b + \cdots + dn_i + \cdots = 0$$

and

$$\epsilon_1\, dn_1 + \epsilon_2\, dn_2 + \cdots + \epsilon_a\, dn_a + \epsilon_b\, dn_b + \cdots + \epsilon_i\, dn_i + \cdots = 0$$

Two of the energy levels have been given special identifying subscripts a and b. These two summations can be written more compactly as

$$dn_a + dn_b + \sum_j dn_j = 0 \tag{A-11}$$

and

$$\epsilon_a\, dn_a + \epsilon_b\, dn_b + \sum_j \epsilon_j\, dn_j = 0 \tag{A-12}$$

where the sum of j terms signifies a summation over all energy levels except the two levels denoted by a and b, which are any two arbitrarily chosen levels. In a similar fashion we may rewrite Eq. (A-8) in the form

$$\frac{\partial \ln W}{\partial n_a}\, dn_a + \frac{\partial \ln W}{\partial n_b}\, dn_b + \sum_j \frac{\partial \ln W}{\partial n_j}\, dn_j = 0 \tag{A-13}$$

Next solve Eqs. (A-11) and (A-12) for dn_a and dn_b. Hence

$$dn_a = \frac{\sum_j \epsilon_j\, dn_j - \epsilon_b \sum_j dn_j}{\epsilon_b - \epsilon_a}$$

and

$$dn_b = \frac{\epsilon_a \sum_j dn_j - \sum_j \epsilon_j\, dn_j}{\epsilon_b - \epsilon_a}$$

Insertion of these two expressions into Eq. (A-13) yields, after rearrangement,

$$\sum_j \left[\frac{\partial \ln W}{\partial n_j} + \left(\frac{\epsilon_a \dfrac{\partial \ln W}{\partial n_b} - \epsilon_b \dfrac{\partial \ln W}{\partial n_a}}{\epsilon_b - \epsilon_a} \right) + \epsilon_j \left(\frac{\dfrac{\partial \ln W}{\partial n_a} - \dfrac{\partial \ln W}{\partial n_b}}{\epsilon_b - \epsilon_a} \right) \right] dn_j = 0 \tag{A-14}$$

In effect, we have used the two constraining equations to eliminate two of the variables in Eq. (A-13). Since two dn_i terms have been arbitrarily eliminated in the above equation, the remaining $(n - 2)\ dn_i$ terms are independent of one another. In our new notation, these remaining terms are indicated by a sum over j terms.

Now note that all the quantities in the two terms in parentheses in Eq. (A-14) refer only to energy levels a and b. Since the choice of these two levels is completely arbitrary, the two terms in parentheses must be constants. If they were not constants, we would get different solutions to Eq. (A-14) depending upon which energy levels were represented by a and b. But there can be only one solution, since we are dealing with the most probable macrostate. We shall represent these constants by $-A$ and $-B$. Consequently, Eq. (A-14) can be written as

$$\sum_j \left(\frac{\partial \ln W_{\text{mp}}}{\partial n_j} - A - B\epsilon_j \right) dn_j = 0$$

where we have reintroduced the subscript "mp" for the most probable macrostate. Now, if the dn_j quantities can be varied independently, the only way for this summation to equal zero is for the coefficients of all the dn_j terms to be zero. These coefficients all have the same form so that

$$\frac{\partial \ln W_{\text{mp}}}{\partial n_j} - A - B\epsilon_j = 0$$

or

$$\frac{\partial \ln W_{\text{mp}}}{\partial n_j} = A + B\epsilon_j \tag{A-15}$$

Although Eq. (A-15) was derived for j terms, the expression actually holds for energy levels a and b as well. This is easily proved by setting the two terms in parentheses in Eq. (A-14) equal to $-A$ and $-B$, respectively, and solving. This proof is left as an exercise. Therefore Eq. (A-15) really applies to $(j + 2)$ levels, or all i energy levels. That is, Eq. (A-15) is a general solution. This equation appears as Eq. (9-11) in the text, and is the desired result.

APPENDIX

A-3

SUPPLEMENTARY TABLES AND FIGURES (METRIC UNITS)

Table A-1M Physical constants and conversion factors

Physical constants	
Avogadro's number	$N_A = 6.023 \times 10^{26}$ atoms/kg · mol
Universal gas constant	$R_u = 0.08205$ L · atm/(g · mol)(°K)
	$= 8.314$ kJ/(kg · mol)(°K)
	$= 0.08314$ bar · m^3/(kg · mol)(°K)
Planck's constant	$h = 6.626 \times 10^{-34}$ J · s/molecule
Boltzmann's constant	$k = 1.380 \times 10^{-23}$ J/(°K)(molecule)
Speed of light	$c = 2.988 \times 10^{10}$ cm/s
Standard gravity	$g = 9.80665$ m/s^2

Conversion factors	
Length	1 cm = 0.3937 in = 10^4 μm = 10^8 Å
	1 km = 0.6215 mi = 3281 ft
Mass	1 kg = 2.205 lb_m
Force	1 N = 1 kg · m/s^2 = 0.2248 lb_f
Pressure	1 bar = 10^5 N/m^2 = 0.9869 atm
	= 100 kilopascals (kPa)
	1 torr = 1 mmHg at 0°C = 1.333 mbar
	= 1.933×10^{-2} psi
	1 mbar = 0.402 in H_2O
Volume	1 L = 0.0353 ft^3 = 0.2642 gal = 61.025 in^3 = 10^{-3} m^3
Density	1 g/cm^3 = 1 kg/L = 62.4 lb_m/ft^3 = 10^3 kg/m^3
Energy	1 J = 1 N · m = 1 V · C
	= 0.7375 ft · lb_f = 10 bar · cm^3 = 0.624×10^{19} eV
	1 kJ = 0.948 Btu = 737.6 ft · lb_f = 10^{-2} bar · m^3
	1 kJ/kg = 0.431 Btu/lb
Power	1 W = 1 J/s = 3.413 Btu/h
	1 kW = 1.3405 hp = 737.3 ft · lb_f/s
Velocity	1 m/s = 2.237 mi/h = 3.60 km/h = 3.281 ft/s

Table A-2M Values of the molar mass (molecular weight) of some common elements and compounds

Substance	Formula	Molar Mass	Substance	Formula	Molar Mass
Argon	Ar	39.94	Water	H_2O	18.02
Aluminum	Al	26.97	Hydrogen peroxide	H_2O_2	34.02
Carbon	C	12.01	Ammonia	NH_3	17.04
Copper	Cu	63.54	Hydroxyl	—OH	17.01
Helium	He	4.003	Methane	CH_4	16.04
Hydrogen	H_2	2.016	Acetylene	C_2H_2	26.04
Iron	Fe	55.85	Ethylene	C_2H_4	28.05
Lead	Pb	207.2	Ethane	C_2H_6	30.07
Mercury	Hg	200.6	Propylene	C_3H_6	42.08
Nitrogen	N_2	28.008	Propane	C_3H_8	44.09
Oxygen	O_2	32.00	*n*-Butane	C_4H_{10}	58.12
Potassium	K	39.096	*n*-Pentane	C_5H_{12}	72.15
Silver	Ag	107.88	*n*-Octane	C_8H_{18}	114.22
Sodium	Na	22.997	Benzene	C_6H_6	78.11
Air		28.97	Methyl alcohol	CH_3OH	32.05
Carbon monoxide	CO	28.01	Ethyl alcohol	C_2H_5OH	46.07
Carbon dioxide	CO_2	44.01	Refrigerant 12	CCl_2F_2	120.92

Table A-3M Critical properties and van der Waals constants

Substance	T_c, °K	P_c, bar	v_c, $\frac{m^3}{kg \cdot mol}$	$Z_c = \frac{P_c v_c}{RT_c}$	van der Waals: a, $bar\left(\frac{m^3}{kg \cdot mol}\right)^2$	van der Waals: b, $\frac{m^3}{kg \cdot mol}$
Acetylene (C_2H_2)	309	62.8	0.112	0.274	4.410	0.0510
Air (equivalent)	133	37.7	0.0829	0.284	1.358	0.0364
Ammonia (NH_3)	406	112.8	0.0723	0.242	4.233	0.0373
Benzene (C_6H_6)	562	49.3	0.256	0.274	18.63	0.1181
n-Butane (C_4H_{10})	425.2	38.0	0.257	0.274	13.80	0.1196
Carbon dioxide (CO_2)	304.2	73.9	0.0941	0.276	3.643	0.0427
Carbon monoxide (CO)	133	35.0	0.0928	0.294	1.463	0.0394
Refrigerant 12 (CCl_2F_2)	385	41.2	0.216	0.278	10.78	0.0998
Ethane (C_2H_6)	305.4	48.8	0.221	0.273	5.575	0.0650
Ethylene (C_2H_4)	283	51.2	0.143	0.284	4.563	0.0574
Helium (He)	5.2	2.3	0.0579	0.300	0.0341	0.0234
Hydrogen (H_2)	33.2	13.0	0.0648	0.304	0.247	0.0265
Methane (CH_4)	190.7	46.4	0.0991	0.290	2.285	0.0427
Nitrogen (N_2)	126.2	33.9	0.0897	0.291	1.361	0.0385
Oxygen (O_2)	154.4	50.5	0.0741	0.290	1.369	0.0315
Propane (C_3H_8)	370	42.7	0.195	0.276	9.315	0.0900
Sulfur dioxide (SO_2)	431	78.7	0.124	0.268	6.837	0.0568
Water (H_2O)	647.3	220.9	0.0558	0.230	5.507	0.0304

Source: Adapted from the data in Table A-3.

Table A-4M Ideal-gas specific-heat data for selected gases, kJ/(kg)(°K)

1. Zero-pressure specific heats for six common gases

Temp. °K	c_p	c_v	k	c_p	c_v	k	c_p	c_v	k	Temp. °K
	Air			Carbon dioxide, CO_2			Carbon monoxide, CO			
250	1.003	0.716	1.401	0.791	0.602	1.314	1.039	0.743	1.400	250
300	1.005	0.718	1.400	0.846	0.657	1.288	1.040	0.744	1.399	300
350	1.008	0.721	1.398	0.895	0.706	1.268	1.043	0.746	1.398	350
400	1.013	0.726	1.395	0.939	0.750	1.252	1.047	0.751	1.395	400
450	1.020	0.733	1.391	0.978	0.790	1.239	1.054	0.757	1.392	450
500	1.029	0.742	1.387	1.014	0.825	1.229	1.063	0.767	1.387	500
550	1.040	0.753	1.381	1.046	0.857	1.220	1.075	0.778	1.382	550
600	1.051	0.764	1.376	1.075	0.886	1.213	1.087	0.790	1.376	600
650	1.063	0.776	1.370	1.102	0.913	1.207	1.100	0.803	1.370	650
700	1.075	0.788	1.364	1.126	0.937	1.202	1.113	0.816	1.364	700
750	1.087	0.800	1.359	1.148	0.959	1.197	1.126	0.829	1.358	750
800	1.099	0.812	1.354	1.169	0.980	1.193	1.139	0.842	1.353	800
900	1.121	0.834	1.344	1.204	1.015	1.186	1.163	0.866	1.343	900
1000	1.142	0.855	1.336	1.234	1.045	1.181	1.185	0.888	1.335	1000
	Hydrogen, H_2			Nitrogen, N_2			Oxygen, O_2			
250	14.051	9.927	1.416	1.039	0.742	1.400	0.913	0.653	1.398	250
300	14.307	10.183	1.405	1.039	0.743	1.400	0.918	0.658	1.395	300
350	14.427	10.302	1.400	1.041	0.744	1.399	0.928	0.668	1.389	350
400	14.476	10.352	1.398	1.044	0.747	1.397	0.941	0.681	1.382	400
450	14.501	10.377	1.398	1.049	0.752	1.395	0.956	0.696	1.373	450
500	14.513	10.389	1.397	1.056	0.759	1.391	0.972	0.712	1.365	500
550	14.530	10.405	1.396	1.065	0.768	1.387	0.988	0.728	1.358	550
600	14.546	10.422	1.396	1.075	0.778	1.382	1.003	0.743	1.350	600
650	14.571	10.447	1.395	1.086	0.789	1.376	1.017	0.758	1.343	650
700	14.604	10.480	1.394	1.098	0.801	1.371	1.031	0.771	1.337	700
750	14.645	10.521	1.392	1.110	0.813	1.365	1.043	0.783	1.332	750
800	14.695	10.570	1.390	1.121	0.825	1.360	1.054	0.794	1.327	800
900	14.822	10.698	1.385	1.145	0.849	1.349	1.074	0.814	1.319	900
1000	14.983	10.859	1.380	1.167	0.870	1.341	1.090	0.830	1.313	1000

Source: Data adapted from "Tables of Thermal Properties of Gases," NBS Circular 564, 1955.

$k = \frac{c_p}{c_v}$ for perfect gas

Table A-4M *(Continued)*

2. Specific-heat data for monatomic gases

Over a wide range of temperatures at low pressures the specific heats c_v and c_p of all monatomic gases are essentially independent of temperature and pressure. In addition, on a molar basis all monatomic gases have the same value for either c_v or c_p in a given set of units. One set of values is

$$c_v = 12.5 \text{ kJ/(kg} \cdot \text{mol)(°K)} \qquad \text{and} \qquad c_p = 20.8 \text{ kJ/(kg} \cdot \text{mol)(°K)}$$

Source: Data adapted from "Tables of Thermal Properties of Gases," NBS Circular 564, 1955.

3. Constant-pressure specific-heat equations for various gases at zero pressure (metric units)

$$\frac{c_p}{R_u} = a + bT + cT^2 + dT^3 + eT^4$$

T is in degrees kelvin, equations valid from 300 to 1000°K

Gas	a	$b \times 10^3$	$c \times 10^6$	$d \times 10^9$	$e \times 10^{12}$
CO	3.710	−1.619	3.692	−2.032	0.240
CO_2	2.401	8.735	−6.607	2.002	
H_2	3.057	2.677	−5.180	5.521	−1.812
H_2O	4.070	−1.108	4.152	−2.964	0.807
O_2	3.626	−1.878	7.056	−6.764	2.156
N_2	3.675	−1.208	2.324	−0.632	−0.226
Air (dry)	3.640	−1.101	2.466	−0.942	
NH_3	3.591	0.494	8.345	−8.383	2.730
NO	4.046	−3.418	7.982	−6.114	1.592
NO_2	3.459	2.065	6.687	−9.556	3.620
SO_2	3.267	5.324	0.684	−5.281	2.559
SO_3	2.578	14.556	−9.176	−0.792	1.971
CH_4	3.826	−3.979	24.558	−22.733	6.963
C_2H_2	1.410	19.057	−24.501	16.391	−4.135
C_2H_4	1.426	11.383	7.989	−16.254	6.749

Source: Adapted from the data in NASA SP-273, U.S. Government Printing Office, Washington, 1971.

Table A-5M Ideal-gas properties of air

T, °K; h, kJ/kg; u, kJ/kg; s^0, kJ/(kg)(°K)

T	h	p_r	u	v_r	s^0	T	h	p_r	u	v_r	s^0
200	199.97	0.3363	142.56	1707.	1.29559	460	462.02	6.245	329.97	211.4	2.13407
210	209.97	0.3987	149.69	1512.	1.34444	470	472.24	6.742	337.32	200.1	2.15604
220	219.97	0.4690	156.82	1346.	1.39105	480	482.49	7.268	344.70	189.5	2.17760
230	230.02	0.5477	164.00	1205.	1.43557	490	492.74	7.824	352.08	179.7	2.19876
240	240.02	0.6355	171.13	1084.	1.47824	500	503.02	8.411	359.49	170.6	2.21952
250	250.05	0.7329	178.28	979.	1.51917	510	513.32	9.031	366.92	162.1	2.23993
260	260.09	0.8405	185.45	887.8	1.55848	520	523.63	9.684	374.36	154.1	2.25997
270	270.11	0.9590	192.60	808.0	1.59634	530	533.98	10.37	381.84	146.7	2.27967
280	280.13	1.0889	199.75	738.0	1.63279	540	544.35	11.10	389.34	139.7	2.29906
285	285.14	1.1584	203.33	706.1	1.65055	550	554.74	11.86	396.86	133.1	2.31809
290	290.16	1.2311	206.91	676.1	1.66802	560	565.17	12.66	404.42	127.0	2.33685
295	295.17	1.3068	210.49	647.9	1.68515	570	575.59	13.50	411.97	121.2	2.35531
300	300.19	1.3860	214.07	621.2	1.70203	580	586.04	14.38	419.55	115.7	2.37348
305	305.22	1.4686	217.67	596.0	1.71865	590	596.52	15.31	427.15	110.6	2.39140
310	310.24	1.5546	221.25	572.3	1.73498	600	607.02	16.28	434.78	105.8	2.40902
315	315.27	1.6442	224.85	549.8	1.75106	610	617.53	17.30	442.42	101.2	2.42644
320	320.29	1.7375	228.42	528.6	1.76690	620	628.07	18.36	450.09	96.92	2.44356
325	325.31	1.8345	232.02	508.4	1.78249	630	638.63	19.84	457.78	92.84	2.46048
330	330.34	1.9352	235.61	489.4	1.79783	640	649.22	20.64	465.50	88.99	2.47716
340	340.42	2.149	242.82	454.1	1.82790	650	659.84	21.86	473.25	85.34	2.49364
350	350.49	2.379	250.02	422.2	1.85708	660	670.47	23.13	481.01	81.89	2.50985
360	360.58	2.626	257.24	393.4	1.88543	670	681.14	24.46	488.81	78.61	2.52589
370	370.67	2.892	264.46	367.2	1.91313	680	691.82	25.85	496.62	75.50	2.54175
380	380.77	3.176	271.69	343.4	1.94001	690	702.52	27.29	504.45	72.56	2.55731
390	390.88	3.481	278.93	321.5	1.96633	700	713.27	28.80	512.33	69.76	2.57277
400	400.98	3.806	286.16	301.6	1.99194	710	724.04	30.38	520.23	67.07	2.58810
410	411.12	4.153	293.43	283.3	2.01699	720	734.82	32.02	528.14	64.53	2.60319
420	421.26	4.522	300.69	266.6	2.04142	730	745.62	33.72	536.07	62.13	2.61803
430	431.43	4.915	307.99	251.1	2.06533	740	756.44	35.50	544.02	59.82	2.63280
440	441.61	5.332	315.30	236.8	2.08870	750	767.29	37.35	551.99	57.63	2.64737
450	451.80	5.775	322.62	223.6	2.11161	760	778.18	39.27	560.01	55.54	2.66176

Table A-5M *(Continued)*

T, °K; h, kJ/kg; u, kJ/kg; $s°$, kJ/(kg)(°K)

T	h	p_r	u	v_r	s^0	T	h	p_r	u	v_r	s^0
780	800.03	43.35	576.12	51.64	2.69013	1360	1467.49	399.1	1077.10	9.780	3.32724
800	821.95	47.75	592.30	48.08	2.71787	1380	1491.44	424.2	1095.26	9.337	3.34474
820	843.98	52.59	608.59	44.84	2.74504	1400	1515.42	450.5	1113.52	8.919	3.36200
840	866.08	57.60	624.95	41.85	2.77170	1420	1539.44	478.0	1131.77	8.526	3.37901
860	888.27	63.09	641.40	39.12	2.79783	1440	1563.51	506.9	1150.13	8.153	3.39586
880	910.56	68.98	657.95	36.61	2.82344	1460	1587.63	537.1	1168.49	7.801	3.41247
900	932.93	75.29	674.58	34.31	2.84856	1480	1611.79	568.8	1186.95	7.468	3.42892
920	955.38	82.05	691.28	32.18	2.87324	1500	1635.97	601.9	1205.41	7.152	3.44516
940	977.92	89.28	708.08	30.22	2.89748	1520	1660.23	636.5	1223.87	6.854	3.46120
960	1000.55	97.00	725.02	28.40	2.92128	1540	1684.51	672.8	1242.43	6.569	3.47712
980	1023.25	105.2	741.98	26.73	2.94468	1560	1708.82	710.5	1260.99	6.301	3.49276
1000	1046.04	114.0	758.94	25.17	2.96770	1580	1733.17	750.0	1279.65	6.046	3.50829
1020	1068.89	123.4	776.10	23.72	2.99034	1600	1757.57	791.2	1298.30	5.804	3.52364
1040	1091.85	133.3	793.36	22.39	3.01260	1620	1782.00	834.1	1316.96	5.574	3.53879
1060	1114.86	143.9	810.62	21.14	3.03449	1640	1806.46	878.9	1335.72	5.355	3.55381
1080	1137.89	155.2	827.88	19.98	3.05608	1660	1830.96	925.6	1354.48	5.147	3.56867
1100	1161.07	167.1	845.33	18.896	3.07732	1680	1855.50	974.2	1373.24	4.949	3.58335
1120	1184.28	179.7	862.79	17.886	3.09825	1700	1880.1	1025	1392.7	4.761	3.5979
1140	1207.57	193.1	880.35	16.946	3.11883	1750	1941.6	1161	1439.8	4.328	3.6336
1160	1230.92	207.2	897.91	16.064	3.13916	1800	2003.3	1310	1487.2	3.944	3.6684
1180	1254.34	222.2	915.57	15.241	3.15916	1850	2065.3	1475	1534.9	3.601	3.7023
1200	1277.79	238.0	933.33	14.470	3.17888	1900	2127.4	1655	1582.6	3.295	3.7354
1220	1301.31	254.7	951.09	13.747	3.19834	1950	2189.7	1852	1630.6	3.022	3.7677
1240	1324.93	272.3	968.95	13.069	3.21751	2000	2252.1	2068	1678.7	2.776	3.7994
1260	1348.55	290.8	986.90	12.435	3.23638	2050	2314.6	2303	1726.8	2.555	3.8303
1280	1372.24	310.4	1004.76	11.835	3.25510	2100	2377.4	2559	1775.3	2.356	3.8605
1300	1395.97	330.9	1022.82	11.275	3.27345	2150	2440.3	2837	1823.8	2.175	3.8901
1320	1419.76	352.5	1040.88	10.747	3.29160	2200	2503.2	3138	1872.4	2.012	3.9191
1340	1443.60	375.3	1058.94	10.247	3.30959	2250	2566.4	3464	1921.3	1.864	3.9474

Source: Adapted from Keenan, J. H., and J. Kaye, "Gas Tables," Wiley, New York, 1945.

Table A-6M Ideal-gas enthalpy, internal energy, and absolute entropy of diatomic nitrogen, N_2

$\Delta h_f = 0$ kJ/kg · mol

T, °K; h and u, kJ/kg · mol; s, kJ/(kg · mol)(°K)

T	h	u	s^0	T	h	u	s^0
0	0	0	0	600	17,563	12,574	212.066
220	6,391	4,562	182.639	610	17,864	12,792	212.564
230	6,683	4,770	183.938	620	18,166	13,011	213.055
240	6,975	4,979	185.180	630	18,468	13,230	213.541
250	7,266	5,188	186.370	640	18,772	13,450	214.018
260	7,558	5,396	187.514	650	19,075	13,671	214.489
270	7,849	5,604	188.614	660	19,380	13,892	214.954
280	8,141	5,813	189.673	670	19,685	14,114	215.413
290	8,432	6,021	190.695	680	19,991	14,337	215.866
298	8,669	6,190	191.502	690	20,297	14,560	216.314
300	8,723	6,229	191.682	700	20,604	14,784	216.756
310	9,014	6,437	192.638	710	20,912	15,008	217.192
320	9,306	6,645	193.562	720	21,220	15,234	217.624
330	9,597	6,853	194.459	730	21,529	15,460	218.059
340	9,888	7,061	195.328	740	21,839	15,686	218.472
350	10,180	7,270	196.173	750	22,149	15,913	218.889
360	10,471	7,478	196.995	760	22,460	16,141	219.301
370	10,763	7,687	197.794	770	22,772	16,370	219.709
380	11,055	7,895	198.572	780	23,085	16,599	220.113
390	11,347	8,104	199.331	790	23,398	16,830	220.512
400	11,640	8,314	200.071	800	23,714	17,061	220.907
410	11,932	8,523	200.794	810	24,027	17,292	221.298
420	12,225	8,733	201.499	820	24,342	17,524	221.684
430	12,518	8,943	202.189	830	24,658	17,757	222.067
440	12,811	9,153	202.863	840	24,974	17,990	222.447
450	13,105	9,363	203.523	850	25,292	18,224	222.822
460	13,399	9,574	204.170	860	25,610	18,459	223.194
470	13,693	9,786	204.803	870	25,928	18,695	223.562
480	13,988	9,997	205.424	880	26,248	18,931	223.927
490	14,285	10,210	206.033	890	26,568	19,168	224.288
500	14,581	10,423	206.630	900	26,890	19,407	224.647
510	14,876	10,635	207.216	910	27,210	19,644	225.002
520	15,172	10,848	207.792	920	27,532	19,883	225.353
530	15,469	11,062	208.358	930	27,854	20,122	225.701
540	15,766	11,277	208.914	940	28,178	20,362	226.047
550	16,064	11,492	209.461	950	28,501	20,603	226.389
560	16,363	11,707	209.999	960	28,826	20,844	226.728
570	16,662	11,923	210.528	970	29,151	21,086	227.064
580	16,962	12,139	211.049	980	29,476	21,328	227.398
590	17,262	12,356	211.562	990	29,803	21,571	227.728

Table A-6M (*Continued*)

T	h	u	s^0	T	h	u	s^0
1000	30,129	21,815	228.057	1760	56,227	41.594	247,396
1020	30,784	22,304	228.706	1780	56,938	42,139	247.798
1040	31,442	22,795	229.344	1800	57,651	42,685	248.195
1060	32,101	23,288	229.973	1820	58,363	43,231	248.589
1080	32,762	23,782	230.591	1840	59,075	43,777	248.979
1100	33,426	24,280	231.199	1860	59,790	44,324	249.365
1120	34,092	24,780	231.799	1880	60,504	44,873	249.748
1140	34,760	25,282	232.391	1900	61,220	45,423	250.128
1160	35,430	25,786	232.973	1920	61,936	45,973	250.502
1180	36,104	26,291	233.549	1940	62,654	46,524	250.874
1200	36,777	26,799	234.115	1960	63,381	47,075	251.242
1220	37,452	27,308	234.673	1980	64,090	47,627	251.607
1240	38,129	27,819	235.223	2000	64,810	48,181	251.969
1260	38,807	28,331	235.766	2050	66,612	49,567	252.858
1280	39,488	28,845	236.302	2100	68,417	50,957	253.726
1300	40,170	29,361	236.831	2150	70,226	52,351	254.578
1320	40,853	29,878	237.353	2200	72,040	53,749	255.412
1340	41,539	30,398	237.867	2250	73,856	55,149	256.227
1360	42,227	30,919	238.376	2300	75,676	56,553	257.027
1380	42,915	31,441	238.878	2350	77,496	57,958	257.810
1400	43,605	31,964	239.375	2400	79,320	59,366	258.580
1420	44,295	32,489	239.865	2450	81,149	60,779	259.332
1440	44,988	33,014	240.350	2500	82,981	62,195	260.073
1460	45,682	33,543	240.827	2550	84,814	63,613	260.799
1480	46,377	34,071	241.301	2600	86,650	65,033	261.512
1500	47,073	34,601	241.768	2650	88,488	66,455	262.213
1520	47,771	35,133	242.228	2700	90,328	67,880	262.902
1540	48,470	35,665	242.685	2750	92,171	69,306	263.577
1560	49,168	36,197	243.137	2800	94,014	70,734	264.241
1580	49,869	36,732	243.585	2850	95,859	72,163	264.895
1600	50,571	37,268	244.028	2900	97,705	73,593	265.538
1620	51,275	37,806	244.464	2950	99,556	75,028	266.170
1640	51,980	38,344	244.896	3000	101,407	76,464	266.793
1660	52,686	38,884	245.324	3050	103,260	77,902	267.404
1680	53,393	39,424	245.747	3100	105,115	79,341	268.007
1700	54,099	39,965	246.166	3150	106,972	80,782	268.601
1720	54,807	40,507	246.580	3200	108,830	82,224	269.186
1740	55,516	41,049	246.990	3250	110,690	83,668	269.763

Source: Based on data from the JANAF Thermochemical Tables, NSRDS-NBS-37, 1971.

Table A-7M Ideal-gas enthalpy, internal energy, and absolute entropy of diatomic oxygen, O_2

$\Delta h_f = 0$ kJ/kg · mol

T, °K; h and u, kJ/kg · mol; s, kJ/(kg · mol)(°K)

T	h	u	s^0	T	h	u	s^0
0	0	0	0	600	17,929	12,940	226.346
220	6,404	4,575	196.171	610	18,250	13,178	226.877
230	6,694	4,782	197.461	620	18,572	13,417	227.400
240	6,984	4,989	198.696	630	18,895	13,657	227.918
250	7,275	5,197	199.885	640	19,219	13,898	228.429
260	7,566	5,405	201.027	650	19,544	14,140	228.932
270	7,858	5,613	202.128	660	19,870	14,383	229.430
280	8,150	5,822	203.191	670	20,197	14,626	229.920
290	8,443	6,032	204.218	680	20,524	14,871	230.405
298	8,682	6,203	205.033	690	20,854	15,116	230.885
300	8,736	6,242	205.213	700	21,184	15,364	231.358
310	9,030	6,453	206.177	710	21,514	15,611	231.827
320	9,325	6,664	207.112	720	21,845	15,859	232.291
330	9,620	6,877	208.020	730	22,177	16,107	232.748
340	9,916	7,090	208.904	740	22,510	16,357	233.201
350	10,213	7,303	209.765	750	22,844	16,607	233.649
360	10,511	7,518	210.604	760	23,178	16,859	234.091
370	10,809	7,733	211.423	770	23,513	17,111	234.528
380	11,109	7,949	212.222	780	23,850	17,364	234.960
390	11,409	8,166	213.002	790	24,186	17,618	235.387
400	11,711	8,384	213.765	800	24,523	17,872	235.810
410	12,012	8,603	214.510	810	24,861	18,126	236.230
420	12,314	8,822	215.241	820	25,199	18,382	236.644
430	12,618	9,043	215.955	830	25,537	18,637	237.055
440	12,923	9,264	216.656	840	25,877	18,893	237.462
450	13,228	9,487	217.342	850	26,218	19,150	237.864
460	13,535	9,710	218.016	860	26,559	19,408	238.264
470	13,842	9,935	218.676	870	26,899	19,666	238.660
480	14,151	10,160	219.326	880	27,242	19,925	239.051
490	14,460	10,386	219.963	890	27,584	20,185	239.439
500	14,770	10,614	220.589	900	27,928	20,445	239.823
510	15,082	10,842	221.206	910	28,272	20,706	240.203
520	15,395	11,071	221.812	920	28,616	20,967	240.580
530	15,708	11,301	222.409	930	28,960	21,228	240.953
540	16,022	11,533	222.997	940	29,306	21,491	241.323
550	16,338	11,765	223.576	950	29,652	21,754	241.689
560	16,654	11,998	224.146	960	29,999	22,017	242.052
570	16,971	12,232	224.708	970	30,345	22,280	242.411
580	17,290	12,467	225.262	980	30,692	22,544	242.768
590	17,609	12,703	225.808	990	31,041	22,809	243.120

Table A-7M (*Continued*)

T	h	u	s^0	T	h	u	s^0
1000	31,389	23,075	243.471	1760	58,880	44,247	263.861
1020	32,088	23,607	244.164	1780	59,624	44,825	264.283
1040	32,789	24,142	244.844	1800	60,371	45,405	264.701
1060	33,490	24,677	245.513	1820	61,118	45,986	265.113
1080	34,194	25,214	246.171	1840	61,866	46,568	265.521
1100	34,899	25,753	246.818	1860	62,616	47,151	265.925
1120	35,606	26,294	247.454	1880	63,365	47,734	266.326
1140	36,314	26,836	248.081	1900	64,116	48,319	266.722
1160	37,023	27,379	248.698	1920	64,868	48,904	267.115
1180	37,734	27,923	249.307	1940	65,620	49,490	267.505
1200	38,447	28,469	249.906	1960	66,374	50,078	267.891
1220	39,162	29,018	250.497	1980	67,127	50,665	268.275
1240	39,877	29,568	251.079	2000	67,881	51,253	268.655
1260	40,594	30,118	251.653	2050	69,772	52,727	269.588
1280	41,312	30,670	252.219	2100	71,668	54,208	270.504
1300	42,033	31,224	252.776	2150	73,573	55,697	271.399
1320	42,753	31,778	253.325	2200	75,484	57,192	272.278
1340	43,475	32,334	253.868	2250	77,397	58,690	273.136
1360	44,198	32,891	254.404	2300	79,316	60,193	273.981
1380	44,923	33,449	254.932	2350	81,243	61,704	274.809
1400	45,648	34,008	255.454	2400	83,174	63,219	275.625
1420	46,374	34,567	255.968	2450	85,112	64,742	276.424
1440	47,102	35,129	256.475	2500	87,057	66,271	277.207
1460	47,831	35,692	256.978	2550	89,004	67,802	277.979
1480	48,561	36,256	257.474	2600	90,956	69,339	278.738
1500	49,292	36,821	257.965	2650	92,916	70,883	279.485
1520	50,024	37,387	258.450	2700	94,881	72,433	280.219
1540	50,756	37,952	258.928	2750	96,852	73,987	280.942
1560	51,490	38,520	259.402	2800	98,826	75,546	281.654
1580	52,224	39,088	259.870	2850	100,808	77,112	282.357
1600	52,961	39,658	260.333	2900	102,793	78,682	283.048
1620	53,696	40,227	260.791	2950	104,785	80,258	283.728
1640	54,434	40,799	261.242	3000	106,780	81,837	284.399
1660	55,172	41,370	261.690	3050	108,778	83,419	285.060
1680	55,912	41,944	262.132	3100	110,784	85,009	285.713
1700	56,652	42,517	262.571	3150	112,795	86,601	286.355
1720	57,394	43,093	263.005	3200	114,809	88,203	286.989
1740	58,136	43,669	263.435	3250	116,827	89,804	287.614

Source: Based on data from the JANAF Thermochemical Tables, NSRDS-NBS-37, 1971.

Table A-8M Ideal-gas enthalpy, internal energy, and absolute entropy of carbon monoxide, CO

$\Delta h_f = -110{,}530$ kJ/kg · mol

T, °K; h and u, kJ/kg · mol; s, kJ/(kg · mol)(°K)

T	h	u	s^0	T	h	u	s^0
0	0	0	0	600	17,611	12,622	218.204
220	6,391	4,562	188.683	610	17,915	12,843	218.708
230	6,683	4,771	189.980	620	18,221	13,066	219.205
240	6,975	4,979	191.221	630	18,527	13,289	219.695
250	7,266	5,188	192.411	640	18,833	13,512	220.179
260	7,558	5,396	193.554	650	19,141	13,736	220.656
270	7,849	5,604	194.654	660	19,449	13,962	221.127
280	8,140	5,812	195.173	670	19,758	14,187	221.592
290	8,432	6,020	196.735	680	20,068	14,414	222.052
298	8,669	6,190	197.543	690	20,378	14,641	222.505
300	8,723	6,229	197.723	700	20,690	14,870	222.953
310	9,014	6,437	198.678	710	21,002	15,099	223.396
320	9,306	6,645	199.603	720	21,315	15,328	223.833
330	9,597	6,854	200.500	730	21,628	15,558	224.265
340	9,889	7,062	201.371	740	21,943	15,789	224.692
350	10,181	7,271	202.217	750	22,258	16,022	225.115
360	10,473	7,480	203.040	760	22,573	16,255	225.533
370	10,765	7,689	203.842	770	22,890	16,488	225.947
380	11,058	7,899	204.622	780	23,208	16,723	226.357
390	11,351	8,108	205.383	790	23,526	16,957	226.762
400	11,644	8,319	206.125	800	23,844	17,193	227.162
410	11,938	8,529	206.850	810	24,164	17,429	227.559
420	12,232	8,740	207.549	820	24,483	17,665	227.952
430	12,526	8,951	208.252	830	24,803	17,902	228.339
440	12,821	9,163	208.929	840	25,124	18,140	228.724
450	13,116	9,375	209.593	850	25,446	18,379	229.106
460	13,412	9,587	210.243	860	25,768	18,617	229.482
470	13,708	9,800	210.880	870	26,091	18,858	229.856
480	14,005	10,014	211.504	880	26,415	19,099	230.227
490	14,302	10,228	212.117	890	26,740	19,341	230.593
500	14,600	10,443	212.719	900	27,066	19,583	230.957
510	14,898	10,658	213.310	910	27,392	19,826	231.317
520	15,197	10,874	213.890	920	27,719	20,070	231.674
530	15,497	11,090	214.460	930	28,046	20,314	232.028
540	15,797	11,307	215.020	940	28,375	20,559	232.379
550	16,097	11,524	215.572	950	28,703	20,805	232.727
560	16,399	11,743	216.115	960	29,033	21,051	233.072
570	16,701	11,961	216.649	970	29,362	21,298	233.413
580	17,003	12,181	217.175	980	29,693	21,545	233.752
590	17,307	12,401	217.693	990	30,024	21,793	234.088

Table A-8M (*Continued*)

T	h	u	s^0	T	h	u	s^0
1000	30,355	22,041	234.421	1760	56,756	42,123	253.991
1020	31,020	22,540	235.079	1780	57,473	42,673	254.398
1040	31,688	23,041	235.728	1800	58,191	43,225	254.797
1060	32,357	23,544	236.364	1820	58,910	43,778	255.194
1080	33,029	24,049	236.992	1840	59,629	44,331	255.587
1100	33,702	24,557	237.609	1860	60,351	44,886	255.976
1120	34,377	25,065	238.217	1880	61,072	45,441	256.361
1140	35,054	25,575	238.817	1900	61,794	45,997	256.743
1160	35,733	26,088	239.407	1920	62,516	46,552	257.122
1180	36,406	26,602	239.989	1940	63,238	47,108	257.497
1200	37,095	27,118	240.663	1960	63,961	47,665	257.868
1220	37,780	27,637	241.128	1980	64,684	48,221	258.236
1240	38,466	28,426	241.686	2000	65,408	48,780	258.600
1260	39,154	28,678	242.236	2050	67,224	50,179	259.494
1280	39,884	29,201	242.780	2100	69,044	51,584	260.370
1300	40,534	29,725	243.316	2150	70,864	52,988	261.226
1320	41,266	30,251	243.844	2200	72,688	54,396	262.065
1340	41,919	30,778	244.366	2250	74,516	55,809	262.887
1360	42,613	31,306	244.880	2300	76,345	57,222	263.692
1380	43,309	31,836	245.388	2350	78,178	58,640	264.480
1400	44,007	32,367	245,889	2400	80,015	60,060	265.253
1420	44,707	32,900	246,385	2450	81,852	61,482	266.012
1440	45,408	33,434	246.876	2500	83,692	62,906	266.755
1460	46,110	33,971	247.360	2550	85,537	64,335	267.485
1480	46,813	34,508	247.839	2600	87,383	65,766	268.202
1500	47,517	35,046	248.312	2650	89,230	67,197	268.905
1520	48,222	35,584	248.778	2700	91,077	68,628	269.596
1540	48,928	36,124	249.240	2750	92,930	70,066	270.285
1560	49,635	36,665	249.695	2800	94,784	71,504	270.943
1580	50,344	37,207	250.147	2850	96,639	72,945	271.602
1600	51,053	37,750	250.592	2900	98,495	74,383	272.249
1620	51,763	38,293	251.033	2950	100,352	75,825	272.884
1640	52,472	38,837	251.470	3000	102,210	77,267	273.508
1660	53,184	39,382	251.901	3050	104.073	78,715	274.123
1680	53,895	39,927	252.329	3100	105,939	80,164	274.730
1700	54,609	40,474	252.751	3150	107,802	81,612	275.326
1720	55,323	41,023	253.169	3200	109,667	83,061	275.914
1740	56,039	41,572	253.582	3250	111,534	84,513	276.494

Source: Based on data from the JANAF Thermochemical Tables, NSRDS-NBS-37, 1971.

Table A-9M Ideal-gas enthalpy, internal energy, and absolute entropy of carbon dioxide, CO_2

$\Delta h_f = -393{,}520$ kJ/kg · mol

T, °K; h and u, kJ/kg · mol; s, kJ/(kg · mol)(°K)

T	h	u	s^0	T	h	u	s^0
0	0	0	0	600	22,280	17,291	243.199
220	6,601	4,772	202.966	610	22,754	17,683	243.983
230	6,938	5,026	204.464	620	23,231	18,076	244.758
240	7,280	5,285	205.920	630	23,709	18,471	245.524
250	7,627	5,548	207.337	640	24,190	18,869	246.282
260	7,979	5,817	208.717	650	24,674	19,270	247.032
270	8,335	6,091	210.062	660	25,160	19,672	247.773
280	8,697	6,369	211.376	670	25,648	20,078	248.507
290	9,063	6,651	212.660	680	26,138	20,484	249.233
298	9,364	6,885	213.685	690	26,631	20,894	249.952
300	9,431	6,939	213.915	700	27,125	21,305	250.663
310	9,807	7,230	215.146	710	27,622	21,719	251.368
320	10,186	7,526	216.351	720	28,121	22,134	252.065
330	10,570	7,826	217.534	730	28,622	22,552	252.755
340	10,959	8,131	218.694	740	29,124	22,972	253.439
350	11,351	8,439	219.831	750	29,629	23,393	254.117
360	11,748	8,752	220.948	760	30,135	23,817	254.787
370	12,148	9,068	222.044	770	30,644	24,242	255.452
380	12,552	9,392	223.122	780	31,154	24,669	256.110
390	12,960	9,718	224.182	790	31,665	25,097	256.762
400	13,372	10,046	225.225	800	32,179	25,527	257.408
410	13,787	10,378	226.250	810	32,694	25,959	258.048
420	14,206	10,714	227.258	820	33,212	26,394	258.682
430	14,628	11,053	228.252	830	33,730	26,829	259.311
440	15,054	11,393	229.230	840	34,251	27,267	259.934
450	15,483	11,742	230.194	850	34,773	27,706	260.551
460	15,916	12,091	231.144	860	35,296	28,125	261.164
470	16,351	12,444	232.080	870	35,821	28,588	261.770
480	16,791	12,800	233.004	880	36,347	29,031	262.371
490	17,232	13,158	233.916	890	36,876	29,476	262.968
500	17,678	13,521	234.814	900	37,405	29,922	263.559
510	18,126	13,885	235.700	910	37,935	30,369	264.146
520	18,576	14,253	236.575	920	38,467	30,818	264.728
530	19,029	14,622	237.439	930	39,000	31,268	265.304
540	19,485	14,996	238.292	940	39,535	31,719	265.877
550	19,945	15,372	239.135	950	40,070	32,171	266.444
560	20,407	15,751	239.962	960	40,607	32,625	267.007
570	20,870	16,131	240.789	970	41,145	33,081	267.566
580	21,337	16,515	241.602	980	41,685	33,537	268.119
590	21,807	16,902	242.405	990	42,226	33,995	268.670

Table A-9M *(Continued)*

T	h	u	s^0	T	h	u	s^0
1000	42,769	34,455	269.215	1760	86,420	71,787	301.543
1020	43,859	35,378	270.293	1780	87,612	72,812	302.271
1040	44,953	36,306	271.354	1800	88,806	73,840	302.884
1060	46,051	37,238	272.400	1820	90,000	74,868	303.544
1080	47,153	38,174	273.430	1840	91,196	75,897	304.198
1100	48,258	39,112	274.445	1860	92,394	76,929	304.845
1120	49,369	40,057	275.444	1880	93,593	77,962	305.487
1140	50,484	41,006	276.430	1900	94,793	78,996	306.122
1160	51,602	41,957	277.403	1920	95,995	80,031	306.751
1180	52,724	42,913	278.362	1940	97,197	81,067	307.374
1200	53,848	43,871	279.307	1960	98,401	82,105	307.992
1220	54,977	44,834	280.238	1980	99,606	83,144	308.604
1240	56,108	45,799	281.158	2000	100,804	84,185	309.210
1260	57,244	46,768	282.066	2050	103,835	86,791	310.701
1280	58,381	47,739	282.962	2100	106,864	89,404	312.160
1300	59,522	48,713	283.847	2150	109,898	92,023	313.589
1320	60,666	49,691	284.722	2200	112,939	94,648	314.988
1340	61,813	50,672	285.586	2250	115,984	97,277	316.356
1360	62,963	51,656	286.439	2300	119,035	99,912	317.695
1380	64,116	52,643	287.283	2350	122,091	102,552	319.011
1400	65,271	53,631	288.106	2400	125,152	105,197	320.302
1420	66,427	54,621	288.934	2450	128,219	107,849	321.566
1440	67,586	55,614	289.743	2500	131,290	110,504	322.808
1460	68,748	56,609	290.542	2550	134,368	113,166	324.026
1480	69,911	57,606	291.333	2600	137,449	115,832	325.222
1500	71,078	58,606	292.114	2650	140,533	118,500	326.396
1520	72,246	59,609	292.888	2700	143,620	121,172	327.549
1540	73,417	60,613	292.654	2750	146,713	123,849	328.684
1560	74,590	61,620	294.411	2800	149,808	126,528	329.800
1580	76,767	62,630	295.161	2850	152,908	129,212	330.896
1600	76,944	63,741	295.901	2900	156,009	131,898	331.975
1620	78,123	64,653	296.632	2950	159,117	134,589	333.037
1640	79,303	65,668	297.356	3000	162,226	137,283	334.084
1660	80,486	66,592	298.072	3050	165,341	139,982	335.114
1680	81,670	67,702	298.781	3100	168,456	142,681	336.126
1700	82,856	68,721	299.482	3150	171,576	145,385	337.124
1720	84,043	69,742	300.177	3200	174,695	148,089	338.109
1740	85,231	70,764	300.863	3250	177,822	150,801	339.069

Source: Based on data from the JANAF Thermochemical Tables, NSRDS-NBS-37, 1971.

Table A-10M Ideal-gas enthalpy, internal energy, and absolute entropy of water, H_2O

$\Delta h_f = -241{,}810$ kJ/kg · mol

T, °K; h and u, kJ/kg · mol; s, kJ/(kg · mol)(°K)

T	h	u	s^0	T	h	u	s^0
0	0	0	0	600	20,402	15,413	212.920
220	7,295	5,466	178.576	610	20,765	15,693	213.529
230	7,628	5,715	180.054	620	21,130	15,975	214.122
240	7,961	5,965	181.471	630	21,495	16,257	214.707
250	8,294	6,215	182.831	640	21,862	16,541	215.285
260	8,627	6,466	184.139	650	22,230	16,826	215.856
270	8,961	6,716	185.399	660	22,600	17,112	216.419
280	9,296	6,968	186.616	670	22,970	17,399	216.976
290	9,631	7,219	187.791	680	23,342	17,688	217.527
298	9,904	7,425	188.720	690	23,714	17,978	218.071
300	9,966	7,472	188.928	700	24,088	18,268	218.610
310	10,302	7,725	190.030	710	24,464	18,561	219.142
320	10,639	7,978	191.098	720	24,840	18,854	219.668
330	10,976	8,232	192.136	730	25,218	19,148	220.189
340	11,314	8,487	193.144	740	25,597	19,444	220.707
350	11,652	8,742	194.125	750	25,977	19,741	221.215
360	11,992	8,998	195.081	760	26,358	20,039	221.720
370	12,331	9,255	196.012	770	26,741	20,339	222.221
380	12,672	9,513	196.920	780	27,125	20,639	222.717
390	13,014	9,771	197.807	790	27,510	20,941	223.207
400	13,356	10,030	198.673	800	27,896	21,245	223.693
410	13,699	10,290	199.521	810	28,284	21,549	224.174
420	14,043	10,551	200.350	820	28,672	21,855	224.651
430	14,388	10,813	201.160	830	29,062	22,162	225.123
440	14,734	11,075	201.955	840	29,454	22,470	225.592
450	15,080	11,339	202.734	850	29,846	22,779	226.057
460	15,428	11,603	203.497	860	30,240	23,090	226.517
470	15,777	11,869	204.247	870	30,635	23,402	226.973
480	16,126	12,135	204.982	880	21,032	23,715	227.426
490	16,477	12,403	205.705	890	31,429	24,029	227.875
500	16,828	12,671	206.413	900	31,828	24,345	228.321
510	17,181	12,940	207.112	910	32,228	24,662	228.763
520	17,534	13,211	207.799	920	32,629	24,980	229.202
530	17,889	13,482	208.475	930	33,032	25,300	229.637
540	18,245	13,755	209.139	940	33,436	25,621	230.070
550	18,601	14,028	209.795	950	33,841	25,943	230.499
560	18,959	14,303	210.440	960	34,247	26,265	230.924
570	19,318	14,579	211.075	970	34,653	26,588	231.347
580	19,678	14,856	211.702	980	35,061	26,913	231.767
590	20,039	15,134	212.320	990	35,472	27,240	232.184

Table A-10M *(Continued)*

T	h	u	s^0	T	h	u	s^0
1000	35,882	27,568	232.597	1760	70,535	55,902	258.151
1020	36,709	28,228	233.415	1780	71,523	56,723	258.708
1040	37,542	28,895	234.223	1800	72,513	57,547	259.262
1060	38,380	29,567	235.020	1820	73,507	58,375	259.811
1080	39,223	30,243	235.806	1840	74,506	59,207	260.357
1100	40,071	30,925	236.584	1860	75,506	60,042	260.898
1120	40,923	31,611	237.352	1880	76,511	60,880	261.436
1140	41,780	32,301	238.110	1900	77,517	61,720	261.969
1160	42,642	32,997	238.859	1920	78,527	62,564	262.497
1180	43,509	33,698	239.600	1940	79,540	63,411	263.022
1200	44,380	34,403	240.333	1960	80,555	64,259	263.542
1220	45,256	35,112	241.057	1980	81,573	65,111	264.059
1240	46,137	35,827	241.773	2000	82,593	65,965	264.571
1260	47,022	36,546	242.482	2050	85,156	68,111	265.838
1280	47,912	37,270	243.183	2100	87,735	70,275	267.081
1300	48,807	38,000	243.877	2150	90,330	72,454	268.301
1320	49,707	38,732	244.564	2200	92,940	74,649	269.500
1340	50,612	39,470	245.243	2250	95,562	76,855	270.679
1360	51,521	40,213	245.915	2300	98,199	79,076	271.839
1380	52,434	40,960	246.582	2350	100,846	81,308	272.978
1400	53,351	41,711	247.241	2400	103,508	83,553	274.098
1420	54,273	42,466	247.895	2450	106,183	85,811	275.201
1440	55,198	43,226	248.543	2500	108,868	88,082	276.286
1460	56,128	43,989	249.185	2550	111,565	90,364	277.354
1480	57,062	44,756	249.820	2600	114,273	92,656	278.407
1500	57,999	45,528	250.450	2650	116,991	94,958	279.441
1520	58,942	46,304	251.074	2700	119,717	97,269	280.462
1540	59,888	47,084	251.693	2750	122,453	99,588	281.464
1560	60,838	17,868	252.305	2800	125,198	101,917	282.453
1580	61,792	48,655	252.912	2850	127,952	104,256	283.429
1600	62,748	49,445	253.513	2900	130,717	106,605	284.390
1620	63,709	50,240	254.111	2950	133,486	108,959	285.338
1640	64,675	51,039	254.703	3000	136,264	111,321	286.273
1660	65,643	51,841	255.290	3050	139,051	113,692	287.194
1680	66,614	52,646	255.873	3100	141,846	116,072	288.102
1700	67,589	53,455	256.450	3150	144,648	118,458	288.999
1720	68,567	54,267	257.022	3200	147,457	120,851	289.884
1740	69,550	55,083	257.589	3250	150,272	123,250	290.756

Source: Based on data from the JANAF Thermochemical Tables, NSRDS-NBS-37, 1971.

Table A-11M Ideal-gas enthalpy, internal energy, and absolute entropy of diatomic hydrogen, H_2

$\Delta h_f = 0$ kJ/kg · mol

T, °K; h and u, kJ/kg · mol; s, kJ/(kg · mol)(°K)

T	h	u	s^0	T	h	u	s^0
0	0	0	0	1440	42,808	30,835	177.410
260	7,370	5,209	126.636	1480	44,091	31,786	178.291
270	7,657	5,412	127.719	1520	45,384	32,746	179.153
280	7,945	5,617	128.765	1560	46,683	33,713	179.995
290	8,233	5,822	129.775	1600	47,990	34,687	180.820
298	8,468	5,989	130.574	1640	49,303	35,668	181.632
300	8,522	6,027	130.754	1680	50,662	36,654	182.428
320	9,100	6,440	132.621	1720	51,947	37,646	183.208
340	9,680	6,853	134.378	1760	53,279	38,645	183.973
360	10,262	7,268	136.039	1800	54,618	39,652	184.724
380	10,843	7,684	137.612	1840	55,962	40,663	185.463
400	11,426	8,100	139.106	1880	57,311	41,680	186.190
420	12,010	8,518	140.529	1920	58,668	42,705	186.904
440	12,594	8,936	141.888	1960	60,031	43,735	187.607
460	13,179	9,355	143.187	2000	61,400	44,771	188.297
480	13,764	9,773	144.432	2050	63,119	46,074	189.148
500	14,350	10,193	145.628	2100	64,847	47,386	189.979
520	14,935	10,611	146.775	2150	66,584	48,708	190.796
560	16,107	11,451	148.945	2200	68,328	50,037	191.598
600	17,280	12,291	150.968	2250	70,080	51,373	192.385
640	18,453	13,133	152.863	2300	71,839	52,716	193.159
680	19,630	13,976	154.645	2350	73,608	54,069	193.921
720	20,807	14,821	156.328	2400	75,383	55,429	194.669
760	21,988	15,669	157.923	2450	77,168	56,798	195.403
800	23,171	16,520	159.440	2500	78,960	58,175	196.125
840	24,359	17,375	160.891	2550	80,755	59,554	196.837
880	25,551	18,235	162.277	2600	82,558	60,941	197.539
920	26,747	19,098	163.607	2650	84,386	62,335	198.229
960	27,948	19,966	164.884	2700	86,186	63,737	198.907
1000	29,154	20,839	166.114	2750	88,008	65,144	199.575
1040	30,364	21,717	167.300	2800	89,838	66,558	200.234
1080	31,580	22,601	168.449	2850	91,671	67,976	200.885
1120	32,802	23,490	169.560	2900	93,512	69,401	201.527
1160	34,028	24,384	170.636	2950	95,358	70,831	202.157
1200	35,262	25,284	171.682	3000	97,211	72,268	202.778
1240	36,502	26,192	172.698	3050	99,065	73,707	203.391
1280	37,749	27,106	173.687	3100	100,926	75,152	203.995
1320	39,002	28,027	174.652	3150	102,793	76,604	204.592
1360	40,263	28,955	175.593	3200	104,667	78,061	205.181
1400	41,530	29,889	176.510	3250	106,545	79,523	205.765

Table A-11M *(Continued)* **Ideal-gas enthalpy, internal energy, and absolute entropy for monatomic oxygen, O**

$\Delta h_f = 249{,}190$ kJ/kg · mol

T	h	u	s^0	T	h	u	s^0
0	0	0	0	2400	50,894	30,940	204.932
298	6,852	4,373	160.944	2450	51,936	31,566	205.362
300	6,892	4,398	161.079	2500	52,979	32,193	205.783
500	11,197	7,040	172.088	2550	54,021	32,820	206.196
1000	21,713	13,398	186.678	2600	55,064	33,447	206.601
1500	32,150	19,679	195.143	2650	56,108	34,075	206.999
1600	34,234	20,931	196.488	2700	57,152	34,703	207.389
1700	36,317	22,183	197.751	2750	58,196	35,332	207.772
1800	38,400	23,434	198.941	2800	59,241	35,961	208.148
1900	40,482	24,685	200.067	2850	60,286	36,590	208.518
2000	42,564	25,935	201.135	2900	61,332	37,220	208.882
2050	43,605	26,560	201.649	2950	62,378	37,851	209.240
2100	44,646	27,186	202.151	3000	63,425	38,482	209.592
2150	45,687	27,811	202.641	3100	65,520	39,746	210.279
2200	46,728	28,436	203.119	3200	67,619	41,013	210.945
2250	47,769	29,062	203.588	3300	69,720	42,283	211.592
2300	48,811	29,688	204.045	3400	71,824	43,556	212.220
2350	49,852	30,314	204.493	3500	73,932	44,832	212.831

Ideal-gas enthalpy, internal energy, and absolute entropy for hydroxyl, OH

$\Delta h_f = 39{,}040$ kJ/kg · mol

T	h	u	s^0	T	h	u	s^0
0	0	0	0	2400	77,015	57,061	248.628
298	9,188	6,709	183.594	2450	78,801	58,431	249.364
300	9,244	6,749	183.779	2500	80,592	59,806	250.088
500	15,181	11,024	198.955	2550	82,388	61,186	250.799
1000	30,123	21,809	219.624	2600	84,189	62,572	251.499
1500	46,046	33,575	232.506	2650	85,995	63,962	252.187
1600	49,358	36,055	234.642	2700	87,806	65,358	252.864
1700	52,706	38,571	236.672	2750	89,622	66,757	253.530
1800	56,089	41,123	238.606	2800	91,442	68,162	254.186
1900	59,505	43,708	240.453	2850	93,266	69,570	254.832
2000	62,952	46,323	242.221	2900	95,095	70,983	255.468
2050	64,687	47,642	243.077	2950	96,927	72,400	256.094
2100	66,428	48,968	243.917	3000	98,763	73,820	256.712
2150	68,177	50,301	244.740	3100	102,447	76,673	257.919
2200	69,932	51,641	245.547	3200	106,145	79,539	259.093
2250	71,694	52,987	246.338	3300	109,855	82,418	260.235
2300	73,462	54,339	247.116	3400	113,578	85,309	261.347
2350	75,236	55,697	247.879	3500	117,312	88,212	262.429

Source: Based on data from the JANAF Thermochemical Tables, NSRDS-NBS-37, 1971.

Table A-12M Properties of saturated water: temperature table

v, cm^3/g; u, kJ/kg; h, kJ/kg; s, kJ/(kg)(°K)

		Specific volume		Internal energy		Enthalpy			Entropy	
Temp. °C T	Press. bars P	Sat. liquid v_f	Sat. vapor v_g	Sat. liquid u_f	Sat. vapor u_g	Sat. liquid h_f	Evap. h_{fg}	Sat. vapor h_g	Sat. liquid s_f	Sat. vapor s_g
0	0.00611	1.0002	206278	−0.03	2375.4	−0.02	2501.4	2501.3	0.0001	9.1565
4	0.00813	1.0001	157232	16.77	2380.9	16.78	2491.9	2508.7	0.0610	9.0514
5	0.00872	1.0001	147120	20.97	2382.3	20.98	2489.6	2510.6	0.0761	9.0257
6	0.00935	1.0001	137734	25.19	2383.6	25.20	2487.2	2512.4	0.0912	9.0003
8	0.01072	1.0002	120917	33.59	2386.4	33.60	2482.5	2516.1	0.1212	8.9501
10	0.01228	1.0004	106379	42.00	2389.2	42.01	2477.7	2519.8	0.1510	8.9008
11	0.01312	1.0004	99857	46.20	2390.5	46.20	2475.4	2521.6	0.1658	8.8765
12	0.01402	1.0005	93784	50.41	2391.9	50.41	2473.0	2523.4	0.1806	8.8524
13	0.01497	1.0007	88124	54.60	2393.3	54.60	2470.7	2525.3	0.1953	8.8285
14	0.01598	1.0008	82848	58.79	2394.7	58.80	2468.3	2527.1	0.2099	8.8048
15	0.01705	1.0009	77926	62.99	2396.1	62.99	2465.9	2528.9	0.2245	8.7814
16	0.01818	1.0011	73333	67.18	2397.4	67.19	2463.6	2530.8	0.2390	8.7582
17	0.01938	1.0012	69044	71.38	2398.8	71.38	2461.2	2532.6	0.2535	8.7351
18	0.02064	1.0014	65038	75.57	2400.2	75.58	2458.8	2534.4	0.2679	8.7123
19	0.02198	1.0016	61293	79.76	2401.6	79.77	2456.5	2536.2	0.2823	8.6897
20	0.02339	1.0018	57791	83.95	2402.9	83.96	2454.1	2538.1	0.2966	8.6672
21	0.02487	1.0020	54514	88.14	2404.3	88.14	2451.8	2539.9	0.3109	8.6450
22	0.02645	1.0022	51447	92.32	2405.7	92.33	2449.4	2541.7	0.3251	8.6229
23	0.02810	1.0024	48574	96.51	2407.0	96.52	2447.0	2543.5	0.3393	8.6011
24	0.02985	1.0027	45883	100.70	2408.4	100.70	2444.7	2545.4	0.3534	8.5794
25	0.03169	1.0029	43360	104.88	2409.8	104.89	2442.3	2547.2	0.3674	8.5580
26	0.03363	1.0032	40994	109.06	2411.1	109.07	2439.9	2549.0	0.3814	8.5367
27	0.03567	1.0035	38774	113.25	2412.5	113.25	2437.6	2550.8	0.3954	8.5156
28	0.03782	1.0037	36690	117.42	2413.9	117.43	2435.2	2552.6	0.4093	8.4946
29	0.04008	1.0040	34733	121.60	2415.2	121.61	2432.8	2554.5	0.4231	8.4739
30	0.04246	1.0043	32894	125.78	2416.6	125.79	2430.5	2556.3	0.4369	8.4533
31	0.04496	1.0046	31165	129.96	2418.0	129.97	2428.1	2558.1	0.4507	8.4329
32	0.04759	1.0050	29540	134.14	2419.3	134.15	2425.7	2559.9	0.4644	8.4127
33	0.05034	1.0053	28011	138.32	2420.7	138.33	2423.4	2561.7	0.4781	8.3927
34	0.05324	1.0056	26571	142.50	2422.0	142.50	2421.0	2563.5	0.4917	8.3728
35	0.05628	1.0060	25216	146.67	2423.4	146.68	2418.6	2565.3	0.5053	8.3531
36	0.05947	1.0063	23940	150.85	2424.7	150.86	2416.2	2567.1	0.5188	8.3336
38	0.06632	1.0071	21602	159.20	2427.4	159.21	2411.5	2570.7	0.5458	8.2950
40	0.07384	1.0078	19523	167.56	2430.1	167.57	2406.7	2574.3	0.5725	8.2570
45	0.09593	1.0099	15258	188.44	2436.8	188.45	2394.8	2583.2	0.6387	8.1648

Table A-12M *(Continued)*

Temp. °C T	Press. bars P	Specific volume: Sat. liquid v_f	Specific volume: Sat. vapor v_g	Internal energy: Sat. liquid u_f	Internal energy: Sat. vapor u_g	Enthalpy: Sat. liquid h_f	Enthalpy: Evap. h_{fg}	Enthalpy: Sat. vapor h_g	Entropy: Sat. liquid s_f	Entropy: Sat. vapor s_g
50	.1235	1.0121	12032	209.32	2443.5	209.33	2382.7	2592.1	.7038	8.0763
55	.1576	1.0146	9568	230.21	2450.1	230.23	2370.7	2600.9	.7679	7.9913
60	.1994	1.0172	7671	251.11	2456.6	251.13	2358.5	2609.6	.8312	7.9096
65	.2503	1.0199	6197	272.02	2463.1	272.06	2346.2	2618.3	.8935	7.8310
70	.3119	1.0228	5042	292.95	2469.6	292.98	2333.8	2626.8	.9549	7.7553
75	.3858	1.0259	4131	313.90	2475.9	313.93	2321.4	2635.3	1.0155	7.6824
80	.4739	1.0291	3407	334.86	2482.2	334.91	2308.8	2643.7	1.0753	7.6122
85	.5783	1.0325	2828	355.84	2488.4	355.90	2296.0	2651.9	1.1343	7.5445
90	.7014	1.0360	2361	376.85	2494.5	376.92	2283.2	2660.1	1.1925	7.4791
95	.8455	1.0397	1982	397.88	2500.6	397.96	2270.2	2668.1	1.2500	7.4159
100	1.014	1.0435	1673.	418.94	2506.5	419.04	2257.0	2676.1	1.3069	7.3549
110	1.433	1.0516	1210.	461.14	2518.1	461.30	2230.2	2691.5	1.4185	7.2387
120	1.985	1.0603	891.9	503.50	2529.3	503.71	2202.6	2706.3	1.5276	7.1296
130	2.701	1.0697	668.5	546.02	2539.9	546.31	2174.2	2720.5	1.6344	7.0269
140	3.613	1.0797	508.9	588.74	2550.0	589.13	2144.7	2733.9	1.7391	6.9299
150	4.758	1.0905	392.8	631.68	2559.5	632.20	2114.3	2746.5	1.8418	6.8379
160	6.178	1.1020	307.1	674.86	2568.4	675.55	2082.6	2758.1	1.9427	6.7502
170	7.917	1.1143	242.8	718.33	2576.5	719.21	2049.5	2768.7	2.0419	6.6663
180	10.02	1.1274	194.1	762.09	2583.7	763.22	2015.0	2778.2	2.1396	6,5857
190	12.54	1.1414	156.5	806.19	2590.0	807.62	1978.8	2786.4	2.2359	6.5079
200	15.54	1.1565	127.4	850.65	2595.3	852.45	1940.7	2793.2	2.3309	6.4323
210	19.06	1.1726	104.4	895.53	2599.5	897.76	1900.7	2798.5	2.4248	6.3585
220	23.18	1.1900	86.19	940.87	2602.4	943.62	1858.5	2802.1	2.5178	6.2861
230	27.95	1.2088	71.58	986.74	2603.9	990.12	1813.8	2804.0	2.6099	6.2146
240	33.44	1.2291	59.76	1033.2	2604.0	1037.3	1766.5	2803.8	2.7015	6.1437
250	39.73	1.2512	50.13	1080.4	2602.4	1085.4	1716.2	2801.5	2.7927	6.0730
260	46.88	1.2755	42.21	1128.4	2599.0	1134.4	1662.5	2796.6	2.8838	6.0019
270	54.99	1.3023	35.64	1177.4	2593.7	1184.5	1605.2	2789.7	2.9751	5.9301
280	64.12	1.3321	30.17	1227.5	2586.1	1236.0	1543.6	2779.6	3.0668	5.8571
290	74.36	1.3656	25.57	1278.9	2576.0	1289.1	1477.1	2766.2	3.1594	5.7821
300	85.81	1.4036	21.67	1332.0	2563.0	1344.0	1404.9	2749.0	3.2534	5.7045
320	112.7	1.4988	15.49	1444.6	2525.5	1461.5	1238.6	2700.1	3.4480	5.5362
340	145.9	1.6379	10.80	1570.3	2464.6	1594.2	1027.9	2622.0	36594	5.3357
360	186.5	1.8925	6.945	1725.2	2351.5	1760.5	720.5	2481.0	3.9147	5.0526
374.14	220.9	3.155	3.155	2029.6	2029.6	2099.3	0	2099.3	4.4298	4.4298

Source: Keenan, J. H., F. G. Keyes, P. G. Hill, and J. G. Moore, "Steam Tables," Wiley, New York, 1969.

Table A-13M Properties of saturated water: pressure table

v, cm^3/g; u, kJ/kg; h kJ/kg; s, kJ/(kg)(°K)

		Specific volume		Internal energy		Enthalpy			Entropy	
Press. bars P	Temp. °C T	Sat. liquid v_f	Sat. vapor v_g	Sat. liquid u_f	Sat. vapor u_g	Sat. liquid h_f	Evap. h_{fg}	Sat. vapor h_g	Sat. liquid s_f	Sat. vapor s_g
.040	28.96	1.0040	34800	121.45	2415.2	121.46	2432.9	2554.4	.4226	8.4746
.060	36.16	1.0064	23739	151.53	2425.0	151.53	2415.9	2567.4	.5210	8.3304
.080	41.51	1.0084	18103	173.87	2432.2	173.88	2403.1	2577.0	.5926	8.2287
0.10	45.81	1.0102	14674	191.82	2437.9	191.83	2392.8	2584.7	.6493	8.1502
0.20	60.06	1.0172	7649	251.38	2456.7	251.40	2358.3	2609.7	.8320	7.9085
0.30	69.10	1.0223	5229.	289.20	2468.4	289.23	2336.1	2625.3	.9439	7.7686
0.40	75.87	1.0265	3993.	317.53	2477.0	317.58	2319.2	2636.8	1.0259	7.6700
0.50	81.33	1.0300	3240.	340.44	2483.9	340.49	2305.4	2645.9	1.0910	7.5939
0.60	85.94	1.0331	2732.	359.79	2489.6	359.86	2293.6	2653.5	1.1453	7.5320
0.70	89.95	1.0360	2365.	376.63	2494.5	376.70	2283.3	2660.0	1.1919	7.4797
0.80	93.50	1.0380	2087.	391.58	2498.8	391.66	2274.1	2665.8	1.2329	7.4346
0.90	96.71	1.0410	1869.	405.06	2502.6	405.15	2265.7	2670.9	1.2695	7.3949
1.00	99.63	1.0432	1694.	417.36	2506.1	417.46	2258.0	2675.5	1.3026	7.3594
1.50	111.4	1.0528	1159.	466.94	2519.7	467.11	2226.5	2693.6	1.4336	7.2233
2.00	120.2	1.0605	885.7	504.49	2529.5	504.70	2201.9	2706.7	1.5301	7.1271
2.50	127.4	1.0672	718.7	535.10	2537.2	535.37	2181.5	2716.9	1.6072	7.0527
3.00	133.6	1.0732	605.8	561.15	2543.6	561.47	2163.8	2725.3	1.6718	6.9919
3.50	138.9	1.0786	524.3	583.95	2546.9	584.33	2148.1	2732.4	1.7275	6.9405
4.00	143.6	1.0836	462.5	604.31	2553.6	604.74	2133.8	2738.6	1.7766	6.8959
4.50	147.9	1.0882	414.0	622.25	2557.6	623.25	2120.7	2743.9	1.8207	6.8565
5.00	151.9	1.0926	374.9	639.68	2561.2	640.23	2108.5	2748.7	1.8607	6.8212
6.00	158.9	1.1006	315.7	669.90	2567.4	670.56	2086.3	2756.8	1.9312	6.7600
7.00	165.0	1.1080	272.9	696.44	2572.5	697.22	2066.3	2763.5	1.9922	6.7080
8.00	170.4	1.1148	240.4	720.22	2576.8	721.11	2048.0	2769.1	2.0462	6.6628
9.00	175.4	1.1212	215.0	741.83	2580.5	742.83	2031.1	2773.9	2.0946	6.6226
10.0	179.9	1.1273	194.4	761.68	2583.6	762.81	2015.3	2778.1	2.1387	6.5863
15.0	198.3	1.1539	131.8	843.16	2594.5	844.84	1947.3	2792.2	2.3150	6.4448
20.0	212.4	1.1767	99.63	906.44	2600.3	908.79	1890.7	2799.5	2.4474	6.3409
25.0	224.0	1.1973	79.98	959.11	2603.1	962.11	1841.0	2803.1	2.5547	6.2575
30.0	233.9	1.2165	66.68	1004.8	2604.1	1008.4	1795.7	2804.2	2.6457	6.1869
35.0	242.6	1.2347	57.07	1045.4	2603.7	1049.8	1753.7	2803.4	2.7253	6.1253
40.0	250.4	1.2522	49.78	1082.3	2602.3	1087.3	1714.1	2801.4	2.7964	6.0701
45.0	257.5	1.2692	44.06	1116.2	2600.1	1121.9	1676.4	2798.3	2.8610	6.0199
50.0	264.0	1.2859	39.44	1147.8	2597.1	1154.2	1640.1	2794.3	2.9202	5.9734
60.0	275.6	1.3187	32.44	1205.4	2589.7	1213.4	1571.0	2784.3	3.0267	5.8892
70.0	285.9	1.3513	27.37	1257.6	2580.5	1267.0	1505.1	2772.1	3.1211	5.8133
80.0	295.1	1.3842	23.52	1305.6	2569.8	1316.6	1441.3	2758.0	3.2068	5.7432
90.0	303.4	1.4178	20.48	1350.5	2557.8	1363.3	1378.9	2742.1	3.2858	5.6772
100.	311.1	1.4524	18.03	1393.0	2544.4	1407.6	1317.1	2724.7	3.3596	5.6141
110.	318.2	1.4886	15.99	1433.7	2529.8	1450.1	1255.5	2705.6	3.4295	5.5527
120.	324.8	1.5267	14.26	1473.0	2513.7	1491.3	1193.6	2684.9	3.4962	5.4924
130.	330.9	1.5671	12.78	1511.1	2496.1	1531.5	1130.7	2662.2	3.5606	5.4323
140.	336.8	1.6107	11.49	1548.6	2476.8	1571.1	1066.5	2637.6	3.6232	5.3717
150.	342.2	1.6581	10.34	1585.6	2455.5	1610.5	1000.0	2610.5	3.6848	5.3098
160.	347.4	1.7107	9.306	1622.7	2431.7	1650.1	930.6	2580.6	3.7461	5.2455
170.	352.4	1.7702	8.364	1660.2	2405.0	1690.3	856.9	2547.2	3.8079	5.1777
180.	357.1	1.8397	7.489	1698.9	2374.3	1732.0	777.1	2509.1	3.8715	5.1044
190.	361.5	1.9243	6.657	1739.9	2338.1	1776.5	688.0	2464.5	3.9388	5.0228
200.	365.8	2.036	5.834	1785.6	2293.0	1826.3	583.4	2409.7	4.0139	4.9269
220.9	374.1	3.155	3.155	2029.6	2029.6	2099.3	0	2099.3	4.4298	4.4298

Source: Keenan et al., "Steam Tables," Wiley, New York, 1969.

Table A-14M Properties of water: superheated-vapor table

v, cm³/g; *u*, kJ/kg; *h*, kJ/kg; *s*, kJ/(kg)(°K)

Temp. °C	*v*	*u*	*h*	*s*	*v*	*u*	*h*	*s*
	0.06 bar (0.006 MPa)(T_{sat} = 36.16°C)				0.35 bar (0.035 MPa) (T_{sat} = 72.69°C)			
Sat.	23739	2425.0	2567.4	8.3304	4526.	2473.0	2631.4	7.7158
80	27132	2487.3	2650.1	8.5804	4625.	2483.7	2645.6	7.7564
120	30219	2544.7	2726.0	8.7840	5163.	2542.4	2723.1	7.9644
160	33302	2602.7	2802.5	8.9693	5696.	2601.2	2800.6	8.1519
200	36383	2661.4	2879.7	9.1398	6228.	2660.4	2878.4	8.3237
240	39462	2721.0	2957.8	9.2982	6758.	2720.3	2956.8	8.4828
280	42540	2781.5	3036.8	9.4464	7287.	2780.9	3036.0	8.6314
320	45618	2843.0	3116.7	9.5859	7815.	2842.5	3116.1	8.7712
360	48696	2905.5	3197.7	9.7180	8344.	2905.1	3197.1	8.9034
400	51774	2969.0	3279.6	9.8435	8872.	2968.6	3279.2	9.0291
440	54851	3033.5	3362.6	9.9633	9400.	3033.2	3362.2	9.1490
500	59467	3132.3	3489.1	10.134	10192.	3132.1	3488.8	9.3194
	0.70 bar (0.07 MPa) (T_{sat} = 89.95°C)				1.0 bar (0.10 MPa) (T_{sat} = 99.63°C)			
Sat.	2365.	2494.5	2660.0	7.4797	1694.	2506.1	2675.5	7.3594
100	2434.	2509.7	2680.0	7.5341	1696.	2506.7	2676.2	7.3614
120	2571.	2539.7	2719.6	7.6375	1793.	2537.3	2716.6	7.4668
160	2841.	2599.4	2798.2	7.8279	1984.	2597.8	2796.2	7.6597
200	3108.	2659.1	2876.7	8.0012	2172.	2658.1	2875.3	7.8343
240	3374.	2719.3	2955.5	8.1611	2359.	2718.5	2954.5	7.9949
280	3640.	2780.2	3035.0	8.3162	2546.	2779.6	3034.2	8.1445
320	3905.	2842.0	3115.3	8.4504	2732.	2841.5	3114.6	8.2849
360	4170.	2904.6	3196.5	8.5828	2917.	2904.2	3195.9	8.4175
400	4434.	2968.2	3278.6	8.7086	3103.	2967.9	3278.2	8.5435
440	4698.	3032.9	3361.8	8.8286	3288.	3032.6	3361.4	8.6636
500	5095.	3131.8	3488.5	8.9991	3565.	3131.6	3488.1	8.8342
	1.5 bars (0.15 MPa) (T_{sat} = 111.37°C)				3.0 bars (0.30 MPa) (T_{sat} = 133.55°C)			
Sat.	1159	2519.7	2693.6	7.2233	606.	2543.6	2725.3	6.9919
120	1188.	2533.3	2711.4	7.2693				
160	1317.	2595.2	2792.8	7.4665	651.	2587.1	2782.3	7.1276
200	1444.	2656.2	2872.9	7.6433	716.	2650.7	2865.5	7.3115
240	1570.	2717.2	2952.7	7.8052	781.	2713.1	2947.3	7.4774
280	1695.	2778.6	3032.8	7.9555	844.	2775.4	3028.6	7.6299
320	1819.	2840.6	3113.5	8.0964	907.	2838.1	3110.1	7.7722
360	1943.	2903.5	3195.0	8.2293	969.	2901.4	3192.2	7.9061
400	2067.	2967.3	3277.4	8.3555	1032.	2965.6	3275.0	8.0330
440	2191.	3032.1	3360.7	8.4757	1094.	3030.6	3358.7	8.1538
500	2376.	3131.2	3487.6	8.6466	1187.	3130.0	3486.0	8.3251
600	2685.	3301.7	3704.3	8.9101	1341.	3300.8	3703.2	8.5892

Table A-14M (*Continued*)

Temp. °C	v	u	h	s	v	u	h	s
	5.0 bars (0.50 MPa) (T_{sat} = 151.86°C)				7.0 bars (0.70 MPa) (T_{sat} = 164.97°C)			
Sat.	374.9	2561.2	2748.7	6.8213	272.9	2572.5	2763.5	6.7080
180	404.5	2609.7	2812.0	6.9656	284.7	2599.8	2799.1	6.7880
200	424.9	2642.9	2855.4	7.0592	299.9	2634.8	2844.8	6.8865
240	464.6	2707.6	2939.9	7.2307	329.2	2701.8	2932.2	7.0641
280	503.4	2771.2	3022.9	7.3865	357.4	2766.9	3017.1	7.2233
320	541.6	2834.7	3105.6	7.5308	385.2	2831.3	3100.9	7.3697
360	579.6	2898.7	3188.4	7.6660	412.6	2895.8	3184.7	7.5063
400	617.3	2963.2	3271.9	7.7938	439.7	2960.9	3268.7	7.6350
440	654.8	3028.6	3356.0	7.9152	466.7	3026.6	3353.3	7.7571
500	710.9	3128.4	3483.9	8.0873	507.0	3126.8	3481.7	7.9299
600	804.1	3299.6	3701.7	8.3522	573.8	3298.5	3700.2	8.1956
700	896.9	3477.5	3925.9	8.5952	640.3	3476.6	3924.8	8.4391
	10.0 bars (1.0 MPa) (T_{sat} = 179.91°C)				15.0 bars (1.5MPa) (T_{sat} = 198.32°C)			
Sat.	194.4	2583.6	2778.1	6.5865	131.8	2594.5	2792.2	6.4448
200	206.0	2621.9	2827.9	6.6940	132.5	2598.1	2796.8	6.4546
240	227.5	2692.9	2920.4	6.8817	148.3	2676.9	2899.3	6.6628
280	248.0	2760.2	3008.2	7.0465	162.7	2748.6	2992.7	6.8381
320	267.8	2826.1	3093.9	7.1962	176.5	2817.1	3081.9	6.9938
360	287.3	2891.6	3178.9	7.3349	189.9	2884.4	3169.2	7.1363
400	306.6	2957.3	3263.9	7.4651	203.0	2951.3	3255.8	7.2690
440	325.7	3023.6	3349.3	7.5883	216.0	3018.5	3342.5	7.3940
500	354.1	3124.4	3478.5	7.7622	235.2	3120.3	3473.1	7.5698
540	372.9	3192.6	3565.6	7.8720	247.8	3189.1	3560.9	7.6805
600	401.1	3296.8	3697.9	8.0290	266.8	3293.9	3694.0	7.8385
640	419.8	3367.4	3787.2	8.1290	279.3	3364.8	3783.8	7.9391
	20.0 bars (2.0 MPa) (T_{sat} = 212.42°C)				30.0 bars (3.0 MPa) (T_{sat} = 233.90°C)			
Sat.	99.6	2600.3	2799.5	6.3409	66.7	2604.1	2804.2	6.1869
240	108.5	2659.6	2876.5	6.4952	68.2	2619.7	2824.3	6.2265
280	120.0	2736.4	2976.4	6.6828	77.1	2709.9	2941.3	6.4462
320	130.8	2807.9	3069.5	6.8452	85.0	2788.4	3043.4	6.6245
360	141.1	2877.0	3159.3	6.9917	92.3	2861.7	3138.7	6.7801
400	151.2	2945.2	3247.6	7.1271	99.4	2932.8	3230.9	6.9212
440	161.1	3013.4	3335.5	7.2540	106.2	3002.9	3321.5	7.0520
500	175.7	3116.2	3467.6	7.4317	116.2	3108.0	3456.5	7.2338
540	185.3	3185.6	3556.1	7.5434	122.7	3178.4	3546.6	7.3474
600	199.6	3290.9	3690.1	7.7024	132.4	3285.0	3682.3	7.5085
640	209.1	3362.2	3780.4	7.8035	138.8	3357.0	3773.5	7.6106
700	223.2	3470.9	3917.4	7.9487	148.4	3466.5	3911.7	7.7571

Table A-14M Properties of superheated water vapor (*Continued*)

Temp. °C	v	u	h	s	v	u	h	s
	40 bars (4.0 MPa) (T_{sat} = 250.40°C)				60 bars (6.0 MPa) (T_{sat} = 275.64°C)			
Sat.	49.78	2602.3	2801.4	6.0701	32.44	2589.7	2784.3	5.8892
280	55.46	2680.0	2901.8	6.2568	33.17	2605.2	2804.2	5.9252
320	61.99	2767.4	3015.4	6.4553	38.76	2720.0	2952.6	6.1846
360	67.88	2845.7	3117.2	6.6215	43.31	2811.2	3071.1	6.3782
400	73.41	2919.9	3213.6	6.7690	47.39	2892.9	3177.2	6.5408
440	78.72	2992.2	3307.1	6.9041	51.22	2970.0	3277.3	6.6853
500	86.43	3099.5	3445.3	7.0901	56.65	3082.2	3422.2	6.8803
540	91.45	3171.1	3536.9	7.2056	60.15	3156.1	3517.0	6.9999
600	98.85	3279.1	3674.4	7.3688	65.25	3266.9	3658.4	7.1677
640	103.7	3351.8	3766.6	7.4720	68.59	3341.0	3752.6	7.2731
700	111.0	3462.1	3905.9	7.6198	73.52	3453.1	3894.1	7.4234
740	115.7	3536.6	3999.6	7.7141	76.77	3528.3	3989.2	7.5190
	80 bars (8.0 MPa) (T_{sat} = 295.06°C)				100 bars (10.0 MPa) (T_{sat} = 311.06°C)			
Sat.	23.52	2569.8	2758.0	5.7432	18.03	2544.4	2724.7	5.6141
320	26.82	2662.7	2877.2	5.9489	19.25	2588.8	2781.3	5.7103
360	30.89	2772.7	3019.8	6.1819	23.31	2729.1	2962.1	6.0060
400	34.32	2863.8	3138.3	6.3634	26.41	2832.4	3096.5	6.2120
440	37.42	2946.7	3246.1	6.5190	29.11	2922.1	3213.2	6.3805
480	40.34	3025.7	3348.4	6.6586	31.60	3005.4	3321.4	6.5282
520	43.13	3102.7	3447.7	6.7871	33.94	3085.6	3425.1	6.6622
560	45.82	3178.7	3545.3	6.9072	36.19	3164.1	3526.0	6.7864
600	48.45	3254.4	3642.0	7.0206	38.37	3241.7	3625.3	6.9029
640	51.02	3330.1	3738.3	7.1283	40.48	3318.9	3723.7	7.0131
700	54.81	3443.9	3882.4	7.2812	43.58	3434.7	3870.5	7.1687
740	57.29	3520.4	3978.7	7.3782	45.60	3512.1	3968.1	7.2670
	120 bars (12.0 MPa) (T_{sat} = 324.75°C)				140 bars (14.0 MPa) (T_{sat} = 336.75°C)			
Sat.	14.26	2513.7	2684.9	5.4924	11.49	2476.8	2637.6	5.3717
360	18.11	2678.4	2895.7	5.8361	14.22	2617.4	2816.5	5.6602
400	21.08	2798.3	3051.3	6.0747	17.22	2760.9	3001.9	5.9448
440	23.55	2896.1	3178.7	6.2586	19.54	2868.6	3142.2	6.1474
480	25.76	2984.4	3293.5	6.4154	21.57	2962.5	3264.5	6.3143
520	27.81	3068.0	3401.8	6.5555	23.43	3049.8	3377.8	6.4610
560	29.77	3149.0	3506.2	6.6840	25.17	3133.6	3486.0	6.5941
600	31.64	3228.7	3608.3	6.8037	26.83	3215.4	3591.1	6.7172
640	33.45	3307.5	3709.0	6.9164	28.43	3296.0	3694.1	6.8326
700	36.10	3425.2	3858.4	7.0749	30.75	3415.7	3846.2	6.9939
740	37.81	3503.7	3957.4	7.1746	32.25	3495.2	3946.7	7.0952

Table A-14M *(Continued)*

Temp. °C	v	u	h	s	v	u	h	s
	160 bars (16.0 MPa) (T_{sat} = 347.44°C)				180 bars (18.0 MPa) (T_{sat} = 357.06°C)			
Sat.	9.31	2431.7	2580.6	5.2455	7.49	2374.3	2509.1	5.1044
360	11.05	2539.0	2715.8	5.4614	8.09	2418.9	2564.5	5.1922
400	14.26	2719.4	2947.6	5.8175	11.90	2672.8	2887.0	5.6887
440	16.52	2839.4	3103.7	6.0429	14.14	2808.2	3062.8	5.9428
480	18.42	2939.7	3234.4	6.2215	15.96	2915.9	3203.2	6.1345
520	20.13	3031.1	3353.3	6.3752	17.57	3011.8	3378.0	6.2960
560	21.72	3117.8	3465.4	6.5132	19.04	3101.7	3444.4	6.4392
600	23.23	3201.8	3573.5	6.6399	20.42	3188.0	3555.6	6.5696
640	24.67	3284.2	3678.9	6.7580	21.74	3272.3	3663.6	6.6905
700	26.74	3406.0	3833.9	6.9224	23.62	3396.3	3821.5	6.8580
740	28.08	3486.7	3935.9	7.0251	24.83	3478.0	3925.0	6.9623
	200 bars (20.0 MPa) (T_{sat} = 365.81°C)				240 bars (24.0 MPa)			
Sat.	5.83	2293.0	2409.7	4.9269				
400	9.94	2619.3	2818.1	5.5540	6.73	2477.8	2639.4	5.2393
440	12.22	2774.9	3019.4	5.8450	9.29	2700.6	2923.4	5.6506
480	13.99	2891.2	3170.8	6.0518	11.00	2838.3	3102.3	5.8950
520	15.51	2992.0	3302.2	6.2218	12.41	2950.5	3248.5	6.0842
560	16.89	3085.2	3423.0	6.3705	13.66	3051.1	3379.0	6.2448
600	18.18	3174.0	3537.6	6.5048	14.81	3145.2	3500.7	6.3875
640	19.40	3260.2	3648.1	6.6286	15.88	3235.5	3616.7	6.5174
700	21.13	3386.4	3809.0	6.7993	17.39	3366.4	3783.8	6.6947
740	22.24	3469.3	3914.1	6.9052	18.35	3451.7	3892.1	6.8038
800	23.85	3592.7	4069.7	7.0544	19.74	3578.0	4051.6	6.9567
	280 bars (28.0 MPa)				320 bars (32.0 MPa)			
400	3.83	2223.5	2330.7	4.7494	2.36	1980.4	2055.9	4.3239
440	7.12	2613.2	2812.6	5.4494	5.44	2509.0	2683.0	5.2327
480	8.85	2780.8	3028.5	5.7446	7.22	2718.1	3949.2	5.5968
520	10.20	3906.8	3192.3	5.9566	8.53	2860.7	3133.7	5.8357
560	11.36	3015.7	3333.7	6.1307	9.63	2979.0	3287.2	6.0246
600	12.41	3115.6	3463.0	6.2823	10.61	3085.3	3424.6	6.1858
640	13.38	3210.3	3584.8	6.4187	11.50	3184.5	3552.5	6.3290
700	14.73	3346.1	3758.4	6.6029	12.73	3325.4	3732.8	6.5203
740	15.58	3433.9	3870.0	6.7153	13.50	3415.9	3847.8	6.6361
800	16.80	3563.1	4033.4	6.8720	14.60	3548.0	4015.1	6.7966
900	18.73	3774.3	4298.8	7.1084	16.33	3762.7	4285.1	7.0372

Source: Keenan, J. H., F. G. Keyes, P. G. Hill, and J. G. Moore, "Steam Tables," Wiley, New York, 1969.

Table A-15M Properties of water: compressed liquid table

v, cm^3/g; u, kJ/kg; h, kJ/kg; s, kJ/(kg)(°K)

Temp. °C	v	u	h	s	v	u	h	s
	25 bars (223.99°C)				50 bars (263.99°C)			
20	1.0006	83.80	86.30	.2961	.9995	83.65	88.65	.2956
40	1.0067	167.25	169.77	.5715	1.0056	166.95	171.97	.5705
80	1.0280	334.29	336.86	1.0737	1.0268	333.72	338.85	1.0720
100	1.0423	418.24	420.85	1.3050	1.0410	417.52	422.72	1.3030
140	1.0784	587.82	590.52	1.7369	1.0768	586.76	592.15	1.7343
180	1.1261	761.16	763.97	2.1375	1.1240	759.63	765.25	2.1341
200	1.1555	849.9	852.8	2.3294	1.1530	848.1	848.1	2.3255
220	1.1898	940.7	943.7	2.5174	1.1866	938.4	944.4	2.5128
Sat.	1.1973	959.1	962.1	2.5546	1.2859	1147.8	1154.2	2.9202
	75 bars (290.59°C)				100 bars (311.06°C)			
20	.9984	83.50	90.99	.2950	.9972	83.36	93.33	.2945
40	1.0045	166.64	174.18	.5696	1.0034	166.35	176.38	.5686
80	1.0256	333.15	340.84	1.0704	1.0245	332.59	342.83	1.0688
100	1.0397	416.81	424.62	1.3011	1.0385	416.12	426.50	1.2992
140	1.0752	585.72	593.78	1.7317	1.0737	584.68	595.42	1.7292
180	1.1219	758.13	766.55	2.1308	1.1199	756.65	767.84	2.1275
220	1.1835	936.2	945.1	2.5083	1.1805	934.1	945.9	2.5039
260	1.2696	1124.4	1134.0	2.8763	1.2645	1121.1	1133.7	2.8699
Sat.	1.3677	1282.0	1292.2	3.1649	1.4524	1393.0	1407.6	3.3596
	150 bars (342.24°C)				200 bars (365.81°C)			
20	.9950	83.06	97.99	.2934	.9928	82.77	102.62	.2923
40	1.0013	165.76	180.78	.5666	.9992	165.17	185.16	.5646
80	1.0222	331.48	346.81	1.0656	1.0199	330.40	350.80	1.0624
100	1.0361	414.74	430.28	1.2955	1.0337	413.39	434.06	1.2917
140	1.0707	582.66	598.72	1.7242	1.0678	580.69	602.04	1.7193
180	1.1159	753.76	770.50	2.1210	1.1120	750.95	773.20	2.1147
220	1.1748	929.9	947.5	2.4953	1.1693	925.9	949.3	2.4870
260	1.2550	1114.6	1133.4	2.8576	1.2462	1108.6	1133.5	2.8459
300	1.3770	1316.6	1337.3	3.2260	1.3596	1306.1	1333.3	3.2071
Sat.	1.6581	1585.6	1610.5	3.6848	2.036	1785.6	1826.3	4.0139
	250 bars				300 bars			
20	.9907	82.47	107.24	.2911	.9886	82.17	111.84	.2899
40	.9971	164.60	189.52	.5626	.9951	164.04	193.89	.5607
100	1.0313	412.08	437.85	1.2881	1.0290	410.78	441.66	1.2844
200	1.1344	834.5	862.8	2.2961	1.1302	831.4	865.3	2.2893
300	1.3442	1296.6	1330.2	3.1900	1.3304	1287.9	1327.8	3.1741

Source: Keenan, J. H., F. G. Keyes, P. G. Hill, and J. G. Moore, "Steam Tables," Wiley, New York, 1969.

Table A-16M Properties of saturated refrigerant 12, $CC1_2F_2$: temperature table

v, cm^3/g; u, kJ/kg; h, kJ/kg; s, kJ/(kg)(°K)

		Specific volume		Internal energy		Enthalpy			Entropy	
Temp. °C T	Press. bar(s) P	Sat. liquid v_f	Sat. vapor v_g	Sat. liquid u_f	Sat. vapor u_g	Sat. liquid h_f	Evap. h_{fg}	Sat. vapor h_g	Sat. liquid s_f	Sat. vapor s_g
−40	0.6417	0.6595	241.91	−0.04	154.07	0	169.59	169.59	0	0.7274
−35	0.8071	0.6656	195.40	4.37	156.13	4.42	167.48	171.90	0.0187	0.7219
−30	1.0041	0.6720	159.38	8.79	158.20	8.86	165.33	174.20	0.0371	0.7170
−28	1.0927	0.6746	147.28	10.58	159.02	10.65	164.46	175.11	0.0444	0.7153
−26	1.1872	0.6773	136.28	12.35	159.84	12.43	163.59	176.02	0.0517	0.7135
−25	1.2368	0.6786	131.17	13.25	160.26	13.33	163.15	176.48	0.0552	0.7126
−24	1.2880	0.6800	126.28	14.13	160.67	14.22	162.71	176.93	0.0589	0.7119
−22	1.3953	0.6827	117.17	15.92	161.48	16.02	161.82	177.83	0.0660	0.7103
−20	1.5093	0.6855	108.85	17.72	162.31	17.82	160.92	178.74	0.0731	0.7087
−18	1.6304	0.6883	101.24	19.51	163.12	19.62	160.01	179.63	0.0802	0.7073
−15	1.8260	.6926	91.02	22.20	164.35	22.33	158.64	180.97	0.0906	0.7051
−10	2.1912	0.7000	76.65	26.72	166.39	26.87	156.31	183.19	0.1080	0.7019
−5	2.6096	0.7078	64.96	31.27	168.42	31.45	153.93	185.37	0.1251	0.6991
0	3.0861	0.7159	55.39	35.83	170.44	36.05	151.48	187.53	0.1420	0.6965
4	3.5124	0.7227	48.95	39.51	172.04	39.76	149.47	189.23	0.1553	0.6946
8	3.9815	0.7297	43.40	43.21	173.63	43.50	147.41	190.91	0.1686	0.6929
12	4.4962	0.7370	38.60	46.93	175.20	47.26	145.30	192.56	0.1817	0.6913
16	5.0591	0.7446	34.42	50.67	176.78	51.05	143.14	194.19	0.1948	0.6898
20	5.6729	0.7525	30.78	54.44	178.32	54.87	140.91	195.78	0.2078	0.6884
24	6.3405	0.7607	27.59	58.25	179.85	58,73	138.61	197.34	0.2207	0.6871

Table A-16M (*Continued*)

		Specific volume		Internal energy		Enthalpy			Entropy	
Temp. °C T	Press. bar(s) P	Sat. liquid v_f	Sat. vapor v_g	Sat. liquid u_f	Sat. vapor u_g	Sat. liquid h_f	Evap. h_{fg}	Sat. vapor h_g	Sat. liquid s_f	Sat. vapor s_g
26	6.6954	0.7650	26.14	60.17	180.61	60.68	137.44	198.11	0.2271	0.6865
28	7.0648	0.7694	24.78	62.09	181.36	62.63	136.24	198.87	0.2335	0.6859
30	7.4490	0.7739	23.51	64.01	182.11	64.59	135.03	199.62	0.2400	0.6853
32	7.8485	0.7785	22.31	65.96	182.85	66.57	133.79	200.36	0.2463	0.6847
34	8.2636	0.7832	21.18	67.90	183.59	68.55	132.53	201.09	0.2527	0.6842
36	8.6948	0.7880	20.12	69.86	184.31	70.55	131.25	201.80	0.2591	0.6836
38	9.1423	0.7929	19.12	71.84	185.03	72.56	129.94	202.51	0.2655	0.6831
40	9.6065	0.7980	18.17	73.82	185.74	74.59	128.61	203.20	0.2718	0.6825
42	10.088	0.8033	17.28	75.82	186.45	76.63	127.25	203.88	0.2782	0.6820
44	10.587	0.8086	16.44	77.82	187.13	78.68	125.87	204.54	0.2845	0.6814
48	11.639	.8199	14.88	81.88	188.51	82.83	123.00	205.83	0.2973	0.6802
52	12.766	.8318	13.49	86.00	189.83	87.06	119.99	207.05	0.3101	0.6791
56	13.972	.8445	12.24	90.18	191.10	91.36	116.84	208.20	0.3229	0.6779
60	15.259	.8581	11.11	94.43	192.31	95.74	113.52	209.26	0.3358	0.6765
112	41.155	1.792	1.79	175.98	175.98	183.35	0	183.35	0.5687	0.5687

Source: Based on data supplied by Freon Products Division, E. I. du Pont de Nemours & Company, 1969.

Table A-17M Properties of saturated refrigerant 12, CCl_2F_2: pressure table

v, cm^3/g; u, kJ/kg; h, kJ/kg; s, kJ/(kg)(°K)

		Specific volume		Internal energy		Enthalpy			Entropy	
Press bar(s) P	Temp. °C T	Sat. liquid v_f	Sat. vapor v_g	Sat. liquid u_f	Sat. vapor u_g	Sat. liquid h_f	Evap. h_{fg}	Sat. vapor h_g	Sat. liquid s_f	Sat. vapor s_g
0.6	−41.42	0.6578	257.5	−1.29	153.49	−1.25	170.19	168.94	−0.0054	0.7290
1.0	−30.10	.6719	160.0	8.71	158.15	8.78	165.37	174.15	0.0368	.7171
1.2	−25.74	.6776	134.9	12.58	159.95	12.66	163.48	176.14	.0526	.7133
1.4	−21.91	.6828	116.8	15.99	161.52	16.09	161.78	177.87	.0663	.7102
1.6	−18.49	.6876	103.1	19.07	162.91	19.18	160.23	179.41	.0784	.7076
1.8	−15.38	.6921	92.25	21.86	164.19	21.98	158.82	180.80	.0893	.7054
2.0	−12.53	.6962	83.54	24.43	165.36	24.57	157.50	182.07	.0992	.7035
2.4	−7.42	.7040	70.33	29.06	167.44	29.23	155.09	184.32	.1168	.7004
2.8	−2.93	.7111	60.76	33.15	169.26	33.35	152.92	186.27	.1321	.6980
3.2	1.11	.7177	53.51	36.85	170.88	37.08	150.92	188.00	.1457	6960
4.0	8.15	.7299	43.21	43.35	173.69	43.64	147.33	190.97	.1691	.6928
5.0	15.60	.7438	34.82	50.30	176.61	50.67	143.35	194.02	.1935	.6899
6.0	22.00	.7566	29.13	56.35	179.09	56.80	139.77	196.57	.2142	.6878
7.0	27.65	.7686	25.01	61.75	181.23	62.29	136.45	198.74	.2324	.6860
8.0	32.74	.7802	21.88	66.68	183.13	67.30	133.33	200.63	.2487	.6845
9.0	37:37	.7914	19.42	71.22	184.81	71.93	130.36	202.29	.2634	.6832
10.0	41.64	.8023	17.44	75.46	186.32	76.26	127.50	203.76	.2770	.6820
12.0	49.31	.8237	14.41	83.22	188.95	84.21	122.03	206.24	.3015	.6799
14.0	56.09	.8448	12.22	90.28	191.11	91.46	116.76	208.22	.3232	.6778
16.0	62.19	.8660	10.54	96.80	192.95	98.19	111.62	209.81	.3329	.6758

Source: Based on data supplied by Freon Products Division, E. I. du Pont de Nemours & Company, 1969.

Table A-18M Properties of superheated refrigerant 12 (CCl_2F_2)
v, cm^3/g; *u*, kJ/kg; *h*, kJ/kg; *s*, kJ/(kg)(°K)

Temp. °C	*v*	*u*	*h*	*s*	*v*	*u*	*h*	*s*
	0.6 bar (0.060 MPa)(T_{sat} = −41.42°C)				1.0 bar (0.10 MPa)(T_{sat} = −30.10°C)			
Sat.	257.5	153.49	168.94	0.7290	160.0	158.15	174.15	0.7171
−40	259.3	154.16	169.72	.7324				
−20	283.8	163.91	180.94	.7785	167.7	163.22	179.99	.7406
0	307.9	174.05	192.52	.8225	182.7	173.50	191.77	.7854
10	319.8	179.26	198.45	.8439	190.0	178.77	197.77	.8070
20	331.7	184.57	204.47	.8647	197.3	184.12	203.85	.8281
30	343.5	189.96	210.57	.8852	204.5	189.57	210.02	.8488
40	355.2	195.46	216.77	.9053	211.7	195.09	216.26	.8691
50	367.0	201.02	223.04	.9251	218.8	200.70	222.58	.8889
60	378.7	206.69	229.41	.9444	226.0	206.38	228.98	.9084
80	402.0	218.25	242.37	.9822	240.1	218.00	242.01	.9464
	1.4 bars (0.14 MPa)(T_{sat} = −21.91°C)				1.8 bars (0.18 MPa)(T_{sat} = −15.38°C)			
Sat.	116.8	161.52	177.87	0.7102	92.2	164.20	180.80	0.7054
−20	117.9	162.50	179.01	.7147				
−10	123.5	167.69	184.97	.7378	92.5	164.39	181.03	.7181
0	128.9	172.94	190.99	.7602	99.1	172.37	190.21	.7408
10	134.3	178.28	197.08	.7821	103.4	177.77	196.38	.7630
20	139.7	183.67	203.23	.8035	107.6	183.23	202.60	.7846
30	144.9	189.17	209.46	.8243	111.8	188.77	208.89	.8057
40	150.2	194.72	215.75	.8447	116.0	194.35	215.23	.8263
50	155.3	200.38	222.12	.8648	120.1	200.02	221.64	.8464
60	160.5	206.08	228.55	.8844	124.1	205.78	228.12	.8662
80	170.7	217.74	241.64	.9225	132.2	217.47	241.27	.9045
100	180.9	229.67	255.00	.9593	140.2	229.45	254.69	.9414
	2.0 bars (0.20 MPa)(T_{sat} = −12.53°C)				2.4 bars (0.24 MPa)(T_{sat} = −7.42°C)			
Sat.	83.5	165.37	182.07	0.7035	70.3	167.45	184.32	0.7004
0	88.6	172.08	189.08	.7325	72.9	171.49	188.99	.7177
10	92.6	177.50	196.02	.7548	76.3	176.98	195.29	.7404
20	96.4	183.00	202.28	.7766	79.6	182.53	201.63	.7624
30	100.2	188.56	208.60	.7978	82.8	188.14	208.01	.7838
40	104.0	194.17	214.97	.8184	86.0	193.80	214.44	.8047
50	107.7	199.86	221.40	.8387	89.2	199.51	220.92	.8251
60	111.4	205.62	227.90	.8585	92.3	205.31	227.46	.8450
80	118.7	217.35	241.09	.8969	98.5	217.07	240.71	.8836
100	125.9	229.35	254.53	.9339	104.5	229.12	254.20	.9208
120	133.1	241.59	268.21	.9696	110.5	241.41	267.93	.9566

Table A-18M *(Continued)*

Temp. °C	v	u	h	s	v	u	h	s
	2.8 bars (0.28 MPa)(T_{sat} = −2.93°C)				3.2 bars (0.32 MPa)(T_{sat} = 1.11°C)			
Sat.	60.76	169.26	186.27	0.6980	53.51	170.88	188.00	0.6960
0	61.66	170.89	188.15	.7049				
10	64.64	176.45	194.55	.7279	55.90	175.90	193.79	0.7167
20	67.55	182.06	200.97	.7502	58.52	181.57	200.30	.7393
30	70.40	187.71	207.42	.7718	61.06	187.28	206.82	.7612
40	73.19	193.42	213.91	.7928	63.55	193.02	213.36	.7824
50	75.94	199.18	220.44	.8134	66.00	198.82	219.94	.8031
60	78.65	205.00	227.02	.8334	68.41	204.68	226.57	.8233
80	83.99	216.82	240.34	.8722	73.14	216.55	239.96	.8623
100	89.24	228.29	253.88	.9095	77.78	228.66	253.55	.8997
120	94.43	241.21	267.65	.9455	82.36	241.00	267.36	.9358
	4.0 bars (0.40 MPa)(T_{sat} = 8.15°C)				5.0 bars (0.50 MPa)(T_{sat} = 15.60°C)			
Sat.	43.21	173.69	190.97	0.6928	34.82	176.61	194.02	0.6899
10	43.63	174.76	192.21	.6972				
20	45.84	180.57	198.91	.7204	35.65	179.26	197.08	0.7004
30	47.97	186.39	205.58	.7428	37.46	185.23	203.96	.7235
40	50.05	192.23	212.25	.7645	39.22	191.20	210.81	.7457
50	52.07	198.11	218.94	.7855	40.91	197.19	217.64	.7672
60	54.06	204.03	225.65	.8060	42.57	203.20	224.48	.7881
80	57.91	216.03	239.19	.8454	45.78	215.32	238.21	.8281
100	61.73	228.20	252.89	.8831	48.89	227.61	252.05	.8662
120	65.46	240.61	266.79	.9194	51.93	240.10	266.06	.9028
140	69.13	253.23	280.88	.9544	54.92	252.77	280.23	.9379
	6.0 bars (0.60 MPa)(T_{sat} = 22.00°C)				7.0 bars (0.70 MPa)(T_{sat} = 27.65°C)			
Sat.	29.13	179.09	196.57	0.6878	25.01	181.23	198.74	0.6860
30	30.42	184.01	202.26	.7068	25.35	182.72	200.46	.6917
40	31.97	190.13	209.31	.7297	26.76	189.00	207.73	.7153
50	33.45	196.23	216.30	.7516	28.10	195.23	214.90	.7378
60	34.89	202.34	223.27	.7729	29.39	201.45	222.02	.7595
80	37.65	214.61	237.20	.8135	31.84	213.88	236.17	.8008
100	40.32	227.01	251.20	.8520	34.19	226.40	250.33	.8398
120	42.91	239.57	265.32	.8889	36.46	239.05	264.57	.8769
140	45.45	252.31	279.58	.9243	38.67	251.85	278.92	.9125
160	47.94	265.25	294.01	.9584	40.85	264.83	293.42	.9468

Table A-18M Properties of superheated refrigerant 12 (CCl_2F_2) (*Continued*)

Temp. °C	v	u	h	s	v	u	h	s
	8.0 bars (0.80 MPa)(T_{sat} = 32.74°C)				9.0 bars (0.90 MPa)(T_{sat} = 37.37°C)			
Sat.	21.88	183.13	200.63	0.6845	19.42	184.81	202.29	0.6832
40	22.83	187.81	206.07	.7021	19.74	186.55	204.32	.6897
50	24.07	194.19	213.45	.7253	20.91	193.10	211.92	.7136
60	25.25	200.52	220.72	.7474	22.01	199.56	219.37	.7363
80	27.48	213.13	235.11	.7894	24.07	212.37	234.03	.7790
100	29.59	225.77	249.44	.8289	26.01	225.13	248.54	.8190
120	31.62	238.51	263.81	.8664	27.85	237.97	263.03	.8569
140	33.59	251.39	278.26	.9022	29.64	250.90	277.58	.8930
160	35.52	264.41	292.83	.9367	31.38	263.99	292.23	.9276
180	37.42	277.60	307.54	.9699	33.09	277.23	307.01	.9609
	10.0 bars (1.0 MPa)(T_{sat} = 41.64°C)				12.0 bars (1.2 MPa)(T_{sat} = 49.31°C)			
Sat.	17.44	186.32	203.76	0.6820	14.41	188.95	206.24	0.6799
50	18.37	191.95	210.32	.7026	14.48	189.43	206.81	.6816
60	19.41	198.56	217.97	.7259	15.46	196.41	214.96	.7065
80	21.34	211.57	232.91	.7695	17.22	209.91	230.57	.7520
100	23.13	224.48	247.61	.8100	18.81	223.13	245.70	.7937
120	24.84	237.41	262.25	.8482	20.30	236.27	260.63	.8326
140	26.47	250.43	276.90	.8845	21.72	249.45	275.51	.8696
160	28.07	263.56	291.63	.9193	23.09	263.70	290.41	.9048
180	29.63	276.84	306.47	.9528	24.43	276.05	305.37	.9385
200	31.16	290.26	321.42	.9851	25.74	289.55	320.44	.9711
	14.0 bars (1.4 MPa)(T_{sat} = 56.09°C)				16.0 bars (1.6 MPa)(T_{sat} = 62.19°C)			
Sat.	12.22	191.11	208.22	0.6778	10.54	192.95	209.81	0.6758
60	12.58	194.00	211.61	.6881				
80	14.25	208.11	228.06	.7360	11.98	206.17	225.34	0.7209
100	15.71	221.70	243.69	.7791	13.37	220.19	241.58	.7656
120	17.05	235.09	258.96	.8189	14.61	233.84	257.22	.8065
140	18.32	248.43	274.08	.8564	15.77	247.38	272.61	.8447
160	19.54	261.80	289.16	.8921	16.86	260.90	287.88	.8808
180	20.71	275.27	304.26	.9262	17.92	274.47	303.14	.9152
200	21.86	288.84	319.44	.9589	18.95	288.11	318.43	.9482
220	22.99	302.51	334.70	.9905	19.96	301.84	333.78	.9800

Source: Based on data supplied by Freon Products Division, E. I. du Pont de Nemours & Company, 1969.

Table A-19M Specific heats of some common liquids and solids

c_p, kJ/(kg)(°C)

A. Liquids

Substance	State	c_p	Substance	State	c_p
Water	1 atm, 273°K	4.217	Benzene	1 atm, 15°C	1.80
	1 atm, 280°K	4.198		1 atm, 65°C	1.92
	1 atm, 300°K	4.179	Glycerin	1 atm, 10°C	2.32
	1 atm, 320°K	4.180		1 atm, 50°C	2.58
	1 atm, 340°K	4.188	Mercury	1 atm, 10°C	0.138
	1 atm, 360°K	4.203		1 atm, 315°C	0.134
	1 atm, 373°K	4.218	Sodium	1 atm, 95°C	1.38
Ammonia	sat., −20°C	4.52		1 atm, 540°C	1.26
	sat., 50°C	5.10	Propane	1 atm, 0°C	2.41
Refrigerant 12	sat., −40°C	0.883	Bismuth	1 atm, 425°C	0.144
	sat., −20°C	0.908		1 atm, 760°C	0.164
	sat., 50°C	1.02	Ethyl alcohol	1 atm, 25°C	2.43

B. Solids

Substance	Temp.	c_p	Substance	Temp.	c_p
Ice	200°K	1.56	Silver	20°C	0.233
	220°K	1.71		200°C	0.243
	240°K	1.86	Lead	−173°C	0.118
	260°K	2.01		−50°C	0.126
	270°K	2.08		27°C	0.129
	273°K	2.11		100°C	0.131
Aluminum	200°K	0.797		200°C	0.136
	250°K	0.859	Copper	−173°C	0.254
	300°K	0.902		−100°C	0.342
	350°K	0.929		−50°C	0.367
	400°K	0.949		0°C	0.381
	450°K	0.973		27°C	0.386
	500°K	0.997		100°C	0.393
Iron	20°K	0.448		200°C	0.403

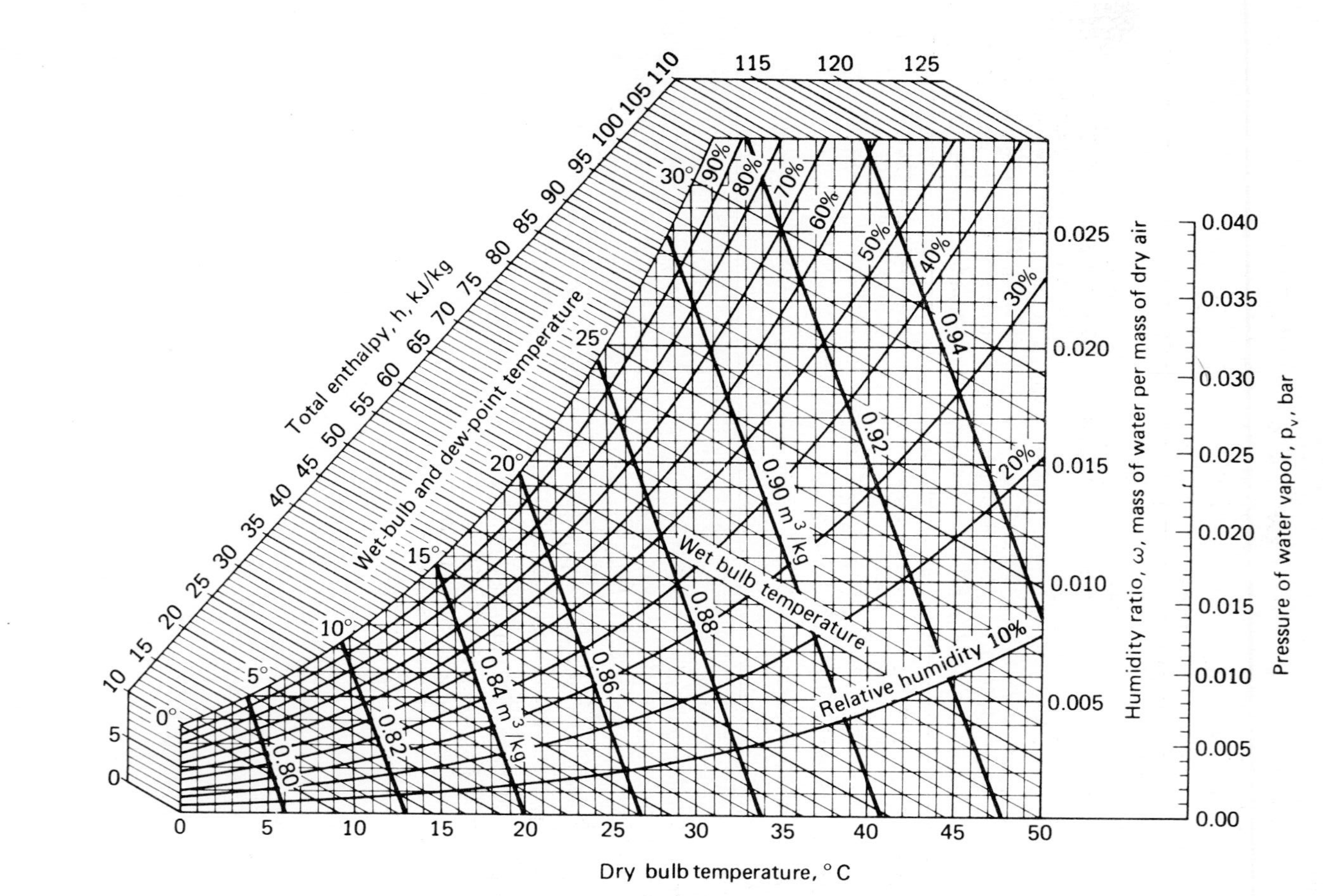

Figure A-20M Psychrometric chart, metric units, barometric pressure 1.01 bar.

Table A-21M Constants for the Benedict-Webb-Rubin and Redlich-Kwong equations of state

1. Benedict-Webb-Rubin; Units are bar(s), $m^3/kg \cdot mol$, and °K

Constants	*n*-Butane, C_4H_{10}	Carbon dioxide, CO_2	Carbon monoxide, CO	Methane, CH_4	Nitrogen, N_2
a	1.9068	0.1386	0.0371	0.0500	0.0254
A_0	10.216	2.7730	1.3587	1.8791	1.0673
b	0.039998	0.007210	0.002632	0.003380	0.002328
B_0	0.12436	0.04991	0.05454	0.04260	0.04074
c	3.205×10^5	1.511×10^4	1.054×10^3	2.578×10^3	7.379×10^2
C_0	1.006×10^6	1.404×10^5	8.673×10^3	2.286×10^4	8.164×10^3
α	1.101×10^{-3}	8.470×10^{-5}	1.350×10^{-4}	1.244×10^{-4}	1.272×10^{-4}
γ	0.0340	0.00539	0.0060	0.0060	0.0053

Source: Cooper, H. W., and J. C. Goldfrank, *Hydrocarbon Processing*, ***46*** (12): 141 (1967).

2. Redlich-Kwong: Units are bar(s), $m^3/kg \cdot mol$, and °K

Substance	a	b
Carbon dioxide, CO_2	64.64	0.02969
Carbon monoxide, CO	17.26	0.02743
Methane, CH_4	32.19	0.02969
Nitrogen, N_2	15.59	0.02681
Oxygen, O_2	17.38	0.02199
Propane, C_3H_8	183.07	0.06269
Refrigerant 12, CCL_2F_2	214.03	0.06913
Sulfur dioxide, SO_2	144.49	0.03939
Water, H_2O	142.64	0.02110

Source: Computed from critical data.

Table A-22M Values of the enthalpy of formation, Gibbs function of formation, absolute entropy, and enthalpy of vaporization at 25°C and 1 atm

Δh_f^0, Δg_f^0, and h_{fg} in kJ/kg · mol and s^0 in kJ/(kg · mol)(°K)

Substance	Formula	Δh_f^0	Δg_f^0	s^0	h_{fg}
Carbon	$C(s)$	0	0	5.74	
Hydrogen	$H_2(g)$	0	0	130.57	
Nitrogen	$N_2(g)$	0	0	191.50	
Oxygen	$O_2(g)$	0	0	205.04	
Carbon monoxide	$CO(g)$	−110,530	−137,150	197.56	
Carbon dioxide	$CO_2(g)$	−393,520	−394,380	213.67	
Water	$H_2O(g)$	−241,820	−228,590	188.72	
Water	$H_2O(l)$	−285,830	−237,180	69.95	44,010
Hydrogen peroxide	$H_2O_2(g)$	−136,310	−105,600	232.63	61,090
Ammonia	$NH_3(g)$	−46,190	−16,590	192.33	
Oxygen	$O(g)$	249,170	231,770	160.95	
Hydrogen	$H(g)$	218,000	203,290	114.61	
Nitrogen	$N(g)$	472,680	455,510	153.19	
Hydroxyl	$OH(g)$	39,460	34,280	183.75	
Methane	$CH_4(g)$	−74,850	−50,790	186.16	
Acetylene (Ethyne)	$C_2H_2(g)$	226,730	209,170	200.85	
Ethylene (Ethene)	$C_2H_4(g)$	52,280	68,120	219.83	
Ethane	$C_2H_6(g)$	−84,680	−32,890	229.49	
Propylene (Propene)	$C_3H_6(g)$	20,410	62,720	266.94	
Propane	$C_3H_8(g)$	−103,850	−23,490	269.91	15,060
n-Butane	$C_4H_{10}(g)$	−126,150	−15,710	310.03	21,060
n-Pentane	$C_5H_{12}(g)$	−146,440	−8,200	348.40	31,410
n-Octane	$C_8H_{18}(g)$	−208,450	17,320	463.67	41,460
n-Octane	$C_8H_{18}(l)$	−249,910	6,610	360.79	
Benzene	$C_6H_6(g)$	82,930	129,660	269.20	33,830
Methyl alcohol	$CH_3OH(g)$	−200,890	−162,140	239.70	37,900
Methyl alcohol	$CH_3OH(l)$	−238,810	−166,290	126.80	
Ethyl alcohol	$C_2H_5OH(g)$	−235,310	−168,570	282.59	42,340
Ethyl alcohol	$C_2H_5OH(l)$	−277,690	−174,890	160.70	
Mercury	$Hg(l)$	0	0	77.24	
Mecuric oxide	$HgO(c)$	−90,210	−58,400	70.45	
Manganese	$Mn(c)$	0	0	31.8	
Manganese dioxide	$MnO_2(c)$	−520,030	−465,180	53.14	
Manganese trioxide	$Mn_2O_3(c)$	−958,970	−881,150	110.5	
Lead	$Pb(c)$	0	0	64.81	
Lead oxide	$PbO_2(c)$	−277,400	−217,360	68.6	
Lead sulfate	$PbSO_4(c)$	−919,940	−813,200	148.57	
Zinc	$Zn(c)$	0	0	41.63	
Zinc oxide	$ZnO(c)$	−348,280	−318,320	43.64	
Sulfuric acid	$H_2SO_4(l)$	−813,990	−690,100	156.90	
Sulfuric acid	(aq, m = 1)	−909,270	−744,630	20.1	
Silver oxide	$Ag_2O(c)$	−31,050	−11,200	121.7	

Sources: From the JANAF Thermochemical Tables, Dow Chemical Co., 1971; *Selected Values of Chemical Thermodynamic Properties*, NBS Tech. Note 270-3, 1968; and *API Research Project 44*, Carnegie Press, 1953.

Table A-23M Thermodynamic properties of potassium

T, °K; P, atm; v, L/kg; h, kJ/kg; s, kJ/(kg)(°K)

A. Saturation temperature table

T	P	v_f	v_g	h_f	h_g	s_f	s_g
900	0.251	1.438	7180	731.0	2739.7	2.6874	4.9175
950	0.447	1.462	4204	771.4	2750.4	2.7313	4.8129
1000	0.753	1.493	2592	812.6	2760.4	2.7732	4.7196

B. Superheat table

P	v	h	s	P	v	h	s
1075°K (1.494 atm)				1200°K (3.860 atm)			
Sat.	1.375	2772.9	4.5977	Sat.	0.573	2795.1	4.4367
1.0	2.120	2813.6	4.7144	3.0	0.759	2831.1	4.5144
0.8	2.684	2830.0	4.7745	2.0	1.178	2874.0	4.6295
1325°K (8.293 atm)				1450°K (15.55 atm)			
Sat.	0.285	2820.4	4.3163	Sat.	0.161	2849.4	4.2255
6.0	0.411	2873.7	4.4162	10.0	0.267	2928.5	4.3599
4.0	0.641	2923.1	4.5316	8.0	0.343	2959.2	4.4233
3.0	0.873	2948.7	4.6078	6.0	0.469	2991.3	4.5011

Source: Data adapted from Naval Research Laboratory Report 6233, 1965, and Air Force Aero Propulsion Laboratory Technical Report 66-104, 1966.

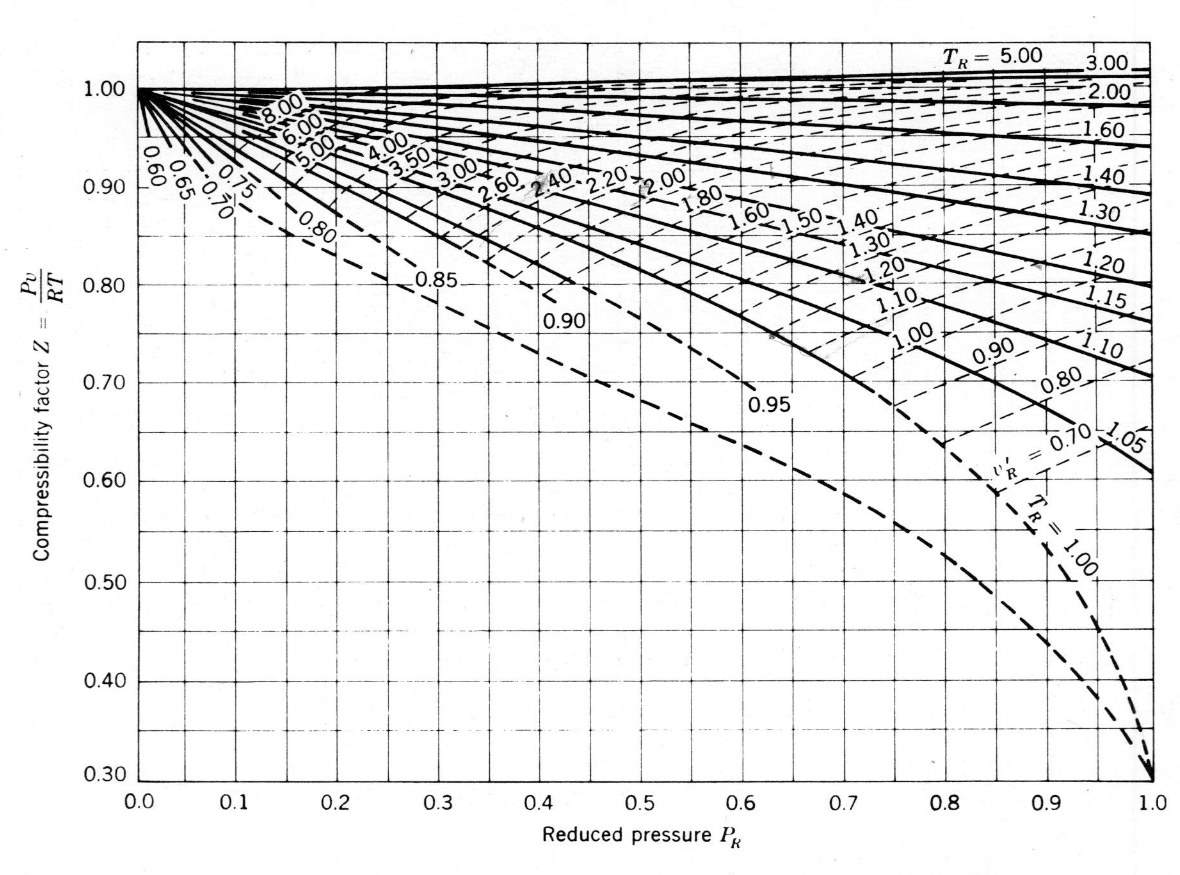

Figure A-24M Generalized compressibility chart [L. C. Nelson and E. F. Obert: Generalized Compressibility Charts, *Chem. Eng.*, **61**:203 (1954).]

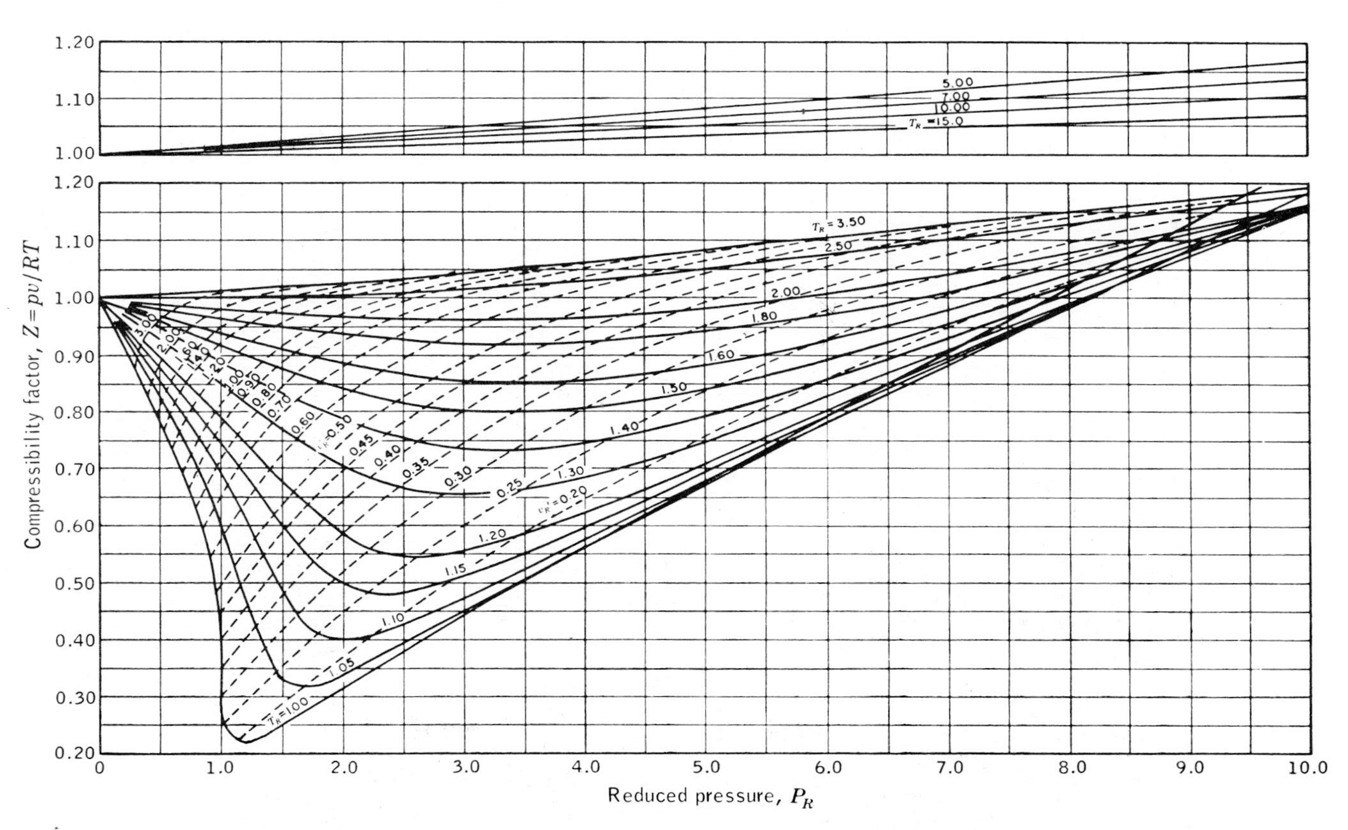

Figure A-25M Compressibility chart, low-pressure range. *Source*: V. M. Faires, "Problems on Thermodynamics," Macmillan, New York, 1962. Data from L. C. Nelson and E. F. Obert, Generalized Compressibility Charts, *Chem. Eng.* **61**:203 (1954).

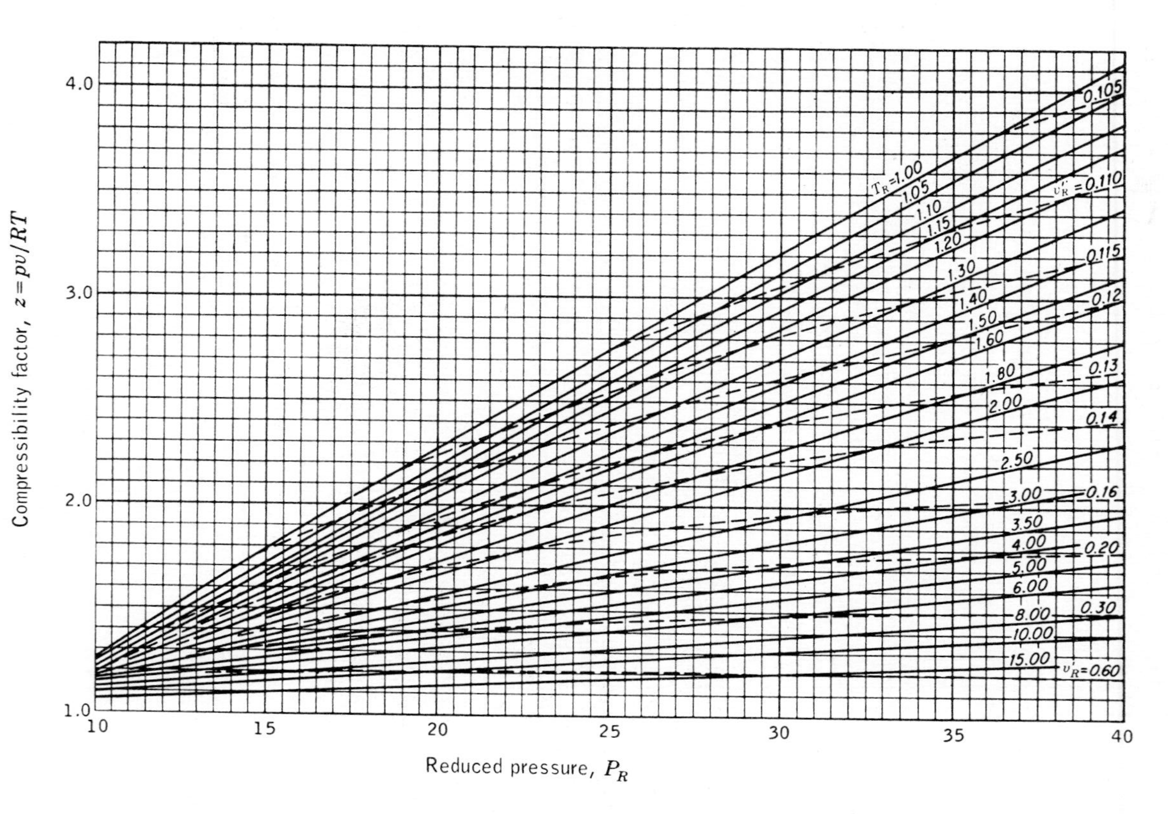

Figure A-26M Compressibility chart, high-pressure range. Adapted from E. F. Obert, "Concepts of Thermodynamics," McGraw-Hill, New York, 1960

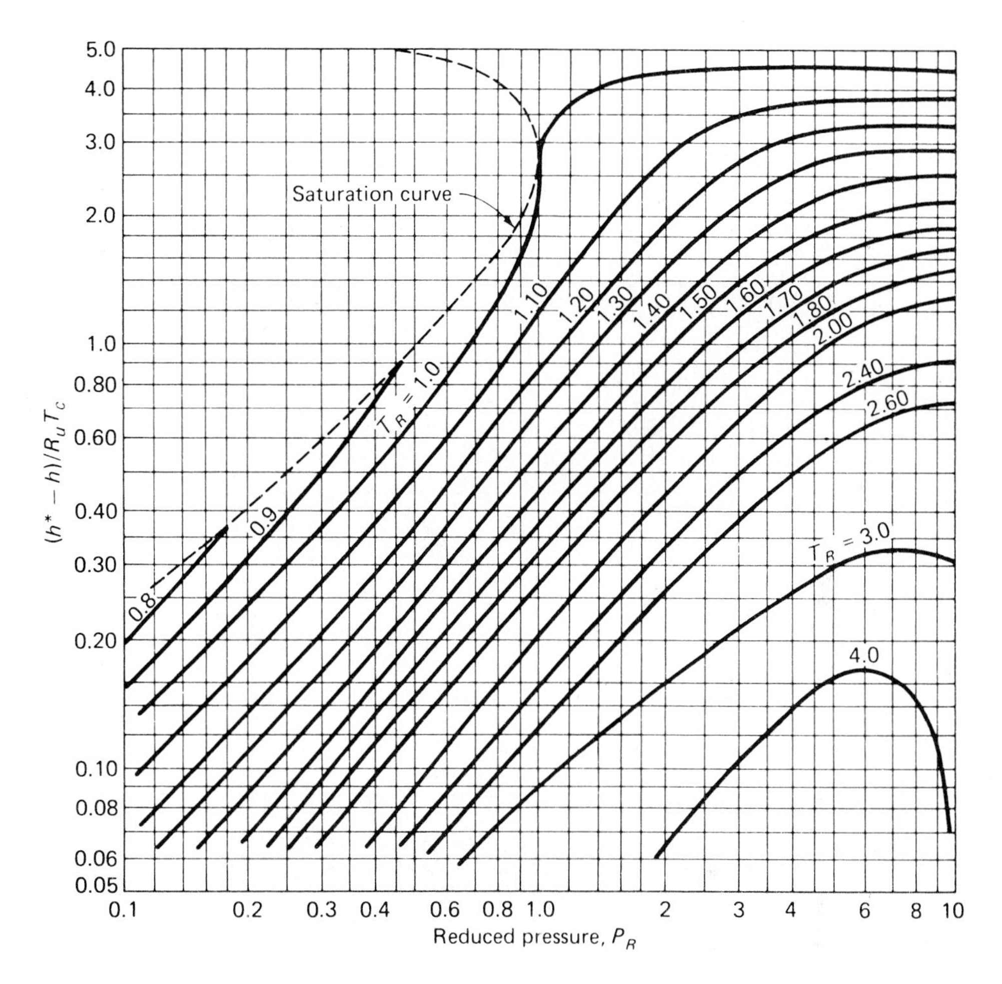

Figure A-27M Generalized enthalpy chart. *Source*: Based on data from A. L. Lydersen, R. A. Greenkorn, and O. A. Hougen, "Engineering Experiment Station Report No. 4," University of Wisconsin, 1955. $T_R = T/T_c$ = reduced temperature, $P_R = P/P_c$ = reduced pressure, T_c = critical temperature, P_c = critical pressure, h^* = enthalpy of an ideal gas, h = enthalpy of an actual gas.

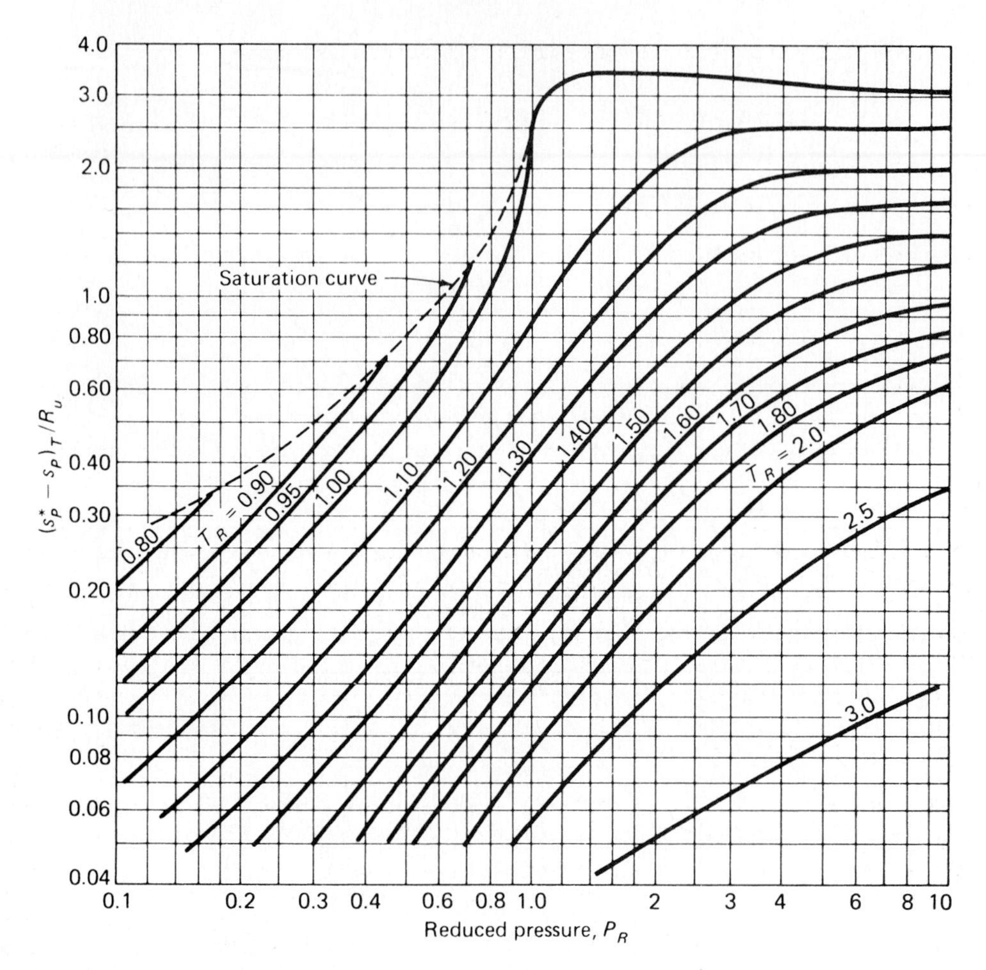

Figure A-28M Generalized entropy chart. *Source:* Based on data from A. L. Lydersen, R. A. Greenkorn, and O. A. Hougen, "Engineering Experiment Station Report No. 4," University of Wisconsin, 1955. $T_R = T/T_c$ = reduced temperature, $P_R = P/P_c$ = reduced pressure, T_c = critical temperature, P_c = critical pressure, s_P^* = entropy of an ideal gas, s_P = entropy of an actual gas.

Table A-29M Logarithms to the base 10 of the equilibrium constant K_p

$$K_p = \frac{(p_E)^{\nu_E}(p_F)^{\nu_F}}{(p_A)^{\nu_A}(p_B)^{\nu_B}} \text{ for the reaction } \nu_A A + \nu_B B \leftrightarrows \nu_E E + \nu_F F$$

Numbered reactions

(1) $H_2 \leftrightarrows 2H$

(2) $O_2 \leftrightarrows 2O$

(3) $N_2 \leftrightarrows 2N$

(4) $\frac{1}{2}O_2 + \frac{1}{2}N_2 \leftrightarrows NO$

(5) $H_2O \leftrightarrows H_2 + \frac{1}{2}O_2$

(6) $H_2O \leftrightarrows OH + \frac{1}{2}H_2$

(7) $CO_2 \leftrightarrows CO + \frac{1}{2}O_2$

(8) $CO_2 + H_2 \leftrightarrows CO + H_2O$

Temp. °K	$\log_{10} K_p$ values for reactions numbered above							
	(1)	(2)	(3)	(4)	(5)	(6)	(7)	(8)
298	−71.224	−81.208	−159.600	−15.171	−40.048	−46.054	−45.066	−5.018
500	−40.316	−45.880	−92.672	−8.783	−22.886	−26.130	−25.025	−2.139
1000	−17.292	−19.614	−43.056	−4.062	−10.062	−11.280	−10.221	−0.159
1200	−13.414	−15.208	−34.754	−3.275	−7.899	−8.811	−7.764	+0.135
1400	−10.630	−12.054	−28.812	−2.712	−6.347	−7.021	−6.014	+0.333
1600	−8.532	−9.684	−24.350	−2.290	−5.180	−5.677	−4.706	+0.474
1700	−7.666	−8.706	−22.512	−2.116	−4.699	−5.124	−4.169	+0.530
1800	−6.896	−7.836	−20.874	−1.962	−4.270	−4.613	−3.693	+0.577
1900	−6.204	−7.058	−19.410	−1.823	−3.886	−4.190	−3.267	+0.619
2000	−5.580	−6.356	−18.092	−1.699	−3.540	−3.776	−2.884	+0.656
2100	−5.016	−5.720	−16.898	−1.586	−3.227	−3.434	−2.539	+0.688
2200	−4.502	−5.142	−15.810	−1.484	−2.942	−3.091	−2.226	+0.716
2300	−4.032	−4.614	−14.818	−1.391	−2.682	−2.809	−1.940	+0.742
2400	−3.600	−4.130	−13.908	−1.305	−2.443	−2.520	−1.679	+0.764
2500	−3.202	−3.684	−13.070	−1.227	−2.224	−2.270	−1.440	+0.784
2600	−2.836	−3.272	−12.298	−1.154	−2.021	−2.038	−1.219	+0.802
2700	−2.494	−2.892	−11.580	−1.087	−1.833	−1.823	−1.015	+0.818
2800	−2.178	−2.536	−10.914	−1.025	−1.658	−1.624	−0.825	+0.833
2900	−1.882	−2.206	−10.294	−0.967	−1.495	−1.438	−0.649	+0.846
3000	−1.606	−1.898	−9.716	−0.913	−1.343	−1.265	−0.485	+0.858
3100	−1.348	−1.610	−9.174	−0.863	−1.201	−1.103	−0.332	+0.869
3200	−1.106	−1.340	−8.664	−0.815	−1.067	−0.951	−0.189	+0.878
3300	−0.878	−1.086	−8.186	−0.771	−0.942	−0.809	−0.054	+0.888
3400	−0.664	−0.846	−7.736	−0.729	−0.824	−0.674	+0.071	+0.895
3500	−0.462	−0.620	−7.312	−0.690	−0.712	−0.547	+0.190	+0.902

Source: Based on data from the JANAF Tables, NSRDS-NBS-37, 1971.

Table A-30M Properties of saturated nitrogen (N_2): temperature and pressure tables

v, L/kg or cm^3/g; u and h, kJ/kg; s, kJ/(kg)(°K)

		Specific volume		Internal energy		Enthalpy			Entropy	
Temp. °K T	Press. bars P	Sat. liquid v_f	Sat. vapor v_g	Sat. liquid u_f	Sat. vapor u_g	Sat. liquid h_f	Evap. h_{fg}	Sat. vapor h_g	Sat. liquid s_f	Sat. vapor s_g
63.15	0.125	1.152	1477.	−50.8	145.6	−50.8	214.9	164.1	2.423	5.826
70.0	0.386	1.191	526.	−37.1	150.2	−37.0	207.5	170.5	2.631	5.595
80.0	1.369	1.257	164.	−16.4	156.4	−16.2	195.1	178.9	2.906	5.345
90.0	3.600	1.340	66.3	4.5	161.1	5.0	180.0	185.0	3.152	5.152
100.0	7.775	1.448	31.3	25.7	163.4	26.8	160.9	187.7	3.376	4.985
110.0	14.67	1.606	16.0	48.3	162.2	50.7	134.9	185.6	3.594	4.820
120.0	25.15	1.908	8.03	76.9	154.1	81.7	92.6	174.3	3.847	4.619
126.25	33.96	3.388	3.39	123.2	123.2	134.7	0	134.7	4.257	4.257
77.24	1.00	1.238	219.1	−22.0	154.9	−21.9	198.7	176.8	2.835	5.407
83.63	2.00	1.285	115.3	−8.9	158.3	−8.6	190.0	181.4	2.998	5.270
91.25	4.00	1.352	60.0	7.1	161.5	7.6	177.9	185.5	3.180	5.130
96.41	6.00	1.406	40.46	18.0	162.9	18.8	168.4	187.2	3.297	5.044
100.41	8.00	1.454	30.37	26.4	163.4	27.6	160.1	187.7	3.384	4.978
103.76	10.00	1.500	24.16	33.9	163.4	35.4	152.2	187.6	3.457	4.924
110.38	15.00	1.614	15.57	49.3	162.0	51.7	133.7	185.4	3.603	4.814
115.56	20.00	1.741	11.04	63.0	159.1	66.5	114.7	181.2	3.726	4.718
119.88	25.00	1.902	8.11	76.5	154.3	81.3	93.3	174.6	3.844	4.622
123.61	30.00	2.152	5.87	91.6	146.3	98.1	65.8	163.9	3.974	4.506
125.63	33.00	2.497	4.550	104.7	137.1	112.9	39.2	152.1	4.088	4.400

Source: Data adapted from A. A. Vasserman, Ya. Z. Kazavchinskii, and V. A. Rabinovich, "Thermophysical Properties of Air and Air Components," Izdatel' stvo Nauka, Moscow, 1966.

Table A-31M Properties of nitrogen (N_2): superheated-vapor table

v, L/kg or cm^3/g; u and h, kJ/kg; s, kJ/(kg)(°K)

Temp. °K	v	u	h	s	v	u	h	s
	1 bar (0.1 MPa) (T_{sat} = 77.2°K)				5 bars (0.5 MPa)(T_{sat} = 94.0°K)			
100	290.8	172.8	201.9	5.691	52.9	167.7	194.1	5.162
150	442.6	210.4	210.4	6.120	86.4	208.1	251.3	5.627
200	592.3	247.8	307.0	6.421	117.5	246.4	305.1	5.937
250	741.4	285.0	359.1	6.654	147.8	284.0	357.9	6.172
300	890.2	322.2	411.2	6.844	177.9	321.4	410.3	6.364
350	1039.	359.4	463.3	7.005	207.9	358.8	462.7	6.525
400	1187.	396.7	515.4	7.144	237.7	396.2	515.0	6.665
	10 bars (1.0 MPa)(T_{sat} = 108.8°K)				20 bars (2.0 MPa)(T_{sat} = 115.6°K)			
150	41.9	205.0	246.9	5.401	19.6	198.4	237.5	5.150
200	58.1	244.6	302.7	5.722	28.5	240.9	297.8	5.498
250	73.6	282.7	356.3	5.962	36.6	280.1	353.2	5.746
300	88.9	320.3	409.2	6.155	44.4	318.3	407.1	5.942
350	104.0	357.9	461.9	6.317	52.1	356.2	460.4	6.107
400	119.0	395.5	514.5	6.457	59.7	394.1	513.5	6.248
450	134.0	433.2	567.2	6.582	67.3	432.0	566.5	6.373
	50 bars (5.0 MPa)				100 bars (10.0 MPa)			
150	5.92	171.5	201.1	4.692	2.39	123.5	147.4	4.218
200	10.7	229.3	282.9	5.169	5.02	208.8	259.0	4.870
250	14.4	272.2	344.1	5.443	7.12	259.3	330.4	5.190
300	17.8	312.4	401.1	5.651	8.95	302.7	392.2	5.416
350	21.0	351.4	456.3	5.821	10.7	343.6	450.3	5.596
400	24.1	390.1	510.7	5.967	12.3	383.6	506.8	5.746
450	27.2	428.7	564.8	6.094	13.9	423.1	562.4	5.877
	150 bars (15.0 MPa)				200 bars (20.0 MPa)			
150	1.95	94.3	134.9	4.064	1.78	96.1	131.6	3.980
200	3.37	190.8	241.3	4.680	2.69	177.2	231.0	4.554
250	4.80	247.1	319.2	5.030	3.73	236.6	311.2	4.913
300	6.09	293.5	384.8	5.269	4.70	285.3	379.3	5.162
350	7.28	336.3	445.5	5.456	5.62	329.5	441.8	5.355
400	8.42	377.5	503.7	5.612	6.49	371.8	501.5	5.515
450	9.52	418.0	560.7	5.746	7.33	413.0	559.6	5.651

Source: Data adapted from A. A. Vasserman, Ya. Z. Kazavchinskii, and V. A. Rabinovich, "Thermophysical Properties of Air and Air Components," Izdatel' stvo Nauka, Moscow, 1966.

APPENDIX

A-4

SUPPLEMENTARY TABLES AND FIGURES (USCS UNITS)

Table A-1 Physical constants and conversion factors

	Physical constants
Avogadro's number	$N_A = 6.023 \times 10^{23}$ atoms/g · mol
Universal gas constant	$R_u = 0.08205$ L · atm/(g · mol)(°K)
	$= 8.314$ J/(g · mol)(°K)
	$= 0.08314$ bar · m^3/(kg · mol)(°K)
	$= 1545$ ft · lb$_f$/(lb · mol)(°R)
	$= 1.986$ Btu/(lb · mol)(°R)
	$= 0.730$ atm · ft^3/(lb · mol)(°R)
	$= 10.73$ psia · ft^3/(lb · mol)(°R)
Planck's constant	$h = 6.626 \times 10^{-34}$ J · s/molecule
Boltzmann's constant	$k = 1.380 \times 10^{-23}$ J/(°K)(molecule)
Speed of light	$c = 2.998 \times 10^{10}$ cm/s
	Conversion factors
Length	1 cm = 0.3937 in = 10^4 μm = 10^8 Å
	1 in = 2.540 cm
	1 ft = 30.48 cm
Mass	1 lb$_m$ = 453.59 g$_m$ = 7000 grains
	1 kg$_m$ = 2.205 lb$_m$
Force	1 lb$_f$ = 32.174 lb$_m$ · ft/s^2
	1 dyne = 1 g$_m$ · cm/s^2
	1 N = 1 kg$_m$ · m/s^2
	1 lb$_f$ = 444,800 dyne = 4.448 N

Table A-1 (*Continued*)

Pressure	1 lb_f/in^2 = 2.036 inHg at 32°F
	1 inHg = 33,864 $dyne/cm^2$ = 0.0334 atm = 0.491 lb_f/in^2
	1 atm = 14.696 lb_f/in^2 = 760 mmHg at 32°F
	= 29.92 inHg at 32°F
	1 bar = 0.9869 atm = 10^5 N/m^2 = 100 kPa
	1 torr = 1 mmHg at 0°C = 10^3 μm Hg
	= 1.933×10^{-2} psi
Volume	1L = 0.0353 ft^3 = 0.2642 gal = 61.025 in^3
	1 ft^3 = 28.316 L = 7.4805 gal = 0.02832 m^3
	1 in^3 = 16.387 cm^3
Density	1 lb_m/ft^3 = 0.01602 g_m/cm^3
Energy	1 Btu = 778.16 ft · lb_f = 252.16 cal = 1055 J
	1 Btu/lb = 2.32 kJ/kg
	1 J = 1 N · m = 1 V · C
	1 eV = 1.602×10^{-19} J
Power	1 W = 1 J/s = 860.42 cal/h = 3.413 Btu/h
	1 hp = 746 W = 550 ft · lb_f/s = 2545 Btu/h
Velocity	1 m/h = 0.447 m/s

Table A-2 Values of the molar mass (molecular weight) of some common elements and compounds

Substance	Formula	Molar mass
Argon	Ar	39.94
Aluminum	Al	26.97
Carbon	C	12.01
Copper	Cu	63.54
Helium	He	4.003
Hydrogen	H_2	2.016
Iron	Fe	55.85
Lead	Pb	207.2
Mercury	Hg	200.6
Nitrogen	N_2	28.008
Oxygen	O_2	32.00
Potassium	K	39.096
Silver	Ag	107.88
Sodium	Na	22.997
Air		28.97
Carbon monoxide	CO	28.01
Carbon dioxide	CO_2	44.01
Water	H_2O	18.02
Hydrogen peroxide	H_2O_2	34.02
Ammonia	NH_3	17.04
Hydroxyl	—OH	17.01
Methane	CH_4	16.04
Acetylene	C_2H_2	26.04
Ethylene	C_2H_4	28.05
Ethane	C_2H_6	30.07
Propylene	C_3H_6	42.08
Propane	C_3H_8	44.09
n-Butane	C_4H_{10}	58.12
n-Pentane	C_5H_{12}	72.15
n-Octane	C_8H_{18}	114.22
Benzene	C_6H_6	78.11
Methyl alcohol	CH_3OH	32.05
Ethyl alcohol	C_2H_5OH	46.07
Refrigerant 12	CCl_2F_2	120.92

Table A-3 Critical properties and van der Waals constants

Substance	T_c, °R	P_c, atm	$\bar{v}_c$, ft³/lb · mol	$Z_c = \frac{p_c v_c}{RT_c}$	van der Waals a, atm(ft³/lb · mol)²	van der Waals b, ft³/lb · mol
Acetylene (C_2H_2)	556	62	1.80	0.274	1121	0.818
Air (equivalent)	239	37.2	1.33	0.284	345.2	0.585
Ammonia (NH_3)	730	111.3	1.16	0.242	1076	0.598
Benzene (C_6H_6)	1013	48.7	4.11	0.274	4736	1.896
n-Butane (C_4H_{10})	765	37.5	4.13	0.274	3508	1.919
Carbon dioxide (CO_2)	548	72.9	1.51	0.276	926	0.686
Carbon monoxide (CO)	239	34.5	1.49	0.294	372	0.632
Refrigerant 12 (CCl_2F_2)	693	39.6	3.43	0.270	2718	1.595
Ethane (C_2H_6)	549	48.2	3.55	0.273	1410	1.041
Ethylene (C_2H_4)	510	50.5	2.29	0.284	1158	0.922
Helium (He)	9.33	2.26	0.93	0.300	8.66	0.376
n-Heptane (C_7H_{16})	972	27	6.86	0.26	7866	3.298
Hydrogen (H_2)	59.8	12.8	1.04	0.304	62.8	0.426
Methane (CH_4)	344	45.8	1.59	0.290	581	0.685
Methyl chloride (CH_3Cl)	749	65.8	2.29	0.276	1917	1.040
Nitrogen (N_2)	227	33.5	1.44	0.291	346	0.618
Nonane (C_9H_{20})	1071	22.86	8.86	0.250		
n-Octane (C_8H_{18})	1025	24.6	7.82	0.258	9601	3.76
Oxygen (O_2)	278	49.8	1.19	0.290	348	0.506
Propane (C_3H_8)	666	42.1	3.13	0.276	2368	1.445
Sulfur dioxide (SO_2)	775	77.7	1.99	0.268	1738	0.911
Water (H_2O)	1165	218.0	0.896	0.230	1400	0.488

Note: Most of the values of T_c and p_c are adapted from Nelson and E. F. Obert, Generalized Compressibility Charts, *Chem. Eng.*, **61**:203 (1954); others from International Critical Tables. Most values of $\bar{v}_c$ are from Physical Tables, Smithsonian Institution, 1954; others from International Critical Tables. The values of Z_c, a, and b are computed from these critical data; $a = 27R^2T_c^2/64P_c$ and $b = RT_c/8P_c$.

Source: Faires, V. M., "Problems on Thermodynamics," 4th ed., Macmillan, New York, 1962.

Table A-4 Ideal-gas specific-heat data for various gases, Btu/(lb)(°F)

1. Zero-pressure specific heats for six common gases

Temp. °F	Air c_p	Air c_v	Air k	Carbon dioxide, CO_2 c_p	CO_2 c_v	CO_2 k	Carbon monoxide, CO c_p	CO c_v	CO k	Temp. °F
40	0.240	0.171	1.401	0.195	0.150	1.300	0.248	0.177	1.400	40
100	0.240	0.172	1.400	0.205	0.160	1.283	0.249	0.178	1.399	100
200	0.241	0.173	1.397	0.217	0.172	1.262	0.249	0.179	1.397	200
300	0.243	0.174	1.394	0.229	0.184	1.246	0.251	0.180	1.394	300
400	0.245	0.176	1.389	0.239	0.193	1.233	0.253	0.182	1.389	400
500	0.248	0.179	1.383	0.247	0.202	1.223	0.256	0.185	1.384	500
600	0.250	0.182	1.377	0.255	0.210	1.215	0.259	0.188	1.377	600
700	0.254	0.185	1.371	0.262	0.217	1.208	0.262	0.191	1.371	700
800	0.257	0.188	1.365	0.269	0.224	1.202	0.266	0.195	1.364	800
900	0.259	0.191	1.358	0.275	0.230	1.197	0.269	0.198	1.357	900
1000	0.263	0.195	1.353	0.280	0.235	1.192	0.273	0.202	1.351	1000
1500	0.276	0.208	1.330	0.298	0.253	1.178	0.287	0.216	1.328	1500
2000	0.286	0.217	1.312	0.312	0.267	1.169	0.297	0.226	1.314	2000

Temp. °F	Hydrogen, H_2 c_p	H_2 c_v	H_2 k	Nitrogen, N_2 c_p	N_2 c_v	N_2 k	Oxygen, O_2 c_p	O_2 c_v	O_2 k	Temp. °F
40	3.397	2.412	1.409	0.248	0.177	1.400	0.219	0.156	1.397	40
100	3.426	2.441	1.404	0.248	0.178	1.399	0.220	0.158	1.394	100
200	3.451	2.466	1.399	0.249	0.178	1.398	0.223	0.161	1.387	200
300	3.461	2.476	1.398	0.250	0.179	1.396	0.226	0.164	1.378	300
400	3.466	2.480	1.397	0.251	0.180	1.393	0.230	0.168	1.368	400
500	3.469	2.484	1.397	0.254	0.183	1.388	0.235	0.173	1.360	500
600	3.473	2.488	1.396	0.256	0.185	1.383	0.239	0.177	1.352	600
700	3.477	2.492	1.395	0.260	0.189	1.377	0.242	0.181	1.344	700
800	3.494	2.509	1.393	0.262	0.191	1.371	0.246	0.184	1.337	800
900	3.502	2.519	1.392	0.265	0.194	1.364	0.249	0.187	1.331	900
1000	3.513	2.528	1.390	0.269	0.198	1.359	0.252	0.190	1.326	1000
1500	3.618	2.633	1.374	0.283	0.212	1.334	0.263	0.201	1.309	1500
2000	3.758	2.773	1.355	0.293	0.222	1.319	0.270	0.208	1.298	2000

Source: Data adapted from "Tables of Thermal Properties of Gases," NBS Circ. 564, 1955.

Table A-4 (*Continued*)

2. Specific-heat data for monatomic gases

Over a wide range of temperatures at low pressures the specific heats c_v and c_p of all monatomic gases are essentially independent of temperature and pressure. In addition, on a mole basis all monatomic gases have the same value for either c_v or c_p in a given set of units. One set of values is

$$c_v = 3.0 \text{ Btu/(lb} \cdot \text{mol)(°F)} \qquad \text{and} \qquad c_p = 5.0 \text{ Btu/(lb} \cdot \text{mol)(°F)}$$

Source: Data adapted from "Tables of Thermal Properties of Gases," NBS Circular 564, 1955.

3. Constant-pressure specific-heat equations for various gases at zero pressure (USCS units)

$$c_p/R_u = a + bT + cT^2 + dT^3 + eT^4$$

T is in °R, equations valid from 540 to 1800°R.

Gas	a	$b \times 10^3$	$c \times 10^6$	$d \times 10^9$	$e \times 10^{12}$
CO	3.710	−1.009	1.140	−0.348	0.0229
CO_2	2.401	4.853	−2.039	0.343	
H_2	3.057	1.487	−1.793	0.947	−0.1726
H_2O	4.070	−0.616	1.281	−0.508	0.0769
O_2	3.626	−1.043	2.178	−1.160	0.2054
N_2	3.675	−0.671	0.717	−0.108	−0.0215
Air (dry)	3.640	−0.617	0.761	−0.162	
NH_3	3.591	0.274	2.576	−1.437	0.2601
NO	4.046	−1.899	2.464	−1.048	0.1517
NO_2	3.459	1.147	2.064	−1.639	0.3448
SO_2	3.267	2.958	0.211	−0.906	0.2438
SO_3	2.578	8.087	−2.832	−0.136	0.1878
CH_4	3.826	−2.211	7.580	−3.898	0.6633
C_2H_2	1.410	10.587	−7.562	2.811	−0.3939
C_2H_4	1.426	6.324	2.466	−2.787	0.6429

Source: Adapted from the data in NASA SP-273, U.S. Government Printing Office, Washington, 1971.

Table A-5 Ideal-gas properties of air

T °R	h Btu/lb	p_r	u Btu/lb	v_r	s^0 Btu/(lb)(°R)	T °R	h Btu/lb	p_r	u Btu/lb	v_r	s^0 Btu/(lb)(°R)
360	85.97	0.3363	61.29	396.6	0.50369	860	206.46	7.149	147.50	44.57	0.71323
380	90.75	0.4061	64.70	346.6	0.51663	880	211.35	7.761	151.02	42.01	0.71886
400	95.53	0.4858	68.11	305.0	0.52890	900	216.26	8.411	154.57	39.64	0.72438
420	100.32	0.5760	71.52	270.1	0.54058	920	221.18	9.102	158.12	37.44	0.72979
440	105.11	0.6776	74.93	240.6	0.55172	940	226.11	9.834	161.68	35.41	0.73509
460	109.90	0.7913	78.36	215.33	0.56235	960	231.06	10.61	165.26	33.52	0.74030
480	114.69	0.9182	81.77	193.65	0.57255	980	236.02	11.43	168.83	31.76	0.74540
500	119.48	1.0590	85.20	174.90	0.58233	1000	240.98	12.30	172.43	30.12	0.75042
520	124.27	1.2147	88.62	158.58	0.59173	1040	250.95	14.18	179.66	27.17	0.76019
537	128.10	1.3593	91.53	146.34	0.59945	1080	260.97	16.28	186.93	24.58	0.76964
540	129.06	1.3860	92.04	144.32	0.60078	1120	271.03	18.60	194.25	22.30	0.77880
560	133.86	1.5742	95.47	131.78	0.60950	1160	281.14	21.18	201.63	20.29	0.78767
580	138.66	1.7800	98.90	120.70	0.61793	1200	291.30	24.01	209.05	18.51	0.79628
600	143.47	2.005	102.34	110.88	0.62607	1240	301.52	27.13	216.53	16.93	0.80466
620	148.28	2.249	105.78	102.12	0.63395	1280	311.79	30.55	224.05	15.52	0.81280
640	153.09	2.514	109.21	94.30	0.64159	1320	322.11	34.31	231.63	14.25	0.82075
660	157.92	2.801	112.67	87.27	0.64902	1360	332.48	38.41	239.25	13.12	0.82848
680	162.73	3.111	116.12	80.96	0.65621	1400	342.90	42.88	246.93	12.10	0.83604
700	167.56	3.446	119.58	75.25	0.66321	1440	353.37	47.75	254.66	11.17	0.84341
720	172.39	3.806	123.04	70.07	0.67002	1480	363.89	53.04	262.44	10.34	0.85062
740	177.23	4.193	126.51	65.38	0.67665	1520	374.47	58.78	270.26	9.578	0.85767
760	182.08	4.607	129.99	61.10	0.68312	1560	385.08	65.00	278.13	8.890	0.86456
780	186.94	5.051	133.47	57.20	0.68942	1600	395.74	71.73	286.06	8.263	0.87130
800	191.81	5.526	136.97	53.63	0.69558	1650	409.13	80.89	296.03	7.556	0.87954
820	196.69	6.033	140.47	50.35	0.70160	1700	422.59	90.95	306.06	6.924	0.88758
840	201.56	6.573	143.98	47.34	0.70747	1750	436.12	101.98	316.16	6.357	0.89542

Table A-5 *(Continued)*

T °R	h Btu/lb	p_r	u Btu/lb	v_r	s^0 Btu/(lb)(°R)	T °R	h Btu/lb	p_r	u Btu/lb	v_r	s^0 Btu/(lb)(°R)
1800	449.71	114.0	326.32	5.847	0.90308	3300	879.02	1418	652.81	.8621	1.07585
1850	463.37	127.2	336.55	5.388	0.91056	3350	893.83	1513	664.20	.8202	1.08031
1900	477.09	141.5	346.85	4.974	0.91788	3400	908.66	1613	675.60	.7807	1.08470
1950	490.88	157.1	357.20	4.598	0.92504	3450	923.52	1719	687.04	.7436	1.08904
2000	504.71	174.0	367.61	4.258	0.93205	3500	938.40	1829	698.48	.7087	1.09332
2050	518.61	192.3	378.08	3.949	0.93891	3550	953.30	1946	709.95	.6759	1.09755
2100	532.55	212.1	388.60	3.667	0.94564	3600	968.21	2068	721.44	.6449	1.10172
2150	546.54	233.5	399.17	3.410	0.95222	3650	983.15	2196	732.95	.6157	1.10584
2200	560.59	256.6	409.78	3.176	0.95919	3700	998.11	2330	744.48	.5882	1.10991
2250	574.69	281.4	420.46	2.961	0.96501	3750	1013.1	2471	756.04	.5621	1.11393
2300	588.82	308.1	431.16	2.765	0.97123	3800	1028.1	2618	767.60	.5376	1.11791
2350	603.00	336.8	441.91	2.585	0.97732	3850	1043.1	2773	779.19	.5143	1.12183
2400	617.22	367.6	452.70	2.419	0.98331	3900	1058.1	2934	790.80	.4923	1.12571
2450	631.48	400.5	463.54	2.266	0.98919	3950	1073.2	3103	802.43	.4715	1.12955
2500	645.78	435.7	474.40	2.125	0.99497	4000	1088.3	3280	814.06	.4518	1.13334
2550	660.12	473.3	485.31	1.996	1.00064	4050	1103.4	3464	825.72	.4331	1.13709
2600	674.49	513.5	496.26	1.876	1.00623	4100	1118.5	3656	837.40	.4154	1.14079
2650	688.90	556.3	507.25	1.765	1.01172	4150	1133.6	3858	849.09	.3985	1.14446
2700	703.35	601.9	518.26	1.662	1.01712	4200	1148.7	4067	860.81	.3826	1.14809
2750	717.83	650.4	529.31	1.566	1.02244	4300	1179.0	4513	884.28	.3529	1.15522
2800	732.33	702.0	540.40	1.478	1.02767	4400	1209.4	4997	907.81	.3262	1.16221
2850	746.88	756.7	551.52	1.395	1.03282	4500	1239.9	5521	931.39	.3019	1.16905
2900	761.45	814.8	562.66	1.318	1.03788	4600	1270.4	6089	955.04	.2799	1.17575
2950	776.05	876.4	573.84	1.247	1.04288	4700	1300.9	6701	978.73	.2598	1.18232
3000	790.68	941.4	585.04	1.180	1.04779	4800	1331.5	7362	1002.5	.2415	1.18876
3050	805.34	1011	596.28	1.118	1.05264	4900	1362.2	8073	1026.3	.2248	1.19508
3100	820.03	1083	607.53	1.060	1.05741	5000	1392.9	8837	1050.1	.2096	1.20129
3150	834.75	1161	618.82	1.006	1.06212	5100	1423.6	9658	1074.0	.1956	1.20738
3200	849.48	1242	630.12	0.955	1.06676	5200	1454.4	10539	1098.0	.1828	1.21336
3250	864.24	1328	641.46	0.907	1.07134	5300	1485.3	11481	1122.0	.1710	1.21923

Source: Data abridged from J. H. Keenan and J. Kaye, "Gas Tables," Wiley, New York, 1945.

Table A-6 Ideal-gas enthalpy, internal energy, and absolute entropy of diatomic nitrogen, N_2

$\Delta h_f = 0$ Btu/lb · mol

h and u, Btu/lb · mol, s, Btu/(lb · mol)(°R)

T, °R	h	u	s^0	T, °R	h	u	s^0
300	2082.0	1486.2	41.695	1080	7551.0	5406.2	50.651
320	2221.0	1585.5	42.143	1100	7695.0	5510.5	50.783
340	2360.0	1684.4	42.564	1120	7839.3	5615.2	50.912
360	2498.9	1784.0	42.962	1140	7984.0	5720.1	51.040
380	2638.0	1883.4	43.337	1160	8129.0	5825.4	51.167
400	2777.0	1982.6	43.694	1180	8274.4	5931.0	51.291
420	2916.1	2082.0	44.034	1200	8420.0	6037.0	51.413
440	3055.1	2181.3	44.357	1220	8566.1	6143.4	51.534
460	3194.1	2280.6	44.665	1240	8712.6	6250.1	51.653
480	3333.1	2379.9	44.962	1260	8859.3	6357.2	51.771
500	3472.2	2479.3	45.246	1280	9006.4	6464.5	51.887
520	3611.3	2578.6	45.519	1300	9153.9	6572.3	51.001
537	3729.5	2663.1	45.743	1320	9301.8	6680.4	52.114
540	3750.3	2678.0	45.781	1340	9450.0	6788.9	52.225
560	3889.5	2777.4	46.034	1360	9598.6	6897.8	52.335
580	4028.7	2876.9	46.278	1380	9747.5	7007.0	52.444
600	4167.9	2976.4	46.514	1400	9896.9	7116.7	52.551
620	4307.1	3075.9	46.742	1420	10046.6	7226.7	52.658
640	4446.4	3175.5	46.964	1440	10196.6	7337.0	52.763
660	4585.8	3275.2	47.178	1460	10347.0	7447.6	52.867
680	4725.3	3374.9	47.386	1480	10497.8	7558.7	52.969
700	4864.9	3474.8	47.588	1500	10648.0	7670.1	53.071
720	5004.5	3574.7	47.785	1520	10800.4	7781.9	53.171
740	5144.3	3674.7	47.977	1540	10952.2	7893.9	53.271
760	5284.1	3774.9	48.164	1560	11104.3	8006.4	53.369
780	5424.2	3875.2	48.345	1580	11256.9	8119.2	53.465
800	5564.4	3975.7	48.522	1600	11409.7	8232.3	53.561
820	5704.7	4076.3	48.696	1620	11562.8	8345.7	53.656
840	5845.3	4177.1	48.865	1640	11716.4	8459.6	53.751
860	5985.9	4278.1	49.031	1660	11870.2	8573.6	53.844
880	6126.9	4379.4	49.193	1680	12024.3	8688.1	53.936
900	6268.1	4480.8	49.352	1700	12178.9	8802.9	54.028
920	6409.6	4582.6	49.507	1720	12333.7	8918.0	54.118
940	6551.2	4684.5	49.659	1740	12488.8	9033.4	54.208
960	6693.1	4786.7	49.808	1760	12644.3	9149.2	54.297
980	6835.4	4889.3	49.955	1780	12800.2	9265.3	54.385
1000	6977.9	4992.0	50.099	1800	12956.3	9381.7	54.472
1020	7120.7	5095.1	50.241	1820	13112.7	9498.4	54.559
1040	7263.8	5198.5	50.380	1840	13269.5	9615.5	54.645
1060	7407.2	5302.2	50.516	1860	13426.5	9732.8	54.729

Table A-6 (*Continued*)

T, °R	h	u	s^0	T, °R	h	u	s^0
1900	13,742	9,968	54.896	3500	27,016	20,065	59.944
1940	14,058	10,205	55.061	3540	27,359	20,329	60.041
1980	14,375	10,443	55.223	3580	27,703	20,593	60.138
2020	14,694	10,682	55.383	3620	28,046	20,858	60.234
2060	15,013	10,923	55.540	3660	28,391	21,122	60.328
2100	15,334	11,164	55.694	3700	28,735	21,387	60.422
2140	15,656	11,406	55.846	3740	29,080	21,653	60.515
2180	15,978	11,649	55.995	3780	29,425	21,919	60.607
2220	16,302	11,893	56.141	3820	29,771	22,185	60.698
2260	16,626	12,138	56.286	3860	30,117	22,451	60.788
2300	16,951	12,384	56.429	3900	30,463	22,718	60.966
2340	17,277	12,630	56.570	3940	30,809	22,985	60.877
2380	17,604	12,878	56.708	3980	31,156	23,252	61.053
2420	17,932	13,126	56.845	4020	31,503	23,520	61.139
2460	18,260	13,375	56.980	4060	31,850	23,788	61.225
2500	18,590	13,625	57.112	4100	32,198	24,056	61.310
2540	18,919	13,875	57.243	4140	32,546	24,324	61.395
2580	19,250	14,127	57.372	4180	32,894	24,593	61.479
2620	19,582	14,379	57.499	4220	33,242	24,862	61.562
2660	19,914	14,631	57.625	4260	33,591	25,131	61.644
2700	20,246	14,885	57.750	4300	33,940	25,401	61.726
2740	20,580	15,139	57.872	4340	34,289	25,670	61.806
2780	20,914	15,393	57.993	4380	34,638	25,940	61.887
2820	21,248	15,648	58.113	4420	34,988	26,210	61.966
2860	21,584	15,905	58.231	4460	35,338	26,481	62.045
2900	21,920	16,161	58.348	4500	35,688	26,751	62.123
2940	22,256	16,417	58.463	4540	36,038	27,022	62.201
2980	22,593	16,675	58.576	4580	36,389	27,293	62.278
3020	22,930	16,933	58.688	4620	36,739	27,565	62.354
3060	23,268	17,192	58.800	4660	37,090	27,836	62.429
3100	23,607	17,451	58.910	4700	37,441	28,108	62.504
3140	23,946	17,710	59.019	4740	37,792	28,379	62.578
3180	24,285	17,970	59.126	4780	38,144	28,651	62.652
3220	24,625	18,231	59.232	4820	38,495	28,924	62.725
3260	24,965	18,491	59.338	4860	38,847	29,196	62.798
3300	25,306	18,753	59.442	4900	39,199	29,468	62.870
3340	25,647	19,014	59.544	5000	40,080	30,151	63.049
3380	25,989	19,277	59.646	5100	40,962	30,834	63.223
3420	26,331	19,539	59.747	5200	41,844	31,518	63.395
3460	26,673	19,802	59.846	5300	42,728	32,203	63.563

Source: Data abridged from J. H. Keenan and J. Kaye, "Gas Tables," Wiley, New York, 1945.

Table A-7 Ideal-gas enthalpy, internal energy, and absolute entropy of diatomic oxygen, O_2

$\Delta h_f = 0$ Btu/lb · mol

h and u, Btu/lb · mol, s, Btu/(lb · mol)(°R)

T, °R	h	u	s^0	T, °R	h	u	s^0
300	2073.5	1477.8	44.927	1080	7696.8	5552.1	54.064
320	2212.6	1577.1	45.375	1100	7850.4	5665.9	54.204
340	2351.7	1676.5	45.797	1120	8004.5	5780.3	54.343
360	2490.8	1775.9	46.195	1140	8159.1	5895.2	54.480
380	2630.0	1875.3	46.571	1160	8314.2	6010.6	54.614
400	2769.1	1974.8	46.927	1180	8469.8	6126.5	54.748
420	2908.3	2074.3	47.267	1200	8625.8	6242.8	54.879
440	3047.5	2173.8	47.591	1220	8782.4	6359.6	55.008
460	3186.9	2273.4	47.900	1240	8939.4	6476.9	55.136
480	3326.5	2373.3	48.198	1260	9096.7	6594.5	55.262
500	3466.2	2473.2	48.483	1280	9254.6	6712.7	55.386
520	3606.1	2573.4	48.757	1300	9412.9	6831.3	55.508
537	3725.1	2658.7	48.982	1320	9571.6	6950.2	55.630
540	3746.2	2673.8	49.021	1340	9730.7	7069.6	55.750
560	3886.6	2774.5	49.276	1360	9890.2	7189.4	55.867
580	4027.3	2875.5	49.522	1380	10050.1	7309.6	55.984
600	4168.3	2976.8	49.762	1400	10210.4	7430.1	56.099
620	4309.7	3078.4	49.993	1420	10371.0	7551.1	56.213
640	4451.4	3180.4	50.218	1440	10532.0	7672.4	56.326
660	4593.5	3282.9	50.437	1460	10693.3	7793.9	56.437
680	4736.2	3385.8	50.650	1480	10855.1	7916.0	56.547
700	4879.3	3489.2	50.858	1500	11017.1	8038.3	56.656
720	5022.9	3593.1	51.059	1520	11179.6	8161.1	56.763
740	5167.0	3697.4	51.257	1540	11342.4	8284.2	56.869
760	5311.4	3802.2	51.450	1560	11505.4	8407.4	56.975
780	5456.4	3907.5	51.638	1580	11668.8	8531.1	57.079
800	5602.0	4013.3	51.821	1600	11832.5	8655.1	57.182
820	5748.1	4119.7	52.002	1620	11996.6	8779.5	57.284
840	5894.8	4226.6	52.179	1640	12160.9	8904.1	57.385
860	6041.9	4334.1	52.352	1660	12325.5	9029.0	57.484
880	6189.6	4442.0	52.522	1680	12490.4	9154.1	57.582
900	6337.9	4550.6	52.688	1700	12655.6	9279.6	57.680
920	6486.7	4659.7	52.852	1720	12821.1	9405.4	57.777
940	6636.1	4769.4	53.012	1740	12986.9	9531.5	57.873
960	6786.0	4879.5	53.170	1760	13153.0	9657.9	57.968
980	6936.4	4990.3	53.326	1780	13319.2	9784.4	58.062
1000	7087.5	5101.6	53.477	1800	13485.8	9911.2	58.155
1020	7238.9	5213.3	53.628	1820	13652.5	10038.2	58.247
1040	7391.0	5325.7	53.775	1840	13819.6	10165.6	58.339
1060	7543.6	5438.6	53.921	1860	13986.8	10293.1	58.428

Table A-7 (*Continued*)

T, °R	*h*	*u*	s^0	*T*, °R	*h*	*u*	s^0
1900	14,322	10,549	58.607	3500	28,273	21,323	63.914
1940	14,658	10,806	58.782	3540	28,633	21,603	64.016
1980	14,995	11,063	58.954	3580	28,994	21,884	64.114
2020	15,333	11,321	59.123	3620	29,354	22,165	64.217
2060	15,672	11,581	59.289	3660	29,716	22,447	64.316
2100	16,011	11,841	59.451	3700	30,078	22,730	64.415
2140	16,351	12,101	59.612	3740	30,440	23,013	64.512
2180	16,692	12,363	59.770	3780	30,803	23,296	64.609
2220	17,036	12,625	59.926	3820	31,166	23,580	64.704
2260	17,376	12,888	60.077	3860	31,529	23,864	64.800
2300	17,719	13,151	60.228	3900	31,894	24,149	64.893
2340	18,062	13,416	60.376	3940	32,258	24,434	64.986
2380	18,407	13,680	60.522	3980	32,623	24,720	65.078
2420	18,572	13,946	60.666	4020	32,989	25,006	65.169
2460	19,097	14,212	60.808	4060	33,355	25,292	65.260
2500	19,443	14,479	60.946	4100	33,722	25,580	65.350
2540	19,790	14,746	61.084	4140	34,089	25,867	64.439
2580	20,138	15,014	61.220	4180	34,456	26,155	65.527
2620	20,485	15,282	61.354	4220	34,824	26,444	65.615
2660	20,834	15,551	61.486	4260	35,192	26,733	65.702
2700	21,183	15,821	61.616	4300	35,561	27,022	65.788
2740	21,533	16,091	61.744	4340	35,930	27,312	65.873
2780	21,883	16,362	61.871	4380	36,300	27,602	65.958
2820	22,232	16,633	61.996	4420	36,670	27,823	66.042
2860	22,584	16,905	62.120	4460	37,041	28,184	66.125
2900	22,936	17,177	62.242	4500	37,412	28,475	66.208
2940	23,288	17,450	62.363	4540	37,783	28,768	66.290
2980	23,641	17,723	62.483	4580	38,155	29,060	66.372
3020	23,994	17,997	62.599	4620	38,528	29,353	66.453
3060	24,348	18,271	62.716	4660	38,900	29,646	66.533
3100	24,703	18,546	62.831	4700	39,274	29,940	66.613
3140	25,057	18,822	62.945	4740	39,647	30,234	66.691
3180	25,413	19,098	63.057	4780	40,021	30,529	66.770
3220	25,769	19,374	63.169	4820	40,396	30,824	66.848
3260	26,175	19,651	63.279	4860	40,771	31,120	66.925
3300	26,412	19,928	63.386	4900	41,146	31,415	67.003
3340	26,839	20,206	63.494	5000	42,086	32,157	67.193
3380	27,197	20,485	63.601	5100	43,021	32,901	67.380
3420	27,555	20,763	63.706	5200	43,974	33,648	67.562
3460	27,914	21,043	63.811	5300	44,922	34,397	67.743

Source: Data abridged from J. H. Keenan and J. Kaye, "Gas Tables," Wiley, New York, 1945.

Table A-8 Ideal-gas enthalpy, internal energy, and absolute entropy of carbon monoxide, CO

$\Delta h_f = -47{,}540$ Btu/lb · mol

h and u, Btu/lb · mol; s, Btu/(lb · mol)(°R)

T, °R	h	u	s^0	T, °R	h	u	s^0
300	2081.9	1486.1	43.223	1080	7571.1	5426.4	52.203
320	2220.9	1585.4	43.672	1100	7716.8	5532.3	52.337
340	2359.9	1684.7	44.093	1120	7862.9	5638.7	52.468
360	2498.8	1783.9	44.490	1140	8009.2	5745.4	52.598
380	2637.9	1883.3	44.866	1160	8156.1	5851.5	52.726
400	2776.9	1982.6	45.223	1180	8303.3	5960.0	52.852
420	2916.0	2081.9	45.563	1200	8450.8	6067.8	52.976
440	3055.0	2181.2	45.886	1220	8598.8	6176.0	53.098
460	3194.0	2280.5	46.194	1240	8747.2	6284.7	53.218
480	3333.0	2379.8	46.491	1260	8896.0	6393.8	53.337
500	3472.1	2479.2	46.775	1280	9045.0	6503.1	53.455
520	3611.2	2578.6	47.048	1300	9194.6	6613.0	53.571
537	3725.1	2663.1	47.272	1320	9344.6	6723.2	53.685
540	3750.3	2677.9	47.310	1340	9494.8	6833.7	53.799
560	3889.5	2777.4	47.563	1360	9645.5	6944.7	53.910
580	4028.7	2876.9	47.807	1380	9796.6	7056.1	54.021
600	4168.0	2976.5	48.044	1400	9948.1	7167.9	54.129
620	4307.4	3076.2	48.272	1420	10100.0	7280.1	54.237
640	4446.9	3175.9	48.494	1440	10252.2	7392.6	54.344
660	4586.6	3275.8	48.709	1460	10404.8	7505.4	54.448
680	4726.2	3375.8	48.917	1480	10557.8	7618.7	54.522
700	4866.0	3475.9	49.120	1500	10711.1	7732.3	54.665
720	5006.1	3576.3	49.317	1520	10864.9	7846.4	54.757
740	5146.4	3676.9	49.509	1540	11019.0	7960.8	54.858
760	5286.8	3777.5	49.697	1560	11173.4	8075.4	54.958
780	5427.4	3878.4	49.880	1580	11328.2	8190.5	55.056
800	5568.2	3979.5	50.058	1600	11483.4	8306.0	55.154
820	5709.4	4081.0	50.232	1620	11638.9	8421.8	55.251
840	5850.7	4182.6	50.402	1640	11794.7	8537.9	55.347
860	5992.3	4284.5	50.569	1660	11950.9	8654.4	55.411
880	6134.2	4386.6	50.732	1680	12107.5	8771.2	55.535
900	6276.4	4489.1	50.892	1700	12264.3	8888.3	55.628
920	6419.0	4592.0	51.048	1720	12421.4	9005.7	55.720
940	6561.7	4695.0	51.202	1740	12579.0	9123.6	55.811
960	6704.9	4798.5	51.353	1760	12736.7	9241.6	55.900
980	6848.4	4902.3	51.501	1780	12894.9	9360.0	55.990
1000	6992.2	5006.3	51.646	1800	13053.2	9478.6	56.078
1020	7136.4	5110.8	51.788	1820	13212.0	9597.7	56.166
1040	7281.0	5215.7	51.929	1840	13371.0	9717.0	56.253
1060	7425.9	5320.9	52.067	1860	13530.2	9836.5	56.339

Table A-8 (*Continued*)

T, °R	h	u	s^0	T, °R	h	u	s^0
1900	13,850	10,077	56.509	3500	27,262	20,311	61.612
1940	14,170	10,318	56.677	3540	27,608	20,576	61.710
1980	14,492	10,560	56.841	3580	27,954	20,844	61.807
2020	14,815	10,803	57.007	3620	28,300	21,111	61.903
2060	15,139	11,048	57.161	3660	28,647	21,378	61.998
2100	15,463	11,293	57.317	3700	28,994	21,646	62.093
2140	15,789	11,539	57.470	3740	29,341	21,914	62.186
2180	16,116	11,787	57.621	3780	29,688	22,182	62.279
2220	16,443	12,035	57.770	3820	30,036	22,450	62.370
2260	16,722	12,284	57.917	3860	30,384	22,719	62.461
2300	17,101	12,534	58.062	3900	30,733	22,988	62.511
2340	17,431	12,784	58.204	3940	31,082	23,257	62.640
2380	17,762	13,035	58.344	3980	31,431	23,527	62.728
2420	18,093	13,287	58.482	4020	31,780	23,797	62.816
2460	18,426	13,541	58.619	4060	32,129	24,067	62.902
2500	18,759	13,794	58.754	4100	32,479	24,337	62.988
2540	19,093	14,048	58.885	4140	32,829	24,608	63.072
2580	19,427	14,303	59.016	4180	33,179	24,878	63.156
2620	19,762	14,559	59.145	4220	33,530	25,149	63.240
2660	20,098	14,815	59.272	4260	33,880	25,421	63.323
2700	20,434	15,072	59.398	4300	34,231	25,692	63.405
2740	20,771	15,330	59.521	4340	34,582	25,934	63.486
2780	21,108	15,588	59.644	4380	34,934	26,235	63.567
2820	21,446	15,846	59.765	4420	35,285	26,508	63.647
2860	21,785	16,105	59.884	4460	35,637	26,780	63.726
2900	22,124	16,365	60.002	4500	35,989	27,052	63.805
2940	22,463	16,225	60.118	4540	36,341	27,325	63.883
2980	22,803	16,885	60.232	4580	36,693	27,598	63.960
3020	23,144	17,146	60.346	4620	37,046	27,871	64.036
3060	23,485	17,408	60.458	4660	37,398	28,144	64.113
3100	23,826	17,670	60.569	4700	37,751	28,417	64.188
3140	24,168	17,932	60.679	4740	38,104	28,691	64.263
3180	24,510	18,195	60.787	4780	38,457	28,965	64.337
3220	24,853	18,458	60.894	4820	38,811	29,239	64.411
3260	25,196	18,722	61.000	4860	39,164	29,513	64.484
3300	25,539	18,986	61.105	4900	39,518	29,787	64.556
3340	25,883	19,250	61.209	5000	40,403	30,473	64.735
3380	26,227	19,515	61.311	5100	41,289	31,161	64.910
3420	26,572	19,780	61.412	5200	42,176	31,849	65.082
3460	26,917	20,045	61.513	5300	43,063	32,538	65.252

Source: Data abridged from J. H. Keenan and J. Kaye, "Gas Tables," Wiley, New York, 1945.

Table A-9 Ideal-gas enthalpy, internal energy, and absolute entropy of carbon dioxide, CO_2

$\Delta h_f = -169{,}290$ Btu/lb · mol

h and u, Btu/lb · mol; s, Btu/(lb · mol)(°R)

T, °R	h	u	s^0	T, °R	h	u	s^0
300	2108.2	1512.4	46.353	1080	9575.8	7431.1	58.072
320	2256.6	1621.1	46.832	1100	9802.6	7618.1	58.281
340	2407.3	1732.1	47.289	1120	10030.6	7806.4	58.485
360	2560.5	1845.6	47.728	1140	10260.1	7996.2	58.689
380	2716.4	1961.8	48.148	1160	10490.6	8187.0	58.889
400	2874.7	2080.4	48.555	1180	10722.3	8379.0	59.088
420	3035.7	2201.7	48.947	1200	10955.3	8572.3	59.283
440	3199.4	2325.6	49.329	1220	11189.4	8766.6	59.477
460	3365.7	2452.2	49.698	1240	11424.6	8962.1	59.668
480	3534.7	2581.5	50.058	1260	11661.0	9158.8	59.858
500	3706.2	2713.3	50.408	1280	11898.4	9356.5	60.044
520	3880.3	2847.7	50.750	1300	12136.9	9555.3	60.229
537	4027.5	2963.8	51.032	1320	12376.4	9755.0	60.412
540	4056.8	2984.4	51.082	1340	12617.0	9955.9	60.593
560	4235.8	3123.7	51.408	1360	12858.5	10157.7	60.772
580	4417.2	3265.4	51.726	1380	13101.0	10360.5	60.949
600	4600.9	3409.4	52.038	1400	13344.7	10564.5	61.124
620	4786.6	3555.6	52.343	1420	13589.1	10769.2	61.298
640	4974.9	3704.0	52.641	1440	13834.5	10974.8	61.469
660	5165.2	3854.6	52.934	1460	14080.8	11181.4	61.639
680	5357.6	4007.2	53.225	1480	14328.0	11388.9	61.800
700	5552.0	4161.9	53.503	1500	14576.0	11597.2	61.974
720	5748.4	4318.6	53.780	1520	14824.9	11806.4	62.138
740	5946.8	4477.3	54.051	1540	15074.7	12016.5	62.302
760	6147.0	4637.9	54.319	1560	15325.3	12227.3	62.464
780	6349.1	4800.1	54.582	1580	15576.7	12439.0	62.624
800	6552.9	4964.2	54.839	1600	15829.0	12651.6	62.783
820	6758.3	5129.9	55.093	1620	16081.9	12864.8	62.939
840	6965.7	5297.6	55.343	1640	16335.7	13078.9	63.095
860	7174.7	5466.9	55.589	1660	16590.2	13293.7	63.250
880	7385.3	5637.7	55.831	1680	16845.5	13509.2	63.403
900	7597.6	5810.3	56.070	1700	17101.4	13725.4	63.555
920	7811.4	5984.4	56.305	1720	17358.1	13942.4	63.704
940	8026.8	6160.1	56.536	1740	17615.5	14160.1	63.853
960	8243.8	6337.4	56.765	1760	17873.5	14378.4	64.001
980	8462.2	6516.1	56.990	1780	18132.2	14597.4	64.147
1000	8682.1	6696.2	57.212	1800	18391.5	14816.9	64.292
1020	8903.4	6877.8	57.432	1820	18651.5	15037.2	64.435
1040	9126.2	7060.9	57.647	1840	18912.2	15258.2	64.578
1060	9350.3	7245.3	57.861	1860	19173.4	15479.7	64.719

Table A-9 (*Continued*)

T, °R	h	u	s^0	T, °R	h	u	s^0
1900	19,698	15,925	64.999	3500	41,965	35,015	73.462
1940	20,224	16,372	65.272	3540	42,543	35,513	73.627
1980	20,753	16,821	65.543	3580	43,121	36,012	73.789
2020	21,284	17,273	65.809	3620	43,701	36,512	73.951
2060	21,818	17,727	66.069	3660	44,280	37,012	74.110
2100	22,353	18,182	66.327	3700	44,861	37,513	74.267
2140	22,890	18,640	66.581	3740	45,442	38,014	74.423
2180	23,429	19,101	66.830	3780	46,023	38,517	74.578
2220	23,970	19,561	67.076	3820	46,605	39,019	74.732
2260	24,512	20,024	67.319	3860	47,188	39.522	74.884
2300	25,056	20,489	67.557	3900	47,771	40,026	75.033
2340	25,602	20,955	67.792	3940	48,355	40,531	75.182
2380	26,150	21,423	68.025	3980	48,939	41,035	75.330
2420	26,699	21,893	68.253	4020	49,524	41,541	75.477
2460	27,249	22,364	68.479	4060	50,109	42,047	75.622
2500	27,801	22,837	68.702	4100	50,695	42,553	75.765
2540	28,355	23,310	68.921	4140	51,282	43,060	75.907
2580	28,910	23,786	69.138	4180	51,868	43,568	76.048
2620	29,465	24,262	69.352	4220	52,456	44,075	76.188
2660	30,023	24,740	69.563	4260	53,044	44,584	76.327
2700	30,581	25,220	69.771	4300	53,632	45,093	76.464
2740	31,141	25,701	69.977	4340	54,221	45,602	76.601
2780	31,702	26,181	70.181	4380	54,810	46,112	76.736
2820	32,264	26,664	70.382	4420	55,400	46,622	76.870
2860	32,827	27,148	70.580	4460	55,990	47,133	77.003
2900	33,392	27,633	70.776	4500	56,581	47,645	77.135
2940	33,957	28,118	70.970	4540	57,172	48,156	77.266
2980	34,523	28,605	71.160	4580	57,764	48,668	77.395
3020	35,090	29,093	71.350	4620	58,356	49,181	77.581
3060	35,659	29,582	71.537	4660	58,948	49,694	77.652
3100	36,228	30,072	71.722	4700	59,541	50,208	77.779
3140	36,798	30,562	71.904	4740	60,134	50,721	77.905
3180	37,369	31,054	72.085	4780	60,728	51,236	78.029
3220	37,941	31,546	72.264	4820	61,322	51,750	78.153
3260	38,513	32,039	72.441	4860	61,916	52,265	78.276
3300	39,087	32,533	72.616	4900	62,511	52,781	78.398
3340	39,661	33,028	72.788	5000	64,000	54,071	78.698
3380	40,236	33,524	72.960	5100	65,491	55,363	78.994
3420	40,812	34,020	73.129	5200	66,984	56,658	79.284
3460	41,388	34,517	73.297	5300	68,471	57,954	79.569

Source: Data abridged from J. H. Keenan and J. Kaye, "Gas Tables," Wiley, New York, 1945.

Table A-10 Ideal-gas enthalpy, internal energy, and absolute entropy of water, H_2O

$\Delta h_f = -104{,}040$ Btu/lb · mol

h and u, Btu/lb · mol; s, Btu/(lb · mol)(°R)

T, °R	h	u	s^0	T, °R	h	u	s^0
300	2367.6	1771.8	40.439	1080	8768.2	6623.5	50.854
320	2526.8	1891.3	40.952	1100	8942.0	6757.5	51.013
340	2686.0	2010.8	41.435	1120	9116.4	6892.2	51.171
360	2845.1	2130.2	41.889	1140	9291.4	7027.5	51.325
380	3004.4	2249.8	42.320	1160	9467.1	7163.5	51.478
400	3163.8	2369.4	42.728	1180	9643.4	7300.1	51.630
420	3323.2	2489.1	43.117	1200	9820.4	7437.4	51.777
440	3482.7	2608.9	43.487	1220	9998.0	7575.2	51.925
460	3642.3	2728.8	43.841	1240	10176.1	7713.6	52.070
480	3802.0	2848.8	44.182	1260	10354.9	7852.7	52.212
500	3962.0	2969.1	44.508	1280	10534.4	7992.5	52.354
520	4122.0	3089.4	44.821	1300	10714.5	8132.9	52.494
537	4258.0	3191.9	45.079	1320	10895.3	8274.0	52.631
540	4282.4	3210.0	45.124	1340	11076.6	8415.5	52.768
560	4442.8	3330.7	45.415	1360	11258.7	8557.9	52.903
580	4603.7	3451.9	45.696	1380	11441.4	8700.9	53.037
600	4764.7	3573.2	45.970	1400	11624.8	8844.6	53.168
620	4926.1	3694.9	46.235	1420	11808.8	8988.9	53.299
640	5087.8	3816.8	46.492	1440	11993.4	9133.8	53.428
660	5250.0	3939.3	46.741	1460	12178.8	9279.4	53.556
680	5412.5	4062.1	46.984	1480	12364.8	9425.7	53.682
700	5575.4	4185.3	47.219	1500	12551.4	9572.7	53.808
720	5738.8	4309.0	47.450	1520	12738.8	9720.3	53.932
740	5902.6	4433.1	47.673	1540	12926.8	9868.6	54.055
760	6066.9	4557.6	47.893	1560	13115.6	10017.6	54.117
780	6231.7	4682.7	48.106	1580	13305.0	10167.3	54.298
800	6396.9	4808.2	48.316	1600	13494.4	10317.6	54.418
820	6562.6	4934.2	48.520	1620	13685.7	10468.6	54.535
840	6728.9	5060.8	48.721	1640	13877.0	10620.2	54.653
860	6895.6	5187.8	48.916	1660	14069.2	10772.7	54.770
880	7062.9	5315.3	49.109	1680	14261.9	10925.6	54.886
900	7230.9	5443.6	49,298	1700	14455.4	11079.4	54.999
920	7399.4	5572.4	49.483	1720	14649.5	11233.8	55.113
940	7568.4	5701.7	49.665	1740	14844.3	11388.9	55.226
960	7738.0	5831.6	49.843	1760	15039.8	11544.7	55.339
980	7908.2	5962.0	50.019	1780	15236.1	11701.2	55.449
1000	8078.9	6093.0	50.191	1800	15433.0	11858.4	55.559
1020	8250.4	6224.8	50.360	1820	15630.6	12016.3	55.668
1040	8422.4	6357.1	50.528	1840	15828.7	12174.7	55.777
1060	8595.0	6490.0	50.693	1860	16027.6	12333.9	55.884

Table A-10 (*Continued*)

T, °R	h	u	s^0	T, °R	h	u	s^0
1900	16,428	12,654	56.097	3500	34,324	27,373	62.876
1940	16,830	12,977	56.307	3540	34,809	27,779	63.015
1980	17,235	13,303	56.514	3580	35,296	28,187	63.153
2020	17,643	13,632	56.719	3620	35,785	28,596	63.288
2060	18,054	13,963	56.920	3660	36,274	29,006	63.423
2100	18,467	14,297	57.119	3700	36,765	29,418	63.557
2140	18,883	14,633	57.315	3740	37,258	29,831	63.690
2180	19,301	14,972	57.509	3780	37,752	30,245	63.821
2220	19,722	15,313	57.701	3820	38,247	30,661	63.952
2260	20,145	15,657	57.889	3860	38,743	31,077	64.082
2300	20,571	16,003	58.077	3900	39,240	31,495	64.210
2340	20,999	16,352	58.261	3940	39,739	31,915	64.338
2380	21,429	16,703	58.445	3980	40,239	32,335	64.465
2420	21,862	17,057	58.625	4020	40,740	32,757	64.591
2460	22,298	17,413	58.803	4060	41,242	33,179	64.715
2500	22,735	17,771	58.980	4100	41,745	33,603	64.839
2540	23,175	18,131	59.155	4140	42,250	34,028	64.962
2580	23,618	18,494	59.328	4180	42,755	34,454	65.084
2620	24,062	18,859	59.500	4220	43,267	34,881	65.204
2660	24,508	19,226	59.669	4260	43,769	35,310	65.325
2700	24,957	19,595	59.837	4300	44,278	35,739	65.444
2740	25,408	19,967	60.003	4340	44,788	36,169	65.563
2780	25,861	20,340	60.167	4380	45,298	36,600	65.680
2820	26,316	20,715	60.330	4420	45,810	37,032	65.797
2860	26,773	21,093	60.490	4460	46,322	37,465	65.913
2900	27,231	21,472	60.650	4500	46,836	37,900	66.028
2940	27,692	21,853	60.809	4540	47,350	38,334	66.142
2980	28,154	22,237	60.965	4580	47,866	38,770	66.255
3020	28,619	22,621	61.120	4620	48,382	39,207	66.368
3060	29,085	23,085	61.274	4660	48,899	39,645	66.480
3100	29,553	23,397	61.426	4700	49,417	40,083	66.591
3140	30,023	23,787	61.577	4740	49,936	40,523	66.701
3180	30,494	24,179	61.727	4780	50,455	40,963	66.811
3220	30,967	24,572	61.874	4820	50,976	41,404	66.920
3260	31,442	24,968	62.022	4860	51,497	41,856	67.028
3300	31,918	25,365	62.167	4900	52,019	42,288	67.135
3340	32,396	25,763	62.312	5000	53,327	43,398	67.401
3380	32,876	26,164	62.454	5100	54,640	44,512	67.662
3420	33,357	26,565	62.597	5200	55,957	45,631	67.918
3460	33,839	26,968	62.738	5300	57,279	46,754	68.172

Source: Data abridged from J. H. Keenan and J. Kaye, "Gas Tables," Wiley, New York, 1945.

Table A-11 Ideal-gas enthalpy, internal energy, and absolute entropy of diatomic hydrogen, H_2

$\Delta h_f = 0$ Btu/lb · mol

h and u, Btu/lb · mol; s, Btu/(lb · mol)(°R)

T, °R	h	u	s^0	T, °R	h	u	s^0
300	2063.5	1467.7	27.337	1400	9673.8	6893.6	37.883
320	2189.4	1553.9	27.742	1500	10381.5	7402.7	38.372
340	2317.2	1642.0	28.130	1600	11092.5	7915.1	38.830
360	2446.8	1731.9	28.501	1700	11807.4	8431.4	39.264
380	2577.8	1823.2	28.856	1800	12526.8	8952.2	39.675
400	2710.2	1915.8	29.195	1900	13250.9	9477.8	40.067
420	2843.7	2009.6	29.520	2000	13980.1	10008.4	40.441
440	2978.1	2104.3	29.833	2100	14714.5	10544.2	40.799
460	3113.5	2200.0	30.133	2200	15454.4	11085.5	41.143
480	3249.4	2296.2	30.424	2300	16199.8	11632.3	41.475
500	3386.1	2393.2	30.703	2400	16950.6	12184.5	41.794
520	3523.2	2490.6	30.972	2500	17707.3	12742.6	42.104
537	3640.3	2573.9	31.194	2600	18469.7	13306.4	42.403
540	3660.9	2588.5	31.232	2700	19237.8	13876.0	42.692
560	3798.8	2686.7	31.482	2800	20011.8	14451.4	42.973
580	3937.1	2785.3	31.724	2900	20791.5	15032.5	43.247
600	4075.6	2884.1	31.959	3000	21576.9	15619.3	43.514
620	4214.3	2983.1	32.187	3100	22367.7	16211.5	43.773
640	4353.1	3082.1	32.407	3200	23164.1	16809.3	44.026
660	4492.1	3181.4	32.621	3300	23965.5	17412.1	44.273
680	4631.1	3280.7	32.829	3400	24771.9	18019.9	44.513
700	4770.2	3380.1	33.031	3500	25582.9	18632.4	44.748
720	4909.5	3479.6	33.226	3600	26398.5	19249.4	44.978
740	5048.8	3579.2	33.417	3700	27218.5	19870.8	45.203
760	5188.1	3678.8	33.603	3800	28042.8	20496.5	45.423
780	5327.6	3778.6	33.784	3900	28871.1	21126.2	45.638
800	5467.1	3878.4	33.961	4000	29703.5	21760.0	45.849
820	5606.7	3978.3	34.134	4100	30539.8	22397.7	46.056
840	5746.3	4078.2	34.302	4200	31379.8	23039.2	46.257
860	5885.9	4178.0	34.466	4300	32223.5	23684.3	46.456
880	6025.6	4278.0	34.627	4400	33070.9	24333.1	46.651
900	6165.3	4378.0	34.784	4500	33921.6	24985.2	46.842
920	6305.1	4478.1	34.938	4600	34775.7	25640.7	47.030
940	6444.9	4578.1	35.087	4700	35633.0	26299.4	47.215
960	6584.7	4678.3	35.235	4800	36493.4	26961.2	47.396
980	6724.6	4778.4	35.379	4900	35356.9	27626.1	47.574
1000	6864.5	4878.6	35.520	5000	38223.3	28294.0	47.749
1100	7564.6	5380.1	36.188	5100	39092.8	28964.9	47.921
1200	8265.8	5882.8	36.798	5200	39965.1	29638.6	48.090
1300	8968.7	6387.1	37.360	5300	40840.2	30315.1	48.257

Source: Data abridged from J. H. Keenan and J. Kaye, "Gas Tables," Wiley, New York, 1945.

Table A-12 Properties of saturated water: temperature table

v, ft^3/lb; u and h, Btu/lb; s, Btu/(lb)(°R)

		Specific volume		Internal energy		Enthalpy			Entropy	
Temp., °F T	Press., psia P	Sat. liquid- v_f	Sat. vapor v_g	Sat. liquid u_f	Sat. vapor u_g	Sat. liquid h_f	Evap. h_{fg}	Sat. vapor h_g	Sat. liquid s_f	Sat. vapor s_g
32	0.0886	0.01602	3305	−.01	1021.2	−.01	1075.4	1075.4	−.00003	2.1870
35	0.0999	0.01602	2948	2.99	1022.2	3.00	1073.7	1076.7	0.00607	2.1764
40	0.1217	0.01602	2445	8.02	1023.9	8.02	1070.9	1078.9	0.01617	2.1592
45	0.1475	0.01602	2037	13.04	1025.5	13.04	1068.1	1081.1	0.02618	2.1423
50	0.1780	0.01602	1704	18.06	1027.2	18.06	1065.2	1083.3	0.03607	2.1259
52	0.1917	0.01603	1589	20.06	1027.8	20.07	1064.1	1084.2	0.04000	2.1195
54	0.2064	0.01603	1482	22.07	1028.5	22.07	1063.0	1085.1	0.04391	2.1131
56	0.2219	0.01603	1383	24.08	1029.1	24.08	1061.9	1085.9	0.04781	2.1068
58	0.2386	0.01603	1292	26.08	1029.8	26.08	1060.7	1086.8	0.05159	2.1005
60	0.2563	0.01604	1207	28.08	1030.4	28.08	1059.6	1087.7	0.05555	2.0943
62	0.2751	0.01604	1129	30.09	1031.1	30.09	1058.5	1088.6	0.05940	2.0882
64	0.2952	0.01604	1056	32.09	1031.8	32.09	1057.3	1089.4	0.06323	2.0821
66	0.3165	0.01604	988.4	34.09	1032.4	34.09	1056.2	1090.3	0.06704	2.0761
68	0.3391	0.01605	925.8	36.09	1033.1	36.09	1055.1	1091.2	0.07084	2.0701
70	0.3632	0.01605	867.7	38.09	1033.7	38.09	1054.0	1092.0	0.07463	2.0642
72	0.3887	0.01606	813.7	40.09	1034.4	40.09	1052.8	1092.9	0.07839	2.0584
74	0.4158	0.01606	763.5	42.09	1035.0	42.09	1051.7	1093.8	0.08215	2.0526
76	0.4446	0.01606	716.8	44.09	1035.7	44.09	1050.6	1094.7	0.08589	2.0469
78	0.4750	0.01607	673.3	46.09	1036.3	46.09	1049.4	1095.5	0.08961	2.0412
80	0.5073	0.01607	632.8	48.08	1037.0	48.09	1048.3	1096.4	0.09332	2.0356
82	0.5414	0.01608	595.0	50.08	1037.6	50.08	1047.2	1097.3	0.09701	2.0300
84	0.5776	0.01608	559.8	52.08	1038.3	52.08	1046.0	1098.1	0.1007	2.0245
86	0.6158	0.01609	527.0	54.08	1038.9	54.08	1044.9	1099.0	0.1044	2.0190
88	0.6562	0.01609	496.3	56.07	1039.6	56.07	1043.8	1099.9	0.1080	2.0136
90	0.6988	0.01610	467.7	58.07	1040.2	58.07	1042.7	1100.7	0.1117	2.0083
92	0.7439	0.01611	440.9	60.06	1040.9	60.06	1041.5	1101.6	0.1153	2.0030
94	0.7914	0.01611	415.9	62.06	1041.5	62.06	1040.4	1102.4	0.1189	1.9977
96	0.8416	0.01612	392.4	64.05	1041.2	64.06	1039.2	1103.3	0.1225	1.9925
98	0.8945	0.01612	370.5	66.05	1042.8	66.05	1038.1	1104.2	0.1261	1.9874
100	0.9503	0.01613	350.0	68.04	1043.5	68.05	1037.0	1105.0	0.1296	1.9822
110	1.276	0.01617	265.1	78.02	1046.7	78.02	1031.3	1109.3	0.1473	1.9574
120	1.695	0.01621	203.0	87.99	1049.9	88.00	1025.5	1113.5	0.1647	1.9336
130	2.225	0.01625	157.2	97.97	1053.0	97.98	1019.8	1117.8	0.1817	1.9109
140	2.892	0.01629	122.9	107.95	1056.2	107.96	1014.0	1121.9	0.1985	1.8892
150	3.722	0.01634	97.0	117.95	1059.3	117.96	1008.1	1126.1	0.2150	1.8684
160	4.745	0.01640	77.2	127.94	1062.3	127.96	1002.2	1130.1	0.2313	1.8484
170	5.996	0.01645	62.0	137.95	1065.4	137.97	996.2	1134.2	0.2473	1.8293
180	7.515	0.01651	50.2	147.97	1068.3	147.99	990.2	1138.2	0.2631	1.8109
190	9.343	0.01657	41.0	158.00	1071.3	158.03	984.1	1142.1	0.2787	1.7932
200	11.529	0.01663	33.6	168.04	1074.2	168.07	977.9	1145.9	0.2940	1.7762

Table A-12 (*Continued*)

Temp. °F T	Press. psia P	Specific volume: Sat. liquid v_f	Specific volume: Sat. vapor v_g	Internal energy: Sat. liquid u_f	Internal energy: Sat. vapor u_g	Enthalpy: Sat. liquid h_f	Enthalpy: Evap. h_{fg}	Enthalpy: Sat. vapor h_g	Entropy: Sat. liquid s_f	Entropy: Sat. vapor s_g
210	14.13	0.01670	27.82	178.1	1077.0	178.1	971.6	1149.7	0.3091	1.7599
212	14.70	0.01672	26.80	180.1	1077.6	180.2	970.3	1150.5	0.3121	1.7567
220	17.19	0.01677	23.15	188.2	1079.8	188.2	965.3	1153.5	0.3241	1.7441
230	20.78	0.01685	19.39	198.3	1082.6	198.3	958.8	1157.1	0.3388	1.7289
240	24.97	0.01692	16.33	208.4	1085.3	208.4	952.3	1160.7	0.3534	1.7143
250	29.82	0.01700	13.83	218.5	1087.9	218.6	945.6	1164.2	0.3677	1.7001
260	35.42	0.01708	11.77	228.6	1090.5	228.8	938.8	1167.6	0.3819	1.6864
270	41.85	0.01717	10.07	238.8	1093.0	239.0	932.0	1170.9	0.3960	1.6731
280	49.18	0.01726	8.65	249.0	1095.4	249.2	924.9	1174.1	0.4099	1.6602
290	57.53	0.01735	7.47	259.3	1097.7	259.4	917.8	1177.2	0.4236	1.6477
300	66.98	0.01745	6.472	269.5	1100.0	269.7	910.4	1180.2	0.4372	1.6356
310	77.64	0.01755	5.632	279.8	1102.1	280.1	903.0	1183.0	0.4507	1.6238
320	89.60	0.01765	4.919	290.1	1104.2	290.4	895.3	1185.8	0.4640	1.6123
330	103.00	0.01776	4.312	300.5	1106.2	300.8	887.5	1188.4	0.4772	1.6010
340	117.93	0.01787	3.792	310.9	1108.0	311.3	879.5	1190.8	0.4903	1.5901
350	134.53	0.01799	3.346	321.4	1109.8	321.8	871.3	1193.1	0.5033	1.5793
360	152.92	0.01811	2.961	331.8	1111.4	332.4	862.9	1195.2	0.5162	1.5688
370	173.23	0.01823	2.628	342.4	1112.9	343.0	854.2	1197.2	0.5289	1.5585
380	195.60	0.01836	2.339	353.0	1114.3	353.6	845.4	1199.0	0.5416	1.5483
390	220.2	0.01850	2.087	363.6	1115.6	364.3	836.2	1200.6	0.5542	1.5383
400	247.1	0.01864	1.866	374.3	1116.6	375.1	826.8	1202.0	0.5667	1.5284
410	276.5	0.01878	1.673	385.0	1117.6	386.0	817.2	1203.1	0.5792	1.5187
420	308.5	0.01894	1.502	395.8	1118.3	396.9	807.2	1204.1	0.5915	1.5091
430	343.3	0.01909	1.352	406.7	1118.9	407.9	796.9	1204.8	0.6038	1.4995
440	381.2	0.01926	1.219	417.6	1119.3	419.0	786.3	1205.3	0.6161	1.4900
450	422.1	0.01943	1.1011	428.6	1119.5	430.2	775.4	1205.6	0.6282	1.4806
460	466.3	0.01961	0.9961	439.7	1119.6	441.4	764.1	1205.5	0.6404	1.4712
470	514.1	0.01980	0.9025	450.9	1119.4	452.8	752.4	1205.2	0.6525	1.4618
480	565.5	0.02000	0.8187	462.2	1118.9	464.3	740.3	1204.6	0.6646	1.4524
490	620.7	0.02021	0.7436	473.6	1118.3	475.9	727.8	1203.7	0.6767	1.4430
500	680.0	0.02043	0.6761	485.1	1117.4	487.7	714.8	1202.5	0.6888	1.4335
520	811.4	0.02091	0.5605	508.5	1114.8	511.7	687.3	1198.9	0.7130	1.4145
540	961.5	0.02145	0.4658	532.6	1111.0	536.4	657.5	1193.8	0.7374	1.3950
560	1131.8	0.02207	0.3877	548.4	1105.8	562.0	625.0	1187.0	0.7620	1.3749
580	1324.3	0.02278	0.3225	583.1	1098.9	588.6	589.3	1178.0	0.7872	1.3540
600	1541.0	0.02363	0.2677	609.9	1090.0	616.7	549.7	1166.4	0.8130	1.3317
620	1784.4	0.02465	0.2209	638.3	1078.5	646.4	505.0	1151.4	0.8398	1.3075
640	2057.1	0.02593	0.1805	668.7	1063.2	678.6	453.4	1131.9	0.8681	1.2803
660	2362	0.02767	0.1446	702.3	1042.3	714.4	391.1	1105.5	0.8990	1.2483
680	2705	0.03032	0.1113	741.7	1011.0	756.9	309.8	1066.7	0.9350	1.2068
700	3090	0.03666	0.0744	801.7	947.7	822.7	167.5	990.2	0.9902	1.1346
705.4	3204	0.05053	0.05053	872.6	872.6	902.5	0	902.5	1.0580	1.0580

Source: Keenan, J. H., F. G. Keyes, P. G. Hill, and J. G. Moore, "Steam Tables," Wiley, New York, 1979.

Table A-13 Properties of saturated water: pressure table

v, ft^3/lb; u and h, Btu/lb; s, Btu/(lb)(°R)

		Specific volume		Internal energy		Enthalpy			Entropy			
Abs. press., psi P	Temp., °F T	Sat. liquid v_f	Sat. vapor v_g	Sat. liquid u_f	Sat. vapor u_g	Sat. liquid h_f	Evap. h_{fg}	Sat. vapor h_g	Sat. liquid s_f	Evap. s_{fg}	Sat. vapor s_g	Abs. press., psi P
0.4	72.84	0.01606	792.0	40.94	1034.7	40.94	1052.3	1093.3	0.0800	1.9760	2.0559	0.4
0.6	85.19	0.01609	540.0	53.26	1038.7	53.27	1045.4	1098.6	0.1029	1.9184	2.0213	0.6
0.8	94.35	0.01611	411.7	62.41	1041.7	62.41	1040.2	1102.6	0.1195	1.8773	1.9968	0.8
1.0	101.70	0.01614	333.6	69.74	1044.0	69.74	1036.0	1105.8	0.1327	1.8453	1.9779	1.0
1.2	107.88	0.01616	280.9	75.90	1046.0	75.90	1032.5	1108.4	0.1436	1.8190	1.9626	1.2
1.5	115.65	0.01619	227.7	83.65	1048.5	83.65	1028.0	1111.7	0.1571	1.7867	1.9438	1.5
2.0	126.04	0.01623	173.75	94.02	1051.8	94.02	1022.1	1116.1	0.1750	1.7448	1.9198	2.0
3.0	141.43	0.01630	118.72	109.38	1056.6	109.39	1013.1	1122.5	0.2009	1.6852	1.8861	3.0
4.0	152.93	0.01636	90.64	120.88	1060.2	120.89	1006.4	1127.3	0.2198	1.6426	1.8624	4.0
5.0	162.21	0.01641	73.53	130.15	1063.0	130.17	1000.9	1131.0	0.2349	1.6093	1.8441	5.0
6.0	170.03	0.01645	61.98	137.98	1065.4	138.00	996.2	1134.2	0.2474	1.5819	1.8292	6.0
7.0	176.82	0.01649	53.65	144.78	1067.4	144.80	992.1	1136.9	0.2581	1.5585	1.8167	7.0
8.0	182.84	0.01653	47.35	150.81	1069.2	150.84	988.4	1139.3	0.2675	1.5383	1.8058	8.0
9.0	188.26	0.01656	42.41	156.25	1070.8	156.27	985.1	1141.4	0.2760	1.5203	1.7963	9.0
10	193.19	0.01659	38.42	161.20	1072.2	161.23	982.1	1143.3	0.2836	1.5041	1.7877	10
14.696	211.99	0.01672	26.80	180.10	1077.6	180.15	970.4	1150.5	0.3121	1.4446	1.7567	14.696
15	213.03	0.01672	26.29	181.14	1077.9	181.19	969.7	1150.9	0.3137	1.4414	1.7551	15
20	227.96	0.01683	20.09	196.19	1082.0	196.26	960.1	1156.4	0.3358	1.3962	1.7320	20
25	240.08	0.01692	16.31	208.44	1085.3	208.52	952.2	1160.7	0.3535	1.3607	1.7142	25
30	250.34	0.01700	13.75	218.84	1088.0	218.93	945.4	1164.3	0.3682	1.3314	1.6996	30
35	259.30	0.01708	11.90	227.93	1090.3	228.04	939.3	1167.4	0.3809	1.3064	1.6873	35
40	267.26	0.01715	10.50	236.03	1092.3	236.16	933.8	1170.0	0.3921	1.2845	1.6767	40
45	274.46	0.01721	9.40	243.37	1094.0	243.51	928.8	1172.3	0.4022	1.2651	1.6673	45
50	281.03	0.01727	8.52	250.08	1095.6	250.24	924.2	1174.4	0.4113	1.2476	1.6589	50
55	287.10	0.01733	7.79	256.28	1097.0	256.46	919.9	1176.3	0.4196	1.2317	1.6513	55

Table A-13 (*Continued*)

		Specific volume		Internal energy		Enthalpy			Entropy			
Abs. press., psi P	Temp., °F T	Sat. liquid v_f	Sat. vapor v_g	Sat. liquid u_f	Sat. vapor u_g	Sat. liquid h_f	Evap. h_{fg}	Sat. vapor h_g	Sat. liquid s_f	Evap. s_{fg}	Sat. vapor s_g	Abs. press., psi P
60	292.73	0.01738	7.177	262.1	1098.3	262.2	915.8	1178.0	0.4273	1.2170	1.6443	60
65	298.00	0.01743	6.647	267.5	1099.5	267.7	911.9	1179.6	0.4345	1.2035	1.6380	65
70	302.96	0.01748	6.209	272.6	1100.6	272.8	908.3	1181.0	0.4412	1.1909	1.6321	70
75	307.63	0.01752	5.818	277.4	1101.6	277.6	904.8	1182.4	0.4475	1.1790	1.6265	75
80	312.07	0.01757	5.474	282.0	1102.6	282.2	901.4	1183.6	0.4534	1.1679	1.6213	80
85	316.29	0.01761	5.170	286.3	1103.5	286.6	898.2	1184.8	0.4591	1.1574	1.6165	85
90	320.31	0.01766	4.898	290.5	1104.3	290.8	895.1	1185.9	0.4644	1.1475	1.6119	90
95	324.16	0.01770	4.654	294.5	1105.0	294.8	892.1	1186.9	0.4695	1.1380	1.6075	95
100	327.86	0.01774	4.434	298.3	1105.8	298.6	889.2	1187.8	0.4744	1.1290	1.6034	100
110	334.82	0.01781	4.051	305.5	1107.1	305.9	883.7	1189.6	0.4836	1.1122	1.5958	110
120	341.30	0.01789	3.730	312.3	1108.3	312.7	878.5	1191.1	0.4920	1.0966	1.5886	120
130	347.37	0.01796	3.457	318.6	1109.4	319.0	873.5	1192.5	0.4999	1.0822	1.5821	130
140	353.08	0.01802	3.221	324.6	1110.3	325.1	868.7	1193.8	0.5073	1.0688	1.5761	140
150	358.48	0.01809	3.016	330.2	1111.2	330.8	864.2	1194.9	0.5142	1.0562	1.5704	150
160	363.60	0.01815	2.836	335.6	1112.0	336.2	859.8	1196.0	0.5208	1.0443	1.5651	160
170	368.47	0.01821	2.676	340.8	1112.7	341.3	855.6	1196.9	0.5270	1.0330	1.5600	170
180	373.13	0.01827	2.553	345.7	1113.4	346.3	851.5	1197.8	0.5329	1.0223	1.5552	180
190	377.59	0.01833	2.405	350.4	1114.0	351.0	847.5	1198.6	0.5386	1.0122	1.5508	190
200	381.86	0.01839	2.289	354.9	1114.6	355.6	843.7	1199.3	0.5440	1.0025	1.5465	200
250	401.04	0.01865	1.845	375.4	1116.7	376.2	825.8	1202.1	0.5680	0.9594	1.5274	250
300	417.43	0.01890	1.544	393.0	1118.2	394.1	809.8	1203.9	0.5883	0.9232	1.5115	300
350	431.82	0.01912	1.327	408.7	1119.0	409.9	795.0	1204.9	0.6060	0.8917	1.4977	350
400	444.70	0.01934	1.162	422.8	1119.5	424.2	781.2	1205.5	0.6218	0.8638	1.4856	400
450	456.39	0.01955	1.033	435.7	1119.6	437.4	768.2	1205.6	0.6360	0.8385	1.4745	450
500	467.13	0.01975	0.928	447.7	1119.4	449.5	755.8	1205.3	0.6490	0.8154	1.4644	500

Table A-13 (*Continued*)

		Specific volume		Internal energy		Enthalpy			Entropy			
Abs. press., psi P	Temp., °F T	Sat. liquid v_f	Sat. vapor v_g	Sat. liquid u_f	Sat. vapor u_g	Sat. liquid h_f	Evap. h_{fg}	Sat. vapor h_g	Sat. liquid s_f	Evap. s_{fg}	Sat. vapor s_g	Abs. press., psi P
550	477.07	0.01994	0.842	458.9	1119.1	460.9	743.9	1204.8	0.6611	0.7941	1.4551	550
600	486.33	0.02013	0.770	469.4	1118.6	471.7	732.4	1204.1	0.6723	0.7742	0.4464	600
700	503.23	0.02051	0.656	488.9	1117.0	491.5	710.5	1202.0	0.6927	0.7378	1.4305	700
800	518.36	0.02087	0.569	506.6	1115.0	509.7	689.6	1199.3	0.7110	0.7050	1.4160	800
900	532.12	0.02123	0.501	523.0	1112.6	526.6	669.5	1196.0	0.7277	0.6750	1.4027	900
1000	544.75	0.02159	0.446	538.4	1109.9	542.4	650.0	1192.4	0.7432	0.6471	1.3903	1000
1100	556.45	0.02195	0.401	552.9	1106.8	557.4	631.0	1188.3	0.7576	0.6209	1.3786	1100
1200	567.37	0.02232	0.362	566.7	1103.5	571.7	612.3	1183.9	0.7712	0.5961	1.3673	1200
1300	577.60	0.02269	0.330	579.9	1099.8	585.4	593.8	1179.2	0.7841	0.5724	1.3565	1300
1400	587.25	0.02307	0.302	592.7	1096.0	598.6	575.5	1174.1	0.7964	0.5497	1.3461	1400
1500	596.39	0.02346	0.277	605.0	1091.8	611.5	557.2	1168.7	0.8082	0.5276	1.3359	1500
1600	605.06	0.02386	0.255	616.9	1087.4	624.0	538.9	1162.9	0.8196	0.5062	1.3258	1600
1700	613.32	0.02428	0.236	628.6	1082.7	636.2	520.6	1156.9	0.8307	0.4852	1.3159	1700
1800	621.21	0.02472	0.218	640.0	1077.7	648.3	502.1	1150.4	0.8414	0.4645	1.3060	1800
1900	628.76	0.02517	0.203	651.3	1072.3	660.1	483.4	1143.5	0.8519	0.4441	1.2961	1900
2000	636.00	0.02565	0.188	662.4	1066.6	671.9	464.4	1136.3	0.8623	0.4238	1.2861	2000
2250	652.90	0.02698	0.157	689.9	1050.6	701.1	414.8	1115.9	0.8876	0.3728	1.2604	2250
2500	668.31	0.02860	0.131	717.7	1031.0	730.9	360.5	1091.4	0.9131	0.3196	1.2327	2500
2750	682.46	0.03077	0.107	747.3	1005.9	763.0	297.4	1060.4	0.9401	0.2604	1.2005	2750
3000	695.52	0.03431	0.084	783.4	968.8	802.5	213.0	1015.5	0.9732	0.1843	1.1575	3000
3203.6	705.44	0.05053	0.0505	872.6	872.6	902.5	0	902.5	1.0580	0	1.0580	3203.6

Source: Keenan, J. H., F. G. Keyes, P. G. Hill, and J. G. Moore, "Steam Tables," Wiley, New York, 1969.

Table A-14 Properties of water: superheated-vapor table

v, ft^3/lb: u and h, Btu/lb; s, Btu/(lb)(°R)

Temp., °F	v	u	h	s	v	u	h	s
	1 psia (T_{sat} = 101.7°F)				5 psia (T_{sat} = 162.2°F)			
Sat.	333.6	1044.0	1105.8	1.9779	73.53	1063.0	1131.0	1.8441
150	362.6	1060.4	1127.5	2.0151				
200	392.5	1077.5	1150.1	2.0508	78.15	1076.0	1148.6	1.8715
250	422.4	1094.7	1172.8	2.0839	84.21	1093.8	1171.7	1.9052
300	452.3	1112.0	1195.7	2.1150	90.24	1111.3	1194.8	1.9367
400	511.9	1147.0	1241.8	2.1720	102.24	1146.6	1241.2	1.9941
500	571.5	1182.8	1288.5	2.2235	114.20	1182.5	1288.2	2.0458
600	631.1	1219.3	1336.1	2.2706	126.15	1219.1	1335.8	2.0930
700	690.7	1256.7	1384.5	2.3142	138.08	1256.5	1384.3	2.1367
800	750.3	1294.4	1433.7	2.3550	150.01	1294.7	1433.5	2.1775
900	809.9	1333.9	1483.8	2.3932	161.94	1333.8	1483.7	2.2158
1000	869.5	1373.9	1534.8	2.4294	173.86	1373.9	1534.7	2.2520
	10 psia (T_{sat} = 193.2°F)				14.7 psia (T_{sat} = 212.0°F)			
Sat.	38.42	1072.2	1143.3	1.7877	26.80	1077.6	1150.5	1.7567
200	38.85	1074.7	1146.6	1.7927				
250	41.95	1092.6	1170.2	1.8272	28.42	1091.5	1168.8	1.7832
300	44.99	1110.4	1193.7	1.8592	30.52	1109.6	1192.6	1.8157
400	51.03	1146.1	1240.5	1.9171	34.67	1145.6	1239.9	1.8741
500	57.04	1182.2	1287.7	1.9690	38.77	1181.8	1287.3	1.9263
600	63.03	1218.9	1335.5	2.0164	42.86	1218.6	1335.2	1.9737
700	69.01	1256.3	1384.0	2.0601	46.93	1256.1	1383.8	2.0175
800	74.98	1294.6	1433.3	2.1009	51.00	1294.4	1433.1	2.0584
900	80.95	1333.7	1483.5	2.1393	55.07	1333.6	1483.4	2.0967
1000	86.91	1373.8	1534.6	2.1755	59.13	1373.7	1534.5	2.1330
1100	92.88	1414.7	1586.6	2.2099	63.19	1414.6	1586.4	2.1674
	20 psia (T_{sat} = 228.0°F)				40 psia (T_{sat} = 267.3°F)			
Sat.	20.09	1082.0	1156.4	1.7320	10.50	1093.3	1170.0	1.6767
250	20.79	1090.3	1167.2	1.7475				
300	22.36	1108.7	1191.5	1.7805	11.04	1105.1	1186.8	1.6993
350	23.90	1126.9	1215.4	1.8110	11.84	1124.2	1211.8	1.7312
400	25.43	1145.1	1239.2	1.8395	12.62	1143.0	1236.4	1.7606
500	28.46	1181.5	1286.8	1.8919	14.16	1180.1	1284.9	1.8140
600	31.47	1218.4	1334.8	1.9395	15.69	1217.3	1333.4	1.8621
700	34.47	1255.9	1383.5	1.9834	17.20	1255.1	1382.4	1.9063
800	37.46	1294.3	1432.9	2.0243	18.70	1293.7	1432.1	1.9474
900	40.45	1333.5	1483.2	2.0627	20.20	1333.0	1482.5	1.9859
1000	43.44	1373.5	1534.3	2.0989	21.70	1373.1	1533.8	2.0223
1100	46.42	1414.5	1586.3	2.1334	23.20	1414.2	1585.9	2.0568

Table A-14 (*Continued*)

Temp., °F	v	u	h	s	v	u	h	s
	60 psia (T_{sat} = 292.7°F)				80 psia (T_{sat} = 312.1°F)			
Sat.	7.17	1098.3	1178.0	1.6444	5.47	1102.6	1183.6	1.6214
300	7.26	1101.3	1181.9	1.6496				
350	7.82	1121.4	1208.2	1.6830	5.80	1118.5	1204.3	1.6476
400	8.35	1140.8	1233.5	1.7134	6.22	1138.5	1230.6	1.6790
500	9.40	1178.6	1283.0	1.7678	7.02	1177.2	1281.1	1.7346
600	10.43	1216.3	1332.1	1.8165	7.79	1215.3	1330.7	1.7838
700	11.44	1254.4	1381.4	1.8609	8.56	1253.6	1380.3	1.8285
800	12.45	1293.0	1431.2	1.9022	9.32	1292.4	1430.4	1.8700
900	13.45	1332.5	1481.8	1.9408	10.08	1332.0	1481.2	1.9087
1000	14.45	1372.7	1533.2	1.9773	10.83	1372.3	1532.6	1.9453
1100	15.45	1413.8	1585.4	2.0119	11.58	1413.5	1584.9	1.9799
1200	16.45	1455.8	1638.5	2.0448	12.33	1455.5	1638.1	2.0130
	100 psia (T_{sat} = 327.8°F)				120 psia (T_{sat} = 341.3°F)			
Sat.	4.434	1105.8	1187.8	1.6034	3.730	1108.3	1191.1	1.5886
350	4.592	1115.4	1200.4	1.6191	3.783	1112.2	1196.2	1.5950
400	4.934	1136.2	1227.5	1.6517	4.079	1133.8	1224.4	1.6288
450	5.265	1156.2	1253.6	1.6812	4.360	1154.3	1251.2	1.6590
500	5.587	1175.7	1279.1	1.7085	4.633	1174.2	1277.1	1.6868
600	6.216	1214.2	1329.3	1.7582	5.164	1213.2	1327.8	1.7371
700	6.834	1252.8	1379.2	1.8033	5.682	1252.0	1378.2	1.7825
800	7.445	1291.8	1429.6	1.8449	6.195	1291.2	1428.7	1.8243
900	8.053	1331.5	1480.5	1.8838	6.703	1330.9	1479.8	1.8633
1000	8.657	1371.9	1532.1	1.9204	7.208	1371.5	1531.5	1.9000
1100	9.260	1413.1	1584.5	1.9551	7.711	1412.8	1584.0	1.9348
1200	9.861	1455.2	1637.7	1.9882	8.213	1454.9	1637.3	1.9679
	140 psia (T_{sat} = 353.1°F)				160 psia (T_{sat} = 363.6°F)			
Sat.	3.221	1110.3	1193.8	1.5761	2.836	1112.0	1196.0	1.5651
400	3.466	1131.4	1221.2	1.6088	3.007	1128.8	1217.8	1.5911
450	3.713	1152.4	1248.6	1.6399	3.228	1150.5	1246.1	1.6230
500	3.952	1172.7	1275.1	1.6682	3.440	1171.2	1273.0	1.6518
550	4.184	1192.5	1300.9	1.6945	3.646	1191.3	1299.2	1.6785
600	4.412	1212.1	1326.4	1.7191	3.848	1211.1	1325.0	1.7034
700	4.860	1251.2	1377.1	1.7648	4.243	1250.4	1376.0	1.7494
800	5.301	1290.5	1427.9	1.8068	4.631	1289.9	1427.0	1.7916
900	5.739	1330.4	1479.1	1.8459	5.015	1329.9	1478.4	1.8308
1000	6.173	1371.0	1531.0	1.8827	5.397	1370.6	1530.4	1.8677
1100	6.605	1412.4	1583.6	1.9176	5.776	1412.1	1583.1	1.9026
1200	7.036	1454.6	1636.9	1.9507	6.154	1454.3	1636.5	1.9358

Table A-14 Properties of water: superheated-vapor table *(Continued)*

Temp., °F	v	u	h	s	v	u	h	s
	180 psia (T_{sat} = 373.1°F)				200 psia (T_{sat} = 381.8°F)			
Sat.	2.533	113.4	1197.8	1.5553	2.289	1114.6	1199.3	1.5464
400	2.648	1126.2	1214.4	1.5749	2.361	1123.5	1210.8	1.5600
450	2.850	1148.5	1243.4	1.6078	2.548	1146.4	1240.7	1.5938
500	3.042	1169.6	1270.9	1.6372	2.724	1168.0	1268.8	1.6239
550	3.228	1190.0	1297.5	1.6642	2.893	1188.7	1295.7	1.6512
600	3.409	1210.0	1323.5	1.6893	3.058	1208.9	1322.1	1.6767
700	3.763	1249.6	1374.9	1.7357	3.379	1248.8	1373.8	1.7234
800	4.110	1289.3	1426.2	1.7781	3.693	1288.6	1425.3	1.7660
900	4.453	1329.4	1477.7	1.8174	4.003	1328.9	1477.1	1.8055
1000	4.793	1370.2	1529.8	1.8545	4.310	1369.8	1529.3	1.8425
1100	5.131	1411.7	1582.6	1.8894	4.615	1411.4	1582.2	1.8776
1200	5.467	1454.0	1636.1	1.9227	4.918	1453.7	1635.7	1.9109
	250 psia (T_{sat} = 401.0°F)				300 psia (T_{sat} = 417.4°F)			
Sat.	1.845	1116.7	1202.1	1.5274	1.544	1118.2	1203.9	1.5115
450	2.002	1141.1	1233.7	1.5632	1.636	1135.4	1226.2	1.5365
500	2.150	1163.8	1263.3	1.5948	1.766	1159.5	1257.5	1.5701
550	2.290	1185.3	1291.3	1.6233	1.888	1181.9	1286.7	1.5997
600	2.426	1206.1	1318.3	1.6494	2.004	1203.2	1314.5	1.6266
700	2.688	1246.7	1371.1	1.6970	2.227	1244.0	1368.3	1.6751
800	2.943	1287.0	1423.2	1.7301	2.442	1285.4	1421.0	1.7187
900	3.193	1327.6	1475.3	1.7799	2.653	1326.3	1473.6	1.7589
1000	3.440	1368.7	1527.9	1.8172	2.860	1367.7	1526.5	1.7964
1100	3.685	1410.5	1581.0	1.8524	3.066	1409.6	1579.8	1.8317
1200	3.929	1453.0	1634.8	1.8858	3.270	1452.2	1633.8	1.8653
1300	4.172	1496.3	1689.3	1.9177	3.473	1495.6	1688.4	1.8973
	350 psia (T_{sat} = 431.8°F)				400 psia (T_{sat} = 444.7°F)			
Sat.	1.327	1119.0	1204.9	1.4978	1.162	1119.5	1205.5	1.4856
450	1.373	1129.2	1218.2	1.5125	1.175	1122.6	1209.6	1.4901
500	1.491	1154.9	1251.5	1.5482	1.284	1150.1	1245.2	1.5282
550	1.600	1178.3	1281.9	1.5790	1.383	1174.6	1277.0	1.5605
600	1.703	1200.3	1310.6	1.6068	1.476	1197.3	1306.6	1.5892
700	1.898	1242.5	1365.4	1.6562	1.650	1240.4	1362.5	1.6397
800	2.085	1283.8	1418.8	1.7004	1.816	1282.1	1416.6	1.6844
900	2.267	1325.0	1471.8	1.7409	1.978	1323.7	1470.1	1.7252
1000	2.446	1366.6	1525.0	1.7787	2.136	1365.5	1523.6	1.7632
1100	2.624	1408.7	1578.6	1.8142	2.292	1407.8	1577.4	1.7989
1200	2.799	1451.5	1632.8	1.8478	2.446	1450.7	1631.8	1.8327
1300	2.974	1495.0	1687.6	1.8799	2.599	1494.3	1686.8	1.8648

Table A-14 (*Continued*)

Temp., °F	v	u	h	s	v	u	h	s
	450 psia (T_{sat} = 456.4°F)				500 psia (T_{sat} = 467.1°F)			
Sat.	1.033	1119.6	1205.6	1.4746	0.928	1119.4	1205.3	1.4645
500	1.123	1145.1	1238.5	1.5097	0.992	1139.7	1231.5	1.4923
550	1.215	1170.7	1271.9	1.5436	1.079	1166.7	1266.6	1.5279
600	1.300	1194.3	1302.5	1.5732	1.158	1191.1	1298.3	1.5585
700	1.458	1238.2	1359.6	1.6248	1.304	1236.0	1356.7	1.6112
800	1.608	1280.5	1414.4	1.6701	1.441	1278.8	1412.1	1.6571
900	1.752	1322.4	1468.3	1.7113	1.572	1321.0	1466.5	1.6987
1000	1.894	1364.4	1522.2	1.7495	1.701	1363.3	1520.7	1.7471
1100	2.034	1406.9	1576.3	1.7853	1.827	1406.0	1575.1	1.7731
1200	2.172	1450.0	1630.8	1.8192	1.952	1449.2	1629.8	1.8072
1300	2.308	1493.7	1685.9	1.8515	2.075	1493.1	1685.1	1.8395
1400	2.444	1538.1	1741.7	1.8823	2.198	1537.6	1741.0	1.8704
	600 psia (T_{sat} = 486.3°F)				700 psia (T_{sat} = 503.2°F)			
Sat.	0.770	1118.6	1204.1	1.4464	0.656	1117.0	1202.0	1.4305
500	0.795	1128.0	1216.2	1.4592				
550	0.875	1158.2	1255.4	1.4990	0.728	1149.0	1243.2	1.4723
600	0.946	1184.5	1289.5	1.5320	0.793	1177.5	1280.2	1.5081
700	1.073	1231.5	1350.6	1.5872	0.907	1226.9	1344.4	1.5661
800	1.190	1275.4	1407.6	1.6343	1.011	1272.0	1402.9	1.6145
900	1.302	1318.4	1462.9	1.6766	1.109	1315.6	1459.3	1.6576
1000	1.411	1361.2	1517.8	1.7155	1.204	1358.9	1514.9	1.6970
1100	1.517	1404.2	1572.7	1.7519	1.296	1402.4	1570.2	1.7337
1200	1.622	1447.7	1627.8	1.7861	1.387	1446.2	1625.8	1.7682
1300	1.726	1491.7	1683.4	1.8186	1.476	1490.4	1681.7	1.8009
1400	1.829	1536.5	1739.5	1.8497	1.565	1535.3	1738.1	1.8321
	800 psia (T_{sat} = 518.3°F)				900 psia (T_{sat} = 532.1°F)			
Sat.	0.569	1115.0	1199.3	1.4160	0.501	1112.6	1196.0	1.4027
550	0.615	1138.8	1229.9	1.4469	0.527	1127.5	1215.2	1.4219
600	0.677	1170.1	1270.4	1.4861	0.587	1162.2	1260.0	1.4652
650	0.732	1197.2	1305.6	1.5186	0.639	1191.1	1297.5	1.4999
700	0.783	1222.1	1338.0	1.5471	0.686	1217.1	1331.4	1.5297
800	0.876	1268.5	1398.2	1.5969	0.772	1264.9	1393.4	1.5810
900	0.964	1312.9	1455.6	1.6408	0.851	1310.1	1451.9	1.6257
1000	1.048	1356.7	1511.9	1.6807	0.927	1354.5	1508.9	1.6662
1100	1.130	1400.5	1567.8	1.7178	1.001	1398.7	1565.4	1.7036
1200	1.210	1444.6	1623.8	1.7526	1.073	1443.0	1621.7	1.7386
1300	1.289	1489.1	1680.0	1.7854	1.144	1487.8	1687.3	1.7717
1400	1.367	1534.2	1736.6	1.8167	1.214	1533.0	1735.1	1.8031

Table A-14 Properties of water: superheated-vapor table (*Continued*)

Temp., °F	v	u	h	s	v	u	h	s
	1000 psia (T_{sat} = 544.7°F)				1200 psia (T_{sat} = 567.4°F)			
Sat.	0.446	1109.0	1192.4	1.3903	0.362	1103.5	1183.9	1.3673
600	0.514	1153.7	1248.8	1.4450	0.402	1134.4	1223.6	1.4054
650	0.564	1184.7	1289.1	1.4822	0.450	1170.9	1270.8	1.4490
700	0.608	1212.0	1324.6	1.5135	0.491	1201.3	1310.2	1.4837
800	0.688	1261.2	1388.5	1.5665	0.562	1253.7	1378.4	1.5402
900	0.761	1307.3	1448.1	1.6120	0.626	1301.5	1440.4	1.5876
1000	0.831	1352.2	1505.9	1.6530	0.685	1347.5	1499.7	1.6297
1100	0.898	1396.8	1562.9	1.6908	0.743	1393.0	1557.9	1.6682
1200	0.963	1441.5	1619.7	1.7261	0.798	1438.3	1615.5	1.7040
1300	1.027	1486.5	1676.5	1.7593	0.853	1483.8	1673.1	1.7377
1400	1.091	1531.9	1733.7	1.7909	0.906	1529.6	1730.7	1.7696
1600	1.215	1624.4	1849.3	1.8499	1.011	1622.6	1847.1	1.8290
	1400 psia (T_{sat} = 587.2°F)				1600 psia (T_{sat} = 605.1°F)			
Sat.	0.302	1096.0	1174.1	1.3461	0.255	1087.4	1162.9	1.3258
600	0.318	1110.9	1193.1	1.3641				
650	0.367	1155.5	1250.5	1.4171	0.303	1137.8	1227.4	1.3852
700	0.406	1189.6	1294.8	1.4562	0.342	1177.0	1278.1	1.4299
800	0.471	1245.8	1367.9	1.5168	0.403	1237.7	1357.0	1.4953
900	0.529	1295.6	1432.5	1.5661	0.466	1289.5	1424.4	1.5468
1000	0.582	1342.8	1493.5	1.6094	0.504	1338.0	1487.1	1.5913
1100	0.632	1389.1	1552.8	1.6487	0.549	1385.2	1547.7	1.6315
1200	0.681	1435.1	1611.4	1.6851	0.592	1431.8	1607.1	1.6684
1300	0.728	1481.1	1669.6	1.7192	0.634	1478.3	1666.1	1.7029
1400	0.774	1527.2	1727.8	1.7513	0.675	1524.9	1724.8	1.7354
1600	0.865	1620.8	1844.8	1.8111	0.755	1619.0	1842.6	1.7955
	1800 psia (T_{sat} = 621.2°F)				2000 psia (T_{sat} = 636.0°F)			
Sat.	0.218	1077.7	1150.4	1.3060	0.188	1066.6	1136.3	1.2861
650	0.251	1117.0	1200.4	1.3517	0.206	1091.1	1167.2	1.3141
700	0.291	1163.1	1259.9	1.4042	0.249	1147.7	1239.8	1.3782
750	0.322	1198.6	1305.9	1.4430	0.280	1187.3	1291.1	1.4216
800	0.350	1229.1	1345.7	1.4753	0.307	1220.1	1333.8	1.4562
900	0.399	1283.2	1416.1	1.5291	0.353	1276.8	1407.6	1.5126
1000	0.443	1333.1	1480.7	1.5749	0.395	1328.1	1474.1	1.5598
1100	0.484	1381.2	1542.5	1.6159	0.433	1377.2	1537.2	1.6017
1200	0.524	1428.5	1602.9	1.6534	0.469	1425.2	1598.6	1.6398
1300	0.561	1475.5	1662.5	1.6883	0.503	1472.7	1659.0	1.6751
1400	0.598	1522.5	1721.8	1.7211	0.537	1520.2	1718.8	1.7082
1600	0.670	1617.2	1840.4	1.7817	0.602	1615.4	1838.2	1.7692

Table A-14 *(Continued)*

Temp., °F	v	u	h	s	v	u	h	s
	2500 psia (T_{sat} = 668.3°F)				3000 psia (T_{sat} = 695.5°F)			
Sat.	0.1306	1031.0	1091.4	1.2327	0.0840	968.8	1015.5	1.1575
700	0.1684	1098.7	1176.6	1.3073	0.0977	1003.9	1058.1	1.1944
750	0.2030	1155.2	1249.1	1.3686	0.1483	1114.7	1197.1	1.3122
800	0.2291	1195.7	1301.7	1.4112	0.1757	1167.6	1265.2	1.3675
900	0.2712	1259.9	1385.4	1.4752	0.2160	1241.8	1361.7	1.4414
1000	0.3069	1315.2	1457.2	1.5262	0.2485	1301.7	1439.6	1.4967
1100	0.3393	1366.8	1523.8	1.5704	0.2772	1356.2	1510.1	1.5434
1200	0.3696	1416.7	1587.7	1.6101	0.3086	1408.0	1576.6	1.5848
1300	0.3984	1465.7	1650.0	1.6465	0.3285	1458.5	1640.9	1.6224
1400	0.4261	1514.2	1711.3	1.6804	0.3524	1508.1	1703.7	1.6571
1500	0.4531	1562.5	1772.1	1.7123	0.3754	1557.3	1765.7	1.6896
1600	0.4795	1610.8	1832.6	1.7424	0.3978	1606.3	1827.1	1.7201
	3500 psia				4000 psia			
650	0.0249	663.5	679.7	0.8630	0.0245	657.7	675.8	0.8574
700	0.0306	759.5	779.3	0.9506	0.0287	742.1	763.4	0.9345
750	0.1046	1058.4	1126.1	1.2440	0.0633	960.7	1007.5	1.1395
800	0.1363	1134.7	1223.0	1.3226	0.1052	1095.0	1172.9	1.2740
900	0.1763	1222.4	1336.5	1.4096	0.1462	1201.5	1309.7	1.3789
1000	0.2066	1287.6	1421.4	1.4699	0.1752	1272.9	1402.6	1.4449
1100	0.2328	1345.2	1496.0	1.5193	0.1995	1333.9	1481.6	1.4973
1200	0.2566	1399.2	1565.3	1.5624	0.2213	1390.1	1553.9	1.5423
1300	0.2787	1451.1	1631.7	1.6012	0.2414	1443.7	1622.4	1.5823
1400	0.2997	1501.9	1696.1	1.6368	0.2603	1495.7	1688.4	1.6188
1500	0.3199	1552.0	1759.2	1.6699	0.2784	1546.7	1752.8	1.6526
1600	0.3395	1601.7	1831.6	1.7010	0.2959	1597.1	1816.1	1.6841
	4400 psia				4800 psia			
650	0.0242	653.6	673.3	0.8535	0.0237	649.8	671.0	0.8499
700	0.0278	732.7	755.3	0.9257	0.0271	725.1	749.1	0.9187
750	0.0415	870.8	904.6	1.0513	0.0352	832.6	863.9	1.0154
800	0.0844	1056.5	1125.3	1.2306	0.0668	1011.2	1070.5	1.1827
900	0.1270	1183.7	1287.1	1.3548	0.1109	1164.8	1263.4	1.3310
1000	0.1552	1260.8	1387.2	1.4260	0.1385	1248.3	1317.4	1.4078
1100	0.1784	1324.7	1469.9	1.4809	0.1608	1315.3	1458.1	1.4653
1200	0.1989	1382.8	1544.7	1.5274	0.1802	1375.4	1535.4	1.5133
1300	0.2176	1437.7	1614.9	1.5685	0.1979	1431.7	1607.4	1.5555
1400	0.2352	1490.7	1682.3	1.6057	0.2143	1485.7	1676.1	1.5934
1500	0.2520	1542.7	1747.6	1.6399	0.2300	1538.2	1742.5	1.6282
1600	0.2681	1593.4	1811.7	1.6718	0.2450	1589.8	1807.4	1.6605

Source: Keenan, J. H., F. G. Keyes, P. G. Hill, and J. G. Moore, "Steam Tables," Wiley, New York, 1969.

Table A-15 Properties of water: compressed liquid table

v, ft^3/lb; u and h, Btu/lb; s, Btu/(lb)(°R)

Temp. °F	500 psia (T_{sat} = 467.1°F)				1000 psia (T_{sat} = 544.7°F)			
	v	u	h	s	v	u	h	s
32	0.015994	0.00	1.49	0.00000	0.015967	0.03	2.99	0.00005
50	0.015998	18.02	19.50	0.03599	0.015972	17.99	20.94	0.03592
100	0.016106	67.87	69.36	0.12932	0.016082	67.70	70.68	0.12901
150	0.016318	117.66	119.17	0.21457	0.016293	117.38	120.40	0.21410
200	0.016608	167.65	169.19	0.29341	0.016580	167.26	170.32	0.29281
300	0.017416	268.92	270.53	0.43641	0.017379	268.24	271.46	0.43552
400	0.018608	373.68	375.40	0.56604	0.018550	372.55	375.98	0.56472
Sat.	0.019748	447.70	449.53	0.64904	0.021591	538.39	542.38	0.74320
	1500 psia (T_{sat} = 596.4°F)				2000 psia (T_{sat} = 636.0°F)			
32	0.015939	0.05	4.47	0.00007	0.015912	0.06	5.95	0.00008
50	0.015946	17.95	22.38	0.03584	0.015920	17.91	23.81	0.03575
100	0.016058	67.53	71.99	0.12870	0.016034	67.37	73.30	0.12839
150	0.016268	117.10	121.62	0.21364	0.016244	116.83	122.84	0.21318
200	0.016554	166.87	171.46	0.29221	0.016527	166.49	172.60	0.29162
300	0.017343	267.58	272.39	0.43463	0.017308	266.93	273.33	0.43376
400	0.018493	371.45	376.59	0.56343	0.018439	370.38	377.21	0.56216
500	0.02024	481.8	487.4	0.6853	0.02014	479.8	487.3	0.6832
Sat.	0.02346	605.0	611.5	0.8082	0.02565	662.4	671.9	0.8623
	3000 psia (T_{sat} = 695.5°F)				4000 psia			
32	0.015859	0.09	8.90	0.00009	0.015807	0.10	11.80	0.00005
50	0.015870	17.84	26.65	0.03555	0.015821	17.76	29.47	0.03534
100	0.015987	67.04	75.91	0.12777	0.015942	66.72	78.52	0.12714
150	0.016196	116.30	125.29	0.21226	0.016150	115.77	127.73	0.21136
200	0.016476	165.74	174.89	0.29046	0.016425	165.02	177.18	0.28931
300	0.017240	265.66	275.23	0.43205	0.017174	264.43	277.15	0.43038
400	0.018334	368.32	378.50	0.55970	0.018235	366.35	379.85	0.55734
500	0.019944	476.2	487.3	0.6794	0.019766	472.9	487.5	0.6758
Sat.	0.034310	783.5	802.5	0.9732				

Source: Keenan, J. H., F. G. Keyes, P. G. Hill, and J. G. Moore, "Steam Tables," Wiley, New York, 1969.

Table A-16 Properties of saturated refrigerant 12 (CCl_2F_2): temperature table

v, ft^3/lb; u, Btu/lb; h, Btu/lb; s, Btu/(lb)(°R)

		Specific volume		Internal energy		Enthalpy			Entropy	
Temp. °F T	Press. psi P	Sat. liquid v_f	Sat. vapor v_g	Sat. liquid u_f	Sat. vapor u_g	Sat. liquid h_f	Evap. h_{fg}	Sat. vapor h_g	Sat. liquid s_f	Sat. vapor s_g
−40	9.308	0.01056	3.8750	−0.02	66.24	0	72.91	72.91	0	0.1737
−30	11.999	0.01067	3.0585	1.93	67.22	2.11	71.90	74.01	0.0050	0.1723
−20	15.267	0.01079	2.4429	4.21	68.21	4.24	70.87	75.11	0.0098	0.1710
−15	17.141	0.01085	2.1924	5.27	68.69	5.30	70.35	75.65	0.0122	0.1704
−10	19.189	0.01091	1.9727	6.33	69.19	6.37	69.82	76.19	0.0146	0.1699
−5	21.422	0.01097	1.7794	7.40	69.67	7.44	69.29	76.73	0.0170	0.1694
0	23.849	0.01103	1.6089	8.47	70.17	8.52	68.75	77.27	0.0193	0.1689
5	26.483	0.01109	1.4580	9.55	70.65	9.60	68.20	77.80	0.0216	0.1684
10	29.335	0.01116	1.3241	10.62	71.15	10.68	67.65	78.33	0.0240	0.1680
20	35.736	0.01130	1.0988	12.79	72.12	12.86	66.52	79.38	0.0285	0.1672
25	39.310	0.01137	1.0039	13.88	72.61	13.96	65.95	79.91	0.0308	0.1668
30	43.148	0.01144	0.9188	14.97	73.08	15.06	65.36	80.42	0.0330	0.1665
40	51.667	0.01159	0.7736	17.16	74.04	17.27	64.16	81.43	0.0375	0.1659
50	61.394	0.01175	0.6554	19.38	74.99	19.51	62.93	82.44	0.0418	0.1653
60	72.433	0.01191	0.5584	21.61	75.92	21.77	61.64	83.41	0.0462	0.1648
70	84.89	0.01209	0.4782	23.86	76.85	24.05	60.31	84.36	0.0505	0.1643
80	98.87	0.01228	0.4114	26.14	77.76	26.37	58.92	85.29	0.0548	0.1639
85	106.47	0.01238	0.3821	27.29	78.20	27.53	58.20	85.73	0.0569	0.1637
90	114.49	0.01248	0.3553	28.45	78.65	28.71	57.46	86.17	0.0590	0.1635
95	122.95	0.01258	0.3306	29.61	79.09	29.90	56.71	86.61	0.0611	0.1633
100	131.86	0.01269	0.3079	30.79	79.51	31.10	55.93	87.03	0.0632	0.1632
105	141.25	0.01281	0.2870	31.98	79.94	32.31	55.13	87.44	0.0653	0.1630
110	151.11	0.01292	0.2677	33.16	80.36	33.53	54.31	87.84	0.0675	0.1628
115	161.47	0.01305	0.2498	34.37	80.76	34.76	53.47	88.23	0.0696	0.1626
120	172.35	0.01317	0.2333	35.59	81.17	36.01	52.60	88.61	0.0717	0.1624
140	221.32	0.01375	0.1780	40.60	82.68	41.16	48.81	89.97	0.0802	0.1616
160	279.82	0.01445	0.1360	45.88	83.96	46.63	44.37	91.00	0.0889	0.1605
180	349.00	0.01536	0.1033	51.57	84.89	52.56	39.00	91.56	0.0980	0.1590
200	430.09	0.01666	0.0767	57.87	85.17	59.20	32.08	91.28	0.1079	0.1565
233.6	596.9	0.0287	0.0287	75.69	75.69	78.86	0	78.86	0.1359	0.1359

Source: Data from Freon Products Division, E. I. du Pont de Nemours & Company, 1956.

Table A-17 Properties of saturated refrigerant 12 (CCl_2F_2): pressure table

v, ft^3/lb; u, Btu/lb; h, Btu/lb; s, Btu/(lb)(°R)

		Specific volume		Internal energy		Enthalpy			Entropy	
Press. psi P	Temp. °F T	Sat. liquid v_f	Sat. vapor v_g	Sat. liquid u_f	Sat. vapor u_g	Sat. liquid h_f	Evap. h_{fg}	Sat. vapor h_g	Sat. liquid s_f	Sat. vapor s_g
5	−62.35	0.0103	6.9069	−4.69	64.04	−4.68	75.11	70.43	−0.0114	0.1776
10	−37.23	0.0106	3.6246	0.56	66.51	0.58	72.64	73.22	0.0014	0.1733
15	−20.75	0.0108	2.4835	4.05	68.13	4.08	70.95	75.03	0.0095	0.1711
20	−8.13	0.0109	1.8977	6.73	69.37	6.77	69.63	76.40	0.0155	0.1697
30	11.11	0.0112	1.2964	10.86	71.25	10.93	67.53	78.45	0.0245	0.1679
40	25.93	0.0114	0.9874	14.08	72.69	14.16	65.84	80.00	0.0312	0.1668
50	38.15	0.0116	0.7982	16.75	73.86	16.86	64.39	81.25	0.0366	0.1660
60	48.64	0.0117	0.6701	19.07	74.86	19.20	63.10	82.30	0.0413	0.1654
70	57.90	0.0119	0.5772	21.13	75.73	21.29	61.92	83.21	0.0453	0.1649
80	66.21	0.0120	0.5068	23.00	76.50	23.18	60.82	84.00	0.0489	0.1645
90	73.79	0.0122	0.4514	24.72	77.20	24.92	59.79	84.71	0.0521	0.1642
100	80.76	0.0123	0.4067	26.31	77.82	26.54	58.81	85.35	0.0551	0.1639
120	93.29	0.0126	0.3389	29.21	78.93	29.49	56.97	86.46	0.0604	0.1634
140	104.35	0.0128	0.2896	31.82	79.89	32.15	55.24	87.39	0.0651	0.1630
160	114.30	0.0130	0.2522	34.21	80.71	34.59	53.59	88.18	0.0693	0.1626
180	123.38	0.0133	0.2228	36.42	81.44	36.86	52.00	88.86	0.0731	0.1623
200	131.74	0.0135	0.1989	38.50	82.08	39.00	50.44	89.44	0.0767	0.1620
220	139.51	0.0137	0.1792	40.48	82.08	41.03	48.90	89.94	0.0816	0.1616
240	146.77	0.0140	0.1625	42.35	83.14	42.97	47.39	90.36	0.0831	0.1613
260	153.60	0.0142	0.1483	44.16	83.58	44.84	45.88	90.72	0.0861	0.1609
280	160.06	0.0145	0.1359	45.90	83.97	46.65	44.36	91.01	0.0890	0.1605
300	166.18	0.0147	0.1251	47.59	84.30	48.41	42.83	91.24	0.0917	0.1601

Source: Data from Freon Products Division, E. I. du Pont de Nemours & Company, 1956.

Table A-18 Properties of superheated refrigerant 12 (CCl_2F_2)

v, ft^3/lb; u, Btu/lb; h, Btu/lb; s, Btu/(lb)(°R)

Temp., °F	v	u	h	s	v	u	h	s
	10 psia (T_{sat} = −37.23°F)				15 psia (T_{sat} = −20.75°F)			
Sat.	3.6246	66.512	73.219	0.1733	2.4835	68.134	75.028	0.1711
0	3.9809	70.879	78.246	0.1847	2.6201	70.629	77.902	0.1775
20	4.1691	73.299	81.014	0.1906	2.7494	73.080	80.712	0.1835
40	4.3556	75.768	83.828	0.1964	2.8770	75.575	83.561	0.1893
60	4.5408	78.286	86.689	0.2020	3.0031	78.115	86.451	0.1950
80	4.7248	80.853	89.596	0.2075	3.1281	80.700	89.383	0.2005
100	4.9079	83.466	92.548	0.2128	3.2521	83.330	92.357	0.2059
120	5.0903	86.126	95.546	0.2181	3.3754	86.004	95.373	0.2112
140	5.2720	88.830	98.586	0.2233	3.4981	88.719	98.429	0.2164
160	5.4533	91.578	101.669	0.2283	3.6202	91.476	101.525	0.2215
180	5.6341	94.367	104.793	0.2333	3.7419	94.274	104.661	0.2265
200	5.8145	97.197	107.957	0.2381	3.8632	97.112	107.835	0.2314
	20 psia (T_{sat} = −8.13°F)				30 psia (T_{sat} = 11.11°F)			
Sat.	1.8977	69.374	76.397	0.1697	1.2964	71.255	78.452	0.1679
20	2.0391	72.856	80.403	0.1783	1.3278	72.394	79.765	0.1707
40	2.1373	75.379	83.289	0.1842	1.3969	74.975	82.730	0.1767
60	2.2340	77.942	86.210	0.1899	1.4644	77.586	85.716	0.1826
80	2.3295	80.546	89.168	0.1955	1.5306	80.232	88.729	0.1883
100	2.4241	83.192	92.164	0.2010	1.5957	82.911	91.770	0.1938
120	2.5179	85.879	95.198	0.2063	1.6600	85.627	94.843	0.1992
140	2.6110	88.607	98.270	0.2115	1.7237	88.379	97.948	0.2045
160	2.7036	91.374	101.380	0.2166	1.7868	91.166	101.086	0.2096
180	2.7957	94.181	104.528	0.2216	1.8494	93.991	104.258	0.2146
200	2.8874	97.026	107.712	0.2265	1.9116	96.852	107.464	0.2196
220	2.9789	99.907	110.932	0.2313	1.9735	99.746	110.702	0.2244
	40 psia (T_{sat} = 25.93°F)				50 psia (T_{sat} = 38.15°F)			
Sat.	0.9874	72.691	80.000	0.1668	0.7982	73.863	81.249	0.1660
40	1.0258	74.555	82.148	0.1711	0.8025	74.115	81.540	0.1666
60	1.0789	77.220	85.206	0.1771	0.8471	76.838	84.676	0.1727
80	1.1306	79.908	88.277	0.1829	0.8903	79.574	87.811	0.1786
100	1.1812	82.624	91.367	0.1885	0.9322	82.328	90.953	0.1843
120	1.2309	85.369	94.480	0.1940	0.9731	85.106	94.110	0.1899
140	1.2798	88.147	97.620	0.1993	1.0133	87.910	97.286	0.1953
160	1.3282	90.957	100.788	0.2045	1.0529	90.743	100.485	0.2005
180	1.3761	93.800	103.985	0.2096	1.0920	93.604	103.708	0.2056
200	1.4236	96.674	107.212	0.2146	1.1307	96.496	106.958	0.2106
220	1.4707	99.583	110.469	0.2194	1.1690	99.419	110.235	0.2155
240	1.5176	102.524	113.757	0.2242	1.2070	102.371	113.539	0.2203

Table A-18 Properties of superheated refrigerant 12 (CCl_2F_2) (*Continued*)

Temp., °F	v	u	h	s	v	u	h	s
	60 psia (T_{sat} = 48.64°F)				70 psia (T_{sat} = 57.90°F)			
Sat.	0.6701	74.859	82.299	0.1654	0.5772	75.729	83.206	0.1649
60	0.6921	76.442	84.126	0.1689	0.5809	76.027	83.552	0.1656
80	0.7296	79.229	87.330	0.1750	0.6146	78.871	86.832	0.1718
100	0.7659	82.024	90.528	0.1808	0.6469	81.712	90.091	0.1777
120	0.8011	84.836	93.731	0.1864	0.6780	84.560	93.343	0.1834
140	0.8355	87.668	96.945	0.1919	0.7084	87.421	96.597	0.1889
160	0.8693	90.524	100.776	0.1972	0.7380	90.302	99.862	0.1943
180	0.9025	93.406	103.427	0.2023	0.7671	93.205	103.141	0.1995
200	0.9353	96.315	106.700	0.2074	0.7957	96.132	106.439	0.2046
220	0.9678	99.252	109.997	0.2123	0.8240	99.083	109.756	0.2095
240	0.9998	102.217	113.319	0.2171	0.8519	102.061	113.096	0.2144
260	1.0318	105.210	116.666	0.2218	0.8796	105.065	116.459	0.2191
	80 psia (T_{sat} = 66.21°F)				90 psia (T_{sat} = 73.79°F)			
Sat.	0.5068	76.500	84.003	0.1645	0.4514	77.194	84.713	0.1642
80	0.5280	78.500	86.316	0.1689	0.4602	78.115	85.779	0.1662
100	0.5573	81.389	89.640	0.1749	0.4875	81.056	89.175	0.1723
120	0.5856	84.276	92.945	0.1807	0.5135	83.984	92.536	0.1782
140	0.6129	87.169	96.242	0.1863	0.5385	86.911	95.879	0.1839
160	0.6394	90.076	99.542	0.1917	0.5627	89.845	99.216	0.1894
180	0.6654	93.000	102.851	0.1970	0.5863	92.793	102.557	0.1947
200	0.6910	95.945	106.174	0.2021	0.6094	95.755	105.905	0.1998
220	0.7161	98.912	109.513	0.2071	0.6321	98.739	109.267	0.2049
240	0.7409	101.904	112.872	0.2119	0.6545	101.743	112.644	0.2098
260	0.7654	104.919	116.251	0.2167	0.6766	104.771	116.040	0.2146
280	0.7898	107.960	119.652	0.2214	0.6985	107.823	119.456	0.2192
	100 psia (T_{sat} = 80.76°F)				120 psia (T_{sat} = 93.29°F)			
Sat.	0.4067	77.824	85.351	0.1639	0.3389	78.933	86.459	0.1634
100	0.4314	80.711	88.694	0.1700	0.3466	79.978	87.675	0.1656
120	0.4556	83.685	92.116	0.1760	0.3684	83.056	91.237	0.1718
140	0.4788	86.647	95.507	0.1817	0.3890	86.098	94.736	0.1778
160	0.5012	89.610	98.884	0.1873	0.4087	89.123	98.199	0.1835
180	0.5229	92.580	102.257	0.1926	0.4277	92.144	101.642	0.1889
200	0.5441	95.564	105.633	0.1978	0.4461	95.170	105.076	0.1942
220	0.5649	98.564	109.018	0.2029	0.4640	98.205	108.509	0.1993
240	0.5854	101.582	112.415	0.2078	0.4816	101.253	111.948	0.2043
260	0.6055	104.622	115.828	0.2126	0.4989	104.317	115.396	0.2092
280	0.6255	107.684	119.258	0.2173	0.5159	107.401	118.857	0.2139
300	0.6452	110.768	122.707	0.2219	0.5327	110.504	122.333	0.2186

Table A-18 (*Continued*)

Temp., °F	v	u	h	s	v	u	h	s
	140 psia (T_{sat} = 104.35°F)				160 psia (T_{sat} = 114.30°F)			
Sat.	0.2896	79.886	87.389	0.1630	0.2522	80.713	88.180	0.1626
120	0.3055	82.382	90.297	0.1681	0.2576	81.656	89.283	0.1645
140	0.3245	85.516	93.923	0.1742	0.2756	84.899	93.059	0.1709
160	0.3423	88.615	97.483	0.1801	0.2922	88.080	96.732	0.1770
180	0.3594	91.692	101.003	0.1857	0.3080	91.221	100.340	0.1827
200	0.3758	94.765	104.501	0.1910	0.3230	94.344	103.907	0.1882
220	0.3918	97.837	107.987	0.1963	0.3375	97.457	107.450	0.1935
240	0.4073	100.918	111.470	0.2013	0.3516	100.570	110.980	0.1986
260	0.4226	104.008	114.956	0.2062	0.3653	103.690	114.506	0.2036
280	0.4375	107.115	118.449	0.2110	0.3787	106.820	118.033	0.2084
300	0.4523	110.235	121.953	0.2157	0.3919	109.964	121.567	0.2131
320	0.4668	113.376	125.470	0.2202	0.4049	113.121	125.109	0.2177
	180 psia (T_{sat} = 123.38°F)				200 psia (T_{sat} = 131.74°F)			
Sat.	0.2228	81.436	88.857	0.1623	0.1989	82.077	89.439	0.1620
140	0.2371	84.238	92.136	0.1678	0.2058	83.521	91.137	0.1648
160	0.2530	87.513	95.940	0.1741	0.2212	86.913	95.100	0.1713
180	0.2678	90.727	99.647	0.1800	0.2354	90.211	98.921	0.1774
200	0.2818	93.904	103.291	0.1856	0.2486	93.451	102.652	0.1831
220	0.2952	97.063	106.896	0.1910	0.2612	96.659	106.325	0.1886
240	0.3081	100.215	110.478	0.1961	0.2732	99.850	109.962	0.1939
260	0.3207	103.364	114.046	0.2012	0.2849	103.032	113.576	0.1990
280	0.3329	106.521	117.610	0.2061	0.2962	106.214	117.178	0.2039
300	0.3449	109.686	121.174	0.2108	0.3073	109.402	120.775	0.2087
320	0.3567	112.863	124.744	0.2155	0.3182	112.598	124.373	0.2134
340	0.3683	116.053	128.321	0.2200	0.3288	115.805	127.974	0.2179
	300 psia (T_{sat} = 166.18°F)				400 psia (T_{sat} = 192.93°F)			
Sat.	0.1251	84.295	91.240	0.1601	0.0856	85.178	91.513	0.1576
180	0.1348	87.071	94.556	0.1654				
200	0.1470	90.816	98.975	0.1722	0.0910	86.982	93.718	0.1609
220	0.1577	94.379	103.136	0.1784	0.1032	91.410	99.046	0.1689
240	0.1676	97.835	107.140	0.1842	0.1130	95.371	103.735	0.1757
260	0.1769	101.225	111.043	0.1897	0.1216	99.102	108.105	0.1818
280	0.1856	104.574	114.879	0.1950	0.1295	102.701	112.286	0.1876
300	0.1940	107.899	118.670	0.2000	0.1368	106.217	116.343	0.1930
320	0.2021	111.208	122.430	0.2049	0.1437	109.680	120.318	0.1981
340	0.2100	114.512	126.171	0.2096	0.1503	113.108	124.235	0.2031
360	0.2177	117.814	129.900	0.2142	0.1567	116.514	128.112	0.2079

Source: Data from Freon Products Division, E. I. du Pont de Nemours & Company, 1956.

Table A-19 Specific heats of some common liquids and solids

c_p, Btu/(lb)(°F)

A. Liquids

Substance	State	c_p	Substance	State	c_p
Water	1 atm, 32°F	1.007	Glycerin	1 atm, 50°F	0.554
	1 atm, 77°F	0.998		1 atm, 120°F	0.617
	1 atm, 212°F	1.007	Bismuth	1 atm, 800°F	0.0345
Ammonia	sat., 0°F	1.08		1atm, 1400°F	0.0393
	sat., 120°F	1.22	Mercury	1 atm, 50°F	0.033
Refrigerant 12	sat., −40°F	0.211		1 atm, 600°F	0.032
	sat., 0°F	0.217	Sodium	1 atm, 200°F	0.33
	sat., 120°F	0.244		1 atm, 1000°F	0.30
Benzene	1 atm, 60°F	0.43	Propane	1 atm, 32°F	0.576
	1 atm, 150°F	0.46			

B. Solids

Substance	T, °F	c_p	Substance	T, °F	c_p
Ice	−100	0.375	Lead	−455	0.0008
	−50	0.424		−435	0.0073
	0	0.471		−150	0.0283
	20	0.491		32	0.0297
	32	0.502		210	0.0320
Aluminum	−150	0.167		570	0.0356
	−100	0.192	Copper	−240	0.0674
	32	0.212		−150	0.0784
	100	0.218		−60	0.0862
	200	0.224		0	0.0893
	300	0.229		100	0.0925
	400	0.235		200	0.0938
	500	0.240		390	0.0963
Iron	68	0.107	Silver	68	0.0558

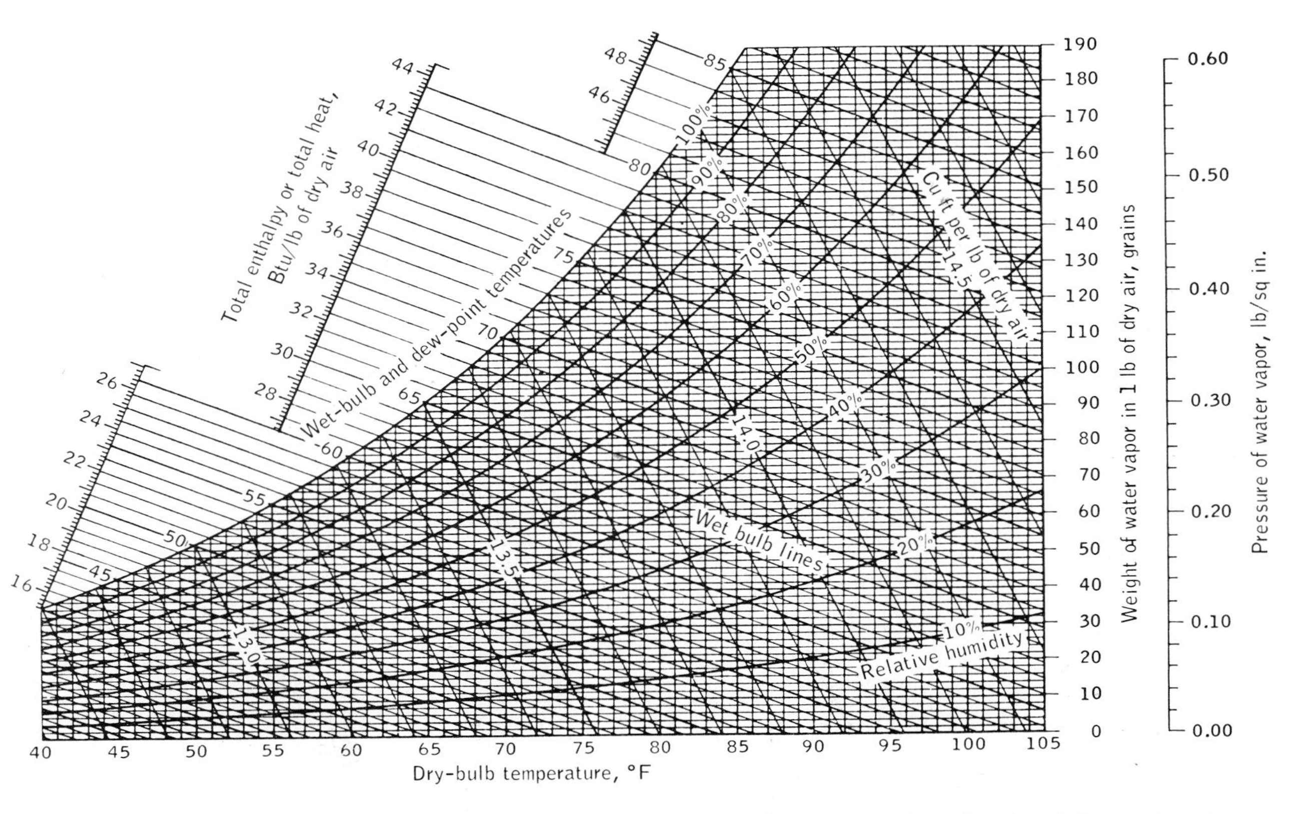

Figure A-20 General Electric psychrometric chart, barometric pressure 14.696 psia. (*Copyright, 1942, by General Electric Company*)

Table A-21 Constants for the Benedict-Webb-Rubin and Redlich-Kwong equations of state

1. Benedict-Webb-Rubin: Units are atm, $ft^3/lb \cdot mol$, and °R

Constants	*n*-Butane C_4H_{10}	Carbon dioxide, CO_2	Carbon monoxide, CO	Methane CH_4	Nitrogen N_2
a	7747	563.1	150.7	203.1	103.2
A_0	2590	703.0	344.5	476.4	270.6
b	10.27	1.852	0.676	0.868	0.598
B_0	1.993	0.7998	0.8740	0.6827	0.6529
c	4.219×10^9	1.989×10^8	1.387×10^7	3.393×10^7	9.713×10^6
C_0	8.263×10^8	1.153×10^8	7.124×10^6	1.878×10^7	6.706×10^6
α	4.531	0.3486	0.5556	0.5120	0.5235
γ	8.732	1.384	1.541	1.541	1.361

Source: Cooper, H. W., and J. C. Goldfrank, *Hydrocarbon Processing*, **46** (12): 141 (1967).

2. Redlich-Kwong: Units are atm, $ft^3/lb \cdot mol$, and °R

Substance	a	b
Carbon dioxide, CO_2	21,970	0.4757
Carbon monoxide, CO	5,870	0.4395
Methane, CH_4	10,930	0.4757
Nitrogen, N_2	5,300	0.4294
Oxygen, O_2	5,900	0.3522
Propane, C_3H_8	62,190	1.0040
Refrigerant 12, CCl_2F_2	72,710	1.1080
Water, H_2O	48,460	0.3381

Source: Computed from critical data.

Table A-22 Values of the enthalpy of formation, Gibbs function of formation, absolute entropy, and enthalpy of vaporization at 77°F and 1 atm

Δh_f^0, Δg_f^0, and h_{fg} in Btu/lb · mol, s^0 in Btu/(lb · mol)(°R)

Substance	Formula	Δh_f^0	Δg_f^0	s^0	h_{fg}
Carbon	$C(s)$	0	0	1.36	
Hydrogen	$H_2(g)$	0	0	31.21	
Nitrogen	$N_2(g)$	0	0	45.77	
Oxygen	$O_2(g)$	0	0	49.00	
Carbon monoxide	$CO(g)$	−47,540	−59,010	47.21	
Carbon dioxide	$CO_2(g)$	−169,300	−169.680	51.07	
Water	$H_2O(g)$	−104,040	−98,350	45.11	
Water	$H_2O(l)$	−122,970	−102,040	16.71	
Hydrogen peroxide	$H_2O_2(g)$	−58,640	−45,430	55.60	
Ammonia	$NH_3(g)$	−19,750	−7,140	45.97	
Methane	$CH_4(g)$	−32,210	−21,860	44.49	
Acetylene	$C_2H_2(g)$	+97,540	+87,990	48.00	
Ethylene	$C_2H_4(g)$	+22,490	+29,306	52.54	
Ethane	$C_2H_6(g)$	−36,420	−14,150	54.85	
Propylene	$C_3H_6(g)$	+8,790	+26,980	63.80	
Propane	$C_3H_8(g)$	−44,680	−10,105	64.51	6,480
n-Butane	$C_4H_{10}(g)$	−54,270	−6,760	74.11	9,060
n-Octane	$C_8H_{18}(g)$	−89,680	+7,110	111.55	
n-Octane	$C_8H_{18}(l)$	−107,530	+2,840	86.23	17,835
Benzene	$C_6H_6(g)$	+35,680	+55,780	64.34	14,550
Methyl alcohol	$CH_3OH(g)$	−86,540	−69,700	57.29	
Methyl alcohol	$CH_3OH(l)$	−102,670	−71,570	30.30	16,090
Ethyl alcohol	$C_2H_5OH(g)$	−101,230	−72,520	67.54	
Ethyl alcohol	$C_2H_5OH(l)$	−119,470	−75,240	38.40	18,220
Oxygen	$O(g)$	+107,210	+99,710	38.47	
Hydrogen	$H(g)$	+93,780	+87,460	27.39	
Nitrogen	$N(g)$	+203,340	+195,970	36.61	
Hydroxyl	$OH(g)$	+16,790	+14,750	43.92	

Sources: From the JANAF Thermochemical Tables, NSRDS-NBS-37, 1971; Selected Values of Chemical Thermodynamic Properties, *NBS Tech. Note* 270-3, 1968; and API Res. Project 44, Carnegie Press, Carnegie Institute of Technology, Pittsburgh, Pa., 1953.

Table A-23 Thermodynamic properties of potassium

T, °R; P, psia; v, ft³/lb; h, Btu/lb; s, Btu/(lb)(°R)

A. Saturation temperature table

T	P	v_f	v_g	h_f	h_g	s_f	s_g
1500	1.523	0.0225	261.13	291.7	1170.8	0.6279	1.2140
1600	3.210	0.0229	130.62	310.4	1176.6	0.6400	1.1813
1700	6.185	0.0234	71.09	329.6	1181.8	0.6516	1.1529
1800	11.060	0.0239	41.49	349.3	1186.6	0.6628	1.1280

B. Superheat table

P	v	h	s	P	v	h	s
2000°R (P_{sat} = 29.58 psia)				2200°R (P_{sat} = 65.83 psia)			
Sat.	16.71	1195.0	1.0867	Sat.	8.01	1203.5	1.0547
20.0	25.58	1213.7	1.1143	50.0	10.89	1220.8	1.0749
14.0	27.34	1225.5	1.1372	40.0	13.89	1232.0	1.0904
8.0	66.73	1237.3	1.1705	30.0	18.91	1243.5	1.1092
2400°R (P_{sat} = 127.7 psia)				2600°R (P_{sat} = 222.9 psia)			
Sat.	4.37	1214.0	1.0303	Sat.	2.63	1224.4	1.0108
110	5.19	1225.2	1.0415	150	4.16	1255.2	1.0398
90	6.50	1238.5	1.0559	130	4.88	1264.3	1.0497
70	8.57	1252.3	1.0732	110	5.87	1273.5	1.0608
50	12.32	1266.5	1.0949	90	7.30	1283.1	1.0738

Source: Air Force Aero Propulsion Laboratory Technical Report 66-104, 1966.

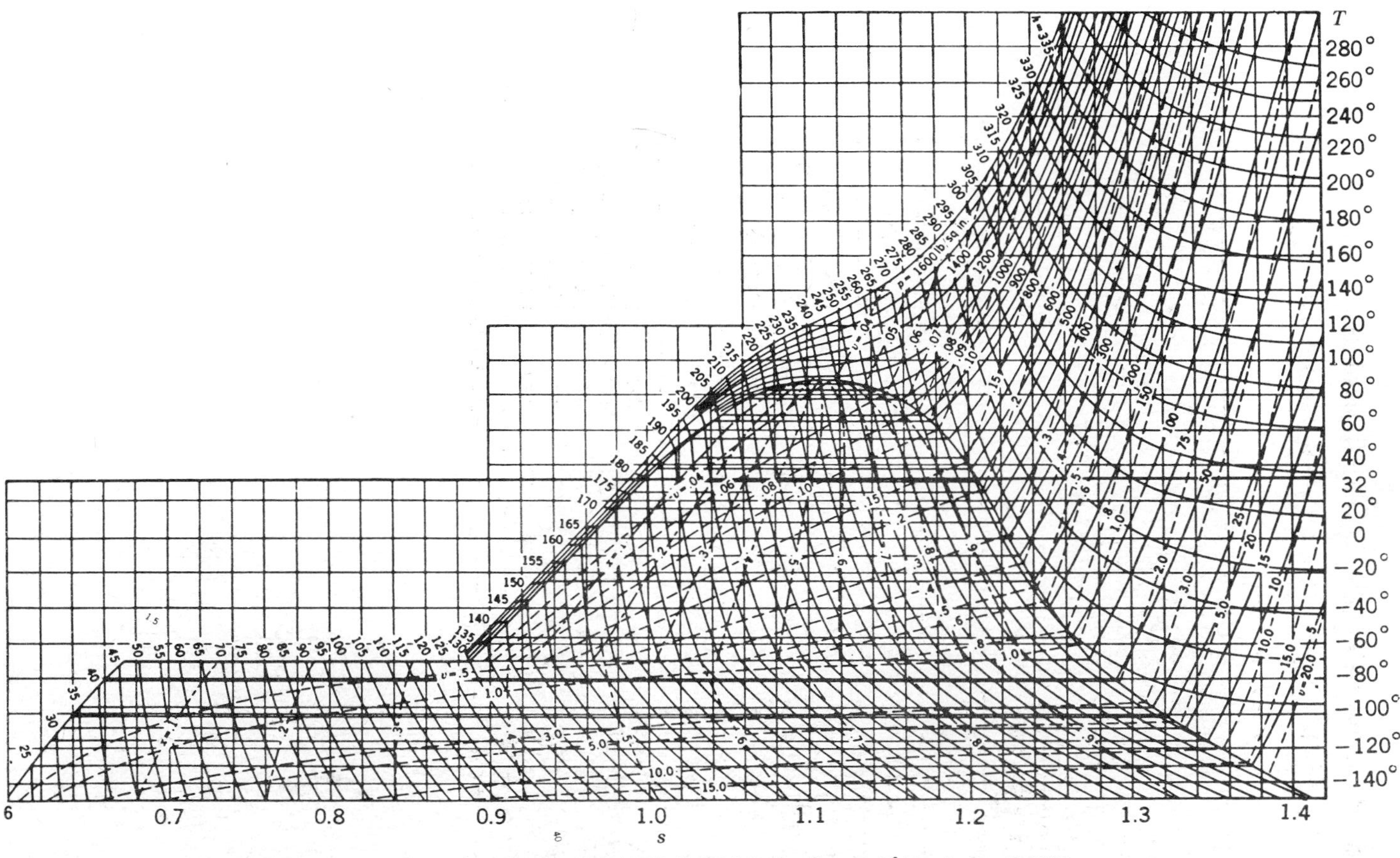

Figure A-24 Temperature-entropy diagram for carbon dioxide (CO_2). T is in °F, h in Btu/lb, v in ft³/lb, s in Btu/(lb)(°R).

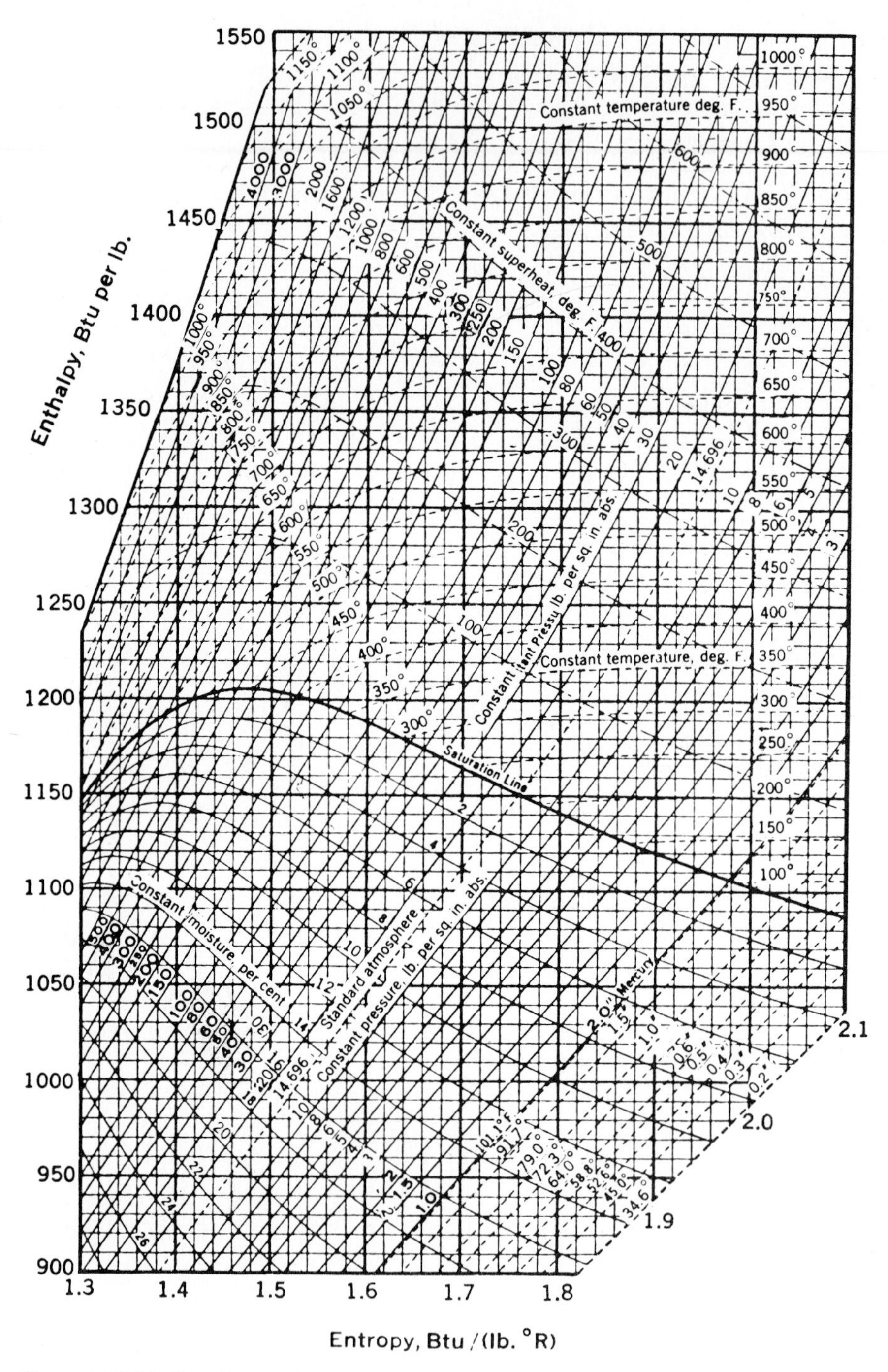

Figure A-25 Mollier diagram for steam. Source: J. H. Keenan and J. Keyes, "Thermodynamic Properties of Steam," Wiley, New York, 1936.

Table A-26 Properties of saturated nitrogen (N_2): temperature table

v, ft^3/lb; u and h, Btu/lb; s, Btu/(lb)(°R)

		Specific volume		Internal energy		Enthalpy			Entropy	
Temp. °R T	Press. psia P	Sat. liquid v_f	Sat. vapor v_g	Sat. liquid u_f	Sat. vapor u_g	Sat. liquid h_f	Evap. h_{fg}	Sat. vapor h_g	Sat. liquid s_f	Sat. vapor s_g
113.7*	1.82	0.0185	23.73	35.32	119.91	35.33	92.59	127.90	0.5802	1.395
120.0	3.34	0.0188	13.56	38.33	120.93	38.34	90.98	129.32	0.6060	1.365
130.0	7.67	0.0193	6.321	43.19	122.47	43.22	88.22	131.44	0.6449	1.324
139.2	14.70	0.0198	3.473	47.71	123.77	47.77	85.45	133.22	0.6785	1.293
140.0	15.46	0.0199	3.315	48.09	123.87	48.15	85.21	133.36	0.6812	1.290
150.0	28.19	0.0205	1.899	53.02	125.11	53.13	81.89	135.02	0.7153	1.262
160.0	47.52	0.0213	1.164	58.00	126.14	58.19	78.19	136.38	0.7474	1.236
170.0	75.18	0.0222	0.750	63.08	126.91	63.39	73.96	137.35	0.7782	1.214
180.0	113.0	0.0233	0.502	68.30	127.36	68.79	69.07	137.86	0.8082	1.192
190.0	162.8	0.0246	0.344	73.75	127.40	74.49	63.28	137.77	0.8378	1.171
200.0	226.9	0.0262	0.239	79.52	126.84	80.62	56.24	136.86	0.8677	1.149
210.0	307.3	0.0285	0.164	85.86	125.33	87.48	47.20	134.68	0.8992	1.124
220.0	406.9	0.0325	0.107	93.57	121.80	96.01	33.85	129.86	0.9363	1.090
226.0	477.9	0.0394	0.071	101.46	115.66	104.95	16.97	121.92	0.9742	1.049
227.2*	493.1	0.051	0.051	108.59	108.59	113.25	0	113.25	1.010	1.010

* 113.7°R is the triple state and 227.2°R is the critical state.

Source: Adapted from National Bureau of Standards Technical Note 648, 1973.

Table A-27 Properties of nitrogen (N_2): superheated-vapor table

v, ft³/lb; u and h, Btu/lb; s, Btu/(lb)(°R)

Temp. °R	v	u	h	s	v	u	h	s
	20 psia (T_{sat} = 144.1°R)				50 psia (T_{sat} = 161.0°R)			
200	3.755	134.83	148.73	1.364	1.454	133.97	147.44	1.295
250	4.740	143.87	161.42	1.421	1.866	143.30	160.58	1.354
300	5.714	152.82	173.98	1.467	2.266	152.40	173.38	1.400
350	6.682	161.74	186.49	1.505	2.659	161.41	186.03	1.439
400	7.647	170.65	198.97	1.539	3.050	170.37	198.61	1.473
450	8.610	179.54	211.4	1.568	3.438	179.30	211.1	1.502
500	9.572	188.42	223.9	1.594	3.826	188.22	223.6	1.529
550	10.53	197.31	236.3	1.618	4.212	197.13	236.1	1.553
	100 psia (T_{sat} = 176.9°R)				200 psia			
200	0.684	132.40	145.07	1.238	0.289	128.33	139.04	1.166
250	0.908	142.31	159.13	1.300	0.427	140.18	156.01	1.243
300	1.116	151.68	172.36	1.349	0.542	150.19	170.25	1.295
350	1.319	160.84	185.26	1.389	0.649	159.69	183.71	1.336
400	1.518	169.90	198.01	1.423	0.752	168.96	196.81	1.371
450	1.715	178.91	210.7	1.452	0.853	178.12	209.7	1.401
500	1.910	187.88	223.3	1.479	0.953	187.20	222.5	1.428
550	2.105	196.83	235.8	1.503	1.052	196.23	235.2	1.453
	500 psia				1000 psia			
250	0.132	131.52	143.78	1.141	0.038	106.30	113.40	0.994
300	0.197	145.25	163.45	1.213	0.083	135.26	150.59	1.131
350	0.247	156.09	178.96	1.261	0.115	149.67	170.95	1.194
400	0.293	166.10	193.23	1.299	0.142	161.23	187.45	1.238
450	0.337	175.73	206.9	1.331	0.166	171.77	202.5	1.274
500	0.379	185.16	220.3	1.359	0.189	181.81	216.8	1.304
550	0.421	194.46	233.4	1.384	0.211	191.56	230.6	1.330
	2000 psia				3000 psia			
250	0.029	95.00	115.61	0.939	0.026	90.47	114.97	0.917
300	0.040	118.41	133.17	1.040	0.032	110.73	128.60	1.003
350	0.055	137.38	157.78	1.116	0.040	128.97	151.34	1.073
400	0.070	151.93	177.78	1.169	0.049	144.47	171.81	1.128
450	0.083	164.27	195.11	1.210	0.058	157.89	190.16	1.171
500	0.096	175.50	211.0	1.244	0.066	169.99	207.0	1.206
550	0.108	186.11	226.0	1.272	0.075	181.28	222.8	1.237

Source: Adapted from National Bureau of Standards Technical Note 648, 1973.

SYMBOLS

A	Area
	Helmholtz function, $A = U - TS$
	Lagrangian undetermined multiplier
a	Specific Helmholtz function, $a = u - Ts$
	Virial coefficient
	Acceleration
B	Lagrangian undetermined multiplier, $B = 1/kT$
b	Virial coefficient
C	Number of components (in Gibbs's phase rule)
	A constant
COP	Coefficient of performance
CV	Control volume
c	Virial coefficient
	Velocity of a particle
	Speed of light
c_v	Specific heat at constant volume, $(\partial u/\partial T)_v$
c_p	Specific heat at constant pressure, $(\partial h/\partial T)_P$
d	An infinitesimal increase in a point function
E	Stored energy
$\mathbf{E}$	Electric field strength
e	Specific stored energy
F	Force
	Variance or degrees of freedom (in phase rule)
F_k	Measurable generalized force
$F_{k,\mathrm{eq}}$	Equilibrium generalized force
G	Gibbs function, $G = H - TS$
ΔG	Gibbs-function change for a unit reaction
ΔG_T^0	Standard-state Gibbs-function change for a unit reaction
g	Local acceleration of gravity
g_c	Gravitational constant in Newton's law, $F = (1/g_c)ma$
g_i	Degeneracy, or statistical weight
H	Enthalpy, $H = U + PV$
$\mathbf{H}$	Magnetic field strength
Δh_R	Enthalpy of reaction
Δh_f	Enthalpy of formation
Δh_c	Enthalpy of combustion
h	Specific enthalpy, $h = u + Pv$
	Planck's constant
I	Irreversibility
KE	Kinetic energy
K_T	Isothermal coefficient of compressibility
K_p	Equilibrium constant for ideal-gas reactions
k	Specific-heat ratio, c_p/c_v
	Boltzmann's constant
k_s	Spring constant
L	Length

M	Mass, as a dimension
	Molar mass, or molecular weight
$\mathbf{M}$	Magnetization per unit volume
MEP	Mean effective pressure
m	Mass of a substance
$\dot{m}$	Mass flow rate
N	Number of moles
N_A	Avogadro's number
n	Number of particles
	Polytropic constant
P	Pressure
	Number of phases (in the phase rule)
$\mathbf{P}$	Polarization
PE	Potential energy
P_m	Permutations
P_R	Reduced pressure
p_r	Relative pressure
p_i	Probability, n_i/n
	Component pressure
p'_i	Partial pressure
Q	Heat interaction
q	Heat interaction per unit mass
Q_A	Available energy
Q_U	Unavailable energy
Q_e	Electric charge
R	Gas constant, R_u/M
R_u	Universal gas constant
r	Distance between masses
	Compression ratio
r_c	Cutoff ratio
r_p	Pressure ratio
S	Entropy
	Surface tension
s	Specific entropy
	Distance
T	Temperature
t	Time
U	Internal energy
u	Specific internal energy
	Velocity of particle in x direction
Δu_R	Internal energy of reaction
V	Volume
	Velocity
	Electrostatic potential
v	Specific volume
	Vibrational quantum number
v_r	Relative volume
W	Work interaction
	Weight
	Thermodynamic probability
w	Specific weight
X_k	Generalized displacement
x	Quality
	Mole fraction
	Cartesian coordinate

y — Mole fraction in the vapor phase
  Cartesian coordinate
Z — Compressibility factor, $Z = Pv/RT$
z — Height
  Cartesian coordinate
  Molecular partition function

Greek symbols and special notation

β — Isobaric compressibility
  Work ratio for a combined cycle
Δ — A finite increase in a point function or property
δ — Symbol for an infinitesimal increase in a path function
ϵ — Energy of a particle
  Strain
$\mathscr{E}$ — emf, electromotive force
η — Efficiency
$\mathscr{F}$ — Faraday constant
θ — Temperature function
μ_{JT} — Joule-Thomson coefficient
μ_i — Chemical potential of the ith component
ν — Fundamental frequency of an oscillator
  Stoichiometric coefficient
$\Delta\nu$ — Change in the stoichiometric coefficients for a unit reaction
ρ — Density
σ — External heat ratio in a combined cycle
  Stress
τ — Time
  Torque
ϕ — Closed system availability per unit mass
  Equivalence ratio
  Relative humidity
$\boldsymbol{\Psi}$ — Time-dependent wave function
  Total stream availability
ψ — Stream availability per unit mass
ω — Angular velocity
  Specific humidity, humidity ratio
$\sum_i x_i$ — The sum $x_1 + x_2 + \cdots + x_n$
$\prod_i x_i$ — The product $x_1 x_2 x_3 \cdots x_n$
$\equiv$ — Identity symbol, used to specify a definition

Subscripts

a — End state of actual process
c — Critical state
f — Saturated-liquid value
fg — Change in value between saturated-liquid and saturated-vapor phases
g — Saturated-vapor value
H — High temperature (as in T_H and Q_H)
L — Low temperature (as in T_L and Q_L)
m — Mixture value
r — Relative value
s — End state of isentropic process

R	Reduced state
v	Vapor state
x	Property value in wet region
σ	System designation
mp	Most probable macrostate

SELECTED PROBLEM ANSWERS (METRIC)

1-3M (*a*) 32.5 percent, (*b*) 0.814

1-6M 70.6 percent

1-9M (*a*) 0.398, (*b*) \$1,412,000, (*c*) \$70,200

1-12M (*a*) 25°C, 1 bar, 0.855 m^3/kg, 3 m^3, 3.51 kg; (*b*) 11.3 N/m^3

1-15M (*a*) 29.6 km, (*b*) 59.1 km, (*c*) 118.2 km

1-18M (*a*) 0.273, (*b*) 0.00367, (*c*) 6.12

1-21M 1035 mb

1-24M −0.65 bar

1-27M (*a*) 476 m, (*b*) 884 m, (*c*) 1572 m

1-30M 296°K

1-33M 509°K

2-3M 878,000 kJ

2-6M (*a*) −7, −11, (*b*) 2, −5

2-9M (*a*) −39,000, (*b*) −26,110, (*c*) −18,000

2-12M (*a*) −49.5 kJ, (*b*) −42.1 kJ

2-15M (*a*) 54,030 N · m, (*b*) 54,530 N · m (input)

2-18M −736,000 N · m

2-21M 1,830 J

2-24M (*a*) $ALE\epsilon^2/2$, (*b*) 31.05 N · m

2-27M (*a*) 9 cm, (*b*) 8 cm

2-30M 0.121 J

2-36M (*a*) 1528, (*b*) 4775

2-39M (*a*) 0.27 kJ, (*b*) 0.72 kJ

2-42M (*a*) 0, 2.20; 10.08, −2.73; 0, −4.23; −5.32, 0; (*b*) 0, 0.168; 0.473, −0.119, 0, −0.137; −0.385, 0

2-45M (*a*) −1500 Nm, (*b*) −15 kJ

2-48M 0.625 m

3-3M (*a*) 269, (*b*) 198

3-6M (*a*) 131, (*b*) 95

3-9M (*a*) 3.3, (*b*) 7.4 kg

3-12M (*a*) 82.45 J/g, (*b*) 38.81, (*c*) 61.3

3-15M (*a*) −9920, (*b*) −5950 kJ/kg · mol

3-18M (*a*) 8235; 8247, 0.7 percent; (*b*) 17,690; 17,694; 1.9 percent; (*c*) 18,810; 18,807; 1.4 percent

3-21M (*a*) 0, 75.1, (*b*) 0, 36.8

3-24M (*a*) 2.0, (*b*) 2.58

3-27M (*a*) 23, −13.2, −13.1; (*b*) −2, 18.24, 18.04

3-30M (*a*) 226, (*b*) 177

3-33M (*a*) 72.2, 72.09; (*b*) 1460, 145.42

3-36M (*a*) 0.137, 0.118; (*b*) 0.0913, 0.0786

3-39M 0.657

3-42M (*a*) 19.7, (*b*) 26.1, (*c*) 28.3 kJ/(kg · mol)(°C)

3-45M 1.30

3-48M 360 (kJ)

4-3M (*a*) 15.54, 2793.2, 2595.3, 1.0; (*b*) 151.2, 3247.6, 2945.2; (*c*) 1.014, 1394, 2158, 0.833; (*d*) 275.6, 2157, 2037, 0.601; (*e*) 1.0752, 593.78, 585.72; (*f*) 99.4, 3230.9, 2932.8; (*g*) 179.9, 175.1, 2576.6, 2401.4; (*h*) 12.54, 156.5, 2590.0, 1.0; (*i*) 1.014, 1267, 2128, 0.759; (*j*) 1.433, 1.0516, 461.14, 0.0; (*k*) 127.4, 1.0672, 535.37, 0.0

4-6M (*a*) 6.3405, 0.7607; (*b*) 22.00, 196.57; (*c*) 29.39, 201.45; (*d*) 1.11, 16.56; (*e*) 0.723, 39.76; (*f*) 1.826, 54.06; (*g*) 50, 193.10; (*h*) 0.70, 11.75; (*i*) 178.32, 30.78; (*j*) 6, 209.31; (*k*) 54.87, 0.7525

4-9M (*a*) 36,412, 36,383; (*b*) 219.0, 206.0; (*c*) 403.2, 401.1; (*d*) 63.86, 55.46; (*e*) 117.0, 115.7

4-12M (*a*) 0.00703, 1.686; (*b*) 0.0911, 1.298

4-15M (*a*) 0.615, (*b*) 0.318, (*c*) 0.134

4-18M (*a*) 15.54 bars, (*b*) 10.45 bars, (*c*) 18°C, (*d*) −389, 408 kJ/kg

4-21M (*a*) 3.94, 0; (*b*) 7.92, 0; (*c*) 15.89, 0

4-24M (*a*) 419.2, 421.18, 0.47; (*b*) 417.12, 421.18, 0.97 percent; (*c*) 413.13, 421.18, 1.95 percent

4-33M (*a*) 1.62, (*b*) −12.5

4-36M 1.25 g

4-39M 181.5 kJ

4-42M −898

4-45M (*a*) 1.366, (*b*) 31.6, (*c*) −27.4

4-48M (*a*) 250.12, 251.47, 0.54 percent; (*b*) 591.46, 595.65, 0.71 percent

4-51M (*a*) 18.5, 14.8, 14.48; (*b*) 16.36, 12.7, 12.58

4-54M (*a*) ~75–78; (*b*) 80

4-57M (*a*) 21.3 bars, (*b*) 31.7 bars

4-60M 70°C

4-63M (*a*) 949.8, 945.1; (*b*) 952.8, 945.9; (*c*) 958.7, 947.5

4-66M 25.72

5-3M (*a*) 1.977×10^6, (*b*) 180

5-6M (*a*) 1.76, (*b*) 3.30

5-9M (*a*) 1.46 bars, (*b*) 385

5-12M (*a*) 0.169, (*b*) 0.346

5-15M (*a*) 70, 121; (*b*) 30.5, 107

5-18M (*a*) 26, (*b*) 5.69

5-21M 8.7/1

5-24M (*a*) −3.75, (*b*) −847, (*c*) 38.83, (*d*) 32,900

5-27M (*a*) 401, (*b*) 2.67

5-30M 96.9

5-33M 13.2

5-36M (*a*) 191, (*b*) 442, (*c*) 154

5-39M (*a*) 20.14, (*b*) 204.5

5-42M (*a*) 88, (*b*) 0.667, (*c*) 0.550

5-45M (*a*) 179.9, (*b*) 200

5-48M 0.108

5-51M (*a*) 6.60, (*b*) 425

5-54M (*a*) 0.326, (*b*) 10.9

6-3M (*a*) 679°C, (*b*) 847°C

6-6M (*a*) 3, (*b*) 240 kJ

6-9M (*a*) 0.335, (*b*) 1,430

6-12M (*a*) 367, (*b*) 256

6-15M no

6-18M −0.870, + 0.870

6-21M (*a*) 315, (*b*) 231

6-24M (*a*) I, (*b*) III, (*c*) IV, (*d*) II

6-27M 97.8

6-30M (*a*) 5, 6; (*b*) 4, 5

6-33M (*a*) 2.03, (*b*) $3.17

6-36M (*a*) 60 kJ, (*b*) 66.6 percent, (*c*) −73, (*d*) 3

6-39M 0.0903

6-42M (*a*) −0.182, 0.357, yes, (*b*) −0.222, 0.357, yes

6-45M 3000 kJ

7-3M (*a*) 0.6820, (*b*) 0.7718, (*c*) 0.4666, (*d*) 0.6814, (*e*) 0.6928, (*f*) 0.7729, (*g*) 0.5208, (*h*) 0.6898, (*i*) 0.7235, (*j*) 0.6836, (*k*) 0.6010

7-6M (*a*) 25.26, (*b*) −0.0857, (*c*) −0.0023, impossible

7-9M (*a*) 14.93, (*b*) 0.4510, (*c*) −110.5, (*d*) 0.0147, irreversible

7-12M (*a*) 0.0133, yes, 31.2; (*b*) 0.0480, yes, 2.33; (*c*) 0.1324, yes, 8.84

7-15M (*a*) 248.3, out, (*b*) −0.6674, (*c*) 0.1603,

7-18M (*a*) 115, (*b*) 32.7, (*c*) −0.315, (*d*) 0.075

7-21M (*a*) −0.00498, (*b*) 0.00777, (*c*) irreversible

7-24M (*a*) −0.0094, (*b*) 5.83, (*c*) 0.0102,

7-27M (*a*) −720, (*b*) −0.0404, (*c*) 0.0417, (*d*) irreversible

7-30M (*a*) −0.0545, (*b*) 0.0467, (*c*) impossible

7-33M (*a*) 6.02, (*b*) −3.48, (*c*) 2.54 kJ/°K

7-36M 0.108

7-39M (*a*) 0, 11.17, (*b*) 8.95, 8.95

7-42M (*a*) −28.20, (*b*) 40.55, (*c*) 12.35

7-45M 0.167 kJ/kg · °K

7-48M 299.5

7-51M (*a*) 88.0, (*b*) 246

7-54M (*a*) 27, (*b*) 513, (*c*) 43.4

7-57M (*a*) 129, (*b*) 2190

7-60M (*a*) 267, (*b*) 36.2, (*c*) 590

7-63M 7.56

7-66M 1.41

7-69M (*a*) 12.5, (*b*) 3.19

7-72M 0.236

7-75M 292
7-78M 52.2
7-81M (*a*) 655, (*b*) 665
7-84M (*a*) 106, (*b*) 0.83, (*c*) 106
7-87M (*a*) 0.771, (*b*) 0.81
7-90M (*a*) 362, (*b*) 0.014
7-93M (*a*) 1.96, (*b*) 1.84
7-96M (*a*) 0.612, (*b*) 0.472
8-3M 147.5
8-6M −348
8-9M (*a*) 27.46, (*b*) 49.5, (*c*) 15.68
8-12M (*a*) 2.50, (*b*) 6.0
8-15M $RT_0\,[1 - (P_0/P) - \ln (P_0/P)]$
8-18M (*a*) 1095, (*b*) 315
8-21M 0.424
8-24M (*a*) 134.7, (*b*) 132.2
8-27M (*a*) −47,700, (*b*) 7800, (*c*) −39,900
8-30M −39.4
8-33M 39.4
8-36M 25.3
8-39M 2.5
8-42M 39,900
9-3M (*a*) 210, (*b*) 105
9-6M (*a*) 0.247, (*b*) 0.0823
9-9M *C*
9-12M 3, 10
9-15M (*a*) 4, (*b*) 3490, (*c*) 2520, (*d*) 5, 3, 2
9-18M (*a*) 8, 4852, (*b*) No
9-21M 2, 3, 4
9-24M −4, 2, 8
9-27M (*a*) 554, 292, 154; (*b*) 333, 333, 333, (*c*) 154, 292, 554
9-30M 231°C
9-33M 7.17 kW, 156°C
9-36M 0.444 kW, 224°C
9-39M −0.862, 0.862 kJ/(°K)(cycle)
10-3M 250
10-6M 79
10-9M (*a*) 274, (*b*) 385°K, 3.09 bars, (*c*) 4.77 kg, (*d*) 3.089 kJ/°K
10-12M (*a*) 19.7 min, (*b*) 25.2 min
10-15M (*a*) 210,000, (*b*) 184,800
10-18M (*a*) 59, (*b*) 0.0954
11-3M (*a*) 0.500 N_2, (*b*) 26.9
11-6M (*a*) 0.522 O_2, (*b*) 108, (*c*) 36.8
11-9M (*a*) 624, (*b*) 0.05, 0.05, (*c*) 1247, (*d*) 0.231
11-12M (*a*) 3/7, (*b*) 3/7

11-15M (*a*) −4690, (*b*) −4685
11-18M 427
11-21M (*a*) −62.3, (*b*) 500
11-24M (*a*) 403, (*b*) 445
11-27M (*a*) 321, (*b*) 327
11-30M (*a*) 442, 3045, (*b*) 431, 3090
11-33M (*a*) 400, (*b*) 6.69
11-36M (*a*) 148.9, (*b*) 3.23, (*c*) 0.3910, 0.6484, (*d*) 0 − 0.248
11-39M (*a*) 252, (*b*) 370 mbar
11-42M (*a*) 78.9 percent, (*b*) 21°C, (*c*) 16.3 g/kg, (*d*) 66.64, (*e*) 0.895
11-45M (*a*) 6.12, (*b*) 5.9, (*c*) 35.63
11-48M (*a*) 35, 8, 19, (*b*) 16, 4, 10, (*c*) 26, 6, 15
11-51M (*a*) 13.34, (*b*) 58.9 percent, (*c*) 61.17
11-54M (*a*) 13.2, (*b*) 59+, (*c*) 61.2
11-57M (*a*) 39 percent, (*b*) 0.0148 bar, (*c*) 51.7 kJ/kg, (*d*) 0.873 m^3/kg, (*e*) 12.5°C
11-60M −6050
11-63M (*a*) −22,400, (*b*) 11,200, (*c*) 4.94
11-66M (*a*) 0.0143 kg/kg da, (*b*) 58.8, (*c*) 14.6
11-69M (*a*) 16.1°C, (*b*) (1) 0.0034, 27.4; (2) 0.0054, 22.5
11-72M (*a*) 0.0052, 9.5, (*b*) 0.0068, 14.0
11-75M (*a*) 24.0, (*b*) 18.3, (*c*) 0.0111, (*d*) 60 percent
11-78M (*a*) 11,790, (*b*) 10,380, (*c*) 259
12-9M (*a*) 20.2, (*b*) 18.9, (*c*) 16.1, (*d*) 16.5, (*e*) 16.0 bars
12-12M (*a*) 134, (*b*) 65.3, (*c*) 76.2, (*d*) 67.7, (*e*) 66 bars
12-15M (*a*) 69.6, (*b*) 68.6, (*c*) 68.7 bars
12-18M (*a*) 0.759, (*b*) 0.76
12-21M 645°K
12-24M 116 percent
12-27M (*a*) 117, (*b*) 106 $cm^3/g \cdot mol$
12-30M 119.5, 113.0 $cm^3/g \cdot mol$
12-33M 2600 $kJ/kg \cdot mol$
12-36M (*a*) 0.71, (*b*) 1.1 m^3
13-9M (*a*) 0.692, 0.698; (*b*) 0.765, 0.769
13-15M (*a*) 0.331, 0.343 $cm^3/(g)(°C)$; (*b*) 0.0816, 0.0822 $cm^3/(g)(°C)$
13-18M $[3a/2vT^{1/2}(v + b)]\, dv$
13-21M (*b*) $R \ln [(v_2 - b)/(v_1 - b)] - (a/T^2)[(1/v_2) - (1/v_1)]$
13-24M $(2aR/T)\, \Delta P$
13-27M 0.3987 percent
13-30M (*a*) 1.248, (*b*) 5.02 percent
13-33M 208 kJ/kg
13-36M (*a*) −1.86, (*b*) −3.70, (*c*) −5.53°C
13-39M (*a*) 71.7, (*b*) 0.479, (*c*) 2.1×10^{-5}
13-42M (*a*) 1.125, (*b*) 1.29, (*c*) 0.708
13-45M (*a*) 1.55, (*b*) 1.63, (*c*) 1.69
13-48M ~650°K

13-51M (*a*) 93, (*b*) 101

13-54M 56°C, 60°C

13-57M (*a*) 1677, (*b*) −5980

13-60M (*a*) 3520 kJ/kg · mol, 227 bars, (*b*) 3120, 175

13-63M (*a*) −80.0, (*b*) −9,080, (*c*) −27,200, (*d*) 18,120

13-66M 58°C, 155 m/s

14-3M (*a*) 28, (*b*) 0.781, (*c*) 12.7, (*d*) 37, (*e*) 64.7

14-6M (*a*) 17.75, (*b*) 0.833, (*c*) 74.6, (*d*) 48, (*e*) 74

14-9M (*a*) – (*d*) 14.62, (*e*) 16.62, (*f*) 15.27, (*g*) 13.69, (*h*) 15.27

14-12M (*a*) 18.2, (*b*) 8.2, (*c*) 52.5, (*d*) 83.5

14-15M (*a*) 16.27, (*b*) 0.115, (*c*) 50

14-18M (*a*) 50, 1.71; (*b*) 54, 2.08

14-21M (*a*) 10.8, (*b*) 13.4, (*c*) 19.6

14-24M (*a*) 16.0, (*b*) 5.0

14-27M 10.0

14-30M (*a*) − 3,450,000, (*b*) − 3,826,000, (*c*) − 1,103,400

14-33M (*a*) 267, (*b*) 331

14-36M 639,000 (out)

14-39M (*a*) 3030, (*b*) 2830, (*c*) 2415

14-42M (*a*) 2265, (*b*) 2015

14-45M (*a*) 2230, (*b*) 2290

14-48M 7.51 (T_f = 3050°K)

14-51M (*a*) 0.0571, 0.2937, 0.0130, 10.77, (*b*) −25,180

14-54M (*a*) − 258.0, 2730, (*b*) 22.4, 2713

14-57M (*a*) −88.0, 2349, (*b*) 188.5, 2330

14-60M (*a*) 165.3, (*b*) 65.6

14-63M (*a*) 1,421,900, −683,500, 683,500, 0.671, (*b*) 1,205,700, −899,700, 899,700, 0.56

14-66M (*a*) 282,990, (*b*) −258,730, (*b*) 258,730

14-69M (*a*) 450,850, (*b*) −249,300, (*c*) 249,300, (*d*) 0.644

15-3M (*a*) (1) −61,100, 201,230, (2) 463,560, (3) −187.1; (*b*) (1) −532,210, −157,800, (2) 216,610, (3) −11.84; (*c*) (1) −642,445, −239,760, (2) 162,925, (3) −7.537

15-6M (*a*) 0.004855, (*b*) 2600

15-9M (*a*) 2490°K, (*b*) 2390°K

15-12M (*a*) 3025°K, (*b*) 2970°K

15-15M (*a*) 1.08 atm, (*b*) 1.96 atm

15-18M (*a*) 0.156 NH_3, (*b*) 0.334 NH_3

15-21M (*a*) 0.939, (*b*) 2.84, (*c*) 2.89

15-24M (*a*) 0.53 CO, 0.47 H_2; (*b*) 0.84 CO, 0.16 H_2

15-27M (*a*) 63.3, (*b*) 75.6

15-30M (*a*) 140,100, (*b*) 123,100

15-33M (*a*) 2765°K

15-36M 0.84

15-39M 0.11 H_2, 0.50 CO, 0.305 O_2

15-42M (*a*) −509,500, −509,340; (*b*) −511,100, −510,600; (*c*) −512,960, −513,430

15-45M 1.60×10^6, 16.5

15-48M (*a*) 94,800, (*b*) 750°K, (*c*) 2.56 atm
16-3M (*b*) 877, 129; (*c*) 0.147, (*d*) 0.75
16-6M (*a*) 34.5 bars; (*b*) 384, (*c*) 0.0830, 1.10; (*d*) 0.616
16-9M (*a*) 581, 291; (*b*) 20, 10; −20, −98; 10, 98; (*c*) 0.154, 0.702
16-12M (*a*) 36.4, 0.144 MPa; (*b*) 36.4, 0.0957 MPa
16-15M (*a*) 17.6, 673; 52.3, 2000; 3.41, 1043; (*b*) 0.510, (*c*) 7.88
16-18M (*a*) 700, 2100, 1077; (*b*) 0.52, (*c*) 0.85
16-21M (*a*) 350, (*b*) 0.52
16-24M (*a*) 1800, (*b*) 0.295, (*c*) 1070
16-27M (*a*) 2.74, (*b*) 0.47, (*c*) 10.4
16-30M (*a*) 506, (*b*) 0.58
16-33M (*a*) 64.2 bars, (*b*) 1300, 2090°K; (*c*) 1040°K, (*d*) 0.576
16-36M (*a*) 252, (*b*) 0.396
16-39M (*a*) 162, 395; (*b*) 0.33, (*c*) 932, (*d*) 206, (*e*) 61.4
16-45M (*a*) 138, (*b*) 0.234
16-48M (*a*) 195, 495; (*b*) 0.218, (*c*) 580, (*d*) 206, (*e*) 64.4
16-51M 36 percent
16-54M (*a*) 0.281, 17 percent, (*b*) 0.307, 23.8 percent
16-57M (*a*) 0.274, (*b*) 0.59, (*c*) 0.837, (*d*) 0.843
16-60M (*a*) 120, −30; (*b*) 117, 0; (*c*) 129, −129
16-63M (*a*) 159, −159; (*b*) 198, −37; (*c*) 208, 0
16-66M (*a*) 234, (*b*) 280
16-69M (*a*) 139.6, (*b*) 142.3
16-72M 174
16-75M (*a*) 185, 0.277; (*b*) 185, 0.329
16-78M (*a*) 160.5, 346, (*b*) 0.37, (*c*) 250
16-81M (*a*) 7.26, 560, 7.26, 1300, 2.71, 1060, 0.60, 737 (*b*) 282, (*c*) 850
16-84M (*a*) 324, (*b*) 0.362, (*c*) 984
16-87M (*a*) 29,500, (*b*) 153, (*c*) 0.468
16-90M (*a*) 2310, (*b*) 645,000
16-93M 20
16-96M (*a*) 0.533, (*b*) 2.68 bars
17-3M (*a*) 0.666, (*b*) 0.428, (*c*) 119.4, (*d*) 51.1
17-6M (*a*) 0.219, 0.417, (*b*) 0.189, 0.403
17-9M (*a*) 2147, 6.777, (*b*) 3138, (*c*) 2943, (*d*) 0.336 (*e*) 6550
17-12M (*a*) 0.891, 0.403, (*b*) 0.866, 0.419
17-15M (*a*) 3320, (*b*) 0.446, (*c*) 39,000, (*d*) 2.12×10^6 kg/h
17-18M (*a*) 0.776, 0.418, 35,700; (*b*) 0.757, 0.436, 34,200
17-21M (*a*) 0.232, 56,800, 0.467; (*b*) 0.229, 56,700, 0.466
17-24M (*a*) 0.0566, 0.1474; (*b*) 1132, 4; (*c*) 0.411
17-27M 0.370
17-30M (*a*) 0.127, (*b*) 3250, (*c*) 0.425
17-33M (*a*) 0.347, (*b*) 0.361, (*c*) 0.356, (*d*) 0.339, (*e*) 0.369 (*f*) 0.359, (*g*) 0.387, (*h*) 0.372
17-36M (*a*) 325, (*b*) 0.128, (*c*) 1510
17-39M (*a*) 2958, 178, 1087, 1057; (*b*) 2422, 3537; (*c*) 0.32, (*d*) 0.562

17-42M (*a*) 0.216, (*b*) 0.186
17-45M (*a*) 1.40, (*b*) 0.472
18-3M (*a*) 5.49, (*b*) 0.253, (*c*) 24.7
18-6M (*a*) 35, (*b*) 5.45, (*c*) 736, (*d*) 3.23
18-9M (*a*) 6.3 percent, (*b*) 9.4 percent
18-12M (*a*) 29, (*b*) 10.3, (*c*) −30.6
18-15M (*a*) 60, (*b*) 2.72, (*c*) 3.32
18-18M (*a*) 8.96, (*b*) 4.54, (*c*) 0.80, (*d*) 4.90, (*e*) 22.2
18-21M (*a*) 137, −69, (*b*) 2.74, (*c*) 159, −60, (*d*) 1.37
18-24M (*a*) −54, −69, (*b*) 2.72, 2.00
18-22M (*a*) 2.65, (*b*) 3.65
18-30M (*a*) 7.96, 10.98, (*b*) 3.62, 4.17, (*c*) 2.26, (*d*) −24.4, (*e*) −22.8
18-33M (*a*) 7.62, (*b*) 1.97, (*c*) 2.17, (*d*) 3.58, (*e*) 3.17
18-36M 2.71, 152
18-39M 3.0, 94.6
19-3M (*a*) −1,467,370, (*b*) 0.941, (*c*) 1.086
19-6M (*a*) −223,900, 0.922, 1.16, (*b*) −214,015, 0.874, 1.11 (*c*) −203,520, 0.825, 1.05
19-9M (*a*) −378,370, (*b*) 1.06, (*c*) 2, (*d*) 1.96, (*e*) 0.171
19-12M (*a*) 850, (*b*) 6.95, (*c*) 285, (*d*) 1031, (*e*) 0.509
19-15M (*a*) 811, (*b*) 7.95, (*c*) 255, (*d*) 1086, (*e*) 0.483
19-18M (*a*) 0.552, 1.17, (*c*) 0.545, 1.31
19-21M (*a*) 1000, 2360, (*b*) 2845 kJ/kg, (*c*) 0.508, (*d*) 0.360

SELECTED PROBLEM ANSWERS (USCS)

1-3 (*a*) 32.4 percent, (*b*) 71.9 percent
1-6 62.3 percent
1-9 (*a*) 0.398, (*b*) $1,321,000, (*c*) $61,920
1-12 (*a*) 70°F, 14.6 psia, 13.4 ft^3/lb_m, 2 ft^3, 0.149 lb, (*b*) 0.072 lb_f/ft^3
1-15 (*a*) 18.3 mi (*b*) 36.5 mi
1-18 (*a*) 5.0, (*b*) 0.20, (*c*) 0.0340
1-21 15.14 psia
1-24 −9.8 psig
1-27 (*a*) 1003, (*b*) 2255, (*c*) 3650 (ft)
1-30 532°R
1-33 892°R
2-3 6.15×10^8 ft · lb_f
2-6 (*a*) 50, −40, 20, (*b*) 50, 40, −70
2-9 (*a*) −26,680, (*b*) −17,840, (*c*) −12,280
2-12 (*a*) −17,100, (*b*) −13,680
2-15 (*a*) 12,250 ft · lb_f/lb_m
2-18 −121,100
2-21 0.362 ft · lb_f
2-24 (*a*) (EAL/2) $(\varepsilon_2^2 - \varepsilon_1^2)$, (*b*) 0.0005 in/in, (*c*) 7.5 ft · lb_f

2-27 (*a*) 7.24 in, (*b*) 8.2 in

2-30 0.121 J

2-36 (*a*) 1750, (*b*) 1750

2-39 (*a*) 0.205, (*b*) 0.956

2-42 1440 Btu

2-45 3940

2-48 −6.8

3-3 (*a*) 173, (*b*) 166

3-6 (*a*) 285, (*b*) 267

3-9 (*a*) 12.45, (*b*) 24.34

3-12 (*a*) 27,300, (*b*) 12,900, (*c*) 19,850

3-15 (*a*) -1.844×10^6, (*b*) -2.212×10^6

3-18 (*a*) 7428, 7425, −1.7; (*b*) 8915, 8916, −1.9; (*c*) 10,444, 10,443, −2.1

3-21 (*a*) 0, −31.9, (*b*) 0, −15.6

3-24 (*a*) 0.967, (*b*) 1.846

3-27 (*a*) 3, −3.5, −3.5, (*b*) 37, −2.31, −2.32

3-30 (*a*) 580, (*b*) 480

3-33 (*a*) 17.2, 17.16, (*b*) 24.1, 24.05

3-36 (*a*) 0.307, 3.92, (*b*) 0.385, 4.91

3-39 0.158

3-42 (*a*) 80, (*b*) 30 psia

3-45 (*a*) 167, (*b*) 0

3-48 −11.1, +18.4, +7.3

4-3 (*a*) 417.43, 1.544, 1118.2, 1.0; (*b*) 1.304, 1356.7, 1236.0; (*c*) 24.97, 12.76, 893.3, 0.781; (*d*) 444.7, 889.8, 838.0, 0.596; (*e*) 0.016554, 171.46, 166.87; (*f*) 550, 3.228, 1297.5; (*g*) 401.04, 1.66, 1119, 1043; (*h*) 66.98, 1180.2, 1100, 1.0; (*i*) 11.529, 30.8, 1066, 0.918; (*j*) 134.53, 0.01799, 321.4, 0.0; (*k*) 312.07, 0.01757, 282.2, 0.0

4-6 (*a*) 98.87, 0.01228, (*b*) 114.3, 88.18, (*c*) 0.6129, 87.169, (*d*) 25.93, 0.304, (*e*) 0.01130, 12.86, (*f*) 29.335, 27.59, (*g*) 160, 89.610, (*h*) 0.70, 0.292, (*i*) 79.51, 0.3079, (*j*) 140, 93,923, (*k*) 19.51, 0.01175

4-9 (*a*) 453, 453.3, (*b*) 2.565, 2.361, (*c*) 4.35, 4.31, (*d*) 0.331, 0.206, (*e*) 0.614, 0.602

4-12 (*a*) 0.00724, 1.677, (*b*) 0.2005, 1.16

4-15 (*a*) 0.366, (*b*) 0.190, (*c*) 0.118

4-18 (*a*) 67.01 psia, (*b*) 58.6 psia, (*c*) 300°F, (*d*) −361, 354.4 Btu/lb

4-21 (*a*) 1.21, 0, (*b*) 3.66, 0, (*c*) 7.33, 0

4-24 (*a*) 100.88, 101.46, 0.57 percent, (*b*) 99.54, 101.46, 1.93 percent, (*c*) 97.62, 101.46, 3.9 percent

4-33 (*a*) 1.38, (*b*) −10.68

4-36 0.0025 lb

4-39 253

4-42 329, out

4-45 (*a*) 0.753 lb, (*b*) 14.0 Btu, (*c*) −7.09 Btu

4-48 (*a*) 99.24, 99.9, 0.67 percent, (*b*) 250.01, 251.64, 0.65 percent

4-51 (*a*) 0.296, 0.243, 0.2371, (*b*) 0.275, 0.227, 0.2212

4-54 (*a*) 100, (*b*) 195, (*c*) 200

4-57 (*a*) 33.8, (*b*) 29.9

4-60 142.3°F

4-63 (*a*) 272.7, 271.46, (*b*) 274.3, 272.4, (*c*) 275.9, 273.33

4-66 88.06

5-3 (*a*) 1.68×10^6, (*b*) 1380

5-6 (*a*) 0.572, (*b*) 10.3

5-9 (*a*) 17.6, (*b*) 863

5-12 (*a*) 0.425, (*b*) 1.25

5-15 (*a*) 172, 506, (*b*) 180, 512

5-18 (*a*) 76.5, (*b*) 12.3

5-21 288

5-24 (*a*) 36, (*b*) 487, (*c*) 8.65×10^6

5-27 (*a*) 500, (*b*) 0.378

5-30 (*a*) 98.7

5-33 −9.9

5-36 (*a*) 547, (*b*) 846, (*c*) 302

5-39 (*a*) 0.395, (*b*) 1.584

5-42 (*a*) 200, (*b*) 5.9, (*c*) 1.62

5-45 (*a*) 327.86, (*b*) 350

5-48 0.100

5-51 (*a*) 26.5, (*b*) 1237

5-54 (*a*) 27.7, (*b*) 3.62

6-3 1100°F

6-6 (*a*) 70, (*b*) 260

6-9 (*a*) 0.333, (*b*) 2850

6-12 (*a*) 630, (*b*) 469

6-15 Yes

6-18 −0.490, +0.490

6-21 (*a*) 530, (*b*) 280

6-24 (*a*) I, (*b*) III, (*c*) IV, (*d*) II

6-27 189

6-30 (*a*) 5, 6, (*b*) 4, 5

6-33 (*a*) 3.15, (*b*) $3.66

6-36 (*a*) 61.1, (*b*) 0.679, (*c*) −58.0, (*d*) 3.41

6-39 0.239

6-42 (*a*) 0.20, −0.10, yes, (*b*) 0.20, −0.125, yes

6-45 3000 Btu

7-3 (*a*) 0.1642, (*b*) 0.1940, (*c*) 0.1189, (*d*) 0.1632, (*e*) 0.0604, (*f*) 0.1873, (*g*) 0.1256, (*h*) 0.0418, (*i*) 0.1749, (*j*) 0.0675, (*k*) 0.1616

7-6 (*a*) −10.24, (*b*) −0.02006, (*c*) 0.01896, (*d*) 0.00110, impossible

7-9 (*a*) 0.5620, (*b*) 0.1273, (*c*) −32.18, (*d*) −0.0019, impossible

7-12 (*a*) 0.0041, yes, 46.5/1, (*b*) 0.0120, yes, 2.44/1, (*c*) 0.0316, yes, 8.85/1

7-15 (*a*) −151, (*b*) −0.1921, (*c*) 0.0928

7-18 (*a*) −49.8, (*b*) 14.2, (*c*) −0.0761, (*d*) 0.0179

7-21 (*a*) −0.00260, (*b*) 0.00410, (*c*) irreversible

7-24 (*a*) −0.0018, (*b*) −2.2, (*c*) 0.0023

7-27 (*a*) −57,100, (*b*) 0.498, (*c*) 1.866, (*d*) irreversible

7-30 (*a*) −0.00668, (*b*) 0.01296, (*c*) irreversible
7-33 (*a*) 0.892, (*b*) −0.524, (*c*) 0.368
7-36 0.016
7-39 (*a*) 0, 0.00580, (*b*) 0.00509, 0.00509
7-42 (*a*) −14.12, (*b*) 18.90, (*c*) 4.78
7-45 0.0333
7-48 (*a*) 97,760, (*b*) 1780
7-51 (*a*) 41.6, (*b*) 567
7-54 (*a*) 233, (*b*) 1385, (*c*) 9.66
7-57 (*a*) 307, (*b*) 867
7-60 (*a*) 500, (*b*) 20.6, (*c*) 984
7-63 22.7
7-66 381
7-69 (*a*) 37.7, (*b*) 1.25
7-72 −10.4
7-75 653
7-78 11.45
7-81 (*a*) 2140, (*b*) 2180 ft/s
7-84 (*a*) 45.4, (*b*) 0.796, (*c*) 16.6
7-87 0.818
7-90 (*a*) 1028 (hp), (*b*) 0.047
7-93 (*a*) 403, (*b*) 361
7-96 (*a*) 1.51, (*b*) 1.23
8-3 (*a*) 445, (*b*) 0.81
8-6 −151
8-9 (*a*) 14.7, (*b*) 121, (*c*) 7.2
8-12 (*a*) 1.60, (*b*) 2.35
8-15 $RT_0[1 - (P_0/P) - \ln (P_0/P)]$
8-18 (*a*) 513 Btu, (*b*) 167 Btu
8-21 0.392
8-24 (*a*) 52.1, (*b*) 42.1
8-27 (*a*) −22,800, (*b*) 400, (*c*) −22,400
8-30 −4.84
8-33 17.1
8-36 23.0
8-39 10.0
8-42 22,400
9-3 (*a*) 210, (*b*) 105
9-6 (*a*) 0.247, (*b*) 0.0823
9-9 C
9-12 3, 10
9-15 (*a*) 4, (*b*) 3490, (*c*) 2520, (*d*) 5, 3, 2
9-18 (*a*) 8, 4852, (*b*) no
9-21 2, 3, 4
9-24 −4, 2, 8

9-27 (*a*) 554, 292, 154; (*b*) 333, 333, 333; (*c*) 154, 292, 554
9-30 231°C
9-33 7.17 kW, 156°C
9-36 0.444 kW, 224°C
9-39 −0.862, 0.862 kJ/(°K)(cycle)
10-3 110 Btu
10-6 180°F
10-9 (*a*) 110, (*b*) 204°F, 36 psia, (*c*) 1.71, (*d*) 0.0683
10-12 (*a*) 7.47, (*b*) 8.56
10-15 40,100
10-18 (*a*) 149, (*b*) 0.060
11-3 (*a*) 50.0 percent N_2, (*b*) 173.8
11-6 (*a*) 52.2 percent O_2, (*b*) 21, (*c*) 36.8
11-9 (*a*) 14.8, (*b*) 5.0, (*c*) 29.5, (*d*) 0.0551
11-12 (*a*) $\frac{3}{7}$, (*b*) $\frac{3}{7}$
11-15 (*a*) −2260, (*b*) −2265
11-18 153
11-21 (*a*) −10.5, (*b*) 75.6
11-24 (*a*) 1110, (*b*) 1228
11-27 (*a*) 572, (*b*) 580
11-30 (*a*) 786, 1260, (*b*) 775, 1324
11-33 (*a*) 700, (*b*) 1.74
11-36 (*a*) 170°F, (*b*) 37.6 psia, (*c*) 36, (*d*) 0.100, (*e*) −0.049
11-39 (*a*) 11.64, (*b*) 4.07
11-42 (*a*) 0.59, (*b*) 65°F, (*c*) 0.0130, (*d*) 33.5, (*e*) 13.97
11-45 (*a*) 44.07, (*b*) 45, (*c*) 23.7
11-48 (*a*) 97, 46, 66, (*b*) 74, 50, 60, (*c*) 90, 60, 70
11-51 (*a*) 0.01456, (*b*) 0.621, (*c*) 35.66
11-54 (*a*) 0.01457, (*b*) 62 percent, (*c*) 35.7
11-57 (*a*) 32 percent, (*b*) 0.22 psia, (*c*) 32.3, (*d*) 14.06, (*e*) 56°F
11-60 2950
11-63 (*a*) 2100, (*b*) 693, (*c*) 0.96
11-66 (*a*) 0.0141, (*b*) 25 Btu/lb, (*c*) 6.3 Btu/lb
11-69 (*a*) 58, (*b*) (1) 0.0046, 69°F, (2) 0.0062, 63°F
11-72 (*a*) 0.00443, 2.4 (*b*) 0.00622, 4.4
11-75 (*a*) 76.9, (*b*) 82.8 gr/lb, (*c*) 59.6 percent
11-78 (*a*) 22,370, (*b*) 19,350, (*c*) 492
12-3 (*a*) 0.362, (*b*) 0.447, (*c*) 0.375, (*d*) 0.368 ft^3/lb
12-6 (*a*) 1165, (*b*) 1005, (*c*) 952, (*d*) 991, (*e*) 1000 psia
12-9 0.54 ft^3/lb
12-12 (*a*) 21, 970, 0.476, (*b*) 2.91 ft^3/lb · mol, (*c*) 9.05 ft^3/lb · mol
12-15 (*a*) 1218, (*b*) 1740
12-18 (*a*) 0.0775, (*b*) 3000
12-21 116.1, 100.2 cm^3/g · mol
12-24 (*a*) 529, (*b*) 871, (*c*) 1675, (*d*) 1000 atm

12-27 (*a*) 5 ft^3, (*b*) 7 ft^3
13-3 (*b*) $T(\partial P/\partial T)_v - P$
13-9 0.394 Btu/(lb)(°F)
13-15 0.00345 ft^3/(lb)(°F)
13-18 $\Delta s = R \ln [(v_2 - b)/(v_1 - b)] - (a/T^2)[(1/v_2) - (1/v_1)]$
13-21 $\Delta T + (B/2)(T_2^2 - T_1^2) + AR\,\Delta P$
13-24 (*a*) 0.82, (*b*) 14 percent
13-27 (*a*) 60.45, (*b*) 71.1, 60.31 Btu/lb
13-30 830 Btu/lb
13-33 Approx. $T_2 = -10°C$
13-36 No
13-39 670
13-42 (*a*) 0.395, (*b*) 0.083, (*c*) −0.0254°F/atm
13-45 4070
13-48 (*a*) 750 Btu/lb · mol, (*b*) −1850
13-51 (*a*) ~5.5 Btu/lb, (*b*) ~5.4 Btu/lb
13-54 (*a*) $[-AT^2/(T - b)^2]\,\Delta P$, (*b*) decrease, (*c*) yes
13-57 2000 hp
13-60 (*a*) 62.4, (*b*) 53.3 Btu/lb
13-63 −0.566, −126, −330, −206
13-66 125°F, 430 ft/s
14-3 (*a*) 28, (*b*) 0.781, (*c*) 12.7, (*d*) 99, (*e*) 62.7
14-6 (*a*) 17.75, (*b*) 0.833, (*c*) 74.6, (*d*) 118, (*e*) 71.4
14-9 (*a*)–(*d*) 231, (*e*) 263, (*f*) 241, (*g*) 216, (*h*) 241
14-12 (*a*) 18.2, (*b*) 8.2, (*c*) 127°F, (*d*) 83.4
14-15 (*a*) 16.27, (*b*) 0.115, (*c*) 121
14-18 (*a*) 124, 1.702, (*b*) 129, 2.07
14-21 (*a*) 10.8, (*b*) 13.4, (*c*) 19.6
14-24 (*a*) 16.0, (*b*) 5.0
14-27 10.0
14-30 (*a*) −1,485,000, (*b*) −1,486,000, (*c*) −474,000
14-33 (*a*) 260, (*b*) 328
14-36 −275,000
14-39 (*a*) 4580, (*b*) 4140, (*c*) 3810
14-42 (*a*) 4080, (*b*) 3625
14-45 (*a*) 4020, (*b*) 4165
14-48 109 ($T_f = 5490°R$)
14-51 (*a*) 0.0571, 0.2937, 0.0130, 10.77; (*b*) 10,890
14-54 (*a*) −61.60, 67.50, (*b*) 5.35, 669.1
14-57 (*a*) −21.0, 580.1, (*b*) 45.1, 573.3
14-60 39.45
14-63 (*a*) 610,270, (*b*) −295,500, (*c*) 295,500, (*d*) 0.674
14-66 (*a*) −121,750, (*b*) −110,340, (*c*) 110,340
14-69 (*a*) 194,400, (*b*) −106,900, (*c*) 106,900, (*d*) 0.645
16-3 (*a*) 377, −50.8; (*b*) 0.135, (*c*) 0.75

16-6 (*a*) 26.1, (*b*) 480, (*c*) 1.16, 12.08, (*d*) 0.60
16-9 (*a*) 1080, 540; (*b*) 20, −20; 0, −94.9; −10, 10; 0, 94.9; (*c*) 5.24, 30.8
16-12 (*a*) 42.2, 26.4 psi, (*b*) 42.2, 17.6 psi
16-15 (*a*) 259, 1208°R; 1030, 4800°R; 68.5, 2555°R; (*b*) 0.504, (*c*) 179
18-18 (*a*) 1260, 3780, 1940°R; (*b*) 0.518, (*c*) 0.854
16-21 (*a*) 243, (*b*) 0.58
16-24 (*a*) 3120, (*b*) 42, (*c*) 441
16-27 (*a*) 0.499, (*b*) 177 psi
16-30 (*a*) 235, 0.61, (*b*) 82.7, 0.36
16-33 (*a*) 1035, (*b*) 2460, 4080, (*c*) 2075, (*d*) 0.57
16-36 (*a*) 100, (*b*) 0.313
16-39 (*a*) 80.1, 187.3, (*b*) 0.385, (*c*) 1515, (*d*) 7400, (*e*) 0.015
16-45 (*a*) 57.2, (*b*) 0.189
16-48 (*a*) 100, 159, (*b*) 0.230, (*c*) 835, (*d*) 7400, (*e*) 0.0139
16-51 32.8
16-54 (*a*) 0.34, 44.2, (*b*) 0.284, 33.3
16-57 (*a*) 0.278, (*b*) 0.56, (*c*) 0.84, (*d*) 0.85
16-60 (*a*) 51.4, −12.5, (*b*) 50, 0, (*c*) 55.1, −55.1
16-63 (*a*) 68.7, −68.7, (*b*) 85.2, −15.7, (*c*) 90.6, 0
16-66 (*a*) 101, (*b*) 122
16-69 (*a*) 59.6, (*b*) 61.1
16-72 75
16.75 94, 0.314
16.78 (*a*) 67.4, 150, (*b*) 0.379, (*c*) 480
16-81 (*a*) 12.93, 519; 103.4, 1025; 103.4, 2300; 37.02, 1860; 8.7 psia, 1260°R; (*b*) 123, (*c*) 2825
16-84 (*a*) 121, (*b*) 55.4, (*c*) 3125
16-87 (*a*) 6160, (*b*) 66.7 (*c*) 0.460
16-90 (*a*) 2730, (*b*) 574,000
16-93 50
16-96 (*a*) 0.533, (*b*) 33.7 psia
17-3 (*a*) 0.59, (*b*) 0.545, (*c*) 45.3, (*d*) 24.7
17-6 (*a*) 21 percent, 0.42, (*b*) 16 percent, 0.396
17-9 (*a*) 931, 1.650, (*b*) 1404, (*c*) 1326, (*d*) 0.355, (*e*) 7260
17-12 (*a*) 0.928, 0.379, (*b*) 0.908, 0.394
17-15 (*a*) 1692, (*b*) 0.440, (*c*) 459,000, (*d*) 31.1×10^6
17-18 (*a*) 0.387, 79,200, (*b*) 0.408, 75,200
17-21 (*a*) 0.242, 645,000, 0.467; (*b*) 0.236, 644,000, 0.465
17-24 (*a*) 0.0605, 0.140, (*b*) 465, 1.5, (*c*) 0.397
17-27 (*a*) 0.434, (*b*) 0.411
17-30 (*a*) 0.122, (*b*) 1365, (*c*) 0.413
17-33 (*a*) 0.327, (*b*) 0.340, (*c*) 0.360, (*d*) 0.338, (*e*) 0.376, (*f*) 0.369, (*g*) 0.382, (*h*) 0.360
17-36 (*a*) 108, (*b*) 0.101, (*c*) 1430
17-39 (*a*) 1233, 83, 428, 424; (*b*) 1051, 430, (*c*) 0.30, (*d*) 0.436
17-42 (*a*) 0.236, (*b*) 0.203
17-45 (*a*) 1.53, (*b*) 0.456

18-3 (*a*) 4.60, (*b*) 0.293, (*c*) 12.38
18-6 (*a*) 93.4, (*b*) 5.71, (*c*) 25.0, (*d*) 4.13
18-9 (*a*) 2.6, (*b*) 2.9
18-12 (*a*) 50, (*b*) −9.6, (*c*) −40
18-15 (*a*) 150, (*b*) 2.60, (*c*) 3.06
18-18 (*a*) 22.2, (*b*) 5.98, 8.01, (*c*) 0.752, (*d*) 3.91, (*e*) 23.4
18-21 (*a*) 278, −95, (*b*) 2.86, (*c*) 315, −79, (*d*) 1.35
18-24 (*a*) −65, −91, (*b*) 2.67, 2.0
18-27 (*a*) 3.15, (*b*) 4.15
18-30 (*a*) 16, (*b*) 80
18-33 (*a*) 17.02, 26.85, (*b*) 5.65, 10.10, (*c*) 1.50, (*d*) −29.7, (*e*) −30.6
18-36 (*a*) 19.3, (*b*) 2.29, (*c*) 5.85, (*d*) 5.05, (*e*) 3.85
18-39 3.15
19-3 (*a*) 390, (*b*) 8.68, (*c*) 139, (*d*) 518, (*e*) 0.508
19-6 (*a*) 345, (*b*) 6.91, (*c*) 108, (*d*) 472, (*e*) 0.512
19.9 (*a*) 1090, 5880, (*b*) 9.0×10^6, (*c*) 0.428, (*d*) 0.337
19-12 78,300

INDEX